Lecture Notes in Computer Science 5717

Commenced Publication in 1973
Founding and Former Series Editors:
Gerhard Goos, Juris Hartmanis, and Jan van Leeuwen

Roberto Moreno-Díaz   Franz Pichler
Alexis Quesada-Arencibia (Eds.)

Computer Aided Systems Theory - EUROCAST 2009

12th International Conference
Las Palmas de Gran Canaria, Spain
February 15-20, 2009
Revised Selected Papers

Springer

Volume Editors

Roberto Moreno-Díaz
Alexis Quesada-Arencibia
Universidad de Las Palmas de Gran Canaria
Instituto Universitario de Ciencias y Tecnologías Cibernéticas
Campus de Tafira
35017, Las Palmas de Gran Canaria, Spain
E-mail: rmoreno@ciber.ulpgc.es, aquesada@dis.ulpgc.es

Franz Pichler
Johannes Kepler University Linz
Institute of Systems Science
Altenbergerstrasse 69
4040 Linz, Austria
E-mail: pichler@cast.uni-linz.ac.at

Library of Congress Control Number: 2009935345

CR Subject Classification (1998): H.1.1, J.1, I.4, I.5.4, I.5, J.2, C.2.1, J.6

LNCS Sublibrary: SL 1 – Theoretical Computer Science and General Issues

ISSN 0302-9743
ISBN-10 3-642-04771-8 Springer Berlin Heidelberg New York
ISBN-13 978-3-642-04771-8 Springer Berlin Heidelberg New York

springer.com

Printed in Germany

Typesetting: Camera-ready by author, data conversion by Scientific Publishing Services, Chennai, India
Printed on acid-free paper SPIN: 12769384 06/3180 5 4 3 2 1 0

Preface

The concept of CAST as Computer Aided Systems Theory was introduced by F. Pichler in the late 1980s to refer to computer theoretical and practical developments as tools for solving problems in system science. It was thought of as the third component (the other two being CAD and CAM) required to complete the path from computer and systems sciences to practical developments in science and engineering.

Franz Pichler, of the University of Linz, organized the first CAST workshop in April 1988, which demonstrated the acceptance of the concepts by the scientific and technical community. Next, the University of Las Palmas de Gran Canaria joined the University of Linz to organize the first international meeting on CAST (Las Palmas, February 1989) under the name EUROCAST'89. This proved to be a very successful gathering of systems theorists, computer scientists and engineers from most European countries, North America and Japan.

It was agreed that EUROCAST international conferences would be organized every two years, alternating between Las Palmas de Gran Canaria and a continental European location. From 2001 the conference has been held exclusively in Las Palmas. Thus, successive EUROCAST meetings took place in Krems (1991), Las Palmas (1993), Innsbruck (1995), Las Palmas (1997), Vienna (1999), Las Palmas (2001), Las Palmas (2003) Las Palmas (2005) and Las Palmas (2007), in addition to an extra-European CAST conference in Ottawa in 1994. Selected papers from those meetings were published as Springer's Lecture Notes in Computer Science volumes 410, 585, 763,1030, 1333, 1798, 2178, 2809, 3643 and 4739 and in several special issues of Cybernetics and Systems: an International journal. EUROCAST and CAST meetings are definitely consolidated, as has been shown by the number and quality of the contributions over the years.

EUROCAST 2009 took place in the Elder Museum of Science and Technology of Las Palmas, February 15–20, and it continued with the approach tested at previous conferences as an international computer-related conference with a true interdisciplinary character. There were different specialized workshops which, on this occasion, were devoted to the following topics: Systems Theory and Simulation: Formal Approaches, chaired by Pichler (Linz), Moreno Díaz (Las Palmas) and Albrecht (Innsbruck); Computation and Simulation in Modelling Biological Systems, chaired by Ricciardi (Napoli); Intelligent Information Processing, chaired by Freire (A Coruña); Applied Formal Verification, chaired by Biere (Linz); Computer Vision and Image Processing, chaired by Sotelo (Madrid); Mobile and Autonomous Systems: Robots and Cars, chaired by García-Rosa and De Pedro (Madrid); Simulation Based System Optimization, chaired by Huemer (Klagenfurt) and Jungwirth (Wels); Signal Processing Methods in Systems Design and Cybernetics, chaired by Astola (Tampere), Moraga (Asturias, Dortmund) and Stankovic (Nis); Polynomial Models in Control System Design, chaired by Kucera and Hromčik (Prague); Heurist Problem Solving, chaired by Affenzeller and Jacak (Hagenberg) and Raidl (Vienna); Simulation and Formal

Methods in Systems Design and Engineering, chaired by Ceska (Brno); and Models of Co-operative Engineering Systems, chaired by Braun (Sydney) and Klempous (Wroclaw).

The 120 papers included in this volume are the result of two successive selection processes: the first for presentation at the conference, and the second for inclusion in this book. The selections were the responsibility of the chairpersons, with the counselling of the International Advisory Committee. The present volume has been divided into 12 chapters corresponding to the workshops.

The event and this volume were possible thanks to the efforts of the chairmen of the workshops in the selection and organization of all the material. The editors would like to express their acknowledgement to all contributors and participants and to the invited speakers, Heinz Schwärtzel from Munich, Miguel A. Sotelo from Madrid and Peter Kopacek from Vienna, for their readiness to collaborate. We would also like to thank the Director of the Elder Museum of Science and Technology, D. Fernando Pérez and the members of the museum. Special thanks are due to the staff of Springer in Heidelberg for their valuable support.

July 2009

Roberto Moreno-Díaz
Franz Pichler
Alexis Quesada-Arencibia

Table of Contents

Systems Theory and Simulation: Formal Approaches

Computation and Simulation in Modelling Biological Systems

Intelligent Information Processing

Applied Formal Verification

Computer Vision and Image Processing

Mobile and Autonomous Systems: Robots and Cars

Simulation Based System Optimization

Signal Processing Methods in Systems Design and Cybernetics

Polynomial Models in Control System Design

Heuristic Problem Solving

Simulation and Formal Methods in Systems Design and Engineering

Models of Co-operative Engineering Systems

Kolmogorov Stream Ciphers

Josef Scharinger

Johannes Kepler University, Institute of Computational Perception,
4040 Linz, Austria
Josef.Scharinger@jku.at

Abstract. Stream ciphers are essential tools for encrypting sensitive data. While having the limitation that a single key may never be used twice, they are often very fast and can offer a valuable alternative to block ciphers in many applications.

In this contribution we describe a novel stream cipher based on discrete Kolmogorov systems. Based on a theorem stating that discrete Kolmogorov systems can provide a perfect permutation operator, we develop a strong generator for pseudo-random bits or bytes. These bits or bytes are then added to the plaintext stream to produce the desired ciphertext stream in a straightforward manner.

1 Introduction and Motivation

Stream ciphers are essential tools for encrypting sensitive data. They are often very fast and can offer a valuable alternative to block ciphers in many applications. However, there is the fundamental limitation that in stream ciphering the same key may never be used twice [10]. But this is not a big problem in practice, since combining a fixed secret key with a varying *nonce* (number used once) that can even be exchanged in plain delivers a different session key for each communication session.

Unfortunately, some of the most popular stream ciphers like A5 [14] or RC4 [8] have been found to be not very secure [3,4,13], particularly when used with nonces as described above.

Due to these shortcomings of existing stream ciphers, this contribution intends to propose a novel stream cipher based on discrete Kolmogorov systems. Based on a theorem stating that discrete Kolmogorov systems can provide a perfect permutation operator, we develop a strong generator for pseudo-random bits or bytes. These bits or bytes are then added to the plaintext stream to produce the desired ciphertext stream in a straightforward manner.

The remainder of this contribution is organized as follows. In section 2 we provide a short introduction to stream ciphers in general and our inspiration RC4 in particular. Section 3 describes and analyzes discrete Kolmogorov systems, followed by section 4 where a novel class of Kolmogorov stream ciphers is introduced. Finally, section 5 summarizes the main ideas presented in this contribution.

R. Moreno-Díaz et al. (Eds.): EUROCAST 2009, LNCS 5717, pp. 1–8, 2009.

2 Stream Ciphers

This contribution intends to introduce a novel stream cipher inspired by the extremely popular stream cipher RC4. Therefore, it seems mandatory to first introduce the concept of stream ciphering in general and our inspiration RC4 in particular.

2.1 Introduction to Stream Ciphers

Stream ciphers like A5 [14] or RC4 [8] are essential tools for encrypting sensitive data. The principle of stream ciphering is simple. At the side of the sender a stream p_i of plaintext bits (bytes) is input to the system and combined via bit-wise exclusive-or operation with a key stream k_i of pseudo-random bits (bytes) generated by a cryptographically strong pseudo-random bit (byte) generator that has an output heavily dependent on the systems key K. The output produced this way is denoted as cipher stream c_i. On the receivers side decryption can proceed in a perfectly analogous way due to $c_i \oplus k_i = (p_i \oplus k_i) \oplus k_i = p_i$.

2.2 RC4

Our novel approach of Kolmogorov stream ciphering is inspired by one of the most popular stream ciphers, RC4 [8]. According to[10], RC4 has a 8×8 S-box $S_0, S_1, \ldots, S_{255}$ where entries are a permutation of the numbers 0 through 255, initialized according to systems key K. RC4 has two counters, i and j, initialized to zero. To generate a random byte B, do the following: $i = (i + 1) \bmod 256$, $j = (j + S_i) \bmod 256$, swap S_i and S_j, $t = (S_i + S_j) \bmod 256$, and finally output $B = S_t$.

Note that RC4 is based on a 8×8 S-box that is continuously permuted (swap S_i and S_j) as output bytes are produced. Unfortunately, this permutation is rather simple and acts in a very local manner, with the consequence that weaknesses have been found in RC4 [4,13].

Due to these shortcomings, this contribution intends to propose a novel stream cipher based on discrete Kolmogorov systems. Contrary to the rather simple permutation mechanism utilized in RC4, permutations generated by chaotic Kolmogorov systems exhibit an outstanding degree of instability and can thus serve as a much stronger generator for pseudo-random bits or bytes.

3 Chaotic Kolmogorov Systems

In our contribution we focus on the class of chaotic Kolmogorov systems [5,6,12]. This class has been of great interest to systems scientists for a long time due to some unique properties amongst which the outstanding degree of instability is particularly remarkable.

3.1 Continuous Kolmogorov Systems

Continuous chaotic Kolmogorov systems act as permutation operators upon the unit square $\mathbb{E}$. Figure 1 is intended to give a notion of the dynamics associated with a specific Kolmogorov system parameterized by the partition $\pi = (\frac{1}{3}, \frac{1}{2}, \frac{1}{6})$. As can be seen, the unit square is first partitioned into three vertical strips according to $\frac{1}{3}, \frac{1}{2}, \frac{1}{6}$. These strips are then stretched to full width in the horizontal and squeezed by the same factor in the vertical direction and finally these transformed strips are stacked atop of each other. After just a few applications (see Fig. 1 from top left to bottom right depicting the initial and the transformed state space after 1, 2, 3, 6 and 9 applications of T_π) this iterated stretching, squeezing and folding achieves excellent mixing of the elements within the state space.

Fig. 1. Illustrating the chaotic and mixing dynamics associated when iterating a Kolmogorov system

Formally this process of stretching, squeezing and folding is specified as follows. Given a partition $\pi = (p_1, p_2, \ldots, p_k)$, $0 < p_i < 1$ and $\sum_{i=1}^{k} p_i = 1$ of the unit interval $\mathbb{U}$ and stretching and squeezing factors defined by $q_i = \frac{1}{p_i}$. Furthermore, let F_i defined by $F_1 = 0$ and $F_i = F_{i-1} + p_{i-1}$ denote the left border of the vertical strip containing the point $(x, y) \in \mathbb{E}$ to transform. Then the continuous Kolmogorov system T_π will move $(x, y) \in [F_i, F_i + p_i) \times [0, 1)$ to the position

$$T_\pi(x, y) = (q_i(x - F_i), \frac{y}{q_i} + F_i). \tag{1}$$

It has been proven [2] that *continuous* Kolmogorov systems T_π guarantee ergodicity, exponential divergence and perfect mixing of the underlying state space for almost all valid choices of parameter π. Note that these properties perfectly match the properties of *confusion and diffusion* (as first defined by *C. Shannon* in [11]) that are so fundamental in cryptography. Our task now is to develop a *discrete* version of Kolmogorov systems that preserves these outstanding properties. That is precisely what will be done in the next subsection.

3.2 Discrete Kolmogorov Systems

In our notation a specific discrete Kolmogorov system for permuting a data block of dimensions $n \times n$ shall be defined by a list $\delta = (n_1, n_2, \ldots, n_k)$, $0 < n_i < n$ and $\sum_{i=1}^{k} n_i = n$ of positive integers that adhere to the restriction that all $n_i \in \delta$ must partition the side length n.

Furthermore let the quantities q_i be defined by $q_i = \frac{n}{n_i}$ and let N_i specified by $N_1 = 0$ and $N_i = N_{i-1} + n_{i-1}$ denote the left border of the vertical strip that contains the point (x, y) to transform.

Then the discrete Kolmogorov system $T_{n,\delta}$ will move the point $(x, y) \in [N_i, N_i + n_i) \times [0, n)$ to the position

$$T_{n,\delta}(x, y) = (q_i(x - N_i) + (y \bmod q_i), (y \operatorname{div} q_i) + N_i). \tag{2}$$

Note that this definition has the nice practical implication that if the side length n is a power of 2, only additions, subtractions and bit-shifts are needed for implementing the transformation.

3.3 Analysis of Discrete Kolmogorov Systems

As detailed in [9], the following theorem can be proven for discrete Kolmogorov systems T_{n,δ_t}:

Theorem 1. *Let the side-length $n = p^m$ be an integral power of a prime p. Then the application of discrete Kolmogorov systems T_{n,δ_t} leads to ergodicity, exponential divergence and mixing provided that at least $4m$ iterations are performed and parameters δ_t used in every round t are chosen independently and at random.*

Obviously, this definitely is assured if at least $4 \log_2 n$ iterations are performed and parameters δ_t used in every round t are chosen independently and at random.

One note of caution is appropriate when interpreting theorem 1. While ergodicity, exponential divergence and the mixing property hold true for almost all points in state space, there are two points that cause problems, namely points $(0, 0)$ and $(n-1, n-1)$ which are fixed points never leaving their position. So to account for this, one should shift the array by some offset after each round, thereby ensuring that ergodicity etc. really holds true for all points in state space.

For any cryptographic system it is always essential to know how many different keys are available to the cryptographic system. In our case of discrete Kolmogorov systems $T_{n,\delta}$ this reduces to the question, how many different lists $\delta = (n_1, n_2, \ldots, n_k)$ of n_i summing up to n do exist when all n_i have to part n?

As detailed in e.g. [1], a computationally feasible answer to this question can be found by a method based on formal power series expansion leading to a simple recursion relation that can be evaluated without any difficulties. Some selected results are given in table 1. To fully appreciate these impressive numbers note that values given express the number of permissible keys for just one round and that the total number of particles in the universe is estimated to be in the range of about 2^{265}.

Table 1. Number of permissible parameters δ for parameterizing the discrete Kolmogorov system $T_{n,\delta}$ for some selected values of n

n	c_n	n	c_n	n	c_n
4	1	8	5	16	55
32	5.271	64	47.350.055	128	$\approx 2^{50}$
256	$\approx 2^{103}$	512	$\approx 2^{209}$	1024	$\approx 2^{418}$

4 Kolmogorov Stream Ciphers

While completely different in the building blocks used, our approach to Kolmogorov stream ciphering is similar to and inspired by RC4 in it's architecture. Recall the following observations from our inspiration RC4 (see subsection 2.2)

- RC4 utilizes an $S-$array $S_0, \ldots, S_{255}$ of bytes
- the sorted S-array is initially permuted according to the key
- pseudo-random generation is realized based on
  - stepwise permutation and output generation solely guided by the current state of the S-array (not at all affected by the key)
  - a very simple permutation mechanism (swap two entries)

Based on these ideas we will now develop our novel approach to Kolmogorov stream ciphering.

4.1 State Space

The key dependent pseudo-random bit (byte) generator we propose is based on discrete Kolmogorov systems. To this end, we utilize a Kolmogorov system with side-length $n = 16$. As depicted in figure 2 this state space is initially filled with bits in a balanced and ordered manner (e.g. left half ones, right half zeros). So this state space allows for up to $\frac{256!}{128! \cdot 128!} > 2^{251}$ different states and discrete Kolmogorov systems offer a perfect choice for permuting this array of bits.

4.2 How to Parameterize One Step

If we are to apply discrete Kolmogorov systems to permute the state array depicted in figure 2, we first have to specify how to parameterize such an application.

Recall from table 1 that there are 55 different valid partitions for side length $n = 16$. So one byte is sufficient to select one out of these valid partitions. For $n = 16$ valid divisors of n obviously are $2^1 = 2$, $2^2 = 4$ and $2^3 = 8$. The basic idea now is to map runs of equal bits onto valid divisors, always taking care of the constraint $\sum_{i=1}^{k} n_i = n$. In order to illustrate how this can be done, consider the following simple example. Let the input byte be 001111010. Then we obtain $n_1 = 2^2 = 4$ (from bits 00), $n_2 = 2^3 = 8$ (use only bits 111 since 2^4 would be too large), $n_3 = 2^1 = 2$ (last bit of group 1111) and $n_4 = 2^1 = 2$ (from bit 0). Note

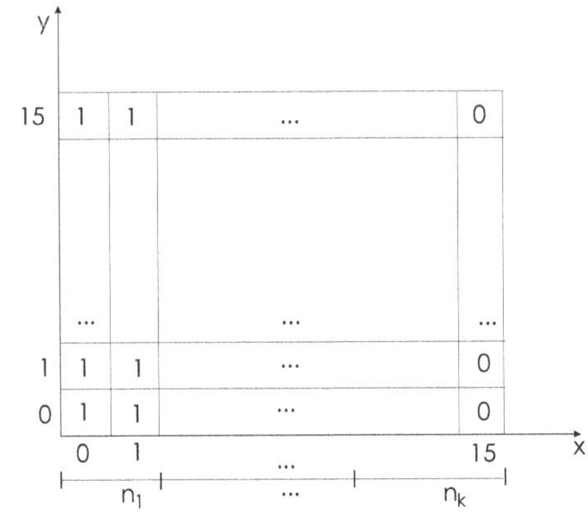

Fig. 2. State space used in proposed Kolmogorov stream ciphering system

that this procedure is guaranteed to always deliver a valid partition of $n = 16$ from one byte in very few steps of computation.

Furthermore, recall from subsection 3.3 that there exist problems with fixed points $(0, 0)$ and $(n-1, n-1)$. As a simple approach a cyclic shift of the 16×16 array allows to solve this problem. So one more byte is needed to parameterize this shift and we end up with the conclusion that 2 bytes are needed in total to parameterize one application of a discrete Kolmogorov system to permute the state array depicted in figure 2.

4.3 Architecture of Proposed Stream Cipher

Now that the state array of the cipher and the way how to parameterize the application of discrete Kolmogorov systems to permute the state array are defined, the architecture of the proposed Kolmogorov stream cipher is straightforward. During key-schedule (warm-up), the initially ordered array is transformed into a perfectly mixed state dependent on the key input to the cipher. During random generation, stepwise permutation and output generation solely guided by the current state of the state array are taking place. The next two subsections will elaborate in more detail on those two phases.

4.4 Key-Schedule

The purpose of this phase is to transform the initially ordered state array into a perfectly mixed state dependent on the key input to the cipher. Recall from theorem 1 that in general the iteration of $4m$ rounds is needed to ensure perfect mixing of an array with side-length $n = 2^m$. So in our particular case of $n = 16 = 2^4$, $4 * 4 = 16$ rounds are needed for perfectly mixing the 16×16 array. Furthermore, recall from subsection 4.2 that 2 bytes are needed for parameterizing one round.

Taking these arguments into consideration, the warm-up phase therefore can proceed quite straightforward (see also figure 3).

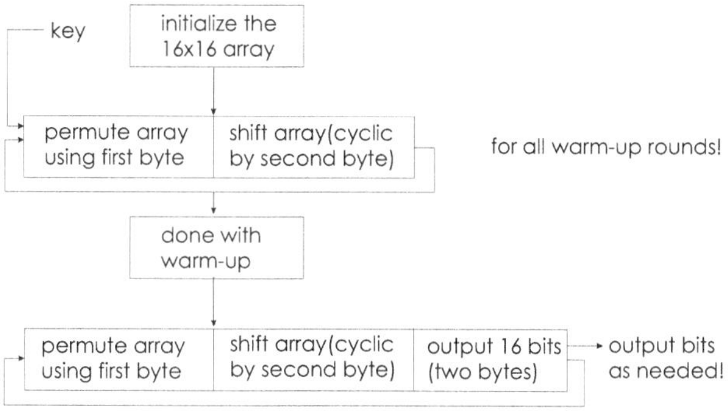

Fig. 3. Architecture of proposed Kolmogorov stream cipher

- take key input to the system (usually as user's secret password concatenated with a running nonce) and apply some strong cryptographic hash function like SHA-256 [7] to obtain (at least) 32 bytes for parameterizing the 16 rounds
- for 16 rounds (16 two-bytes groups): permute array by Kolmogorov permutation parameterized by first byte and cyclically shift array by the value that is represented by the second byte

This way the state array is transformed from a perfectly ordered state to a perfectly chaotic state in a key-dependent manner.

4.5 Random Generation

The purpose of this phase is to realize stepwise permutation and output generation solely guided by the current state of the state array.

Let us first turn to the permutation step based on the current state. According to subsection 4.2, we need two bytes to parameterize one step. Our approach to generate these 16 bits is simple. First calculate column parity bits $cp(x) = XOR_{y=0}^{15} statearray(x, y)$ for $x = 0, \ldots, 15$. Then utilize $cp(x)$ for $x = 0, \ldots, 7$ (first byte) to permute the state array by the Kolmogorov permutation parameterized by this byte and $cp(x)$ for $x = 8, \ldots, 15$ (second byte) to cyclically shift the array by the value that is represented by this second byte (see also figure 3).

Finally, output generation based on the current state proceeds just along the same line. Just calculate row parity bits $rp(y) = XOR_{x=0}^{15} statearray(x, y)$ for $y = 0, \ldots, 15$ and then output this 16 bits value for XORing with the plaintext.

5 Conclusion

In this contribution a remarkably simple key stream generator for Kolmogorov stream ciphering systems has been described. Based on a theorem stating that discrete Kolmogorov systems can provide a perfect permutation operator, we have developed a strong generator for pseudo-random bits or bytes. These bits or bytes can then be added to the plaintext stream to produce the desired ciphertext stream in a straightforward manner.

References

1. Aigner, M.: Kombinatorik. Springer, Heidelberg (1975)
2. Arnold, V.I., Avez, A.: Ergodic Problems of Classical Mechanics. W.A. Benjamin, New York (1968)
3. Biryukov, A., Shamir, A., Wagner, D.: Real time cryptanalysis of A5/1 on a PC. In: Fast Software Encryption Workshop (2000)
4. Fluhrer, S.R., Mantin, I., Shamir, A.: Weaknesses in the key scheduling algorithm of RC4. In: Selected Areas in Cryptography, pp. 1–24 (2001)
5. Goldstein, S., Misra, B., Courbage, M.: On intrinsic randomness of dynamical systems. Journal of Statistical Physics 25(1), 111–126 (1981)
6. Moser, J.: Stable and Random Motions in Dynamical Systems. Princeton University Press, Princeton (1973)
7. NIST. Secure hash standard (SHS). FIPS 180-2 (August 2002)
8. Robshaw, M.J.B.: Stream ciphers. Technical report, RSA Laboratories (July 1995)
9. Scharinger, J.: An excellent permutation operator for cryptographic applications. In: Moreno Díaz, R., Pichler, F., Quesada Arencibia, A. (eds.) EUROCAST 2005. LNCS, vol. 3643, pp. 317–326. Springer, Heidelberg (2005)
10. Schneier, B.: Applied Cryptography. Addison-Wesley, Reading (1996)
11. Shannon, C.E.: Communication theory of secure systems. Bell System Technical Journal 28(4), 656–715 (1949)
12. Shields, P.: The Theory of Bernoulli Shifts. The University of Chicago Press, Chicago (1973)
13. Tews, E., Weinmann, R.-P., Pyshkin, A.: Breaking 104 bit wep in less than 60 seconds. Cryptology ePrint Archive, Report 2007/120 (2007), `http://eprint.iacr.org/`
14. Xu, S.B., He, D.K., Wang, X.M.: An implementation of the GSM general data encryption algorithm A5. In: Chinacrypt 1994, pp. 287–291 (1994)

Morphotronic System (Theory)

Germano Resconi[1] and Zenon Chaczko[2]

[1] Dept. of Mathematics and Physics, Catholic University, Brescia, I-25121
[2] FEIT, University of Technology Sydney, NSW, Australia
resconi@numerica.it, zenon@eng.uts.edu.au

Abstract. The Morphotronic approach postulates a significant improvement to traditional system design thinking based on the Turing Machine model. The paper presents a range of important concepts and definitions supporting this proposition. The Morphotronic system represents an abstract universe of the objects. This universe of objects has two interpretations as in the case of the voltages and currents in the electrical circuit. For the space of the voltages the objects are the voltages at edges of the electrical circuit. For the current space of the currents the objects are the currents in any edge. The dimension of the object space is equal to the number of edges in the electrical circuit. Such a space allows dual interpretation of the current and voltages. Other possible dual variables can be used in the morphotronic system as forces and the fluxes in mechanics or dissipative thermodynamics, in a general way the dual interpretation of the object space will be denoted as causes and effects. The morphogenetic system can be modelled by samples of the causes and effects. The morphotronic system with the samples generates the algorithm to implement the purpose in the system. Providing that the samples of the effect and the purpose denote a virtual cause, the vector E can be computed so that it represents the effective origin of the causes inside the purpose map. With the *cause-effect* rule the effective causes can be computed obtaining results that are coherent with the samples. Providing that the virtual cause is given by purpose the effective causes can be generated in agreement with the samples. The described algorithm is denoted as the projection operator that transforms a virtual cause (purpose) into an effective cause.

Keywords: Morphotronics, Turing machine, Dissipative thermodynamics.

1 Introduction

Traditionally software is built for a specific purpose where its actions are represented as a set of instructions to be executed in the context of the required task. Software uses a standard programming language that contains the syntactic rules which aggregate a set of instructions. The Turing Machine (TM) is used as a conceptual model for constructing software system that is based on the above principles to realise the purpose. However, nature provides many examples where the purpose is obtained without taking into consideration of the TM model. The recent work in the domains of

R. Moreno-Díaz et al. (Eds.): EUROCAST 2009, LNCS 5717, pp. 9–16, 2009.

complex network systems[5], biomimetic middleware systems[3], constructal theory [1], immuno-computing [10], holographic computing, morphic computing [7,8,9], quantum computers [6], DNA computing, secondary sound sources, General system logical theory, Neural Networks and Group Theory [4] and several other areas of study suggest a change is required to the traditional approaches based on the Turing Machine model. The proposed morphotronic computation model offers a radical change of perspective. Firstly, by beginning to state the purpose as the initial point for defining the process of computation so that the purpose becomes the conceptual input to the program or computer machine; and secondly, the local machine state can be ignored when generating the context and its rules as resources to located and allocate the computational component(s).

2 Morphotronic System and Samples of Cause and Effect

In Morphotronic system the cause is considered as input and effect as output, where inputs are p strings with q values; a similar case is with outputs; Matrix Z represents the transformation (Fig.1).

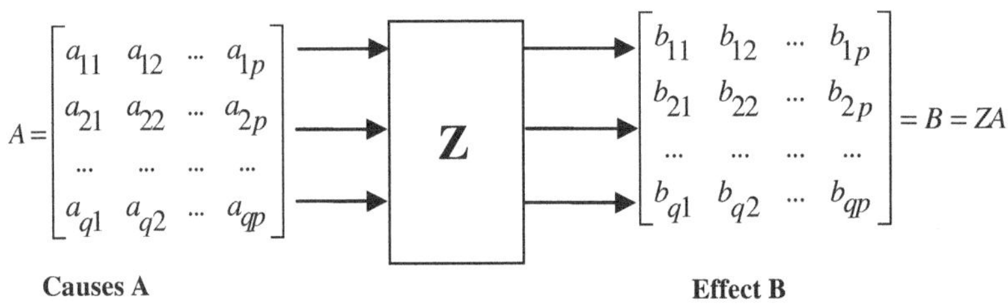

Fig. 1. Morphotronics as the *cause and effect* system (MIMO system)

The matrix A is the matrices of the p samples for q inputs of the causes

$$A = \begin{bmatrix} a_{11} & a_{12} & \dots & a_{1p} \\ a_{21} & a_{22} & \dots & a_{2p} \\ \dots & \dots & \dots & \dots \\ a_{q1} & a_{q2} & \dots & a_{qp} \end{bmatrix}, \quad B = ZA = \begin{bmatrix} b_{11} & b_{12} & \dots & b_{1p} \\ b_{21} & b_{22} & \dots & b_{2p} \\ \dots & \dots & \dots & \dots \\ b_{q1} & b_{q2} & \dots & b_{qp} \end{bmatrix}$$

B and A expressions we can be connected by the matrix:

$Z = B(A^T A)^{-1} A^T$, thus ZA can be denoted as: $ZA = B(A^T A)^{-1} A^T A = B$

Given A and B a simple loop can be generated as depicted in Fig.2

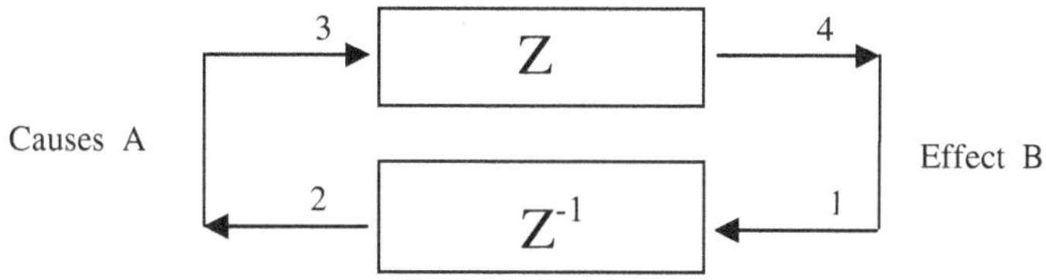

Fig. 2. A loop of *Causes* and *Effects* matrices

For the definition of Z in the previous case the following equation can be derived:

$Z = B\,(A^T A)^{-1} A^T, \quad Z^{-1} = A\,(B^T B)^{-1} B^T, \quad Z A = B \quad \text{and} \quad Z^{-1} B = A$

In the *Causes* and *Effects* loop (Fig.2) the Projector Operator Q can be calculated as:

$Q = Z Z^{-1} = B\,(A^T A)^{-1} A^T A\,(B^T B)^{-1} B^T = B\,(B^T B)^{-1} B^T$

where, $Q B = B$, and $Q^2 = B\,(B^T B)^{-1} B^T B\,(B^T B)^{-1} B^T = B\,(B^T B)^{-1} B^T = Q$

Alternatively, the projection operator Q can also be denoted in as:

$Q = Z^{-1} Z = A\,(B^T B)^{-1} B^T B\,(A^T A)^{-1} A^T = A\,(A^T A)^{-1} A^T$, where $QA = A$, and
$Q^2 = A\,(A^T A)^{-1} A^T A\,(A^T A)^{-1} A^T = A\,(A^T A)^{-1} A^T = Q$

thus for Z another representation can be obtained:

$Z = B\,(B^T A)^{-1} B^T, \quad Z^{-1} = A\,(A^T B)^{-1} A^T$ and
$Q = Z^{-1} Z = A\,(A^T B)^{-1} A^T B\,(B^T A)^{1} B^T = A\,(B^T A)^{-1} B^T$
Where $Q A = A$ and $Q^2 = A\,(B^T A)^{-1} B^T A\,(B^T A)^{-1} B^T = A\,(B^T A)^{-1} B^T = Q$
We have also: $Q = Z Z^{-1} = B\,(B^T A)^{-1} B^T A\,(A^T B)^{-1} A^T = A\,(A^T B)^{-1} A^T$
where, $Q B = B$ and $Q^2 = A\,(B^T A)^{-1} B^T A\,(B^T A)^{-1} B^T = A\,(B^T A)^{-1} B^T = Q$.

3 Geometric Image of the Projection Operator

Let the matrice *Y* be expressed as a product of *H* and *W* matrices:

$$Y = HW = \begin{bmatrix} h_{11} & h_{12} & \dots & h_{1p} \\ h_{21} & h_{22} & \dots & h_{2p} \\ \dots & \dots & \dots & \dots \\ h_{q1} & h_{q2} & \dots & h_{qp} \end{bmatrix} \begin{bmatrix} w_1 \\ w_2 \\ \dots \\ w_p \end{bmatrix} = \begin{bmatrix} y_1 \\ y_2 \\ \dots \\ y_q \end{bmatrix}$$

and

$$Y = w_1 \begin{bmatrix} h_{11} \\ h_{21} \\ \dots \\ h_{q1} \end{bmatrix} + w_2 \begin{bmatrix} h_{12} \\ h_{22} \\ \dots \\ h_{q2} \end{bmatrix} + \dots\dots + w_p \begin{bmatrix} h_{1p} \\ h_{2p} \\ \dots \\ h_{qp} \end{bmatrix} = w_1 H_{\alpha,1} + w_2 H_{\alpha,2} + \dots + w_p H_{\alpha,p}$$

The Objects A_k and attributes H_J can be geometrically represented in 3D as Colon Vectors in Matrix H with the coordinates set by the vector W (Fig. 3)

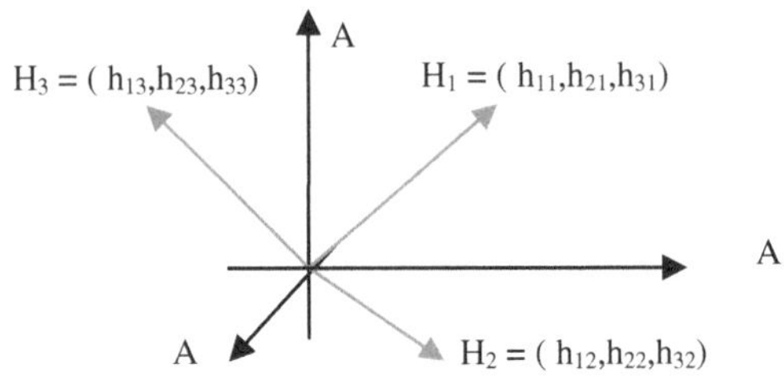

Fig. 3. Objects A_k and Attributes H_j as Colon Vectors in Matrix H

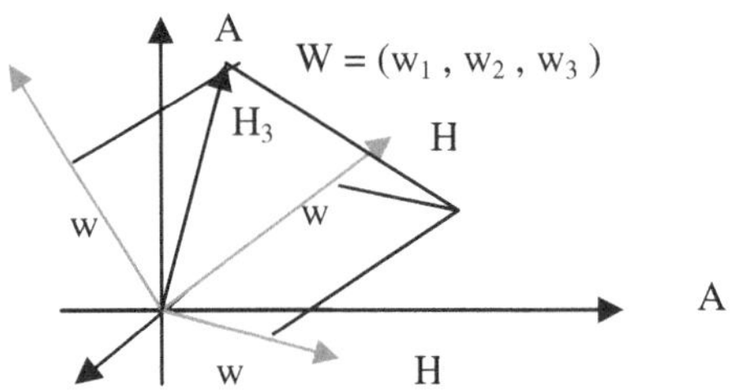

Fig. 4. Objects A_k and Attributes H_j represented by column vectors in matrix H

The vector $W = (w_1, w_2, w_3)$ has three components w_1, w_2, w_3 on the reference given by three vectors H_1, H_2, H_3 as shown in Fig. 4.

3.1 Projection Operator Q and Weights W

Considering that $Y = HW$, and because H is a rectangular matrix, then the pseudo inverse matrix can be expressed as:

$$W = (H^T H)^{-1} H^T Y \text{ , } HW = H(H^T H)^{-1} H^T Y$$

$$For \;\; Y = HW \text{ , } HW = H(H^T H)^{-1} H^T HW = HW = Y$$

For $g = H^T H$, we can write by index notation in tensor calculus

$$W = x^h = g^{-1} x_k = (H^T H)^{-1} H^T Y$$

Where $W = x^h , H^T X = x_k$ in tensor calculus g is the metric tensor and

$$g = H^T H = \sum_k h_{kj} h_{ki} = h^k_j h_{ki}$$

The covariant components of X are: $x_k = H^T X = h_k^j A_j$, where X_j is a component of X in the space of the object. The contravariant components of X are: $x^i = g^{-1} x_j = g^{i,j} x_j = W$

When H is a square matrix then Y can be obtained as:

$$Y = HW = h_{j,i} x^i = A_j = X$$

When H is a rectangular matrix with q > p, we have

$$QX = HW = H(H^T H)^{-1} H^T X = h_{j,i} x^i = y_j$$

With the property: $Q^2 X = H(H^T H)^{-1} H^T H(H^T H)^{-1} H^T X = H(H^T H)^{-1} H^T X = QX$

A geometric image of the Q projection operator in three dimensional object space and with two dimensional attribute space is shown in Fig. 5.

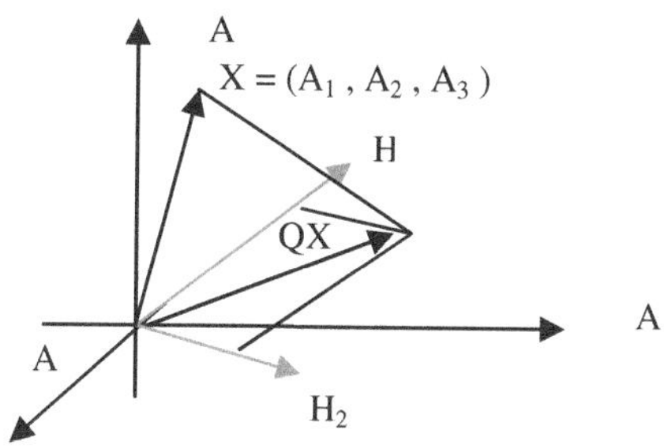

Fig. 5. An example of the Projection Operator Q

3.2 Projection Operator and Lagrangian Function Minimum Condition

We separate the projection operator in two parts one is the source of the force E and the other is the source of the flux J. The two parts are:

$J = x^h = (B^T A)^{-1} B^T X$ *and* $E = B^T X = x_k$, therefore the Lagrangian form is:

$$L = x^h g x^h = J^T g J + \lambda(E - B^T QX) = J^T g J + \lambda(E - B^T A J) = J^T g J + \lambda^T (E - g J)$$

and the derivative for minimum condition is: $\dfrac{dL}{dJ} = 2 g J - \lambda g = 0$

The solution is $\lambda = \dfrac{J}{2}$ when substituting in L it can be denoted:

$$L = J^T g J + \frac{1}{2} J^T (E - g J) = \frac{1}{2} J^T g J + \frac{1}{2} J^T E$$

Thus thc dcfinition of E and J can be expressed as:

$$L = \frac{1}{2} J^T g J + \frac{1}{2} J^T E = \frac{1}{2} J^T g J + \frac{1}{2} J^T g J = J^T g J = J^T E = x^h x_h$$

The invariant $L = J^T g J$, thus the min. value is obtained under the constraint $E = g J$.

4 Symmetry in Morphotronic System

Given one dimensional space, the symmetric points $P_1 = 1$ and $P_2 = -1$ can be considered as a sample for symmetry so thus the matrix of samples can be denoted as:

$H = \begin{bmatrix} 1 \\ -1 \end{bmatrix}$ For the transformation U A = -A we have $UH = \begin{bmatrix} -1 \\ 1 \end{bmatrix} = \begin{bmatrix} 1 \\ -1 \end{bmatrix}$

where the space of objects is a 2D space and H is one vector in this space. The purpose can be defined a set of vectors that are functions of the x parameter.

$X(x) = \begin{bmatrix} f(x) \\ f(-x) \end{bmatrix}$, X has this symmetry $X(Ux) = \begin{bmatrix} f(Ux) \\ f(U(-x)) \end{bmatrix} = \begin{bmatrix} f(-x) \\ f(x) \end{bmatrix}$

Now the weights can be computed as follows:

$$W(x) = (H^T H)^{-1} H^T X = (H^T H)^{-1} H^T \begin{bmatrix} f(x) \\ f(-x) \end{bmatrix} = \frac{1}{2}(f(x) - f(-x))$$

For which we have $W(Ux) = \frac{1}{2}(f(Ux) - f(-Ux)) = -W(x) = UW(x)$

W(x) is a basis function for the group G of transformations: $G = (U^2 = 1, U)$, since the transformation of the space by U, W(Ux) is equal to the transformation U of the function W(x). Thus the following group is denoted as:

$G\,W(x) = (U^2 W(x) = W(x), UW(x) = W(Ux))$ and

$$G(x) = Y(x) = HW(x) = \begin{bmatrix} 1 \\ -1 \end{bmatrix} W(x) = \begin{bmatrix} W(x) \\ -W(x) \end{bmatrix} = \begin{bmatrix} \frac{1}{2}(f(x) - f(-x) \\ -\frac{1}{2}(f(x) - f(-x) \end{bmatrix}$$

also $$Y(Ux) = \begin{bmatrix} W(Ux) \\ -W(Ux) \end{bmatrix} = UY(x) = U \begin{bmatrix} \frac{1}{2}(f(x) - f(-x) \\ -\frac{1}{2}(f(x) - f(-x) \end{bmatrix}$$

It is known that $UH = U\begin{bmatrix} 1 \\ -1 \end{bmatrix} = \begin{bmatrix} -1 \\ 1 \end{bmatrix}$ and that Y changes in the similar way

$$Y(Ux) = U\begin{bmatrix} \frac{1}{2}(f(x) - f(-x) \\ -\frac{1}{2}(f(x) - f(-x) \end{bmatrix} = \begin{bmatrix} -\frac{1}{2}(f(x) - f(-x) \\ \frac{1}{2}(f(x) - f(-x) \end{bmatrix}$$

hence the vector Y change in the same manner as H, although at the same time the following equations can be defined:

$$X(Ux) = \begin{bmatrix} f(Ux) \\ -f(Ux) \end{bmatrix} = \begin{bmatrix} f(-x) \\ -f(x) \end{bmatrix}, \text{ and}$$

$$Y(Ux) = \begin{bmatrix} \frac{1}{2}(f(Ux) - f(-Ux) \\ -\frac{1}{2}(f(Ux) - f(-Ux) \end{bmatrix} = \begin{bmatrix} -\frac{1}{2}(f(x) - f(-x) \\ \frac{1}{2}(f(x) - f(-x) \end{bmatrix} = UY(x)$$

Y has the symmetry of H and X at the same time. The orthogonal function is given as:

$$F(x) = (1-Q)X(x) = \begin{bmatrix} f(x) \\ f(-x) \end{bmatrix} - \begin{bmatrix} \frac{1}{2}(f(x) - f(-x)) \\ -\frac{1}{2}(f(x) - f(-x)) \end{bmatrix} = \begin{bmatrix} \frac{1}{2}(f(x) + f(-x)) \\ \frac{1}{2}(f(x) + f(-x)) \end{bmatrix} \text{ and}$$

$$QF(x) = Q(1-Q)X(x) = (Q - Q^2)X(x) = (Q - Q)X(x) = 0$$

Considering that X = F(x), Q X = 0 then the projection of F on the samples H is equal to zero. In F (x) it is impossible to introduce the symmetry of H because in F the two components are equal. Thanks to the following property: QX + (1–Q) X = X any set of functions $X(x) = \begin{bmatrix} f(x) \\ f(-x) \end{bmatrix}$ is decomposed into:

$$X(x) = \begin{bmatrix} f(x) \\ f(-x) \end{bmatrix} = \begin{bmatrix} G_1(x) + F_1(x) \\ G_2(x) + F_2(x) \end{bmatrix}$$

A deeper asymmetry in QX for which the set of function G(X) included in X(x) is actually lost. Therefore when only Y(x) is known it is not possible to return to X(x). The metric can be computed as:

$$L(x) = J(x)^T \text{g J}(x) = \frac{1}{2}(f(x) - f(-x))2\frac{1}{2}(f(x) - f(-x)) = \frac{1}{2}(f(x) - f(-x))^2$$

and L(-x) = L(x) hence for a symmetric change of the variable x , L(x) is invariant.

5 Conclusion

A complex system can be perceived as a network made of connectors and components with variables such as time, connector length, the inverse of transmission speed or velocity. In such a network the Extreme principle can be used to compute the length of the connectors and to obtain the minimum path. In Morphotronics the local description is substituted with the global description. This resembles the Lagrange approach and minimum action in physics, where the variables are: position, mass and velocity. In thermodynamics, far from equilibrium, a force equals the difference of temperature, thermal resistance, and thermal flux. Similarly when applying the Kirchhoff's 1st law in electrical circuits for fixed voltages and resistances - the currents for which dissipation assumes the min. value can be obtained [11]. The global rule (allometric laws) can be defined as minimum velocity of entropy production [2]. An ideal (perfect) communication is embedded in a context where global rules reside as a minimum condition or an invariant. The theory of Morphotronics offers a new class of computation for finding optimal communication that is coherent with the ideal communication model inside a context described by its rules. Morphotronics theory uses Non-Euclidean geometry offering a powerful modelling tool to shape the context and define the projection operators for an ideal network. The morphotronic approach can represent complex systems in such domains as: biology, chemistry, computing, telecommunication, transportation and business.

References

1. Bejan, A.: Shape and Structure, from Engineering to Nature. Cambridge University Press, Cambridge (2000)
2. Bruers, S.: Classification and discussion of macroscopic entropy production principles, ar Xiv: cond-mat0604482v3 (cond-mat.stat-mech) (May 2, 2007)
3. Chaczko, Z.: Autopoietics of Biomimetic Middleware System, private correspondence (November 2007)
4. Cotton, F.A.: Chemical Application of Group Theory. Wiley & Sopns Inc., New York (1971)
5. Newman, M.E.J.: The Structure and Function of Complex Networks. Santa Fe Institute Publication (2004)
6. Perus, M., Bischof, H., Caulfield, H.J., Loo, C.K.: Quantum–Implementable Selective Reconstruction of High-resolution Images. Applied Optics 43(33) (November 20, 2004)
7. Resconi, G., Nikravesh, M.: Morphic Computing: Concepts and Foundation. In: Nikravesh, M., Zadeh, L.A., Kacprzyk, J. (eds.) Forging the new Frontieres: Fuzzy Pioneers I. Studies in Fuzziness and Soft Computing. Springer, Heidelberg (2007)
8. Resconi, G., Nikravesh, M.: Morphic Computing: Quantum and Field. In: Nikravesh, M., Zadeh, L.A., Kacprzyk, J. (eds.) Forging the new Frontieres: Fuzzy Pioneers II. Studies in Fuzziness and Soft Computing. Springer, Heidelberg (2007)
9. Resconi, G.: The Morphogenetic Systems in Risk Analysis. In: Proceeding of the 1st International Conference on Risk Analysis and Crisis Response, Shangai, China, September 25-26, pp. 161–165 (2007)
10. Tarakanov, A.O., et al.: Immunocomputing: Principles and Applications. Springer, Heidelberg (2003)
11. Zupanovic, P., Juretic, D.: The chemical Cycle Kinetics close to the Equilibrium State and Elecrrical Circuit Analogy. Croatia Chemical Acta CCACAA 77(4), 561–571 (2004)

Knowledge Discovery in Databases Using Multivalued Array Algebra

Margaret Miró-Julià

Departament de Ciències Matemàtiques i Informàtica
Universitat de les Illes Balears
07122 Palma de Mallorca, Spain
margaret.miro@uib.es

Abstract. In the past, the way of turning data into knowledge relied on manual analysis and interpretation. Nowadays computational techniques are used in order to extract knowledge from data. Changing data into knowledge is not a straightforward task. Data is generally disorganized, contains useless details and may be incomplete. Knowledge is the opposite, organized but expressed using a poorer language, which might even be imprecise or vague.

Knowledge Discovery in Databases is the nontrivial process of identifying valid, novel, potentially useful, and understandable patterns in data. The Multivalued Array Algebra does not handle raw data, it handles declarative descriptions of the data by means of a multivalued language. This paper proposes and addresses the use of the Multivalued Array Algebra for Knowledge Discovery in Databases.

1 Knowledge Discovery in Databases

In the past, the way of turning data into knowledge relied on manual analysis and interpretation. The classical approach to data analysis depended on one or more analysts becoming familiar with the data and interpreting the data to the users. This manual form of data treatment was slow, expensive and highly subjective. As the volume of data grew, this manual data analysis became totally impractical. Nowadays, the use of computers has allowed humans to collect more data than that we can interpret. The obvious step is to use computational techniques in order to extract knowledge from data.

Computers store and exploit knowledge, at least that is one of the aims of the Artificial Intelligence and the Information Systems research fields. However, the problem is to understand what knowledge means, to find ways of representing knowledge, and to extract useful information from stored data. Changing data into knowledge is not a straightforward task. Data is generally disorganized, contains useless details and may be incomplete. Knowledge is the opposite, organized but expressed using a poorer language, which might even be imprecise or vague.

Knowledge Discovery in Databases is the nontrivial process of identifying valid, novel, potentially useful, and understandable patterns in data [1]. It is aimed towards the discovery of methods, techniques and tools that support analysts in the process of discovering useful information and knowledge in databases.

R. Moreno-Díaz et al. (Eds.): EUROCAST 2009, LNCS 5717, pp. 17–24, 2009.

The Knowledge Discovery in Databases process involves several steps that are summarized in Figure 1.

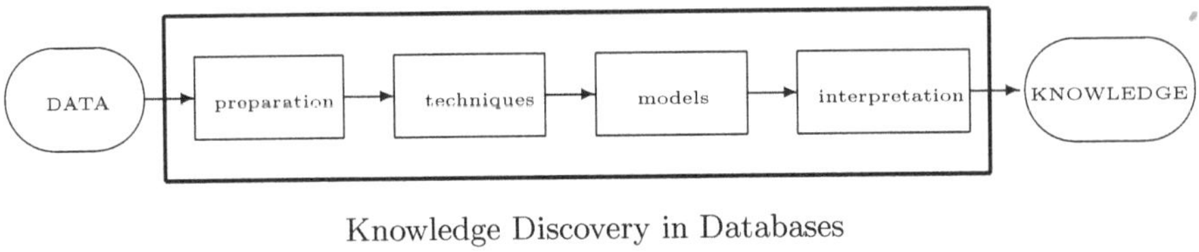

Fig. 1. An overview of the Knowledge Discovery in Databases process

The starting point of the Knowledge Discovery in Databases process is the data. This data has to be prepared, this step includes the selection, preprocessing, subsampling and transformation of the data. At this time, useful features with which to represent the data must be found. Once the data is transformed, data mining techniques are used and new models are found. These mined patterns are then interpreted and evaluated and converted into useful knowledge.

2 Conceptual Knowledge Discovery

Many real-world knowledge discovery tasks are too complex to be found by simply applying a learning or data mining algorithm. Therefore Knowledge Discovery in Databases may not be strong enough to extract useful knowledge due to the fact that it relies heavily on the know-how of the analyzing expert.

Concepts are necessary for expressing human knowledge. Therefore, the process of discovering knowledge in databases benefits from a comprehensive formalization of concepts. Formal Concept Analysis [2] provides such a formalization by introducing the formal concepts. A formal concept is defined as the pair (D_i, R_i), where $D_i \subseteq D$, the extension, is a set of objects that exhibit a set of attribute values $R_i \subseteq R$, the intension, and all these values only apply to those objects [3]. The set of all concepts together with an order relation form a complete lattice, called the concept lattice.

The development of Conceptual Knowledge Discovery in Databases is based on the mathematical theory of Formal Concept Analysis and aims to develop methods and procedures that allow the analysis of given data by examination and visualization of their conceptual structure [4]. Knowledge is discovered in interaction with the data during an iterative process using Conceptual Data Analysis techniques that are guided by theoretical preconceptions and declared purposes of the expert.

3 Multivalued Array Algebra

The Multivalued Array Algebra introduced in [5] does not handle raw data, it handles declarative descriptions of the data by means of a multivalued language.

This language, based on arrays, allows a multivalued description of the knowledge contained in a data table, by means of array expressions. Declarative expressions from a multivalued data table can be obtained using arrays and declarative expressions can be transformed by application of algebraic techniques.

3.1 Multivalued Object Attribute Table

The knowledge of an environment can be descriptive and can be expressed in declarative form by means of a language. The objects that form the environment are elements of the domain and can be described in terms of the values of the characteristics or attributes they possesses.

Definition 1. *Let $D = \{d_1, d_2, \ldots, d_i, \ldots, d_m\}$ be an ordered set called domain, of elements d_i representing the m objects, let $R = \{r_g, \ldots, r_c, \ldots, r_a\}$ be a set of the g multivalued attributes or properties of the objects. The set of values of attribute r_c is represented by $C = \{[c_{n_c}], \ldots, [c_j], \ldots, [c_1]\}$. The elements of set C, $[c_j]$, are called 1-spec-sets since the elements are defined by means of one specification. An Object Attribute Table (OAT) is a table whose rows represent the objects, and whose columns represent the attributes of these objects. Each element $[c_i]$ represents the value of attribute r_c that corresponds to object d_i as is shown in Table 1.*

Table 1. Object Attribute Table

	r_g	$\ldots$	r_c	$\ldots$	r_a
d_1	$[g_1]$	$\ldots$	$[c_1]$	$\ldots$	$[a_1]$
d_2	$[g_2]$	$\ldots$	$[c_2]$	$\ldots$	$[a_2]$
$\vdots$	$\vdots$	$\ddots$	$\vdots$	$\ddots$	$\vdots$
d_i	$[g_i]$	$\ldots$	$[c_i]$	$\ldots$	$[a_i]$
$\vdots$	$\vdots$	$\ddots$	$\vdots$	$\ddots$	$\vdots$
d_m	$[g_m]$	$\ldots$	$[c_m]$	$\ldots$	$[a_m]$

An Object Attribute Table (OAT) represents the complete itemized description of a given specified domain. Each line of the table describes an object by means of the values of its features, therefore it may be considered that the description of element d_i is the ordered sequence of the corresponding attribute values.

3.2 Multivalued Language

In order to handle the descriptions that rise from a multivalued OAT, where attributes take values from a given set, a multivalued language is needed.

The set of all subsets of a given set C (the power set of C), $\rho(C)$, constitutes a Boolean algebra $< \rho(C), \cup, \cap, \widehat{\ }, \emptyset, C >$. If a symbolic representation or a description of subsets is considered, there is a parallel Boolean algebra $< \mathcal{S}_c, +, \cdot, \hat{\ }, \vee_c, \wedge_c >$ defined on the set $\mathcal{S}_c$ of all possible symbols representing subsets of C. The zero of this algebra is $\vee_c$ (the symbol representing the empty set). The identity is $\wedge_c$ (the symbol representing set C). Set $\mathcal{S}_c = \{c_1, c_2, \ldots, c_k\}$ is formed by all possible symbols describing subsets of C using the octal code introduced in [6]. In this paper, the expression $c_i \looparrowright C_i$ may be read as "c_i is the symbol describing subset C_i".

Regular set operations can also be described using multivalued language. The symbolic notation used for complement ($\widehat{\ }$), union ($\cup$) and intersection ($\cap$) of subsets are:

$$\hat{c_h} \looparrowright \widehat{C_h} \qquad c_h + c_k \looparrowright C_h \cup C_k \qquad c_h \cdot c_k \looparrowright C_h \cap C_k$$

Furthermore, operation tables using octal code can easily be constructed.

3.3 Multivalued Arrays

All the concepts, operations and special elements introduced above make reference to only one set of values, that is, one attribute. A data table has more than one attribute. Let's consider g sets $G, \ldots, B$ and A, the elements of each of these sets are the 1-spec-sets (one specification). A g-spec-set, $[g_k, \ldots, b_j, a_i]$, is a chain ordered description of g specifications, one from set G, $\ldots$, one from set B and one from set A. Each spec-set represents itself and all possible permutations.

The cross product $G \otimes \cdots \otimes B \otimes A$ is the set of all possible g-spec-sets formed by one element of G, $\ldots$, one element of B and one element of A. The set of all possible g-spec-sets induced by sets G, $\ldots$, B and A is called the universe and every subset of the universe is called a subuniverse.

It is important to mention that the cross product is not the cartesian product. A g-spec-set represents itself and all possible permutations whereas the elements of the cartesian product are different if the order in which they are written varies.

The basic element of the Multivalued Algebra is the array. An array is a description of those subuniverses (subsets of g-spec-sets) that can be written as a cross product.

Definition 2. *Given sets G, $\ldots$, B, A, let $G_i \subseteq G$, $\ldots$, $B_i \subseteq B$, $A_i \subseteq A$, an array $|t_i| = |g_i, \ldots, b_i, a_i|$ is the symbolic representation of the cross product $G_i \otimes \ldots \otimes B_i \otimes A_i$ where $g_i \looparrowright G_i$, $\ldots$, $b_i \looparrowright B_i$ and $a_i \looparrowright A_i$.*

$$|t_i| = |g_i, \ldots, b_i, a_i| \looparrowright G_i \otimes \cdots \otimes B_i \otimes A_i$$

Arrays are symbolic representations of subuniverses, 2-dimensional (two attributes) arrays are represented graphically in Fig. 2.

The zero Array. There is an array that deserves special consideration: the zero array. The zero array $\bigvee$ is the symbol that describes the empty subuniverse:

$$\bigvee = |\vee_g, \ldots, \vee_b, \vee_a| \looparrowright \emptyset$$

In the development of the array theory the following theorem was proven in [7].

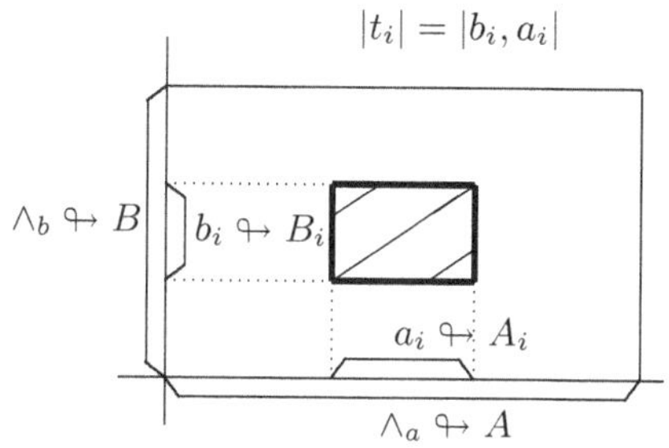

Fig. 2. Two dimensional arrays

Theorem 1. *An array with a $\vee$ component is equal to $\bigvee$*

$$\forall b \quad |g_i, \ldots, \vee_b, a_i| = \bigvee$$

where:

$$\rightsquigarrow |g_i, \ldots, \vee_b, a_i| \looparrowright G_i \otimes \cdots \otimes \emptyset_B \otimes A_i$$

This theorem gives rise to some interesting questions. Even though the cross product is not the cartesian product it inherits an undesirable property: the cartesian product of a set by the empty set is the empty set. If an OAT is considered, just because there is a missing piece of information can we say that we have no information at all?

In [8], the zero array is singled out and its interpretation analyzed. There is not a unique zero array. There are levels of zero arrays depending on the number of $\vee$ components in the array. They are called the n-order projection arrays. If two attributes are considered, there is one 2-order projection array and two 1-order projection array. These projection arrays are related to Wille's formal concepts and can be used in Conceptual Knowledge Discovery in Databases. These projection arrays describe the intension of a formal concept.

Array Operations. Since the arrays describe subuniverses (subsets of spec-sets), regular set operations may be performed with them. Let $|t_i| = |g_i, \ldots, b_i, a_i| \looparrowright G_i \otimes \cdots \otimes B_i \otimes A_i$ and $|t_j| = |g_j, \ldots, b_j, a_j| \looparrowright G_j \otimes \cdots \otimes B_j \otimes A_j$ be two arrays, the following operations are introduced:

- $\sim$ complement (symbolic representation of the complement of a subuniverse respect to the universe):

$$\sim |t_i| \looparrowright \quad \sim (G_i \otimes \ldots \otimes B_i \otimes A_i)$$

- $\ddagger$ sum (symbolic representation of the union of two subuniverses):

$$|t_i| \ddagger |t_j| \looparrowright (G_i \otimes \ldots \otimes B_i \otimes A_i) \cup (G_j \otimes \ldots \otimes B_j \otimes A_j)$$

- $\circ$ product (symbolic representation of the intersection of two subuniverses):

$$|t_i| \circ |t_j| \looparrowright (G_i \otimes \ldots \otimes B_i \otimes A_i) \cap (G_j \otimes \ldots \otimes B_j \otimes A_j)$$

The $\circ$ product of two continuous arrays is a continuous array.

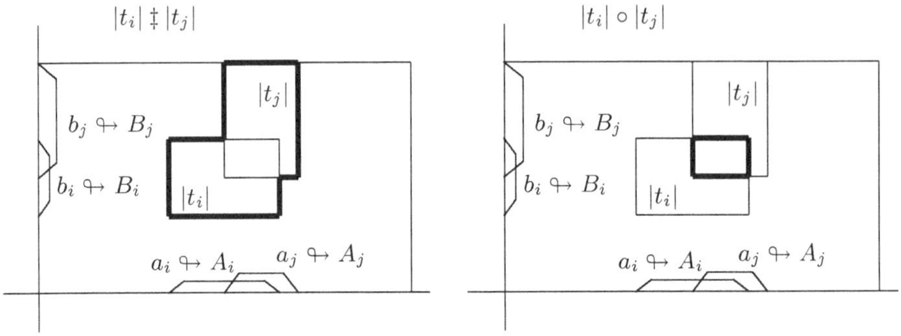

Fig. 3. Two-dimensional ‡ sum and ∘ product of arrays

All the results obtained by use of operations $\sim$, ‡ and ∘ on arrays are symbolic representations of subuniverses. If only two attributes are considered, these operations can be represented graphically as shown in Fig. 3.

3.4 Array Expressions

Subuniverses can be symbolically represented by arrays or by algebraic expressions of arrays. An expression E_i is a symbolic representation of a subuniverse U_i, it represents the reality described by an OAT. Array expressions can be used to describe any subuniverse.

Definition 3. *Any combination of arrays using operations $\sim$, ‡ and ∘ (well formed formula) is called an expression E_i.*

$$E_i = \sim |t_i| \ddagger |t_j| \circ |t_k| \ldots \looparrowright U_i$$

Expressions represent subuniverses, therefore an order relation that symbolically represents set inclusion may be introduced: $E_i \preceq E_j \looparrowright U_i \subseteq U_j$. This order relation has been studied in [7] and has been used to find simplified equivalent expressions.

Definition 4. *An expression E_i is called an array expression if it is written as a ‡ sum of arrays.*

$$E_i = |t_z| \ddagger \cdots \ddagger |t_y| \ddagger \cdots \ddagger |t_x| \looparrowright U_i$$

An array expression in 2 dimensions is shown in Fig. 4.

3.5 Minimum Expression

The same OAT can be described by different expressions. The number of arrays appearing in the expression is not necessarily minimal. Algorithmic techniques can be used to find a minimal declarative expression.

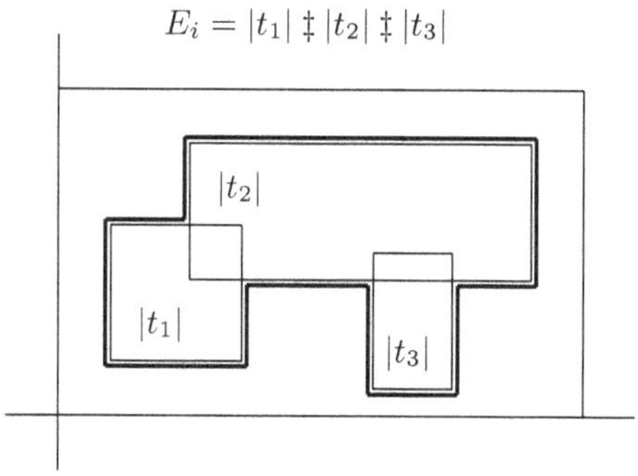

Fig. 4. Two-dimensional array expression

Definition 5. *Given a set of arrays* $A = \{|t_1|, |t_2|, \ldots, |t_n|\}$ *an array* $|t_i|$ *is said to be covered by* A *or is called a covered array respect to* A *if and only if*

$$|t_i| \preceq |t_1| \ddagger |t_2| \ddagger \cdots \ddagger |t_n|$$

An array that is not a covered array respect to A *is called an uncovered array respect to* A.

Definition 6. *Given an array expression* $E = |t_1| \ddagger \cdots \ddagger |t_i| \ddagger \cdots \ddagger |t_k| \ddagger \cdots \ddagger |t_z|$ *an array* $|t_i|$ *of the expression is called a redundant array respect to* E *if and only if*

$$|t_i| \preceq |t_1| \ddagger \cdots \ddagger |t_{i-1}| \ddagger |t_{i+1}| \ddagger \cdots \ddagger |t_z|$$

In other words, $|t_i|$ *is covered by* $A - \{|t_i|\} = \{|t_1|, \ldots, |t_{i-1}|, |t_{i+1}|, \ldots, |t_z|\}$. *An array that is not a redundant array respect to* E *is said to be an essential array respect to* E.

Given an expression E, a minimum expression of E is formed by all the essential arrays of the expression and some of the redundant arrays. Let E_f be the array expression of all the essential arrays. A minimum expression is formed by all the essential arrays respect to E and those redundant arrays respect to E that are essential respect to E_f.

4 Conclusions and Future Work

This paper proposes and addresses the use of the multivalued Array Algebra for Knowledge Discovery in Databases. This algebra handles declarative descriptions of the multivalued data and allows a multivalued description of the knowledge contained in a data table by means of array expressions. Furthermore, the multivalued Array Algebra provides the methods and techniques necessary to the find all the array concepts of a context and construct the array lattice without identifying the extensional part of the concept by means of the n-order projection arrays

An Object Attribute Table can be described by array expressions, that is by descriptions of the data in terms of only the attribute values. From the many array expressions describing an OAT, minimum expressions can be found. This minimum expression is not unique. The ordering of the arrays in the expression and the order in which calculations are performed can change the outcome.

Acknowledgements

This work has been partially supported by the Dirección General de Investigación del Ministerio de Educación, Ciencia y Tecnología through the TIN2007-67993 project.

References

1. Fayyed, U., Piatestsky-Shapiro, G., Smyth, P.: From Data Mining to Knowledge Discovery in Databases. American Association for Artificial Intelligence, AI Magazine Fall 96, 37–54 (1996)
2. Ganter, B., Wille, R.: Formal Concept Analysis. Mathematical Foundations. Springer, Heidelberg (1999)
3. Wille, R.: Restructuring Lattice Theory: an Approach based on Hierarchies of Concepts. In: Ordered Sets, pp. 445–470. Reidel Publishing Company (1982)
4. Hereth, J., Stumme, G., Wille, R., Wille, U.: Conceptual Knowledge Discovery – a Human-Centered Approach. Applied Artificial Intelligence 17(3), 281–302 (2003)
5. Miró-Julià, M., Fiol-Roig, G.: An algebra for the treatment of multivalued information systems. In: Perales, F.J., Campilho, A.C., Pérez, N., Sanfeliu, A. (eds.) IbPRIA 2003. LNCS, vol. 2652, pp. 556–563. Springer, Heidelberg (2003)
6. Miró-Julià, M.: A New Approach for Handling Information Systems. In: Isaias, P., Kommers, P., McPherson, M. (eds.) Proceedings of the IADIS Intenational Conference. e-Society, vol. 1, pp. 549–556 (2004)
7. Miró-Julià, M.: A Contribution to Multivalued Systems. PhD thesis, Universitat de les Illes Balears (2000)
8. Miró-Julià, M.: The Zero Array: A Twilight Zone. In: Moreno-Díaz Jr., R., Pichler, F. (eds.) EUROCAST 2003. LNCS, vol. 2809, pp. 92–103. Springer, Heidelberg (2003)

Local Space-Time Systems Simulation of Linear and Non-linear Retinal Processes

Roberto Moreno-Díaz[1], Arminda Moreno-Díaz[2], and Gabriel de Blasio[1]

[1] Instituto Universitario de Ciencias y Tecnologías Cibernéticas
Universidad de Las Palmas de Gran Canaria
rmoreno@ciber.ulpgc.es
gdeblasio@dis.ulpgc.es
[2] School of Computer Science. Madrid Technical University
amoreno@fi.upm.es

1 General

The realization of local space-time models of retinal processes can be achieved by means of available typical dynamical systems simulation tools (like *Simulink*) because only a very small number of parallel channels is needed. In short, the aim is to simulate both the time and space dimensions as delay chains, where the travelling signals are available at different points of the delay chain to interact among them. These models provide an interesting and fruitful insight into the neurophysiological processes.

For the discrete case, this is equivalent to modelling a generalized shift register, which provides for the transformation of space into time. Thus, a spatial contrast of illumination (e.g. a local spatial pulse of light) is equivalent to a time pulse. For the temporal local modelling (linear and non-linear ON, OFF and ON-OFF neurons) we consider the modelling based on the interaction of fast and retarded processes. Spatial local contrast detectors are modelled by a similar mechanism [1], [2], [3].

2 Postsynaptic Inertial Delay Models

The simplest inertial delaying state equation for one state variable, x_i, is

$$\dot{x}_i = a_i(-x_i + u) \tag{1}$$

where a_i is a positive constant (representing the speed of the delay) and u is the external input. Its effect is similar to that of a RC integrating circuit.

2.1 Models Based in Two Non-linearities

Consider two processes having state variables x_1 and x_2, delayed at different "speeds" a_1 and a_2, each following equation 1. The main postsynaptic interaction of both processes consists in the linear combination of the signals x_1 and x_2

R. Moreno-Díaz et al. (Eds.): EUROCAST 2009, LNCS 5717, pp. 25–32, 2009.

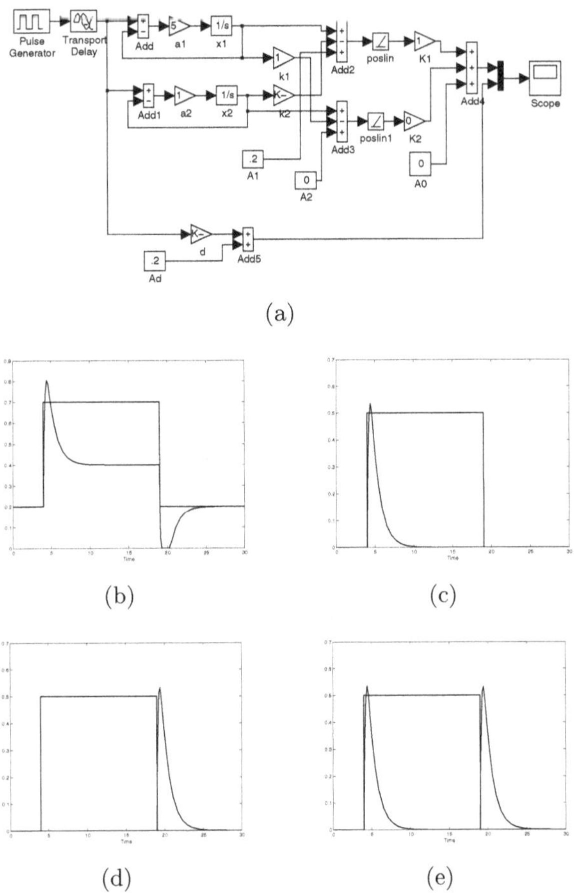

Fig. 1. (a) Simulation for the general postsynaptic two non-linearities model. Several responses for different values of the parameters: (b) Sustained ON, (c) Non sustained ON, (d) Pure OFF and (e) Non-linear ON-OFF.

resulting from said processes, followed by a non-linearity similar to a half wave rectification.

That is, we have the following general postsynaptic two non-linearities model, where it is supposed that x_1 is a fast signal and x_2 is a retarded signal ($a_1 > a_2$)

$$A_c = K_1 \cdot Pos(x_1 - k_2 x_2 + A_1) + K_2 \cdot Pos(x_2 - k_1 x_1 + A_2) + A_0 \tag{2}$$

where A_c is the total postsynaptic local activity contributing to the firing of a retinal cell; K_1 and K_2 are two positive factors which determine the nature of the process: for $K_1 \neq 0$ and $K_2 = 0$, we have an ON process. For $K_1 = 0$, $K_2 \neq 0$, an OFF process. For $K_1 \neq 0$ and $K_2 \neq 0$, an ON-OFF non-linear process is obtained, with relative weights K_1 and K_2 respectively. Finally, k_1 and k_2 are factors which determine the existence or not of an ON and/or OFF sustained response. Thus, if $k_1 = k_2 = 0$ there is no sustained response. A_0, A_1, A_2 are sustained activities due to the background illumination.

The general diagram for the model is shown in figure 1(a). The two local non-linearities acting on the linear interaction of the fast and retarded processes are weighted and added to generate the total activity A_c, that is recorded in the scope. The stimulation with a long pulse of local light (long enough to allow for stationary regime) shall produce the corresponding responses in the model for the various situations. Thus figure 1(b) is the sustained ON response, for $K_1 = 1$, $K_2 = 0$, $k_1 = 1$, $k_2 = 0.8$ and parameters $A_0 = 0$, $A_1 = 0.2$, $A_2 = 0$ and the display parameters $A_d = 0.2$ and gain $d = 0.5$. Figure 1(c) is the non-sustained ON response, for $k_1 = k_2 = 1$, $K_1 = 1$, $K_2 = 0$, and all the rest of parameters set to zero. Figure 1(d) is the pure OFF response for $k_1 = k_2 = 1$, $K_1 = 0$, $K_2 = 1$ and all the rest of parameters set to zero. Finally, the non-linear ON-OFF response is illustrated in figure 1(e), for $k_1 = k_2 = K_1 = K_2 = 1$ and the rest set to zero.

2.2 Models Based in Temporal Center-Periphery

The typical time phenomena which characterize retinal cells correspond to the detection of "edge" in time, both positive (ON effects) and negative (OFF effects). A classical edge detector in space, which corresponds to a band pass filter, is the geometrical spatial structure of center-periphery. The realization of this structure for inertial delay models requieres at least three inertial delay variables, two corresponding to the one dimensional periphery and one corresponding to the center.

If x_1 and x_3 are the state variables of the periphery and x_2 the one of the center, the linear center-periphery (postsynaptic type) of formulation gives, for the activity of the cell

$$A_c = Pos\{k_1 x_1 + k_2 x_2 + k_3 x_3\} + A_0 \tag{3}$$

For inhibitory center and excitatory surround, which corresponds to the linear ON-OFF case, we have

$$A_c = Pos\{k_1 x_1 - k_2 x_2 + k_3 x_3\} + A_0 \tag{4}$$

where k_1, k_2 and k_3 are positive. Note that the weights k_1, k_2, k_3 represent the "time kernel" of the linear part of the operation of the cell.

The general model for this case is constructed in a way similar to section 2.1. Delay parameters are $a_1 = 2$, $a_2 = 0.4$ and $a_3 = 0.2$, for state variables x_1, x_2, x_3. The kernel factors are $k_1 = k_3 = 1$ and $k_2 = -2$. Figure 2(a) shows the ON-OFF response to a long pulse of stimulus. Notice the typical properties of the inertial delay models for center-periphery: the ON response is stronger and shorter than the OFF response, although they may carry the same energy. Also, because the nature of the linearity of the kernel, the OFF response appears delayed with respect to the end of the pulse stimulus by an amount equal to the length of the ON response.

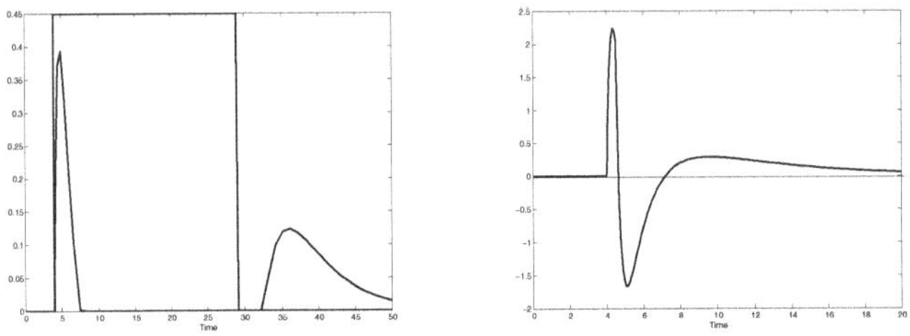

(a) (b)

Fig. 2. (a) The ON-OFF response for a long pulse of stimulus. (b) Impulsive response of the linear part of the system, where time central (inhibitory) and time peripheral (excitatory) areas are apparent.

3 Postsynaptic Transport Delay Models

These models correspond to the generation of retarded signals x_i by means of the usually available transport delay module. In this case, the expressions for processes x_i will be given by the operator D (transport delay)

$$x_i = D_i(v) = v(t - \Delta) \qquad (5)$$

where v is a low pass filtered version of the input stimulus. Δ is a time delay.

The case of two rectifying non-linearities corresponds to a signal $x = v$ and its delayed version, Dx, interacting in an "exclusive OR" fashion, which provokes an activity A_c

$$A_c = K_1 \cdot Pos\{x - k_1 Dx + A_1) + K_2 \cdot Pos(Dx - k_2 x + A_2\} + A_0 \qquad (6)$$

Again A_1, A_2 and A_0 are factors which determine the existence of a response to background illumination.

The model corresponding to this formulation proceeds similarly as before. Figure 3(a) shows the corresponding output for a sustained ON effect with background illumination. Parameters values are $K_1 = 1$, $K_2 = 0$, $k_1 = 0.8$, $k_2 = 1$ and $A_1 = 0.3$, $A_2 = A_0 = 0$, $a = 3$, $\Delta = 3$. Figure 3(b) illustrates de output of the model for a non-linear ON-OFF type of response. Parameters are now $K_1 = 1$, $K_2 = 0.8$, $k_1 = k_2 = 1$, $A_1 = A_2 = A_0 = 0$.

A chain of transport delays will provide for a shift-register type of effect, so that a temporal center-periphery structure can be modelled. If $v = x_1$ and $x_2 = Dx_1$, $x_3 = Dx_2$, the activity corresponding to the central (for x_2)-periphery (for x_1 and x_3) structure produces a total activity

$$A_c = Pos\{k_1 x_1 + k_2 x_2 + k_3 x_3\} + A_0 \qquad (7)$$

For an inhibitory negative center ($k_2 < 0$) and excitatory periphery ($k_1 > 0$, $k_2 > 0$), the linear ON-OFF type of response results. For $k_1 + k_2 + k_3 = 0$, there

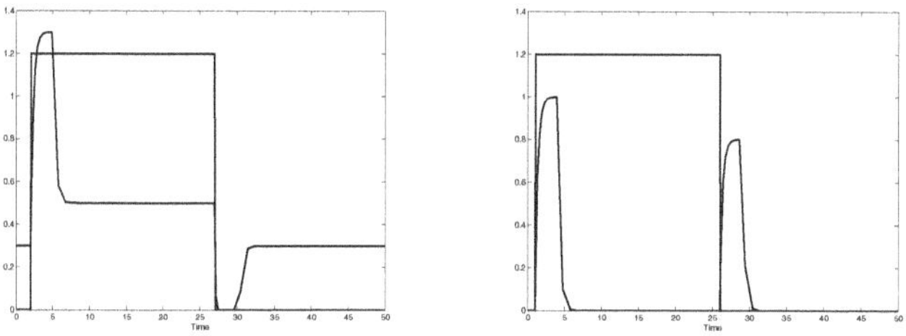

(a) (b)

Fig. 3. Responses of the general postsynaptic transport delay model of two rectifying non-linearities for different values of the parameters: (a) Sustained ON with background illumination, (b) Non-linear ON-OFF

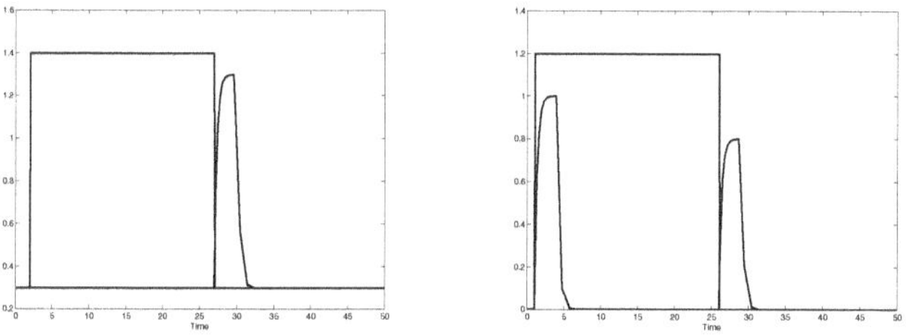

(a) (b)

Fig. 4. Temporal center-periphery transport delay model. Responses: (a) OFF and (b) ON-OFF.

is no sustained response. A_0 provides for the output to background stimulus, as before.

Figure 4(a) shows the OFF output response, corresponding to the linear kernel values $k_1 = 1,\ k_2 = -1,\ k_3 = 0,\ A_0 = 0.3,\ a = 3,\ \Delta = 3$. Figure 4(b) shows the ON-OFF response for parameters values $k_1 = 1,\ k_2 = -2,\ k_3 = 1,\ A_0 = 0.3,\ a = 3,\ \Delta = 3$.

4 Presynaptic Inhibition Models

The inhibition previous to the synapsis admits two types of non-linear formulations, divisional or shunting inhibition and exponential inhibition [4], [5]. For the first case, if signals x_1 and x_2 correspond to fast and delayed processes, then mutual inhibition provides the total activity

$$A_c = Pos\left\{\frac{K_1 x_1}{1 + k_2 x_2} + \frac{K_2 x_2}{1 + k_1 x_1} - \theta\right\} \tag{8}$$

where a threshold θ have been included.

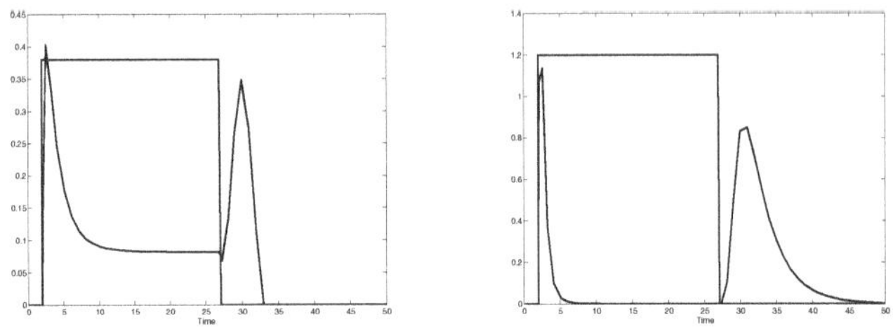

(a) (b)

Fig. 5. (a) Presynaptic divisive inhibition model ON-sustained-OFF response. (b) Presynaptic exponential inhibition model ON-OFF response.

Figure 5(a) shows the output of the model for the following parameters: $a_1 = 1$, $a_2 = 0.3$, $k_1 = k_2 = 10$, $K_1 = 13$, $K_2 = 1$ and a threshold $\theta = 0.6$. This corresponds to an ON-sustained OFF type of response, as it is shown in figure 5(a). Other types of response are for different values of the parameters, as it was the case for the previous models.

The activity for the exponential mutual inhibition of fast and retarded signals x_1 and x_2 is given by

$$A_c = K_1 x_1 \cdot exp\left(-k_2 x_2\right) + K_2 x_2 \cdot exp\left(-k_1 x_1\right) \tag{9}$$

Figure 5(b) shows a typical ON-OFF process, corresponding to parameter values $a_1 = 1$, $a_2 = 0.3$, $k_1 = k_2 = 10$, $K_1 = 13$, $K_2 = 1$.

In both cases, because the non-linearities required by the mutual inhibitions, there is no significant delay between the edges of the stimulus and the apparition of the ON and OFF responses.

5 Generation of Spike Trains to Model Ganglion Cells Outputs

The ganglion cell coding into spike trains of the slow signals processed, and represented by the activity A_c in previous sections, can be modelled and approximated in a variety of ways. Typical membrane simulations use voltage controlled oscillators (VCO) [6]. We develop here a simple spike generator model based on a VCO followed by a signal shaping subsystem, which is stimulated by the corresponding activities according to previous slow potential models.

The basic sinusoidal VCO is a second order non-linear system having state equations

$$\dot{z}_1 = z_2; \; \dot{z}_2 = -k^2(v) z_1 \tag{10}$$

where $k^2(v)$ is a positive parameter, function of the input voltage $v(t)$ and z_1, z_2 are the state variables.

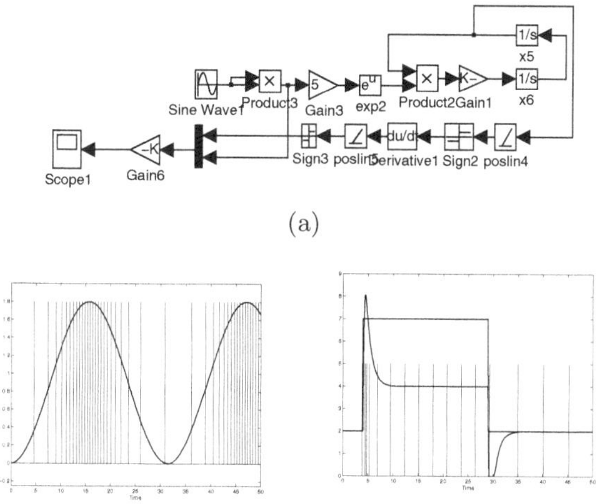

Fig. 6. (a) Simulation that models the spike generator subsystem. (b) Coded output for a positive sinusoidal signal. (c) Spike coding of a sustained-ON response with background illumination.

For a slowly varying input voltage and initial states $z_1(0) = 1,\ z_2(0) = 0$, the solution is approximately given by

$$z_1(t) = cos[k(v)t];\ z_2(t) = -k(v)sin[k(v)t] \tag{11}$$

The signal shaping subsystem (SSS) processes the almost cosinusoidal output of the VCO, $z_1(t)$, to provide for an equivalent spike-frequency signal, which is the code of $k(v)$. In the models $w^2(v)$ is obtained by emphasizing the signal by an $exp(v)$ function.

The complete spike generator is shown in figure 6(a); Figure 6(b) shows the coded output for a signal $v = sin(t)$. Figure 6(c) shows the ganglion spike response corresponding to the sustained ON with background illumination.

6 Modelling the Equivalence between Time and Space

As pointed before, a chain of shift registers transform a time signal into a space one. This can be used to model one dimensional space processes. For example, the one dimensional center-periphery contrast detector corresponds to the structure shown in figure 7(a), where the corresponding spatial kernel has the weights $(-1,\ -1,\ 2,\ 2,\ -1,\ -1)$. The double contrast (edge) detection is shown in figure 7(b), which is typical of spatial center-periphery structures. Other spatial local operations (kernels) are simulated in a similar way.

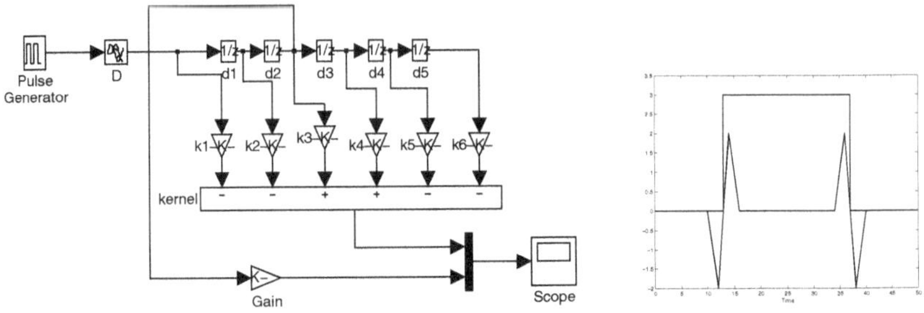

(a) (b)

Fig. 7. (a) Simulation to model one-dimensional space process. (b) Double contrast (edge) detector.

Acknowledgments

This work is supported in part by Spanish Ministry of Science and Innovation (MICINN) under grant TIN2008-06796-C04-02.

References

1. Moreno-Díaz, R., de Blasio, G., Moreno-Díaz, A.: A Framework for Modelling Competitive and Cooperative Computation in Retinal Processing. In: Ricciardi, L.M., Buonocuore, A., Pirozzi, E. (eds.) Collective Dynamics: Topics on Competition and Cooperation in the Biosciences, pp. 88–97. American Institute of Physics, New York (2008)
2. Troy, J.B., Shou, T.: The Receptive Fields of Cat Retinal Ganglion Cells in Phisiological and Pathological States: Where We Are After Half a Century of Research. Progress in Retinal and Eye Research 21, 263–302 (2002)
3. Moreno-Díaz, R., Rubio, E.: A Model for Non-linear Processing in Cat's Retina. Biological Cybernetics 37, 25–31 (1980)
4. Lettvin, J.Y.: Form-Function Relations in Neurons. Research Laboratory of Electronics, MIT Quaterly Progress Report, 333–335 (June 1962)
5. Schipperheyn, J.J.: Contrast detection in frog's retina. Acta Physiol. Pharmacol. Neerlandica 13, 231–277 (1965)
6. Hoppensteadt, F.C.: An Introduction to the Mathematics of Neurons. Cambridge University Press, Cambridge (1997)

Analytical Representation of Intrinsic Directionality in Retinal Cells

Gabriel de Blasio, Roberto Moreno-Díaz jr., and Roberto Moreno-Díaz

Instituto Universitario de Ciencias y Tecnologías Cibernéticas
Universidad de Las Palmas de Gran Canaria
{gdeblasio,rmorenoj,rmoreno}@dis.ulpgc.es

1 General

Directional sensitivity to local stimuli by retinal ganglion cells are related to processes which probably are located at the Inner Plexiform Layer of the retina, at the ganglion cells dendrites and it is the result of at least two mechanisms. First, at the ganglion dendrites, either by postsynaptic inhibition from amacrines or by presynaptic inhibition of bipolar synapses, also by amacrines. Second, there seems to be an "intrinsic" amacrine directionality by the so called "starbust" amacrines which is itself emphasized by amacrine-amacrine interaction and then transmitted, by inhibition, to presynaptic ganglia connections [1], [2], [3].

When considered globally and not in the details of the synaptic connections, directional selectivity has in general the "intrinsic" character found in starbust amacrines. In fact, directional selectivity has to be always associated to the existence of an asymmetry in location of two processes, one fast and excitatory and a second inhibitory and long lasting, the preferred direction being indicated by the arrow that points from the site of the fast excitatory process to that of the slow inhibitory one.

The sensitivity of retinal cells in a variety of situations has been shown to be well represented by the so called Newton Filters in two dimensions [4], [5], [6] which can be generalized to the continuum to two variable Hermite functions, or to cuasi-Hermite functionals.

For receptive fields having rotational symmetry (center and center-surround structures), even Hermite functionals on r^2, $(r^2 = x^2 + y^2)$, are the appropriate representation tool. For other possible receptive field configurations, the corresponding Hermite functionals are of a mixed nature.

2 Directional Selectivity in Rotational Receptive Fields

Directional selectivity will intrinsically appear whenever there are distinct excitatory and inhibitory areas of a global receptive field having strong different or long latencies.

A series of experimental studies in the last years [1], [2], [7] have shown that a particular type of amacrine cells, the starburst amacrine cells, which receive signals from earlier retinal neurons, not only supply the inhibitory signals provoking

R. Moreno-Díaz et al. (Eds.): EUROCAST 2009, LNCS 5717, pp. 33–40, 2009.

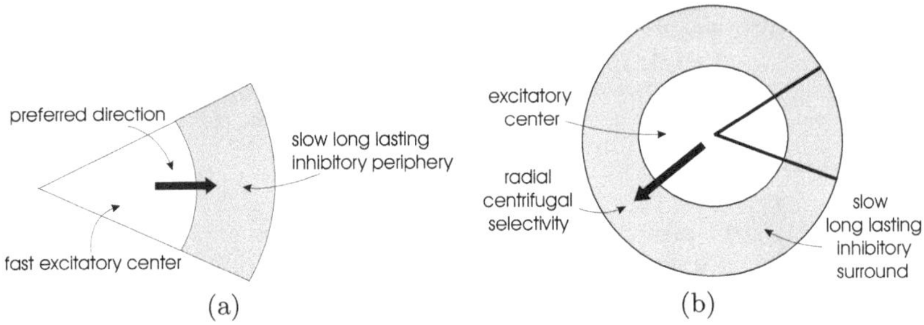

Fig. 1. Intrinsic dendrite directional selectivity appears as centrifugal selectivity for sectors which correspond to areas of fast excitatory and slow inhibitory sectorial zones (a). This is rotated to generate the total amacrine field (b).

directionality in ganglion cells but that they are directional sensitive themselves. The reported motion selectivity appears by sectors, and the optimum direction is always centrifugal.

The first obvious explanation for the existence of this centrifugal motion selectivity is to assume for each sector an asymmetric structure, the inhibitory "external" part of the field providing for slow, longer lasting effect, as shown in figure 1(a). This inhibition is probably coming from neighbouring amacrines. Summing up all sectors of the starburst cell one obtains a central excitatory fast response area surrounded by a slow longer lasting inhibitory ring, a structure which is rather classical in many of the retinal cells (figure 1(b)).

Notice that the centrifugal directional selectivity of the whole cell is a straightforward intrinsic consequence of the existence of a fast-excitatory center surrounded by a slow, longer lasting inhibitory periphery. Thus it would be expected that any retinal cell having this structure will show overall centrifugal motion selectivity. The effect can be demonstrated by using typical "mexican hat" weight profiles. It has been shown that plausible weighting functions for retinal cells, consequence of the operation of excitatory and inhibitory microprocesses, are representable by Newton filters that can be generalized to the continuum as Hermite functions of different order. For rotationally symmetric (radial) fields, the weighting Hermite profile of order n is given by

$$W_n(r) = \pm \frac{d^n}{dr^n} \left(exp\left[-r^2/2\right] \right) \tag{1}$$

where r is the distance to the center and the orden n indicates the number of inhibitory layers in the microstructure [4], [8].

The second order Hermite profile is precisely the "mexican hat", that since David Marr's Laplacian of Gaussians [9] has been accepted as higher plausible center-periphery "visual filter" rather than the alternative difference of Gaussians [10].

For R_0 the radius of the excitatory center, the radial second order Hermite profile is

$$W_2(r) = \left(1 - r^2/R_0^2\right) \cdot exp\left[-r^2/2R_0^2\right] \tag{2}$$

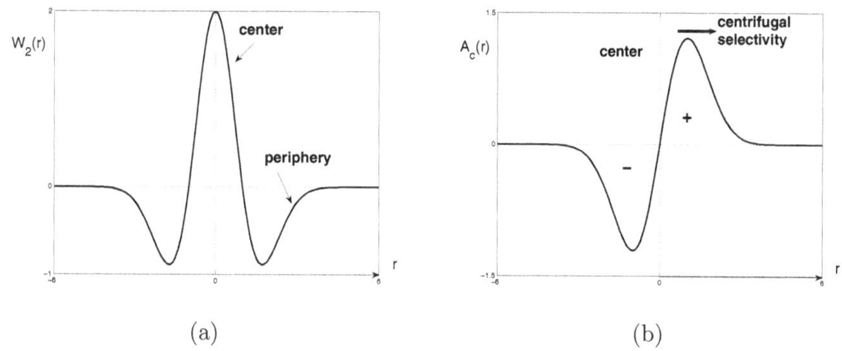

Fig. 2. (a) Center-periphery structure of receptive field, corresponding to the second order Hermite profile. (b) Cumulative activity from the field in (a) for long lasting excitatory-inhibitory effects, for a local stimulus moving diametrally from left to right. Note that excitatory activity only appears during centrifugal motion.

with $r^2 = x^2 + y^2$, which is shown in figure 2(a). For a local stimulus crossing diametrally the receptive field of the cell, and for relatively long lasting depolarizing-hyperpolarizing actions, the cumulative activity of the cell, when the stimulus reaches point r, is proportional to the integral, from the left border $(-\infty)$ to r, of $W_2(r)$, that is

$$A_c(r) = \int_{-\infty}^{r} W_2(r)\,dr = r \cdot exp\big[-r^2/2R_0^2\big] \tag{3}$$

which is the Hermite profile of order 1. $A_c(r)$ is shown in figure 2(b). It can be seen there that the cumulative activity is excitatory only during the centrifugal motion of the stimulus (right lobule of figure), even if the motion is started at the center, because the cumulative activity is null there.

Notice that the above structure that has been assumed for starburst amacrine cells, can, in principle, correspond to a whole ganglion cell (the so called macrokernels). In this case the cell will show a centrifugal directional selectivity, in addition to other local ON-OFF, contrast detector properties. Long ago, classical centripetally directional selective retinal ganglion cells (besides being locally ON-OFF) have been described for Group 2 ganglion cells in frogs [11]. It would correspond to an inhibitory-center, excitatory-surround situation, that is to $-W_2(r)$.

For macrokernels of ganglion cells, and perhaps for the intrinsic directionality of starburst amacrines, there is a possibility of a "multiple ring" structure of alternating excitatory-inhibitory zones [12], [13], provoking peculiar directional selectivity. The corresponding kernels can be represented by higher even Hermite functions. Figure 3(a) shows the fourth Hermite profile corresponding to the even radial kernel

$$W_4(r) = \big(r^4 - 6r^2 + 3\big) \cdot exp\big[-r^2/2\big] \tag{4}$$

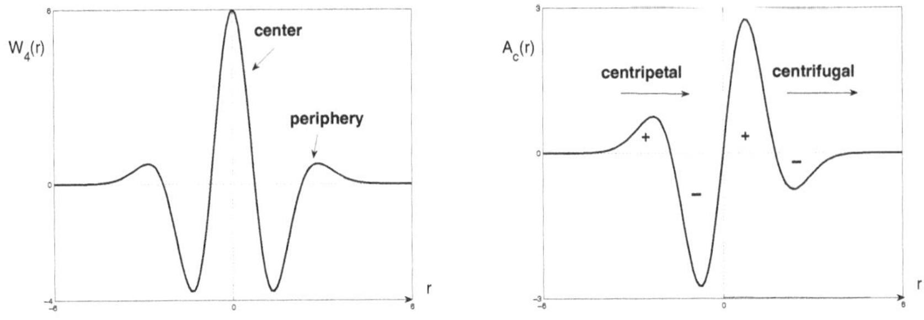

(a) (b)

Fig. 3. (a) A double ring receptive field, represented by a Hermite profile of order 4. Note the central excitatory area, surrounded by a first inhibitory ring and a second excitatory one. (b) The cumulative activity for a local stimulus moving diametrally from left to right, provokes first a sensitivity to centripetal motion and then a centrifugal selectivity.

which consists of a central excitatory zone, surrounded by an inhibitory ring plus an extra far excitatory ring area. The cumulative activity A_c for a diametrally crossing stimulus is

$$A_c(r) = \left(r^3 - 3r\right) \cdot exp\left[-r^2/2\right] \tag{5}$$

which is the third order Hermite profile, shown in figure 3(b). Notice that for said stimulus, there is centripetal selectivity at the first moment, to became centrifugal selectivity as the stimulus leave the receptive field.

3 Non Rotational Directional Selectivity

When the kernel or weighting function is not of rotational symmetry there is a variety of potential directionality properties, always depending on the latencies of the excitatory and inhibitory areas. The preferred direction points from the excitatory to the inhibitory zones. The simplest case which corresponds to a single inhibitory layer in the x direction for the Newton Filters representation, is the Hermite bidimensional kernel of order 1, given by

$$H_x = -\frac{\partial}{\partial x}[exp\,(-r^2/2)] = x \cdot exp\,(-r^2/2)$$

This kernel is represented in figure 4, where the preferred direction is as indicated.

Zonal directional selectivity appears when there are inhibitory layers in the x and/or y directions of the microstructure represented by the bidimensional discrete Newton Filters [4]. In the continuum generalization, each inhibitory microstructure in one direction corresponds to a partial derivative, giving rise to the corresponding two dimensional Hermite kernel.

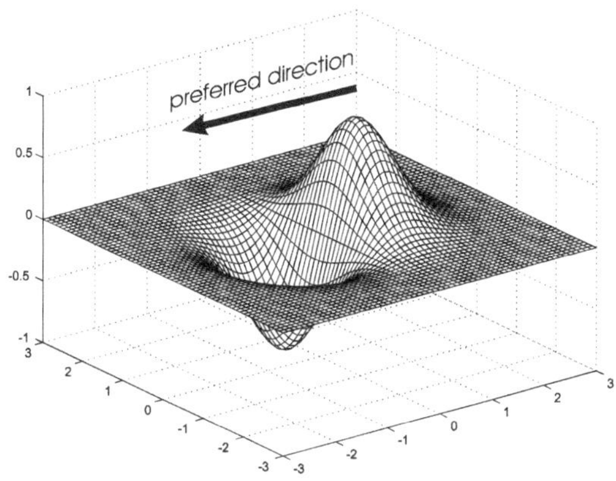

Fig. 4. Directional selectivity represented by a first order bidimensional Hermite kernel, corresponding to one microstructure inhibition in the x direction

For example, for one inhibitory layer in each direction x and y of the microstructure, it results the kernel:

$$H_{xy} = \frac{\partial^2 H_0}{\partial x \partial y} = xy \cdot exp\left(-r^2/2\right)$$

This kernel is represented in figure 5. Favored directions are from excitatory to inhibitory zones. Notice that there is a null direction normal to a optimum detection one, as it is indicated in said figure.

As an additional illustration of the potentials of two dimensional Newton Filters and the corresponding continuum Hermite kernel, consider the case of two inhibitory layers in the x direction and two in the y direction. It corresponds to the bidimensional Hermite kernel of order four:

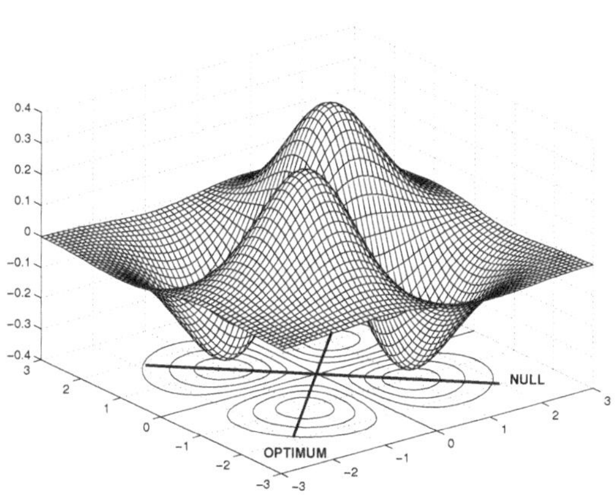

Fig. 5. Second order Hermite kernel showing normal null-optimum directions

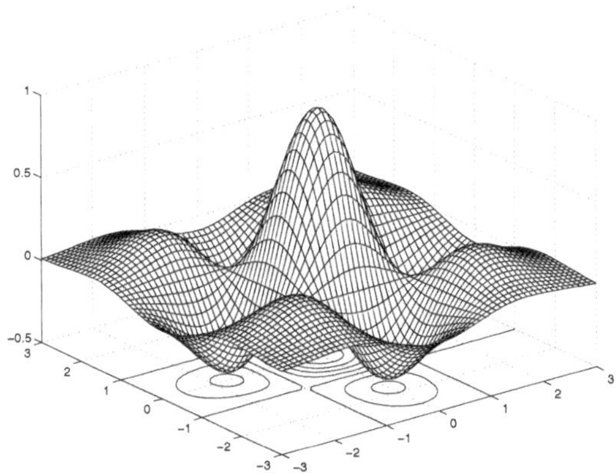

Fig. 6. Fourth order Hermite kernel with a ring of alternating excitatory and inhibitory zones

$$H_{xxyy} = \frac{\partial^4 H_0}{\partial^2 x \partial^2 y} = (x^2 - 1)(y^2 - 1) \cdot exp\,(-r^2/2)$$

This kernel is shown in figure 6, where it can be remarked the existence of a centre excitatory area surrounded by a ring having alternating excitatory and inhibitory zones. There are four favored directions and four which are centrifugal.

4 Quasi-Hermite Receptive Fields

Hermite functions are a very appropriate analytical representation for sensorial neuronal receptive fields, but in many cases, only in a simple qualitative way, excluding non-linearities.

However, their microstructural substrate, provided by discrete Newton Filters, point to convenient generalizations to the continuum to cover situations like the "widening" of the inhibitory periphery in center-periphery cuasi-linear neurons [14]; the non linear interactions between excitatory and inhibitory subfields and the existence of sustained linear and non linear responses.

These refinements, although do not change the qualitative nature of the directionality due to asymmetric Hermite kernels, provide for better approaches to the experimental results.

Widening of the inhibitory ring may be a consequence of the lowering of the resolution from center to periphery, which in the case of the Hermite kernel representation would be a type of widening the scale for the r coordinate as going from center to periphery. Figure 7 illustrates this effect for an expansion of the type $r' = r^{1.3}$, where the widening of the inhibitory ring is observed.

Other non linearities are due to the interaction of excitatory and inhibitory components of signals, which can also give rise to sustained responses under constant uniform stimulation. This is the case for a local realistic rectifying non linearity of the form:

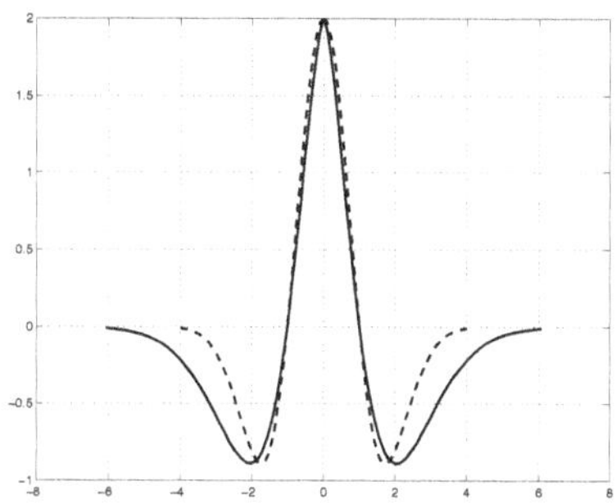

Fig. 7. Widening of the inhibitory ring as a consequence of the lowering of the resolution from center to periphery by a factor $r' = r^{1.3}$

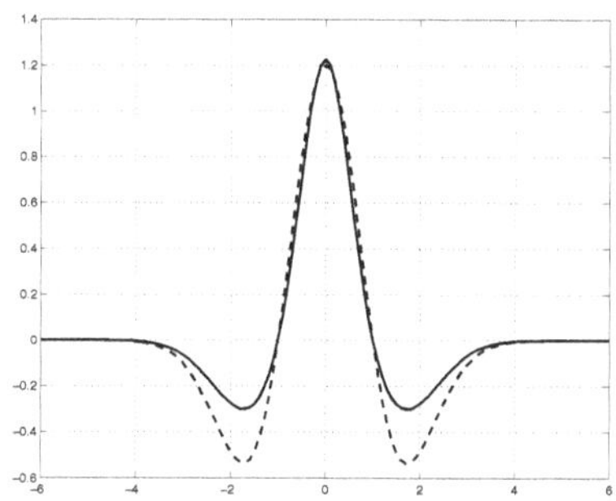

Fig. 8. Decreasing of the amplitude for the inhibitory ring in the case of the H_2 kernel

$$F[H(x,y)] = [exp(kH) - 1], \quad k < 1$$

Figure 8 illustrates the changes in the kernel for the mexican hat H_2 and $k = 0.4$. Notice the decrease in amplitude of the inhibitory ring, resulting in a non-zero positive value for the mean value of kernel F.

References

1. Fried, S.I., Masland, R.H.: Image Processing: How the Retina Detects the Direction of Image Motion. Current Biology 17(2), R63–R66 (2007)
2. Taylor, W.R., He, S., Levick, W.R., Vaney, D.I.: Dendritic Computation of Direction Selectivity by Retinal Ganglion Cells. Science 289, 2347–2350 (2000)
3. Fried, S.I., Münch, T.A.: Mechanisms and Circuitry Underlying Directional Selectivity. Nature 40, 411–413 (2002)
4. Moreno-Díaz Jr., R.: Computación Paralela y Distribuida: Relaciones Estructura-Función en Retinas. Ph.D. Thesis, Universidad de Las Palmas de G.C (1993)
5. Moreno-Diaz, R., de Blasio, G.: Systems and Computational Tools for Neuronal Retinal Models. In: Moreno-Díaz Jr., R., Pichler, F. (eds.) EUROCAST 2003. LNCS, vol. 2809, pp. 494–505. Springer, Heidelberg (2004)

6. Moreno-Díaz, R., de Blasio, G., Moreno-Díaz, A.: A Framework for Modelling Competitive and Cooperative Computation in Retinal Processing. In: Ricciardi, L.M., Buonocuore, A., Pirozzi, E. (eds.) Collective Dynamics: Topics on Competition and Cooperation in the Biosciences, pp. 88–97. American Institute of Physics, New York (2008)
7. Yoshida, K., Watanabe, D., Ishikane, H., Tachibana, M., Pastan, I., Nakanishi, S.: A Key Role of Starburst Amacrine Cells in Originating Retinal Directional Selectivity and Optokinetic Eye Movement. Neuron 30, 771–780 (2001)
8. Moreno-Diaz, R., de Blasio, G.: Systems Methods in Visual Modelling. Sys. Anal. Model Simul. 43, 1159–1171 (2003)
9. Marr, D., Vision, W.H.: Freeman and Company, San Fransisco (1982)
10. Rodieck, R.W., Stone, J.: Response of Cat Retinal Ganglion Cells to Moving Visual Patterns. J. Neurophysiol. 28, 819–832 (1965)
11. Lettvin, J.T., Maturana, H.R., McCulloch, W.S., Pitts, W.H.: What the Frog's Eye Tells the Frog's Brain? Proc. of the I.R.E. 47(11), 1940–1951 (1959)
12. Hammond, P.: Contrasts in Spatial Organization of Receptive Fields at Geniculate and Retinal Levels: Centre Surround and Outer Surround. J. Physiol. 228, 115–137 (1973)
13. Li, C.Y., Zhou, Y.X., Pei, X., Qiu, F.T., Tang, C.Q., Xu, X.Z.: Extensive Disinhibitory Region Beyond the Classical Receptive Field of Cat Retinal Ganglion Cells. Vision Res. 32(2), 219–228 (1992)
14. Troy, J.B., Shou, T.: The Receptive Fields of Cat Retinal Ganglion Cells in Phisiological and Pathological States: Where We Are After Half a Century of Research. Progress in Retinal and Eye Research 21, 263–302 (2002)

Linear Complexity Measures for Multi-valued Cryptographic Data Streams by Application of the Rissanen Partial Realization Method

Franz Pichler*

c/o Institute of Formal Models and Verification, Johannes Kepler University Linz
A 4040 Linz, Austria
franz.pichler@jku.at

Abstract. Jorma Rissanen developed in his papers [1],[2] a method to compute recursively for a matrix-valued data stream S of finite length the associated minimal linear system Σ=(F,G,H) which has S as its impulse response. The method of Rissanen is based on the fundamental algebraic theory of linear systems realization as developed earlier by the fundamental research in mathematical systems theory by the work of Rudolf Kalman [3],[4],[5]. In our presentation we show how the Rissanen method of Hankel matrix decomposition can be applied to measure the linear complexity profile of vector-valued cryptographic data streams as it is applied in stream cipher testing. Our method generalizes the well known Massey-Berlekamp algorithm which is applied in testing scalar-valued data streams. For this reason we call it the "Rissanen algorithm". Although the author has been familiar already for a long time with the realization theory of Kalman and contributed to the topic earlier [6], only recently the reported applicability in cryptographic testing of pseudorandom sequences has been found. The result presented here proves that results of mathematical systems theory and automata theory, which were developed nearly half a century ago by Rudolf Kalman, Jorma Rissanen, Michael Arbib and others are until today of scientific interest and can successfully be applied to solve engineering problem of todays interest. Jochinger [7] gives a report on the software implementation of the Rissanen Method of recursive Hankel matrix decomposition and the effective computation of partial linear systems, following [1] and [2]. A more detailed presentation of the topic discussed here, which includes also the discussion of the theory of linear systems realization, has been given earlier by Pichler [8].

1 Linear Complexity Measures in Cryptography

An important task in the design of stream cipher devices is to measure the linear complexity profile of the pseudo random sequences which are used in Vernam-like cryptographic systems for mixing with the plaintext data stream. In the following we give a short description of this task of taking measurements.

* Professor Emeritus (Systems Theory).

R. Moreno-Díaz et al. (Eds.): EUROCAST 2009, LNCS 5717, pp. 41–46, 2009.

Let $SM=S(0),S(1),...,S(M-1)$ denote a random sequence of length M with values in $GF(q)^p$. The goal is to compute to *SM* an associated autonomous linear state machine ALFSM=(F,H) of minimal dimension which generates *SM* from a certain initial state $x(0)$.We then say ALFSM „realizes“ SM. Let n(M) denote the dimension of the state space of the ALFSM which realizes *SM*. The function $L:N_0 \rightarrow N_0$ which is given by $L(M):=n(M)$ is called the linear complexity profile of the sequence $S = S(0),S(1),S(2),....$ The function *L* is not decreasing. It stays constant from a point *M* on if $n(M)=n(M+1)$. Only in this case we have full knowledge of *L*. In cryptography it is desired to be able to compute $L(M)$ for very long sequences *SM*.

The Massey-Berlekamp algorithmus [9] computes for scalar sequences *S* with values in $GF(q)$ the linear complexity profile *L*. Since it is desired to compute the linear complexity profile for a extremely long interval the computation has to be computational effective by a recursive procedure to compute $L(M+1)$ from $L(M)$ and $S(M)$. The Massey-Berlekamp algorithm fulfills this.

The method used is based on polynomial presentation of sequences by the *D*-transform (Laurent expansion of sequences) which is common in shift register theory. As a result the Massey-Berlekamp algorithm computes a realizing ALFSM which turns out in this case to be a autonomous linear feedback shift register ALFSR of minimal length $n(M)$. Our goal is to provide for the computation of the linear complexity profile of vector-valued sequences *SM* over $GF(q)$ a computational effective method. To reach this we discuss in the following shortly the method of linear realization as developed originally by the work of Rudolf Kalman [3],[4],[5].

2 Computation of a Linear System Realization

The algebraic theory of linear systems realizations deals with determination of a minimal linear system $\Sigma=(F,G,H)$ for a given (observed) impulse response $A:T \rightarrow \boldsymbol{M}(p \times m)$.T denotes the time scale and $\boldsymbol{M}(p \times m)$ is the set of matrices of size $p \times m$ over a field $\boldsymbol{K}$. The impulse response A can be interpreted as a multiple I/O experiment on Σ.

If $T=\boldsymbol{R}$ (continuous time) and $\boldsymbol{K}=\boldsymbol{R}$, then Σ is a linear differential system

if $T=\boldsymbol{Z}$ (discrete time) and $\boldsymbol{K}=\boldsymbol{R}$, Σ is a linear difference system

if $T=\boldsymbol{Z}$ (discrete time),and $\boldsymbol{K}=GF(q)$,we have for Σ a linear finite state machine LFSM.

In the digital world of cryptography the interest has the case of linear finite state machines LFSM=(F,G,H). In this case the impulse response A of Σ is given by $A=A(0),A(1),...$ where $A(k)$ are matrices with elements in $GF(q)$.

It can be shown that the impulse response A of a discrete time linear system $\Sigma=(F,G,H)$ can generally be expressed by

$$A=(HG,HFG,HF^2G,HF^3G, ...) \tag{1}$$

The linear system realization problem has the goal to determine a minimal linear system(F,G,H) which meets equation (1). For a solution of the linear realization problem we have at first to determine the state space Q of Σ. Let $f: U \rightarrow Y$ denote the

„zero state" I/O function of Σ which assigns to each input function u with finite support („input word") the associated output function $y=f(u)$ („output word").Then the state space Q of Σ can be constructed by the quotient space

$$Q=U/ker(f) \tag{2}$$

3 Linear Partial Realizations

In case of the linear partial realization problem we have the task to determine $\Sigma=(F,G,H)$ which generates a impulse response of finite length M which is given by

$$AM=(A(0),A(1),...,A(M-1)) \tag{3}$$

In this case the I/O function $f(N):U(N)\rightarrow Y(N)$ is restricted to the set $U(N)$ of input functions u and to the set $Y(N)$ of output functions y of length N, respectively. The Hankel matrix $H(N)$ of size $N\times N$ associated to (3) shows the values $f(N)(eik)$ of the I/O function $f(N)$ for the set of basis vectors eik $(i=0,1,...,N\text{-}1,\ k=1,2,...,m)$ of $U(N)$.

$$H(N)=\begin{bmatrix} A(0) & A(1) & A(2) & \dots & A(N-1) \\ A(1) & A(2) & A(3) & \dots & A(N) \\ \dots & \dots & \dots & \dots & \dots \\ \dots & \dots & \dots & \dots & \dots \\ A(N-1) & A(N) & \dots & \dots & A(M-1) \end{bmatrix} \tag{4}$$

It can be shown that the dimension $n(M)$ of the state space Q of the LFSM which realizes AM can be determined by $n(M)=rank\ H(N)$ under the condition that we have $rank\ H(N)=rank\ H(N+1)$.

In the following we give the steps which are necessary to compute a partial linear realization. We follow in our discussion the method which is given by the textbook of Padulo-Arbib [10].

Since $Q=U(N)/ker(f(N))$ we derive from rank $H(N)=n$ the result $Q=\boldsymbol{K}^n$

Let $X=\{x1,x2,x3,...,xn\}$ denote a basis for Q which we chose by taking n unit input functions $e1,e2,...,en$ which correspond to n linear independent rows $f(ei)$ of $H(N)$.For any input function u we denote by $[u]$ the representation of u by a column vector in the basis X. Then $xi=[ei]$ for $i=1,2,...,n$. It can be shown that the matrices F,G,H of the partial realization Σ can be determined by

$$\begin{gathered} F=[[e10],[e20],...,[en0]] \\ G=[[e1],[e2],...,[em]] \\ H=[f(N)(e1)(0),f(N)(e2)(0),...,f(N)(en)(0)] \end{gathered} \tag{5}$$

($ei0$ denotes the input function of length $N+1$ which is given by concatenation of ei with 0).

4 Determination of Linear Complexity Measures by Partial Linear Realization

The solution of the linear realization problem allows a solution of the cryptanalytic problem of section 1 in the following way: For $m=1$ (scalar input) the finite length impulse response AM of a LFSM is a vector-valued sequence $AM=(A(0),A(1),...,A(M-1))$ with $A(k)\in GF(q)^p$. We consider AM as identical to a partial pseudorandom stream SM by $AM=SM$. If $\Sigma=(F,G,H)$ is for AM the solution of the linear partial realization problem then (F,H) generates from the initial state $x(0)=Ge$ the finite length sequence SM as output (e denotes the unit input word of length 1 which is given by $e=1$). Since G is in the case $m=1$ a column vector of length n, we get $x(0)=G1=G$. We see that the application of the method of linear partial realization determines for a given p-valued finite sequence SM the minimal ALFSM$=(F,H)$ and the initial state $x(0)=G$ of the ALFSM.

This result seems is to our knowledge new in public cryptologic research. To get this result it was only necessary to know the result of Kalman´s realization theory and to consider the (trivial) fact that the impulse response AM is defined only from time 1 on. By time-invariance of a linear system Σ and by the fact that $\Sigma=(F,G,H)$ operates with zero-input as the corresponding autonomous linear System ALFSM$=(F,H)$ the above solution is rather trivial.

5 Effective Computation of Linear Partial Realizations by the Method of Rissanen

Jorma Rissanen (IBM research CA and Stanford University, now with MDL-Research, Tampere, Finland, expert in „Statistical Modeling) developed in the late 1960´s for the computation of partial realizations an effective method to compute recursively stepwise by the length M of an observed impulse response AM the associated linear partial realization $\Sigma M=(FM,GM,HM)$ and also its dimension $n(M)$.The computation of $\Sigma M+1$ depends by the method of Rissanen only on the result ΣM and the last value $A(M)$ of $AM+1$. For the case of a scalar impulse response this was shown by Rissanen in the paper [1]. For the more general matrix-valued case of impulse response, that is the case of multivariable I/O system Σ, this method is discussed by Rissanen-Kailath [2].

For the author no implementation of the Rissanen method was available. Dominik Jochinger implemented by the request of the author the method which is reported in [7]. Since the application in cryptography was the main goal, the implementation was done for finite fields $\boldsymbol{K}=GF(q)$ such that a computed linear system $\Sigma=(F,G,H)$ becomes a linear finite state machine LFSM. Examples of computation prove the necessary effectiveness which is required for determining linear complexity measures for realistic stream cipher data. In the future the Rissanen method will be included in the already existing “Crypto Workbench” of Dominik Jochinger and will serve there for applications in cryptography.

6 Generalization of the Massey-Berlekamp Algorithm

Since the Rissanen method of partial linear realization is computational effective, which compares to the Massey-Berlekamp algorithm, it can be applied for the computation of the linear complexity profile L of a vector-valued random sequence $S=S(0),S(1),S(2),\ldots$For the case of cryptographic testing as discussed in section 4 we call the Rissanen method Rissanen algorithm. The Rissanen algorithm generalizes for vector-valued sequences of data the Massey-Berlekamp algorithm. The above result has been reported in more detail in an earlier paper [8].

7 Conclusion and Final Remarks

Modern cryptographic devices for fast stream cipher systems need pseudo random generators which generate vector-valued (Byte-oriented) sequences to be ⊕-mixed with the plain text data to get the cipher text. For getting the linear complexity profile of sequences by cryptographic testing the Rissanen algorithm can be applied. The Rissanen algorithm computes to a given vector-valued digital signal for „windows" of length M (overlapping or non-overlapping) an associated set of parameters which are given by the matrices F,G,H of the linear system Σ which allow a reconstruction of the signal. Other applications in data compression, signal classification and data coding seem to be possible. Australian outback-patents by Pichler-Kookaburra are in sincere consideration.

Of systems theoretical interest would be the determination of a basis for the state space Q of the computed LFSM=(F,G,H) such that F becomes rational canonical form, which means that the LFSM consists of a parallel composition of autonomous LFSR´s.

This paper proves that „classical" topics of mathematical systems theory, such as the „Algebraic Theory of Linear Systems" as developed by Rudolf Kalman and others have after nearly fifty years still the „power" to lead to new applications of current interest. The „linear realization method", which is a part of the theory, was originally developed for applications in control theory. I myself was interested in research and teaching linear systems realization some time ago [6],[12]. Today this method seems to be rather neglected in academic education.

By the family visit of the author during the (wet) summer 2008 at Forresters Beach, Central Coast, Australia, and with the valuable support of my fellow L. Kookaburra there, the here reported application was found [11]. As a work which comes close to our results the paper [13] should be mentioned. There in connection with the Massey-Berlekamp algorithm reference to realization theory is given, however the authors seem to have not seen the possibility of a generalization to vector-valued sequences.

I would like to thank Professor Jorma Rissanen, Tampere, Finland, for his interest in the paper and my former student Dominik Jochinger for his cooperation in implementing the Rissanen method.

References

1. Rissanen, J.: Recursive Identification of Linear Systems. SIAM Journal on Control 9(3), 420–430 (1971)
2. Rissanen, J., Kailath, T.: Partial Realization of Random Systems Automatica, vol. 8, pp. 389–396. Pergamon Press, Oxford (1972)
3. Kalman, R.E.: Mathematical description of linear dynamical systems. SIAM Journal on Control, 152–192 (1963)
4. Kalman, R.E., Falb, P.L., Arbib, M.A.: Topics in Mathematical Systems Theory. In: Algebraic theory of linear systems, ch. 10. McGraw Hill, New York (1969)
5. Ho, B.L., Kalman, R.E.: Effective construction of linear state-variable models from input/output functions, Regelungstechnik, Oldenbourg, pp. 545–548 (1966)
6. Pichler, F.: General Dynamical Systems: Construction and Realization. In: Mathematical Systems Theory-Udine 1975. Lecture Notes in Economics and Mathematical Systems, pp. 393–408. Springer, Berlin (1976)
7. Jochinger, D.: A Software Implementation of the Rissanen Method for partial linear systems realization. In: Moreno-Díaz, R., et al. (eds.) EUROCAST 2009. LNCS, vol. 5717, pp. 47–52. Springer, Heidelberg (2009)
8. Pichler, F.: Effective Computation of Cryptanalytic Measures for Stream Cipher Data by the Rissanen Algorithmus. Revista de la Accademia. Canaria de Ciences, XIX (Núms. 1-2), pp. 9-22 (2007)
9. Massey, J.: Shift register synthesis and BCH decoding. IEEE Trans. on Information Theory IT-15, 122–127 (1967)
10. Padulo, L., Arbib, M.: System Theory. An Unified Approach to Continuous and Discrete Systems. Hemisphere Publishing Corporation, Washington D.C (1974)
11. Pichler, F., Kookaburra, L.: Forresters Beach Notes, Forresters Beach, Central Coast, NSW, Australia (manuscript) (February 2008)
12. Pichler, F.: Realisierung linearer Input-Output Prozesse I: Diskrete Prozesse. Technischer Bericht SYS-PED 1, Lehrkanzel für Systemtheorie, Universität Linz, Dezember, 35 pages (1974) (in German)
13. Jonckheere, E., Ma, C.: A Simple Hankel Interpretation of the Massey-Berlekamp Algorithm. Linear Algebra and its Applications, 65–76 (1989)

A Software Implementation of the Rissanen Method for Partial Linear Systems Realization

Dominik Jochinger

Johannes Kepler University Linz, Altenberger Str. 69, A-4040 Linz
dominik.jochinger@jku.at

Abstract. This work deals with the software solution for calculating the minimal partial realization of a discrete multi-variable linear system. The Rissanen method for the computation of the partial linear system is proposed. This method is based on a recursive Hankel matrix decomposition. It is implemented in the JAVA programming language to determine the linear system $\Sigma=(F,G,H)$ over a finite field $K=GF(q)$. The method is illustrated by some cryptological experiments. The Rissanen method generalizes the Massey-Berlekamp algorithm.

1 Introduction

The systems-theoretical problem of effective computing a linear system (F, G, H) from an infinite sequence given by the impulse response was first considered by Ho and Kalman in [1]. We consider here "partial realization" to compute for a given finite sequence $A(N)=(A_0, A_1, A_2, \ldots, A_{N-1})$ as given by the impulse response the minimal linear system (F, G, H) which generates $A(N)$. An efficient recursive method for partial linear systems realization method for the scalar case has been derived earlier by the paper of Rissanen [2]. An extension of the method to multi-variable sequences was presented by Rissanen and Kailath in [3]. In this paper we present a software implementation of the Rissanen method for partial realization of linear finite state machines. To our knowledge, such an implementation is currently not available in open sources (e.g. Mathematica, Matlab and others).

2 Hankel Matrix Decomposition

The main software component of the method is the factorization of a Hankel matrix A.

A Hankel matrix, named after Hermann Hankel, is a square matrix with constant skew diagonals. An example for a 4×4 Hankel matrix with entries in $GF(2)$ is given in Fig. 1.

The input of the Rissanen method is the Hankel matrix $A(m,N)$ of size N derived from a finite impulse response sequence of length $m+N-1$.

The Rissanen algorithm is based on a factorization of the Hankel matrix $A(m,N)$ of the following type:

R. Moreno-Díaz et al. (Eds.): EUROCAST 2009, LNCS 5717, pp. 47–52, 2009.

$$A(m,N) = P(m,m).Q(m,N),\ N \geq m,\ rank(A) \geq m-1 \tag{1}$$

Where *P(m,m)* is a lower triangular matrix with 1's on the diagonal.

$$\begin{bmatrix} A_1 & A_2 & & A_N \\ A_2 & \cdot^{\cdot^{\cdot}} & & \\ & & \cdot^{\cdot^{\cdot}} & \\ A_m & & & A_{m+N-1} \end{bmatrix} = \begin{bmatrix} 1 & 0 & 0 & 0 \\ p_{21} & 1 & 0 & 0 \\ \dots & & 1 & 0 \\ p_{m1} & \dots & p_{m,m-1} & 1 \end{bmatrix} \cdot \begin{bmatrix} q_{11} & q_{12} & \dots & q_{1N} \\ q_{21} & q_{22} & & q_{2N} \\ \dots & & & \dots \\ q_{m1} & q_{m2} & \dots & q_{m,N} \end{bmatrix} \tag{2}$$

The problem of factorization of the Hankel matrix *A* is reduced to recursively solving algebraic equations, where in each step a single row for each matrix *P* and *Q* is computed.

A detailed explanation of the factorization algorithm is given in [2] and also in [4]. The three matrices *F*, *G*, *H* which form the minimal partial realization can be determined by the matrices *P* and *Q* of the factorization algorithm [2],[4]. For a detailed description of the algebraic theory of linear realization of partial realization we advise also to consult [5].

$$A(4,4) = \begin{bmatrix} 0 & 1 & 0 & 1 \\ 1 & 0 & 1 & 0 \\ 0 & 1 & 0 & 0 \\ 1 & 0 & 0 & 1 \end{bmatrix}$$

Fig. 1. Hankel matrix representation over GF(2)

3 Rissanen Method for PQ-Decomposition

The recursive Rissanen method determines how *N* has to be increased at each step and for which values of *N* the representation needs to be recalculated. The extension for the multi-variable case as proposed in [3] is also part of our software implementation. In this case the impulse response consists of *p x q*-matrices. The flow diagram of Rissanen method as given in [3] is shown in Figure 2.

The goal is, to compute from the given partial impulse response *A* of length *M* the minimal linear system LS=*(F, G, H)* which generates *A*. *(F, G, H)* is called the linear realization of *A*.

The algorithmic steps can be described for short as follows:

An impulse response of length *N* is given. The algorithm starts with the smallest possible Hankel matrix. In each iteration the method tries to find a decomposition of the Hankel matrix. If the last row of *Q* is zero, just the column of the Hankel matrix is increased by one. Otherwise another row and column is added to the Hankel matrix, and the decomposition is recalculated.

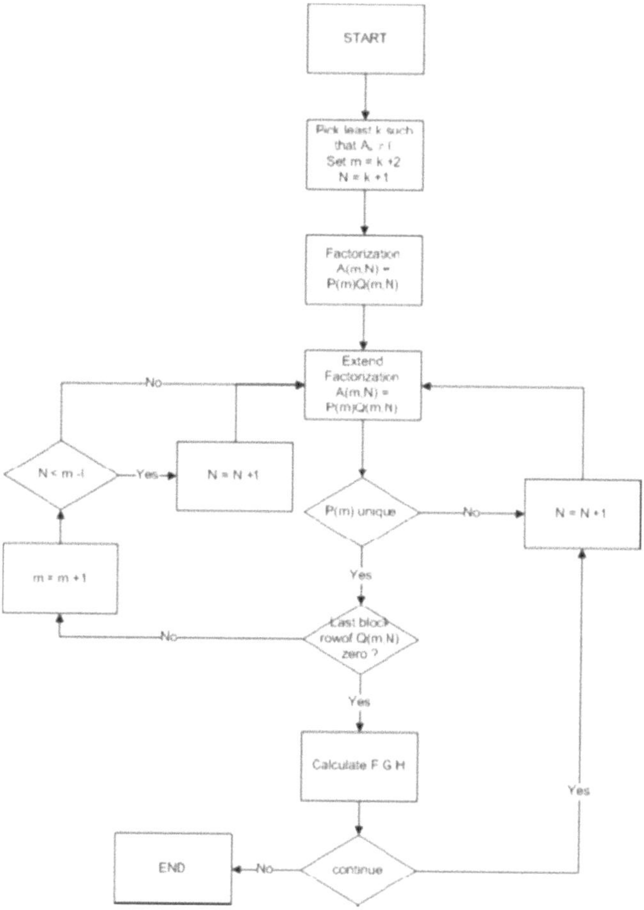

Fig. 2. Flowchart of the Rissanen method [3]

4 Experimental Results for Applications in Cryptology

In [5], [6] Pichler has shown how to apply the Rissanen method to determine the linear complexity profile of a vector-valued random sequence.

The Rissanen method has been implemented in the object-oriented programming language JAVA in order to perform some cryptological experiments.

We show the results obtained by applying the proposed Rissanen method to different finite sequences generated by pseudo random generators. For each experiment the linear complexity profile is graphed by plotting the dimension of the computed matrix *F*.

All tests were performed on an Intel Quad Core 2Ghz computer.

In the first experiment we determined the minimal partial realization for a scalar valued output of an MLFSR over *GF(2)* of length 5 with its characteristic polynomial:

$$x^5 + x^2 + 1 \tag{3}$$

The maximum period of a MLFSR generator of this length is 2^5-1.

The scalar values of the impulse response of the MLFSR are:

1001101001000001010110...

Applying the Rissanen method to the impulse response gives the following solution:

$$F=\begin{bmatrix}0&1&0&0&0\\0&1&1&0&0\\1&1&0&1&0\\0&1&0&1&1\\0&0&0&1&0\end{bmatrix},G=\begin{bmatrix}1\\0\\0\\0\\0\end{bmatrix},H=\begin{bmatrix}1&0&0&0&0\end{bmatrix} \quad (4)$$

Dimension of the matrix F is the linear complexity.

Verification of the result is given as

$$\begin{aligned}&A_1=H\cdot F^0\cdot G=1\\&A_2=H\cdot F^1\cdot G=0\\:\ &A_3=H\cdot F^2\cdot G=0\\&A_4=H\cdot F^3\cdot G=1\\&A_5=H\cdot F^4\cdot G=1\\&A_6=H\cdot F^5\cdot G=0\\&\vdots\\&A_{16}=H\cdot F^{15}\cdot G=1\end{aligned} \quad (5)$$

Fig.3 shows a linear complexity profile obtained form an impulse response of a LFSR of length 5.

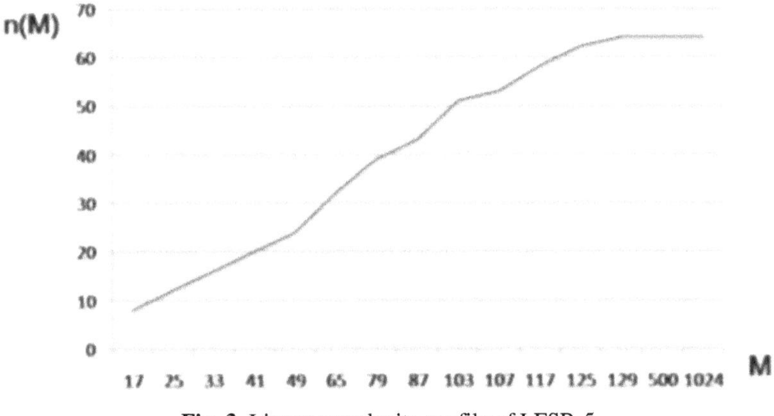

Fig. 3. Linear complexity profile of LFSR-5

To demonstrate the Rissanen method for vector valued impulse response A we generate a vector valued output, as follows:

$$\begin{bmatrix}1\\0\\1\\0\end{bmatrix},\begin{bmatrix}0\\0\\0\\0\end{bmatrix},\begin{bmatrix}0\\0\\0\\1\end{bmatrix},\begin{bmatrix}1\\1\\0\\0\end{bmatrix},\begin{bmatrix}0\\1\\0\\0\end{bmatrix},\begin{bmatrix}1\\1\\0\\1\end{bmatrix},\begin{bmatrix}1\\0\\1\\1\end{bmatrix},\begin{bmatrix}1\\1\\1\\0\end{bmatrix},\begin{bmatrix}1\\0\\1\\0\end{bmatrix},\begin{bmatrix}0\\0\\1\\1\end{bmatrix},\begin{bmatrix}0\\0\\0\\1\end{bmatrix},\begin{bmatrix}0\\0\\0\\1\end{bmatrix},\begin{bmatrix}1\\1\\0\\1\end{bmatrix},\begin{bmatrix}0\\0\\0\\1\end{bmatrix},\begin{bmatrix}0\\0\\0\\1\end{bmatrix},\begin{bmatrix}0\\0\\1\\0\end{bmatrix},\ldots \quad (6)$$

We have q=1 and p=4. Using the Rissanen method and by considering A as a random sequence S, we obtain the following linear complexity profile:

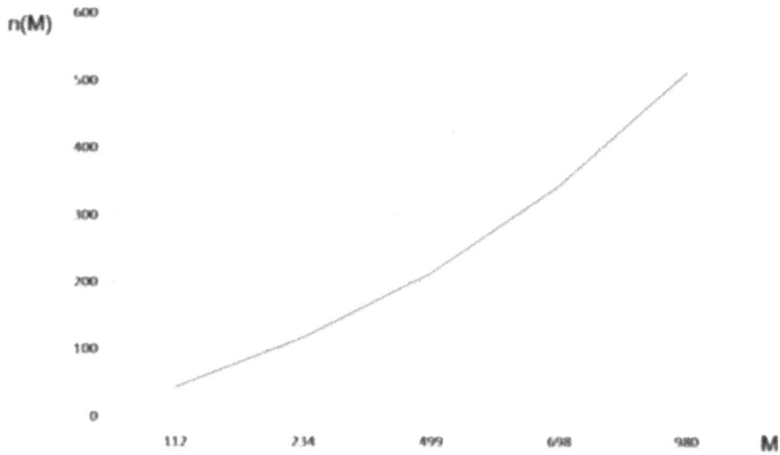

Fig. 4. Linear complexity profile of vector valued data stream

The computation of the partial minimal realization $\Sigma(20)=(F(20),G(20),H(20))$ for the partial impulse response $A(20)$ of length 20 has the following result:

$$F=\begin{bmatrix}
0&0&1&1&1&0&0&0&0&0&0&0&0&0&0&0\\
0&1&0&1&1&1&0&0&0&0&0&0&0&0&0&0\\
0&1&1&0&0&0&1&0&0&0&0&0&0&0&0&0\\
0&0&1&0&1&1&1&1&0&0&0&0&0&0&0&0\\
0&0&0&0&0&0&0&0&1&0&0&0&0&0&0&0\\
0&0&0&0&0&0&0&0&0&1&0&0&0&0&0&0\\
0&0&1&0&0&0&1&0&0&1&1&0&0&0&0&0\\
1&1&1&0&1&1&1&0&1&0&0&1&0&0&0&0\\
0&0&0&0&1&0&0&0&0&0&1&1&1&0&0&0\\
0&0&0&0&1&1&1&0&0&1&0&1&1&1&0&0\\
0&0&0&0&0&1&0&0&1&0&0&0&1&0&1&0\\
0&0&0&0&1&0&0&0&0&0&0&0&1&0&1&1\\
0&0&0&0&0&0&0&0&0&0&0&1&1&1&1&0\\
0&0&0&0&0&0&0&0&0&0&0&0&0&0&1&0\\
0&0&0&0&0&0&0&0&0&0&0&0&1&0&1&0\\
0&0&0&0&0&0&0&0&0&0&0&0&0&1&0&1
\end{bmatrix}
\qquad
G=\begin{bmatrix}1\\0\\0\\0\\0\\0\\0\\0\\0\\0\\0\\0\\0\\0\\0\\0\end{bmatrix}$$

$$H=\begin{bmatrix}
1&0&0&0&0&0&0&0&0&0&0&0&0&0&0&0\\
0&1&0&0&0&0&0&0&0&0&0&0&0&0&0&0\\
1&1&1&0&0&0&0&0&0&0&0&0&0&0&0&0\\
0&0&0&1&0&0&0&0&0&0&0&0&0&0&0&0
\end{bmatrix}$$

Fig. 5. (F,G,H) at M = 20

Observe the tubular form of the matrix F.

The autonomous linear finite state machine *(F, H)* generates from its initial state *x(0)=G* the partial random sequence *S(20)*.

5 Conclusion

The calculation of the minimal partial realization of a discrete multi-variable linear system has been formulated and solved in a software implementation. The extension for vector-valued inputs as proposed in [3] is also part of the software implementation.

It was demonstrated that the Rissanen method works also for linear systems over a finite field *K=GF(q)*.

As reported in [5], [6] the Rissanen method for computing the partial linear systems realization by its effective software implementation has recently found an important application in cryptology for the cryptological evaluation of random sequences. It extends the Massey-Berlekamp algorithm to vector-valued sequences.

In the future, we will investigate further applications using the software implementation of the Rissanen method.

References

1. Ho, B.L., Kalman, R.E.: Effective construction of linear state-variable models from input/output functions, pp. 545–548. Regelungstechnik, Oldenbourg (1966)
2. Rissanen, J.: Recursive Identification of Linear Systems. SIAM Journal on Control 9(3), 420–430 (1971)
3. Rissanen, J., Kailath, T.: Partial Realization of Random Systems Automatica, vol. 8, pp. 389–396. Pergamon Press, Oxford (1972)
4. Padulo, L., Arbib, M.: System Theory. In: An Unified Approach to Continuous and Discrete Systems. Hemisphere Publishing Corporation, Washington (1974)
5. Pichler, F.: Effective Computation of Cryptanalytic Measures for Stream Cipher Data by the Rissanen Algorithmus. Revista de la Accademia. Canaria de Ciences XIX (Núms. 1-2), 9–22 (2007)
6. Pichler, F.: Linear Complexity Measures for Multi-valued Cryptographic Data Streams by Application of the Rissanen Partial Realization Method. In: To be presented at EUROCAST 2009, Las Palmas (February 2009)

New Frontiers in the Validation of Simulation Models–Structural Dominance Analysis

Markus Schwaninger and Stefan Groesser

University of St. Gallen, Institute of Management,
Dufourstrasse 40a, 9000 St. Gallen, Switzerland
markus.schwaninger@unisg.ch

Abstract. Building better models is crucial for coping with complexity in general, and for the management of organizations in particular. This paper discusses the epistemological aspects of model validation for the achievement of high-quality models. Then it provides an overview of validation methods. The logic of validation is demonstrated by introducing the Structural Dominance Test as a means for testing the correspondence of the structural dominance between model and reality.

Keywords: Modeling, Simulation, Validation, Validity, Model Quality, System Dynamics, Structural Dominance Analysis.

1 Introduction

Validation is the process by which the correspondence between model system and real system is systematically enhanced. It consists in gradually building confidence in the usefulness of a model by applying validation tests. In principle, validation pervades all phases of the modeling process, and, in addition, reaches into the phases of model use and implementation.

In this contribution, the issue of validation is addressed with respect to dynamic models of social systems. Special reference will be made to System Dynamics models. System Dynamics is a methodology for the modeling and simulation of complex, dynamic systems [3]. It is particularly adequate for modeling socio-technical systems, such as private and public enterprises, communities, etc. They are structured as meshes of interconnected feedback loops. Causal relationships, delays, and closed-loop structures are characteristic of System Dynamics models.

Simulation is a way of experimenting with mathematical models to gain insights and then employ them to improve the real system under study. Validity in this context consists of a stringent correspondence between an abstract model system and a concrete "real" system. We will concentrate on the crucial philosophical underpinnings and a canon of methods for model validation.

2 Philosophy of Model Validation

One of the frequent convictions about science is the obsessive idea that proofs are the touchstone of the validity of both theories and models. To orientate model validation,

R. Moreno-Díaz et al. (Eds.): EUROCAST 2009, LNCS 5717, pp. 53–62, 2009.

we follow a different rationale. We argue for the adoption of the philosophical position of critical rationalism, a philosophical position founded by Karl R. Popper.

Critical rationalism posits that, in the social domain, theories can never be definitely proved, but can only reach greater or lesser levels of truth. Scientific proofs are confined to the realm of the formal sciences, namely logic and mathematics [12, 13]. As Popper demonstrates, all theories are provisional. As a consequence, the main criterion for the assessment of a theory or model's truth status is falsification (see also: [17]).

Popper's refutationist concept (as opposed to a verificationist concept) of theory-testing implies both an evolutionist perspective and an empiricist stance. The evolutionist perspective is primary because it welcomes the challenges posed to a theory, since the attempts at falsification lead to an evolutionary process: Successful falsification efforts result in revisions and improvements of the theory. Correspondingly, empiricism is paramount in the social sciences, because the main source for the refutation of a theory is empirical evidence. However, falsification can also be grounded in logical arguments where empirical evidence cannot be obtained.

3 Validation Methods

For the enhancement of model validity, a considerable set of qualitative and quantitative tests has been developed. We give an overview of the types of tests developed for System Dynamics models. Three domains of tests are expounded: the model-related context, model structure, and model behavior.[1]

3.1 Tests about Model-Related Context

These tests deal with aspects related to the situation in which the model is to be developed and embedded. They imply meta-level decisions which have to be taken in the first place, before engaging in model-building. Applied ex-post-facto, i.e., after modeling, they allow for assessing the utility of the modeling endeavor as such.

Examples: Issue Identification Test, Adequacy of Methodology Test, System Configuration Test, System Improvement Test.

3.2 Tests of Model Structure

Tests of model structure refer to the "nuts and bolts" of System Dynamics modeling, i.e., to the concepts and interrelationships which represent the real system. Model structure tests - direct and indirect - aim to increase confidence in the structure of the theory created to assess the behavior mode of interest. The model structure can be assessed by means of either direct or indirect inspection. Tests of model structure assess if the logic of the model is attuned to the corresponding structure in the real world.

Examples of Direct Structure Tests: Structure Examination Test, Parameter Examination Test, Direct Extreme Condition Test, Boundary Adequacy Structure Test, Dimensional Consistency Test.

[1] For a detailed description of these tests, see: [18] as well as literature quoted therein.

Examples of Indirect Structure Tests: Indirect Extreme Condition Test, Behavior Sensitivity Test, Integration Error Test, Boundary Adequacy Behavior Test / Boundary Adequacy Policy Test, Loop Dominance Test.

3.3 Tests of Model Behavior

Tests of model behavior are empirical and compare simulation outcomes with data from the real system under study. On that basis, inferences about the adequacy of the model can be made. The empirical data can either be historical or refer to reasonable expectations about possible future developments.

Examples: Behavior Reproduction Tests, Behavior Anticipation Tests, Family Member Test, Surprise Behavior Test, Turing Test.

4 Structural Dominance Test

As an example, we will now demonstrate a new test about the model structure, called the Structural Dominance Test (SDT). In principle, SDT evaluates the relative partial influence of individual feedback loops on the behavior of chosen variables, and compares the result to the real system. Structural dominance signifies which particular piece of model structure is dominant, i.e., most influential for the behavior over a certain period of time. SDT is based on Structural Dominance Analysis [6]. Structural Dominance Analysis is a field of System Dynamics with a three-decade history. Only today are the approaches mature enough to sustain frequent use. Groesser conceptualizes the use of SDA for model validation purposes, namely the SDT [4].

4.1 Axiom of the Structural Dominance Test: Feedback Structure Creates a System's Behaviour

The dominant structure of a feedback system is the cause for the behaviour of the system [3]. The system's behaviour over time can, in principle, be partitioned into such time intervals that for each interval one of three behaviour patterns can be recognized: linear growth, exponential growth, or logarithmic growth [2]. These behaviours are created by the dominant structures. Complex phenomena are generated by the interaction of a multitude of structural feedback loops. This kind of analysis of dynamic complex feedback models is a daily challenge for modellers – even for advanced system dynamicists. The analysis of structural dominance supports modellers in determining which part of the model structure is mainly responsible for creating the system behaviour. On this basis rests the topic of model validation by means of the Structural Dominance Test.

In order to transfer the general notion of the structural dominance analysis to a validation test, the basic elements for the structural analysis of dynamic systems have to be elaborated. These are the concepts of *dominant feedback loop*, the *polarity of a feedback loop* and *the change of dominant structure.*

4.2 Basic Concepts for Structural Dominance Analysis

4.2.1 Dominant Structure

The approaches available for the analysis of a model's dominant structure use the concept of a *dominant feedback loop* as their basic element. According to Richardson and Pugh, "a dominant loop is a loop that is primarily responsible for the model behaviour over some time interval" [14: 231]. A feedback loop is a chain of causal effects which includes as a constitutive property at least one state variable[2]. Richardson's and Pugh's definition of a dominant feedback loop is valid for relatively small structures (=first or second order models, see [15]). Higher order systems result in more complex interactions among a greater number of feedback loops. In such models, it is likely that multiple feedback loops are simultaneously responsible for causing the system's behaviour; we term this the dominance of a feedback loop cluster.

4.2.2 Polarity of a Feedback Loop

The polarity of a feedback loop indicates the direction of the resulting behaviour of a feedback loop as the consequence of a change in any variable of the respective loop. Given an increase in a variable's value by one unit ($dx_{t=0} = 1$), a feedback loop with a positive (self-reinforcing) polarity further amplifies the initial increase, i.e., $dx_{t=1} > 1$, $(dx/dt)/dx > 0$. A negative (balancing) feedback loop reduces an initial increase accordingly, i.e., $dx_{t=1} < 1$, $(dx/dt)/dx < 0$. Expressed in mathematical terms: the polarity of a feedback loop is the signum of the loop (Equation 1). The signum is the multiplication of the partial derivatives of all bi-variate causal relationships within that feedback loop. The dominant polarity of the feedback structure can be inferred by comparing the polarity of feedback loops with the model's dominant patterns in each time interval [2].

$$\operatorname{sgn}\left(\frac{\partial\left(\frac{\partial x_1}{\partial t}\right)}{\partial x_n} * \frac{\partial\left(\frac{\partial x_2}{\partial t}\right)}{\partial x_1} * \frac{\partial\left(\frac{\partial x_3}{\partial t}\right)}{\partial x_2} * \cdots \frac{\partial\left(\frac{\partial x_n}{\partial t}\right)}{\partial x_{n-1}}\right)$$

Equation 1. Definition of the polarity of a feedback loop

The notation of dominant polarity is bi-unique in systems with only one feedback loop (=simple system). In a system with two or more feedback loops, all feedback loops which contain the target variable as an element must be taken into account to calculate the dominant polarity.[3]

---

[2] State variables are system variables, which represent accumulations over time.

[3] Even though the identification and calculation of the dominant polarity according to the procedure established by Richardson [15] are of limited convenience for multi-loop systems, they are helpful to illustrate the basic logic of the approach. The simultaneous consideration of multiple feedback loops becomes increasingly impossible for the analyst. The authors estimate that it is impossible to mentally cope with the dynamic complexity of more than 3-5 feedback loops.

4.2.3 Change in Dominant Structure

A change in dominant structure marks the point in time at which the dominance of one feedback loop is neutralized by the dominance of another. This change is identifiable by an alteration of the signum of the feedback structure [15]. The example in Equation 2 shows that the state variable (x) and the modelled system context (plainly represented by the intercept a and the slope b) determine the dominant polarity and the position in time [x=a/b] of the change in the dominant polarity. Taking the example from Equation 2, the reinforcing structure (positive signum) dominates for relatively small values of x; correspondingly, this changes the dominance to a balancing structure (negative signum) for relatively large values of x.

$$\text{Dominant Polarity} = \text{sgn}(a - bx) = \begin{cases} + \text{ if } x < a/b \\ - \text{ if } x > a/b \end{cases} \quad ; \text{with } a, b > 0;$$

Equation 2. The change of the signum (*sgn*) indicates the change of the dominant polarity. The example shows a linear function with the intercept a and the slope b.

The interpretation of the change of the dominant structure depends on the degree of complexity of the considered structure. In simple systems with two feedback loops with opposing polarities, a change of the polarity displays a change in the dominance of the feedback loops. In more complex systems, this interpretation is no longer valid. A change in the dominant structure can no longer be directly inferred from individual feedback loops. In a system containing more than three feedback loops only the polarity of a cluster of loops can be calculated. If a change of polarity occurs, the whole feedback structure then has the opposing polarity; individual feedback loops, which might be responsible for the dominant behaviour, cannot be detected. It might be, for instance, that within a given time interval, during which no change of the signum takes place, two positive feedback loops dominate each other in sequence. Obviously, the dominance of the individual loops shifts over time. This sequential dominance, however, cannot be recognized by the indicator of the dominant polarity. Diagnosing changes in structural dominance within a time interval that is continuously dominated by either positive or negative loops requires mathematical Eigenvalue methods for the exact analytical determination of the structural dominance over time [5]. Several mathematical approaches of structural dominance analyses have emerged over the past decade ([6], [7], [8], [10], [11]). An experimental approach to the analysis of dominant structure has emerged early in the field of System Dynamics [2]. However, this approach lacks the mathematical rigour of the Eigenvalue-analytical methods.

4.3 Validation Logic

The validation logic of the structural dominance test corresponds to the general logic of any other validation effort: the analyst – or a group of participants in a group modelling project – compares the model assumptions, model inputs, as well as model outcomes with the perceived counterparts of the modelled section of reality (Figure 1). *The evaluation of the correspondence of the model and the modelled part of reality is (always) based on the available empirical material.* This fundamental validation logic will now be applied to the analysis of the structural dominance of a model. Thereby, the unique contribution of the structural dominance test will become more vivid.

Modelling with System Dynamics tries to elicit the basic system structure with a special focus on the following properties: feedback structure of the system, dynamic behaviour of the system, and change of the dominant system structure over time. System Dynamics analysts explore the reality based on these three criteria. As a result, they manifest their insights in their mental models about dynamic systems [1] (Figure 1). The major difference between the informal presentation of the reality in the mental model and the reality itself is that the perception of reality is imperfect given the perceptual apparatus of the observer. Subject-related distortions can be reduced by group-based modelling practices and validation methods [19].

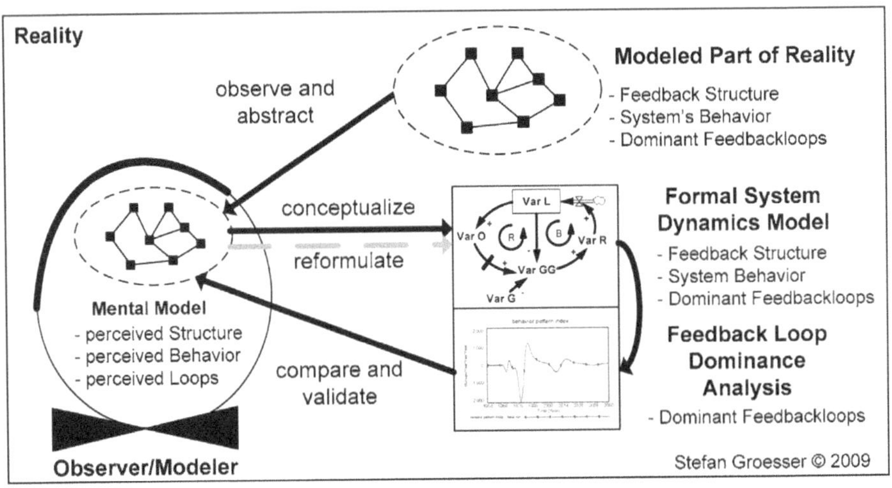

Fig. 1. Association of the modelling of the reality, the mental model of the observer, and the formal simulation model as well as the validation process. Validation by means of Structural Dominance Analysis supports the modeller in identifying the dominant feedback structure of the formal model, and enables evaluation of the correspondence between the formal model and the modelled part of reality.

The conceptualization and creation of a formal model is fully grounded only in the mental models of the analysts. The formal model should incorporate the structure of the feedback system, the system's dynamic behaviour, and the change of its dominant structures over time. Validation efforts in all possible validation domains [18] attempt to ensure the correspondence of the formal model and its counterparts in reality. The dominance analysis focuses on the comparison of the dominant feedback structures in the simulation model and reality. Thereby, it becomes possible to test their correspondence. The following decision logic is applied:

$$\text{Structural Dominance}_{\text{formal}} \approx \text{Structural Dominance}_{\text{real}} \begin{Bmatrix} \text{true} \rightarrow \text{Abort Validation} \\ \text{false} \rightarrow \text{Reformulation} \end{Bmatrix}$$

If the dominant structure of the formal model corresponds to the dominant structure of the perceived part of reality over time, no further validation is required – the potential of the structural dominance test is then exhausted. However, if the structures do not correspond, it must be that either the formal model does not reflect reality adequately or that the mental model is not an adequate representation of reality. In either case, it is appropriate to reconsider both the formal and the mental models. Thereby, immense possibilities for substantial learning effects open up. The tracing of the dominant structures in the formal model supports locating the discrepancy between model and reality.

In the following, the principle of the structural dominance test is briefly applied. A full application and analysis is provided in [4]. The example is based on a simplified version of Meadow's commodity cycle theory [9]. At its core, it consists of two interacting feedback mechanisms (B1 and B2) which are mutually regulated by the commodity price (Figure 2). The price stimulates capacity expansion which is introduced into the market with a medium-term time lag. Demand-side reactions to price changes are short-term oriented. The interaction of both goal-seeking feedback loops (B1 and B2) creates an oscillatory behaviour of the indicator variable 'price' (see Figure 3, dotted line 'price').

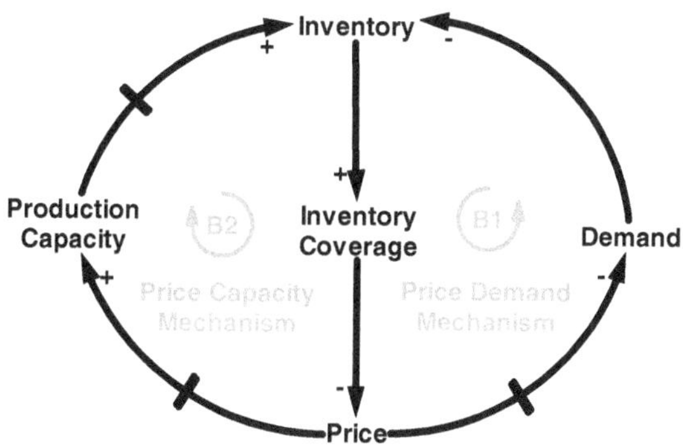

Fig. 2. Commodity cycle theory of Meadows [9]

Figure 3 plots the Behaviour Pattern Index (BPI; solid line; for a detailed definition of the BPI: see [2]). The BPI is an operationalization of both the concept of dominant polarity (see 4.2.) and the pattern of growth (linear, exponential, and logarithmic) of the variable of interest. For positive values of the BPI, the commodity price (dotted line) exhibits an exponential-reinforcing growth pattern (e. g., for the time intervals $ti_{p1} = [4..6]$[4], $ti_{p2} = [15..18]$); for negative values of the BPI, the price follows a logarithmic, i.e., goal-seeking, behaviour (for example, $ti_{n1} = [0..4]$, $ti_{n2} = [6..15]$). An additional demand step of 30% of the regular demand (for 10 months) disturbs the model initial equilibrium at $p_{commodity} = 50$ [USD].

For the excess demand (+30% for t>10 months), the equilibrium price is now at $p_{commodity} = 55$ [USD]; the equilibrium demand at 675 [units]. In the short-term, the higher demand meets a relatively fixed production capacity which almost immediately results in higher commodity prices. Consequently, the demand per capita is reduced towards the equilibrium demand; this is shown by the logarithmic growth pattern

[4] *p* stands for positive, *n* for negative values of BPI.

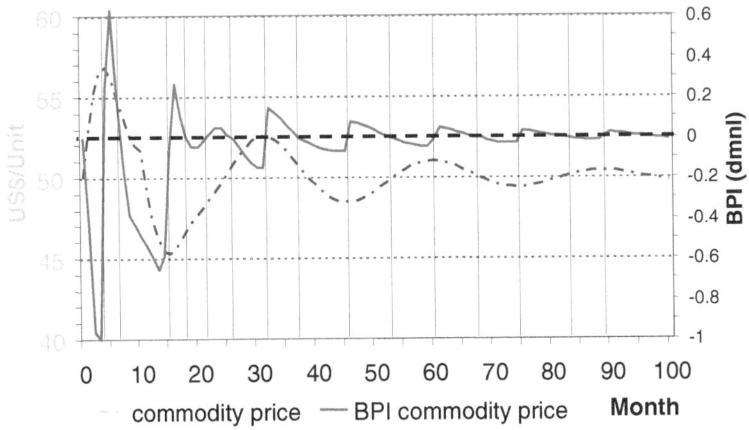

Fig. 3. Dynamic behaviour of the price in Meadow's commodity cycle theory [9]. The Behaviour Pattern Index (BPI) is one possibility for operationalizing an indicator for loop dominance analysis [2].

[$0<t<4$] of the BPI. Due to the time delay in the per-capita demand adjustment, the system undershoots the equilibrium demand leading to a brief reinforced correction process in the time interval from $4<t<6$. After a delay for building, additional production capacity enters the market, leading to further decline in the commodity price. In terms of the model structure, capacity expansion (B2) now dominates the development of the price; B1 is still active. For $t>10$, the excess demand is reduced; a different equilibrium price and demand is now established. Both loops try to achieve their implicit goals; hence, the main dynamics of the BPI show logarithmic growth. Several spikes of exponential growth tendencies occur (e.g., $15<t<18$), which indicate a reinforcing correction process of the demand as explained above.

The example chosen is a well-tested theory about commodity prices. At first, it might be counter-intuitive that a balancing feedback loop can generate a reinforcing growth behaviour for short time periods. However, a balancing loop which shows lagged action can exhibit transitional reinforcing growth dynamics [16].

The application of the SDT does not indicate a falsification of the model; the model's change in structural dominance seems to correspond to reality. In this case, the SDT cannot invalidate the model.

5 Conclusion

Validation is a rich and well-defined process by which the confidence in a model is gradually enhanced, with the help of the application of a battery of tests. Validity is always a matter of degree, never an absolute property.

By this demonstration of the Structural Dominance Test, we have made a case for a new approach which extends the frontiers of model validation. More generally, we have used the description of SDT as an exemplar of the logic underlying the validation process.

Simulation based on formal dynamic models is likely to become ever more important for research and practice. It will continue to support scholars and managers at all levels in research, decision-making, and policy design. The more that models are relied upon in all these areas, the greater the importance of their quality. Therefore, model validation is one of the major challenges that we face in advancing modeling and simulation.

References

1. Doyle, J.K., Ford, D.N.: Mental Models Concepts for System Dynamics Research. System Dynamics Review 14(1), 3–29 (1998)
2. Ford, D.N.: A Behavioral Approach to Feedback Loop Dominance Analysis. System Dynamics Review 15(1), 3–36 (1999)
3. Forrester, J.W.: Industrial Dynamics. Productivity Press, Cambridge (1961)
4. Groesser, S.N.: Validität und Qualität von Simulationsmodellen: Strukturdominanz Analyse angewendet als Validierungsmethode. Discussion Paper No. 57. University of St. Gallen, Institute of Management (2009)
5. Güneralp, B.: Towards Coherent Loop Dominance Analysis: Progress in Eigenvalue Elasticity Analysis. System Dynamics Review 22(3), 263–289 (2006)
6. Kampmann, C.E., Oliva, R.: Loop Eigenvalue Elasticity Analysis: Three Case Studies. System Dynamics Review 22(2), 141–162 (2006)
7. Kampmann, C.E., Oliva, R.: Structural Dominance Analysis and Theory Building in System Dynamics. Systems Research and Behavioral Science 25(4), 505–519 (2008)
8. Kampmann, C.E., Oliva, R.: System Dynamics: Analytical Methods for Structural Dominance Analysis. In: Encyclopedia of Complexity and Systems Science. Springer, Berlin (2009)
9. Meadows, D.L.: Dynamics of Commodity Production Cycles. Wright-Allen Press, Cambridge (1970)
10. Mojahedzadeh, M., Andersen, D.F., Richardson, G.P.: Using Digest to Implement the Pathway Participation Method for Detecting Influential System Structure. System Dynamics Review 20(1), 1–20 (2004)
11. Mojtahedzadeh, M.: Do Parallel Lines Meet? How Can Pathway Participation Metrics and Eigenvalue Analysis Produce Similar Results? System Dynamics Review 24(4), 451–478 (2008)
12. Popper, K.R.: The Logic of Scientific Discovery. Hutchinson, London (1959)
13. Popper, K.R.: Objective Knowledge: An Evolutionary Approach. Clarendon Press, Oxford (1972)
14. Richardson, G.P., Pugh III, A.L.: Introduction to System Dynamics Modeling with DYNAMO. Productivity Press, Cambridge (1981)
15. Richardson, G.P.: Loop Polarity, Loop Dominance, and the Concept of Dominant Polarity. System Dynamics Review (Wiley 1984) 11(1), 67–88 (1995)
16. Saleh, M., Davidsen, P.: Extending Eigenvalue Analysis to Nonlinear Models via Incorporating Higher Order Terms of Taylor Series Expansion. In: Proceedings of the 2007 International Conference of the System Dynamics Society. The System Dynamics Society, Boston (2007)

17. Schwaninger, M., Groesser, S.: System Dynamics as Model-based Theory-building. Research and Behavioral Science 25(4), 447–465 (2008)
18. Schwaninger, M., Groesser, S.N.: System Dynamics Modeling: Validation for Quality Assurance. In: Encyclopedia of Complexity and System Science. Springer, Berlin (2009)
19. Vennix, J.A.M.: Group Model Building: Facilitating Team Learning Using System Dynamics. Wiley, Chichester (1996)

Optimizing the Hardware Usage of Parallel FSMs

Rainer Findenig, Florian Eibensteiner, and Markus Pfaff

FH Hagenberg, Hardware/Software Systems Engineering,
Softwarepark 11, 4232 Hagenberg, Austria
{rainer.findenig,florian.eibensteiner,markus.pfaff}@fh-hagenberg.at

Abstract. Hardware design is traditionally done by modeling finite state machines (FSMs). In this paper, we present how a basic round-robing scheduling mechanism, well-known from operating systems, can be applied to a design that needs several identical FSMs running (quasi) in parallel.

This approach allows exploiting the classical trade-off between chip area and operating frequency to severely cut down the hardware resources needed to implement the FSMs by increasing the operating frequency of the design. We additionally show that, in a system-on-a-chip design using only a single clock domain, the design's overall operating frequency is dependent on the processor's frequency, making especially low-speed communication cores already clocked faster than needed. This means that with regard to the design's frequency, our approach may come at no additional cost.

Keywords: FSM, Scheduling, Serialization, Resource Sharing.

1 Introduction

Modern system-on-a-chip designs often feature a considerable amount of IP cores. Communication and control tasks with hard real-time requirements can effortlessly be implemented in hardware, thereby alleviating the software from several timing constraints [1].

For many applications, it is preferable to implement a single IP core several times. The application at hand, for example, currently features a single system-on-a-chip (SoC) and 18 LIN nodes, all of which have to be controlled in real-time by the SoC. While LIN is a perfect choice when it comes to hardware costs, a single LIN bus does not provide enough throughput for our application. Since the nodes need not communicate between each other, this problem can easily be overcome by designing several independent LIN busses, and implementing a master for each of them in the central SoC. This parallel implementation, however, obviously severely affects the amount of hardware resources needed.

Additionally, in the application at hand, the SoC uses a single clock domain. The system clock is therefore dependent on the internal processor's frequency, which is several times higher than the frequency necessary for the LIN cores. This leads to the approach presented in this paper: we show how applying traditional

R. Moreno-Díaz et al. (Eds.): EUROCAST 2009, LNCS 5717, pp. 63–68, 2009.

scheduling algorithms can mitigate the additional hardware usage, introduced by implementing several identical automata in parallel, by exploiting the classical trade-off between chip area and operating frequency [2,4].

1.1 Definitions

In this paper, we define an input/output-automaton (IO-automaton) as a six-tuple $A = (Q, q_0, \Sigma, \delta, O, \lambda)$, with the components being the following:

- Q is a finite set of states,
- $q_0 \in Q$ is the initial state,
- Σ is a finite set of input symbols,
- $\delta : Q \times \Sigma \rightarrow Q$ is the (total) transition function,
- O is a finite set of output symbols, and
- $\lambda : Q \times \Sigma \rightarrow O$ is the (total) output function.

For an input word $w = w_0 w_1 w_2 \ldots$ with $w_i \in \Sigma$, the automaton produces a trace q of states ($q = q_0 q_1 q_2 \ldots$ and $q_i \in Q$) and an output word o of output symbols ($o = o_0 o_1 o_2 \ldots$ with $o_i \in O$), where $q_{i+1} = \delta(q_i, w_i)$ and $o_i = \lambda(q_i, w_i)$.

2 Implementation and Scheduling

When implementing several automata in parallel, one can think of the input and output symbols as tuples. For n automata in parallel, an input symbol is a tuple $w_i = (w_i^0, w_i^1, \ldots, w_i^{n-1}) \in \Sigma^0 \times \Sigma^1 \times \ldots \times \Sigma^{n-1}$, with w_i^j being the i-th input symbol to automaton j and Σ^j being the input alphabet of automaton j. Similarly, the produced output $o_i = (o_i^0, o_i^1, \ldots, o_i^n) \in O^0 \times O^1 \times \ldots O^{n-1}$. This is shown in Fig. 1.

Our contribution is based on a round-robin scheduling algorithm with static time slices of exactly one cycle. This allows a system running at $f_n = n \cdot f_a$ (or, obviously, $f_n \geq n \cdot f_a$), with f_a being the maximal execution speed required for a single automaton, to compute the next state and output of n automata one after the other instead of in parallel, while still meeting the timing requirements.

Note that, while in the following we are focusing on the input-/output-behavior, the same is true for the states of the automata.

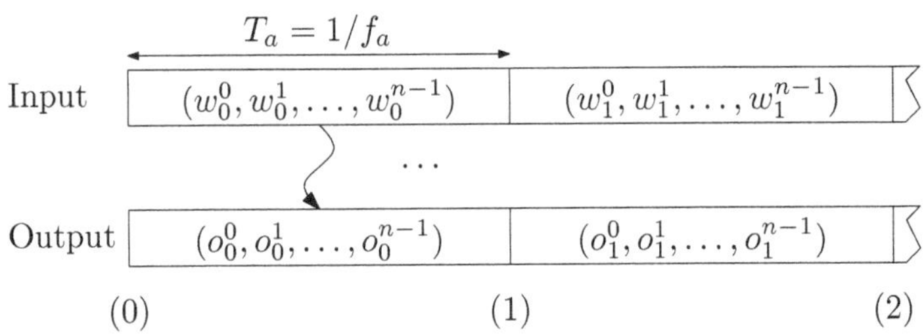

Fig. 1. Parallel implementation of the automata

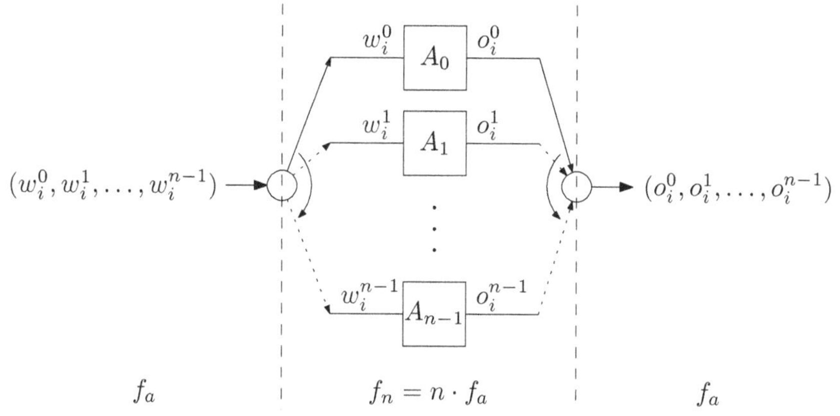

Fig. 2. Parallel to serial conversion of the automata

The switching between the automata, which forms the scheduler in our approach, is shown in Fig. 2. The input word w is applied with the frequency f_a (left part of Fig. 2), taken by a commutator, and it's components are applied to the n automata consecutively in a round-robin fashion with the frequency f_n (middle). After the automaton has calculated its output symbol, it is collected and stored by another commutator (right part), where it can be sampled with f_a again.

This leads to an input-/output-behavior as shown in Fig. 3. If the output of the second commutator is sampled at the same time as the original implementation would have been (eg. at timestamp (1) for w_0), it is obvious that the output of both the serial and the parallel version is identical.

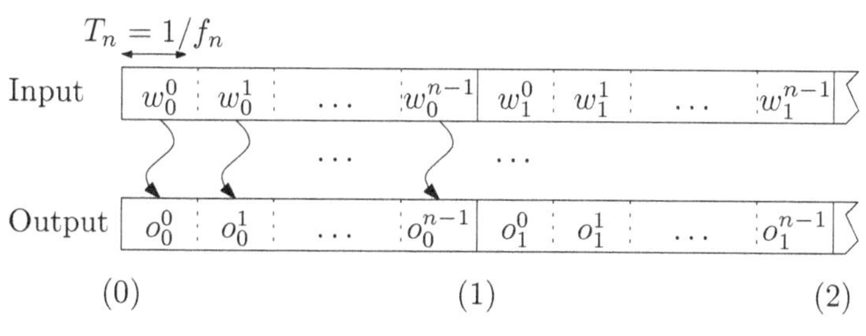

Fig. 3. Serialized implementation of the automata

3 Optimized Implementation

While we have shown that it is possible to serialize the calculation of the automata, there was no benefit achieved by doing so. When implementing several *identical* automata, we can return to the automaton's definition to optimize the implementation.

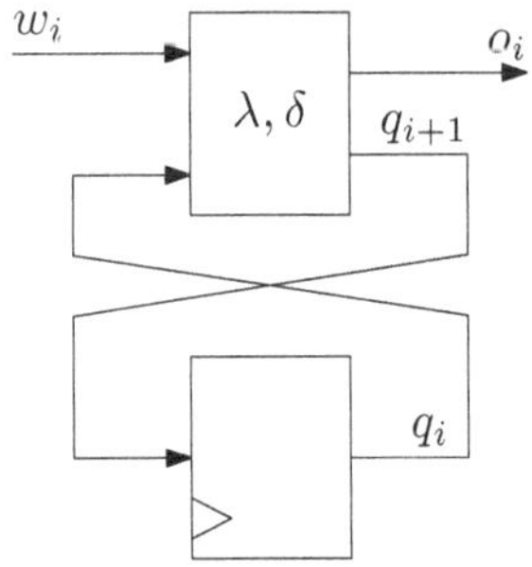

Fig. 4. Basic structure of a single IO-automaton when implemented in hardware. For simplicity, δ and λ are shown combined to block.

Implementing an IO-automaton $A = (Q, q_0, \Sigma, \delta, O, \lambda)$ in hardware results in a structure as shown in Fig. 4. As can be seen, an IO-automaton is largely defined by its transition function δ and its output function λ—the sets Q, Σ, and O are defined implicitly by those functions and q_0 is the register's reset value.

Obviously, the combinational logic for each automaton is only used while the automaton is calculating a new state and output. Due to the serialization shown before, it is guaranteed that no two automata are doing so at the same time; therefore, all automata can safely share *one* implementation of δ and λ. Instead of placing the commutator before the *automata*, as shown in Fig. 2, it can be placed before the *register*, creating a structure as shown in Fig. 5. This obviously greatly reduces the amount of hardware necessary for implementing several automata, as the combinational logic need only be available one single time.

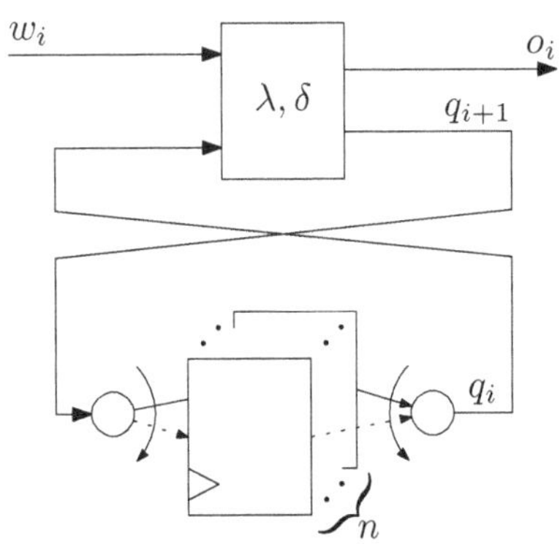

Fig. 5. Optimized implementation of n automata

3.1 State Storage in RAMs

The second commutator can be implemented as a hierarchical multiplexer structure with the depth of $d \geq \lceil \mathrm{ld}(n) \rceil$ (using two-input multiplexers and n being the amount of automata implemented). This, naturally, creates an additional hardware overhead, but has an even greater impact on the critical path.

Using dedicated RAMs, as, for example, available in nearly all modern FPGAs, this problem can easily be addressed. Selecting a state from the RAM is only a matter of addressing it correctly, no additional multiplexer structure is needed. Hence, when using a RAM, the two commutators can be reduced to a single mod-n-counter that generates the respective addresses [3].

4 Results

The presented approach has been verified using a simple automaton for recognizing a regular language over ASCII input symbols. A single instance of this automaton needed 4349 μm in a 0.35μm ASIC technology, as can be seen from Fig. 6. Implementing 100 instances in parallel obviously leads to 100 times the area. Using our approach and implementing all 100 instances sequentially using a straightforward scheduler consisting of a counter and a multiplexer, 46% area can be saved. Since the overhead introduced by the scheduler is considerable, further research will be required to optimize the implementation of the approach. Such approaches have, for example, been presented in [3].

Implementation	Area Comb	Area Seq	Total	%
Single	2948	1401	4349	1
100x, parallel	$2948 \cdot 100$	$1401 \cdot 100$	434900	100
100x, scheduled	$2948 + \underbrace{93420}_{\text{Scheduler}}$	$1401 \cdot 100$	236468	54

Fig. 6. Area requirements of a single automaton and 100 instances in parallel as well as scheduled

Fig. 7 shows the result achieved when implementing 18 LIN master cores in parallel as well as when utilizing our sequential and scheduled approach. Again, while the overhead is considerable, the total area consumption is reduced by more than 22%. The difference in both the combinational and sequential area requirements of the scheduled version when compared to the single implementation can be attributed to different synthesis optimizations.

Implementation	Area Comb	Area Seq	Total	%
Single	30084	33269	63353	5.6
18x, parallel	$30084 \cdot 18$	$33269 \cdot 18$	1140354	100.0
18x, scheduled	$30230 + \underbrace{231754}_{\text{Scheduler}}$	$34343 \cdot 18$	880158	77.2

Fig. 7. Area requirements of a single LIN master core and 18 instances in parallel as well as scheduled

5 Conclusion

In this paper, we presented a simple scheduling-based approach for sharing resources between several automata.

We showed how our serialization approach can be used for several *identical* parallel state machines with $\lambda_0 = \lambda_1 = \ldots = \lambda_{n-1}$ and $\delta_0 = \delta_1 = \ldots = \delta_{n-1}$. This can, however, be partly generalized for *similar* combinational functions. Assume that the functions λ_i can be decomposed to be $\lambda_i(\lambda_{i,0}, \lambda_{i,1}, \ldots)$ of "sub-functions" $\lambda_{i,j}$. Obviously, if two automata i and i' share sub-functions, ie. $\lambda_{i,j} = \lambda_{i',j'}$ for any $i \neq i'$, the combinational logic calculating $\lambda_{i,j}$ can just as well be shared between the automata i and i'. This obviously is also applicable to the functions δ_i and to any combinations of $\lambda_{i,j}$ and $\delta_{i',j'}$, as long as $i \neq i'$.

Additionally, we assumed that the shared resources, ie. the logic implementing δ and λ, are scheduled in a round-robin manner with time slices of exactly one cycle. This makes the relation between both implementations, as shown in Sec. 2, obvious. Other scheduling algorithms, such as earliest deadline first, are possible, tough, and may, depending on the application, increase the system's performance with regard to real-time constraints.

References

1. Eibensteiner, F., Findenig, R., Tossold, J., Kubinger, W., Langer, J., Pfaff, M.: Embedded robotic solution: Integrating robotics interfaces with a high-level CPU in a system-on-a-chip. In: Moreno Díaz, R., Pichler, F., Quesada Arencibia, A. (eds.) EUROCAST 2007. LNCS, vol. 4739, pp. 1017–1024. Springer, Heidelberg (2007)
2. Flynn, M.J.: Area – time – power and design effort: the basic tradeoffs in application specific systems. In: Proc. 16th IEEE International Conference on Application-Specific Systems, Architecture Processors, 2005. ASAP 2005, July 2005, pp. 3–6 (2005)
3. Hofstätter, M., Mayer, G., Findenig, R., Eibensteiner, F., Pfaff, M.: RAM basiertes Entwurfskonzept für flächenoptimierte Multi-IP-Core-Designs. In: Tagungsband des 3. Forschungsforums der österreichischen Fachhochschulen (2009)
4. Ullman, J.D.: Computational Aspects of VLSI. Computer Science Press, Rockville (1984)

SynPSL: Behavioral Synthesis of PSL Assertions

Florian Eibensteiner, Rainer Findenig, and Markus Pfaff

FH Hagenberg, Hardware/Software Systems Engineering,
Softwarepark 11, 4232 Hagenberg, Austria
{florian.eibensteiner,rainer.findenig,markus.pfaff}@fh-hagenberg.at

Abstract. The effort of verifying state-of-the-art hardware designs undeviatingly increases with the complexity of those designs. The design's state space, directly related to its complexity, grows exponentially, while the computational performance for verifying the design grows only linearly. This so-called verification gap can, for example, be met by using methods such as assertion-based verification (ABV), which can be used for both specifying the system's properties as well as verifying the relating implementation during simulation phase.

In this paper, we present an open-source tool which generates synthesizable HDL code from assertions specified in the Property Specification Language (PSL). This is done by first reducing the PSL formulas into base cases, called PSL_{min}, and then generating automata which can be transformed to synthesizable HDL code and therefore into hardware.

Keywords: PSL, Assertion-based Verification, Synthesis.

1 Introduction

The complexity of modern state-of-the-art hardware designs becomes more and more of a problem for the verification engineers, because they can hardly keep up with this complexity. This is especially the case if only traditional verification methods such as testbenches are used to check the conformance of a highly sophisticated design. Therefore, the semiconductor industry increasingly turns to approaches such as assertion based verification (ABV) to achieve a better and faster way of verification.

As the name suggests, ABV is based on assertions, which are used to check the properties of a system. Those properties and therefore assertions expressing them are usually defined during concept phase and further on are used for formal and functional verification from model level down to register transfer level (RTL).

At the same time, other methods of speeding up the verification process, such as emulation and prototyping, are gaining the industry's interest. Such systems allow sourcing out parts of the design under verification (DUV) to hardware, which can greatly increase the simulation speed [8].

As long as assertions cannot be synthesized, however, those two techniques are obviously mutually exclusive. Therefore, in this paper we present a tool that can bridge this gap by transforming PSL assertions into synthesizable HDL code.

R. Moreno-Díaz et al. (Eds.): EUROCAST 2009, LNCS 5717, pp. 69–74, 2009.

1.1 Assertion Based Verification

As mentioned, ABV allows the definition of system properties in a certain language such as PSL or SystemVerilog Assertions (SVA). Assertions based on these properties can be checked by a verification tool during functional verification and therefore provide an extension of the traditional testbench based functional verification.

Approaches based on a traditional testbenches verifying a DUV in a black-box manner, because they are using a predefined input stimuli and checking the correctness of the output. Therefore, errors not appearing at the output within the simulation cycle will not be detected. The usage of conventional assertions as already available in hardware description languages, could partially solve this problem, but complex sequences, especially timing constraints, can not be modeled by these kind of assertions.

However, ABV allows the definition of such complex system requirements, as early as in concept phase. Therefore, not only the output but also the internals of a system can be verified and so ABV facilitates a white-box approach for verification.

1.2 Property Specification Language

PSL itself consists of a foundation language (FL) and an optional branching extension (OBE). The latter is is mainly used in formal verification, therefore we will focus on the FL. FL is a linear temporal logic, which means that only linear traces are considered and therefore system properties can be described only on a single trace [4]. Furthermore, FL is composed of Linear Temporal Logic (LTL) and Sequential Extended Regular Expressions (SEREs) that offer a higher expressiveness: properties written in LTL can be rewritten in SEREs, but not the other way around [4].

As their name suggests, SEREs and conventional regular expressions are very similar. The main difference is that regular expressions are applied to single characters and SEREs works with Boolean expressions, as will be shown later.

Consider the following example:

$$\textbf{assert always } (\{\textit{request}\} \;\; |{=}{>} \; (\{\textit{ack};\ \textit{grant};\ \textit{data}\} \;|\; \{\textit{nack}\})) \; @ \; \textbf{rising\_edge}(\textit{clk});$$

Informally, this assertion requires the antecedent, the signal request, to be followed either by the sequence *ack, grant, data* or the single expression *nack*, otherwise the assertion will fail. The operator @ specifies the clock expression, which controls when the formula is evaluated [1]. In the considered example, the assertion will be checked at every rising edge of *clk*.

2 Implementation

The described tool basically consists of three blocks structured in a form well known from compiler design: an input parser to read a PSL assertion and convert

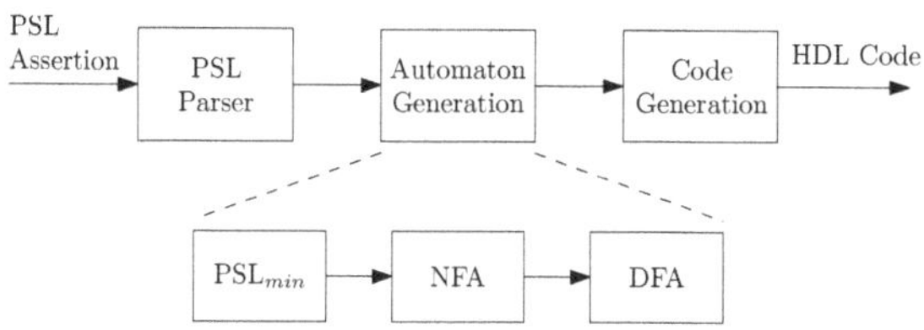

Fig. 1. Blockdiagram of the internal structure of SynPSL

it to an internal representation, automata generation algorithms [2,5,6] to build a corresponding automaton, and finally a code generation stage which outputs synthesizable HDL code implementing the automaton (Fig. 1).

The most sophisticated part is the automaton generation (middle block), where in a first step the assertion is reduced to base cases defined in PSL$_{min}$ [6]; this reduces the tool's complexity since automata generation algorithms need only be provided for those cases. A nondeterministic finite automaton (NFA) is constructed, and, for the implementation in hardware, converted to a deterministic finite automaton (DFA). In a final step, the DFA is translated into HDL code which can be synthesized afterwards.

2.1 PSL$_{min}$

PSL assertions can be rewritten to a relatively small set of base cases defined in PSL$_{min}$ [2,6]. The transformation allows for easier implementation of the tool since the complexity of the automata generation algorithms be reduced. As a result, in a first step the syntax tree, generated by the PSL parser, will be rewritten into its base cases shown in Fig. 2.

- boolean HDL expressions
- $\{b\}$
- $\{r_1;r_2\}$
- $r[*0]$
- $r_1 \mid r_2$
- b
- p **abort** b
- r |-> p
- p **until** b
- **eventually**! r
- **prev**(b, c)
- $r[*]$
- $\{r_1{:}r_2\}$
- $r[*l$ **to** $h]$
- r_1 && r_2
- r
- p_1 && p_2
- r |=> p
- $r!$

Fig. 2. Rewriting rules for reducing PSL to PSL$_{min}$

The rewriting rules to simplify a PSL assertion are beyond the scope of this work and can be found in [1] and a, in some cases, for our tool more efficient approach is introduced in [6].

2.2 Automata Representation of PSL_{min}

In order to get a model of a PSL assertion which can be synthesized and represented in hardware, a deterministic automaton must be constructed. Therefore a automaton is recursively assembled from the assertion rewritten in PSL_{min} [6].

Consider, for example, from the assertion **assert always** ($\{a\}$ |=> $\{b[*];\ c;\ d\}$): intuitively, the automaton shown in Fig. 3(a) can be constructed where the antecedent $\{a\}$ corresponds to the left part and the postcondition is represented by the right part of the automaton. Also, consider the trace shown in figure 3(b).

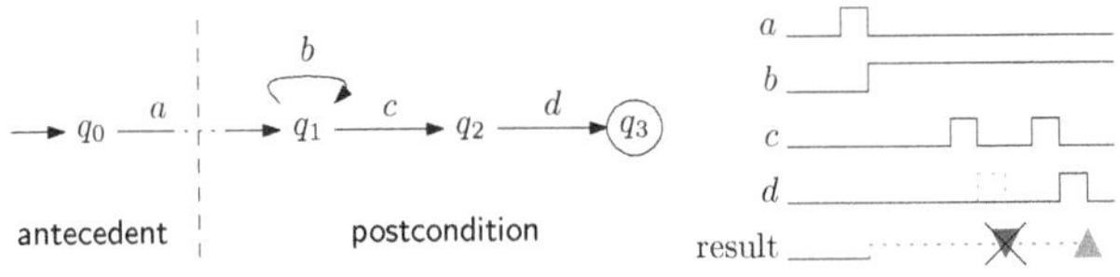

Fig. 3. Gernerated automata

If the automaton left the state q_1 at the first occurrence of c, it would detect the missing d an will report an error. This is incorrect, though: As can be seen in the trace, b and c can occur simultaneously: the automaton has to guess whether to stay in q_1 or go to q_2, which means the automaton is nondeterministic. This can be attributed to the difference between a regular expression's semantic and its language [9].

Therefore, a determinization algorithm to convert the NFA into a DFA, which was introduced in detail in [6] and is, to a large extent, based on the work by Ruah et al. [9], is applied: First, all Boolean expressions appearing in the outgoing transitions of S, where S is be a set of states of the NFA, must be determined:

$$conditions(S) = \{l | (q_1, l, q_2) \in outgoing(S)\}.$$

For an implementation in hardware it is necessary to know the successor state for all possible combinations of input symbols. The powerset of the conditions in S provides these combinations, and every

$$P \in \mathcal{P}(conditions(S))$$

corresponds to a transition in the resulting DFA. In the last step, all states these transitions lead to can be calculated:

$$succ(P, S) = \begin{cases} \{q_{\text{sink}}\} & \text{if } P = \emptyset \\ \{q_2 \in Q | (q_1, l, q_2) \in outgoing(S), l \in P\} & \text{otherwise.} \end{cases}$$

P contains all input symbols which must be true in order for the next state to be $succ(P, S)$, but it contains no information about the other input symbols5. To ensure that they are false, the full version of P is calculated by

$$det\_edge\_cond(P, S) = \left(\bigwedge_{l \in P}\right) \wedge \left(\bigwedge_{l \in conditions(S) \setminus P}\right).$$

Applying this algorithm to the before mentioned example, where the state q_1 was nondeterministic, will lead to a deterministic result shown in Fig. 4.

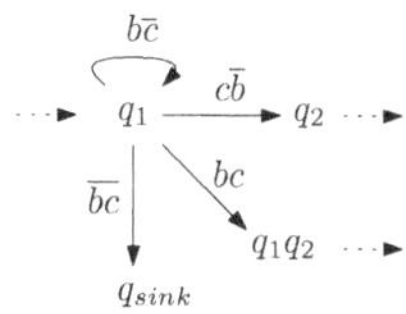

Fig. 4. Result of determinisation of state q_1

Addtionally, the PSL LRM denotes that the semantic of the **always** operator is a synonym of LTL's operator **G** [1], and $\pi \models \mathbf{G}f$ iff, for all $i \geq 1, \pi^i \models f$, where π is a sequence of states $q_0q_1q_2\ldots$ and π^i is the suffix starting in q_i [3]. Therefore, the **always** operator requires the assertion is evaluated in every cycle which possibly leads to multiple concurrent runs. In other words, a pipelined evaluation of the assertion is needed.

To this end, the automaton is implemented with an unconditional self loop in the initial state and one flip-flop per state. This is equivalent to an implicit subset construction.

3 Results

The results shown in Fig. 5, where FFs is the number of flip-flops, ALUTs the amount of logic cells and f_{max} is the maximal clock frequency, were achieved by synthesizing the given assertions for an Altera Stratix-III FPGA.

Obviously, the amount of hardware resources needed for one assertion is marginal. However, in a design, there may are thousands of assertions, depending on the complexity of the DUV, therefore the hardware overhead need for

Assertion	FFs	ALUTs	f_{max} in MHz
assert always ({a[*2];b} |-> {c[*3]; b **and** c; c});	8	8	1114.43
assert always ({a} |-> {b[*]; c});	4	5	1428.57
assert always ({a} |-> ({b[*]; c}));	4	5	1428.57
assert always ({a} |=> ({{b; c}[*]; d}));	6	6	1193.32
assert always ({a; b} |=> {b; c[*2]});	6	5	1324.50
assert always ({a} |-> {{b; c}[*]; d});	2	2	1390.82
assert always ({a **and** b} |-> {c[->]});	9	9	1210.65
assert always ({a **and** b} |-> {c[->4]});	6	7	1436.78
assert always ({a} |-> {b[=1]; c});	4	3	1459.85
assert always ({a} |=> (**next_event**(b)(c)));	8	7	1379.31

Fig. 5. Synthesis results of PSL assertions

monitoring my be substantial. This leads to a tradeoff between observability and chip area consumption. Due to the high operating frequency f_{max} of the assertion checkers, the critical path of a design is unlikely to be influenced. This can be attributed to the hardware structure of the assertion checkers: even for complex assertions the needed combinational logic is quite simple.

4 Conclusion

In this paper, we presented an open-source tool which allows implementing PSL assertions in hardware by generating synthesizable HDL code from PSL formulas. The implementation of assertions in hardware provides several possibilities in emulation, post-silicon debugging, and non-intrusive observation of real-time systems during runtime [7].

We outlined an algorithm to convert nondeterministic automata, arising from the semantic of regular expressions, to deterministic automata. Furthermore, an approach was shown to allow a pipelined evaluation of assertions. As shown in the results of synthesis, the hardware overhead for a small number of assertion checkers is negligible and the critical path is unlikely to be influenced.

References

1. Accellera: Property Specification Language Reference Manual (September 2004)
2. Boulé, M., Zilic, Z.: Efficient automata-based assertion-checker synthesis of PSL properties. In: Proceedings of the 2006 IEEE International High Level Design Validation and Test Workshop (HLDVT 2006), pp. 69–76 (2006)
3. Clarke, E.M., Grumberg, O., Peled, D.A.: Model Checking. MIT Press, Cambridge (1999)
4. Eisner, C., Fisman, D.: A Practical Introduction to PSL. Springer Science+Business Media LLC, 223 Spring Street, New York, NY 10013, USA (2006)
5. Abarbanel, Y., et al.: FoCs: Automatic generation of simulation checkers from formal specifications. In: Emerson, E.A., Sistla, A.P. (eds.) CAV 2000. LNCS, vol. 1855, pp. 538–542. Springer, Heidelberg (2000)
6. Findenig, R.: Behavioral synthesis of PSL assertions. Master's thesis, Upper Austrian University of Applied Sciences, Hagenberg (July 2007)
7. Findenig, R., Eibensteiner, F., Pfaff, M.: Echtzeit-Überwachung von Hard- und Software mit PSL. In: Tagungsband Austrochip 2008, October 2008, pp. 44–47 (2008)
8. Pfaff, M.: Verfahren zur beschleunigten systemsimulation mit vhdl durch integration von externen hardware/software-komponenten. In: Reihe, C. (ed.) Schriften der Johannes Kepler Universität Linz, Technik und Naturwissenschaften. Universitätsverlag Rudolf Trauner (1999)
9. Ruah, S., Fisman, D., Ben-David, S.: Automata construction for on-the-fly model checking PSL safety simple subset. Technical Report H-0234, IBM Haifa Research Lab, Haifa (April 2005)

Learning Autonomous Helicopter Flight with Evolutionary Reinforcement Learning*

José Antonio Martín H.[1] and Javier de Lope[2]

[1] Dep. Sistemas Informáticos y Computación, Universidad Complutense de Madrid
jamartinh@fdi.ucm.es
[2] Dept. Applied Intelligent Systems, Universidad Politécnica de Madrid
javier.delope@upm.es

Abstract. In this paper we present a method to obtain a near optimal neuro-controller for the autonomous helicopter flight by means of an ad hoc evolutionary reinforcement learning method. The method presented here was developed for the Second Annual Reinforcement Learning Competition (RL2008) held in Helsinki-Finland. The present work uses a Helicopter Hovering simulator created in the Stanford University that simulates a Radio Control XCell Tempest helicopter in the flight regime close to hover. The objective of the controller is to hover the helicopter by manipulating four continuous control actions based on a 12-dimensional state space.

Keywords: Reinforcement Learning, Evolutionary Computation, Autonomous Helicopter.

1 Introduction

Helicopters have complex and noisy dynamics that makes difficult to create simple and suitable controllers. We have previously shown [1] the viability of modeling the dynamics of an autonomous helicopter using computer vision techniques and artificial neural networks (ANNs).

In this paper we present a method to obtain a near optimal neuro-controller for the autonomous helicopter flight by means of an ad hoc evolutionary reinforcement learning (ERL) method [2]. The method presented here was developed for the Second Annual Reinforcement Learning Competition (RL2008) held in Helsinki-Finland. The first place was obtained by Rogier Koppejan [3] using a similar but without online learning preventing, in a more aggressive strategy, any possible helicopter crashing.

The Helicopter Hovering competition-track was created by Pieter Abbeel, Adam Coates and Andrew Y. Ng from Stanford University [4]. The competition environment simulates an XCell Tempest helicopter in the flight regime close to hover. The agent's objective is to hover the helicopter by manipulating four

* This work has been partially funded by the Spanish Ministry of Science and Technology, project DPI2006-15346-C03-02.

R. Moreno-Díaz et al. (Eds.): EUROCAST 2009, LNCS 5717, pp. 75–82, 2009.

continuous control inputs based on a 12-dimensional state space. one of the added difficulties for the competition was the addition of simulated wind affecting the dynamics of the helicopter flight.

Also there is another competition planned for the 2009 in which new parameters will be modified for the helicopter simulator.

The main advantages of our approach are:

- Initial feasible solutions based on the base-line controller.
- Very fast convergence to reasonable good controllers.
- Very efficient execution time trough the use of ANNs.
- Convergence to deeply optimal values trough ad hoc adaptation procedures.
- Ad hoc exploration degree control to prevent dangerous exploratory maneuvers.

2 Common Evolutionary Optimization Framework vs. Real-Time Interaction Evolutionary Reinforcement Learning

There are many performance measures about the behavior of an evolutionary algorithm. In general the best fitness is used as the performance of the optimization algorithm since not success exploratory trials are ignored. But, in the interactive case such failed exploratory trials cannot be ignored since they have been actually performed and has influenced the behavior of the system. So, in this case the performance measure is the summation of all the fitness of all individuals or alternatively the mean fitness, which is also generally used in standard evolutionary optimization. These facts characterize a problem type that we call "Real-time interaction Evolutionary Reinforcement Learning" that requires of special solution methods different from the more traditional ERL (e.g. [2]) yielding thus to a more systemic approach when the entire population of the evolutionary algorithm has to be considered as part of the solution.

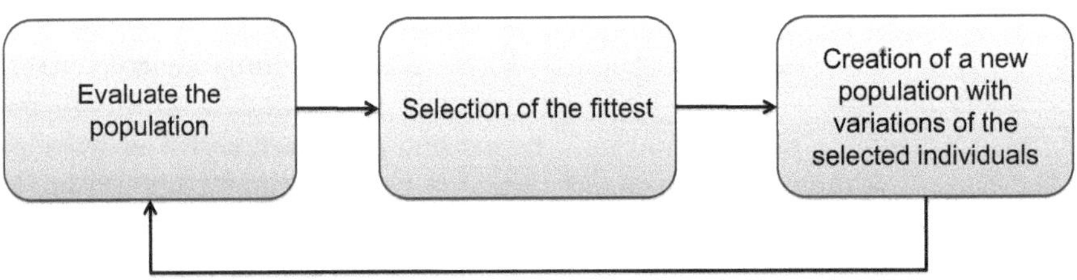

Fig. 1. Common evolutionary optimization framework

The common evolutionary optimization framework (fig. 1) is mainly characterized by the following features:

1. The Evolutionary algorithm has absolute control over the evaluation cycle and can evaluate individuals when it decides.

2. There is no need for real-time operation between evaluations.
3. There is *no penalty for exploration*, indeed exploration is the main purpose of an Evolutionary Algorithm.

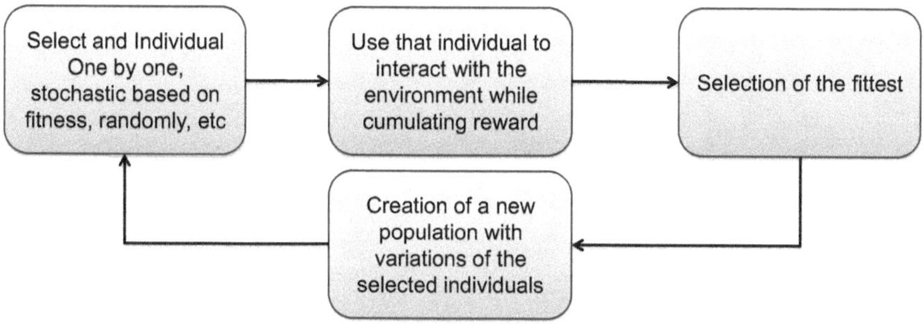

Fig. 2. Real-time interaction Evolutionary Reinforcement Learning

Instead, Real-time interaction ERL (fig. 2) is mainly characterized by the following features:

1. The Evolutionary algorithm has no control over the evaluation cycle and evaluate individuals on demand.
2. There is a need for real-time operation between evaluations.
3. The nature of the operation is interactive and the goal is generally to obtain high reward, so very well balance between exploration and exploitation is needed.
4. There is *penalty for exploration* when low reward is achieved.

3 The Helicopter Hovering Problem

The competition environment simulates an XCell Tempest helicopter (fig. 3) in the flight regime close to hover. The agent's objective is to hover the helicopter by manipulating four continuous control inputs based on a 12-dimensional state space.

The observation space is 12 dimensional continuous valued defined as follows:

u**:** forward velocity.
v**:** sideways velocity (to the right).
w**:** downward velocity.
x_{err}**:** helicopter x-coord position − desired x-coord position.
y_{err}**:** helicopter y-coord position − desired y-coord position.
z_{err}**:** helicopter z-coord position − desired z-coord position.
ϕ**:** angular rate around helicopter's x axis.

Fig. 3. A picture of the real XCell Tempest RC Helicopter used in [4]

θ: angular rate around helicopter's y axis.
ω: angular rate around helicopter's z axis.
p: angular velocity helicopter's x axis.
q: angular velocity around helicopter's y axis.
r: angular velocity around helicopter's z axis.

The action space is defined as a 4 dimensional continuous valued vector:

aileron: longitudinal (front-back) cyclic pitch.
elevator: latitudinal (left-right) cyclic pitch.
rudder: main rotor collective pitch.
coll: tail rotor collective pitch.

The main difficulty in solving this problem is what we call the "Helicopter Paradox".

The competition simulator[1] is set up to run for six thousand (6000) time steps, and each simulation step is 0.1 seconds, thus giving runs of ten minutes (although the simulator runs faster than real-time). If the simulator enters a terminal state (the helicopter crashes) before six thousand time steps. The crashing event is determined by a set of constraints imposed to the simulator such as:

velocity: the velocity along any of the main axes exceeds $5m/s$.
position: any of the true position measures x, y, z (not observable by the controller) is off by more than $20m$.
angular rate: the angular rate in any of the helicopter's axes is more than 30 deg from the target orientation.

[1] After the RL2008 competition the software of the simulator was made public in the following address: `http://rl-competition.googlecode.com`

angular velocity: the angular velocity around any of the helicopter's axes exceeds $4\pi rad/s$.

When the helicopter crashes a large negative reward is given, corresponding to getting the most negative reward achievable for the remaining time.

Thus the main problem is that: *for obtaining a high reward, high exploration rates are needed, but high exploration rates can cause a system crash yielding a large negative reward.*

So, free, global and wide exploration, which is the main purpose of an Evolutionary Algorithm is, in this case, dangerous and in some sense opposed to get higher rewards.

4 Evolutionary Reinforcement Learning Approach

The selected approach to solve the Helicopter Hovering Problem is the evolution of neuro-controllers (e.g. [5]).

We designed our evolutionary algorithm following a systemic methodology, that is, it is necessary to consider all the population of the evolutionary algorithm as part of the solution. We must recall that a system is a collection of *interacting interrelated* elements.

So, in a population of an evolutionary algorithm from a system point of view:

1. There is a degree of interrelation or integration of all the elements.
2. There is a degree of interaction between the elements.

Thus, we can design and take effective control of an evolutionary system by controlling these two variables.

The degree of interrelation or integration in the system will be mainly controlled by the evolutionary operators for producing offspring, such as the mutation and matting mechanisms, while the interaction between the elements, apart from the mutation and matting mechanisms, have an additional mechanism consisting on a competitive and selectional process.

4.1 Evolutionary Algorithm Description

Each genotype is a vector whose length is determined by three fixed variables:

1. The number of input nodes $(ni + bias)$
2. The number of nodes in the hidden layer (nh)
3. The number of output nodes of the multi-layer perceptron (no)

Thus, the genotype will be formed by all the weights of each layer, say:

1. The weights of the input layer:

$$|w_i| = (ni + 1) \times nh$$

Algorithm 1. The EvaluateNN function to evaluate a Neuro-Controller

function EvaluateNN (inputs, genotype)
  $w_i \leftarrow$ genotype$[1...(ni + 1) \times nh]$
  $w_o \leftarrow$ genotype$[((ni + 1) \times nh + 1) \; ...N]$
  $a_i \leftarrow [inputs; 1]$ —1 for bias node
  $a_h \leftarrow \tanh(a_i \cdot w_i)$
  $a_o \leftarrow \tanh(a_h \cdot w_o)$
return a_o

2. The weights of the output layer:

$$|w_o| = nh \times no$$

Hence, a genotype will be a vector of real numbers of cardinality:

$$N = |w_i| + |w_o|$$

Thus, for constructing an individual's phenotype, that is, its respective artificial neural network, we need just to extract the corresponding segments of the genotype vector in order to create the weights matrix of each layer. In order to evaluate the behavior of any evolved controller we can then apply the Algorithm 1 to each genotype:

4.2 Evolutionary Mechanism

While in traditional evolutionary approaches the algorithm has complete control over de optimization process and generally implements an loop of evaluation over all individuals of the population, in interactive evolutionary reinforcement learning such a loop can not be implemented. Instead, the results of the evaluation of all individuals is recorded when they are evaluated. We used an exhaustive exploration policy, that is evaluating all the individuals one by one, and recorded the fitness for all individuals. After each complete evaluation of the population the evolutionary operators (i.e. selection, matting and mutation) are applied to the population taking into account its respective fitness.

We have carefully designed this operators in order to maintain learning but taking care about the dangerous effect of producing non-viable offspring (i.e. crashing controllers). A controlled "Chained Weighted Matting" mechanism and a "Low Deviation Gaussian Mutation" where implemented.

Chained Weighted Matting. Weighted matting (fig. 4) is an effective way of assuring offspring viability since weights can be set in order to produce individuals very close to one of its parents. This matting strategy assures the interrelation of the system since this produces a chain of correlations that propagates over the entire population. We selected for the weighted matting the best 21 (from 1 to 20 vs. 2 to 21) individuals.

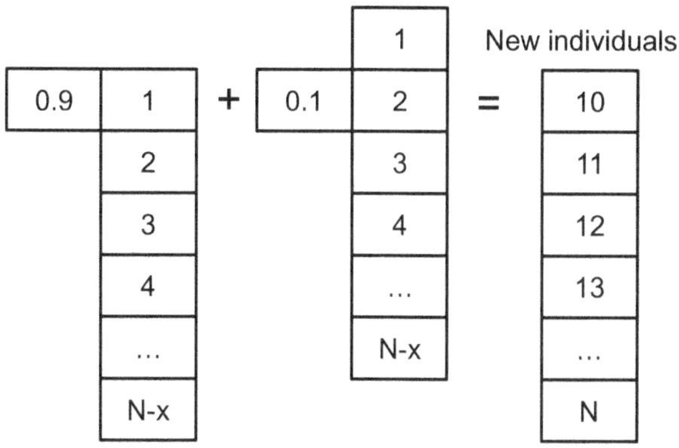

Fig. 4. Chained Weighted Matting

Low deviation Gaussian Mutation. A strategy for producing viable offspring with high probability is to produce new individuals as Gaussian mutations (fig. 5), with very low standard deviation, from the best individuals. This is achieved by generating a matrix of random normal distributed numbers with zero mean and a very low standard deviation and then adding up this matrix to the selected individuals to be mutated. We, again, selected the best 20 individuals to produce mutated individuals.

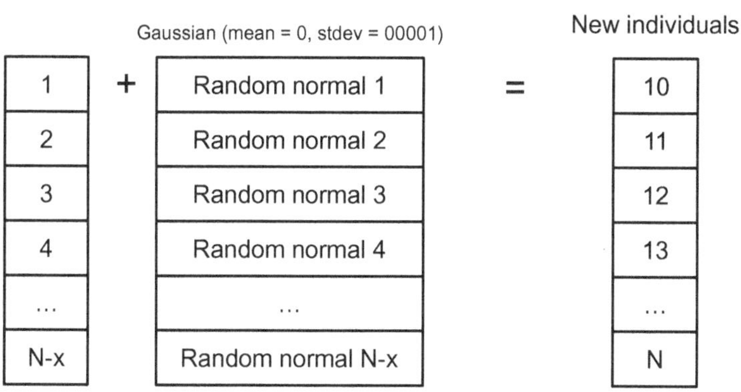

Fig. 5. Low deviation Gaussian Mutation

5 Conclusion and Further Work

We have presented a method to obtain near optimal neuro-controllers for the autonomous helicopter flight by means of an ad hoc evolutionary reinforcement learning method. Our method has been designed to deal with a kind of problem that we called "Real-time interaction Evolutionary Reinforcement Learning" that requires special solution methods different from more traditional ERL (e.g. [2]).

Our proposed method was based on a systemic approach that consider the entire population of the evolutionary algorithm as the solution as a whole and not only the best individual but also this approach is different from the classical distinction between Michigan vs. Pittsburgh style evolutionary algorithms.

We where able to produce highly efficient Helicopter Controllers without crashes and we where able to continue learning during execution time which is a feature known as online learning (learning while in operation mode). Also, we obtained the second best mark during the proving phase on the RL2008 Competition with no statistical difference with the first place which means that the algorithm performs well in comparison with other approaches.

Our further work will include the control of a physical helicopter in a controlled ambient (indoor) and also outdoor for trying special maneuvers beyond hovering.

References

1. de Lope Asiaín, J., Martín, J.J.S., José Antonio Martin, H.: Helicopter flight dynamics using soft computing models. In: Moreno Díaz, R., Pichler, F., Quesada-Arencibia, A. (eds.) EUROCAST 2007. LNCS, vol. 4739, pp. 621–628. Springer, Heidelberg (2007)
2. Moriarty, D.E., Schultz, A.C., Grefenstette, J.J.: Reinforcement learning through evolutionary computation (1999)
3. Koppejan, R., Whiteson, S.: Neuroevolutionary reinforcement learning for generalized helicopter control. In: GECCO 2009: Proceedings of the Genetic and Evolutionary Computation Conference (to appear, July 2009)
4. Ng, A.Y., Kim, H.J., Jordan, M.I., Sastry, S.: Autonomous helicopter flight via reinforcement learning. In: Thrun, S., Saul, L.K., Schölkopf, B. (eds.) NIPS. MIT Press, Cambridge (2003)
5. Jose Antonio Martin, H., de Lope, J., Santos, M.: Evolution of neuro-controllers for multi-link robots. In: Innovations in Hybrid Intelligent Systems. Advances in Soft Computing, vol. 44, pp. 175–182 (2008)

Designing Communication Space in Wireless Sensor Network Based on Relational Attempt

Jan Nikodem

The Institute of Computer Engineering, Control and Robotics
Wrocław University of Technology
11/17 Janiszewskiego Street, 50-372 Wrocław, Poland
jan.nikodem@pwr.wroc.pl

Abstract. In this paper, we describe results of designing a communication space in WSN. To achieve this goal we propose notional system which can convey sophisticated WSN reality onto some mathematical abstraction. We concentrate on three of them; neighborhood, communication space and relations. Consequently our work concentrate on developing the formal methods and techniques necessary to model and evaluate communication space in the network. Proposed approach pointed that: neighborhoods are more useful than clusters because they provide more communication connections, communication space is better than routing paths because it spreads evenly energy losses, relations work better than functions because of topology based properties.

Keywords: wireless sensors network, relational systems.

1 Problem Formulation and Related Work

The fact that compound problems can be solved by partitioning them into small pieces that are simpler and even easier to investigate, was noticed a long time ago. The most fascinating in this issue is, how to make such partitioning, that subsequent process of partial solutions looks up and next their composition gives us a decent result.

WSN communication management in WSN is undoubtedly a complex issue, due to the vast number of network components, abundance of potential communication junctions and their changeability in time and a space. We cannot forget about common transmission medium utilization, which introduces collisions, arbitration and bandwidth consumption, and of course, a limited range of wireless communication, and that carry with it the necessity of a strict cooperation during information pass on for long distances.

There is a huge number of papers considered communication aspects in WSN, related mainly to clustering and routing problems. On the one hand, scientists have discussed self-configuring of sensors [1], self-management [2,14,18], adaptive clustering [1,7,19] or concept of adjustable autonomy [4]. On the other hand, there are papers, which discussed bio-inspired ideas and tend to isolate some aspects of the natural world for computer emulation. Authors [3] have shown that

R. Moreno-Díaz et al. (Eds.): EUROCAST 2009, LNCS 5717, pp. 83–90, 2009.

the communication topology of some biological, social and technological networks is neither completely regular nor completely random but stays somehow in between these two extreme cases. It is worth to mention papers [19,18,2] devoted to self-organizing protocols using both random and deterministic elements.

In order to manage communication activities effectively, one has to address the problems of sensor network organization and the subsequent reorganization and maintenance. It would be desirable that the initial sensor network organization take advantage of the basic physical attitudes and topological characteristics. The initial communication structure must be reorganized repeatedly to adapt to the changing environment and varied network traffic. The decision about reorganization is taken globally and this process is realized mainly in deterministic manner [19]. Based on functional description, nodes perform algorithms (we *control* this process) and as a result, we obtain new routing paths.

2 Modeling Node's Neighborhood

Processes of clustering and routing path selection are experientially grounded in wireless network communication. Partitioning a complex problem into simple subproblems allows to use less sophisticated tools and leads to efficient methods and algorithms. On this foundation, it is intended to develop mostly hierarchical, either proactive or reactive methods of managing communication activity in WSN. To achieve this goal it is required to transform notional system which can convey sophisticated WSN reality onto some mathematical abstraction.

Let $Map(X,Y)$ denotes a set of mapping functions from set X onto set Y (i.e. surjection) and let $Sub(X)$ denotes a family of all subsets of X. Using $Map(X,Y)$ and $Sub(X)$ enables to define the neighborhood $\mathcal{N}$ as follows:

$$\mathcal{N} \in Map(Nodes, Sub(Nodes)). \tag{1}$$

Neighborhood of node k is thus defined as:

$$\mathcal{N}(k)_{|k \in Nodes} := \{y \in Nodes \mid y\, \mathcal{R}_{\mathcal{N}}\, k\}, \tag{2}$$

whereas $\mathcal{N}(S)$

$$\mathcal{N}(S)_{|S \subset Nodes} := \{y \in Nodes \mid (\exists x \in S)(y\, \mathcal{R}_{\mathcal{N}}\, x)\}. \tag{3}$$

defines neighborhood of all nodes from the set S. The $\mathcal{N}(k)$ abstraction is a very flexible tool that allows to model WSN communication activity. It can be used to model neighborhoods which can be defined in various ways. We can point out some communication properties either essential or background for our algorithm, but usually communication range is used since it is essential to the operation of the nodes and the whole WSN. Radio link range is thus an example of $\mathcal{R}_{\mathcal{N}}$ ("is in the range") relation.

Routing algorithms take advantages of a concept of neighborhood and often partitioning a set of WSN nodes onto subsets called clusters. It is so, since clusters have several advantages:

- allow to build hierarchical structures with cluster heads and regular nodes,
- reduce mesh communication, place restrictions on regular nodes activity within cluster,
- increase efficiency of multi-hop communication since only cluster heads are responsible for message routing.

Any clustering algorithm decides whether particular node becomes the cluster head or a regular one. Nodes can communicate only with a small set of its neighbors and its activity is restricted to this set of nodes. As a consequence a specific types of neighborhood are created; clusters as well as routing trees and/or routing graphs.

Let us define cluster C as a mapping function

$$C \in \{Map(Nodes, Sub(Nodes))\} \tag{4}$$

where

$$C(k)_{|_{k \in Nodes}} := \{y \in Nodes \mid y\, R_C\, k\,\} \tag{5}$$

and k is a cluster's main node (cluster head).

Based on clustering relation R_C we can build clusters which are both pairwise disjoint and collectively exhaustive with respect to the set $Nodes$. Formally, clusters are indexed family of sets $\mathcal{C} = \{C_i \mid i \in I\}$ for which following properties hold:

$$(\forall i \in I)(C_i \neq \emptyset) \;\wedge\; \bigcup C_i = Nodes \tag{6}$$

and

$$(\forall i, j \in I \mid i \neq j)(C_i \bigcap C_j = \emptyset) \tag{7}$$

Conditions (6) and (7) imply that:

$$(\forall y \in Nodes)(\exists!\, i \in I \mid\; y \in C_i) \tag{8}$$

where $\exists!$ means "exists exactly one".

The formulas (6),(7) describe clusters from global (network) point of view and advantages of such approach are evident and clearly seen on figure 1. While the right part of 1 illustrates neighborhoods defined in terms of communication range (marked with circles), the left side illustrates clusters and routing trees There is no doubt that cardinality of neighborhood set and its extensive overlapping shows clusterization more attractive than neighborhood. Moreover, notations used in eq. (6) and (7) simplifies and clear our view of Wireless Sensor Network. The cardinality of clusters is significantly lower because cluster brings in hierarchy of nodes that also simplifies and differentiates both; the data collection processes (from regular node towards cluster head, inside the cluster) and data routing (between cluster heads only, towards base station). Clusterization (6),(7) allows to restrict a set of possible communication channels to multi-hop routing paths on a cluster head level only. Regular nodes (all which are not a cluster head) must communicate only with its cluster head. As a result we obtain clear and well determined situation (see left side of fig.1). At least, ordering and

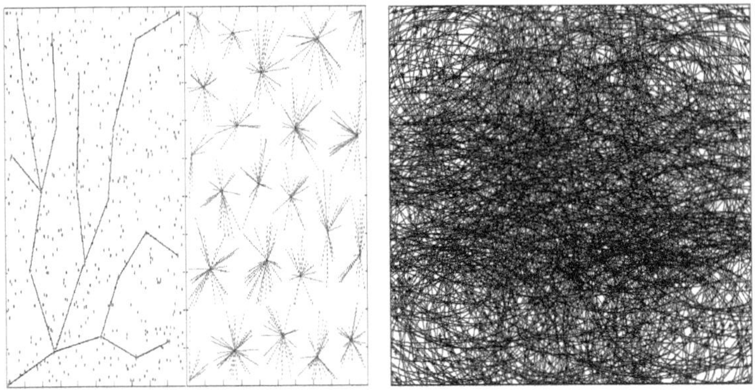

Fig. 1. Clusters vs. neighborhoods in WSN (globally)

hierarchization of WSN nodes are real advantages of clusterization and routing processes.

Apart from advantages clusterisation has drawback too. When comparing (1) with (4) and (5) it can be realised that clusters are in fact restricted neighborhoods but restriction leads to limitations. In order show what are the limitations of clusterisation we construct a neighborhood abstraction based on (6) and contradiction of corollary (7)

$$(\forall i \in I)(N_i \neq \emptyset) \ \wedge \ \bigcup N_i = Nodes \tag{9}$$

$$(\forall i, j \in I \mid i \neq j)(N_i \bigcap N_j \neq \emptyset). \tag{10}$$

Conditions (9) and (10) imply that:

$$(\forall y \in Nodes)(\exists^{\sim} i \in I \mid y \in \bigcap N_i \neq \emptyset). \tag{11}$$

where $\exists^{\sim}$ means "exist as many as feasible" Considering clusterization process from the local point of view (fig.2), radical restriction of node's communication capabilities is its main disadvantage. Any node y in the network has only one (via cluster head) communication path towards base station whereas there are other nodes in its neighborhood $N(y)$ (fig.2.a). Each of these neighbors $k \in N(y)$ can provide communication services (fig.2.b) and some nodes $x \in N(y)$ can do this with similar efficiency as cluster head can, based on its routing path (fig.2.c). Thus, any of these nodes can realize next hop in communication path towards base station.

3 Relational Approach to Communication Activity

Recognising that functions are not sufficient for describing complex and distributed systems, we have introduced (following by [6]) the general presumption

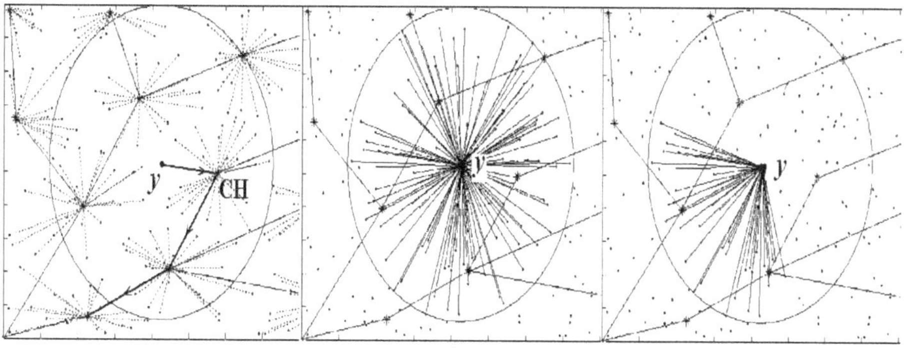

Fig. 2. Clusters vs. neighborhoods in node y (locally)

of the existence of three binary relations. These three relations; collision ($\varkappa$), subordination (π) and tolerance (ϑ) are defined on the set of action (Act) and describe communication activities in WSN. Because binary relation $\mathcal{R}$ on a set A can be represented as a set of ordered pairs $< x, y >$ where $x, y \in A$ and x, y are in relation $\mathcal{R}$ (which we denote $x\mathcal{R}y$) thus we can define relation as:

$$\mathcal{R} = \{< x, y >| \, x, y \in A, x\mathcal{R}y\}. \tag{12}$$

Based on this definition one may interpret a relation as a set such that $\mathcal{R} \subset A \times A$. We may also define the converse of relation $\mathcal{R}$ as:

$$\mathcal{R}^{-1} = \{< x, y >| \, x, y \in A, y\mathcal{R}x\}. \tag{13}$$

Based on these definitions we can write down two basic properties of $\varkappa, \pi$ and ϑ relations [6]:

$$\pi \cup \vartheta \cup \varkappa \subset Act \times Act \neq \emptyset, \tag{14}$$

and

$$\iota \cup (\pi \circ \pi) \subset \pi, \tag{15}$$

where ι is a identity relation on the set Act. Formula (14) states that all three relations are binary on non-empty set of $Actions$. Formula (15) states that subordination is reflexive ($\iota \subset \pi$) and transitive ($\pi \circ \pi \subset \pi$). Further

$$\pi \cup \vartheta^{-1} \cup (\vartheta \circ \pi) \subset \vartheta \tag{16}$$

means that:

- subordination implies tolerance – if π holds for some $x, y \in Act$ then ϑ also holds for these,
- tolerance is reflexive – if $x\vartheta y \Rightarrow y\vartheta x$,
- subordinated action tolerate all actions tolerated by the dominant – if $(x\pi y \wedge y\vartheta z) \Rightarrow x\vartheta z$.

For collision relation we have that

$$\varkappa^{-1} \cup \{\pi \circ \varkappa\} \subset \varkappa \subset \vartheta' \tag{17}$$

where ϑ' is the complement of ϑ:

$$\vartheta' = \{< x, y >\in X \times Y \mid < x, y >\notin \vartheta\}. \tag{18}$$

Formula (17) states that collision is symmetric ($\varkappa^{-1} \subset \varkappa$), disjoint to tolerance ($\varkappa^{-1} \subset \vartheta'$) and subordinated action must be in collision with any action being in collision with dominant ($(\pi \circ \varkappa\} \subset \varkappa$).

Paper [10] presents step by step how to use in simulation algorithms, relations π, ϑ and $\varkappa$ in order to model spatial communication. Subordination π is responsible for multihop path generation. A growing intensity quotient of π results in extension of different multihop paths in w communication space. π is responsible for a set of pontifex (elements joined different paths). Tolerance ϑ is responsible for range of communication space. A bigger intensity quotient of ϑ widens communication space and extends possibility of parallel paths. However collision $\varkappa$ allows to form surface restrictions for the communication space. Relational approach provides us with good tool for profile communication space. Using this tool, it is possible to design required properties of communication space. It is possible to profile communication space narrow or wide (ϑ), to obstruct selected area ($\varkappa$) and other point as especially recommended freeways for information flow (π).

4 Concluding Remarks

The global description of WSN communication activity presented by formulas (6),(7) (for clusters) or (9),(10) (for neighborhoods) are suitable for simulators. When communication activity is design in real network, we should write programs for each node. In such situation the most useful are corollars (8),(11) because they describe requirements individually for each node. WSN is real distributed system in which the dilemma of holographic principle (the whole not only contains its parts, but also is contained by each part) is up to date. Novel approach proposed in this paper is distinctly different from mentioned above. Neighborhood abstraction and relational attempt [6,8,9], based on more abstract areas of mathematics (set theory), considers three relations: subordination π , tolerance ϑ and collision $\varkappa$. Using these relations, we perform initial organization, but there is no necessity to perform any reorganization during a network lifetime. Each node selects element of routing path, based on local balance between π , ϑ and $\varkappa$, so the final result is not specified by a concrete function but is governed by relational principles at nodes vicinity. Based on relational description nodes perform actions (we *design/shape* this process) and as a result we obtain new element of routing paths. Because of using both random and deterministic elements, the resulted routing path is unrepeatable.

Proposed approach allows an interaction between sensor's vicinity during the routing path selection. When sensor detects changes in his vicinity or detects deviations from the normal neighbor's behavior it has many different choices of next element of routing path and thus reduces the impact of a failure in any single component.

A developed mathematical model applies to the distributed systems. Currently by its utilization, we can model such dichotomical aspects of complex system as locality vs. globality, cooperation vs. autonomy or adaptation vs. immunity. In the future, we plan to study a migration of Pareto front inside and between neighborhoods and designing in complex system such abstraction as tactics, strategy and politics as a choice of intensity quotients of subordination, tolerance and collision relations.

References

1. Cerpa, A., Estrin, D.: ASCENT: Adaptive Self-Configuring Sensor Networks Topologies. IEEE Transactions on Mobile Computing 3(3) (July-September 2004)
2. Chevallay, C., Van Dyck, R.E., Hall, T.A.: Self-organization Protocols for Wireless Sensor Networks. In: Thirty Sixth Conference on Information Sciences and Systems (March 2002)
3. Cohn, A.G., Bennett, B., Gooday, J.M., Gotts, N.M.: Representing and Reasoning with Qalitative Spatial Relations about Regions. In: Cohn, A.G., Bennett, B., Gooday, J.M., Gotts, N.M. (eds.) Spatial and Temporal Reasoning, pp. 97–134. Kulwer, Dordrecht (1997)
4. Crandall, J.W., Goodrich, M.A.: Experiments in adjustable autonomy. In: IEEE International Conference on Systems, Man, and Cybernetics, Tucson, USA, vol. 3, pp. 1624–1629 (2001)
5. Chaczko, Z., Ahmad, F.: Wireless Sensor Network Based System for Fire Endangered Areas. In: ICITA 2005, Sydney (2005)
6. Jaroń, J.: Systemic Prolegomena to Theoretical Cybernetics, Scient. Papers of Inst. of Techn. Cybernetics, Wrocław Techn. Univ., Wrocław, vol. 45 (1978)
7. Lin, C.R., Gerla, M.: Adaptive Clustering for Mobile Wireless Networks. IEEE Journal On Selected Areas In Communications 15(7) (September 1997)
8. Nikodem, J.: Autonomy and Cooperation as Factors of Dependability in Wireless Sensor Network. In: Proceedings of the Conference in Dependability of Computer Systems, DepCoS - RELCOMEX 2008, Szklarska Poreba, Poland, June 2008, pp. 406–413 (2008)
9. Nikodem, J., Klempous, R., Chaczko, Z.: Modelling of immune functions in a wireless sensors network. In: Giovanni, C.S. (ed.) W: The 20th European Modeling and Simulation Symposium. EMSS 2008, Italy (2008)
10. Nikodem, J.: Relational Approach Towards Feasibility Performance for Routing Algorithms in Wireless Sensor Network. In: Proceedings of the Conference in Dependability of Computer Systems, DepCoS - RELCOMEX 2009, Szklarska Poreba, Poland, June 2008 (2009) (in printing)
11. Pichler, F.: Modeling Complex Systems by Multi-Agent Holarchies. In: Kopacek, P., Moreno-Díaz, R., Pichler, F. (eds.) EUROCAST 1999. LNCS, vol. 1798, pp. 154–168. Springer, Heidelberg (2000)
12. Scerri, P., Pynadath, D., Tambe, M.: Towards Adjustable Autonomy for the Real World. Journal of Artificial Intelligence Research 17 (2003)
13. Schillo, M.: Self-organization and adjustable autonomy: Two sides of the same medal? Connection Science 14(4), 345–359 (2003)

14. Sohrabi, K., Gao, J., Ailawadhi, V., Pottie, G.J.: Protocols for Self-Organization of a Wireless Sensor Network. IEEE Personal Communications (October 2000)
15. Su, P., Feng, D.: The Design of an Artificial Immune System. In: Int. Conf. on Networking, Systems and Mobile Communications and Learning Technologies (2006)
16. Vaidya, D., Peng, J., Yang, L., Rozenblit, J.W.: A Framework for Sensor Management in Wireless and Heterogeneous Sensor Network. In: ECBS 2005, 12th IEEE International Conference on the Engineering of Computer-Based Systems, Greenbelt, USA, April 4-7, pp. 155–162 (2005)
17. Veeramachaneni, K., Osadciw, L.: Dynamic Particle Swarm Optimizer for Information Fusion in Non Stationary Sensor Networks. In: IEEE Swarm Intelligence Symposium, Indianapolis, USA (2006)
18. Veyseh, M., Wei, B., Mir, N.F.: An Information Management Protocol to Control Routing and Clustering in Sensor Networks. Journal of Computing and Information Technology - CIT 13(1), 53–68 (2005)
19. Younis, O., Fahmy, S.: HEED: A Hybrid, Energy-Efficient,Distributed Clustering Approach for Ad Hoc Sensor Networks. IEEE Transactions on Mobile Computing 3(4) (October-December 2004)

Boundary Scan Security Enhancements for a Cryptographic Hardware

Maciej Nikodem

Wrocław University of Technology,
The Institute of Computer Engineering, Control and Robotics
Wybrzeże Wyspiańskieg 27, 50-370 Wrocław, Poland

Abstract. Boundary scan (JTAG) is a powerful testing scheme that is widely used in nowadays circuits to maintain and verify operation of the hardware. However, JTAG is not used in cryptographic hardware since it may be used to compromise security of the implemented cryptographic algorithm. This paper analyses different solutions proposed to overcome the threat of such attacks, presents requirements that have to be satisfied in order to construct effective security solution, and presents novel proposal that improves security of the boundary scan.

Keywords: boundary scan, IEEE 1149, side-channel attacks, countermeasures.

1 Introduction

Scan based Design-for-Test (DFT) is a powerful testing scheme that enables to verify correctness of device operation during its whole life time. Large potential of this scheme results from the ability to control and observe device behaviour in detail. These properties are desired for most hardware but not for cryptographic devices where controllability and observability may simplify attacks and thus led to security leaks [3,10]. Although, there is no unambiguous information about how danger to cryptographic hardware test circuits may be [8], it is a common practice to test cryptographic devices only during manufacturing process. Afterwards, before leaving the factory, all test outputs and/or connections embedded in device are destroyed, turning it into an untestable black-box. Destroying test circuits and/or connections is a method to preserve security of the cryptographic hardware. However, impossibility to test the cryptographic hardware is also a drawback, especially from user's point of view. This is due to existence of a subliminal channels that may be embedded in hardware and leak user's secret information to the manufacturer. Verifying whether device implements subliminal channel based on input-output analysis is very difficult or even impossible.

2 Related Work

Since both testability and security of cryptographic circuits is required thus there is a need to construct secure DFT scheme. Several proposals how to bring

R. Moreno-Díaz et al. (Eds.): EUROCAST 2009, LNCS 5717, pp. 91–97, 2009.

together these two properties have been proposed in literature [4,5,7,9,11]. Most of them can be described as an extension to the boundary scan (BS) technique which is a IEEE 1149 standard [12]. It is so, since BS introduce relatively small implementation complexity, enables to test the device during whole life cycle and gives the possibility to adjust test vectors, reducing costs of manufacture and service.

Sophisticated solution was proposed by Hély et al. [5] who proposed to substitute the standard BS with a dynamic BS that reconfigures randomly every clock cycle. Reconfiguration process is controlled by a control circuit with a pseudo-random bit generator (PRBG) which defines the order of flip-flops in the BS. Reconfiguration happens all the time causing the BS structure to be unknown to everyone except dedicated user. On the other hand dedicated user may enter the secret key that configures BS in a predefined manner that allows to test the device easily.

Mirror key registers (MKRs) are another proposal presented in [11]. Purpose of MKRs is to separate the secret key, stored inside the device, from the BS. Although this proposal is both sound and effective it address specific situation when the BS runs around a secret key register. Additionally using MKRs does not ensures security against all types of the side channel attacks – the attacker may benefit from complete input-output data analysis and use this knowledge to attack the device.

Solution proposed by Gomółkiewicz et al. [4] takes advantage of message authentication code based on cyclic redundancy check (CRC-MAC). CRC-MAC was selected since it can be simply implemented in hardware with linear feedback shift register (LFSR) and integrated to the boundary scan. Precisely, Gomółkiewicz et al. proposed to connect CRC-MAC serially at the end of the scan chain and use it to calculate the hash value of the test response. Together with additional counter the whole solution ensures that only the hash value outputs the scan chain giving no information about exact test response to the user. Testability in this proposal requires that user is given either the correct hash value for a given test vector or knows initial vector of the LFSR. Given the correct hash value one can only compare it with hash output from the scan chain. If there is a difference between these two it means that the device operates incorrectly. On the other if the hash output is correct then with high probability device operates correctly. However, due to error masking during computation of CRC-MAC, it may happen that computed hash value is correct while test response is incorrect due to errors. If the user is given the initial vector of the LFSR then he may use computed hash to trace back the computation of CRC-MAC and determine the test response to a given test input. Solution proposed in [4] also lowers the throughput of the boundary scan, reducing the maximal clock frequency due to cascades of EXOR gates in the feedback of LFSR.

Interesting approach was presented by Lee et al. [7] who join the test key with test vector so that the resulting test input is dedicated to a particular device (i.e. device with this key embedded). This proposal eliminates the need to modify the TAP (Test Access Port) since there is no dedicated instruction to load the test

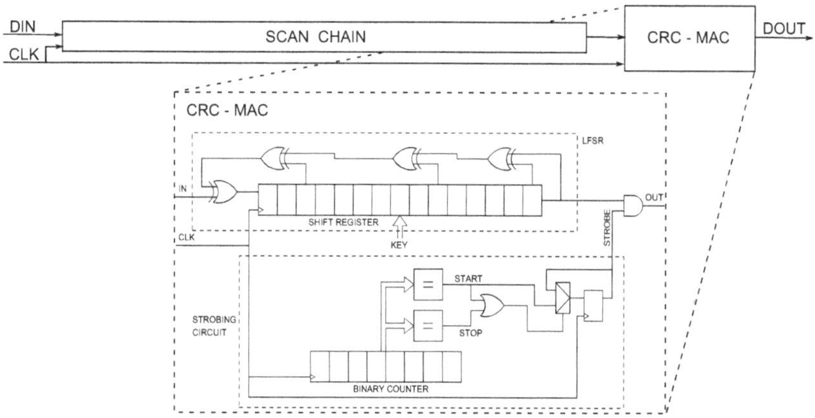

Fig. 1. Structure of CRC-MAC architecture for securing boundary scan [4]

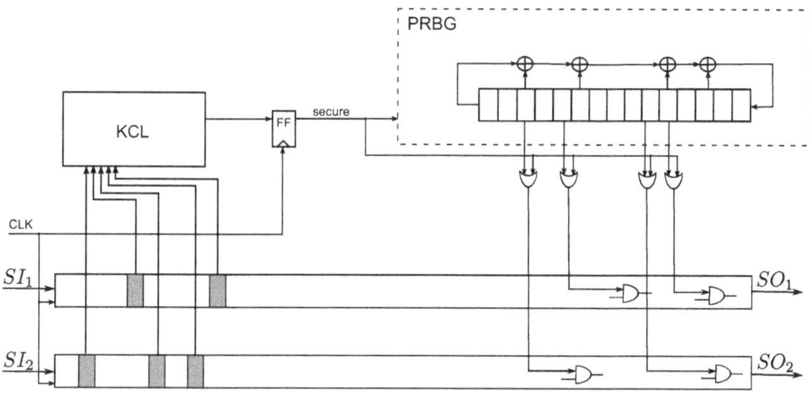

Fig. 2. Overview of secure scan solution proposed by Lee [7]

key (as it was in [5,11]) but key is loaded along with the test vector. However, this solution sounds it is susceptible to attacks since every single bit of the test input belongs either to the key or the test vector. If the attacker gets several test inputs for a particular device then he may compare them and decide which bits surely belong to the key. It is so since positions of key bits within the test input are constant so these bits will be the same for all the test inputs while bits of the test vector may change. Moreover, even given only one correct test input for a given device, the attacker can determine position of the key and its value. To verify whether a particular bit belongs to the key or the test vector, the attacker simply flips this bit and tests the device twice. Because security solution proposed by [7] generates random output if the key is incorrect, therefore if the attacker gets two different test outputs then he knows that bit of the key was flipped. If both test outputs are the same then the bit flipped belongs to the test vector. Repeating this test for every single bit of the test input gives information about key and its position within the test input.

There are several other proposals that aim to incorporate logic circuits, protocols or simple cryptographic algorithms in order to ensure that only entitled users are able to test the device [4,5,7,11]. Unfortunately neither sound nor effective solution has been proposed so far – some proposals yields large implementation overhead [5], ensure security against specific types of attack [11], limit the testability [4] or can be easily evaded [7].

In our paper we analyse existing improvements of BS security that attempt to determine properties of effective secure design-for-test scheme for cryptographic hardware. We also present new proposal that utilises simple cryptographic unit that gives ability to test the device to dedicated users only. Our proposal can be easily integrated with the standard BS and some of the previous proposals (e.g. [4]). It ensures security through output secrecy based on secret key embedded in the device and enables to maintain both security and testability by proper choice of parameters.

3 Boundary Scan for Cryptographic Hardware

There are several properties that one has to keep in mind when designing a secure boundary scan. First of all it has to be remembered that primary aim of using BS is to ensure testability. It means that user has to be able to use BS in order to verify whether the device operates correctly or not. Therefore any security enhancement should not affect the testability. If it does (like proposal [4]) then it should be kept as small as possible.

The purpose of securing the BS is to restrict observability of the device under test for an unauthorized user. At the same time user should be allowed to verify device operation correctness. Ability to verify device operation is crucial to security since sometimes even single erroneous execution of a cryptographic algorithm may leak secrets. This requirement states that the manufacturer has to be able to equip user with test input-output pairs that will allow him to verify operation of the device. Simultaneously user has to be prevented from making experiments by introducing his own test inputs and analyzing resulting test outputs.

Concerning implementation details one has to keep in mind that boundary scan is stream-oriented circuit so any improvements should be stream-oriented too for ease of integration. Therefore it is advisable to use stream-oriented cryptographic algorithms rather than block-oriented ones. That is why Gomółkiewicz et al. [4] proposed to use CRC-MAC instead of block hash functions (e.g. MD5, SHA1).

3.1 Our Proposal

Nevertheless security drawbacks of Lee et al. [7] proposal we think that incorporating the key into test vector is a promising idea. However, key checking logic

(KLC) and random response network (RRN) proposed in [7] are two sub circuits that need to be modified, since they are responsible for threats presented in previous section. KCL is a simple circuit that verifies whether predefined bits of the test input compose the correct key. If not then KCL outputs *secure* signal to notify the RRN that the key is incorrect and the test output has to be randomly modified in order to prevent attacks. Specific fact about the KCL is that it verifies only the predefined bits of the test input causing any changes in the remaining bits to go undetected. When KCL detects changes in the test key the RRN network is responsible for randomizing the output of the BS in order to prevent analysis of the test output. RRN circuit composes of pseudo random bit generator (PRBG) that determines the way BS output is modified. Unfortunately, this is not done in key dependent manner but key simply turns the RRN on and off depending on whether it was correct or not.

Analysis of previous security proposals for BS lead us to following observations:

1. user should not be able to play with the key,
2. the security countermeasures has to be key dependent,
3. any change to the test input has to cause changes in the test output.

In order achieve first goal we propose to split the key into two separate parts:

- user dependent part,
- device dependent part.

User part of the key is incorporated into test input, similarly like it was proposed by Lee et al. [7]. On the other hand, device dependent part of the key is specific to a particular device and is inaccessible to the user. Purpose to split the key into two parts is to use the device depend part as a seed for a modified PRBG while user dependent part is used to modify its operation. In this way we achieve second goal – PRBG operates in a key dependent manner. Third goal can be achieved if we allow selected bits of the scan chain to affect the operation of the PRBG. Moreover, connecting these bits directly to the PRBG without any key checking sub circuit ensures that all bits of the test input will affect operation of the PRBG. Additional *Scan out* signal strobes the PRBG influence on the scan chain, so it influences test output only when data is shifted out of the boundary scan. This is additional signal which is not in JTAG standard but can be easily derived based on the states of test access port (TAP) [12].

3.2 Operation and Security of the Proposal

Since our proposal takes advantages of device dependent test key thus every test input-output pair is dedicated to a particular device. Precisely, test inputs are universal and can be used to test different devices. On the other hand test output depends on the feedback of the LFSR, that PRBG is composed of, and device dependent key that initializes shift register. Using the same test input on different devices gives different outputs and so to test the device one has to

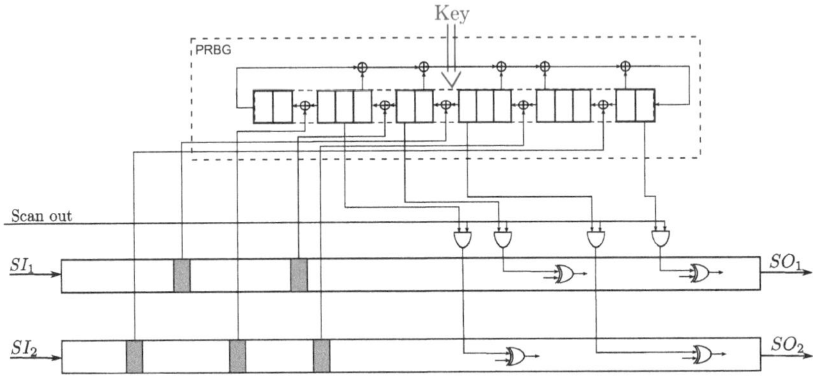

Fig. 3. Overview of proposed secure scan solution

know both test input and correct test output. This allows user to verify whether device operates correctly by simply testing the device with test input–output pairs given from the manufacturer.

While user can test the device he has no information about the exact test response generated by the circuit. He doesn't know the internal structure of the scan chain, has no information on how pseudo random bits affect the test output and what is their value. Therefore even given the correct input–output pairs he cannot trace output vector back to determine the exact test response.

Proposed modification to the standard boundary scan also helps to protect against scan-based cryptanalysis since it is very difficult for the attacker to decide what is the test response for a given input vector. Despite the fact that the attacker may control the device and the test input freely the proposed countermeasure minimises observability if the secret key is not known. On the other hand the attacker may repeat testing using the same test input several times in a row in order to get enough output data to reconstruct the structure of the LFSR. To minimise the threat of such attack it may be advisable to reset the LFSR before shifting in next test input. Therefore in our proposal the device dependent key is loaded into the LFSR when the device is run in a test mode and TAP goes into data shift state.

There are several other cryptographic attacks that one may try to implement against proposed solution such as timing attacks, fault attacks, SPA / DPA and / or SEMA / DEMA ([1,2,6]). While defense against such attacks is out of scope of this paper, it seems that due to extremely simple construction of the proposal (single LFSR) at least some of those techniques (eg. timing attack, SPA and SEMA) are rather infeasible.

4 Conclusions and Open Problems

Securing boundary scan is a challenging task that is investigated by many researchers. This is due to its significant importance to the security of modern cryptographic devices. This papers tries to sum up the most important results presented so far, states the most relevant aspects of a secure JTAG architecture and proposes a novel solution to improve security of boundary scan. Our proposal yields small implementation overhead and can be easily extended with additional countermeasures to further improve security level. Since security of the proposal depends on the parametres of the PRBG used, therefore, future work will focus on determining the best security parameters and full integration with IEEE 1149 standard.

References

1. Biham, E., Dunkelman, O.: Cryptanalysis of the A5/1 GSM Stream Cipher. In: Roy, B., Okamoto, E. (eds.) INDOCRYPT 2000. LNCS, vol. 1977, pp. 43–51. Springer, Heidelberg (2000)
2. Gandolfi, K., Mourtel, C., Olivier, F.: Electromagnetic Analysis: Concrete Results. In: Koç, Ç.K., Naccache, D., Paar, C. (eds.) CHES 2001. LNCS, vol. 2162, pp. 251–261. Springer, Heidelberg (2001)
3. Goering, R.: Scan Design Called Portal for Hackers, EE Times (October 2004), `http://www.eetimes.com/news/latest/showArticle.jhtml?articleID=51200146`
4. Gomółkiewicz, M., Tomczak, T., Nikodem, M.: Low-cost and Universal Secure Scan: a Design-for-Test Architecture for Crypto Chips. In: International Conference on Dependability of Computer Systems 2006, May 25-27, pp. 282–288 (2006)
5. Hély, D., Flotters, M.-L., Bancel, F., Rouzeyre, B., Bérard, N.: Scan Design and Secure Chip. In: Proceedings of the International On-Line Testing Symposium, 10th IEEE (IOLTS 2004), July 12-14, p. 219 (2004)
6. Kocher, P.C., Jaffe, J., Jun, B.: Differential power analysis. In: Wiener, M. (ed.) CRYPTO 1999. LNCS, vol. 1666, pp. 388–397. Springer, Heidelberg (1999)
7. Lee, J., Tehranipoor, M., Plusquellic, J.: A Low-Cost Solution for Protecting IPs Against Scan-Based Side-Channel Attacks. IEEE Trans. Dependable Sec. Comput. 4(4), 325–336 (2007)
8. Santos, L., Rela, M.Z.: Constraints on the Use of Boundary-Scan for Fault Injection. Dependable Computing, 39–55 (2003)
9. Sengar, G., Mukhopadhyay, D., Chowdhury, D.R.: Secured Flipped Scan Chain Model for Crypto-architecture. IEEE Trans. on CAD of Integrated Circuits and Systems 26(7), 1331–1339 (2007)
10. Yang, B., Wu, K., Karri, R.: Scan Based Side Channel Attack on Data Encryption Standard, Cryptology ePrint Archive: Report 2004/083 (2004)
11. Yang, B., Wu, K., Karri, R.: Secure scan: a design-for-test architecture for crypto chips. In: DAC 2005: Proceedings of the 42nd annual conference on Design automation, San Diego, California, USA, pp. 135–140. ACM Press, New York (2005)
12. IEEE Standard Test Access Port and Boundary-Scan Architecture, June 14. IEEE Computer Society, New York (2001) (reaffirmed 26 March 2008)

Automated Design of Totally Self-Checking Sequential Circuits

Jerzy Greblicki and Jerzy Kotowski

Institute of Computer Engineering, Control and Robotics
Wrocław University of Technology
11/17 Janiszewskiego St., 50-372 Wrocław, Poland
{jerzy.greblicki,jerzy.kotowski}@pwr.wroc.pl

Abstract. In this paper methods of designing of a class of highly reliable digital circuit - Totally Self Checking Sequential Machines are presented. The main problem in TSC sequential machines (TSC SM) designing is synthesis TSC functional excitation circuit. Formal condition of ST property for both AND-OR and AND-AND-OR structures are presented. The description of design methodology of TSC SM is presented. Owing to our methods we can design TSC circuits in a fully automatic way.

Keywords: Fault tolerant systems, totally self-checking circuits, sequential circuits.

1 Introduction

Recently we observe, in some areas, that great stress is laid on reliability of VLSI digital circuits, i.e., correctness, continuity, and safety of such circuits. We consider the most difficult problem of reliable sequential circuits - Totally Self Checking (TSC) circuits. In TSC circuits the first error caused by any single fault is detected. The main problem in TSC sequential machines (SM) designing is synthesis TSC functional excitation circuit. Unfortunately, there are not many automated methods for designing such circuits. In literature, problems of TSC SM using unordered codes was considered in [2]. Basically, three properties of TSC SMs should be considered: FS, ST, and CD. Definitions of properties are presented in literature in [3,4,8]. It was shown in [1] that any inverter free combinational circuit using m/n codes on its inputs and outputs is FS; this can be easily generalized to include any other unordered code as well. Diaz [2] formulated conditions for self-testing (ST) property verification for circuits with disjoint products and inputs, internal states and outputs, all encoded with m/n code. Diaz also showed method of circuits modification for ST. Unfortunately, a modified circuit is not code disjoint (CD). Criteria given by Diaz were generalized by Piestrak [8] for circuits using shared logic. In [2,8], circuits using only m/n and double rail codes are considered. In [3,5,4], we show that it is possible to construct a ST circuit without modification of circuit functioning by adding an extra shared AND gates layer (modification of [8]). The method can be applied in TSC/CD circuit designing because it doesn't affect CD property for

R. Moreno-Díaz et al. (Eds.): EUROCAST 2009, LNCS 5717, pp. 98–105, 2009.

non-code input vectors. Authors of almost all methods do no give any automatic method of ST. In addition they claim [6,7], that encoding inputs, internal states, and outputs by unordered codes and then constructing a circuit as inverter free leads to ST circuits. Unfortunately, this is not true in most cases. Jha and Wang [6] presented a method for designing both a circuit and software package which doesn't guarantee 100% of fault coverage.

In literature two architectures of TSC SM are considered: TST/STC and TSC/CD. For both structures problem of design on TSC/SM leads to problem of design of combination circuit **H** (transition circuit of SM) as a ST and CD for TSC/CD circuits and as a ST for TSC/STC circuits. The approaches presented here will apply the model TSC/CD, wherein the CD property is assured by CD of functional circuit. The STC is placed only on output interface of a SM. We propose theorems to verify self testing (ST). CD property veryfication methods were presented by us in [3] . In [3] we also consider TST/STC circuits, where CD is provided by STC checkers on all interfaces. We limit our considerations to Mealy SM and we assume that inputs, internal states and outputs are encoded using any unordered codes. Set of faults that we are considering are limited to single $stuck - at$ 1 or 0.

Unordered Codes and Inverter-Free Circuits

Definition 1. *Let X and Y be two binary n-tuples. We say that X covers Y (written $Y \leq X$) if and only if X has 1's everywhere Y has 1's. If neither $Y \leq X$ nor $X \leq Y$, then we say X and Y are unordered (written $X \not\subset Y$).*

Definition 2. *A set of binary n-tuples C is called unordered code if for every $X, Y \in C$, $X \neq Y$ implies $X \not\subset Y$.*

A code C is able to detect all unidirectional errors if and only if it is unordered [9]. The functions of a circuit **H** with unordered input code space C_{IN} can be derived as follows.

Definition 3. *We say that* a word X_i of an unordered code C_{IN} corresponds to an implicant m_i, *written $X_i \leftrightarrow m_i$, if m_i is a product of uncomplemented variables x_k which correspond to 1's in X_i.*

For instance, for $X = (001101)$ we have $X \leftrightarrow x_3x_4x_6$, since X has 1's on the bits x_3, x_4, and x_6.

The functions z_j of a circuit **H** can be expressed as

$$z_j = \sum_{m_i \in M_j} m_i, \quad j = \{1, 2, \ldots, s\}, \tag{1}$$

where: $\sum$ denotes logic OR and M_j is the set of all implicants m_i which correspond to the inputs $X_i \in C_{IN}$ for which z_j is 1. Implicants m_i corresponds to first layer of AND gates.

2 Formal Conditions for Self-testing Property of AND-OR Circuits

Assuming that the initial functions (1) do not contain identical AND gates feeding the same outputs (the circuit **H** is irredundant), only s/1 faults on the inputs of AND gates need to be considered. All other s/1 faults other than those on the inputs of AND gates are detected by the latter tests as well due to dominance relations. Any s/0 fault of the AND gate that implements m_j is detected by the codeword $X_j \leftrightarrow m_j$. The same X_j detects the s/0 fault on the output of the OR gate fed by m_j.

Let us note: $z^*(m_l)$- set of z_i functions containing implicant m_l, $m_j(x^*)$- set of x variables that acure in m_j, $m_j(x_k)$- divider of $m_j$by variable x_k calculated by applying $x_k = 1$ in m_j, $H(X_l)$-function of circuit **H**, $H(X_l, f)$- function of circuit **H** in presence of fault f.

Theorem 1. *The circuit* ***H*** *that implements (1) is ST for all single s/1 faults of input lines x_k of AND gates, if and only if*

$$((\forall m_j \in M)\,(\forall x_k \in m_j(x^*))\,(\exists m_l \in M \,\|, m_j(x_k) \subset m_l \text{ and } z^*(m_j) \neq z^*(m_l)) \,. \tag{2}$$

Proof. **Sufficiency** Consider an $s/1$ fault of line $x_k \in m_j(x^*)$ at the input of AND gate implementing a product $m_j \leftrightarrow X_j$. Moreover, assume that condition of Theorem (1) holds:

$$(\exists m_l \in M, \|, m_j(x_k) \subset m_l \text{ and } z^*(m_j) \neq z^*(m_l)) \,. \tag{3}$$

In a fault free circuit for input codeword X_j, only $m_j = 1$, $m_j \leftrightarrow X_j$. The $s/1$ fault of x_k can be tested by input codeword $X_l \leftrightarrow m_l$, such that $m_j(x_k) \subset m_l$. However in the presence of this fault not only $m_l = 1$ but also $m_j(x_k) = 1$ that is $m_j = m_j(x_k) \cdot x_k = 1$ (the $0 \rightarrow 1$ error). Because the circuit **H** is inverter free $H(X_l, f) = (X_l) \vee H(X_j)$ and $H(X_l) \subset H(X_l, f)$ and $H(X_j) \subset H(X_l, f)$ and $H(X_l) \neq H(X_j)$, (by assumption that $z^*(m_k) \neq z^*(m_l)$), the $0 \rightarrow 1$ error can be observed on at least one output. ■

Necessity. Let as assume that the circuit is ST for the $s/1$ fault of line x_k in the implicant m_j. It means that there exists an input codeword $X_l \leftrightarrow m_l$ which is the test codeword for this fault. It can be easily noticed that X_l covers the sub implicant $m_j(x_k)$, owing to that in the presence $s/1$ fault of line x_k not only $m_l = 1$ but also $m_j = 1$. Moreover since the sets $z^*(m_j)$ and $z^*(m_l)$ are different ($z^*(m_j) \neq z^*(m_l)$), that guarantees that $H(X_l, f) \notin C_{OUT}$, hence the conditions of Theorem 1 holds. ■

Theorem 1 serves as a formal algebraic tool to verify that a given 2-level AND-OR circuit **H** realized using complete products is ST w.r.t. s/1 faults. If Theorem 1 does not hold for some x_k in m_j, it indicates that this circuit is not ST for the $x_k/1$ fault on the input of m_j.

3 Formal Conditions for Self-testing Property of AND-AND-OR Circuits

Unfortunately in 2-layer AND-OR circuits implementing function (1) untestable $s/1$ faults may occur on inputs of AND gates (such input lines can be detested by checking condition of Theorem 1). It is possible to modified AND-OR structure by adding an extra layer of AND gates, such modification may eliminate untestable lines.

Let assume that, implicants m_i and m_j exists with at least two common variables $p_u = x_{j_1}...x_{j_a}$ $(a \geq 2)$ where $m_j = \underbrace{(x_{j_1}...x_{j_a})}_{p_u} x_{j_{a+1}}...x_{j_c}$ and $m_i = \underbrace{(x_{j_1}...x_{j_a})}_{p_u} x_{i_{a+1}}...x_{i_c}$. Moreover let assume that sub implicant $p_u$contain variables, that are not testable in m_j but are testable in implicant m_i . It is possible to eliminate an untestable fault in m_j by implementing m_j in 2-layers of AND gates. First layer generates shared sub implicant p_u, second layer generates implicants m_i and m_j using p_u from first layer of AND gates. Formally we write: $p_u = x_{j_1}...x_{j_a}$ and $m_j = p_u x_{j_{a+1}}...x_{j_c}$ and $m_i = p_u x_{i_{a+1}}...x_{i_c}$.

For example let us consider 2 output functions z_1 and z_2 with untested fault $s/1$ (marked by $\downarrow$): $z_1 = ... + x_1 x_2 \overset{\downarrow}{x_3} + ...$ and $z_2 = ... + x_2 x_3 x_4 + ...$ Let us note that sub implicant $p_u = x_2 x_3$ is common for implicants $x_1 x_2 x_3$ and $x_2 x_3 x_4$, hence:

$$p_u = x_2 x_3$$
$$z_1 = ... + x_1 p_u + ...$$
$$z_2 = ... + p_u x_4 + ...$$

The untestable fault of input line x_3 of gate m_{123} can be tested by m_{234}, by fulfilling some other conditions presented below.

Let p_u, $u \subset \{1, r\}$, is a sub implicant of $m_i = (x_{j_1}...x_{ja})p_1...p_r$. and P - denote set of all sub implicants p_u in circuit **H** (P is related to all AND gates from first shared layer) in **H**. Let $m_i(p^*) = \{p_1, ..., p_r\}$ denote set of all sub implicants in m_i. Divider $m_j(p_u)$ of implicant m_i by $p_u (p_u \subset m_i)$ is calculated by applying $p_u = 1$ in implicant m_i.

Theorem 2. *If in 3-layer (AND-AND-OR) combinational circuit **H** implementing function (1), using shared implicants of first layer of AND gates, for every $m_j \in M$ the following conditions are fulfilled :*

$$1) \quad (\forall x_k \in m_j(x^*))(\exists m_l \mid m_j(x_k) \subset m_l) \, and \, z^*(m_j) \neq z^*(m_l),$$
$$2) \quad (\forall p_u \in m_j(p^*))(\exists m_l \mid m_j(p_u) \subset m_l) \, and \, z^*(m_j) \neq z^*(m_l),$$

and for every $p_u \in P$ the following condition is fulfilled

$$3) \quad (\forall x_k \in p_u(x^*)(\exists m_r, m_l, m_r \in M, m_l \in M \mid p_u \in m_l(p^*) \; and \; m_l(x_k) \subset m_r \; and \; z^*(m_l) \neq z^*(m_r)),$$

*then circuits **H** is self testing for every single $s/1$ fault.*

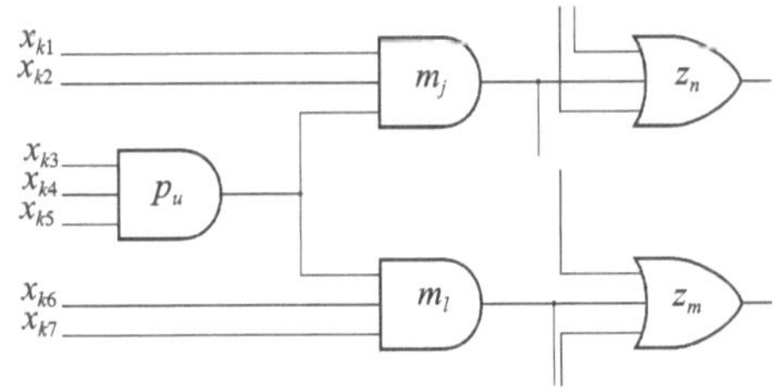

Fig. 1. Fragment of AND-AND-OR circuit

Proof. Let us consider fragment of example structure from figure 1. Condition 1 of theorem 2 corresponds to testability of x_k primary input lines of second layer of AND gates (implementing m_j i m_l). Condition 2 corresponds to testability of lines connecting first and second layer of AND gates. Condition 3 corresponds to testability of primary inputs of first layer of AND gates (calculating p_u).

Proof of condition 1 of Theorem 2 is analogous to proof of Theorem 1. If condition 2 holds then there exists an input code word that tests such a fault. Potential test is such $X_l \leftrightarrow m_l$, that: 1. cause error on faulty $s/1$ line p_u and 2. will propagate this error to the output of circuit. This condition is fulfilled for codeword $X_l \leftrightarrow m_l$ such that, $m_j(p_u) \subset m_l$ and $z^*(m_l) \neq z^*(m_j)$. Let us note that in a non faulty circuit with X_l codeword on inputs, we have: $p_u = 0$ and we note than becouse $m_j(p_u) \subset m_l$ also $m_j(p_u) = 1$ in light of this $m_j = m_j(p_u) \cdot p_u = 0$ and $m_l = 1$. However in presence of fault $p_u/1$ with X_l codeword on inputs, we have $m_j = 1$ becouse: $p_u = 1$ coused by fault $s/1$ and also $m_j(p_u) = 1$, hence $m_j = m_j(p_u) \cdot p_u = 1$ and $m_l = 1$. Because sets $z^*(m_j)$ i $z^*(m_l)$ are not equal , an error $0 \to 1$ of line p_u will propagate to to output of circuit.

Condition 3 guarantees ST of faults of primary inputs of first layer of AND gates (calculating sub implicants p_u). Proof of this condition is similar to proof of Theorem 1.

Let us consider an example set of functions (an excitation circuit of sample SM [4]). Analysis of ST property, by Theorem 1, of example circuit implemented as AND-OR,showed 3 untestable lines (marked $\downarrow$):

$$z_1 = 145 + 157 + 247 + 256 + 3\overset{\downarrow}{4}5 + 326 + 35\overset{\downarrow}{6} + 35\overset{\downarrow}{7}$$

$$z_2 = 146 + 147 + 156 + 245 + 246 + 257 + 3\overset{\downarrow}{4}5 + 347 + 35\overset{\downarrow}{6} + 35\overset{\downarrow}{7}$$

$$z_3 = 145 + 146 + 147 + 256 + 257 + 346$$

$$z_4 = 156 + 157 + 245 + 246 + 247 + 347$$

Notation: 145 denotes implicant $m_{145} = x_1 x_4 x_5$.

We modified circuit with untestable faults to AND-AND-OR structure and proved with Theorem 2 that such a circuit is ST. The modified circuit is presented below:

$p_{34} = 34 \quad p_{56} = 56 \quad p_{57} = 57$

$$z_1 = 145 + 1p_{57} + 247 + 2p_{56} + p_{34}5 + 326 + 3p_{56} + 3p_{57}$$
$$z_2 = 146 + 147 + 1p_{56} + 245 + 246 + 2p_{57} + p_{34}5 + p_{34}7 + 3p_{56} + 3p_{57}$$
$$z_3 = 145 + 146 + 147 + 2p_{56} + 2p_{57} + p_{34}6$$
$$z_4 = 1p_{56} + 1p_{57} + 245 + 246 + 247 + p_{34}7$$

Modified circuit is ST for all single *stuck − at* faults.

4 Design of TSC/STC Circuits

Our analysis of benchmark circuits showed that the conditions of Theorem 1 and Theorem 2 do not hold for some circuits. In [3] we presented extended version of method from [2] which minimize inverter free combinational circuit using any unordered code and guarantee the ST property of minimized circuit and makes posible design of TSC/STC SM. Below we present an example of usage of this method.

Let us consider an example circuit from section 3. It can be noticed that the sub implicant $p_{16} = x_1x_6$ belongs to implicants m_{146} and m_{156}, moreover both of them belong to z_2 and p_{16} is sub implicant of only m_{146} and m_{156}. Assuming that during normal operation only codewords from code C_{IN} will appear at the circuit input. It means that only two input codewords X_{146} and X_{156} cover p_{16}. So we can assume that $m_{146}(p_{16}) = 1$ and $m_{156}(p_{16}) = 1$ in implicants from sum z_2. In this way, we modify these functions z_2 as follows: $z_2 = 16 + 147 + 245 + 246 + 257 + 345 + 347 + 356 + 357$.

We use same method to minimize function (1). The minimized function of example circuits from section 3 is shown below:

$$z_1 = 3 + 145 + 256 + 157$$
$$z_2 = 16 + 27 + 24 + 35 + 147$$
$$z_3 = 14 + 346 + 256 + 257$$
$$z_4 = 24 + 347 + 156 + 157$$

Detailed description of this method was presented by us in [3].

5 CAD Software

Owing to the theorems presented above, we design an algorithm presented in figure 2. Circuit specification is verified and detected errors are corrected. Next, the structure of the circuit is verified by checking a graph of transitions. The aim is to find unreachable nodes or nodes without return, and/or constant input/output lines. Any specification error makes the synthesis impossible. In the next step input, internal state, and output codes are constructed. The library of STC circuits is searched for STC for selected codes. If the search is successful,

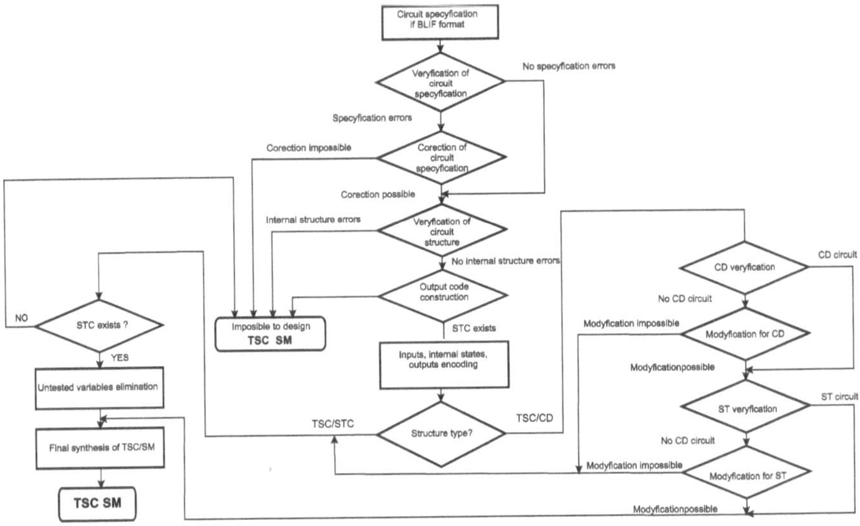

Fig. 2. Algorithm of TSC/SM CAD software

the ST property of STC is checked. In the next stage, a structure TSC/CD of TSC/STC is chosen. For TSC/CD circuits, ST and CD properties are verified. If necessary, a circuit is modified for ST and/or CD in the way mentioned above and presented [3] in details. If a ST and CD circuit is constructed, synthesis of TSC/CD SM is completed. In the other case, TSC/STC SM is synthesized. In TSC/STC SM designing, we have to guarantee ST for all used STC. Then, circuit **H** is synthesized and all untested variables are eliminated with methods presented in [3] . Our algorithms and methods for self-checking SM were implemented in a software package TSCSM CAD (Totally Self-Checking SM CAD). The software was implemented in C++.

6 Numerical Results

The efficiency of proposed algorithms were tested on a set of benchmark circuits (ISCAS' 89). Two types of TSC SM were constructed: TSC/CD and TSC/STC. For both of them, circuit structure were verified for errors making synthesis impossible. We use a novel method for internal states encoding which guarantees ST of Self Testing Checker (STC) circuit for internal states code, presented by us in [5]. Output code was constructed with algorithm proposed in [7]. In Table 1 results are presented. Complexities of designed circuits are given in number of literals calculated in SIS software. We designed circuits for several encoding and for both structures TSC/CD (with internal AND-AND-OR structure) and TSC/STC. Due to our theorems properties of circuits where we verified. For almost all circuits, we designed circuits that are smaller than duplicated circuits. We would like to note that all circuits were designed in a fully automatic way.

Table 1. Complexities of designed TSC SM's

Circuit	Inputs encoding: 2-rail				Inputs encoding: Berger				Circuit
	Int. states.: $1/n$		Int. states.: $2/n$		Int. states.: $1/n$		Int. states.: $2/n$		duplicated
	TSC CD	TSC STC	TSC CD	TSC STC	TSC CD	TSC STC	TSC CD	TSC STC	
bbara	233	212		264	348	243		273	188
bbtas	118	104	121	94					100
beecount	163	113	172	117	145	134	140	142	136
cse		676		706		609		662	517
dk14	344	340		344	326	364		359	295
dk15	266	250	269	263	263	254	251	274	257
dk16	514	506	468	488					715
dk27	103	113		112	103				84
dvram		604		523		748		697	843
mc	191	172	204	174	208	205	231	219	129
planet		1247		1245					1851

7 Conclusions

Circuits designed with the package are fully testable for all single faults. Moreover, we successfully designed CD circuits in a fully automatic way. No such method is known to the authors.

References

1. Diaz, M., et al.: Unified design of self-checking and fail-safe combinational circuits and sequential machines. IEEE Trans. Comput. C-28, 276–281 (1979)
2. Diaz, M., de Souza, J.M.: Design of self-checking microprogrammed controls. In: Digest of Papers 5th Int. FTCS, Paris, France, June 1975, pp. 137–142 (1975)
3. Greblicki, J.W.: Synthesis of sequential circuits using unordered codes. PhD thesis, Wroclaw University of Technology, Wroclaw (in polish) (October 2003)
4. Greblicki, J.W., Piestrak, S.J.: Design of totally self-checking code-disjoint synchronous sequential circuits. In: Hlavicka, J., Maehle, E., Pataricza, A. (eds.) EDDC 1999. LNCS, vol. 1667, pp. 250–266. Springer, Heidelberg (1999)
5. Greblicki, J.W.: CAD software for designing of totally self checking sequential circuits. In: DepCoS - RELCOMEX 2006, pp. 289–296. IEEE Comp. Society Press, Los Alamitos (2006)
6. Jha, N.K., Wang, S.-J.: Design and synthesis of self-checking VLSI circuits. IEEE Transactions on Computer–Aided Design of Integrated Circuits 12, 878–887 (1993)
7. Lai, C.-S., Wey, C.-L.: An efficient output function partitioning algorithm reducing hardware overhead in self-checking circuits and systems. In: Proceedings 35th Midwest Symp. Circuits System, pp. 1538–1541 (1992)
8. Piestrak, S.J.: PLA implementation of totally self-checking circuits using m-out-of-n codes. In: Port Chester, N.Y. (ed.) Proceedings ICCD 1985, International Conference on Computer Design: VLSI in Computers, October 1-3, pp. 777–781 (1985)
9. Smith, J.E.: The design of totally self-checking check circuits for a class of unordered codes. J. Des. Autom. Fault-Tolerant Comput. 2, 321–342 (1977)

A General Purpouse Control System

Adrián Peñate-Sánchez, Alexis Quesada-Arencibia, and Roberto Moreno-Díaz jr.

Institute for Cybernetics, University of Las Palmas de Gran Canaria, Las Palmas de G.C., Las Palmas, España

Abstract. Industrial control applications are developed as ad hoc solutions. These solutions require specific programming and installation, and very often it takes large economic investments. We have developed a system that makes a special effort trying to offer versatility and a solution to a widespread of industrial installations. Our system understands an industrial plant as a compound of sensors, actuators and processes. The user will define its own processes, sensors and actuators. It will also be in his hand the design of the plants visualization diagram. In order to achieve this generalness it is of essence to offer a user interface as friendly as possible. Via a very intuitive user interface the implemented functionalities are far more powerful because the user understand them and makes use of them.

Keywords: User friendly, industrial control, web application.

1 Introduction

Most of the industrial control applications we have encountered are specific solutions for each kind of installations. For example, two desalination plants property of the same company, and with very similar characteristics, use the same industrial control solution. The problem with this is that each plant requires a customized application that needs to be modified by the provider, and then, it requires a specialist to go to each plant and install its customized solution.

A specific solution for each plant involves huge costs, for the company that requires the control solution as well as for the developer. Not all industrial installations can overtake these costs, leaving them without the functionalities a control solution can offer.

We have opened a line of research trying to develop an industrial control application that doesn't need to be customized or installed by an expert, and that, with time will come to give a low cost solution for industrial plants. In this paper we will show the system we have developed, but we will emphasize the assets that make our solution user friendly and versatile.

2 System Structure

The system includes different components such as a rule based system that defines and controls the processes that take place in the system; a sensor manager and an actuator manager that take care of defining sensors and actuators in the system to be

R. Moreno-Díaz et al. (Eds.): EUROCAST 2009, LNCS 5717, pp. 106–112, 2009.

used by the rule based system. It also includes different assets that help the user to monitor the state of the plant. Examples of these is the use of charts to visualize the historical data of the plant, the fact that the system can be used in any location by using a web browser, and the plant diagram editor where the user designs its own visual image of the plant where he can view the sensors and actuators actual state in a more intuitive manner. Further detail of the system can be found in [1], and [2].

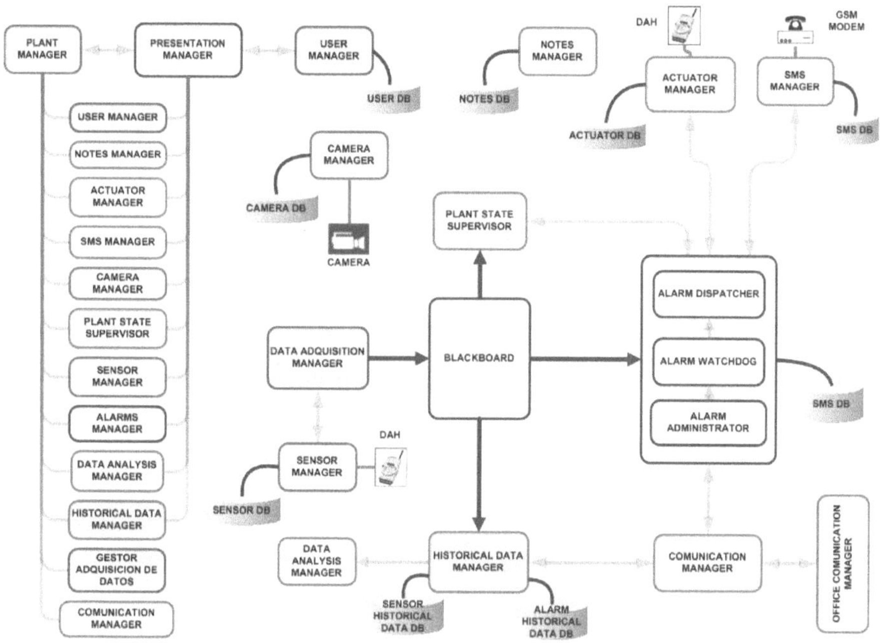

Fig. 1. Modular architecture. It can be seen in this picture the different modules of the plant architecture, the interaction with each other and their access to their specific databases.

3 Generic Functionality

We wanted to create a system that achieved easily a "build it yourself" philosophy, for this to come true we identified two main conditions:

- Connecting sensors and actuators. It had to be possible to modify the set of components of the system at all time. In order to be generic it needed to be able of changing over time.
- Generic control definition. We needed to make an abstraction of the different industrial plants that we could face. The common variables in control had to be identified in order to create a generic system.

All the functionalities we have equipped the system with have been designed to be as general as possible. A user can define its own control rules, and can use any kind of sensor or actuator, from any provider, as long as it applies to the 4-20 mA input or output standard.

3.1 Connecting Different Kinds of Sensors and Actuators

To introduce different kinds of sensors and actuators in the system, we have made use of the Modbus protocol. The Modbus protocol is a standard TCP based protocol that defines an interface, which abstracts us from the fact of dealing with different kinds of components. This way the user will have to introduce the data without thinking about how to connect the hardware, the only thing that will be required will be the IP direction where the data acquisition hardware is.

The Modbus protocol establishes a server-client connection exchanging data between each side's registers.

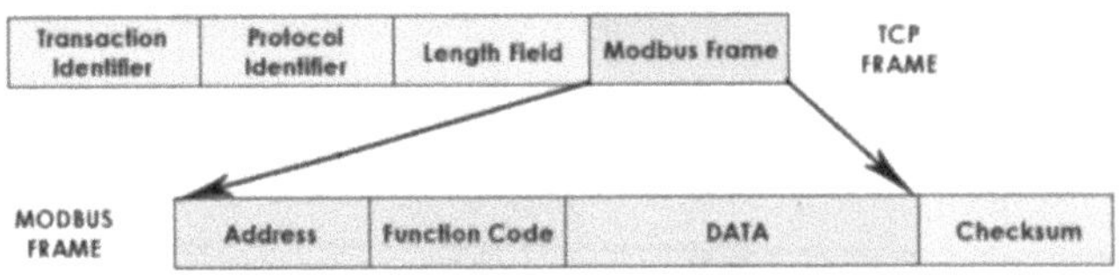

Fig. 2. Modbus encapsulation

By using this interface with our hardware we have the possibility of using any sensor or actuator that uses the markets standards. Studying the market we found out that the vast majority of components came in the standards we use.

3.2 Generic Control Definition

A control process in an industrial plant is a compound of sensors, actuators and control logic. Understanding that all control processes are composed of the same elements sets the basis to define a generic control capable of giving an answer to all kinds of plants.

Each plant will have its own set of actuators and sensors, and its own set of control rules. Rules need to be created by the user because he is the one who carries the specific knowledge of each plant. We have introduced a rule based engine that enables the user to define rules with comparisons between sensors, comparisons with constants, with time or other alarms.

Our system is also capable of changing, depending on the requirements of the plant. The rule set that controls the plant can change at any time without need of stopping the plant.

4 User Interface

The user interface has been defined to be as simple as possible. It has been carefully designed to take the user through the least menus as possible. In order for the user not to get muddled up with the application, it needs to be organized in a coherent way. Most of the functionalities are grouped by context and linked when necessary. It is also very flexible when it comes to modify a process or a component, you don't have to redefine the whole process or even the whole rule, and you just have to change the parameters you want.

The system will be defined by the user at the beginning by introducing its sensors, actuators, users, processes, phone numbers, design of the plants diagram, etc. These settings in all cases are intended to be done and modified by the user and not by an expert, to achieve this we have made use of a very easy to use interface.

The definition of the most complex tasks has been implemented using Wizards. Wizards divide complex tasks in small simple ones, this way it guides the user through the process one step at a time.

Fig. 3. Sensor connection Wizard

4.1. Sensor and Actuator Definition

The user will introduce de IP direction and the port where the component is connected. All the process will be guided by a Wizard; it will ask the information to establish the connection and afterwards it will provide the option of testing the component, either by reading its value or activating or deactivating it depending of the kind of component.

In Fig. 4 you can see one of the forms that the user will fill, in this case to connect a sensor. We only need to select the sensor and the port of the hardware where we want to connect it.

4.2 Control Definition

If it was going to be the user the one to define the control over the plant, it had to be hidden from him the technical details it involved. To define rules that control the system we have introduced a Wizard that guides the user through the process. To get the logic propositions we make simple questions, such as: which sensor?, should the sensor be greater or less than what value?, how much time should it pass to consider it true?. At the end of the process the user will have, without knowing, defined a logical predicate that will be the rule that will be monitored by the engine.

In Fig. 4 you can see one of the forms that the user will fill, in this case to introduce an action in the event of the rule being triggered.

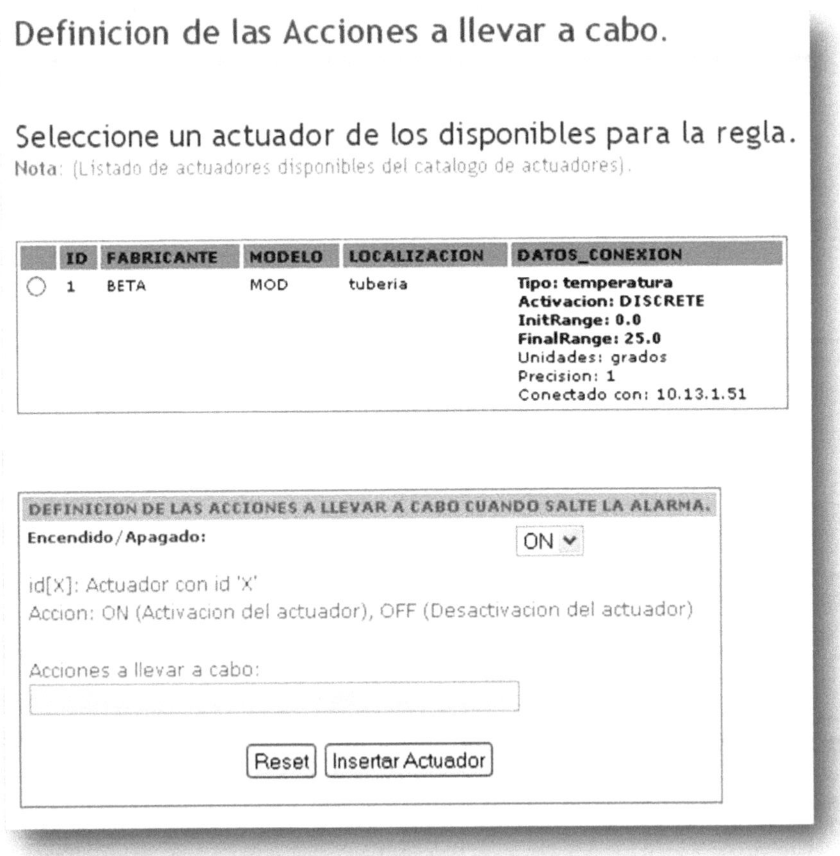

Fig. 4. Alarm definition Wizard. Action definition.

All the changes are also updated in real time, when something is modified you don't have to refresh the system, it automatically introduces the changes into the rule based system. One of the principal advantages is that the system is web based. This allows users to access the system from any place without any particular requirements, and with full functionality.

4.3. Understanding What Happens in the Plant

For a control system to be effective the user needs to know what is happening in the plant. Special effort has to be made in showing this information in the easiest manner. For this purpose we have introduced the possibility of creating our own diagram of the plant. In Fig. 5 you can see an example of a diagram created from scratch in the application that is monitoring a part of a plant.

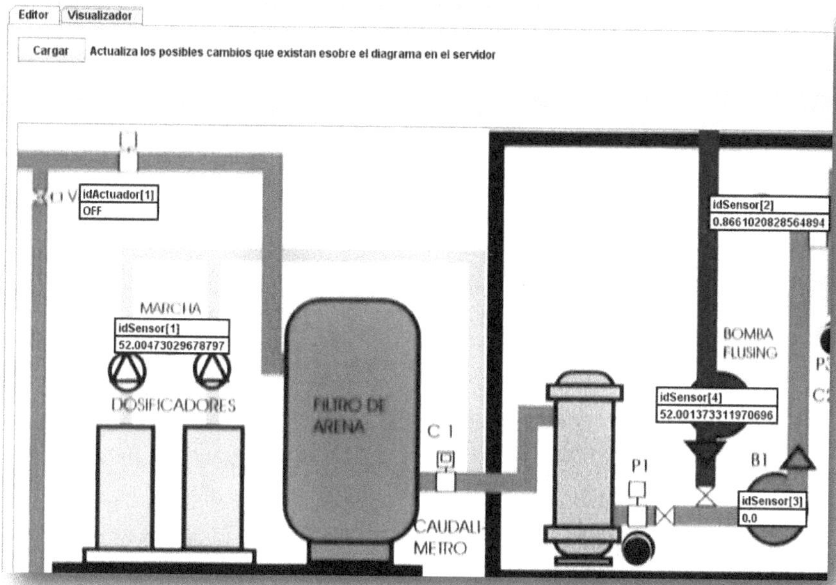

Fig. 5. Plant Visualization diagram

If something is happening in the system we will receive text messages in our mobile phone alerting us. This is a web based application, so when we receive the SMS alerting us, we can connect ourselves to the plant from any place with an internet connection and access the control system.

The system also includes the possibility of visualizing the historical data with charts that are created dynamically. This is a very important asset because it gives data another dimension.

5 Conclusions

As it can be seen the assets that make our system generic are non specificity and the capacity of change over time. We are capable of defining a customized control solution for each kind of plant we encounter.

As for easy-to-use, we have chopped everything into simple pieces and hidden the technical details from the user. This way the customized control solution can be achieved by a user lacking technical knowledge about automated industrial control.

We have already introduced our system in two real contexts: desalination plants and student residences. The results have been quite satisfactory; users get to develop their control system quite easily. We have found out that it takes time to introduce all initial data into the system. We intend to introduce the system in as many contexts as possible to get further feedback in order to improve the system.

References

1. Peñate-Sánchez, A., Solinis-Camalich, I., Quesada-Arencibia, A., Rodríguez-Rodríguez, J.C.: A general system for monitoring and controlling via internet. In: Proceedings of the Tenth International Conference on Enterprise Information Systems - ICEIS 2008, pp. 267–270. INSTICC, Institute for Systems and Technologies of Information, Control and Communication (2008)
2. Peñate-Sánchez, A., Solinis-Camalich, I., Quesada-Arencibia, A., Rodríguez-Rodríguez, J.C.: A Web Monitoring Application Based on JAVA technologies – ICOMP 2008, pp. 154–160. CSREA Press (2008)

On the First Exit Time Problem for a Gompertz-Type Tumor Growth*

G. Albano[1] and V. Giorno[2]

[1] Dip. di Scienze Economiche e Statistiche, Università di Salerno, Via Ponte don Melillo, Fisciano (SA), Italy
pialbano@unisa.it

[2] Dipartimento di Matematica e Informatica, Università di Salerno, Via Ponte don Melillo, Fisciano (SA), Italy
giorno@unisa.it

Abstract. A stochastic model describing tumor growth based on Gompertz law is considered. We pay attention on the tumor size at time detection. We assume the initial state as a random variable since it may suffer from errors due to measurement and diagnostics. The aim of the present work is to study the first exit time problem for the resulting stochastic process. A numerical analysis is also performed for particular choices of the initial distribution.

1 Introduction and Background

Many of the proposed models in literature to describe the dynamics of tumor growth are based on Gompertz growth ([3], [5], [6], [7], [9]), indeed it seems to be particularly consistent with the evidence of tumor growth. Recently, to include random fluctuations in experimental data, in [1] a stochastic model generalizing Gompertz growth has also been considered. In particular, in [1] and [2] tumor size is described by the following stochastic differential equation:

$$d\,X_C(t) = \{[\alpha - C(t)]X_C(t) - \beta\, X_C(t)\, \ln X_C(t)\}dt + \sigma X_C(t)\, d\,W(t), \tag{1}$$

$$X_C(t_0) = x_0 \; a.s. \tag{2}$$

where α, β and σ are positive constants representing the growth, death rates and the width of random fluctuations, respectively. The deterministic function $C(t)$ describes the tumor regression rate due to an antiangiogenetic therapy. Here $W(t)$ is a standard Brownian motion and x_0 represents tumor size at the time when the illness is diagnosed.

The transition probability density function (pdf) for the process defined in (1) with the initial condition (2) is:

$$f_C(x,t|x_0,t_0) = \frac{1}{x\,\sqrt{2\pi V(t-t_0)}} \exp\Big\{ -\frac{[\ln x + d(t) - M(t|x_0,t_0)]^2}{2\,V(t-t_0)} \Big\} \tag{3}$$

* Work performed under partial support by G.N.C.S.-INdAM and by Regione Campania.

R. Moreno-Díaz et al. (Eds.): EUROCAST 2009, LNCS 5717, pp. 113–120, 2009.

where

$$M(t|x_0,t_0) = \frac{\alpha - \sigma^2/2}{\beta}\left(1 - e^{-\beta(t-t_0)}\right) + \ln x_0\, e^{-\beta(t-t_0)} + d(t_0)e^{-\beta(t-t_0)}, \tag{4}$$

$$V(t) = \frac{\sigma^2}{2\beta}\left(1 - e^{-2\beta t}\right)$$

are the mean and the variance of (3) respectively, and

$$d(t) = \varphi(t)\exp(-\beta t), \qquad \varphi(t) = \int^t C(\tau)\exp(\beta\tau)\, d\tau. \tag{5}$$

Moreover, two real boundaries S_1 and S_2 ($S_1 < S_2$), representing the "recovery level" and the "carrying capacity" respectively, are introduced in the model. In this way, to analyse the evolution of $X_C(t)$ corresponds to study the first exit time (FET) of $X_C(t)$ from the real interval $(S_1,\ S_2)$. The diffusion process $X_C(t)$ is time-non-homogeneous, so standard procedures to analyse this kind of process do not exist in literature. However, the transformation

$$y = \ln x + d(t)\ , \qquad y_0 = \ln x_0 + d(t_0) \tag{6}$$

leads $X_C(t)$ to an Ornstein-Uhlenbeck (OU) process $Y(t)$ defined in $(-\infty,\infty)$ with infinitesimal moments:

$$B_1(x) = \left(\alpha - \frac{\sigma^2}{2}\right) - \beta x, \qquad B_2(x) = \sigma^2. \tag{7}$$

Making use of (6), the FET problem of $X_C(t)$ from (S_1, S_2) is changed into the FET problem of $Y(t)$ from $(\bar{S}_1(t),\ \bar{S}_2(t))$, where

$$\bar{S}_1(t) = \ln S_1 + d(t), \qquad \bar{S}_2(t) = \ln S_2 + d(t). \tag{8}$$

More precisely, assuming $Y(t_0) = y_0$, con $y_0 \in (\bar{S}_1(t_0),\ \bar{S}_2(t_0))$, we define the following random variables (r.v.'s):

$$\begin{aligned}
T^- &= \inf_{t\geq t_0}\ \{t\ :\ Y(t) < \bar{S}_1(t);\ Y(\theta) < \bar{S}_2(\theta), \forall\ \theta \in (t_0,t)\}, \quad Y(t_0) = y_0,\\
T^+ &= \inf_{t\geq t_0}\ \{t\ :\ Y(t) > \bar{S}_2(t);\ Y(\theta) > \bar{S}_1(\theta), \forall\ \theta \in (t_0,t)\}, \quad Y(t_0) = y_0,\\
T &= \inf\ \{T^-,\ T^+\},
\end{aligned}$$

and the respective pdf's

$$\gamma^-(t|y_0,t_0) = \frac{\partial}{\partial t}\, P(T^- < t), \quad \gamma^+(t|y_0,t_0) = \frac{\partial}{\partial t}\, P(T^+ < t) \tag{9}$$

and

$$\gamma(t|y_0,t_0) = \frac{\partial P(T<t)}{\partial t} = \gamma^-(t|y_0,t_0) + \gamma^+(t|y_0,t_0). \tag{10}$$

Since an analytical form for γ^-, γ^+ and γ does not exist in literature, we made use of a numerical approach as suggested in [4]. Precisely, the functions γ^- and γ^+ are solution of the second-kind Volterra integral equations system:

$$\begin{aligned}\gamma^-(t|\, y_0,t_0) &= 2\Psi_1(\bar{S}_1(t),t|\, y_0,t_0) - 2\int_{t_0}^{t}\{\gamma^-(\theta|\, y_0,t_0)\Psi_1(\bar{S}_1(t),t|\, \bar{S}_1(\theta),\theta) \\ &\quad + \gamma^+(\theta|\, y_0,t_0)\Psi_1(\bar{S}_1(t),t|\, \bar{S}_2(\theta),\theta)\}\, d\theta \\ \gamma^+(t|\, y_0,t_0) &= -2\Psi_2(\bar{S}_2(t),t|\, x_0,t_0) + 2\int_{t_0}^{t}\{\gamma^-(\theta|\, y_0,t_0)\Psi_2(\bar{S}_2(t),t|\, \bar{S}_1(\theta),\theta) \\ &\quad + \gamma^+(\theta|\, y_0,t_0)\Psi_2(\bar{S}_2(t),t|\, \bar{S}_2(\theta),\theta)\}\, d\theta\end{aligned} \tag{11}$$

where

$$\begin{aligned}\Psi_i(\bar{S}_i(t),t|\, z,\theta) = \frac{1}{2}\Bigg\{ & C(t) - \alpha + \frac{\sigma^2}{2} + \beta \ln S_i - \frac{2\beta}{\sigma^2}\left[1 - \mathrm{e}^{-2\beta\,(t-\theta)}\right]^{-1} \\ & \times\left[- \ln\, S_i + d(t) - \frac{\alpha - \sigma^2/2}{\beta}\left(1 - \mathrm{e}^{-\beta\,(t-\theta)}\right) + z\mathrm{e}^{-\beta\,(t-\theta)}\right]\Bigg\} \\ & \times f_{OU}[\ln\, S_i(t) + d(t), t \mid z, \theta],\end{aligned} \tag{12}$$

and f_{OU} is the transition pdf for the OU process (7):

$$f_{OU}(x,t|z,\theta) = \frac{1}{\sqrt{2\pi V_{OU}(t-\theta)}}\exp\left\{-\frac{[x - M_{OU}(t-\theta|z)]^2}{2\, V_{OU}(t-\theta)}\right\} \tag{13}$$

with

$$M_{OU}(t|z) = \frac{\alpha - \sigma^2/2}{\beta}\left(1 - e^{-\beta t}\right) + z\, e^{-\beta t}, \qquad V_{OU}(t) = \frac{\sigma^2}{2\beta}\left(1 - e^{-2\beta t}\right).$$

We point out that in the model defined in (1) and (2) $X_C(t_0) = x_0$ is a fixed real value. However, since, in our context, $X_C(t_0)$ represents the tumor size at the time detection of the illness, it is reasonable to assume fluctuations in this parameter. In this direction, in the following section, we extend the model in [1] to the case in which the initial value of the process $X_C(t)$ is a r.v.. In Section 3 we consider the mean FET of $X_C(t)$ from the interval (S_1, S_2) in the presence of a therapy with a fixed intensity.

2 The Model

Let X_0 be an absolutely continuous r.v. with pdf $g_{X_0}(x)$ non zero in a real interval (a, b) with $S_1 < a < b < S_2$. We assume that condition (2) is substitute by the following:

$$X_C(t_0) = X_0 \ a.s. \tag{14}$$

Making use of (6) we obtain the initial condition for the OU process $Y(t)$. So $Y(t_0) = Y_0$ a.s. is a r.v. with pdf $g_{Y_0}(y_0) = g_{X_0}(e^{y_0 - d(t_0)})e^{y_0-d(t_0)}$. With this assumption, from (9) we can define the following functions:

$$\hat{\gamma}^-(t \mid t_0) = \int_{\ln a + d(t_0)}^{\ln b + d(t_0)} e^{y_0 - d(t_0)} \, g_{X_0}(e^{y_0-d(t_0)}) \gamma^-(t \mid y_0, t_0) \, dy_0,$$

$$\hat{\gamma}^+(t \mid t_0) = \int_{\ln a + d(t_0)}^{\ln b + d(t_0)} e^{y_0 - d(t_0)} \, g_{X_0}(e^{y_0-d(t_0)}) \gamma^+(t \mid y_0, t_0) \, dy_0. \tag{15}$$

Taking the mean value of both sides in (11), we obtain:

$$\hat{\gamma}^-(t \mid t_0) = 2\hat{\Psi}_1(\bar{S}_1(t), t \mid t_0) - 2 \int_{t_0}^{t} \{\hat{\gamma}^-(\theta \mid t_0) \Psi_1(\bar{S}_1(t), t \mid \bar{S}_1(\theta), \theta) + \\ + \hat{\gamma}^+(\theta \mid t_0) \Psi_1(\bar{S}_1(t), t \mid \bar{S}_2(\theta), \theta)\} \, d\theta$$

$$\hat{\gamma}^+(t \mid t_0) = -2\hat{\Psi}_2(\bar{S}_2(t), t \mid t_0) + 2 \int_{t_0}^{t} \{\hat{\gamma}^-(\theta \mid t_0) \Psi_2(\bar{S}_2(t), t \mid \bar{S}_1(\theta), \theta) + \\ + \hat{\gamma}^+(\theta \mid t_0) \Psi_2(\bar{S}_2(t), t \mid \bar{S}_2(\theta), \theta)\} \, d\theta \tag{16}$$

with

$$\hat{\Psi}_i(\tilde{S}_i(t), t \mid t_0) = \int_{\ln a + d(t_0)}^{\ln b + d(t_0)} e^{y_0 - d(t_0)} \, g_{X_0}(e^{y_0-d(t_0)}) \Psi_i(\tilde{S}_i(t), t \mid y_0, t_0) \, dy_0 \qquad (i = 1, 2). \tag{17}$$

The system (16) consists of two Volterra-integral equations with non singular kernels whose solution, obtained via a numerical procedure, will give the functions $\hat{\gamma}^-$ and $\hat{\gamma}^+$.

In the following we consider some numerical evaluations of (16) for particular distributions of the initial value. As suggested in [6], we assume the following expression for $C(t)$:

$$C(t) = C_0 \ln(e + \xi \, t)$$

where e is the Neper constant, $C_0 = 1 \, year^{-1}$ and $\xi = 5 \, year^{-1}$. Furthermore, the values for the boundaries S_1 and S_2 are $S_1 = 1$, $S_2 = 9.3438 \cdot 10^8$. Finally, we choose $\alpha = 6.46 \, year^{-1}$, $\beta = 0.314 \, year^{-1}$, $\sigma^2 = 1 \, year^{-1}$, corresponding to a parathyroid tumor with mean age of 19.6 *years* (cf. [8]).

2.1 Uniform Initial Distribution

Assume that X_0 is an uniform r.v. in (a, b) with $S_1 < a < b < S_2$. Recalling (6) one has:

$$g_{Y_0}(y) = \begin{cases} \dfrac{\exp[y - d(t_0)]}{b - a} & y \in \big(\ln a + d(t_0), \ln b + d(t_0)\big) \\ 0 & \text{otherwise.} \end{cases} \tag{18}$$

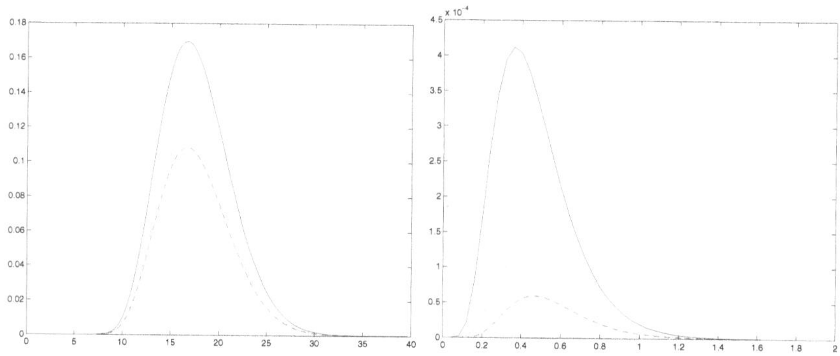

Fig. 1. The densities $\hat{\gamma}^-(t \mid 0)$ (on the left) and $\hat{\gamma}^+(t \mid 0)$ (on the right) are plotted for X_0 uniformly distributed in the interval $(10^8, 1.148 \cdot 10^8)$ (dashed line), $(0.5 \cdot 10^8, 1.648 \cdot 10^8)$ (dotted line) and $(10^7, 2.048 \cdot 10^8)$ (solid line)

According with [8], we choose in the algorithm (16)

$$\mathrm{E}X_0 = \frac{b+a}{2} = 1.074 \cdot 10^8.$$

In Figure 1 the densities $\hat{\gamma}^-(t \mid 0)$ (on the left) and $\hat{\gamma}^+(t \mid 0)$ (on the right) are plotted for X_0 uniformly distributed in the interval $(10^8, 1.148 \cdot 10^8)$ (dashed line), $(0.5 \cdot 10^8, 1.648 \cdot 10^8)$ (dotted line) and $(10^7, 2.048 \cdot 10^8)$ (solid line). We note that $\hat{\gamma}^-(t \mid 0)$ and $\hat{\gamma}^+(t \mid 0)$ increase as the interval (a, b) becomes larger and larger.

2.2 Gaussian Distribution Normalized in (a, b)

Now we consider X_0 having a normal distribution restricted to a real interval (a, b), i.e.

$$g_{X_0}(x) = \frac{A}{\sqrt{2\pi v^2}} \exp\left\{ -\frac{(x-\mu)^2}{2\,v^2} \right\} \qquad (\mu \in \mathbf{R},\ v > 0) \tag{19}$$

with

$$A = \left[\Phi\left(\frac{b-\mu}{v}\right) - \Phi\left(\frac{a-\mu}{v}\right) \right]^{-1},$$

where $\Phi(\cdot)$ is the normal cumulative distribution function. In Figure 2 the densities $\hat{\gamma}^-(t \mid 0)$ (on the left) and $\hat{\gamma}^+(t \mid 0)$ (on the right) are plotted for X_0 distributed according to (19) with $\mu = (a+b)/2 = 1.074 \cdot 10^8$ and $v^2 = 1$. The widths of the interval (a, b) are chosen as in the uniform case. Comparing Figures 1 and 2 we can observe that the shape of both the curves, $\hat{\gamma}^-$ and $\hat{\gamma}^+$, is similar in the uniform and Gaussian cases.

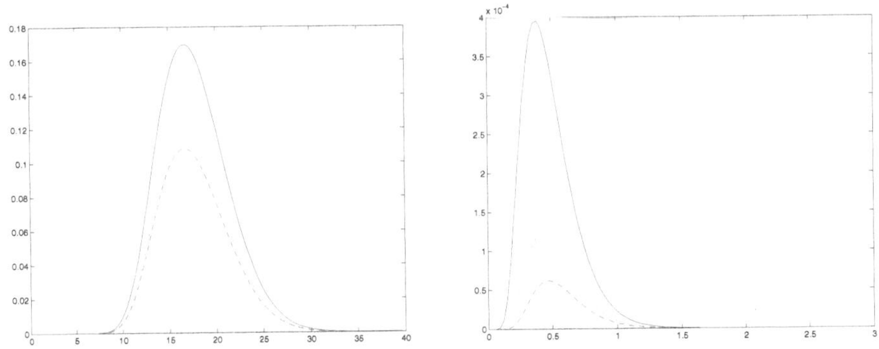

Fig. 2. $\hat{\gamma}^-(t \mid t_0)$ (on the left) and $\hat{\gamma}^+(t \mid t_0)$ (on the right) are plotted for X_0 distributed as in (19) in the interval $(10^8, 1.148 \cdot 10^8)$ (dashed line), $(0.5 \cdot 10^8, 1.648 \cdot 10^8)$ (dotted line) and $(10^7, 2.048 \cdot 10^8)$ (solid line)

3 Mean First Exit Time

In this section, we consider the case of a therapy with a constant intensity, i.e. $C(t) = C_0$. It can be handled as the no therapy case setting $\alpha - C_0 = \delta$. In this way, $X_C(t)$ is a time-homogeneous process and it can be transformed in the OU process

$$D_1(x) = \delta - \sigma^2/2 - \beta x, \qquad D_2(x) = \sigma^2. \tag{20}$$

The FET problem for $X_C(t)$ through S_1 and S_2 becomes the FET problem for the process (20) through $\bar{S}_1 = \ln S_1$ and $\bar{S}_2 = \ln S_2$. In this case (cf. [1]) we can also evaluate the mean FET. Indeed, denoting by

$$P^-(y_0) = \frac{\mathrm{Erfi}\left[\dfrac{\bar{S}_2\beta - \alpha}{\sqrt{\sigma^2\beta}}\right] - \mathrm{Erfi}\left[\dfrac{y_0\beta - \alpha}{\sqrt{\sigma^2\beta}}\right]}{\mathrm{Erfi}\left[\dfrac{\bar{S}_2\beta - \alpha}{\sqrt{\sigma^2\beta}}\right] - \mathrm{Erfi}\left[\dfrac{\bar{S}_1\beta - \alpha}{\sqrt{\sigma^2\beta}}\right]} \quad \text{(FET probability through } S_1\text{)},$$

$$P^+(y_0) = \frac{\mathrm{Erfi}\left[\dfrac{y_0\beta - \alpha}{\sqrt{\sigma^2\beta}}\right] - \mathrm{Erfi}\left[\dfrac{\bar{S}_1\beta - \alpha}{\sqrt{\sigma^2\beta}}\right]}{\mathrm{Erfi}\left[\dfrac{\bar{S}_2\beta - \alpha}{\sqrt{\sigma^2\beta}}\right] - \mathrm{Erfi}\left[\dfrac{\bar{S}_1\beta - \alpha}{\sqrt{\sigma^2\beta}}\right]} \quad \text{(FET probability through } S_2\text{)},$$

we have:

$$\mathrm{E}(T \mid y_0) = \int_0^\infty t\,\gamma(t \mid y_0)\,dt = P^+(y_0)\big[l(\bar{S}_2) - g(\bar{S}_2)\big] + P^-(y_0)\big[l(\bar{S}_1) - g(\bar{S}_1)\big] \tag{21}$$

with

$$l(x) = \frac{\pi}{2\beta}\mathrm{Erfi}\left[\frac{\beta x - \alpha}{\sqrt{\sigma^2\beta}}\right]\left\{\mathrm{Erf}\left[\frac{\beta x - \alpha}{\sqrt{\sigma^2\beta}}\right] - \mathrm{Erf}\left[\frac{\beta x_0 - \alpha}{\sqrt{\sigma^2\beta}}\right]\right\}, \tag{22}$$

$$g(x) = \frac{1}{\sigma^2\beta^2}\left\{(\beta x-\alpha)^2\ {}_2F_2\left[1,1;3/2,2;-\frac{(\beta x-\alpha)^2}{\sigma^2\beta}\right]\right\}$$
$$-(\beta x_0-\alpha)^2\ {}_2F_2\left[1,1;3/2,2;-\frac{(\beta x_0-\alpha)^2}{\sigma^2\beta}\right]\Bigg\} \tag{23}$$

where

$${}_2F_2\big[1,1;3/2,2;z\big] = \sum_{k=0}^{\infty}\frac{(2z)^k}{(k+1)(2k+1)!!} \tag{24}$$

is the generalized hypergeometric function. Under the condition (14) we can immediately write:

$$\mathrm{E}(T) = \int_{\ln a}^{\ln b} e^y\, g_{X_0}(e^y)\,\mathrm{E}(T \mid y)\,dy. \tag{25}$$

Then, the mean FET can be obtained from (25) specifying the pdf of the initial state $Y(t)$. In the same way, we can obtain the FET probability of $Y(t)$ through $\bar{S}_1$ and $\bar{S}_2$:

$$\hat{P}^- = \int_{\ln a}^{\ln b} e^y\, g_{X_0}(e^y)\,\hat{P}^-(y)\,dy, \qquad P^+ = \int_{\ln a}^{\ln b} e^y\, g_{X_0}(e^y)\,P^+(y)\,dy.$$

Table 1 lists the probability P^- and the mean FET for the process (20) through the $\bar{S}_1 = \ln(S_1) = 0$ and $\bar{S}_2 = \ln(S_2) = 20.6554$ when the initial state X_0 is uniformly distributed in the intervals $(10^8, 1.148\cdot 10^8)$ and $(10^7, 2.048\cdot 10^8)$. We can note that the FET probability $\hat{P}^-$ monotonically increases as $\delta = \alpha - C_0$ decreases. This means that for aggressive therapies the probability to reach the recovery level becomes larger and larger. The mean FET is an unimodal function

Table 1. FET probability through $\bar{S}_1$ and mean FET of the process (20) for different values of δ for the initial state X_0 uniformly distributed in the interval (a, b)

δ	P^-	E(T)	$P^-(x_0)$	E(T)
	$a = 10^8$ $b = 1.148\cdot 10^8$	$a = 10^8$ $b = 1.148\cdot 10^8$	$a = 10^7$ $b = 2.048\cdot 10^8$	$a = 10^7$ $b = 1.148\cdot 10^8$
6.46	$2.49000\cdot 10^{-48}$	12.56953	$2.91708\cdot 10^{-48}$	12.75055
6.06	$1.88015\cdot 10^{-41}$	58.42840	$1.90708\cdot 10^{-41}$	58.45927
5.66	$1.71518\cdot 10^{-34}$	706.49063	$1.70788\cdot 10^{-34}$	703.07736
5.26	$1.75559\cdot 10^{-27}$	25572.28392	$1.74716\cdot 10^{-27}$	25449.21482
4.86	$1.91743\cdot 10^{-20}$	$2.74948\cdot 10^{6}$	$1.90939\cdot 10^{-20}$	$2.73794\cdot 10^{6}$
4.46	$2.17099\cdot 10^{-13}$	$8.56665\cdot 10^{8}$	$2.16249\cdot 10^{-13}$	$8.53312\cdot 10^{8}$
4.06	$2.50528\cdot 10^{-6}$	$7.62035\cdot 10^{11}$	$2.49572\cdot 10^{-6}$	$7.59124\cdot 10^{11}$
3.66	0.96680	$6.36792\cdot 10^{13}$	0.96313	$6.34377\cdot 10^{13}$
3.26	0.99999	$4.02340\cdot 10^{10}$	0.99620	$4.00818\cdot 10^{10}$
2.86	0.99999	$6.97889\cdot 10^{7}$	0.99621	$6.95250\cdot 10^{7}$

of δ. The maximum of E(T) can be interpreted by remarking that for small values of δ the FET through $\bar{S}_1$ is preferred. Instead, when δ increases the exit from the upper boundary $\bar{S}_2$ becomes more likely. Furthermore, comparing column 2 with 4 and column 3 with 5, we can observe that when the interval (a, b) is changed the order of the involved quantities is kept.

4 Conclusions

In this work we have considered a stochastic model generalizing Gompertz growth for describing monoclonal tumor in the presence of a time-dependent therapy. We have assumed that the tumor size at time detection is a r.v.. The FET problem has been investigated for uniform and Gaussian initial distribution. Furthermore, in the presence of a therapy with fixed dosage the mean FET and the probability to reach the "recovery level" have been analysed. From Table 1 we can observe that therapies able to reduce by 6% the tumor growth rate (corresponding to $\delta = 6.06$), though not implying complete recovery, leads to a 58 years mean FET.

References

1. Albano, G., Giorno, V.: A stochastic model in tumor growth. J. Theor. Biol. 242(2), 229–236 (2006)
2. Albano, G., Giorno, V.: Towards a stochastic two-compartment model in tumor growth. Scientiae Mathematicae Japonicae 67(2), 305–318, e2008, 15-28 (2008)
3. Albano, G., Giorno, V., Saturnino, C.: A Prey-Predator Model for Immune Response and Drug Resistance in Tumor Growth. In: Moreno Díaz, R., Pichler, F., Quesada Arencibia, A. (eds.) EUROCAST 2007. LNCS, vol. 4739, pp. 171–178. Springer, Heidelberg (2007)
4. Buonocore, A., Giorno, V., Nobile, A.G., Ricciardi, L.M.: On the two-boundary first-crossing-time problem for diffusion processes. J. Appl. Prob. 27, 102–114 (1990)
5. Castorina, P., Zappalà, D.: Tumor Gompertzian growth by cellular energetic balance. Physica A 365(2), 473–480 (2004)
6. de Vladar, H.P., Gonzalez, J.A.: Dynamic response of cancer under the influence of immunological activity and therapy. J. Theor. Biol. 227(3), 335–348 (2004)
7. Kozusko, F., Bajzer, Z.: Combining Gompertzian growth and cell population dynamics. Mathematical Biosciences 185, 153–167 (2003)
8. Parfitt, A.M., Fyhrie, D.P.: Gompertzian growth curves in parathyroid tumours: further evidence for the set-point hypothesis. Cell Prolif. 30, 341–349 (1997)
9. Stepanova, N.: Course of the immune reaction during the development of a malignant tumor. Biophysics 24, 917–923 (1980)

A Neuronal Model with Excitatory and Inhibitory Inputs Governed by a Birth-Death Process

Antonio Di Crescenzo and Barbara Martinucci

Dipartimento di Matematica e Informatica, Università di Salerno
Via Ponte don Melillo, I-84084 Fisciano (SA), Italy
{adicrescenzo,bmartinucci}@unisa.it

Abstract. A stochastic model for the firing activity of a neuronal unit has been recently proposed in [4]. It includes the decay effect of the membrane potential in the absence of stimuli, and the occurrence of excitatory inputs driven by a Poisson process. In order to add the effects of inhibitory stimuli, we now propose a Stein-type model based on a suitable exponential transformation of a bilateral birth-death process on $\mathbb{Z}$ and characterized by state-dependent nonlinear birth and death rates. We perform an analysis of the probability distribution of the stochastic process describing the membrane potential and make use of a simulation-based approach to obtain some results on the firing density.

1 Introduction

One of the first models for the description of the membrane potential in single neuronal units is due to Stein [13], who proposed a leaky-integrate-and-fire model characterized by the linear summation of excitatory and inhibitory synaptic inputs driven by time-homogeneous Poisson processes. This is the starting point of many more complex stochastic neuronal models.

A Stein-type model has been studied recently in [4], where the dynamics of neuronal membrane potential is described by the stochastic process

$$\widetilde{V}(t) = v_0\, \mathrm{e}^{-\nu t+\widetilde{X}(t)}, \quad t > 0, \qquad \widetilde{V}(0) = v_0 > 0, \tag{1}$$

where $\nu > 0$ is the decay rate to the resting level in absence of stimuli, and $\widetilde{X}(t)$ is a suitable compound Poisson process. Differently from Stein model, (1) includes a multiplicative state-dependent effect, since the membrane depolarizations are random and depend on the voltage level at the stimulus time. However, since the sample-paths of $\widetilde{X}(t)$ are non-decreasing, this model allows only for the effects of excitatory inputs. In this paper we aim to modify (1) by including the effects of inhibitory inputs. The new model is described in Section 2, where we obtain the probability distribution of the stochastic process $V(t)$ that describes the membrane potential, and discuss some symmetry properties. The mean and the variance of $V(t)$ are also studied. Section 3 is devoted to analyze the firing

R. Moreno-Díaz et al. (Eds.): EUROCAST 2009, LNCS 5717, pp. 121–128, 2009.

activity of the model. We develop a simulation approach in order to estimate the firing density, the mean and the coefficient of variation of the firing times. In particular, the behavior of the firing density and the role of the involved parameters are investigated.

2 The Model

Let us now introduce a new state-dependent neuronal model based on a continuous-time stochastic process $\{V(t); t \geq 0\}$ that describes the neuronal membrane potential. We assume that

$$V(t) = v_0\, \mathrm{e}^{-\nu t + \gamma X(t)}, \quad t > 0, \qquad V(0) = v_0 > 0, \tag{2}$$

where $X(t)$ is a bilateral birth-death process on $\mathbb{Z}$, with initial state $X(0) = 0$, and characterized by birth and death rates

$$\lambda_n = \lambda\, \frac{1 + c\left(\frac{\mu}{\lambda}\right)^{n+1}}{1 + c\left(\frac{\mu}{\lambda}\right)^{n}}, \qquad \mu_n = \mu\, \frac{1 + c\left(\frac{\mu}{\lambda}\right)^{n-1}}{1 + c\left(\frac{\mu}{\lambda}\right)^{n}}, \qquad n \in \mathbb{Z}, \tag{3}$$

with $\nu, \gamma, c, \lambda, \mu > 0$. We remark that the sum of birth and death rates is independent on n; indeed, from (3) we have $\lambda_n + \mu_n = \lambda + \mu$. Moreover, we note that $\lambda_{n-1}\, \mu_n = \lambda\, \mu$. Straightforward calculations thus show that when $c = 1$ and $\lambda + \mu = 1$, process $X(t)$ identifies with the nonlinear birth-death process studied by Hongler and Parthasarathy [7].

Examples of simulated sample-paths of $V(t)$ are shown in Figure 1 for different choices of c. We point out that neuronal models involving birth-death processes have been already employed in the past (see, for instance Giorno *et al.* [6] and Ricciardi *et al.* [12]) and that the condition of state-dependent rates is often assumed in investigations related to biological systems (see, for instance, Pokora and Lánský [8] where the number of activated olfactory receptor neurons is

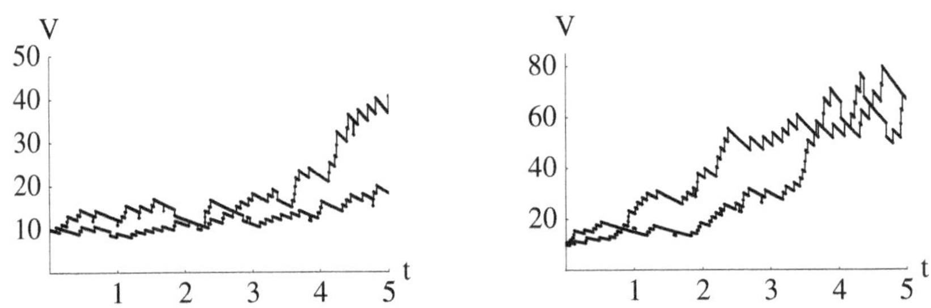

Fig. 1. Two simulated sample-paths of $V(t)$ for c = 0.1 (left-hand side) and c = 0.01 (right-hand side), with $\nu = 0.5$, $\gamma = 0.1$, $\lambda = 8$, $\mu = 1$

described by a state-dependent birth-death process). See also Ricciardi [11] for a comprehensive review on birth-death processes in stochastic population models.

Process $X(t)$ describes the difference between the numbers of excitatory and inhibitory inputs arrived in $[0,t]$. Indeed, births (deaths) of $X(t)$ correspond to upward (downward) jumps of the sample-paths of $V(t)$, produced by excitatory (inhibitory) inputs. Denoting by t_n the arrival time of the n-th stimulus, from (2) we have

$$V(t_n) - V(t_n^-) = w\, V(t_n^-), \qquad n = 1, 2, \dots \tag{4}$$

where

$$w = \begin{cases} e^{\gamma} - 1 & \text{if the stimulus is excitatory,} \\ e^{-\gamma} - 1 & \text{if the stimulus is inhibitory.} \end{cases}$$

The effect of excitatory (inhibitory) stimuli is thus state-dependent. Indeed, the amplitude of the jump performed by the neuronal membrane potential at time t_n is equal to the level of the potential attained just before t_n multiplied by a positive (negative) factor. We point out that the ratio $V(t_n)/V(t_n^-)$ in model (1) is random, whereas in the present model it is constant (see (4)).

We note that the excitatory and inhibitory stimuli arrival rates λ_n and μ_n are respectively increasing and decreasing in n, both approaching constant values as $|n| \to +\infty$. Indeed, from (3) we have

$$\lim_{n\to+\infty} \lambda_n = \lim_{n\to-\infty} \mu_n = \max\{\lambda, \mu\},$$

$$\lim_{n\to-\infty} \lambda_n = \lim_{n\to+\infty} \mu_n = \min\{\lambda, \mu\}.$$

The above mentioned properties of the rates are suitable to describe a physiological feature by which the excitatory (inhibitory) inputs prevail over the inhibitory (excitatory) ones when the membrane potential approaches the firing threshold. Moreover, a similar feature occurs if c is close to 0 and if $\lambda > \mu$. Indeed, in this case λ_n is decreasing and μ_n is increasing in $c > 0$ with

$$\lim_{c\to 0} \lambda_n = \lim_{c\to+\infty} \mu_n = \lambda, \qquad \lim_{c\to 0} \mu_n = \lim_{c\to+\infty} \lambda_n = \mu.$$

We note that $X(t)$ is a Markov process which is *similar* to a bilateral birth-death process with constant rates λ and μ, in the sense that the ratio of their transition functions is time independent (see Di Crescenzo [2], Pollett [9] and references therein for various results on *similar* processes). Other properties of $X(t)$ are given in [3].

From (2) we have that $\{V(t); t \geq 0\}$ is a stochastic process with discrete state-space $\{v_0\, e^{-\nu t+\gamma n},\ n \in \mathbb{Z}\}$ and probability distribution (see Section 4 of [3])

$$\begin{aligned} p(\lambda,\mu,c,t,n) &:= P\{V(t) = v_0\, e^{-\nu t+\gamma n} \mid V(0) = v_0\} \\ &= \frac{e^{-(\lambda+\mu)t}}{1+c} \left[\left(\frac{\mu}{\lambda}\right)^{-\frac{n}{2}} + c \left(\frac{\mu}{\lambda}\right)^{\frac{n}{2}} \right] I_n\left(2t\sqrt{\lambda\mu}\right), \qquad t > 0, \end{aligned} \tag{5}$$

where

$$I_n\left(2t\sqrt{\lambda\mu}\right) = \sum_{i=0}^{\infty} \frac{\left(t\sqrt{\lambda\mu}\right)^{n+2i}}{i!\,(n+i)!}$$

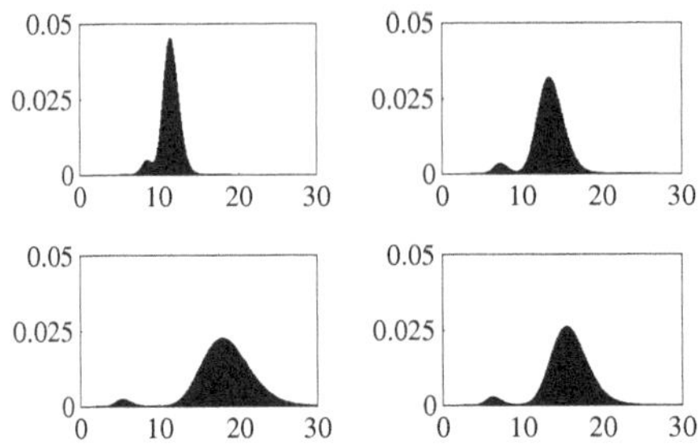

Fig. 2. Distribution $p(\lambda, \mu, c, t, n)$ for $n = 0, 1, \ldots, 30$, with $v_0 = 10$, $\lambda = 8$, $\mu = 5$, $\gamma = 0.01$, $\nu = 0.001$, $c = 0.1$ and $t = 5, 10, 15, 20$ (clockwise from the top)

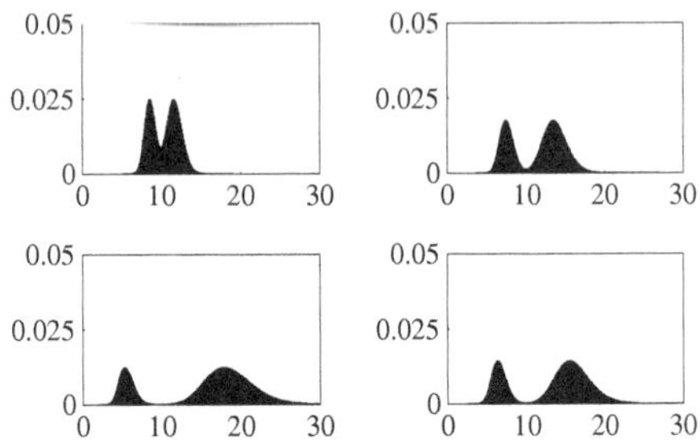

Fig. 3. Same as Figure 2, with $c = 1$

denotes the modified Bessel function of the first kind. We note that the terms in square brackets of Eq. (5) can be expressed as the following combination of hyperbolic functions:

$$(c+1)\cosh\left(\frac{n}{2}\log\frac{\mu}{\lambda}\right) + (c-1)\sinh\left(\frac{n}{2}\log\frac{\mu}{\lambda}\right),$$

or equivalently in terms of the generalized hyperbolic cosine function (cf. Eq. (2.2) of Ren and Zhang [10]).

Due to (5), the probability distribution of $V(t)$ exhibits the following symmetry properties, for all $n \in \mathbb{Z}$:

$$\begin{aligned} p(\lambda, \mu, c, t, n) &= p(\mu, \lambda, \frac{1}{c}, t, n), \\ p(\lambda, \mu, c, t, n) &= p(\mu, \lambda, c, t, -n), \\ p(\lambda, \mu, c, t, n) &= p(\lambda, \mu, \frac{1}{c}, t, -n). \end{aligned}$$

Note that $p(\lambda, \mu, c, t, n)$ is symmetric with respect to n for $c = 1$, whereas, in the case $\lambda > \mu$, it is positive (negative) skewed for $0 < c < 1$ ($c > 1$). Figures 2 and 3 show some plots of $p(\lambda, \mu, c, t, n)$ in the case $\lambda > \mu$, for different choices of the parameters. The probability distribution exhibits a bimodal shape with the well located at the initial value v_0. In the case $\lambda > \mu$, if c approaches 0^+ then the right-hand peak increases. Moreover, as t increases the probability mass spreads over the n-axis.

In the following proposition we obtain the mean and the variance of the stochastic process $V(t)$.

Proposition 1. *For all $t > 0$ we have*

$$E[V(t) \mid V(0) = v_0] = \frac{v_0\, \mathrm{e}^{-(\lambda+\mu+\nu)t}}{1+c} \left\{\exp\left[t\left(\lambda \mathrm{e}^{\gamma} + \mu \mathrm{e}^{-\gamma}\right)\right] + c \exp\left[t\left(\mu \mathrm{e}^{\gamma} + \lambda \mathrm{e}^{-\gamma}\right)\right]\right\} \tag{6}$$

and

$$
\begin{aligned}
&Var\left[V(t)\,|\,V(0)=v_0\right] \\
&\quad = \frac{v_0^2\,\mathrm{e}^{-(2\nu+\lambda+\mu)t}}{1+c}\left\{\exp\left[t\left(\lambda \mathrm{e}^{2\gamma}+\mu \mathrm{e}^{-2\gamma}\right)\right]+c\exp\left[t\left(\mu \mathrm{e}^{2\gamma}+\lambda \mathrm{e}^{-2\gamma}\right)\right]\right. \\
&\quad \left. -\frac{\mathrm{e}^{-(\lambda+\mu)t}}{1+c}\left\{\exp\left[t\left(\lambda \mathrm{e}^{\gamma}+\mu \mathrm{e}^{-\gamma}\right)\right]+c\exp\left[t\left(\mu \mathrm{e}^{\gamma}+\lambda \mathrm{e}^{-\gamma}\right)\right]\right\}^2\right\}. \qquad (7)
\end{aligned}
$$

Proof. Aiming to obtain the mean and the variance of $V(t)$ let us evaluate the moment generating function of $X(t)$. Being

$$
\begin{aligned}
E\left[\mathrm{e}^{sX(t)}\,|\,X(0)=0\right] &= \sum_{n\in\mathbb{Z}} \mathrm{e}^{sn} P\{X(t)=n\,|\,X(0)=0\} \\
&= \frac{\mathrm{e}^{-(\lambda+\mu)t}}{1+c}\sum_{n\in\mathbb{Z}}\left[\left(\mathrm{e}^s\sqrt{\frac{\lambda}{\mu}}\right)^n + c\left(\mathrm{e}^s\sqrt{\frac{\mu}{\lambda}}\right)^n\right] I_n\left(2t\sqrt{\lambda\mu}\right).
\end{aligned}
$$

Hence, recalling that (see Eq. 9.6.33 of [1])

$$
\sum_{k=-\infty}^{+\infty} t^k I_k(z) = \mathrm{e}^{\frac{1}{2}z(t+1/t)}, \qquad t\neq 0,
$$

we obtain

$$
E\left[\mathrm{e}^{sX(t)}\,|\,X(0)=0\right] = \frac{\mathrm{e}^{-(\lambda+\mu)t}}{1+c}
$$

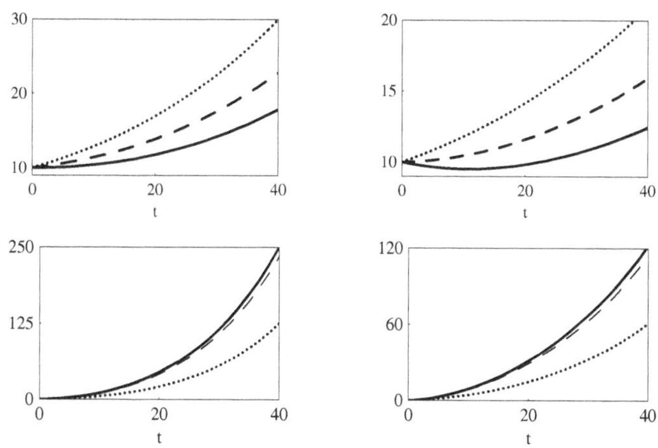

Fig. 4. First row: $E\left[V(t)\,|\,V(0)=v_0\right]$ for $\lambda=8, \mu=5, \gamma=0.01$ and $c=0.1, 0.5, 1$ (from top to bottom) with $\nu=0.001$ (left-hand side) and $\nu=0.01$ (right-hand side). Second row: $Var\left[V(t)\,|\,V(0)=v_0\right]$ for the same choice of parameters (from bottom to top).

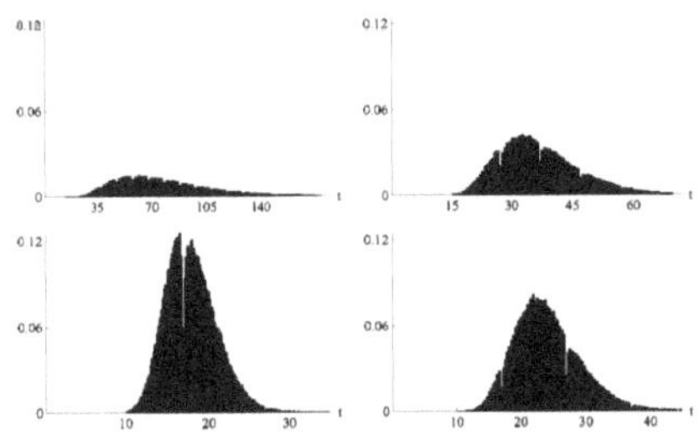

Fig. 5. Histograms of simulated firing times for $\nu = 0.001$, $v_0 = 10$, $\beta = 20$, $c = 0.1$, $\lambda = 6$, $\gamma = 0.01$, with (a) $\mu = 5$, (b) $\mu = 4$, (c) $\mu = 3$, (d) $\mu = 2$

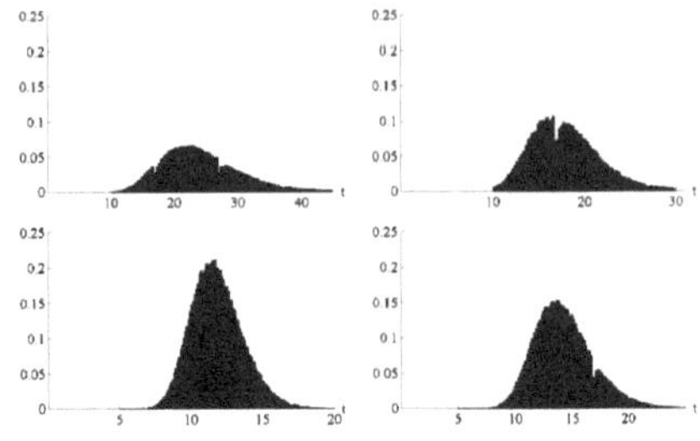

Fig. 6. Same as Figure 5 for $\lambda = 8$

$$\times \left\{ \exp\left[t\sqrt{\lambda\mu}\left(\mathrm{e}^{s}\sqrt{\frac{\lambda}{\mu}} + \mathrm{e}^{-s}\sqrt{\frac{\mu}{\lambda}} \right) \right] + c\exp\left[t\sqrt{\lambda\mu}\left(\mathrm{e}^{s}\sqrt{\frac{\mu}{\lambda}} + \mathrm{e}^{-s}\sqrt{\frac{\lambda}{\mu}} \right) \right] \right\}$$

$$= \frac{\mathrm{e}^{-(\lambda+\mu)t}}{1+c}\left\{ \exp\left[t\left(\lambda \mathrm{e}^{s} + \mu \mathrm{e}^{-s} \right) \right] + c\exp\left[t\left(\mu \mathrm{e}^{s} + \lambda \mathrm{e}^{-s} \right) \right] \right\}. \tag{8}$$

Due to (2), Eq. (6) follows from (8). Similarly, Eq. (7) can be obtained making use of (6) and (8).

Figure 4 shows some plots of the mean and the variance of $V(t)$ for various choices of c and ν. We remark that the mean value (6) is decreasing in $c > 0$ if $\lambda > \mu$, and it is not necessarily monotonic in t. For instance, each of the following sets of parameters provides sufficient conditions such that $E\left[V(t) \,|\, V(0) = v_0\right]$ is increasing in $t > 0$:

(i) $\mu < \lambda < \mu \mathrm{e}^{\gamma}$, $0 < \nu < \mu(\mathrm{e}^{\gamma} - 1) - \lambda(1 - \mathrm{e}^{-\gamma})$;

(ii) $\lambda > \mu \mathrm{e}^{\gamma}$, $0 < \nu < \lambda(\mathrm{e}^{\gamma}-1) - \mu(1-\mathrm{e}^{-\gamma})$, $\dfrac{c\left[\nu + \lambda(1 - \mathrm{e}^{-\gamma}) - \mu(\mathrm{e}^{\gamma} - 1)\right]}{\left[\lambda(\mathrm{e}^{\gamma} - 1) - \mu(1 - \mathrm{e}^{-\gamma}) - \nu\right]} < 1.$

3 Firing Activity

The main interest in neuronal models relies in the properties of the first-passage time of $V(t)$ through the firing threshold, which identifies with the firing time. The latter is described as the first-passage time of $V(t)$ through the constant firing level β:

$$T_V^{(\beta)} = \inf\{t \geq 0 : V(t) > \beta\}, \qquad \beta > v_0. \tag{9}$$

Hence, $T_V^{(\beta)}$ is the random time between the instant when the membrane potential resets to v_0 and the instant when an action potential is generated due to a crossing of the firing threshold β. See [12] for a comprehensive review of theoretical and algorithmic approaches to first-passage-time problems with applications to biological modeling.

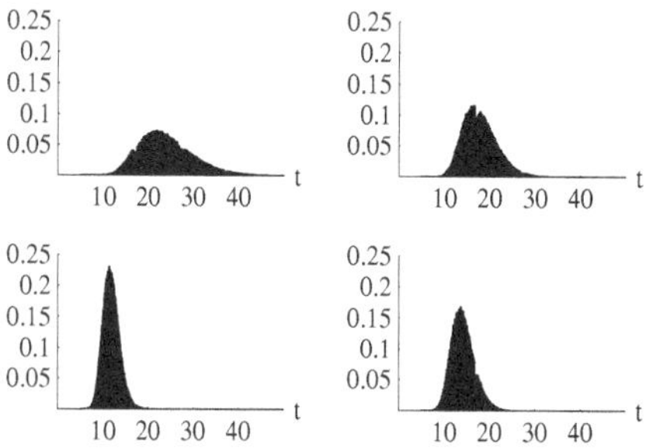

Fig. 7. Same as Figure 6 for $c = 0.01$

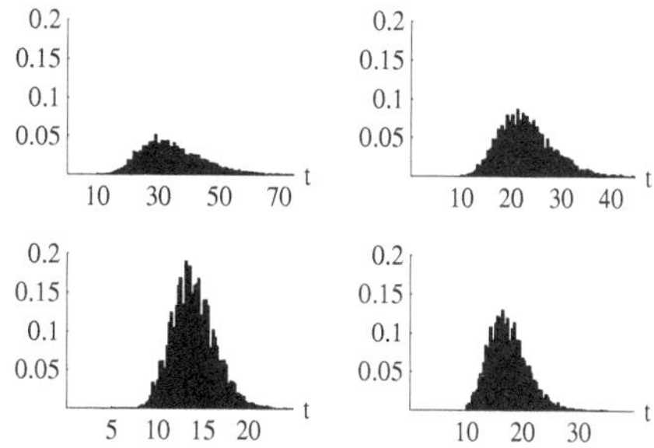

Fig. 8. Same as Figure 6 for $\nu = 0.01$

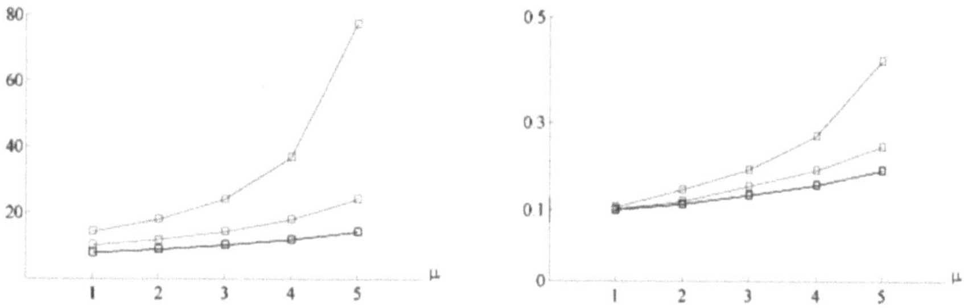

Fig. 9. On the left: mean of the firing time for $\lambda = 6, 8, 10$ (from top to bottom); on the right: CV of the firing time for the same choice of λ. Other parameters are: $\nu = 0.001$, $v_0 = 10$, $\beta = 20$, $\gamma = 0.01$ and $c = 0.1$.

Since closed-form expressions of the firing density appear to be unavailable, we adopt a Monte-Carlo simulation approach in order to study the firing activity of model (2). Figures 5÷8 show various histograms of the simulated firing times obtained by means of 2×10^5 simulations. The first-passage-time problem of $V(t)$ through a constant boundary can be seen as an analogous problem for $X(t)$ in the presence of a linearly increasing boundary. Hence, the discrete nature of $X(t)$ and the form of the boundary justify the non-smooth shape of the firing density (see Figures 5÷8). This is more pronounced when ν is large. Moreover, specific analysis of the simulated histograms suggests that this feature is strictly related also to the difference $\lambda - \mu$. Furthermore, simulation results show that the firing probability is increasing as $c \to 0^+$, and that the firing densities have similar shape for different values of c. Future developments of the present research will be oriented to develop an analytical approach to study the firing density.

Figure 8 shows the mean and the coefficient of variation of the firing time for different choices of λ and μ. It suggests that both the mean and the coefficient of variation are decreasing in $\lambda - \mu$, and take large values as $\lambda - \mu$ approaches zero. We remark that for different choices of λ and μ the computational approach provides a coefficient of variation of the firing time smaller that 0.5, this suggesting that the model is characterized by regular firing patterns.

Acknowledgments

This work has been partially supported by Regione Campania and G.N.C.S.-INdAM.

References

1. Abramowitz, M., Stegun, I.A.: Handbook of Mathematical Functions with Formulas, Graphs, and Mathematical Tables. Dover, New York (1992)
2. Di Crescenzo, A.: On certain transformation properties of birth-and-death processes. In: Trappl, R. (ed.) Cybernetics and Systems 1994, vol. 1, pp. 839–846. World Scientific, Singapore (1994)
3. Di Crescenzo, A.: On some transformations of bilateral birth-and-death processes with applications to first passage time evaluations. In: SITA 1994 – Proc. 17th Symp. Inf. Theory Appl., Hiroshima, pp. 739–742 (1994), `http://arXiv.org/pdf/0803.1413v1`
4. Di Crescenzo, A., Martinucci, B.: Analysis of a stochastic neuronal model with excitatory inputs and state-dependent effects. Math. Biosci. 209, 547–563 (2007)
5. Di Crescenzo, A., Martinucci, B.: A first-passage-time problem for symmetric and similar two-dimensional birth-death processes. Stoch. Models 24, 451–469 (2008)
6. Giorno, V., Lánský, P., Nobile, A.G., Ricciardi, L.M.: Diffusion approximation and first-passage-time problem for a model neuron. III. A birth-and-death process approach. Biol. Cybernet. 58, 387–404 (1988)
7. Hongler, M.O., Parthasarathy, P.R.: On a super-diffusive, nonlinear birth and death process. Phys. Lett. A 372, 3360–3362 (2008)
8. Pokora, O., Lánský, P.: Statistical approach in search for optimal signal in simple olfactory neuronal models. Math. Biosci. 214, 100–108 (2008)
9. Pollett, P.K.: Similar Markov chains. In: Daley, D.J. (ed.) Probability, Statistics and Seismology. A Festschrift for David Vere-Jones; J. Appl. Prob. 38A, 53–65 (2001)
10. Ren, Y.J., Zhang, H.Q.: New generalized hyperbolic functions and auto-Bäcklund transformation to find new exact solutions of the (2+1)-dimensional NNV equation. Phys. Lett. A 357, 438–448 (2006)
11. Ricciardi, L.M.: Stochastic population theory: birth and death processes. In: Hallam, T.G., Levin, S.A. (eds.) Biomathematics. Mathematical Ecology, vol. 17, pp. 155–190. Springer, Heidelberg (1986)
12. Ricciardi, L.M., Di Crescenzo, A., Giorno, V., Nobile, A.G.: An outline of theoretical and algorithmic approaches to first passage time problems with applications to biological modeling. Math. Japonica 50, 247–322 (1999)
13. Stein, R.B.: A theoretical analysis of neuronal variability. Biophys. J. 5, 173–194 (1965)

Diffusion Processes Subject to Catastrophes*

Roberta di Cesare, Virginia Giorno, and Amelia G. Nobile

Dipartimento di Matematica e Informatica, Università di Salerno,
Via Ponte don Melillo, Fisciano (SA), Italy
{rdicesare,giorno,nobile}@unisa.it

Abstract. The aim of the present paper is to provide some quantitative informations on the role of catastrophes in diffusion models. Analytical and computational results for the Wiener and for the Ornstein-Uhlenbeck processes are determined.

1 Introduction

During the last three decades great attention has been paid in the literature to the description of biological, physical and engineering systems subject to various types of catastrophes. The usual framework is that the system evolves according to the dynamics of some continuous-time Markov chain and it is influenced by catastrophes that occur at exponential rate ξ and reduce instantaneously to zero the state of the system (cf. [2], [3], [6], [7], [8], [9], and references therein). These works are concerned with various quantities of interest, such as the transient and the stationary probabilities, the time of extinction and the first occurrence time of effective catastrophe. The results obtained for continuous-time Markov chains suggest the possibility of deriving corresponding results for the analogous diffusion models subject to catastrophic events.

In Section 2 some general results for the transient and steady-state probability density functions (pdf's) in the presence of catastrophes will be obtained. Furthermore, in Section 3 and 4 the Wiener and Ornstein-Uhlenbeck processes with catastrophes will be considered.

2 Diffusion Processes with Catastrophes

Let $\{\widetilde{X}(t), t \geq 0\}$ be a regular one-dimensional time-homogeneous diffusion process with drift $A_1(x)$ and infinitesimal variance $A_2(x)$ restricted to the interval $I \equiv [0, r_2)$ by a reflecting boundary in zero state. For this process we denote with $\widetilde{r}(x,t|x_0) := dP\{\widetilde{X}(t) < x|\widetilde{X}(0) = x_0\}/dx$ the transition pdf, with $0 \leq x_0 < r_2$. We construct a new stochastic process $X(t)$ defined in I as follows. Starting at the state x_0 at the initial time $t = 0$, the process $X(t)$ evolves according to the process $\widetilde{X}(t)$ until a catastrophe occurs. In our approach the catastrophes occur

* Work performed under partial support by G.N.C.S.- INdAM and by Campania Region.

R. Moreno-Díaz et al. (Eds.): EUROCAST 2009, LNCS 5717, pp. 129–136, 2009.

randomly, and are exponentially distributed with mean $1/\xi$ ($\xi > 0$). The effect of a catastrophe is to reset the state of $X(t)$ to zero, after which the process evolves like $\widetilde{X}(t)$, until a new catastrophe occurs. Hence, the zero state of $X(t)$ may be reached either as the natural evolution of the process $\widetilde{X}(t)$ or as the effect of a catastrophe occurrence.

Denoting by $r(x,t|x_0) := dP\{X(t) < x|X(0) = x_0\}/dx$ the transition pdf of the process $X(t)$, one has:

$$r(x,t|x_0) = e^{-\xi t}\,\widetilde{r}(x,t|x_0) + \xi \int_0^t e^{-\xi\tau}\, r(x,t-\tau|0)\,d\tau \qquad (x, x_0 \geq 0). \tag{1}$$

The first term on the right-hand side of (1) shows that there are no catastrophes until the time t but there could be some reflections each time the zero state is reached, matching the behavior of the $\widetilde{X}(t)$. Furthermore, the second term shows that at the time $\tau \in (0,t)$ the process may reach the zero state due to the occurrence of a catastrophe and then, the process starts at the zero state and evolves according to $X(t)$. In the sequel with the notation

$$\varphi_\lambda(x|x_0) = \int_0^{+\infty} e^{-\lambda t}\,\varphi(x,t|x_0)\,dt \qquad (x, x_0 \in I, \lambda > 0) \tag{2}$$

we denote the Laplace transform of the function $\varphi(x,t|x_0)$. From (1) one obtains:

$$r_\lambda(x|0) = \frac{\lambda+\xi}{\lambda}\,\widetilde{r}_{\lambda+\xi}(x|0) \qquad (x \geq 0, \lambda > 0), \tag{3}$$

$$r_\lambda(x|x_0) = \widetilde{r}_{\lambda+\xi}(x|x_0) + \frac{\xi}{\lambda+\xi}\,r_\lambda(x|0) \qquad (x \geq 0, x_0 > 0, \lambda > 0). \tag{4}$$

Making use of (3) in (4) one has:

$$r_\lambda(x|x_0) = \widetilde{r}_{\lambda+\xi}(x|x_0) + \frac{\xi}{\lambda}\,\widetilde{r}_{\lambda+\xi}(x|0) \qquad (x, x_0 \geq 0, \lambda > 0). \tag{5}$$

Taking the inverse Laplace transform of (5), we note that the transition pdf of $X(t)$ can be expressed in terms of the transition density of $\widetilde{X}(t)$ as follows:

$$r(x,t|x_0) = e^{-\xi t}\,\widetilde{r}(x,t|x_0) + \xi \int_0^t e^{-\xi\tau}\,\widetilde{r}(x,\tau|0)\,d\tau \qquad (x, x_0 \geq 0). \tag{6}$$

The process $X(t)$ always possesses a steady-state density, and from (6) one has

$$W(x) := \lim_{t\to+\infty} r(x,t|x_0) = \xi \int_0^{+\infty} e^{-\xi\tau}\,\widetilde{r}(x,\tau|0)\,d\tau = \xi\,\widetilde{r}_\xi(x|0) \quad (x \geq 0), \tag{7}$$

where $\widetilde{r}_\xi(x|0)$ is the Laplace transform of $\widetilde{r}(x,t|0)$. For $n = 1, 2, \ldots$ we denote by

$$\widetilde{M}_n(t|x_0) = \int_0^{+\infty} x^n\,\widetilde{r}(x,t|x_0)\,dx, \quad M_n(t|x_0) = \int_0^{+\infty} x^n\,r(x,t|x_0)\,dx$$

the nth-order conditional moments of the processes $\widetilde{X}(t)$ and $X(t)$, respectively. By virtue of (6), one obtains:

$$M_n(t|x_0) = e^{-\xi t}\,\widetilde{M}_n(t|x_0) + \xi \int_0^t e^{-\xi\tau}\,\widetilde{M}_n(\tau|0)\,d\tau \qquad (n = 1, 2, \ldots). \tag{8}$$

Let now $\alpha(x,t|x_0)$ be the transition pdf in the presence of an absorbing boundary at 0. For the process $X(t)$ one has:

$$\alpha(x,t|x_0) = e^{-\xi t}\,\widetilde{\alpha}(x,t|x_0) \qquad (x, x_0 > 0), \tag{9}$$

where $\widetilde{\alpha}(x,t|x_0)$ is the corresponding density for the process $\widetilde{X}(t)$ without catastrophes. Equation (9) shows that until the time t no catastrophe occurs and that the process $\widetilde{X}(t)$ never reached the state 0 before the time t. Furthermore, for the process $X(t)$ we consider the random variable T_{x_0} representing the first-visit time (FVT) to 0 starting from x_0, and we denote by $g(0,t|x_0) := dP(T_{x_0} < t)/dt$ the related pdf. Since

$$\int_0^{+\infty} \alpha(z,t|x_0)\,dz + \int_0^t g(0,\tau|x_0)\,d\tau = 1 \qquad (x_0 > 0),$$

by virtue of (9) for $x_0 > 0$ one has:

$$\int_0^t g(0,\tau|x_0)\,d\tau = 1 - e^{-\xi t}\int_0^{+\infty} \widetilde{\alpha}(z,t|x_0)\,dz = 1 - e^{-\xi t}\left[1 - \int_0^t \widetilde{g}(0,\tau|x_0)\,d\tau\right], \tag{10}$$

where $\widetilde{g}(0,\tau|x_0)$ is the first-passage time (FPT) pdf from x_0 to 0 for the process $\widetilde{X}(t)$. Hence, from (10) it follows that

$$g(0,t|x_0) = e^{-\xi t}\,\widetilde{g}(0,t|x_0) + \xi\, e^{-\xi t}\left[1 - \int_0^t \widetilde{g}(0,\tau|x_0)\,d\tau\right] \qquad (x_0 > 0), \tag{11}$$

so that the first visit to 0 is a sure event for the process $X(t)$.

The densities $r(x,t|x_0)$, $\alpha(x,t|x_0)$ and $g(0,t|x_0)$ are related by the following relation:

$$r(x,t|x_0) = \alpha(x,t|x_0) + \int_0^t g(0,\tau|x_0)\,r(x,t|0,\tau)\,d\tau, \qquad (x, x_0 \geq 0). \tag{12}$$

Indeed, for the process $\widetilde{X}(t)$ one has:

$$\widetilde{r}(x,t|x_0) = \widetilde{\alpha}(x,t|x_0) + \int_0^t \widetilde{g}(0,\tau|x_0)\,\widetilde{r}(x,t|0,\tau)\,d\tau \qquad (x, x_0 \geq 0), \tag{13}$$

so that taking the Laplace transform of (9), (11) and (13) one obtains:

$$\begin{aligned}
&\alpha_\lambda(x|x_0) = \widetilde{\alpha}_{\lambda+\xi}(x|x_0) \quad (x, x_0 > 0),\\
&g_\lambda(0|x_0) = \frac{\lambda}{\lambda+\xi}\,\widetilde{g}_{\lambda+\xi}(0|x_0) + \frac{\xi}{\lambda+\xi} \qquad (x_0 > 0),\\
&\widetilde{r}_\lambda(x|x_0) = \widetilde{\alpha}_{\lambda+\xi}(x|x_0) + \widetilde{r}_{\lambda+\xi}(x|0)\,\widetilde{g}_{\lambda+\xi}(0|x_0) \qquad (x, x_0 \geq 0).
\end{aligned} \tag{14}$$

Making use of (14) and (3) in (5) one has:

$$\begin{aligned} r_\lambda(x|x_0) &= \big[\widetilde{\alpha}_{\lambda+\xi}(x|x_0) + \widetilde{r}_{\lambda+\xi}(x|0)\,\widetilde{g}_{\lambda+\xi}(0|x_0)\big] + \frac{\xi}{\lambda}\,\widetilde{r}_{\lambda+\xi}(x|0) \\ &= \widetilde{\alpha}_{\lambda+\xi}(x|x_0) + r_\lambda(x|0)\left[\frac{\lambda}{\lambda+\xi}\,\widetilde{g}_{\lambda+\xi}(0|x_0) + \frac{\xi}{\lambda+\xi}\right] \\ &= \alpha_\lambda(x|x_0) + g_\lambda(0|x_0)\,r_\lambda(x|0) \qquad (x, x_0 \ge 0), \end{aligned} \tag{15}$$

so that, taking the inverse Laplace transform, Eq. (12) follows.

3 Wiener Process with Catastrophes

Let $\widetilde{X}(t)$ be a Wiener process restricted to the interval $I \equiv [0, +\infty)$ by a reflecting boundary in zero state, characterized by drift and infinitesimal variance:

$$A_1(x) = \mu, \qquad A_2(x) = \sigma^2 \qquad (\mu \in \mathbb{R}, \sigma > 0). \tag{16}$$

As well-known the transition pdf of $\widetilde{X}(t)$ is given by (cf. [1]):

$$\begin{aligned} \widetilde{r}(x,t|x_0) = &\frac{1}{\sigma\sqrt{2\pi t}}\left[\exp\left\{-\frac{(x-x_0-\mu t)^2}{2\sigma^2 t}\right\} + \exp\left\{\frac{-4\mu t x_0 - (x+x_0-\mu t)^2}{2\sigma^2 t}\right\}\right] \\ &-\frac{\mu}{\sigma^2}\exp\left\{\frac{2\mu x}{\sigma^2}\right\}\mathrm{Erfc}\left(\frac{x+x_0+\mu t}{\sigma\sqrt{2t}}\right) \qquad (x, x_0 \ge 0), \end{aligned} \tag{17}$$

where $\mathrm{Erfc}(x) = (2/\sqrt{\pi})\int_x^{+\infty} e^{-z^2}\,dz$ denotes the complementary error function. Making use of (17) in (6), for the Wiener process in the presence of catastrophes we obtain:

$$\begin{aligned} r(x,t|x_0) = &\, e^{-\xi t}\,\widetilde{r}(x,t|x_0) + \frac{\mu}{\sigma^2}\,e^{-\xi t}\exp\left\{\frac{2\mu x}{\sigma^2}\right\}\mathrm{Erfc}\left(\frac{x+\mu t}{\sigma\sqrt{2t}}\right) \\ &+\frac{\sqrt{\mu^2+2\sigma^2\xi}-\mu}{2\sigma^2}\exp\left\{-\frac{\sqrt{\mu^2+2\sigma^2\xi}-\mu}{\sigma^2}\,x\right\}\mathrm{Erfc}\left(\frac{x-t\sqrt{\mu^2+2\sigma^2\xi}}{\sigma\sqrt{2t}}\right) \\ &-\frac{\sqrt{\mu^2+2\sigma^2\xi}+\mu}{2\sigma^2}\exp\left\{\frac{\sqrt{\mu^2+2\sigma^2\xi}+\mu}{\sigma^2}\,x\right\}\mathrm{Erfc}\left(\frac{x+t\sqrt{\mu^2+2\sigma^2\xi}}{\sigma\sqrt{2t}}\right) \\ &\qquad (x, x_0 \ge 0), \end{aligned} \tag{18}$$

that for $x_0 = 0$ identifies with Eq. (38) in [2]. The steady state density of the Wiener process with catastrophes is an exponential density. Indeed, taking the limit as $t \to +\infty$ in (18), one has:

$$W(x) = \frac{\sqrt{\mu^2+2\sigma^2\xi}-\mu}{\sigma^2}\exp\left\{-\frac{\sqrt{\mu^2+2\sigma^2\xi}-\mu}{\sigma^2}\,x\right\} \qquad (x \ge 0). \tag{19}$$

Setting $\xi = 0$, if $\mu < 0$ from (19) we obtain the steady-state density of the Wiener process without catastrophes. In Fig. 1 we compare the conditional pdf (18) with the steady state density (19). Since (cf. Eq. (3.12) in [4]):

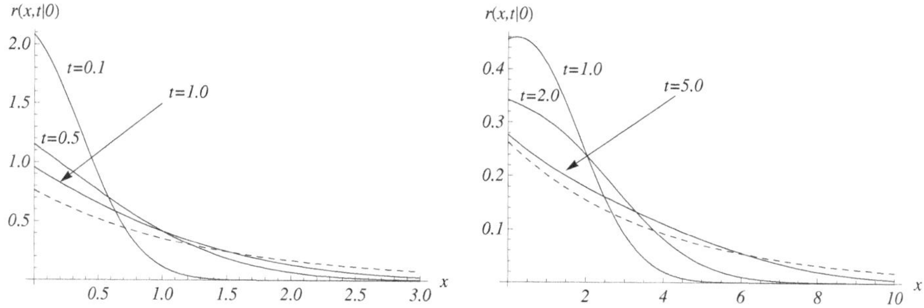

Fig. 1. For the Wiener process with $x_0 = 0$, $\sigma^2 = 2$ and $\xi = 0.2$, the conditional pdf $r(x,t|0)$ is plotted as function of x for $\mu = -0.5$ on the left and $\mu = 0.5$ on the right. The dashed curve indicates the steady state density $W(x)$.

$$\widetilde{M}_1(t|x_0) = \frac{x_0 + \mu\, t}{2}\left[1 + \mathrm{Erf}\Big(\frac{x_0 + \mu\, t}{\sigma\sqrt{2\,t}}\Big)\right] + \sigma\sqrt{\frac{t}{2\,\pi}}\,\exp\Big\{-\frac{(x_0 + \mu\, t)^2}{2\,\sigma^2\, t}\Big\}$$
$$-\frac{\sigma^2}{4\,\mu}\,\mathrm{Erfc}\Big(\frac{x_0 + \mu\, t}{\sigma\sqrt{2\,t}}\Big) + \frac{\sigma^2}{4\,\mu}\,\exp\Big\{-\frac{2\,\mu\, x_0}{\sigma^2}\Big\}\,\mathrm{Erfc}\Big(\frac{x_0 - \mu\, t}{\sigma\sqrt{2\,t}}\Big) \quad (x_0 \geq 0),$$

with $\mathrm{Erf}(x) = 1 - \mathrm{Erfc}(x)$, making use of (8) the conditional mean of the Wiener process with catastrophes for $x_0 \geq 0$ is:

$$M_1(t|x_0) = e^{-\xi\, t}\,\widetilde{M}_1(t|x_0) + \frac{\mu}{2\,\xi} + \frac{\sqrt{\mu^2 + 2\,\sigma^2\,\xi}}{2\,\xi}\,\mathrm{Erf}\left(\frac{\sqrt{(\mu^2 + 2\,\sigma^2\,\xi)\,t}}{\sigma\sqrt{2}}\right)$$
$$-\frac{e^{-\xi\, t}}{2}\left[\mu\, t + \frac{\mu}{\xi} + \Big(\mu\, t + \frac{\mu}{\xi} + \frac{\sigma^2}{\mu}\Big)\mathrm{Erf}\Big(\frac{\mu\sqrt{t}}{\sigma\sqrt{2}}\Big) + \frac{\sigma\sqrt{2\,t}}{\sqrt{\pi}}\,\exp\Big\{-\frac{\mu^2 t}{2\,\sigma^2}\Big\}\right]. \quad (20)$$

We note that as t increases $\widetilde{M}_1(t|x_0)$ admits an asymptotic limit if and only $\mu < 0$, given by $\sigma^2/(2|\mu|)$. Instead, $M_1(t|x_0)$ possesses an asymptotic limit given by $(\mu + \sqrt{\mu^2 + 2\,\sigma^2\,\xi})/(2\,\xi)$ as t increases (see Fig. 2).

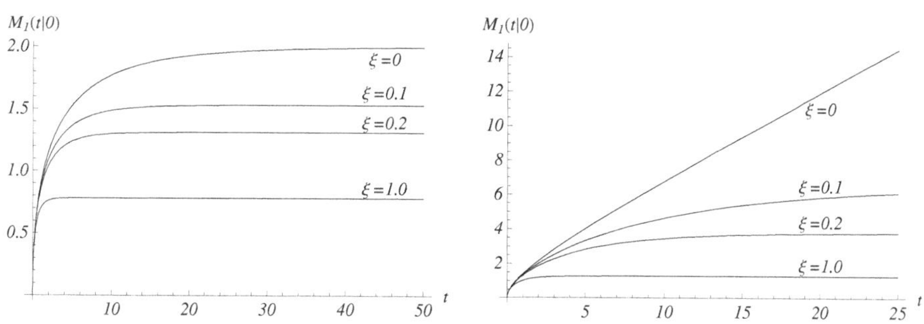

Fig. 2. Conditional mean (20) is plotted as function of t for $\mu = -0.5$ on the left and $\mu = 0.5$ on the right for the Wiener process with $x_0 = 0$ and $\sigma^2 = 2$

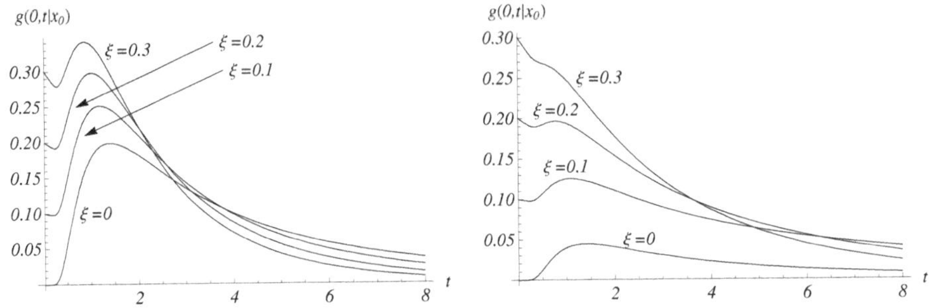

Fig. 3. FPT pdf ($\xi = 0$) and FVT pdf ($\xi \neq 0$) are plotted as function of t for $\mu = -0.5$ on the left and $\mu = 0.5$ on the right for the Wiener process with $x_0 = 3$ and $\sigma^2 = 2$

Starting from the transition pdf in the presence of an absorbing boundary in 0 for the Wiener process (cf. [1]), by virtue of (9) for $x, x_0 > 0$ one obtains:

$$\alpha(x,t|x_0) = \frac{e^{-\xi t}}{\sigma\sqrt{2\pi t}}\left[\exp\Big\{-\frac{(x-x_0-\mu t)^2}{2\sigma^2 t}\Big\} - \exp\Big\{\frac{-4\mu t x_0 - (x+x_0-\mu t)^2}{2\sigma^2 t}\Big\}\right]$$

Recalling that

$$\widetilde{g}(0,t|x_0) = \frac{x_0}{\sigma\sqrt{2\pi t^3}}\exp\Big\{-\frac{(x_0+\mu t)^2}{2\sigma^2 t}\Big\},$$

$$\int_0^t \widetilde{g}(0,\tau|x_0)\,d\tau = \frac{1}{2}\,\mathrm{Erfc}\Big(\frac{x_0+\mu t}{\sigma\sqrt{2t}}\Big) + \frac{1}{2}\exp\Big\{-\frac{2\mu x_0}{\sigma^2}\Big\}\,\mathrm{Erfc}\Big(\frac{x_0-\mu t}{\sigma\sqrt{2t}}\Big),$$

with $x_0 > 0$, from (11) one has:

$$g(0,t|x_0) = e^{-\xi t}\,\widetilde{g}(0,t|x_0) + \xi\, e^{-\xi t}\left[1 - \frac{1}{2}\,\mathrm{Erfc}\Big(\frac{x_0+\mu t}{\sigma\sqrt{2t}}\Big) - \frac{1}{2}\exp\Big\{-\frac{2\mu x_0}{\sigma^2}\Big\}\,\mathrm{Erfc}\Big(\frac{x_0-\mu t}{\sigma\sqrt{2t}}\Big)\right] \qquad (x_0 > 0). \tag{21}$$

For the Wiener process, in Fig. 3 the FPT pdf ($\xi = 0$) is compared with the FVT pdf (21) for $\xi = 0.1, 0.2, 0.3$.

4 Ornstein-Uhlenbeck Process with Catastrophes

Let $\widetilde{X}(t)$ be an Ornstein-Uhlenbeck (OU) process restricted to the interval $I \equiv [0, +\infty)$ by a reflecting boundary in zero state, characterized by drift and infinitesimal variance:

$$A_1(x) = \eta\, x, \qquad A_2(x) = \sigma^2 \qquad (\eta \in \mathbb{R}, \sigma > 0). \tag{22}$$

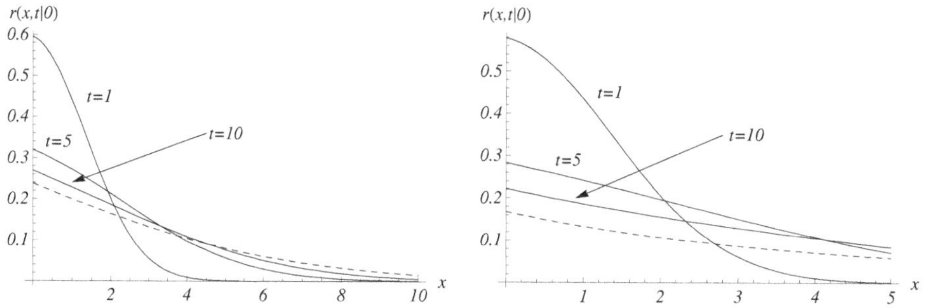

Fig. 4. For the OU process with $x_0 = 0$, $\sigma^2 = 2$ and $\xi = 0.04$ the conditional pdf $r(x,t|0)$ is plotted as function of x for $\eta = -0.03$ on the left and $\eta = 0.03$ on the right. The dashed curve indicates the steady state density $W(x)$.

Since the transition pdf of $\widetilde{X}(t)$ is given by (cf. [4]):

$$\widetilde{r}(x,t|x_0)=\sqrt{\frac{\eta}{\sigma^2\pi\,(e^{2\eta t}-1)}}\left[\exp\left\{-\frac{\eta\,(x-x_0\,e^{\eta t})^2}{\sigma^2\,(e^{2\eta t}-1)}\right\}+\exp\left\{-\frac{\eta\,(x+x_0\,e^{\eta t})^2}{\sigma^2\,(e^{2\eta t}-1)}\right\}\right] \tag{23}$$

with $x, x_0 \geq 0$, making use of (6) the transition pdf of the OU process in the presence of catastrophes can be obtained via a numerical procedure. Furthermore, by virtue of (7), for $x \geq 0$ one is led to the steady-state density of $X(t)$:

$$W(x)=\begin{cases}\dfrac{\sqrt{|\eta|}\;2^{\xi/(2|\eta|)+1}}{\sigma\sqrt{\pi}}\,\Gamma\Big(\dfrac{\xi}{2|\eta|}+1\Big)\exp\Big\{-\dfrac{|\eta|\,x^2}{2\sigma^2}\Big\}\,D_{-\xi/|\eta|}\Big(\dfrac{\sqrt{2|\eta|}}{\sigma}\,x\Big), & \eta<0\\[2ex] \dfrac{\xi\,2^{\xi/(2\eta)+1/2}}{\sigma\sqrt{\pi\,\eta}}\,\Gamma\Big(\dfrac{\xi}{2\eta}+\dfrac{1}{2}\Big)\exp\Big\{\dfrac{\eta\,x^2}{2\sigma^2}\Big\}\,D_{-\xi/\eta-1}\Big(\dfrac{\sqrt{2\eta}}{\sigma}\,x\Big), & \eta>0,\end{cases} \tag{24}$$

where $D_\nu(z)$ denotes the parabolic cylinder function (cf. [5]). Note that if $\eta < 0$, setting $\xi = 0$ in (24) and recalling that $D_0(z) = e^{-z^2/4}$, we obtain the steady-state density of $\widetilde{X}(t)$. In Fig. 4 we compare the steady state density (24) with the conditional pdf $r(x,t|0)$, numerically obtained by means of (6) and (23).

The transition pdf of $X(t)$ in the presence of an absorbing boundary in zero can be evaluated recalling (9); indeed, for $x, x_0 > 0$ one has:

$$\alpha(x,t|x_0)=\frac{e^{-\xi t}}{\sigma\sqrt{\pi}}\sqrt{\frac{\eta}{e^{2\eta t}-1}}\left[\exp\left\{-\frac{\eta\,(x-x_0\,e^{\eta t})^2}{\sigma^2\,(e^{2\eta t}-1)}\right\}-\exp\left\{-\frac{\eta\,(x+x_0\,e^{\eta t})^2}{\sigma^2\,(e^{2\eta t}-1)}\right\}\right]. \tag{25}$$

Furthermore, since

$$\widetilde{g}(0,t|x_0) = \frac{2\,x_0\,e^{\eta t}}{\sigma\sqrt{\pi}}\left(\frac{\eta}{e^{2\eta t}-1}\right)^{3/2}\exp\left\{-\frac{\eta\,x_0^2\,e^{2\eta t}}{\sigma^2\,(e^{2\eta t}-1)}\right\} \qquad (x_0>0),$$

$$\int_0^t \widetilde{g}(0,\tau|x_0)\,d\tau = 1-\mathrm{Erf}\left(\frac{x_0}{\sigma}\sqrt{\frac{\eta}{1-e^{-2\eta t}}}\right) \qquad (x_0>0),$$

from (11) one obtains:

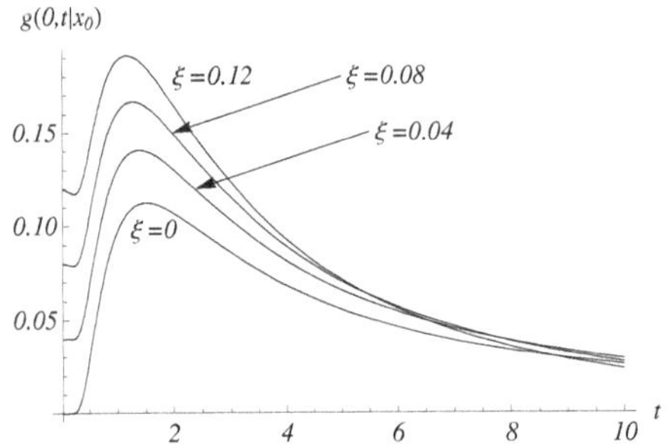

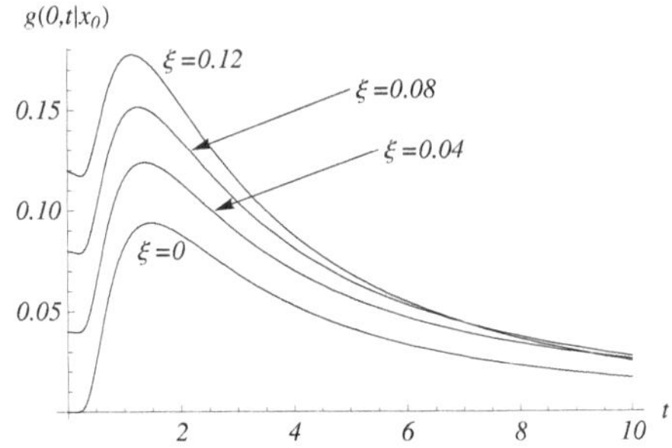

Fig. 5. FPT pdf ($\xi = 0$) and FVT pdf ($\xi \neq 0$) are plotted as function of t for $\eta = -0.03$ on the left and $\eta = 0.03$ on the right for the OU process with $x_0 = 3$ and $\sigma^2 = 2$

$$g(0,t|x_0) = e^{-\xi t}\,\widetilde{g}(0,t|x_0) + \xi\, e^{-\xi t}\,\mathrm{Erf}\Big(\frac{x_0}{\sigma}\sqrt{\frac{\eta}{1-e^{-2\eta t}}}\Big) \qquad (x_0 > 0). \quad (26)$$

For the OU process, in Fig. 5 the FPT pdf $\widetilde{g}(0,t|x_0)$ is compared with the first visit time pdf (26) for different choices of ξ.

Further details, as well as some extensions of the above results, will be the object of future works.

References

1. Cox, D.R., Miller, H.D.: The Theory of Stochastic Processes. Methuen, London (1970)
2. Di Crescenzo, A., Giorno, V., Nobile, A.G., Ricciardi, L.M.: On the $M/M/1$ queue with catastrophes and its continuous approximation. Queueing Systems 43, 329–347 (2003)
3. Di Crescenzo, A., Giorno, V., Nobile, A.G., Ricciardi, L.M.: A note on birth-death processes with catastrophes. Stat. Prob. Letters 78, 2248–2257 (2008)
4. Giorno, V., Nobile, A.G., Ricciardi, L.M.: On some diffusion approximations to queueing systems. Adv. Appl. Prob. 18, 991–1014 (1986)
5. Gradshteyn, I.S., Ryzhik, I.M.: Table of Integrals, Series, and Products. Academic Press, London (2007)
6. Pakes, A.G.: Killing and resurrection of Markov processes. Comm. Stat. - Stoch. Mod. 13(2), 255–269 (1997)
7. Saura, A., Giorno, V., Nobile, A.G.: Prendiville Stochastic Growth Model in the Presence of Catastrophes. In: Trappl, R. (ed.) Cybernetics and Systems 2004, pp. 151–156. Austrian Society for Cybernetics Studies, Vienna (2004)
8. Saura, A., Giorno, V., Nobile, A.G.: Loss system in the presence of catastrophes. In: Trappl, R. (ed.) Cybernetics and Systems 2008, pp. 261–266. Austrian Society for Cybernetics Studies, Vienna (2008)
9. Swift, R.J.: Transient probabilities for a simple birth-death-immigration processes under the influence of total catastrophes. Int. J. Math. Sci. 25, 689–692 (2001)

Automatic System Identification of Tissue Abnormalities Based on 2D B–Mode Ultrasound Images

Víctor D. Díaz-Suárez[1], Carlos M. Travieso[2], Javier González-Fernández[1,3], Miguel A. Ferrer[2], Luis Gómez[1,3], and Jesús B. Alonso[2]

[1] Center for Technology in Medicine (CTM), University of Las Palmas de Gran Canaria (ULPGC), Campus Universitario de Tafira, Telecomunicación, Pabellón B. E-35017, Las Palmas de Gran Canaria, Spain
{victor,jgonzalez,lgomez}@ctm.ulpgc.es

[2] Signals and Communications Department, Technological Center for Innovation on Communication (CeTIC, ULPGC), Campus Universitario de Tafira, Telecomunicación, Pabellón B. E-35017, Las Palmas de G.C., Spain
{ctravieso,mferrer,jalonso}@dsc.ulpgc.es

[3] Department of Electronic and Automatic Engineering, Campus Universitario de Tafira, Telecomunicación, Pabellón B. , E-35017, Las Palmas de G.C., Spain
{jgonzalezf,lgomez}@diea.ulpgc.es

Abstract. A neural network with characteristic parameters to recognize abnormalities in ultrasound images acquired from echographic tissue-mimicking materials is proposed. The neural network has been implemented in MATLAB and it can be used in real time to assist the clinical diagnoses in the early phases. The parameters are extracted from a database of B-mode ultrasound images. After training and testing the network, using a statistically significative set of experimental data and a non-commercial phantom, results show that the proposal can be successfully applied to efficiently deal with this problem.

Keywords: Ultrasound, tissue-mimicking phantom, B-mode imaging database, Higuchi Fractal dimension, Feed-forward neural network.

1 Introduction

Detecting tissue abnormalities from B-mode ultrasound images is a major problem to be addressed to help in the early medical diagnoses phases. B-mode Ultrasound images, which are extensively used mainly in ambulatory explorations , manifest what is named as speckle noise [1], making that the detection of tumor (specially cancer tumors) cannot be accomplished in a fast and an objective manner.

We remark that is in the very early phases of a tumor (cancer, specially) when it must be detected in order to proceed with the medical protocols. Several approaches regarding this problem can be found in the clinical image processing (see for instance [2] for a review on elastography) and an important effort has been carried out focused on reducing image noise or even exploring new ultrasound scan methods. In this paper we do not research on those areas but we focus on the fast, objective and, what is more important, efficient detection of tumors in the very early phases.

R. Moreno-Díaz et al. (Eds.): EUROCAST 2009, LNCS 5717, pp. 137–142, 2009.

A classical feed-forward neural network with characteristic parameters to recognize abnormalities in ultrasound images acquired from tissue-mimicking materials (phantoms) is proposed in this paper. A neural network (artificial neural network, ANN) is a computational paradigm that differs substantially from those based on the standard von Neumann architecture [3]. Neural networks have been successfully applied to many problems during the last two decades (graph problems, pattern recognition) and even to well established NP problems such as TSP (Travelling Salesman Problem).

As a difference from other heuristic techniques widely applied in artificial intelligence (Simulate Annealing or Tabu Search), feature recognition suits naturally well with the neural network schemes, due that feature recognition ANNs learn from experience (a software simulated experience) instead of being formally programmed with rigid or flexible rules as in conventional artificial intelligence.

Neural networks learning has its roots in the statistical theory [4] and it can be assessed that the quality of the results to get depend mainly on the quality of the statistics estimators to be used. In this first approach, we apply only a set of basic estimators which seems to work well with the problem. We are not aware of any research similar to this involving a neural network. We also present a new low-cost and easy to use material (based on solution of agar, propanol and formaldehyde) to make phantoms which reproduce the desired human tissue echographic properties. An inclusion to account for the lesion to detect is embedded in the phantom (the size of this phantom is 12 cm long, 10 cm wide and 5 cm high and the inclusion is 1 cm in diameter size).

This paper is organized as follows: first (section II) we present the neural network designed methodology, that is, the database and the proposed parametrization system. Next section (section III) deals with the preparation of the phantom and finally, we present and discuss the main results.

2 Database and Parameterization System

A database with a total of 280 images was created using a custom-made ultrasound phantom with a 12 MHz linear probe (LA12-9, 80% bandwidth) of an ultrasound scanner (Ultrasonix ES500 RP, BA,Canada) (see figure 2). 140 B-mode images were obtained insonifying the phantom so that a hyperechoic inclusion, mimicking abnormal tissue, was positioned in different random locations.

Water-based hydrocolloids containing diverse acoustic scatterers have been shown to have appropriate ultrasonic characteristics [5]. Mean elastic modulus for human abdominal tissue was estimated to be approximately 25 KPa. In this manner, the amount of gel powder (Gelatin Gold DC, 200 Bloom) was calculated according to [6]:

$$E_{gel} = 0.003 \times C^{2.09} \tag{1}$$

where E_{gel} is the desired Young's modulus of the gel in KPa, and C is the gel concentration in grams per liter. Thus, a solution of de-ionized water (500 ml.) with 10 g/l. of agar powder and 75 g/l. of pre-hydrated gelatin was heated and stirred until the temperature reached 60 ◦C. It was then placed into a water bath for cooling until the temperature decreased to 45 ◦C. Subsequently 80 ml/l. of n-propanol and 4 ml/l. of 35-40% formaldehyde were added to the mixture. Formaldehyde was added to increase cross-linking

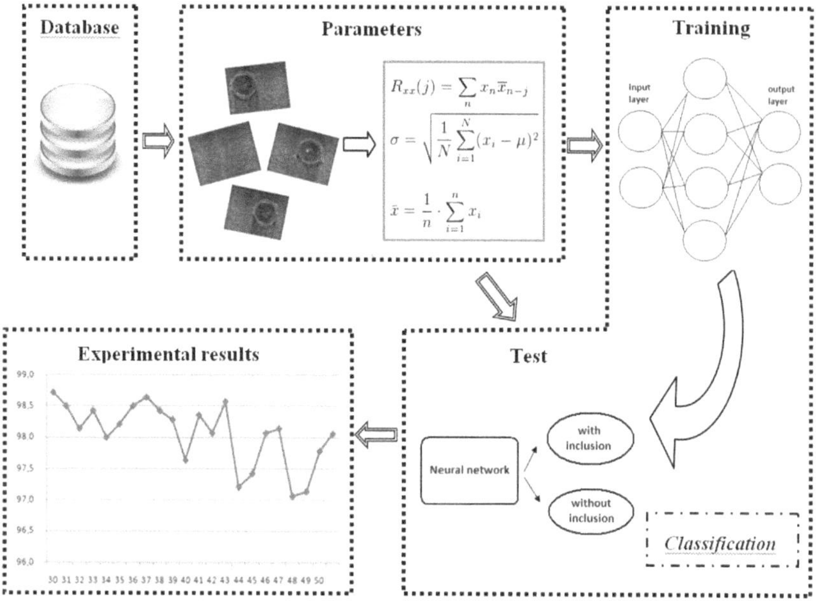

Fig. 1. Block diagram for the implemented system

among collagen fibers and raise the melting point of the gel [6], whereas n-propanol concentration increases sound speed [5]. At that time, cellulose powder was added in order to obtain appropriate scattering and absorption properties according to [5, 6]. The solution was then poured into the phantom container holding a hyperechoic polyurethane inclusion, and let to rest in a refrigerator at 5°C for approximately 12 hours.

In order to obtain a set of local parameters, each 640x480 image is segmented into 100x100 pixels wide, 10% overlapping windows. The first order statistics estimators selected to load the network were the mean, the standard deviation and the autocorrelation, yielding a total of 35x3 parameters. These parameters on each grid are calculated to train the neural network.

Acquisition depth was 4 cm. Another set of 140 B-mode images was acquired insonifying regions without the inclusion. Table 1 presents characteristics from our database.

Table 1. Database created from US Image

Total number of images	280
Number of images without abnormalities	140
Number of images with abnormalities	140
Image sizes	12-17 Kbytes
Digital format	Gray scale – 8bits – jpeg

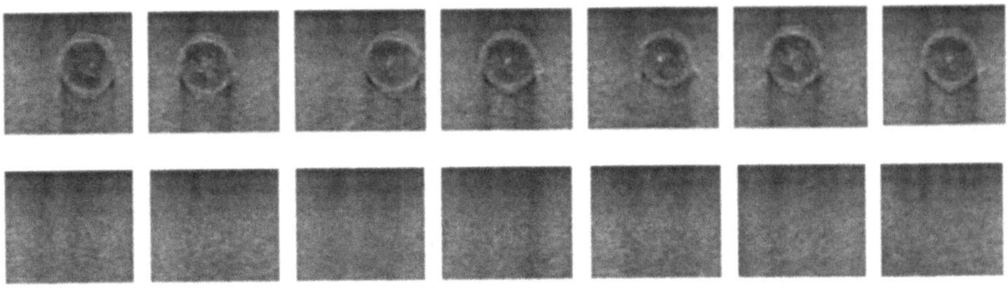

Fig. 2. Images with and without abnormalities from the database

3 Classification System Based on Neural Network

In recent years several classification systems have been implemented using different techniques, as Neural Networks (NN). The Neural Networks techniques are widely well known on applications for pattern recognition.

An Artificial Neural Network (ANN) is an information processing paradigm that is inspired by the way biological nervous systems, such as the brain, process information [7]. The key element of this paradigm is the novel structure of the information processing system. It is composed of a large number of highly interconnected processing elements (neurones) working in unison to solve specific problems. ANNs, like people, learn by example. An ANN is configured for a specific application, such as pattern recognition or data classification, through a learning process. Learning in biological systems involves adjustments to the synaptic connections that exist between the neurones. This is true of ANNs as well.

One of the simplest ANN is the so called perceptron that consist of a simple layer that establishes its correspondence with a rule of discrimination between classes based on the linear discriminator. However, it is possible to define a decision for non linearly separable classes, using multilayer perceptron (MLP). This kind of NN has feedforward propagation from input layer to output layer, having one or more layers as hidden layers. These additional layers contain hidden neurons or nodes, which are directly connected between the input and output layers [7, 8]. This inter-connected is called sinapsis (see figure 3).

Each neuron is associated with a weight and biases. These weights and biases have been adjusted in the training process for each connection of the network, in order to make their suitable values for the classification task between the different classes.

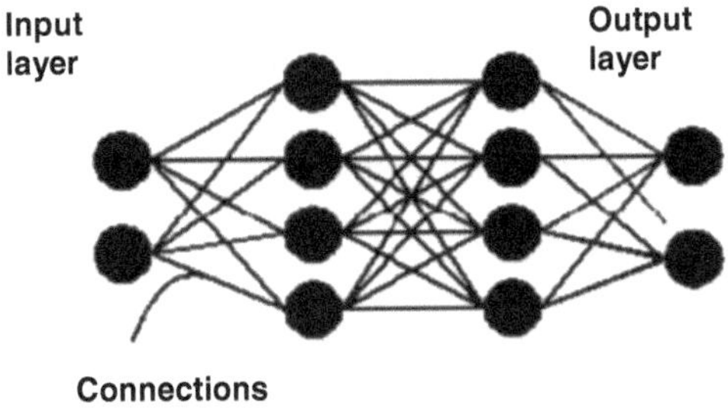

Fig. 3. Structure of layer for a Neural Network classifier

In this work a FeedForward NN has been used, which is defined on layers: one input layer, one hidden layer and one output layer; where the output of a neuron feeds to all neurons of the next layer. This Feed-forward Network is trained with the Perceptron Multilayer (MLP) using Back-propagation algorithm [7]. Some neurons are defined with a non-lineal threshold. This non-linear aspect is necessary in order to achieve the minimization of the error gradient. We remark that this aspect is as well the critic point of a MLP-NN and must be taken in consideration in the neural layers implementation through the design phases.

4 Experiments and Results

Images were automatically classified as containing abnormalities or not. The extracted parameters from the images were used to make a supervised classification with a neural network with perceptron topology. Finally, back-propagation method and a feed-forward network have been used [7].

Once the system has been implemented, 50% from our database has been used for training and the other 50% has been used for testing. Each experiment has been repeated 10 times. For each experiment, training and test samples randomly have been chosen from our database, but with a condition per experiment; samples used for training mode never can be used for test mode, therefore, this system uses independent samples. It is important to indicate that the training phase time for our implemented neural approach was around 10 minutes in a Pentium IV, 2 GHz, 3 MB RAM and the execution time to get the results were practically instantaneous for a simple given image (using the same hardware). Therefore for test phase, this application is considered in real time. The system has been implemented in MATLAB [9] and special care has been taken in consideration during its design to speed-up the runs.

Figure 4 shows the averaged success rates and the standard deviation. The best result exhibits a 98.36% of success rate with a standard deviation of 0.89 and it has been obtained applying a 41 neurons in the hidden layer.

From the analysis of the quality of the results, it can be concluded that the proposed methodology works well for detecting tissue abnormalities.

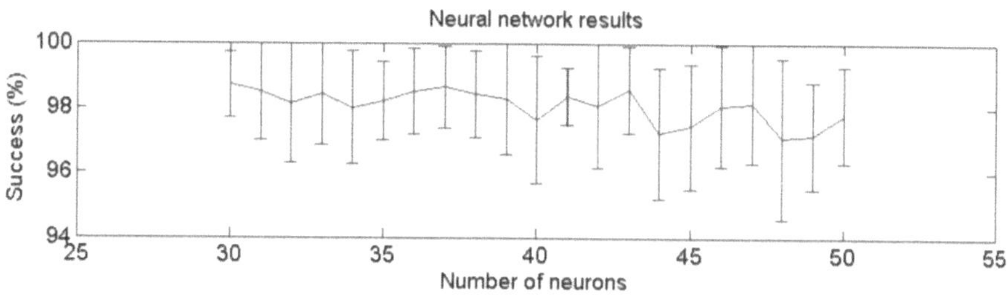

Fig. 4. Success rates for detecting tissues abnormalities using a neural network classifier

5 Conclusions

At summary, a simple and robust system has been implemented in order to detect abnormalities in ultrasound images (phantoms), using an artificial neural network as classifier. This system has built a supervised classification, and after training process, this system achieves an average success rate of 98.36% with a low computation cost.

For future works, we are working to improve the extraction of parameters introducing other transforms (Higuchi transform) to further characterize the abnormalities and lead to an overall network optimization. We also want to explore the application of new statistical estimators based on Bayesian models. Moreover, we will test these algorithms on real application by a medical supervised process.

References

1. Tao, Z., Tagare, H.D., Beaty, J.D.: Evaluation of Four Probability Distribution Models for Speckle in Clinical Cardiac Ultrasound Images. IEEE Transactions on Medical Imaging 25(11), 1483–1491 (2006)
2. Varghese, T., Ophir, J., Konofagou, E., Kallel, F., Righetti, R.: Tradeoffs in Elastographic Imaging. Ultrasonic Imaging 23, 216–248 (2001)
3. Reilly, D.L., Cooper, L.N., Elbaum, C.: A Neural Model for Category Learning. Biological Cybernetics 45, 35–41 (1982)
4. Muller, P., Insua, D.R.: Issues in Bayesian Analysis of Neural Network Models. Neural Computation 10, 571–592 (1995)
5. Madsen, E.L., Zagzebski, J.A., Banjavie, R.A., Jutila, R.E.: Tissue Mimicking Materials for Ultrasound Phantoms. Med. Phys. 5, 391–394 (1978)
6. Hall, T.J., Bilgen, M., et al.: Phantom Materials for Elastography. IEEE Transactions on Ultrasonics, Ferroelectrics and Frecuency Control 44(6), 1355–1365 (1997)
7. Bishop, C.M.: Neural Networks for Pattern Recognition. Oxford University Press, Oxford (1995)
8. Hush, D.R., Horne, B.G.: Progress in Supervised Neural Networks. IEEE Signal Processing Magazine 10(1), 8–39 (1993)
9. MATLAB: The Mathworks, Inc. (2008)

Vision–An Essay from a Computational View Point

José Luís S. Da Fonseca, José Barahona da Fonseca,
and Isabel Barahona da Fonseca

Abstract. Vision is considered form a theoretical view point stressing a traditional view concerning the 3D problem. Proprioceptive variables as well as cross correlation and regression analysis are considered. It is proposed a semantic model of logical processes based on the Theory of Artificial Neural Netwoks and a calculus of Signification and Intention developed by J. S. Fonseca and J. Mira y Mira in 1970. First, the cogency of semantic proposal is tested using a poem by Fernando Pessoa. Finally, it is shown that a visual counterpart of the approach has been used by Piet Mondrian.

Keywords: vision, computation, signification, intention, logigcs.

1 Introduction

The problem of understanding visual processing is much wider and complex than what can be said in a single scientific paper.

Our aim is to propose a mathematical model that expresses the viewpoint of the psycophysiologist more than the ideas of the systems engineer.

We take most ideas from human sciences, namely psychology, logics, linguistics as well as from strategies that have been traditionally used in painting which aims to the representation of reality as this task is understood in Western culture.

2 Two Dimensions versus Three Dimensions

Our eyes receive photographic like information in curve plane surfaces – the two Retinas. Let us consider the two eyes, with their rotation centre along their vertical axis which correspond to a coordinated movement of both eyes which exerted by extrinsic muscles which depend on the command by Common Oculomotor Nerve, Pathetic Cranial Nerves and indirectly from the Optic Nerve through information sent from Retina as well as from Visual Cortex to both Colliculi Superiores in the Mesencephalon, in the Brain Stem. It is traditionally thought by medicine students that one of the most important clues for the calculation of distance is the information obtained from ocular extrinsic muscles through neuro-muscular proprioceptive receptors. Let us consider what we know a priori: Od (interocular distance d1) and θ (convergent angle θ relative to Od is obtanined for the case of an isosceles triangle from $\theta = 180° - 2\,\alpha$. We have immediately $h^2 = (½d')^2 + d''^2$, where h is the hypotenusa and d'' the perpendicular from target point to d', that is the distance from target to the interocular axis. We have immediately $d'' = \alpha/\sin\alpha$ or alternatively $d'' = \sqrt{[(\alpha/\cos\alpha)^2 - (½d')^2]}$ in which d'' is the only quantity unknown to the Central Nervous System.

R. Moreno-Díaz et al. (Eds.): EUROCAST 2009, LNCS 5717, pp. 143–151, 2009.

For the general case of a target whose projection lies outside d' we will have three distances: d', d'' and d''', respectively, d' = β'/ cos β', d'' = β''/cos β'' and d''' = β'/ sin β' and further more $d'^2 = d'''^2 + (d + v)^2$ in which, now d is the interocular distance and v is the extension of the interocular distance to the point in which the target point projects itself in to the interocular axis.

We may conclude that effectively our program of expressing distance in terms or known values of variables has been fulfilled.

If we scan a volume using a meridian and parallel strategy, we obtain a net which describes perfectly the visible surface of the object – therefore we have volume. From 2 D we obtain 3 D.

3 Visual Message

We may now put the question about the exactitude of our measurement: our method is reliable but its resolution is coarse. Remember now that we have two curved planar surfaces, which in principle occupy symmetric positions concerning the longitudinal axis of the eye globes. When an object is fixated, an absolute coincidence must exist, in normal conditions, concerning the representation of the target and of neighbourhood areas.

As a matter of fact, maximal two-dimensional correlation, C (x,y), occurs when in each Retina the target projection and its neighbourhood are focused in each fovea. $C(x,y) = ½T \int T \int T f(x,y).f'(x+\tau,y+\tau')dx.dy.$ in which C(x,y) is the correlation value, f1 and f2 are two dimensional periodic functions of time τ is a variable that shifts one function over the other to find the maximal value of C(x,y)- In the visual cortex of the brain, namely in associative areas, distance is represented and calculated as it results from single visual cell records in external temporal cortex of the brain.

Elementary statistics provide immediately an answer to our question. Distance can be calculated with high precision using correlation measurements:

The greatest value of C provides information about the space orientation or H as in the preceding example and the problem is again solved. We may comment that we are in presence of "redundancy of potential command" as it was proposed in the sixties by Warren Sturgis McCulloch (McCulloch, W. S., 1985).

4 Texture

We may now ask a further question, namely which are the best indicator variables that allow a calculus with maximum precision of distance. The answer is immediate: taking distance as a dependent variable, the regression of other random variables allows us to calculate which variables contribute more significantly to solve our problem. Multiple regression analysis answer this question. If we want to simplify matters, we may perform a multivariate analysis, namely cluster analyses or else factor analyses by principal components.

5 Colour

Cubist painters like and first of all Paul Cézanne use black and white graining and colour saturation gradients to express volumes. According to De Valois (de Valois, R. L., 1960) our eyes perform their operations using receptive areas in the brain organized according to a particular topology: red centre -green periphery; green centre - red periphery; blue centre - yellow periphery; yellow centre - blue periphery and finally white centre- black periphery, and black center - white periphery. For the sake of simplicity we leave aside in this moment Edwin Land (Land, E. H., 1959) results. In this view, colour plays a role very similar to texture, providing reliable indicator variables.

6 Canonical Representations

Suppose that a very famous movie star, for instance Marilyn Monroe, if she were alive, comes to visit Instituto Bento da Rocha Cabral and is willing to collaborate in an investigation. Then we ask her to stand in front of us and we scan her body with meridians and parallels in a frontal perspective. Thus we obtain a 3 D virtual sculpture of Marilyn. Thereafter we asked her to rotate 45°, 90°, 135°, 180°. The first position is 0°. If we want to have planar representations or Marilyn, we use projective groups of transformations or Euclidian planes and we obtain similitude transformations. If we want positions other than canonical positions, we will obtain invariants using affine transformations.

7 Economy of Means

When linguistics faced the problem of phonemes being expressed by a potentially infinite class of equivalence of allophones, structuralistic linguists like Roman Jakobson (Jakobson, R., 1962) proposed the following approach: phonemes are characterized by invariants concerning their possession of characteristic distinctive qualities.

Using this information as a metaphor, Marilyn would be perfectly described by a set of predicates, namely: a) colour of the hair; b) distance from the top of the head to the top of her face; c) distance from this point to eyebrow; d) thickness of eyebrow; e) form and colour of the eye; f) maximal length and height of the nose; g) distance from the base of the nose to the tangent of the superior contour of the upper lip; h) thickness of the superior lip and thickness if the inferior lip; i) distance from the inferior lip to the extreme position of the face; j) ratio between horizontal and vertical maximum diameters of the face; k) ratio of the diameter passing through the extreme of the ears and the extremes of the skin of the face; l) distance between the two extremes of the eye corners; m) width of the nose; n) length of the mouth; o) form and position of the teeth. The face of Marilyn can be so described by fifteen predicates leaving aside, for the moment, the possibility of using again Factor Analyses.

8 "The Number That Has Been Is Odd"

Consider the initial condition:

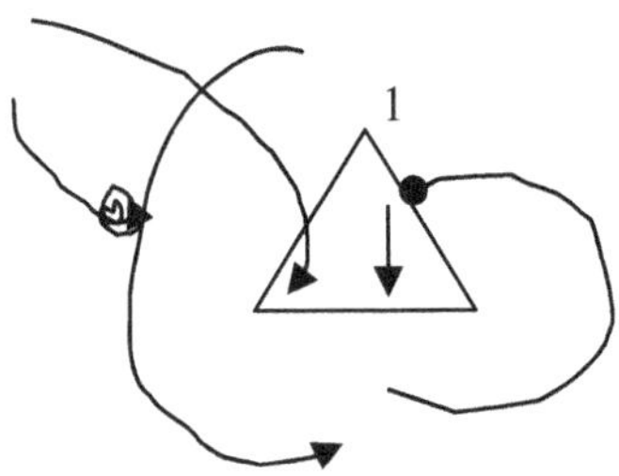

and suppose that 1 is introduced at the input. One is odd. The next time instant neuron will be active. "The number that has been is odd." At t+2 if the input is 1 ⟶ pre-synaptic inhibition cuts the feedback from the neuron to itself then thereafter the neuron is silent "the number that has been is not odd, the neuron stays silent". Next one at the input produces again activity in the neuron and thereafter. "The number that has been is odd." As a matter of fact, to the second one corresponds number 3 – the number that has been is odd. The only interest of this neuronal semantics is that it provides an initial understanding about the way conventional meaning is conveyed in neural functioning of the C.N.S. These methods require a more delicate logical treatment as we did in "Signification and Intention" in 1970 where we furthermore introduced a neuronal network approach to Husserl's Phenomenology.

9 Dictionary

Finally, what is expressed in semantics can also be expressed in a linguistic declarative system. It is elementary to build a dictionary that translates propositions into sentences in plain English.

$$S(t+1)=X(t) + MS(t)$$
$$Y(t+1)= X(t) + NS(t)$$

In which X is the input of the network, Y is its output, S is the internal state of the network, M is the decision rule that assigns a truth value to the next internal state of the network and corresponds to the decision making rules used to attribute a signification to the internal state, M. Furthermore M is the matrix that expresses the decision rules that given an input and a state of the net real produce an output Y, that is, N is the intention of the net. The signification is given by $\alpha \equiv$ dist (x (t) + MS (t)), reference $\leq \Delta$ for the case of signification and β Y (t+1), reference' $< \gamma$ for the case of intention. Our (da Fonseca, J. S., 1970)approach has some similarity with both Wittgenstein's (Wittgenstein, L., 1961)and Searle's (Searle, J., 1989) proposals for semantic meaning has being defined by a social set of rules that I use conventionally in a practice.

We propose hereby, we think a "case signification logics" and case intention logic, as we will discuss extensively.

10 Measure of Information Carried by Propositions

Let us consider now a channel in which the information that is sent by the encoder and decoded by the receiver concerns the representation of logical propositions. Suppose that we have a logical relationship or between propositions x and y xUy - - - xyU~x~y U ~xy equivalent to x U y - - - ~(x~y).

Let us suppose that symbols x and y considered isolated occur with stationary probability P(x)=P(y)=p and P(x=0)=P(y=0)=1-p=q. We further consider that signals x=1 and y=1 are corrupted by noise in a way such that the conditional probability of P(x=1/x=1)=p' and P(x=0/x=1)=q', p'+q'=1 and the same holds true for y; the conditional probability of P(x=0/x=0)=q'' and P(x=1/x=0)=p''. We can now calculate both the joint probabilities and the mean entropy for relevant cases: a) isolated symbols p and q for the case of a pair of symbols pq and /H_1=-p log_2 p – q log_2 q. Let us suppose, for the sake of simplicity that p=p'=p''=1/2.

We have immediately **H**1=1. In the particular case we are considering in what concerns the preservation of information concerning the logical relationship errors that convert mean terms of the normal disjunctive form into other mean terms for which the logical relationship holds true are not considered as corruption of the propositional message.

P1=P(xy/xy)+P(x~y / xy)+P(~xy / xy)=p^2(p'^2+p'q'+q' p')
P2=P(x~y/x~y)+P(xy/x~y)+P(~xy/x~y)= pq (p'q''+p' p''+q'q'')
and finally
P3=P(~xy/~xy)+P(xy/~xy)+P(x~y/~xy)=qp(q'' p' + p'' p' + p'' p')
<H1>=P1 + P2 +P3
<H2>=(x U y)=-P1 log_2P1 – P2 log_2 P2 – P3 log_2 P3= 2.25

Finally let us consider those cases in which noise corrupts information about the proposition:

P(~x~y/xy)+P(~x~y/x~y)+P(~x~y/~xy)=p_2 q'^2 + pq q' q'' + qp q'' q'

And the amount of information that is corrupted is given by

<H3>=-P4 log_2 P4= 0.75

As a single mean term is sufficient to make the logical relationship hold true, the amount of redundancy is R=(P1+P2+P3) –P_i, i=1..3.

Finally, the specification of a propositional connective implies, in our case, three mean terms in succession , which are represented by three couples of binary digits.

As the order of mean terms is irrelevant, we have six permutations that are equivalent and represent an implicit type of redundancy.

11 Signification of Visual Propositions

The interpretation of the signification of the activity of the neuron which characterizes odd numbers implies a conception of the activity of the neural operator as a process of decision-making that takes into account a set of rules which is used to make the attribution of a sense to the symbols that impinge on its input.

The algebraic structure that is required to define an exact structure for a logic of signification and intention will now be described.

At a first level we have a field of elementary entities included in the set {0,1}.

We specify the operation of addition and multiplication modulo(2).

This is a Galois Field (2) which includes a Boolean algebra once

$\bar{x} \equiv x \oplus 1$ in which $\oplus$ stands for exclusive 'or'.

A useful relationship is $x \oplus x \equiv 0$. If we wish a dimensional characterization of any qualitative Boolean variable x, we may use Pseudo-Boolean equations and inequalities with integer coefficients associated to Boolean variables.

Over the field of scalars, of elementary qualities, we define n-tuples, vectors and matrices which form a complete additive group and an incomplete multiplicative group as well as operations which involve scalars and n-tuples. Therefore, we have an algebraic structure of a ring. We also define a Hamming distance between n-tuples.

Signification and Intention are now described within a vectorial space by vectors and matrices and relationships between them.

Again, the signification of a concept is specified by the set of decision rules expressed in this vectorial space, relatively to a reference.

Identity is defined as: dist. MX, Ref $\leq \Phi$ in which M represents the decision criteria, X the vector of elementary qualities, Ref is the corresponding reference expression, and Φ the threshold for decision.

Our aim is to deal with the problem of signification and intention as far as visual data are concerned. As a first step we will, nevertheless, first consider the case of signification and intention carried by expressions belonging to the declarative system.

Let us first consider the poem of Fernando Pessoa, which in Portuguese and in a strict English translation is, respectively:

Com que ânsia tão raiva	With anxiety so anger
Eu quero aquele outrora	I want that other time ago
Eu era feliz? Não sei	Was I happy? I don´t know
Fui-o outrora agora	I was it another time ago now

We will, as a first step, introduce the concept of concatenation, the operation which links successive components of a sentence. When we make a formal representation of a colloquial phrase, our matrices

$${}^{t}_{i,j}X^{k,l,m,n\ldots}_{p,q,\ldots,s}$$

in wich t symbolises the level in the Russel -Whithead theory of types, i,j a specific component defined by the line and the column, k,l,m,n... are the total number of lines and columns in p,q …,s denote other operators which contribute together with M to the contextual definition of a given concept. This tensor representation is from a connotative view point equivalent to the usual phrase – markers, or a component of it, for the sake of completenessused by Noam Chomsky to denote syntactic structures. (Chomsky, N., 1956)

Returning to our poetical example, in the formal representation we propose, it becames

with what | $R_{1.}\bar{x}_1 \oplus$

anxiety | emotion
feeling
motivation
expectancy
negative evaluation
visceral perception
sensory-motor component

which corresponds to the vector $\bar{x}_2$ the total expression for this component being $M_2 \bar{x}_2$

Note that we are omitting, for the sake of simplicity, and because it is irrelevant for our first example, indices m,n,p,q…,t.

so anger | association
dimension
emotion
motivation
negative evaluation of a situation
visceral component
behavioral expression
to which corresponds $M_3\bar{x}_3$

I want | subject of action
statement of intention
subjective representation
plan of action
…
$M_4\bar{x}_4$

That other time ago | specificity
time interval
past
specification of difference
…
$M_5\bar{x}_5$

Was I happy? | state of agent
Question
Introduction of two opposite statements

Relative to other time
past
reference to a preceding operator
emotion
feeling
motivation
pleasant
...

I don't know | behavioral component
agent
state of agent
negative
state of consciousness
subjective
(reference to other operators)
...
$M_7\bar{x}_7$

I was it another time ago | agent
state of the agent
past (reference to other operators)
...
$M_8\bar{x}_8$

now | time
conflictual reference
contextual connotation
alteration of signification
resolution of the contradiction implied by 'I don't know'
assimilation to 'another time ago' creating a new time concept
element for the thematic sequence ABA, by assimilation of B to A modified
$M_9\bar{x}_9$

Finally, we remark that concatenation is a complex operation which besides its mathematical meaning involves the denotation of the role of a particular component operator in the architecture of a phrase operator or even of a global text operator.

If we consider some paintings by Piet Mondrian, it is immediately evident that instead of trying an immediate representation of nature, he has tried a conceptual representation using visual categories. This is what we might call the concept within the percept which is not only present in Art but also in our current daily Weltanschaung.

Here we must distinguish between the iconic characteristics of perception and, on the other hand, the concept within the percept. Mondrian, like many other painters, tries to transpose reality to his representations, using a set of visual attributes. Visual attributes and concepts are not reducible to the linguistic declarative system of communication. In the case of Piet Mondrian, it is nevertheless apparent an attempt to convey the signification of what is represented by means of relationships defined over visual attributes or more elementary visual concepts.

This approach is very similar to our attempt to decode the poem of Fernando Pessoa.

To conclude, we present 'Woman' and 'Tree' from Piet Mondrian to render explicit the reasons behind our argument.

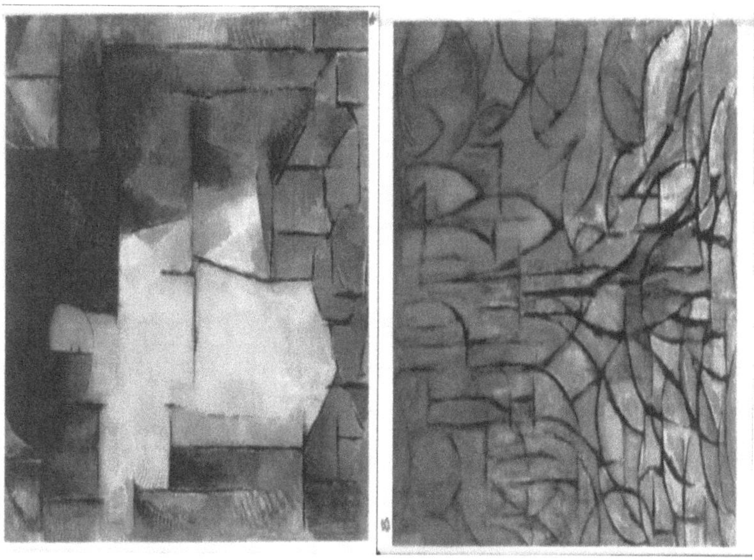

References

1. Chomsky, N.: Syntactic Structures. The Hague, Mouton (1956)
2. Da Fonseca, J.S., Mira, J.: A calculus of Signification and Intention. In: Da Fonseca, Mira, J. (eds.) Signification and Intention, pp. 5–12. Faculdade de Medicina de Lisboa, Da Fonseca (1970)
3. Jakobson, R.: The phonemic concept or distinctive features. In: Proceed. of the 4th Int. Cong. of Phonetic Sc. Helsinki, Mouton. The Hague (1962)
4. Land, E.H.: Experiments in colour vision, p. 84. Scientific American (1959)
5. McCulloch, W.S.: Embodiment of the Mind. MIT Press, Cambridge (1985)
6. De Valois, R.L.: Colour vision mechanism in the monkey. J Gen. Physiol. 43(supp. II), 115 (1960)
7. Searle, J.: Speech Acts, London (1969)
8. Von Helmhotz, H.: Helmholtz's treatise on Physiological Optics. Dover, New York (1962)
9. Wittgenstein, L.: Tratado Lógico-Filisófico – Investigações Filosóficas, Fundação Calouste Gulbenkian, Lisboa (1987)

On a Generalized Leaky Integrate–and–Fire Model for Single Neuron Activity

Aniello Buonocore[1], Luigia Caputo[2], Enrica Pirozzi[1], and Luigi M. Ricciardi[1]

[1] Dipartimento di Matematica e Applicazioni, Università di Napoli Federico II
Via Cintia, 80126 Napoli, Italy
{aniello.buonocore,enrica.pirozzi,luigi.ricciardi}@unina.it
[2] Dipartimento di Matematica, Università di Torino
Via Carlo Alberto 10, 10123 Torino, Italy
luigia.caputo@unito.it

Abstract. Motivated by some experimental results of [13], the standard stochastic Leaky Integrate-and-Fire model for single neuron firing activity is generalized in a way to include evolutionary instantaneous time constant and resting potential. The main features of the ensuing Gauss-diffusion process are disclosed by making use of a space-time transformation leading to the Ornstein-Uhlenbeck process. On the grounds of simulations of the time course of the membrane potential, we are led to conclude that our generalized model well accounts for a variety of experimental recordings that appear to indicate that the standard model is inadequate to reproduce statistically reliable features of spike trains generated by certain types of cortical neurons.

Keywords: Neuron firing, LIF model, diffusion process.

1 Introduction

Leaky Integrate-and-Fire (LIF) models for describing neuron's firing activity under the effect of constant or periodic stimuli as the first-passage-time (FPT) for a diffusion process through time-varying boundaries have received a noteworthy attention in recent times [10], [11]. From a mathematical point of view, the encountered difficulties are mainly of a two-fold kinds: The form of the equations in the unknown FPT probability density function (pdf), and the general lack of closed-form solutions for biologically significant thresholds. Since the available closed-form results are scarce and fragmentary, attention has been generally paid on devising algorithms and numerical procedures able to allow one to evaluate the FPT pdf's of interest via implementation on digital computers, often in need of long computation times and large memory storage space. New input and vigor towards the FPT pdf determination for diffusion processes were originated by Durbin's seminal paper [4] in which an ingenuous algorithm to compute the FPT pdf for the Wiener process through a continuous time-dependent boundary was proposed. However, in general, especially within the neurophysiological context, it is necessary to determine, or to evaluate, FPT pdf's for Gauss-Markov (GM)

R. Moreno-Díaz et al. (Eds.): EUROCAST 2009, LNCS 5717, pp. 152–158, 2009.

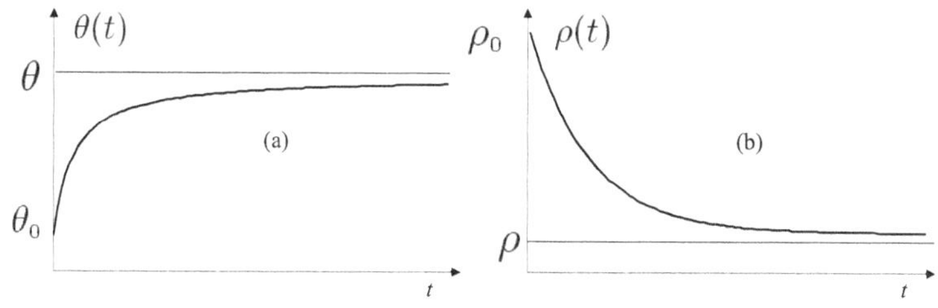

Fig. 1. Qualitative time courses of the decay constant $\theta(t)$ and of the resting potential for a generalized LIF model

processes possessing an equilibrium distribution, which allows to achieve certain useful results ([7], [8]). However, it is fair to claim that very little is known in such an area and that some subtle questions have been overlooked.

Motivated by biological problems, and especially by those related to LIF neuronal models, the FPT problem for Gauss-diffusion processes was recently taken up afresh by us [2] both via space-time transformations to the Ornstein-Uhlenbeck (OU) process and in algorithmic terms.

Specifically, after stating the essential features of Gauss-Markov process in a fashion useful for such aims, a convenient definition of FPT for such processes was introduced. In particular, it was pointed out that a constant can be determined that leads to the simplest space-time transformation. The obtained results have been applied to the neuronal LIF models. Since even for the simplest kind of LIF models analytical solutions of neurobiological interest are not available, we has to rely on numerical approximations and on time asymptotically analytical approximations.

The forthcoming considerations assume that the reader is familiar with the basic ideas underlying single neuron's activity modeling and, in particular, with the LIF model as described, for instance, in [5], [9] and [14]. Hence, we shall proceed directly to the formulation of a generalized version of it. This has been motivated by certain results in [13] aiming to analyze the role of the conductances of the ionic channels of neuronal membrane during the time intervals elapsing between pairs of successive action potentials (spikes releases). Full details will be the object of a forthcoming paper [3]. Here we limit ourselves to mentioning that the purpose was to probe experimentally the conjecture that resistance and driving force of the neuronal membrane exhibits well-defined typical time changes during such time intervals, a conjecture that appears to be confirmed in [13]. Hence, the membrane time constant will be viewed as a time dependent function $\theta(t)$, and the resting potential will be modeled as a time dependent function $\rho(t)$. Both functions are assumed to tend asymptotically to the constant values θ and ρ of the standard LIF model as qualitatively indicated in Figure 1.

We remark that the above experimental findings provide a rational to enrich the standard LIF model with the inclusion of some form of refractoriness, which is not accounted for by the classical model.

2 The Generalized LIF Model

On the grounds of the mentioned evidences, we re-write Equation (3) of [13] within a stochastic context, as the following stochastic differential equation:

$$dV(t) = -\left[\frac{V(t) - \rho(t)}{\theta(t)} - \mu\right] dt + \sqrt{\frac{\sigma^2}{\varsigma^2}} dW. \tag{1}$$

Here $V(t)$ stands for the membrane potential of the neuron and $W(t)$ is the standard Wiener process with $\varsigma^2 = 1(\mathrm{mV})^2/\mathrm{ms}$. In order to achieve some preliminary results without resorting to unduly complicated technicalities, here we shall assume that the decay constant $\theta(t)$ and the resting potential $\rho(t)$ are as sketched in Figure 2.

The solution to equation (1) is a Gauss-diffusion process (see, for instance [1]). Hence, quantitative information on the FPT pdf through the neuronal firing threshold can be obtained via the above mentioned theory and related computational algorithms. Note that the solution to Eq. (1) is a diffusion process $\{V(t),\ t \in [t_0, +\infty[\}$ having drift and infinitesimal variance

$$A_{1,\mathrm{V}}(v,t) = -\frac{v}{\theta(t)} + \frac{\rho(t)}{\theta(t)} + \mu, \tag{2}$$

$$A_{2,\mathrm{V}}(v,t) = A_2(t) \equiv \sigma^2. \tag{3}$$

In conclusion, the time course of the membrane potential is represented by the Gaussian process specified by mean $m_\mathrm{V}(t) := \mathbb{E}[V(t)]$ and autocovariance $c_\mathrm{V}(\tau,t) := \mathbb{E}\{[V(\tau) - m_\mathrm{V}(\tau)][V(t) - m_\mathrm{V}(t)]\}$ that (see, for instance, [6]) is the product of two functions: $u_\mathrm{V}(\tau)$ and $v_\mathrm{V}(t)$, for all pairs (τ, t) such that $\tau \leq t$. In addition, functions $m(t)$, $u(t)$ and $v(t)$ satisfy system (41) in [2]. Henceforth, without loss of generality we shall take $t_0 = 0$ and, for the sake of concreteness, shall assume

$$\theta(t) := \begin{cases} \theta_0 + \frac{\theta - \theta_0}{t^*} t, \ 0 \leq t \leq t^*; \\ \theta, \qquad t \geq t^*; \end{cases} \quad \text{and} \quad \rho(t) := \begin{cases} \rho_0 - \frac{\rho_0}{t^*} t, \ 0 \leq t \leq t^*; \\ \rho = 0, \quad t \geq t^* \end{cases} \tag{4}$$

with $0 < \theta_0 < \theta$ and $t^* > 0$. With such a choice of $\theta(t)$ and $\rho(t)$ it is possible to prove that $\{V(t),\ t \in [t_0, +\infty[\}$, the solution to Eq. (1) with initial condition $V(0) = v_0$ w.p. 1, is characterized by

$$m_\mathrm{V}(t) = \begin{cases} (v_0 - v_1)\left(1 + r\frac{t}{\theta_0}\right)^{-1/r} + \frac{1}{r+1} t \left(r\mu - \frac{\rho_0}{t^*}\right) + v_1, \ 0 \leq t \leq t^*; \\ \left(\frac{\theta_0}{\theta}\right)^{1/r} \left[v_0 + v_2 - \mu\theta \left(\frac{\theta}{\theta_0}\right)^{1/r}\right] e^{-(t-t^*)/\theta} + \mu\theta, \ t \geq t^*; \end{cases} \tag{5}$$

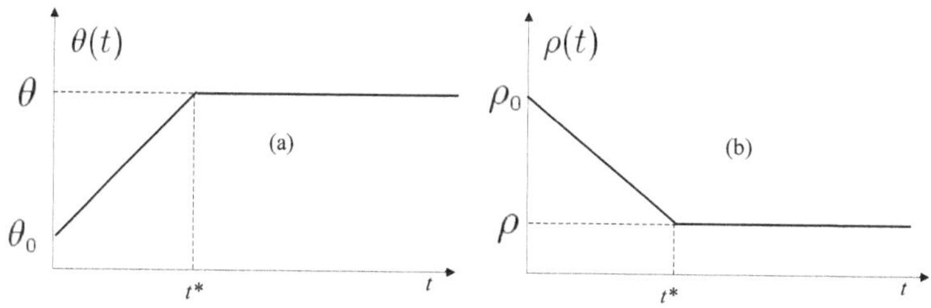

Fig. 2. The assumed decay constant $\theta(t)$ and resting potential $\rho(t)$. Ordinates θ and ρ indicate the constants of the standard LIF model, and t^* is a characteristic time to be specified in the model implementations.

and by the following functions

$$u_V(t) = \begin{cases} \sigma \frac{\theta_0}{r+2}\left[\left(1+r\frac{t}{\theta_0}\right)^{1+1/r} - \left(1+r\frac{t}{\theta_0}\right)^{-1/r}\right], 0 \le t \le t^*; \\ \sigma\left(\frac{\theta_0}{\theta}\right)^{1/r}\frac{\theta}{2}\left\{-\frac{1}{r+2}\left[\left(\frac{\theta}{\theta_0}\right)^{2/r} r + 2\frac{\theta_0}{\theta}\right] \times \right. \\ \left. e^{-(t-t^*)/\theta} + \left(\frac{\theta}{\theta_0}\right)^{2/r} e^{(t-t^*)/\theta}\right\}, \qquad t \ge t^*; \end{cases} \tag{6}$$

$$v_V(t) = \begin{cases} \sigma\left(1+r\frac{t}{\theta_0}\right)^{-1/r}, \quad 0 \le t \le t^*; \\ \sigma\left(\frac{\theta_0}{\theta}\right)^{1/r} e^{-(t-t^*)/\theta}, t \ge t^*; \end{cases} \tag{7}$$

where we have set

$$r := \frac{\theta-\theta_0}{t^*}, \quad v_1 := \rho_0 + \frac{\theta_0}{r+1}\left(\frac{\rho_0}{t^*}+\mu\right),$$
$$v_2 := \frac{1}{r+1}\left(\frac{\theta}{\theta_0}\right)^{1/r}\left(\frac{\theta}{t^*}\rho_0 + \mu\theta\right) - v_1. \tag{8}$$

The conditional probability density function (pdf) of $\{V(t),\ t \in [0,+\infty[\}$ is Gaussian with mean and variance

$$M_V(t|y,\tau)(t) = m_V(t) - \frac{v_V(t)}{v_V(\tau)}[y - m_V(\tau)], \tag{9}$$

$$D_V^2(t|\tau) = \frac{v_V(t)}{v_V(\tau)}\left[u_V(t)v_V(\tau) - u_V(\tau)v_V(t)\right]. \tag{10}$$

For $t \to +\infty$ they yield $\mu\theta$ and $\sigma^2\theta/2$, respectively.

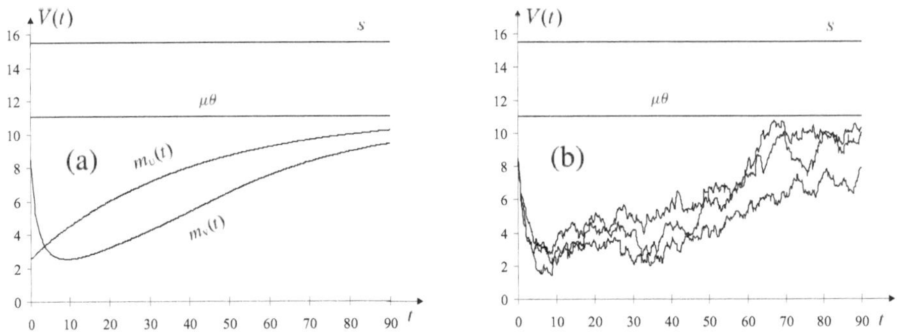

Fig. 3. In (a) means of $\{V(t),\ t \in [0,+\infty[\}$ and $\{U(t),\ t \in [0,+\infty[\}$ are plotted, while (b) shows simulated sample paths of the generalized LIF model. Here parameters are chosen as follows: $\theta = 38.8$ ms, $\mu = 0.285$ mV/ms, $\sigma^2 = 0.1824$ mV^2/ms, $S_V \equiv S = 15.48$ mV, $\theta_0 = 2$ ms, $\rho_0 = 0$ mV, $t^* = 50$ ms. For the generalized LIF model we have taken $v_0 = 8.5$ mV, whereas $v_0 = 2.5$ mV in the case of the standard LIF model.

It is useful to remark explicitly that the GM $\{V(t),\ t \in [0,+\infty[\}$ can be related to the Ornstein-Uhlenbeck (OU) process $\{U(t),\ t \in [0,+\infty[\}$ having zero mean and autocovariance

$$c_U(\tau,t) := \frac{\varsigma^2\vartheta}{2}\left(e^{\tau/\vartheta} - e^{-\tau/\vartheta}\right) e^{-t/\vartheta}, \qquad \tau \le t \tag{11}$$

and such that $U(0) = 0$ w.p. 1. Indeed, processes $\{V(t),\ t \in [0,+\infty[\}$ and $\{U(t),\ t \in [0,+\infty[\}$ are related as follows (see [2]):

$$V(t) = m_V(t) + \varphi(t)\, U[\varrho(t)], \qquad t \in [0,+\infty[\tag{12}$$

where, for $t \ge t^*$, $\varphi(t)$ and $\varrho(t)$ become

$$\varphi(t) = \sqrt{\frac{\sigma^2\theta}{\varsigma^2\vartheta}},$$
$$\varrho(t) = \frac{\vartheta}{2}\ln\left[\frac{r+2}{r+2\left(\frac{\theta_0}{\theta}\right)^{1+2/r}}\right] + \frac{\vartheta}{\theta}(t-t^*),$$

respectively.

3 Some Quantitative Features

The substantial qualitative and quantitative diversities of the predictions obtained via the standard LIF model and its generalization considered in the present paper can be pinpointed by studying the features of the firing pdf. In our generalized LIF model, this is represented by the pdf of the first-passage

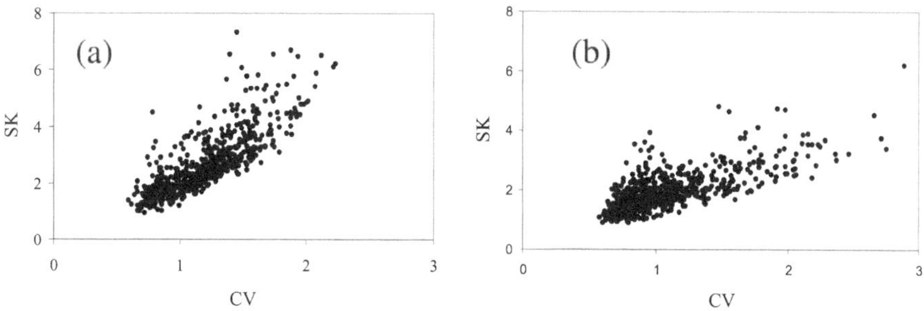

Fig. 4. Plots of the 611 points having coordinates (CV,SK): (a) refers to $\{V(t),\ t \in [0,+\infty[\}$ whereas (b) refers to $\{U(t),\ t \in [0,+\infty[\}$. In (a) $\theta_0 = 10$ ms and ρ_0 has been randomly chosen in $(0, 9.75)$ mV having assumed a parabolic distribution. In (b) a similar choice has been made for v_0. The remaining parameters are $S = 10$ mV, $\theta = 20$ ms, $\mu = 0.4$ mV/ms and $\sigma^2 = 0.2$ mV^2/ms.

time (FPT) of $\{V(t),\ t \in [0,+\infty[\}$ through the firing threshold $S_{\rm V}(t)$, namely by the function

$$g_{\rm V}[S_{\rm V}(t),t] := \frac{d}{dt}\mathrm{Prob}\,(\mathcal{T} \le t) \tag{13}$$

where

$$\mathcal{T} := \inf\{t > t_0 : V(t) \ge S_{\rm V}(t)\} \tag{14}$$

is random variable describing the firing time. In the case of the standard LIF model, analogous definitions hold with the proviso that in (13) and in (14) letter V must be changed to U.

The comparisons can then be performed by implementing the computational methods described for instance in [2], or by relying on simulation procedures based on repeatedly solved the respective stochastic differential equations by suitably discretizing meshes. While referring to [3] for computational discussions, here we limit ourselves to pointing out an evident distinctive feature of our generalized LIF model with the above choices of $\theta(t)$ and $\rho(t)$ with respect to the standard one. The obvious diversity of the time courses of their respective mean values, analytically determined as well as simulated in the generalized case, is shown in Figure 3. Figure 4 witnesses instead an a priory unexpected pattern diversity of the two models. Namely, with a motivation related to experimental findings of [12], Figure 4 shows clouds of 611 points, each point having coordinates (CV, SK), where CV denotes the coefficient of variation and SK the skewness. The coordinates of each point have been estimated from a sample of size 100 of neuronal firing trains. Note that (a) is characterized by an overall shape that much better than (b) reproduces the experimental evidence (see reference [12]). The conclusion, strongly supported by a large mass of computations and simulations, is that our generalized LIF model is suitable to account for the experimental evidences on the ground of which the authors of [12] were led to reject the standard LIF model.

References

1. Arnold, L.: Stochastic Differential Equations: Theory and Applications. Wiley and Sons, Chichester (1974)
2. Buonocore, A., Caputo, L., Pirozzi, E., Ricciardi, L.M.: The First Passage Time Problem for Gauss-Diffusion Processes: Algorithmic Approaches and Applications to LIF Neuronal Model. Methodol. Comput. Appl., doi:10.1007/s11009-009-9132-8
3. Buonocore, A., Caputo, L., Pirozzi, E., Ricciardi, L.M.: In preparation
4. Durbin, J.: Boundary-crossing probabilities for the Brownian motion and Poisson processes and techniques for computing the power of the Kolmogorov-Smirnov test. J. Appl. Prob. 8, 431–453 (1971)
5. Lánský, P.: Sources of periodical force in noisy integrate-and-fire models of neuronal dynamics. Physical Review E 55(2), 2040–2043 (1997)
6. Mehr, C.B., McFadden, J.A.: Certain properties of Gaussian processes and their first passage time. J. R. Statist. Soc. B 27, 505–522 (1965)
7. Nobile, A.G., Pirozzi, E., Ricciardi, L.M.: On the estimation of first-passage time densities for a class of Gauss-Markov processes. In: Moreno Díaz, R., Pichler, F., Quesada Arencibia, A. (eds.) EUROCAST 2007. LNCS, vol. 4739, pp. 146–153. Springer, Heidelberg (2007)
8. Nobile, A.G., Pirozzi, E., Ricciardi, L.M.: Asymptotics and evaluations of FPT densities through varying boundaries for Gauss-Markov processe. SCMJ 67(2), 241–266 (2008)
9. Ricciardi, L.M.: Diffusion processes and related topics in biology. Springer, Berlin (1977)
10. Schindler, M., Talkner, P., Hänggi, P.: Firing Times Statistics for Driven Neuron Models: Analytic Expressions versus Numerics. Physical Review Letters 93(4), 048102-1–048102-4 (2004)
11. Schindler, M., Talkner, P., Hänggi, P.: Escape rates in periodically driven Markov processes. Physica A 351, 40–50 (2005)
12. Shinomoto, S., Sakai, Y., Funahashi, S.: The Ornstein-Uhlenbeck process does not reproduce spiking statistics of neurons in prefrontal cortex. Neural Computation 11, 935–951 (1999)
13. Stevens, C.F., Zador, A.M.: Novel Integrate-and-fire-like Model of Repetitive Firing in Cortical Neurons. In: Proceedings of the 5th Joint Symposium on Neural Computation, UCSD, La Jolla, CA (1998)
14. Tuckwell, H.: Introduction to Theoretical Neurobiology. Cambridge University Press, Cambridge (1988)

Mathematical and Computational Modeling of Neurons and Neuronal Ensembles

Andreas Schierwagen

Institute for Computer Science, Intelligent Systems Department,
University of Leipzig, Leipzig, Germany
schierwa@informatik.uni-leipzig.de
http://www.informatik.uni-leipzig.de/~schierwa

Abstract. In Computational Neuroscience, mathematical and computational modeling are differentiated. In this paper, both kinds of modeling are considered. In particular, modeling approaches to signal generation and processing in single neurons (i.e., membrane excitation dynamics, spike propagation, and dendritic integration) and to spatiotemporal activity patterns in neuronal ensembles are discussed.

Keywords: neurons, neuronal ensembles, mathematical, computational, modeling.

1 Introduction

Two kinds of modeling are differentiated in Computational Neuroscience (CNS), mathematical and computational modeling. The first emerges from applying mathematics to neuroscience in the way that is standard in science. This conventional modeling concentrates on the analysis of structure and dynamics of the brain and its parts. Thus it is confined to the nervous system itself. The other kind, computational modeling, is concerned with function, which in the classical AI tradition means, information representation, processing and manipulation, learning and decision-making. Alternative directions think of function rather in terms of behavior, e.g. visual scene analysis, sensomotor coordination, reaching and grasping, and navigation. Here modelers are forced to deal with an organism embedded in an environment. There are several features common to both modeling kinds: they require an abstraction process by which presumably negligible details are eliminated. Any instance of a computational model is a mathematical model, and the two are normally closely meshed with one another in particular research subjects. In principle, by this meshing computational constraints could guide the appropriate design of mathematical models, but these techniques are not yet widely used. In the following, both kinds of modeling in CNS are considered. In particular, modeling approaches to signal generation and processing in single neurons and to spatiotemporal activity patterns in neuronal ensembles are discussed.

R. Moreno-Díaz et al. (Eds.): EUROCAST 2009, LNCS 5717, pp. 159–166, 2009.

2 Single Neuron Modeling

At the level of single neuron modeling, the cable theory of signal spread in passive dendrites, the Hodgkin-Huxley model relating action potential, ionic conductances and membrane particles, and the compartmental modeling approach to complex branched neurons represent the "working horses" of CNS [1]. Two types of complexity must be dealt with: the intricate interplay of active conductances underlying the complex neuronal excitation dynamics, and the elaborate dendritic morphology that allows neurons to receive and process inputs from many other neurons (e.g. [2]). Their specific morphology is used to classify neurons[1] (Fig. 1).

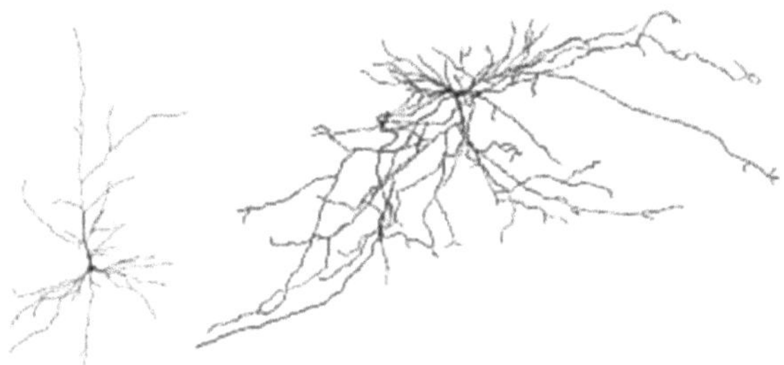

Fig. 1. Examples of rendered neuron morphologies. Left: pyramidal neuron of the mouse somatosensory cortex [4], Right: wide-field neuron of the cat superior colliculus [5].

2.1 Mathematical Models

Membrane excitation dynamics. The evolution of the membrane potential of a neuron is described by a system of coupled, non-linear ordinary differential equations, such as the Hodgkin-Huxley model[6]. The Hodgkin-Huxley model is well beyond analytical solution; fortunately, dynamical systems theory provides insights into how neuron activity is shaped by individual neuronal parameters, such as the maximal conductance of a particular membrane current. For instance, in [7], the functional role of the repolarizing ionic currents in the periodic activity of nerve membranes was analyzed. The Hodgkin-Huxley equations can be driven into repetitive activity by a maintained depolarizing current I, a decrease in the maximal K^+–conductance g_K, or by moving the Nernst potential V_K for K^+ in the depolarizing direction. In all these cases large amplitude periodic solutions are obtained (Fig. 2, left). Changing two parameters gives rise to bifurcation curves in the parameter plane. For the parameters I and V_K a region of multiple equilibria is found. The three equilibrium solutions occur in the interior of the solid curves in Fig. 2 (right) where the dashed curve represents the Hopf bifurcation.

[1] For a review of our approaches to morphological quantification and mathematical modeling of neuron growth and structure, see [3].

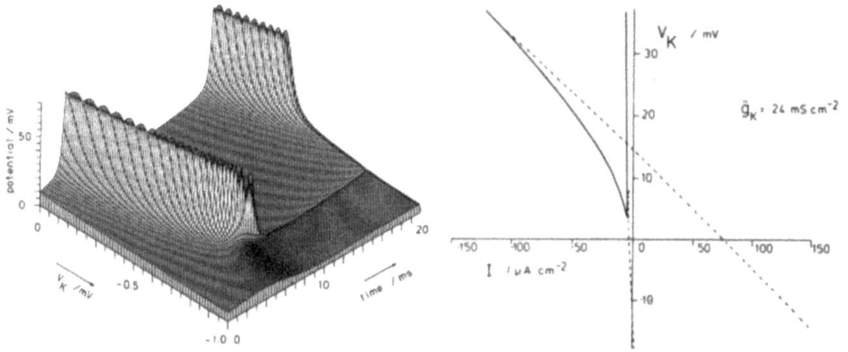

Fig. 2. Left: Numerical solutions of the standard Hodgkin-Huxley membrane equations as V_K, the Nernst potential for K^+, is changed.. Right: Bifurcation curve in the $I - V_K$ plane at a specific value of g_K [7]. See text for details.

Models of dendritic integration. Models of the integrative neuron function differ in the extent to which the branching geometry of the dendrites is considered. Originally, the cable equation describing passive membrane voltage changes and current spread was assumed to be linear[2], and the complex morphology was reduced to a single, electrically *equivalent cylinder* (EC) [8]. It turned out, however, that most of the restrictions imposed by Rall's EC model are not fulfilled in reality [9], and the model has lost the favorite state for the interpretation and prediction of experiments. In order to overcome this, two alternatives have been followed: a) to modify the EC model and b) to implement a compartmental model.

In [10], the main types of passive dendritic cable models, including the *equivalent cable* (ECa) model were presented. The ECa model is based on the cable equation with spatially varying parameters. While this equation can be solved in general only numerically, we were able to formulate a branching condition (comprising the idealized geometry necessary for the EC model) under which analytical solutions can be deduced (Fig. 3, left), and branching patterns found in dendritic neurons could be analytically treated [11,12].

Based on experimental data of several types of neurons, *compartment models* have been used for exploring intraneuronal signal processing, and to analyze the impact of dendritic morphology and non-uniform ion channel distribution on neuron function. These models can be employed in two ways: to solve the *inverse problem* (i.e., to determine membrane parameters) and to do *forward calculations* [13]. The inverse problem was shown to be ill-posed, i.e. parameter estimation is not unique. Using physiological restrictions, several admissible parameter combinations can be determined. In the forward calculations, a model

[2] Now it is unquestionable that many if not most dendrites are not passive but active, and thus the nonlinear cable equation or a corresponding compartmental model must be used.

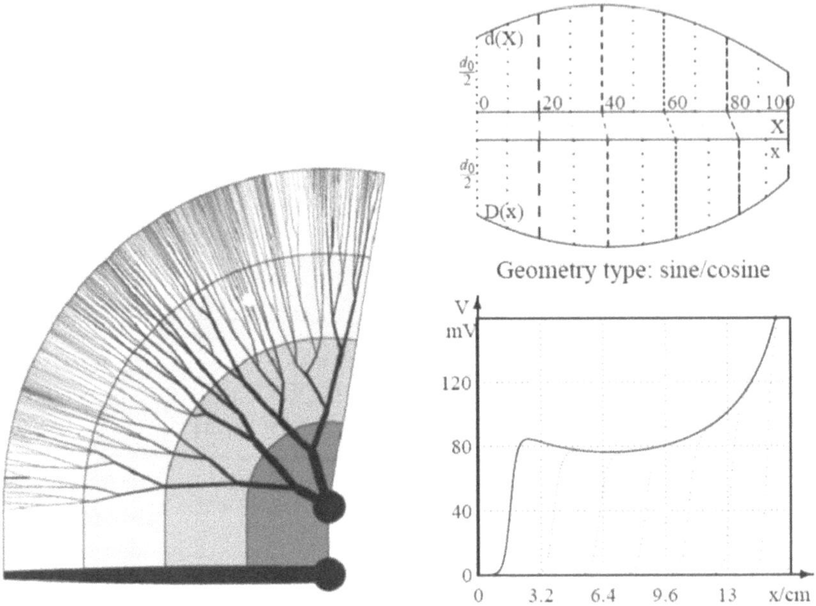

Fig. 3. Left: Equivalent cable model. The reduction of a neuron with two stem dendrites to an equivalent cable is illustrated. The lines connect points of equal electrotonic distance on the dendrites and on the nonuniform equivalent cable with sine/cosine profile. Right: Propagation of excitation front along a neurite of the sine/cosine geometry type. Displayed is the neurite diameter function in anatomical and electrotonic space (top), and snapshots of an excitation front moving leftwards (bottom). Notice that amplitude changes with diameter. After [12].

with fitted parameters is applied to calculate functional characteristics of, e.g., passive dendrites such as attenuation, delay and time window of synaptic input signals. In this way, clues for the possible function of the neurons studied can be derived. For example, neurons from superior colliculus (a part of the midbrain) could be differentiated w.r.t. to computing function as *coincidence detectors* and *integrators*, respectively [5] (see below, paragraph 2.2).

In the study [4], we employed compartmental modeling to perform a comparative electrotonic analysis of two samples of cortical pyramidal neurons, one from wildtype, and the other from transgenic mice. While anatomical dendritic trees of transgenic pyramidal neurons were significantly enlarged, the statistical analysis of the sample morphoelectrotonic (MET) dendrograms revealed that the transgenic neurons scaled in a MET-conserving mode. This means, the dimensions of their dendritic trees changed in such a way that direction- and frequency-dependent signal propagation in the passive neuron models was little affected!

The nerve conduction model. For the conduction of nerve impulses along a neurite (dendrite or axon), the model involves a nonlinear parabolic partial differential equation (PDE) such as those of Hodgkin and Huxley [6] or FitzHugh and Nagumo ([14]). Its solutions mimic membrane potential and current as a function of time and distance along the neurite.

In the standard case of neurites with uniform electrical and geometric properties, a *traveling wave* solution exists, and an explicit velocity–diameter relationship for nerve fibers can be given. Experimental effects which could not be explained with this theory include blocking of impulse conduction and changes of action potential (AP) shape in regions of nonuniform axon geometries (for review, see [15]). Using the FitzHugh-Nagumo PDE, we could show that AP propagation in a non-uniform neurite is equivalent (under certain variable transformations) to the homogeneous case [14,16]. The transformation conditions determine six specific neurite geometries enabling exact solutions, including diameter profiles of the type power, exponential and sine/cosine function. For these inhomogeneous neurite geometries, explicit formulas were derived reproducing the observed relationship between neurite geometry type and AP shape, velocity and frequency [14,16] (see Fig. 3, right).

2.2 Single-Neuron Computations

Computing in computer science means implementing an algorithm, that is, a sequence of simple computational steps that map the input to the output. Based on this model, ways are searched for decomposing into such simpler building blocks the very complex mapping done by a neuron, and for determining the *units of computation.* At the level of single neurons, literally each structural part, i.e. dendrites, spines, cell body and axon, has been considered as possible functional units[3]. For example, models of dendritic neurons have been used to implement Boolean logical operations, to compute the movement direction of a stimulus, and to simulate coincidence detection in auditory neurons (see [18]) for review). The possible computational functions of axons have been less studied. The nonlinear interactions (observed and modeled, see above, paragraph 2.1) of action potentials at regions of changing axon geometry could serve computational functions. E.g., a reduction of spike frequency at branching points [15]) could be exploited in brain networks using rate coding.

Summing up, there is no doubt that single neurons dispose of a range of mechanisms that could be used to implement elementary computations. Proving that neurons, dendrites etc. do really a specific computation is not possible, even if this has been claimed. In the cases mentioned (and in general) only indirect evidence is available or can be expected, due to the problems that inhere in the computational approach itself [19].

[3] The underlying concept of *decompositional brain analysis* has been critiziced in [17]. There I concluded that in complex systems like the brain, structural and functional components generally do not match up one-to-one.

3 Modeling Neural Ensembles

The distributed activity of neural ensembles, i.e. large populations of neurons, in the form of, e.g. oscillations and traveling waves, is known to play an important role in the nervous system. A common starting point for analyzing the large-scale dynamics of cortex is to treat nerve tissue as a continuous two-dimensional medium, so-called neural fields.

3.1 Neural Fields and Their Dynamics

The work of Amari [20] has provided a categorization of the dynamics of one-dimensional, homogeneous neural fields with symmetrical lateral coupling functions. In one-layer fields, five types of dynamics were proved to exist, which are in general multi-stable. Among them are stationary, localized excitation regions, often referred to as *bumps*, and several modes of interaction of excitation regions. Two-layer fields admit oscillatory and traveling wave solutions. These results transfer to two-dimensional neural fields, but new types of dynamics appear [21,22,23].

Using computer simulations [24], we found that inhomogeneous neural fields with asymmetrical coupling functions can produce stable bumps moving on the neural field (Fig. 4).

3.2 Analog Computations in Neural Fields

The rich dynamic behavior of neural fields has been successfully employed to realize *analog computations*. This has been based on the idea of mapping a particular problem to be solved onto the dynamics of a neural field. The problem solution can be obtained then by following the spatiotemporal field evolution.

In general, homogeneous, symmetrically connected networks have been used as models of neural computations. However, biological neural networks have asymmetrical connections, at the very least because of the separation between excitatory and inhibitory neurons in the brain. It has been shown that the distinctly different dynamical behaviors they present can make the asymmetrical networks computationally beneficial [24]. In [25], we proposed a neural field model of dynamic control of fast orienting eye movements (saccades). The model realizes the short-term memory of target location using a homogeneous field with symmetrical couplings, and the dynamic motor error coding via the hill-shift effect in an inhomogeneous field with asymmetrical couplings. The different schemes of lateral coupling have been chosen in general agreement with experimental findings. Fig. 4 shows the modeled *hill-shift effect* as found in the superior colliculus of the cat.

From a general point of view, the interpretation in terms of computations of the activity patterns appearing in neural fields can be easier achieved, as compared with single neurons. This is due to the experimentally available techniques which can be used to demonstrate correlations between recorded activity

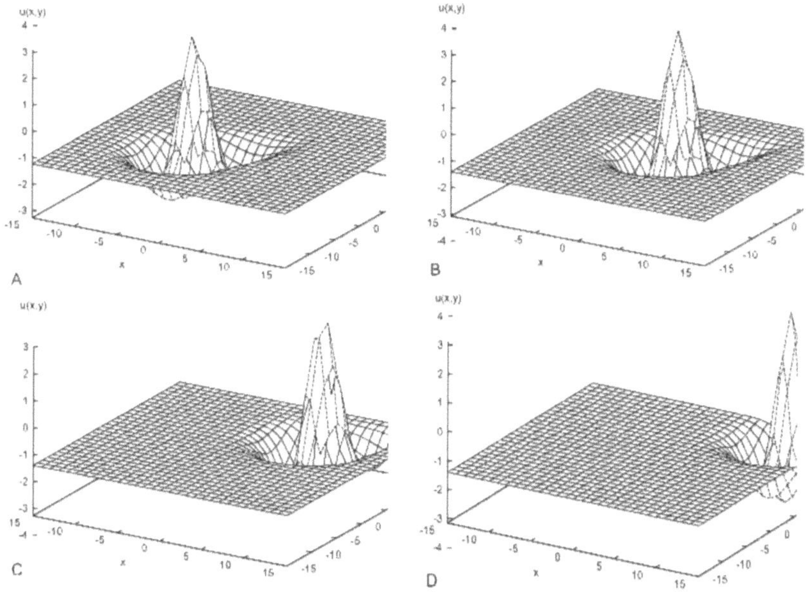

Fig. 4. Hill-shift effect in an inhomogeneous field with asymmetrical couplings. The bump moves from start location (A) via (B), (C) to location (D), see [24].

patterns and behavior of the animal[4]. Because of the nature of the *modeling relation* [17] for complex systems like the brain, an ultimate proof that a particular computation produces a particular behavior is not possible.

References

1. Koch, C., Segev, I. (eds.): Methods in Neuronal Modeling. From Ions to Networks. MIT Press, Cambridge (1998)
2. Schierwagen, A.: Growth, structure and dynamics of real neurons: Model studies and experimental results. Biomed. Biochim. Acta 49, 709–722 (1990)
3. Schierwagen, A.: Neuronal morphology: Shape characteristics and models. Neirofiziologiya/Neurophysiology 40, 366–372 (2008)
4. Schierwagen, A., Alpár, A., Gärtner, U.: Scaling properties of pyramidal neurons in mice neocortex. Mathematical Biosciences 207, 352–364 (2007)
5. Schierwagen, A., Claus, C.: Dendritic morphology and signal delay in superior colliculus neurons. Neurocomputing 38-40, 343–350 (2001)
6. Hodgkin, A.L., Huxley, A.F.: A quantitative description of membrane current and its application to conduction and excitation in nerve. J. Physiol. 117, 500–544 (1952)

[4] At the level single neurons or below, such correlations with animal behavior cannot be seriously expected.

7. Holden, A.V., Muhamad, M.A., Schierwagen, A.K.: Repolarizing currents and periodic activity in nerve membrane. J. Theor. Neurobiol. 4, 61–71 (1985)
8. Rall, W.: Theory of physiological properties of dendrites. Ann. N.Y. Acad. Sci. 96, 1071–1092 (1962)
9. Schierwagen, A.: Segmental cable modelling of electrotonic transfer properties of deep superior colliculus neurons in the cat. J. Hirnforsch 27, 679–690 (1986)
10. Schierwagen, A.: Exploring the computational capabilities of single neurons by continuous cable modelling. Prog. Brain Research 102, 151–168 (1994)
11. Schierwagen, A.: A non-uniform equivalent cable model of membrane voltage changes in a passive dendritic tree. J. theor. Biol. 141, 159–180 (1989)
12. Ohme, M., Schierwagen, A.: An equivalent cable model for neuronal trees with active membrane. Biol. Cybernet. 78, 227–243 (1998)
13. Schierwagen, A.: Identification problems in distributed parameter neuron models. Automatica 26, 739–755 (1990)
14. Schierwagen, A.: Impulse propagation in nonuniform fibres: analytical treatment. In: Fusco, D., Jeffrey, A. (eds.) Nonlinear Waves and Dissipative Effects. Pitman Research Notes in Mathematics, vol. 227, pp. 133–140. Longman, London (1991)
15. Waxman, S.G., Kocsis, J.D., Stys, P.K. (eds.): The Axon. Structure, Function and Pathophysiology. Oxford University Press, Oxford (1995)
16. Schierwagen, A., Ohme, M.: A Model for the Propagation of Action Potentials in Nonuniform Axons. AIP Proc. 1028, 98–112 (2008)
17. Schierwagen, A.: Brain Complexity: Anlysis, Models and Limits of Understanding. In: Mira, J., et al. (eds.) IWINAC 2009, Part I. LNCS, vol. 5601, pp. 195–204. Springer, Heidelberg (2009)
18. London, M., Häusser, M.: Dendritic computation. Ann. Rev. Neurosci. 28, 503–532 (2005)
19. Schierwagen, A.: Brain Organization and Computation. In: Mira, J., Álvarez, J.R. (eds.) IWINAC 2007. LNCS, vol. 4527, pp. 31–40. Springer, Heidelberg (2007)
20. Amari, S.: Dynamics of pattern formation in lateral-inhibition type neural fields. Biol. Cybernet. 27, 77–87 (1977)
21. Wellner, J., Schierwagen, A.: Cellular-Automata-like Simulations of Dynamic Neural Fields. In: Holcombe, M., Paton, R.C. (eds.) Information Processing in Cells and Tissues, pp. 295–304. Plenum, New York (1998)
22. Werner, H., Schierwagen, A.: Categorization of the dynamics in two-dimensional neural fields. In: Abstracts of the 9th Annual Computational Neuroscience Meeting, Brugge, Belgium, July 16-21, p. 156 (2000)
23. Werner, H., Richter, T.: Circular stationary solutions in two-dimensional neural fields. Biol. Cybern. 85, 211–217 (2001)
24. Schierwagen, A., Werner, H.: Analog computations with mapped neural fields. In: Trappl, R. (ed.) Cybernetics and Systems 1996, pp. 1084–1089. Austrian Society for Cybernetic Studies, Vienna (1996)
25. Schierwagen, A., Werner, H.: Fast orienting movements to visual targets: Neural field model of dynamic gaze control. In: 6th European Symposium on Artificial Neural Networks - ESANN 1998, pp. 91–98. D-facto publications, Brussels (1998)

The Foldl Operator as a Coequalizer Using Coq*

Antonio Blanco, Enrique Freire, Jose Luis Freire, and Javier Paris

Departamento de Computacion, Facultade de Informatica, Universidade da Coruna
{blanco,efreire,freire,javierparis}@udc.es

Abstract. In the present work a Coq based approach is taken to characterize the *foldl* using a functorial structure from which an inductive type is determined. With μ_F being an initial F–algebra and (B, θ) another F–algebra, two F–algebras with support $B \times \mu_F$ are constructed and then coequalized. This coequalization morphism allows the definition of *foldl* structurally.

After examining some significant examples we propose the following methodology to define a *foldl* operator. Let F be a polynomial endofunctor and (μ_F, in_F) its initial algebra. We define two F–algebras with support $B \times \mu_F$, and $h_1, h_2 : F(B \times \mu_F) \rightarrow B \times \mu_F$ constructed such that in one of them the argument of the initial type is syntactically (structurally) lower than that in the other. Then, $foldl : B \times \mu_F \rightarrow B$ can be defined as a specific morphism that coequalizes them $(h_1; foldl = h_2; foldl)$.

For an initial F–algebra with distinguished element (as in the case of lists), *foldl* is a coequalizer of h_1 and h_2.

The proofs are performed using the Coq proof system. In this context, constructive approach stress that existence is constructive, therefore with a computational content.

Detailed structure for proofs in Coq is included but interactive code is omited. See the section 4 for a repository with the complete source code.

1 Introduction

Research on categorical representation of programs and program transformation using some form of representing them in a type theory has been done for years. Some of this research has led to PoplMark Challenge [13] forum which deserves special mention. There, automated proof assistans, like Coq, are used as they ofer the hope of significantly easing the involved proofs.

In that context, and following pioneering papers [7], [8], [12], [2] and [14] our paper seeks to provide a methodology for building the *foldl* operator.

In [7] a categorical characterization of when a morphism is a catamorphism, i.e. is a *foldr*, is given. In [14] is detected, and fix it, that this characterization is not constructive in the sense that there is a function h satisfying the conditions demanded in [7] but such that the morphism g needed to say that h is *foldr g*, is not computable. Although we do not deal with this problem, we characterize the

* Partially supported by Xunta de Galicia PGIDIT07TIC005105PR and MEC TIN2005-08986.

R. Moreno-Díaz et al. (Eds.): EUROCAST 2009, LNCS 5717, pp. 167–176, 2009.

foldl following also a categorical construction that, since it deals with specific cases, do not raise the general problem of non computability.

Instead of having a catamorphism as in the *foldr* case, we discovered that *foldl* is a coequalizer.

There are some proposals to define *foldl* on lists and trees using Prolog and Haskell but without any verification and scheduling function directly without using any previous characterization. Deserves special mention the work [10] where, using an invariant concept, a charaterization of *foldl* as a colimit is given for lists and natural numbers.

For a good introduction to the system Coq see [3].

1.1 Categorical Context

If $\mathbf{C}$ is a category and $F : \mathbf{C} \to \mathbf{C}$ a cocontinuous endofunctor there exists an initial algebra, $(\mu F, in_F)$ [1], where μ_F is a fixed point for the functor F. So, there is an object $\mu F \in \mathbf{C}$ (called inductive type) and an isomorphism $in_F : F(\mu F) \to \mu F$ such that, for each F-algebra (A, θ), there is a single morphism, denoted as $foldr\ \theta : \mu F \to A$, satisfying

$$in_F; (foldr\ \theta) = F(foldr\ \theta); \theta.$$

Given $f : A \to B$ a homomorphism between the F-algebras (A, θ), (B, ξ) the initial character guarantees the known fusion theorem

$$(foldr\ \theta); f = foldr\ \xi.$$

From basic facts of Category theory, that

$$in_F; (foldr\ \theta; f) = (in_F; foldr\ \theta); f = (F(foldr\ \theta); \theta); f = F(foldr\ \theta); (\theta; f) = F(foldr\ \theta); (Ff; \xi) = (F(foldr\ \theta); Ff); \xi = F(foldr\ \theta; f); \xi.$$

However, $foldr\ \xi$ is, by definition, the only morphism that satisfies $in_F; (foldr \xi) = F(foldr\ \xi); \xi$ hence, proving the indicated result.

There also exists a functional with similar semantics, *foldl*, but it has not been studied as extensively. This functional, given its tail-recursive character, is widely used due to its efficiency (at least for strict evaluation). However, while *foldr* has a catamorphism version for any inductive type, *foldl* is limited nearly exclusively to the lists type.

In the present work *foldl* is characterized using a functorial structure on which the inductive type is determined. Since (μ_F, in_F) is an initial F–algebra and (B, θ) another F–algebra, two F–algebras with $\mu_F \times B$ support are constructed and coequalized. This coequalization morphism allows the definition of the *foldl* structurally.

In our approach, the Coq system is used to build the proofs. The constructions and the proofs have been formalized in this type theory system [3].

In section 2 *foldl* is studied for lists ($foldl_{list}$) and the fusion theorem for $foldl_{list}$ is proven as a corollary of its categorical characterization.

Thereafter, more examples of inductive types will be analyzed following the same pattern.

2 *Foldl* and the Type of Lists

Let **C** be a category with terminal object $*$ and finite products and finite coproducts. We know that the type ($List\ A$) of lists of type $A \in \mathbf{C}$ can be constructed as a fixed point of the functor $L_A : \mathbf{C} \to \mathbf{C}$ given by $L_A(B) = * + A \times B$. This gives an isomorphism induced on the coproduct by the projections $[\,]$ and $::$

$$([\,], ::) : * + A \times (List\ A) \to List\ A,$$

where $[\,]$ represents the empty list and $::$ means adding an element to a list.

Furthermore, the type $List\ A$ is an initial object in the category of L_A–algebras. That is to say, given any L_A–algebra (B, ϵ) (with $\epsilon = (b_0, f) : * + A \times B \to B$ for some $b0 \in B$ and $f : A \times B \to B$), there exists a unique morphism $foldr\ b0\ f : List\ A \to B$ such that for every $x \in A$ and $xs \in List\ A$

$$\begin{aligned} foldr_{list}\ f\ b_0\ [\,] &= b_0 \\ foldr_{list}\ f\ b_0\ x :: xs &= f(x, foldr_{list}\ f\ b_0\ xs). \end{aligned}$$

In this form the function is not tail–recursive.

A purely tail–recursive operator $foldl_{list}$ can be specified as:

$$\begin{aligned} foldl_{list}\ f\ b_0\ [\,] &= b_0 & (1) \\ foldl_{list}\ f\ b_0\ x :: xs &= foldl_{list}\ f\ f(x, b_0)\ xs. & (2) \end{aligned}$$

although, $foldr_{list}\ f\ b_0$ and $foldl_{list}\ f\ b_0$ do not compute the same function in general.

2.1 $foldl_{list}$ as a Coequalizer

We now characterize this construction in a categorical framework as follows.

Theorem 1. *Given* $(A, List\,A, [\,], ::)$ *and any* L_A*-algebra* (B, ϵ) *as above, let us define functions* $h_1, h_2 : * + A \times B \times List\ A \to B \times List\ A$ *by*

$$\begin{aligned} h_1(x, b, xs) &= (b, x :: xs) & (3) \\ h_2(x, b, xs) &= (f(x, b), xs) \text{ and} & (4) \\ h_1(*) = h_2(*) &= (b_0, [\,]). & (5) \end{aligned}$$

Then, $foldl_{list}$ *is the coequalizer of* h_1, h_2.

We give a proof directly in Coq. For simplicity, we use the curryfication, that is, the isomorphism $\mathbf{Set}(A \times B, C) \simeq \mathbf{Set}(A, C^B)$ is applied.

Note that the second argument on the right side of (4) is structurally less than that in (3). This allows to proceed in the proof process by induction on the inductive type of lists.

```
Section foldl_coeq.
Require Export List.
Variables A B:Set.
Variable f:A → B → B.
Variable b0:B.

Fixpoint foldl_list (b:B) (l:list A){struct l}:B:=
match l with
  nil ⇒ b
  |cons x xs ⇒ foldl_list (f x b) xs
end.

Implicit Arguments foldl_list.

Definition h1:=fun (a:A) ⇒ fun (b:B) ⇒ fun (ls:list A) ⇒ (pair b (cons a ls)).
Definition h2 :=fun (a:A) ⇒ fun (b:B) ⇒ fun (ls:list A) ⇒ ((f a b),ls).
```

The next lemma proves that $foldl_{list}$ coequalizes h_1 and h_2

```
Lemma one: ∀ a:A, ∀ b:B, ∀ ls:list A,
(foldl_list (fst (h1 a b ls)) (snd (h1 a b ls)))=(foldl_list (fst (h2 a b ls ))(snd
(h2 a b ls))).
```

Now, using induction we prove its initial character for the coequalizers of h_1 and h_2. For this, we consider a pair (C, h) where $h : B \times List\ A \rightarrow C$ such that $h_1; h = h_2; h$. We prove that $\exists f' : B \rightarrow C$ such that $foldl_{list}; f' = h$ and also that $\forall g' : B \rightarrow C$ such that $\forall ls \in list\ A$ and $\forall b \in B$, $g'(foldl_{list}\ b\, ls) = h\, b\, ls$. Then, we verify that $\forall b \in B, g'\, b = f'\, b$

```
Variable C:Set.

Lemma coequalizer: ∀ h:B→list A→C, (∀ b:B, ∀ a:A, ∀ ls:list A,
(h (fst (h1 a b ls))(snd (h1 a b ls)) )=(h (fst (h2 a b ls))(snd (h2 a b ls)) ))
   → ∃f':B→C | (∀ ls:list A,∀ b:B, f'(foldl_list b ls)=h b ls)
                    ∧ (∀ (g':B→C),(∀ ls:list A, ∀ b:B, g' (foldl_list b ls)=h b ls)
                        → ∀ b:B,g' b=f' b) .
End foldl_coeq.
```

2.2 Optimization

As a direct consequence of 1, we obtain the following new optimization result for $foldl_{list}$.

A proof of the analogous Fusion theorem for *foldr* in a similar context can be found in [5].

Theorem 2. *Considering a homomorphism of L_A-algebras $\theta : (B, (b_0, f)) \rightarrow (C, (c_0, g))$, that is, for each $a \in A$, $b \in B$,*

$$\theta(b_0) = c_0; \tag{6}$$
$$\theta(f(a, b)) = g(a, \theta(b))). \tag{7}$$

Then, for any list ls, $\theta(foldl_{list}\ f\ b_0\ ls) = foldl_{list}\ g\ c_0\ ls$.

The structure of the proof in Coq using the previous result `coequalizer` in 2.1 follows

Section *fusion.*
Variables *A B C:Set.*
Variable *f:A → B → B g:A → C → C theta:B → C.*
Implicit *Arguments foldl_list.*

Hypothesis *F1:∀ a:A, ∀ b:B, theta (f a b)=g a (theta b).*

Definition *unif:=fun (b:B) ⇒ fun (ls:(list A)) ⇒ foldl_list g (theta b) ls .*

Lemma *UNIF: ∀ b:B, ∀ a:A,∀ ls:list A,*
(unif (fst (h1 A B a b ls)) (snd (h1 A B a b ls)))=
(unif (fst (h2 A B f a b ls)) (snd (h2 A B f a b ls))).

Lemma *unification: ∃K1:B → C | (∀ ls:list A,∀ b:B,*
K1 (foldl_list f b ls)=unif b ls)
∧ (∀ (g1:B→C),(∀ ls:list A, ∀ b:B,g1 (foldl_list f b ls)=unif b ls)
→ ∀ b:B,g1 b=K1 b) .

Definition *K:= proj1_sig unification.*

Note that this object K provides the computational content of the proof.

Lemma *prefusion: ∀ ls:list A, ∀ b:B, theta (foldl_list f b ls)=*
foldl_list g (theta b) ls.

Theorem *fusion: ∀ ls:list A, ∀ b:B, ∀ c:C, theta b=c*
→ (∀ a:A, ∀ b:B,theta (f a b)=g a (theta b))
→ theta (foldl_list f b ls)=foldl_list g c ls.
End *fusion.*

3 Extension of *Foldl* Operator to Other Inductive Types

In the previous section we characterized categorically the functional *foldl* for the type of lists. One might ask now if this can be applied to other types, and if so, what approach should be taken.

Let us analyze some examples illustrating how our approach fits well and creates a good methodology.

Example 1. Considering the functor $N : \mathbf{Set} \to \mathbf{Set}$ defined as $N(B) = (\,) + B$. The initial N-algebra associated with this functor is the natural numbers type `nat` $(\mathbf{N}, (O, S))$. Let $(B, \epsilon = (b_0, f))$ be another N-algebra.

As a specialization of a natural generalization of the usual *foldr* we can define:

$$foldr_{nat}\ f\ b_0\ 0 = b_0$$
$$foldr_{nat}\ f\ b_0\ (Sn) = f\ (foldr_{nat}\ f\ b_0\ n).$$

It is easy to see that for each $n \in \mathbf{N}$,

$$foldr_{nat}\ f\ b_0\ n = f^n(b_0).$$

Now, following the same pattern as in 2.1 we can define $foldl_{nat}$ as follows. Let us consider the functions $h_1,\ h_2 : (\,) + B \times \mathbf{N} \to B \times \mathbf{N}$:

$$\begin{aligned} h_1(\,) = (b_0, 0) &= h_2(\,) \\ h_1(b, n) &= (b, S(n)) \\ h_2(b, n) &= (f(b), n) \end{aligned}$$

where $n \in \mathbf{N},\ b \in B$. We define $foldl_{nat} : B \times \mathbf{N} \to B$, as the coequalizer of h_1, h_2. Therefore, for all $n \in \mathbf{N},\ b \in B$ we must have:

$$\begin{aligned} foldl_{nat}(b, S(n)) &= foldl_{nat}(f(b), n) \\ foldl_{nat}(b, 0) &= b. \end{aligned}$$

This tail–recursive operator coincides with the $foldr_{nat}$ operator. The proof using Coq is available at `http://www.dc.fi.udc.es/staff/freire/publications/publications.shtml`

Example 2. Consider A as a fixed set. Let the functor $S_A : \mathbf{Set} \to \mathbf{Set}$ defined as $S_A(B) = A + B$. The initial S_A-algebra associated with this functor is $(\mathbf{S}, [L, I])$, that coincides with the type $\mathbf{N} \times A$ given that for each $a \in A$ one can identify (n, a) with $I^n(a)$ where $I^0(a) = L(a)$ and $I^k(a) = I(I^{k-1}(a))$ if $k > 0$.

If we take $(B, [f, g] : A + B \to B)$ another S_A-algebra, then the *foldr* operator is defined as follows:

$$\begin{aligned} foldr\ f\ g\ L(a) &= f(a); \\ foldr\ f\ g\ I(s) &= g(foldr\ f\ g\ s). \end{aligned}$$

Hence,

$$foldr\ f\ g\ I^n(L(a)) = g^n(f(a)). \tag{8}$$

For the characterization of *foldl*, we define the functions $h_1,\ h_2 : A + \mathbf{S} \times B \to \mathbf{S} \times B$:

$$\begin{aligned} h_1(a) = h_2(a) &= (f(a), L(a)) \\ h_1(b, s) &= (b, I(s)) \\ h_2(b, s) &= (g(b), s) \end{aligned}$$

where $a \in A,\ s \in \mathbf{S},\ b \in B$. Therefore

$$\begin{aligned} foldl\ f\ g\ b\ L(a) &= b; \\ foldl\ f\ g\ b\ I(s) &= (foldl\ f\ g\ (g\ b)\ s) \end{aligned}$$

The formalization using Coq that *foldl* coequalizes (h_1, h_2) and that foldr is a particular case of foldl is easy and is available at `http://www.dc.fi.udc.es/staff/freire/publications/publications.shtml`

Now, let us analyze what occurs with the type of trees, proposed by Naish in [12].

Example 3. The functor $T_A : \mathbf{Set} \to \mathbf{Set}$ defined as $T_A(B) = A + B \times B$ has as initial object $Tree\ A$, the binary tree types whose leaves are type A, with the isomorphism $[L, T] : A + (Tree\ A) \times (Tree\ A) \to Tree\ A$.

To follow our pattern, if $(B, [f, g])$ is a T_A-algebra, we want to construct morphisms

$$h_1,\ h_2 : A + (Tree\ A) \times (Tree\ A) \times B \times B \to (Tree\ A) \times B.$$

to obtain our *foldl* as a morphism that coequalized them.

It seems natural to define, for each $a \in A$

$$h_1(a) = h_2(a) = (L(a), f(a)).$$

How do we define $h_1(l, r, b_1, b_2)$, $h_2(l, r, b_1, b_2)$? Let us assume a given function $h : B \times Tree\ A \to B$. Then, we define

$$h_1(l, r, b_1, b_2) = (T(l, r), g(b_1, b_2))$$
$$h_2(l, r, b_1, b_2) = (r, h(g(g(b_1, b_2), g(b_1, b_2)), l))$$

Let us consider $\xi(f, g, h) : Tree\ A \times B \to B$ a morphism coequalizing h_1 and h_2. the following equation must hold

$$\xi(f, g, h)(T(l, r), b) = \xi(f, g, h)(l, h(r, b)).$$

Because we do not want to explicitly attach the function h we will use one of type $Tree\ A \to B$. If we fix an element $b \in B$ the actual function ξ could take the role of function h, which would allow us to obtain the following characterization of *foldl* :

$$foldl\ f\ b_0\ (L\ a) = f\ a$$
$$foldl\ f\ b_0\ (T(l, r))) = foldl\ f\ b_0\ (foldl\ f\ (g\ b_0\ b_0\ r)\ l$$

For example, in the work of Naish and Sterling [12] the following version of the function is defined *foldl*: (we write it in Haskell as it appears in the original) for the type of binary trees:

```
data Bt a = Leaf a | Tree (Bt a) (Bt a)
foldltbt :: (a -> a) -> (b -> a -> a) -> (Bt b) -> a -> a
foldlbt f g (Leaf x) m = g x m
foldlbt f g (Tree l r) m =
        foldlbt f g r (foldlbt f g l (f m))
```

We observe that this is an adaptation of the conventional operator *foldr*, since it does not originate from any general pattern. However, we find that this morphism is actually a particular case of a coequalization morphism of (h_1, h_2) defined by taking as h precisely this morphism.

The structure of the proof, using Coq, that this morphism is indeed a particular case of the technique just described, is as follows:

```
Section foldl_Tree.
Variable A:Set B:Set.
Variable f:A → B g:B → B → B.

Definition k:=fun a:A⇒fun b:B⇒g (f a) b.

Inductive Tree (A:Set):Set :=
    Leaf:A → Tree A
  | Bin:Tree A → Tree A → Tree A.


Fixpoint foldl_Naish (b:B)(l:Tree A){struct l}:B:=
match l with
    Leaf a ⇒ (k a b)
  | Bin l r ⇒ foldl_Naish (foldl_Naish (g b b) l) r
end.

Definition h1 := fun (l r:Tree A) ⇒ fun (b1 b2:B) ⇒ (Bin A l r,g b1 b2 ).
Definition h2 := fun (l r:Tree A) ⇒ fun (b1 b2:B) ⇒
                                (r,foldl_Naish (g (g b1 b2) (g b1 b2)) l).


Lemma one: ∀ a:A, ∀ b1 b2:B, ∀ lt1 lt2:Tree A,
(foldl_Naish
        (snd (h1 lt1 lt2 b1 b2)) (fst (h1 lt1 lt2 b1 b2)))=
(foldl_Naish (snd (h2 lt1 lt2 b1 b2)) (fst (h2 lt1 lt2 b1 b2)) ).
End foldl_Tree.
```

Example 4. In this last example, we will see how to apply the coequalization technique in the type $Rose(A)$. This type is the final coalgebra of the endofunctor $F_A(X) = A \times List(X)$. In [4] the authors prove that $Rose(A)$ is the support of the initial F_A–algebra.

To optimize the construction of tail–recursive functions for this data type we will use the $foldl_{list}$ studied in section 2 of this work. Here we rename it as $foldl'_{list}$ just to make this part more readable.

```
Section foldl_List.
Require Export List.
Variable A:Set B:Set f:B→A→B.

Fixpoint foldl_list' (b:B)(l:list A){struct l}:B:=
match l with
nil ⇒ b
|cons a ls ⇒foldl_list' (f b a) ls
end.

End foldl_List.
Section foldl_Rose.
Require Export List.
```

```
Inductive Rose (A:Set):Set :=
 RCons: A → list (Rose A) → Rose A.

Variable A B:Set .
Variable eps:A→B→B.

Fixpoint foldl_Rose (b:B) (r:Rose A){struct r}:B:=
match r with
  RCons a rs ⇒ match rs with
      nil ⇒ eps a b
    | r::rs'⇒ (foldl_Rose (foldl_list' (Rose A) B
                              (foldl_Rose) (eps a b) rs') r)
     end
end.
End foldl_Rose.

Section CoeqRose.
Variable A B:Set.
Variable eps: A→ B → B.
Implicit Arguments foldl_Rose.

Definition h1 :=fun (a:A) ⇒ fun (rs:list (Rose A))
     ⇒ fun (xs: B) ⇒ (RCons A a rs,eps a xs).

Definition h2 :=fun (a:A) ⇒ fun (rs:list (Rose A))
     ⇒ match rs with
         nil ⇒ fun (xs:B) ⇒ (RCons A a nil,eps a xs)
       | cons r rs' ⇒ match rs' with
             nil ⇒ fun (xs: B) ⇒ (r , eps a (eps a xs))
           | r'::rs"⇒ fun (xs: B) ⇒ (r ,(foldl_list' (Rose A) B
                        (foldl_Rose A B eps) (eps a (eps a xs)) rs'))
                         end
         end.

Lemma ONE: ∀ rs:list (Rose A), ∀ a:A,∀ xs:B,
((foldl_Rose eps) (snd (h1 a rs xs)) (fst (h1 a rs xs)) )=
 ((foldl_Rose eps) (snd (h2 a rs xs)) (fst (h2 a rs xs))).
End CoeqRose.
```

4 Conclusions

In this paper we present a methodology for defining the operator *foldl* based on a characterization as a coequalizer (Theorem 1). This provides a guide to be followed in each particular case. This methodology is illustrated with a variety of examples. When a distinguished element exists in the type, as in Lists or Nat, then we prove all the universal properties defining a coequalizer. But, in other cases we just found the coigualization property.

Also, a new fusion theorem for *foldl* is obtained in 2.

In the spirit of the forum PoplMark Challenge [13], the use of a theory of types as Coq allows to obtain a correct operator regarding the specification and on the other hand, using Coq also for the categorical context, provides proofs, in many cases automatic, of the corresponding theorems.

The Coq tool coqdoc has been used to format the Coq code.

Full source Coq code is availabe at `http://www.dc.fi.udc.es/staff/freire/publications/publications.shtml`

References

[1] Arbib, M.A., Manes, E.G.: Algebraic Approaches to Program Semantics. The AKM Series in Theoretical Computer Science. Springer, Heidelberg (1986)

[2] Belleannée, C., Brisset, P., Ridoux, O.: A Pragmatic Reconstruction of λ–Prolog. The Journal of Logic Programming (1994)

[3] Bertot, Y., Castéran, P.: Interactive Theorem Proving and Program Development. EATCS Series. Springer, Heidelberg (2004)

[4] Freire, J.L., Blanco, A.: On the abstraction process. In: Brain Processes, Theories and Models. MIT Press, Cambridge (1996)

[5] Freire Nistal, J.L., Blanco Ferro, A., Freire Brañas, J.E., Sánchez Penas, J.J.: EUROCAST 2001. LNCS, vol. 2178, pp. 583–596. Springer, Heidelberg (2001)

[6] Freire, J.L., Freire, E., Blanco, A.: On Recursive Functions and Well–Founded Relations in the Calculus of Constructions. In: Moreno Díaz, R., Pichler, F., Quesada Arencibia, A. (eds.) EUROCAST 2005. LNCS, vol. 3643, pp. 69–80. Springer, Heidelberg (2005)

[7] Gibbons, J., Hutton, G.: Proof methods for corecursive programs. Fundamenta Informaticae Special Issue on Program Transformation 66(4), 353–366 (2005)

[8] Gibbons, J., Hutton, G., Altenkirch, T.: When is a function a fold or an unfold? Electronic Notes in Theoretical Computer Science 44(1) (2001)

[9] Greiner, J.: Programming with Inductive and Co-Inductive Types, Technical report, School of Computer Science, Carnegie Mellon University, Pittsburgh (1992)

[10] Jay, C.B.: Tail Recursion through Universal Invariants, Technical report, University of Edinburgh (1993)

[11] Jacobs, B., Rutten, J.: A Tutorial on (Co)Algebras and (Co)Induction. EATCS Bulletin 62 (1997)

[12] Naish, L., Sterling, L.: A Higher Order Reconstruction of Stepwise Enhancement. In: Fuchs, N.E. (ed.) LOPSTR 1997. LNCS, vol. 1463, pp. 245–262. Springer, Heidelberg (1998)

[13] `http://alliance.seas.upenn.edu/~plclub/cgi-bin/poplmark`

[14] Weber, T., Caldwell, J.: Constructively Characterizing Fold and Unfold. In: Bruynooghe, M. (ed.) LOPSTR 2003. LNCS, vol. 3018, pp. 110–127. Springer, Heidelberg (2003)

Algorithm for Testing the Leibniz Algebra Structure

José Manuel Casas[1], Manuel A. Insua[2], Manuel Ladra[2], and Susana Ladra[3]

[1] Universidad de Vigo, 36005 Pontevedra, Spain
jmcasas@uvigo.es
[2] Universidad de Santiago, E-15782 Santiago, Spain
avelino.insua@gmail.com,
manuel.ladra@usc.es
http://web.usc.es/~mladra
[3] Universidad de A Coruña, 15071 A Coruña, Spain
sladra@udc.es

Abstract. Given a basis of a vector space V over a field $\mathbb{K}$ and a multiplication table which defines a bilinear map on V, we develop a computer program on Mathematica which checks if the bilinear map satisfies the Leibniz identity, that is, if the multiplication table endows V with a Leibniz algebra structure. In case of a positive answer, the program informs whether the structure corresponds to a Lie algebra or not, that is, if the bilinear map is skew-symmetric or not.

The algorithm is based on the computation of a Gröbner basis of an ideal, which is employed in the construction of the universal enveloping algebra of a Leibniz algebra. Finally, we describe a program in the NCAlgebra package which permits the construction of Gröbner bases in non commutative algebras.

1 Introduction

A classical problem in Lie algebras theory is to know how many different (up to isomorphisms) finite-dimensional Lie algebras are for each dimension [11,15].

The classical methods to obtain the classifications essentially solve the system of equations given by the bracket laws, that is, for a Lie algebra $\mathfrak{g}$ over a field $\mathbb{K}$ with basis $\{a_1, \ldots, a_n\}$, the bracket is completely determined by the scalars $c_{ij}^k \in \mathbb{K}$ such that

$$[a_i, a_j] = \sum_{k=1}^{n} c_{ij}^k a_k \qquad (1)$$

so that the Lie algebra structure is determined by means of the computation of the structure constants c_{ij}^k. In order to reduce the system given by (1) for $i, j \in \{1, 2, \ldots, n\}$, different invariants as center, derived algebra, nilindex, nilradical, Levi subalgebra, Cartan subalgebra, etc., are used. Nevertheless, a new approach by using Gröbner bases techniques is available [9,10].

On the other hand, in 1993, J.-L. Loday [17] introduced a non-skew symmetric generalization of Lie algebras, the so called Leibniz algebras. They are $\mathbb{K}$-vector

R. Moreno-Díaz et al. (Eds.): EUROCAST 2009, LNCS 5717, pp. 177–186, 2009.

spaces $\mathfrak{g}$ endowed with a bilinear map $[-,-]:\mathfrak{g}\times\mathfrak{g}\to\mathfrak{g}$ such that the Leibniz relation holds

$$[x,[y,z]]=[[x,y],z]-[[x,z],y], \quad \text{for all } \ x,y,z\in\mathfrak{g}\,. \tag{2}$$

When the bracket satisfies $[x,x]=0$ for all $x\in\mathfrak{g}$, then the Leibniz identity (2) becomes the Jacobi identity, and so a Leibniz algebra is a Lie algebra. From the beginning, the classification problem of finite-dimensional Leibniz algebras is present in a lot of papers [1,2,3,4,5,6,8]. Nevertheless, the space of solutions of the system given by the structure constants becomes very hard to compute, especially for dimensions greater than 3 because the system has new equations coming from the non-skew symmetry of the bracket. In these situations, the literature only collects the classification of specific classes of algebras (solvable, nilpotent, filiform, etc.). In order to simplify the problem, new techniques applying Gröbner bases methods are developed [14]. However, the space of solutions is usually huge and two kind of problems can occur in applications:

1. Given a multiplication table of an algebra for a fixed dimension n, how can we know if it corresponds to a Leibniz algebra structure among the ones obtained by the classification?
2. Since the classification provides a lot of isomorphism classes, how can we be sure that the classification is well done?

The aim of the present paper is to give an answer to the following question: Given a multiplication table of an algebra for a fixed dimension n, how can we know if it corresponds to a Leibniz algebra structure? We present a computer program in NCAlgebra [12] (a package running under Mathematica which permits the construction of Gröbner bases in non commutative algebras) that implements the algorithm to test if a multiplication table for a fixed dimension n corresponds to a Leibniz algebra structure. In case of a positive answer, the program distinguishes between Lie and non-Lie algebras. The algorithm is based on the computation of a Gröbner basis of an ideal, which is employed in the construction of the universal enveloping algebra UL($\mathfrak{g}$) [19] of a Leibniz algebra $\mathfrak{g}$ over a field $\mathbb{K}$ with basis $\{a_1,a_2,\ldots,a_n\}$.

In order to do this, we consider the ideal $J=<\{-\Phi(r_{[a_i,a_j]})+x_ix_j-x_jx_i,-\Phi(l_{[a_i,a_j]})+y_ix_j-x_jy_i, i,j=1,\ldots,n\}>$ of the free associative non-commutative unitary algebra $\mathbb{K}<y_1,\ldots,y_n,x_1,\ldots,x_n>$, where $\Phi:T(\mathfrak{g}^l\oplus\mathfrak{g}^r)\to\mathbb{K}<y_1,\ldots,y_n,x_1,\ldots,x_n>$ is the isomorphism given by $\Phi(l_{a_i})=y_i,\Phi(r_{a_i})=x_i$, being $\mathfrak{g}^l$ and $\mathfrak{g}^r$ two copies of $\mathfrak{g}$, and an element $x\in\mathfrak{g}$ corresponds to the elements l_x and r_x in the left and right copies, respectively.

Then $(\mathfrak{g},[-,-])$ is a Leibniz algebra if and only if the Gröbner basis corresponding to the ideal J with respect to any monomial order does not contain linear polynomials in the variables $y_1,\ldots,y_n$.

We also obtain that $(\mathfrak{g},[-,-])$ is a Lie algebra if and only if the Gröbner basis with respect to any monomial order of the ideal $J'=<\{-\Phi(r_{[a_i,a_j]})+x_ix_j-x_jx_i, i,j=1,\ldots,n\}>$ does not contain linear polynomials in the variables $x_1,\ldots,x_n$.

2 On Leibniz Algebras

Definition 1. *A Leibniz algebra* $\mathfrak{g}$ *is a* $\mathbb{K}$*-vector space equipped with a bilinear map* $[-,-] : \mathfrak{g} \times \mathfrak{g} \to \mathfrak{g}$ *satisfying the Leibniz identity*

$$[x,[y,z]] = [[x,y],z] - [[x,z],y], \quad \textit{for all} \;\; x,y,z \in \mathfrak{g}\,. \tag{2}$$

When the bracket satisfies $[x,x] = 0$ for all $x \in \mathfrak{g}$, then the Leibniz identity (2) becomes the Jacobi identity; so a Leibniz algebra is a Lie algebra. Hence, there is a canonical inclusion functor from the category **Lie** of Lie algebras to the category **Leib** of Leibniz algebras. This functor has as left adjoint the Liezation functor which assigns to a Leibniz algebra $\mathfrak{g}$ the Lie algebra $\mathfrak{g}_{\mathrm{Lie}} = \mathfrak{g}/\mathfrak{g}^{\mathrm{ann}}$, where $\mathfrak{g}^{\mathrm{ann}} = \langle\{[x,x], x \in \mathfrak{g}\}\rangle$.

Example 1.

1. Lie algebras.
2. Let A be a K-associative algebra equipped with a K-linear map $D : A \to A$ satisfying

$$D(a(Db)) = DaDb = D((Da)b), \quad \text{for all} \;\; a,b \in A\,. \tag{3}$$

   Then A with the bracket $[a,b] = aDb - Dba$ is a Leibniz algebra.
   If $D = \mathrm{Id}$, we obtain the Lie algebra structure associated to an associative algebra. If D is an idempotent algebra endomorphism ($D^2 = D$) or D is a derivation of square zero ($D^2 = 0$), then D satisfies equation (3) and the bracket gives rise to a structure of non-Lie Leibniz algebra.
3. Let D be a dialgebra [18]. Then $(D,[-,-])$ is a Leibniz algebra with respect to the bracket defined by $[x,y] = x \dashv y - y \vdash x, x,y \in D$.
4. Let $\mathfrak{g}$ be a differential Lie algebra, then $(\mathfrak{g},[-,-]_d)$ with $[x,y]_d := [x,dy]$ is a non-Lie Leibniz algebra.

For a Leibniz algebra $\mathfrak{g}$, we consider two copies of $\mathfrak{g}$, left and right, denoted by $\mathfrak{g}^l$ and $\mathfrak{g}^r$, respectively. For an element $x \in \mathfrak{g}$, we denote by l_x and r_x the corresponding elements in the left and right copies, respectively. The universal enveloping algebra of $\mathfrak{g}$ was defined in [19] as

$$\mathrm{UL}(\mathfrak{g}) := T(\mathfrak{g}^l \oplus \mathfrak{g}^r)/I$$

where $T(V)$ is the tensor algebra on V and I is the two-sided ideal spanned by the following relations:
i) $r_{[x,y]} - (r_x r_y - r_y r_x)$
ii) $l_{[x,y]} - (l_x r_y - r_y l_x)$
iii) $(r_y + l_y)l_x$.
Moreover, [19, Theorem (2.3)] establishes that the category of representations (resp. co-representations) of the Leibniz algebra $\mathfrak{g}$ is equivalent to the category of right (resp. left) modules over $\mathrm{UL}(\mathfrak{g})$. After this, [16, Corollary 1.4] establishes

the equivalence between the categories of representations and co-representations of a Leibniz algebra.

Let $\mathfrak{g}$ be a finite-dimensional Leibniz algebra with basis $\{a_1, \ldots, a_n\}$. There is an isomorphism of algebras

$$\Phi : T(\mathfrak{g}^l \oplus \mathfrak{g}^r) \to \mathbb{K} < y_1, \ldots, y_n, x_1, \ldots, x_n >$$

given by $\Phi(l_{a_i}) = y_i, \Phi(r_{a_i}) = x_i$, where $\mathbb{K} < y_1, \ldots, y_n, x_1, \ldots, x_n >$ denotes the free associative non-commutative unitary $\mathbb{K}$-algebra of polynomials.

Having in mind the inclusion $\mathfrak{g} \hookrightarrow T(\mathfrak{g}^l \oplus \mathfrak{g}^r)$ and the isomorphism Φ, we obtain that

$$\mathrm{UL}(\mathfrak{g}) \cong \frac{\mathbb{K} < y_1, \ldots, y_n, x_1, \ldots, x_n >}{< -\Phi(r_{[a_i,a_j]}) + x_i x_j - x_j x_i, -\Phi(l_{[a_i,a_j]}) + y_i x_j - x_j y_i, (x_i + y_i) y_j >},$$

and hence we can use the theory of Gröbner bases on $\mathbb{K} < y_1, \ldots, y_n, x_1, \ldots, x_n >$ [20] to obtain results about the structure of $\mathfrak{g}$.

Let $\prec$ be a given monomial order on the noncommutative polynomial ring $\mathbb{K}<X>$. For an arbitrary polynomial $p \in \mathbb{K}<X>$, we will use $lm(p)$ to denote the leading monomial of p.

Definition 2. *Let I be a two-sided ideal of $\mathbb{K}<X>$. A subset $\{0\} \subsetneq G \subset I$ is called a* Gröbner basis *for I if for every $0 \neq f \in I$, there exists $g \in G$, such that $lm(g)$ is a factor of $lm(f)$.*

Lemma 1 (Diamond Lemma [7]). *Let $\prec$ be a monomial order on $\mathbb{K}<X>$ and let $G = \{g_1, \ldots, g_m\}$ be a set of generators of an ideal I in $\mathbb{K}<X>$. If all the overlap relations involving members of G reduce to zero modulo G, then G is a Gröbner basis for I.*

We note that the overlap relations are the noncommutative version of the S-polynomials.

Now, we consider the ideal $J =< \{g_{ij} = -\Phi(r_{[a_i,a_j]}) + x_i x_j - x_j x_i, h_{ij} = -\Phi(l_{[a_i,a_j]}) + y_i x_j - x_j y_i, i,j = 1, \ldots, n\} >$. For $i,j,k \in \{1, \ldots, n\}$, let be $P_{ijk} = h_{ij} x_k + y_i g_{kj}$.

Lemma 2. *$P_{ijk} \to_J -\Phi(l_{[[a_i,a_j],a_k]}) - \Phi(l_{[a_i,[a_k,a_j]]}) + \Phi(l_{[[a_i,a_k],a_j]})$ for all $i,j,k \in \{1, \ldots, n\}$.*

Proof. $P_{ijk} = h_{ij} x_k + y_i g_{kj} = -\Phi(l_{[a_i,a_j]}) x_k - x_j y_i x_k - y_i \Phi(r_{[a_k,a_j]}) + y_i x_k x_j \to_J -\Phi(l_{[a_i,a_j]}) x_k - x_j y_i x_k - y_i \Phi(r_{[a_k,a_j]}) + \Phi(l_{[a_i,a_k]}) x_j + x_k y_i x_j \to_J -\Phi(l_{[a_i,a_j]}) x_k - y_i \Phi(r_{[a_k,a_j]}) + \Phi(l_{[a_i,a_k]}) x_j + x_k y_i x_j - x_j \Phi(l_{[a_i,a_k]}) - x_j x_k y_i \to_J -\Phi(l_{[a_i,a_j]}) x_k - y_i \Phi(r_{[a_k,a_j]}) + \Phi(l_{[a_i,a_k]}) x_j + x_k y_i x_j - x_j \Phi(l_{[a_i,a_k]}) + \Phi(r_{[a_k,a_j]}) y_i - x_k x_j y_i \to_J -\Phi(l_{[a_i,a_j]}) x_k - y_i \Phi(r_{[a_k,a_j]}) + \Phi(l_{[a_i,a_k]}) x_j + x_k \Phi(l_{[a_i,a_j]}) - x_j \Phi(l_{[a_i,a_k]}) + \Phi(r_{[a_k,a_j]}) y_i$.

Let $[a_k, a_j] = \alpha_1^{kj} a_1 + \cdots + \alpha_n^{kj} a_n$.

$\Phi(r_{[a_k,a_j]}) y_i - y_i \Phi(r_{[a_k,a_j]}) = (\alpha_1^{kj} x_1 + \cdots + \alpha_n^{kj} x_n) y_i - y_i (\alpha_1^{kj} x_1 + \cdots + \alpha_n^{kj} x_n) = \alpha_1^{kj} (x_1 y_i - y_i x_1) + \cdots + \alpha_n^{kj} (x_n y_i - y_i x_n) \to_J -\alpha_1^{kj} \Phi(l_{[a_i,a_1]}) - \cdots - \alpha_n^{kj} \Phi(l_{[a_i,a_n]}) = -\Phi(l_{[a_i,[a_k,a_j]]})$.

$\Phi(l_{[a_i,a_j]})x_k - x_k\Phi(l_{[a_i,a_j]}) = (\alpha_1^{ij}y_1+\cdots+\alpha_n^{ij}y_n)x_k - x_k(\alpha_1^{ij}y_1+\cdots+\alpha_n^{ij}y_n) = \alpha_1^{ij}(y_1x_k - x_ky_1)+\cdots+\alpha_n^{ij}(y_nx_k - x_ky_n) \to_J \alpha_1^{ij}\Phi(l_{[a_1,a_k]})+\cdots+\alpha_n^{ij}\Phi(l_{[a_n,a_k]}) = \Phi(l_{[[a_i,a_j],a_k]})$. □

Theorem 1. *Let* $\mathfrak{g}$ *be a finite-dimensional* $\mathbb{K}$*-vector space with basis* $\{a_1,\ldots,a_n\}$ *together with a bilinear map* $[-,-] : \mathfrak{g}\times\mathfrak{g}\to\mathfrak{g}$.
$(\mathfrak{g},[-,-])$ *is a Leibniz algebra if and only if the Gröbner basis* $G = \{g_1,\ldots,g_t\}$ *corresponding to the ideal* J *with respect to any monomial order does not contain linear polynomials* $g_j(y_1,\ldots,y_n)$.

Proof. If $(\mathfrak{g},[-,-])$ is a Leibniz algebra, then $P_{ijk}=0$ for all $i,j,k\in\{1,\ldots,n\}$ because of the Leibniz identity (2).

On the other hand, if $(\mathfrak{g},[-,-])$ is not a Leibniz algebra, then there exist $i,j,k\in\{1,\ldots,n\}$ such that $[[a_i,a_j],a_k]+[a_i,[a_k,a_j]]-[[a_i,a_k],a_j]\neq 0$, so J contains a degree 1 polynomial on the variables $y_1,\ldots,y_n$. □

Example 2. Let $(\mathfrak{g} =< a_1,a_2,a_3 >_{\mathbb{K}},[-,-])$ be the vector space such that

$$[a_1,a_3]=a_2,\ [a_2,a_3]=a_1+a_2,\ [a_3,a_3]=a_3,\ \text{ and 0 in other case.}$$

In this case, $J =< -x_1x_2+x_2x_1, -x_1x_3+x_3x_1+x_2, x_1x_2-x_2x_1, -x_2x_3+x_3x_2+x_1+x_2, x_1x_3-x_3x_1, x_2x_3-x_3x_2, x_3, x_1y_1-y_1x_1, x_2y_1-y_1x_2, x_3y_1-y_1x_3+y_2, x_1y_2-y_2x_1, x_2y_2-y_2x_2, x_3y_2-y_2x_3+y_1+y_2, x_1y_3-y_3x_1, x_2y_3-y_3x_2, x_3y_3-y_3x_3+y_3 >\subset \mathbb{K}<y_1,y_2,y_3,x_1,x_2,x_3>$.

The Gröbner basis of J with respect to degree lexicographical ordering with $y_3>y_2>y_1>x_3>x_2>x_1$ is $\{y_3,y_2,y_1,x_3,x_2,x_1\}$.

Therefore, $\mathfrak{g}$ is not a Leibniz algebra.

According to Theorem 1, we can know if a finite-dimensional algebra is Leibniz or not. In case of positive answer, it is natural to ask if the Leibniz algebra is a Lie algebra or a non-Lie Leibniz algebra.

As it is well-known [13]

$$\mathfrak{g}^{\mathrm{ann}} =< \{[a_i,a_i], i=1,\ldots,n\}\cup\{[a_i,a_j]+[a_j,a_i], i,j=1,\ldots,n\} >,$$

and having in mind the identification

$$\begin{array}{ccccc} \mathfrak{g} & \hookrightarrow & T(\mathfrak{g}^l\oplus\mathfrak{g}^r) & \overset{\Phi}{\underset{\simeq}{\longrightarrow}} & \mathbb{K}<x_1,\ldots,x_n,y_1,\ldots,y_n> \\ a_i & \mapsto & r_{a_i} & \mapsto & x_i \end{array}$$

then the generators of $\mathfrak{g}^{\mathrm{ann}}$ can be written as

$$\{\Phi(r_{[a_i,a_i]}), i=1,\ldots,n\}\cup\{\Phi(r_{[a_i,a_j]})+\Phi(r_{[a_j,a_i]}),\ i,j=1,\ldots,n\}$$

and these generators belongs to J, since

$$g_{ii}=\Phi(r_{[a_i,a_i]}),\ i=1,\ldots,n\,,$$

$$-g_{ij}-g_{ji}=\Phi(r_{[a_i,a_j]})+\Phi(r_{[a_j,a_i]}),\ i,j=1,\ldots,n\,.$$

Consequently, if we compute the Gröbner basis G with respect to any monomial order for the ideal J, we can obtain two possible results:

1. There are some linear polynomials $g_j(y_1, \dots, y_n)$ in G, so $\mathfrak{g}$ is not a Leibniz algebra.
2. There are not any linear polynomials $g_j(y_1, \dots, y_n)$ in G, so $\mathfrak{g}$ is a Leibniz algebra, and:
   (a) If there are some linear polynomials $g_i(x_1, \dots, x_n)$ in G, then $\mathfrak{g}^{\mathrm{ann}} \neq 0$ and so $\mathfrak{g}$ is a non Lie-Leibniz algebra.
   (b) In other case $\mathfrak{g}$ is a Lie algebra.

Example 3. Let $(\mathfrak{g} =< a_1, a_2, a_3 >_{\mathbb{K}}, [-,-])$ be the vector space such that

$$[a_1, a_3] = a_2,\ [a_2, a_3] = a_1 + a_2,\ [a_3, a_3] = a_1, \text{ and 0 in other case.}$$

In this case, $J =< -x_1x_2 + x_2x_1, -x_1x_3 + x_3x_1 + x_2, x_1x_2 - x_2x_1, -x_2x_3 + x_3x_2 + x_1 + x_2, x_1x_3 - x_3x_1, x_2x_3 - x_3x_2, x_1, x_1y_1 - y_1x_1, x_2y_1 - y_1x_2, x_3y_1 - y_1x_3 + y_2, x_1y_2 - y_2x_1, x_2y_2 - y_2x_2, x_3y_2 - y_2x_3 + y_1 + y_2, x_1y_3 - y_3x_1, x_2y_3 - y_3x_2, x_3y_3 - y_3x_3 + y_1 > \subset \mathbb{K}<y_1, y_2, y_3, x_1, x_2, x_3>$.

The Gröbner basis of J with respect to degree lexicographical ordering with $y_3 > y_2 > y_1 > x_3 > x_2 > x_1$ is $\{x_2, x_1, -x_3y_1 + y_1x_3 - y_2, -x_3y_2 + y_2x_3 - y_1 - y_2, -x_3y_3 + y_3x_3 - y_1\}$.

Therefore, $\mathfrak{g}$ is a Leibniz algebra and not a Lie algebra.

Example 4. Let $(\mathfrak{g} =< a_1, a_2, a_3, a_4 >_{\mathbb{K}}, [-,-])$ be the vector space such that

$$[a_4, a_1] = a_1, [a_1, a_4] = -a_1, [a_4, a_2] = a_2, [a_2, a_4] = -a_2, [a_4, a_3] = a_3, [a_3, a_4] = -a_3,$$

and 0 in other case.

In this case, $J =< -x_1x_2 + x_2x_1, -x_1x_3 + x_3x_1, -x_1x_4 + x_4x_1 - x_1, x_1x_2 - x_2x_1, -x_2x_3 + x_3x_2, -x_2x_4 + x_4x_2 - x_2, x_1x_3 - x_3x_1, x_2x_3 - x_3x_2, -x_3x_4 + x_4x_3 - x_3, x_1x_4 - x_4x_1 + x_1, x_2x_4 - x_4x_2 + x_2, x_3x_4 - x_4x_3 + x_3, x_1y_1 - y_1x_1, x_2y_1 - y_1x_2, x_3y_1 - y_1x_3, x_4y_1 - y_1x_4 - y_1, x_1y_2 - y_2x_1, x_2y_2 - y_2x_2, x_3y_2 - y_2x_3, x_4y_2 - y_2x_4 - y_2, x_1y_3 - y_3x_1, x_2y_3 - y_3x_2, x_3y_3 - y_3x_3, x_4y_3 - y_3x_4 - y_3, x_1y_4 - y_4x_1 + y_1, x_2y_4 - y_4x_2 + y_2, x_3y_4 - y_4x_3 + y_3, x_4y_4 - y_4x_4 > \subset \mathbb{K}<y_1, y_2, y_3, y_4, x_1, x_2, x_3, x_4>$.

The Gröbner basis of J with respect to degree lexicographical ordering with $y_4 > y_3 > y_2 > y_1 > x_4 > x_3 > x_2 > x_1$ is $\{-x_1x_2 + x_2x_1, -x_1x_3 + x_3x_1, -x_1x_4 + x_4x_1 - x_1, -x_2x_3 + x_3x_2, -x_2x_4 + x_4x_2 - x_2, -x_3x_4 + x_4x_3 - x_3, -x_1y_1 + y_1x_1, -x_2y_1 + y_1x_2, -x_3y_1 + y_1x_3, -x_4y_1 + y_1x_4 + y_1, -x_1y_2 + y_2x_1, -x_2y_2 + y_2x_2, -x_3y_2 + y_2x_3, -x_4y_2 + y_2x_4 + y_2, -x_1y_3 + y_3x_1, -x_2y_3 + y_3x_2, -x_3y_3 + y_3x_3, -x_4y_3 + y_3x_4 + y_3, -x_1y_4 + y_4x_1 - y_1, -x_2y_4 + y_4x_2 - y_2, -x_3y_4 + y_4x_3 - y_3, -x_4y_4 + y_4x_4\}$.

Therefore, $\mathfrak{g}$ is a Lie algebra.

We can also solve the dichotomy of knowing if a $\mathbb{K}$-vector space with a given multiplication table is a Lie algebra by means of the following ideal

$$J' =< \{g_{ij} = -\Phi(r_{[a_i,a_j]}) + x_ix_j - x_jx_i, i, j = 1, \dots, n\} > .$$

For $i, j, k \in \{1, \dots, n\}$, let be $Q_{ijk} = x_jg_{ki} - g_{ij}x_k$.

Lemma 3. *$Q_{ijk} \to_{J'} -\Phi(r_{[a_k,[a_i,a_j]]}) + \Phi(r_{[[a_k,a_i],a_j]}) - \Phi(r_{[[a_k,a_j],a_i]})$ for all $i, j, k \in \{1, \dots, n\}$.*

Proof. $x_j g_{ki} - g_{ij} x_k = \Phi(r_{[a_i,a_j]})x_k - x_i x_j x_k - x_j \Phi(r_{[a_k,a_i]}) + x_j x_k x_i \rightarrow_{g_{kj}x_i} \Phi(r_{[a_i,a_j]})x_k - x_i x_j x_k - x_j \Phi(r_{[a_k,a_i]}) - \Phi(r_{[a_k,a_j]})x_i + x_k x_j x_i \rightarrow_{-x_i g_{kj}} \Phi(r_{[a_i,a_j]})x_k - x_j \Phi(r_{[a_k,a_i]}) - \Phi(r_{[a_k,a_j]})x_i + x_k x_j x_i + x_i \Phi(r_{[a_k,a_j]}) - x_i x_k x_j \rightarrow_{-g_{ki}x_j} \Phi(r_{[a_i,a_j]})x_k - x_j \Phi(r_{[a_k,a_i]}) - \Phi(r_{[a_k,a_j]})x_i + x_k x_j x_i + x_i \Phi(r_{[a_k,a_j]}) + \Phi(r_{[a_k,a_i]})x_j - x_k x_i x_j = \Phi(r_{[a_i,a_j]})x_k + x_k(x_j x_i - x_i x_j) + \Phi(r_{[a_k,a_i]})x_j - x_j \Phi(r_{[a_k,a_i]}) - \Phi(r_{[a_k,a_j]})x_i + x_i \Phi(r_{[a_k,a_j]}) \rightarrow_{J'} \Phi(r_{[a_i,a_j]})x_k - x_k \Phi(r_{[a_i,a_j]}) + \Phi(r_{[a_k,a_i]})x_j - x_j \Phi(r_{[a_k,a_i]}) - \Phi(r_{[a_k,a_j]})x_i + x_i \Phi(r_{[a_k,a_j]})$.

Let $[a_i, a_j] = \alpha_1^{ij} a_1 + \cdots + \alpha_n^{ij} a_n$.

$\Phi(r_{[a_i,a_j]})x_k - x_k \Phi(r_{[a_i,a_j]}) = (\alpha_1^{ij} x_1 + \cdots + \alpha_n^{ij} x_n)x_k - x_k(\alpha_1^{ij} x_1 + \cdots + \alpha_n^{ij} x_n) = \alpha_1^{ij}(x_1 x_k - x_k x_1) + \cdots + \alpha_{k-1}^{ij}(x_{k-1}x_k - x_k x_{k-1}) + \alpha_k^{ij} \cdot 0 + \alpha_{k+1}^{ij}(x_{k+1}x_k - x_k x_{k+1}) + \cdots + \alpha_n^{ij}(x_n x_k - x_k x_n) \rightarrow_{J'} \alpha_1^{ij}\Phi(r_{[a_1,a_k]}) + \cdots + \alpha_{k-1}^{ij}\Phi(r_{[a_{k-1},a_k]}) - \alpha_{k+1}^{ij}\Phi(r_{[a_k,a_{k+1}]}) - \cdots - \alpha_n^{ij}\Phi(r_{[a_k,a_n]}) \rightarrow_{J'} \alpha_1^{ij}\Phi(r_{[a_1,a_k]}) + \ldots + \alpha_{k-1}^{ij}\Phi(r_{[a_{k-1},a_k]}) + \alpha_{k+1}^{ij}\Phi(r_{[a_{k+1},a_k]}) - \cdots - \alpha_n^{ij}\Phi(r_{[a_n,a_k]}) = \Phi(r_{[[a_i,a_j],a_k]}) - \alpha_k^{ij}\Phi(r_{[a_k,a_k]}) \rightarrow_{J'} \Phi(r_{[[a_i,a_j],a_k]})$, or $\Phi(r_{[a_i,a_j]})x_k - x_k \Phi(r_{[a_i,a_j]}) \rightarrow_{J'} -\Phi(r_{[a_k,[a_i,a_j]]})$. □

Theorem 2. *Let $\mathfrak{g}$ be a finite-dimensional $\mathbb{K}$-vector space with basis $\{a_1, \ldots, a_n\}$ together with a bilinear map $[-,-] : \mathfrak{g} \times \mathfrak{g} \to \mathfrak{g}$.*
$(\mathfrak{g}, [-,-])$ is a Lie algebra if and only if the Gröbner basis $G = \{g_1, \ldots, g_t\}$ corresponding to the ideal J' with respect to any monomial order does not contain linear polynomials $g_j(x_1, \ldots, x_n)$.

Example 5. Let $(\mathfrak{g} =< a_1, a_2, a_3, a_4 >_{\mathbb{K}}, [-,-])$ be the vector space such that

$$[a_4, a_1] = a_2, \quad [a_1, a_4] = -a_2, \quad [a_4, a_2] = a_3, \quad [a_2, a_4] = -a_3,$$
$$[a_4, a_3] = a_1 + a_2, \quad [a_3, a_4] = -a_1 - a_2, \quad \text{and} \quad 0 \quad \text{in other case} .$$

In this case, $J' =< -x_1x_2 + x_2x_1, -x_1x_3 + x_3x_1, -x_1x_4 + x_4x_1 - x_2, x_1x_2 - x_2x_1, -x_2x_3 + x_3x_2, -x_2x_4 + x_4x_2 - x_3, x_1x_3 - x_3x_1, x_2x_3 - x_3x_2, -x_3x_4 + x_4x_3 - x_1 - x_2, x_1x_4 - x_4x_1 + x_2, x_2x_4 - x_4x_2 + x_3, x_3x_4 - x_4x_3 + x_1 + x_2 > \subset \mathbb{K}<x_1, x_2, x_3, x_4>$.

The Gröbner basis of J' with respect to degree lexicographical ordering with $x_4 > x_3 > x_2 > x_1$ is $\{-x_1x_2 + x_2x_1, -x_1x_3 + x_3x_1, -x_1x_4 + x_4x_1 - x_2, -x_2x_3 + x_3x_2, -x_2x_4 + x_4x_2 - x_3, -x_3x_4 + x_4x_3 - x_1 - x_2\}$.

Therefore, $\mathfrak{g}$ is a Lie algebra.

3 Computer Program

In this section we describe a program in NCAlgebra [12] (a package running under Mathematica) that implements the algorithms discussed in the previous section. The program computes the reduced Gröbner basis of the ideal J which determines if the introduced multiplication table of $\mathfrak{g}$ corresponds to a Leibniz algebra or not. In case of positive answer, the program decides whether the algebra is a Lie or non-Lie Leibniz algebra. The Mathematica code together with some examples are available in `http://web.usc.es/~mladra/research.html`.

```
##########################################################################
   (* This program tests if an introduced multiplication table
  corresponds to a Lie algebra, a non-Lie Leibniz algebra or an
 algebra which has not Leibniz algebra structure. To run this code
      properly it is necessary to load the NCGB package*)
##########################################################################
```

(* Let $\mathfrak{g} =< a_1, \ldots, a_n >$ be an algebra of dimension n *)
(* Insert the Bracket represented by Bracket$[\{i1, i2\}] := \{\lambda_1, \ldots, \lambda_n\}$ where $[a_{i1}, a_{i2}] = \lambda_1 a_1 + \cdots + \lambda_n a_n$. In Example 2, e.g., Bracket[{2, 3}] := {1,1,0} *)

```
LeibnizQ[n_]:=Module[{G,A,lengA,Variabs,BaseG,varaux,rvarsx,lvarsy},

    (* First of all we construct the generators of the ideal *)

    G={};
    A=Tuples[Table[i,{i,1,n}],2];
    lengA=Length[A];

    Do[
        G=Join[
          G,{Bracket[A[[i]]].Table[x[i],{i,1,n}]-(x[A[[i,1]]]** x[A[[i,2]]]-
                x[A[[i,2]]]**x[A[[i,1]]])}]
      ,{i,1,lengA}];

    Do[
        G=Join[
          G,{Bracket[A[[i]]].Table[y[i],{i,1,n}]-(y[A[[i,1]]]** x[A[[i,2]]]-
                x[A[[i,2]]]**y[A[[i,1]]])}]
      ,{i,1,lengA}];

    rvarsx=Table[x[i],{i,1,n}];
    lvarsy=Table[y[i],{i,1,n}];
    Variabs=Join[rvarsx,lvarsy];

    (* Now we compute a Gröbner Basis of G *)

    SetNonCommutative@@Variabs;

    SetMonomialOrder[Variabs] ;

    BaseG=NCMakeGB[G,15];

    (* Finally, we check the Gröbner Basis *)

    varaux=Select[BaseG,Union[Variables[#],lvarsy]===lvarsy&];
```

```
  If[varaux==={},

     (* g is a Leibniz algebra but, is g a Lie algebra? *)

    varaux=Select[BaseG,Union[Variables[#],rvarsx]===rvarsx&];

    If[varaux==={},

       Print["g is a Lie algebra"];
       ,
      Print["g is not a Lie algebra but g is a Leibniz algebra"];

      ];
     ,
   Print["g is not a Leibniz algebra"];
   ];
]
```

Acknowledgements

First and third authors were supported by Ministerio de Educación y Ciencia, Grant MTM2006-15338-C02-02 (European FEDER support included) and by Xunta de Galicia, Grant PGIDIT06PXIB371128PR.

References

1. Albeverio, S., Ayupov, S.A., Omirov, B.A.: On nilpotent and simple Leibniz algebras. Comm. Algebra 33(1), 159–172 (2005)
2. Albeverio, S., Omirov, B.A., Rakhimov, I.S.: Varieties of nilpotent complex Leibniz algebras of dimension less than five. Comm. Algebra 33(5), 1575–1585 (2005)
3. Albeverio, S., Omirov, B.A., Rakhimov, I.S.: Classification of 4-dimensional nilpotent complex Leibniz algebras. Extracta Math. 21(3), 197–210 (2006)
4. Ayupov, S.A., Omirov, B.A.: On Leibniz algebras. Algebra and Operator Theory (Tashkent, 1997), pp. 1–12. Kluwer Acad. Publ., Dordrecht (1997)
5. Ayupov, S.A., Omirov, B.A.: On a description of irreducible component in the set of nilpotent Leibniz algebras containing the algebra of maximal nilindex, and classification of graded filiform Leibniz algebras. In: Computer algebra in scientific computing (Samarkand, 2000), pp. 21–34. Springer, Berlin (2000)
6. Ayupov, S.A., Omirov, B.A.: On some classes of nilpotent Leibniz algebras. Siberian Math. J. 42(1), 15–24 (2001)
7. Bergman, G.M.: The diamond lemma for ring theory. Adv. in Math. 29, 178–218 (1978)
8. Cuvier, C.: Algèbres de Leibnitz: définitions, propriétés. Ann. Sci. École Norm. Sup. 27(1) (4), 1–45 (1994)
9. de Graaf, W.: Lie algebras: theory and algorithms. North-Holland Publishing Co., Amsterdam (2000)

10. de Graaf, W.: Classification of solvable Lie algebras. Experiment. Math. 14, 15–25 (2005)
11. Goze, M., Khakimdjanov, Y.: Nilpotent Lie algebras. Mathematics and its Applications, vol. 361. Kluwer Acad. Publ., Dordrecht (1996)
12. Helton, J.W., Miller, R.L., Stankus, M.: NCAlgebra: A Mathematica package for doing noncommuting algebra (1996), `http://math.ucsd.edu/~ncalg`
13. Insua, M.A.: Varias perspectivas sobre las bases de Gröbner: forma normal de Smith, algoritmo de Berlekamp y álgebras de Leibniz. Ph. D. Thesis (2005)
14. Insua, M.A., Ladra, M.: Gröbner bases in universal enveloping algebras of Leibniz algebras. J. Symbolic Comput. 44, 517–526 (2009)
15. Jacobson, N.: Lie algebras. Interscience Publ., New York (1962)
16. Kurdiani, R.: Cohomology of Lie algebras in the tensor category of linear maps. Comm. Algebra 27(10), 5033–5048 (1999)
17. Loday, J.-L.: Une version non commutative des algèbres de Lie: les algèbres de Leibniz. Enseign. Math. 39(2), 269–293 (1993)
18. Loday, J.-L.: Algèbres ayant deux opérations associatives (digèbres). C. R. Acad. Sci. Paris Sér. I Math. 321(2), 141–146 (1995)
19. Loday, J.-L., Pirashvili, T.: Universal enveloping algebras of Leibniz algebras and (co)homology. Math. Ann. 296, 139–158 (1993)
20. Mora, T.: An introduction to commutative and noncommutative Gröbner bases. Theor. Comp. Sci. 134, 131–173 (1994)

Automatic Drusen Detection from Digital Retinal Images: AMD Prevention

B. Remeseiro, N. Barreira, D. Calvo, M. Ortega, and M.G. Penedo

VARPA Group, Dept. of Computer Science, Univ. A Coruña, Spain
{bremeseiro,nbarreira,dcalvo,mortega,mgpenedo}@udc.es

Abstract. The age-related macular degeneration (AMD) is the main cause of blindness among people over 50 years in developed countries and there are 150 million people affected worlwide. This disease can lead to severe loss central vision and adversely affect the patient's quality of life. The appearance of drusen is associated with the early AMD, so we proposed a top-down methodology to detect drusen in initial stages to prevent AMD. The proposed methodology has several stages where the key issues are the detection and characterization of suspect areas. We test our method with a set of 1280 × 1024 images, obtaining a system with a high sensitivity in the localization of drusen, not just fake injuries.

Keywords: drusen, AMD, retinal images, template matching, normalized cross correlation, region growing.

1 Introduction

The age-related macular degeneration (AMD) [1] is a degenerative eye disease that affects the central vision. This kind of vision is needed to perform daily tasks such as reading, sewing or driving. The AMD causes significant visual impact to the center of the retina, the macula, and therefore the center of the visual field. In 1995, an international classification was proposed by Bird, using color eye fundus images. The AMD was defined as a degenerative disease that affects people over 50 years old and has two stages: early and late AMD. The former is characterized by the presence of drusen and pigment epithelium abnormalities. The later includes the late, atrophic and humid injuries.

In this sense, early detection of drusen is useful in the diagnose and treatment of patients that suffer AMD. Therefore, the development of a screening system based on drusen could prevent the AMD. Drusen, in their early stages, are circular, small and white structures which can be observed in retinal images as Fig. 1 shows.

There are some works in the literature that try to detect drusen, but none of them is focused on initial stages. For example, the work proposed by Sbeh *et al.* [2] tries to segment drusen using an adaptative algorithm based on morphological operations. Rapantzikos and Zervakis [3] developed a segmentation technique called HALT (Histogram-based Adaptive Local Thresholding) with the aim of detecting drusen in eye fundus images by extracting useful information. In 2003,

R. Moreno-Díaz et al. (Eds.): EUROCAST 2009, LNCS 5717, pp. 187–194, 2009.

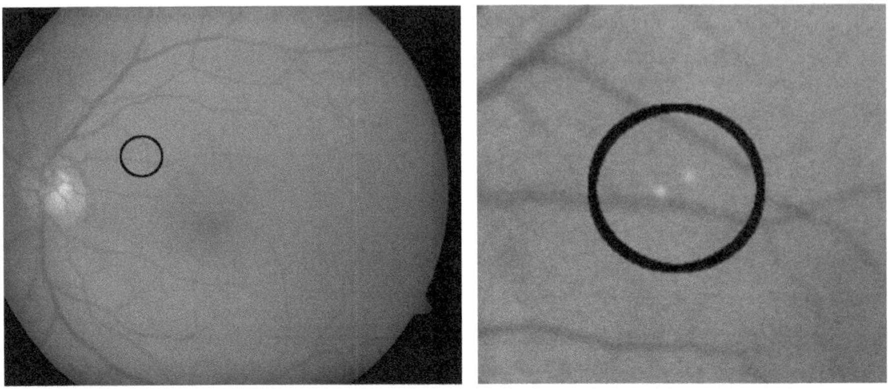

(a) (b)

Fig. 1. Examples of drusen: (a) shows a retinal image with two drusen in the macular area and (b) shows the zoom in this area (contrast enhanced)

Brandon and Hoover [4] proposed a multilevel algorithm to detect drusen in retinal images without human supervision. One year later, the work proposed by Mora *et al.* [5] uses classical image processing techniques to detect and model drusen. In 2006, Garg *et al.* [6] proposed two different methods to detect, count and segment drusen in retinal images, without human interaction or supervision. Both of them use morphological characteristics of drusen, such as texture and their 3D profiles. Another work, the proposed by Niemeijer *et al.* [7], presents a system that allows to detect exudates and cotton-wool spots in color fundus images and distinguish them from drusen. All mentioned works have one thing in common: they detect drusen at any stage, but they do not provide results on their performance in initial stages. However, only drusen detected at early stages can be used to prevent AMD.

Thus this work is focused on the automatic detection and characterization of drusen in early stages. We propose a top-down methodology to detect circular diffuse spot with a maximun diameter of $125\mu m$, using techniques such as template matching and region growing. This methodology can be integrated in a screening system for AMD diagnose.

This paper is organised as follows: in section 2 a description of the five stages methodology is presented. Section 3 shows the experimental results and validation obtained using a set of retinal images provided by ophthalmologists. Finally, section 4 provides some discussion and conclusion.

2 Methodology

The proposed methodology consists of five stages (see Fig. 2). The first stage involves the acquisition of the retinal image. The second stage entails the extraction of the green channel of the colour image. In the third stage, the search area

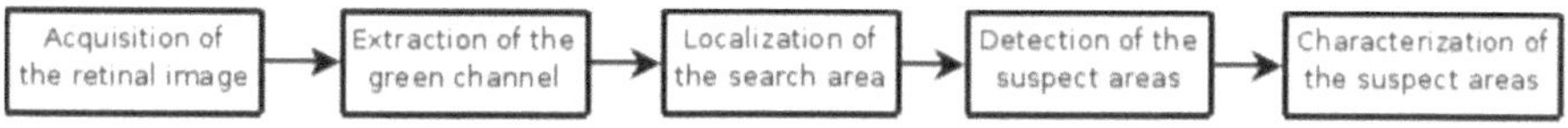

Fig. 2. Methodology general chart

is restricted to the inside of the ETDRS (Early Treatment Diabetic Retinophaty Study) protocol grille. The fourth stage tries to localize the areas of the image which are suspected of being drusen using the template matching technique. Finally, the suspect areas are segmented using the region growing technique and filtered to rule out false lesions. In the following sections, all these stages will be explained in detail.

2.1 Acquisition of the Retinal Image

The acquisition of the image is the first step towards the drusen detection. All of the images used in this work have been acquired with the FF 450^{plus} *Fundus Camera*, a 2 Mpx camera. They are colour fundus images, in PPM format and their resolution is 1280×1024 pixels.

2.2 Extraction of the Green Channel

The green channel of the colour image contains the most of the image information since its contrast is greater than the contrast of the other RGB channels. This is due to the optical characteristics of the eye and the nature of the cameras. the blue channel of the image contains little information while the red channel is too saturated. For this reason, the green channel of the image is extracted and it will be used in next stages. Other works also use the green channel according to the same reasoning [4,6,7].

2.3 Localization of the Search Area

The ETDRS [8] is a standard protocol that studies the diabetic retinopathy. The ETDRS protocol grille was initially created to divide the central retina in different areas to the treatment of diabetic people. Nowadays, ophthalmologists use it in other pathologies such as AMD.

The drusen outside the grille correspond to a very peripheral area of the vision. In this area, there is neither vision in detail nor colour vision, so the presence of drusen outside the grille has a negligible impact on the visual field. In addition, all of the images are focused on the macula so the peripheral drusen may appear blurred and deformed, distorting, consequently, their analysis.

Therefore, the proposed system will detect the drusen inside the grille, so that this area has been called *search area*. The grille (see Fig. 3) consists of three concentric circumferences focused on the macula. The *search area* is limited to the area occupied by the grille in order to focus the system on the area of interest. As a result, the drusen are searched in a circumference of 7.2 mm diameter and

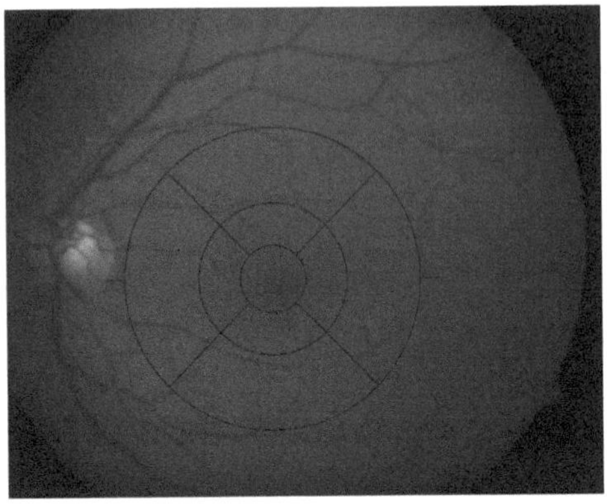

Fig. 3. ETDRS protocol grille over a retinal image

centered on the macula. The idea proposed by Mariño *et al.* [9] was used to center the circumference on the macula.

2.4 Detection of the Suspect Areas

The detection of the suspect areas is one of the key stages in the proposed methodology. The goal is to identify the regions of the image that might be drusen. It is intended to achieve the fullest possible detection, which means high sensitivity. The technique used is the template matching [10]. Its adaptation to the suspect area problem entails the creation of a template that represents a drusen and the search for parts of the retinography that resemble the template.

The similarity measurement used is the normalized cross correlation [10], so the output image will have pixels with values between -1 and 1. A threshold δ is selected to determine which are the suspect areas.

Drusen have a circular shape with fuzzy edges and a whitish colour. Their intensity is variable, but always higher than the surrounding retinal tissue. Regarding the size, we only consider those drusen with a maximum diameter of $125\mu m$.

Due to the drusen characteristics, two different templates were tested: circular templates and gaussian templates. As drusen have different sizes, a multiscale approach was used. The experimental results obtained with four test images proved that the most suitable configuration includes two gaussian templates with radius 3 and 4 and square window sizes 9 and 15, respectively. The threshold was set to $\delta = 0.35$. Figure 4 shows the results which were obtained in a retinal image after applying this stage, using the above-mentioned parameters.

2.5 Characterization of the Suspect Areas

In the previous stage, all the suspect areas, this is, the candidate areas to contain drusen, were identified. The goal was to get a high sensitivity despite of

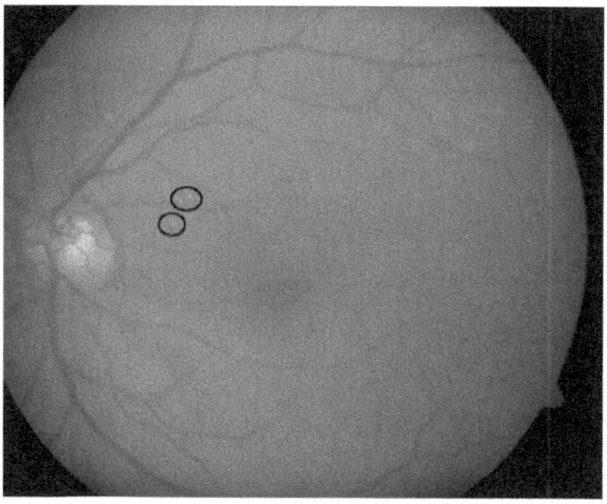

Fig. 4. Results after the detection of the suspect areas. Two drusen were detected (upper circle), which means 100% sensitivity, and one false positive was included in the set of suspect areas (lower circle).

the number of false positives. In this stage, the areas previously detected are analyzed to determine if they are drusen. This way, the number of false positives is reduced.

This stage has two important steps: the segmentation of the suspect areas, to achieve a good fit of the candidate regions, and the region filtering, to reduce the number of false positives.

The goal of the segmentation process is to distinguish the different regions the suspect areas contain. In order to achieve a good fit of the candidate areas, the technique used is region growing [11]. This technique involves three steps: the selection of the center of mass or seed associated with each region, the definition of a criterion to include a pixel in a region and, finally, the creation of a stopping criterion to finish the segmentation.

In our case, the seed for each suspect area is the point of maximum correlation for each region:

$$\forall R_i, S_i = p_j / corr(p_j) = max\{corr(p_k), \forall p_k \in R_i\}, i = 1 \ldots N \; . \tag{1}$$

where R_i is the ith region of the N suspect areas of this stage, S_i is the seed of the R_i region and p_j is the jth pixel of the R_i which correlation value is the maximum of this region.

Moreover, a pixel is added to a region if it exceeds a threshold ϑ and is neighbor of another pixel that belongs to that region. Since lighting is not constant throughout the retina this threshold is computed for each suspect area using the next equation:

$$\vartheta(x,y) = I_{bg}(x,y) - \alpha(I_{bg}(x,y) - I(x,y)) \; . \tag{2}$$

where I is the input image, I_{bg} is the input image after applying a median filter and α is a weighting variable, with values between 0 and 1. In this work $\alpha = 0.6$.

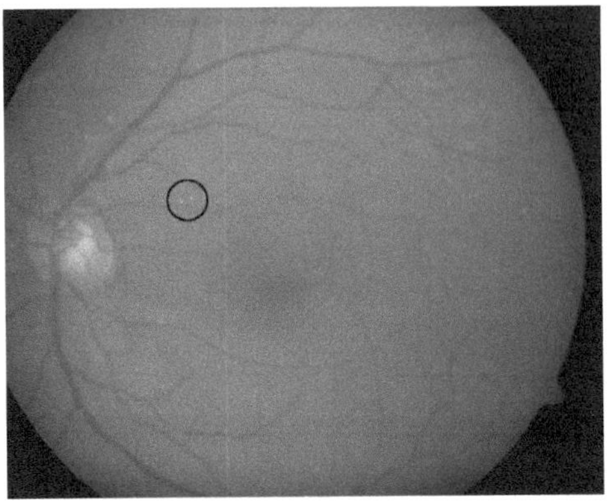

Fig. 5. Results after characterizing the suspect areas: two drusen detected (inside the circle), which means 100% sensitiviy, and no false positive

The process finishes when no more pixels can be added to any existing region or if the region exceeds the maximum size $\zeta = 150$ pixels.

After the segmentation process, we have a vector which contains all the candidate areas. This vector is processed to analyze the candidate areas and do the region filtering process. In this case, four properties were studied to reduce the number of false positives: size, circularity, intensity and correlation mean. The first two do not work because the segmented structures are very tiny. Also, the third one does not work due to the high variability in the tonality of the images. This way, the correlation mean is analyzed in order to rule out false lesions from the suspect areas.

The idea is to create a correlation mean filter to eliminate the candidates which pixels do not show continuity with respect to their correlation value. The average of the correlation values of the pixels in each region is computed as follows:

$$\forall R_i, \nu(R_i) = \frac{1}{m}\sum_{j=1}^{m} corr(p_j), p_j \in R_i, i = 1 \ldots N \ . \qquad (3)$$

where R_i is the ith region of the N segmented areas, $\nu(R_i)$ is the correlation average value of R_i and p_j is the jth pixel of the m pixels belonging to this group.

Then, the candidates which average value does not exceed a threshold ϱ are eliminated. We have set $\varrho = 0.35$ after several experiments with four test images. Figure 5 shows the results which were obtained in the same retinal image than in the previous stage, after applying the segmentation process and the region filtering process, using the above-mentioned parameters.

3 Results

The proposed methodology was tested with a set of 1280×1024 images in PPM format obtained with a FF 450^{plus} *Fundus Camera*. Two different experiments have been used to prove the accuracy of the proposed methodology.

The first experiment consists of two test benchs. The first bench has four images with 11 drusen in initial stages marked by ophtalmologists and the second one has five healthy retinal images. We had obtained a sensitivity of 82% whereas the number of false positives is close to 0 (see Table 1).

Table 1. Final results for the proposed methodology in the first experiment

Bench 1					Bench 2
TP	FN	FP	Average FP	Sensitivity	Average FP
9	2	0	0	82%	0.6

The second experiment arises due to the problem to evaluate the method, because of the shortage of images showing drusen in incipient stages. This experiment consists of two test benchs too. The first bench has four images with 11 drusen in initial stages and 13 drusen added artificially. The second bench has five healthy retinal images with 16 drusen added artificially. The method to add drusen artificially is based on inserting real drusen in retinal images by means of a cloning process. The same process is used by specialists. We had obtained results similar to the previous experiment (see Table 2).

Table 2. Final results for the proposed methodology in the second experiment

Bench 1			
Artificial drusen	Real drusen	Average FP	Sensitivity
13	11	0	83%

Bench 2			
Artificial drusen	Real drusen	Average FP	Sensitivity
16	0	0.6	87%

These results can not be compared with previous work since there is no previous work devoted to detect drusen in initial stages. Also, it is too difficult to do a thorough testing process because of the shortage of images. Anyway, we have got a high sensitivity, more than 82%, and the number of false positives is practically zero.

4 Conclusions and Future Research

In this work a method for the detection of drusen in initial stages has been presented, to support ophthalmologists in the prevention of the AMD. This method does a first detection of suspect areas and a later classification of them. We have developed a system that is able to automatically detect drusen in retinal images with a high sensitivity (over 80%) and without hardly detecting false lesions. The proposed system could be integrated into a screening system to prevent the AMD.

This system could be improved in several ways. First, new drusen properties could be used to create new filters to eliminate spurious injuries. Furthermore, we could use aditional information by means of other kind of images, such as OCT images. It would be very important to create a database with images containing drusen marked by specialists in order to do a more exhaustive testing process.

References

1. Age-Related Macular Degeneration (AMD), National Eye Institute, `http://www.nei.nih.gov/health/maculardegen/index.asp`
2. Sbeh, Z.B., Cohen, L.D., Mimoun, G., Coscas, G., Soubrane, G.: An adaptive contrast method for segmentation of drusen. In: ICIP 1997: Proceedings of the 1997 International Conference on Image Processing (ICIP 1997), vol. 3, 1, p. 255. IEEE Computer Society, Washington (1997)
3. Rapantzikos, K., Zervakis, M.: Nonlinear enhancement and segmentation Algorithm for the Detection of Age-related Macular Degeneration (AMD) in Human Eye's Retina. In: Proceedings of ICIP 2001, Thessaloniki, Greece (October 2001)
4. Brandon, L., Hoover, A.: Drusen Detection in a Retinal Image Using Multi-level Analysis. In: Ellis, R.E., Peters, T.M. (eds.) MICCAI 2003. LNCS, vol. 2878, pp. 618–625. Springer, Heidelberg (2003)
5. Mora, A., Vieira, P., Fonseca, J.: Drusen Deposits on Retina Images: Detection and Modeling. In: MEDSIP 2004, Malta (2004)
6. Garg, S., Sivasway, J., Joshi, G.D.: Automatic Drusen Detection from Colour Retinal Images. In: Proc. of Indian Conference on Medical Informatics and Telemedicine (ICMIT), Kharagpur, pp. 84–88 (2006)
7. Niemeijer, M., van Ginneken, B., Russel, S., Suttorp-Schulten, M., Abramoff, M.: Automated detection and differentiation of Drusen, exudates, and cotton-wool spots in digital color fundus photographs for early diagnosis of Diabetic Retinopathy. Investigative Ophthalmology & Visual Science 48, 2260–2267 (2007)
8. Early Treatment Diabetic Retinopathy Study (ETDRS), National Eye Institute, `http://www.nei.nih.gov/neitrials/viewStudyWeb.aspx?id=53`
9. Mariño, C., Pena, S., Penedo, M.G., Rouco, J., Barja, J.M.: Macula precise localization using digital retinal angiographies. In: ICCOMP 2007: Proceedings of the 11th WSEAS International Conference on Computers, Stevens Point, Wisconsin, USA, pp. 601–607. World Scientific and Engineering Academy and Society (WSEAS), Singapore (2007)
10. Russ, J.C.: The image processing handbook, 3rd edn. CRC Press, Inc., Boca Raton (1999)
11. González, R., Woods, R.: Digital image processing (1992)

A Study of Extracting Knowledge from Guideline Documents

M. Taboada[1], M. Meizoso[1], D. Martínez[2], and S. Tellado[2]

[1] Dpto. de Electrónica e Computación,
Universidad de Santiago de Compostela, 15782 Santiago de Compostela, Spain
maria.taboada@usc.es, maria.meizoso@usc.es
[2] Dpto. de Física Aplicada,
Universidad de Santiago de Compostela, 27002 Lugo, Spain
diego.martinez@usc.es, serafin.tellado@usc.es

Abstract. Guideline documents offer a rich repository of information on clinical decisions, actions and prescriptions. However, clinicians do not use them as much as expected since health care organisations started to develop them. One alternative to promote the use of guidelines is to automatically select the relevant information at the point of care. But, extracting knowledge from a guideline document is an arduous and complex task. In this paper, we propose to apply the methodology CommonKADS in the analysis phase of a clinical practice guideline, with the aim of systematizing knowledge acquisition, providing a methodological support that helps to detect and document all the transformations from natural language to the structured representation of a knowledge model. When forcing to the knowledge engineer to keep these transformations, the knowledge modelling becomes more gradual.

Keywords: CommonKADs methodology, clinical practice guideline, knowledge modelling, ontologies.

1 Introduction

Guideline documents offer a rich repository of information on clinical decisions, actions and prescriptions. So, they can play an important role in the daily practice of medicine [1]. However, clinicians do not use them as much as expected since health care organisations started to develop them. The obstacles found by clinicians when they try to access relevant information also paralyses the dissemination of the information included in guidelines. There is an overload of medical information and clinicians do not have the required time to select the relevant information at the point of care. One alternative to promote the use of guidelines is to implement these as decision support systems generating medical advice and monitoring the clinician actions [2]. With this goal in mind, many Knowledge Engineering methodologies and representation languages have been proposed and successfully applied to the medical domain. CommonKADS [3] is an example of a methodology oriented to develop, in a structured way, knowledge-intensive systems. Asbru, EON, GLIF, Guide, Prodigy and PROForma, among others, are well-known examples of languages proposed to formally represent clinical practice guidelines [2].

R. Moreno-Díaz et al. (Eds.): EUROCAST 2009, LNCS 5717, pp. 195–202, 2009.

The codification of a clinical guideline directly in these representation languages is an arduous and complex task, and the final resulting model is illegible for medical specialists, making difficult its validation. To overcome these drawbacks, at least two types of solutions have emerged. The first one involves to describe guidelines in a higher level of abstraction. For example, Clercq et al. [4] have proposed to model guidelines by combining typical components of a knowledge model (such as, ontologies and problem-solving methods) with guideline representation primitives. Another approach was presented by Vollebregt et al. [5]. They carried out a study on re-engineering a realistic system comparing the general purpose AI methodology CommonKADS [3] and the special purpose methodology PROforma [6].

The second solution can be found in more recent works, focused on reducing the existing gap between natural language and formal guideline representations. In [7,8], the authors focus on keeping traces of the transformation process from natural language to formal representation, with the aim of facilitating validation and maintenance.

In this paper, we propose to apply CommonKADS in the analysis phase, with the aim of systematizing knowledge acquisition, providing a methodological support that helps to detect and document all the transformations from natural language to the structured representation of a knowledge model. When forcing to the knowledge engineer to keep these transformations, the knowledge modelling becomes more gradual.

The structure of the article is the following one. We begin presenting a case-study and the main stages in the construction of a knowledge model. We detail the set of required activities for each stage, including examples of the actual case-study we performed. In section 3, we discuss the results of our work. Finally, we present the related work in section 4 and the conclusions in section 5.

2 An Approach to Model Knowledge from Guideline Documents

This section presents an approach to model knowledge from guideline documents in the domain of medical diagnosis. It is an extension of the construction process of CommonKADS, which allows knowledge engineer to keep the transformation paths during knowledge modelling. All examples we will use in this section come from our case-study: the knowledge modelling on medical diagnosis in Conjunctivitis from the textual documentation provided by a clinical practice guideline[1].

Our approach includes three main stages in the construction of a knowledge model: *1. Preprocessing steps*, where all the pieces that will be handled during the modelling process are specified; 2. *Knowledge components recognition*, where explicit correspondences between textual knowledge and the different types of knowledge pieces are set, at the same time as the knowledge model is constructed; *3. Validate knowledge*, where gaps in the guideline are detected.

2.1 Preprocessing Stage

In this stage, all the pieces that will be handled during the modelling process are specified. This stage includes three activities.

[1] http://www.aao.org/aao/education/library/ppp

Identify External Resources for Entity Recognition. This activity involves the identification and selection of the external resources that can improve the entity recognition stage: the Unified Medical Language System[2] (UMLS), one of the most important publicly available resources in the biomedical domain; the Semantic Network, a small hierarchy of semantic concepts used to classify the concepts in the UMLS; and the OpenNLP[3], a natural language processing tool to parse the biomedical entities.

Identify Types of Knowledge. This activity identified the following types of knowledge:

- *Noun phrases representing medical entities*. Examples are *diagnosis*, *cause*, *visual function* or *discharge*.
- *Vague* expressions that reflect the state of medical entities that can be measured or quantified. For example, an expression like *decrease of the visual acuity* or *Rapid development of severe hyperpurulent conjunctivitis*.
- *Generic linguistics expressions*, including
  - Verbs that describe structures or parts. For example, in the expression *The patient population includes individuals of all ages who present with symptoms suggestive of conjunctivitis, such as red eye or discharge*, the verb *include* refers to the set of patients, which the clinical practice guideline aims at.
  - Verbs that describe medical actions. For example, the set of goals in the guideline include clinical actions, such as *Establish the diagnosis of conjunctivitis* or *Establish appropriate therapy*.
  - Decisions or causalities. An example is the expression *Questions about the following elements of the patient history may elicit helpful information: ...*, that expresses an indication on what information to compile during the interrogation of the patient.

Establish Correspondences between Textual Knowledge. This activity involves the revision of documents on the part of clinical experts with the purpose of grouping and relating noncontiguous textual portions that make reference to the same knowledge. We have used the DELTA tool[4], as it provides an easy way to establish links between textual portions and create an XML file with these links.

2.2 Knowledge Component Recognition

In this stage, the correspondences between textual knowledge and the different types of knowledge components are set, at the same time the knowledge components are recognized.

Transform Textual Terms into Standard Terms. Noun phrases included in texts must be extracted and replaced by standard terms. The entities representing symptoms, signs, procedures, diseases, etc. are extracted by implementing an NLP-based entity recognition algorithm, which uses the OpenNLP and the Metathesaurus as external resources.

---
[2] http://http://www.nlm.nih.gov/research/umls/

[3] http://opennlp.sourceforge.net/

[4] http://www.ifs.tuwien.ac.at/~votruba/DELTA/

The tool OpenNLP annotates each NP with part-of-speech tags. The algorithm works in the following phases:

1. Identify noun phrases (NPs) in the selected parts of the guideline document. This syntactic analysis relies on the OpenNLP tools.
2. Extract the NPs from tokenized textual portions.
3. Map each NP to one or more UMLS concepts.
4. For each NP that does not occur in the Metathesaurus, rebuild the NP by removing an adjacent word and go to 3 until the NP only contains the head name.

In Table 1, we show an example of three medical entities identified in the document. In some cases, only one medical entity is recovered from the Metathesaurus. For example, *photophobia* and *purulent conjunctivitis* map to the Metathesaurus through a single match, *Photophobia* and *Bacterial Conjunctivitis*, respectively. However, many NPs map to two or more medical entities, leading to ambiguous mappings (*diagnosis* in Table 1). In addition, some NPs correspond to medical entities, but the complete NP found in the text has more precision than the corresponding concept included in the Metathesaurus. For example, *recurrent purulent conjunctivitis* cannot be mapped directly to a Metathesaurus concept. However, *purulent conjunctivitis* is mapped to *Bacterial conjunctivitis*. Our algorithm manages this situation by firstly identifying *recurrent* as the concept *Recurrence* and secondly, searching the core-block *purulent conjunctivitis*.

Table 1. Examples of medical entities extracted from the document

Textual Term	**Type of Match**	**Cardinality of Match**	**Medical Entity**	**Semantic Category**
Diagnosis	Complete	MULTIPLE	-Diagnosis -Diagnosis Classification -Diagnosis Aspect -Diagnosis Study	Health Care Activity Classification Qualitative Concept Research Activity
Photophobia	Complete	SINGLE	Photophobia	Sign or Symptom
[Recurrent] Purulent Conjunctivitis	Partial	SINGLE	Bacterial Conjunctivitis	Disease or Syndrome

Transform Standard Terms into Knowledge Components. The extracted concepts from the Metathesaurus can have one or more assigned semantic types, which provides a track about the knowledge component that the concept is referred to. For example, *Diagnosis (CUI: C0011900)* is a *Diagnostic Procedure* and *Therapeutic procedure (CUI: C0087111)* is a *Therapeutic and preventive procedure*. So, they can be modeled as stereotypical tasks. On the other hand, *Discharge from eye (CUI: C0423006)* is a *Sign or Symptom* and *Duration (CUI: C0449238)* is a *Temporal Concept*. So, they correspond to domain concepts.

Transform Generic Linguistic Expressions into Knowledge Components. Three types of linguistic expressions are identified and transformed:

- Verbs describing structures or parts are transformed into domain standard concepts or relations. For example, in the expression *The initial eye examination includes measurement of visual acuity, external examination, and slit-lamp biomicroscopy*, the verb *include* describes the parts that an ocular exploration must consist of. It is, therefore, a relation among components of the domain ontology.
- Verbs describing actions are transformed into PSMs, inferences or transfer functions. For example, in the expression describing the goal *Establish the diagnosis of conjunctivitis, differentiating it from other causes of red eye*, the noun *diagnosis* indicates a stereotypical task.
- Decisions or causalities (*if ..., but ..., it would have to be ..*, etc.) are transformed into tasks or PSMs. For example, in the description *The external examination should include the following elements: ...*, the verb *include* describes the parts that, in ideal conditions of work, an external ocular exploration must consist of. It is, therefore, a relation among components of the domain ontology. But, the clarification *should include* also indicates that the provided list is too exhaustive and the doctor must decide what elements are the most important for each patient. So, the selection of the parts will be dynamic and depend on the particular data of the patient. Therefore, this expression matches an assessment task, consisting of collecting additional data, based on the current diagnosis hypotheses as well as the costs implied in the external examination.

2.3 Validate Knowledge

This stage includes two activities:

1. *Validate*, as much as possible, the carried out *knowledge transformations*. An important technique is to create test cases of real scenes of the medical diagnosis task and simulate them.
2. *Detect knowledge gaps*. From the validation stage, we will be able to detect whether the knowledge model is complete. In our example, we have found that the clinical practice guideline includes knowledge on some dimensions characterizing the conjunctivitis diagnosis, such as differential, etiologic and multiple fault diagnosis. Nevertheless, it does not contemplate specific knowledge on the PSMs to be used in:
   - Generation of diagnostic hypotheses from the patient symptoms.
   - Decision of how to carry out the ocular exploration.
   - Evaluation of diagnostic hypotheses, in order to confirm or reject them.

3 Results

A CommonKADS knowledge model consists of three layers of knowledge: 1) *domain knowledge* describing concepts, attributes, relationships, etc. in the application domain; 2) *inference knowledge* representing the reasoning steps and the role the domain knowledge plays in the reasoning process; 3) *task knowledge* describing the goals to be achieved with the use of the knowledge and the strategies to be employed for realizing the goals. To model a guideline in CommonKADS, firstly, we identified the three layers of knowledge. The clinical experts collaborating in the project revised the document and grouped portions of text in two initial parts: general objectives and knowledge about the disease model and treatment. The general objectives are mainly located in the *orientation* part of the guideline (*purpose* and *goals*) and the disease and treatment knowledge is located in tables. So, we assumed the rest of the text in the guideline would bridge the gap between general goals and the specific domain knowledge. However, in many cases, the guideline does not specify the strategies to be employed for realizing the goals with the provided domain knowledge. For example, the guideline does not include information on how the diagnostic hypotheses should be generated from the patient symptoms or how the ocular exploration should be carried out. As a result, we will need to acquire this knowledge from the experts in a future.

Once we had divided the guideline text into two parts, firstly we focused in modelling general objectives. The guideline describes these by generic linguistic expressions including verbs, such as *Preserve visual function* or *Establish of diagnosis of Conjunctivitis*. These expressions are grouped by two flat lists in the guideline: one list with four expressions for purpose and another list with seven expressions for goals. Then, we revised the text again, looking for expressions like these in other parts of the guideline. We found out some of these expressions outside the *orientation* part of the guideline: a flat list of three expressions in the *care process* part (named *patient outcome criteria*) and another three expressions embedded in isolated paragraphs. This latter case was the most difficult to detect, as we needed to identify it into the text and extract it. Next, we transformed the set of textual terms into standard terms, using the utilities of the UMLS server. Table 2 shows some of these expressions transformed to standard terms and grouped with other text in the guideline. We can summarize the main results of using a standard thesaurus to model expressions describing goals and intentions in three aspects:

1. **The UMLS Metathesaurus includes many of the objectives described in natural language in the guideline**. Surprisingly, we found many of these expressions as standard concepts in the Metathesaurus. For example, expressions in the guideline such as *Minimize the spread of infectious disease* (third row in Table 2) or *Educate and engage the patient in the management of the disease* (second row in Table 2) could be replaced by the standard terms *Minimize opportunities for transmission of infection* and *Educate the patient*, respectively. In total, we found 12 of the 17 expressions on goals and intentions in the Metathesaurus. We detected that for only 1 of the 5 not found expressions the guideline contains some knowledge to describe it and for the remaining 4 expressions there is a complete lack of knowledge. An example is the purpose *Preserve visual function* (last row in Table 2). There is no standard term for this expression (first column in Table 2) and the guideline does not provide knowledge on how the purpose can be reached (fourth column in Table 2).
2. **The UMLS Metathesaurus provides a means of disambiguating the linguistic expressions on goals and intentions**. For example, the guideline text distinguishes between purpose and goals. But, the semantic difference between them is too small. Revising the meaning of the standard UMLS concepts, we have found that some expressions on goals and intentions were equivalent. For example, the third row

Table 2. An example showing several expressions of the guideline linked to standard terms and grouped with other chunks of text in the guideline

Standard Term	Chunk	Location	Grouped Chunks
Differential Diagnosis (Diagnostic Procedure)	"Establish the diagnosis of conjunctivitis, differentiating it from other causes of red eye"	Goals	Diagnosis (pages 8-11) Risk factors (Pages 4-5) Natural History (Pages 6-7)
Educate the patient (Educational Activity)	"Educate and engage the patient in the management" of the disease"	Goals	Prevention and early detection (Paragraph 4)
Minimize opportunities for transmission of infection (Therapeutic or Preventive procedure)	"Minimize the spread of infection disease" "Prevent the spread of .."	Purpose Goals	Prevention and early detection (paragraph 5)
Early Diagnosis (Diagnostic Procedure)	"Early detection of conjunctivitis"	Embedded expression (page 7)	Prevention and early detection (paragraphs 2-4)
	"Preserve visual function" "Restoring or maintaning normal visual function"	Purpose Patient outcome criteria	

in Table 2 shows an example of a goal and a purpose that were linked to the same standard concept. In addition, all these expressions correspond to standard concepts expressing procedures of health care activity. So, they were modelled as stereotypical tasks in CommonKADS.

3. **The UMLS Semantic Network provides a structured organization of the objectives described by flat lists in the guideline**. Initially, we did really think that the knowledge in the guideline was well-organized and that was not the case. The use of a standard terminology has been very useful in this analysis-phase to organize the objectives of the guideline.

Each task groups and relates noncontiguous chunks. For this purpose, we used the DELTA tool, which allowed us to structure chunks. However, once the text was well-structured using DELTA links, medical experts detected the need of revising the new structure of guideline but in natural language. This option, the generation of the new organization of the text, is not provided automatically by DELTA. We think that this facility could enhance the process of authoring guidelines.

On the other hand, modelling the inference and the domain layer was more difficult. In the first case, the guideline text does not provide knowledge enough, so we will need to acquire it in a future in order to complete the model. In the second case, CommonKADS does not provide primitives to model temporal constraints, so we will need to follow some language to formalize these parts of knowledge.

4 Conclusions

In this paper, we have applied CommonKADS in the analysis phase of a clinical practice guideline, with the aim of systematizing knowledge acquisition. Our proposal extends the knowledge model construction of CommonKADS, as it provides a methodological support that helps to detect and document all the transformations from natural language to the structured representation of a knowledge model. When forcing to the knowledge engineer to keep these transformations, the knowledge modelling becomes more gradual. In addition, we have provided a limited number of transformations in each stage, so from the validation stage, we are able to detect what parts of the knowledge model are complete and what are missing in the guideline.

Acknowledgements. This work has been funded by the Ministerio de Educación y Ciencia, through the national research project HYGIA (TIN2006-15453-C04-02).

References

1. Grimshaw, J.M., Russel, I.T.: Effects of clinical guidelines on medical practice: a systematic review of rigorous evaluation. Lancet 342, 1317–1322 (1993)
2. de Clercq, P.A., Blom, J.A., Korsten, H.H.M., Hasman, A.: Approaches for creating computer-interpretable guidelines that facilitate decision support. Artificial Intelligence in Medicine 31(1), 1–27 (2004)
3. Schreiber, G., Akkermans, H., Anjewierden, A., de Hoog, R., Shadbolt, N., Van de Velde, W., Wielinga, W.: Knowledge Engineering and Management. In: The CommonKADS Methodology. MIT Press, Cambridge (1999)
4. de Clercq, P., Hasman, A., Blom, J., Korsten, H.: The application of ontologies for the development of shareable guidelines. Artificial Intelligence in Medicine 22, 1–22 (2001)
5. Vollebregt, A., ten Teije, A., van Harmelen, F., van der Lei, J., Mosseveld, M.: A study of PROforma, a development methodology for clinical procedures. Artificial Intelligence in Medicine 17, 195–221 (1999)
6. Sutton, D., Fox, J.: The Syntax and Semantics of the PROforma guideline modelling language. J. Am. Med. Inform. Assoc. 10(5), 433–443 (2003)
7. Svatek, V., Ruzicka, M.: Step-by-step formalisation of medical guideline content. International Journal of Medical Informatics 70(2-3), 329–335 (2003)
8. Votruba, P., Miksch, S., Kosara, R.: Proc. of MEDINFO 2004. In: Fieschi, et al. (eds.) Facilitating Knowledge Maintenance of Clinica Guidelines and Protocols. IOS Press, Amsterdam (2004)

Modelling Differential Structures in Proof Assistants: The Graded Case*

Jesús Aransay and César Domínguez

Departamento de Matemáticas y Computación, Universidad de La Rioja
Edificio Vives, Luis de Ulloa s/n, E-26004 Logroño, La Rioja, Spain
{jesus-maria.aransay,cesar.dominguez}@unirioja.es

Abstract. In this work we propose a representation of graded algebraic structures and morphisms over them appearing in the field of Homological Algebra in the proof assistants Isabelle and Coq. We provide particular instances of these representations in both systems showing the correctness of the representation. Moreover the adequacy of such representations is illustrated by developing a formal proof of the Trivial Perturbation Lemma in both systems.

1 Introduction

Sergeraert's ideas on effective homology gave rise to the symbolic computation system Kenzo [7]. This system has produced remarkable results in the field of Homological Algebra, some of them previously unknown. A formal study of this system was simultaneously proposed, from which some results have already been obtained regarding the algebraic specification of the structures involved in the computations [9,5,6]. A formal study of certain crucial algorithms of the Kenzo system with theorem proving assistants was also tackled. One of these algorithms is a result named *Basic Perturbation Lemma*. A formal proof of this result in the proof assistant Isabelle [12] in the case of non-graded structures was presented in [1]. The result for graded structures can be seen as a generalization of the non-graded version. In that paper, the problem of the implementation of the degree was omitted since the proof itself does not depend on any relevant property of such structures. Coquand and Spiwack are using a type theory approach to formalize that result [4].

As a continuation of the previous work [1], we consider in this paper the representation of graded structures in the proof assistants Isabelle and Coq [11]. The problem of choosing appropriate representations of algebraic structures before formalizing proofs is well-known, and it has provided ideas to enhance proof assistants with additional tools (for instance, see [8]). A graded structure can be thought of as a family of structures indexed by the integers. Accordingly, a graded structure morphism between two graded structures will be a family of morphisms between algebraic structures belonging to each graded structure.

* This work has been partially supported by the Spanish Government, project MTM2006-06513.

R. Moreno-Díaz et al. (Eds.): EUROCAST 2009, LNCS 5717, pp. 203–210, 2009.

The algebraic structures appearing in the Basic Perturbation Lemma are chain complexes, perturbations and reductions. Roughly speaking, a *chain complex* is defined as a graded module together with a graded module endomorphism (called differential), where the differential satisfies a nilpotency condition. A *perturbation* over a chain complex is a module endomorphism of that chain complex, which produces a chain complex when added to the original differential. A *reduction* is a triple of morphisms between a pair of chain complexes (usually called top and bottom) satisfying some special requirements. Reductions are relevant since they preserve the homology of chain complexes. The Basic Perturbation Lemma states that given a reduction and a perturbation of the top chain complex, a new reduction can be obtained. This reduction can be algorithmically calculated and thus a way to compute the homology of the top chain complex is provided. In a similar way, the *Trivial Perturbation Lemma* builds a new reduction from a reduction and a perturbation of the bottom chain complex.

A working representation of the previous concepts (*i.e.*, graded structures and graded structure morphisms) in a proof assistant has to be sound, but also needs to be useful. The first feature is shown by providing instances of the representations. The second feature can be shown by formally proving some results with them. In this work, we propose a formalization of these graded structures in the proof assistants Isabelle and Coq, provide instances of them and formally prove the Trivial Perturbation Lemma. The representations in both proof assistants have subtleties that depend on the underlying logic and the previous built-in definitions of the systems. They allow for a comparison between these systems. Some other comparisons of proof assistants are also worth noting [13,2], although they focus on problems different from the one presented here.

The paper is organized as follows. In Section 2, we introduce the Homological Algebra definitions required to state the Trivial Perturbation Lemma. In Sections 3 and 4, we present a formalization of the previous concepts in Isabelle and Coq. Finally, in Section 5, we briefly compare both representations and explore some further applications.

2 Mathematical Definitions

The following definitions have been obtained from [10].

Definition 1. *Given a ring R, a* graded module *M is a family of left R-modules $(M_n)_{n \in \mathbb{Z}}$.*

Definition 2. *Given a pair of graded modules M and M', a* graded module morphism *f of degree k between them is a family of module morphisms $(f_n)_{n \in \mathbb{Z}}$ such that $f_n \colon M_n \to M'_{n+k}$ for all $n \in \mathbb{Z}$.*

Definition 3. *Given a graded module M, a* differential *$(d_n)_{n \in \mathbb{Z}}$ is a family of module endomorphisms of M of degree -1 such that $d_{n-1} \circ d_n = 0_{\operatorname{Hom} M_n\, M_{n-2}}$ for all $n \in \mathbb{Z}$.*

From the previous definitions, the notion of chain complex can be stated as follows.

Definition 4. *A* chain complex *is a family of pairs* $(M_n, d_n)_{n\in\mathbb{Z}}$ *where* $(M_n)_{n\in\mathbb{Z}}$ *is a graded module and* $(d_n)_{n\in\mathbb{Z}}$ *is a differential.*

Definition 5. *Given a pair of chain complexes* $(M_n, d_n)_{n\in\mathbb{Z}}$ *and* $(M'_n, d'_n)_{n\in\mathbb{Z}}$*, a* chain complex morphism *between them is a family of module morphisms* $(f_n)_{n\in\mathbb{Z}}$ *of degree* 0 *between* $(M_n)_{n\in\mathbb{Z}}$ *and* $(M'_n)_{n\in\mathbb{Z}}$*, such that* $d'_n \circ f_n = f_{n-1} \circ d_n$ *for all* n *in* $\mathbb{Z}$.

Based on the previous definitions, the notions of reduction and perturbation can be introduced.

Definition 6. *Given a pair of chain complexes* $M = (M_n, d_n)_{n\in\mathbb{Z}}$ *and* $M' = (M'_n, d'_n)_{n\in\mathbb{Z}}$*, a* reduction *from* M *to* M' *is a triple of morphisms* (f, g, h) *where* f *is a chain complex morphism from* M *to* M'*,* g *is a chain complex morphism from* M' *to* M *and* h *is a module endomorphism of degree* $+1$ *of* M *(called* homotopy operator*), which satisfy the following properties (for all* n *in* $\mathbb{Z}$*):*

(1) $f_n \circ g_n = \mathrm{id}_{M'_n}$ *(2)* $f_{n+1} \circ h_n = 0_{\mathrm{Hom}\, M_n\, M'_{n+1}}$
(3) $h_n \circ g_n = 0_{\mathrm{Hom}\, M'_n\, M_{n+1}}$ *(4)* $h_{n+1} \circ h_n = 0_{\mathrm{Hom}\, M_n\, M_{n+2}}$
(5) $g_n \circ f_n + d_{n+1} \circ h_n + h_{n-1} \circ d_n = \mathrm{id}_{M_n}$

Definition 7. *Given a chain complex* $(M_n, d_n)_{n\in\mathbb{Z}}$*, a* perturbation *is a module endomorphism* δ *of degree* -1 *of* $(M_n)_{n\in\mathbb{Z}}$ *such that* $(M_n, d_n + \delta_n)_{n\in\mathbb{Z}}$ *is a chain complex (i.e.,* $(d_n + \delta_n)_{n\in\mathbb{Z}}$ *is also a differential).*

Finally, the Trivial Perturbation Lemma is stated as follows.

Theorem 1. (Trivial Perturbation Lemma) *Given a pair of chain complexes* $(M_n, d_n)_{n\in\mathbb{Z}}$ *and* $(M'_n, d'_n)_{n\in\mathbb{Z}}$*, a reduction* (f, g, h) *from* $(M_n, d_n)_{n\in\mathbb{Z}}$ *to* $(M'_n, d'_n)_{n\in\mathbb{Z}}$*, and a perturbation* δ' *of* $(M'_n, d'_n)_{n\in\mathbb{Z}}$*, then a new reduction from* $(M_n, d_n + g_{n-1} \circ \delta'_n \circ f_n)_{n\in\mathbb{Z}}$ *to* $(M'_n, d'_n + \delta'_n)_{n\in\mathbb{Z}}$ *is given by means of* (f, g, h)*.*

The previous result differs from the Basic Perturbation Lemma because that result demands a perturbation of the top chain complex (*i.e.*, $(M_n, d_n)_{n\in\mathbb{Z}}$) that satisfies an additional requirement, called local nilpotency condition, and also because the output reduction contains a pointwise defined (finite) series.

3 Implementation in Isabelle/HOL

Isabelle [12] is a generic proof assistant with respect to the logics that can be implemented on top of it. Its meta-logic is based on Church's simple type theory. From this variety of logics, HOL (Higher-Order Logic) is widely used due to its simplicity and expressive power[1]. Type variables (`'a`) and constructors for total

[1] In the following we will refer to Isabelle/HOL as simply Isabelle.

functions (⇒) and *pairs* (×) are the main tools used to represent the definitions in Section 2. In this work, we use the definitions available in the standard Isabelle distribution up to rings and modules over them. These structures are defined by means of record types (record types are a reformulation of *pair* types where fields are labeled).

Definitions in Isabelle are usually introduced by means of a *type definition* together with a *proposition* describing the properties of the objects defined. Thus, according to Definition 1, a representation of graded modules can be given in Isabelle by means of the following definition:

```
definition
graded-R-module :: ('a ring) ⇒ (int ⇒ ('a, 'b) module) ⇒ bool
 where graded-R-module R f = ∀n::int. module R (f n)
```

It should be noted that `module` denotes a record type constructor (in which the different fields represent the operations of the algebraic structure and also an explicit representation of the domain in an additional field named `carrier`) in the type definition, as well as a predicate stating the usual properties of such algebraic structure (closedness of binary operations, associativity, commutativity, and so on) in the proposition.

It is also relevant to observe that in the previous definition the modules belonging to the graded structure being defined share a common underlying type (`'b`); dependent types would allow us to define a different type for each integer number, but they are not available in the HOL type system.

As a consequence of this limitation, we can consider the following question. Should it be possible to write down expressions such as $x_n +_{M_n} x_{n+1}$, with $x_n \in M_n$ and $x_{n+1} \in M_{n+1}$? In our proposed representation, the answer is yes, since every module shares a common underlying type. Thus, it remains up to the user to ensure the correctness of the definitions and expressions provided to the system (type checking will not directly reject the previous expression).

From the previous definition and the one of module morphism (given by a predicate `hom-module`), we define now graded group morphisms (see Definition 2) between graded groups as follows:

```
definition
graded-group-hom :: ('a ring) ⇒
  (int ⇒ ('a, 'b) module) ⇒
  (int ⇒ ('a, 'd) module) ⇒ int ⇒
  (int  ⇒ ('b ⇒ 'd)) set
 where graded-group-hom R M M' k =
 {h. (∀n::int. (h n) ∈ hom-module R (M n) (M'(n + k)))}
```

The type definition corresponds again to a family of module morphisms *indexed by the integer numbers*, each of them of degree k. There is one subtlety in the previous definition of morphisms. The definition of `hom-module` has to satisfy the following predicate:

```
definition
completion :: 'a ring ⇒ 'b ring ⇒ ('a ⇒ 'b) set
 where completion M M' = {f. ∀x. x ∉ carrier M⟶f x = zero M'}
```

This property maps every point out of a function domain of definition (represented by the `carrier` of the source ring) to a distinguished point (in this case, the additive unit of the target ring, but another distinguished point could be used). Thanks to this property, distributivity of addition *w.r.t.* composition of morphisms can be later proved, which would not be provable with a generic representation.

Definitions of chain complexes (Definition 4), reductions (Definition 6) and perturbations (Definition 7) are produced from the previous ones. Concrete instances of such algebraic structures are also defined and proved in the system.

With the previous definitions, the statement of the TPL (Theorem 1) is now:

```
lemma TPL
 assumes reduction R M diff M' diff' f g h
 and δ ∈ perturbation R M' diff'
 shows reduction R M (diff⊕R M M -1(g⊙-1(δ⊙0f)))
                     M'(diff'⊕R M' M' -1δ)
                     f g h
```

In the previous statement, it can be seen how each operation over graded module morphisms must contain explicit information on the degree of the morphisms being operated. For instance, when composing δ and f, and being f of degree 0, the notation $\delta \odot_0 f$ corresponds with $\delta_n \circ f_n$ for each n in $\mathbb{Z}$, whereas $g \odot_{-1} (\delta \odot_0 f)$ represents $g_{n-1} \circ (\delta_n \circ f_n)$. Accordingly, the addition of graded morphisms must be defined over the target module. Dependent types, also in this case, would help to avoid the need of explicitly passing as a parameter information on the degree of morphisms being operated.

The result can be formally proven in Isabelle by applying mainly equational reasoning over morphisms, following closely the style of a *paper and pencil* proof. An arbitrary degree is fixed, and then proofs are carried out over module morphisms of such degree. The main properties used are distributivity laws, and the ones in the definitions of reduction, perturbation and differential.

4 Implementation in Coq

Coq [11] is a proof assistant based on a very expressive variation of typed λ-calculus called Calculus of Inductive Constructions [3]. Although the standard library of Coq does not include the basic algebraic structures, different representations of them can be found in the literature. For instance, L. Pottier's Coq development published in the users' contributions in [11] that includes modules (over a ring) and module morphisms. They consist on records called `Module` and `Module_hom`, respectively. Besides, further constructions such as, for instance, the addition or the composition of module morphisms are defined and represented using the infix notation `[+h]` or `[oh]`, respectively (see [11] for a detailed description).

Using the previous development, a graded R-module, where R is a given ring, can be represented in Coq through a function type `GradedModule:= Z -> Module R` (where Z is the type for integers). The ring could be parameterized in the construction (and indeed, this will be done by the section mechanism of Coq). Then, given a graded module `GM:GradedModule R`, a graded module endomorphism of degree −1 is implemented with the *dependent type* `GrdMod_hom_1:= forall i:Z, Module_hom(GM i)(GM(i-1))`. It is worth mentioning that using dependent types we obtain that every module in the family has a different type (depending on their degree). Now, using the following definition of the nilpotence property, `Definition Nilpotence(g:Module_hom B C)(f:Module_hom A B):= forall a:A, ((g[oh]f) a)[=]Zero`, the chain complex structure can be defined as a record:

```
Record ChainComplex: Type:=
 {GrdMod:> GradedModule R;
  Diff: GrdMod_hom_1 GrdMod;
  NilpotenceDiff: forall i:Z, (Nilpotence (Diff (i-1))(Diff i))}.
```

Some remarks on the above definition are required. First, the base of an algebraic structure is a *Setoid*, *i.e.* a set with an equality (`[=]` in this case). Second, records in Coq consist on labeled fields in which the type of one label may depend on a previously defined label. This allows to include the representation of the domain and operations in the record, as well as the properties of an algebraic structure. Finally, the `GrdMod` component is declared (by the annotation `:>`) to be a *coercion function*. This means that the type checker will insert this function over a chain complex when a graded module is required. These techniques are extensively used in the implementation of algebraic structures [11].

In a similar way, given two chain complexes `CC1 CC2:ChainComplex R`, a chain complex morphism `ChainComplex_hom` is represented as a record with a family of module morphisms `GrdMod_hom:> forall i:Z, Module_hom (CC1 i)(CC2 i)` which commutes with the differentials of the chain complexes. A homotopy operator is defined as a family of module morphisms `HomotopyOperator:= forall i:Z, Module_hom(CC1 i)(CC1(i+1))`.

With these definitions, it is possible to formalize the notions of reduction and perturbation. For instance, the notion of reduction is again a record structure:

```
Record Reduction:Type:=
 {topCC: ChainComplex R;
  bottomCC: ChainComplex R;
  f_t_b: ChainComplex_hom topCC bottomCC;
  g_b_t: ChainComplex_hom bottomCC topCC;
  h_t_t: HomotopyOperator topCC;
  rp1: forall (i:Z)(a:(bottomCC i)),((f_t_b i)[oh](g_b_t i))a[=]a;
  rp2: forall (i:Z)(a:(topCC i)),((f_t_b(i+1))[oh](h_t_t i))a[=]Zero;
  rp3: forall (i:Z)(a:(bottomCC i)),((h_t_t i)[oh](g_b_t i))a[=]Zero;
  rp4: forall (i:Z)(a:(topCC i)),((h_t_t(i+1))[oh](h_t_t i))a[=]Zero;
  rp5: homotopy_operator_prop f_t_b g_b_t h_t_t}.
```

The rp5 property contains the following homotopy_operator_prop definition.

```
Definition homotopy_operator_prop(f:ChainComplex_hom C1 C2)
  (g:ChainComplex_hom C2 C1)(h:HomotopyOperator C1):=
  forall (i:Z)(a:(C1 i)),
  ((transition_h C1 (eqpm i)[oh]((Diff C1 (i+1))[oh](h i)))[+h]
   (((transition_h C1 (eqmp i)[oh](h (i-1))[oh](Diff C1 i)))[+h]
    ((g i)[oh](f i))))a[=]a.
```

In the previous definition, a type problem appears. The composition `((Diff C1 (i+1))[oh](h i)))` has as underlying type `(C1 i) -> (C1 (i+1)-1)`, the composition `(h (i-1))[oh](Diff C1 i)` has type `(C1 i) -> (C1 (i-1)+1)`, and the composition `((g i)[oh](f i))` has type `(C1 i) -> (C1 i)`. The problem is that `(C1 (i+1)-1)` and `(C1 (i-1)+1)` are not convertible types in Coq which is required in order to make the morphism addition, and that they are either not convertible to `(C1 i)`. The `transition_h` module morphism is defined to transform a type in an equal (but not directly convertible) type. This *type transformation* is essentially an identity and allows to obtain the required type. More precisely, this module morphism is defined using the functional type `transition: forall i j:Z, eq i j -> CC1 i -> CC1 j` and the equalities `eqpm: forall i:Z, i+1-1=i` and `eqmp: forall i:Z, i-1+1=i`.

Instances of the these data structures are included. For instance, it is not difficult to build the chain complex with the integers in each degree and null morphisms as differential. Finally, the Trivial Perturbation Lemma is obtained as a definition in which, given a reduction `r: Reduction R` and a perturbation `p: Perturbation (bottomCC r)`, it is possible to build a new instance of the reduction record. In this construction, new domains and operations of the new reduction are built and the properties of the reduction structure on these new components are obtained as lemmas. For instance, the new top chain complex is defined through the differential: `new_topCC_diff:=fun i:Z => (Diff (topCC r) i)[+h](g_b_t r (i-1))[oh](p i)[oh](f_t_b r i))`.

5 Comparison of Both Approaches and Conclusions

In this paper, a representation of the graded structures appearing in the Basic Perturbation Lemma in two different proof assistants (Isabelle/HOL and Coq) has been proposed. A representation of morphisms over that graded structures has been also introduced. Additionally, instances of the representations have been provided in both systems, and a formal proof of the Trivial Perturbation Lemma has been obtained.

The formalization obtained in the proof assistants has subtleties that depend directly on the type systems, the underlying logics, the specific style and the previous built-in definitions of the systems that allow for a brief comparison between both proof assistants.

The richer type theory underlying Coq allows to build a precise specification of graded structures. It uses dependent types that allow to assign a different type to each structure in the graded structure. Thus, type checking prevents the user, for instance, from writing down expressions where operations defined on a degree operate over elements of other degrees. On the contrary, and also as a consequence of the Coq type system, it is required to include type transformations in order to work with equal but not convertible types.

The Isabelle formalization has been carried out without using dependent types. A unique type is used to represent each graded structure. Therefore, it does not require the insertion of type conversions, although, for instance, type checking cannot be used to ensure the correctness of the expressions used. Another relevant feature of the representation chosen is the explicit use of carrier sets to represent the module domain. This feature increases the expressivity of our approach, but also forces us to impose some restrictions (the use of completions) on the representation of morphisms.

Further work would be needed with both proof assistants to develop more formal proofs (as, for instance, the Basic Perturbation Lemma), enrich the hierarchy of graded structures, and also compare the possibilities of the systems.

References

1. Aransay, J., Ballarin, C., Rubio, J.: A Mechanized Proof of the Basic Perturbation Lemma. Journal of Automated Reasoning 40(4), 271–292 (2008)
2. Aransay, J., Domínguez, C.: A Case-Study in Algebraic Manipulation Using Mechanised Reasoning Tools. To appear in International Journal of Computer Mathematics, doi:10.1080/00207160802676604
3. Coquand, T., Huet, G.: The Calculus of Constructions. Information and Computation 76, 95–120 (1988)
4. Coquand, T., Spiwack, A.: Towards Constructive Homological Algebra in Type Theory. In: Kauers, M., Kerber, M., Miner, R., Windsteiger, W. (eds.) MKM/CALCULEMUS 2007. LNCS (LNAI), vol. 4573, pp. 40–54. Springer, Heidelberg (2007)
5. Domínguez, C., Lambán, L., Rubio, J.: Object-Oriented Institutions to Specify Symbolic Computation Systems. Rairo - Theoretical Informatics and Applications 41, 191–214 (2007)
6. Domínguez, C., Rubio, J., Sergeraert, F.: Modelling Inheritance as Coercion in the Kenzo System. Journal of Universal Computer Science 12(12), 1701–1730 (2006)
7. The Kenzo Program (1999), `http://www-fourier.ujf-grenoble.fr/~sergerar/Kenzo`
8. Gonthier, G., Mahboubi, A., Rideau, L., Tassi, E., Théry, L.: A Modular Formalisation of Finite Group Theory. In: Schneider, K., Brandt, J. (eds.) TPHOLs 2007. LNCS, vol. 4732, pp. 86–101. Springer, Heidelberg (2007)
9. Lambán, L., Pascual, V., Rubio, J.: An Object-Oriented Interpretation of the EAT System. Applicable Algebra in Engineering, Communication and Computing 14(3), 187–215 (2003)
10. Rubio, J., Sergeraert, F.: Constructive Algebraic Topology. Bulletin Sciences Mathématiques 126, 389–412 (2002)
11. The Coq Proof Assistant (2009), `http://coq.inria.fr`
12. The Isabelle Proof Assistant (2009), `http://isabelle.in.tum.de`
13. Wiedijk, F. (ed.): The Seventeen Provers of the World, Foreword by Dana S. Scott. LNCS (LNAI), vol. 3600. Springer, Heidelberg (2006)

Vascular Landmark Detection in Retinal Images

M. Ortega, J. Rouco, J. Novo, and M.G. Penedo

VARPA Group, Department of Computer Science, University of A Coruña, Spain
{mortega,jrouco,jnovo,mgpenedo}@udc.es

Abstract. This paper describes a methodology for the detection of landmark points in the retinal vascular tree using eye fundus images. The procedure is fully automatic and is based in modified order filters, morphological operators and local analysis along the vascular tree. The results show a detection rate of 90% approx. using VARIA, a retinal image database designed to test techniques of retinal processing in heterogeneous conditions.

Keywords: Landmark points, retinal vascular tree, feature extraction.

1 Introduction

The analysis of retinal vascular tree can lead to detection and diagnosis of several problems related to vision. Also, it is possible to define a person by his/her retinal tree. The segmentation of the vessels is an useful procedure but it is a poor characterisation by itself if no additional information is obtained. The vascular tree is a 3D-structure being usually analysed by means of 2D images. The proper detection of landmark points in the structure adds more information regarding the vascular tree and allows to use them as reference points for registering images. Detecting vascular tree feature points is a complex task particularly due to the complexity of the vessel structure whose illumination and size is highly heterogeneous both between images and between regions from the same image.

Many methods for extracting information from the retina vessel tree can be found in the literature, but authors usually limit their work to a two dimensional extraction of the information. An analysis of the third dimension, depth, is needed. In the bibliography there are some works that try to solve this problem. For instance, the work proposed by Ali Can *et al.* [1] tries to solve the problem in difficult images using the central vessel line to detect and classify the feature points. Other methods, like the proposed by Chia-Ling Tsai *et al.* [2], use vessel segments intersections as seeds to track vessel centre lines and classify feature points according to intersection angles. The work proposed by Enrico Grisan *et al.* [3] extracts the structure using a vessel tracking based method needing a previous step before detecting feature points to fix the loss of connectivity in the intersections.

In this paper a landmark detection procedure is described starting from a segmented image [4]. The landmark points of interest in the vascular tree are the bifurcations, crossovers and endpoints in the vessels [5].

R. Moreno-Díaz et al. (Eds.): EUROCAST 2009, LNCS 5717, pp. 211–217, 2009.

The paper is organised as follows: in section 2 a description of the initial detection method is presented. Section 3 describes a filtering process taking place after the initial detection. Section 4 shows the experimental results and validation obtained using standard retinal image databases. Finally, section 5 provides some discussion and conclusions.

2 Feature Point Extraction

The goal in this first stage is to detect the feature points of the retinal vessel tree. This detection implies an analysis of the vascular structure. The first step is to perform a segmentation of the vascular tree. In this approach it has been used a technique with a particularly high sensitivity and specificity at classifying points as vessel or non vessel points, discussed in [4]. As discussed before, properties are not constant along all the structure, like vessel width, that decreases as the branch level of the structure becomes deeper. To unify this property a method able to reduce vessel width to one pixel without changing either vessel direction or connectivity is needed. The skeleton is the structure that meets all these properties.

However, the results of the segmentation process force a previous preprocessing step before the skeletonization. Fig.1 (a) shows gaps inside the vessels in the segmented image that would give a wrong skeleton structure if the next step is applied to images with this problem. A vessel with this problem in the segmented image would produce two parallel vessels in the skeletonized image (one for each border of the gap) creating false feature points, as shown in Fig.1 (b).

To avoid these false positive feature points it is necessary to "fill" the gaps inside the vessels. To perform this task, a dilation process is applied making the lateral vessel borders grow towards the center filling the mentioned gaps. The dilation process is done using a modified median filter. As in this case the filter is applied to a binary image the result central pixel value will be the most repeated value in the original window. In order to avoid an erosion when the filter is applied to the external border of vessels, the result value will only be

(a) (b)

Fig. 1. Segmentation problems, creating gaps inside the vessels. Subfigure (a) shows the segmentation problem with inside vessel gaps colored in red. Subfigure (b) shows the skeleton of a vessel with gaps, false feature points are marked in red.

set if it is a vessel pixel. To "fill" as much white gaps as possible the dilation process is applied in a iterative way, this is, dilation is applied to the previous dilation result N times. The value of N must be big enough to fill as much gaps as possible and, at the same time, small enough to avoid merging not connected vessels. The value of N depends on the spatial resolution of the images used, with the images used in this work (768x584) it was determined empirically that optimal values for N were around 4. The iterative process is shown in Fig.2.

Usually, skeletonization goal is to represent global objects properties with reducing the original image as much as possible. The skeleton, as stated before, expresses the structural connectivity of the objects with a width of one pixel. The basic method to obtain the skeleton is thinning, an iterative technique that erases pixels of the borders with, at least, one background neighbor if this erasing does not change the connectivity. The skeleton is defined by the medial axis function (MAF)[6], defined this as the set of points center of the maximum radius circles that fit inside the object. Calculating directly the MAF is a very expensive task and thus template based methods are used due to its versatility and effectiveness. In this work the Stentiford thinning method [7] is used. This method uses four templates (one for each of the four different borders of the objects) erasing the pixels only when the template matches and the connectivity is not affected. Fig.2(d) shows the results obtained with this approach.

As defined previously, feature points are landmarks in the vessel tree where several vessels appear together in the 2D representation. This allows to locate

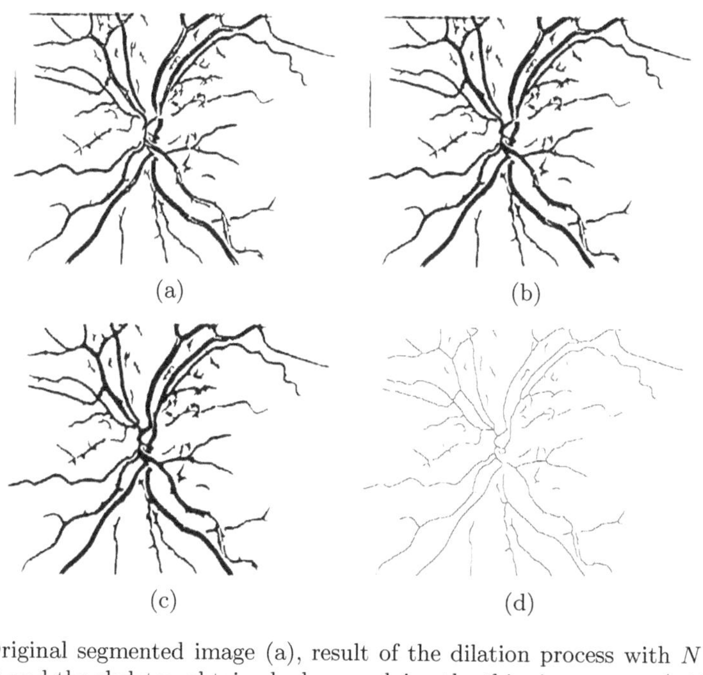

(a) (b) (c) (d)

Fig. 2. Original segmented image (a), result of the dilation process with $N = 2$ (b), $N = 4$ (c) and the skeleton obtained when applying the thinning process (with $N = 4$)

the feature points in the vessel tree using local information along it. This information is obtained with the analysis of the neighbors of each point. This way, the intersection number, $I(v)$, is calculated for each point, v, of the structure as showed in Eq.1, where the $N_i(v)$ are the neighbors of the analyzed point, v, named clockwise consecutively.

$$I(v) = \frac{1}{2}\left(\sum_{i=1}^{8} |N_i(v) - N_{i+1}(v)|\right) \tag{1}$$

According to its intersection number each point will be marked as,

- Vessel end point if $I(v) = 1$
- Vessel internal point if $I(v) = 2$
- Vessel bifurcation or crossover if $I(v) > 2$

In this approach, points are labelled as feature points when their intersection number $I(v)$ is greater than two corresponding to bifurcations or crossovers.

The problem in this detection is that not all the points are real points, this is, not every point detected exists in the real image due to the small branches that the skeletonization process creates in the border of the vessels 3. In the next section, a filtering process is depicted to deal with that particular issue.

3 Feature Points Filtering

The skeleton of the retinal vascular tree, as shown before, is obtained from a segmented image through a thinning process that erases the pixels from the borders towards the vessel center without affecting the connectivity. To adapt this structure to a correct point detection, it is necessary to erase the branches that do not actually belong to the retinal tree but its appearance is due to small waves in the borders of the vessels. The process to remove the spurious branches is performed following the next methodology:

The points previously detected are divided into two sets,

- C_1: Set of points labelled as vessel end points. ($I(v) = 1$)
- C_2: Set of points labelled as bifurcation or crossover. ($I(v) > 2$)

With these two sets, the extraction algorithm is as follows,

1. A point, $c \in C_1$ is taken as initial point (seed).
2. Starting in c, the vessel is tracked following the direction of neighbor pixels. Note that every pixel has only one predecessor and one successor.
3. When a previously labelled point, v, is found,
   (a) If $v \in C_1$, the segment is labelled as an independent segment and the process ends.
   (b) If $v \in C_2$, the segment is labelled as a branch and the process ends.

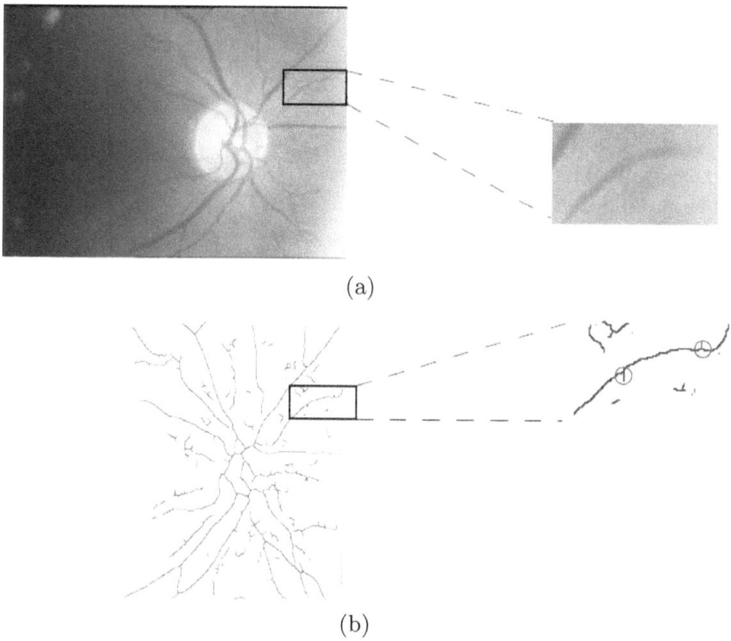

Fig. 3. Example of branches appearing after skeletonization process. (a) A region is zoomed in the original image and (b) associated skeleton where circles surround branches not corresponding to any real vessel.

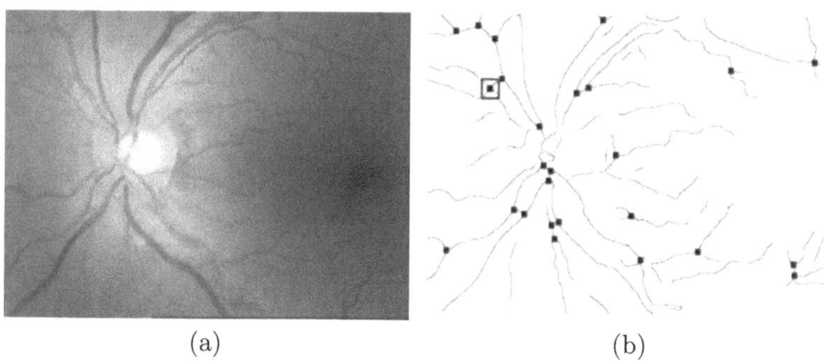

Fig. 4. Example of feature points extracted from original image with the vessel segmentation approach. *(a)* Original Image. *(b)* Feature points marked over the image after the pruning of branches. Again, spurious points are signalled. Squares surround pairs of points corresponding to the same crossover (detected as two bifurcations). The same heuristics than in the crease approach may be followed to avoid those problems.

Once obtained all the segments labelled as branches, and defined each of them by its final points (initial and end point), the internal points and its length, the pruning task consists of an analysis of all the branches deleting the ones shorter than the established threshold (ζ). Erasing a branch implies erasing the intersections associated to it, removing that particular intersection point from the list of feature points.

The chosen value for ζ is given by the own origin of the false branches, the ones due to small undulations in vessel borders. So, ζ is the maximum vessel width expected in the image. Fig.4 shows an example of final feature points extracted with this approach.

4 Experiments and Results

The images used for the experiments were extracted from the [8] database. The images have been acquired over a span of several years with a TopCon NW-100 model non-mydriatic retinal camera and are optic disc centered with a resolution of 768x584. These images have a high variability in contrast and illumination allowing the system to be tested in quite hard conditions, simulating a more realistic environment. The different conditions are also due to the fact that different experts with different illumination configurations on the camera acquired the images. 50 images were randomly selected from VARIA database. In order to make execution times more reliable, the tests were repeated 100 times for each image and method in a random order.

The efficacy of the system will be measured in terms of precision and recall. Also efficiency (computation time) and accuracy (distance from the real feature point location to the system obtained location for that point) are computed.

Table 1 shows the best results obtained with and without running the spurious point filtering algorithm (branch pruning) for the segmentation approach. The filtering of points improves the specificity even more significantly than the crease

Table 1. Performance of the feature points extraction using metrics of efficacy, efficiency and accuracy. First row represents the results obtained without applying filtering of branches and the second row applying the filtering.

	Efficacy		Efficiency			Accuracy
	Recall	Precision	T_{avg}	T_{min}	T_{max}	Deviation
No filtering	93.2%	71.7%	4.14s	1.84s	5.93s	4.77px
Filtering	90.5%	99.2%	4.56s	2.41s	6.11s	3.69px

Table 2. Parameters configuration for the feature point extraction using the vessel segmentation approach

Parameter	Description	Value
N	Number of dilations to fill holes in the segmented vessel tree	4
ζ	Threshold to prune tree branches	15 pixels

case due to the nature itself of the skeletonization. In fact, the filtering stage with this approach is unavoidable because otherwise the precision would be too low.

Table 2 shows the best values for the parameters using the same set of images.

5 Conclusions

In this work a method for the detection of the feature points of the retinal vascular tree using several image processing techniques has been presented. The detection of these points is crucial because it allows to increase the information about the retinal vascular structure. Having the feature points of the tree enables an objective analysis of the diseases that cause modifications in the vascular morphology or facilitates the retinal recognition. As a conclusion, the presented work is able to be applied in many other domains such as authentication or medical tasks.

In line with this this line of work, a goal based in these results is to perform a reliable classification of the feature points to enhance the retinal vessel tree data available.

Acknowledgements

This paper has been partly funded by the Xunta de Galicia through the grant contracts PGIDIT06TIC10502PR.

References

1. Can, A., Shen, H., Turner, J., Tanenbaum, H., Roysam, B.: Rapid automated tracing and feature extraction from retinal fundus images using direct exploratory algorithms. IEEE Transactions on Information Technology in Biomedicine 3(2), 125–138 (1999)
2. Tsai, C.L., Stewart, C., Tanenbaum, H., Roysam, B.: Model-based method for improving the accuracy and repeatability of estimating vascular bifurcations and crossovers from retinal fundus images. IEEE Transactions on Information Technology in Biomedicine 8(2), 122–130 (2004)
3. Grisan, E., Pesce, A., Giani, A., Foracchia, M., Ruggeri, A.: A new tracking system for the robust extraction of retinal vessel structure. In: 26th Annual International Conference of the IEEE Engineering in Medicine and Biology Society, IEMBS 2004, September 2004, vol. 1, 3, pp. 1620–1623 (2004)
4. Condurache, A.P., Aach, T.: Vessel segmentation in angiograms using hysteresis thresholding. In: Proceedings of the Ninth IAPR conference on Machine Vision Applications 2005, vol. 1, pp. 269–272 (2005)
5. Bevilacqua, V., Cambo, S., Cariello, L., Mastronardi, G.: A combined method to detect retinal fundus features. In: ECEACDA (2005)
6. Blum, H.: A Transformation for Extracting New Descriptors of Shape. In: Wathen-Dunn, W. (ed.) Models for the Perception of Speech and Visual Form, pp. 362–380. MIT Press, Cambridge (1967)
7. Stentiford, F.W.M., Mortimer, R.G.: Some new heuristics for thinning binary handprinted characters for ocr. IEEE Transactions on Systems, Man, and Cybernetics 13(1), 81–84 (1983)
8. VARIA: Varpa retinal images for authentication, `http://www.varpa.es/varia.html`

Web Applications: A Proposal to Improve Response Time and Its Application to MOODLE

David Horat and Alexis Quesada Arencibia

Instituto Universitario de Ciencias y Tecnologías Cibernéticas
Universidad de Las Palmas de Gran Canaria
Las Palmas, E35017, Spain
david.horat@gmail.com, aquesada@dis.ulpgc.es

Abstract. This paper covers some of the most advanced optimization techniques for web servers and web applications applied to a Modular Object Oriented Distance Learning Environment based on PHP 5 and Apache 2.

Keywords: web, optimization, web application, Moodle, PHP, Apache, javascript, HTTP, DNS, CSS, XHTML, HTML, minification, cookies.

1 Introduction

Response time in web applications is one of the most important issues today. The whole stack of intermediate software we use today to deliver web interfaces (Apache, MySQL, PHP 5, .NET, Java 5 Enterprise Edition, XHTML, Javascript, ...) added to the increasing popularity of web services make web applications slow (Fig. 1). Consider some of the most used web applications out there such as Google Search, Yahoo Search, Facebook, MySpace, etc. They all need huge amounts of infrastructures behind them and they really understand why optimization is important. Most of the methods we talk about here are used by these giants or are being considered at the moment of writing this paper. There are also teams on all these companies that address this problem and share their information, such as the Yahoo Exceptional Performance Team [1], as well as books [2] and papers that try to solve some of the problems stated above.

To access all these services, users make use of web browsers, which are the ones in charge of rendering the final screens and interacting with the user. They need to interpret heavy scripting languages such as HTML, XHTML, CSS and Javascript and they use old technologies such as HTTP. So from a users´ perspective and for usability reasons, web applications should be delivered as fast as possible and they should be structured in such a way that web browsers can render them fast enough so the user can begin to use it in the least time.

R. Moreno-Díaz et al. (Eds.): EUROCAST 2009, LNCS 5717, pp. 218–225, 2009.

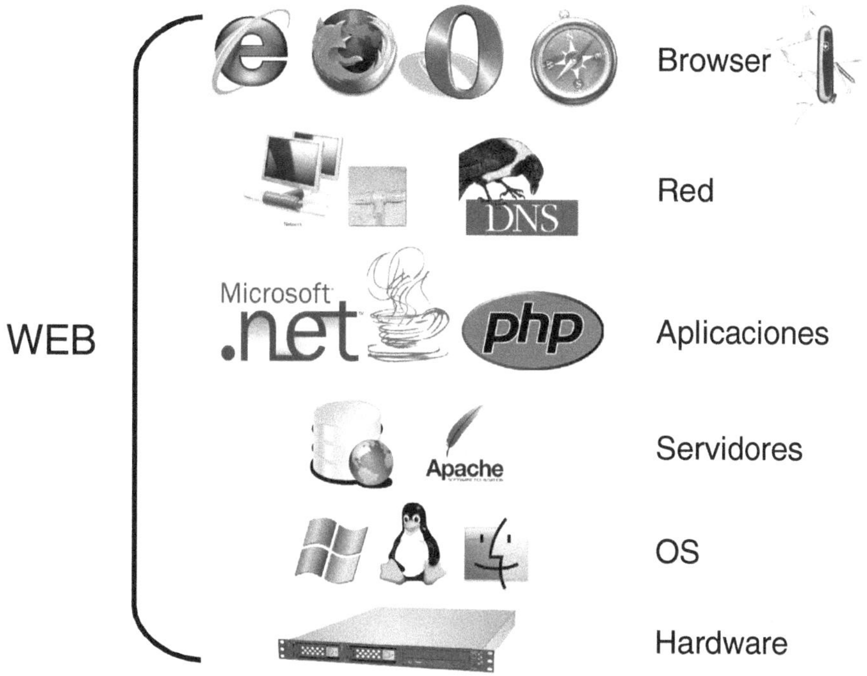

Fig. 1. Stack of technologies for Web Applications

2 Web Server and Network Technologies Methods

2.1 Reduce HTTP Requests

HTTP [3] is a heavy TCP/IP protocol and each HTTP header is around 1-2 KB [2]. On the other hand, around 80% of user´s response time is due to the download of all components regarding the web page that the user requested [1]. Thus, reducing the number of HTTP requests will reduce the number of HTTP headers, reducing the amount of information to download and, in the end, reducing the user´s response time (Fig. 2).

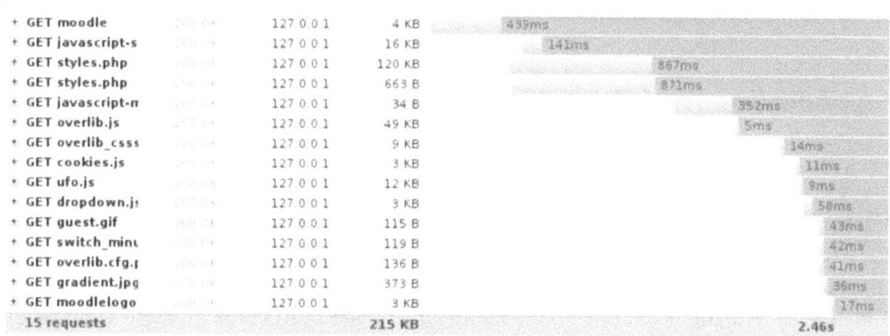

Fig. 2. Download time diagram

2.2 Use a Content Delivery Network

User´s geographical proximity to a web server reduces latency, thus increasing response time. If we distribute our web components along several servers distributed geographically, the user will get all the information needed faster. Since most of the download time is due to static components and they are much easier to distribute, we fully recommend this option.

2.3 Add an Expires Header

When a user visits a page for the first time, he has to download all components. In the next visits, reusage of components depends on caching policies and expire headers of each component. If a component is reused, it doesn't need to be downloaded again until it expires. Thus, adding an expires header to each component is highly recommended. Generally, adding a 1 year ahead expire header is enough [1], although this value can be tailored for each component depending on its usage.

2.4 Autocompress Components

Upon the appearance of the protocol HTTP 1.1 [3], we have the possibility to compress components in both deflate[8] and gzip [5] format. Gzip is the most common one and has the best ratio of compression.

The recommended components to compress are the ones that uses plan text, such as HTML [9], scripts [15], CSS [7], XML [10], JSON [11], RSS [12], ATOM [13] and SVG [14]. Components that are already compressed, such as the most common image formats, ZIP and PDF, are not recommended to be compressed again.

2.5 Deactivate ETags

ETags or Entity Tags specified in the protocol HTTP 1.1 [3] used to ensure that a cached component is exactly the one it is stored in the server. It is an improvement over the classical "if-modified-since" header which uses the last modification timestamp. ETags are strings that identify uniquely each component and its implementation depends on the web server software. Each ETag varies on each physical web server, thus making it unusable on web server clusters.

If we know that ETags are not going to help us in our application and we prefer to rely on the "if-modified-since" method, we recommend to deactivate them.

3 Application Architecture Methods

3.1 Avoid Redirections

HTTP redirections [3] are achieved using state codes 301 (moved permanently) and 302 (found). The main problem is that they have a great impact on response time due to the fact that each redirection requires two extra HTTP messages. Thus we recommend not to use them except when using the Post/Redirect/Get design pattern [15]. There are several solutions to avoid its usage: use server aliases, use mod_rewrite in Apache for changes in the same server or use CNAME [4] in the DNS if the redirection is a change of domain.

3.2 Flush the Buffer Soon

When a user request a page, the generation of the entire page within the server can take from 200 to 500 ms. in average. Thus, it is a good practice to send parts of the semi-generated page so the browser can begin downloading components earlier.

The best place to send a part of the page is just after the head [9] because it has the biggest density of external components, usually CSS and script files.

3.3 Reduce DOM Elements

Access to the DOM [6] is relatively slow. There is a big difference between acceding to 100 or to 1000 elements. Thus, we should try to reduce the number. In order to do so, we can ask ourselves a few questions:

- Do we use tables for design? This is disregarded.
- Do we use <div> tags as line breaks? This is disregarded.
- Do our tags are semantically appropriate? Review them all!

3.4 Reduce Number of iframes

Iframes [9] are tags that let us embed HTML documents in other HTML documents. They have several advantages, such as letting us embed third-party content, the embedded pages have a security sandbox regarding its parent and they allow to download scripts in parallel. Unfortunately, it has even bigger drawbacks, such as the long time needed to download a full HTML page, the fact that it can cause problems with scripts, that it is not semantically correct as it has been specified by W3C and that is prone to security attacks [16], such as the "Italian Job" [17], based on the "iframe attack" [18]. Thus, the use of iframes is disregarded.

3.5 Reduce Cookie Size

An HTTP cookie [3] is a data fragment which is stored in our disk accessible by a concrete webpage. Cookies are used for several reasons, such as storing authentication information, site personalization, local data storage, etc. Cookies have a deep impact in time response [1] because a 3 KB generates a delay of 78 ms compared to the 1 ms. of delay that generates one with 500 bytes [1]. Thus, it is highly recommended to use cookies below 500 bytes, which, indeed is what happens with most of the big Internet services out there such as Amazon, Google, Yahoo, CNN, YouTube, MSN, eBay or MySpace.

3.6 Use Domains without Cookies for Components

When a browser requests a static component, it also downloads the cookie associated to the domain that contains it. But only dynamic pages use the cookie. Thus, it is better to use domains without cookies for static components. Several Internet services are already using this method: Amazon uses "images-amazon.com", Yahoo uses "yimg.com" and YouTube uses "ytimg.com".

4 Javascript and CSS Methods

4.1 CSS at the Top

The importance of giving the user visual feedback about the progress has been already studied [19]. When a stylesheet is placed in the body [3], it will block partial rendering in most browsers until it has been completely downloaded. Thus, it is highly recommended that all CSS files are placed in the head using the <link> tag [3].

4.2 Scripts at the Bottom

HTTP 1.1 Protocol Specification [3] suggests that browsers should not download more than two components from the same domain at the same time. Unfortunately, most browsers do not downloaded more than one script from all domains at the same time. This fact is a big blocker or parallel downloads. Thus, it is highly recommended that all scripts are placed at the bottom, if possible, so they are the last components to be downloaded.

4.3 Avoid CSS Expressions

CSS expressions are powerful and dangerous way of modifying CSS properties dynamically. Some of their current problems are that they are not part of CSS standard, they are only supported by Internet Explorer and they are reevaluated each time there is a change in the page view, such as moving it up and down or mouse actions. Thus, this technique is not recommended.

4.4 GET Requests in AJAX

The HTTP protocol [3] specifies that the GET method should be used to request information, while the POST method should be used to send information. The facts are that the GET method is faster, it encodes all the parameters in the URL but it can only hold 2 KB of data. Since AJAX requires the maximum speed, we recommend to use the GET method not only if we want to request data, but also when we want to send data below 2 KB and it is not sensitive information.

5 Other Components Methods

5.1 Externalize Component Files

At this point, some people may think that in order to apply some of the methods above, it will better to embed all static files, such as images, CSS and scripts, into an HTML file. The fact is that we embed them, the benefits of caching and reusing the same components in different HTML files will be lost, and caching is a major benefit. On the other hand, if we find a concrete set of static information that is going to be used in a single HTML file, then we can definitely embed it. Thus, the general recommendation is to externalize all component files except in very rare occasions.

5.2 Minification of Plain Text Files

Minification is a technique used to reduce the size of scripts and CSS files erasing strings that do not modify their workflow and functionality. Some of these strings could be comments, white spaces, line breaks and tabs. There are already some tools that help us achieve this technique, such as JSMin [20] and YUI Compressor [21]. In our tests, we have been able to reduce up to 15% of a full application, counting images and all files, just by minifiying its CSS and script files with this technique.

5.3 Preload Components

The objective of preloading is to use the spare time while a user is watching a page to download components in advance that will be used by the user in the future. There are basically three different techniques:

- Unconditional preload: We download every possible file that the user will download in the future.
- Conditional preload: We download just the files that we know the user will need in the future using contextual and historical information.
- New design preload: If we are going to make heavy changes in our service in the future, it is interesting to let people download files that will be used in the new design, thus not overloading the server when the new design becomes available.

All these techniques are recommended.

5.4 Optimize Images

When images are ready, there are several ways to optimize them. We can use the most appropriate format, such as JPG for photos, GIF or PNG for drawings with few colors or SVG for vectorial drawings. We should try several options before deciding which one. Some of the parameters we can play with are the size, compressions rate or color pallet. Be careful not to make browsers resize images dynamically because it is a waste of time and bandwidth. To automate some of these process we can use several tools, such as ImageMagick [22], Pngcrush [23], jpegtran [24] or GIMP [25].

5.5 Reduce Favicon

Favicon, also known as page icon, is an image associated to a web page or domain that identifies it. Traditionally it was used as an static file called "favicon.ico" in the root of a domain. This is not an official standard [26]. Nevertheless, W3C has made several suggestions about it [26]: It should be an official supported image format such as ICO, GIF or PNG; it should be 16x16 or 32x32 with color-depth between 8 and 24 bits; and it should be inserted in the head of the document with the <link> tag [3]. Now icons are associated to a web page and not to a domain. Since browsers request this special kind of file with the old method if the new method is not used, it is heavily recommended to have always one. Moreover, since it will be requested a lot, it should be small, preferably under 1 KB. ImageMagick [22] can help us achieve this.

6 Optimization in Moodle

In this paper, we have also studied the impact of all the methods shown above in a Modular Object Oriented Distance Learning Environment called Moodle [27]. Moodle is Course Management System written in PHP, which uses XHTML 1.0 Strict, CSS level 2 and Javascript for its web User Interface. Our conclusions are:

Table 1. State and comments of the optimization methods studied within Moodle

Name	State	Comments
2.1 Reduce HTTP Requests	FIXED	Combine javascript files.
2.2 Use a Content Delivery Network	NOK	Architectural changes needed.
2.3 Add an expires header	FIXED	Web server and code changes.
2.4 Autocompress components.	FIXED	Web server configuration.
2.5 Deactivate ETags	FIXED	Web server configuration.
3.1 Avoid redirections	OK	Ok.
3.2 Flush the buffer soon	FIXED	Code changes needed.
3.3 Reduce DOM elements	OK	Ok.
3.4 Reduce number of iframes	NOK	Architectural changes needed.
3.5 Reduce cookie size	OK	Ok.
3.6 Use domains without cookies	NOK	Architectural changes needed.
4.1 CSS at the top	OK	Ok.
4.2 Scripts at the bottom	FIXED	Modification of themes.
4.3 Avoid CSS expressions	OK	They are not used.
4.4 GET requests in AJAX	OK	Ok.
5.1 Externalize component files	FIXED	Javascript modification needed.
5.2 Minification of plan text files	FIXED	Bash script created.
5.3 Preload components	FIXED	Unconditional preload applied.
5.4 Optimize images	OK	Ok.
5.5 Reduce favicon	OK	Ok.

7 Conclusions

In this paper we have studied several methods to optimize the response time in web applications and we have studied the application of those methods in Moodle. We have seen the importance of getting deep into protocol specifications, common usage and browser implementations.

Moodle seems to be fit in many of the methods studied here, but to comply with all of them, it will need some heavy architectural changes. In the meantime, there are several methods that could be applied with light changes of code and web server configuration.

Big Internet service providers will keep on improving these kind of methods as more and more users connect to them, trying to reduce the load on their severs and to improve User Experience, which is nowadays a determining quality factor.

References

1. Yahoo Exceptional Performance Team, http://developer.yahoo.com/performance/
2. Souders, S.: High Performance Web Sites. Editorial O'Really (2007)
3. HTTP/1.1 Protocol Specification, http://tools.ietf.org/HTML/rfc2616
4. Domain Names – Implementation and Specification, http://www.ietf.org/rfc/rfc1035.txt
5. RFC 1952, http://www.ietf.org/rfc/rfc1952.txt
6. Document Object Model Specification, http://www.w3.org/DOM/
7. CSS Level 2 Specification, http://www.w3.org/TR/CSS2/
8. RFC 1951, http://www.ietf.org/rfc/rfc1951.txt
9. HTML 4.01 Protocol Specification, http://www.w3.org/TR/html4/
10. XML Protocol Specification, http://www.w3.org/XML/
11. JSON Protocol Specification, http://www.json.org/
12. RSS Protocol Specification, http://www.rssboard.org/
13. ATOM Protocol Specification, http://tools.ietf.org/HTML/rfc4287
14. SVG Protocol Specification, http://www.w3.org/Graphics/SVG/
15. Post/Redirect/Get design pattern, http://en.wikipedia.org/wiki/Post/Redirect/Get
16. Article Hackers expand massive IFRAME attack to prime sites, http://www.pcworld.idg.com.au/index.PHP/id;271828304;fp;2;fpid;1
17. Article 'Italian job' Web attack hits 10,000 sites, http://www.networkworld.com/news/2007/061907-italian-job-web-attack.HTML
18. Article What's an IFrame attack and why should I care?, http://www.guardian.co.uk/technology/2008/apr/03/security.google
19. Article "Response time", http://www.useit.com/papers/responsetime.HTML
20. Project JSMin, http://crockford.com/javascript/jsmin
21. Project YUI Compressor, http://developer.yahoo.com/yui/compressor/
22. Project ImageMagick, http://www.imagemagick.org
23. Project Pngcrush, http://pmt.sourceforge.net/pngcrush/
24. Project JPEG image compression library, http://www.ijg.org/
25. Project GIMP, http://www.gimp.org/
26. Article "How to add a favicon" from W3C, http://www.w3.org/2005/10/howto-favicon
27. Project Moodle, http://moodle.org

Functional Disambiguation Using the Syntactic Structures Algorithm for Each Functional Interpretation for Spanish Language

Octavio Santana Suárez, José Rafael Pérez Aguiar,
Idafen Santana Pérez, and Rubén Quesada López

Department of Computer Science, University of Las Palmas de Gran Canaria, 35017
Las Palmas de Gran Canaria, Spain
{osantana,jperez,isantana,rquesada}@dis.ulpgc.es

Abstract. This paper presents a disambiguation method that diminishes the functional combinations of the words of a sentence taking into account the context in which they appear. This process uses an algorithm which does the syntactic analysis of every functional combination of the sentece. In order to control this analysis, a grammar with restrictions has been developed to model the valid syntactic structures of the Spanish language. The main target of our algorithm is the separation between the disambiguation method and the grammar which governs it.

Keywords: Functional Disambiguation, Syntactic Analysis, Computational Linguistics, Natural Language Processing.

1 Introduction

One of the main problems we need to raise at the Automatic Text Analysis in Spanish is the high number of combinations which emerges because of the ambiguation grade of words which configure the language. The fact that each word could raise many grammar functions makes an excessive number of combinations by sentence.

With the goal of minimizing this problem, a disambiguation algorithm has been developed which is based on the syntactic analysis of each combination of sentence, starting of an input grammar and the morphological analyser by [5] and [6]. There will be two goals in the developed work, one is to minimize the number of interpretations of the sentence, through the study of syntactic structures, and the other one is to obtain the syntactic analysis trees of the sentence.

Furthermore, it develops an algorithm which works independently from the grammar rules, which allows to obtain a tool with a high ease maintainability and it permits to be used by linguistic users without computing skills.

2 XML Grammar

An important point before developing the processes of the disambiguation is to formalize the grammar which will model the syntactic valid structures of the Spanish language.

R. Moreno-Díaz et al. (Eds.): EUROCAST 2009, LNCS 5717, pp. 226–230, 2009.

Our grammar will be formed by rules *generated symbol: list of generator symbols*. In addition, a series of requirements associated with every rule will appear. They will permit to introduce a dependence of the context at the moment of the rule application. That means that a rule could be applied when at least one of its requirements is checked.

Formally, the rules of our grammar will follow the following structure:

<Regla N="valor_N" NR="valor_NR" Simbolo="nombre símbolo generado">
<Generatriz Simbolo="nombre símbolo generatriz_1" Posicion="1">
</Generatriz>
<Generatriz Simbolo="nombre símbolo generatriz_2" Posicion="2">
</Generatriz>
...
<Requisito>
</Requisito>
...
</Regla>

On the other way, the requirements of the rule will be as following:

<Requisito R="valor_R" Frecuencia="normal/baja" Tipo="validacion/rechazo">
<Condiciones>...</Condiciones>
<Concordancias>...</Concordancias>
<Herencia>...</Herencia>
</Requisito>

The *<Condiciones>* tag will accommodate the conditions which each generator symbol should achieve individually. The *<Concordancias>* tag will accommodate all restrictions that several generators will verify coordinately. If the content of these structures is checked, we could say that the requirement has been verified, and it could begin a new process: generating the new symbol. This generated symbol could accommodate some imposed or extracted information (from the generator symbols), all this being specified inside the *<Herencia>* tag.

There will be some special symbols in the grammar that will represent a complete and valid Spanish sentence. This kind of symbols will be called root symbols and will be tagged by some information specified inside the *<Herencia>* tag of the rules which generate them.

3 Functional Disambiguation by Syntactic Analysis Algorithm

Taking into account the different functional categories of every word that take part in the sentence, returned by a morphological analyser, we proceed to discard those categories that do not satisfy the syntactic restrictions specified in the input grammar.

Considering the following sample sentence: *Lo hemos conseguido*, its morphological analysis returns this result:

Table 1. Morphological analysis of the sentence *Lo hemos conseguido*

Lo	hemos	conseguido
article	verb	verb
personal pronoun		adjective
noun		

This analysis produces six functional combinations for the sentence:

(*Lo*) article (*hemos*) verb (*conseguido*) verb
(*Lo*) article (*hemos*) verb (*conseguido*) adjective
(*Lo*) personal pronoun (*hemos*) verb (*conseguido*) verb
(*Lo*) personal pronoun (*hemos*) verb (*conseguido*) adjective
(*Lo*) noun (*hemos*) verb (*conseguido*) verb
(*Lo*) noun (*hemos*) verb (*conseguido*) adjective

The disambiguation algorithm tries to do the syntactic analysis of every combination. By doing this, the number of interpretations will be reduced to only that ones which allow to obtain, at least, one valid syntax tree. In the case of the sample sentence, five combinations will be discarded and only one remains:

(*Lo*) personal pronoun (*hemos*) verb (*conseguido*) verb

This combination is the only one which returns a valid syntax tree and, because of this, it is the only combination that verifies the structures included in the XML grammar.

In order to get the valid combinations for the sentence, an iterative process is performed to achieve a bottom up syntax tree creation. This process will work with a set of node structures, being the content of each node a grammar symbol and some extra morphological and control information.

Firstly, starting from an initial set of leaf nodes, which are the interpretations of each word, the input grammar is inspected to extract those rules whose generator symbols match a subset of these initial nodes. These rules are studied in order to check if this subset of nodes satisfies at least one of their requirements. If they do, a new node will be created containing the generated symbol of the rule and the information specified inside its <*Herencia*> tag. This new node will be added to the set and will be taken into account as a possible generator symbol in the future rule applications.

This process will be repeated over the growing set of nodes until any new rule cannot be applied. In that moment, those nodes that contain any root symbol of the grammar and which cover the whole sentence will be the syntax solutions, hence, their bottom leaf nodes will be the valid combinations for the sentence.

4 Results

The developed algorithm has been tested over a group of 7000 representative Spanish sentences. The overall results were these:

Table 2. Overall results

Overall results	
Number of initial combinations	205.69
Number of final combinations	3.65
Disambiguation percentage (%)	86.63
Number of syntax trees	14.97
Number of nodes	19155.17
Number of useful nodes	123.15
Disambiguation time (sec.)	4.73
Total time (sec.)	4.87

This method generates all the possible functional combinations using the different functional categories of every word in the sentence, returning 205.69 initial combinations. Once applied the algorithm, an average number of 3.65 final combinations per sentence is obtained, which means a disambiguation percentage of 86.63%. These combinations create 14.97 valid syntax trees using the rules that are included in the grammar.

If we focus on the memory usage, 19155.17 nodes are generated, 123.15 of these are useful nodes, i.e. they take part in a valid syntax tree. The average time for the disambiguation is 4.73 seconds per sentence, and 4.87 seconds if we consider the previous morphological analysis. This test has been executed in a Pentium IV @ 2,80 GHz and 2 GB RAM.

5 Conclusions

In conclusion, the present paper exposes a method capable of carrying out the functional disambiguation by means of the use of the Syntactic Analysis, being based exclusively on the grammatical structures codified as external rules. With all, a high percentage of disambiguation is obtained in a reasonable time of execution, and moreover, a totally independent operation from the input grammar.

The obtained results are slightly lower than the ones obtained by [2] and [8]. However, this circumstance is due to several optimization techniques used in that PhD (such as verb number limitation, elimination of non-common interpretations, and so on), whose usage was avoided in this algorithm in order to provide a more general solution.

Moreover, this research could be used in the future to build new applications that go beyond the computational linguistics field, resolving the rest of Spanish ambiguities or, holding on the obtained results, which will be able to carry out other procedures such as automatic translation or analysis of searching requests. Generally speaking, all applications that aim to resolve Natural Language Processing problems could use the techniques exposed in this paper.

References

1. Chomsky, N.: Syntactic Structures. The Hague, Mouton (1957)
2. Losada, L.: Automatización del análisis sintáctico del español. Phd thesis, University of Las Palmas de Gran Canaria (2002)
3. Quesada, J.: Un modelo robusto y eficiente para el análisis sintáctico de lenguajes naturales mediante árboles múltiples virtuales. Centro Informático Científico de Andalucía, CICA (1996)
4. Quesada, J.: El algoritmo SCP de análisis sintáctico mediante propagación de restricciones. Phd thesis, University of Sevilla (1997)
5. Santana, O., Pérez, J., Hernández, Z., Carreras, F., Rodríguez, G.: FLAVER: Flexionador y lematizador automático de formas verbales. Lingüística Española Actual XIX 2, 229–282 (1997)
6. Santana, O., Pérez, J., Duque, J., Hernández, Z., Rodríguez, G.: FLANOM: Flexionador y lematizador automático de formas nominales. Lingüística Española Actual XXI 2, 253–297 (1999)
7. Santana, O., Pérez, J., Losada, L., Carreras, F.: Hacia la desambiguación funcional automática en Español. Procesamiento del Lenguaje Natural 1(3), 1–15 (2002)
8. Santana, O., Pérez, J., Losada, L., Carreras, F.: Bases para la desambiguación estructural de árboles de representación sintáctica. Procesamiento del Lenguaje Natural 32, 43–65 (2004)

On Similarity in Case-Based Reasoning for Structural Health Monitoring

Reinhard Stumptner, Bernhard Freudenthaler, and Josef Küng

FAW – Institute for Applied Knowledge Processing, University of Linz,
Altenbergerstr. 69, 4040 Linz, Austria
{rstumptner,bfreudenthaler,jkueng}@faw.jku.at

Abstract. This contribution deals with the importance of similarity in Case-based Reasoning and the application for problems in the field of Structural Health Monitoring. Case-based Reasoning and different methodologies for the retrieval of knowledge stored in a case base are presented. Furthermore, different approaches for object representation and applicable similarity measures including indexing techniques to reduce the number of required distance-calculations are introduced, particularly the M-tree approach. Finally, a prototype for a Case-based Decision Support System for Structural Health Monitoring is described.

Keywords: Case-based Reasoning, Indexing, Similarity, Structural Health Monitoring.

1 Introduction

Case-based Reasoning (CBR) is a problem solving method, which uses already known problems and solutions of a certain domain to solve new problems [1]. Thus, new problems should be solved by reusing solutions of similar problems/situations which are retrieved by means of so-called similarity measures. An already known problem with an appropriate solution defines a case which is stored in the case base. The aim of CBR is to reuse already existing knowledge for similar problems and not to develop a solution for any new problem. Consequently cost-effective and rapid solutions are probable.

Similarity is one of the most important fundamentals in CBR. According to [18] the aim of retrieving similar cases "is to retrieve the most useful previous cases towards the optimal resolution of a new case and to ignore those previous cases that are irrelevant." A case mostly consists of a set of attribute - value pairs which are called descriptors. Descriptors of a new case are compared with descriptors of cases in the case base and the most similar cases are filtered out. These cases can be ranked according to the similarity values to the new case. Three retrieval methodologies to support the matching of previous cases are proposed [18]:

- Similarity assessment along descriptors: The descriptors of a new case are compared with the descriptors of previous cases in the case base.

R. Moreno-Díaz et al. (Eds.): EUROCAST 2009, LNCS 5717, pp. 231–238, 2009.

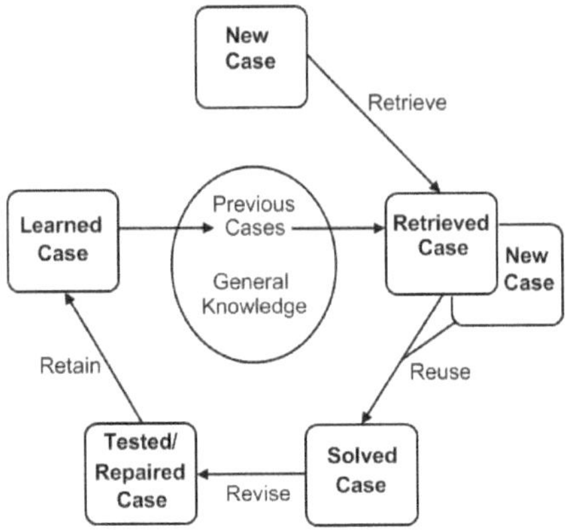

Fig. 1. From Problem to a new Case. (According to [1]).

- Contextual determination of importance of descriptors: Descriptors can have different influence on calculating similarities between cases, what can be expressed by user-defined weights for instance.
- Acquisition and application of validity constraints: Methods to get validity constraints to prevent over-generalization.

To calculate the similarity between cases mostly not only quantitative features have to be considered, but also qualitative aspects. These qualitative aspects can be handled using weights or other abstractions [3]. Many kinds of similarity measures are described in literature ([3], [4], [21], [26]) and most of them have special areas of application. Raw data has to be analysed carefully to decide, which similarity measure (incl. indexing methodologies) would be the best for a certain domain/problem. Similarity measures can provide numeric values which represent the similarity or the relation between different cases. A certain number of cases (1...n) with the highest similarity values can be presented to the user in addition with some kind of distance to the initial problem. Such feature can make a Decision Support System more transparent and would support the acceptance by users.

Monitoring of structures like it is suggested by the Structural Health Monitoring (SHM) community became more and more important. Our infrastructure is aging and consequently has to be checked to guarantee safety and to predict life time, especially of critical structures like bridges, monuments, industrial plants and so on.

Hereby, different methodologies of expressing similarity are shown in conjunction with an indexing technique.

2 Structural Health Monitoring

There already exist many areas of application where CBR is utilized successfully. CBR techniques even are used for methodologies like flexible querying [24] for instance. Actually computer support of processes in Structural Health Monitoring has become a topic of growing interest for civil engineers. In particular they have to deal with the question "Is a structure (e.g. a bridge) still safe?" or the like.

Main goals of Structural Health Monitoring (SHM) are damage identification and lifetime prediction for the civil infrastructure defining damages as changes of materialistic or geometric properties. According to [12] CBR can be used to monitor bridges and support the engineers in making decisions relying on the interpretation of measurement data. Due to the fact that each bridge has its individual dynamic parameters, the interpretation and analysis of raw data done by human beings is very time-consuming and consequently cost-intensive. Furthermore, the interpretation process is subjective, as experts have different levels of experience and use different concepts for interpreting measurement data. Taking these problems into consideration, CBR should be used to develop an intelligent Decision Support System to disburden and support engineers in interpreting measurement results of structures ([12], [13]).

3 Similarity Measures

According to [5], the three main approaches of computing similarities between object representations are feature-based, geometric or structural approaches.

- **Feature-based approach:** Objects are represented by attribute-value pairs (descriptors). Thereby, the descriptors of a new case are compared regarding commonality and difference with the descriptors of already known cases in the case base.
- **Geometric approach:** Objects are represented by points in an n-dimensional space. A new case is more similar to a case in the case base the smaller the distance between them is.
- **Structural approach:** Objects are represented by nodes in a graph-like arrangement and associations between objects are represented by the edges of the graph. Similarity measuring in structural approaches is based on graph matching algorithms.

Hybrid and extended forms of these approaches are also possible and usual ([4] for instance). In most cases data can be transferred to an n-dimensional (metric) space more or less appropriately. It is well known that the comparison of cases and consequently the provision of indicators for similarities between cases are main principles especially of CBR's retrieve phase [19] and mostly are sticking points of the design of a Case-based System. In the following only a few approaches for similarity measures can be introduced.

Searching in general always was a very popular data processing operation. While at first models (algorithms, index structures) for exact-match searches were developed, the search paradigms soon changed and requirements like "calculate the similarity between objects x and y" were arising. Similar objects can be described as the set of objects, being in some sense "near" to each other. The metric space notation satisfies the needs of representations and abstractions in this connection. There are different types of similarity queries but the most popular ones are "similarity range" and "nearest neighbour search". The similarity search in the metric space can be seen as a kind of ranking of objects in respect to a query object and to a certain similarity.

Before some similarity/distance measures will be described, two types of similarity queries (maybe the most important ones) are introduced. The first example for a similarity query is the range query *R*. It is based on a query object *o* and on a radius *r*. From the whole set of objects *O*, all objects within this range o_{res} are returned.

$$R(o,r) = \{o_{res} \in O, d(o,o_{res}) \leq r\} \tag{1}$$

If the search radius of a range query is zero, a so-called point query, then it is equal to the exact-match search. The following figure tries to visualise the principle of the range query.

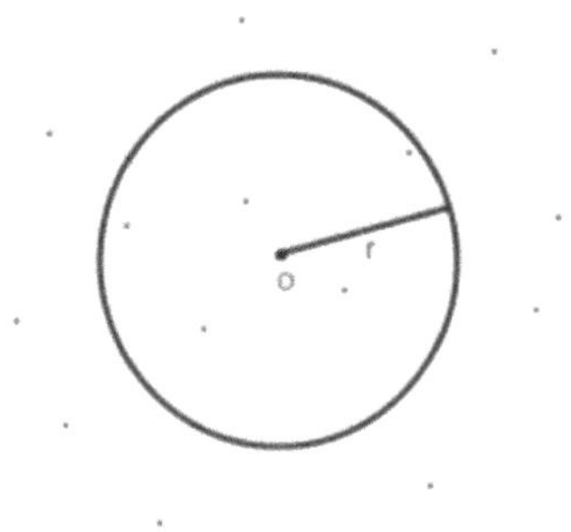

Fig. 2. Range Query

Another type of similarity query is the nearest neighbour query whereby a certain number (*k*) of nearest objects regarding a certain query object should be provided.

$$\begin{aligned} kNN(o) = \{NN \subseteq O, length(NN) = \\ k \wedge \forall_{x \in NN, y \in O \wedge y \notin NN} (d(o,x) \leq d(o,y))\} \end{aligned} \tag{2}$$

Especially for CBR this query-type is very important. A nearest neighbour search can provide a set of the most similar cases regarding a new case or "query case". It should be mentioned that there exist further similarity queries which cannot be described in this contribution.

In general the similarity between two objects (cases) *X* and *Y* can be represented by their "inverse standardised distance" *1-d(X,Y)*. There are numerous different methods to calculate this distance. The distance function of the metric space expresses the nearness between objects. Generally there are two types of distance functions, discrete (set of possible values) and continuous (mostly standardised).

There exist many approaches dealing with similarity measurement. However, for any kind of attributes (quantitative, qualitative, graph...) a method to express distances/similarities has to be defined and finally the similarity regarding all different attributes has to be represented by a numerical value. A popular discrete measure for similarity between many kinds of objects is the Hamming Similarity. It measures the minimum number of substitutions which are required to transform one object into the other [26]. The Hamming Similarity is a very general measure, nevertheless not useful for all kinds of data sources. More specialised measures in certain cases may provide more adequate results.

A distance measure, which concerning its main principles is very similar to the Hamming or Edit Distance, is the Tree Edit Distance [26] (structural approach of object representation). This measure describes the cost to transform a “source tree” into a “target tree”. There is a set of predefined allowed operations (insert/delete node) to perform the transformation, having different weights depending on the level of a tree where they are carried out because operations near to the root for instance may be more cost-intensive (depending on the domain of an application) than somewhere near to the leaves. The needed steps to perform the transformation describe the distance between two certain tree structures. This distance measure could be used for the comparison of tree-like documents like XML files for instance.

The Euclidean Distance is the most intuitive example for a continuous distance measure.

$$d(X,Y) = \sqrt{\sum_{i=1}^{n}(x_i - y_i)^2} \tag{3}$$

But, there exist many other possibilities depending on any eventual preconditions. A possibility to bring some semantic information to this distance measure is the Weighted Euclidean Distance [26].

$$d(X,Y) = \sqrt{\sum_{i=1}^{n} w_i (x_i - y_i)^2} \tag{4}$$

Each dimension has a predefined weight w_i. The weights describe the importance of the dimensions in the metric space which can be seen as a deformation. In a CBR-System this deformation consequently models the relevance of certain attributes when retrieving similar cases from the case base.

Finally it should be mentioned, that in some cases it may be more advantageous to predefine distances between certain sets of objects/attributes in the form of a matrix for instance. Sometimes, especially when the number of possible values of an attribute is fixed, this number is not too high and the calculation of distances is a very complicate task and requires deep expert’s knowledge, then it may be better to predefine all possible distances with respect to the experience of an expert.

3.1 Indexing

The output of CBR's retrieve phase usually is the result of a nearest neighbour search in the space formed by the cases of the case base. To improve the performance of this phase it is important to use suitable indexing methodologies in respect to the object representations and to the similarity measure. In case of geometric object representation the usage of a metric index structure, like the M-tree [7] or related models [23] or extensions like the M^+-tree [27], are efficient methods to improve the runtime performance of the case-retrieval phase. The M-tree grows dynamically, even frequent inserting and deleting of objects is not a very problematic performance problem. Due to this fact and because it is open for any similarity metric the M-tree is an efficient runtime improvement for Case-based Decision Support for SHM. The M-tree is growing bottom-up and each node has a certain capacity. If an overflow occurs, the concerned node is divided and two new nodes are created due to a certain split-policy. The tree only contains so-called routing objects, but in the leaf nodes there are references to the data-objects stored. Each parent object of the tree in some way bounds all underlying nodes using a covering radius and divides the metric space into ball-like regions. The improvement of the runtime-performance of searching is achieved by selecting regions according to a search radius and to the covering radius of sub-trees of the structure.

The distance function ($d(o_1, o_2)$) between two objects o of a M-tree has to have the following characteristics [7]:

$$\begin{aligned}
&1)\ d(o_x,o_y)=d(o_y,o_x)\\
&2)\ d(o_x,o_x)=0\\
&3)\ d(o_x,o_y)>0\,for(o_x\neq o_y)\\
&4)\ d(o_x,o_y)\leq d(o_x,o_z)+d(o_z,o_y)
\end{aligned}\tag{5}$$

The distance between two objects has to be symmetric (1), the distance of an object to itself has to be zero (2), a distance cannot be negative (3) and the triangle inequality has to be valid (4).

Due to the fact that the number of attributes forming objects/cases probably is numerous, the costs of calculations of distances are correspondingly high. The M-tree model in average reduces the number of required distance-calculations by approximately 40% [7] and so this model is a good improvement of similarity calculations in metric spaces.

4 Conclusion

Due to an increasing age of many partly critical structures, like bridges for instance, Structural Health Monitoring gains more and more in importance and interest. In this contribution a possible application of Case-based Reasoning for the field of "Structural Health Monitoring" is introduced whereby similarity measurement and indexing is focused on. Research was done within a project called "safePipes" [8] funded by the European Union. Within this project a prototype for a Case-based Decision Support System was developed which should support the engineer in interpreting measurement data taken from different types of structures. Experiments with different kinds of test-data verified that Case-based Reasoning suits well to problems in this domain of Structural Health Monitoring and especially the (fully) automated interpretation of measurements from simple structures (e.g. pipes, ...) is a very promising research topic.

References

1. Aamodt, A., Plaza, E.: Case-Based Reasoning: Foundational Issues, Methodological Variation and System Approaches. In: AI Communications, pp. 39–59. IOS Press, Amsterdam (1994)
2. Armengol, E., Plaza, E.: Similarity Assessment for Relational CBR. In: Aha, D.W., Watson, I. (eds.) ICCBR 2001. LNCS (LNAI), vol. 2080, pp. 44–58. Springer, Heidelberg (2001)
3. Beierle, C., Kern-Isberner, G.: Methoden wissensbasierter Systeme: Grundlagen, Algorithmen, Anwendungen, vol. 2. Auflage. Vieweg, Wiesbaden (2003)
4. Bergmann, R., Stahl, A.: Similarity Measures for Object-Oriented Case Representations. In: Smyth, B., Cunningham, P. (eds.) EWCBR 1998. LNCS (LNAI), vol. 1488, pp. 25–36. Springer, Heidelberg (1998)
5. Bridge, D.G.: Defining and Combining Symmetric and Asymmetric Similarity Measures. In: Smyth, B., Cunningham, P. (eds.) EWCBR 1998. LNCS (LNAI), vol. 1488, pp. 52–63. Springer, Heidelberg (1998)
6. Ciaccia, P., Patella, M.: Searching in Metric Spaces with User-Defined and Approximate Distances. ACM Transactions on Database Systems 27(4), 398–437 (2002)
7. Ciaccia, P., Patella, M., Zezula, P.: M-tree: An Efficient Access Method for Similarity Search in Metric Spaces. In: Proceedings of 23rd VLDB Conference, pp. 426–435 (1997)
8. European Communities, SAFE PIPES, Safety Assessment and Lifetime Management of Industrial Piping Systems, `http://cordis.europa.eu/fetch?CALLER=FP6_PROJ&ACTION=D&DOC=1&CAT=PROJ&QUERY=1206532260841&RCN=75379`
9. Falkman, G.: Similarity Measures for Structured Representations: A Definitional Approach. In: Blanzieri, E., Portinale, L. (eds.) EWCBR 2000. LNCS (LNAI), vol. 1898, pp. 380–392. Springer, Heidelberg (2000)
10. Finnie, G., Sun, Z.: Similarity and Metrics in Case-Based Reasoning. International Journal of Intelligent Systems 17, 273–287 (2002)
11. Freudenthaler, B.: Case-based Reasoning (CBR): Grundlagen und ausgewählte Anwendungsgebiete des fallbasierten Schließens. VDM Verlag Dr. Müller, Saarbrücken (2008)
12. Freudenthaler, B., Gutenbrunner, G., Stumptner, R., Küng, J.: Case-based Decision Support for Bridge Monitoring. In: Proceedings of the Third International Multi-Conference on Computing in the Global Information Technology, ICCGI 2008 (2008)
13. Freudenthaler, B., Stumptner, R., Forstner, E., Küng, J.: Case-based Reasoning for Structural Health Monitoring. In: Uhl, T., Ostachowicz, W., Holnicki-Szulc, J. (eds.) Proceedings of the Fourth European Workshop on Structural Health Monitoring. DEStech Publications, Lancaster (2008)
14. Gierl, L., Bull, M., Schmidt, R.: CBR in Medicine. In: Lenz, M., Bartsch-Spörl, B., Burkhard, H.D., Wess, S. (eds.) Case-Based Reasoning Technology: From Foundations to Applications. LNCS (LNAI), vol. 1400, pp. 273–297. Springer, Heidelberg (1998)
15. Heit, E.: Features of similarity and category-based induction. In: Proceedings of the Interdisciplinary Workshop on Similarity and Categorisation, pp. 115–121 (1997)
16. Lenz, M., Hübner, A., Kunze, M.: Textual CBR. In: Lenz, M., Bartsch-Spörl, B., Burkhard, H.D., Wess, S. (eds.) Case-Based Reasoning Technology: From Foundations to Applications, pp. 115–137. Springer, Heidelberg (1998)

17. Montani, S., Portinale, L.: Case Based Representation and Retrieval with Time Dependent Features. In: Muñoz-Ávila, H., Ricci, F. (eds.) ICCBR 2005. LNCS (LNAI), vol. 3620, pp. 353–367. Springer, Heidelberg (2005)
18. Montazemi, A.R., Gupta, K.M.: A framework for retrieval in case-based reasoning systems. Annals of Operations Research 72, 51–73 (1997)
19. Perner, P.: Are case-based reasoning and dissimilarity-based classification two sides of the same coin? In: Engineering Applications of Artificial Intelligence, vol. 15, pp. 193–203 (2002)
20. Puppe, F., Gappa, U., Poeck, K., Bamberger, S.: Wissensbasierte Diagnose- und Informationssysteme: Mit Anwendungen des Expertensystem-Shell-Baukastens D3. Springer, Heidelberg (1996)
21. Spertus, E., Sahami, M., Buyukkokten, O.: Evaluating Similarity Measures: A Large-Scale Study in the Orkut Social Network. In: KDD 2005: The Eleventh ACM SIGKDD International Conference on Knowledge Discovery and Data Mining (2005)
22. Stolpmann, M., Wess, S.: Optimierung der Kundenbeziehung mit CBR-Systemen: Intelligente Systeme für E-Commerce und Support, 1. Auflage. Addison-Wesley, Bonn (1999)
23. Tasan, M., Ozsoyoglu, Z.M.: Improvements in Distance-Based Indexing. In: Proceedings of the 16th International Conference on Scientific and Statistical Database Management – SSDBM 2004, pp. 161–170 (2004)
24. de Tre, G., Matthé, T., KordJamshidi, P., Demoor, M.: On the Use of Case Based Reasoning Techniques in Flexible Querying (invited paper). In: Wagner, R., Revell, N., Pernul, G. (eds.) DEXA 2007. LNCS, vol. 4653, pp. 345–349. Springer, Heidelberg (2007)
25. Wilke, W., Lenz, M., Wess, S.: Intelligent Sales Support with CBR. In: Lenz, M., Bartsch-Spörl, B., Burkhard, H.D., Wess, S. (eds.) Case-Based Reasoning Technology: From Foundations to Applications, pp. 91–113. Springer, Heidelberg (1998)
26. Zezula, P., Amato, G., Dohnal, V., Batko, M.: Similarity Search: The Metric Space Approach. Springer, New York (2006)
27. Zhou, X., Wang, G., Xu Yu, J., Yu, G.: M+-tree: A New Dynamical Multidimensional Index for Metric Spaces. In: Proceedings of the 14th Australasian database conference – ADC 2003, vol. 17, pp. 161–168 (2003)

A Distributed System for Massive Generation of Synthetic Video Using GPUs*

Javier Paris, Victor Gulías, and Carlos Abalde

Departamento de Computacion, Facultade de Informatica, Universidade da Coruna
{javierparis,gulias,cabalde}@udc.es

Abstract. Audio-visual interactive content is very common nowadays. It has many applications in very different fields, from videogames to visualization of scientific data. However, there are environments such as digital television in which the delivery of interactive content is of interest, but are limited by the shortcomings of the players. For example, in cable TV environments users access content through a set-top box, which is usually very limited in computing power due to cost, power consumption and the need to keep a moderate size. Furthermore, set-top boxes do not usually have specific hardware for graphics processing (GPU, Graphics Processing Unit) desirable for high quality interactive content, but rather are optimized for real time decoding of video in hardware (usually Mpeg-2, in very recent ones h.264). In this work we describe a distributed system for the creation of synthetic content and its encoding to digital video to send it to the clients. The most important features to provide are scalability and fault tolerance, in order to support a large number of concurrent users with an uninterrupted service.

1 Introduction

Digital Television has been growing over the past decade. It is distributed by satellite, cable, terrestrial, and more recently, in mobile phone networks. Due to the spreading of its use there are now lots of devices which can decode digital video in real time. Most of these devices have scarce computing power for other tasks because of design requirements such as low cost for the set-top boxes used in satellite, cable and terrestrial TV or small size and low power consumption in mobile TV. However, as decoding digital video is computationally intensive, they usually include some kind of hardware decoder to accelerate the process.

Due to the appeal that interactive content has to the users, Digital video broadcasters would like to deliver interactive content to their clients. The problem is that Interactive content is most usually delivered in formats adequate for computers and the user accesses the services through devices which, as we said before, are limited in computing power. This devices rarely have a CPU capable of running the programs to generate content with the quality that users expect.

* Partially Supported by Partially supported by Xunta de Galicia PGIDIT07TIC005105PR and MEC TIN2005-08986.

R. Moreno-Díaz et al. (Eds.): EUROCAST 2009, LNCS 5717, pp. 239–246, 2009.

The cost of providing all the users of a broadcaster with an adequate device to access interactive content is not feasible (Broadcasters in cable networks usually select one set-top box over another because of differences in price of as little as 20€).

To address this lack of computational power without unreasonable costs we propose to move the generation of the synthetic content to the server side. This makes it possible to build a system sized to fit the expected usage of the system, instead of having to provide all the potential users of the system with a suitable device. The content can be delivered to the clients in a format which, as was said before, all the devices have special hardware to decode: digital video.

In this work we present a system which generates interactive synthetic content in a cluster of servers, encodes it into digital video, and streams it to a client device with low computing power but special hardware to decode video. The rest of the paper is structured as follows: In section 2, we present the global design of the system, and how the video generation is approached. In section 4, we show how to scale the system and provide fault tolerance using distributed techniques. Next we show in section 3 how the user interaction is going to be managed. Finally, we present our conclusions in section 5.

2 System Design

The problem addressed in this work is the massive distribution of complex interactive content in a digital television network. In this environment there will be a large amount of users who will access the system through some digital television distribution network (for the purposes of this work, the distribution network is not specified, but it is supposed to be any that has a reasonable return channel, such as the Internet, or a DVB-C or DVB-H network). The interactive content must be generated in real time, which poses a challenge. A typical application for the generation of interactive content in a computer (by far the most common interactive content application) can be seen in figure 1. The user provides input commands to the application, which generates content[1] which generates content that is displayed in the screen. The main goal of this system is to move the application to a server and distribute the content as video, so that it can be used in simple devices such as mobile phones or set-top boxes.

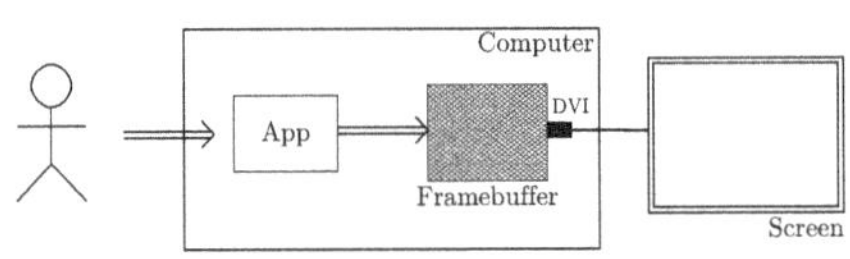

Fig. 1. Structure of an usual interactive synthetic application

[1] For simplicity, only video content is shown in the figures. However, audio is also considered in this work.

2.1 Hardware Encoder

The simplest approach to this problem is to connect a hardware encoder to the DVI output of the computer. This approach is shown in figure 2. There are several problems to this approach:

- Managing user input. In a computer the user controls the application directly, but in this environment the user input must be somehow carried back to the server where the content is being generated. This problem affects all the remote generation approches, and will be addressed more in detail in section 3.
- Scalability. The biggest problem is that this approach requires a hardware encoder for each concurrent user in the system. The cost of a professional quality encoder is approximately 4000€, which is clearly too expensive, as it would be cheaper to equip all potential users with high cost devices.

However, this approach provides an implementation reference. Any system developed must provide at least the same features as this, but trying to reduce the cost.

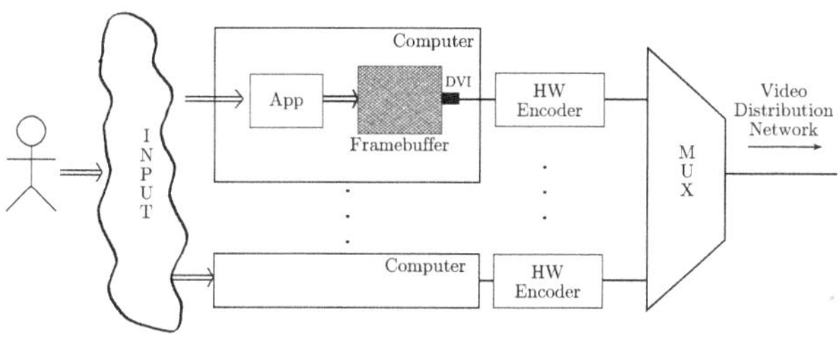

Fig. 2. Initial approach to a remote interactive content delivery

2.2 Software Encoder

Using a software encoder instead of the hardware encoder mitigates the cost problem. Figure 3 shows the architecture of this solution. The software encoder makes it easier to run more than one application in each computer. Although this was also possible using hardware encoders the limiting factor in that case is the cost of the encoder. Adding several GPUs to each computer to have several DVI outputs would only barely reduce the cost.

Running more than one application at a time in each server is possible because the video to be generated does not require high definition. The devices used to display the content will be televisions and mobile phones, which have very low resolutions. This reduces the strain on the CPU and makes it possible to serve more than one client with each server. However, the scalability of this solution is limited by the CPU performance. Real time encoding is a time consuming operation, and the applications which generate synthetic interactive content are usually heavy. All this places a limit in the performance of this approach.

However, this approach is easy to implement and relatively inexpensive, so it is adequate for initial tests of unrelated parts of the architecture, such as user interaction management, or the application checkpointing service.

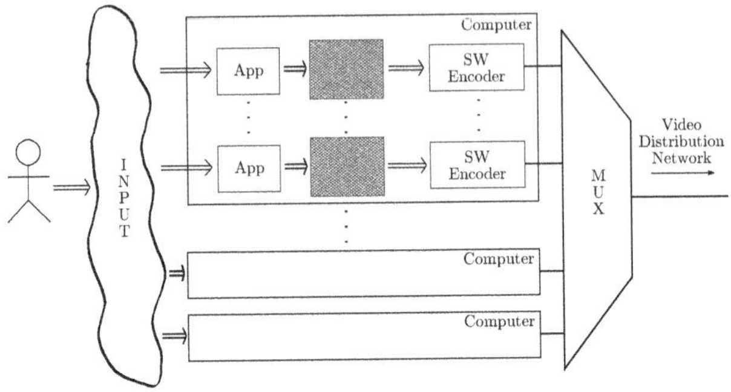

Fig. 3. Remote interactive content using software encoders

2.3 GPU Encoding

The next approach is to try to encode the video inside the GPU. The GPU architecture is well suited for video encoding (see [4]), as video encoding is a highly parallel task which GPUs are very good at. While GPUs can be expensive they are much cheaper than a hardware encoder, and at the same time are capable of supporting a larger number of simultaneous users than using software encoders. Furthermore, each server may host more than one GPU (using PCI-E cards), which reduces the costs. The CPU is going to be the limiting factor again, but in this case the only task it has is to run the applications, as the video encoding will be done in the GPUs.

Another interesting feature is that using GPUs the framebuffer will be placed in the Video Card RAM. The advantage is that, as the video generation takes place there, there is no need to move framebuffer data through the PCIe bus. When the encoding is complete, the video will go through the bus, but encoded video will have a much lower bitrate than raw framebuffer data.

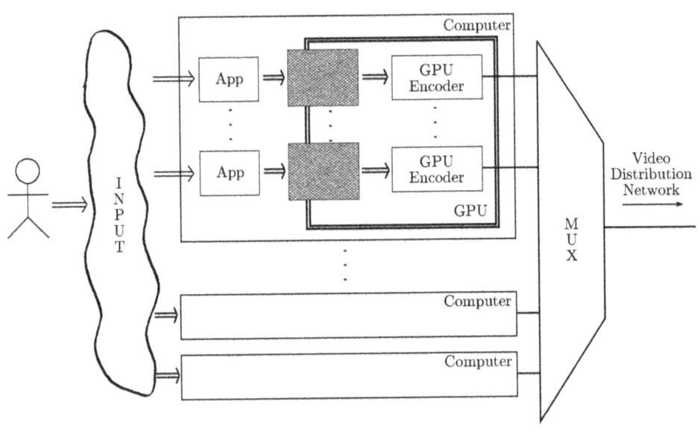

Fig. 4. GPU Encoding

The number of users that the system can support with this approach is much larger than with any of the previous ones, but it will ultimately be limited by the demands of the interactive applications.

2.4 Generic Encoding

As more than one approach will be explored, it is desirable to be able to adapt the system to several different ways of generating the video. The idea is to create two generic APIs that will be implemented by all the encoding modules, so that they are interchangeable (see figure 5).

The design of a generic API is not easy due to the very different nature of the encoding approaches used. As an example, in the GPU technique it is likely that part of the frame producer runs in the GPU, whereas in the software encoder all will be done in the CPU. Similarly, the hardware encoder will require a different consumer than the software and GPU-based encoders.

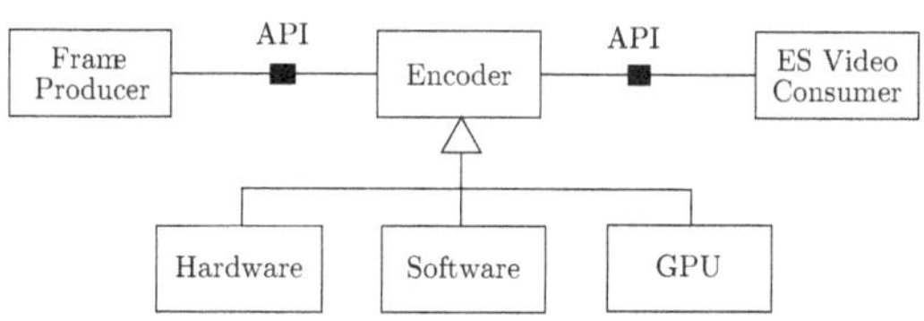

Fig. 5. Encoding Architecture

3 User Interaction

User interaction is a critical part of any interactive system. It provides the user with the capacity to change the behaviour of the system. In most interactive systems, the user is interacting with the same computer that runs the application. In this system the application runs in a server separate from the user. Furthermore, the content is being delivered as video to devices like mobile phones or set-top boxes. Any interaction mechanism will have to be ad-hoc for the device, and use whatever controls are available to the user (the keypad in the phone, or the remote control in the set-top box).

In the Cell phone, the control mechanism would be through a program run on the phone which sent commands through the telephone network. In the set-top box, the remote commands would be translated to requests either through the DVB network, or through the Internet using a different communication channel.

Something to be considered in this system is the response time. Video is going to be encoded to MPEG-2, because its simplicity makes it easier to implement in a GPU. However, encoding to MPEG-2 imposes a delay of about 1 second, so the interactions of the user will have at least that delay before he can see any effect on his screen. This limits the kind of interactive applications which can

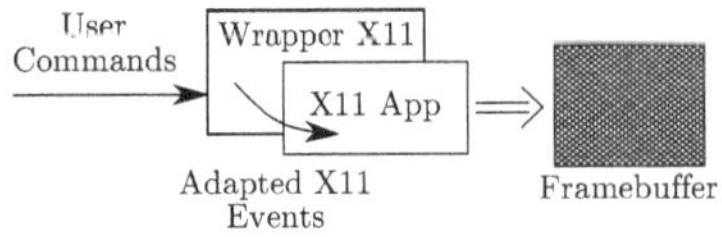

Fig. 6. Adaptation of a Common Application

be used in the system. We hope to adopt lower latency codecs in the future to mitigate this problem.

The system must provide a generic API for the different mechanism so that user input arrives through a standard interface to the applications regardless of the user device.

It is also desirable to be able to use standard interactive applications in the system without having to create specific ones. For this purpose, an adaptor for X11 applications (Figure 6) has been designed. This adaptor translates the input commands from the standard API to X11 events, so that the applications may receive the remote commands as normal X11 input events. As the system takes the application output from the framebuffer, this makes it possible to run unmodified X11 applications.

4 Distributing the Problem

The system described in this work is intended for environments like a cable-tv or a cell phone network, with a large number of simultaneous users (see [2]). It is also likely to be a paid-for service, so uninterrupted service is desirable. As it has been shown in previous figures, the approach used is to design a distributed system in which there will be a certain amount of servers (depending on the expected load) with several GPUs to serve the incoming clients.

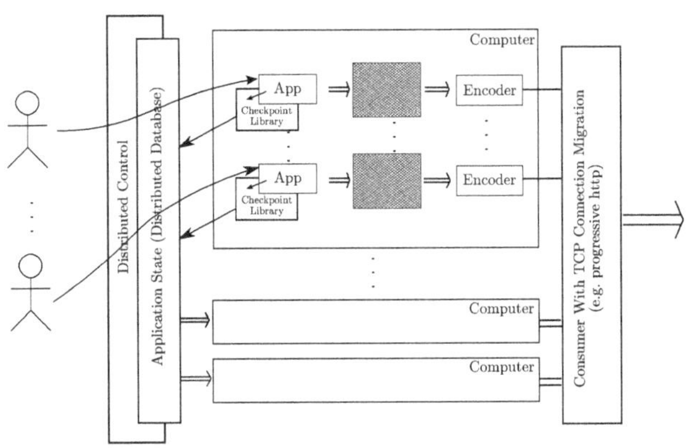

Fig. 7. System Overview

In the previous sections we have described the necessary components for a single server, from dealing with the user input to the encoding framework. In this section we will describe the distributed components (see figure 7) that bind all the servers in the system together:

- The distributed control that schedules incoming requests and manages the return channel with the user input.
- The checkpointing system used for fault tolerance.
- The video consumer that distributes the generated video to the clients.

4.1 Distributed Control

The distributed control has several responsibilities:

- Assign incoming requests from clients to one of the free resources (computer+encoder) in the distributed system.
- Monitor the servers in the systems for failures, and act whenever there is one. This includes recovering the applications that were running in the failed server if it is possible.
- Manage the input from the users and redirect it to the right application.

The control itself is distributed as an Erlang[1] distributed application, so that there are no single points of failure.

4.2 Checkpointing

One of the design goals of the system is that it is capable of delivering uninterrupted operation. To mitigate the effects of the failure of a server there is a checkpointing API for applications to save a representation of their state (a memento). This memento is stored in a distributed database. In case of failure in one of the server, the applications which had their state saved in the database are restarted in other servers. This service is optional for applications, if they choose not to support it the only drawback is that they will not be restarted in case of failure.

Depending on the video transmission network this restarting mechanism may be coupled with a fault tolerant TCP/IP stack[3] which can recover TCP connections in other nodes (for example, if the video is transmitted over the Internet using progressive HTTP).

4.3 Consumer

The consumer is the component that translates the generated video (which is a MPEG2 elementary stream) into something suitable for transmission. Depending on the network in which it will be transmitted, it can be a MPEG2 transport stream (in a DVB-C network), an UDP or TCP packet with a MPEG2TS inside (in the Internet), or any other packet type.

5 Conclusions

We have presented the design for an interactive content distribution system. The design presented is able to provide uninterrupted operation, deliver video over different transmission networks, and bring remote interactivity to the user in a transparent way.

The distributed design of the system brings several advantages. Distributed systems are usually a cheaper alternative to high cost monolithic systems. It is also easier to provide fault tolerance in a distributed system because of the nature of the system. They tend to be highly scalable as well. This features have been included in this design as there is no single point of failure, and the tasks of the distributed control are not heavyweight.

References

1. Armstrong, J.: Programming Erlang: Software for a Concurrent World. Pragmatic Bookshelf (2007)
2. Gulias, V.M., Barreiro, M., Freire, J.L.: Vodka: Developing a video-on-demand server using distributed functional programming. J. Funct. Program 15(3), 403–430 (2005)
3. Paris, J., Gulias, V.M., Valderruten, A.: Developing a functional tcp/ip stack oriented towards tcp connection replication. In: Proceedings of the 2005 IFIP/ACM Latin American Networking Conference (2005)
4. Thompson, C.J., Hahn, S., Oskin, M.: Using modern graphics architectures for general-purpose computing: a framework and analysis. In: MICRO 35: Proceedings of the 35th annual ACM/IEEE international symposium on Microarchitecture, pp. 306–317. IEEE Computer Society Press, Los Alamitos (2002)

Using a Rank Fusion Technique to Improve Shot Boundary Detection Effectiveness

M. Eduardo Ares and Álvaro Barreiro

IRLab, Department of Computer Science, University of A Coruña,
Campus de Elviña s/n, 15071, A Coruña, Spain
{maresb,barreiro}@udc.es
http://www.dc.fi.udc.es/irlab

Abstract. Achieving high effectiveness in Shot Boundary Detection (SBD) paves the way for high-level analysis of the video (keyframe extraction, story segmentation, etc.), which makes this step very important. Thus, the SBD problem has been extensively addressed, and many approaches have been proposed. As these approaches have their own different strengths and weaknesses, merging the outcomes of different detectors in order to obtain a better detector comes naturally. In this paper we propose an approach to SBD which takes into account the outcomes of two shot boundary detectors, using a rank fusion technique. This new detector is tested with videos from the TRECVid initiative, finding that it outperforms the two original methods. Moreover, the computation of this merging method is very fast, so it is very attractive for an operational environment.

1 Introduction

Nowadays, multimedia (and specially video) information has achieved a great importance in our society. Arguably, the best example of this situation is television, which has risen to a privileged position among the mass media, rivalling in followers and influence with newspapers. Besides, in the last few years, a great number of video repositories have appeared. Those popular webpages host an ever-growing collection of videos, letting users upload, search and browse them.

Consequently, in the last two decades there have been great efforts to develop techniques to perform automated video analysis and retrieval, like those available for text. Maybe the most significant proof of this trend was the concern that TREC [1] (Text Retrieval Conference) showed in that subject. That interest led in the first place to the creation of the TREC Video Track which in 2003 was turned into a initiative of its own, the TREC Video Retrieval Evaluation (TRECVid)[2], being the showcase of the new developments and advancements on automated video processing.

Almost all techniques developed for video analysis take the *shot* as retrieval unit. A shot is a video sequence which has been taken with the same camera without cuts in between, and is widely regarded as the minimal meaningful

R. Moreno-Díaz et al. (Eds.): EUROCAST 2009, LNCS 5717, pp. 247–254, 2009.

portion of a video, hence its importance to video analysis. Even when shots are not the outcome of the system (for instance, in TV News retrieval systems, where the retrieval unit is the *story*), the shots are used as building blocks of those bigger retrieval units. Thus, segmenting a video stream into shots is a very important process, whose effectiveness has a huge impact on the performance of the whole system. This process is called Shot Boundary Detection (SBD).

The boundary between shots can be categorised in two types, attending to the abruptness of the the transition: *hard cuts*, in which the transition happens instantly, and *gradual transitions*, in which the transition spans some frames. The detectors use the expected similarity between frames of the same shot, characterising the frontiers between shots as frames that bear a low visual similarity with previous ones. This similarity between frames is measured with features related with their visual content, being the main difference among SBD methods the choice of these features and the way they are used. Furthermore, the techniques used to detect one type of transition may (and often will) not be useful to detect the other type, which makes SBD even harder.

Due to its importance and difficulty, the SBD problem has been extensively addressed. The existing works have proposed a lot of features to measure the similarity of the frames: the plain absolute difference between pixels, the similarity between colour histograms[3], motion-compensated pixel differences[4], similarity between edge images[5]... The researches have also proposed many ways of using these features, which range from the simplest approaches, based on thresholding one feature, to more complex approaches, such as adaptive thresholding[3,6], statistical modelling[4], combining several features using rules[7,8] and machine learning techniques [9]. This intense research in the field and the great results that have been achieved have led to some authors to deem the SBD problem as almost resolved[10]. However, and precisely due to the problem's importance to the whole video retrieval process, there are still some aspects worth studying.

In this paper we propose an approach to SBD based on using a voting technique (the *Borda count*, which has been successfully used in metasearch to merge ranks[11]) to merge the outcomes of different detectors. Since each Shot Boundary Detector has its own strengths and weaknesses, creating a detector that improves their performance merging them comes naturally, making possible to reach higher effectiveness avoiding the need for a laborious parameter tuning. And even when the base performance is very good, a little improvement, whatever small it may be, is important. Furthermore, the merging method proposed is very fast and does not need any parameter tuning. Its simplicity and speed makes it very attractive for operational environments.

This work is focused on testing the feasibility of this method trying to merge the results of two simple shot boundary detectors to detect hard cuts, due to their relative simplicity compared with gradual transitions. The experimentation carried out showed that the aggregated method outperforms the methods it aggregates, providing an improvement which ranges from 3.5% to 15.6%.

Next, section 2 details the approach we are proposing; section 3 specifies how the evaluation was carried out; section 4 lists the results obtained in that evaluation and section 5 concludes and discusses the future work.

2 Details of the Approach

The approach to SBD proposed in this paper merges the outcomes of two detectors (a block matching detector and an edge-based detector) using a ranking merging method (the Borda count). These detectors were chosen because they are two well-known detectors, which are relatively simple and whose strengths are complementary.

2.1 Detector 1: Block Matching Detector

This detector uses a motion compensation approach similar to the one proposed in [4]. The next steps are followed when trying to compare frames f_i and f_{i+1}:

1. The frames' size is reduced (or reduced versions are taken directly from the video stream, using the DC coefficients).
2. The frames are splitted in n non-overlapping blocks
3. The best match for each block b_j of f_{i+1} is searched in f_i:
   - The search area for the best match is the region of f_i composed by the block of f_i which is located in the same position as b_j and the pixels which lay in a certain neighbourhood of that block.
   - The difference between two blocks b and b' is measured in terms of the square differences between their pixels in all colour channels

$$\text{block difference}(b, b') = \sum_{x,y,c \in \text{channels}} (b(x,y,c) - b'(x,y,c))^2 \qquad (1)$$

   - For each b_j, the best matching block is the candidate which minimises this difference value.
4. The difference value between f_i and f_{i+1} is the sum of the difference values (1) between each b_j and its best matching candidate.

2.2 Detector 2: Edge-Based Detector

This detector is an implementation of one of the detectors proposed in [5]. It uses an approach based on comparing the edges detected in the frames. In order to compare two frames f_i and f_{i+1} the next steps are followed:

1. Both images are resized to half their size and converted to greyscale.
2. A Sobel edge detector is applied to both frames.
3. The edge images e_i and e_{i+1} are created from the results of step 2, taking the points whose edge intensity is greater than a certain threshold t as 1 (*edge*) and the others as 0 (*non-edge*).
4. A dilation morphological operation is applied to e_i and e_{i+1} ($\overline{e}_i$ and $\overline{e}_{i+1}$).

5. The ratios of "entering" (edge pixels in f_{i+1} which are not so in f_i) and "exiting" (edge pixels in f_i which are not so in f_{i+1}) edge pixels are calculated:

$$p_{in} = \frac{\sum_{x,y} e_{i+1}(x,y)\bar{e}_i(x,y)}{\sum_{x,y} e_{i+1}(x,y)}; p_{out} = \frac{\sum_{x,y} e_i(x,y)\bar{e}_{i+1}(x,y)}{\sum_{x,y} e_i(x,y)} \quad (2)$$

6. The difference value between f_i and f_{i+1} is the maximum of p_{in} and p_{out}.

In both detectors, the steps are applied all along the video to every pair of consecutive frames.

2.3 Rank Aggregation: Borda Count

The Borda count is a voting technique which has been successfully used in metasearch to merge document rankings [11]. We will use the simplest version of Borda count, where all voters are equal and their opinions are given the same weight. Given a set of n candidates and k voters, it follows the next steps:

1. each voter creates a ranking of the n candidates
2. each voter assigns n votes to the first candidate in its ranking, $n-1$ votes to the second, and so on.
3. the votes for each candidate are added
4. the aggregated ranking is created according to the scores calculated in 3.

The system we propose in this paper uses the Borda count in order to merge the outcomes of the two presented shot boundary detectors. Given a video composed by n frames (from f_1 to f_n), the *candidates* are the $n-1$ possible pairs composed by consecutive frames. Thus, the candidate c_1 would be composed by f_1 and f_2, c_2 would be composed by f_2 and f_3 and so on until c_{n-1}, which would be composed by f_{n-1} and f_n. On the other hand, the *voters* are the shot boundary detectors. For each candidate, the detectors calculate the difference value between the frames the candidate is composed of. Then, each detector ranks the pairs of frames according to this difference value, and each candidate is given a number of votes depending of its position in this ranking, where candidates with higher difference values are ranked higher. These votes are the only output of each detector. In other words, the difference values used by each detector are not considered further, avoiding the need for the normalisation of these scores (this is one of the most important advantages of Borda count). Once the votes of the two detectors are calculated, and as it was explained above, they are totalled. These sums are the outcome of the system, and the new difference values.

2.4 Cut Detection

The methods presented output a list of difference values for the transitions between each pair of frames. Once these values are calculated, a criterion has to be defined on how to use them to detect the cuts. A lot of criteria have been proposed, ranging from simple thresholding to adaptive approaches.

3 Evaluation

To test the feasibility of an approach to SBD based on ranking aggregation we will compare the effectiveness of the method explained in Sect. 2 with the two detectors (the Motion compensation based detector and the Edge based detector) on their own. As we are proposing and testing the first sketches of a new approach to combining evidences in SBD we have focused solely on the hard cuts, because of their simplicity compared with gradual transitions. Moreover, the hard cuts are the most common transition type in almost all video genres, so a good hard cut detection is mandatory for a successful SBD system. A proof of their importance are the ratios of transition types in the videos used in the SBD task of TRECVid. Since its inception in 2003 (and also in the TREC track) the hard cuts were the most common transition type, reaching in 2007 (the last year this task was considered) a ratio of 89.5%[10]. In that edition, two of the fifteen participating groups (U. Sheffield and U. Brno) focused exclusively in hard cuts.

In order to assess the effectiveness of our approach we have used a collection of five videos used in the TRECVid in years 2001 and 2005, which are in the public domain (Table 1). The human annotated ground truth marking the hard cuts present in these videos was obtained from the TRECVid page.

To measure the effectiveness of the detectors we have used precision (ratio of cuts which have been correctly detected) and recall (portion of the proper cuts on the video which have been detected), which are used in the TRECVid SBD task[10]. So as to summarise these two metrics we have used the F-measure (3), which gives the same importance to precision and recall.

$$\text{F-Measure} = \frac{2 \times \text{precision} \times \text{recall}}{\text{precision} + \text{recall}} \tag{3}$$

The detectors presented in Sects. 2.1 and 2.2 depend on some parameters. After testing several different settings, we chose the ones which showed the best results. For the block matching detector the reduction ratio of the frames was set to $\frac{1}{8}$ of the original size. The block size was set to 4×4 pixels, and the zone to search was 12 pixels. On the other hand, for the edge based detector the threshold of the intensity of the edges to detect the edges (t) was set to 200, and the structuring element used in the dilation step was a square of 3×3 pixels.

In the experiments the detectors explained in Subsects. 2.1, 2.2 and 2.3 were applied to the video collection. In the aggregated method a transition between a

Table 1. Video collection

Identifier	Video name	Length (frames)	Hard cuts
`anni005`	NASA 25th Anniversary Show, Segment 5	11364	38
`anni009`	NASA 25th Anniversary Show, Segment 9	12307	38
`NASAConnect-HT`	NASA Connect - Hidden	50823	143
`NASAConnect-AO`	NASA Connect - Ancient Observatories	51290	67
`NASADT18`	NASA Destination Tomorrow 18	51299	135

pair of frames was labelled as *cut* if and only if its number of votes was greater than a certain threshold t, which was the same for the whole video. In the other approaches the thresholding was made on the ranking position of the difference value of the pair of frames. For these methods, a threshold of t means that a transition was labelled as *cut* if and only if its difference value was above t difference values or more corresponding to other transitions of the video. Note that this has the same effect as thresholding the number of votes which that candidate would be given by that method in the aggregated approach. We have chosen this technique due to its homogeneity and simplicity, avoiding external factors which could influence the effectiveness of the system and the tuning of a more complex method. The best threshold for each method was set manually, testing a set of thresholds. For the aggregated method the thresholds tested ranged from 99.90% to 99.00% of the votes of the most voted candidate, while in the block matching and the edge based approaches the thresholds tested ranged from 99.90% to 99.00% of the total number of transitions. Then, the threshold which reached the best F-Measure was selected. This methodology was chosen in order to test the effectiveness of the proposed rank aggregation technique, regardless of the threshold calculating method. Obviously, this approach is not suitable for real world applications, where the threshold must be calculated automatically.

4 Results

We present in Table 2 the results of the three methods tested (block matching, edge based and aggregated) for each video in the collection. The results shown are the F-Measures obtained with the best threshold (see Table 3) and the improvement achieved by the aggregated method over the best performing of the two plain shot boundary detectors.

4.1 Discussion

In all videos the aggregated method we propose in this paper outperforms the methods it is composed of. Even though this was the expected behaviour, there are some aspects worth remarking:

First, and even though the effectiveness of the aggregated method is obviously dependant on the effectiveness of the methods merged, it provides an improvement of the performance which ranges from 3.4% to 15.6%. This improvement

Table 2. F-measure for each method and improvement achieved over the best performing detector

Video	anni005	anni009	NasaConnect-HT	NasaConnect-AO	NASADT18
Block matching	0.904	0.880	0.818	0.773	0.692
Edge	0.932	0.640	0.761	0.255	0.591
Borda count	0.961	0.911	0.930	0.887	0.800
Improvement	3.1%	3.5%	5.0%	11.4%	15.6%

Table 3. Cut ratios and Best thresholds for all videos

Video	Cuts ratio	Best threshold
anni005	00.33%	99.60%
anni009	00.31%	99.30%
NasaConnect-HT	00.29%	99.60%
NasaConnect-AO	00.13%	99.60%
NASADT18	00.26%	99.60%

is higher in the videos where the two simple methods perform worse. It should be also noted that this improvement of the results is achieved in all videos, regardless of the difference of effectiveness between the proposed detectors. In the results of NasaConnect-AO it can be noted how the outcomes of a detector which performs fairly well and another one which performs very poorly are merged without degrading the performance of the first one, and in fact improving it. This shows the robustness of the merging method chosen.

Moreover, as shown in Table 3, the best threshold for the aggregated method is stable along the collection (almost always the 99.60% of the maximum value), regardless of the varying cut ratio. This fact should be taken into account when devising a strategy to calculate automatically the threshold.

In order to assess the statistical significance of the results we have performed a Wilcoxon test. The hypothesis were H_0: our method does not outperform the individual detectors and H_1: our method is better than the individual detectors. The test showed that the aggregated method is significantly better than the best of the others detectors in each video with a p-value < 0.05.

5 Conclusions and Future Work

In a general level, we should note that, as it has been previously stated, the objective of this work was to test the feasibility of a method to merge the outcomes of different shot boundary detectors based on the Borda count. According to the results presented in Sect. 4, its feasibility is proved. It is also worth remarking the great improvement achieved by a method which merges blindly the outcomes of other detectors, despite not having previous training or being imbued with domain knowledge. It should be also noted the simplicity and the low computational cost of the merging method: once the difference values are calculated, the time invested in merging them is negligible. Also, we are avoiding the need for a normalisation of the difference values and the problems that would be associated with that process, such as modelling of the outcomes of each method.

Future work should be aimed in two main directions: testing the behaviour of this method when trying to detect gradual transitions and developing a method to set automatically the threshold of the detection. Also, the method should be tested against a bigger video collection and with more detectors.

Moreover, and centring in the method itself, there are two main questions still worth addressing. The first one is trying to devise a way to avoid having to wait until the whole video is processed to have results. This could be a problem if there are restrictions of time (i.e. a cut should be detected in no more than a certain time since it happens) and specially if we are trying to segment long videos or if the detectors we are trying to merge are computationally expensive.

The second aspect that we think worth studying is using a weighted Borda count approach, which would enable us to weight more the votes of those detectors which seem more reliable.

Acknowledgements. This work is co-funded by *Ministerio de Ciencia e Innovación*, FEDER and *Xunta de Galicia* under projects TIN2008-06566-C04-04 and 07SIN005206PR. M. Eduardo Ares also wants to acknowledge the support of the FPU programme of *Ministerio de Educación* of Spanish Government.

References

1. Voorhees, E.M., Harman, D.K.: TREC: Experiment and Evaluation in Information Retrieval (Digital Libraries and Electronic Publishing). MIT Press, Cambridge (2005)
2. Smeaton, A.F., Over, P., Kraaij, W.: Evaluation campaigns and TRECVid. In: Proceedings of MIR 2006, pp. 321–330 (2006)
3. O'Toole, C., Smeaton, A.F., Murphy, N., Marlow., S.: Evaluation of automatic shot boundary detection on a large video test suite. In: Proceedings of CIR 1999 (1999)
4. Hanjalic, A.: Shot-boundary detection: unraveled and resolved? IEEE TCSV 12(2), 90–105 (2002)
5. Smeaton, A.F., Gilvarry, J., Gormley, G., Tobin, B., Marlow, S., Murphy., N.: An evaluation of alternative techniques for automatic detection of shot boundaries in digital video. In: Proceedings of IMVIP 1999 (1999)
6. Yeo, B.L., Liu, B.: Rapid scene analysis on compressed video. IEEE TCSV 5(6), 533–544 (1995)
7. Browne, P., Smeaton, A.F., Murphy, N., O'Connor, N., Marlow, S., Berrut, C.: Evaluating and combining digital video shot boundary detection algorithms. In: Proceedings of IMVIP 2000, pp. 93–100 (2000)
8. Liu, Z., Gibbon, D., Zavesky, E., Shahraray, B., Haffner, P.: A fast, comprehensive shot boundary determination system. In: Proceedings of IEEE ICME 2007, pp. 1487–1490 (2007)
9. Matsumoto, K., Naito, M., Hoashi, K., Sugaya, F.: SVM-based shot boundary detection with a novel feature. In: Proceedings of IEEE ICME 2006, pp. 1837–1840 (2006)
10. Over, P., Awad, G., Kraaij, W., Smeaton., A.F.: TRECVID 2007 overview. In: TRECVid 2007 - Text REtrieval Conference TRECVid Workshop (2007)
11. Aslam, J.A., Montague, M.: Models for metasearch. In: Proceedings of SIGIR 2001, pp. 276–284 (2001)

Step-Guided Clinical Workflow Fulfilment Measure for Clinical Guidelines*

Jose M. Juarez[1], Patricia Martinez[1], Manuel Campos[2], and Jose Palma[1]

[1] Dept. of Information and Communication Engineering, Universidad de Murcia, Spain
jmjuarez@um.es, patricia.martinez@carm.es, jtpalma@um.es
[2] Dept. of Languages and Systems, Universidad de Murcia, Spain
manuelcampos@um.es

Abstract. Health-care quality control is a challenge that medical services and their information systems will deal with in the following years. Clinical Practice Guidelines are a structured set of recommendations for physicians to solve a medical problem. The correct fulfilment of a guideline seems to be a good indicator of health-care quality. However, the guideline fulfilment checking is an open clinical problem that requires collaborative efforts. In this sense, clinical workflows have demonstrated to be an effective approach to partially model a clinical guideline. Moreover, some efforts have been done in consistency checking and debugging management in order to obtain a correct description of the clinical processes. However, the clinical practice not always strictly fulfils clinical workflows, since patients require personalised care and unexpected situations occur. Therefore, in order to obtain a workflow fulfilment degree, it seems reasonable to compare *a posteriori* the evolution of the patient records and the clinical workflow, providing a flexible fulfilment measure. In this work we present a general framework to classify and study the development of these measures, and we propose a workflow fulfilment function, illustrating its suitability in the clinical domain by an example of a real medical problem.

1 Introduction

A Clinical Practise Guideline (CPG) is a set of recommendations/rules developed in some systematic way in order to help professionals and patients during the decision-making process concerning an appropriate health-care pathway, by means of opportune diagnosis and therapy choices, on specific health problems or clinical conditions [4]. The common use of CPG has demonstrated to improve the health-care processes and to reduce their cost [1].

Moreover, the correct use of CPG for each patient can be considered a good quality indicator for health-care processes. The key question is how to check and quantify the correct application of the CPG. Despite the efforts done in the medical field to check the CPG fulfilment [1], this is still an open clinical problem that requires collaborative efforts. In this sense, advancements in medical computer-based systems have meant a sounded impact in the medical activity, such as the Health Information System (HIS),

* This work was partially supported by the Spanish MEC under the national projects TIN2006-15460-C04-01, PET 2006-0406 and PET2007-0033.

R. Moreno-Díaz et al. (Eds.): EUROCAST 2009, LNCS 5717, pp. 255–262, 2009.

the Electronic Health Records (EHRs), but also CPGs. Therefore, it seems reasonable to consider that this task (to measure health-care quality) could be partially solved by computer methods, providing quality indicators based only in the recorded data within EHRs and the HIS.

Main efforts on Computerized CPG are focused on the quality improvement and the optimisation of financial and human resources, and patients' care. The basic principles for the assitential quality through such tools consist of 1) reducing the variability during resources utilization, 2) improving the efficiency of such resources, and 3) reducing the cost of the whole process. The first step in order to produce Computerized CPG is the modelling process. On one hand, specific languages have been proposed for modelling CPG such as GLIF3, ASBRU or GLARE.

On the other hand, workflow models are useful formalisms to model some aspects of the CPG. Workflows are collections of organised tasks to carry out some process made by software systems, groups of people, or the combination of both [8]. Workflow Management Systems (WfMS) provide support for modelling, executing and monitoring the workflows. Traditional WfMS states an effective strategy to assist information processes driven by the workflow, especially in industry. Few efforts have been done in checking the workflow consistency on modelling and handling unexpected actions [5,6]. These proposals can be considered and *a priori* approach for checking the correct execution of the workflow, since any kind of variation from the standard information process must be specified. However, due to the complexity of the medical knowledge (imprecise, incomplete, and multidisciplinary), clinical environments must allow a flexible performance of the workflow (when represents a CPG). For instance, physicians must solve unexpected medical problems of a patient daily. In other words, the activities done by physicians (registered in the EHR) could not strictly follow the workflow. Therefore, it seems reasonable to consider that the difference between the workflow and the activities executed can be measured *a posteriori*, quantifying the accomplishment of the CPG for a patient.

This paper is a preliminary work on the analysis of the fulfilment of CPGs represented by Clinical Workflows based on *retrospective* checking of the EHR. The remainder of this paper is organized as follows. In Section 2 we address the clinical workflow principles. In Section 3 we present a general framework to classify workflow fulfilment approaches, we propose a workflow fulfilment measure and we present a practical example in the medical domain. Section 4 draws a conclusion and future works.

2 Clinical Workflow

Workflows are traditionally dedicated to the representation of business tasks, but there is an increasing interest in the medical field to represent CPG by workflow models, named Clinical Workflows [9,3].

In short, we define a workflow as:

$$W = < \mathbb{T}, \mathbb{C}, \mathbb{E} > \tag{1}$$

$$\mathbb{T} = \{< taskname, id >\} \tag{2}$$

$$\mathbb{C} \in \wp(\zeta) \tag{3}$$

$$\mathbb{E} \subseteq (\mathbb{T} \times \mathbb{T}) \cup (\mathbb{T} \times \mathbb{C}) \cup (\mathbb{C} \times \mathbb{T}) \cup (\mathbb{C} \times \mathbb{C}) \tag{4}$$

where $\mathbb{T}$ is the set of task that can be performed, $\wp$ is the power set, $\mathbb{C}$ is the set of control nodes (characterized by $\zeta = \{BE, EN, XORs, XORj, ORs, ORj, ANDs, ANDj\}$) and $\mathbb{E}$ is the set of edges that state the precedence between tasks, control nodes or both.

In medicine, another key concept is the pathway, a set of the taken care of plans characterized by a procedure applied to a clinical scenario in the time [7]. In other words, a pathway is the subset of tasks and split-paths in the Clinical Workflow followed by the medical team for a concrete patient. When a pathway is selected, the workflow can be executed.

A workflow case (C), the execution of a workflow, is the set of performed tasks of a workflow (named case tasks, $ct \in CT$). A case task is a tuple $< taskName, temp\_exec >$ where $taskName$ is a task label, and $temp\_exec$ is defined by the timestamps t_b, t_e, describing the beginning and ending time of when the performed task occurred.

3 Measuring Fulfilment of Clinical Workflows

From a methodological perspective, the workflow fulfilment checking is a complex issue that could be interpreted by several ways and it could be composed of different tasks. In this section we propose a framework to analyse the possible alternatives to develop a workflow fulfilment checking. Then, we propose a fulfilment measure for clinical workflows and we illustrate it with a real piece of CPG represented by a workflow schema.

3.1 Strategies of Workflow Fulfilment Measures

There is not a consensus about the meaning of the concept *fulfilment measure* since it could indicate if a case fulfils workflow or not (classification), it could identify the concrete parts of workflow that is being fulfilled (to diagnose), or it could establish a measurement of the fulfilment degree (quantification). Moreover, different methods could be applied to solve this tasks and the workflow languages play an essential role since they state the flow control of the tasks (e.g. control nodes or edges).

In order to characterise the alternatives for the development of workflow fulfilment measures, we identify three dimensions (see Figure 1): technique, problem and task.

Technique Dimension establishes the different ways to solve the problem. *Step-Guided* approach attempts to find the decisions made for each condition when the workflow case was performed, checking step by step the path of the workflow schema followed by the case. Unlike *Step-Guided*, *Pathway* approach requires a pre-processing step that calculates all possible (without loops) pathways of the workflow schema. Then, the fulfilment checking means to find the pathway followed by the case. *Brute-force* is similar to pathway approach but considering even the execution of tasks in loops (useful only when the number of loops is known beforehand).

Problem Dimension raises the different elements that we are going to find in workflow and how treat them, depending on the workflow language.

In this work, we chose YAWL (Yet Another Workflow Language) [11]. This workflow language is based on the patterns of workflow with high level Petri Nets, but it

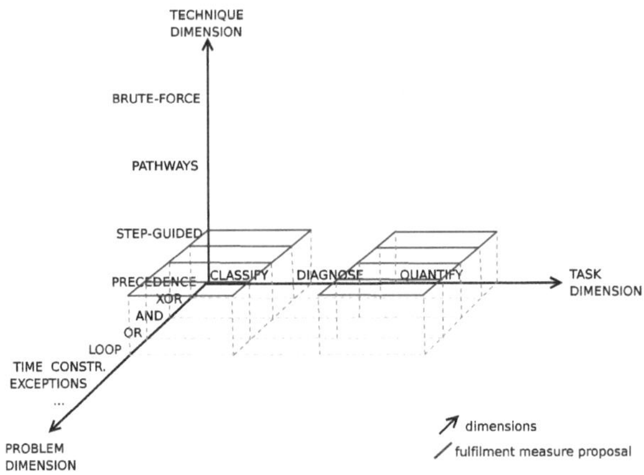

Fig. 1. Framework

extends this type of networks with additional characteristics for a better representation of workflows. Therefore, we enumerate (but not limited to) some of the most commonly used flow-control components: begin, end, or, xor, and, precedence edge, loops, temporal constraints and exceptions.

Task Dimension concerns the nature of the fulfilment measure. We consider three possible tasks. *Classify* is the process of deciding if the case belongs to the set of cases that strictly fulfils the workflow schema. A more general approach could state the pertenence degree to this set, *quantifying* the degree of fulfilment. Finally, other approach could not only obtain a fulfilment degree, but also to diagnose it, that is, to identify which elements of the workflow schema are not fulfilled by the case.

3.2 Fulfilment Measure Proposal for Clinical Workflows

In this work, we present a workflow fulfilment measure based on similarity techniques. Our proposal uses different similarity functions, depending on the connector nodes of the workflow, to quantify the fulfilment degree.

In order to simplify the computational model, we assume that workflow schema does not have nested connectors. However its extension for more complex schemata could be obtained using recursive calls of the functions proposed.

Similarity Measures. Similarity, taken from the cognitive psychology idea, is a key concept of many Artificial Intelligence approaches (e.g. case-based reasoning). In general, a similarity function is a normalised binary function that tries to quantify how similar two elements are. Therefore, it seems reasonable to think that the fulfilment measure of a workflow could be based on the similarity degree between parts of the workflow schema (composed by tasks) and the workflow case (composed by executed tasks).

In the following, we propose some basic functions (named SIM) to measure the degree of similarity between a workflow case and primitive subworkflow schemata.

For the sake of simplicity, we describe these functions in incrementally complexity, describing how to obtain the similarity degree depending on the control nodes.

Firstly, we define how a case task ($ct \in CT$) and the tasks of a workflow schema ($t \in \mathbb{T}$) are compared.

$$SIM_{task}(ct, w) = \begin{cases} 1 \text{ if } ct \in \mathbb{T} \\ 0 \text{ if } ct \notin \mathbb{T} \end{cases} \quad (5)$$

That is, $SIM_{task} = 1$ if ct is a task of w that was executed.

The following expression defines how to check fulfilment of a sequential workflow (i.e. no control nodes but begin and end).

$$SIM_{tot}(c, w) = \frac{2 * \sum_{i=1}^{|c|} SIM_{task}(ct_i, w)}{|\mathbb{T}| + |c|} \quad (6)$$

where $|c|$ and $|\mathbb{T}|$ are the cardinalities of the case and the workflow task set respectively and ct_i is the i-th task of the case.

In the following, we describe how to measure the fulfilment of a *and*-branched workflow.

$$SIM_{and}(c, w) = \frac{\sum_{i=1}^{n} SIM_{part}(c, w_i)}{n} \quad (7)$$

$$SIM_{part}(c, w_i) = \frac{\sum_{j=1}^{|c|} SIM_{task}(ct_{i,j}, w_i)}{|\mathbb{T}_i|} \quad (8)$$

where n is the number of subworkflow branches of the *and* control node, $|\mathbb{T}_i|$ is the task cardinality of a subworkflow branch w_i, and $ct_{i,j}$ is its j-th case task.

In order to check the fulfilment of *xor* split, we define:

$$SIM_{xor}(c, w) = MAX(\{SIM_{tot}(c, w_n), \forall w_n\}) \quad (9)$$

where w_n is the n-th subworkflow branch of the *xor* control node.

The following expression defines how to check fulfilment of a *or*-branched workflow.

$$SIM_{or}(c, w) = \begin{cases} 1 & \text{if } (\exists |w_n| = |c|) \wedge (SIM_{tot}(c, w_n) = 1) \\ SIM_{AND}(c, w) & \text{other case} \end{cases} \quad (10)$$

where w_n is the n-th subworkflow branch of the *or* control node.

Finally, the following definition states how to measure the fulfilment of a *loop* in a workflow.

$$SIM_{loop} = \frac{\sum_{i=1}^{n} SIM_{tot}(c_i, w)}{n} \quad (11)$$

where n is the number of loops performed, while c_i are the tasks performed in the i-th iteration of the loop.

Fulfilment Measure Proposal. We propose a fulfilment measure that can be characterised as: guided (technique dimension), quantifier (dimension task), and covers precedence, or, and, xor, and loop control nodes (problem dimension). Figure 1 depicts the dimensions covered by our proposal.

We present $FULFIL\!-\!CHECK$ (see Algorithm 2), a workflow fulfilment measure based on the $NEXT$ (see Algorithm 1) function that obtains the pieces of the workflow schema that the case followed (guided strategy).

Algorithm 1. NEXT function

Function *NEXT*(c, w) **return** $(C \times W)$
  $con_i = firstControl(w)$
  {Copy tasks and edges from w until the next connector}
  $w' = copy(\forall task_i, E_i)\ until\ con_{i+1}$ {built a subworkflow of w}
  {built a subcase of C}
  last $task_j \in \mathbb{T}_{w'}$ that $mach(c, task_j)$
  $ct_j = match(C, task_j)$
  $c' =< ct_1, \ldots, ct_j > \ ct_i \in c$
  return (c', w')

Given $w =< \mathbb{T}, \mathbb{C}, \mathbb{E} >$, $firstTask$ and $firstControl$ functions returns the first elements of $\mathbb{T}$ and $\mathbb{C}$ respectively. Given a workflow task $task_i$ and a case c, $match$ function returns the case task $ct \in c$ as the result of the execution of $task_i$.

Algorithm 2. The fulfilment measure proposed

Function *FULFIL-CHECK*(c, w) **return** $[0, 1]$
  $c' = \emptyset,\ w' = \emptyset$
  $overall = 0,\ n = 0$
  while $NEXT(c, w) \neq \emptyset$ **do**
    $(c', w') = NEXT(c, w)$
    $con' = firstControl(w')$
    if $con' = \emptyset$ **then**
      $tot = SIM_{tot}(c', w')$
    else
      case of (con') {depending on connector type}
      AND: $tot = SIM_{and}(c', w')$
      OR: $tot = SIM_{or}(c', w')$
      XOR: $tot = SIM_{xor}(c', w')$
      $LOOP$: $tot = SIM_{loop}(c', w')$
    end if
    $c = c \backslash c'$
    $w = w \backslash w'$
    $n{+}{+}$
    $overall = overall + tot$
  end while
  return $(overall/n)$

This work is a preliminary study focused on the methodological and the similarity aspects. Note that $NEXT$ function is a simplistic approach only suitable for simple executions, and real situations require a more sophisticated version. Nevertheless, modifications on function $NEXT$ do not affect the rest of the proposal.

3.3 Example in the Medical Domain

We ilustrate the application in the medical domain of the fulfilment measure proposed for a real CPG. In particular, a portion of a real CPG have been modelled (see Figure 2) for the treatment of cerebral vasoespasms of patients affected by Subarachnoid Haemorrhage (SAH) [10].

Therefore, physicians could follow part of this Clinical Workflow (w_{SAH}) for patients affected by SAH. This could be represented by the following examples: $Case\_a$ = <(T1, (0,11)), (T2,(28,148)), (T3,(150,152)), (T4, (170,175))>, $Case\_b$ = <(T2, (0,121)), (T1,(125,138)), (T6, (142,148))>, and $Case\_c$ = <(T2, (0,124)), (T4,(128,134)), (T1, (139,145))>. Then, the $FULFIL-CHECK$ function is used to compute the fulfilment degree of w_{SAH} of cases. Table 1 summarizes the fulfilment calculi. Note that Algorithm 2 iterations depends on $NEXT$ function. For the sake of simplicity, in Figure 2 pointed rectangles (labeled $NEXT(i)$) represents the loops executed by Algorithm 2 (loops $NEXT(-)$ are not considered).

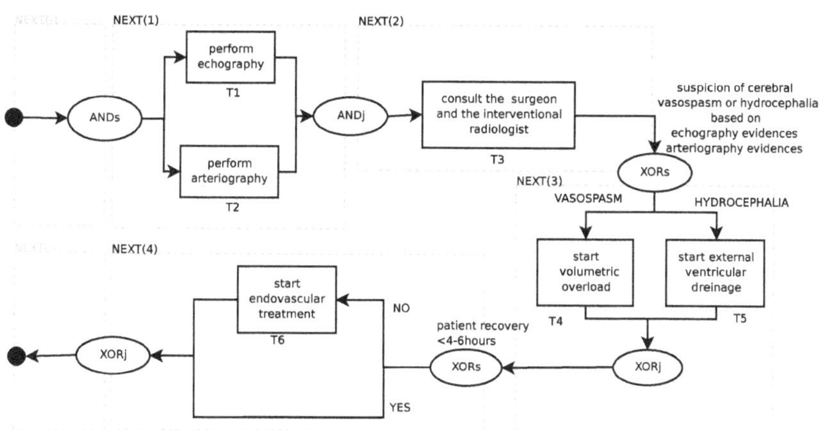

Fig. 2. Clinical Workflow for the treatment of SAH (w_{SAH})

Table 1. Fulfilment degree of w_{SAH}

$Case$	$NEXT_1$	$NEXT_2$	$NEXT_3$	$NEXT_4$	n	overall	Result
$Case\_a$	$SIM_{and} = 1$	$SIM_{tot} = 1$	$SIM_{xor} = 1$	$SIM_{xor} = 1$	4	4	1.0
$Case\_b$	$SIM_{and} = 1$	$SIM_{tot} = 0$	$SIM_{xor} = 0$	$SIM_{xor} = 1$	4	2	0.5
$Case\_c$	$SIM_{and} = 1$	$SIM_{tot} = 0$	$SIM_{xor} = 0$	$SIM_{xor} = 0$	4	1	0.25

4 Conclusions and Future Works

In this work we deal with the correct application of CPGs using Clinical Workflows, as a health-care quality indicator, by measuring the fulfilment of their tasks. Thus, the main results presented are: (1) to present a general workflow model, (2) to state a framework to classify and analyse the development of workflow fulfilment measures; and (3) to propose a method to quantify the fulfilment based on similarity techniques, illustrating its application by the use of part of the Subarachnoid Haemorrhage guideline.

As far as we are concerned, most efforts in workflow modelling focus on the consistent definition of the workflow schema. In [5] the authors states a set of correctness issues. The work described in [6] deals with the management of unexpected situations and handling exceptions. However, our workflow fulfilment measure assumes that Clinical Workflow represents a CPG, i.e. the set of recommendations to follow by the physician team, and therefore the workflow schema and the workflow case could differ. Inspired in the Case-Based Reasoning approach, in [2] a similarity measure is proposed to face the fulfilment of the temporal aspects of CPGs. Unlike that proposal, in this work we focus on the fulfilment of the alternative paths, instead of the temporal facet.

Future works focus on the development of methods for techniques proposed in the framework presented, the extension of the proposed checking method for real life cases (composed workflow schema, $NEXT$ function and weighted pathways).

References

1. Basinski, A.: Evaluation of clinical guidelines. Can Med. Assoc. J. 153(11), 1575–1581 (1995)
2. Combi, C., Gozzi, M., Juarez, J.M., Marin, R., Oliboni, B.: Temporal similarity measures for querying clinical workflows. In: Artificial Intelligence in Medicine (in press, 2009)
3. Combi, C., Gozzi, M., Juarez, J.M., Oliboni, B., Pozzi, G.: Conceptual modeling of temporal clinical workflows. In: Proceedings of the 14th International Symposium on Temporal Representation and Reasoning. TIME 2007, pp. 70–81. IEEE Computer Society Press, Los Alamitos (2007)
4. Field, M.J., Lohr, K.N.: Guidelines for clinical practice: from development to use (1992)
5. Kamath, M., Ramamritham, K.: Correctness issues in workflow management. Distributed Systems Engineering 3(4), 213–221 (1996)
6. Kumar, A., Wainer, J.: Meta workflows as a control and coordination mechanism for exception handling in workflow systems. Decision Support Systems 40(1), 80–105 (2005)
7. Lin, F., Chou, S., Pan, S., Chen, Y.: Mining time dependency patterns in clinical pathway. International Journal of Medical Informatics 62(1), 11–25 (2001)
8. Mentzas, G., Halaris, C., Kavadias, S.: Modelling business processes with workflow systems: an evaluation of alternative approaches. International Journal of Information Management 21(1), 123–135 (2001)
9. Quaglini, S., Ciccarese, P.: Models for guideline representation. Neurological Sciences 27 (2006)
10. Singer, R.J., Ogilvy, C.S., Rordorf, G.: Treatment of subarachnoid hemorrhage. up to date. 9.1(vs 14.1) (2006)
11. van der Aalst, W.M.P., ter Hofstede, A.H.M.: Yawl: Yet another workflow language. Information Systems 30(4), 245–275 (2005)

Debugging and Verification of Multi-Agent Systems*

Clara Benac Earle and Lars-Åke Fredlund

Babel group
DLSIIS, Facultad de Informática,
Universidad Politécnica de Madrid
{cbenac,fredlund}@fi.upm.es

Abstract. Multi-agent systems are systems composed of multiple interacting autonomous agents forming complex systems. Verifying multi-agent systems is a challenging task due to their dynamic nature, and the complex interactions between agents. In this paper, we propose the use of the McErlang model checker as a testing tool, as it affords precise control of the scheduling of agents, and provides convenient access to the internal states and actions of the agents. We illustrate the suitability of the approach by discussing our experiences in applying this verification technique to RoboCup teams. The experiments we conducted discovered a number of bugs in two such teams.

1 Introduction

It is commonly understood that a multi-agent system (MAS) is a system composed of multiple interacting autonomous agents with either diverging information or diverging interests, or both. Wooldridge and Jennings [14] identify the following properties of an agent:

- Autonomy: an agent is capable of operating without external intervention.
- Social ability: an agent is capable of interacting with other agents and/or with its environment.
- Reactivity: an agent is capable of responding to external changes.
- Pro-activity: an agent is capable of behaving accordingly to its goals.

Besides, rationality (defined as the ability of an agent to act consistently with its goals) is often assumed when reasoning about agents. Agents typically interact with other agents in some environment to form more complex structures. Examples of MAS are online auctions, e-commerce and RoboCup [5].

As multi-agent systems grow in complexity, verifying that they satisfy their design requirements becomes a challenging task. Verification is even more crucial in multi-agent systems, which are intrinsically more complex than traditional

* This work has been partially supported by the FP7-ICT-2007-1 Objective 1.2. IST number 215868 (ProTest project).

R. Moreno-Díaz et al. (Eds.): EUROCAST 2009, LNCS 5717, pp. 263–270, 2009.

distributed systems: by definition, MAS are employed to capture high level properties of large, autonomous systems. Moreover, agents in MAS are often highly autonomous and out of direct human control. Hence, in such scenarios, in-depth verification can save time and money, and improve security [11].

One method that is often advocated to verify such MAS is model-checking. However, in performing model-checking on MAS two main issues arise: i) a model needs to be constructed, and ii) the state space is bound to grow too large. In this paper we propose an approach to the checking of correctness properties in multi-agent systems that involves simulation, in particular we use the McErlang tool [8] to verify properties of RoboCup teams. The RoboCup teams we verify are written in the Erlang programming language, a functional programming language with built-in support for process concurrency and message passing, which is seen as a good platform for developing multi-agent systems. As the McErlang model checker supports full Erlang, the task of constructing an accurate model for the purpose of model checking is considerably eased. While it might still be necessary to abstract the original multi-agent system to reduce its complexity, and thus to enable effective model checking, the result of the abstraction step can be represented in the same programming language as the original design.

The use of tool support in the task of verifying multi-agent systems is recently attracting significant interest from the agent community. In [3], for example, a variant of the abstract agent-oriented programming language AgentSpeak, AgentSpeak(F), is proposed. By translating AgentSpeak(F) programs into Promela or Java, properties written in LTL can be model-checked with SPIN or the Java Path Finder [13], a general purpose model checker for Java. A difference between their approach and ours is that AgentSpeak is based on the BDI agent architecture while we do not consider any specific agent architecture. In [9] a combination of UML statecharts and hybrid automata was proposed for modeling multi-agent systems, and the method was applied to the task of model checking agents of the RoboCup rescue simulation league. In [4] a trace based approach is used to study a complex agent scenario.

The rest of the paper is organized as follows: in Sect. 2 we introduce the most important features of the Erlang programming language for programming multi-agent systems, and Sect. 3 describes the McErlang model checking tool. Next, in Sect. 4 we examine how a number of complex multi-agent systems (RoboCup soccer teams) are verified. Finally Sect. 5 discusses the conclusions of the verification effort, and further research directions.

2 The Erlang Programming Language

Erlang [1] is a dynamically typed functional programming language originating from Ericsson, a Swedish telecommunication company. In 1998 Erlang was released as Open Source. Today several commercially available products developed by Ericsson as well as other companies (Yahoo, T-Mobile, Facebook, Amazon) are at least partly programmed in Erlang, an example is the AXD 301 ATM switch [2]. In the last couple of years the language has gained a lot of attention because of its concurrency oriented programming paradigm which matches well the multi-core processor hardware architecture that is becoming increasingly common. In addition the programming paradigm is a good match for the underlying assumptions for many multi-agent systems frameworks, making an Erlang system an excellent implementation platform for multi-agent systems.

Erlang provides a functional core language, which is enriched with the concept of *processes*. Processes are isolated from each other, communicating only via asynchronous message passing. Erlang processes have a unique name, a process identifier, and messages sent to the process are stored in the *process mailbox*.

On top of the process concept the distribution layer of Erlang is implemented in terms of *nodes*. A node (e.g. a host runtime system) provides a distribution mechanism where processes are mapped onto different nodes. Both intra-node and inter-node communication are implemented, and the distribution aspect is mostly transparent; meaning that a process communicates in the same way with a local and a distributed process. A typical Erlang application is organized in many, relatively small, source components. At runtime the components execute as a dynamically varying number of processes running parallel (Erlang processes are very lightweight, and it is not uncommon having systems with several hundred thousand simultaneous processes.)

Handling a large number of processes easily turns into an unmanageable task, and therefore Erlang programmers mostly work with higher-level language components. The OTP component library [12] offers industrially proven design patterns such as: a generic server component (for client-server communication), a finite state machine component, generic TCP/IP communication, and a supervisor component for structuring fault-tolerant systems. The Erlang programming system is nowadays often referred to as Erlang/OTP, to stress the benefit to distributed application programming that the OTP library provides.

3 The McErlang Tool

In this section we introduce the McErlang[8] model checker [8], which has been used to verify a number of distributed Erlang applications [8,7]. The model checker has recently been released as open source, and is available for downloading at `http://babel.ls.fi.upm.es/trac/McErlang/`.

The input to the model checker is an Erlang program, together with a special call-back module (called a *monitor*) also written in Erlang, which specifies the behavioral property to be checked (implementing either a safety automaton or a Büchi automaton). The output can be either a positive answer saying that the property holds, or a negative one together with a counterexample (typically an execution trace leading to a faulty program state).

The main idea behind McErlang is to re-use as much of a normal Erlang programming language implementation as possible, but adding a model checking capability. To achieve this, the tool replaces the part of the Erlang runtime system which implements concurrency and message passing, while still using the runtime system for the evaluation of the sequential part of the input programs.

The model checker has a complex internal state in which the current state of the runtime system is represented. The structure that is maintained by the model checker records the state of all alive processes (their process identifiers, mailboxes, computation state, etc). Moreover the global state kept by the model checker runtime system includes a structure to record process links, information about registered process identifiers, etc.

The model checker implements full linear-temporal logic (LTL) checking. Correctness properties are represented as Büchi automatons which are checked using various on–the–fly model checking and simulation algorithms. The task of such an automaton, or monitor, is to check whether the correctness property holds of the combination of the new program state, a sequence of actions (side effects) that occurred during the computation of the new program state from the old one, and the old monitor state. Actions are, for example, the sending of a message from one process to another. If the monitor deems that new program state, and the associated actions, are acceptable in its current state, the monitor should returns an new monitor state. Correctness properties can be implemented, therefore, as finite state machines where depending on the monitor state, actions leading to new states are accepted or not. Such correctness properties have full access to the internal state of the program run (including message queues, state of processes, and so on).

McErlang has built-in support for some Erlang/OTP component behaviours that are used in almost all serious Erlang programs such as the supervisor component (for implementing fault-tolerant applications) and the generic server component (implementing a client–server component). The presence of such high-level components in the model checker significantly reduces the gap between original program and the verifiable model, compared to other model checkers.

In addition to the usual model checking algorithms for checking safety and liveness properties, the McErlang tool implements a simulation algorithm, whereby instead of exploring the whole state space of an application only a single execution branch is followed. Which execution to branch to follow is by default a random choice, however finer control can be exercised by specifying either a custom scheduler, or refining the safety monitor module above, which in addition to checking safety properties can mark certain states an "uninteresting", preventing the model checker to examine them and instead choosing an alternative next state during the simulation.

The checking of the multi-agent systems has required a number of changes to the McErlang model checker, including supporting the simulation of real-time systems. Many multi-agent systems are highly time dependent, and have to respond in a timely fashion to information sent. Moreover the model checker had to be "opened up to the outside world". In many agents frameworks commands are communicated to agents using TCP or UDP sockets. To support sending UDP commands from McErlang was trivial, whereas receiving messages was more difficult. The solution was to program a new Erlang process, constantly listening for incoming UDP messages. This (real) Erlang process keeps a map of virtual (simulated) Erlang processes to which incoming messages should be resent (using the virtual message communication mechanism). Thus virtual processes wanting to receive UDP messages on a certain UDP port communicates this to the UDP Erlang process, which in turn starts receiving and forwarding incoming messages on behalf of the virtual process.

Notable is also that when the application has been opened up to the outside world, no longer is absence of enabled transitions necessarily a reason to halt the simulation run. Enabled timers has to be taken into account, as well as the possibility that the environment may later send messages to a simulated process.

There are a number of advantages by using McErlang as a simulation tool compared to using traditional testing frameworks:

- correctness properties can be more elegantly and compactly be expressed as automatons rather than sets of tests,
- the tool provides detailed control of the scheduling of processes, and delivery of messages (which a traditional runtime system does not provide at all). Testing a multi-agent system under different assumptions regarding processes scheduling can often reveal errors that are difficult to reproduce using normal testing procedures,
- no or very little source code modification is necessary to interpret testing outcome (i.e., as all the agents states, and all the processes implementing an agent – can be inspected, there is generally little need to export extra information from an agent),
- since we are using an untyped functional programming language (Erlang) we can treat programs (e.g., pending function calls, sent messages, etc) as data, and analyze such data using powerful data abstraction functions. Moreover we can often reuse functions and data structures used in the program itself to formulate correctness properties.

4 A Case Study

The RoboCup Soccer Simulator, the soccer server [5], is a research and educational tool for multi-agent systems and artificial intelligence. It enables two teams of eleven simulated autonomous players to play soccer (football). A match is carried out in client/server style: the server provides a virtual field and simulates all movements of a ball and the players, and each client controls the movements of one player. Communication is done via UDP/IP sockets, enabling players to be written in any programming language that supports UDP communication.

The IT-university of Gothenburg has been organizing local RoboCup competitions for their students as part of a course for undergraduate students enrolled in a software engineering and management program. Students were asked to developed in groups a RoboCup soccer simulation team in Erlang to play against teams developed by other groups. We have taken a number of such teams as a starting point for a case study on verifying complex multi-agent systems.

Seen as a verification task, checking properties of a RoboCup team is very challenging. A team consists of eleven to a large extent independently acting agents with complex internal states, that cooperate to solve a common task in real-time. Unfortunately the hostile environment, i.e., the opponent team, tries to sabotage task solving. Moreover, the setting is not static, the number of opponents will vary, and in addition the soccer simulation server contains random features[1] that will alter the outcome of various actions of the agents.

To apply model checking techniques to such a verification problem one would have to construct, with substantial effort, a simplified model of the soccer server, the agents (for both teams) and their environment. Even so the real state space would be huge, and model checking would be unlikely to cover more than a very tiny fragment of that state space. For this reason we decided upon a different verification strategy: to use the McErlang model checker for executing the agents, and to formulate correctness properties to check as monitors, but instead of model checking a team we used the model checker as a testing/simulation environment. What we lose by not performing a complete verification, which could anyway never be complete due to the abstractions needed to obtain a verifiable model, we hope to gain by checking the actual source code of the agents.

We use the monitor concept of McErlang to check properties of a RoboCup team programmed in Erlang during games with opponents. Monitors to check correctness properties of the team are written in Erlang as well, and have full access to the state of all agents (players), message in communication channels, and so on. That is, by using the McErlang tool instead of a normal testing framework, we gain instant access to the complete state of the underlying distributed system on top of which the multi-agent system runs, and are thus able to formulate correctness properties over the whole multi-agent system state in a concise manner. In a traditional testing framework we would instead have had to partition the "implementation" of the correctness property over the different nodes comprising the distributed system, and implement communication between the different parts of the correctness property implementation, a decidedly non-trivial task.

However, the states of player agents may of course not reflect reality, as they may have incorrect or simply insufficient knowledge of the state of the game. Clearly to determine whether a property holds, in general we need access to the state of the soccer server as well. As the server is not written in Erlang, McErlang does not have direct access to its internal state. However, by programming a "Coach agent" in Erlang[2], that repeatedly gets truthful and complete situational information from the soccer server (e.g., ball position, and the position and movement of all players), we gain access, using the McErlang tool, to the complete simulation state.

In case a property violation is detected by a monitor, the complete trace of the simulation up to that point, including the states and actions of all agents and the coach, are available for further analysis in the McErlang debugger.

The experimental setup is depicted in Fig. 1; note that there is no direct communication between agents comprising a team.

To check a team in varying situations the opposition teams were chosen with care. To evaluate defensive play we matched the team to check against good teams from previous international robocup competitions. Concretely such teams

[1] e.g. reporting all positional information to an agent with a possible slight error.

[2] The coach interface is provided by the soccer simulation server.

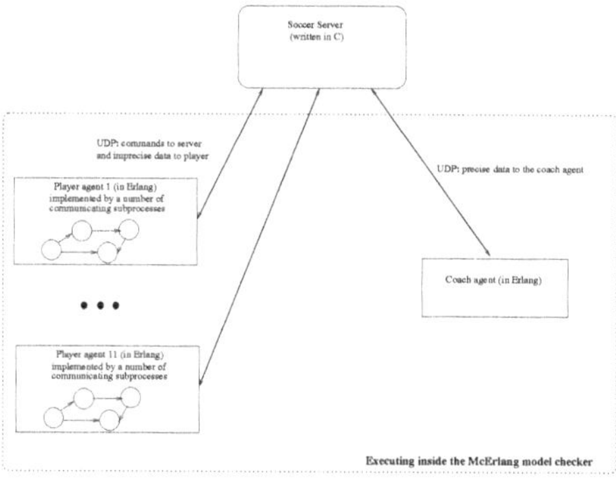

Fig. 1. RoboCup verification setup

include `fcportugal2004` and `tokyotech2004`, both from the 2004 international robocup competition. For evaluating offensive play a particularly bad student team was selected as an opponent. Finally, to evaluate the team in a more fluctuating situation we played the team against itself. All games were repeated multiple times, to increase the coverage of the verification experiment.

By formulating properties regarding a number of obvious correctness criteria for football play (respecting playing field boundaries etc), and also a number of properties that concern the inner logic of the agents, we were able to find several bugs in two RoboCup teams written in Erlang [6].

5 Conclusions

The analysis and verification of multi-agent systems is not an easy task due to their dynamic nature, and the complex interactions between agents. One method that is often advocated to verify such systems is model checking. However, in performing model-checking on multi-agent systems two main issues arise: i) a model needs to be constructed, and ii) the state space is bound to grow too large. In this paper we propose an alternative approach to the verification of properties in multi-agent systems by means of testing using the simulation capability of McErlang, a model checking tool developed by us. While the approach is then more similar to traditional testing, we gain a number of advantages: (i) using McErlang we have full control over the scheduling of processes and delivery of messages, and can thus test the teams under challenging environmental conditions, (ii) McErlang provides full access to the complete state of the multi-agent system under test, and (iii) McErlang provides a rich language for formulating correctness properties on a higher abstraction level, instead of having to write a large set of lower-level traditional tests.

We illustrate the suitability of the approach by discussing our experiences in applying this verification technique to RoboCup teams. The experiments we conducted discovered a number of bugs in two such teams.

One important direction for future research concerns the integration of simulation (testing) with model checking. The idea is to run a game between the team we want to verify against an opponent team in simulation mode, and when a state deemed interesting is seen, we switch the analysis mode to model checking. Thus, in this way one can simulate part of the state space until a certain point, then model checking can be applied to a smaller portion of the state space. Early results are promising, but point to a need to implement partial-order reductions [10] for more efficient model checking of multi-agent systems comprising a large (but mostly independent) number of processes.

References

1. Armstrong, J., Virding, R., Wikström, C., Williams, M.: Concurrent Programming in Erlang. Prentice-Hall, Englewood Cliffs (1996)
2. Blau, S., Rooth, J.: AXD 301 - a new generation ATM switching system. Ericsson Review 1, 10–17 (1998)
3. Bordini, R.H., Fisher, M., Visser, W., Wooldridge, M.: Verifying multi-agent programs by model checking. Autonomous Agents and Multi-Agent Systems 12(2), 239–256 (2006)
4. Bosse, T., Lam, D.N., Barber, K.S.: Automated analysis and verification of agent behavior. In: AAMAS 2006: Proceedings of the fifth international joint conference on Autonomous agents and multiagent systems, pp. 1317–1319. ACM, New York (2006)
5. Chen, M., Dorer, K., Foroughi, E., Heintz, F., Huang, Z., Kapetanakis, S., Kostiadis, K., Kummeneje, J., Murray, J., Noda, I., Obst, O., Riley, P., Steffens, T., Wang, Y., Yin, X.: RoboCup Soccer Server. Manual for Soccer Server Version 7.07 and later (obtainable from `sserver.sf.net`) (2003)
6. Earle, C.B., Fredlund, L., Iglesias, J., Ledezma, A.: Verifying robocup teams. Electronic Notes in Theoretical Computer Science 5348/2009, 34–48 (2008)
7. Fredlund, L., Penas, J.S.: Model checking a VoD server using McErlang. In: Proceedings of the 2007 Eurocast conference (February 2007)
8. Fredlund, L., Svensson, H.: McErlang: a model checker for a distributed programming language. In: Proceedings of the 2007 ACM SIGPLAN International Conference on Functional Programming (2007)
9. Furbach, U., Murray, J., Schmidsberger, F., Stolzenburg, F.: Model Checking Hybrid Multiagent Systems for the RoboCup. In: Visser, U., Ribeiro, F., Ohashi, T., Dellaert, F. (eds.) RoboCup 2007: Robot Soccer World Cup XI. LNCS (LNAI), vol. 5001, pp. 262–269. Springer, Heidelberg (2008)
10. Peled, D.: Combining partial order reductions with on-the-fly model-checking. Formal Methods in System Design 8(1) (1996)
11. Raimondi, F.: Model Checking Multi-Agent Systems. PhD thesis, University College London, UK (2006)
12. Torstendahl, S.: Open telecom platform. Ericsson Review 1 (1997)
13. Visser, W., Havelund, K., Brat, G., Park, S.: Java pathfinder - second generation of a java model checker (2000)
14. Wooldridge, M., Jennings, N.R.: Intelligent agents: Theory and practice. Knowledge Engineering Review 10(2), 115–152 (1995)

Easing the Definition of N–Ary Relations for Supporting Spatio–Temporal Models in OWL

Alberto G. Salguero, Cecilia Delgado, and Francisco Araque

Dpt. of Computer Languages and Systems
University of Granada, Granada (Andalucía), Spain
{agsh,cdelgado,faraque}@ugr.es

Abstract. There are many issues to overcome when integrating different data sources due to the number of variables that are involved in the integration phase. However we are interested in the integration of temporal and spatial information due to the nature of modern Information Systems. We have previously developed a model, called STOWL, which is a spatio-temporal extension of OWL. We use this model as the common model for defining the schemes of the data sources in order to ease their integration. This paper presents the part of STOWL which has to do with the definition of n-ary relations.

Keywords: Ontology, data sources, data integration, OWL.

1 Introduction

OWL is the language adopted by the W3C for defining ontologies and supporting the Semantic Web. A growing number of Semantic Web applications, which can interact between them and cooperate to find better solutions, is being developed [2], [5], [6]. It is a common belief that Semantic Web technology would significantly impact the use of the Web, essentially in terms of increased task delegation to intelligent software agents, and a subsequent amelioration of the information overload effect [3].

This work focuses on the integration of spatial information. This kind of information has grown in importance in recent years due to the proliferation of GIS-based applications and the global positioning system (GPS). Increasingly, companies rely on such information to develop their business or enhance their productivity. The work is restricted to spatio-temporal information for reducing complexity in terms of integration possibilities: it is easier to perform the integration if the working environment is considerably reduced. It would be impractical to propose a model which considers all aspects of the real world. Moreover, we believe that this type of information is general enough to solve most of the problems that we can find nowadays. It could be possible to develop a financial-oriented model but the number of supported situations would be fewer than using a more general model. The problem is that OWL does not support some of the desired characteristics of a spatio-temporal data model.

One of these missing characteristics is the possibility of defining n-ary relations between concepts. It is common when dealing with geographic information which evolves in time to mark the instant when a measurement has been made, but using

R. Moreno-Díaz et al. (Eds.): EUROCAST 2009, LNCS 5717, pp. 271–278, 2009.

OWL only binary relations can be defined. Although the W3C on its own has described how to represent n-ary relations using OWL the resulting code is long and not easily understandable by a person.

In this paper an extension of OWL is presented in order to support the definition of n-ary relations in the context of geographic information systems. We call this extension STOWL [1].

The remaining part of this paper is organized as follows. In the following section the differences between OWL and STOWL are illustrated; in section 3 our model is presented; finally, section 4 summarizes the conclusions of this paper.

2 Differences between OWL and STOWL Ontologies

STOWL is the name of the common data model which is proposed in this work for describing the schemes of the different data sources. In this point, the differences between OWL and STOWL are explained.

The use of a data model based on ontologies is proposed as a common data model to deal with the data sources schemes integration. Although it is not the first time the ontology model has been proposed for this purpose [5], [6], [7], [8], to our knowledge, this is the first time an ontology language has been used to improve the Information System data refreshment process design.

Both languages allow the description of ontologies based on the OWL ontology. The OWL language can be used for describing spatio-temporal data repositories. The main problem is that this language lacks some desired features which makes difficult to express certain type of knowledge which is common when dealing with spatial and temporal information:

- *Description of exhaustive decompositions.* As well as in the DAML+OIL ontology (the OWL ontology derive from), in the OWL ontology is possible to express exhaustive decompositions. For this is used the primitive *owl:oneOf.* In spite of that, it is important to note that the classes described using this primitive are exhaustive decompositions of instances of other classes. The problem is that, in certain cases, it is necessary to describe exhaustive decompositions of classes, not specific instances.
- *Description of partitions.* Unlike the DAM+OIL model, in OWL is not possible to express partitions of concepts. This is because, in the case it is needed, a partition can be expressed combining the primitives *owl:disjoint With* and *owl:unionOf.* Although valid, the code needed for describing partitions with this approach is long and complex to follow.
- *N-ary relations.* OWL only allows the definition of binary relations. In case of higher arity relations need to be defined, the W3C suggests the creation of artificial classes (or concepts) for representing those relations. As well as with the previously described properties, although it is possible to express n-ary relations using the OWL language the resultant code is difficult to follow. The STOWL language has been defined to solve this issue.
- *N-ary functions.* OWL has not been designed for supporting functions. In case of necessity, the primitive *owl:FunctionalProperty* can be used as a mechanism for defining binary relations. It is not possible to express functions in

OWL with higher arity. It is therefore not possible to define layered information, a common type of data managed by the geographical information systems (terrain height, terrain usage…).

- *Formal axioms.* None of the markup-based languages for describing ontologies, including OWL, support the definition of formal axioms.
- *Rules.* They are a special type of formal axioms, so they cannot be expressed in OWL. They can be used for inferring new knowledge. Geographic information system can use this rules for performing complex queries.
- *Integrity constraints.* The integrity constraints make the maintenance of semantic consistency of data easier. OWL only considers the one type of integrity constraint: the functional dependency, which can be expressed by mean of the primitive *owl:FunctionalProperty.* There are other types of integrity constraints which should be addressed: unique integrity constraint, assertions… The integrity constraints are related to the quality of data.

The rest of this work is focused in how the OWL language has been extended in order to consider n-ary relations.

3 Extending OWL with N-Ary Relations

STOWL is build on top of OWL language. Most of its features rely on OWL features On the other hand, there are spatio-temporal characteristics which can not be expressed in the OWL language because to the OWL ontology does not support those desired features. In this case the modification of the OWL ontology should be made in order to incorporate those features and the OWL language, consequently, has also to be modified in order to reflect those changes (giving STOWL as result).

In the former situation STOWL can be seen as a software layer that transform the new defined spatio-temporal features, which can be expressed straightforwardly in STOWL, in a more complex and equivalent definition in OWL. Due to this transformation the results continue being an ontology description expressed in OWL, so all the OWL tools can be used as usual (reasoners, editors...). This is the case of the proposed extension in this paper. N-ary relations in STOWL are transformed to concepts in OWL that represent those relations.

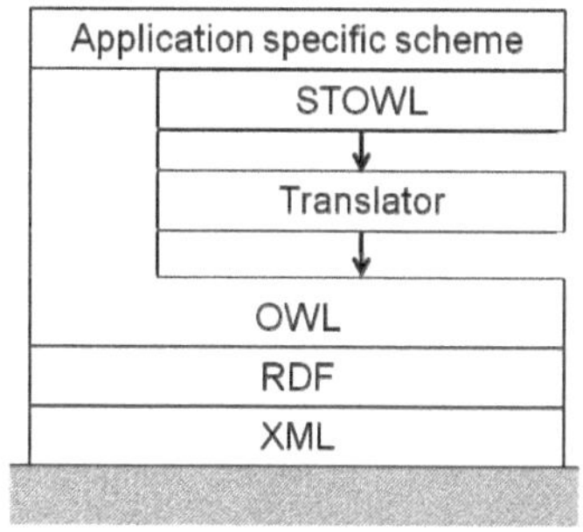

Fig. 1. STOWL functional diagram

3.1 Defining N-Ary Relations in OWL

N-ary relations cannot be expressed directly in OWL. The binary relations are the relations with the higher arity which can be defined using OWL. The W3C, being aware of this problem, has proposed some solutions in order to solve this issue. Their usage depends on the specific situation and the intended meaning for the n-ary relation [4]:

- Additional attributes describing relations.
- Different subjects about the same relation.
- N-ary relation with no distinguished participant.

In the former situation, for describing a relation is necessary to add an attribute to it. This is the case when, for instance, the definition of a relation for tracking the position of a person or any object is needed. In OWL is possible to specify that a person is located at some location but it is not possible to specify that the person were located at some location at some instant. For this usage pattern the W3C suggest the creation of a new artificial concept to represents the ternary relation (figure 2).

Although valid, this solution has some important drawbacks. One of those drawbacks is, for instance, the necessity of creating instances of the introduced artificial concepts representing the n-ary relations. Those instances are also artificial and do not provide useful knowledge from the specific problem point of view. Furthermore, as it can be seen in figure 3, its implementation could be relatively complex, even when the implicated concepts are relatively few (as in this case). From the developer point of view, its usage is difficult and error-prone. Much of the effort is dedicated to handling with the artificial concepts and instances and how they are related with the rest of the elements of the scheme (the really important elements).

The second of the usage patterns (different subjects about the same relation) is resolved in the same way as the former. This kind of relation appears when the relation to be represented has ranges of complex objects with multiple relations involved.

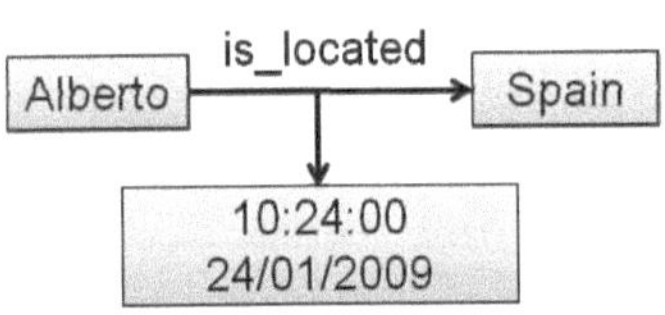

Fig. 2a. Example of ternary relation

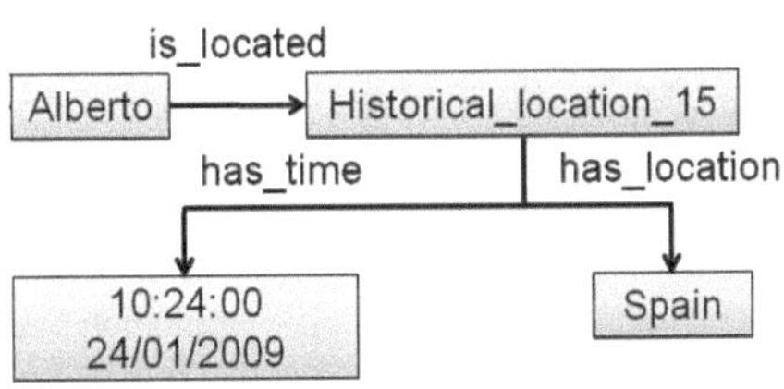

Fig. 2b. Example of artificial concept created for representing the ternary relation of figure 2a

On the other hand, the latter usage pattern (n-ary relation with no distinguished participant) is slightly different. In this case, there is no a main concept in the relation. In spite of this, the solution is equivalent to the former usage patterns: create an artificial concept which represents the n-ary relation.

```
<owl:Class rdf:ID="Person">
<rdfs:subClassOf>
        <owl:Restriction>
                <owl:onProperty>
                        <owl:ObjectProperty rdf:about="#is_located"/>
                </owl:onProperty>
                <owl:allValuesFrom>
                        <owl:Class rdf:about="# Historical_location"/>
                </owl:allValuesFrom>
        </owl:Restriction>
</rdfs:subClassOf>
</owl:Class>
...
<owl:Class rdf:ID="Historical_location">
<rdfs:subClassOf>
        <owl:Restriction>
                <owl:someValuesFrom rdf:resource="#Location"/>
                <owl:onProperty>
                        <owl:FunctionalProperty rdf:about="#has_location"/>
                </owl:onProperty>
        </owl:Restriction>
</rdfs:subClassOf>
<rdfs:subClassOf>
        <owl:Restriction>
                <owl:allValuesFrom rdf:resource="#TimePeriod"/>
                <owl:onProperty>
                        <owl:FunctionalProperty rdf:about="#has_time"/>
                </owl:onProperty>
        </owl:Restriction>
</rdfs:subClassOf>
</owl:Class>
```

Fig. 3. Implementation of the example in figure 2b using OWL

There are some considerations which have to be taken into account when introducing a new concept as a relation:

- It is necessary to implement some kind of method for assigning *meaningful names* to instances of properties or to the artificial concepts used to represent instances of n-ary relations.
- Creating a class to represent an n-ary relation limits the use of many OWL constructs and creates a *maintenance problem.*
- Defining *inverse properties* with n-ary relations, using any of the patterns above, requires more work than with binary relations.

In a spatio-temporal environment it is very common to find such kind of relations and their management could results very difficult. It is very common, for instance, to find in any spatial-based IS a ternary relation such as “connect” (representing that two locations are connected by a segment). It would require the definition of a new concept representing the relation and two new properties relating that concept and two of the three related concepts. This operation can be performed when there are few n-ary relations and instances defined in the ontology but it is very difficult to maintain an ontology when the number of such kind of relations increases. This is precisely the case of geographical information. On the other hand, only one relation (and no new concepts) should be defined in order to represent the same situation with STOWL, capable of defining n-ary relations. As explained previously, STOWL is a software layer which tries to overcome the drawbacks stated before.

Following are detailed how OWL has been extended in order to consider n-ary relations.

3.2 STOWL Abstract Syntax

The first step consists on extends OWL in order to support n-ary relations. Actually, we consider that the best option for accomplishing this task is to directly modify RDFS, which OWL relies on. The changes introduced have to do with the possibility of defining multiple range values for the relations. They consisted in the definition of *n* subclasses of the class *rdfs:Range* (figure 4), being *n* the desired maximum arity. The changes are declared in a new namespace called RDFSN.

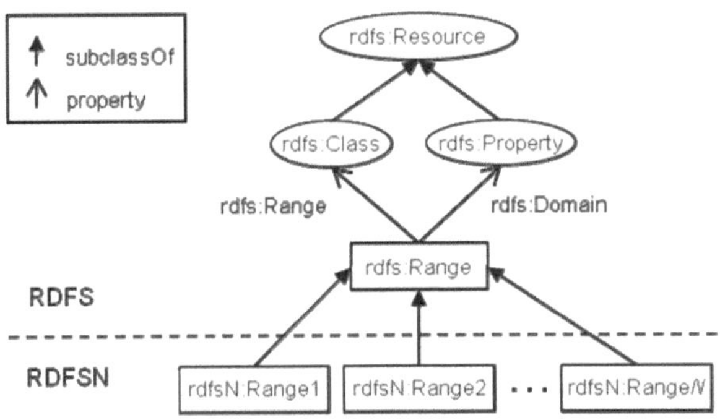

Fig. 4. Definition of RDFSN as an extension of RDFS for supporting properties with multiple ranges

The OWL abstract syntax does not have to be modified in order to consider multiple ranges. As illustrated in figure 5, the OWL abstract syntax is designed to allow multi-ranges properties.

```
axiom ::= 'DatatypeProperty(' datavaluedPropertyID ['Deprecated'] { annotation }
            { 'super(' datavaluedPropertyID ')' } ['Functional']
            { 'domain(' classID' ')' } { 'range(' dataRange ')' } ')'
        | 'ObjectProperty(' individualvaluedPropertyID ['Deprecated'] { annotation }
            { 'super(' individualvaluedPropertyID ')' }
            [ 'inverseOf(' individualvaluedPropertyID ')' ] [ 'Symmetric' ]
            [ 'Functional' | 'InverseFunctional' | 'Functional' 'InverseFunctional'
              | 'Transitive' ]
            { 'domain(' classID ')' } { 'range(' classID ')' } ')'
```

Fig. 5. OWL abstract syntax relating to relations range description

3.3 STOWL RDF Syntax

Although the OWL abstract syntax allows the definition of properties with multiple ranges, the RDF model does not allow such kind of knowledge to be expressed. This is basically the reason why the subclasses of the concept *rdfs:Range* have been introduced in the previous section. When needed, they can be used to relate more than two classes in the same relation. The solution consists on replacing the property *rdfsn:Range* by as many as *rdfsn:RangeX* properties as concepts involved in the relation. The property *is_located* of figure 2, for instance, is defined in STOWL using two derived classes of *rdfs:Range*, as illustrated in figure 6.

```
<owl:ObjectProperty rdf:ID="is_located">
        <rdfs:domain rdf:resource="#Person"/>
        <rdfsn:range1 rdf:resource="#Location"/>
        <rdfsn:range2 rdf:resource="#Time"/>
</owl:ObjectProperty >
...
```

Fig. 6. Definition of property *is_located* in STOWL using n-ary relations

When defining restrictions about the n-ary relations is necessary to specify what of the concepts are involved. This can be performed, as illustrated in figure 7, adding the *rdfsn:RangeX* attribute to the property description to indicate the affected concept.

```
<owl:Class rdf:ID="Person">
<rdfs:subClassOf>
        <owl:Restriction>
                <owl:onProperty>
                        <owl:ObjectProperty rdf:about="#is_located" rdfs:range="#range1"/>
                </owl:onProperty>
                <owl:allValuesFrom>
                        <owl:Class rdf:about="#Location"/>
                </owl:allValuesFrom>
        </owl:Restriction>
</rdfs:subClassOf>
</owl:Class>
```

Fig. 7. Example of definition of property restrictions on n-ary relations in STOWL

As result it is obtained the possibility of defining n-ary relations straightforwardly, without having to manually create artificial classes or instances. As well as with the case of the restrictions, the unique difference with respect to OWL when defining instances are that it is necessary to specify the range of the property value. Figure 8 illustrated how the example in figure 2 can be expressed easily in STOWL without having to create artificial classes and instances.

```
<Person rdf:ID="Alberto">
        <is_located rdf:resource="Spain" rdf:range="range1">
        <is_located rdf:resource="10:24:00 24/01/2009" rdf:range="range2">
</Person>
```

Fig. 8. Example of definition of restrictions about n-ary relations in STOWL

4 Conclusions

In this paper we have presented our work relating to the definition of n-ary relations in OWL. Furthermore, we have described the elements in STOWL, a spatio-temporal extension of OWL we have developed in order to ease the integration of different data sources schemes, which have to do with the possibility of defining n-ary relations.

We have justified why this kind of knowledge is important when dealing with spatio-temporal information and presented the difficulties for implementing it directly in OWL. Although the W3C provides some valid solutions to overcome this issue they require a big effort. STOWL is a software layer on top of OWL which eases the definition of n-ary relations.

Acknowledgements

This work has been supported by the Andalucía Research Program under project GR2007/07-2 and by the Spanish Research Program under projects EA-2007-0228 and TIN2005-09098-C05-03.

References

1. Salguero, A., Araque, F., Delgado, C.: Using ontology meta data for data warehousing. In: Filipe, J., Cordeiro, J. (eds.) ICEIS 2008. LNBIP, vol. 19, pp. 28–35. Springer, Heidelberg (2009)
2. Kolas, D., Dean, M., Hebeler, J.: Geospatial semantic Web: architecture of ontologies. In: Rodríguez, M.A., Cruz, I., Levashkin, S., Egenhofer, M.J. (eds.) GeoS 2005. LNCS, vol. 3799, pp. 183–194. Springer, Heidelberg (2005)
3. Lytras, M.D., García, R.: Semantic Web applications: a framework for industry and business exploitation – What is needed for the adoption of the Semantic Web from the market and industry. International Journal of Knowledge and Learning 4(1), 93–108 (2008)
4. Hayes, P., Welty, C.: Defining N-ary Relations on the Semantic Web. W3C Working Group Note April 12 (2006), `http://www.w3.org/TR/swbp-n-aryRelations`
5. Goudos, S.K., Peristeras, V., Tarabanis, K.A.: Semantic Web Application for Public Administration using OWL for Public Domain Data Knowledge Representation. WSEAS Transactions on Information Science & Applicattions 4(4), 725–730 (2007)
6. Yang, S.Y.: How Does Ontology help Web Information Management Processing. WSEAS Transactions on Computers 5(9), 1843–1850 (2006)
7. Skotas, D., Simitsis, A.: Ontology-Based Conceptual Design of ETL Processes for Both Structured and Semi-Structured Data. International Journal on Semantic Web and Information Systems 3(4), 1–24 (2006)
8. Ale, M.A., Gerarduzzi, C., Chiotti, O., Galli, M.R.: Organizational Knowledge Sources Integration through an Ontology-Based Approach: The Onto-DOM Architecture. In: Lytras, M.D., Carroll, J.M., Damiani, E., Tennyson, R.D. (eds.) WSKS 2008. LNCS (LNAI), vol. 5288, pp. 441–450. Springer, Heidelberg (2008)

Separation of Transitions, Actions, and Exceptions in Model-Based Testing

Cyrille Artho

Research Center for Information Security (RCIS), AIST, Tokyo, Japan

Abstract. Model-based testing generates test cases from a high-level model. Current models employ extensions to finite-state machines. This work proposes a separation of transitions in the model and their corresponding actions in the target implementation, and also includes special treatment of exceptional states.

1 Introduction

Software testing entails execution of a system under test (SUT) or parts thereof. A series of stimuli (inputs) is fed to the SUT, which responds with a series of observable events (outputs). The oldest form of testing consists of executing a system manually, with a human administering the input and observing the output. In the last few decades, various techniques have been developed to automate testing, and to impose some rigor by systematic selection of test data and measurement of the effectiveness of tests [5,6].

For testing on a smaller scale, unit testing has become a widely accepted and used way of testing relatively small, self-contained units in software [4]. Unit testing automates test execution and the verification of the test output. In this way, once a test is written, it can be re-used throughout the life time of a system. The continuous use of a test suite throughout a product cycle is called *regression testing*. Unit tests automate regression testing, provided the interface (input specification) or the output format of a system does not change.

Despite its success, unit testing has the drawbacks that test creation often involves writing large parts of low-level code. Furthermore, unit tests require maintenance whenever the interface or output of the SUT undergoes a change. It is therefore desirable to automate test creation, while providing a more abstract view of inputs and outputs. The aspiration is to reduce the amount of low-level test code to be written, while also decoupling test behavior from the detailed format of data. In this way, the data format can be updated while the test model can be (mostly) retained.

Model-based testing (MBT) is a technology that automates creation of test cases in certain domains [3,8]. Test code is generated from the model by specialized tools [8] or by leveraging model transformation tools [3] from the domain of model-driven architecture (MDA) [7]. MDA represents an approach where problem-specific features (such as the description of system behavior) are represented by a domain-specific language. Standardized tools then transform the domain-specific language into the desired target format, such that an executable system is obtained. In this paper, the domain-specific language entails a description of the system behavior as a finite-state machine (FSM). This description is then transformed into Java code.

R. Moreno-Díaz et al. (Eds.): EUROCAST 2009, LNCS 5717, pp. 279–286, 2009.

Prior to this work, to our knowledge, openly available MBT tools for Java lacked the flexibility of the approach that is presented here. This paper makes the following contributions:

1. The existing test model used in ModelJUnit [8] is replaced with a high-level model that minimizes the amount of code to be written, while being more flexible.
2. Our new model separates two orthogonal concerns, model transitions and implementation actions, presenting a conceptually cleaner solution.
3. Exceptional behavior, a feature that is often used in modern programming languages, can be modeled naturally and expediently with our architecture.

The rest of this paper is organized as follows: Section 2 introduces a running example that shows how the model handles different aspects of the problem. Problems with the existing approach are detailed in Section 3. Our proposed architecture is described in Section 4. Section 5 concludes and outlines future work.

2 Example

2.1 Elevator System

The example that is used throughout this paper describes the possible state space of an elevator that has two pairs of doors: one at the front, another one at the rear. The elevator has been inspired by an elevator from Yoyogi station of the Oedo metro subway line in Tokyo. The elevator ranges over five floors, from the street-level entrance on the second floor, to the lower entrance on the first floor (reachable through a passage from a different train station), to the underground floors of the actual metro station. While the first and second basement level exist, they are not reachable by the elevator, and in general inaccessible to subway commuters, even by stairs.[1]

The model has to reflect the range of permitted configurations and operations that change the system state. Figure 1 shows the state space of all permitted elevator states. Each state has its own label, while transitions are labeled with the action required to reach the next state from the current state. This example does not include transitions leading to error states; such an addition will be described later.

2.2 Test Cases Using JUnit

A simple property that has to hold for the given elevator example is that a if it moves down a number of floors, it will be back at the starting position after having moved up again the same number of floors. If the doors open and close in between, the property still has to hold. A few unit tests verifying this property are shown in Figure 2. As can

[1] Construction of the Toei Oedo line started in 1991, when most of Tokyo city and its metro network were already built. This made it necessary to construct the Oedo line more than 40 m below ground, requiring the elevator to skip a couple of levels before arriving at the ticket gate on the third underground floor. On that floor, the rear doors open. This design makes it more convenient to use the elevator, as people to not have to back out of the elevator through the same door they entered in, but they can instead use the opposite door in front of them.

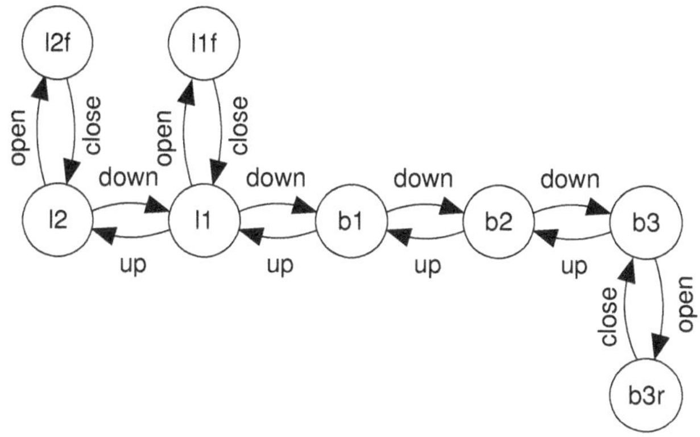

Fig. 1. Example of an elevator with two doors, which are only allowed to open on certain floors. On levels 1 and 2 (states l_1 and l_2), the front doors are allowed to open, changing the configuration to state l_{1f} and l_{2f}, respectively. On the first and second basement floor (b_1 and b_2), the doors must remain closed. On the third basement floor, the rear doors are allowed to open, which is reflected by transition $b_3 \rightarrow b_{3r}$.

```
@Test void test1() {
  pos = l2;
  down();
  up();
  assert(pos == l2);
}

@Test void test2() {
  pos = l2;
  down();
  open();
  close();
  up();
  assert(pos == l2);
}
```

```
@Test void test3() {
  pos = l2;
  down();
  down();
  down();
  down();
  open();
  close();
  up();
  up();
  up();
  up();
  assert(pos == l2);
}
```

Fig. 2. Code example of unit tests using JUnit

be easily seen, much of the test code is simple and repetitive. Yet, only a small part of the possible state space is covered.[2] While better coverage can be achieved in principle, it is unlikely to happen in practice, as too much code has to be written. Model-based testing aims to replace the manual work required to generate a high coverage of the possible state space, while keeping the task of modeling the system simple.

[2] For instance, the possibility of moving to the bottom floor without opening the doors is not tested; opening and closing the doors twice before moving on is not tested; and opening the doors on the first floor before moving to the bottom floor is not tested either.

3 MBT Using ModelJUnit

3.1 Tool Architecture

ModelJUnit is an openly available test case generation tool [8], using an extended finite state machine (EFSM) as input. An EFSM is a Java implementation of a finite-state machine, including two extensions: First, a custom state comparison function can provide a more refined or more abstract behavior than a pure FSM. Second, transitions of an EFSM include their mapping to concrete actions of the system under test (SUT).

ModelJUnit searches the graph of an EFSM at run-time. As new transitions are explored, corresponding actions in the SUT are executed. Results are verified by using JUnit assertions [4]. Specification of the entire model as a Java program allows one artifact to describe all system states, transitions, and corresponding actions of the SUT. However, states and transitions have to be encoded in Java. Each transition requires two methods: a *guard* specifying when a transition is enabled, and an *action* method specifying the successor state and the corresponding action of the SUT. In ModelJUnit, each transition/action pair requires about six lines of repetitive yet error-prone code.

3.2 Elevator System Described in ModelJUnit

As described above, ModelJUnit explores a specifically structured Java program. The program has to contain an initialization method, a number of guard conditions, and actions that update the model state and execute test actions. The tasks of exploring the model, measuring and reporting model coverage, are then done by ModelJUnit. Figure 3 shows an excerpt of a finite-state machine encoded for ModelJUnit. As can be seen, the specification of the state space and the initial state is rather simple, but parts relating to the SUT (lines 2 and 8) are interleaved with model code. This interleaving of model and implementation code continues throughout each guard method (lines 12–14) and action method (lines 16–19). Out of these seven lines, only one line, the call to `down()`, corresponds to a concrete test action. The remainder of the code is quite repetitive and can be fully expressed by the graphical finite-state machine from Figure 1.

4 Proposed Architecture

4.1 Separation of the FSM from System Actions

We propose a separation of the behavioral model and the program code. The state space of our model is described by a conventional FSM, using the "dot" file format from graphviz [1]. This format is concise, human-readable, and supported by visualization and editing tools. A transition can be specified in a single line of text of form `pre -> post [ label = "action" ]`. A label corresponds to an action in the SUT. The same action may be executed in different transitions.[3]

Most methods of a SUT require arguments that cannot be constructed trivially from a high-level model such as an FSM. Instead, actions are delegated to a bridge class written in Java. The bridge class implements test actions and describes how parameters are constructed and verified.

[3] The same effect would be achieved in an existing ModelJUnit model by either duplicating code or by writing complex transition guards and post-state conditions.

```java
public class ElevatorFSM implements FsmModel {
  private ElevatorController elevator;

  enum State { l2, l2f, l1, l1f, b1, b2, b3, b3r }
  private State state;

  public void reset(boolean testing) {
    elevator = new ElevatorController();
    state = State.l2;
  }

  public boolean ActionL2L1Guard() {
    return state == State.l2;
  }

  public @Action void actionL2L1() {
    elevator.down();
    state = State.l1;
  }
}
```

Fig. 3. Part of the elevator model for ModelJUnit

4.2 Elevator Model Using Our Architecture

Figure 4 shows how our model splits the concerns into two parts: The FSM, encoded in the dot format, and the bridge class called *ElevatorImpl,* which contains the application-specific test code. Note that the entire state space of the model is conveniently managed in the FSM, so our Java code contains no guard or state variables. At the same time, any implementation-specific code is removed from the FSM.

4.3 Model Annotations

Libraries may contain redundant interfaces ("convenience methods") as shorthands. It is desirable to test all of these methods, yet the FSM would be cluttered by the inclusion of the full set of redundant methods. We chose to use annotations of FSM transitions to describe cases where a single FSM transition covers a set of actions. Annotated transitions are internally expanded to the full set of methods before the test model is explored. Thereby, interface variants are tested by selecting a random variant each time the transition is executed. A set of methods that only read data without changing the system state can also be represented by one FSM action and an annotation. In this way, the fact that no access method actually modifies the system state can be tested against.

4.4 Exceptional Behavior

Finally, a given action may cause an exception, depending on the state of the SUT. Exception annotations specify states where an exception is expected to occur. As an example, take an extension of the elevator from the ticket gate level (b_3) to the tracks (b_4). Customers are required to use the ticket gate for the metro, so they are not authorized to

```
digraph Elevator {
  init -> l2;
  l2   -> l1 [ label = "down" ];
  l1   -> b1 [ label = "down" ];
  b1   -> b2 [ label = "down" ];
  b2   -> b3 [ label = "down" ];
  b3   -> b2 [ label = "up" ];
  b2   -> b1 [ label = "up" ];
  b1   -> l1   [ label = "up" ];
  l1   -> l2   [ label = "up" ];
  l2   -> l2f  [ label = "open" ];
  l2f  -> l2   [ label = "close" ];
  l1   -> l1f  [ label = "open" ];
  l1f  -> l1   [ label = "close" ];
  b3   -> b3r  [ label = "open" ];
  b3r  -> b3   [ label = "close" ]; }
```

```
public class ElevatorImpl {
  private ElevatorController elevator;

  public void init() {
    elevator = new ElevatorController();
  }

  public void down() {
    elevator.down();
  }
}
```

Fig. 4. The FSM of the elevator model in dot format (top) and a part of the bridge to its implementation in Java (bottom)

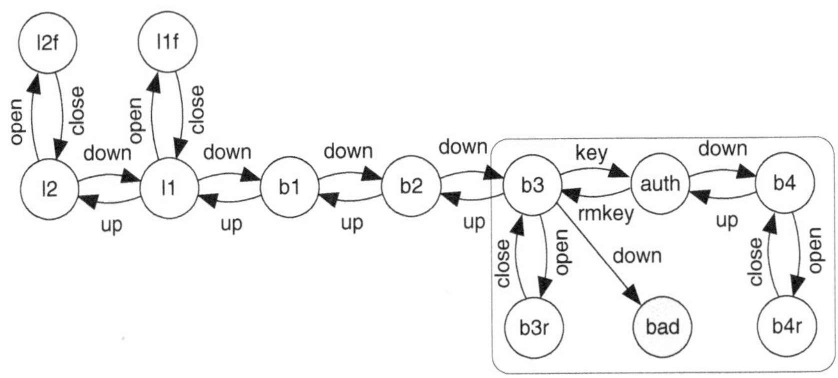

Fig. 5. Extended model of the elevator. Assume there is a forth basement level, which is only allowed to be accessed by authorized personnel. Authorization is granted when a key is inserted when in basement level 3 (b_3). If the "down" action is executed without authorization, an exception should be thrown, as indicated by state *bad*.

access level b_4 directly through the elevator. Assume that, for maintenance, the elevator extends to b_4. If a key is inserted in b_3, the elevator changes to an authorized state, from which access to the lowest level is given. Otherwise, any attempt to reach b_4 should be indicated by an exception, leading to a "bad" state.[4] Figure 5 shows the extended FSM.

[4] In the Tokyo Metro stations, this problem is usually solved by having a separate elevator from the ticket gates down to the tracks below.

The code necessary in Java to verify the presence or absence of an exception is quite lengthy. In Java, so-called *checked exceptions* are defined by methods that declare that they may throw an exception [2]. Each call to such a method has to be guarded against possible exceptions by using a `try/catch` block. In our model, annotations of FSM states specify which states correspond to an exceptional state. All transitions to that state are expected to throw precisely that type of exception. Equivalent actions leading to non-exceptional states may not throw any exception. As the two cases differ only slightly, the code in Figure 6 can be easily generated automatically.

```
public void actionb3bad() {
  boolean exceptionOccurred = false;
  try {
    impl.down();
  }
  catch (IllegalStateException e) {
    exceptionOccurred = true;
  }
  assert(exceptionOccurred);
}
```

Fig. 6. Code that verifies the presence of an exception

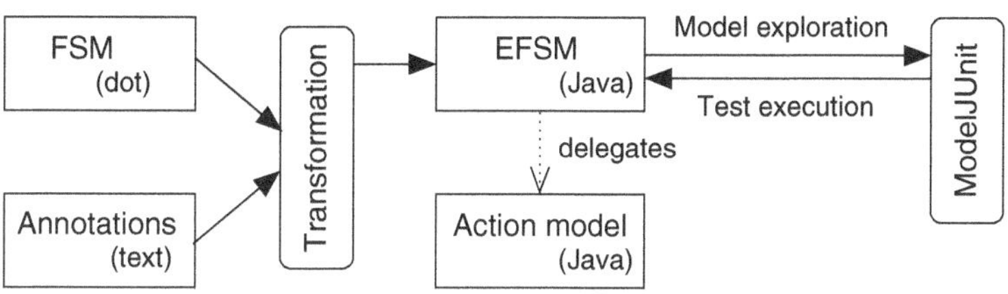

Fig. 7. Architecture of system to generate the ModelJUnit model

4.5 Tool Architecture

For use with ModelJUnit, the FSM is expanded into its Java representation (see Figure 7). The generated code includes `try/catch` blocks for verification of the presence or absence of exceptions. Transformation of the FSM can be automated either by leveraging existing model transformation tools [7] or by extending ModelJUnit with a parser for the additional file formats. The former approach requires additional tools but is independent of the programming language and unit test library.

5 Conclusions and Future Work

Current tools for model-based testing do not completely separate all features. Specifically, different transitions in an abstract model may correspond to the same action in the implementation. Separation of these two artifacts leads to a more concise model.

Inclusion of exceptional states in the model further increases the amount of code that can be generated automatically, making the model more expressive and maintainable.

Future work includes exploring the possibility of tying state invariant code, rather than just exceptions, to states in the model. This would be done in a way that is equivalent to how actions are tied to transitions. Furthermore, in our architecture, the ModelJUnit tool only serves to explore the graph of the model. The transformation step shown in Figure 7 could therefore be subsumed by direct execution of the action model, without going through ModelJUnit. Independence of ModelJUnit would have the added benefit of being able to implement features that ModelJUnit does not support, such as non-deterministic transitions.

References

1. Gansner, E., North, S.: An open graph visualization system and its applications. Software – Practice and Experience 30, 1203–1233 (1999)
2. Gosling, J., Joy, B., Steele, G., Bracha, G.: The Java Language Specification, 3rd edn. Addison-Wesley, Reading (2005)
3. Javed, A., Strooper, P., Watson, G.: Automated generation of test cases using model-driven architecture. In: Proc. 2nd Int. Workshop on Automation of Software Test (AST 2007), p. 3. IEEE Computer Society, Washington (2007)
4. Link, J., Fröhlich, P.: Unit Testing in Java: How Tests Drive the Code. Morgan Kaufmann Publishers, Inc., San Francisco (2003)
5. Myers, G.: Art of Software Testing. John Wiley & Sons, Inc., Chichester (1979)
6. Peled, D.: Software Reliability Methods. Springer, Heidelberg (2001)
7. Poole, J.: Model-driven architecture: Vision, standards and emerging technologies. In: Workshop on Metamodeling and Adaptive Object Models, Budapest, Hungary (2001)
8. Utting, M., Legeard, B.: Practical Model-Based Testing: A Tools Approach. Morgan Kaufmann Publishers, Inc., San Francisco (2006)

Automatic Test Generation for Coverage Analysis Using CBMC

Damiano Angeletti, Enrico Giunchiglia, Massimo Narizzano, Alessandra Puddu*, and Salvatore Sabina

DIST, University of Genova,
Via all'Opera Pia 13 16145 Genova, Italy
{enrico.giunchiglia,massimo.narizzano,alessandra.puddu}@unige.it
Ansaldo STS
Via Paolo Mantovani, 3 - 16151 Genova, Italy
{damiano.angeletti,salvatore.sabina}@ansaldo-sts.com

Abstract. Testing is the most used technique for software verification: it is easy to use and even if no error is found, it can release a set of tests certifying the (partial) correctness of the compiled system. Moreover, in order to increase the confidence of the correctness of the compiled system, it is often required that the provided set of tests covers 100% of the code. This requirement, however, substantially increases the costs associated to the testing phase, since it may involve the manual generation of tests. In this paper we show how to use a Bounded Model Checker for C programs (CBMC) as an automatic test generator for the Coverage Analysis, and we show how its use can substantially reduce the costs of the testing phase.

1 Introduction

BMC has been successfully used in the last decade to formally verify finite systems, such as sequential circuits and protocols, see e.g. [3]. The key idea of BMC is to build a propositional formula whose models correspond to program traces (of bounded length) that violate some given properties and check the resulting formulae for satisfiability. Recently BMC has been also applied to the Software Verification. In [1] the authors built a Bounded Model Checker, CBMC, to reason about low-level ANSI-C programs, checking safety properties such as the correctness of pointer constructs, array bounds, and user-provided assertions. Testing is the most used technique for software verification: it is easy to use and even if no error is found, it can release a set of tests certifying the (partial) correctness of the compiled system. The most used technique for testing generation is random testing since it is automatic and simple to apply. However, it does not ensure an extensive test of the code: since it merely relies on probability it has quite low chance in finding semantically small faults. Thus, in order to increase the confidence of the correctness of the compiled system, it is often required

* Partially supported by a Ph.D. grant (2007-2009) financed by Ansaldo STS.

R. Moreno-Díaz et al. (Eds.): EUROCAST 2009, LNCS 5717, pp. 287–294, 2009.

that the provided set of tests covers 100% of the code. This requirement, however, substantially increases the costs associated to the testing phase, since it may involve the manual generation of tests. In this paper we show how to use a Bounded Model Checker for C programs (CBMC) as an automatic test generator for the Coverage Analysis. We experimented our methodology on a subset of modules of the ERTMS/ETCS source code, an industrial system for the control of the traffic railway, provided by Ansaldo STS. Using CBMC we were able to verify 5 different modules of the ERTMS, i.e we were able to generate a set of test covering the 100% of the code for each function in each module. The use of CBMC for test generation led to a dramatic increase in the productivity of the entire Software Development process by substantially reducing the costs of the testing phase.

The use of BMC for automatic test generation is not completely new in the field of verification of circuits and microprocessor design [12], [16]. There are also some related work that apply symbolic execution to Software Testing. For example, *Pex* [13], [14] is a tool for Automatic Test generation, developed at Microsoft Research. It helps developers to write PUTs (Parametrized Unit Tests) [9] in .NET language. For each PUT, Pex uses dynamic test-generation techniques to compute a set of input values that exercises all the statements and assertions in the analyzed program. Pex is not completely automatic (PUTs are written by hand) and does not guarantee the 100% of decision coverage. Another approach is presented in *DART* (Directed Automated Random Testing) [7], where interface extraction, random testing, and dynamic analysis are combined during the execution of a random test, in order to systematically drive the generation of the new tests along alternative program paths. DART uses Constraint Solving techniques it does not guarantee 100% of decision coverage. Another approach, similar to DART and Pex is presented in a tool called KLEE [10]. KLEE explores a path along the control flow graph of the program, collecting its path constraint. Then the path constraint is passed to a constraint solver that returns an assignment to the input variables, if any, which makes true the path constraint. KLEE will continue until all the statements of the program are covered by the set of tests computed. Also KLEE uses Constraint Solving techniques and it covers all the statements (lines of code), while we guarantee decision coverage. Another similar approach is presented in TestEra [11], a tool for automatic test generation for Java Programs. Given a method in Java (sourcecode or bytecode), a formal specification of the pre- and post-conditions of that method and a bound limit on the size of the test cases to be generated, TestEra automatically generates all nonisomorphic test inputs up to the given bound. Specifications are first order logic formulae. TestEra is not fully automated and it does not guarantee the 100% of decision coverage.

The paper is structured as follows: first we present the usage of CBMC as a test generator and then we show how to use it for Coverage Analysis. Finally we present some experimental results.

2 Automatic Test Generation Using CBMC

CBMC [1] is a Bounded Model Checker of ANSI-C programs, allowing the verification of safety properties such as the correctness of pointer constructs, array bounds, and user-provided assertions. Given a property and a piece of code, CBMC, in case that the program violate the property, will return an error-trace, i.e. an assignment to the input variables. In more details, CBMC, after a small manipulation of the original code such as function calls replacement and loop unwinding, it rewrites the code and the property into an equivalent program in Single Static Assignment (SSA) form [5] that is an Intermediate Representation where each variable is assigned exactly once. In figure 1 is presented an example of SSA transformation, for more details about SSA transformation, please refer to [1]. After the SSA transformation, the resulting program is translated into a Boolean formula in Conjunctive Normal Form (CNF) and a SAT solver, like chaff [4] or MiniSat [8], is applied. If the SAT solver returns false then the property holds, otherwise the property does not hold. If the property does no hold the SAT solver will return an assignment to the input variables. Since a test is an assignment to the input variables causing the execution of a portion of code, or violating a property, we take advantage of the error trace returned by CBMC to automatic generate a test. However in order to generate a test, CBMC needs to violate a property, such as an user-provided assertions: in our framework we insert into the code an assertions that must be violated and CBMC will return an assignment to the input variables violating it. In figure 1, on the left is presented a piece of code where an assertion is introduced, line 5 on the left: if the $assert(a \neq 0)$ is violated then an assignment to the input variable a is returned, for example $a = -1$. It is not always possible to write a code dependent property that can be violated, instead we use a code independet property that is always violated: such a property in our framework is $assert(0)$. So in the code we insert

```
int FUT(int a)
0  int r = i = 0
1  while i < max do
2    g++
3    if i > 0 then
4      a++
5      assert(a ≠ 0)
6      r = r + (g+2)/a
7    else
8      r = r + g + i
9    i++
10 r = r * 2
11 return r
```

```
int FUT(int a)
0  int r = i = 0
1  if i < max then
2    g++
3    if i > 0 then
4      a++
5      assert(a ≠ 0)
6      r = r + (g+2)/a
7    else
8      r = r + g + i
9    i++
10 r = r * 2
```

$$
\begin{aligned}
C := \;& r_0 = 0 \wedge \\
& i_0 = 0 \wedge \\
& g_1 = g_0 + 1 \wedge \\
& a_1 = a_0 + 1 \wedge \\
& r_1 = r_0 + \tfrac{g_1+2}{a_1} \wedge \\
& r_2 = r_0 + g_0 + i_0 \wedge \\
& i_1 = i_0 + 1 \wedge \\
& r_3 = i_0 > 0 ? r_1 : r_2 \wedge \\
& r_4 = i_0 < max_0 ? r_3 : r_0 \wedge \\
& r_5 = r_4 * 2 \wedge \\
P := \;& a_1 \neq 0
\end{aligned}
$$

Fig. 1. Example of SSA transformation; Left: A generic function written in a subset of the ANSI-C; Center: Unwinding with $k = 1$; Right: SSA form transformation of the program on the left an $assert(0)$ and running CBMC we obtain an assignment to the input variables, i.e a test, causing the execution of the assert: the test is automatic generated.

3 Coverage Analysis Using CBMC

The CENELEC EN50128 [6] is a part of a group of European standards for the development of Railway applications. It concentrates on the methods which need to be used in order to provide software which meets the demands for safety integrity. The European standards have identified techniques and measures for 5 levels of software safety integrity (*sil*) where 0 is the minimum level and 4 the highest level. The railway system requires *sil* 4, meaning that the system verification should be done producing a set of test having 100% of code coverage. In particular Ansaldo STS, that produce software that has to meet the CENELEC requirements, it has adopted as coverage criteria the branch coverage, that means that all the decision in a function under test have to be executed. This means that for covering the following decision

if $a == 0$ **then**
    $block_1$
else
    $block_2$

we need at least two test, one covering $a == 0$ and one covering $a \neq 0$. In the previous section we show that adding an $assert(0)$ in a particular line of the code and running CBMC on that code, it will return an assignment to the input variables exercising(covering) the portion of code containing the assert. So in this case, following the branch coverage criteria, we would like to automatic generate two different test, one exercising the $block_1$ and another one exercising the $block_2$. In order have CBMC generating the two tests, we have to insert two $assert(0)$: one before $block_1$ and the other before $block_2$. Following this idea we insert in the code as many assert as we need to cover all the decisions.

This is the idea behind the methodology sketched in Figure 2 and explained in details below:

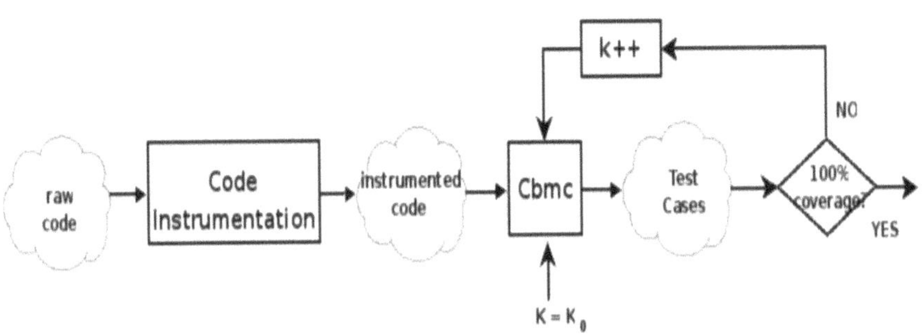

Fig. 2. Testing Process with CBMC

1. *Code Instrumentation.* CBMC requires that each function called in the function under test is completely defined. Nevertheless a tester can not always have the access to each function, for example to the functions defined in modules which are not available yet. So, for each missing function we have to create a *stub*, i.e. a function returning an appropriate random value. Moreover, it is also necessary for CBMC to create a *main* function that calls the function under test, and sets all the input values. Finally, in CBMC we insert an *assert*(0) for each branch predicate and one for its negation. Unfortunally they can not be inserted all together since during the generation phase CBMC will stop at the first assertion violated. An alternative could be to insert an assert at a time generating one file for each assertion introduced. However, in this way many files will be generated, one for each test, causing difficulties in the management of the test cases. So, if we want to generate n tests we have to introduce n macros like:

```
#if defined(ASSERT_i)
assert(0)
#endif
```

where $i = 1..n$, and n is the number of branches of the program. For a clearer code we used pseudo code where each ASSERT_i inside the function under test means that it is replaced by the proper macro. Figure 3 presents the instrumented version of the code in Figure 1. Notice that inside the main function each input variable is initialized with a random value returned by a non deterministic function, that is necessary, otherwise CBMC will set each (global) input variable to 0.

```
#ifdef ASSERT_1
  assert(0)
# endif
#ifdef ASSERT_2
  assert(0)
# endif
#ifdef ASSERT_3
  assert(0)
# endif
#ifdef ASSERT_4
  assert(0)
# endif
#ifdef ASSERT_5
  assert(0)
# endif
#ifdef ASSERT_6
  assert(0)
# endif
#ifdef ASSERT_7
  assert(0)
# endif

int MAIN(int argc, char * argv[ ])
    max = NONDET_INT()
    g = NONDET_INT()
    int a = NONDET_INT()
    return fut(a)
```

```
int FUT(int a)
          ASSERT_1
s0        int r = i = 0
b0, b1    while i < max do
          ASSERT_2
s1          g++
b2          if i > 0 then
          ASSERT_3
s2            a++
b4            if a ≠ 0 then
          ASSERT_4
s3              r = r + (g+2)/a
b5            else
          ASSERT_5
b3          else
          ASSERT_6
s4            r = r + g + i
s5          i++
          ASSERT_7
s6        r = r * 2
s7        return r
```

Fig. 3. An example of instrumented function

2. *Test Generation.* Given the instrumented code CBMC is run n times with a fixed k, where n is the number of assertions added to the code and for each run an assertion is activated at a time. From each run a counter-example is produced and from which a test is extracted.
3. *Coverage Analysis.* Adding an assertion for each branch ensures that CBMC will generate enough tests covering the 100% of the branches. However if k is not correctly fixed then some branches can not be covered. For example, looking at Figure 1 on the left, if k is set to 1, then the statements s_2, s_3 and s_4 can not be covered. In order to prevent this behavior, a coverage analysis process with Cantata is executed on the test-set and if the branch covered is less than 100% then k is incremented and the testing generation phase is executed again; otherwise the process will stop.

4 Experimental Analysis and Conclusions

We applied our framework to an industrial project developed by Ansaldo STS, the European Rail Traffic Management System [2]. ERTMS is an EU "major European industrial project" to enhance cross-border interoperability and signalling procurement by creating a single Europe-wide standard for railway signalling. Ansaldo STS as part of the European Project produces the European Vital Computer (EVC) software, a fail-safe system which supervises and controls the speed profiles using the information received during the navigation transmitted to the train. Following the CENELEC standards Ansaldo STS needs to provide a certificate of the integrity level required, i.e. it has to provide a set of tests covering the 100% of the branches. In order to simplify the readability, the Ansaldo STS implementation of the EVC is developed into different modules of fixed size. In our experimental analysis we took 5 interconnected modules of the EVC and we applied the automatic test generation strategy seen in the section above. The five modules under test contain 73 different functions presenting no recursive calls and the max iterations of each loop is known a priori. For industrial copyright purpose, we just omit each module's name, substituting them with m_i, where i is a number between 1 and 5. The five modules contain more than 10000 lines of code, while the entire EVC project contains more than 100000 lines of code. Table 1 shows the results of the automatic test generation on the 5 modules of the EVC. The table is divided in 3 parts each one including 2 columns: the

Table 1. Experimental analysis on the five modules of the EVC

		Cbmc		Ansaldo STS	
Modules	#functions	#tests	time(s)	#tests	time(s)
m_1	19	148	1212	64	57600
m_2	7	47	1444	26	23400
m_3	13	193	6256	80	72000
m_4	18	184	1308	110	99000
m_5	16	185	1667	105	94500
Total	73	757	11887	385	346500

first two columns describe some parameters of the modules, module's name and number of function in the module respectively. The second group of two columns represents the number of tests generated by CBMC and the time spent during the generation phase. The last group of two columns represents the number of tests generated by Ansaldo STS and the estimation of the time spent during the generation, taking into account an average of fifteen minutes for each test. Each line of the table represents a module, while the last line represents the total. As it can be seen for each module the number of tests automatically generated by CBMC is close to the double with respect to the ones generated manually by Ansaldo STS, but on the other hand the time spent to generate the test manually is lower of an order of magnitude than the time spent by the automatic generation. Looking at the total we can see that CBMC will generate almost 757 tests for the coverage of 73 functions in less than 13 minutes, while the test manually generated are only 385, but the time spent is more than six and a half hours: in practice, CBMC can generate tests to cover 5 different modules in the time that a single test is manually generated. Table 1 shows also that CBMC will never generate less tests than the ones generated by Ansaldo STS. This is mainly due to the fact that CBMC generates tests without any reasoning or a priori knowledge: Indeed a person, with a priori knowledge also on the semantic of the program, has a clear advantage. For example, looking at Figure 1 on the left, CBMC will generate a test for each different branch, even if some of them can be avoided, such as, for instance, the exploration of b_0 and b_1 that is enforced by the traversal of the branches b_3 and b_4. So, in this case some tests are subsumed by a (set of) test in the test-set: a smarter tester may construct a more complex test-set that minimize this phenomenon.

5 Conclusions

In this paper we have shown how (C)BMC can be successfully used for the automatic generation of a set of tests covering the 100% of branches. Our experiments report that the use of CBMC led to a dramatic increase in the productivity of the entire Software Development process, by substantially reducing the time spent, and consequently, the costs of the testing phase. To the best of our knowledge, this is the first time that BMC techniques have been used for coverage analysis of Safety-Critical Software in an Industrial setting. These results demonstrate the maturity of the Bounded Model Checking technique for automatic test generation in industry.

References

1. Clarke, E., Kroening, D., Lerda, F.: A Tool for Checking ANSI-C Program. In: Tools and Algorithms for the Construction and Analysis of Systems, pp. 168–176
2. ERTMS: The official Website, `http://www.ertms.com/`
3. Biere, A., Cimatti, A., Clarke, E.M., Strichman, O., Zhu, Y.: Bounded Model Checking. Advances inComputers 58

4. Moskewicz, W.M., Madigan, C.F., Zhao, Y., Zhang, L., Malik, S.: Chaff:Engineering an Efficient SAT Solver. In: Proceedings of the 38th Design Automation Conference, pp. 530–535
5. Cytron, R., Ferrante, J., Rosen, B.K., Wegman, M.N., Zadeck, F.K.: Efficiently-Computing Static Single Assignment Form and the Control DependenceGraph. ACM Transactions on Programming Languages and Systems 13(4), 451–490
6. European Committee forElectrotechnical Standardization. Railway Applications - Communication, signalling and processing systems - Software forrailway control and protection systems, `http://www.cenelec.eu`
7. Godefroid, P., Klarlund, N., Sen, K.: DART: Directed automated random testing. In: Proceedings of the ACM SIGPLAN Conference on Programming Language Design and Implementation, pp. 213–223
8. Een, N., Sorensson, N.: An ExtensibleSAT-solver. In: Satisfiability Workshop, pp. 502–518
9. Tillmann, N., Schulte, W.: Parameterized unit tests. In: Proceedings of the 10th European Software Engineering Conference Held Jointly with 13th ACM SIGSOFT international Symposium on Foundations of Software Engineering, pp. 253–261
10. Cadar, C., Dunbar, D., Engler, D.R.: KLEE: Unassisted and Automatic Generation of High-Coverage Tests for Complex Systems Programs. In: OSDI, pp. 209–224
11. Khurshid, S., Marinov, D.: TestEra: Specification-based testing of java programs using sat. Automated Software Engg., 403–434
12. Chockler, H., Kupferman, O., Kurshan, R.P., Vardi, M.Y.: A Practical Approach to Coverage in Model Checking. In: Berry, G., Comon, H., Finkel, A. (eds.) CAV 2001. LNCS, vol. 2102, pp. 66–78. Springer, Heidelberg (2001)
13. Tillmann, N., de Halleux, J.: Pex White Box Test Generation for.NET. In: Beckert, B., Hähnle, R. (eds.) TAP 2008. LNCS, vol. 4966, pp. 134–153. Springer, Heidelberg (2008)
14. de Halleux, J., Tillmann, N.: Parameterized unit testing with pex. Tests and Proofs, 171–181 (2008), `http://dx.doi.org/10.1007/978-3-540-79124-9_12`
15. Jackson, D., Shlyakhter, I., Sridharan, M.: A micromodularity mechanism. In: Proceedings of the 8th European Software Engineering Conference Held Jointly with 9th ACM SIGSOFT international Symposium on Foundations of Software Engineering, pp. 62–73
16. Vedula, V.M.: Hdl Slicing for Verification and Test, available electronically from `http://hdl.handle.net/2152/1033`

Self-healing Assurance Based on Bounded Model Checking*

Vendula Hrubá, Bohuslav Křena, and Tomáš Vojnar

Faculty of Information Technology, Brno University of Technology
Božetěchova 2, Brno, CZ-612 66, Czech Republic
{ihruba,vojnar,krena}@fit.vutbr.cz

Abstract. This paper presents an approach of using bounded model checking for healing assurance within a framework for self-healing of concurrent Java programs. In this framework, dynamic (i.e., runtime) analysis is used to detect possible data races for which some pre-defined healing strategy may subsequently be applied. Before applying such a strategy, it is desirable to confirm that the detected possible error is indeed a real error and that the suggested healing strategy will solve it without introducing an even worse problem (namely, a deadlock). For this purpose, we suggest bounded model checking to be applied in the neighbourhood of the state in which the possible error is detected. In order to make this possible, we record certain points in the trace leading to the suspicious state within a run of the tested system, and then replay the trace in the chosen model checker (in particular, Java PathFinder) using its state space exploration capabilities to navigate between the recorded points.

1 Introduction

Despite the constantly growing pressure on quality of software applications, many software bugs still appear in the field. This happens in spite of the ongoing development and applications of various software engineering techniques targeted at developing reliable software as well as of various automated techniques for verification of software, including different formal techniques. This situation has recently become a motivation for developing various *self-healing technologies*, such as those considered in [4], whose task is to catch bugs remaining in software at the runtime and, moreover, to try to automatically correct them.

One class of bugs that may be treated via self-healing are *concurrency-related bugs*, e.g., *data races* caused by missing or improper synchronisation. A data race is, in particular, a situation when two threads access the same shared variable, at least once for writing, and there is no synchronization between the accesses. Such bugs may be detected at runtime using various *dynamic analyses* (e.g., [8,2,5]). A self-healing action can then be implemented by *influencing the scheduling* or by supplying some *additional synchronisation* [4]. However, self-healing actions can be unsuccessful or even

* The work was supported by the European FP6 project IST-035157 SHADOWS, the Czech Science Foundation by the project 102/07/0322, and the Czech Ministry of Education by the project MSM 0021630528.

R. Moreno-Díaz et al. (Eds.): EUROCAST 2009, LNCS 5717, pp. 295–303, 2009.

introduce new bugs to the healed application (such as a deadlock). In addition, they can also cause a significant drop down in the efficiency of the application.

Therefore, it is desirable to accompany self-healing by a possibility to check whether the detected bugs are real bugs requiring some self-healing action to be applied and subsequently whether the automatically proposed healing action will remove the encountered bug while not causing other problems considered to be even worse. We denote such checks as *self-healing assurance* and, in this paper, we discuss our practical experience with the use of *bounded model checking* for implementing the first phase of healing assurance (namely checking whether a detected possible bug is a real one) in the context of self-healing of *data races* in concurrent Java programs.

In particular, we assume dynamic analysis to be performed over a running Java program, indicating a possible data race. Subsequently, we want to use bounded model checking to confirm that a real data race was detected and, even more, that it is the so called *true data race*, i.e., that the two unsynchronized concurrent read/write or write/write actions, which were found by the dynamic analysis, can happen in both orders.

In order to allow for bounded model checking in the neighbourhood of the detected error state, we *record* certain points in the trace leading to this state within a run of the tested system (using the ConTest [3] infrastructure that we also use to implement the Eraser+ and AtomRace dynamic analyses [6] that we concretely employ to detect data races). This partially recorded trace (or some similar trace with the same control points as the recorded ones) is subsequently *replayed* in the chosen model checker (in our case, Java PathFinder [7]).

Despite we have not yet implemented and practically tested the second phase of healing assurance, namely checking whether the chosen healing action is effective and not harmful, the framework we have built provides a solid starting point for implementing such checks.

Plan of the Paper. The paper is organized as follows. In Section 2, we briefly recall some features of the Java PathFinder model checker that we use in our work. In Section 3, we describe how we get to the problematic state in whose neighbourhood we want to perform bounded model checking, which is then discussed in Section 4. Section 5 gives some experimental data before we conclude in Section 6.

2 Java PathFinder

Java PathFinder (JPF) [7] is an explicit-state model checker for Java bytecode, implemented over a specialized Java virtual machine. JPF provides several *state space search strategies* like depth-first search (DFS) or breadth-first search (BFS) as well as a listener-based mechanism allowing new user-specific strategies to be defined. We exploit this feature in our work to implement a directed run through the state space controlled by trying to follow a (partially) recorded suspicious run of a given program.

JPF implements several techniques for *state space reduction*, including *partial order reduction* which is based on the observation that a large number of theoretically possible thread interleavings are irrelevant for the properties being checked in concurrent programs. JPF automatically detects which schedules are relevant when checking a certain property. Moreover, JPF provides a mechanism to influence the mechanism of choosing the relevant schedules. This is important for us because, on one hand, we cannot avoid partial order reduction not to significantly increase the number of explored states, and, on the other hand, the implicit partial order reduction of JPF may interfere with our intention to reconstruct a certain specific run of the given program.

3 Getting to the Given State

In order to be able to use JPF for bounded model checking with the above described aims, we need to get JPF into some close predecessor of the state of the tested system in which a possible data race was detected by dynamic analysis. From such a predecessor state, we can then start a systematic (bounded) state space search. Starting the bounded model checking from some close predecessor of the suspicious state (instead of this state itself) is desirable as after some problematic behaviour has already happened, it may not be possible to see it happening again.

To get JPF into the desired state, one could think of *saving the suspicious state* from the regular Java virtual machine (JVM) in which the monitored application is running (together with the dynamic analyser attached to it) and subsequently *restoring the state* in the special JVM of JPF. Such an approach would not slow down the monitored application during a regular run by a need to continuously record some information, it would not require any extra space during the regular run, and it would not require JPF to spend time by going through many predecessor states before getting to the suspicious state. However, there are significant obstacles associated with this approach. A major disadvantage is that bounded model checking would have to be started from the suspicious state, and not from some of its predecessors, which is not what we need. Furthermore, from an implementation point of view, it is technically very complicated to reproduce a state (including existing threads, their stacks, the heap, etc.) saved from one JVM in another JVM.

Therefore, to get JPF into some close predecessor of the suspicious state, we have decided to *record some points in the trace* executed by the monitored program and then *replay this trace* in JPF, using its state search to navigate between the saved points. In this case, once we reach the suspicious state, we can get back to some of its predecessors and then start a systematic bounded state space search. Moreover, the approach is relatively easy to implement. Of course, the recording slows down the monitored application, requires potentially a lot of space for storing the recorded data, requires some time to replay the trace, and if only some points in the trace are recorded (to save space and time during the recording), there is no guarantee that we really get into the suspicious state in JPF. However, according to our opinion, the disadvantages of this approach are still less significant than its advantages, and the approach also provides a lot of space for further improvements in the future (e.g., via using various heuristics about what to record, etc.).

We next give some more details on our implementation of recording a trace of an application running on a regular JVM and on replaying it in JPF.

3.1 Recording a Trace Using ConTest

We have implemented trace recording in the ConTest infrastructure [3], over which we have also implemented two dynamic analyses for detecting possible data races—Eraser+ [4], a slightly optimized version of the original Eraser algorithm [8] (which, e.g., takes into account the join synchronization in Java, etc.), and AtomRace [5]. ConTest was used to implement the healing actions [4] based on influencing the scheduler or based on adding new locks too. ConTest provides a heuristic noise injection mechanism (to increase probability of manifestation of concurrency-related bugs) and a listener architecture implemented via Java bytecode instrumentation. We use the listener architecture for implementing the dynamic analyses, healing actions as well as for recording some points in the run of the monitored application.

The special listener that we have implemented for recording information about the trace followed by the monitored application records several selected events—in particular, (1) *thread beginning*, (2) *thread end*, and (3) *entering a basic block* of the bytecode. For each event, the appropriate thread identifier, the Java source code line, and the number of instruction generated from this line that causes the encountered event are recorded. We call the recorded events *control points* and they are subsequently used to navigate JPF to the detected suspicious state (or, more precisely, some of its close predecessors).

3.2 Replaying a Trace in JPF

When replaying in JPF a trace recorded with the help of ConTest from a running application, we are driven by the sequence of recorded control points. Given the next recorded control point that the replaying has to go through, one has to solve a problem with *identifying which instruction to be executed in JPF corresponds to this control point.* This is not easy because JPF does not provide instruction numbering compatible with that of ConTest. That is why we do not run the non-instrumented bytecode in JPF, but the instrumented one, and we exploit the following fact: Each original instruction corresponding to a recorded control point is preceded in the instrumented code by some additional instructions which are accompanied with information about which line of the source code and which of the instructions generated from it have been instrumented. When executing the instrumented code in JPF, we get this information and we may easily detect when an instruction corresponding to a recorded control point is to be executed by JPF.

On the other hand, simply running the instrumented code in JPF would mean that *JPF's state space generation would examine also the internals of ConTest*, which would significantly worsen the state space explosion. To cope with this situation, we modify the instrumented bytecode when it is loaded to JPF as follows: Using JPF's standard means, we mark the instructions that have been added into the code by the ConTest's instrumentation as instructions that should not be executed during the state space exploration—they are, however, still recognized during the state space generation, and in case they correspond to a recorded event, we get this information through a special listener that we prepared for this purpose. Moreover, ConTest not only adds some instructions, but also replaces some original bytecode instructions[1] by a call of some ConTest code, which then eventually executes the original instruction. In this case, we modify the bytecode when it is loaded in such a way that we put back the original instruction.

Another problem, which has to be resolved when replaying a recorded trace in JPF, is the possibility that the order in which particular events happen in the recorded trace may be *incompatible* with the way threads are scheduled in JPF when the *partial order reduction* is applied—namely, the recorded trace may not appear in the reduced state space even when we fully explore it. At the same time, we cannot afford to rule out partial order reduction because without it, too many different interleavings of the non-recorded events would have to be explored within replaying the recorded trace.

To cope with this problem, we have modified the JPF's depth-first state space search in such a way that each instruction belonging among those that yield control points is checked against the recorded sequence of these points. If the instruction to be executed by JPF matches with the recorded control point at the current position in the trace, we let JPF behave as usual. If the recorded instruction and the one chosen by JPF differ, we *override the default behaviour* of JPF and its partial order reduction.

The overriding takes into account that in JPF, we can distinguish states where a single successor action within the currently active thread is considered[2], and states where several possible actions within possibly multiple (though not necessarily all) threads are explored using the so called choice generator. If the difference between the recorded action and the one chosen by JPF happens in a state where JPF intends to explore a single outgoing transition only (as shown in Figure 1(a)[3]), we add a choice generator to the encountered state and force JPF to iterate through actions of all threads ready to run. Out of them, the one that corresponds to the recorded control point is executed[4] as illustrated in Figure 1(b). When the difference between the recorded action and the one that JPF wants to execute happens in a state where JPF is about to explore several possible continuations, we force it to follow the recorded one. If the recorded one is not among those selected with respect to the rules of partial order reduction, we again force JPF to iterate through all firable actions and to follow the recorded one.

Once we finish executing the entire recorded trace, we back-track some predefined number of steps and then start systematic bounded model checking as described below. Note that it may happen that we, in fact, followed a different trace than the one that was followed by the application, but these two traces at least agree on the recorded sequence of key events, and so there is a hope that we got really close to the problematic state and we will hit it within the subsequent bounded model checking.

---

[1] This applies, e.g., for instructions `wait`, `notify`, `join`, etc.—these instructions are not among those corresponding to the recorded control points.

[2] This happens when partial order reduction allows JPF to keep exploring just the behaviour of the currently active thread (despite there may be more runnable threads), and, moreover, the active thread behaves in a deterministic way.

[3] The big circles depict states where JPF considers exploring more different outgoing transitions, and the small circles correspond to states where a single outgoing transition is only considered. In the recorded trace, $t1$–$t3$ are threads, $bBBi$ are instructions corresponding to the beginning of a basic block.

[4] Due to the properties of partial order reduction, this must be possible.

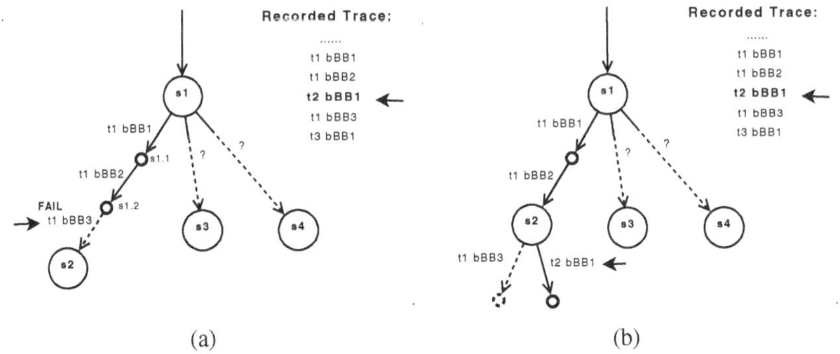

Fig. 1. Replaying a trace in JPF

4 Bounded Model Checking

Model checking automatically checks whether a system satisfies a specification via a systematic exploration of its state space. A crucial problem with applying it to real software is the state explosion problem. To cope with this problem in our setting, we use bounded model checking in which we *limit the depth of the search*. This is motivated by an assumption that to show that a suspected concurrency bug is real or to show that a healing action is unsafe, it often suffices to explore a bounded neighbourhood of the suspected state. This approach is not complete, but a full state space exploration is anyway usually impossible.

In our particular setting of checking whether a *suspected data race is real and even more a true race*, we, use the `PreciseDataRace` listener [1] available in JPF. This listener is based on the observation that two read/write or write/write actions form a true race iff there is a state from which they are both enabled. We try to find such a state in the neighbourhood of the state into which we got by replaying the recorded trace suspected by our dynamic analysis to lead to a data race.

The `PreciseDataRace` listener is parametrized by the variable over which the true race is to be sought. We know this variable from the preceding dynamic analysis. A problem with the default behaviour of the listener is that it looks for the true data race from the very beginning of the state space generation. In our case, however, its application within replaying the recorded trace would mean just a loss of time and space. Therefore, we have implemented a mechanism which allows an activation of the `PreciseDataRace` listener (as well as other similar listeners) only once a suitable starting point in the state space is reached.

For us, the suitable starting point is reached once we replay the entire recorded trace and *backtrack a user-defined number of steps* from the state reached at the end of the replaying. For backtracking, the `backtrack()` method of JPF is used taking an advantage of the fact that the trace was replayed using the depth-first search strategy of JPF (albeit with the above described modifications) for which this method is available.

The depth of the bounded state search is pre-defined by the user, and the search is performed under the standard partial order reduction (without the modifications used when replaying the recorded trace). As we have already mentioned, it may happen that the replaying followed a trace which has the same control points, but is not identical with the original trace followed by the monitored application. Then, a true race needs not be detected within the bounded model checking even if it really happens in the application. To increase the probability that an existing error will be found, we may proceed as follows. After bounded model checking is not successful in detecting a true race, we can backtrack through the replayed trace and find another trace (or several further traces) with the same sequence of control points as the recorded ones are and try bounded model checking at the end of this trace (these traces) too.

5 Experiments

We have so far tested our first prototype implementation of the described approach on a small-sized Java application containing a true data race. In particular, we considered a simple system of selling flight tickets. The system uses concurrent threads representing ticket sellers that are modifying—without a proper synchronization—a shared variable representing the number of available seats. The application contains 3 classes, 159 lines of code, and we were running 4 seller threads.

Table 1 shows results of our experiments done on a notebook with Intel Centrino Duo at 1.73GHz and 1GB of RAM running Java version 1.5 and JPF version 4.1 under Fedora Linux 10, kernel version 2.6.27.19. For running the experiments, we have recorded 100 traces from the considered application. This set of the traces was then used in each of the experiments corresponding to the particular lines of the table. The average number of recorded control points was 51. In all cases, we backtracked 5 steps after the end of replaying a trace.

Table 1. Results of experimenting with a prototype implementation of the proposed approach

BMC max depth	max (avg) num. of replaying attempts	num. of states	time of replaying	time of BMC	error depth	success of detection
15	1 (1)	2607	0.58	3.54	12.4	0.74
25	1 (1)	4384	0.58	6.34	20.3	0.91
15	5 (1.9)	7485	0.64	10.27	12.2	0.84
25	5 (1.3)	10589	0.65	15.52	20.3	0.94
15	no limit (5.35)	19334	1.89	28.41	12.2	1

The first column of Table 1 indicates the allowed maximum depth of bounded model checking. The second column gives the allowed maximum number of different tries of replaying a trace and the number of tries that was on average used. On the last line, there was no limit on the allowed number of tries of replaying the recorded trace. The third column gives the total number of generated states (within the replaying phase as well as in bounded model checking). The 4th and 5th columns give the time in seconds needed for replaying the trace and for the subsequent bounded model checking phase.

Thc 6th column gives the depth at which the error was found on average in bounded model checking. Finally, the last column gives the percentage of cases in which the true race was successfully detected. We see that we easily get a very high success ratio in the considered case study.

6 Conclusion

We have described our experience with implementing healing assurance for self-healing of data races in Java programs via bounded model checking in JPF. The implemented healing assurance is used to check that a suspected data race detected by dynamic analysis is in fact a true data race, which allows for avoiding costly healing actions in cases when they are not needed. In order to be able to start bounded model checking in a neighbourhood of a suspicious state found by dynamic analysis, we record certain events in the trace leading to the suspected error state during the run of the monitored application and then use this information to navigate JPF to some predecessor of the problematic state.

The proposed framework provides also a basis for the further needed phase of healing assurance, namely using bounded model checking to check that if we heal the data race by adding new locks, a deadlock will not be caused at least in the neighbourhood of the detected problematic state (full model checking is usually impossibly due to its too high cost).

In the future, we need to improve the implementation of our prototype and perform experiments with larger case studies. Next, we can explore multiple interesting issues such as some refined heuristics for choosing program actions to be recorded (possibly combined with recording some aspects of the visited states to help navigating through the state space), advanced heuristics for navigating through long partial traces, more advanced notions of bounding the state space search, and/or combining it with some specialised state space reduction strategies.

References

1. van Breugel, F., Kulikov, S., Shafiei, N., Visser, W.: Detecting Data Races with Java PathFinder. A draft paper. University of York, Canada (2008), `http://www.cse.yorku.ca/~franck/research/drafts/race.pdf`
2. Elmas, T., Qadeer, S., Tasiran, S.: Goldilocks: A Race and Transaction-aware Java Runtime. In: Proc. of PLDI 2007. ACM, New York (2007)
3. IBM R&D Labs. ConTest—A Tool for Testing Multi-threaded Java Applications (2008), `http://www.haifa.ibm.com/projects/verification/contest`
4. Křena, B., Letko, Z., Tzoref, R., Ur, S., Vojnar, T.: Healing Data Races On-The-Fly. In: Proc. of PADTAD 2007. ACM, New York (2007)
5. Letko, Z., Vojnar, T., Křena, B.: AtomRace: Data Race and Atomicity Violation Detector and Healer. In: Proc. of PADTAD 2008. ACM, New York (2008)
6. Křena, B., Letko, Z., Nir-Buchbinder, Y., Tzoref-Brill, R., Ur, S., Vojnar, T.: A Concurrency Testing Tool and its Plug-ins for Dynamic Analysis and Runtime Healing. Technical report FIT-TR-2009-01, FIT BUT (2009)

7. Robust Software Engineering Group, NASA Ames Research Center. Java PathFinder (2008), http://javapathfinder.sourceforge.net
8. Savage, S., Burrows, M., Nelson, G., Sobalvarro, P., Anderson, T.: Eraser: A Dynamic Data Race Detector for Multi-threaded Programs. In: Proc. of SOSP 1997. ACM, New York (1997)

Effective Bit-Width and Under-Approximation

Robert Brummayer and Armin Biere

Institute for Formal Models and Verification
Johannes Kepler University Linz, Austria

Abstract. Recently, it has been proposed to use approximation techniques in the context of decision procedures for the quantifier-free theory of fixed-size bit-vectors. We discuss existing and novel variants of under-approximation techniques. Under-approximations produce smaller models and may reduce solving time significantly. We propose a new technique that allows early termination of an under-approximation refinement loop, although the original formula is unsatisfiable. Moreover, we show how over-approximation and under-approximation techniques can be combined. Finally, we evaluate the effectiveness of our approach on array and bit-vector benchmarks of the SMT library.

1 Introduction

The problem of Satisfiability Modulo Theories (SMT) is to decide satisfiability of logical formulas expressed in a combination of first-order theories. SMT solvers are used in many applications, e.g. optimization, scheduling, verification, and test case generation. For text books about SMT see [4,16].

The quantifier-free theory of fixed-size bit-vectors plays an important role in specifying and verifying software and hardware systems. Modelling programs and digital circuits on the bit-vector level, e.g. addition of fixed-size bit-vectors with two's complement arithmetic, allows bit-precise and exact reasoning. Not taking modular arithmetic into account, i.e. using natural numbers instead of bit-vectors, may lead to unsound verification results, e.g. bugs caused by overflows may not be detected.

The theory of bit-vectors can be combined with the theory of arrays in order to model memory in a bit-precise way. This enables reasoning about pointers and pointer arithmetic which is important in software verification. Even reasoning about assembler programs on a symbolic processor is possible [7].

Typically, a system and its properties are represented by formulas. These formulas are combined to one verification formula which is checked by an SMT solver. The solver tries to satisfy the formula in order to find a counter-example where the system violates a property. If the formula is satisfiable, most modern SMT solvers can generate a model of the formula. A model can be used to construct a concrete execution of the system that leads to a property violation.

R. Moreno-Díaz et al. (Eds.): EUROCAST 2009, LNCS 5717, pp. 304–311, 2009.

2 Formula Approximation Techniques

The main motivation of most approximation techniques is to speed-up decision procedures. While over-approximation technique tend to speed-up unsatisfiable formulas, under-approximation techniques tend to speed-up satisfiable formulas. In order to remain sound and complete, approximation techniques are typically combined with a refinement loop. Recently, approximation techniques are also used in the context of decision procedures for bit-vectors [8,15].

Over-approximation techniques are in the spirit of the Counter Example Guided Abstraction Refinement Framework [10] (CEGAR) and are typically used in lazy SMT approaches [17]. For example, over-approximation techniques are used to decide complex array formulas in [5,13]. In the rest of this paper we will focus on under-approximation techniques.

The basic idea of under-approximation techniques in the context of bit-vectors is to restrict individual bits of bit-vectors. While such domain restrictions typically lead to a smaller search space and a speed-up for satisfiable formulas, it additionally produces "smaller" models which means that the domain of the variables are smaller. If a small model can be found, it must also be a model of the original formula. Small models are beneficial for diagnosis, for instance if the model is directly analyzed by users for debugging. Furthermore, in the area of test case generation, small models lead to test cases with reduced test data size.

One way of using over-approximations and under-approximations in bit-vector logic has been pioneered in [8]. In the context of under-approximation, the m most significant bits of variables are additionally restricted. The remaining n least significant bits are not concerned by the under-approximation and remain variable. In the rest of this paper we call n the *effective bit-width*.

in [8] it is proposed to use an under-approximation technique which corresponds to sign-extension. Let n be the effective bit-width. The m most-significant bits of a variable are forced to be equal to the nth least significant bit, the last effective bit. This technique reduces the domains of variables and leads to smaller models where bit-vectors are interpreted in the context of two's complement. An example with an effective bit width of four is shown left in Fig. 1.

It is also possible to force the m most significant bits to zero resp. one which we call *zero-extension* resp. *one-extension*. Zero-extension has been suggested in [8]. While zero-extension leads to smaller models where bit-vectors are interpreted in an unsigned context, one-extension is beneficial if a formula has small models with negative values. Examples with an effective bit-width of four are shown in Fig. 1 resp. Fig. 2. In [14] zero-extension and one-extension were used for bounded model checking of embedded software.

We propose an additional under-approximation technique that *partitions bits* of individual bit-vectors into equivalence classes. All bits in one class are forced to have the same value. An example is shown Fig. 2. The under-approximation refinement increases the number of classes per variable, or splits individual classes. The idea of this technique is that only some individual bits of the vector are important to satisfy the formula. Therefore, the other bits can be forced to the same value in order to reduce the search space.

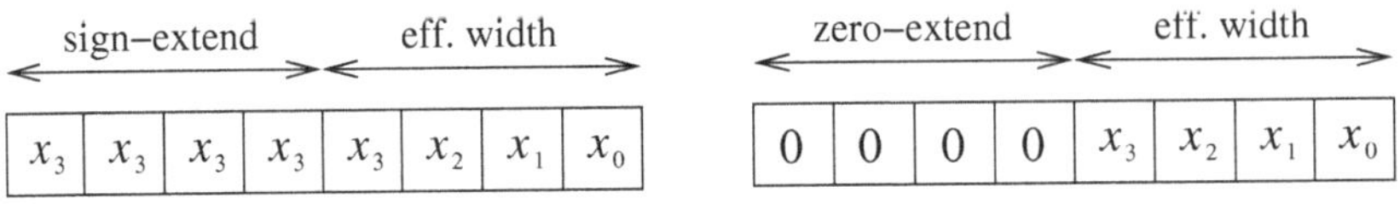

Fig. 1. Under-approximation techniques: Sign-extension is shown left and zero-extension is shown right. The effective bit-width is four in both examples.

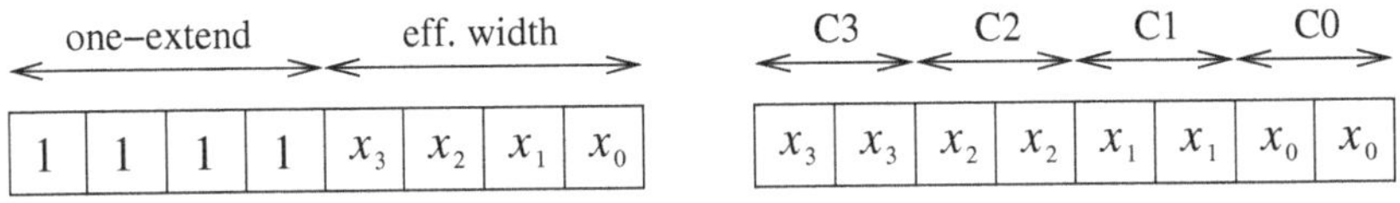

Fig. 2. Under-approximation techniques: One-extension is shown left and class splitting is shown right. The effective bit-width resp. number of classes is four.

3 Under-Approximation Refinement on CNF Layer

We propose to perform under-approximations with the help of additional clauses on the CNF layer. The original formula is translated to CNF once. In each refinement iteration i we perform an under-approximation by adding new clauses to the SAT solver incrementally and also use assumptions as in [9].

First, we introduce a fresh boolean under-approximation variable e. Then, we perform an under-approximation by adding new clauses. Let n be the effective bit width of a bit-vector variable v of bit-width w. To perform a sign-extending under-approximation we add the following clauses:

$$\bigwedge_{i=n}^{w-1} ((v_{n-1} \vee \overline{v_i} \vee \bar{e}) \wedge (\overline{v_{n-1}} \vee v_i \vee \bar{e}))$$

Finally, we assume e to enable the under-approximation.

For example, let v be a bit-vector variable with bit-width eight and an effective bit-width of six. We add the following clauses to perform a sign-extending under-approximation:

$$(v_5 \vee \overline{v_6} \vee \bar{e}) \wedge (\overline{v_5} \vee v_6 \vee \bar{e}) \wedge (v_5 \vee \overline{v_7} \vee \bar{e}) \wedge (\overline{v_5} \vee v_7 \vee \bar{e})$$

Assuming e enforces $v_5 = v_6 = v_7$.

Zero-extension can be encoded as follows. Again, let n be the effective bit width of a bit-vector variable of bit-width w. We add the following clauses:

$$\bigwedge_{i=n}^{w-1} (\overline{v_i} \vee \bar{e})$$

One-extension can be encoded analogously.

If we have to refine our approximation, we add the unit clause $\bar{e}$ in order to disable the current approximation. This gives the SAT solver also the opportunity to recycle clauses. Then, a refined under-approximation is performed with the help of another fresh boolean under-approximation variable.

4 Refinement Strategies

Generally, the effective bit-width can be used as a metric of approximation. Typically, the effective bit-width is initialized to one, i.e. the domain of a bit-vector variable is restricted to $\{-1, 0\}$ in a signed resp. $\{0, 1\}$ in an unsigned context. During the refinement the effective bit width is increased. In order to avoid too many refinement loops, the effective bit-width is typically doubled in each iteration. Traditionally, in the worst case, e.g. if the original formula is unsatisfiable, the effective bit-width reaches the original bit-width.

With the proposed refinement on the CNF layer, two main refinement strategies are possible which we call *global* and *local*. On the one hand, local refinement strategies maintain one fresh boolean under-approximation variable e for each bit-vector in each refinement. The benefit is a precise refinement as we can ask the SAT solver if it has used the respective e to derive unsatisfiability. Only those under-approximations that have been used need to be refined. However, in the worst case we have to introduce $k \cdot r$ fresh variables, where k is the number of bit-vector variables and r is the maximum number of refinements.

On the other hand, the global refinement strategy maintains exactly one under-approximation variable for all bit-vector variables. The benefit is less overhead, as we need only r additional boolean variables, where r is the number of refinements. However, the refinement is imprecise.

5 Early Unsat Termination

Traditional under-approximation techniques perform the under-approximation outside the SAT solver. The CNF is generated from scratch in each refinement iteration. This makes it impossible to find out whether the current under-approximation has been responsible for deriving unsatisfiability or not. In the worst case, the original formula is unsatisfiable and we have the additional overhead of the under-approximation refinement, which is slower than solving the original formula up front.

The proposed under-approximation refinement on the CNF layer enables the decision procedure to terminate earlier, even if the original formula is unsatisfiable. If the under-approximated formula is unsatisfiable, then we can use the under-approximation variables to ask the SAT solver which under-approximations have been used to derive unsatisfiability. If no under-approximation variables have been used, then we can conclude that the original formula is unsatisfiable, and terminate. The early unsat technique is shown in Fig. 3.

Furthermore, the refinement on the CNF layer allows the SAT solver to keep learned conflict clauses over refinement iterations. This would be impossible if the CNF was generated on scratch in each refinement iteration.

In a first implementation we let the SAT solver generate unsat cores modulo assumptions, but simply recording those assumptions [9] that were used in deriving the empty clause is enough, much faster and easier to implement, both on the side of the SAT solver and on the side of the SMT solver.

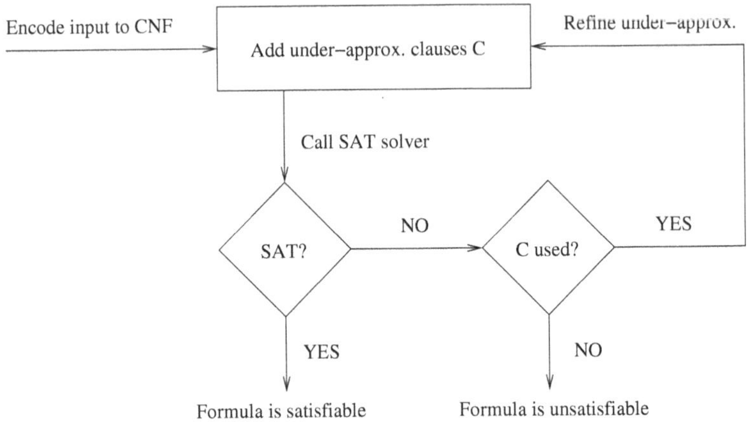

Fig. 3. Early Unsat Termination

6 Combining Approximation Techniques

Figure 4 shows how over-approximation and under-approximation techniques can be combined to solve complex SMT formulas. We consider the quantifier-free theory of arrays combined with the quantifier-free theory of bit-vectors. The idea is to use over-approximation techniques for the array part [5,13] and under-approximation techniques for the bit-vector part.

First of all, we perform an over-approximation by replacing reads by fresh bit-vector variables. Then, we translate the bit-vector part of the formula to CNF and add a set C of under-approximation clauses. In each iteration we call the SAT solver. Depending on the result we have to perform an additional check. On the one hand, if the result is *satisfiable* we have to check if the current model σ respects the theory of arrays. If not, we have to refine our over-approximation with a lemma on demand [11,12,2]. Otherwise, we can terminate with the model σ and the result *satisfiable*. On the other hand, if the result of the SAT solver is *unsatisfiable* we have to check if the current set of under-approximation clauses has been used. If not, we can terminate with the result *unsatisfiable*. Otherwise, we disable the current under-approximation and continue with a refined approximation.

7 Experiments

We implemented the presented approximation techniques in our SMT solver Boolector [6]. Boolector implements a decision procedure for the quantifier-free theory of fixed-size bit-vectors combined with the quantifier-free extensional theory of arrays [5]. It is the winner of the last SMT competition in 2008 [1] in the bit-vector category (`QF_BV`) and also in the division of bit-vectors with arrays (`QF_AUFBV`). Moreover, Boolector can be used as word-level bounded model checker for synchronous hardware and software systems [7]. For our experiments we used Boolector 1.0.

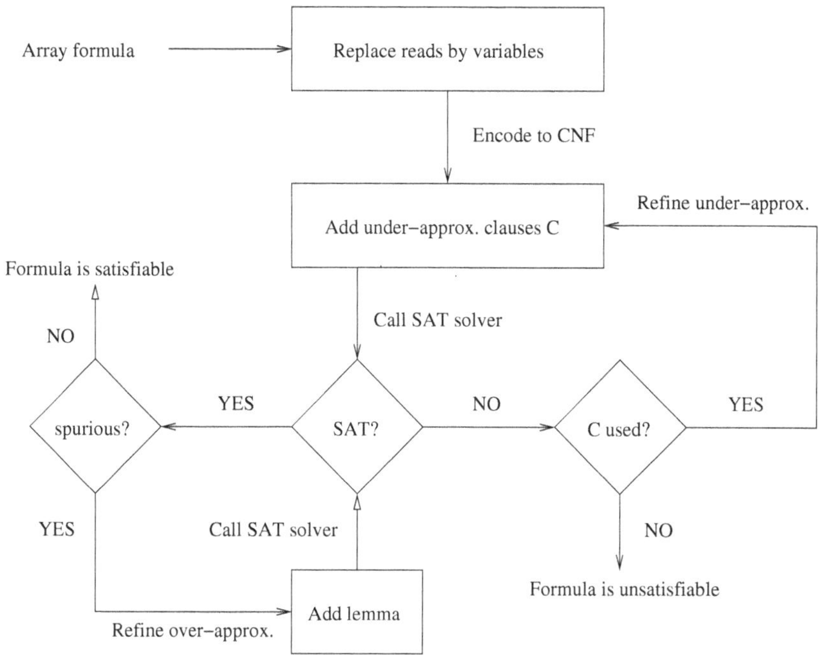

Fig. 4. Combining over-approximation and under-approximation techniques

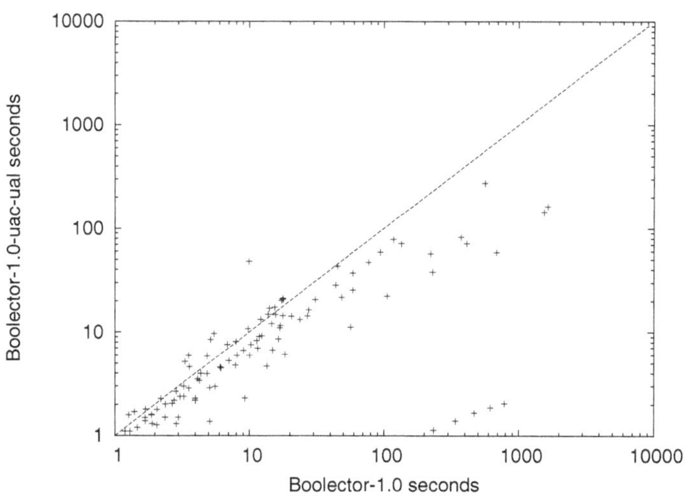

Fig. 5. Boolector (x-axis) vs. Boolector with class under-approximation and local refinement strategy (y-axis). Benchmarks are from QF_AUFBF and are all satisfiable.

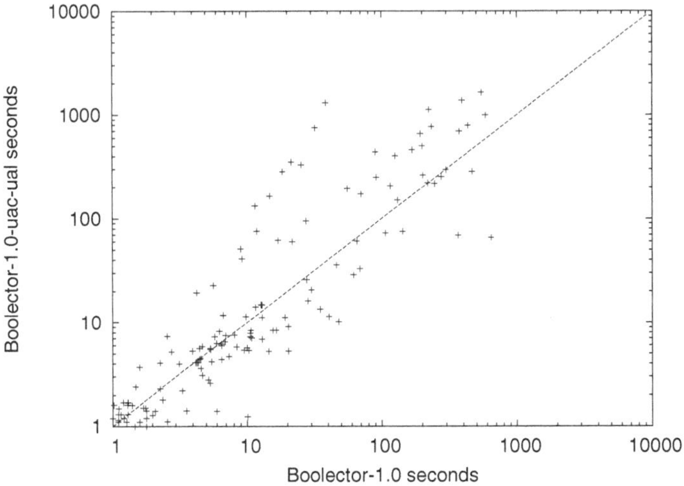

Fig. 6. Boolector (x-axis) vs. Boolector with class under-approximation and local refinement strategy (y-axis). Benchmarks are from QF_AUFBV and are all unsatisfiable.

The results of our experiments are shown in Fig. 5 and Fig. 6. We used benchmarks from QF_AUFBF of the SMT library [3] (June 1st, 2008). As expected, under-approximation techniques speed up satisfiable instances, but slow down unsatisfiable instances. We summarize further observations.

First, the under-approximation by classes technique performs as good as the under-approximation by sign-extension technique on the satisfiable instances of QF_AUFBF. However, it performs worse on the unsatisfiable benchmarks. Second, the average ratio effective bit-width / original bit-width is 16% (without `egt` examples) on satisfiable and 27% on unsatisfiable benchmarks, which corresponds to an impressive reduction of 84% resp. 73%. Third, early unsat termination occurs in 1553 from 3552 unsat cases. Finally, the global refinement strategy seems to be a good approximation of the local strategy. It is much easier to implement, is often as good as the local strategy, and should be sufficient in most cases.

8 Conclusion

We presented formula approximation techniques, in particular for bit-vector. We discussed different techniques and refinement strategies and showed how they can be implemented on the CNF layer which enables further optimizations like early unsat termination. Finally, we showed how under-approximation techniques for bit-vectors can be combined with over-approximation techniques for arrays and evaluated the effectiveness of our approach on benchmarks from the SMT library.

Formula approximation techniques help to handle complex and hard formulas. Under-approximation techniques speed up decision procedures in the context of falsification and generate small models with restricted domains.

References

1. Barrett, C., Deters, M., Oliveras, A., Stump, A.: SMT-Comp (2008), www.smtcomp.org
2. Barrett, C., Dill, D., Stump, A.: Checking Satisfiability of First-Order Formulas by Incremental Translation to SAT. In: Brinksma, E., Larsen, K.G. (eds.) CAV 2002. LNCS, vol. 2404, p. 236. Springer, Heidelberg (2002)
3. Barrett, C., Ranise, S., Stump, A., Tinelli, C.: The Satisfiability Modulo Theories Library (SMT-LIB) (June 2008), www.SMT-LIB.org
4. Bradley, A., Manna, Z.: The Calculus of Computation: Decision Procedures with Applications to Verification. Springer, Heidelberg (2007)
5. Brummayer, R., Biere, A.: Lemmas on Demand for the Extensional Theory of Arrays. In: Proc. SMT 2008, ACM Press, New York (2008)
6. Brummayer, R., Biere, A.: Boolector: An Efficient SMT Solver for Bit-Vectors and Arrays. In: Kowalewski, S., Philippou, A. (eds.) TACAS 2009. LNCS, vol. 5505. Springer, Heidelberg (2009)
7. Brummayer, R., Biere, A., Lonsing, F.: BTOR: Bit-Precise Modelling of Word-Level Problems for Model Checking. In: Proc. BPR 2008. ACM Press, New York (2008)
8. Bryant, R.E., Kroening, D., Ouaknine, J., Seshia, S., Strichman, O., Brady, B.: Deciding Bit-Vector Arithmetic with Abstraction. Software Tools for Technology Transfer, STTT (2009)
9. Claessen, K., Sörensson, N.: New Techniques that Improve MACE-style Finite Model Finding. In: CADE-19, Workshop W4, Model Computation – Principles, Algorithms, Applications (2003)
10. Clarke, E., Grumberg, O., Jha, S., Lu, Y., Veith, H.: Counterexample-Guided Abstraction Refinement for Symbolic Model Checking. Journal of the ACM, JACM (2003)
11. de Moura, L., Rueß, H.: Lemmas on Demand for Satisfiability Solvers. In: Proc. SAT 2002. Springer, Heidelberg (2002)
12. Flanagan, C., Joshi, R., Saxe, J.: Theorem Proving Using Lazy Proof Explication. In: Hunt Jr., W.A., Somenzi, F. (eds.) CAV 2003. LNCS, vol. 2725, pp. 355–367. Springer, Heidelberg (2003)
13. Ganesh, V.: Decision Procedures for Bit-Vectors, Arrays and Integers. PhD thesis, Computer Science Department, Stanford University (2007)
14. He, N., Hsiao, M.: Bounded Model Checking of Embedded Software in Wireless Cognitive Radio Systems. In: Proc. ICCD. IEEE, Los Alamitos (2007)
15. He, N., Hsiao, M.: A new Testability Guided Abstraction to Solving Bit-Vector Formula. In: Proc. BPR 2008. ACM, New York (2008)
16. Kroening, D., Strichman, O.: Decision Procedures: An algorithmic Point of View. Springer, Heidelberg (2008)
17. Sebastiani, R.: Lazy Satisfiability Modulo Theories. Journal on Satisfiability, Boolean Modeling and Computation (JSAT) 3 (2007)

Observable Runtime Behavior for Defects Indicated by Automated Static Analysis

Klaus Wolfmaier[1], Rudolf Ramler[1], Gabor Guta[1], and Heinz Dobler[2]

[1] Software Competence Center Hagenberg, 4232 Hagenberg, Softwarepark 21, Austria
{Klaus.Wolfmaier,Rudolf.Ramler,Gabor.Guta}@scch.at
[2] University of Applied Sciences, Softwarepark 11, 4232 Hagenberg, Austria
heinz.dobler@fh-hagenberg.at

Abstract. For the efficient and effective use of automated static analysis of software systems it is crucial to know what kind of errors can be detected and how seriously a reported problem can or should be taken. In the study conducted for this paper we applied a widely used tool (*PC-lint*) for automated static analysis (ASA) to check C++ code fragments from student exercises. The goal of this research was to discover which types of defects can be identified by automated static analysis. In this paper we present our findings; furthermore the results from classifying the defects are set in relation to detection rules and severity levels provided by ASA, in order to derive insights for calibrating ASA tools in a specific application context.

Keywords: Automated Static Analysis, Defect Classification.

1 Introduction

Automated static analysis (ASA) is widely used in today's software development practice. With the support of ASA, software engineers reveal defects in their software and analyze code quality. The value and effectiveness of this approach has been discussed and documented by researchers and practitioners (e.g., [1], [8], [12]). However, there is still little evidence available on which types of defects typically observed in industrial contexts can be detected efficiently and why.

Figure 1 illustrates possible defect types by means of three alternative ways of processing all elements of an array in C or C++. Instead of using the condition $i<N$ (Figure 1a), the programmer may continue the loop as long as $i<=N$ or $i<N$-1. In the first case (Figure 1b) the programmer most likely has a misconception about the correct index of the last element and introduces a defect that will produce a failure possibly observable as "runtime error". In the second case (Figure 1c) the for loop does not iterate over all elements of the array. In this case no runtime error can be observed. This simple anomaly may however indicate several different types of defects, e.g.: an algorithmic fault according to Barr's defect classification [2], when the intention was to process the last element in a different way, but the complete modification is missing.

R. Moreno-Díaz et al. (Eds.): EUROCAST 2009, LNCS 5717, pp. 312–319, 2009.

`int a[N], i;`
`for (i=0, i<N; ++i)`
`    a[i]=f(i);`
	`int a[N], i;`
	`for (i=0, **i<=N**; ++i)`
	`    a[i]=f(i);`
		`int a[N], i;`
		`for (i=0, **i<N-1**; ++i)`
		`    a[i]=f(i);`

Fig. 1a. Correct **Fig. 1b.** Defect type 1 **Fig. 1c.** Defect type 2

The goal of this paper is to investigate which of these defects types can be identified by ASA. Furthermore, the results from classifying the defects are set in relation to detection rules and severity levels provided by ASA, in order to derive recommendations for calibrating ASA in a specific project context. Specifically, we investigate if all types of defects can be identified by ASA.

The closest research work to our approach is [12]: In this paper the effectiveness of ASA on large C++ projects is discussed. The work to [12] also investigated which defect types are identified according to the orthogonal defect classification (ODC, see [3]). In another study (see [1]) the messages produced by the tool *FindBugs* are analyzed and classified by their practical relevance. We also investigated previous work in the field of defect classification and static analysis tools. Defect classification systems were created to identify different types of defects. These classification systems help us making a quantitative evaluation of the software development process. Probably the oldest classification system has been proposed by Knuth in [9]. The different classification systems are created with different intent: Knuth's or Barr's classifications in [2] focus on the personal development process; others like IBM's ODC already mentioned above (see [3]) or Hewlett Packard's Defect Origins [7] focus on aspects important in large scale software projects. For the scope of this work, we have created a specific classification system indicating the observability and, thus, the immediate impact of a defect. The classification system is described in Section 2.

A broad range of ASA tools has been developed with different capabilities, e.g., in terms of the approach used to detect defects, soundness, completeness, etc. (see [4]). ASA tools can be categorized in two main classes:

(1) *Rule checkers* are looking for violations of coding standards defined by rules. We consider bug pattern detectors as a kind of style checker.

(2) *Semantic analysis tools* use a semantic-based approach, i.e., data flow analysis, constraint based techniques, or abstract interpretation to identify defects.

In [6] and [11] the capabilities of C++ and Java tools are discussed, respectively. For the scope of our work presented in this paper, we used *PC-lint*[1] as ASA tool, as it is used widely in industry.

This paper is structured as follows: Section 2 outlines the steps of our experiment. Section 3 presents the results of the experiment, and Section 4 discusses these results. Section 5 concludes this paper and mentions possible further work.

---

[1] http://www.gimpel.com

2 Study Description

For this work, we evaluated all the messages produced by the selected ASA tool (PC-lint) when applied to student exercises as described in [5]. A message generated by the tool contains three kinds of information: the *message code* which references a defect detection rule, the *line number* for the line in which the defect has been located, and the message *severity level* such as error, warning, info, or note. Administrative messages produced by the tool were omitted from the analysis, as these messages do not indicate possible defects. The remaining messages (especially errors and warnings) were manually investigated by code inspection, in order to judge whether these messages correctly indicate a defect or by mistake a so called false positive.

The identified defects were then classified according to the following three defect types:

- **DT_1 - "Run-time defect":** This type of defect may result in a runtime error when the defective code is executed. Thus, these defects can be revealed by executing the program with certain inputs in certain environments. Thereby, the runtime behavior of the program is unspecified or non-deterministic according to the language semantics or the runtime environment. The language may specify that a runtime error must occur (e.g., Java). This category can be further refined in the sense of the preciseness of the model. Typical models can be built from (1) the standard of the programming language, (2) the language interpretation of the compiler, and (3) the semantics of the runtime environment.
- **DT_2 – "Semantic defect":** This type of defect may result in incorrect program behavior at runtime. If the program is deterministic, these defects can be revealed through a test case that compares the actual behavior with the expected behavior. More formally expressed, the program does not meet with the functional (behavioral) requirements, i.e., the program is not (totally) correct according to its formal specification. This defect type cannot be detected efficiently by ASA tools, because typically no formal specification of the behavior is available. The only possibility to find this kind of defects is the usage of some heuristics, which corresponds to guessing according to empirical experiences.
- **DT_3 – "Deferred defect":** This type of defect has no immediate effect observable at runtime, but has a negative effect on future development activities. For example, improper structured code reduces the readability and new defects are more easily introduced during maintenance work. The program may be (totally) correct according to its formal specification, but fails to meet with other quality requirements such as maintainability or effectiveness. This defect type can be detected efficiently by ASA tools, but it is hard to judge the value when fixing this kind of defects.

The evaluation of the ASA messages was done by analyzing available test results for all student exercises and by a code review conducted by the authors. We used ASA results produced by the tool PC-lint, version 8.0u. We repeated the analysis with version 8.0w and 9.0b, but did not find significant differences in the results.

3 Results from Automated Static Analysis

This section shows three selected code samples for the three different defect types described in Section 2 that were found during analysis.

Figure 2 shows a runtime defect and the related ASA message. The variable `h` may contain an arbitrary value according to the variable `line`. If `h` cannot be parsed to a valid number by the function `atoi` or `atoi` returns 0, then the value of the variable `t` will be 0. In this case the function `processPRel` will produce a *division by 0* error.

```
BTA-13.cpp, 108, 414, Warning, "Possible division by 0 [Ref.: file BTA-13.cpp: lines 54, 108]"

  66   void Teacher::processInput(string line) {
...
  98     h = "";
  99     for(j; j < i; j++) h += line[j];
 100     t = atoi(h.c_str());
 101
 102     processPRel();
 103
 104   }
...
 107   void Teacher::processPRel(){
 108     pRel =  (static_cast<double>(p)) /
                ((static_cast<double>(t)) * 3);
 109 }
```

Fig. 2. Runtime defect

Figure 3 shows a semantic defect. The method `ReadDataFrom` gets a filename as parameter that is never used and the filename for data input is hardcoded in line 111.

```
BTA-23.cpp, 194, 715, Info, "Symbol 'source' (line 108) not referenced"
BTA-23.cpp, 108, 830, Info, "Location cited in prior message"

108    int bta::ReadDataFrom(char* source) {
109
110      // opening file
111    ifstream data("levasy.txt");
...
194    } // end
```

Fig. 3. Semantic defect

Figure 4 shows a deferred defect. The parameter `line` could be declared `const`, which would provide the intention of the parameter more clearly [10] as this parameter is just expected to be read (input parameter). Constant declaration prevents the programmer from modifying this parameter accidentally later on during maintenance. And in the `const` case even a reference could have been used (`const string &`) to improve efficiency.

```
BTA-13.cpp, 60, 952, Note, "Parameter 'line' (line 58) could be declared const
                                --- Eff. C++ 3rd Ed. item 3"
BTA-13.cpp, 58, 830, Info, "Location cited in prior message"

void Teacher::processInput(string line){
  int i, j;
  string h;

  i = 0;

  while((i < line.length()) && (line[i] != ' ')) i++;
...
}
```

Fig. 4. Deferred defect

4 Discussion

This section discusses the results applying descriptive statistics for the messages of ASA and for the classification of these messages. Figure 5 depicts the distribution of defects in the analyzed student's solutions for the exercise (named *BTA-xx*), stacked by defect type as described in Section 2.

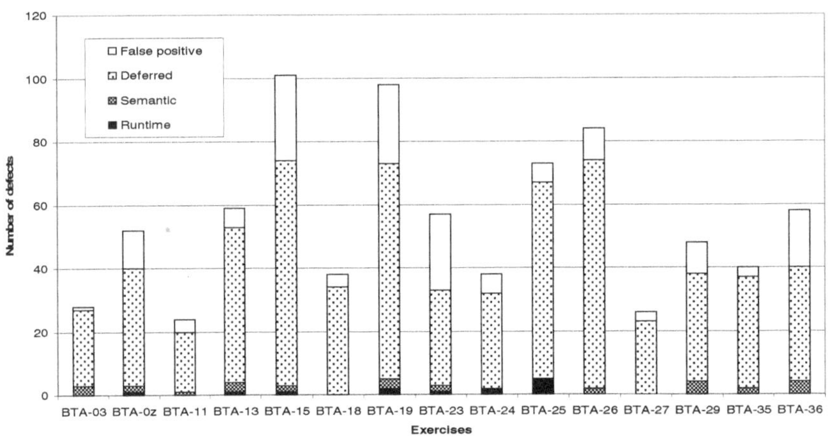

Fig. 5. Defects per Exercise

The number of ASA messages per solution range from 24 to 101, with an average of 54.93 messages. The analyzed solutions had 203 lines of C++ code on average. From Figure 5 one can see that 19.3 percent of the messages were classified as 'False Positive'. The remaining messages are 75.73 percent classified as 'Deferred Defect', 3.4 percent as 'Semantic Defect', and 1.7 percent as 'Runtime Defect'.

Figure 5 shows that different students are likely to produce different solutions in terms of involved defects although all solutions are based on the same requirements and use the same technology. Furthermore, Figure 2 depicts that ASA has identified only a small number of defects (runtime defects + semantic defects = 4.98 percent) that trigger immediate action to fix compared to the total number of messages produced.

From approximately 1,000 available message codes in PC-lint, only 57 different were reported in our analysis. This is due to the fact that the analyzed solutions are rather small and many of the code constructs analyzed by PC-lint have not been used in the implemented algorithms. Furthermore, the six most frequently reported message codes together sum up to 53.22 percent of all messages reported. The average number of messages per message code is 14.

Figure 6 shows the mapping of defect types to ASA message codes. The figure only contains those message codes that were mapped to more than one defect type.

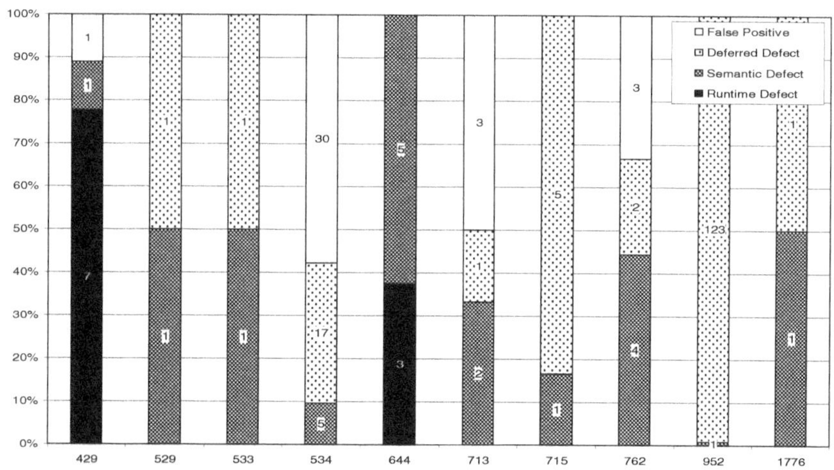

Fig. 6. Mapping of Message Codes to Defect Types

Figure 6 depicts those 10 message codes out of the 57 reported codes that were mapped to two different defect types. None of the message codes was mapped to all three defect types. Mapping of one message code to two different defect types might be due to the fact that the distinction especially between semantic defect and deferred defect is fuzzy. However, there is still a strong indication that mapping of a single code to different defect types also depends on the implementation context, i.e., the same message code in one context might be a deferred defect while it maps to a semantic defect in an other context.

Figure 7 presents the number of messages for each severity level as defined by the ASA tool, stacked by defect type as described in Section 2.

As Figure 7 shows, the severity level for the messages provided by the static analysis tool does not exactly map to the selected defect classification, i.e., not all 'Runtime Defects' are indicated by messages classified as 'Warning'. Yet, defects that trigger immediate action to fix such as 'Runtime Defects' are more likely to be found among messages with a higher severity level. Also, the ratio of false positives differs for the different severity levels. The ratio decreases for lower severity levels whereas the absolute number of false positives increases, compared to a higher severity level.

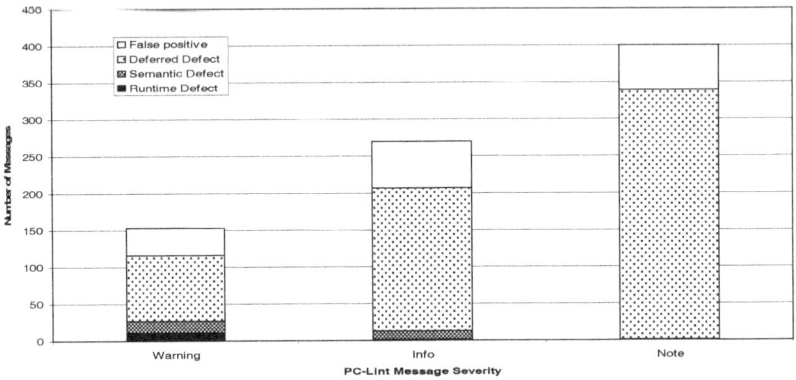

Fig. 7. Number of messages in different severity categories

5 Findings and Future Work

The manual classification of the indicated defects detected with the ASA tool PC-lint resulted in the following findings:

(1) There are message codes that are not solely classified as runtime defect as the classification is context dependent, e.g., an uninitialized variable may lead to a runtime error if it is used as a pointer or an array index while it may "just" lead to an incorrect calculation in the case of an integer variable.

(2) ASA results contain few messages that are classified as semantic defects. In many cases, these message were also classified as deferred defects due to the fact that a clear distinction between semantic defect and deferred defect in our classification is not always possible. A large number (75.73 percent) of messages is classified as deferred defect.

(3) Each ASA message code can be assigned to a defect type according to the classification of the messages with that message code. We can classify only those message codes where at least one message is classified non false positive. After classifying message codes this way, we can calculate the ratio of false positives and non false positive messages for each message code. It can be observed that message codes classified as semantic defect have the highest false positive rate in average compared to the average of the other defect types and messages codes classified as deferred defects have the lowest average false positive rate. The reason for the low false positive rate for deferred defects is the very simple rules that are used to identify those defects.

In summary, our findings support the claim that ASA tools are capable of identifying defects in the code. However, we have to point two major threats to validity in our work. (1) We base our findings on the results of a single ASA tool, namely PC-lint. Several other tools are available which promise to deliver better performance [6] and, thus, may produce different results. (2) The student exercises we used in our study are of limited size and may reflect only a fraction of the typical defects found in large-scale software development projects.

Therefore, for future work, we plan to validate our results on a larger code base and with analysis results form different tools. Furthermore, it seems to be promising to repeat the study also for different languages, e.g., Java, which may be amenable for different types of defects.

References

1. Ayewah, N., Pugh, W., Morgenthaler, J.D., Penix, J., Zhou, Y.: Evaluating static analysis defect warnings on production software. In: PASTE 2007: Proceedings of the 7th ACM SIGPLAN-SIGSOFT workshop on Program analysis for software tools and engineering, pp. 1–8. ACM, New York (2007)
2. Barr, A.: Find the Bug: A Book of Incorrect Programs. Addison-Wesley Professional, Reading (2004)
3. Chillarege, R., Bhandari, I., Chaar, J., Halliday, M., Moebus, D., Ray, B., Wong, M.: Orthogonal Defect Classification: a Concept for In-process Measurements. IEEE Transactions on Software Engineering 18(11), 943–956 (1992)
4. Cousot, P., Cousot, R., Feret, J., Mauborgne, L., Miné, A., Monniaux, D., Rival, X.: Varieties of Static Analyzers: A Comparison with ASTRÉE. In: First IEEE & IFIP International Symposium on Theoretical Aspects of Software Engineering, TASE 2007, Shanghai, China, pp. 3–17 (2007)
5. Dobler, H., Ramler, R., Wolfmaier, K.: A Study of Tool Support for the Evaluation of Student's Solutions of Exercises in Programming. In: EuroCAST (2007)
6. Emanuelsson, P., Nilsson, U.: A Comparative Study of Industrial Static Analysis Tools (extended version), Technical report, Linkoping University, Linköping University Electronic Press (2008)
7. Huber, J. T.: A Comparison of IBM's Orthogonal Defect Classification to Hewlett Packard's Defect Origins, Types, and Modes (1989), `http://www.stickyminds.com/sitewide.asp?Function=edetail&ObjectType=ART&ObjectId=2883`
8. Jaspan, C., Chen, I., Sharma, A.: Understanding the value of program analysis tools. In: OOPSLA 2007: Companion to the 22nd ACM SIGPLAN conference on Object oriented programming systems and applications companion, pp. 963–970. ACM, New York (2007)
9. Knuth, D.E.: The errors of TEX. Softw. Pract. Exper. 19(7), 607–685 (1989)
10. Meyers, S.: Effective C++: 55 Specific Ways to Improve Your Programs and Designs, 3rd edn. Addison-Wesley, Reading (2005)
11. Rutar, N., Almazan, C.B., Foster, J.S.: A Comparison of Bug Finding Tools for Java. In: ISSRE 2004: Proceedings of the 15th International Symposium on Software Reliability Engineering, pp. 245–256. IEEE Computer Society, Washington (2004)
12. Zheng, J., Williams, L., Nagappan, N., Snipes, W., Hudepohl, J.P., Vouk, M.A.: On the value of static analysis for fault detection in software. IEEE Transactions on Software Engineering 32(4), 240–253 (2006)

Real-Time Vision-Based Vehicle Detection for Rear-End Collision Mitigation Systems

D. Balcones, D.F. Llorca, M.A. Sotelo, M. Gavilán, S. Álvarez, I. Parra, and M. Ocaña

Department of Electronics, University of Alcalá, Alcalá de Henares, Madrid, Spain
{llorca,sotelo,parra,miguel.gavilan}@depeca.uah.es

Abstract. This paper describes a real-time vision-based system that detects vehicles approaching from the rear in order to anticipate possible rear-end collisions. A camera mounted on the rear of the vehicle provides images which are analysed by means of computer vision techniques. The detection of candidates is carried out using the top-hat transform in combination with intensity and edge-based symmetries. The candidates are classified by using a Support Vector Machine-based classifier (SVM) with Histograms of Oriented Gradients (HOG features). Finally, the position of each vehicle is tracked using a Kalman filter and template matching techniques. The proposed system is tested using image data collected in real traffic conditions.

1 Introduction

The rear-end collisions are one of the most common types of automobile accidents. A rear facing camera mounted on the rear of the vehicle can provide an important number of driving assistance functions such as collision warning systems that will alert the driver of an impending collision or pre-crash systems (seat belt pretensioning, intelligent headrest, etc.). Accordingly, the work presented in this paper is directly related with the automotive industry.

The system is divided in four main blocks: rear-lane detection, candidates selection, single-frame classification and multi-frame validation and tracking. The global overview of the system is depicted in Figure 1.

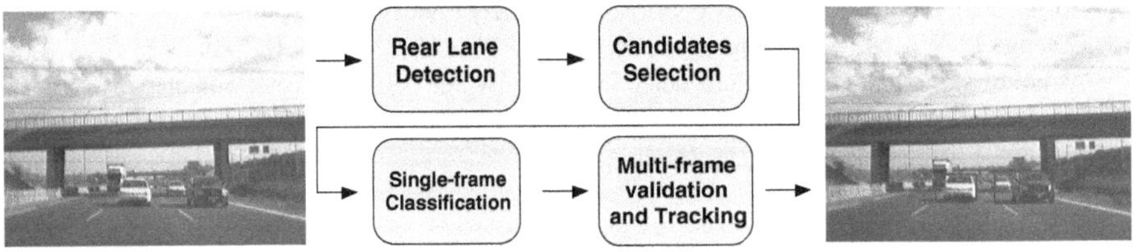

Fig. 1. Global overview of the rear-end vehicle detection system

R. Moreno-Díaz et al. (Eds.): EUROCAST 2009, LNCS 5717, pp. 320–325, 2009.

2 System Description

2.1 Rear-Lane Detection

This stage is carried out by using a Lane Departure Warning (LDW) system previously developed by the authors [1]. The LDW system has been adapted in order to deal with rear conditions. We use this system in combination with flat-world assumption, fixed camera pitch and camera height, so that, the search space is drastically restricted.

2.2 Candidates Selection

Candidates are selected in a three-stage process. Firstly, the vehicle contact point is searched by means of the white top-hat transformation. This operator allows the detection of contrasted objects on non-uniform backgrounds [2]. In our case it enhances the boundary between the vehicles and the road. Horizontal contact points are pre-selected if the number of white top-hat features is greater than a threshold. This process is applied from bottom to top for each detected lane. Then, candidates are pre-selected if the entropy of Canny points is high enough for a region defined by means of perspective constraints and prior knowledge of target objects (see Figure 2).

Fig. 2. From left to right: original image; contact point detection on white top-hat image; candidate pre-selected with high entropy of canny points

In a second step, gray level, vertical edges and horizontal edges symmetries are obtained, so that, candidates will only pass to the next stage if their symmetries values are greater than a threshold (see Figure 3). Symmetry axis are linearly combined to obtain the final position of the candidate. Finally, a weighted variable is defined as a function of the entropy of Canny points, the three symmetry values and the distance to the host vehicle. We use this variable to apply a non-maximum suppression process per lane which removes overlapped candidates. An example of this process is depicted in Figure 4.

2.3 Single-Frame Classification

The selected candidates are classified by means of a SVM classifier with RBF kernel, in combination with HOG features [3]. All candidates are resized to a

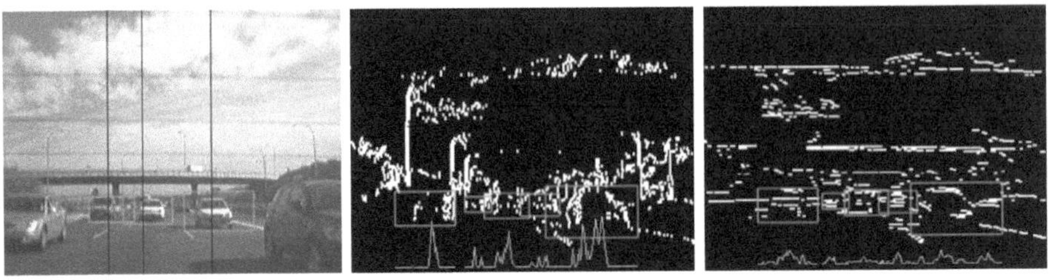

Fig. 3. From left to right: gray-level symmetries, vertical edges symmetries and horizontal edges symmetries

Fig. 4. Left: overlapped candidates. Right: non-maximum suppression results.

fixed size of 64x64 pixels to facilitate the features extraction process. The SVM classifier is trained with 2000 samples and tested with 1000 samples (1/1 positive/negative ratio). Figure 5 depicts some positives and negatives examples of the training and test data sets. The distance to the hyperplane is defined for a detection rate (DR) of 92% and a false positve rate (FPR) of 32%. We have to note that these numbers are defined in a single-frame fashion, so that, they will be improved in subsequent stages.

Fig. 5. Upper row: positive samples (vehicles). Lower row: negative samples.

2.4 Multi-frame Validation and Tracking

The position of the vehicles in the image is tracked by using a Kalman Filter. Data association problem is solved by means of a linear combination of the

Fig. 6. Left: non-rigid grid template. Right: tracking results.

Euclidean distance and the Zero mean Normalized Cross Correlation (ZNCC). Once a vehicle is detected and tracked during a consecutive number of frames, SVM classification is stopped and a non-rigid grid-based matching technique is used until the vehicle disappears from the scene (see Figure 6). Global matching usually fails with close vehiclcs in lateral lanes. This approach allows to achieve a considerable reduction in the computational cost.

3 Results

The algorithm was implemented on a PC on-board a real automobile. Three different test sequences have been recorded in real traffic conditions with a total duration of 240sec and a traffic density of 1.5 vehicles/frame on average. As we can see in Table 1 the system achieves a detection rate of 92.2% with 1 false positive per minute on average.

Table 1. Overall results

	Duration (sec)	**Number of vehicles**	**Detention Rate (%)**	**False Positives per minute**	**Vehicle density (per frame)**
Sequence 1	115	16	87.5%	1.56	1.58
Sequence 2	65	15	93.33%	0.92	1.778
Sequence 3	54	2	100.0%	0.0	0.645

Regarding the computational cost aspects, in Table 2 we can see that candidates selection and single-frame classification stages are the most time consuming parts of the system. Once the system has a certain knowledge of a candidate of being a vehicle, the proposed approach starts to run really fast since the non-rigid grid-based tracking stage has a low computational cost (less than 3ms). The overall system runs in real-time (23 fps).

The output of the system in a real experiment is depicted on the left side of Figure 7. The distance of each rear-vehicle with regard to the camera is showed on the upper-right corner of the bounding boxes. On the right side of Figure 7

Table 2. Computational cost of the different parts of the system using a Pentium IV 2.8 GHz with 512MB of RAM. The system runs at 23 fps.

	Rear LDW	**Candidates selection**	**Single-frame classification**	**Multi-frame tracking**	**Non-rigid grid-based tracking**
Comp. cost (ms)	5	17.74	16.2	1.9	2.7
Percentage	11.4%	40.7%	37.2%	4.4%	6.2%

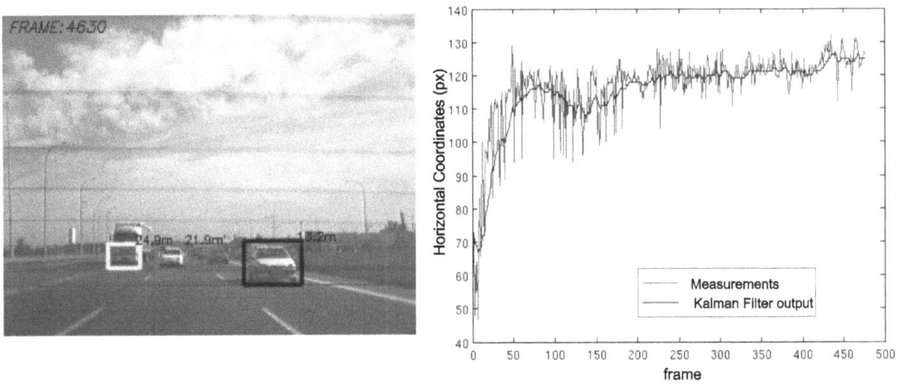

Fig. 7. Left: detected vehicles on a test sequence. Right: filtered y-position of the car located in the middle lane.

the measurement of the y-position in pixel coordinates and its corresponding filtered value for the car located in the middle lane are provided.

At present, the output of the rear-vehicle detection system is being used in combination with Blind Spot Detection (BSD) system developed by the authors [4], resulting in the the so-called Panoramic BSD (see Figure 8).

Fig. 8. Panoramic Blind Spot Detection system (Panoramic BSD)

4 Conclusions and Future Works

This paper presented a real-time vision-based system that detects vehicles approaching from the rear in order to anticipate possible rear-end collisions. The search space is drastically reduced thanks to the rear LDW system which automatically detects the lanes of the road. Candidates are robustly selected using top-hat features and entropy of Canny points in combination with gray-level, horizontal edges and vertical edges symmetries. A single-frame SVM classifier is trained and used with HOG features. This step rejects a considerable amount of false detections. Vehicles are then tracked by using a Kalman filter. Once a vehicle has been classified and tracked during a considerable number of frames, the system performs a non-rigid grid-based matching which decreases the overall computational cost.

The algorithm was implemented on a PC on-board a real automobile. In experiments on datasets captured from a moving vehicle in real traffic conditions, the system achieves a detection rate of 92.2% with 1 false positive per minute on average. Future work involves the detection of vehicles in nighttime conditions by incorporating an infrared camera.

Acknowledgments

This work has been funded by Research Project CCG08-UAH/TIC-3572 (University of Alcalá and CAM).

References

1. Sotelo, M.A., Nuevo, J., Bergasa, L.M., Ocaña, M., Parra, I., Fernández, D.: Road Vehicle Recognition in Monocular Images. In: Proceedings of IEEE ISIE, pp. 1471–1476 (2005)
2. Derong, Y., Yuanyuan, Z., Dongguo, L.: Fast Computation of Multiscale Morphological Operations for Local Contrast Enhancement. In: Proceedings of Engineering in Medicine and Biology 27th Annual Conference (2005)
3. Dalal, N., Triggs, B.: Histograms of Oriented Gradients for Human Detection. In: Proceedings of IEEE CVPR, pp. 886–893 (2005)
4. Sotelo, M.A., Barriga, J., Fernández, D., Parra, I., Naranjo, J.E., Marrón, M., Álvarez, S., Gavilán, M.: Vision-Based Blind Spot Detection Using Optical Flow. In: Moreno Díaz, R., Pichler, F., Quesada Arencibia, A. (eds.) EUROCAST 2007. LNCS, vol. 4739, pp. 1113–1118. Springer, Heidelberg (2007)

Real–Time Hierarchical GPS Aided Visual SLAM on Urban Environments

David Schleicher, Luis M. Bergasa, Manuel Ocaña, Rafael Barea, and Elena López

Department of Electronics, University of Alcala, Alcalá de Henares, 28805 Madrid, Spain
dsg68818@telefonica.net,
{bergasa,barea,elena,mocana}@depeca.uah.es

Abstract. In this paper we present a new real-time hierarchical (topological/metric) Visual SLAM system focusing on the localization of a vehicle in large-scale outdoor urban environments. It is exclusively based on the visual information provided by both a low-cost wide-angle stereo camera and a low-cost GPS. Our approach divides the whole map into local sub-maps identified by the so-called fingerprint (reference poses). At the sub-map level (Low Level SLAM), 3D sequential mapping of natural landmarks and the vehicle location/orientation are obtained using a top-down Bayesian method to model the dynamic behavior. A higher topological level (High Level SLAM) based on references poses has been added to reduce the global accumulated drift, keeping real-time constraints. Using this hierarchical strategy, we keep local consistency of the metric sub-maps, by mean of the EKF, and global consistency by using the topological map and the MultiLevel Relaxation (MLR) algorithm. GPS measurements are integrated at both levels, improving global estimation. Some experimental results for different large-scale urban environments are presented, showing an almost constant processing time.

Keywords: SLAM, Intelligent Vehicles, Computer Vision, Real-Time.

1 Introduction

The interest in Visual SLAM has grown tremendously in recent years as cameras have become much more inexpensive than lasers, and also provide texture rich information about scene elements at practically any distance from the camera. Currently, the main goal in SLAM research is to apply consistent, robust and efficient methods for large-scale environments in real-time. On the other hand, one of the most popular sensors in outdoor navigation is the GPS. However, their standalone information is not always as accurate as needed, especially on urban environments, mainly due to satellites occlusion because of high buildings, tunnels, etc. One of the most popular methods to solve the SLAM problem is the Extended Kalman Filter (EKF) and more recently FastSLAM [1]. The first one has the covariance matrix growing problem while the second one discretizes the problem by using particle filters. Both of them are limited, in terms of computing time, when the environment becomes larger. To cope with that

R. Moreno-Díaz et al. (Eds.): EUROCAST 2009, LNCS 5717, pp. 326–333, 2009.

issue, two different approaches have been developed that try to divide the whole map into smaller ones in a hierarchical way. The original idea of having a set of sub-maps with uncertain relations dates back to [2] and [3]. The first approach introduces a high metric level over pieces of the metric map in the so-called *Metric-Metric* approach [4] [5]. The second one is referred as the *Toplogical-Metric* one, which adds a high topological level over the metric sub-maps [6] [7] [8]. A third alternative to face the large scale SLAM problem is to use only *topological* maps without sub-maps associated to their vertex [9] [10]. These maps lack the details of the environments but they can achieve good results for certain applications.

Our final goal is the autonomous outdoor navigation of a vehicle in large-scale environments where GPS signal does not exist or it is not reliable (tunnels, urban areas with tall buildings, mountainous forested environments, etc). Our approach defines a *Low Level SLAM*, where the system uses stereo vision to feed an EKF to create local sub-maps which are expressed in local coordinates relative to some reference frames (*fingerprints*). Local poses are periodically fused with GPS measurements by using (1) (2). The only output used from the low level is the relation of the final vehicle frame (current fingerprint) relative to the reference vehicle frame (previous fingerprint). Over this low level a *High Level SLAM* is defined, where fingerprints uncertain relations are stored in a graph of relations defining stochastic constraints on the reference vehicle frames (fingerprints), as shown on Fig. 1. GPS is also added there as an absolute constraint on such a frame. This graph of relations is fed into the MultiLevel Relaxation (MLR) algorithm [11], which computes the least square estimate for the graph. Unfortunately, current implementation of MLR does not provide covariance information for this estimate. So, to derive uncertainty information, our approach implements another procedure in parallel. The algorithm exploits that uncertain metrical relations can be compounded by (3). So to obtain uncertainty information about a reference vehicle pose (fingerprint), the shortest path in the above mentioned graph is taken, where the different relations from local maps are compounded. To detect loop closing, some of the fingerprints add visual information to the pose that helps to identify previously visited places. These kind of fingerprints are called SIFT fingerprints because they are based on SIFT features (*Scale Invariant Feature Transform).* In case of *long-term* GPS signal lost, at the time of signal recovering, vehicle pose is corrected and the global map is optimized by mean of the MLR as well.

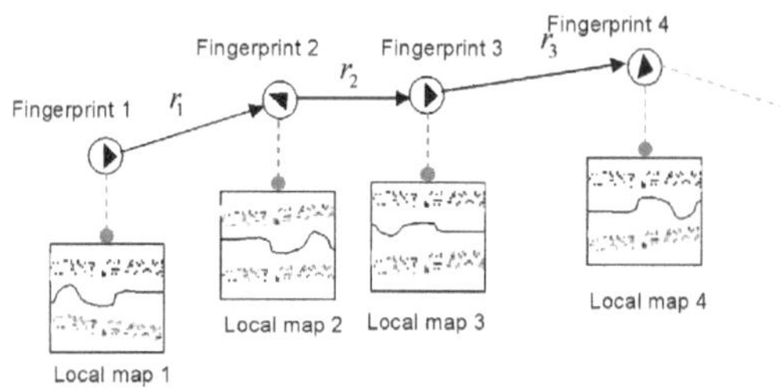

Fig. 1. General architecture of our two hierarchical levels SLAM. Each sub-map has an associated fingerprint.

2 Low Level SLAM

This level is inspired on A. Davison monocular approach [12], however it has been modified for a stereo implementation as detailed in [13]. The low level state vector for the EKF is defined as $X_l = (X_v \quad Y_1 \quad Y_2 \quad \cdots)^T$, which is composed by the vehicle state vector $X_v = (X_{rob} \quad q_{rob} \quad v_{rob} \quad \omega)^T$ plus all local landmarks on the sub-map Y_i. Landmarks are identified by their corresponding features, which on this implementation are defined by the whole set of pixels of the patch. On this equation, X_{rob} is the 3D position of the vehicle relative to the local frame, $q_{rob} = (q_0 \quad q_x \quad q_y \quad q_z)^T$ is the orientation quaternion, v_{rob} is the linear speed and ω is the angular speed. For clarity reasons the sub-map notation is omitted.

Each time a new GPS reading $X_{GPS} = (x_{GPS} \quad y_{GPS})^T$ is available, which under normal conditions occur at 1s period, we proceed to fuse it with our visual estimation by applying a two-dimensional statistical approach based on Bayes Rule and Kalman filters, as shown in (1). Here, $X_{\Pr ob}$ and $P^0_{\Pr ob}$ are the 2D vehicle global position and global covariance respectively.

$$X^{fusion} = X_{\Pr ob} + P^0_{\Pr ob}\left(P^0_{\Pr ob} + P_{GPS}\right)^{-1}\left(X_{GPS} - X_{\Pr ob}\right) \tag{1}$$

GPS uncertainty P_{GPS} is obtained as a function of the HDOP (Horizontal Dilution Of Probability), containing the variable error provided by the GPS, and the UERE (User Equivalent Range Error), covering the estimated constant errors along time. In the same way, the following estimated covariance is calculated by mean of equation (2).

$$P^{fusion} = P^0_{\Pr ob} - P^0_{\Pr ob}\left(P^0_{\Pr ob} + P_{GPS}\right)^{-1} P^0_{\Pr ob} \tag{2}$$

3 High Level SLAM

Our SLAM implementation adds an additional topological level, called high level SLAM, to the explained low level SLAM in order to keep global map consistency with almost constant processing time. This goal is achieved by using the MLR algorithm over the reference poses. Therefore, the global map is divided into local sub-maps referenced by the mentioned fingerprints, one by one. There are two different classes of fingerprints: *Ordinary Fingerprints* and *SIFT fingerprints*. The first ones are denoted as $FP = \{fp_l | l \in 0 \ldots L\}$. Their only purpose is to store the vehicle reference pose $X^{fp_l}_{rob}$ and local covariance $P^{fp_l}_{rob}$ relative to the previous one, i.e., the reference frame of the current sub-map. The sub-map size, after experimental testing, is limited to 10 m of covered path. SIFT fingerprints are a sub-set of the first ones, denoted as $SF = \{sf_q \in FP | q \in 0 \ldots Q, Q < L\}$. Their additional functionality is to store the visual appearance of the environment at the moment of being obtained. That is covered by the definition of a set of *SIFT features* associated to the fingerprint, which identifies the place at that time. These fingerprints are taken only under the condition of having a significant change on the vehicle trajectory, defined by a maximum angular speed increase γ_{max} followed by a minimum decreasing γ_{min}, both experimentally obtained. When a new SIFT fingerprint is taken, it is matched with the previously acquired SIFT fingerprints within an uncertainty search region. This region is obtained from the vehicle global covariance P^0_{rob} because it keeps the global uncertainty information of the vehicle. If the matching is positive, it means that the vehicle is in a previously visited place and a *loop closing* is identified. Then, the MLR algorithm is applied in order to determine the maximum likelihood estimate of all nodes poses. Finally, nodes corrections are transmitted to theirs associated sub-maps. When a new fingerprint is created, an associated sub-map is created as well. Each of the old sub-maps defines the pose $X^{fp_{l-1}}_{fp_l}$ and covariance $P^{fp_{l-1}}_{fp_l}$ of a fingerprint relative to the previous node. The current sub-map defines the vehicle pose $X^{fp_l}_{rob}$ and covariance $P^{fp_l}_{rob}$ relative to the previous node. Then, the global pose of the vehicle is computed by compounding these relations with uncertainty using the equation $X^0_{rob} = X^0_{fp_l} \oplus X^{fp_l}_{rob}$, where X^0_{rob} and $X^0_{fp_l}$ define the vehicle and previous reference absolute poses respectively. Due to the need of being aware about the current global uncertainty at any time, we need to maintain P^0_{rob} updated (see Fig. 2). We calculate it by using the *coupling summation formula* (described in [6]), obtained from the *compounding* operation, in a recursive way: first, to obtain P^0_{rob} we need to evaluate (3); second, to obtain the global covariance of the current fingerprint $P^0_{fp_l}$, we must apply (3) again, but this time to the previous fingerprint, repeating it until we reach the first fingerprint, where $P^0_{fp_l} = P^{fp_0}_{fp_l}$ can be directly solved.

$$P^0_{rob} = \frac{\partial X^0_{rob}}{\partial X^0_{fp_l}} \cdot P^0_{fp_l} \cdot \left(\frac{\partial X^0_{rob}}{\partial X^0_{fp_l}}\right)^T + \frac{\partial X^0_{rob}}{\partial X^{fp_l}_{rob}} \cdot P^{fp_l}_{rob} \cdot \left(\frac{\partial X^0_{rob}}{\partial X^{fp_l}_{rob}}\right)^T \quad (3)$$

3.1 Loop Closing and Map Correction

Our system identifies a specific place using the SIFT fingerprints. These fingerprints, in addition to the vehicle pose, are composed by a number of SIFT [14] landmarks distributed across the reference image and characterize the visual appearance of the image, allowing loop closing detection as explained in [15]. Once a loop-closing has been detected, the whole map must be corrected according to the old place recognized. To do that, we use the MLR algorithm [11], which has proved to show a high efficiency in terms of computation cost and map complexity. The purpose of this algorithm is to assign a globally consistent set of Cartesian coordinates to the fingerprints of the graph based on local, inconsistent measurements, by trying to maximize the total likelihood of all measurements. The MLR inputs are the relative poses and covariances of the fingerprints. As outputs MLR returns the most *likely* set of reference poses, i.e., the set already corrected $X_M = \left(Xc^0_{fp_1} \quad Xc^0_{fp_2} \quad \cdots \quad Xc^0_{fp_L}\right)^T$. The MLR algorithm manages only 2D information, therefore we need to obtain the 2D related fingerprint pose $X^{fp_l}_{2D} = \left(x_{2D} \quad y_{2D} \quad \theta_{2D}\right)^T$ and covariance $P^{fp_l}_{2D}$ from $X^{fp_{l-1}}_{fp_l}$ and $P^{fp_{l-1}}_{fp_l}$. Then the corresponding corrected fingerprints X_M are obtained, assuming flat terrain. To calculate the global vehicle uncertainty P^0_{rob} after closing a loop, there is a situation where one fingerprint has relations with more than one additional fingerprint, as occurs, for example, to sf_3 (see Fig. 2). To calculate the current P^0_{rob} we apply the recursive coupling summation formula (3) to the shortest possible path from the first fingerprint to the current position, which leads to the lowest P^0_{rob}. Being aware of the current global uncertainty is important in order to increase the fingerprints search process efficiency because the number of matched SIFT fingerprints will be lower.

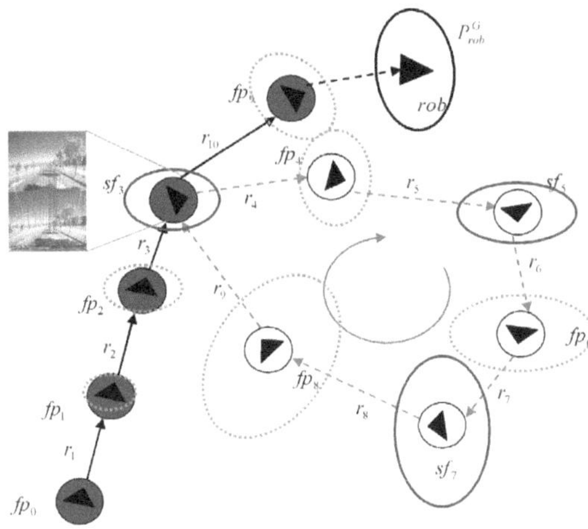

Fig. 2. Representation of the vehicle global uncertainties P^0_{rob}, increasing along the vehicle path at each of the reference poses. Solid red lines represent vehicle global uncertainties at SIFT fingerprints places. Numbers represent each fingerprint. Graph also shows an example of shorter path selection for global uncertainty calculation after a loop-closing situation.

The last step is to transfer the correction performed on the High level SLAM into the Low level SLAM. This is implicitly done by the transformation of each sub-map reference frame, i.e., all the landmarks within each sub-map will be moved according to their corresponding reference frame. At this level, GPS data fusion is taken into account only at *long term* signal loose, which is usually the case within tunnels or in urban areas with high buildings. In that case, the state correction implies a global map correction that concerns mainly the section where the GPS signal was unavailable. Because GPS uncertainty is global, when GPS signal is available, fusion is carried out on global coordinates and nodes are introduced within the graph as global relations, i.e., MLR algorithm is fed with the global pose $X^0_{fp_l}$ and covariance $P^0_{fp_l}$ of each fingerprint.

4 Results

In order to test the behaviour of our system several video sequences were collected from a commercial car manually driven in large urban areas, covering more than 20 km. The employed cameras for the stereo pair were the Unibrain Fire-i IEEE1394 with additional wide-angle lens, which provide a field of view of around 100º horizontal and vertical with a resolution of 320x240. The baseline of the stereo camera was 40 cm. Both cameras were synchronized at the time of commanding the start of transmission. The cameras were mounted inside the car on the top of the windscreen and near the rear-view mirror. We used a low-cost standard GPS, the GlobalSat BU-353 USB. To evaluate the performance of our system we compared our results with a ground truth reference, obtained with an RTK-GPS Maxor GGDT, with an estimated accuracy of 2 cm. Part of the path covered by the vehicle is shown on Fig. 3. The average speed of the car was around 30 km/h. The complete covered path was 3.17 km long. It contained 5 loops inside, taking 8520 low level landmarks and 281 nodes. More landmarks are located on high buildings areas, while GPS signal has more strength in open-spaced areas providing better location estimation. This shows that both sensors complement each other, providing good estimations for different situations. The Euclidean error relative to the ground truth of both the standard GPS and our combined SLAM implementation is depicted in Fig. 4. We obtain an average error of around 4 m and a reasonably low error at the moments of total GPS loose. This error is compared to the global uncertainty covariances for each node using the Euclidean formula applied to the X and Z components as well, showing nearly consistent error estimates. As expected, uncertainty monotonically grows on GPS unavailable sections due to the relative measurements provided by the visual sensor. Fig. 3 depicts the estimation of our combined SLAM system and the standard GPS alone compared to the ground truth. In spite of the increased estimation error on some segments at the beginning of the path, as shown on Fig. 4, we still have a relatively accurate estimation to be able to locate the vehicle. Respect to the processing time, the real-time implementation imposes a time constraint, which shall not exceed 33 ms for a 30 frames per second capturing rate.

Table 1. Processing times

Low level SLAM processing times		High level SLAM processing times (parallelized).	
Number of features / frame	5	Number of features	8520
		Number of nodes	281
Filter step	Time		Time
Measurements	3 ms	Fingerprint matches	3 s
Filter update	5 ms	Loop closing + graphic representation time	1 s + 10s
Feature initializations	7 ms		
GPS processing (1s sampling period)	4ms		

On Table 1 we show the average processing times for some of the most important tasks in the process. Low Level SLAM tasks are limited in regards of time consuming due to the limited sub-map size. High Level SLAM tasks slightly increase over time, but as they do not belong to the continuous self-locating process carried out by the Low Level SLAM they can be calculated apart in a parallel process. Therefore, the total processing time is proved to remain below the real time constraint within all our testing environments.

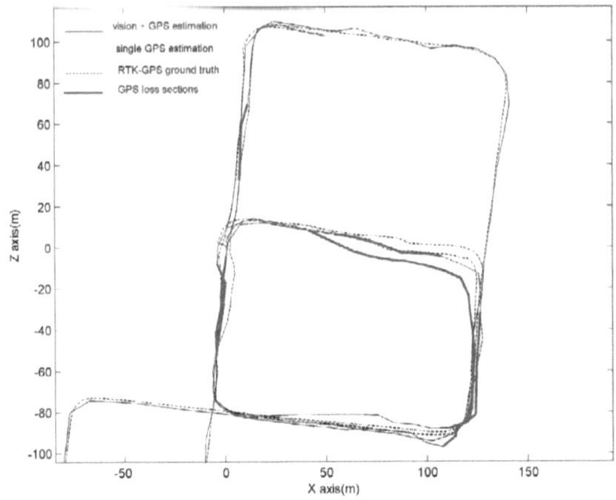

Fig. 3. Path estimation using only a standard low-cost GPS (dotted line), our SLAM method by means of vision and GPS (solid line), and the ground truth (dashed line). Thick red lines indicate path sections where GPS was unavailable.

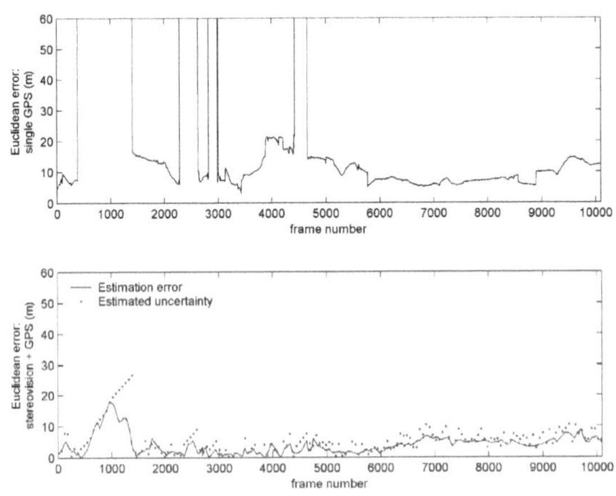

Fig. 4. Euclidean distance error ($\varepsilon = \sqrt{X^2 + Z^2}$) using standard single GPS (up) and our combined SLAM system (down). Global covariances uncertainties for each node are shown as well.

5 Conclusions

In this paper we have presented a two levels (topological/metric) hierarchical SLAM that allows self-locating a vehicle in a large-scale urban environment using a low-cost wide-angle stereo camera and a standard low-cost GPS as sensors. We have shown the positioning improvements of our system regarding to use a simple standard GPS, opening the possibility to improve current vehicle navigation systems. One limitation of our system is that flat terrain is assumed for matching the 2D map of the topological level with the 3D maps of the metric one.

As future work, we plan to generalize the MLR algorithm in order to manage 3D characteristics, as well as to replace the Low Level SLAM by Visual Odometry.

References

1. Montemerlo, M.: FastSLAM: A factored solution to the simultaneous localization and mapping problem with unknown data association. Ph.D. thesis, Carnegie Mellon Univeristy (2003)
2. Durrant-Whyte, H.F.: Uncertain geometry in robotics. IEEE Trans. Robotics 4(1), 23–31 (1988)
3. Smith, R., Self, M., Cheeseman, P.: Estimating Uncertain Spatial Relationships in Robotics. Autonomous Robot Vehicles, 167–193 (1988)
4. Piníes, P., Tardós, J.D.: Scalable SLAM building conditionally independent local maps. IROS (2007)
5. Frese, U.: Treemap: An O(log n) algorithm for indoor simultaneous localization and mapping. Autonomous Robots, 103–122 (2006)
6. Bailey, T.: Mobile robot localisation and mapping in extensive outdoor environments. PhD Thesis, University of Sydney (2002)
7. Bosse, M., Newman, P., Leonard, J., Teller, S.: An Atlas Framework for Scalable Mapping. In: ICRA, pp. 1899–1906 (2003)
8. Eade, E., Drummond, T.: Monocular SLAM as a Graph of Coalesced Observations. In: ICCV, pp. 1–8 (2007)
9. Andreasson, H., Duckett, T., Lilienthal, A.: Mini-SLAM: minimalistic visual SLAM in large-scale environments based on a new interpretation of image similarity. In: ICRA (2007)
10. Cummins, M., Newman, P.: Probabilistic Appearance Based Navigation and Loop Closing. In: IEEE International Conference on Robotics and Automation, pp. 2042–2048 (2007)
11. Frese, U., Larsson, P., Duckett, T.: A multilevel relaxation algorithm for simultaneous localization and mapping. IEEE Transactions on Robotics 21(2), 196–207 (2005)
12. Davison, A.J.: Real-time simultaneous localisation and mapping with a single camera. In: ICCV (2003)
13. Schleicher, D., Bergasa, L.M., Lopez, E., Ocaña, M.: Real-Time simultaneous localization and mapping using a wide-angle stereo camera and adaptive patches. In: IROS (2006)
14. Lowe, D.G.: Object Recognition from Local Scale-invariant Features. In: International Conference on Computer Vision, pp. 1150–1157 (1999)
15. Schleicher, D., Bergasa, L.M., Barea, R., Lopez, E., Ocaña, M., Nuevo, J.: Real-Time wide-angle stereo visual SLAM on large environments using SIFT features correction. In: IROS (2007)

Tomographic Image Reconstruction Using Abstractions

J.A. Alvarez and J. Roca

Dept. Arquitectura de Computadores y Eca.
Universidad de Almería
jaberme@ual.es

Abstract. New trends in High Performance Computing Architectures are arising an old concept, *concurrence*. But concurrence is not parallelism. To afford parallelism in the new computing arena, parallel applications need to consider this or just being completely rewritten in such a way that parallelism can be expressed by means of concurrence. Abstractions may help to keep performance on the new, and also on the legacy, platforms. This paper shows how abstractions may play an important role when used to model the problem.

Keywords: Abstractions, High Level Parallel Constructions, Concurrence, Threads, Object Orientation.

1 Introduction

Abstractions have shown to be an effective tool for bridging the gap between human and computer problem conception. The dizzy evolution experimented by processors hampers software from taking full advantage of hardware improvements. It is here where abstractions become an useful tool. Clusters of processors are a well known platform where highly demanding resources problems, in our case the 3D tomographic image reconstruction problem[1] can be solved efficiently using parallelism. A high percentage of the scientific community has adopted a computational model based on monolithic processes that communicate via messages. For incoming architectures, where more than one core do exist per chip, different programming models are recommended. As there are more cores per chip, the clock frequency is reduced. Therefore a program built under the traditional approach cannot experiment the expected upturn in performance and scalability. Multicores are parallel processors with shared memory. The programming techniques advised for these platforms are mainly multithreaded[2] in order to have all cores running, but it is still based on the monolithic process model, see Figure 1. New abstractions are needed in order to maintain performance gains in HPC [3]. The Object Oriented programming paradigm offers a flexible mean to describe how computations should be carried out. The model that this paradigm follows is inherently parallel, see Figure 1 right. This work shows how using objects, and threads embedding them, a better computational model can

R. Moreno-Díaz et al. (Eds.): EUROCAST 2009, LNCS 5717, pp. 334–341, 2009.

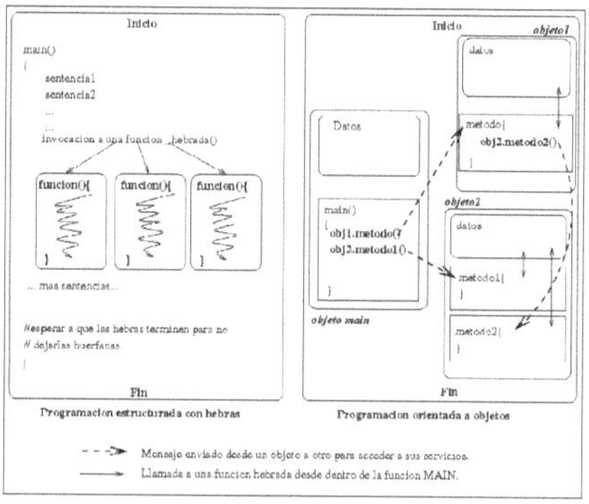

Fig. 1. Threaded vs. OO programming model

be used, for both parallel platforms, clusters of computers and multicore processors. In section 2 the problem of the iterative reconstruction is presented. Section 3 studies the coarse grain approach. Section 4 shows the finer grain approach on both, clusters and multicore. Finally, section 5 summarizes the conclusions.

2 Iterative Image Reconstruction Methods

Series expansion reconstruction methods assume that a 3D object, or function f, can be approximated by a linear combination of a finite set of known and fixed basis functions, with density x_j. The aim is to estimate the unknowns, x_j. These methods are based on an image formation model where the measurements depend linearly on the object in such a way that $y_i = \sum_{j=1}^{J} l_{i,j} \cdot x_j$, where y_i denotes the ith measurement of f and $l_{i,j}$ the value of the ith projection of the jth basis function. Under those assumptions, the image reconstruction problem can be modeled as the inverse problem of estimating the x_j's from the y_i's by solving the system of linear equations aforementioned. Assuming that the whole set of equations in the linear system may be subdivided into B blocks, a generalized version of component averaging methods, *BICAV*[1], can be described. The processing of all the equations in one of the blocks produces a new estimate. All blocks are processed in one iteration of the algorithm. This technique produces iterations which converge to a weighted least squares solution of the system.A volume can be considered made up of 2D slices. The use of the spherically symmetric volume elements (blobs) [4], makes slices interdependent because of blob's overlapping nature. The amount of communications is proportional to the number of blocks and iterations. Reconstruction yields better results as the number of blocks is increased. The main drawback of iterative methods are their high computational requirements. These demands can be faced by means of parallel computing and efficient reconstruction methods with fast convergence. The parallel iterative reconstruction method has been implemented following the Single Program Multiple Data (SPMD) approach [5]. Two different levels of abstractions were used: coarse grain abstractions based on user level threads that embeds MPI processes, useful approach for cluster computing; and a finer grain approach based on object orientation, useful for multicore and clustered platforms.

3 Abstracting MPI Processes

Traditionally, scientific codes have been designed under the assumption of having a single flow of control per process. Parallelism was achieved by connecting these processes via message passing libs such as MPI. Object Orientation, in such cases, can be applied without having to rewrite the code by embedding a process into an object wrapper hosted by a user-level thread. Obviously, this is not a truly Object Oriented approach but a *coarse grain approach* instead. On the other hand multithreaded programming provides a way to divide a program into entities that run concurrently. AMPI [6] allows to abstract MPI processes into user level threads (*virtual processors*), allowing more than one active flow of control within a process. Using objects (Figure 1), and threads embedding them, can result into a better computational model [7], for clusters and multicore processors. Concurrence using AMPI offers the advantage of having more virtual processors than physical processors. Therefore more than one virtual processor can coexist in a physical processor, efficiently.

Our experiments consisted in a varying number of abstracted processes per processor. Efficiency was defined on the idle time computed per processor. Experiments underlined the gain obtained by the multithreaded implementation of our algorithm compared to the MPI version of the reconstruction. Scaling tests were carried out on both versions varying the number of threads/processor and the number of processors, for both versions AMPI and MPI. Here, K defines the number of blocks. This scenario harms the MPI version whereas AMPI is expected to keep good performance. Two test volumes were used, a 256x256x256 and a 512x512x512 voxels volume. All experiences were performed on *Vermeer*, our research cluster(*32* computing nodes with two Pentium IV xeon 3.06 Ghz with 512KB L2 Cache and a 2GB sdram). The relative difference between cpu and wall times, using the higher K value, for AMPI wall and cpu times are alike, which means that cpu was mostly in use, in contrast to the MPI version in which differences turn out to be significant. It can be said that for the multithreaded version, the concurrence is seized at maximum. In Figure 2 wall times for MPI and AMPI (128 virtual processors) versions are shown, for several numbers of blocks K and for 256 and 512 volume sizes. It can be observed that below a threshold of K=64 both versions seem to behave similarly, showing slight improvement in AMPI. But above that threshold, and as K increases, AMPI behaves better than MPI especially for more than 16 processors. AMPI seems to be getting benefits from the hidden concurrence. AMPI keeps its speedup almost linear. Thread switching is succeeding in maintaining an optimal speedup. For non-dedicated clusters, concurrence, if exploited correctly, can play an important role for performance. This criteria is implemented as a complement to the load balancing strategy in [8]. AMPI offers a threaded framework in which latencies

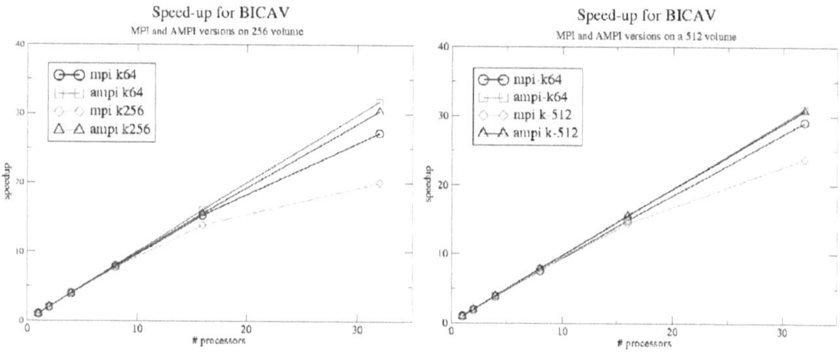

(a) 256 volume (b) 512 volume

Fig. 2. Speedup for 256 dim (left) and 512 dim (right) volume reconstruction

due communications can be handled. A parallel application can be abstracted embedding processes into AMPI threads.

4 A Finer Grain Implementation Based on Objects

One of the key concepts offered by the O.O. paradigm is the encapsulation, concept that when used, exploits the caches [7], without applying further techniques. Using O.O., together with threads, on parallel platforms such as clusters or on implicit parallel architectures such as multicores can result in performance improvements. Ideally a number of threads embeding computational objects can be active at a time in a single core. But concurrence is not parallelism nevertheless on the forthcoming architectures both concepts are closely related [9]. Concurrent execution can improve performance in three fundamental ways: it can hide latency (aspect exploited in section 3), it can reduce latency or it can increase throughput. Using concurrence to hide latency is highly problem-specific requiring a parallel algorithm for the task at hand. Using concurrence to reduce latency requires that a unit of work be long enough as to pay for the costs of coordinating multiple computing elements, so this is also on the problem. When problems resists to parallelization or have no appreciable latency to hide, the third way that concurrent execution can improve performance is to increase the throughput of the system. Instead of using parallel logic to make a single operation faster, one can employ multiple concurrent executions of sequential logic to accommodate more simultaneous work. It is important to note that a system using concurrence needs not to be multithreaded. Rather, those components that share no state (i.e. objects) can be entirely sequential. The sharing in the system can be then offloaded to components explicitly designed around parallel execution on shared state, which can ideally be reduced to those elements already known to operate well in concurrent environments. Migrating scientific applications to these new environments should be done with a maximum transparency. To achieve this, techniques for automatic parallelization must be pushed to their limits and current shared memory programming paradigms must be considered with respect to their ease of use, performance and ability to support a wide range of programming needs. Nevertheless the need of supporting *code migration* by means of explicit parallelism is unavoidable, therefore the ability to achieve expressiveness is required when porting. Considering both approaches, the need for a concurrence platform [10] and the need for more expressiveness [11], then frameworks like Charm++ [12] (a C++ based language that relies on concurrent objects and asynchronous message-driven execution model) may be considered as an alternative. Other alternatives are pthreads and *OpenMP*. Whereas pthreads often requires major reorganization of the program's structure, the insertion of *OpenMP* directives is straightforward. However current *OpenMP* does not yet provide features for the expression of locality and modularity that may be needed for multicore enabled applications. Therefore, given the potentially prohibitive cost of manually parallelization using a low level programming model [10], [11] it is imperative that programming models and environments be provided that offer a reasonably straightforward means of adapting existing code and creating future parallel programs. Automated parallelization is hard to achieve, in fact, it is on research from many years now but advances are in progress, nevertheless object orientation offers a direct way of expressing interactions. Parallel programming models designed for HPC such as MPI may be implemented on a multicore architecture but do not seem to be an appropriate vehicle for migrating mainstream applications [11]. Unfortunately, some of these models require major code reorganizations, in particular, those that offer a *local* view of computation such as MPI, expect the entire program to be rewritten to construct the local code and hence do not permit incremental parallelization. They are likely to waste memory. The drawbacks of memory hierarchies may be experienced without the ability to exploit any possible benefits of cache sharing.

Figure 3 shows the fine grain implementation using Charm++ for the image reconstruction problem. Its objects are computing entities based on the active objects model [13]. The first attempt to port the application into the multicore platform considers the preservation of the iterative nature of the problem. This

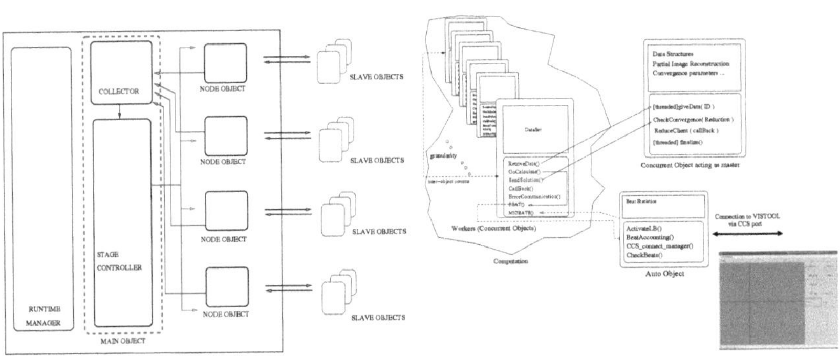

(a) Model used (b) Detail of objects

Fig. 3. Object Oriented implementation for de Image Reconstruction Problem implementation only changes the architecture of the problem to include the concept of object. In it, a concurrent object controls the reconstruction following a master / slave model. This object is in charge of administrative tasks such as creating the data structures, building the worker objects and controlling the convergence to the solution. Worker objects should contact the master concurrent object to get all the information they need to work. All the communication, either at the cluster or at the multicore, is carried out by means of method invocation, in contrast to the MPI implementation that uses real message passing. The mission of each worker object is to iteratively reconstruct a portion of the volume. A number of objects may be active at a processor to show how the concurrence can be exploited to achieve a faster convergence. Automatic adaptativity is under implementation using special automated objects (see *auto* object) that drive the load balancing, a visual tool is also provided for inspecting this object's reactions. Figure 3(a) shows the implementation of a High Level Parallel Composition where objects follow the Farm Pattern[14], where there exists a group of concurrent objects that can work in parallel under the directions of a master objec. The model selected to control the concurrence was MAXPAR [15]. Results obtained encouraged us to explore this model in actual architectures. In the proposed model, the runtime has as a built-in scheduler (useful for the buffered messages queue, global and local nondeterminisms). Fast context switching is achieved thanks to the user level threads (quickthreads) package that offers a suitable wrap for embedding objects. The runtime manager entity is provided by the Charm++ framework. Then, the main object is the one that controls how the program advances. The node objects are placed one per core -or processor- to avoid the master having to communicate with each slave object. Slaves simply do their task using their concurrence capabilities, independently from the processor they are located. Figure 4 shows the behaviour of the reconstruction in MPI and in Charm++ when using a cluster and a multicore system, in the x-axis the legend means # of processors in the cluster and # of instances (processes) in the multicore. In this case the volume was a smaller one. K was set to the maximum in both, MPI and Charm++, versions. Charm++ tests were always launched with four worker objects per processor in the cluster platform.

As Figure 4 shows, Charm++ version shows a better behavior than its MPI counterpart. Using a cluster means that MPI is the owner of the processor. The main harm that this scenario suffers from is the network latencies. Communicating two MPI processes is costly. Current chipsets are extremely optimized but even in these cases thousands of clock cicles are jettisoned waiting for communications to complete. It is a fact, in cluster platforms, non-blocking communications or the alternative developed in [17] are used to alleviate cited latencies. As Figure 4 shows, the Charm++ version improves the reconstruction. This is due a better granularity and concurrence, both helps in a hiding latencies efficiently. Also the encapsulation characteristic of Object Orientation improves the cache usage[16],[7] helping a faster convergence to the solution than the provided by MPI. Nevertheless when running the application in the multicore (Intel Core 2 Quad Q6600), the MPI version reaches a point (8 processes which means two processes per core) where the cache contention (shared by pairs of cores) affects

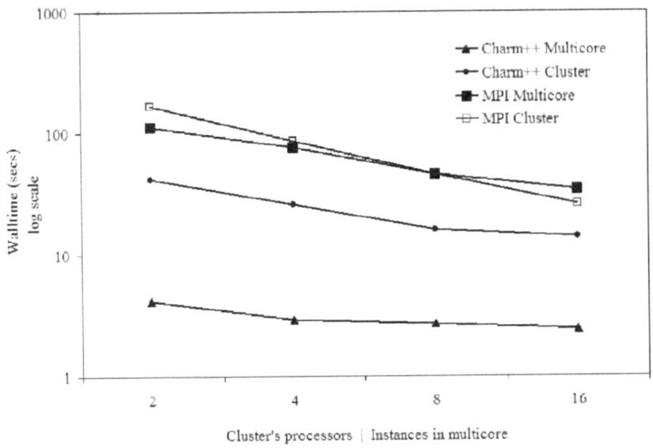

Fig. 4. Execution on a Cluster and on a Multicore of MPI and OO versions

dramatically the performance. Also one may note that MPI is based on passing messages and although the network remains untouched, when sharing memory this message passing is translated into a copy from private to shared memory where conflicts also exist. The fact that the MPI multicore version is not very much better than the clustered one is that the frequency per core in the multicore is lower and the frequency may vary to prevent wasting power. Despite, the object oriented version exploits cache reducing contention by means of data encapsulation. Also, communications between objects is done using method invocation, which assures some kind of transactional access to memory due to the fact that concurrent methods are atomic. In addition, the ability to switch context quickly makes the object oriented version a strong competitor.

5 Conclusions

Parallel programming can be difficult. It moves the programmer closer to the hardware and further from their application space or problem. Abstractions do a good job on smoothing the landing. For clusters platforms do exist abstractions based on threads embedding processes that, with few coding transformations can be translated in performance benefits. When referring to multicore platforms, it is proven that the legacy computing model seems inappropriate. We trusted on abstractions to better express the parallelism we needed (using object orientation). Results showed that using concurrent active objects and a proper platform abstraction, performance improvements can be achieved.

Acknowledgments

This work has been funded by grants **TIN2005-00447** and **TIN2008-01117** (Spanish Ministry of Science and Innovation). **P06-TIC-01426, P08-TIC-3518** (Junta de Andalucía), and by ERDF.

References

1. Fernández, J.J., Lawrence, A.F., Roca, J., Garca, I., Ellisman, M.H., Carazo, J.M.: High Performance Computing in Electron Microscope Tomography of Complex Biological Structures, vol. 2565 (2003)
2. Whitney, J.: Living in a Multi-Core World: Tips for Developers, `http://developer.amd.com/pages/621200628.aspx`
3. McCandless, B., Lumsdaine, A.: The Role of Abstractions in High-Performance Computing. In: Proceedings of the Scientific Computing in Object-Oriented Parallel Environments, pp. 201–210. Springer, Heidelberg
4. Matej, S., Lewitt, R., Herman, G.: Practical considerations for 3-D image reconstruction using spherically symmetric volume elements. IEEE Trans. Med. Imag. 15, 68–78 (1996)
5. Fernández, J., Lawrence, A.F., Roca, J., García, I., Ellisman, M.H., Carazo, J.M.: High performance electron tomography of complex biological specimens. Journal of Structural Biology 138, 6–20 (2002)
6. Huang, C., Lawlor, O., Kale, L.V.: Adaptive MPI. In: Rauchwerger, L. (ed.) LCPC 2003. LNCS, vol. 2958, pp. 306–322. Springer, Heidelberg (2004)
7. Bassetti, F., Davis, K., Marathe, M.: Improving Cache Utilization of Linear Relaxation Methods: Theory and Practice. In: Matsuoka, S., Tholburn, M. (eds.) ISCOPE 1999. LNCS, vol. 1732, pp. 25–36. Springer, Heidelberg (1999)
8. Álvarez, J.A., Roca, J., Fernández, J.J.: A Load Balancing Framework in Multithreaded Tomographic Reconstruction. In: Proceedings of the Intl. Parco Conference, pp. 165–172. John von Newmann Institute for Computing (2007)
9. Cantril, B., Bonwick, J.: Real-World concurrence. ACM Queue, 17–25 (September 2008)
10. Leiserson, C.: The Case for a Concurrence Platform. Dr Dobbs Journal (November 2008)
11. Chapman, B., Huang, L.: Enhacing OpenMP and Its Implementation for Programming Multicore Systems. In: Proceedings of the International Conference Parco 2007, pp. 3–18 (2007)
12. Kale, L., Krishnan, S.: Charm++: a portable concurrent object oriented system based on c++. In: Proceedings of the eighth annual conference on Object-oriented programming systems, languages, and applications, pp. 91–108. ACM Press, New York (1993)
13. Lavender, R.G., Schmidt, D.C.: Active object: an object behavioral pattern for concurrent programming Book: Pattern languages of program design, vol. 2, pp. 483–499. Addison-Wesley Longman Publishing Co., Inc., Amsterdam (1996)
14. Lopes, M.R., Capel Tuñon, M.: An Approach to Structured Parallel Programming Based on a Composition of Parallel Objects. In: CONIELECOMP 2006: Proceedings of the 16th International Conference on Electronics, Communications and Computers, pages 42. IEEE Computer Society, Washington (2006)
15. Corradi, A., Leonardi, L.: Concurrency within objects: layered approach. Inf. Softw. Technol. 33(6), 403–412 (1991)
16. Veldhuizen, T.L., Jernigan, M. (eds.): Will C++ Be Faster than Fortran? Proceedings of the Scientific Computing in Object-Oriented Parallel Environments, pp. 49–56. Springer, Heidelberg (1997); 3-540-63827-X
17. Sievert, O., Casanova, H.: A Simple MPI Process Swapping Architecture for Iterative Applications. International Journal of High Performance Computing Applications 18(3), 341–352 (2004)

Unsupervised Clustering Using Diffusion Maps for Local Shape Modelling

Daniel Valdes-Amaro and Abhir Bhalerao

Department of Computer Science,
The University of Warwick, Coventry, UK, CV4 7AL
{dvaldes,abhir}@dcs.warwick.ac.uk
http://www.dcs.warwick.ac.uk/research/improc/

Abstract. Understanding the biological variability of anatomical objects is essential for statistical shape analysis and to distinguish between healthy and pathological structures. Statistical Shape Modelling (SSM) can be used to analyse the shapes of sub-structures aiming to describe their variation across individual objects and between groups of them [1]. However, when the shapes exhibit self-similarity or are intrinsically fractal, such as often encountered in biomedical problems, global shape models result in highly non-linear shape spaces and it can be difficult to determine a compact set of modes of variation. In this work, we present a method for *local* shape modelling and analysis that uses Diffusion Maps [2] for non-linear, spectral clustering to build a set of linear shape spaces for such analysis. The method uses a curvature scale-space (CSS) description of shape to partition them into sets of self-similar parts and these are then linearly mixed to more compactly model the global shape.

Keywords: Shape, Statistical Shape Modelling, Local Shape Models, Curvature Scale Space, Diffusion Maps, brain contours.

1 Introduction

Study of variability in natural objects has been a topic of research for many years. Different approaches have been used in this type of analysis, but since shape is one of the most important features of human perception it is natural to asses the variation in terms of it. In medicine, fundamental features of the brain structure or function (in health and disease) are revealed by digital imaging, but due to the complexity of the human brain, quantitative analysis is a challenging area of research [3]. Computational Anatomy is a discipline where the objective is to create algorithmic tools to help in the analysis of biological and anatomical structures, in particular, brain substructure. Identification of structural brain changes is associated with different neuro-degenerative diseases, so identifying such variation can bring valuable information in the diagnosis and treatment of many pathologies [4].

Statistical Shape Modelling (SSM) can be used to analyse the shapes of sub-structures aiming to describe their variation across individual objects and

R. Moreno-Díaz et al. (Eds.): EUROCAST 2009, LNCS 5717, pp. 342–349, 2009.

between groups of them [1]. *Active Shape Models* deal well with problems such as the size of the training set and the homology (point-to-point correspondences), but its important to point out that these models are successful in modelling large scale variations, but they struggle with the finer shape details. Shen *et al.* [5], presents a deformable model for segmentation and definition of point correspondences in brain images using an adaptive-focus deformable statistical model based on affine-invariant attribute vectors, minimisation of an energy function and PCA. Worthy of mention are the many techniques created to model the surface of the brain like [6] where a spherical topology mapping and topology correction are used to map accurately the cortex. Shape modelling has shown that shape variation can be successfully modelled as in [7], in which an approach for shape representation that utilises medial representations derived from a spherical harmonics boundary description to study Hippocampus schizophrenia described. Xue *et al.* [8] proposed an automatic segmentation algorithm for neonatal brain MRI using a knowledge based approach to identify and reduce the MLPV in an EM-MRF segmentation scheme.

When the shapes exhibit self-similarity or are intrinsically fractal, such as often encountered in biomedical problems, global shape models results in highly non-linear shape spaces and it can be difficult to determine a compact set of modes of variation. Depending on the application, one approach is to combine rigid shape models together with parametric non-linear deformation, but this can result in far too many degrees of freedom being used. Others have reported on the use of hierarchical analysis of such shapes. In this work, we present a method for *local* shape modelling and analysis that uses Diffusion Maps [2] for non-linear, spectral clustering to build a set of linear shape spaces for such analysis. The method uses a curvature scale-space (CSS) description of shape to partition shapes into sets of self-similar parts [9] and these are then linearly mixed to more compactly model the global shape. We present results on a set of leaves and brain contours to asses the veracity of the method.

2 Method

Our proposed local shape modelling method consists of creating a pose independent shape space where the local contour variability can be measured. The model is based on a Point Distribution Model [1] where each shape obtained from an image (figure 1-(a)) is represented by a set of labelled points (figure 1-(b)):

$$c = \{x_1, y_1, \ldots, x_k, y_k\} \tag{1}$$

Using the consistency of curvature extrema points of the Curvature Scale Space (figure 1-(c)) we proceed to derive a set of local partitions from each contour (figure 1-(d)) . The later will be explained in detail in the next section. Once that a set of meaningful partitions is ready to be analysed we need to compare the shapes, but in order to do this they need to be aligned with respect to a set of axes. Hence an affine alignment is performed over the set to eliminate pose variation (figure 1-(e)). As we select a reference shape, we need to

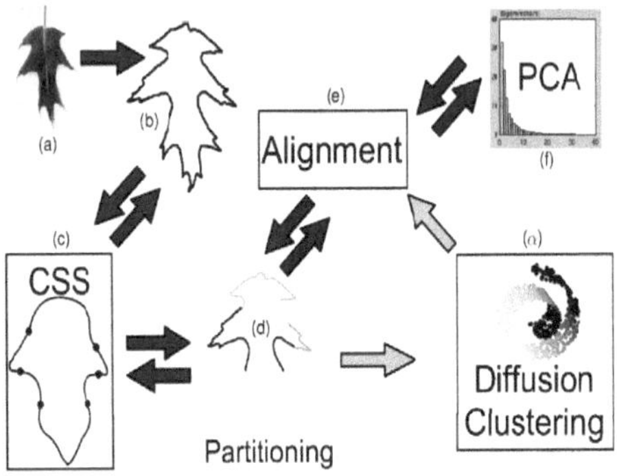

Fig. 1. Process for our local shape model: (a) The source image is processed so the contour (b) can be drawn out of it. Then the CSS process is applied in order to obtain the partitions of the contour (d). From here there are two flows, first we can proceed with the alignment (e) and finish with the PCA analysis (f) or we can proceed to (α) find a low-dimensional embedding of the sub-manifold using Diffusion Maps and k-means clustering to identify a set of local shape models.

align each shape of the set with this one. Rigid body transformation parameters can be used to transform the points from any shape to the reference frame.

Here, we introduce the use of spectral clustering to build a set of linear shape spaces for such analysis (figure 1-(α)). We use the method described in [10] for extracting the intrinsic parameters of multiple shape classes in an unsupervised manner, where the method is based on learning the global structure of the shape manifolds [2].

2.1 Curvature Scale Space Zero-Crossings

Curvature Scale Space is a technique for object representation, invariant under *pose* variations and based on the scale space representation. To build the CSS representation the curve needs to be considered as a parametric vector equation $\Gamma(t) = (x(t), y(t))$, then a series of *evolved versions* of $\Gamma(t)$ are produced by increasing the scale parameter, σ, from 0 to ∞. Every new evolved version is defined as $\Gamma_\sigma = (X(t,\sigma), Y(t,\sigma))$, where

$$X(t,\sigma) = x(t) \circledast g(t,\sigma)\ ,\ Y(t,\sigma) = y(t) \circledast g(t,\sigma). \tag{2}$$

Here, $\circledast$ denotes the convolution operator and $g(t,\sigma)$ is a Gaussian of width σ. Since the CSS representation contains curvature zero-crossings or extrema points from the evolved version of the input curve, these are calculated directly from any Γ_σ by:

$$k(t) = \frac{\dot{X}(t,\sigma)\ddot{Y}(t,\sigma) - \dot{Y}(t,\sigma)\ddot{X}(t,\sigma)}{(\dot{X}(t,\sigma)^2 + \dot{Y}(t,\sigma)^2)^{3/2}} \tag{3}$$

The final step is the construction of the CSS image, but only the generation of evolved versions of the curve and the locations of the curvature zero-crossings are relevant for this work, for further details see [11]. The generation of evolved versions of the curve, produces a set of zero-crossings of the second derivative where there is a change on the curvature of the contour. These points provide a basic but efficient way to create meaningful partitions contours that exhibit self-similar variation (Figure 2).

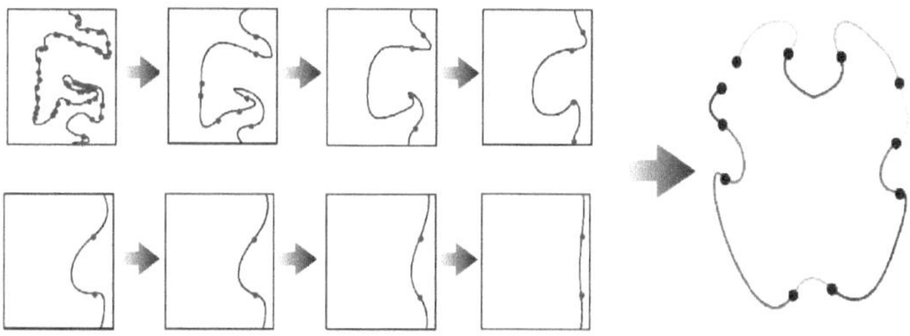

Fig. 2. CSS evolution of a white-matter brain contour. At some appropriate level of smoothing, a set of meaningful partitions can be identified. Pairs of zero-crossings (red points) are used to search and rank local parts on the original shape.

2.2 Shape Representation

In the context of shape analysis is desirable to use a shape representation invariant to translation, rotation, and scale transformations. For this purpose Fourier descriptors (FD) where chosen to represent the shapes since they are effective for many problems of pattern classification and computer vision.

Let us regard any shape as a contour (closed curve) represented as a set of boundary points as in equation 1, then the *Centroidal distance* function is defined as the distance from the boundary points from the centroid of the shape: $r(i) = \sqrt{(x_i - \overline{x})^2 + (y_i - \overline{y})^2}$ where $\overline{x}$ and $\overline{y}$ denotes the coordinates of the centroid of the object. Then, the distance vector $r = \{r(1), r(2), \ldots, r(N)\}$ is transformed into the frequency domain using FFT. Now the feature vector $\mathbf{f}$ is derived as follows:

$$\mathbf{f} = \left(\frac{|F_1|}{|F_0|}, \frac{|F_2|}{|F_0|}, \cdots, \frac{|F_{N/2}|}{|F_0|} \right) \tag{4}$$

here $|F_i|$ denotes the *ith* Fourier coefficient and $|F_0|$ the *DC* component. In the last equation, due to the the fact that the centroidal distance function is real valued, only half of the FDs is needed to index the shape, as well, taking the magnitudes of the coefficients yields rotation invariance and scale invariance is obtained by dividing them by the DC component.

2.3 Shape Manifolds and Diffusion Maps

Like most manifold learning methods, the first step of diffusion maps is to define the feature vectors, hence $\Omega = \{\mathbf{f}_1, \mathbf{f}_2, \ldots, \mathbf{f}_n\}$ (where n denotes the total of shapes) can be regarded as the set of feature vectors that corresponds to our data set of various shapes. Then as Ω regards these feature vectors as the nodes of the adjacency graph G such that $G = (\Omega, W)$, where W, the similarity matrix between $\mathbf{f}_i$ and $\mathbf{f}_j$ can be computed using the Gaussian kernel of width ε:

$$w(\mathbf{f}_i, \mathbf{f}_j) = e^{\frac{-||\mathbf{f}_i - \mathbf{f}_j||^2}{2\varepsilon}} \tag{5}$$

The graph G with weights W represents our knowledge of the local geometry of the set. Next, a Markov random walk is defined on this graph, and the degree of node $d(\mathbf{f}_i)$ of node $\mathbf{f}_i$ is expressed:

$$d(\mathbf{f}_i) = \sum_{z \in \Omega} w(\mathbf{f}_i, z) \tag{6}$$

Now, if P is defined as an $n \times n$ matrix whose entries are given by:

$$p_{ij} = \frac{w(\mathbf{f}_i, \mathbf{f}_j)}{d(\mathbf{f}_i)} \tag{7}$$

then $p(x, y)$ can be viewed as the transition kernel of a Markov chain on V.

As P contains geometric information about the data set, the transitions that it defines directly reflect the local geometry defined by the immediate neighbours of each node in the graph of the data. In other words, $p(i, j)$ represents the probability of transition from node i to node j in one time step [2]. Running the Markov chain forward is equivalent to computing powers of the operator P. For this computation, in theory, the eigenvalues and eigenvectors of P can be used, but instead, these objects can be directly employed in order to characterise the geometry of the data set. Hence, it is possible to define the family of diffusion maps $\{\Psi_t\}_{t \in \mathbb{N}}$ given by:

$$\Psi_t(x) \triangleq \begin{pmatrix} \lambda_1^t \psi_1(x) \\ \lambda_2^t \psi_2(x) \\ \vdots \\ \lambda_{s(\delta,t)}^t \psi_{s(\delta,t)} \end{pmatrix}. \tag{8}$$

Each component of $\psi_t(x)$ is termed *diffusion coordinate*. The mapping $\Psi_t : V \longrightarrow \mathbb{R}^{s(\delta,t)}$ embeds the data set into an Euclidean space of $s(\delta, t)$ dimensions.

3 Experimental Results

Our first experiments use closed contours. The data set was the same data set as in [10], six different shape classes from the Kimia database of object silhouettes. The classes are: carriage (20 shapes), dog (49 shapes), rat (20 shapes), fish (32 shapes), hand (16 shapes) and horse (20 shapes) for a total of 157 samples (Figure 3-(a)). The next data set used for experimental evaluation had partitions from whole contours using the CSS zero-crossings (equation 3) obtained from a

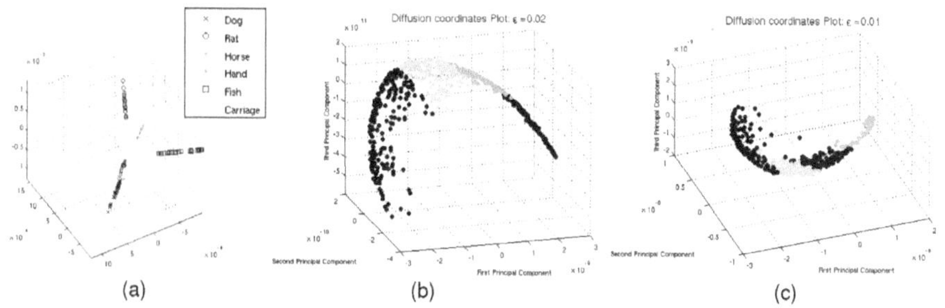

Fig. 3. Diffusion maps coordinates plots, results for contour shapes: (a) Kimia data set of six different classes of shapes: carriage, dog, rat, fish, hand and horse (b) 662 partitions from 50 leaf contours of the class Quercus Kelloggii and (c) 501 partitions from 60 white matter contours

set of leaf images. First, we used a set of 50 shapes of the leaf class Quercus Kelloggii, that generate approximately 600 leaf partitions, each partition was represented by 128 points. Corresponding results are presented in figure 3-(b). The brain contour data set for the following results was from McGill University's BrainWeb data of 20 anatomical models of normal brains [12]. In this case, our data set consisted in 501 partitions coming from 60 white matter contours and figure 3-(c) exhibit the results for this.

4 Evaluation and Discussion

The main contribution of this work is the creation of a method that obtains a set of meaningful shapes, meaning with that this that is possible to find local parts that are similar and localised according to a non-supervised the novel spectral clustering technique of Diffusion Maps. Furthermore, another objective of this idea of generating ordered sets of partitions from contours is to establish a way of determining meaningful local sets of shapes. The spectral clustering is effective in discovering the non-linear manifold in the non-linear shape spaces. The combination of CSS and diffusion maps are a way to map from self-similar contours to a piece-wise shape description. We used spectral-clustering to build a set of 4 local (linear) shape models and then measure the reconstruction error for each model against a single (global) SSM. Figure 4 shows the cluster variation for the local versus global models and demonstrates the better compactness of the 4 classes over the global model.

The method has a number of applications in shape modelling of natural shapes, such as in biology and medical imaging. Elsewhere [9], we have described the latter shape model and with it, a simple windowing and blending technique which allows the modelled parts to be reconstructed back into the original global shape useful for visual feedback (figure 5).

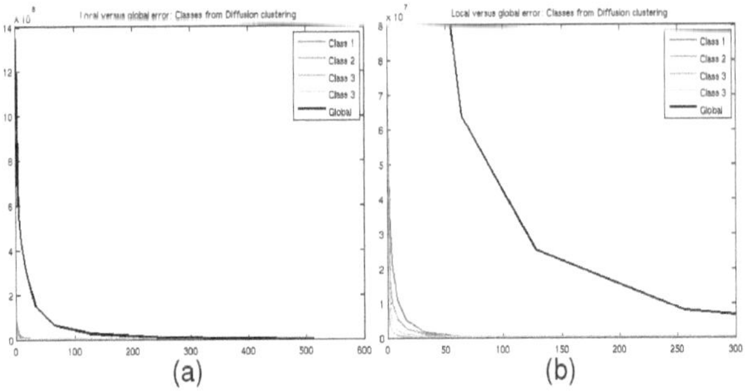

Fig. 4. Reconstruction log-error plot for the different clusters and for the global shape model in black

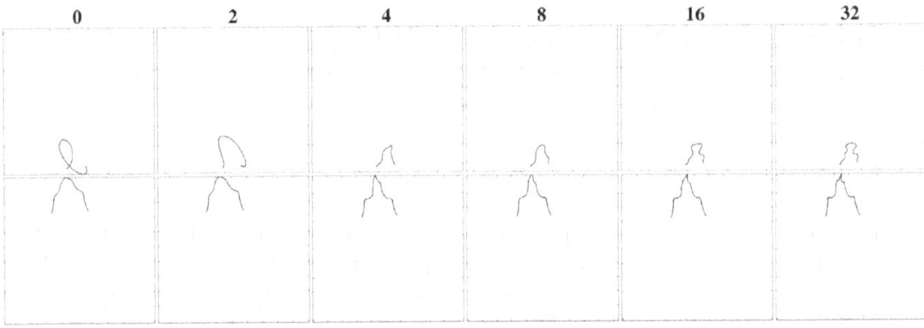

Fig. 5. Reconstruction of the chosen set of shapes, by added a sequence of principal modes of variation: 0, 2, 4, 8, 16, 32. First row corresponds to a white matter contour and second to a leaf from the class Quercus Kelloggii. The modelled partitions are blended back into a smooth scale of the CSS, Γ^{σ} defocussing the general, irrelevant shape variations for the purposes of visualisation.

The results presented here are illustrative and further validation is necessary. The method needs to be extended to surfaces to be properly validated with clinical data but it is not clear now how the local partitioning and the clustering could be easily extended to surface patches. We believe that this model could have useful application in brain morphometrics and computational anatomy. To be more precise, the provided method could be adapted for clinical diagnosis software for assessing changes in local shape variation of anatomical structures, such as, white/gray matter. Finally, the spectral clustering might be adapted for the problem of image database retrieval where the objective can be to discover images which contain objects similar to query objects, in this case brain sections.

Acknowledgements. The first author thanks the Mexican National Research Council for Science and Technology (CONACyT) for the research grant given for the support of his PhD studies.

References

1. Cootes, T.F., Taylor, C.J., Cooper, D.H., Graham, J.: Training models of shape from sets of examples. In: Proc. British Machine Vision Conference, pp. 266–275. Springer, Heidelberg (1992)
2. Lafon, S., Lee, A.B.: Diffusion maps and coarse-graining: A unified framework for dimensionality reduction, graph partitioning, and data set parameterization. IEEE Transactions Pattern Analysis and Machine Intelligence 28(9), 1393–1403 (2006)
3. Grenander, U., Miller, M.I.: Computational anatomy: An emerging discipline. Quarterly of Applied Mathematics 56, 617–694 (1998)
4. Ashburner, J., Csernansky, J.G., Davatzikos, C., Fox, N.C., Frisoni, G.B., Thompson, P.M.: Computer-assisted imaging to assess brain structure in healthy and diseased brains. Lancet Neurology 2, 79–88 (2003)
5. Shen, D., Herskovits, E.H., Davatzikos, C.: An adaptive-focus statistical shape model for segmentation and shape modeling of 3-d brain structures. IEEE Transactions on Medical Imaging 202, 257–270 (2001)
6. Fischl, B., Liu, A., Dale, A.M.: Automated manifold surgery: Constructing geometrically accurate and topologically correct models of the human cerebral cortex. IEEE Transactions on Medical Imaging 2, 70–80 (2001)
7. Styner, M., Gerig, G., Lieberman, J., Jones, D., Weinberger, D.: Statistical shape analysis of neuroanatomical structures based on medial models. Medical Image Analysis 7, 207–220 (2003)
8. Xue, H., Srinivasan, L., Jiang, S., Rutherford, M., Edwards, A.D., Rueckert, D., Hajnal, J.V.: Automatic segmentation and reconstruction of the cortex from neonatal mri. Neuroimage 38, 461–477 (2007)
9. Valdes-Amaro, D., Bhalerao, A.: Local Shape Modelling for Brain Morphometry using Curvature Scale Space. In: McKenna, S., Hoey, J. (eds.) Proceedings of the 12th Annual Conference on Medical Image Understanding and Analysis 2008, July 2008, pp. 64–68. British Machine Vision Association (2008)
10. Rajpoot, N.M., Arif, M., Bhalerao, A.H.: Unsupervised learning of shape manifolds. In: British Machine Vision Conference, pp. 312–321 (2007)
11. Mokhtarian, F., Bober, M.: Curvature Scale Space Representation: Theory, Applications, and MPEG-7 Standardization. Kluwer Academic Publishers, Norwell (2003)
12. Aubert-Broche, B., Griffin, M., Pike, G.B., Evans, A.C., Collins, D.L.: Twenty new digital brain phantoms for creation of validation image data bases. IEEE Transactions on Medical Imaging 25, 1410–14163 (2006)

Sensibility Analysis of an Object Movement Forecast Approximation in Real Image Sequences

J.L. Crespo, P. Bernardos, and E. Mora

Department of Applied Mathematics and Computer Sciences, University of Cantabria, Avda. de los Castros s/n 39005 Santander, Spain
{crespoj,bernardp,morae}@unican.es

Abstract. The objective of this paper is to analyse the influence of the different parameters used for an overall approach to forecasting a future position of the mobile objects of an image sequence after processing the previous images to it. Our approximation uses classical techniques such as optical flow to extract object's trajectories and velocities and autoregressive algorithms to build the predictive model. Applications to outdoor scenarios are possible, for videos where stationary cameras are used and moving objects follow an affine displacement field. In this work, traffic sequences with different meteorological conditions are studied.

Workshop: Computer Vision and Image Processing.

1 Introduction

The movement analysis of objects in temporal image sequences is an important topic with applications in several areas such as navigation and tracking field. Most of the works found in the literature are focused on the tracking of a single object where additional information is also available in several cases.

Recent progress has been made in this field. For example, in [1] the forecasting of an image is successfully obtained; being their possible moving objects identified and located and their positions predicted.

The proposed method in [1] makes use of computer vision techniques such as optical flow to obtain inter-frame displacements and the Box-Jenkins' linear autoregressive algorithms to build the predictive model.

Its application is suitable to real image sequences in which stationary cameras are used, moving objects hold their shape approximately fixed during the sequence and change their position with time, such as several recorded traffic sequences with different meteorological conditions found in [2].

The aim of this work is to study the sensitivity and effect of the different parameters used within the object movement forecast approximation given in [1].

The paper is organized as follows: in Section 2 the prediction method is briefly described, Section 3 presents the results obtained for the three traffic intersection sequences and finally, some conclusions are resumed in Section 4.

R. Moreno-Díaz et al. (Eds.): EUROCAST 2009, LNCS 5717, pp. 350–357, 2009.

2 Prediction Method

Here, the four steps that the prediction method can be split in, are briefly described. The complete description can be found in [1]. The objective of the method as a whole is to take an image sequence with moving objects in it as input, and to forecast the image obtained in a certain future time as output. The following steps are distinguished:

- Pixels' velocity calculation. In the first step, pixel displacements for subsequent images in the sequence are calculated. Optical flow techniques according to the differential method based on spatial gradients proposed by [3] are followed. Although overlapping problems are not addressed, short disruptions in the velocity calculation are permitted by using time domain filtering.
- Moving object identification. The sequence background is obtained as the temporal median of such sequence. Next, the moving areas are pre-identified as the ones that substantially differ from the background. Isolated pixels are removed and then each moving object is identified as a set of 8-connected pixels.
- Velocity assignation to identified objects. In this step, a coherent motion field to every identified block is assigned. For that purpose, the pixels' velocities in each block are fitted to an affine field.
- Predictive model building. Finally, the previously obtained velocities are adjusted to a Box-Jenkins' autoregressive linear model in order to generate the prediction.

Every step in the previous computation introduces noise, and for this reason, some filtering is required along the process. A sensibility analysis of the different filtering parameters is developed in this context. Prediction validity in time is also studied.

3 Results

We will focus on the width of two Gaussian filters: the one used to smooth the image sequence, which improves the computational response of the sequence, and the one used in the Lucas-Kanade optical flow algorithm.

The first one serves a dual purpose: first, it reduces the amount of noise and other troublesome high frequency components; second, it spreads the intensity values around the object borders, hence alleviating the aperture problem, and allowing for more pixels where optical flow can be detected. It is sometimes called the noise scale; we will address it as the pre-filter width.

The second is related to the assumption of constant local velocity inherent to the Lucas-Kanade algorithm. It measures the size of the region where the method assumes similar velocity. It is sometimes called the integration scale; we will address it as the uniform region width.

Regarding the pixel velocity calculation performed by the Lucas-Kanade method, only the image pixels with a minimum eigenvalue significant are valid to calculate their velocity, the other pixels being excluded from the calculus. This minimum eigenvalue will also be varied in the following.

Finally, when correcting pixel velocities with the objects to which they belong, noise points, having numerical values similar to the background must be filtered. We will address it as the noise filter threshold. The decreasing of this parameter together with a previously high precision background removing, have permitted the recovering of the dark vehicles present in the images, which is an important issue.

In the following, we take as a basis the results shown in [1] for three scenarios, available freely from [2]: partly clear, heavy fog and heavy snowfall. In all of them, a pre-filter width of 5 pixels and uniform region width of 15 pixels were used. The minimum eigenvalue reliable for pixel velocity calculation was imposed to 1 and a noise filter threshold value of 10 was considered, which means a 4% of the grey level range. Let us focus first on the partly clear scenario. The fourth previous filtering parameters are varied here in order to explore the validity limits of our approximation. A set of selected cases is represented in Figures 1-12 to see the visual changes of the predictions. The noise filter threshold is 10 in all of them, because smaller values like 5 lead to the inclusion of much noise in the movement, thus producing a cut and paste effect with large portions of the image. Considering bigger values like 15 or 20 have no significant effect in the forecast image.

In the next table we summarize our tests, referring to the pictures where the impact of the changes is reflected.

Table 1. Values of the parameters: pre-filter width, uniform region width (both expressed in pixels) and minimum eigenvalue reliable for pixel velocity calculation, for the construction of the forecast image represented in the corresponding figure, in the partly clear scenario

Pre-filter width	Uniform region width	Minimum eigenvalue	Figure number
5	30	0.75	1, 8, 9, 11
Not used	3	1	3
2	15	1	4
15	15	1	5
5	5	1	6
5	90	1	7

Figure 1 shows our best prediction for this sequence, to be compared with the real image represented in Figure 2. It improves the result obtained in [1] in the sense that dark vehicles are also detected here thanks to a high precision background removing, although now vehicle positions are in general slightly retarded.

Reasonable results are also obtained when predicting images 30 and 40 frames after the last known image (see Figures 9 and 11, to be compared with the real images of Figures 10 and 12, respectively), even if dark vehicles are worse predicted for the increasing of time. However, for 45 frames later the last known image, see Figure 8, the forecast vehicles are distorted and/or present significant noise around their positions.

Regarding Figures 3 to 7, bad predictions are obtained in all cases. Thus, valid pre-filter width is around [5 – 10]. Within this range, vehicle shapes are maintained in the movement thanks to the good prediction of their edges movement. Smaller values out of the previous range tend to produce only fragments of vehicles because their contours are wrongly calculated whereas bigger values tend to loose vehicle movement. Moreover, valid uniform region width is around [15- 30]. Smaller values tend to produce only fragments of vehicles again and to have significant noise around their positions. Bigger values tend to move vehicles together with a halo around them, which therefore retard their predicted positions.

Fig. 1. Forecast image 20 frames after the 193th image of the sequence

Fig. 2. Real 213th image of the sequence

Fig. 3. The same as in Fig. 1 but with no pre-filtering and uniform region width of 3

Fig. 4. The same as in Fig. 1 but with a pre-filter width of 2 and uniform region width of 15

Fig. 5. The same as in Fig. 1 but with a pre-filter width of 15 and uniform region width of 15

Fig. 6. The same as in Fig. 1 but width a pre-filter width of 5 and uniform region width of 5

Fig. 7. The same as in Fig. 1 but with a pre-filter width of 5 and uniform region width of 90

Fig. 8. Forecast image 45 frames after the 193th image of the sequence

Fig. 9. Forecast image 30 frames after the 193th image of the sequence

Fig. 10. Real 223th image of the sequence

Fig. 11. Forecast image 40 frames after the 193th image of the sequence

Fig. 12. Real 233th image of the sequence

Now, the same previous study is made on the heavy fog and heavy snowfall scenarios. A set of selected cases is represented in Figures 13 – 20 and their filter parameter values are summarized in Table 2.

Figure 13 shows the best result obtained for the heavy fog sequence where only small improvement is got with respect to that obtained in [1], due to the better definition of the vehicles on the left hand side of the image. Comparing with the real image of Figure 14, we see that the black car on the right side is still undetected because of the heavy fog presented in the image.

When predicting 40 frames after the last known image, vehicles are all fuzzy although well positioned, as can be seen in Figure 15, comparing with the real image of Figure 16.

For the snowfall sequence, as the vehicle positions were already well predicted in [1], the improvement now is coming from the detection of the black car appearing in the bottom left part of the image, see Figure 17 compared with the real image of Figure 18. For this purpose, different noise filter thresholds are used: 10 for everything clearer than the background and 5 for everything darker than the background. Nevertheless, this black car becomes only a spot when the prediction interval increases, as can be seen in Figure 19, which represents 40 frames later the last known image.

Similar results are produced here for uniform region width from 15 to 10 pixels.

Table 2. Values of the parameters: pre-filter width, uniform region width (both expressed in pixels), minimum eigenvalue reliable for pixel velocity calculation and noise filter threshold, for the construction of the forecast image represented in the corresponding figure, in the heavy fog and heavy snowfall scenarios

Pre-filter width	Uniform region width	Minimum eigenvalue	Noise filter threshold	Figure number
5	10	0.5	5	13, 15
5	15	1	10, 5	17, 19

Fig. 13. Forecast image 20 frames after the 100th image of the sequence

Fig. 14. Real 120th image of the sequence

Fig. 15. Forecast image 40 frames after the 100th image of the sequence

Fig. 16. Real 140th image of the sequence

Fig. 17. Forecast image 20 frames after the 100th image of the sequence

Fig. 18. Real 120th image of the sequence

Fig. 19. Forecast image 40 frames after the 100th image of the sequence

Fig. 20. Real 140th image of the sequence

4 Conclusions

In order to avoid the numerical noise produced in every step of the prediction method presented here, some filtering is required along the process. A sensibility analysis of the different filtering parameters is developed in this context. Prediction validity in time is also studied.

Two Gaussian filters are varied for this purpose: the one used to smooth the image sequence, which improves the computational response of the sequence, and the one used in the Lucas-Kanade optical flow algorithm.

The minimum eigenvalue significant to calculate pixel velocity and the noise filter threshold are also varied. The decreasing of this last parameter together with a previously high precision background removing, have permitted the recovering of the dark vehicles present in the images, which is an important issue.

The presence of all the previous filters have shown to be essential for good predictions. Variations are permitted in general up to a 50% factor, being the chosen intervals critical for the prediction accuracy.

Regarding prediction validity in time, the upper limit seems to be about 40 frames later the last known image, although it depends on the atmospheric conditions sequence and the colour of the vehicles present in the scene.

Acknowledgments

The authors are deeply grateful to the Spanish Interministerial Board for Science and Technology (Comisión Interministerial de Ciencia y Tecnología CICYT) which supported Project TIC2002-01306, for the opportunity given to beginning research in this field.

References

1. Crespo, J.L., Zorrilla, M., Bernardos, P., Mora, E.: Moving objects forecast in image sequences using autoregressive algorithms. The Visual Computer 25, 309–323 (2009), http://dx.doi.org/10.1007/s00371-008-0270-8
2. Group Prof. Dr. H.-H. Nagel, Institut fuer Algorithmen und Kognitive Systeme, Fakultaet fuer Informatik Universitaet Karlsruhe (TH). Traffic intersection sequence, http://i21www.ira.uka.de/imagesequences/
3. Barron, J.L., Fleet, D.J., Beauchemin, S.: Performance of Optical Flow Techniques. Int. J. Computer Vis. 12(1), 43–77 (1994)

Angular Contour Parameterization for Signature Identification

Juan Carlos Briceño[1], Carlos M. Travieso[2], Miguel A. Ferrer[2], Jesús B. Alonso[2], and Francisco Vargas[3]

[1] Computer Science Department, University of Costa Rica
Sede "Rodrigo Facio Brenes", Montes de Oca, Post-Code 2060, San José, Costa rica
[2] Department of Signals and Communications, Technological Centre for Innovation on Communication (CeTIC), University of Las Palmas de Gran Canaria, Campus de Universitario de Tafira, Ed. de Telecomunicación, Pabellón B. 35017, Las Palmas de G.C., Spain
[3] Departamento de Ingeniería Electrónica, Universidad de Antioquia, Colombia
juancarlos.briceno@ecci.ucr.ac.cr,
{ctravieso,mferrer,jalonso}@dsc.ulpgc.es, jfvargas@udea.edu.co

Abstract. This present work presents a parameterization system based on angles from signature edge (2D-shape) for off-line signature identification. We have used three different classifiers, the Nearest Neighbor classifier (K-NN), Neural Networks (NN) and Hidden Markov Models (HMM). Our off-line database has 800 writers with 24 samples per each writer; in total, 19200 images have been used in our experiments. We have got a success rate of 84.64%, applying as classifier Hidden Markov Model, and only used the information from this edge detection method.

Keywords: Signature identification, Biometric system, Edge parameterization, Handwritten Writing, Document analysis, Classification system, Pattern Recognition.

1 Introduction

Signature identification is not only a popular research area in the field of pattern recognition, document processing and security biometric applications [1], but also plays an important role in many applications concerned e.g. with security, access control, or financial and contractual matters [2]. Plamondon and Srihari [3] note that automatic signature verification systems occupy a very specific niche among other automatic identification systems: "On the one hand, they differ from systems based on the possession of something (key, card, etc.) or the knowledge of something (passwords, personal information, etc.), because they rely on a specific, well learned gesture. On the other hand, they also differ from systems based on the biometric properties of an individual (finger prints, voice prints, retinal prints, etc.), because the signature is still the most socially and legally accepted means of personal identification."

The signature image to recognize (in off-line systems) can be considered like a space-time signal due to geometric and sequential characteristics. Generally, known recognition and classification methods are based on geometric parameters extraction, and their classification by linear or non-linear classifiers [4].

R. Moreno-Díaz et al. (Eds.): EUROCAST 2009, LNCS 5717, pp. 358–365, 2009.

Generally the signature classification methods are divided into two kinds: on-line systems and off-line systems. Handwriting recognition in off-line systems is more difficult than in on-line systems as a lot of dynamic information is lost. Hence, online signature verification is generally more successful [5]. Nevertheless, off-line systems have a significant advantage in that they do not require access to special processing devices when the signatures are produced. In fact, if the accuracy of the identification promoted greatly, the off-line method has much more practical application areas than that of the on-line one. Moreover, on-line systems have two problems: we can not recognize an already made signature (we need the signer) and, the electronic equipment is more expensive than the off-line systems.

In off-line classification methods, the signature is written on a sheet of paper and afterwards scanned. Subsequently, from the scanned image, the usual step is to parameterize its geometric as a previous stage to their recognition by a Neural Network based classifier or others [6], [7], [8], [9], and [10].

In order to use the off-line system and alleviate their drawbacks, we propose a new parameterization method of scanned signatures. Concisely, the off-line signature classifier proposed acquire the signatures by a scanner and after their parameterization (which include noise filtering, binarization, thinning and vectorization) as a sequence, it is recognized by a Hidden Markov Models (HMM) classifier, which provides a good probabilistic representation of sequences having large variations [11]. Then, we have into account both the geometric structure and the sequential information. This procedure neglects the temporal information of the signature. In order to alleviate this disadvantage, this paper proposes to use signature parameters with spatio-temporal information and its classification by a classifier being able to cope with spatio-temporal problems as HMM [11]. This information can be added to other kinds of parameters in order to improve the success and to give more discriminate information.

2 Database and Its Pre-processing

The GPDS-800 signature corpus contains 24 genuine signatures of 800 individual. So, there are 800 x 24 = 18400 genuine signatures. For the acquisition of our database we use a scanned resolution of 300 dpi, and its quantification was 8 bits in gray-scale. The building of our database was taken in just one session. The signers filled up a form with 24 boxes of different size. The whole process of signing was accomplished under the supervision of an operator.

After its building, the samples were pre-processed. In particular, the stage includes noise reduction and outline detection. Firstly, images were binarized by Otsu's method. After this step, the noise reduction was applied. Noise is due to the scanned process, and it was applied mathematical morphology [12] in order to remove it. Finally, that step was the most complex due to the difference between signatures. The signature is filling and connected each part, then, its edge was found. Tools of mathematical morphology have been used to compute it.

3 Parameterization System

For the purpose of this study we have only considered the signature shape. Border characterization by (x,y) positions of perimeter pixels, has been achieved first by a process of shadowing (black shape over white background), filtering of isolated points, and finally automatic perimeter points location *x= line y=row* coding, by a point to point continuous follow procedure. In the end, we have a perimeter description of $\{(x_i, y_i)|\ i = 1,\ldots,n\}$ points location description representing a one pixel wide stroke, closed border of a signature shape.

Data compression, size regularization and critical control point selection of perimeters description are achieved by a structuring procedure. This procedure is based on the idea that a one pixel stroke on a black an white image may be described as a graph G_f of a one dimensional trajectory application *f*, if we have preservation of a correct sequencing definition or monotonic behavior on the x ordinate. That is,

$$G_f = \{(x_i, y_i) \mid y_i = f(x_i), i = 1,\ldots,n\}, \tag{1}$$

where ordinate points x_i, of the *f* stroke must be such that: $x_i < x_{i+1}$ or $x_{i+1} < x_i$ for $i = 1,\ldots,n-1$. Afterwards, considering the complete perimeter *G* we define its description relation *F* as the partial definition of piece like 1-D trajectory applications f_j (with graphs G_j) preserving monotonic behavior. That is $G = \bigcup_{j\in J} G_j$, where G_j is a set of positional points, a piece of border for convenient set of index *J* and J_j.

$$G_j = \{(x_\alpha, y_\alpha) \mid y_\alpha = f_j(x_\alpha), \alpha \in J_j\} \tag{2}$$

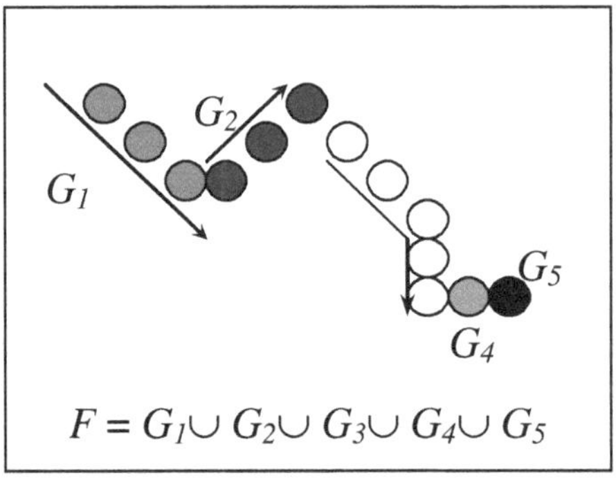

Fig. 1. Example of an F relation decomposed in graphs with a correct function description

The restriction trajectory applications (or coded pieces of border) $f_j = F_{|\{x_\alpha | \alpha\in J_j\}}$ are in such a way that the next point following the last of G_j is the first of G_{j+1}. Accordingly G_j graphs are correct f_j trajectory applications description. To avoid G_j reduced to one pixel, we preserve only the first point of constant *x* ordinate series. Note that the structure of *G* border by partial graph descriptions G_j account for abrupt direction changes on the perimeter description. After building up the G_j we select all the *n* first points of each G_j, *j*= 1,…,*n* and for an arbitrary constant number $p \geq n$ we complete the perimeter points description by *k*= *n*-*p* points, chosen uniformly distributed for each G_j and proportionally to its size.

In order to perform a rotational, scale size and origin reference free coding; we perform an angle transformation for the positional point border coded as before. For a given coded border of n positional control points $G=\{X_i=(x_i,y_i)\mid i=1,...,n\}$, let C_0 be its central point, and let be β_i and α_i the angles referred by C_0 and X_i, $\beta_i = \text{angle}(C_0, X_i, X_{i+1})$ and $\alpha_i = \text{angle}(X_i, C_0, X_{i+1})$. Then the sequence of (x_i,y_i) $i=1,\ldots,n$ positional points are then transformed in sequence of (α_i,β_i) $i = 1,\ldots,n-1$ angular origin free representations points. Note that the choice of the start point X_1 and the C_0 points account for scale and signature shape rotation, as well as geometrical properties of triangular similarities make such sequence of signature shape coding, size and location free (see figure 2).

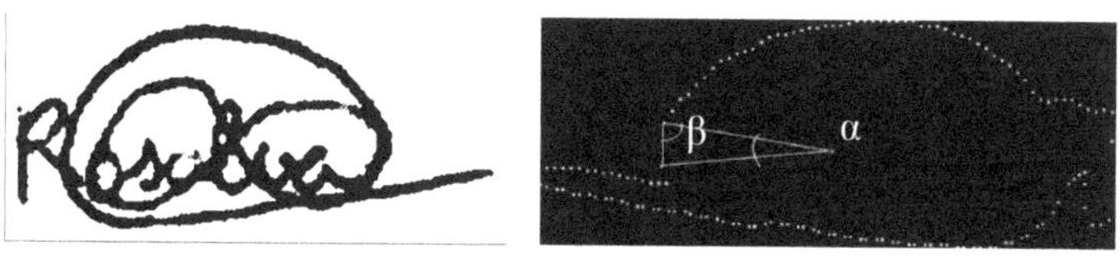

Fig. 2. Illustration example of signature coding, and its angle transformed

Besides, other classical methods have been implemented in order to compare them. In particular, we have used the sweeping of signature edge with polar coordinates for each angle, in total 360 points, the same length than the previous method. But for a same angle, we can found different options, the nearby or far point, or the combination of them. It has been used for our experiments.

4 Classification System

In this present work, three different classifiers have been used, K- nearby neighbor classifier (K-NN) [13], Neural Networks (NN) [13] and Hidden Markov Models (HMM) [14].

4.1 Hidden Markov Models

In this case, a discrete HMM [14] is chosen to model each signer's features vectors. This avoids making assumptions on the form of the underlying distribution, especially when there is a limited amount of data, as is the case in signature identification.

A signature is modeled by two left-to-right discrete HMM, one per p_t sequence and another one for the c_t sequence. The number of states in each signer's HMMs signature is sweept from 20 to 140. The topology only authorizes transitions between each state to itself and to its immediate right-hand neighbors. The classification (evaluation), decoding, and training problems are solved with the Forward-Backward algorithm, the Viterbi algorithm, and the Baum-Welch algorithm. The initialization method is the equal-occupancy method [15]. The K-Means algorithm is used during training to create the multilabeling VQ used by us, which makes a soft decision about which code words are closest to the input vector [16]. Therefore, our VQ generates an output vector whose components indicate the relative closeness of the 10 closest code words to the input.

Additionally, our VQ permits multiple observations training because incoherence can arise if one assumes the independence property among the different components of the input D-dimensional feature vector [17]. The components of the feature vector pt are considered independent, so the two sets of observation symbols contain a certain number of symbols. Therefore, we only have considered a independent group for both angles: [αi , βi]. Experimentally, we have determined 32 symbols for that group.

Once trained the two HMMs that model each signature, the final score is obtained averaging their likelihoods. The verification process accept the signature if the average likelihood is greater than a threshold.

4.2 Neural Network

In recent years several classification systems have been implemented using different techniques, such as Neural Networks. The widely used Neural Networks techniques are very well known in pattern recognition applications.

An Artificial Neural Network (ANN) is an information processing paradigm that is inspired by the way biological nervous systems, such as the brain, process information. The key element of this paradigm is the novel structure of the information processing system. It is composed of a large number of highly interconnected processing elements (neurones) working in unison to solve specific problems. ANNs, like people, learn by example. An ANN is configured for a specific application, such as pattern recognition or data classification, through a learning process. Learning in biological systems involves adjustments to the synaptic connections that exist between the neurones. This is true of ANNs as well.

One of the simplest ANN is the so called perceptron that consist of a simple layer that establishes its correspondence with a rule of discrimination between classes based on the linear discriminator. However, it is possible to define discriminations for non-linearly separable classes using multilayer perceptrons that are networks without refreshing (feed-forward) and with one or more layers of nodes between the input layer and the output layer. These additional layers (the so called hidden layers) contain hidden neurons or nodes, are directly connected to the input and output layer [13] [18].

A neural network multilayer perceptron (NN-MLP) of one hidden layer had been used in this work. Each neuron is associated with weights and biases. These weights and biases are set to each connections of the network and are obtained from training in order to make their values suitable for the classification task between the different classes.

In particular and for our experiments, we have used a Multilayer Perceptron (MLP) Feed-Forward with Back-Propagation training algorithm with only one hidden layer of several different neurons (nodes), obtained empirically in each case. The number of input neurons fits in with the number of edge points, and the number of output neurons with the number of signers.

4.3 K – Nearest Neighbor

This algorithm is one of the simplest methods of classification. All in all, the portion of code of the algorithm stores the data that we present. When we want to do a prediction on a new one vector of characteristics, the KNN algorithm finds nearest vector (according to some metric distance) from the training vectors to the new vector, and predicts the new class. The KNN is a method of learning based on the most next neighbor. This algorithm calculates the similarity among the sample of test and the samples of training, considering the k-nearest vectors in the training, finding the class that more is seemed. To find the vector of more seemed training (not alone in distance), we use the method of the majority of votes.

The degree of similarity between two samples is the distance among them, based on a metric distance. In our simulations we have used the distance Euclidean.

Be t a sample with n characteristic represented by the vector of characteristics $< v_1(t), v_2(), \ldots , v_n(t) >$, where the term $v_i(t)$ is the value of the characteristic one i of the sample t. Therefore the distance among two samples t_i and t_j are $d(t_i, t_j)$, where:

$$d(t_i, t_j) = \sqrt{\sum_{m=1}^{n} (v_m(t_i) - v_m(t_j))^2} . \tag{3}$$

5 Experiments and Results

The classification of the previous classical method is done by K- Nearest Neighbor (KNN) [13], Neural Networks (NN) [13] and Hidden Markov Models (HMM) classifiers, and it is compared with angular parameterization method [14] using HMM. All experiments have been repeated in five times, and the success rates are shown with mean and standard deviation. A supervised classification has been built, and therefore we have two modes, training and test modes. The 50% of the samples (12 samples) have been used for training mode, and the rest of test mode (12 samples). In each experiment, the signature samples for each mode randomly have been chosen.

Experiments have been based on parameters calculated from the signature edge. Therefore, we have achieved our results varying some parameters from the proposed system; in particular, the number of HMM states (between 20 and 140 states) and for neural networks, the number of hidden layer neurons. In table I can be seen the success rates for the classical method.

Table 1. Results for classical method of contour detection

Classifier	Type of contour	Mean ± Standard Deviation
KNN	Nearby	60.77% ± 11.78
	Far	60.90% ± 11.58
	Combination	61.37% ± 12.04
NN	Nearby	76.77% ± 0.17
	Far	76.44% ± 0.17
	Combination	78.24% ± 0.48
HMM	Nearby	75.23% ± 0.78
	Far	77.94% ± 0.61
	Combination	79.18% ± 0.54

The best results are found with HMM classifier, and therefore, it will be compared with our proposal based on angular parameterization of signature edge. The comparison can be seen in table 2.

Table 2. Results of contour detection, with HMM classifier

Classifier	Type of contour	Mean ± Standard Deviation
HMM	Classical method	79.18% ± 0.54
HMM	Method proposed	84.64% ± 0.20

After this second experiment, it clearly shown that our proposed method achieves better results than classical method. Experimentally, the number of HMM states has been 35 and using 32 symbols.

6 Conclusions

In this paper is presented an edge detection method based on angles, therefore, their parameter are rotation, translation and scale invariant. After our experiments, it can be observed that our method improve the results versus classical methods. The experiments are done with three different classifiers for classical parameterization, but the best results are found with HMM, therefore, our method was implemented with that classifier. For 800 off-line signature classes, we have achieved a success of 84.64%.

Acknowledgment

This work has been supported by Canary Government with Mobility Funds in 2008. F. Vargas is supported by the high level scholarships program, Programme Alßan No. E05D049748CO.

References

1. Ross, A., Jain, A.K.: Multimodal Biometrics: An Overview. In: Proceedings XII European Signal Processing Conference, pp. 1221–1224 (2004)
2. Rigoll, A., Kosmala, A.: A Systematic Comparison between On-Line and Off-Line Methods for Signature Verification with Hidden Markov Models. In: 14th International Conference on Pattern Recognition, vol. II, pp. 1755–1757 (1998)
3. Plamondon, R., Srihari, S.N.: On-line and off-line handwriting recognition: a comprehensive survey. IEEE Transactions on Pattern Analysis and Machine Intelligence 22(1), 63–84 (2000)
4. Ferrer, M.A., Alonso, J.B., Travieso, C.M.: Offline geometric parameters for automatic signature verification using fixed-point arithmetic. IEEE Transactions on pattern analysis and machine intelligence 27(6), 993–997 (2005)
5. Plamondon, R., Srihari, S.N.: Online and Off-LineHandwriting Verification: A Comprehensive Survey. IEEE. Transactions on Pattern Analysis & Machine Intelligence 22(1) (2000)

6. Edson, J., Justino, R., El Yacoubi, A., Bortolozzi, F., Sabourin, R.: An Off-Line Signature Verification System Using HMM and Graphometric Features. In: Proceedings Fourth International Workshop Document Analysis Systems, pp. 21–222 (2000)
7. Kovari, B., Kertesz, Z., Major, A.: Off-Line Signature Verification Based on Feature Matching. In: 11th International Conference on Intelligent Engineering Systems, pp. 93–97 (2007)
8. Nguyen, V., Blumenstein, M., Muthukkumarasamy, V., Leedham, G.: Off-line Signature Verification Using Enhanced Modified Direction Features in Conjunction with Neural Classifiers and Support Vector Machines. In: Ninth International Conference on Document Analysis and Recognition, vol. 2, pp. 734–738 (2007)
9. Tian, W., Qiao, Y.: Off-line Chinese Signature Verification based on Optimal Matching of Projection Profiles. In: The Sixth World Congress on Intelligent Control and Automation, vol. 2, pp. 10240–10244 (2006)
10. Chen, S., Srihari, S.: A New Off-line Signature Verification Method based on Graph. In: 18th International Conference on Pattern Recognition, vol. 2, pp. 869–872 (2006)
11. Rabiner, L.R.: A Tutorial on Hidden Markov Models and Selected Applications in Speech Recognition. In: Proceedings of the IEEE, vol. 77, pp. 257–286 (1989)
12. González, R., Woods, R. (eds.): Digital Image Processing. Prentice Hall, Englewood Cliffs (2002)
13. Bishop, C.M.: Neural Networks for Pattern Recognition. Oxford University Press, Oxford (1995)
14. Rabiner, L., Juang, B.: Fundamentals of Speech Recognition. Prentice-Hall, Englewood Cliffs (1993)
15. Ferrer, M.A., Alonso, I., Travieso, C.: Influence of Initialization and Stop Criteria on HMM Based Recognizers. Electronics Letters of IEE 36, 1165–1166 (2000)
16. Hernando, J., Nadeu, C., Mariño, J.B.: Speech Recognition in a Noisy Environment Based on LP of the One-Sided Autocorrelation Sequence and Robust Similarity Measuring Techniques. Speech Communication 21, 17–31 (1997)
17. Xiaolin, L., Parizeau, M., Plamondon, R.: Training Hidden Markov Models with Multiple Observations—A Combinatorial Method. IEEE Transactions on Pattern Analysis and Machine Intelligence 22(4), 371–377 (2000)
18. Hush, D.R., Horne, B.G.: Progress in supervised neural networks. IEEE Signal Processing Magazine 10(1), 8–39 (1993)

Image Sequences Noise Reduction: An Optical Flow Based Approach*

Roman Dudek, Carmelo Cuenca, and Francisca Quintana

Departamento de Informática y Sistemas,
Campus Universitario de Tafira, Las Palmas GC, Spain
roman@idecnet.com, {ccuenca,fquintana}@dis.ulpgc.es

Abstract. We present an optical flow based method for noise reduction in image sequences. To prevent artefacts caused by optical flow imperfections, we propose a method to estimate these imperfections. We use the estimation to adaptively choose either a temporal or a spatial based noise reduction algorithm to be applied in different image zones. Our results have shown that an important noise reduction can be achieved with the proposed method, without the drawbacks of the simpler methods. The method has provided important noise reductions even with complex image sequences.

1 Noise in Film

Either we use digital or chemical resources for shooting image sequences, an important amount of noise is introduced due to the physical nature of the shooting process. This noise needs to be frequently reduced to make further digital processing of the film image sequences. Although many people consider that they miss something important if noise is removed from the grainy film look of the movie theater projections, noise is a technical problem that must be reduced because of their interferences with the digital postproduction process. The noise can be easily added after the image processing, if desired so, for artistic reasons. Multiple methods were proposed and used for the noise reduction.

2 Usual Methods of Noise Reduction and Our Proposed Method

The most obvious approach to reduce image noise is by suppressing the high frequencies in the image spectrum characteristic, for example with Gaussian filtering. While this technique reduces the image noise, also some important details of the image would be removed. Adaptive methods have also been developed, aimed at distinguishing for example the image edges and do not blur across them. However, every such method has its limitations: For example, the edge

* This work has been partially supported by The Spanish Ministry of Science and Innovation under contract TIN2007-60625 and by FEDER funds.

R. Moreno-Díaz et al. (Eds.): EUROCAST 2009, LNCS 5717, pp. 366–373, 2009.

detecting method would blur any detailed textures with not enough contrast to be classified as edges. Another approach is based on averaging multiple adjacent images of an image sequence. However, this method only works well for static scenes as any moving object in the scene creates ghost-like trails.

In our work, we propose and test a noise reduction method based on averaging the image pixel values along the sequence time. As in the simple frame averaging method described above, we will use the sequence time consistency to distinguish noise from actual image. However, to extend the method's applicability to moving objects, we will use the optical flow (OF) to track each pixel's actual position in previous and following images.

3 Noise Reduction Based on Image Averaging and Estimation of the Noise Reduction

The noise of each image pixel is caused by the accumulation of different phenomena, like thermal noise and random particle distribution[1]. All these phenomena usually have a Gaussian distribution, so we can consider that the noise itself would have a Gaussian distribution. Experimental data confirms this assumption.

Let us suppose the value of a given pixel belonging to the image I_n to be the sum of an ideal signal S_n and a noise N_n with Gaussian distribution and zero mean value:

$$I_n = S_n + N_n(0, \sigma^2)$$

As previously mentioned, there is no relation between noise patterns in consecutive frames. The noise distribution for a set of pixels, each one belonging to a different frame, can be considered purely random. On the other hand, the signal value S should be constant in case of static scenes, or in cases of dynamic scenes if we account for movements using OF information, as we will show later.

In our proposed method, we calculate the average for each pixel across n frames,

$$I_{average} = \frac{\sum_{i=1}^{n} S + N_i(0, \sigma^2)}{n} = S + \frac{\sum_{i=1}^{n} N_i(0, \sigma^2)}{n}$$

We can show that while the signal value S is preserved, the dispersion of the averaged noise amplitude is reduced by a factor of $\sqrt{n}$, the number of summed values. However, as the noise energy is proportional to the square of the amplitude, the noise energy is reduced by a factor of n.

4 Optical Flow Based Noise Reduction

In our process, we first calculate the OF fields[2][3][4] for each pair of consecutive images of the sequence, for both forward and backward directions of the OF. Once the OF fields are calculated, we perform the actual image averaging: for each pixel of a given image, we track the corresponding pixel coordinates in certain number of images preceding and following the current image in the sequence[5][6]. The coordinates are calculated recursively, following the track of the pixel along the sequence.

Notice that only the (x_0, y_0) coordinates are integer values. The (x_n, y_n) pairs are typically non-integer, so interpolation is used in order to obtain the $u_n(x_n, y_n)$ and $v_n(x_n, y_n)$ values.

Then, the resulting pixel value is calculated as the average value of the tracked pixels:

$$I_{average}(x, y) = \frac{\sum_{n=-r}^{r} I(x_n, y_n)}{2r + 1} \quad (1)$$

Again, interpolation is used to obtain the image values for non-integer coordinates (x_n, y_n).

Selecting the value of r (number of frames we use for averaging) requires a compromise as increasing its value increases the grade of noise reduction but it also increases the demands on the OF precision, as inaccuracies in the fields would accumulate while tracking individual pixels along more frames. Our tests show that the best results are obtained by averaging 3 to 7 frames, i.e., $r \in [1, 3]$.

4.1 Treating Zones of Large Errors in OF

The main limitations of the method are the precision of the OF fields and the occlusions or large changes in illumination in the image sequence, which can prevent us from following a given pixel position over long sequences of images.

As a first measure to minimize these problems, during the calculation of the noise reduced current frame, we use a group of images consisting of both the previous and the following frames, with the current frame in the center of the group. This leads to a lower error accumulation than, for example, only using frames which are previous to the current one.

Even with perfect OF fields available, the time averaging method will fail in zones of occlusions. Practically, the OF obtained by an estimation method will contain certain errors, with problematic zones containing large errors. Experiments show that ignoring such errors can cause important artifacts in zones of the scene where the OF field is incorrect (or even undefined in case of occlusions or transparencies). We developed a method to detect such zones and threat them differently.

4.2 Detecting Zones of Large Errors in OF

A good detection of the OF validity is needed in order to distinguish where OF based method is applicable and where intra-frame filtering should be used instead.

We can assume that where the OF vectors got "lost", and could not correctly follow the movements in the scene, the values of pixels that we find in the neighboring frames using these OF vectors will differ. To estimate these OF errors for each result pixel, we propose to calculate the dispersion (medium square error) of the values used for the calculation of the average value of each result pixel:

$$E(x,y) = \frac{\sum_{n=-r}^{+r} (I_n(x_n, y_n) - I_{average}(x,y))^2}{2r+1} \tag{2}$$

Unfortunately, considering that the number of averaged values in the above sums is relatively low (3 to 7), the error measure $E(x, y)$ itself contains a large amount of noise.

To analyze the properties of $E(x, y)$, we can express it as a sum of the "OF Error" $E_{OF}(x, y)$, caused only by the OF imperfections, and a noise component $N_{error}(x, y)$:

$$E(x,y) = E_{OF}(x,y) + N_{error}(x,y) \tag{3}$$

The amplitude of $N_{error}(x, y)$ can be derived from the amplitude of the noise in the source images. Let the individual source image pixels be considered the sum of the "Signal" and a "Noise":

$$I_n(x,y) = S_n(x,y) + N_n(x,y) \tag{4}$$

Let us suppose that the OF vectors were perfect, so the $S_n(x, y)$ are equal for each n from $[-r, +r]$, and $E_{OF}(x, y) = 0$. Then,

$$N_{error}(x,y) = E(x,y) = \frac{\sum_{n=-r}^{r} (I_n(x_n, y_n) - I_{average}(x,y))^2}{2r+1} \tag{5}$$

The value $N_{error}(x, y)$ would converge to a certain constant c for large values of r, with a high number of averaged frames. This constant could be used as a threshold to decide the validity of the OF fields for a certain pixel: If the error $E(x, y)$ is similar to c, the pixel was likely correctly followed. If the $E_{OF}(x, y)$ is much higher than c, the OF is likely invalid for this pixel, and time averaging should not be used. We use simple spatial Gaussian filtering of the image for the given pixel instead:

$$I_{result} = \begin{cases} (g \circ I_0)(x,y)) \text{ for} E(x,y) > c \\ I_{average}(x,y) \text{ for} E(x,y) \leq c \end{cases} \tag{6}$$

However, as we need to use only low values of r ($r \in [1, 3]$), the $E(x, y)$ values do not converge enough to the actual noise level of the sequence. There is an important noise component in our $E(x, y)$ itself, making impractical such direct decision per pixel. In practice, the random dispersion of the total error measure $E(x, y)$ is frequently larger than the $E_{OF}(x, y)$ we are actually trying to detect, so neighboring pixels would be frequently randomly misclassified due to this noise component.

To make a better classification, we need to reduce the randomness in the $E(x, y)$ error measure field. We propose to carry out a spatial averaging of the $E(x, y)$ field. This spatial averaging, added to the temporal averaging used to create the $E(x, y)$ itself, can largely reduce the randomness of the $E(x, y)$ error measure, while not affecting much the $E_{OF}(x, y)$ component that we are trying to detect, supposing that it is locally smooth anyway. We used Gaussian filter for this averaging operation, to create a filtered error measure $E'(x, y)$

$$E'(x, y) = (g \circ E)(x, y)) \tag{7}$$

The modified algorithm to obtain the final result will use $E'(x, y)$ instead of $E(x, y)$:

$$I_{result} = \begin{cases} (g \circ I_0)(x, y)) \text{ for} E'(x, y) > c \\ I_{average}(x, y) \text{ for} E'(x, y) \leq c \end{cases} \tag{8}$$

Using $E'(x, y)$, our threshold classification method provides a much better detection of the problematic zones. However, the application of a thereshold classification causes some visible artifacts along the edges of the zones where the decision changes. In order to reduce such artifacts, we propose to create a thin transition zone using a clamped blending equation instead of a thresholding:

$$p = clamp((E'(x, y) - c) * s) \tag{9}$$

where s is a user defined constant and $clamp()$ is a function limiting p to the range $[0, 1]$. Then, the final result is obtained by using a blending equation instead of using a threshold:

$$I_{result}(x, y) = p * (g \circ I_0)(x, y)) + (1 - p) * I_{average}(x, y) \tag{10}$$

In our test application, the constant s is a user defined value adjusted in such a way that a transition zone of only a few pixels wide will be created around OF error zones. Similarly, c is adjusted by the user in order to correctly detect the zones of OF errors, while ignoring errors too small to cause visible artifacts in the result. The adjustment of other parameters, like the Gaussian filtering radius, depend on the resolution and noise level of the used image sequence. However, once these values are adjusted, they seem to be constant for all shots from the same original negative film roll, so only few adjustments need to be done for each film project.

5 Results

For our tests we used a variety of digitalized image sequences, originally shot on negative film. We obtained important noise reductions without observably suppressing any detail in the scene.

Figure 1 shows a sample image from the original image sequence. This image is a part of a sequence filmed intentionally on $8mm$ celluloid film, to obtain obvious film look even when broadcast over standard PAL television. We have chosen this material in order to to make observable the necessary details in a printed form of this document. We can observe important noise caused by film grain, specially in the flat background behind the musician.

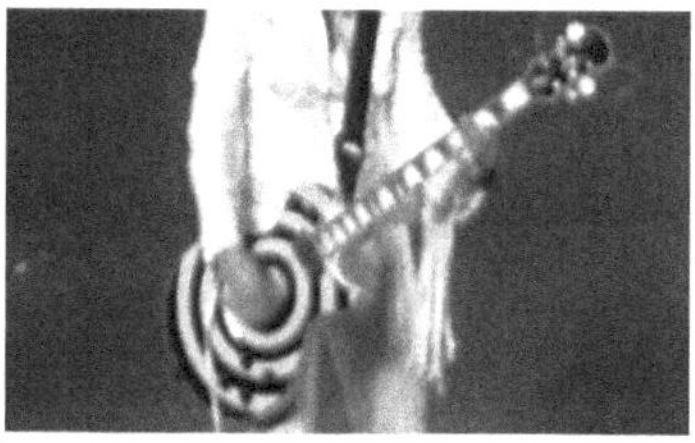

Fig. 1. Sample image from the original sequence

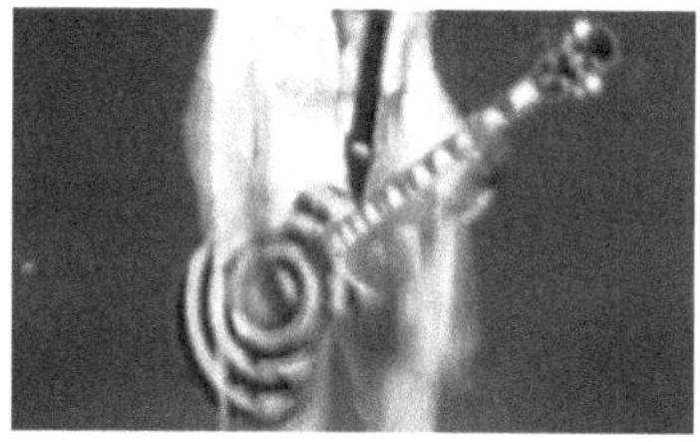

Fig. 2. Sample image after simple frame averaging

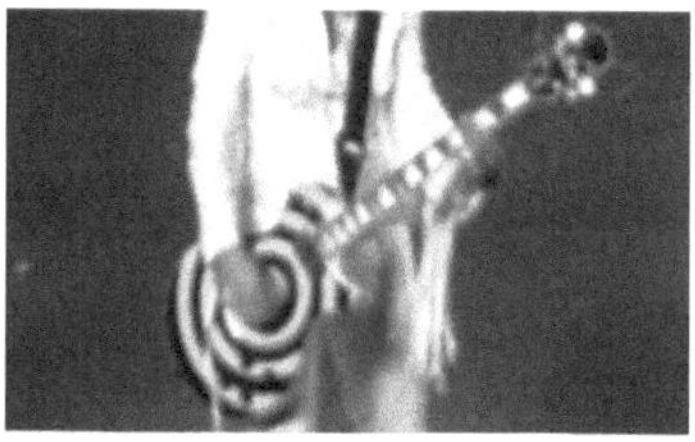

Fig. 3. Sample image after applying a simple Gaussian filter

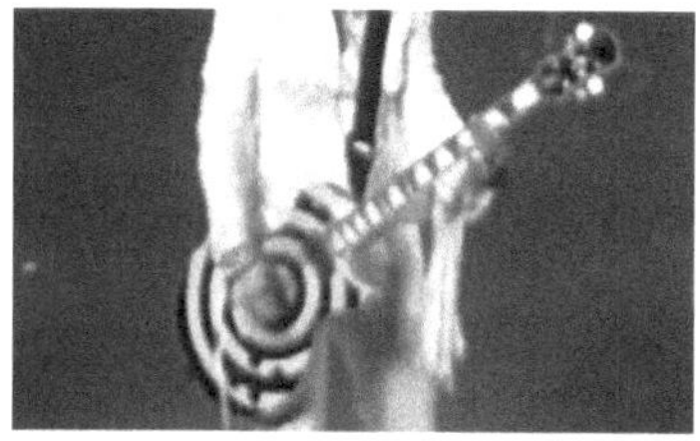

Fig. 4. Sample image after averaging using OF based, with pixel following over 5 frames

In Figure 2 we can see the result obtained after a simple frame averaging of five frames. We can see that the image detail is lost due to the motion trails created around any object in movement. These trails can be clearly observed on the right elbow of the musician. The noise level is generally reduced, as can be obviously observed in the flat background, for example. However image stays sharp only in the few parts where there was no movement in the range of the five frames used.

After applying a simple Gaussian filtering to the image we obtain the Figure 3. The Gaussian filter radius was set just large enough to provide a visually similar noise reduction as it would be obtained by averaging five frames of the sequence. We can see that there are no trails around the shoulder as in Figure 2, but edges are visibly blurred. Compare the black shoulder belt edges, for example. Some image detail is visibly lost.

Figure 4 shows the frame averaging of 5 frames using OF based tracking of the pixel positions. We can look at that the noise is reduced. Neither there are no trails of simple frame averaging, nor there is the detail loss of the spatial Gaussian filtering. However, artifacts can be found in some zones. For example, see the hand hitting the strings of the guitar: some black strips, not present in the original image, can be observed over the hand. This type of artefact is characteristic for the zones where the OF filed calculation failed due to some reason, because of a large advancing occlusion in this case.

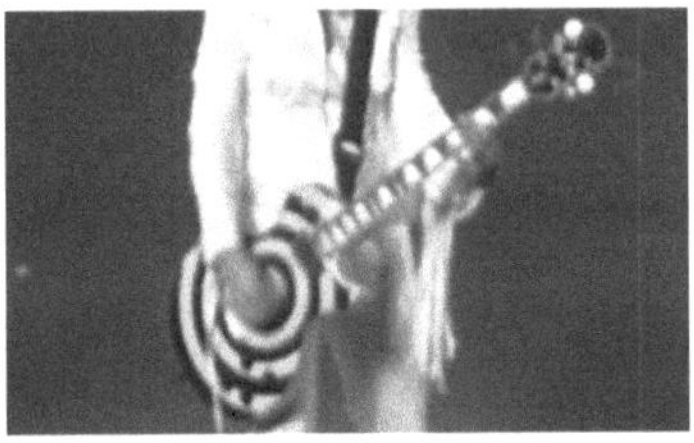

Fig. 5. Sample image using our adaptive algorithm. A simple, hard threshold decision is used in this case.

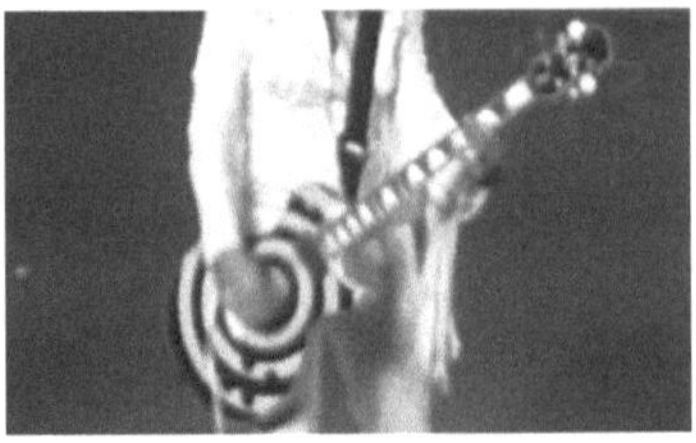

Fig. 6. Sample image using our adaptive algorithm. Improved threshold is used.

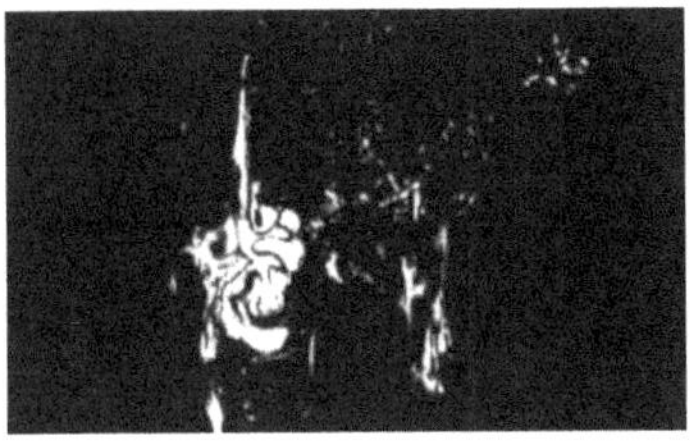

Fig. 7. Mask image resulting from a simple threshold applied to detect errors in OF

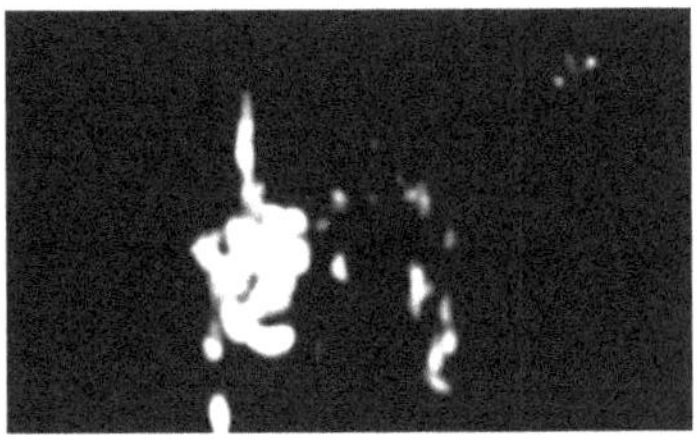

Fig. 8. Mask image resulting from the improved threshold, using a filtered error measure and smooth transitions

Figure 5 is the result of our adaptive algorithm based on the detection of large errors in the OF. The OF corrected frame averaging is used where errors are small. Gaussian filtering is used where OF errors are large. Virtually no artefacts can be observed here, however the hard threshold decision based on unfiltered error measure can be observable in the form of visible spots in some scenes. In this case, a spot is hardly observable on the fingers of the hand hitting the strings.

In Figure 6 improved threshold is used, with filtered error measure and smooth transition between the zones. It is very difficult to find any visible artefacts anymore, anywhere in the image. Some parts of the image, like the hand hitting the strings, may look blurry. However, comparing with the unfiltered original image, and you will find that the hand was blurry in the original too, due to non-zero exposure time of the frame. The resulting image preserves, even makes clearer, any detail present in the original sequence, while reducing the noise in the same time.

Figure 7 shows the mask resulting from a simple threshold applied to detect errors in OF. White pixels represent the zones of large error, where Gaussian filtering image will be used. Black pixels represent the zones of apparently correct OF, where the preferable OF corrected time averaging will be used. Observe the zone of the hand striking the strings: While almost the whole zone is white, detected as containing large errors, there are some unexpected, spurious strips of black pixels. These false negatives (zones of error detected as correct) cause the spots described above.

Figure 8 shows the mask resulting from improved threshold method, using filtered error measure and smooth transition. We can see that the spurious black strips in the zone of the moving hand were eliminated due to the error measure filtering. Softening the outlines, using a soft threshold instead of a hard one, further reduces the visibility of any spot artefacts left.

6 Conclusions and Future Work

In this work we have presented an adaptive optical flow based noise reduction method, combining the advantages of both spatial and temporal filtering methods. We proposed and tested a robust tresholding method to detect flaws in the OF, an choose the prefferable method for each image zone, with good transitioning on the borders between the zones. We have shown that an important noise reduction can be achieved with the proposed method, without the drawbacks of the simpler methods. While the method requires some manual parameter adjusting, virtually in all practical test cases we were able to achieve visually artefact-less noise reduction.

In our future work, we will try to detect different cases of problematic zones, and test alternative methods for improving the results in such zones. Occlusions could be detected and handled explicitly in order to stop tracking a pixel in images where that pixel is not present anymore, while still using lower number of samples for averaging. This could be combined with a smaller amount of spatial filtering to complement the noise reduction already achieved by partial frame averaging, instead of doing a complete fallback to spatial filtering.

Also the natural blurring in zones of movement is anamorphic: There is a loss of resolution in direction of the movement, but not in the direction perpendicular to it. As a consequence, using an anamorphic filtering instead of symmetric Gaussian filter would likely improve the results.

References

1. Minelly, S., Curley, A., Giaccone, P., Jones, G.A.: Reducing chromatic grain noise in film sequences (1998)
2. Álvarez, L., Weickert, J., Sánchez, J.: Reliable Estimation of Dense Optical Flow Fields with Large Displacements. International Journal of Computer Vision 39(1), 41–56 (2000)
3. Beauchemin, S.S., Barron, J.L.: The Computation of Optical Flow. ACM Computing Surveys 27(3), 433–467 (1995)
4. Horn, B., Schunk, B.: Determining Optical Flow, AI Memo 572. Massachusetts Institue of Technology (1980)
5. Jiang, Z., Wong, T.-T., Bao, H.: Practical Super-Resolution from Dynamic Video Sequenes. In: Proceedings of IEEE Computer Vision and Pattern Recognition 2003 (CVPR 2003), Madison, Wisconsin, USA, June 16-22 (2003)
6. Borman, S., Stevenson, R.L.: Spatial Resolution Enhancement of Low-Resolution Image Sequences: A Comprehensive Review with Directions for Future Research. Department of Electrical Engineering, University of Notre Dame (1998)

From Industrial to Ubiqitous Robots

Peter Kopacek

Intelligent Handling and Robotics – IHRT
Vienna University of Technology, Austria
Favoritenstr. 9-11/325 A6, A-1040 Vienna, Austria
kopacek@ihrt.tuwien.ac.at

Abstract. Robotics is a very fast growing field especially in the last years. In the late seventies the first industrial applications of stationary unintelligent industrial robots were realised. Begin of the 90`s a new generation of mobile, intelligent, cooperative robots grows up. This new generation opens new applications areas like in construction, in agriculture, in the food industry, in the household, for medical and rehabilitation applications, in the entertainment industry as well as for leisure and hobby. Current developing trends are humanoid robots and robots supporting humans in every day life. In the future probably ubiquitous robots will support us.

Keywords: Industrial Robots, AGV`s, Humanoid Robots, Ubiquitous Robots.

1 Introduction

Industrial robots have been widely applied in many fields to increase productivity and flexibility and to help workers from physically heavy and dangerous tasks.

Definition according to ISO 8373: A manipulating industrial robot is an automatically controlled, reprogrammable, multipurpose manipulator programmable in three or more axes which may be either fixed in place or mobile for use in industrial automation applications.

From similar aspects the need on robots in service sectors - like robots in hospitals, in households, in amusement parks - is rapidly increasing.

Definition: A service robot is a robot which operates semi- or fully autonomously to perform services useful to well- being of the humans and equipment, excluding manufacturing operations.

Cheap and accurate sensors with a high reliability are the basis for „intelligent“ robots. These intelligent robots can be used for conventional as well as complex applications. Furthermore new applications not only in industry are possible.

There are three "starting" points for the development of intelligent robots (Fig.1): Conventional, stationary industrial robots; mobile, unintelligent platforms (AGV`s) and Walking mechanisms [1].

Stationary industrial robots are equipped with external sensors for "intelligent" operations e.g assembly and disassembly, fuelling cars… and are "intelligent" robots.

R. Moreno-Díaz et al. (Eds.): EUROCAST 2009, LNCS 5717, pp. 374–382, 2009.

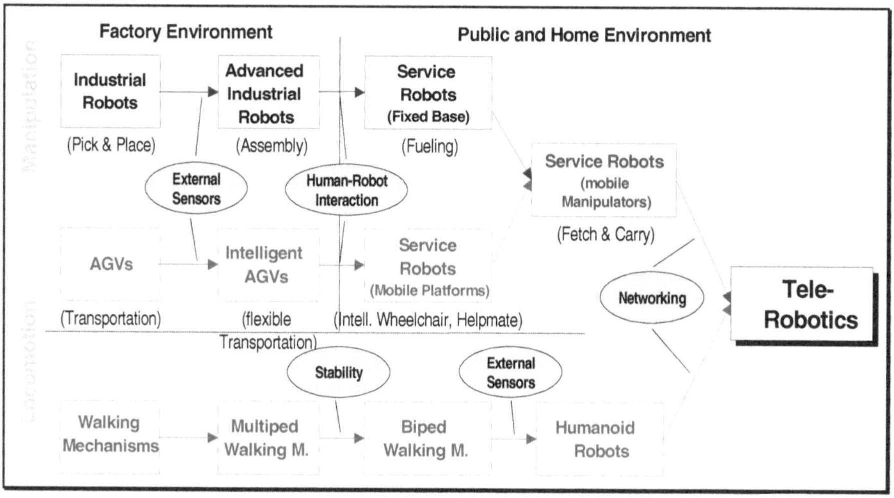

Fig. 1. From Industrial to Service Robots [2]

Partially intelligent mobile platforms "Autonomous Guided Vehicles – AGV`s" are available since some years and are introduced in industry. Equipped with additional external sensors (Intelligent Autonomous Guided Vehicles – Intelligent AGV`s) are currently slowly introduced in industry and cover a broad application field.

Walking machines or mechanisms are well known since some decades. Usually they have 4 to 6 legs (multiped) and only in some cases 2 legs (biped). Walking on two legs is from the view point of control engineering a very complex (nonlinear) stability problem. Biped walking machines equipped with external sensors are the basis for "humanoid" robots. Some prototypes of such robots are available today.

In addition these intelligent robots – especially mobile platforms and humanoid robots - are able to work together on a common task in a cooperative way. The goal is so called "Multi Agent Systems – MAS". A MAS consists of a distinct number of robots (agents), equipped with different devices e.g. arms, lifts, tools, gripping devices ... and a host computer. A MAS has to carry out a whole task e.g. assemble a car. The host computer divides the whole task in a number of subtasks (e.g. assembling of wheels, windows, brakes ...) as long as all this subtasks can be carried out by at least one agent. The agents will fulfil their subtasks in cooperative way until the whole task is solved.

One of the newest application areas of service robots is the field of entertainment, leisure and hobby because people have more and more free time. In addition modern information technologies lead to loneliness of the humans (tele-working, tele-banking, tele-shopping, and others). Therefore service robots will become a real "partner" of humans in the nearest future. One dream of the scientists is the "personal" robot. In 5, 10 or 15 years everybody should have at least one of such a robot because the term personal robot is derived from personal computer and the price should be equal.

2 Development Trends

Fig.2 shows possible development trends in robotics. We are now on the way from unintelligent industrial robots via intelligent industrial robots to intelligent mobile – including humanoid – robots to the third generation "advanced" robots able to interact and work symbiotically with us.

21th century robots will be used in all areas of modern life. The major challenges are:

- To develop robotic systems that can sense and interact useful with the humans.
- To design robotic systems able to perform complex tasks with a high degree of autonomy.

In the same way as mobile phones and laptops have changed our daily habit, robots are poised to become a part of our everyday life. The robot systems of the next decades will be human assistants, helping people do what they want to do in a natural ad intuitive manner. These assistants will include: Robot co-workers in the workplace; robot assistants for service professionals; robot companions in the home; robot servants and playmates; robot agents for security and space.

The role of these robots of the future could improved by embedding them into emerging IT environments characterised by a growing spread of ubiquitous computing and communications and of ad-hoc networks of sensors forming what has been termed "ambient intelligence".

Current available robots are fare away from this vision of the 3rd generation being able to understand their environments, their goals and their own capabilities or to learn from own experiences.

As the number of humanoids increases, the collective population of humanoids will learn, develop and perhaps eventually reproduce themselves more effectively. Unlike cars or televisions that improve along a linear, highly controlled trajectory, humanoids

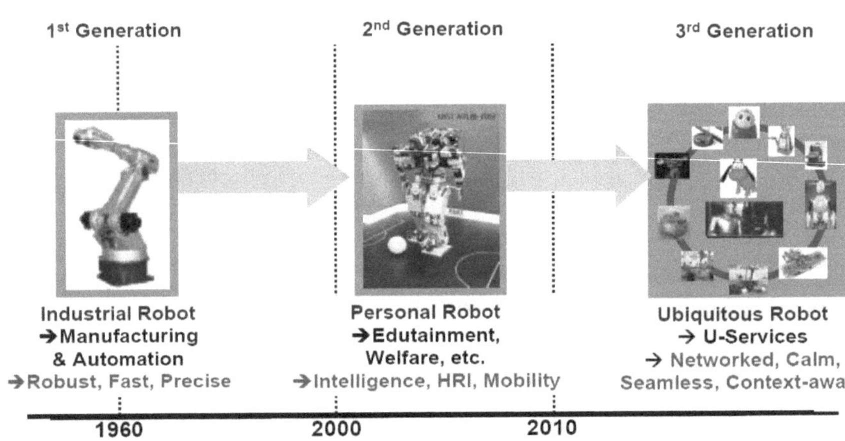

Fig. 2. Development trends in robotics [3]

will be the ultimate in self-accelerating technology. Likewise, robotics is a self-enabling technology. Robotic tools will make the humanoids we ourselves could never make. Once we have a large population of self-motivated agents attending to separate tasks, these agents will negotiate, exchanging tasks and resources in mutually beneficial ways. Humanoids will comprise a new distributed infrastructure not only of information, but real-world action. As a given task arises, humanoids will place bids, often partnering with other humanoids to get the job done. Humanoids will not only share workload and resources, but will also evolve by passing host-independent, modular code.

As robots become more pervasive, they will, like automobiles, become increasingly complex. Already, some robots are comprised of millions of parts. Those skeptical of humanoid research often point to the high price tags of today's humanoids. If fast, cheap, rapid manufacture of robots is to occur, it will be necessary to remove humans from the design and manufacturing process. Through mutation and recombination, the genetic algorithm might modify bar length, split bars, or connect neurons to various components as it propels generations of increasingly fit robots. Finally, the robots are fabricated automatically by a machine that prints the robots, layer by layer, out of plastic.

In an ubiquitous era we [3] will be living in a world where all objects such as electronic appliances are networked to each other and a robot will provide us with various services by any device through any network, at any place anytime. This robot is defined as a ubiquitous robot, Ubibot, which incorporates three forms of robots: software robot (Sobot), embedded robot (Embot) and mobile robot (Mobot). The Ubibot is following the paradigm shift of computer technology. The paradigm shift of robotics is motivated by ubiquitous computing and the evolution of computer technology in terms of the relationship between the technology and humans.

The basic concepts of ubiquitous computing include the characteristics, such as every device should be networked; user interfaces should operate calmly and seamlessly; computers should be accessible at anytime and at any place; and ubiquitous devices should provide services suitable to the specific situation. Computer technology has been evolving from the mainframe era, where a large elaborate computer system was shared by many terminals, through the personal computer era, where a human uses a computer as a stand-alone or networked system, in a work or home environment, to the ubiquitous computing era, where a human uses various networked computers simultaneously, which pervade their environment unobtrusively.

3 Examples of "Advanced" Robots

In the following some realised examples for this new robot generation are shortly described. Special emphasis is on the new headline: Cost Oriented Automation (COA).

3.1 Robots for Landmine Detection

An example for a MAS is robot swarms for landmine detection, removal and destroying [4]. According to current estimates, more than 100.000.000 anti-personnel and other landmines have been laid in different parts of the world. A similar number exists in stockpiles and it is estimated that about two million new ones are being laid each year. According to recent estimates, mines and other unexploded ordnance are killing between 500 and 800 people, and maiming 2.000 others per month.

Landmines are usually very simple devices which are readily manufactured anywhere. There are two basic types of mines:

- anti-vehicle or anti-tank (AT) mines and
- anti-personnel (AP) mines.

AT mines are comparatively large (0.8 – 4 kg explosive), usually laid in unsealed roads or potholes, and detonate which a vehicle drives over one. They are typically activated by force (>100 kg), magnetic influence or remote control. AP mines are much smaller (80-250g explosive, 7-15cm diameter) and are usually activated by force (3-20kg) or tripwires. There are approximately 800 different types with different designs and actuation mechanisms.

Currently demining is carried mostly by human deminers. Because this is very dangerous we developed a prototype of a demining robot [4]. It consists of a platform and a metal detection sensor. This robot is equipped with an internal micro controller as well as internal sonar sensors, position speed encoders and a battery pack for network-independent and autonomous operation. An addressable I/O bus allows the installation of 16 additional sensors or devices like grippers. Furthermore, two RS-232 serial ports, five A/D ports and PSU controllers are accessible via server software. With appropriate software a tele-operation could also be achieved.

The robots base-weight is about 9kg with an ability to carry 30kg. Overall-dimensions of the basic robot setup are about (length/width/height) 55x50x50cm. With the mounted mine detector search head and telescopic pole the length increases up to 120cm.

A commercially available mine detecting set – produced in Austria - is attached on the robot basic-platform. This device is intended to detect land mines with a very small metal content (1.5g) 10cm below the surface of the ground and in fresh or salt water. The overall weight of the mounted sensor components is about 2.5kg.

When an object is detected a tone is released with its intensity and pitch depending on size, shape, depth under ground level and metal content of the object. For very tiny metal objects the tone is higher near the inner ring of the search head than in the middle. When searching for large metal objects, the continuous tone automatically changes to a pulsed tone whereas the pulse rate of the tone will be highest when the search head is immediately above the object. Outdoor tests with this robot were carried out. All functions could be validated only high grass could influence the sonar-sensors of the robot.

This prototype of a six-wheeled robot (HUMI – Robot for Humanitarian Demining) based on the Ackermann Geometry for movement in rough terrain is shown in Fig. 3.

The ultimate target to be reached would be a robot that possesses faculties approaching that of human beings - autonomous robot agents. Leaving such an ideal robot as a goal for the future, intermediate robots that only satisfy a limited selection of the most requisite functions should still find good use in human society. Among the faculties cited above, mobility is the most indispensable feature for a service robot.

Fig. 3. HUMI at Outdoor Tests

3.2 Roby Space

The concept of solar power from the space (SPS) was proposed in 1968. The basic idea of this concept is the generation of emission-free solar energy by means of solar cells from outer space and the transmission of energy to the earth using microwave or laser beam. Because of high launch cost the structure - consisting of solar cells as well as microwave transmitters - should be light weight. Instead of the conventional rigid structures a new concept (Furoshiki Concept) of a large membrane or a mesh structure was proposed. Next step to be realized is the transport of solar panels and microwave transmitters on this mesh structure [5].

The main purpose of this project was the development of mobile mini robots that place solar cells and transmitters on the net structure to build a solar power plant based on the Furoshiki net concept. A sounding rocket launches four satellites (one mother satellite and three daughter satellites), robots, net, solar panel and the microwave transmitters in the orbit. Approximately 60 seconds after launch the rocket reaches an altitude of 60 km. The mother satellite and three daughter-satellites build the Furoshiki net. Robots transport solar cells and microwave transmitters on the net structure (Fig.4). In the frame work of the project a feasibility study was done to verify the performance of the Furoshiki net as well as the crawling robots.

The requirements on the robot are the limited maximum size (10 x 10 x 5 cm), a simple mechanical construction, miniaturized electronics, robustness, "low cost", and independence of the mesh's dimension (from 3 x 3cm to 5 x 5cm). The weight of the robot plays an important role. Even the launching cost per kilogram is very high. Another point to be considered is that in case the robot is too heavy, the satellite can not produce enough net tension. For a free movement the moving and holding mechanism of the robot should be well designed. Other difficulties are the vibration and shock during launching of the rocket. The robot should pass the vibration and shock tests up to 40 g. Last but not least the working environment of the robot is in outer space – 200 km over the earth. The high/low temperature, the radiation as well as the vacuum and others should be considered in the design phase.

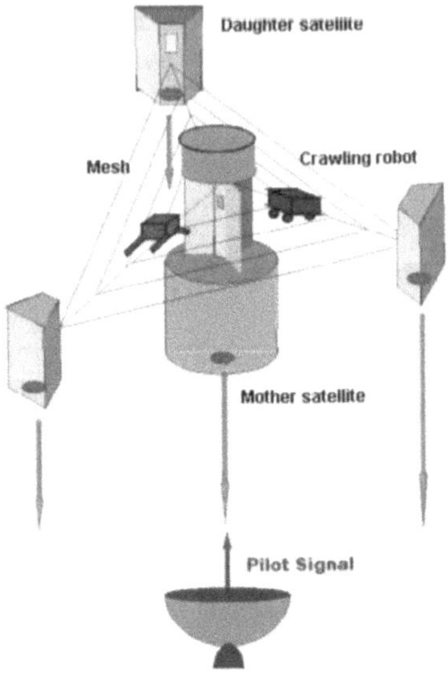

Fig. 4. Solar power plant by means of Furoshiki Satellite [5]

The robot (Fig.5) consists of two parts – the upper part has two active driven belts, the lower one two passive driven belts [6]. Magnetic forces push them together. The special surface between the parts prevents the lower part from moving away.

The advantage of this construction is the very low friction between mesh and robot during operation. There are no high sliding forces between the passive and the active driven belts of the upper and lower part as well as the mesh.

Fig. 5. Robot fixed on the mesh ready for action

The robots passed following tests:

- Microgravity tests during parabolic flights in January and March 2005 in Japan
- Vibration and shock tests in May 2005 at the ESA Mechanical Systems Laboratory, The Netherlands
- Mechanical verification tests, June 2005 in Japan

After the tests and technical updates of the systems a sounding rocket S310 with two RobySpace-Junior were launched on January 22, 2006 at the Uchinoura Space Center, Japan. One of two robots worked well. It crawled on the net with small resistance – the robot had a constant high voltage level and moved with constant velocity for more than 30 seconds. For the other robot we have to wait for the results from the telemetry data.

Three cameras in the satellite delivered the video signals. The video which sent the satellite showed one of robots which moved on the net. According to European Space Agency (ESA) and Japan Aerospace Exploration Agency (JAXA) the experiment completed successfully.

3.3 "Archie" – A Humanoid Robot

A "cost oriented" two legged robot called ARCHIE is currently in development in Austria [7]. The goal is to create a humanoid robot, which can act like a human. This robot should be able to support humans in everyday life; at the working place, in household and for leisure and hobby.

Therefore Archie has a head, a torso, two arms, two hands and two legs and will have the following features:

1. Height: 120 cm
2. Weight: less than 40kg
3. Operation time: minimum 2hrs
4. Walking speed: minimum 1m/s
5. Degrees of freedom: minimum 24
6. "On board" intelligence
7. Hands with three fingers (one fixed, two with three DOFs)
8. Capable to cooperate with other robots to form a humanoid Multi Agent System (MAS) or a "Robot Swarm".
9. Reasonable low selling price – using commercially available standard components.

Archie will be equipped with sensors for measuring distances and to create primitive maps, for temperature, acceleration, pressure and force for feeling and social behaviour, two CMOS-camera-modules for stereoscopic looking, two small microphones for stereoscopic hearing and one loud speaker to communicate with humans in natural language.

The control system is realised by a network of processing nodes (distributed system), each consisting of relative simple and cheap microcontrollers with the necessary interface elements. According to the currently available technologies the main CPU is for example a PGA module, one processor for image processing and audio control and one microcontroller for each structural component.

The upper part is currently in the final test phase.

4 Summary

Robotics is currently a very fast growing field not only in science and industrial application. In the last time more and more mass medias (TV, broadcast, journals, newspapers) are interested in this field because a broader public is in favour to get familiar with these new " intelligent machines". It is a first step for the realisation of the old dream of humans to have a robot available looking like a human. In the nearest future such robots or the next generation – ubiquitous robots - will be available for a reasonable price.

In this contribution some examples of currently available robots under the headline COA (Cost Oriented Automation) were presented. It is possible that small research teams are able to develop such robots in a reasonable time.

Acknowledgements. These projects were supported by the "European Space agency – Advanced Concept team" under contract ESTEC/Contract No.18178/04/NL/MV-Furoshiki Net Mobility Concept, by the Austrian Space Applications Programme (ASAP) under contract "Roby Space"and by the Austrian Ministry for " Transportation, Innovation and Technology – BMVIT " under contract BMVIT, GZ 603.034/0045- III/15/2007.

References

1. Coiffet, P.: New Role of Robotics in the Next Century. In: 7th Intl.Workshop on Robotics in Alpe-Adria-Danube-Region RAAD 1998, Smolenice Castle, Slovakia, June 1998, pp. 261–266 (1998)
2. Kopacek, P.: Advances in Robotics. In: Moreno Díaz, R., Pichler, F., Quesada Arencibia, A. (eds.) EUROCAST 2005. LNCS, vol. 3643, pp. 549–558. Springer, Heidelberg (2005)
3. Kim, J.W.: Humanoid Robots. In: Course Material for the Summer School, KAIST, Taejon, Korea (2006)
4. Silberbauer, L.: A new 6 wheeled robot for humanitarian demining. In: EURON/IARP International Workshop on Robotics for Risky Interventions and Surveillance of the Environment, Spain (2008)
5. Kaya, N., Iwashita, M., Nakasuka, S., Summerer, L., Mankins, J.: Rocket Experiment on Construction of Huge Transmitting Antenna for the SPS using Furoshiki Satellite System with Robots. In: 4th International Conference on Solar Power from Space SPS 2004 Together with The 5th International Conference on Wireless Power Transmission WPT 5, Granada, Spain, pp. 231–236 (2004)
6. Kopacek, P., Putz, B., Han, M.W.: Development of Mobile Mini Robots for Space Application. In: 37th International Symposium on Robotics ISR 2006 and 4th German Conference on Robotics ROBOTIK 2006, p. 209. VDI Publishing Company (2006)
7. Baltes, J., Byagowi, A., Anderson, J., Kopacek, P.: A Teen Sized Humanoid Robot – Archie. Will be published

WiFi Localization System Using Fuzzy Rule-Based Classification

José M. Alonso[1], Manuel Ocaña[2], Miguel A. Sotelo[2], Luis M. Bergasa[2], and Luis Magdalena[1]

[1] European Centre for Soft Computing, Mieres (Asturias), Spain
{jose.alonso,luis.magdalena}@softcomputing.es
[2] Department of Electronics, University of Alcalá (Madrid), Spain
{mocana,sotelo,bergasa}@depeca.uah.es

Abstract. The framework of this paper is robot localization inside buildings using WiFi signal strength measure. This localization is usually made up of two phases: training and estimation stages. In the former the WiFi signal strength of all visible Access Points (APs) are collected and stored in a database or Wifi map, while in the latter the signal strengths received from all APs at a certain position are compared with the WiFi map to estimate the robot location. This work proposes the use of Fuzzy Rule-based Classification in order to obtain the robot position during the estimation stage, after a short training stage where only a few significant WiFi measures are needed. As a result, the proposed method is easily adaptable to new environments where triangulation algorithms can not be applied since the AP physical location is unknown. It has been tested in a real environment using our own robotic platform. Experimental results are better than those achieved by other classical methods.

1 Introduction

WiFi localization systems take advantage of the boom in wireless networks over the last few years. The wireless networks have become a critical component of the networking infrastructure and are available in most corporate environments (universities, airports, train stations, tribunals, hospitals, etc), and in many commercial buildings (cafes, restaurants, cinemas, shopping centres, etc).

In the literature, we can find multiples systems proposed and successfully deployed to find the pose (position and orientation) of a robot from its physical sensors. These systems are based on: infrared sensors [1], computer vision [2], ultrasonic sensors [3], laser [4] or radio frequency (RF) [5] [6]. Within the last group we can find localization systems that use WiFi signal strength measure.

These WiFi systems are attractive for indoor environments where traditional techniques, such as Global Positioning System (GPS) [7], fail. One of the main advantages of these systems is that they do not need to add any extra hardware in the environment. They use the signal strength measure of the wireless communication network established by the WiFi.

The signal strength depends on the distance and obstacles between APs and the robot. Moreover, the system needs more than one base stations or AP to

R. Moreno-Díaz et al. (Eds.): EUROCAST 2009, LNCS 5717, pp. 383–390, 2009.

measure the distance from them to the device. In [8] they use these measures to apply a triangulation algorithm to infer the estimated position.

Unfortunately, in indoor environments, the WiFi channel is very noisy and the RF signal can suffer from reflection, diffraction and multipath effect, which makes the signal strength a complex function of distance [5]. To solve this problem, it can be used a priori WiFi map, which represents the signal strength of each AP at certain points in the area of interest [9] [10] [11] [12].

These systems work in two phases: training and estimation of the position. During the first phase, a WiFi map is built while in the estimation phase, the vector of samples received from each access point is compared with the WiFi map and the "nearest" match is returned as the estimated robot location.

Fuzzy Logic (FL) introduced by Zadeh [13] is acknowledged for both its well-known ability for linguistic concept modeling and its use in system identification. The semantic expressivity of fuzzy logic, using linguistic variables [14] and linguistic rules [15], is quite close to expert natural language. In addition, being universal approximators [16], fuzzy inference systems (FIS) are able to perform non-linear mappings between inputs and output. FL is especially useful to handle problems where the available information is vague. This is the typical situation regarding WiFi localization where measures normally yield incomplete or distorted data.

In this paper we use Fuzzy Classification in the estimation stage to obtain the estimated robot position. Such classification obtains several benefits over the classical methods. The most significant advantages are: (1) The robustness of the built systems which are able to deal with the intrinsic uncertainty of indoor environments; and (2) the adaptability to new environments where AP location is indeterminate.

The rest of the paper is organized as follows: Section 2 provides a description of the proposed Fuzzy Classification system. Section 3 shows the implementation and some experimental results, as well as a description of the used test bed. Finally, the conclusions and future work are described in Section 4.

2 Description of the Fuzzy Classification System

In this section we provide a brief description of the Fuzzy Rule-based Classification system. It was designed and built using Knowledge Base Configuration Tool (KBCT) [17] a free software tool which implements the Highly Interpretable Linguistic Knowledge (HILK) methodology [18]. This new methodology focuses on building interpretable fuzzy classifiers, i.e., classifiers easily understandable by human beings. Applying machine learning techniques it is able to extract useful pieces of knowledge from data sets. In addition, knowledge automatically extracted form data is represented by means of linguistic variables and rules under the fuzzy logic formalism. Rules are of form **If** *condition* **Then** *conclusion*, where both condition and conclusion use linguistic terms. For instance, **If** *Signal received from APi is High* **and** *Signal received from APj is Low* **Then** *The robot is close to Position k*. The semantic expressivity of fuzzy logic makes easier the knowledge extraction and representation phase. In addition, it lets us combine under the same formalism knowledge extracted from data and knowledge described by an expert[1] in natural language.

In classical logic only two crisp values are admissible (0/1, false/true, negative/positive, etc). This is a strong limitation in order to deal with real-world complex problems where there are many important details which are usually vague. In the real world things are not so simple as black and white but there is a continuous scale of grays. To cope with this problem FL is a useful tool. Working with FL everything has a membership degree. For instance, the same person can be considered more or less tall or short depending on the context: 1.80 is tall in Spain but it is not so tall in Sweden.

This information is represented by membership functions like the ones in Figure 1. As it can be seen the same value xi is partially *Low* (0.22) and *Medium* (0.78), but the addition of both membership degrees equals one. This kind of partitions is called strong fuzzy partitions (SFPs) [19] and they are the best ones from an interpretability point of view. By default we use SFPs of seven linguistic terms.

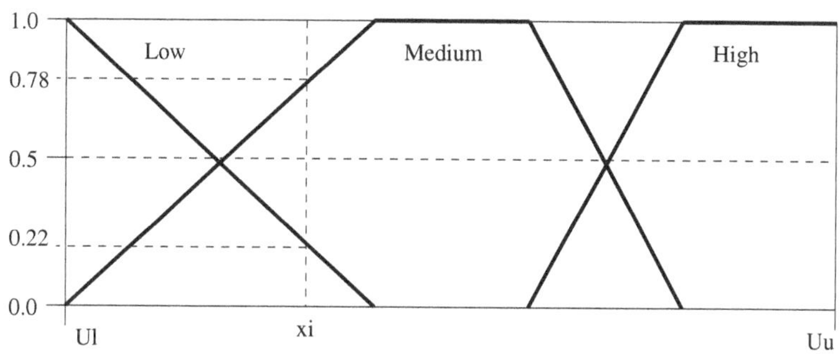

Fig. 1. A strong fuzzy partition with three linguistic terms

Once the fuzzy partitions of the input variables are defined they are used in linguistic rules of form:

$$\textbf{If} \quad \underbrace{\underbrace{X_a \text{ is } A_a^i}_{Partial\ Premise\ P_a} \quad \textbf{AND} \ldots \textbf{AND} \quad \underbrace{X_z \text{ is } A_z^j}_{Partial\ Premise\ P_z}}_{Premise} \quad \textbf{Then} \quad \underbrace{Y \text{ is } C^n}_{Conclusion}$$

On the one hand, rule premises are made up of tuples (*input variable, linguistic term*) where X_a is the name of the input variable a, while A_a^i represents the label i of such variable. Notice that the absence of an input variable in a rule means that the variable is not considered in the evaluation of that rule. On the other

[1] An expert is a person who has a deep knowledge about the problem under study. It is usually an expert domain but not a fuzzy expert.

hand, C^n is one of the possible output classes, i.e., one position in the case of WiFi localization.

Regarding the rule generation from data, there are lots of methods in the fuzzy literature [20]. However, keeping in mind the interpretability goal we have chosen Fuzzy Decision Tree (FDT) [21], a fuzzy version of the popular decision trees defined by Quinlan [22]. Notice that our implementation of FDT is able to build quite general rules with the interpretable partitions previously defined.

Then, a simplification procedure is carried out on the whole fuzzy knowledge base with the aim of removing redundancies and getting still more compact and understandable partitions and rules.

Finally, the output of the fuzzy classifier will be one position along with an activation degree computed as the result of a fuzzy inference that takes into account all defined inputs and rules[2]. Such activation degree can be understood as a degree of confidence on the system output. Notice that several output classes can be activated since several fuzzy rules can be fired at the same time by the same input vector. The activation degrees of the different classes can be used in order to make an interpolation among several positions. For instance, if the system output says that the robot is in position A with degree 0.2 and in position B with degree 0.8, it can be concluded that it is located someway between A and B but closer to B.

3 Implementation and Results

The robot used in the experimentation is called Sancho3. It is shown in Figure 2 and it was developed in the European Centre for Soft Computing (ECSC[3]). This robot is based on a modular architecture whose first version was designed in the Technical University of Madrid (UPM[4]). It has the following configuration: Linux Debian 5.0 Lenny operating system, Orinoco PCMCIA Silver wireless card, wireless tools v.28, two ultrasound sensors mounted over servos and one AXIS 213 pan-tilt-zoom camera.

The Test-Bed environment was established on the ECSC. The layout of this zone is shown in Figure 3. It has a surface of 60m x 20m, with 8 different rooms, including offices, labs, bathrooms, storerooms and meeting rooms. Six APs are available at the whole environment.

For simplicity, the tests were achieved in the main corridor. This was discretized into 16 nodes placed at the positions indicated in Figure 3. Sancho3 was placed at each node and 1000 signal strength samples were collected from all APs. These samples contain the signal and noise levels expressed in dBm and they have been used for both purposes, to train and to test the proposed method.

For each position, we computed the mean and the deviation of the corresponding signal and noise values for each AP. Then, we constructed two tables, one

[2] Please refer to the cited literature for a complete description.
[3] http://www.softcomputing.es
[4] http://www.upm.es

Fig. 2. Real prototype used in the experimentation

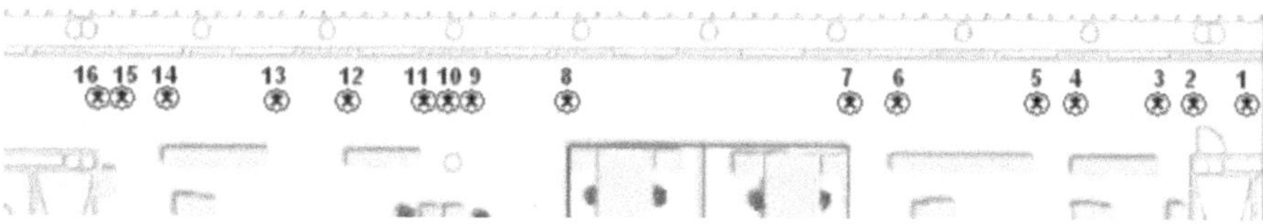

Fig. 3. Test-bed. European Centre for Soft Computing

for training and the other for testing. These tables contain tuples of the form: $(pos, \overline{S_{AP1}}, \sigma_{S_{AP1}}, \overline{N_{AP1}}, \sigma_{N_{AP1}}, ..., \overline{S_{APi}}, \sigma_{S_{APi}}, \overline{N_{APi}}, \sigma_{N_{APi}})$, where *pos* is the environment position and i is the number of APs. The training data were used to automatically generate the partitions and rules of the Fuzzy Classification system.

In addition, the same data were used to compare our method with a classical localization method called Nearest Neighbour (NN) [5]. It obtains the location by means of computing the Euclidean distance from the tuple received at a certain position and the tuples stored in the training table. The lowest distance indicates the estimated position.

The methods have been tested using different number of samples, both in training and test phase. The best classification rate was 60.16% for the NN method and 99.2% for FC, these were obtained with 60 samples in the training and test stages. The results are shown in Table 1.

Also, we have tested the classification rate when the samples taken in the training and test stage were different. It is important to note that the maximum

Table 1. Comparison of classification methods regarding training data

Samples	Data	Errors (NN)	Classification rate (NN)	Errors (FC)	Classification rate (FC)
12	1328	732	44.88	17	98.72
28	560	275	50.89	5	99.1
60	256	102	60.16	2	99.22

acquisition frequency of the WiFi interface is 4Hz, then to take 60 samples it is needed to spend 15 seconds at the same place. We have reduced the samples from 60 to 4 with the aim of checking the classification rate of both methods, Figure 4 shows these results. As it can be seen in this figure, the FC (on the right picture of the figure) maintains a good classification rate even when the samples taken are 12 and 4 in the training and test stages. As a result, the FC yields robust and simple solutions. In the worst case, the classification rate is around 70 % for a FC trained with groups of 60 samples when it is tested regarding groups made up of only 4 samples (the robot only spends 1 second to capture them). In addition, the best classification rate achieved by NN method (on the left picture of the figure) is lower than the worst one obtained by FC.

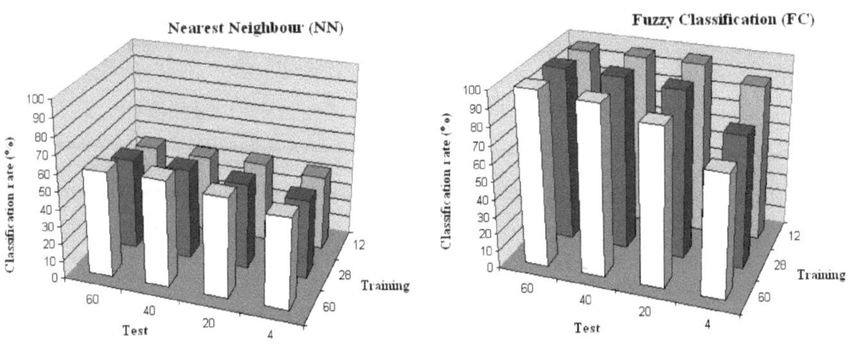

Fig. 4. Comparison of classification rates

Finally, Table 2 gives an idea on the complexity of the fuzzy classifiers built for this problem. They are more easily interpretable (because of the smaller number of rules, inputs, etc.) when the number of samples is increased. However, if the number of samples grows then the acquisition time is increased. In consequence, the system design has to be made carefully looking for a good trade-off (depending on the application) between number of samples (acquisition time), classification rate, and interpretability of the model.

Table 2. Complexity of fuzzy classifiers

Samples	Rules	Inputs	Linguistic Terms
12	77	16	91
28	35	13	57
60	25	10	42

4 Conclusions and Future Works

In this work we have presented a WiFi localization system based on Fuzzy Classification. We demonstrate that it is useful and robust to localize the robot in real conditions.

The classification rate of our method improves the ratings of other classical methods like Nearest Neighbour. This rate is maintained even when we take only a few samples.

In the near future, we have the intention of using this system in other environments to test the applicability of the method. Also we want to add new data sources provided by the robot, such as actions and ultrasound observations to improve the classification rate.

Acknowledgement. This work has been funded by grant S-0505/DPI/000176 (Robocity2030 Project) from the Science Department of Community of Madrid, TIN2008-06890-C02-01 (CWPIE Project) from the Spanish Ministry of Science and Technology (MCyT) and CCG08-UAH/DPI-3919 (SISLOPEWI Project) from the Community of Madrid and University of Alcala.

References

1. Want, R., Hopper, A., Falco, V., Gibbons, J.: The active badge location system. ACM Transactions on Information Systems 10, 91–102 (1992)
2. Krumm, J., Harris, S., Meyers, B., Brumitt, B., Hale, M., Shafer, S.: Multi-camera multi-person tracking for easy living. In: Proc. of 3rd IEEE International Workshop on Visual Surveillance, pp. 3–10 (2002)
3. Priyantha, N., Chakraborthy, A., Balakrishnan, H.: The cricket location support system. In: Proc. of the 6th ACM MobiCom, pp. 155–164 (2002)
4. Barber, R., Mata, M., Boada, M., Armingol, J., Salichs, M.: A perception system based on laser information for mobile robot topologic navigation. In: Proc. of 28th Annual Conference of the IEEE Industrial Electronics Society, pp. 2779–2784 (2002)
5. Bahl, P., Padmanabhan, V.: Radar: A, in-building rf-based user location and tracking system. In: Proc. of the IEEE Infocom, pp. 775–784 (2000)
6. LaMarca, A., et al.: Place lab: Device positioning using radio beacons in the wild. In: Gellersen, H.-W., Want, R., Schmidt, A. (eds.) PERVASIVE 2005. LNCS, vol. 3468, pp. 116–133. Springer, Heidelberg (2005)

7. Enge, P., Misra, P.: Special issue on gps: The global positioning system. In: Proc. of the IEEE, vol. 87, pp. 3–172 (1999)
8. Serrano, O., Cañas, J., Matellán, V., Rodero, L.: Robot localization using wifi signal without intensity map. In: Proc. of the V Workshop Agentes Físicos (WAF 2004), pp. 79–88 (2004)
9. Howard, A., Siddiqi, S., Sukhatme, G.: An experimental study of localization using wireless ethernet. In: Proc. of the International Conference on Field and Service Robotics (2003)
10. Ladd, A., Bekris, K., Rudys, A., Marceu, G., Kavraki, L., Wallach, D.: Robotics-based location sensing using wireless ethernet. In: Proc. of the MOBICOM 2002 (2002)
11. Youssef, M., Agrawala, A., Shankar, A.: Wlan location determination via clustering and probability distributions. In: Proc. of the IEEE PerCom 2003 (2003)
12. Sotelo, M.A., Ocaña, M., Bergasa, L.M., Flores, R., Marrón, M., García, M.A.: Low level controller for a pomdp based on wifi observations. Robot. Auton. Syst. 55(2), 132–145 (2007)
13. Zadeh, L.A.: Fuzzy sets. Information and Control 8, 338–353 (1965)
14. Zadeh, L.A.: The concept of a linguistic variable and its application to approximate reasoning. Parts I, II, and III. Information Sciences 8, 8, 9, 199–249, 301–357, 43–80 (1975)
15. Mamdani, E.H.: Application of fuzzy logic to approximate reasoning using linguistic systems. IEEE Transactions on Computers 26(12), 1182–1191 (1977)
16. Wang, L.X.: Fuzzy systems are universal approximators. In: First IEEE Conference on Fuzzy Systems, San Diego, pp. 1163–1169 (1992)
17. Alonso, J.M., Guillaume, S., Magdalena, L.: Kbct: A knowledge management tool for fuzzy inference systems. In: Free software under GPL license (2003), `http://www.mat.upm.es/projects/advocate/kbct.htm`
18. Alonso, J.M., Magdalena, L., Guillaume, S.: HILK: A new methodology for designing highly interpretable linguistic knowledge bases using the fuzzy logic formalism. International Journal of Intelligent Systems 23(7), 761–794 (2008)
19. Ruspini, E.H.: A new approach to clustering. Information and Control 15(1), 22–32 (1969)
20. Hüllermeier, E.: Fuzzy methods in machine learning and data mining: Status and prospects. Fuzzy Sets and Systems 156, 387–406 (2005)
21. Ichihashi, H., Shirai, T., Nagasaka, K., Miyoshi, T.: Neuro-fuzzy ID3: A method of inducing fuzzy decision trees with linear programming for maximizing entropy and an algebraic method for incremental learning. Fuzzy Sets and Systems 81, 157–167 (1996)
22. Quinlan, J.R.: Induction of decision trees. Machine Learning 1, 81–106 (1986)

Vehicle Detection Based on Laser Radar

Fernando Garcia[1], Pietro Cerri[2], Alberto Broggi[2],
Jose Maria Armingol[1], and Arturo de la Escalera[1]

[1] Intelligent Systems Lab. Universidad Carlos III de Madrid, Spain
{fegarcia,escalera,armingol}@ing.uc3m.es
www.uc3m.es/islab
[2] VisLab.Universitá degli Studi di Parma, Italy
{cerri,broggi}@vislab.it
www.vislab.it

Abstract. This paper describes the detection of moving obstacles using laser radar in road environments. This application is designed to be implemented in further research on data fusion technologies. The developed application uses only a laser radar which provides information to sort objects according to their shape and movement. The subsequent detection and classification provide higher level tracking.

Keywords: ADAS, Intelligent Vehicles, Data Fusion, Laser Radar.

1 Introduction

Over the years, countless efforts have been made to decrease the number of casualties on roads. In the recent years most of these efforts have focused on developing technologies that both help and warn drivers in the event of hazardous situations. Due to recent advances in information technologies, new applications can be developed to prevent these situations. In this context, the lack of cheap and reliable sensors underlines the need to use different sensors at the same time in order to provide a reliable and accurate application.

A possible set of sensors used in data fusion applications are computer vision and radar. The reason for using these sensors is that radar provides a reliable source of possible detections in the surroundings; on the other hand, data provided by vision sensors allow the different objects detected by the laser to be classified. The application presented in this paper focuses on the detection of moving obstacles, mainly vehicles, using a laser radar integrated in the bumper of a test vehicle. Detection based on laser radar is also very useful in regions where no visual information is available and the estimations have to be done with only the laser data.

1.1 State of the Art

While most of the applications developed fuses frequency radar information and visual information[1],[2] and [3], in recent years laser radar are becoming more

R. Moreno-Díaz et al. (Eds.): EUROCAST 2009, LNCS 5717, pp. 391–397, 2009.

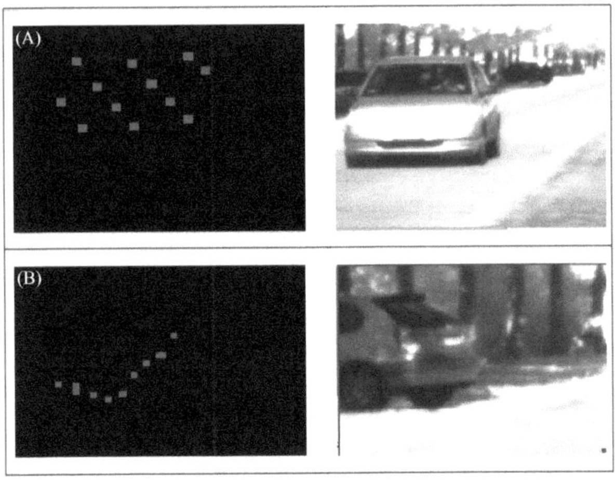

Fig. 1. Different patterns given by the laser. (A) Moving Car.(B) Parked Car.

popular [4] due to their lower cost. Laser radar also give more complete and reliable information.

The methods differ mainly in the level in which fusion is done. Low level classification converts unprocessed data from several sensors into a new set of raw data to be processed [2]. Medium level fusion methods get different patterns given by the sensors (corners, shapes, colors, movements ...) and performs the classification according to them. High level fusion methods performs different classification for each sensor, and at the final stage, all classifications are checked to provide a more reliable detection [5].

The application presented focuses on the detection of moving obstacles, mainly cars, using a laser radar mounted in a vehicle bumper. Once the detection is performed it creates a log of the movement of the object. So after some time, the application is able to give an estimation about which kind of obstacle is more likely to be. This research is a first stage of a further fusion method, based on laser radar and visual information. Typically these fusion methods has some parts of the images where data is not provided by both sensors, thus the detection has to be done using only one them. Thanks to this research, an estimation of the obstacles in the surroundings can be perform, with only the information provided by the radar.

In section 2 laser radar behavior is explained. Section 3 focuses on the algorithm developed for this application. Finally, section 4 presents some results, conclusions and gives future steps .

2 Laser Radar Behavior

The laser radar used is the SICK LMS 211. It gives a 2D reading of 100°around the vehicle with a 0.25° resolution. To achieve this it performs 4 scans

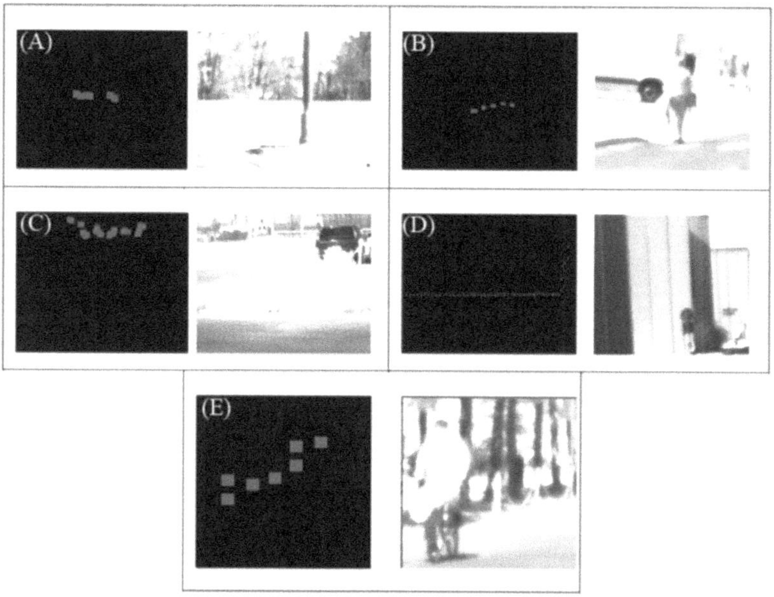

Fig. 2. Different patterns given by the laser. (A) Lamppost. (B) Pedestrian. (C) Bushes. (D) Building. (E) Bicycle.

independently which give 4 sets of spots with 1° of resolution on each spot. When a moving obstacle is found, the four scans performed by the laser for a single detection appear with a variation which is proportional to the speed and direction of the detected object and the test vehicle.Hence, performing egomotion correction moving obstacles show a special pattern proportional to their movement.

A typical problem when dealing with this applications is that moving obstacles pattern given by the laser radar is very similar to other patterns that can lead to mistakes, so false positives are common. Typically false positives appears dealing with pitch movements and bushes as it is shown in Fig. 2. To avoid those false positives, filtering the detected moving obstacles is mandatory. Finally a tracking system is very useful in order to record the movement of the object along the environment. This tracking systems, also allows to detect moving obstacles even when they are not detected (false negatives), also a record of the movement is useful to predict their future position.

3 Algorithm Explanation

3.1 Data Acquisition and Egomotion Correction

The application receives the information from the laser and corrects it according to egomotion information. The egomotion is provided by the test vehicle via

CAN-BUS. This information allows to correct the difference between the four consecutive scans due to test vehicle movement. After the egomotion correction, moving obstacles still appear with a variation which is proportional to the speed and direction of the detected object only (see fig. 1.A); thus by taking advantage of this behavior, it is possible to know whether an obstacle is moving or not.

3.2 Low Level Classification

The sets of points are separated, according to the distance between them, into different obstacles and their shapes are estimated. The shape is estimated by merging the points for each obstacle into a polyline that defines its shape. A low level classification is made taking into account the shape of the obstacles, moving obstacles may be differentiated thanks to their serrated shape. The sets of obstacles that are differentiated are:

- L **Shape Obstacle.** Obstacle with the specific L shape that typically is given by parked car.
- **Possible Pedestrian.** For obstacles with a small size that could represent wether pedestrian or typical road object that could be detected, like lamp-post or signs.
- **Fixed Obstacle.** Obstacles that does not fit in all the rest possible obstacles are labeled as fixed obstacles.
- **Road Border.** When the size and position of the obstacle is suitable to fix in this kind of obstacle it is labeled as road border. Also a high number of possible pedestrian obstacles parallel to the movement of the car which could mean that there is railguards or milestones in the road are labeled as road borders.
- **Moving Obstacle.** This is the main part of the algorithm due to their special and interesting shape.

3.3 Moving Obstacle Detection

The serrated shape that presents moving obstacles and the divergence between the points, make possible to detected them and their trajectory and speed estimated. Some constraints in the speeds or movements (impossible speeds, accelerations and shapes that does not match with cars shape) are mandatory to avoid false positives.

Where T=13msecs, which is the period of a single scanner rotation.

The speed can be computed three times each complete scan, so it is also possible to calculate the acceleration of the moving obstacle by calculating the difference in the calculated speeds. These information makes possible differentiate between real moving cars and false positives. The acceleration and speed calculated are not accurate due to the 0,25° divergence between two rotations, but precise enough to avoid those false positives.

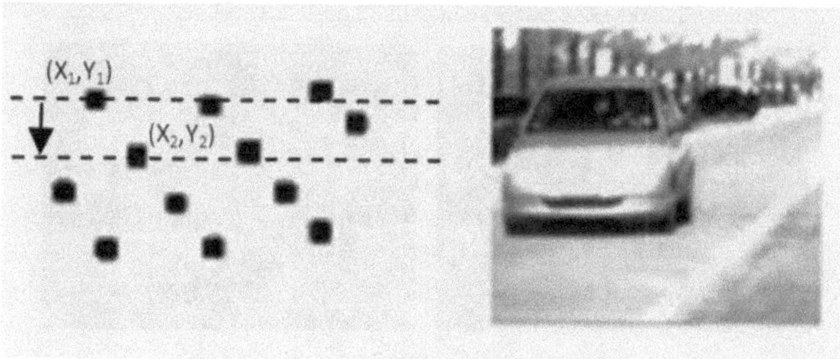

Fig. 3. Divergence between points can be used to detect false positives

$$v = \frac{\sqrt{(x_1 - x_2)^2 + (y_1 - y_2)^2}}{T}$$

3.4 Tracking Stage

After the low level classification, a tracking stage is performed in order to detect similar objects in the following laser scan. This tracking is done with both the information provided by the low level detection in the current scan and the previous tracking.

The tracking stage computes the speed of the car and calculates the position for the next scan. The moving obstacle is searched within a given window which is proportional to the size and shape of the moving obstacle. If a moving obstacle is detected, the speed is actualized according to the difference in the position of the last two frames and the low level classification is stored. This way the estimated speed in this stage is more accurate than lower level speed estimation which has the problem of the 0.25° divergence for each rotation. After some consecutive moving obstacle detections a higher level estimation is done using a voting scheme based in the latest low level estimations. The longer the obstacle is detected, the more reliable the classification.

Low level false positives are one of the most challenging situations in this stage. Tracking algorithm has to deal with these detections and avoid tracking them. Impossible movements and big size changes are detected and labeled as false positives,thus tracking is not performed for these obstacles.

4 Results, Conclusions and Future Steps

4.1 Results

Test has been performed under real conditions to check the reliability of the system.

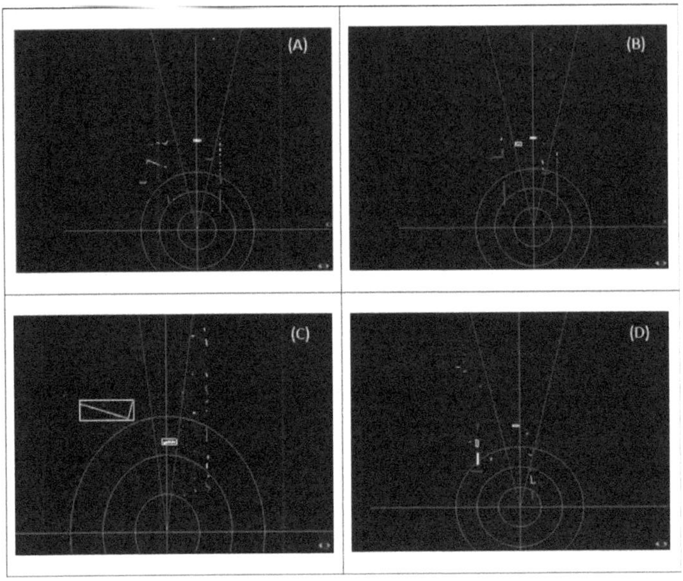

Fig. 4. Results. (A) Real traffic with a car ahead. (B) Two cars in opposite directions. (C) A truck and a bus in a crossroad. (D) A car ahead and two false positives.

Low Level Detection Test. two different Test were performed, the first where perform with the test vehicle stopped. the main purpose of the test was to check the viability of the algorithm avoiding possible problems due to egomotion correction. The results obtained were 80% positive detections for a single car in an approach movement for a distance closer to 48 meters, and 36 meters if the car is separating from the laser. In closer distances the detection percentage is higher. The results were very positive since this percentage could be increased by adding the tracking stage which, once the first detection is performed, can track the vehicle even when it is not detected.

Moving test were performed and egomotion correction included in the application. The results obtained were very similar to previous results. The test showed that egomotion corrections plays an important role in the detections of the surroundings. While the data provided by the test vehicle related to the velocity resulted reliable, the yaw angle were not so accurate. This problem lead to some false positives when lateral movement were involved, mainly curves. Also pitch movements when breaking or accelerating caused some false positives.

Complete Algorithm Test. Tracking resulted to be very useful since once the vehicle were detected in could be tracked until it disappears. The main problems in this stage were caused by low level false positives which should be avoided.

4.2 Conclusions

It has been proved that vehicle detection can be done using only laser radar information. Information provided by this applications are not limited to shapes, it also can be detected speeds, accelerations and movement. A tracking stage can help to avoid missdetections. Finally, it also has been proved that, when dealing with real situations, egomotion correction plays an important role and need to very accurate to avoid false positives.

4.3 Future Steps

The applications represents the first step of a complete fusion algorithm which includes visual and laser radar information. Future step fuses this application with visual information algorithms. Visual information can avoid some false positive due egomotion correction errors, by checking the detections using visual information. Egomotion correction has to be improved.

Acknowledgements

This application has been developed by members of VisLab [6] and Intelligent System Lab [7] as an Integrated Action between the two groups. The system has been tested on recorded sequences provided by the test platform "Grandeur" [4]. This research has been developed in the scope of scanning laser radar-based pedestrian detection project funded by Mando (South Korea).

This work also was supported in part by the Spanish Government through the CICYT projects VISVIA (Grant TRA2007-67786-C02-02) and POCIMA (Grant TRA2007-67374-C02-01).

References

1. Ofer, A.S., Mano, O., Stein, G.P., Kumon, H., Tamatsu, Y., Shashua, A.: Solid or not solid: Vision for radar target validation. In: IEEE Intelligent Vehicles Symposium Proceedings, pp. 819–824 (2004)
2. Steux, B., Laurgeau, C., Salesse, L., Wautier, D.: Fade: a vehicle detection and tracking system featuring monocular color vision and radar data fusion, June 2002, vol. 2, pp. 632–639 (2002)
3. Alessandretti, G., Broggi, A., Cerri, P.: Vehicle and guard rail detection using radar and vision data fusion. IEEE Transactions on Intelligent Transportation Systems 8(1), 95–105 (2007)
4. Broggi, A., Cerri, P., Ghidoni, S., Grisleri, P., Jung, H.G.: Localization and Analysis of Critical Areas in Urban Scenarios. In: Procs. IEEE Intelligent Vehicles Symposium 2008, Eindhoven, Netherlands, June 2008, pp. 1074–1079 (2008)
5. Hofmann, U., Rieder, A., Dickmanns, E.D.: Radar and vision data fusion for hybrid adaptive cruise control on highways. In: Schiele, B., Sagerer, G. (eds.) ICVS 2001. LNCS, vol. 2095, pp. 125–138. Springer, Heidelberg (2001)
6. `http://www.vislab.it`
7. `http://www.uc3m.es/islab`

Biomimetic Controller for Situated Robots Based on State-Driven Behaviour

Gerhard Hoefer and Manfred Mauerkirchner

UDA: Austrian Partner of International Universities, A-4060 Leonding, Austria
g.hoefer@htl-leonding.ac.at
m.mauerkirchner@htl-leonding.ac.at

Abstract. This work introduces the internally and externally grounded core structure of intelligent behaviour in complete robots. It integrates reactive as well as emotional subsystems into a multi-agent network in order to realize emergent intentional behaviour. The architecture of this controller is structured in four layers that interact in a decentralized and reciprocal way. Of special importance is the internal grounding which is based on monitoring internal processes of the robotic system. The integration of these internal states enables the robot to response differently to the same stimulus pattern at different times. Emotional behaviour may be regarded as a bundle of concerted activities that facilitate the successful survival in a potentially unpredictable environment. Additionally, an extension for the controller that may help to achieve long-ranging and aim-oriented behaviour is outlined. Rational behaving robots should be based on grounded cognition.

1 Introduction

In this section recent controller architectures as well as their pros and cons are outlineded. This should motivate the evolution of the structure of the proposed controller architecture.

The symbolic controller architecture is based on input dependent processing of information and symbolic representations [1]. It is realized as a sense-model-plan-act cycle and makes extensive use of representational structures and large-scale computational processes. A major drawback is the frame of reference problem which leads to problems in achieving real time interaction [2].

The reactive controller architecture is based on the rejection of the symbol-based sense-model-plan-act cycle. Behaviour is produced by layered competences that directly interact with the world which is regarded as its own best model [3]. This direct interaction is referred to as sensory-motor coordination [4]. Each competence is designed to realize a distinct task achieving behaviour but is only able to exhibit reflex-like reactive behaviour. Major drawbacks are limited cognitive abilities due to the lack of representational structures and the fact that sophisticated applications may become very complex.

A hybrid architecture like the Three-Layer-Architecture [5] incorporates reactive, executing, and planning layers in order to overcome the problems evident

R. Moreno-Díaz et al. (Eds.): EUROCAST 2009, LNCS 5717, pp. 398–405, 2009.

in both controller architectures introduced above. But we would like to argue that this hybrid architecture only very roughly integrates the paradigms of both architectures. There is no genuine bridging of the methodical gap that separates the participating layers because the symbol structures of the planning layer are not grounded in the reactive layer. Representational processes if really necessary should be based on action centered and modal structures [6][7]. Therefore, this hardly reconcilable separation in reactive and symbolic subsystems should be given up in favour of one unified controller architecture based on mutual interaction of agents coupled by a network.

Because of the grounding problem evident in the hybrid approach a biomimetic architecture should be introduced that resembles the way behaviour is generated in biological entities [8]. The architecture of our proposed controller system biomimetically resembles the general structure evident in all biological controller structures. Autonomous agents are embedded in a network structure and mutually modulate their activity according to actual internal states and external constraints [9]. Behaviour may be regarded as the activity of an overall multi-agent system exhibiting emergent properties [4].

This multi-agent based controller is tightly coupled to the physical body of the robot, the structure of its environment, and even its social context. Behaviour can no longer be seen as limited to pure cognition [10]. Therefore, the basic structure of the proposed architecture must stick to the design principles for complete robots. Such complete robots must behave in an autonomous, self sufficient, embodied, and situated way [4]. Accordingly, the complexity of the environment is resembled in the complexity of the sensory-motor competences coordinated by the reactive controller architecture. The dynamical interaction of the controller, the body, and the environment may be modelled as one overall dynamical system [11][10]. But the proposed controller architecture will not exclusively stick to that approach since we would like to argue in favour of a complementary architecture that integrates dynamical system theory and discrete agent modules [6].

Complete robots are not only grounded in the environment via sensory-motor coordination but should also be grounded in their interior physical and motor system. Behaviour is based on both external and internal grounding processes [12][13]. The current strength of internal state variables drives and modulates the purely reactive competences. Internal drives are able to initiate and maintain behaviour without receiving stimuli from the current environment. The state dependent modulation of reactive behaviour enables different response to the same external stimulus at different times.

2 Principle Units of a Biomimetic Controller

The realisation of a biomimetic controller architecture depends largely on a reasonable structuring of the overall system taking into account the interaction of controller, body, and environment as outlined in the previous section. Our proposed controller architecture is based on the biologically motivated idea of structuring it into two interacting principle units. Therefore, the overall controller architecture is comprised by one ratiomorph apparatus and one rational apparatus, respectively [14]. In order to realize the internal state system we propose an emotionally augmented version of the ratiomorph apparatus since the original version deals mainly with purely epistemic ideas. Generally, the ratiomorph apparatus realises internally and externally grounded behaviour whereas the rational apparatus should be able to guide the execution of complex tasks that depend on reasoning and planning.

2.1 Ratiomorph Apparatus

The proposed ratiomorph apparatus is made up by one externally grounded subsystem and one internally grounded subsystem which both are realized as distinct subsets of the multi-agent system introduced in section 1. The externally grounded subsystem comprises reactive behaviour exerted by its competences whereas the internally grounded subsystem comprises emotional behaviour based on internal states. Both subsystems must be interacting very closely since motivational and emotional processes are intertwined with sensory-motor competences in order to generate intentional behaviour [12]. The motivational and emotional components of the ratiomorph apparatus are arranged in interacting layers [15].

Background emotions represent the current state of interior processes of the robot induced by its metabolism, reflexes, or damage. These internal states represent the current activities of visceral and musculo-skeletal processes which are sampled by various detector systems and represented in brainstem structures especially the tectum and the dorsal tegmentum. They are dynamically represented by a pattern structure of internal states values. Background emotions influence emotional behaviour as basal dispositions and motivations which act as contextual background that modulates the emotional response to certain external stimulus patterns triggering basic emotions [12].

Seeking and rewarding is directed by appetence detectors. They are located in the hypothalamus as well as the ventral tegmentum and detect deviations from the homeostasis of mainly metabolic processes which are manifested in bodily needs like e.g. hunger or thirst. Seeking behaviour is the primary basic emotion which is modulated by detected appetence signals which increases the activity of the motor system above its tonic levels [13]. Additionally, the basically undirected seeking behaviour is supported by the selection of appropriate preafferences which tune the multimodal sensory system in order to facilitate the detection of affording signals from the current environment [9]. The seeking system is reciprocally coupled to the rewarding system which initiates behaviour that might finally restore the state of homeostasis inside the organism [13]. In such cases the rewarding system triggers value based learning by sending neuromodulatory signals to all relevant parts of the brain [9]. The seeking system is made up by brain structures located in hypothalamus and the ventral tegmentum whereas the rewarding system is made by brain structures located in the hypothalamus and the septum [13].

Basic emotions are induced by the lack of spontaneous reward or by the active detection of an emotionally connotated stimulus patterns [15]. Basic emotional behaviour may be regarded as the spontaneous read out of a appropriate set of activities triggered by the presence of such patterns. Consequently, they act as special purpose subsystems that modulate the entire behaviour of the organism. Each basic emotion initiates a bundle of concerted activities that characteristically modulate its interior processes and sensory-motor competences. Well characterized basic emotions are fear [16], rage [13], panic [13], and disgust [17]. Basic emotions are generated by limbic brain structures like the amygdala and the antorior gyrus cinguli [15]. The basal ganglia modulates the current output of the limbic system on the motor system.

2.2 Rational Apparatus

Long-ranging and aim-oriented behaviour is achieved by extending the controller architecture with the rational apparatus. This apparatus steadily interacts with the ratiomorph one in order to carry out full blown cognitive tasks like reasoning and planning [14]. Rational processes determine the overall strategy that the organism should execute in order to achieve such tasks and guide their overall execution. Ratiomorph processes realize the actual tactic of executing the current plan in unpredictable environmental situations and in return modify the overall strategy.

The frontal lobe is the core structure of the rational apparatus and involved in all kinds of cognitive processes [18]. Grounded cognition is based on the reasonable assumption that cognitive tasks are typically grounded in bodily states, situated action, and re-enactive simulation processes [7]. Therefore, rational and ratiomorph apparatus interact circularly via distinct interface structures. At the moment only such basic concepts of the rational apparatus may be proposed but they will be worked out more precisely in recent future.

3 Realization of the Ratiomorph Apparatus

In this section the focus is on the ratiomorph apparatus as the core structure of our proposed controller. All participating agent modules will be introduced and their dynamic interaction will be outlined. The overall controller architecture is shown in Fig.1.

3.1 Realization of Reactive Behaviour

Exteroceptive pathways monitor the surrounding environment of the robot. Various sensory modalities serve as input to the different grounded sensory-motor competences. These input signals are transformed by each competence into motor signals that control the actuators. Additionally, these various modalities are integrated by an agent biomimetically equivalent to the activity of associative cortices in order to construct multimodal input schemes that may characterize situations triggering emotional behaviour. These input schemes are typically realized without the use of amodal physical symbols by the Multimodal Integration agent.

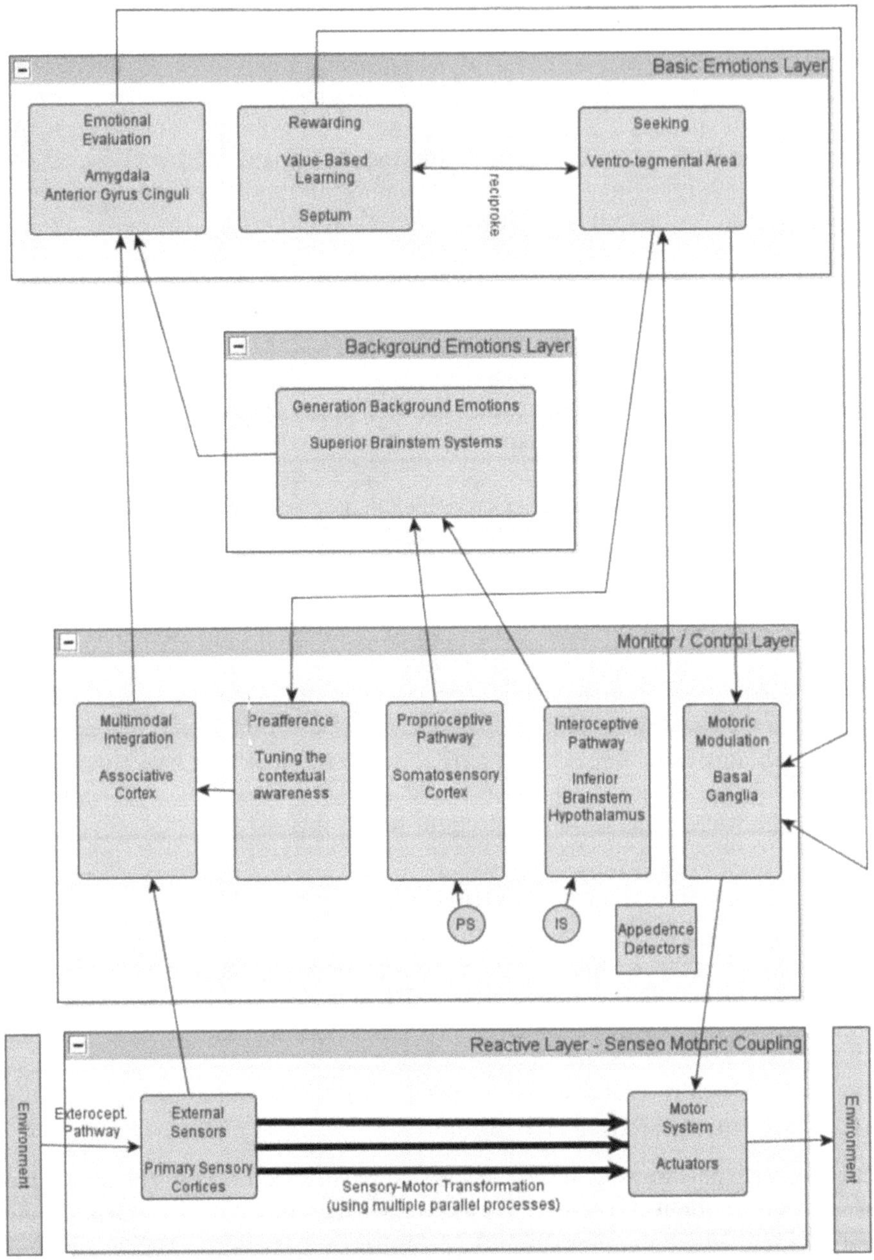

Fig. 1. Structure of Ratiomorph Apparatus (see section 3)

3.2 Realization of Seeking Behaviour

Seeking behaviour is based on and driven by characteristic state values originating from inside the physical system of the robot. Current state values from processes maintaining the operability of the robot are sampled by an internal sensory system made up by specialized appetence detectors. For example, one type of appetence detectors has to sample the remaining amount of energy available for the motor system. The pattern of state values indicates the overall distance of the physical system of the robot from its homeostasis and serves as generator of basal motivations that drive the seeking behaviour.

This undirected seeking process should enable the possible detection of signal patterns out in the environment that may lead to satisfy the actual needs of the robot. Therefore, the Seeking agent selects a seeking strategy according to the current basal motivations and modulates it on the actual motor output of the actuators of all competences The modulation process is carried out by the Motor Modulation agent biomimetically equivalent to the basal ganglia.

The detection process is supported by preafferences that tunes the Multimodal Integration agent torwards the successful detection of suitable external signal patterns. The Preafferences agent acts as a basic memory system that associates suitable multimodal input schemes to the actual pattern of appetences. Optionally, collary discharges may also tune the exteroceptive sensory system to potentially useful patterns of stimuli. In summary, the activity of the Preafferences agent directs the robot more effectively towards situations that may help to restore its state of homeostasis.

The Rewarding agent provides stereotypical and preprogrammed activities that may secure the reestablishment of the state of homeostasis. It is reciprocally coupled to the Seeking agent and is maximally activated if this homeostasis is going to be regained. Therefore, the rewarding system is the principle structure of value based learning.

3.3 Realization of Emotional Behaviour

The input schemes of both the internal and the external grounding process serve as input data for valuation processes that generate basic emotional responses which modulate the coordinated activities of the various behaviourial competences of the reactive subsystem.

Internal physical and motor processes that may trigger background emotions are monitored by a network of sensors placed inside the robot. These sensors sample current values of these processes and transmit them to the Background Emotions agent via two distinct pathways: The interoceptive pathway transmits raw data from sensors (IS) that monitor internal processes like e.g. the internal distribution of temperature or damage of components. The proprioceptive pathway transmits data sampled by kinesthetic sensors (PS) which montor the current state of all important components of the motor system like e.g. joints and motors. The Background Emotions agent transforms the sampled raw data into a pattern of state values that serve as input scheme representing internal processes in a non-symbolic way. The current multimodal input is checked by the Emotional Evaluation agent for stimuli that may trigger basic emotional behaviour in context to the actual background emotions. The output pattern of this agent initiates of a bundle of coordinated activities: Priming signals for motor activities are transmitted to the Motor Modulation agent which modulates the current activities of reactive competences in order to realize the appropriate emotional response. Additionally, internal processes may also be modulated appropriately. In summary, this emotional response may help the robot to behave more effective in its environment.

4 Conclusions

The augmented ratiomorph apparatus and the supplementary rational apparatus characterise distinct stages in the evolution of biological as well as artificial controller structures.

The separation in these two principle units is not a forced diversification but just a matter of functionally structuring the overall multi-agent architecture. The augmented ratiomorph apparatus comprises the senory-motor competences of the reactive subsystem and the layered emotional subsystem as well as a basic multimodal integration and a simple memory system based on preafferences. This basic architecture may be regarded as the core brain structure of intentionality in vertebrates [9]. The various layers of the emotional subsystems are utilized to enrich the behaviourial spectrum of complete robots:

Background emotions display the inner states of robot and therefore render its basal dispositions and motivations visible by characteristic modulation of surface structures. The interaction and communication of robots with humans may be based on such basal emotional expressions.
Reciprocally coupled seeking and rewarding behaviour serves as the base for value based learning. Complete robots have to be equipped with such an onboard mechanism that allows them to monitor and evaluate its current behaviour [4].

Basic emotions act as a bundle of coordinated activities that achieve a more effective and successful interaction with potentially unpredictable events in the current environment of the robot. This kind of behaviour is based on inner states represented by background emotions and consequently offers a much wider range of behaviourial responses.

Contrary to the proposals of symbol based modelling of emotional processes [19] our modelling of basic emotions in the augmented ratiomorph apparatus is realized without the use of amodal physical symbol structures. It does not compute emotions like production systems. Emotional behaviour is realized by decentralized subsystems of the multi-agent network in a contextual appropriate way.

This network connects agents that take part in reactive as well as emotional behaviour. Therefore, it is evident that emotional behaviour is closely intertwined with reactive types of behaviour and it may seem that a functional separation is somewhat artificial. Our proposed controller architecture is based on a trade-off between a modular agent structure exhibiting basic representational abilities and the modulation of the coupling network sticking to the principles of dynamic system theory. It should enable emergent intentional behaviour based on the controller, the body, and the current environment of the complete robot.

References

1. Newell, H., Simon, H.A.: Human Problem Solving. Prentice Hall, Englewood Cliffs (1972)
2. Copeland, J.: Artificial Intelligence: A Philosophical Introduction. Blackwell, Malden (1993)
3. Brooks, R.A.: Intelligence without Representation. Artificial Intelligence (1991)
4. Pfeiffer, R., Scheier, C.: Understanding Intelligence. MIT Press, Cambridge (2000)
5. Gat, E.: On Three-Layer Architecture. AAAI Press, Menlo Park (1998)
6. Clark, A.: Being There. MIT Press, Cambridge (1997)
7. Barsalou, L.W.: Grounded Cognition. Annu. Rev. Psycol. 59, 617–645 (2008)
8. Ayers, J., et al.: Neurotechnology for Biomimetic Robots. MIT Press, Cambridge (2002)
9. Freeman, W.: How Brain Make up Their Minds. Columbia University Press (2001)
10. Beer, R.D.: The Dynamics of Active Categorical Perception in an Evolved Model Agent. Adaptive Behaviour (2003)
11. Kelso, J.A.S.: Dynamic Patterns. MIT Press, Cambridge (1995)
12. Damasio, A.R.: Decartes Error, Harcourt (1994)
13. Panksepp, J.: Affective Neuroscience. Oxford University Press, Oxford (1998)
14. Riedl, R.: Strukturen der Komplexität. Springer, Heidelberg (2000)
15. Damasio, A.R.: Looking for Spinoza, Harcourt (2003)
16. LeDoux, J.: The Emotional Brain, Simon and Schuster (1996)
17. Rozin, P., et al.: Disgust. Guilford Press, New York (2000)
18. Stuss, D.T., Knight, R.T.: Principles of Frontal Lobe Function. Oxford University Press, Oxford (2002)
19. Arbib, M.A., Fellous, J.: Emotions: from Brains to Robots. Trends in Cognitive Science (2004)

Supporting Information Services for Travellers of Public Transport by Road

Carmelo R. García, Ricardo Pérez, Álvaro Lorenz,
Francisco Alayón, and Gabino Padrón

Universidad de Las Palmas de Gran Canaria
Dpto. Informática y Sistemas
Edificio de Informática y Matemáticas
Campus Universitario de Tafira, 35017 Las Palmas
{rgarcia,rperez,falayon,gpadron}@dis.ulpgc.es

Abstract. In this work we describe the main aspects of an information system for travelers of public transport by road, specifically its architecture and functionalities. In the paper we explain how the ubiquitous computing paradigm has been applied to achieve the system goals. Relevant properties of the system are: Its distributed architecture, its capacity to work using local communications infrastructures such as Bluetooth and Wifi, the interaction with the travelers is made using their own mobile devices' communications, such as: cellular phones, PDAs, etc., etc.

Keywords: Information Systems, Public Transport, Ubiquitous computing.

1 Introduction

An efficient public transport system affects the quality of life of citizens; this assertion is accepted by all authorities concerned (local, regional, national and international). To improve public transport there are two types of measures: hard measures; these consist of tax and regulatory measures, for example raising the taxes on fuels, the speed limit, etc. and soft measures; consisting of developing of services in order to improve and encourage the use of public transport, an example of such services is intelligent transport systems. The system described in this paper falls into the second category of initiatives and is located specifically in the context of information systems for travelers. It is specially designed to facilitate access to public transport networks to groups of people with special needs such as disable people, tourists, etc.

The first section in this paper situate the system in the context of public transport, specifically in the context of the intelligent transport system and in the context of the information systems for travelers of public transport, emphasizing the importance of this kind of system to improve the life quality of the citizens. Following the main goals and requirements will be described. Next we'll

R. Moreno-Díaz et al. (Eds.): EUROCAST 2009, LNCS 5717, pp. 406–412, 2009.

explain a general description of the system. This description will begin with a general vision of the system structure, the executing scheme of the services will be described too; in this executing scheme the context processing plays a important role. And finally the architecture of virtual device used by travelers to interact with the system will be presented. The last but one section of this paper will be dedicated to present an example of service, specifically a payment system that uses the mobile phones of the travelers as payment support. At the end, the main conclusions will be exposed.

2 Information Services on Public Transport Context

The main goal of the public transport information systems is to improve the quality of service offered to the passengers. For this, these systems provide a range of information services that are designed to make it: easier to use, more attractive and more accessible. Specifically, for a traveler information system to fulfill the main objective is to provide quality information, or useful information in an appropriate format and available at the right time, it is necessary that these systems have the ability to operate in the different travelers environments, in the context of ubiquitous computing parading this requirement is called context awareness. The fulfillment of this requirement is the main scientific and technological challenge for these systems to solve. In the bibliography we can find examples of such kind of systems: Systems for pedestrians that inform about the route followed by pedestrian in cities, using this kind of system we can know how to arrive to a place of the city and information about places near to the route followed. Examples of this type of system are: Cybeguide [1], Hypermedia Tour Guide [2] and Gulliver's Genie [3]. Systems for train public transport; these inform about the time of departures and arrives in station., an example of this kind of system is the ODIN system [4]. Systems for bus public transport; using geographic information they inform about the time tables of bus stops near to the user. The Bus Cacher system [5] and Stopman system [6] are cases of this type of system. Systems for intermodal transport; these inform to user how to move in a intermodal public transport network, using different transport modes: pedestrian, bus and train. An example of this type of system is the MUMS system [7].

These systems differ in: the mode of transport in operation, the way to address the context awareness and the infrastructure required, especially with regard to mobile communications. In general, a large-scale implementation, which collects a full set of travel-related information in real-time, and disseminates such information in a context aware manner, is not possible within the existing information infrastructure in the public transport networks.

The main goal of the system described in this paper is to improve the quality of service offered to the public passenger transport. For this, the system provides a range of information services that are designed to make it: easier to use, more attractive and more accessible. From a functional point of view, it is structured in three subsystems: timetable information subsystem, route information subsystem and payment subsystem. The *figure* 1 shows us the relationship between these subsystem and the goals mentioned below.

	Usability	Accessibility	Attractiveness
Timetable information subsystem		•	•
Route information subsystem	•		•
Payment subsystem	•	•	•

Fig. 1. Subsystems-Goals Relations

In order to fulfill these goals, it must be: flexible; in order to permit the integrations the information technology advances, scalable; in order to permit the integrations of new functionalities and accessible; in order to permit the friendly interaction of the travel with the system in different physical and logical contexts. Especially important are the specific scheme of interaction for group of persons with specific requirements such as handicapped people, elderly people, tourist, etc. Finally we must comment that all these characteristics must be achieved with an effective operational cost and to fulfill this requirement the mobile communications plays an important role.

3 System Description

3.1 General Vision

Our system is based on ubiquitous computing model [8]. The reason of this election is that the ultimate goal of this computing model is to develop computer systems that meet their goals adapting to every user and every environment in which the user moves. So mobility, context awareness and nature interaction with the user are principles of this model. Our design model follows a server-client model. The servers are available in different places of the public transport network. Each server offers a particular service. The traveler can use the information services using client applications and these client application run on the user's mobile communication devices. Each client is associated with a particular service. But the services could be related. For example, a payment service may request information from the information service routes.

3.2 System Structure

The system has a distributed architecture, the elements are provided by the infrastructure of the transportation company and the users of the system. All the services are always carried out the following common scheme:

1. Context identification achieved by client.
2. Context information request achieved by client and attended by the server.
3. Data transformation achieved by server and client.
4. Transaction confirmation achieved by client.

The complexity of each step depends on the specific service. For example, services where the security plays a critic role, for example payment systems, the steps two, three and four are implemented using encryption mechanisms, but in a service about route information for travellers such mechanisms are not required. Using this executing scheme our system not only provides information services to the travellers, others actors of the transport company, who work in mobility, are benefited from its operation, such as for example: maintenance staff and operation control staff. This is because the system provides different spaces of data, that we name data context, associated with different information subsystems [9].

Conceptually all the functionalities of the system are structured in a three layers. The bottom layer, called basic infrastructure services, has the responsibility of providing the basic functionalities; these are the related with positioning and time. The processes of this level always run in the infrastructure of the transportation company. The second level, called extended service provider, is responsible for providing the different areas of data, data contexts, and it also provides mechanisms for access to these data spaces. Like the previous level, all the processes of this level always run in the infrastructure of the transportation company and never run on the user devices. Finally, at the third level we have the user application level; at this level are all running information services for system users. The processes of this level run on the infrastructure of the transport corporation or on the mobile devices of the users, for example on the cellular phones or PDAs of the travelers.

In order to provide these services regardless of the type of mobile device user and the type of information service to provide, the system introduce the concept of virtual user device, this virtual machine has a layered architecture of five levels (*Figure* 2). The bottom level, called Physical Level, includes to all the necessary resources hardware so that any service can be carried out, consists of three types: processor resource, storage and communication. The second level is called Virtual Device Level; in this level an abstraction is made from physical level providing a common extended machine for the execution of all the programs taking part in the execution of any service. This extended machine is formed by four units: archive's unit, security unit, communications unit and unit of administration of software. The following level is the Common Services; it is responsible to provided all the common functions required to the execution of applications. The set of functions is grouped in two categories: primitives for the accomplishment of basic operations and handling of data, and primitives for communicating and

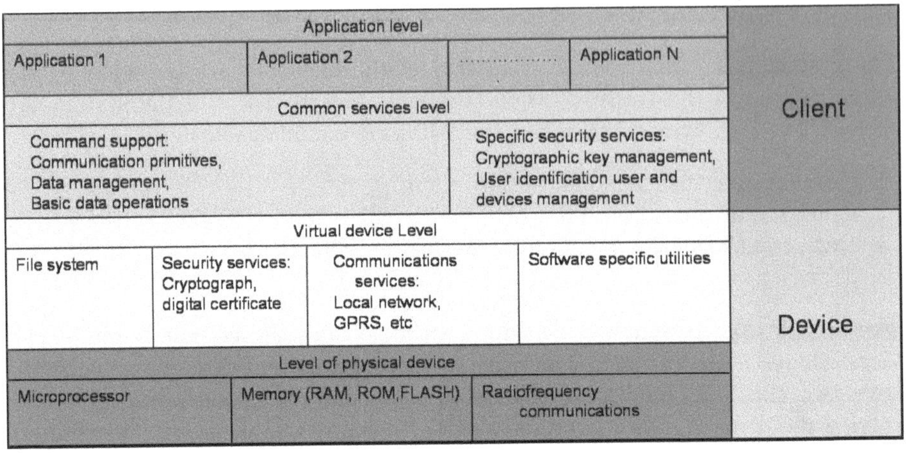

Fig. 2. Architecture of virtual machine

security of the applications. In the upper level is the Application Level; in this level are developed the different applications associated with each service. Each application running on this level, each of these applications is associated to a information service offered by the system to the user. The archives managed by each application are divided in two types: the data files and the archives of orders.

From a point of view of the technology used to implement this model of virtual user device, the system is based on the use of local technologies for mobile communications (IEEE 802.11 and Bluetooth) and platforms for software development: Java 2 Micro Edition (J2ME) for users applications that run on their mobile phones, Java Standard Edition (J2SE), the JSR-82 and Api Bluecove for interaction between the applications of the user application level and the level of service provider to the user by Bluetooth.

4 An Example of Information Service: A Payment System

We have developed a prototype of payment service using Bluetooth technology and a mobile phone as user's device. The payment model is inspired in magnetic cards and contactless cards. In this model the data that describe the pre payment and the trips that have been contracted with the transport company is stored into devices.

The first step is the detection of the payment service by the client application. The application inside the server device explores the environment finding client connections. In this prototype, the user select the server from a list that application client has detected. This system allows with a little increase of information interchange that this step can be performed without user action. This facility is possible because of the client application can obtain information from the servers automatically and taking account the user's preferences, the client application can discover the appropriate vehicle.

The action of user is needed to confirm that he desires go up to vehicle. When the server is selected, the connection is initiated. The connection goes through an authentication process; in this process the server must confirm some requirements of the client applications. Realized the authentication, the server ask for information about user to client application. This information identifies to the user and his characteristics related with the payment service: origin and destiny contracted, the number of trips that the user can do, the cost of the trip, etc. In the server side several verifications must be made, for example: credit balance of virtual card, distance of the trip, expired date, etc. Once all the checks carried out successfully, the balance of travel of the card user is decremented by the server. Finally, a data file is returned to the client. The data of this file have been modified to reflect that the user has used a trip. The transaction is reflected in a record in a file that is stored also in user's device. The trace of use of the service is stored in this transaction file. Also, the server send a ticket to client as a receipt that user can show to inspector.

5 Conclusions

As a conclusion, we can say that this system illustrates the validity of the model of ubiquitous computing to automate production processes of enterprises, providing attractive solutions, both, from a technological point of view and the effective costs. Specifically, we have presented an information system for public transport travelers based on this computer model. Relevant properties of the system are:Distributed architecture. It run on different contexts of the transport corporation: stations, bus stops and vehicles.Travelers can interact with the system using their mobile devices. It is designed to facilitate access to public transport for groups of people with special needs, such as handicapped people, elderly , tourist, etc. Finally, in order to achieve a cost effective ,the interaction with the user is performed using local mobile communication infrastructures such as Bluetooth.

References

1. Abowd, G., et al.: Cyberguide: a mobile context-aware tour guide. Mobile Networking 3(5), 421–433 (1997)
2. Bellotti, F., et al.: User Testing a Hypermedia Tour Guide. IEEE Pervasive computing, mobile and ubiquitous systems 1(2), 33–41 (2002)
3. O'Grady, M., O'Hare, G., Sas, C.: Mobile agents for mobile tourists: a user evaluation of Gulliver's Genie. Interacting with Computers 17(4), 343–366 (2005)
4. ODIN: geOgraphic Distributed INformation.European Commission (2000), http://www.odinproject.org

5. Bertolotto, M., et al.: Bus Catcher: a Context Sensitive Prototype System for Public Transportation Users. In: Proceedings of the Third International Conference on Web Information Systems Engineering (Workshops). IEEE Computer Society, Los Alamitos (2002)
6. Turunen, M., et al.: Mobile Speech-based and Multimodal Public Transport Information Services. In: 2006 Workshop on Speech in Mobile and Pervasive Environments (2006)
7. Hurtig, T.: A mobile multimodal dialogue system for public transportation navigation evaluated. In: Proceedings of the 8th conference on Human-computer interaction with mobile devices and services. ACM International Conference Proceeding Series, vol. 159, pp. 251–254 (2006)
8. Weiser, M.: The computer for the 21st century. IEEE Pervasive computing, mobile and ubiquitous systems 1(1), 18–25 (2002); reprinted with permission Copyright 1991 by Scientific American Inc.
9. García, C., Alayón, F., Caraballo, J., Pérez, R., Padrón, G.: PSPRT: A case of Pervasive System for Public Road Transport. In: Moreno Díaz, R., Pichler, F., Quesada Arencibia, A. (eds.) EUROCAST 2007. LNCS, vol. 4739, pp. 1126–1133. Springer, Heidelberg (2007)

Applying Reinforcement Learning to Multi-robot System Behavior Coordination

Yolanda Sanz, Javier de Lope, and Darío Maravall

Perception for Computer and Robots
Universidad Politécnica de Madrid
Campus de Montegancedo, 28660, Madrid, Spain
yolanda.sanz.sanchez@gmail.com, javier.delope@upm.es,
dmaravall@fi.upm.es

Abstract. We have applied ANLAGIS to a coordination problem Multi-robot Systems, specifically the storage of a set of elements is in the warehouses. We have combined ANLAGIS along with Reinforcement Learning for each of the behaviors that stem from this task, besides drawing up the coordination of these behaviors in order to perform the task in a satisfactory way.

Keywords: Multi-robots systems, cooperative systems, coordination, learning systems.

1 Introduction

Machine Learning allows to create applications that are able to acquire and integrate new knowledge in order to resolve new arised problem. This feature makes very interesting this kind of systems against whose that only can solve problem that are taken into account during the developing stage.

For instance, if you want to build a vision system that is capable of recognizing a set of faces, it would be impossible to program it by hand. Thanks to Machine Learning it is allowed to build a model through a set of examples to learn the goal of recognition. At other times, it will need systems capable of adapting to the environment where they are. You may also have an -application in support of analysing information, extracting knowledge from an automatic way through a set of examples, giving a series of patterns.

2 ANLAGIS and Reinforcement Learning

In this work, we are adapting ANLAGIS [2] for solving a problem of coordination in a multi-robot systems framework by including a simple model of Reinforcement Learning as the Q-learning.

ANLAGIS is divided in two parts; the first one consists of the state variable granulation of the problem, and the second one to choose the best action or

R. Moreno-Díaz et al. (Eds.): EUROCAST 2009, LNCS 5717, pp. 413–420, 2009.

output for each state that forms of the aforementioned granulation. In this part, the choice of action is realised applying Reinforcement Learning.

Reinforcement Learning [3] is included within the area of the Artificial Intelligence known like Machine Learning, in which the models or systems learn by observing the environment. Fundamental components that take part in learning are on the one hand the agent, it makes up the learning subjet; and the other hand the environment, it is the responsible for giving the rewards and the punishments which are received by agent. The environment must be clearly-defined by a set of states, and the agent can choose one and only one action of a limited set of actions. For each step, the agent must be choose an action by observing the state of the environment, and it must be carry it out. After this, the agent receives a reward from the environment.

There exist several reinforcement learning algorithms, but we are considering Q-learning [4] in this work. The Q-learning function is defined by the expression:

$$Q(s,a)_{t+1} = Q(s,a)_t + \alpha[r + \gamma \max_a Q^{\Pi}(s_{t+1},a)_t - Q(s,a)_t] \tag{1}$$

which shows the expected value of reward to take action a from the state s. The discount factor γ is used to distinguish between episodic tasks or no-episodic tasks. An episodic task has a final state, then the value γ is always the unity, and a no-episodic task has not a final state then the value γ is a number in the range $0 \leq \gamma < 1$

The policy selects an action a from a finite set of possible actions from a state s. There are a lot of policies to apply, for instance, ε greedy policy:

$$f(n) = \begin{cases} 1-\varepsilon & \text{a=arg}\max Q^{\Pi}(s_{t+1},a') \\ \varepsilon & \text{otherwise} \end{cases} \tag{2}$$

Note that it is selected through a determined probability, the action whose estimated value is the maximum.

Once it has reached the optimum function of Q, the optimum policy is always choose the possible actions for a state between has higher value. The Q-function that is mapping between states and actions, can be represented by a table. The initial values are random or arbitrary.

3 Task Description

The problem considered in this paper is as follows: the environment where robots live with other robots is an environment of N warehouses with the same size, and a dock that is bigger than the warehouses. For experimentation purpose are considering an environment with N=4 warehouses. In the dock, there is a finite set of disks of four different colors, as shown in Fig 1a). The task to be accomplished by the robots is to store the disks with same color in one of the rooms available, so that the final situation of each room must contain all the disk with the same color, see Fig 1b). The color to be in a storage room is set up when the first disk is placed in it, and from that moment on, all the disks with the same color must be stored in that room.

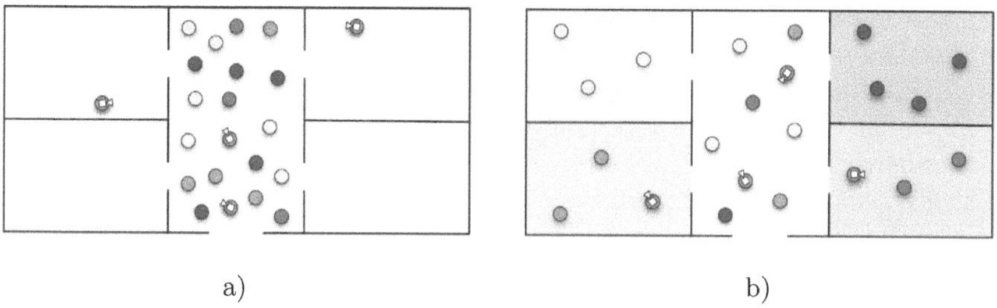

Fig. 1. Location of robots and disks. *(a)* Initially, the robots and disks are in the dock *(b)* Finally, each disks is located in its corresponding room.

The robots team consists of M robots, (for this experiment we are also considering M = 4). The robots are conventional Pioneer P2AT [5]. Each one of them is equipped with a camera, besides all sonars. As we previously said, all robots are located in the dock, the disks are also located in the dock, as shown in Figure 1b. There is no coordination between the robots, they work independent way.

Behaviors associated with this task, which will be later on, are all developed using the technique of Reinforcement Learning, but these behaviors could have developed by using other techniques as for example Evolutionary Computation. Coordinating such behaviors we have used ANLAGIS along with Q-Learning.

4 System of Behavior Coordination

As we said previously, the behaviors were developed through Reinforcement Learning. The behaviors for this task are "avoid obstacles", "grasp", "release", "go warehouse", "go dock", the latter two are based on a conventional wall following plus a specific finishing condition. The diagram of the coordination of behavior shown in Fig 2. It follows such behavior.

- Avoid obstacles: The states of the problem are characterised by the front sonars of the robot. Each state is classified in terms of how objects are positioned with respect to the robot's sonars, so we find that the robot can detect an object on the right or/and to the left and/or forehead.

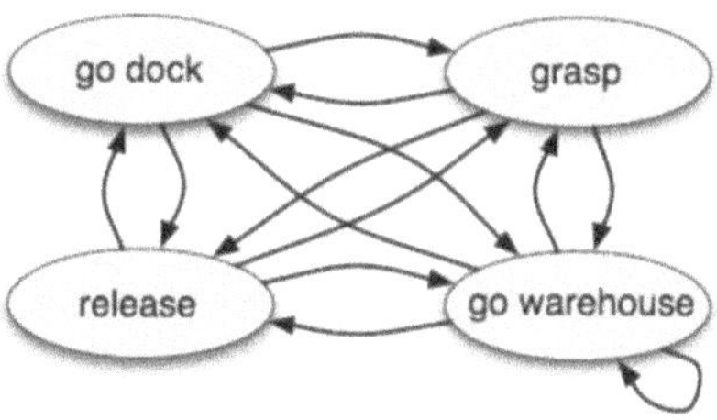

Fig. 2. Diagram of all possible combinations between behaviors

The sonars are classified into three groups, the group left that corresponds to the left sensors of robot (1,0,15), the front panel that contains front sonars (2,3,4,5), and finally the right group, which is formed by the right sonars of the robot (6,7,8). The corresponding group of sonars is activated, according to where the object is. Therefore, at a given time single group can be activated, or the two of them, or three or none of them . When an object is in vision range of some one of three sonar group (left, right, or front), the corresponding bit to the group is turned on, either the left or the right, or the frontal respectively. At this point, the total number of states is 8, that is, in 8 states are defined all the states in which the robot can be found at a given time.

Four actions are determined, "turn right", "turn left", "go straight" or "turn on itself", being the minimum of necessary actions for the robot can learn to avoid all the obstacles of the environment.

It set up a reward as follows: if the robot collides with an obstacle them, the reward is negative −3, and for any other state of the environment the reward will be 0.

$$f(n) = \begin{cases} -3 & \text{the robot collides} \\ 0 & \text{in any other state} \end{cases} \tag{3}$$

- Go dock: The sonar system, specifically the left side sonar, namely, the left group (1,0,15), are responsible for setting the states of robot. These states are basically two state. The first one it follows the left wall, and the second one the robot loses the left wall, this happens when the range of left sonars does not sense the left wall. Using established thresholds, these sonars group are actived at the time that the robot is parallel to the left wall.

  Actions for this behavior are "go to straight", and "rotate counterclockwise.

  The reward is always rewarding to follow wall, and when the wall disappears is rewarded linearly as the group left sonar ever closer to the threshold.

$$f(n) = \begin{cases} +3 & \text{if the robot is following wall} \\ 5/x & \text{in any other state} \end{cases} \tag{4}$$

  where x is the average of the distance that marked the left sonars group.
- Go warehouse: It is exactly like the previous behavior, but it is necessary to differentiate them to know when you are in a warehouse or in the dock. This will set a pattern by size of the rooms so that this pattern is repeated in the four warehouse and did not match up the dock. It is a pattern that is detected with the average of the sonar.
- Grasp the disk: To the behavior of the perception system used is the blobfinder, The blobfinder is used to color blob detectors such as the ACTS vision system. At the very moment that the blobfinder of the robot detects the disk, the robot get close to disk maximizing the size of the image (number of pixels) that is getting through the blobfinder, so the robot can reach it.

To this behavior, the first state that can be defined is the final state, which occurs when the robot has grasped the target, that is a disk in this case. it must be also seen clearly another state that is to detect some disk in the range of its sonars", namely, "the robot see a disk". This state is insufficient by itself, since the robot can see the disk on the right hand side or the left hand side or the centre, and none of these three states are equal. If the robot is seeing the disk in front of it then simply the robot the robot follows a straight path, but on the contrary, if the robot is detecting the disk on the right hand side or the left hand side, and it continues a straight path then the robot is going away disk.

Therefore it is necessary to divide the state "to detect disk" into tree states: the first one is "to detect disk on the right hand side", the second one is "to detect disk on the left hand side' , and the last one "to detect disk in front". Since it must be generated different actions for each of those states, the three states must be separated. Another state necessary is "the loss of the disk", that happens when disk is not detected in range of the sonars. This state is not exactly a state, because the robot can not discriminate between different directions in which the disk has been lost, namely, the robot can give up detecting the disk on right hand side or left hand side. Hence, the target loss state must be separated in two states: the loss target on the right hand side and the loss target on the left hand side. Now if the robot loses the target, the robot will learn to get back it.

Before indicating the actions to this behavior, saying that the robot will always carry a constant linear speed, its angular speed is the only thing that varies, so the robot can move to the right or left when the robot needed it. Instead there are three actions, move to the right, move to the left, continue going to straight on, in the latter case the angular velocity would be 0.

The reward was set as +10 if "the robot grasps the disk", $+3/x$ if the robot detects the disk in the range of its sonars, where x is the distance that marked sonars, -1 if the robot gives up detecting the disk either by right or left, and 0 for any other state.

$$f(n) = \begin{cases} +10 & \text{the robot grasps the disk} \\ 3/x & \text{the robot detects the disk in the range of its sonars} \\ -1 & \text{the robot gives up detecting the disk either by right or left} \\ 0/x & \text{in any other state} \end{cases} \tag{5}$$

- Release disk: This behavior by itself is resolved by the architecture of ANLAGIS, as if the robot is on a warehouse,(the warehouse is detected by robot with its sonar system), and in this warehouse there is not any disks or there is a disk or more of the same color that the robot carries it, the robot just leaves the disk in that warehouse.

4.1 State Variable

In this section, the state variables are defined to implement ANLAGIS. This state variables are the inputs in the granulation layer of the ANLAGIS system.

The state variables represents the current state of a robot, namely, where the robot is located, for instance in a warehouse or a dock, if the robot carries a disk or not.

Our environment has five rooms that are a dock and four warehouses, there are four disks (yellow, green, blue, red). There is a state variable that defines the room where the robot is. If the robot is in a dock this state variable is equal 0, and if the robot in some warehouse, the value of this state variable is in range [1,4]. Another state variable is the color of disk that the robot carry. The robot can carry a disk of the four possible color or the robot can not carry any disk as well. The last state variable is the color of warehouse that can be either yellow or green or blue or red, and the warehouse can be white if the warehouse is empty.

The granulation of the state variables is not necessary because of the state variables already represent discrete values. Summarizing, we have $5 \times 5 \times 5 = 125$ neurons at the *ballungen* layer and four possible output actions, the actions are the behavior described above. These state and actions make up the table that will be implemented by Q-Learning.

4.2 Rewards

The rewards that are defined in this section, are used to learn the coordination of behavior. Such rewards are detailed next:

If the robot R_i is in one of the warehouses (W_i) and the color of such warehouse (CW_j) is equal than the color of disk that the robot carries (CD_k) or simply the warehouse is empty, namely, there is not any disk in warehouse, then if robot releases the disk in such warehouses W_i the reward is $+3$. On the other hand, if the robot is in dock, and it does not carry disk, and it grasps any disk, then the reward is $+1$. The rewards is the same if the robot is in the warehouse and it has not disk, and it goes to the dock. If the robot is in one of the warehouses (W_i) and the color of such warehouse is different than the color of disk that robot carries then the reward is $+2$, if the robot goes to another warehouse. Finally, in any other state, the reward would be -2.

5 Experimental Results

The results are focused on being more interesting, as it develops in the coordination of behavior. The evolution of the learning system is described in Fig 3, represents the reward obtained for each episode. It can see at the beginning of the learning phase, when the rewards gained vary per episode, which is exploring the knowledge of the system to improve the coordination of behavior. Once the policy is reached Pi^* optimal control (episode 400), the coordination of behavior is stabilised following the same structure as shown in Fig 4.

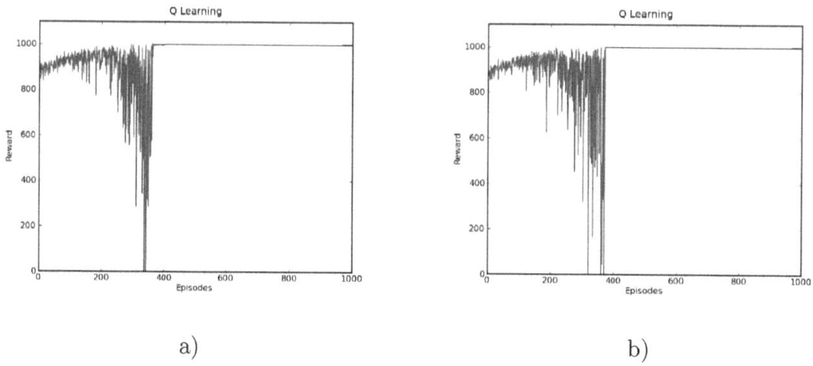

Fig. 3. Learning Curves of the experiment. Reward per episode. *(a)* *(b)*

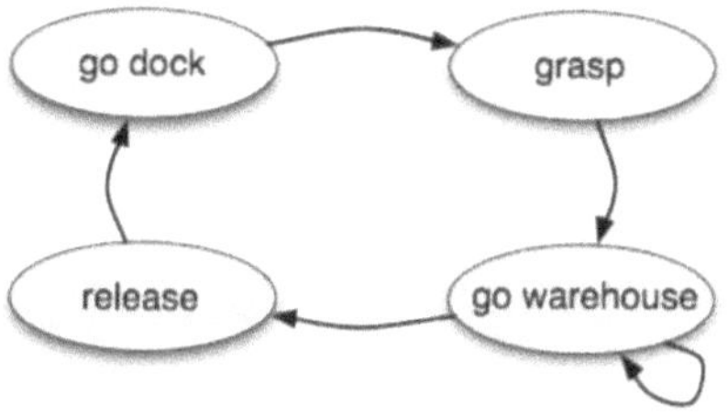

Fig. 4. Diagram of the coordination of behavior

6 Conclusions and Further Work

We have applied ANLAGIS to a problem of coordination a robots team for carring out a cooperative task. The considered task has as goal to store disks with different colors in warehouses. Initially as the robots as the disks are in a special room that we called it dock and the warehouses do not have an specific assigned color, it must be resolved autonomously by the robots.

To the granulation stage, the general ANLAGIS guidelines have been applied. Moreover, the proposed experiment already has discrete states which facilitates the process. For the learning stage, we have successfully applied a conventional Q-learning algorithm due to the applicability restrictions are met. It was the very first time that a Q-learning approach is considered in the ANLAGIS context.

It would be interesting to use an evolutionary computation approach in order to obtain the relationship between the *ballungen* and output layers rather than Reinforcement Learning (i.e. the equivalent to a Q-function or Q-table in the current approach). It would be interesting to compare the convergence speed by using both methods.

Acknowledgements

This work has been partially funded by the Spanish Ministry of Science and Technology, project: DPI2006-15346-C03-02.

References

1. Mahadevan, S., Connell, J.: Automatic Programming of Behavior-based Robots Using Reinforcement Learning. Artificial Intelligence 55, 311–365 (1992)
2. Maravall, D., de Lope, J.: Neuro Granular Networks with Self-learning Stochastic Connections:Fusion of Neuro Granular Networks and Learning Automata Theory. In: Koppen, M., et al. (eds.) ICONIP 2008, Part I. LNCS, vol. 5506, pp. 1025–1032. Springer, Heidelberg (2009)
3. Sutton, R., Barto, A.: Reinforcement Learning: An Introduction. MIT Press, Cambridge (1998)
4. Watkins, C., Dayan, P.: Technical note Q-learning. Machine Learning 8, 279 (1992)
5. `http://www.activmedia.com/`

Safe Crossroads via Vehicle to Vehicle Communication

Javier Alonso, Vicente Milanés, Enrique Onieva, Joshué Perez, and Ricardo García

Instituto de Automática Industrial - CSIC La Poveda, Arganda del Rey,
28500 Madrid, Spain
{jalonso,vmilanes,onieva,jperez,ricardo}@iai.csic.es

Abstract. Driving through crossroads is one of the most dangerous maneuvers. The European community goal of reducing the vehicle accident rate will require a reduction of accidents in crossroads. This paper presents a method of cooperation among vehicles getting into crossroads in order to avoid accidents.

Keywords: Autonomous Driving, Crossroads, Vehicle to Vehicle Communication, Accident Reduction.

1 Introduction

The safety mechanism for automatic driving in a crossroad must be highly reliable and accurate. This paper is going to introduce a method that allows automatic driving in crossroads. This method tries to emphasize in the safety issues instead of trying to optimize the traffic congestions in the crossroads. The experimental results were obtained in our experimental zone and also in the Cybercars2 project final demonstration at Place du Verdun in La Rochelle, France.

The problem to solve is how to merge two flows of cars in a one way crossing. This is a simplified case of study, but can be considered as a starting point for a further research. When a vehicle arrives to a crossing there is always the possibility of an accident, either because of a driving error or because of a misperception of the environment. In this paper we attempt a method to increase the knowledge available to the cars, so accidents can be avoided. This implies cooperation among vehicles and between vehicles and infrastructure. Even if the presentation is about automated driving it is obvious that it can also become an Advance Driven Assistance System to help the drivers.

There are a lot of difficulties to fix for achieving the automatic driving in crossroads. Most of them come from the communications step. Once the car has all the information needed to take a decision it is relatively easy to take a robust decision that allows all the cars to travel safely.

Even with a well tested communications technology, there are lots of problems that can happen in a real car, real world application. In most of our cities there are important buildings that are protected from terrorist attacks, and to do so, there are some radio interferences. The buildings also shade the satellite signals and make signal reflections. The wire connections suffer the vibrations of onboard equipment and sometimes brake. And so on…

R. Moreno-Díaz et al. (Eds.): EUROCAST 2009, LNCS 5717, pp. 421–428, 2009.

2 Automatic Driving though Crossroads

The first step in order to achieve automatic driving is to be able to follow a path. In this project it is performed by storing a map describing a trajectory and using a GPS to monitor the actual position of the vehicle. The comparison of these data permits identify the distance between the actual position and the target one thus permitting to act upon the car controls in order to correct the actual trajectory [1].

This work is about vehicle cooperation. We think the solution of the problems of traffic needs the cooperation among vehicles. In this work the vehicles share information to permit other vehicles to cooperate. Every vehicle will broadcast its own position, as well as its velocity and some more data so other vehicles can take their own decisions.

The vehicle cooperation is not enough. We think the infrastructure is to be part of the solution. Some information the infrastructure has, might be necessary for the orderly driving in the crossroad, for example a command to stop because of an accident blocking circulation. We are actually proposing that the information messages that imply a modification of the driving behaviour (because an accident, rain, etc…) in road be actually sent to the cars directly.

The vehicles have to know in advance the trajectory so the crossroads positions can be determined. This is achieved through a map containing the information about the driving area, be it a nation, or, more likely, a thematic park. In this way the analysis of the map permits the identification of the next crossroad in the path, what on its turn permits to identify which messages from other vehicles have to be analyzed in order to determine the course of action in the crossroad.

After the number of vehicles coming into the crossroads has been determined, the decision taking algorithm will permit to adopt a decision on whether to stop or to proceed, and in this case whether to maintain the speed or to reduce it. This algorithm can be as simple as stopping if there is some vehicle coming from the right side into the crossroad, or as complicate as to introduce rights of way and determination of time other vehicles will take to get into the crossroad in order to adapt their speed and not to have to stop. In this paper we only deal with stopping if a car comes from the right side.

3 Experiments

These experiments have been carried out in La Rochelle as a part of the Cybercars 2 final demonstration involving TNO (Netherlands Organization for Applied Scientific Research), INRIA (Institut National de Recherche en Informatique et Automatique) and CSIC (Consejo Superior de Investigaciones Cientificas) cars. Two vehicles start moving in a one way eight shaped loop (8). The starting positions and speeds have been selected to ensure that both vehicles will arrive to the cross section of the loop at the same time but arriving from different directions. The vehicle which has a car coming from his right has to stop and let the other car to continue and cross the intersection.

To be able to perform this maneuver each car needs the position and speed of the other car. The accuracy and reliability of that information is really important for the safety of the maneuver, so some other information is added to the communication package. A timestamp has been added to ensure that the information is not too old. And the DGPS quality is also added to the comm. package to be able to stop the experiment if the accuracy of the DGPS position measurements goes down.

In addition to this in-code safety measures, some light displays have been added to be able to check whether communications are still alive, and to inform of the intersection state. So, the car occupants will be able to know if the intersection is occupied by any other car and also if his car is going to stop before the intersection entrance to let pass the car coming from the right.

3.1 Intersection Scenario

TNO and CSIC cars head the same intersection. The starting places are shown in figure 1. The CSIC car stops at intersection to let TNO car passes because TNO car comes from its right side. The TNO car arrives at the intersection and crosses it without stopping because of the precedence to the right rule.

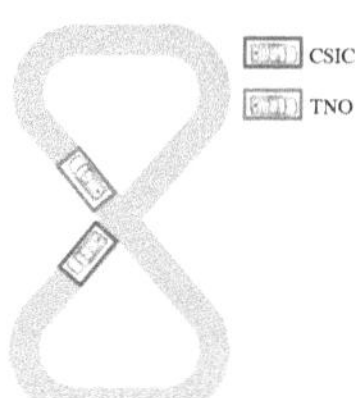

Fig. 1. Starting positions of La Rochelle intersection scenario

3.2 Algorithm: Taking Decisions Based on Time Calculations

Once the car is near the crossroad (at a distance that allows it to stop before entering in the intersection) it should:

1. Check whether there is another car in the intersection.

- If the intersection is free, then it can proceed to step 2).
- If there is another car in the intersection, but its direction is the lane, it can also proceed to step 2, "ACC" and "Stop & Go" guarantee that the safe distance will be kept.
- If there is another car in the intersection and it is not in the same lane, it must stop before getting into the crossroad.

2. Check whether there is another vehicle in the lane that leads to the intersection.

- If there is another car waiting to enter in the intersection and it is on the lane that comes to the intersection from the right, it must stop and let the waiting car to resume its way.

- If there is another car coming into the intersection from the right side and the expected time to arrive to the intersection is equal or less than the time to cross the intersection, it must stop and let the coming car pass.
- If there are not cars in the intersection, or they are at a safe distance to the right, it can proceed and cross the intersection.

3.3 Procedure: Take Decisions Based on Time Calculations

The implementation of this procedure has been done in Matlab for TNO's vehicle and also in C++ code for CSIC's vehicle.

3.4 La Rochelle Tests

These tests were made as CiberCars2 final demo for European Commission reviewers. The cooperation among heterogeneous vehicles was proved. ACC and "Stop and Go" manoeuvres were tested in conjunction with INRIA y Robosoft. And crossroads cooperative manoeuvres were tested in conjunction with TNO. The results of this experiment have a great added value because they were made in a real urban environment. The localization of Verdun square in the middle of La Rochelle was really convenient for the dissemination aspects of the project, but it represents a difficult challenge for the communications and also for the use of GPS systems.

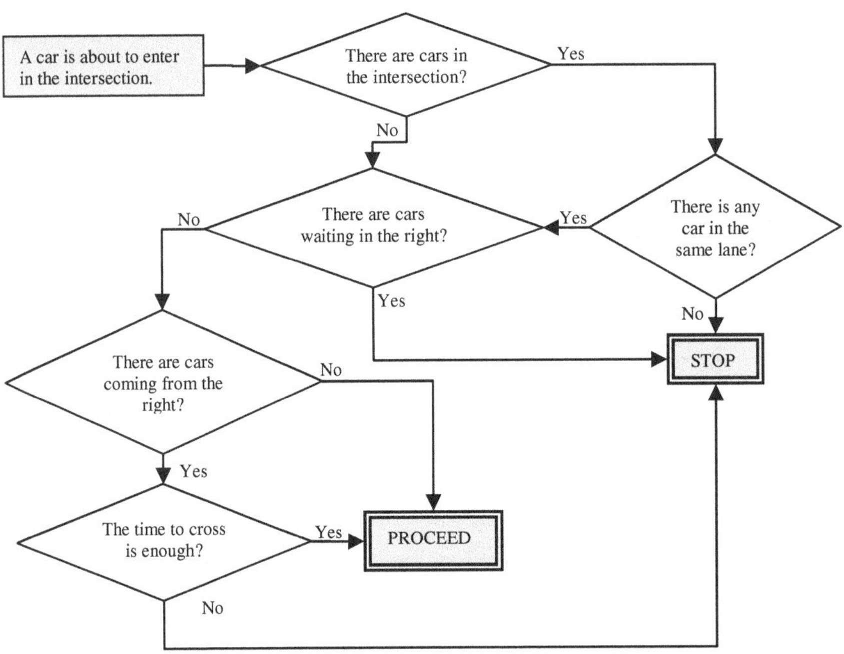

Fig. 2. Intersections decision flow chart based on time

Fig. 3. Verdun square in La Rochelle, France

The buildings might block the signal from near the horizon satellites, make signals to rebound and decrease the quality of the signals. The security devices in the government buildings and the high concentration of electronic aerial signals also interfere in the communications. But, even with so many difficulties the final test was successful.

3.5 Results

The data stored by CSIC car during the CC2 final demonstration are shown in figure 4. The signals shown are:

- Occupation signal: dark blue line, it's 1 when there is another car at the intersection or in the lane that arrives to the intersection from the right side. Otherwise it's 0.
- Speed signal: pink line, it's measured in km/h. It goes from 8 to 10 km/h (goal speed was 9 km/h, a really low speed for a gas car).
- Distance to stop signal: yellow line, it's the distance from the car GPS receiver to the selected point to stop the car before it enters to the intersection.
- DGPS quality signal: cyan line, it shows the GPS quality, when it drops from 4 to 2, the DGPS is not reliable and the car must drive using inertial sensors and odometry [2].

Figure 4 shows the performance of the car's behavior during two loops. The experiment was designed to force the cars to arrive to the cross at the same time twice. The car reference speed is lowered once the system detects a big discrepancy between the DGPS and the estimated position. Once the DGPs quality is recovered the speed reference is set to 10 Km/ hour again.

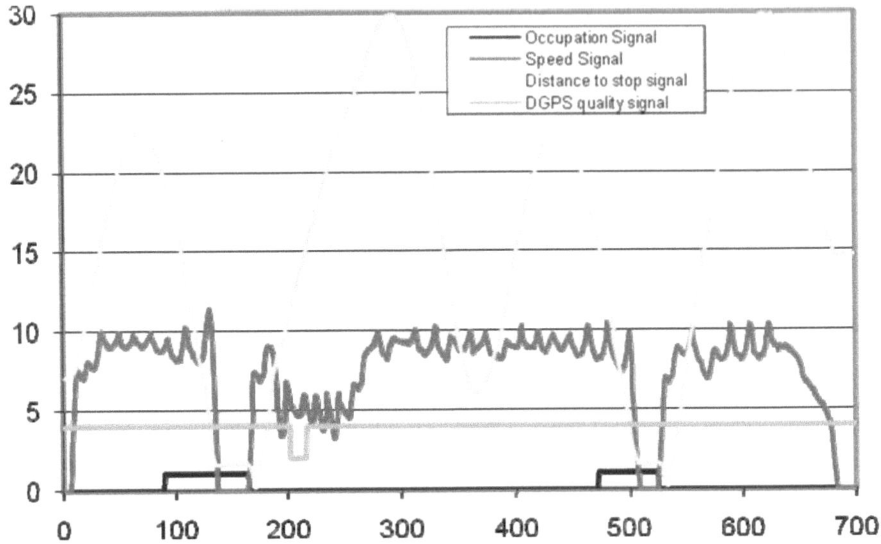

Fig. 4. La Rochelle intersection results

3.5.1 First loop

Figure 5 shows the positions of the CSIC car and the received positions of the TNO car by communications. The data shown start and end at the same time. The CSIC car stops until TNO car crosses the intersection.

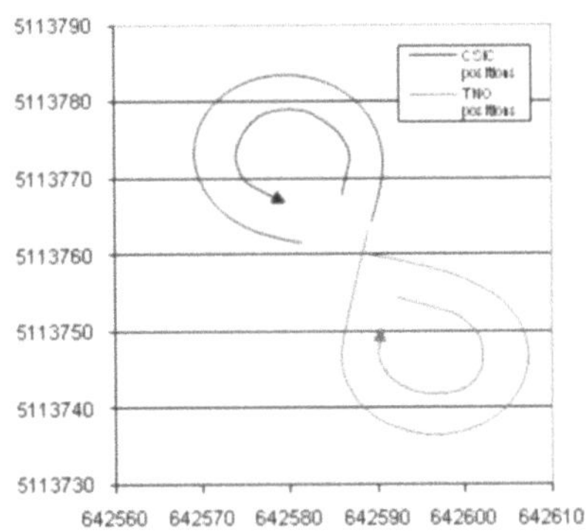

Fig. 5. Vehicle trajectories until CSIC car stops during the first loop of La Rochelle demo

Figure 6 shows the occupation signal turning from 0 to 1 since second 9 (90x100ms), but it has no influence on CSIC car behavior until second 13,2. That's the point where the distance to the stop falls below 3 meters. Then it slows down and uses the precision breaking system to stop the car 1,5 meters before the stop point (this distance was chosen to provide extra safety). Once the TNO vehicle leaves the intersection the occupation signal returns to 0 and CSIC car can resume its route.

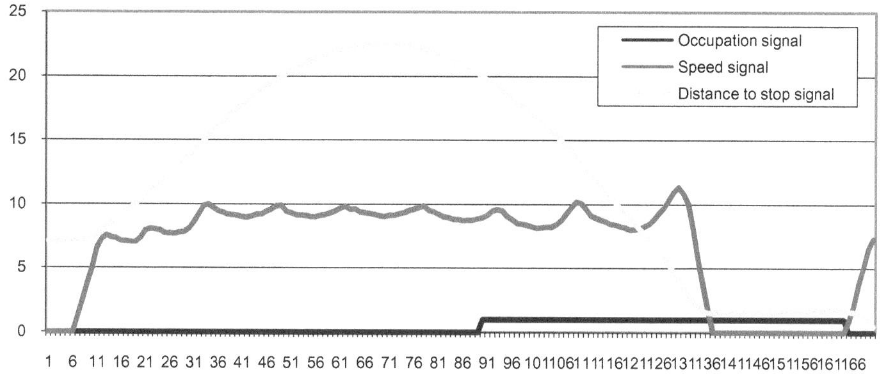

Fig. 6. Occupation, speed and distance to the stop for the first loop of La Rochelle demo

3.5.2 Second loop

The results shown in figure 7 were recorded after a complete loop. The speed difference between both cars is not enough to make the CSIC car reach the intersection and cross it safely, so it must stop again and let TNO car pass. Figure 7 shows the positions of CSIC and TNO car and occupation, speed and distance to stop point signals from CSIC car.

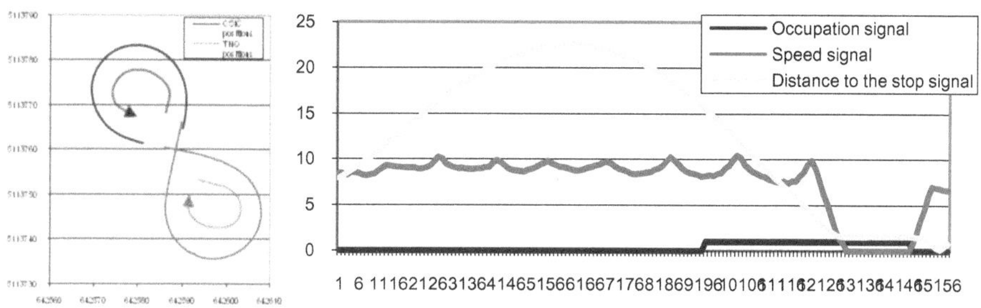

Fig. 7. Results for the second loop of La Rochelle demo

The results from both loops are more or less the same (figure 6 and figure 7). The small differences come from initial positions, speeds, etc... It's also seen how the occupation signal turns on and how the CSIC car stops when the distance to the stop point becomes lower than 3 meters.

4 Conclusions

Automatic driving is far for becoming a fact in cars in the near future, but we consider it will eventually come. The experiment described in this paper shows that it is possible to drive autonomously in simple crossroads or intersections by sharing information among vehicles.

Acknowledgments

This work has been made thanks to the projects: TRANSITO (TRA 2008-06602-C03-01), ENVITE (MFOM T7/2006), and CYBERCARS-2: Close Communications for Cooperation between Cybercars (IST-2004-028062).

References

1. Milanés, V., Naranjo, J.E., González, C., Alonso, J., de Pedro, J.: Autonomous vehicle based in cooperative GPS and inertial systems. Robotica 26, 627–633 (2008)
2. Naranjo, J.E., González, C., de Pedro, T., García, R., Alonso, J., Sotelo, M.A., Fernández, D.: AUTOPIA Architecture for Automatic Driving and Maneuvering. In: IEEE Intelligent Transportation Systems Conference, Toronto, Canada, pp. 1220–1225 (2006)

Cooperation Enforcement Schemes in Vehicular Ad-Hoc Networks*

C. Hernández-Goya, P. Caballero-Gil, J. Molina-Gil, and C. Caballero-Gil

Department of Statistics, Operations Research and Computing,
University of La Laguna, 38271 La Laguna, Tenerife, Spain
{mchgoya,pcaballe}@ull.es,
{jezabelmiriam,candido.caballero}@gmail.com

Abstract. Vehicular Ad-hoc NETworks (VANETs) will provide many interesting services in the near future. One of the most promising is commercial application. In such a case, there will be necessary to motivate drivers to cooperate and contribute to packet forwarding in Vehicle-TO-Vehicle and Vehicle-TO-Roadside communications. This paper examines the problem, analyzes the drawbacks of known schemes, and proposes a new secure incentive scheme to stimulate cooperation in VANETs.

Keywords: Cooperation, Vehicular Ad-Hoc Network, VANET.

1 Introduction

A Vehicular Ad-hoc Network (VANET) is a special type of ad-hoc network used to provide communications between On-Board Units (OBUs) in nearby vehicles, and between OBUs in vehicles and Road-Side Units (RSUs), which are fixed equipment located on the road [14].

The main goal of VANETs is the prevention of road accidents and traffic jams, with a direct effect on safety, efficiency and comfort in everyday road travel. However, their structure will also allow taking advantage of other Added-Value Services such as: advertising support [6], request/provide information about nearby companies, access to Internet, etc. Consequently, we can say that the main objective of VANETs is the deployment of different applications related to the design of an efficient and reliable Intelligent Transportation System.

In particular, this paper deals with the topic of Inter-Vehicle Communication when the systems in a VANET do not rely on RSUs, and consequently constitute a Mobile Ad-hoc Network (MANET).

The main advantage of VANETs is that they do not need an expensive infrastructure. However, their major drawback is the comparatively complex networking management system and security protocols that are required. This difficulty is mainly due to some specific characteristics of VANETs that allow differentiating them from the rest of MANETs such as their hybrid architecture, high

* Research supported by the Spanish Ministry of Education and Science and the European FEDER Fund under TIN2008-02236/TSI Project, and by the Agencia Canaria de Investigación, Innovación y Sociedad de la Información under PI2007/005 Project.

R. Moreno-Díaz et al. (Eds.): EUROCAST 2009, LNCS 5717, pp. 429–436, 2009.

mobility, dynamic topology, scalability problems, and intermittent and unpredictable communications. Consequently, these features have to be taken into account when designing any management service or security protocol.

In order to bring VANETs to their full potential, appropriate schemes to stimulate cooperation need to be developed according to the specific properties and potential applications of VANETs. Many incentive schemes to stimulate cooperation in ad-hoc networks may be found in the bibliography [5] [9] [15] [10] [11]. Some authors have made first approaches to the topic of cooperation in VANETs [3] [4] [12] [13]. Related to the proposal here described, Buttyan and Hubaux proposed in [1] and [2] the use of virtual credit in incentive schemes to stimulate packet forwarding. Also, Li et al. discussed some unique characteristics of the incentive schemes for VANETs in [7] [8] and proposed a receipt counting reward scheme that focuses on the incentive for spraying. However, the receipt counting scheme proposed there has a serious overspending problem. Based on the specific characteristics of VANETs, a more comprehensive weighted rewarding method is proposed here.

In particular, the proposed scheme is based on incentives where the behavior of a node is rewarded depending on its level of involvement in the routing process. Schemes based on reputation were here discarded due to the high mobility of nodes in VANETs, which makes infeasible to maintain historical information about peers behavior.

Note that an important problem that must be dealt with in rewarding incentive schemes is the possibility for selfish or malicious users in the vehicles to exaggerate their contribution in order to get more rewards. In our proposal, we assign different possible incentives to vehicles according to their contribution in packet forwarding, in an effort to achieve fairness and provide stimulation for participation. Our scheme utilizes a weighted rewarding component to decide the specific incentive in each case so they help to keep the packet forwarding attractive to the potential intermediate vehicles.

This paper is organized as follows. Section 2 contains basic definitions related to the forwarding process in VANETs and describes in detail the proposal. The paper ends with the Section of conclusions and open questions.

2 Forwarding Trees in VANETs

Figure 1 shows a typical packet forwarding process in VANETs, called *Forwarding Tree*. In such a figure several important features of routing in VANETs are represented:

1. The root node corresponds to the source vehicle that first sprays the message.
2. Each intermediate vehicle corresponds to one node in the tree.
3. Each node ignores those packets that it had previously received. Consequently, every vehicle is present just once in every forwarding tree.
4. Each link in the tree corresponds to an encounter in the vehicular network, which is associated with a timestamp and the spatial coordinates indicating the position of the vehicles.

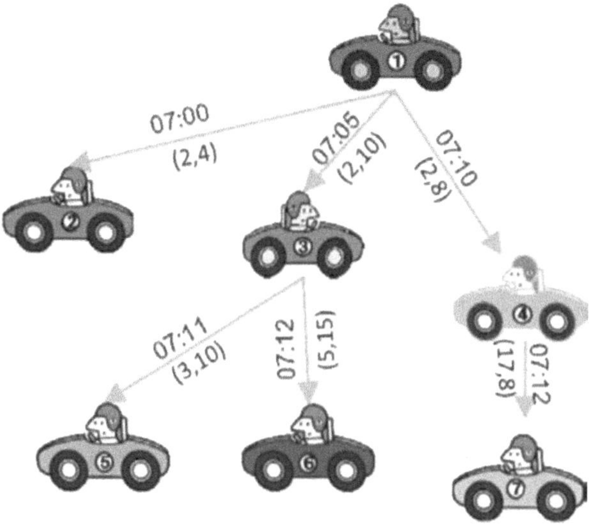

Fig. 1. Packet forwarding

According to the store-and-carry paradigm [7] [8], if an intermediate vehicle stores a packet for a long time or actively sprays the packet to other vehicles, the packet will be either more likely to reach the intended destination, or to arrive to more destinations, depending on the specific goal of the routing. Therefore, by simply combining storage time and number of sprays, we can define a useful contribution metric for the intermediate vehicles. In order to stimulate intermediate vehicles to contribute more, the source vehicle should reward each intermediate vehicle according to its contribution.

Initially, the contribution C_i to packet forwarding of a node i during the forwarding process may be modeled as a linear convex combination balancing numbers of forwarding f_i and the period the packet is stored t_i:

$$C_i = \alpha t_i + (1 - \alpha) f_i.$$

However, this basic model implies a constant share reward R which is promised for the source node to each intermediate node. This model may cause an over-spending problem because the source vehicle cannot guess in advance the total reward since the number of nodes in the tree cannot be predicted easily. Such a problem might be solved maintaining constant the total reward and calculating the reward associated to each intermediate node R_i after the packet reaches the destination according to the following formula:

$$R_i = \frac{R \cdot C_i}{C} \text{ where } C = \sum_i C_i$$

When the packet reaches the destination, each node i that participated in the forwarding should report its contribution C_i to the source. The final contribution C is calculated through the sum of the partial contribution of each node in the forwarding tree. Each intermediate node will receive R_i as reward for forwarding.

This model cannot be considered neither a good solution because selfish nodes might prefer keeping the packet rather than retransmitting it since they do not know in advance how much they can earn for forwarding and/or they might prefer not to share the reward. It happens when an intermediate node forwards the packet to a non final node because its proportional reward might decrease.

Our proposal tries to solve this problem. We propose a new function in which three parameters are used. These parameters may be interesting both for the source node and/or for the forwarding node. In particular we consider the following notation to describe the parameters for the computation of rewards:

- Packet delivery deadline T_j.
- Period t_{ij} that packet j is stored by node i.
- Distance d_{ij} between source and destination nodes when the packet j is relayed by node i.
- Maximum distance D_j where the information in the packet j is considered interesting by the receivers.
- Number of forwardings f_{ij} of package j by node i.

Each of the parameters considered in this convex function has a balancing factor, represented by α_1,α_2 and α_3. The value that is assigned to each α_i depends on on the relevance that the source node prefers to assign to each component represented in the contribution function:

$$C_{ij} = \alpha_1 T_j(1 - e^{-t_{ij}}) + \alpha_2 f_{ij} + \alpha_3(-D_j(1 - e^{-d_{ij}}) + D_j) \text{where} \sum_{k=1}^{3} \alpha_k = 1.$$

In the next subsections each part of this function will be detailed, and the justification why they are needed and the repercussion they will have in the contribution function will be given.

2.1 Time

As discussed above, the time is one of the most important parameters when trying to assure that a packet reaches the intended destination. If a vehicle stores a packet for a long time, it could forward the package to more vehicles. However, this parameter could produce a selfish behavior because a node could prefer not to forward it and in this way not to share the final reward with potential forwarding nodes. This effect is avoided by considering in the metric here proposed the component associated to the following formula:

$$T_j(1 - e^{-t_{ij}}).$$

This function corresponds to the Stokes formula, which has a characteristic asymptotical behavior. This function is intended to set a maximum time T_j that a node should store one packet. Note that the value of contribution increases when time increases. When t_{ij} reaches the threshold T_j, the growth of contribution stops. In this way, the selfish behavior can be avoided because if the time threshold is set properly, the vehicles that retransmit the packet before

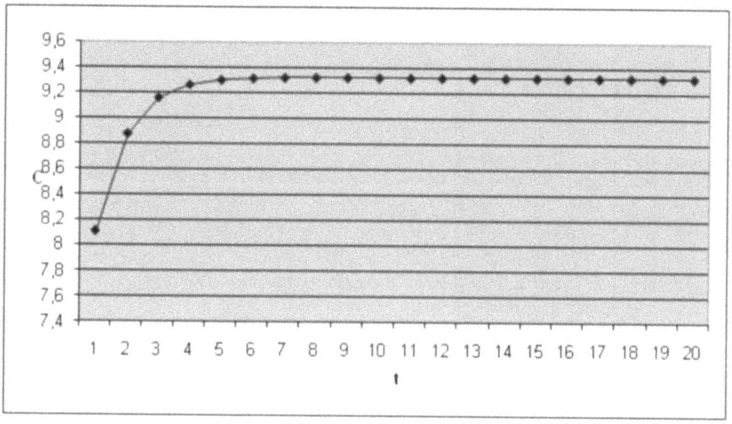

Fig. 2. Contribution versus time

deadline will have increased their contribution. The maximum time T_j has been set to 10 in this example and the three balancing factors have exactly the same values. The remaining parameters have been determined so they do not affect to the final value.

Note that the value of contribution increases when the time increases. When t_{ij} reaches the threshold T_j, the contribution increase stops. In this way, both selfish behavior and forwarding after deadline are discouraged because vehicles that retransmit the packet before deadline will have their contribution increased.

2.2 Forwarding

The second term in the proposed contribution metric is related to the ultimate goal of our work. It deals with measuring the forwarding of packets by each intermediate node. This process is quite simple. It has not any restriction such as maximum or minimum possible values. It consists of increasing the contribution of node i to relay the packet j:

$$f_{ij}.$$

In order to check how this term affects to the final contribution, the same process of previous section has been followed. The other parameters have been defined as constant so they do not affect to the final value. The result is shown in Figure 3.

As shown in Figure 3, the behavior in this case is quite simple: the more collaborate the vehicles in forwarding a packet, the bigger their final contribution is. In the proposed function, this parameter is the one that increases the contribution faster. For this reason the balancing factor for this parameter must be higher than the other two factors in order to encourage the forwarding of packets.

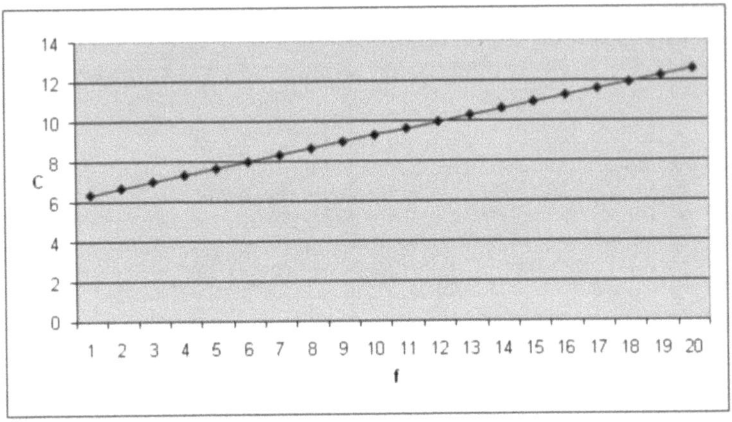

Fig. 3. Contribution versus forwarding

2.3 Distance

The evaluation of the effect of distance in the share rewarding process is the goal of the third term of the contribution function. This has been developed thinking that information generated in a determined point is not interesting out of a radius distance. With this idea in mind, when the vehicles go too far from the source of the original packet, this value decreases.

For example, if we talk about an accident in Madrid city center, it has not sense that the message reaches Alcalá de Henares. Another situation where this idea is applicable is where the information is sent by a commercial or restaurant. These situations are represented in Figure 4.

This term is similar to the one related to time commented in subsection 2.1. The goal is to obtain a function which asymptotical behavior tends to zero when distance is near to D_j. The value D_j is established by the source node. The expression that models this behavior is:

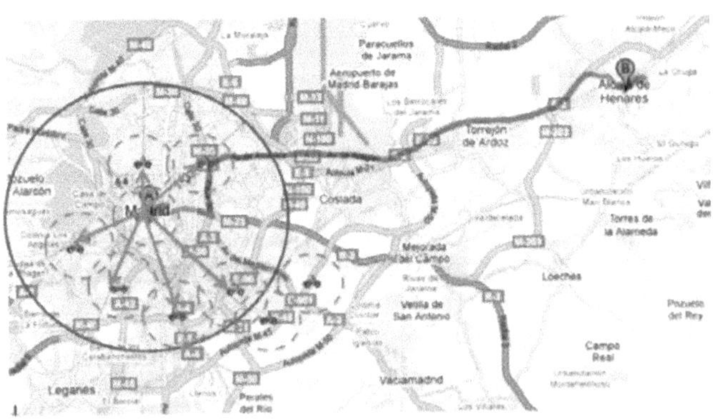

Fig. 4. Radio

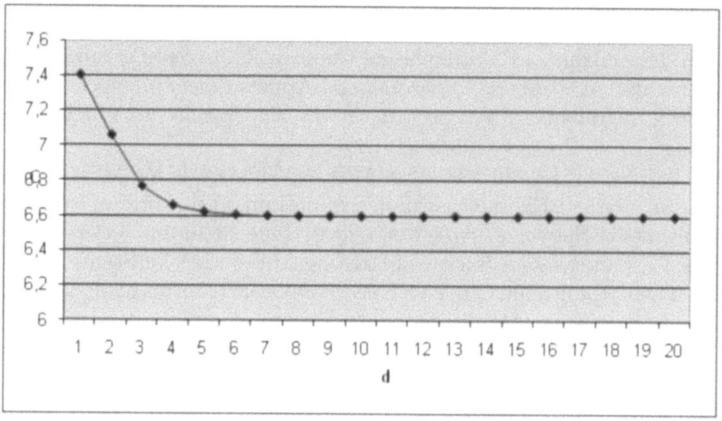

Fig. 5. Contribution versus distance

$$-D_j(1 - e^{-d_{ij}}) + D_j.$$

In order to appreciate better the characteristics of the contribution of this component to the contribution metric we have followed the same process of the previous sections. Time and retransmission have been defined as constants values that do not affect to the final contribution while distance is a variable. In Figure 5 the behavior of the contribution function is showed.

In this case the graphic clearly shows that when the vehicle moves away from the source, its contribution decreases. When it reaches certain point, this value is 0, so that the vehicle will not get any benefit if it retransmits the packet outside the set radio.

3 Conclusions and Open Questions

In this paper we have seen that a simple adaptation of known cooperation enforcement schemes defined originally for MANETs is not adequate to incentivize cooperation in VANETs. Consequently, we have proposed a new scheme where incentives are defined by a convex function that depends on different parameters. We have designed a metric for contribution according to the characteristics of VANETs and to parameters that are important both for source node and for enforcing cooperation among nodes. We conclude from our study that when designing these methods for distributing a reward, the parameters to be taken into account should be carefully assessed according to the network conditions.

Since this is a work in progress, many open questions exist such as the simulation of the new approach using Network Simulator NS2 on different scenarios and network conditions so that nodes will have different transmission ranges, like in the real world. Another open question is the analysis of how can data associated to traffic and weather conditions can be used in order to improve the efficiency of the proposal.

References

1. Buttyan, L., Hubaux, J.P.: Stimulating Cooperation in Self-Organizing Mobile Ad Hoc Networks. ACM Mobile Networks and Applications 8(5) (October 2003)
2. Buttyan, L., Hubaux, J.P.: Security and Cooperation in Wireless Networks. Cambridge Univ. Press, Cambridge (2007)
3. Dotzer, F., Fischer, L., Magiera, P.: VARS: A Vehicle Ad-Hoc Network Reputation System. In: Sixth IEEE International Symposium on a World of Wireless Mobile and Multimedia Networks, WoWMoM 2005, June 13-16, pp. 454–456 (2005)
4. Fonseca, E., Festag, A.: A Survey of Existing Approaches for Secure Ad Hoc Routing and Their Applicability to VANETS, Technical Report NLE-PR-2006-19, NEC Network Laboratories (March 2006)
5. Ho, Y.H., Ho, A.H., Hamza-Lup, G.L., Hua, K.A.: Cooperation Enforcement in Vehicular Networks. In: International Conference on Communication Theory, Reliability, and Quality of Service, CTRQ 2008, June 29-July 5, pp. 7–12 (2008)
6. Lee, S., Pan, G., Park, J., Gerla, M., Lu, S.: Secure Incentives for Commercial Ad Dissemination in Vehicular Networks. In: MobiHoc 2007, Canada, September 9-14 (2007)
7. Li, F., Wu, J.: A Winning-Probability-based Incentive Scheme in Vehicular Networks. In: Proc. of IEEE International Conference on Network Protocols (ICNP), poster abstract (2008)
8. Li, F., Wu, J.: FRAME: An Innovative Incentive Scheme in Vehicular Networks. In: Proc. of IEEE International Conference on Communications, ICC (2009)
9. Liu, P., Zang, W.: Incentive-based modeling and inference of attacker intent, objectives, and strategies. In: Proc. of the 10th ACM Computer and Communications Security Conference (CCS 2003), Washington, DC, October 2003, pp. 179–189 (2003)
10. Shastry, N., Adve, R.S.: Stimulating cooperative diversity in wireless ad hoc networks through pricing. In: IEEE Int. Conf. on Communications (June 2006)
11. Srinivasan, V., Nuggehalli, P., Rao, R.R.: Cooperation in Ad Hoc Networks. In: Proc. of Infocom, San Francisco, CA (2003)
12. Wang, Z., Chigan, C.: Countermeasure Uncooperative Behaviors with Dynamic Trust-Token in VANETs. In: IEEE International Conference on Communications, ICC 2007, June 24-28, pp. 3959–3964 (2007)
13. Wang, Z., Chigan, C.: Cooperation Enhancement for Message Transmission in VANETs. Wireless Personal Communications: An International Journal 43(1), 141–156 (2007)
14. Yousefi, S., Mousavi, M.S., Fathy, M.: Vehicular Ad Hoc Networks (VANETs): Challenges and Perspectives. In: 6th Int. conference on ITS Telecommunications (ITST 2006), China (2006)
15. Zhong, S., et al.: Sprite: A Simple, Cheat-Proof, Credit-Based System for Mobile Ad-Hoc Networks. In: Proceedings of IEEE Infocom, San Francisco, CA, USA (2003)

Cooperative and Competitive Behaviors in a Multi-robot System for Surveillance Tasks

Yadira Quiñonez, Javier de Lope, and Darío Maravall

Perception for Computer and Robots
Universidad Politécnica de Madrid
Campus de Montegancedo, 28660, Madrid, Spain
ay.quinonez@alumnos.upm.es, javier.delope@upm.es,
dmaravall@fi.upm.es

Abstract. In this paper we present a control architecture for multi-robot systems in dynamic environments, where the low level behaviors are obtained through artificial neural networks and evolutionary algorithms to achieve collaborative behaviors in a multi-robot system. As an example, we have cooperative tasks establishing a surveillance scenario stressing cooperation and competition between them.

Keywords: Multi-robot systems, Reactive Behaviors, Artificial Neural Network, Evolutionary Algorithms, Surveillance, Neurocontrollers.

1 Introduction

In recent years, research on control of Multi-Robot Systems (MRS) has attracted the attention of several researchers from the scientific community due to the advantages provided by these systems with respect to a single robot. A MRS is composed by a series of robots that interact with each other to achieve a common goal; some typical applications are execution of a complex task beyond the limits of a single robot, tasks that cover a region and tasks that require redundancy. The main advantages of MRS with regard to a single robot are that can perform more efficiently, possible to increase fault tolerance, distributed sensing and actuating and ability to complete a task more quickly.

A MRS is used to increase the effectiveness of the system, that is, a MRS can perform complex tasks in less time and with higher quality. The control of a MRS is not an easy task. The robots have to communicate, exchange information or interact in some way to achieve a common task. This requires taking into account some features such as: the type of control, system's typologies, communication, etc. During these years, the scientific community has developed some research progress in cooperative robotics with respect to mechanisms for coordination and communication [1]. Dudek *et al.* [2] present a taxonomy for multi-agent robotic systems, where proposed a classification based on the size of the team, communication parameters (communication range, bandwidth and topology), the reconfigurability of the team, the processing capacity of each member and the team composition (homogeneous vs. heterogeneous robots). Farinelli *et al.* [3] propose

R. Moreno-Díaz et al. (Eds.): EUROCAST 2009, LNCS 5717, pp. 437–444, 2009.

a classification based on different levels of coordination (unaware, aware but not coordinated, weakly coordinated, strongly coordinated systems) and introduce a classification based on the coordination dimensions (cooperation, knowledge, coordination, organization) and system dimensions (communication, team composition, system architectures and team size). Finally, a taxonomy based on coordination mechanisms and on multi-robot task allocation is presented in [4].

2 Multi-robot System for Surveillance

As we have previously commented a MRS has several advantages over a single robot, however, achieve the communication and coordination in a dynamic environment between them is not an easy task. There are many problems that need to be considered in a dynamic environment, for example, multiple moving objects, various obstacles, team members, among others. All this makes more difficult to achieve coordination between robots.

Currently one of the main interests of the international community is design strategies for communication and coordination between robots, we present an architecture of communication and coordination for MRS in dynamic environments, which allows robots to modify their behavior to cope with the environmental changes or actions performed by other robots, in order to obtain cooperative behavior that allows them to achieve a common goal. For this purpose we use a robotic device simulator (*Player*) and a multi-robot simulation in 2D (*Stage*) [5].

In the general outline of the MRS here proposed, a set of behaviors associated with each member of a robot team with a common goal is considered: to follow and catch a robot, or flee to avoid being catched. To achieve collaborative and competitive behaviors in a MRS in an unknown environment, we have established a surveillance scenario for illustrating the proposed control architecture: the red robots must patrol and detect the blue robots in an office-like environment (see Fig.1). The objective of red robots is to work coordinately in order to catch the blue robots (collaborative), meanwhile the goal of blue robots is to avoid be caught by any member of red robots (competitive).

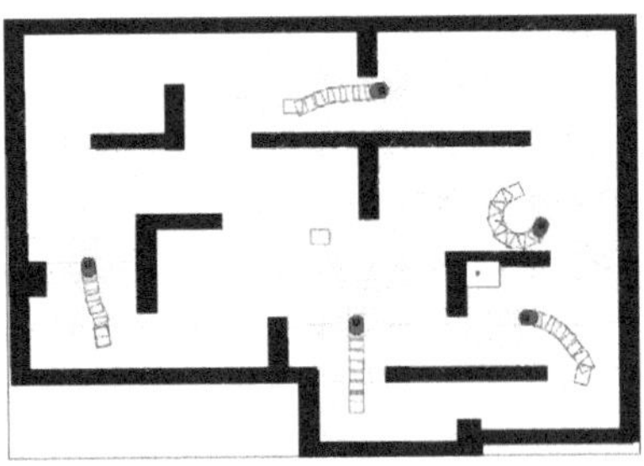

Fig. 1. Multi-robot system

3 Basic Behaviors Description

We have defined several low level behaviors to achieve the goal. We have divided them in navigation and communication behaviors as follows:

3.1 Navigation Behaviors

- *Searching robot*: This behavior is only available for the red robots. It makes that the robots wander around the environment, looking for the blue robots (see Fig.2 a). It also makes that the robot releases 'pucks' or special indirect communication marks. The pucks can be greens (an intruder is been detected) or pinks (that zone is currently explored and it is empty).
- *Avoiding obstacle*: This obstacle avoidance behavior generates a trajectory in which each robot dynamically avoids the physical contact with other elements of the environment as obstacles, walls, or other robots. It uses the sonar information in order to determine the elements positions and generate a turn speed to avoid the possible contact (see Fig.2 b).
- *Unblocking*: While the robots are navigating by the environment, they can be blocked for the obstacles in the environment. An unblocking behavior has been defined in order to solve it. Basically, the behavior makes that the robot goes to the opposite side in which the obstacle has been detected by the sonar (see Fig.2 c).
- *Following robot*: This behavior makes that red robots follow to the blue robots when they can be perceived through the camera. A tracking process based on the robot color is performed and some moments are calculated. The behavior tries to maintain the horizontal position coordinate of the followed robot in the center of the image and also tries to increase the blob area associated with the detected robot (see Fig.2 d).
- *Avoiding robot*: This is a complementary behavior of robot following. It uses a similar scheme but in this case the behavior tries to generate a trajectory that maintains the horizontal position coordinate of the avoided robot far from the center of the image. When the blue robots detect red robots avoid going to a different address to not be caught.

3.2 Communication Behaviors

- *Releasing puck*: The communication between red robots is indirectly made by means of 'pucks'. When a red robot detects a blue robot, it releases a green puck and when a red robot does not detect a blue robot in a room (see Fig.3), it releases a pink puck.
- *Following puck*: This behavior makes that red robots follow the puck when it can be perceived through the camera. When the red robots detect green pucks interpret them as a signal and they will go to that zone (see Fig.3).
- *Avoiding puck*: When the red robots detect pink pucks avoid them because interpret them as a signal that area has been explored and they will explore different rooms.

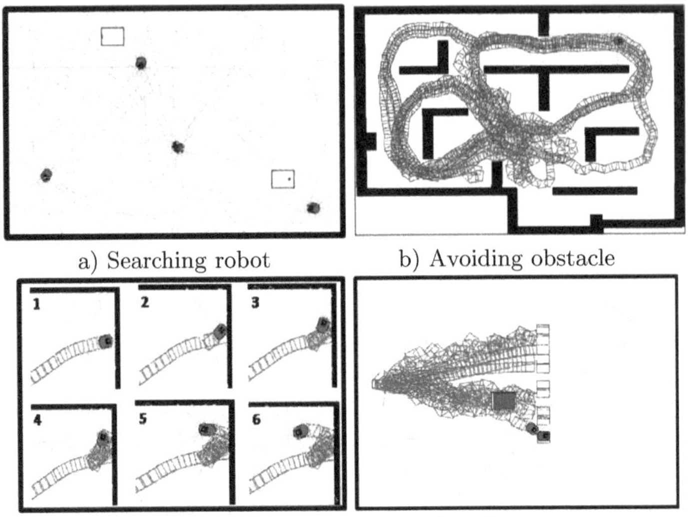

a) Searching robot b) Avoiding obstacle

c) Unblocking d) Following

Fig. 2. Navigation behaviors

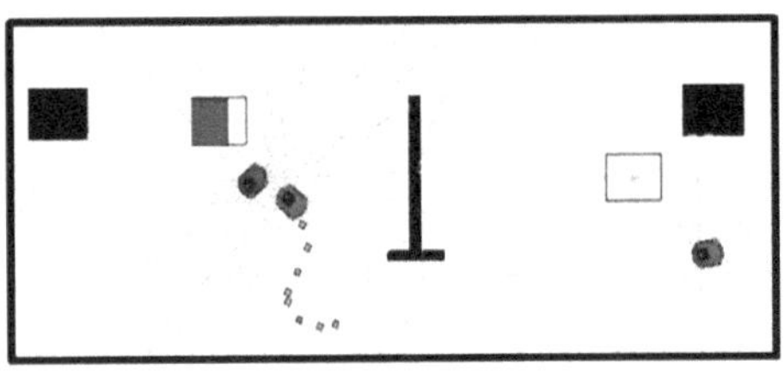

Fig. 3. Communication behaviors

4 Control Architecture

We have defined an architecture for controlling MRS based on behavior. We considered the taxonomy proposed in [3] to classify the coordination approaches in MRS. Our MRS consists of multiple mobile robots, with a model of distributed control where robots are completely autonomous in the decision process [6]. Team composition is homogeneous, that is, robots do not have any advantage or disadvantage with regard to the other. Indirect communication is based on the observed physical environment, where any alteration or modification is interpreted as a communicative act; this paradigm is known as "cooperation without communication" [7]. The architecture is reactive since it is based on behaviors and each individual robot reacts to the changing environment to reorganize its own task.

Fig. 4a and Fig. 4b depict the control architecture for red robot behaviors and blue robot behavior, respectively.

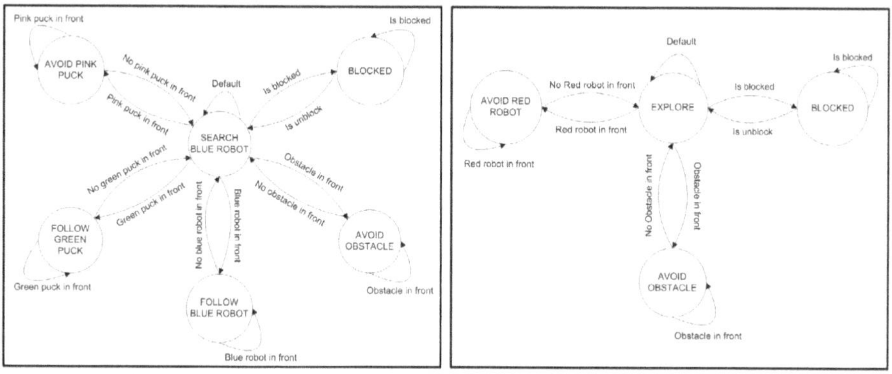

a) Red robots behaviors b) Blue robots behaviors

Fig. 4. Control arquitecture

5 Evolving Neurocontrollers for Low Level Behaviors

Evolutionary Algorithms (EA) are techniques usually employed for searching and optimizing based on natural processes of evolution, in order to solve complex problems using computer models. Currently, there are different EA such as Evolutionary Strategies, Genetic Algorithms, Evolutionary Programming, Genetic Programing, Differential Evolution, Cultural Algorithms, Coevolution [8]. Despite being developed independently, they share the goal of imitating natural evolutionary processes.

Each behavior is modeled by means of a artificial neural network (ANN), whose input parameters are the reading of the sensors and the x position of the robot; and the output parameters correspond to the robot's rotation speed, considering the forward speed as constant for all behaviors.

The ANNs are evolved through genetic algorithms. Fig. 5 shows a flowchart describing the general structure of EA. First, it initiates with a population that is randomly generated; after this, we evaluated and sorted the fitness of each individual of the population, then we select the best individuals randomly, where individuals with better fitness will have more probability of being selected. Then, we apply the mutation operator to each of the individuals selected to generate a new individual and renew the population. The algorithm converges when the best individual of the population satisfies the solution to the problem. The following parameters were considered to the behaviors evolution: the number of generations, the number of individuals, the number of steps and the mutation probability.

We have also considered different ways to evaluate the fitness from the individuals, for example, placing a robot in a specific position and placing it in a random position. Moreover, we have used elitism where the best individuals are copied to a new population without being mutated.

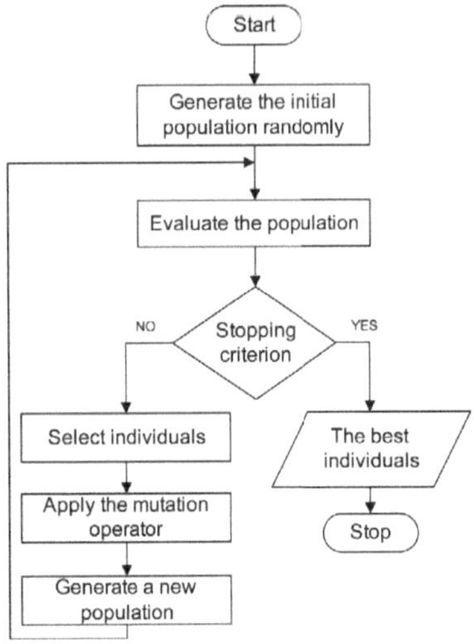

Fig. 5. Evolutionary algorithm designed to evolve low level behaviors

6 Experimental Results

Usually around 40–60 generations were needed for getting a controller that satisfied the requirements; the populations were composed by 10–40 individuals. In training we have made several experiments with elitism and without elitism changing the number of generations, the number of individuals and the mutation probability. The fitness functions depend on each behavior. For avoiding behaviors the fitness function is as follows:

$$f = \frac{steps}{MaxSteps}$$

where *steps* is the number of steps the robot was without collide and *MaxSteps* is the maximum number of steps that a robot can take without collide.

For Following behaviors the fitness function is as follows:

$$f = log\left(1 + \frac{d}{steps}\right)$$

where d is the distance between the robot and the goal, *steps* is the number of steps it takes to reach a goal. The *logarithmic function* is to potentiate the small changes between the numbers of steps and we add 1 to the logarithmic function in order to have no negative results.

The results obtained by 40 generations with population of 20 individuals, mutation factor of 0.3, 0.5 and 0.7 is shows in Fig. 6. In Fig. 6a we present the results of the performance without elitism. It can be seen that the algorithm makes a random search, however, only with mutation factor of 0.3 favorable

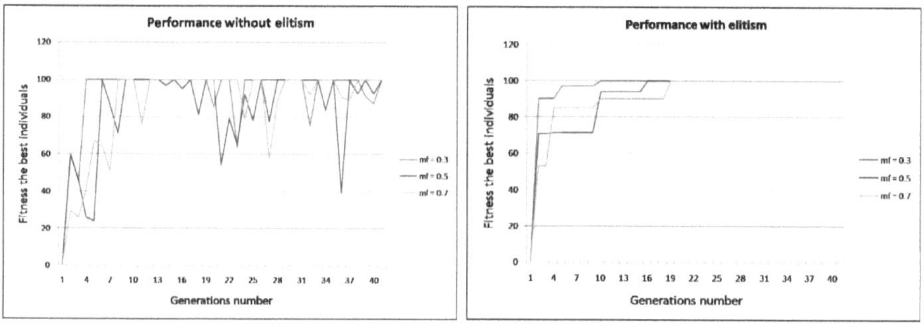

a) Performance without elitism b) Performance with elitism

Fig. 6. Results

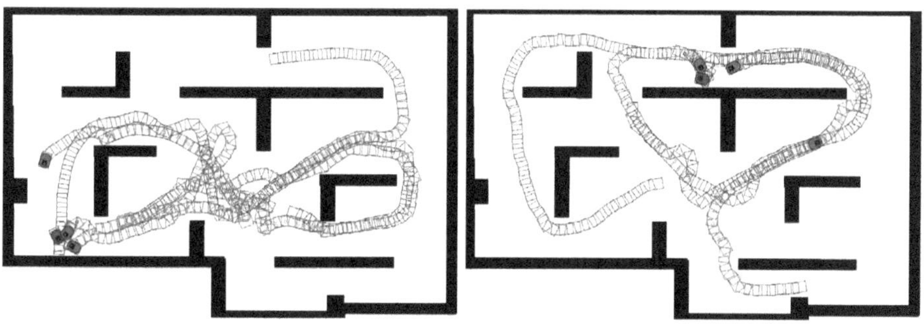

Fig. 7. Simulation multi-robot system

results were obtained. We compare the results obtained by using elitism in Fig. 6b. It can be seen that elitism is maintaining the quality of the population and with mutation factor of 0.3 the algorithm converges faster.

Cooperative and competitive behaviors were obtained through a combination of low-level behaviors through control architecture proposal. Fig. 7 shows the results obtained with the MRS, it can be seen how the red robots team communicate and collaborate among themselves to achieve objective.

7 Conclusion and Further Work

Multi-robot systems in dynamic environments require a well-structured control architecture which allows robots to modify their behavior to cope with the environmental changes or actions performed by other robots. To achieve collaborative behavior between members of a system, it needs a combination of behaviors associated with each robot.

It has been verified that the use of soft-computing techniques like the evolutionary neurocontrollers are appropriate for developing low level robot behaviors.

Moreover, these techniques allow obtaining families of controllers just by modifying the fitness function.

The use of indirect communication brings new possibilities. This multi-robot system has strategies of indirect communication; therefore, it reacts to any changes you have in the environment.

Taking into account that the interaction and coordination between real robots is complicated, it would be interesting to put into practice the results obtained in physical robots. Then, it is necessary to consider the use of other techniques for better results. Currently, we are starting to work with learning classifier system.

Acknowledgments

This work has been partially funded by the Spanish Ministry of Science and Technology, project: DPI2006-15346-C03-02.

References

1. Ge, S.S., Lewis, F.L., Dekker, M., Jones, C., Mataric, M.J.: Behavior-Based Coordination in Multi-Robot Systems. In: Autonomous Mobile Robots: Sensing, Control, Decision-Making, and Applications (2005)
2. Dudek, G., Jenkin, M.R.M., Wilkes, D.: A Taxonomy for Multi-Agent Robotics. Autonomous Robots 3(4), 375–397 (1996)
3. Farinelli, A., Farinelli, R., Iocchi, L., Nardi, D.: Multirobot systems: A classification focused on coordination. IEEE Transactions on Systems Man and Cybernetics Part B (Cybernetics) 34(5), 2015–2028 (2004)
4. Gerkey, B.P., Mataric, M.J.: A formal framework for the study of task allocation in multi-robot systems. International Journal of Robotics Research 23(9), 939–954 (2004)
5. The Player and Stage Project, `http://playerstage.sourceforge.net`
6. McMillen, C., Veloso, M.: Distributed, Play-Based Coordination for Robot Teams in Dynamic Environments. In: Lakemeyer, G., Sklar, E., Sorrenti, D.G., Takahashi, T. (eds.) RoboCup 2006: Robot Soccer World Cup X. LNCS (LNAI), vol. 4434, pp. 483–490. Springer, Heidelberg (2007)
7. Khamis, A.M., Kamel, M.S., Salichs, M.A.: Cooperation: Concepts and General Typology. In: IEEE International Conference on Systems, Man and Cybernetics, pp. 1499–1505 (2006)
8. Engelbrecht, A.: Computational Intelligence: An Introduction, 2nd edn. Halsted Press, New York (2007)

Control Action Continuity on Situation-Based Obstacle Avoidance*

D. Hernandez, J. Cabrera, A. Dominguez, and J. Isern

SIANI, ULPGC, Spain
dhernandez@iusiani.ulpgc.es

Abstract. This work is related to the analysis of reactive obstacle avoidance in general, and specifically to ND algorithms family. Contrary to many previous methods, the ND approach is not aimed at devising a general motion law; instead, it operates over a reduced set of possible situations that are treated by a particular motion law. The big earning of this idea is that it eases the design of control, as now motion laws are specific to every identifiable situation. However, it also raises new issues as nothing guarantees the control action continuity when the diagnostic changes. In this paper a modification of the ND approach, along with experimental results, is presented in order to improve this aspect of the method.

1 Introduction

The study and design of obstacle avoidance algorithms has been in the research agenda of researchers in the mobile robots field for nearly thirty years and very different approaches have been presented. However, the problem of moving autonomously a robot on its environment in a safe and agile manner remains an open challenge.

Some obstacle avoidance methods employ the local map to create an artificial potential field [4] that should drive the robot towards the goal. Algorithms based on potential fields show up important instabilities and are not suitable for navigating narrow passages. Borenstein and Koren [2] developed a different approach, reducing the local path planning to the search of open passages in a polar histogram that is built from the local map. This method has been improved to take into consideration the robot kinematic constraints and a global path planner [10].

Other approaches, like the curvature velocity method [9] or the dynamic window based methods [3] [1] have been proposed to take into consideration the robots kinematics and dynamic constraints in order to guarantee that a planned local trajectory is physically feasible, a question commonly neglected in previous methods.

* This work has been partially supported by Canary Islands Government via F.E.D.E.R. funds (project PI2007/039).

R. Moreno-Díaz et al. (Eds.): EUROCAST 2009, LNCS 5717, pp. 445–452, 2009.

An aspect that all these methods have in common is that the different situations that an agent must face while navigating in its environment are finally treated using a single motion law, where factors as the width of the selected open passage, its angular position relative to the actual direction of motion or the relative position of the goal, are weighted in a control law to derive the commanded linear and angular velocities.

A different approach, termed Nearness Diagram (ND) Navigation, is presented by Minguez et al in [7]. This method uses a "divide and conquer" strategy to classify the local navigation problem into *situations*, treating then each one using a particular motion law. The big earning of this approach is that it eases the design of motion laws, as these are now specific to every possible situation. However, this solution also raises new issues as nothing guarantees the smoothness of motion commands when the situation changes. Concretely, nothing in the ND (and its descendants [5] [8]) approach precludes rapid changes in the identification of the current situation that have as a consequence large modifications of commanded velocities. These unnecessary and drastic changes in the commanded actions seriously degrade the coherence and agility of the agent movement. In this paper a modification of the ND approach will be proposed and evaluated to improve this aspect of the method.

2 The Nearness Diagram (ND) Algorithm

In the following section we will briefly introduce the ND (Nearness Diagram) algorithm for avoiding obstacles, specifically its ND+ version [6]. The ND is a reactive algorithm that uses, as input information, sensory data provided by a laser range finder which produces periodically a data scan. Based on this sensory information the algorithm decides which action should be carried out by the robot in order to reach a goal position avoiding collision with obstacles.

The algorithm operation is divided in three sequential phases:

1. **Calculation of ND Diagrams.** In the first phase two polar diagrams are generated, PND and RND, from sensory data. The first of these diagrams, PND, represents the nearness of the obstacles from the central position of the robot, the second one, RND, represents the nearness of the obstacles from the robots bounds.
2. **Selection of a Navigable Valley.** Using the PND diagram obtained in the previous phase, the algorithm looks for discontinuities in the diagram. A discontinuity is defined as the difference between two polar values such that it is wide enough to allow the robot to pass through the obstacle points given by the polar values, and a "valley" is formed between two adjacent discontinuities found in the diagram. The valley which is closer to the goal in angular terms will be selected first. Next a *navigability test* is applied on it, if the test is positive, the valley is selected, if not the algorithm proceeds equally with the next valley available, until it founds one navigable valley.
3. **Situated Actions.** Based on the navigable valley selected in the previous phase and the RND diagram calculated in the first phase the algorithm

diagnoses six different situations in ND+ (five situations in ND) in order to generate an action aimed at driving the robot to the goal while avoiding obstacles. In turn, those situated actions are classified into two groups.

- **High Safety (HS) Situations.** When there are no obstacles near the robot closer than a given distance, called *security distance*, which defines a *security zone* surrounding the robot, it is said that the robot is in *High Safety*. The algorithm considers three types of high safety situations: *HSWR* (*wide* navigable valley), *HSNR* (*narrow* navigable valley) and *HSGR* (the goal is inside the navigable valley).
- **Low Safety (LS) Situations.** When there are obstacles inside the security zone, it is said that the robot is in *low safety*. Three situations are considered in low safety situations: *LS1* (obstacles only on one side of the robot), *LS2* (obstacles on both sides of the robot) and *LSGR* (the goal is inside the navigable valley).

Based on the previous taxonomy of situations the algorithm determines which actions should be taken in order to avoid obstacles, if any, at the same time that the robot is driven to the goal. The corresponding decision tree is presented in Fig. 1, extracted from the original paper, for clarification purposes.

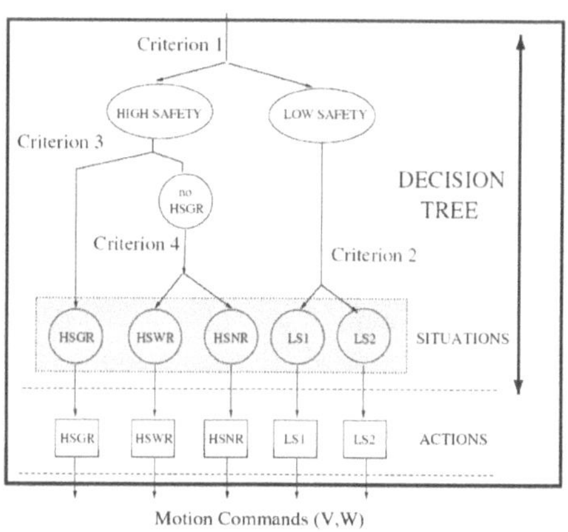

Fig. 1. The ND decision tree for situation identification

3 Proposed Alternative

As commented in the introduction, the ND+ algorithm results that we have obtained both from Player/Stage simulations and real world experiments show frequently sharp oscillations in the commanded and measured angular velocities, especially when medium to high maximum limits are allowed. In most cases, these abrupt changes were not motivated by the selection of a different valley from the ND diagram, they are consequence of some transitions between different situations.

This result is not completely unexpected, as there is not any explicit mechanism in the control strategy that can guarantee continuity or smoothness between successive control actions generated under the aforementioned circumstances. In fact, ND+ algorithm is an extension of a previous version that tries to address control action continuity as one of their objectives. However, this analysis was only developed for a subset of possible situation transitions, the most frequent in the kind of "narrow navigation" problems considered, or, as termed by the authors "troublesome scenarios". In other related work [6], authors make a reference to this issue indicating that they have used a hysteresis mechanism to reduce the discontinuity effect, although no additional details are given.

The origin of the problem can be located in the decision tree used to label the current robot situation. Independently of the implementation, a border line is created so slight sensor data fluctuations could lead to different situation selection. We consider, however, that the problem can be alleviated if a kind of membership function would be taken into account when a new command is to be produced from the estimated robot situation. The approximation proposed in this paper consists in defining a measure of situation certainty in order to determine how trustful the action command generated by the correspondent control rule is. Once computed, this certainty degree is used to calculate the final commanded angular velocity for the robot weighting the value proposed from the situated action and the current angular velocity.

3.1 Certainty Estimation

According to the decision tree used for determining the current situation the following factors are considered: HS_F (High/low security distance factor), GC_F (Goal contained in valley factor) and NV_F (Narrow/wide valley factor). In a first step, the modified algorithm maps the factors involved in the selection of the current situation to a 0-1 range.

Once the factors have been mapped, the situation certainty can be obtained using some evidence fusion method. In our tests, simple minimum or multiplicative rules, following the algorithm decision tree, have produced adequate results. Specifically, the multiplicative rule has been applied in the experiments presented in this paper.

3.2 Performance Evaluation

Although visual estimation is normally sufficient for preliminary analysis of the results, a simple benchmark measure has been defined and used in this work for a more objective quantitative comparison. This measure contains the following factors: time ($TimeF$), curvature ($CurvF$), translational acceleration ($TAccelF$) and rotational acceleration ($RAccelF$). The ideal trajectory is considered to be a straight line joining the initial and goal points, following a trapezoidal velocity profile, using maximum acceleration and velocities. This ideal case will return a value of 1.0 for each factor.

4 Experiments

Some experiments have been conducted in order to evaluate the effect of the proposed solution on the robot behaviour. Two versions of ND+, with and without situation certainty evaluation, have been implemented on the Player/Stage environment to compare their performance. The experiments use a simplified obstacle configuration to facilitate the illustration of the results.

4.1 Scenario 1

Figure 2 shows the obstacle configuration used for the first experiment, including the robot initial position and its trajectory to the final (labeled with time stamps) position using the standard version of the ND+ algorithm. The plot in figure 4 depicts the commanded angular velocities and the situation transitions along the robot path from the origin to the goal. It is possible to verify how certain situation transition have practically no effect on command continuity, for example around the 25-30 seconds interval; while others induce a more noticeable distortion, for example between 30 and 35 seconds.

Finally, figures 5 and 6 include the robot velocities and accelerations corresponding to the execution of the standard ND+ algorithm. Here, the discontinuities observed in the commanded angular velocity graph finally show up as high robot acceleration values.

The figures are now repeated for the modified version of the algorithm: commands and situation transitions in figure 7, velocities in figure 8 and accelerations in figure 9. The results show how the smoothed commands generated by the certainty-based version are responsible for the lower acceleration values measured, without affecting significantly trajectory safety or elapsed time from robot start until the goal is reached. The visual impression of improvement is corroborated by the benchmark measure, that gives better values for the modified version of the algorithm.

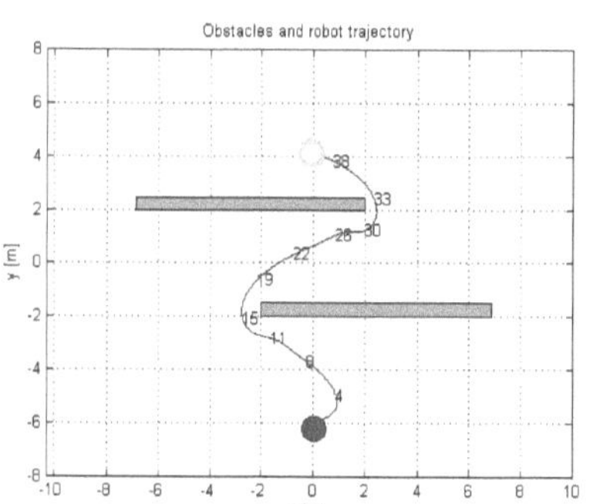

Fig. 2. Scenario 1: setup and trajectory using the standard algorithm

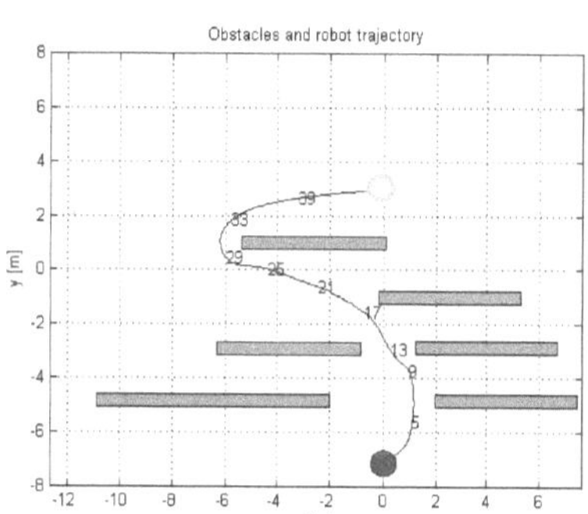

Fig. 3. Scenario 2: setup and trajectory using the standard algorithm

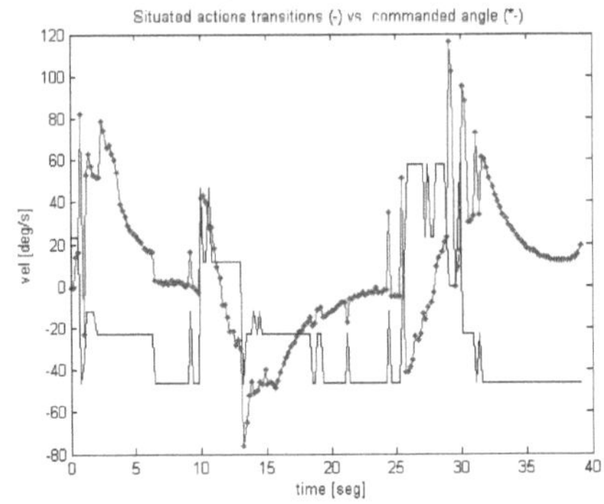

Fig. 4. Situation transitions and commands from standard algorithm (scenario 1)

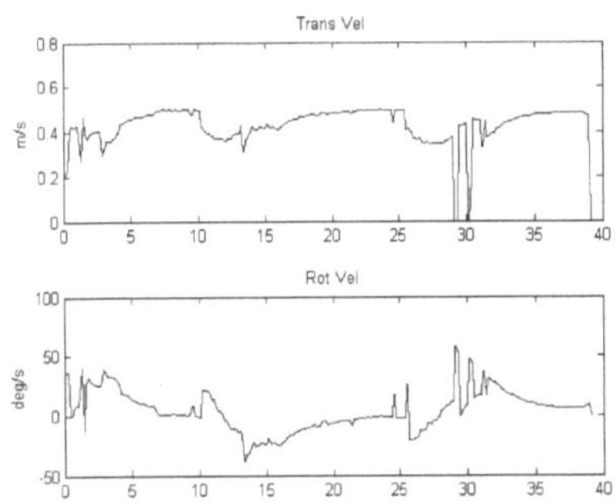

Fig. 5. Robot velocities from standard algorithm (scenario 1)

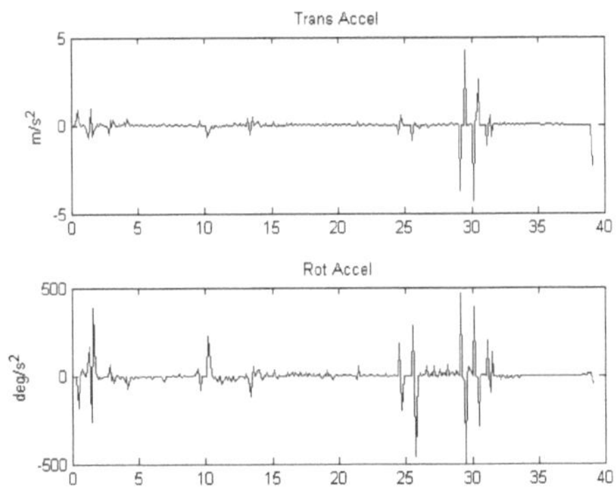

Fig. 6. Robot accelerations from standard algorithm (scenario 1)

As can be observed from the results, some discontinuities, although attenuated, are still present. These occurrences are associated with new transitions where the identified situation has a high certainty level, and must be preserved in order to avoid robot collisions or missed targets.

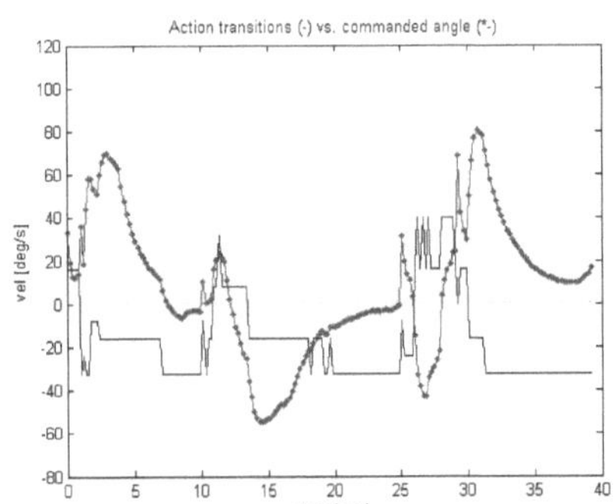

Fig. 7. Situation transitions and commands from modified algorithm (scenario 1)

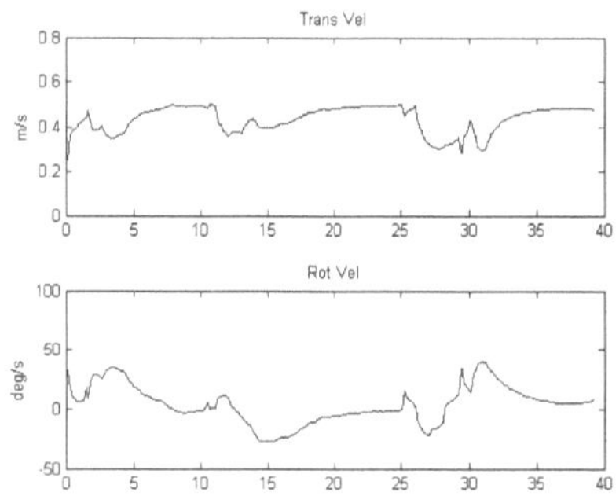

Fig. 8. Robot velocities from modified algorithm (scenario 1)

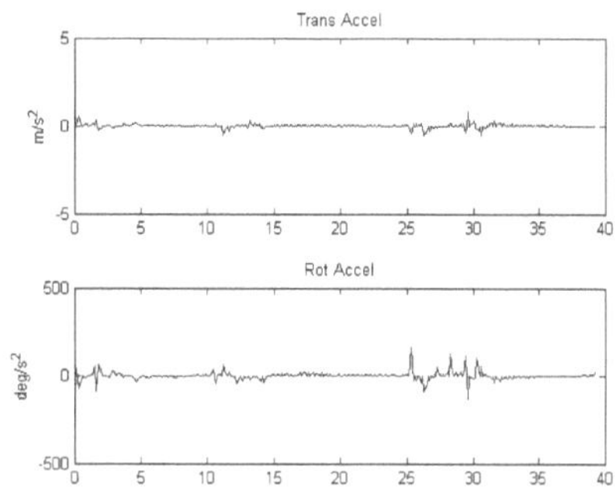

Fig. 9. Robot accelerations from modified algorithm (scenario 1)

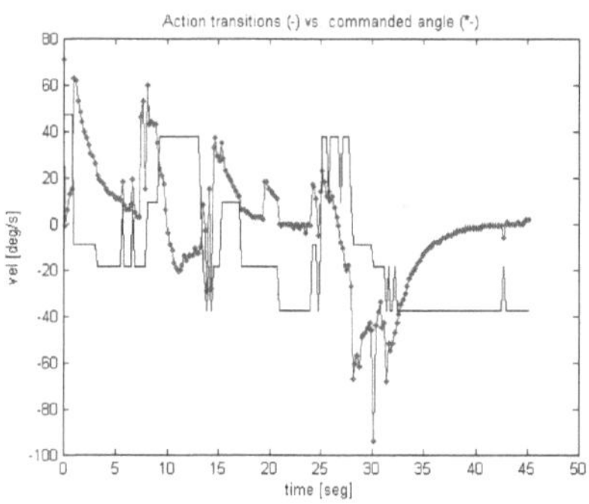

Fig. 10. Situation transitions and commands from standard algorithm (scenario 2)

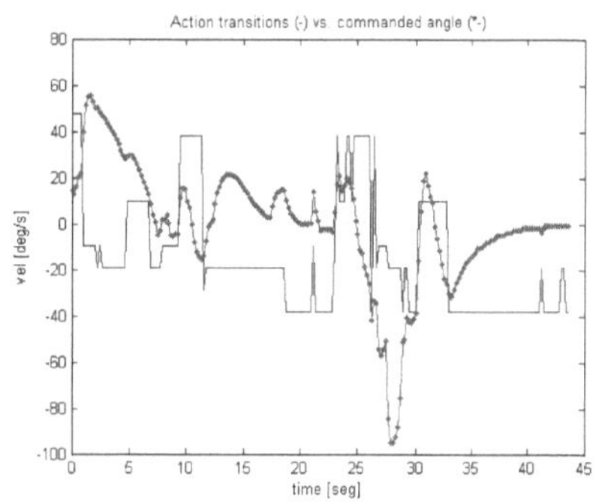

Fig. 11. Situation transitions and commands from modified algorithm (scenario 2)

4.2 Scenario 2

Using a more complicated scenario (figure 3), similar results have been obtained. The figures 10 and 11 resume the collected data for this second case.

5 Conclusion

In this paper a certainty-based method has been applied to reactive obstacle avoidance in order to improve control action continuity. The ND algorithm has been analyzed under this perspective, concluding that control action smoothness cannot be guaranteed under high velocities and rapid situation changes. The experiments have demonstrated that the modified algorithm presented in this paper shows a better stability without compromising reactivity. A benchmarking function has been defined to provide and objective measure for comparison.

References

1. Arras, K.O., Persson, J., Tomatis, N., Siegwart, R.: Real-time obstacle avoidance for polygonal robots with a reduced dynamic window. In: Proc. Int. Conf. on Robotics and Automation, pp. 3050–3055 (2002)
2. Borenstein, J., Koren, Y.: The vector field histogram - fast obstacle avoidance for mobile robots. IEEE Transactions on Robotics and Automation 7(3), 278–288 (1991)
3. Brock, O., Khatib, O.: High-speed navigation using the global dynamic window approach. In: Proc. Int. Conf. on Robotics and Automation, pp. 341–346 (1999)
4. Khatib, M., Chatila, R.: An extended potential field approach for mobile robot sensor-based motions. In: Proc. of International Conference on Intelligent Autonomous Systems (IAS 1995), pp. 490–496 (1995)
5. Minguez, J., Montesano, L., Montano, L.: Extending reactive collision avoidance methods to consider any vehicle shape and the kinematics and dynamic constraints. International Journal of Advanced Robotic Systems 3(1), 85–91 (2006)
6. Minguez, J., Osuna, J., Montano, L.: A "divide and conquer" strategy based on situations to achieve reactive collision avoidance in troublesome scenarios. In: IEEE International Conference on Robotics and Automation, ICRA 2004. Proceedings, April 26-May 1, vol. 4, pp. 3855–3862 (2004)
7. Minguez, J., Montano, L.: Nearness diagram (ND) navigation: collision avoidance in troublesome scenarios. IEEE Transactions on Robotics and Automation 20(1), 45–59 (2004)
8. Minguez, J., Montano, L.: Extending reactive collision avoidance methods to consider any vehicle shape and the kinematics and dynamic constraints. IEEE Transactions on Robotics (2008)
9. Simmons, R.: The curvature-velocity method for local obstacle avoidance. In: International Conference on Robotics and Automation, April 1996, pp. 3375–3382 (1996)
10. Ulrich, I., Borenstein, J.: VFH*: Local obstacle avoidance with look-ahead verification. In: Proc. IEEE Int. Conf. Robotics and Automation, April 2000, pp. 2505–2511 (2000)

Traffic Signals in Traffic Circles: Simulation and Optimization Based Efficiency Study

Javier J. Sánchez Medina, Manuel J. Galán Moreno,
Moisés Díaz Cabrera, and Enrique Rubio Royo

Innovation Center for Information Society (CICEI)
University of Las Palmas de Gran Canaria, Spain
javier.sanchez.medina@gmail.com, mgalan@dmat.ulpgc.es
moisdc@gmail.com, erubio@polaris.ulpgc.es
http://www.cicei.com

1 Introduction

Traffic Circles are frequently used in cities, to control vehicular traffic at intersections. As said in [1], their main advantages can be the provision of an adequate throughput and the improvement of user safety, by slower vehicle speeds and reducing traffic conflicts.

However, when such intersections suffer from overloading, there exists a tendency to include traffic lights inside traffic circles. By the current work our aim is to explore, using a very simplified test case, a feasible methodology to determine whether or not it is useful to use traffic signals within roundabouts.

We have simulated a generic traffic circle including a set of traffic signals placed in it. For every traffic signal we have considered only two states: "red" and "intermittent yellow". Considering the Spain's driving rules, an intermittent yellow traffic signal means the same control command to drivers like if there were not any traffic light at all –"Go on, Carefully!". A traffic signal in a permanent intermittent yellow state could be considered to be removed without any consequence for traffic management.

For this scenario we have performed an optimization of traffic signal state sequences, using a Genetic Algorithm. We have carried out several tests in order to think about the suitability of the proposed methodology to that problem. Although results are not conclusive, this methodology seems to be useful for traffic managers or planners.

2 Methodology

In previously published works ([2], [3], [4], [5], [6], [7], and [8]) we described a traffic signals optimization architecture. The methodology consists on the combination of three techniques: a Genetic Algorithm (GA) as Non-Deterministic Optimization Technique, a Cellular Automata (CA) based Traffic Simulator inside the evaluation routine of the GA, and a Beowulf Cluster as a MIMD multicomputer. Through this section we will give the key aspects of the Genetic Algorithm optimization application to a generic traffic circle intersection.

R. Moreno-Díaz et al. (Eds.): EUROCAST 2009, LNCS 5717, pp. 453–460, 2009.

2.1 Microsimulation

Traffic Simulation is known to be a very complex task. Nowadays, microscopic simulators are widely used. One of the main reasons for this is that they can model the discrete dynamics that arise from the interaction among individual vehicles [9]. Cellular Automata are usually faster than any other traffic microsimulator [10], and, as stated in [11] *"the computational requirements are rather low with respect to both storage and computation time making it possible to simulate large traffic networks on personal computers"*.

We have developed a traffic model based on the SK[1] model [12] and the SchCh[2] model [13]. The SchCh model is a combination of a highway traffic model [14] and a very simple city traffic model [15]. The SK model adds the "smooth braking" to avoid abrupt speed changes. We decided to base our model on the SK model due to its better results for all the tests shown in [16].

Based on the Cellular Automata Model we have developed a non-linear model for simulating traffic behavior. The basic structure is the one used in Cellular Automata. However, in our case, we add a new level of complexity by creating two new abstractions: "Paths" and "Vehicles".

"Paths" are overlapping subsets included in the Cellular Automata set. There is one "path" for every origin-destination pair. To do this, every "path" has a collection of positions and, for each one of them, there exists an array of allowed "reachable" positions. This idea is illustrated in Figure 1.

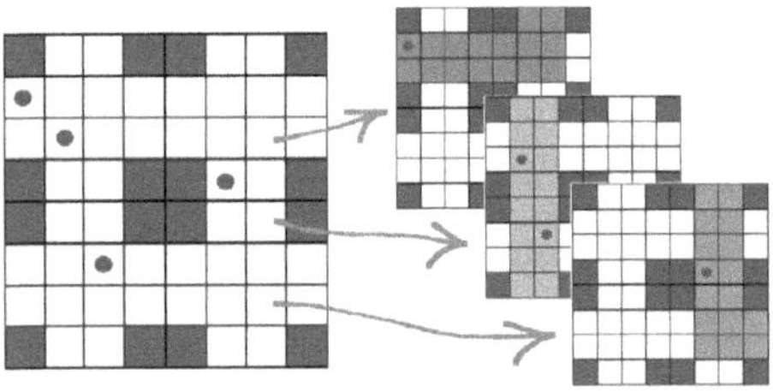

Fig. 1. Paths in our Improved Cellular Automata Model

"Vehicles" consists of an array of structures, each one of them having the following properties:

1. Position: Situation at the Cellular Automata. Note that every cell may be occupied by only one vehicle.
2. Speed: the current speed of a vehicle. It means the number of cells it moves over at every time step.
3. Path: In our model, every vehicle is related to a "path".

---

[1] Stephan Krauss, the author.
[2] Andreas Schadschneider and Debashish Chowdhury, the authors.

These are the rules applied to every vehicle:

1. A vehicle ought to accelerate up to the maximum speed allowed if it has no obstacle in its way (another vehicle, or a red traffic sign). It will accelerate at a rate of 1 point per time step, every time step.
2. If a vehicle can reach an occupied position, it will have to reduce its speed and will occupy the free position just behind the preceding vehicle.
3. If a vehicle has a red traffic signal ahead, it will stop.
4. Smooth Braking: Once the vehicle position is updated, then the vehicle speed is updated too. To do this, the number of free positions from the current position ahead is taken into account. If there is not enough free space for the vehicle to move forward on the next time step going at its current speed (hypothetically, since in the next time step the traffic situation may change), it will reduce its speed in one unit.
5. Multiple Lanes: When a vehicle is trying to move on, or update its speed, it is allowed to consider positions on other parallel lanes. For every origin-destination couple (path), at every point there exists a list of possible "next" positions. The first considered is the one straight ahead, if this one is not available, there may be more possible positions in parallel lanes that will need to be considered. Of course, this list of possible "next" positions is created taking the basic Spanish Highway code into account.

By means of these rules we can have lots of different paths and vehicles running in the same network. This model may be seen as a set of N_{paths} traditional Cellular Automata networks working in parallel over the same physical grid.

2.2 Genetic Algorithm

Chromosome Encoding. In fig. 2 we want to illustrate the chromosome designed for this work. In that figure "N Traffic Signals" means the number of traffic signals contained in the traffic circle under study. "N Stages" means the number of states or stages for every traffic signal cycle. A priori it is known that every traffic signal has a fixed number of stages or states to be in, which is the same number for all of them. All traffic signals are synchronized. Hence, we have a fixed period for the whole intersection.

The chromosome encoding consists of the state of every traffic light, at every step of the fixed period. There are allowed only two possible states: Red – encoded as '1' – and Intermittent Yellow – encoded as '0'. We have set it like this in order to simulate two real world traffic control commands: 'Stop' (Red Light) and 'Pass with caution' (Intermittent Yellow). The chosen state pair makes sense if one notes that with the Intermittent Yellow state one have a 'no traffic signal'-like situation.

The initial population of the GA is created at random.

Selection Strategy. We have chosen a Truncation and Elitism combination as selection strategy. This means that at every generation a reduced group of individuals — in our case, the best two individuals — is cloned to the next

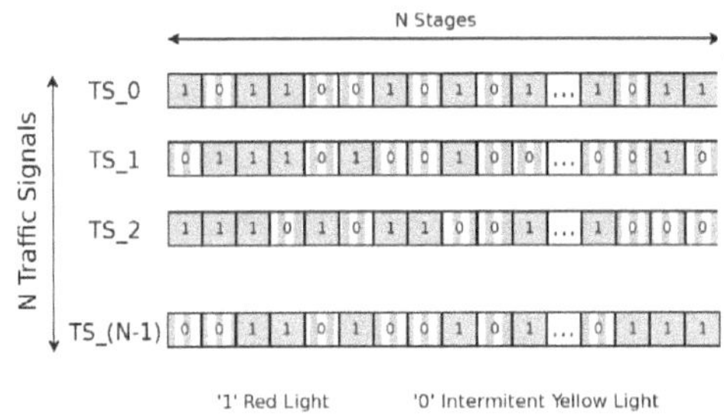

Fig. 2. Chromosome

Fig. 3. Generic T. Circle

generation. The remainder of the next generation is created by crossing the individuals from a best fitness subset – usually 66 percent of the whole population.

Crossover Operator. We have used a standard Two Point Crossover operator. At random points – for a pair of *parent* chromosomes – it selects two random points, cuts them at these positions into three pieces and then interchanges the central chunk.

Mutation Operator. When an individual is chosen to be mutated – according to the mutation probability – the value stored at a randomly chosen position of its chromosome is changed.

The mutation probability is not fixed. It starts with a high mutation probability that will progressively decrease until reaching probability values near to the inverse of the population size at the end of the planned number of generations.

Fitness Function. In this research we have used a single variable sampled from the traffic simulation as fitness function: the number of vehicles that left the network during the traffic simulation carried out.

2.3 Cluster

The Architecture of our system is based on a Beowulf Cluster, due to its price/per formance relationship and the possibility of employing Open Source Software on it. This is also a very scalable MIMD computer, a very desirable feature in order to solve all sorts and scales of traffic problems.

For this research project we set up a eight node cluster, each node consisting of an AMD Opteron64. The nodes were connected through a Gigabit Ethernet Backbone. Every node had the same hardware, except the master node which had an extra Gigabit Ethernet network card for "out world" connection.

Every node had installed CentOS — Kernel 2.6.9-78.0.13.ELsmp. For parallel programming the installation of Open MPI (openmpi-1.2.9-1) was also necessary.

In our application there were two kinds of processes, namely *master* and *slave* processes. There was only one master process running at each test. At every generation it sends the chromosomes (MPI_Send) to slave processes, receives the evaluation results (MPI_Recv) and creates the next population. Slave processes are inside an endless loop, waiting to receive a new chromosome (MPI_Recv). Then, they evaluate it and send the evaluation result (MPI_Send) to the master process.

GA – Chromosome Encoding. In the fig. 2 we want to illustrate the chromosome designed for this work. In this example "N Traffic Signs" means the number of traffic signals contained in the traffic circle under study. "N Stages" means the number of states or stages for every traffic signal cycle.

GA – Optimization Criterion. For this work the GA target is to maximize the absolute number of vehicles that left the traffic network once the simulation finishes.

3 Test Results

We have designed a set of 5 test cases with the hypothetical traffic network of figure 3 varying the initial occupancy of the network from 0% up to 100%. These are the parameters used for every test case:

- GA population size: 200 individuals.
- GA generations: 200 generations.
- Simulation time: 2000 time steps.
- Traffic input: 6 vehicles per minute at each traffic source.

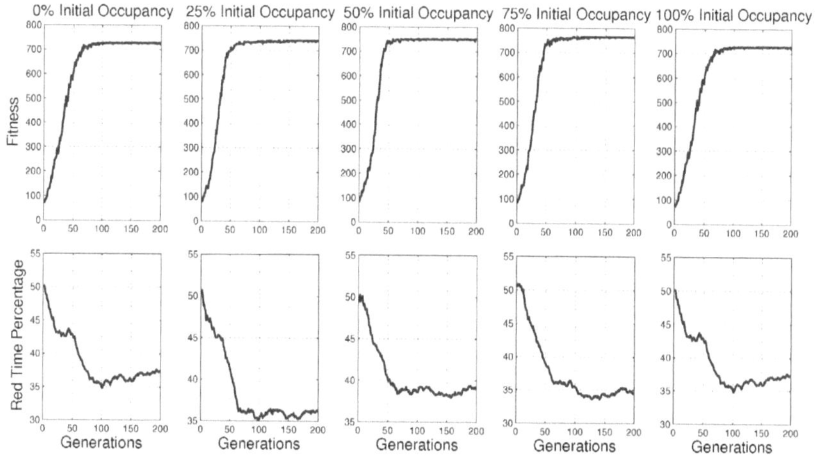

Fig. 4. Fitness and Avg. Red Time Evolution for the 5 Test Cases

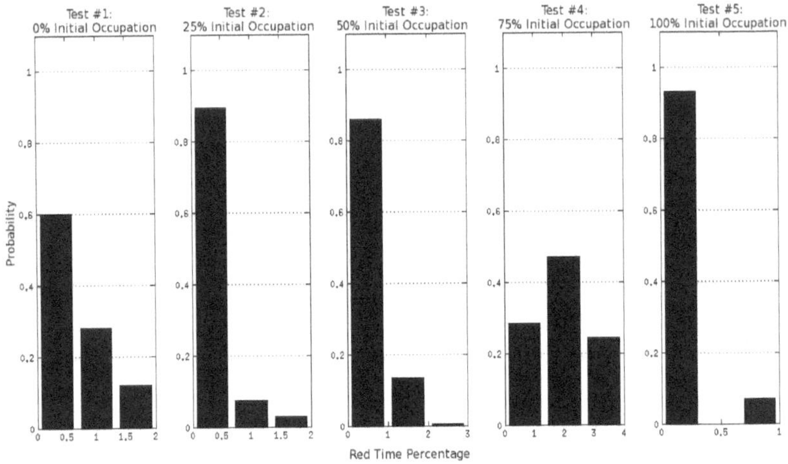

Fig. 5. Red Time Percentage Histograms

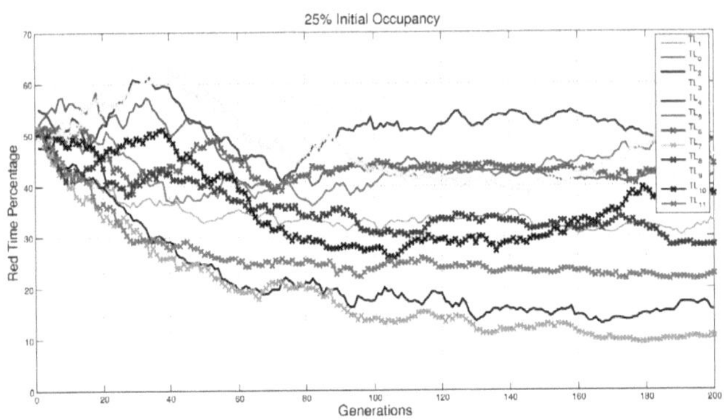

Fig. 6. Traffic Lights Red Time Percentage Evolution with Initial Occupancy of 25%

In figure 4, in the first row, it is represented the average fitness evolution for every test case. In the second row it is displayed the mean Red Time percentage evolution for every test case. It seems to exist a tendency to a reduction of the red time state for all cases, as the fitness increases.

In figure 5 it is represented a three marks histogram for every test case. In the horizontal axis Red Time Percentages values are represented. In the vertical axis relative frequency is represented.

Finally, in figures 6 and 7 for two of the tested cases – 25% and 75% initial occupancy – it is displayed the evolution of mean Red Time of every traffic signal as the GA evolves itself. In that examples one may observe that some traffic lights tend to have a reduced Red Time percentage at the end of the optimization.

Fig. 7. Traffic Lights Red Time Percentage Evolution with Initial Occupancy of 75%

4 Conclusions and Future Research Plans

Through this research we have delivered a set of initial ideas about how to deal with the placing of traffic lights withing roundabouts. Although results are very preliminary, using the simplification of only two possible states – red and intermittent yellow – it seems to exist a tendency towards not including traffic lights withing the rotary.

The presented methodology, although it is quite in a draft state yet, may be useful for practitioners and traffic networks designers.

As for future research we do realize that we need to develop a whole set of tests in order to consistently determine what traffic lights should be removed from the initial layout. First, we should variate the input of traffic including all possible scenarios – peak hour traffic, low density, etc. Besides that, it should be carried out multiple executions of the genetic algorithm, at least 30, for the sake of consistency. After that, traffic lights without a significant portion of red time, can be withdrawn from the traffic network.

References

1. Hossain, M.: Capacity estimation of traffic circles under mixed traffic conditions using micro-simulation technique. Transportation Research Part A: Policy and Practice (1999)
2. Sánchez, J., Galán, M., Rubio, E.: Genetic Algorithms and Cellular Automata: A New Architecture for Traffic Light Cycles Optimization. In: Proceedings of The Congress on Evolutionary Computation 2004 (CEC 2004), vol. 2, pp. 1668–1674 (2004)
3. Sánchez, J., Galán, M., Rubio, E.: Bit Level Versus Gene Level Crossover in a Traffic Modeling Environment. In: Proceedings of the International Conference on Computational Intelligence for Modelling Control and Automation - CIMCA 2005, vol. I, pp. 1190–1195 (2005)

4. Sánchez, J., Galán, M., Rubio, E.: Stochastic vs deterministic traffic simulator. Comparative study for its use within a traffic light cycles optimization architecture. In: Mira, J., Álvarez, J.R. (eds.) IWINAC 2005. LNCS, vol. 3562, pp. 622–631. Springer, Heidelberg (2005)
5. Sánchez, J.J., Galán, M.J., Rubio, E.: A visual and statistical study of a real world traffic optimization problem. In: Topping, B.H.V., Montero, G., Montenegro, R. (eds.) Proceedings of the Fifth International Conference on Engineering Computational Technology, Stirlingshire, United Kingdom, paper 147. Civil-Comp Press (2006)
6. Sánchez, J.J., Galán, M.J., Rubio, E.: Study of Correlation Among Several Traffic Parameters Using Evolutionary Algorithms: Traffic Flow, Greenhouse Emissions and Network Occupancy. In: Moreno Díaz, R., Pichler, F., Quesada Arencibia, A. (eds.) EUROCAST 2007. LNCS, vol. 4739, pp. 1134–1141. Springer, Heidelberg (2007)
7. Sánchez, J.J., Galán, M.J., Rubio, E.: Applying a traffic lights evolutionary optimization technique to a real case: "las ramblas" area in santa cruz de tenerife. IEEE Transactions on Evolutionary Computation (2008)
8. Sanchez-Medina, J.J., Galán, M.J., de Ugarte, N.A., Rubio, E.: Simulation times vs. network size in a genetic algorithm based urban traffic optimization architecture. In: GEM, pp. 255–261 (2008)
9. Benjaafar, S., Dooley, K., Setyawan, W.: Cellular Automata for Traffic Flow Modeling. Technical Report CTS 97-09, Intelligent Transportation Systems Institute (1997)
10. Nagel, K., Schleicher, A.: Microscopic traffic modeling on parallel high performance computers. Parallel Comput. 20(1), 125–146 (1994)
11. Cremer, M., Ludwig, J.: A fast simulation model for traffic flow on the basis of Boolean operations. Mathematics and Computers in Simulation (28), 297–303 (1986)
12. Krauss, S., Wagner, P., Gawron, C.: Metastable states in a microscopic model of traffic flow. Phys. Rev. E 55, 5597–5605 (1997)
13. Schadschneider, A., Chowdhury, D., Brockfeld, E., Klauck, K., Santen, L., Zittartz, J.: A new cellular automata model for city traffic. In: Traffic and Granular Flow 1999: Social, Traffic, and Granular Dynamics. Springer, Heidelberg (1999)
14. Nagel, K., Schreckenberg, M.: A Cellular Automaton Model for Freeway Traffic. Journal de Physique I France 2(33), 2221–2229 (1992)
15. Biham, O., Middleton, A.A., Levine, D.: Self-organization and a dynamical transition in traffic-flow models. Phys. Rev. A 46(10), R6124–R6127 (1992)
16. Brockfeld, E., Kühne, R., Skabardonis, A., Wagner, P.: Towards Benchmarking Microscopic Traffic Flow Models. Transportation Research Record (1852), 124–129 (2003)

Integrated System and Network Simulation of a 5.8 GHz Local Positioning System

Ralf Mosshammer[1], Ralf Eickhoff[2], Mario Huemer[3], and Robert Weigel[1]

[1] Institute for Electronics Engineering,
University of Erlangen-Nuremberg,
Cauerstrasse 9, 91058 Erlangen, Germany
[2] Chair for Circuit Design and Network Theory,
Dresden University of Technology,
Barkhausenbau 257, Helmholtzstraße 18, 01069 Dresden, Germany
[3] Chair of Embedded Systems,
NES Klagenfurt University
Lakeside B02b, 9020 Klagenfurt, Austria
{mosshammer,weigel}@lfte.de, ralf.eickhoff@tu-dresden.de,
mario.huemer@uni-klu.ac.at

Abstract. In this work, performance aspects of a high-precision radiolocation system are evaluated. An integrated simulation environment is introduced, which allows the co-evaluation of both physical layer parameters, such as measurement error, as well as network-related performance figures like the time-to-fix and throughput. Simulation results for different strategies and algorithms are presented and analyzed for their practicability.

1 Introduction

Wireless radiolocation systems have experienced a considerable gain in interest from both industrial and academical sectors in recent years. A number of competing technologies vie to cover a huge application space, reaching from simple recreational or consumer-oriented uses to advanced industrial automation [1].

The vastly different requirements have spawned a number of unique technological approaches. A promising technology in the area of high-precision specialized systems is secondary radar based on the Frequency Modulated Continuous Wave (FMCW) principle [2]. A number of systems in this class are deployed for a variety of applications [3,4].

The platform introduced in this work was developed in the course of the EU-endorsed RESOLUTION project, which aims for an integrated system platform for mobile radiolocation and communication, with the pronounced goals of low cost and reconfigurability for a large number of applications [5].

In this work, we present an integrated analysis of this platform, covering both aspects of the physical layer (PHY) in section 2 as well as the Medium Access Control sublayer (MAC) in section 3. Performance figures obtained with an integrated simulation framework are shown in section 4, and conclusions drawn in section 5.

R. Moreno-Díaz et al. (Eds.): EUROCAST 2009, LNCS 5717, pp. 461–468, 2009.

2 System Overview

The RESOLUTION system is based on the exchange of FMCW ramps of the form

$$s_{\mathrm{TX}}(t) = \cos\big((\omega_0 + \mu t)t\big), \tag{1}$$

where ω_0 is the offset angular frequency, and μ is the ramp steepness, which is proportional to the ratio of bandwidth B to ramp period T. The system is designed to operate in the free ISM band at 5.8 GHz, which allows for a B of 150 MHz. The signal is affected by multipath propagation and noise. It reaches the receiver and is mixed with a locally generated copy, resulting in a baseband term

$$s_{\mathrm{Mix}}(t) = \sum_{i=1}^{N_c} \alpha_i \cos\big((\mu\tau_i)t + \phi\big) + n(t). \tag{2}$$

where N_c is the total number of path components, α_i and τ_i path-specific amplitudes and delays, ϕ a fixed phase term, and $n(t)$ a white Gaussian noise term. A frequency-domain analysis by means of Fast Fourier Transform (FFT) yields

$$S_{\mathrm{RX}}(f) = \sum_{i=1}^{N_c} \bar{\alpha}_i \,\mathrm{F}\big(\pi T(f - \mu\tau_i)\big) + N(f). \tag{3}$$

Here, $\bar{\alpha}_i$ is the modified amplitude of the ith component, T the observation length and $\mathrm{F}(\cdot)$ the Fourier transform of the chosen window function. Due to the limited bandwidth, the overlap of multipath components in the frequency domain impedes exact line of sight path detection, an effect which has been elaborated upon in [5,6]. Even in the absence of multipath, the distance measurement is affected by a Gaussian error term of the form

$$\epsilon_{1\mathrm{D}} = |d - d_m| = N(0, \sigma_\epsilon^2), \tag{4}$$

where d and d_m are the actual and measured distances and σ_ϵ the standard deviation of the error, given by the thermal noise floor plus the noise figures of the transmitter and receiver and the limited resolution of the FFT. In most applications, the two-dimensional (radial) error will be of interest. It is given by

$$\epsilon_{2\mathrm{D}} = \sqrt{(x - x_m)^2 + (y - y_m)^2}, \tag{5}$$

with x, y and x_m, y_m being the real and measured distances in x and y direction. With respect to (4), it takes on a Rayleigh distribution under the approximative assumption that the standard deviations in the x and y coordinates are identical. Thus,

$$\epsilon_{2\mathrm{D}} = \sqrt{N(0,\sigma^2)^2 + N(0,\sigma^2)^2} = \mathrm{Rayleigh}(\sigma). \tag{6}$$

The ramp exchange takes place between a mobile station to be located (henceforth MS, for brevity) and fixed infrastructure installments called base stations (BS), which are based on the same hardware architecture, shown in Fig. 1. The

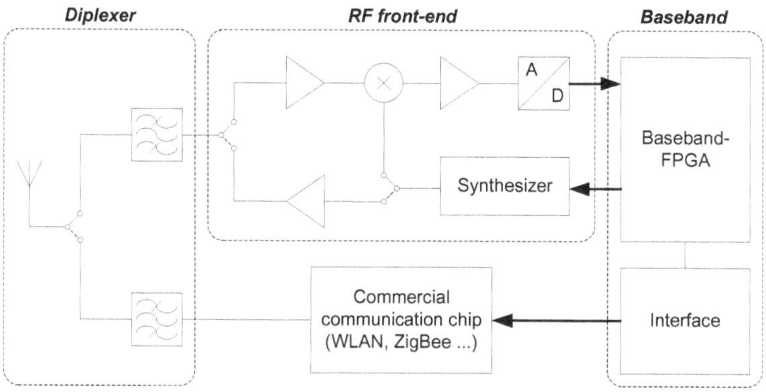

Fig. 1. Block diagram of the integrated RESOLUTION communication and radiolocation hardware

design is centered around a digitally controlled, fully integrated fractional-N PLL synthesizer for ramp generation with extremely precise offsets [7]. The synthesizer feeds a power amplifier for ramp transmission, or alternatively serves as a local oscillator signal in the receive stage. As a hybrid communication and positioning system, the RESOLUTION platform features an interchangeable communications subsystem. Current configurations implement either IEEE 802.11 WLAN (for high-bandwidth communications) or IEEE 802.16 ZigBee standards.

In response to the wide variety of applications that the platform should support, several protocol topologies can be optionally be implemented with the platform. They require no change in the hardware, though for down- or uplink-only scenarios, the respective transmitter and receiver parts can be powered down to save energy.

A solution which shows some implementational elegance is the Roundtrip Time of Flight (RToF) topology. Following a synchronization ramp by the mobile, the BS answer sequentially with their own ramps, allowing the MS to calculate its position. The roles can be reversed to eschew the need for baseband algorithms in the MS.

The timing of the position measurement is as follows:

1. The mobile sends out a ramp at $t_{0,\mathrm{MS}}$ which reaches a base station after the time of flight $\tau_{\mathrm{MS}_i \rightarrow \mathrm{BS}_j}$. There, it is mixed with an internal ramp generated at $t_{0,\mathrm{BS}}$.
2. The base station generates a new ramp after a known offset time T_j and presynchronizes it with the previously calculated time difference, yielding the time instant $T_j + t_{0,\mathrm{MS}} + \tau_{\mathrm{MS}_i \rightarrow \mathrm{BS}_j}$.
3. The mobile receives the synchronized ramp after an additional time of flight $\tau_{\mathrm{MS}_i \rightarrow \mathrm{BS}_j}$ and mixes with an internal ramp generated at $t_{0,\mathrm{MS}} + T_j$, yielding $2\tau_{\mathrm{MS}_i \rightarrow \mathrm{BS}_j}$.

Another protocol option is to have the synchronization/measurement exchange take place sequentially between mobile and base station. This would be necessary to facilitate the use of error estimation/repeat-request techniques, which are discussed in later sections.

The sequential and blocking nature of the base station access leads to channel contention: mobiles which post requests for position to the base station must be deferred if the infrastructure is busy. This mandates Medium Access Control (MAC) schemes, which are discussed in the next section.

3 MAC Layer Proposals

Classical approaches from communications to handle shared channel access are multiplexing/division in any domain, such time, frequency or code (TDMA, FDMA and CDMA, respectively).

Such solutions are sometimes impractical, especially for ad-hoc and dynamic configurations which are often encountered in wireless sensor networks and positioning networks, where there is high volatility and often no chance to assign static frequencies or time slots to MS.

If static access schemes are not feasible to implement for any of the reasons outlined above, *random* or *dynamic channel access* ensues. This is generally taken to mean that all MS share the system band and peruse it at random times. The channel thus becomes a shared resource, and, invariably, contention situations arise. In this case, MAC layer algorithms must be installed to handle concurrent access, or MS will "overshout" each others signals.

Investigation of MAC algorithms targets several specific performance figures.

Most importantly, the total *time-to-fix* encompasses several specific metrics dealing with delays in the positioning process. The average time-to-fix is the total time it takes from the instant the position is requested to when it is finally provided.

In literature, the time-to-fix or waiting time is usually given as

$$W_i = D_i + S_i, \tag{7}$$

where D_i is the delay experienced by the ith terminal until channel contention situations are resolved (*queue time*), and S_i the actual time it takes for service completion [8]. The figures

$$d = \lim_{n \to \infty} \frac{\sum_{i=1}^{n} D_i}{n} \tag{8}$$

and

$$w = \lim_{n \to \infty} \frac{\sum_{i=1}^{n} W_i}{n} \tag{9}$$

describe the *steady-state average delay* and *steady-state average waiting time*. In the following, (9) is called *request latency*, as it describes the total time a client waits for the position request to complete.

Closely related is the classical term *throughput*. In communications networks, it describes the number of (successful) packet transmissions per time unit, summed over all nodes. In this application, it is defined as the number of successfully completed position calculations per second.

For the RToF configuration of the positioning system, two MAC strategies have been under investigation, detailed below.

3.1 Controlled ALOHA

In allusion to the archaic contention scheme from communications, the MS which are rejected from the infrastructure simply enter an exponential backoff interval, which is parameterized by a mean value which is usually lower than the expected mean request intervals of the mobiles.

When the backoff time expires, the mobiles simply repeat their request, which may again be rejected.

3.2 FIFO

An alternative approach to handle collision situations is the installation of a central queue which handles rejected MS according to a First-In First-Out (FIFO) principle. When an MS is rejected, it is told to stand by and wait for acknowledge. Internally, the infrastructure stores a unique identifier associated with the MS in the FIFO. When the infrastructure completes the current position request, it starts processing the FIFO. This process is not interruptible, e.g., by new positioning requests. As long as there are nodes in the FIFO, queue processing has priority and new requests will be deferred to the queue.

4 Simulation Results

To evaluate the system performance with the MAC algorithms described in the above section, a discrete event simulation framework has been implemented.

In addition to selection of various timing- and protocol-related parameters, this framework also allows integration of a single-MS system simulation. This allows to evaluate the effect of repeat-request algorithms and diversity acquisition on the position error and MAC performance figures.

Fig. 2 shows the mean request latency for both access strategies and various mean backoff times T_{BO}. In all simulations, the request times are assumed to be a Poisson process with a mean request interval of 2 s.

The FIFO strategy has a clear advantage over C-ALOHA at low MS densities, where the queue remains small and turnaround times quick. The nearly linear latency increase of C-ALOHA compensates for this when densities increase. Obviously, lower backoff times improve the latency. However, this comes at the cost of an increasing number of rejected MS, which is directly proportional to the energy consumption: each reject means a wasted communication between the MS and the infrastructure.

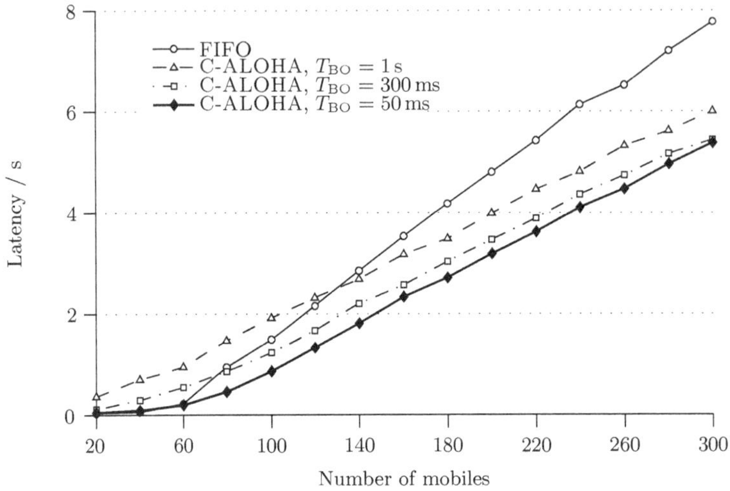

Fig. 2. Mean request latency for both MAC strategies and various C-ALOHA backoff times

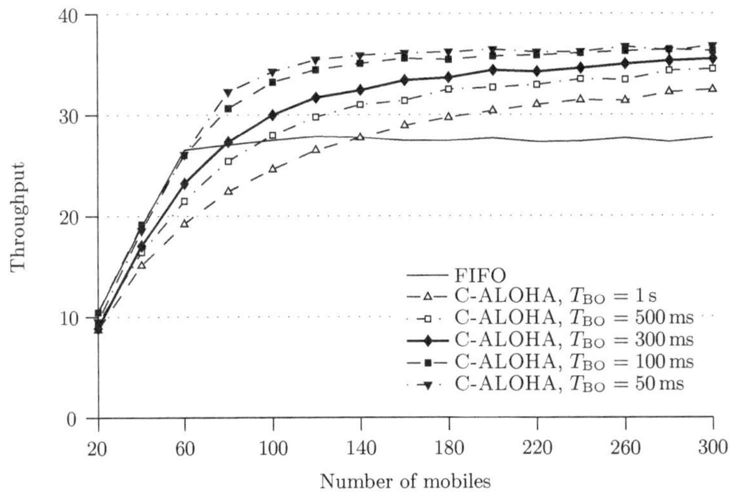

Fig. 3. Total throughput (successful requests per second) for both MAC strategies and various C-ALOHA backoff times

The improving performance of lower backoff times is also evident in Fig. 3. Here, it is notable that the throughput of FIFO settles to a significantly lower value than C-ALOHA. The reason for this is that there is an extra communications step required in FIFO, namely the infrastructure signaling the MS that it can leave the queue.

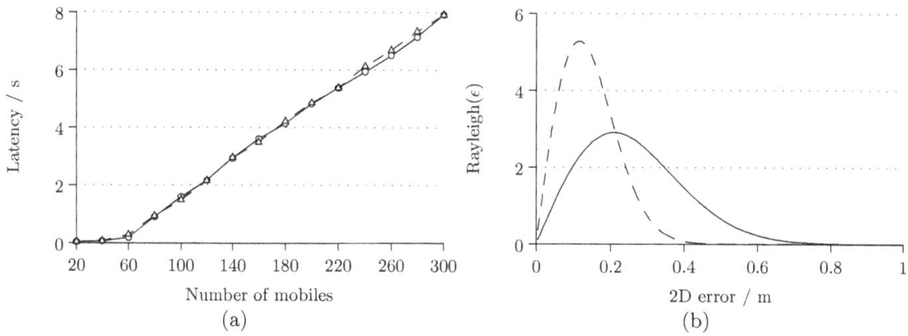

Fig. 4. Performance figures for regular (solid line) and repeat-request (dashed line) positioning. The FIFO access strategy was applied. (a) Latency (b) 2-D Error.

Fig. 4 illustrates the capabilities of the integrated system simulation. Here, an automated repeat-request algorithm is installed.

The idea is that especially in multipath environments, statistical outliers in the distance estimation might severely aggravate the final position error. This might be mitigated by excluding particularly bad distance estimates from the final position calculation.

Estimating the quality of a distance measurement is no trivial task. Ideas for this include calculation of the Rice factor of the channel profile [9], using Neural Networks trained to indoor channel profiles [10], sensor fusion techniques and tracking.

The exact implementation details of such an algorithm are beyond the scope of this work. However, it can reasonably assumed that a good algorithm will estimate the distance error with a Gaussian fault term, thus balancing false positive and false negatives for purposes of latency and error estimation.

Fig. 4 shows error pdfs following (6) for regular and repeat-request simulations with an indoor channel model described in [11]. It can readily be seen that there is a significant performance increase. The latency plot in Fig. 4 documents the fact that there are only few strong statistical outliers responsible for the detrimental effect of multipath on the position fix.

5 Summary

In this paper, system aspects of a novel, high-precision radiolocation system operating in the 5.8 GHz ISM band were presented, with a particular focus on MAC layer aspects.

Of the two access strategies presented, C-ALOHA and FIFO, none can be said to exhibit a clear advantage over the other. For low-complexity operations where energy consumption is no primary concern, C-ALOHA can be deployed with minimal software adaption. The installation of one or several queues, however, yields increased control over the network.

Integration of system simulation into the network layer allows to analyse algorithms for not only their effect on error performance, but also on MAC performance figures. The automated repeat-request algorithm presented here was discovered to have little impact on request latency, but to improve position estimation performance drastically.

References

1. Vossiek, M., Wiebking, L., Gulden, P., Wieghardt, J., Hoffmann, C., Heide, P.: Wireless local positioning. IEEE Microwave Magazine 4(4), 77–86 (2003)
2. Skolnik, M.I.: Radar Handbook. McGraw-Hill Book Co., New York (1990)
3. Stelzer, A., Pourvoyeur, K., Fischer, A.: Concept and application of LPM - a novel 3-d local position measurement system. IEEE Transactions on Microwave Theory and Techniques 52, 2664–2669 (2004)
4. Fuentes Michel, J.C., Millner, H., Vossiek, M.: A novell wireless forklift positioning system for indoor and outdoor use. In: 5th Workshop on Positioning, Navigation and Communication, WPNC 2008, Hannover, Germany, March 27, pp. 219–227 (2008)
5. Mosshammer, R., Huemer, M., Szumny, R., Kurek, K., Hüttner, J., Gierlich, R.: A 5.8 GHz local positioning and communication system. In: IEEE (ed.) Proceedings (Conference CD-ROM) of the IEEE International Microwave Symposium (IMS 2007), June 2007, pp. 1237–1240 (2007)
6. Mosshammer, R., Frank, F., Huemer, M.: Neural network based path detection for an FMCW positioning system. In: Moreno Díaz, R., Pichler, F., Quesada Arencibia, A. (eds.) EUROCAST 2007. LNCS, vol. 4739, pp. 928–935. Springer, Heidelberg (2007)
7. Eickhoff, R., Ellinger, F., Ußmüller, T., Spiegel, S.: A Highly-Integrated Fractional-N Synthesiser for FMCW Radar. In: Proceedings of the International IEEE Conference on Microwaves, Communications, Antennas and Electronic System (COMCAS), Tel-Aviv, Israel (May 2008)
8. Law, A.M.: Simulation Modeling and Analysis, 4th edn. McGraw-Hill, New York (2007)
9. Schmid, A., Neubauer, A.: Channel estimation technique for positioning accuracy improvement in multipath propagation scenarios. In: Proceedings of the ION GNSS 17th International Technical Meeting of the Satellite Division (2004)
10. Götz, A.: Detektionsalgorithmik und Mehrwegekompensationsverfahren für Local-Positioning-Radar-Systeme. Master's thesis, University of Erlangen-Nuremberg (May 2008)
11. Kozlowski, S., Kurek, K., Szumny, R., Modelski, J.: Statistical modelling of wideband propagation channel in an indoor environment. In: 17th International Conference on Microwaves, Radar and Wireless Communications, MIKON 2008, May 19-21, pp. 1–4 (2008)

Simulation Based Optimization of Vertex Packing Decoding Algorithms

Michael Lunglmayr[1], Jens Berkmann[2], and Mario Huemer[1]

[1] Klagenfurt University, Embedded Systems and Signal Processing Group
{michael.lunglmayr,mario.huemer}@uni-klu.ac.at
[2] Infineon Technologies AG, Munich
jens.berkmann@infineon.com

Abstract. Low Density Parity Check (LDPC) codes are considered in many future communication systems for error correction coding. Optimal decoding of LDPC codes is usually too costly to be done in practice. For this reason, sub-optimal algorithms are used. A state-of-the-art algorithm for decoding of LDPC codes is called belief propagation (BP). For short LDPC codes and for codes with an implementation efficient structure, the performance of this algorithm can be far from optimum.

We present a graphical model for representing the decoding problem, called configuration graph. We show the construction of a configuration graph and describe how the decoding problem can be represented as maximum weighted vertex problem (VP) on a configuration graph. We describe decoding approaches utilizing this representation and show the improvements in terms of decoding performance as well as the complexity/performance trade-offs possible with these algorithms.

Keywords: Channel coding, low-density parity-check (LDPC) codes, vertex packing, performance simulation.

1 Introduction

Error correction coding is a widely used tool in digital communications. It allows utilizing redundant information added at the transmitter to correct transmission errors at the receiver. Low Density Parity Check (LDPC) [1] codes are considered in many future communication systems for error correction coding, e.g. WLAN 802.11n [2]. Optimal decoding approaches of LDPC codes are usually too costly to be done in practice. A state-of-the-art algorithm for decoding of LDPC codes is called belief propagation (BP) [3]. Despite the sub-optimality of BP decoding, especially for very long and randomly structured LDPC codes, it can show a very good decoding performance [4]. On the other hand, for short LDPC codes and for LDPC codes with an implementation efficient structure, the performance of this algorithm can be far from optimum [5].

In this work we show new decoding algorithms based on a graphical representation of the decoding problem called configuration graph. We first describe LDPC codes. Then we show the construction of a configuration graph and algorithms for decoding derived from that graphical representation. We present

R. Moreno-Díaz et al. (Eds.): EUROCAST 2009, LNCS 5717, pp. 469–476, 2009.

simulation results demonstrating the performance gains achievable with these algorithms.

2 LDPC Codes

In this work we will only consider binary LDPC codes. LDPC codes are linear block codes with a sparse $m \times n$ parity check matrix $\mathbf{H}$. This parity check matrix $\mathbf{H}$ can be used to define a code. A binary linear block code $\mathbb{C}$ is defined by the (column) vectors

$$\mathbb{C} = \{\mathbf{x} \in \{0,1\}^n | \mathbf{H}\mathbf{x} = \mathbf{0}\}. \quad (1)$$

The elements $H_{i,j}$ of $\mathbf{H}$ are also either 0 or 1 (all calculations are done in the binary field). Every row $\mathbf{h}_i^T$ of $\mathbf{H}$ corresponds to a so-called party check. We say that $\mathbf{x}$ satisfies parity check i, if $\mathbf{h}_i^T\mathbf{x} = 0$. Clearly, a vector $\mathbf{x}$ has to satisfy all m parity checks of $\mathbf{H}$ to be a codeword. Further, we say that a bit j of $\mathbf{x}$ participates to parity check i, if the element $H_{i,j} = 1$. It can be easily shown that a parity check i can only be satisfied if the number of ones of the participating bits is even.

Let J_i be an ascendingly ordered list of the indices of the bits participating to parity check i. Let $|J_i|$ be the number of elements in the list J_i. Because parity check i can only be satisfied when an even number of its participating bits is one, there are $2^{|J_i|-1}$ assignments of those bits, satisfying parity check i. Such an assignment $\mathbf{c}^{(ik)}$ will be called a *local codeword*. We represent the local codewords also as ordered lists. The jth position of $\mathbf{c}^{(ik)}$ defines the value of the bit with the index of the jth position of J_i. Tab. 1 shows the local codewords of an example code with the parity check matrix

$$\mathbf{H} = \begin{bmatrix} 1\,0\,1\,1\,0 \\ 0\,1\,1\,0\,1 \end{bmatrix}. \quad (2)$$

A codeword can be represented by a set P of local codewords, if the elements of the set are chosen according to the following conditions.

1. Exactly one local codeword per parity check is included in P.
2. The local codewords are consistent. This means for all $c^{(ik)}, c^{(lo)} \in P$, if a bit j participates to parity check i and l, the corresponding values for bit j in the local codewords $c^{(ik)}$ and $c^{(lo)}$ must be equal, respectively.

Table 1. Local codewords for example code

Parity check 1	Parity check 2
$J_1 = (1,3,4)$	$J_2 = (2,3,5)$
$\mathbf{c}^{(11)} = (0,0,0)$	$\mathbf{c}^{(21)} = (0,0,0)$
$\mathbf{c}^{(12)} = (0,1,1)$	$\mathbf{c}^{(22)} = (0,1,1)$
$\mathbf{c}^{(13)} = (1,0,1)$	$\mathbf{c}^{(23)} = (1,0,1)$
$\mathbf{c}^{(14)} = (1,1,0)$	$\mathbf{c}^{(24)} = (1,1,0)$

Following these conditions, a codeword $\mathbf{c} = [c_b]$ can be built using a set P according to the following principle: we assign the value of the jth element of the local codeword $\mathbf{c}^{(ik)} \in P$ to the corresponding bit c_b of the jth index (in this case b) of J_i.

Here condition 1 guarantees that for every parity check there must be a local codeword in P and condition 2 guarantees that the assignment of the bits is consistent (the same value is assigned to the common bits of the local codewords).

For the example code, the set $\{\mathbf{c}^{(11)}, \mathbf{c}^{(23)}\}$ represents a codeword (the codeword $[0\ 1\ 0\ 0\ 1]^T$). Contrary, the set $\{\mathbf{c}^{(11)}, \mathbf{c}^{(22)}\}$ does not define a codeword because in $\mathbf{c}^{(11)}$ the bit 3 of a codeword is defined as 0 but in $\mathbf{c}^{(22)}$ it is defined as 1. So, bit 3 of the codeword would not be clearly defined with the set $\{\mathbf{c}^{(11)}, \mathbf{c}^{(22)}\}$.

The consistency between the local codewords is naturally maintained in equation (1) and in a configuration graph defined in the following section. We will see later how we can use this representation for decoding of LDPC codes.

3 Configuration Graphs

For construction of a configuration graph we will assume that the parity check matrix has the structure $\mathbf{H} = [\mathbf{D}\ \mathbf{I}]$. This structure guarantees that decoding can be represented as a so-called vertex packing problem on a configuration graph as proofed in [6]. Parity check matrices of every other form can be brought to this desired structure by elementary matrix row operations.

A configuration graph $G_c = (V, E)$ is constructed by assigning the set of vertices to all local codewords of all parity checks: $V = \{c^{(ik)}\}_{i=1\ldots m,\ k=1\ldots 2^{|J_i|-1}}$. The set of edges of the configuration graph is found by connecting the vertices according to the following rule:

Two vertices $c^{(ik)}$ and $c^{(lo)}$, $c^{(ik)} \neq c^{(lo)}$ of a configuration graph are connected: $\{c^{(ik)}, c^{(lo)}\} \in E$, if and only if, a bit with index b participates to parity check i and to parity check l, and the value of bit b in $c^{(ik)}$ is different than the value of bit b in $c^{(lo)}$. At maximum one connection is allowed between two vertices of the graph.

Fig. 1 shows the configuration graph for the example code. As described before, a codeword can be represented by a set of m local codewords, one per parity check, where the local codewords are consistent to each other.

A configuration graph naturally represents this consistency property: as proofed in [6] a set of m vertices of the configuration graph, where no vertex of the set is connected to any other vertex of the set represents a codeword. Such a set of vertices of a graph, where no vertex is connected to any other vertex of the set is called a *vertex packing* [7].

When assigning weights w_i to every vertex v_i of a graph G, the *(weighted) vertex packing problem* is to find those vertex packing with the maximum weight sum (i.e. the *costs* of a vertex packing)

$$\hat{P} = \underset{P \in \mathcal{P}_G}{\operatorname{argmax}} \sum_{i: v_i \in P} w_i, \qquad (3)$$

with $\mathcal{P}_G$ as the set of all vertex packings on G.

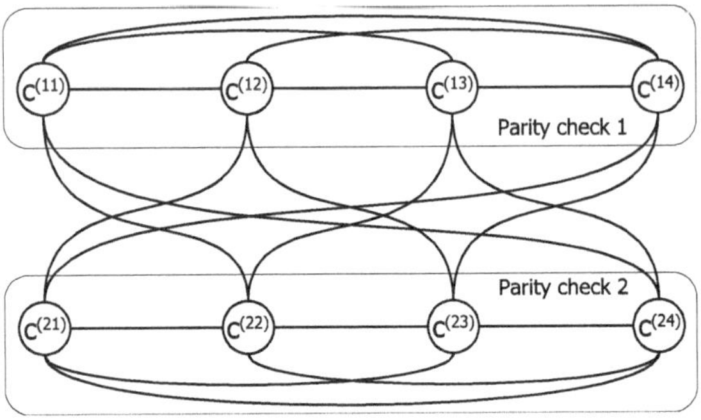

Fig. 1. Configuration graph for the example parity check matrix

It is shown in [6] that with adequate weights of the vertices, decoding of an LDPC code can be represented as weighted vertex packing problem.

Unfortunately, the complexity of solving the weighted vertex packing (similar other optimal decoding approaches) is too high to be done in a communications receiver for practically used codes. For this reason, we present sub-optimal approaches for the weighted vertex packing problem based on greedy algorithms [8] in the next section. We will show, that especially when being combined with belief propagation decoding, the performance of such a combined decoder can be significantly better than the performance of a single BP decoder.

4 Greedy Algorithms for Decoding

Although there is no guarantee that a maximum weighted vertex packing is found with greedy algorithms, they show a good performance-complexity trade-off in practice. The basic structure of these algorithms is described in *GreedyBasic*.

Algorithm. GreedyBasic

1: $P \Leftarrow \emptyset$
2: **for each** parity check i **do**
3: Add the vertex $c^{(ik)}$ from parity check i to P – which is not connected to any vertex already in P – with maximal $f(c^{(ik)})$.
4: **end for**

Here $f(c^{(ik)})$ represents cost function of a vertex. In its most basic form, the costs of a vertex can be only the weights w_{ik} of the vertices. But when using such a simple cost function, the shortsightedness of the greedy algorithm may often prevent finding the optimal solution. To improve this behavior, a more advanced cost function does not only consider the costs of a vertex $c^{(ik)}$ itself, but also the potential of other vertices that could be added (while maintaining a vertex packing) after adding $c^{(ik)}$ to P (another point of view would be to consider the costs of those vertices that are prevented to be added after adding $c^{(ik)}$). Considering this potential of vertices can improve the performance of the greedy algorithm but will also increase its computational complexity.

In the following we will use two algorithms: GreedySimple and GreedyLookAhead. GreedySimple will only use the weights of the vertices as costs, while GreedyLookAhead also estimates the potential of a vertex. This is done by accumulating the costs of a vertex packing built by GreedyLookAhead for the still unprocessed parity checks. The calculation of this potential is shown in the next algorithm:

Algorithm. Calculating $h(c^{(ik)})$

Input: w_{ik}, P
Output: $h(c^{(ik)})$

```
1: P' ⇐ P ∪ {c^(ik)}, h(c^(ik)) ⇐ 0
2: for each parity check p with no vertex already in P' do
3:     Add the next maximal consistent vertex c'^(pj) to P'
4:     h(c^(ik)) ⇐ h(c^(ik)) + w'_pj
5: end for
```

The costs $f(c^{(ik)})$ are then calculated as $w_{ik} + h(c^{(ik)})$.

As described before, the number of vertices of a parity check scales exponentially with the number of participating bits. For practically used codes, several hundreds of bits may participate to a parity check. This makes it infeasible in practice to evaluate all vertices of a parity check. But it has been shown in [6] that the vertex with maximum weight that is still consistent (not connected) to all vertices of a given vertex packing, can be found with a very low effort. Considering the complexity of evaluating all vertices of a parity check, only the maximum vertex and those local codewords (vertices) obtained by flipping up to b bits are evaluated for GreedyLookAhead. This means when j bits participate to parity check i, only $\binom{j-1}{b}$ vertices (only assignments with an even number of ones are to be considered) are evaluated for parity check i, instead of 2^{j-1}.

When analyzing the structure of GreedyBasic, one can see that the order of processing the parity checks, has an impact on the performance of the algorithm. Therefore, for a practical implementation of vertex packing decoding, the greedy algorithms are called t times, each time with a random processing order. The number of processing orders used for decoding scales the decoding performance as well as the decoding complexity. This parameter can be adjusted until a performance complexity trade-off is found. In addition, when using GreedyLookAhead, the number of evaluated neighbors of a maximum vertex, i.e. by flipping b bits, represents an additional parameter for scaling the complexity as well as the performance. In Sect. 6 we will show simulation results allowing to select the least complex parameters for a desired decoding performance.

5 Combined Decoding Approaches

We observed that for LDPC codes, the vertex packing decoding (VPD) algorithms based on greedy approaches, as described before, perform worse in terms of word error rate, compared to belief propagation decoding [3] – the standard approach for decoding LDPC codes. But when being performed *in addition* to BP decoding, the decoding performance of a combined BP/VPD decoding was significantly better than when using BP decoding only. Fig. 2 shows the principle of such a combined decoding concept. The BP decoder uses the samples

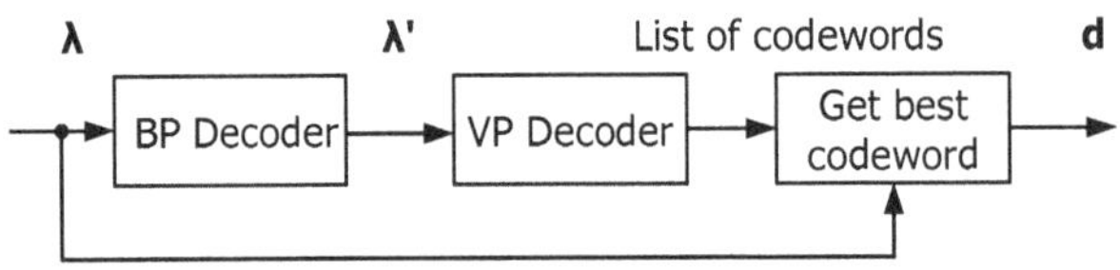

Fig. 2. Combined VP/BP decoder

$\boldsymbol{\lambda}$ of a received signal to calculates its decoding output $\boldsymbol{\lambda}'$. For a more detailed description on the BP decoding algorithm we refer to [3]. The BP decoding output is used by the VPD algorithm (i.e. GreedySimple or GreedyLookAhead) to calculate a vertex packing that corresponds to a codeword. As described before, multiple processing orders of the parity check are used for decoding. Every run of a greedy algorithm with a different processing order potentially results in a different codeword. From this list of output codewords, the best according to the maximum likelihood metric [9] is chosen.

6 Simulation Results

In Fig. 3 we show simulation results in terms of word error rate (WER) for an example code, the code 96.3.963 from [10]. In the legend of this figure, the numbers after the names of the greedy variants specify the number of random processing orders used for decoding. For GreedyLookAhead, the numbers after "flip" specifies how many bits are flipped to evaluate the neighbors of the maximum vertex of a parity check. For the example code, the combined decoder with GreedyLookAhead and evaluating the neighbors of a maximum vertex by flipping up to two bits performs up to 0.6dB better than when using BP only. Interestingly, the performance when using GreedySimple with 2000 processing orders is nearly the same when using GreedyLookAhead with 100 processing orders and evaluating the neighbors of a maximum vertex obtained by flipping one bit. But a complexity comparison reveals that in this case GreedySimple with 2000 processing orders needs about 20 times less he effort than GreedyLookAhead with 100 processing orders. On the other hand, even when increasing the number of processing orders for GreedySimple, the performance was always worse than when using GreedyLookAhead with 100 processing orders and flipping up

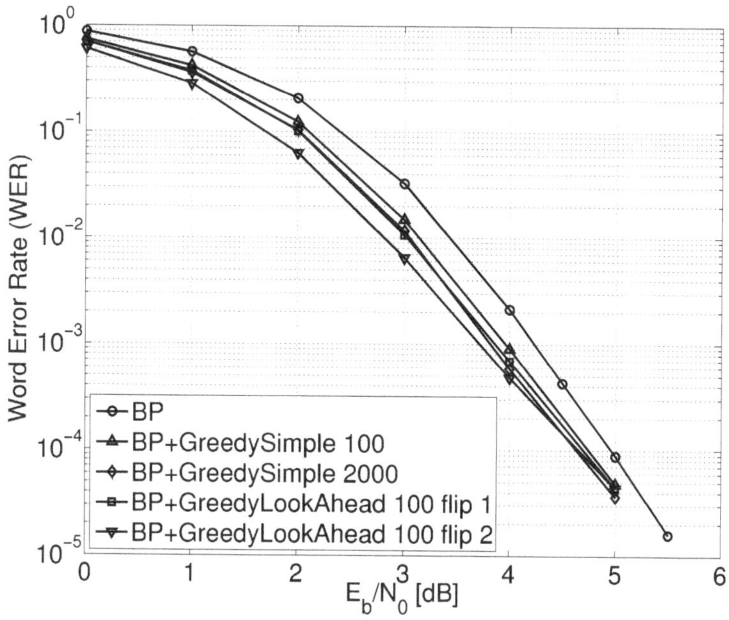

Fig. 3. Simulation results for example LDPC code

to two bits. But it has to be noticed that the complexity of GreedyLookAhead with 100 processing orders and flipping up to two bits is about 500 times higher than the complexity of GreedySimple with 2000 processing orders. For practical application one has to ponder, if the performance of GreedySimple (e.g. with 2000 processing orders) is sufficient. If not, then the algorithm GreedyLookAhead with a higher complexity is to be used. But the results show that with the presented algorithms, the performance-complexity trade-off can be easily adjusted.

7 Conclusion

We presented a new graphical model for representing the decoding problem, called configuration graph. We explained the construction of a configuration graph for binary LDPC codes and explained how the decoding problem can be represented as maximum weighted vertex problem (VP) on a configuration graph. We described decoding approaches based on this graphical representation and showed that especially in combination with BP decoding the decoding performance of this approach is significantly better than when using BP only. We presented simulation results and discussed how a selection of the algorithms' parameters based on simulation results can be performed considering the different complexities of the algorithms.

References

1. Gallager, R.G.: Low-Density Parity-Check Codes. MIT Press, Cambridge (1963)
2. Joint Proposal: High throughput extension to the 802.11 Standard: PHY. IEEE 802.11-05/1102r4 (2006)
3. Kschischang, F., Frey, B., Loeliger, H.: Factor graphs and the sum-product algorithm. IEEE Transactions on Information Theory 51(3), 954–972 (2005)
4. Chung, S.Y., Forney, G.D., Richardson, T.J., Urbanke, R.: On the Design of Low-Density Parity-Check Codes within 0.0045 dB of the Shannon Limit. IEEE Communications Letters 5(2) (2001)
5. Etzion, T., Trachtenberg, A., Vardy, A.: Which Codes Have Cycle-Free Tanner Graphs. IEEE Transactions on Information Theory 45(6), 2173–2181 (1999)
6. Lunglmayr, M., Berkmann, J., Huemer, M.: Vertex Packing Decoding. In: Proc. International Conference on Communications (ICC), Dresden, Germany (2009)
7. Nemhauser, G., Trotter, L.: Vertex Packings: Structural Properties and Algorithms. Mathematical Programming 8(1), 232–248 (1975)
8. Michalewicz, Z., Fogel, D.: How to Solve It: Modern Heuristics, 2nd edn. Springer, Heidelberg (2004)
9. Bossert, M.: Channel Coding for Telecommunications. Wiley, Chichester (1999)
10. MacKay, D.: Encyclopedia of Sparse Graph Codes (hypertext archive), `http://www.inference.phy.cam.ac.uk/mackay/codes/data.html`

Diversity Order of Spatial Multiplexing with Transmit Antenna Correlation Based Precoding

Christian Hofbauer[1,*], Yann Lebrun[2], Valéry Ramon[2], André Bourdoux[2], François Horlin[3], and Mario Huemer[1]

[1] Klagenfurt University, Embedded Systems and Signal Processing, 9020 Klagenfurt, Austria
[2] IMEC, Wireless Group, 3001 Leuven, Belgium
[3] ULB, Université Libre de Bruxelles, 1050 Brussels, Belgium
{chris.hofbauer,mario.huemer}@uni-klu.ac.at
{yann.lebrun,valery.ramon,andre.bourdoux}@imec.be
francois.horlin@ulb.ac.be

Abstract. Spatial Multiplexing (SM) is an effective means for enhancing the transmission data rate in Multiple-Input Multiple-Output (MIMO) systems, particularly when used in combination with precoding techniques. However, it is not always obvious to connect the performance of such a system to its number of data streams and antennas. In this paper, the diversity order of a SM MIMO system using a Minimum Mean Square Error (MMSE) receiver is analytically calculated when a precoder based on transmit antenna correlation is included at the transmitter. It is shown that it is given by $d = N_r - N_s + 1$ where N_r is the number of receive antennas and N_s is the number of data streams. This result is confirmed by simulations. Although enabling a Signal-to-Noise Ratio (SNR) gain, such a precoder is thus not able to improve the diversity order with respect to a non-precoded SM MIMO system.

1 Introduction

The presence of multiple antennas at both the transmitter and the receiver in communication systems can provide a considerable gain in terms of capacity, coverage and link reliability. Thanks to these benefits, Multiple-Input Multiple-Output (MIMO) techniques appear in emerging wireless standards. While MIMO techniques already improve the performance when the receiver alone knows the channel, the achievable gain can be further enhanced when the transmitter has knowledge of the statistical characteristics (for example the mean or the covariance matrix) of the channel [1]. In that case, precoding exploits these slow-varying properties of the channel to perform signal shaping before transmission. While transmit antenna correlation based precoding improves the performance of a MIMO system, it is not straightforward to link its performance in terms

* Christian Hofbauer has been funded by the European Regional Development Fund and the Carinthian Economic Promotion Fund (KWF) under grant 20214/15935/23108.

R. Moreno-Díaz et al. (Eds.): EUROCAST 2009, LNCS 5717, pp. 477–484, 2009.

of diversity order and Signal-to-Noise Ratio (SNR) gain to the system configuration (number of streams N_s, number of transmit antennas N_t and number of receive antennas N_r).

In [2], Gore and al. derive the diversity order per transmit stream in a spatial multiplexing (SM) MIMO system using a Zero Forcing (ZF) receiver. A transmit antenna selection is done based on the knowledge of the channel statistics in order to maximize the average throughput or minimize the average probability of error. They show that the presence of transmit antenna correlation does not impact the diversity order on each stream. Although the antenna selection algorithm proposed in [2] can be seen as a specific type of precoder, the results do not apply when more general precoding matrices are used. The goal of this paper is to derive the diversity order of a SM MIMO system including a Minimum Mean Square Error (MMSE) receiver and a transmit antenna correlation based precoding. We show that it is equal to $d = N_r - N_s + 1$ where N_r is the number of receive antennas and N_s is the number of data streams. This result is confirmed by simulations. We also demonstrate that such precoding schemes bring no diversity gain but only SNR gain compared to non-precoded systems.

The outline of this paper is as follows. Section 2 introduces the system model. In section 3, we provide the calculation of the diversity order of a SM MIMO system in which a precoder based on transmit antenna correlation is used. Simulations in section 4 confirm the validity of this calculation and section 5 concludes our work.

The following notation is used in this paper. The vectors and matrices are in boldface letters, vectors are denoted by lower-case and matrices by capital letters. The superscripts $(\cdot)^T$, $(\cdot)^\dagger$, $(\cdot)^*$ and $(\cdot)^H$ denote the transpose, pseudo-inverse, conjugate and Hermitian operators, respectively. $E[\cdot]$ is the expectation operator, $\mathbf{X}_{ij}$ is the (i,j) element of the matrix $\mathbf{X}$, $\mathbb{C}^{N\times 1}$ denotes the set of complex vectors of size $(N \times 1)$. $\mathbb{C}^{N\times M}$ denotes the set of complex matrices of size $(N \times M)$ and $\mathbf{x} \sim CN(\mathbf{0}, \mathbf{R})$ is the vector of zero-mean Gaussian distributed complex elements with covariance matrix $\mathbf{R}$. The matrix $\mathbf{0}_{i,i}$ is a zero matrix of dimension $(i \times i)$ and $\mathbf{I}_i$ is an identity matrix of dimension $(i \times i)$. $\det(\mathbf{X})$ denotes the determinant of the matrix $\mathbf{X}$ and $\mathrm{etr}(\mathbf{X})$ is the exponential of the trace of $\mathbf{X}$, $e^{\mathrm{tr}(\mathbf{X})}$.

2 System Model

We consider a communication system with N_t transmit antennas and N_r receive antennas. We define N_s as the number of transmit streams, with $N_s \leq \min(N_r, N_t)$. We denote by $\mathbf{H}$ the $N_r \times N_t$ MIMO channel matrix with Rayleigh fading coefficients (flat fading). We assume that $\mathbf{H}$ is fixed during the transmission of a data block and changes independently to a new realization in the next block (quasi-static fading). We suppose also that $\mathrm{rank}(\mathbf{H}) \geq N_s$. At the channel output, the received vector $\mathbf{y} \in \mathbb{C}^{N_r \times 1}$ can be expressed as

$$\mathbf{y} = \mathbf{H}\mathbf{F}\mathbf{x} + \mathbf{n} \tag{1}$$

where $\mathbf{x} \in \mathbb{C}^{N_s \times 1}$ denotes the transmit symbol vector, $\mathbf{F} \in \mathbb{C}^{N_t \times N_s}$ is the precoding matrix and $\mathbf{n} \in \mathbb{C}^{N_r \times 1}$ is the zero-mean circularly symmetric complex additive white Gaussian noise with noise variance N_0.

In this paper, we assume the presence of correlation between antennas at the transmitter whereas the receive antennas are assumed to be decorrelated. Physically, this could correspond to a non-obstructed transmitter situated on a high location with waves coming from the same direction (small angular spread) while the receiver is situated in a rich scattering environment [3]. Based on these assumptions, we can express the channel matrix as $\mathbf{H} = \mathbf{H}_w \mathbf{R}_t^{1/2}$, where $\mathbf{H}_w$ denotes a $N_r \times N_t$ white zero-mean circularly symmetric complex Gaussian (ZMCSCG) channel matrix with independent and identical distributed (i.i.d.) elements of unit variance and $\mathbf{R}_t^{1/2}$ is an $N_t \times N_t$ matrix that represents the "square root" of the transmit antenna correlation matrix $\mathbf{R}_t = \left(\mathbf{R}_t^{1/2}\right)^H \mathbf{R}_t^{1/2}$. We model $\mathbf{R}_t$ by applying the exponential correlation model $\mathbf{R}_{t_{ij}} = \rho^{\left|i-j\right|}$, whereas $\rho = 0$ means no correlation and $\rho = 1$ full correlation, respectively. As we assume no receive antenna correlation, the columns $\mathbf{h}_1 \dots \mathbf{h}_{N_r}$ of $\mathbf{H}^H$ can be assumed to be i.i.d. as $\mathbf{h}_j \sim CN(\mathbf{0}, \mathbf{R}_t)$.

Our precoder $\mathbf{F}$ is designed for minimizing the Mean Square Error (MSE criterion) and is based on the correlation between the transmit antennas [4]. The derivations presented in this paper are also valid for different criteria (for example the BER and SNR). We aim at minimizing the MSE

$$\text{MSE} = E\left[\text{tr}\left((\hat{\mathbf{x}} - \mathbf{x})(\hat{\mathbf{x}} - \mathbf{x})^H\right)\right] = E\left[\text{tr}\left(\mathbf{I}_{N_s} + \gamma_0 \mathbf{F}^H \mathbf{H}^H \mathbf{H} \mathbf{F}\right)^{-1}\right] \quad (2)$$

subject to the power constraint $\text{tr}\left(\mathbf{F}\mathbf{F}^H\right) \leq 1$, assuming $E\left[\mathbf{x}\mathbf{x}^H\right] = \mathbf{I}_{N_s}$ and $\gamma_0 = 1/N_0$. Using the Jensen's inequality to bring the expectation operator inside the matrix inversion in (2) and following the derivations made in [4], the optimal precoder turns out to be $\mathbf{F} = \mathbf{U}_t \mathbf{\Lambda}_F$ with $\mathbf{R}_t = \mathbf{U}_t \mathbf{\Lambda}_t \mathbf{U}_t^H$. $\mathbf{\Lambda}_F$ is a diagonal matrix that implements an inverse water pouring policy.

3 Calculation of the Diversity Order

Our target is to assess how the precoder $\mathbf{F}$ affects the symbol error behavior of the streams, especially with respect to the diversity order. The diversity order d determines the negative slope of the symbol error curves, and the authors in [5] define it as

$$\lim_{SNR \to \infty} \frac{\log_{10} P_e(SNR)}{\log_{10} SNR} = -d \quad (3)$$

where $P_e(SNR)$ denotes the average error probability as a function of the SNR. We target the high SNR region, as d in (3) is derived for $SNR \to \infty$. We can therefore replace the MMSE receiver in our calculations by the simpler definition of the Zero Forcing (ZF) receiver. Hence, after processing by the receiver, the estimated symbol vector is given by

$$\hat{\mathbf{x}} = (\mathbf{H}\mathbf{F})^{\dagger} \mathbf{H}\mathbf{F}\mathbf{x} + (\mathbf{H}\mathbf{F})^{\dagger} \mathbf{n} = \mathbf{x} + (\mathbf{H}\mathbf{F})^{\dagger} \mathbf{n} \quad (4)$$

where $(\mathbf{HF})^\dagger$ denotes the pseudo-inverse of $\mathbf{HF}$. The symbol stream error vector $\mathbf{e}$ is thus $(\mathbf{HF})^\dagger\mathbf{n}$, leading to a SNR on the kth stream of

$$\gamma_k = \frac{E\left[\mathbf{x}_k\mathbf{x}_k^H\right]}{E\left[\mathbf{e}\mathbf{e}^H\right]_{kk}} = \frac{1}{N_0\left[(\mathbf{HF})^\dagger\left((\mathbf{HF})^\dagger\right)^H\right]_{kk}} = \frac{\gamma_0}{[(\mathbf{F}^H\mathbf{H}^H\mathbf{HF})^{-1}]_{kk}}. \tag{5}$$

The remainder of this section is organized as follows. In section 3.1, the characteristic function of $\left(\mathbf{F}^H\mathbf{H}^H\mathbf{HF}\right)$ will be derived. This will enable the calculation, in section 3.2, of the probability density function (pdf) of $\left(\mathbf{F}^H\mathbf{H}^H\mathbf{HF}\right)$. Once we have achieved this, we can easily determine the pdf of γ_k, and finally use these results in order to formulate a symbol error bound. This bound will unveil the impact of $\mathbf{F}$ on the error behavior.

3.1 Derivation of the Characteristic Function of $\mathbf{F}^H\mathbf{H}^H\mathbf{HF}$

The characteristic function for a matrix-valued random variable $\mathbf{A}$ is defined as $\varphi(\mathbf{\Theta}) = E[\text{etr}(i\mathbf{A}\mathbf{\Theta})]$ where $\mathbf{\Theta} \in \mathbb{C}^{N_s\times N_s}$ is Hermitian with $\text{rank}(\mathbf{\Theta}) = N_s$ and i denotes the complex number subject to $i^2 = -1$ [6]. Letting $\mathbf{A} := \mathbf{F}^H\mathbf{H}^H\mathbf{HF}$ and taking into account that the trace of a product is invariant under cyclic permutations of the matrices in this product, the characteristic function of $\mathbf{F}^H\mathbf{H}^H\mathbf{HF}$ is given by

$$\varphi(\mathbf{\Theta}) = E\left[\text{etr}\left(i\mathbf{H}^H\mathbf{HF}\mathbf{\Theta}\mathbf{F}^H\right)\right]. \tag{6}$$

We know that the columns $\mathbf{h}_1 \ldots \mathbf{h}_{N_r}$ of $\mathbf{H}^H$ are i.i.d. as $\mathbf{h}_j \sim CN(\mathbf{0}, \mathbf{R}_t)$. Additionally taking into account that the trace and sum functions can be swapped, it follows that

$$\varphi(\mathbf{\Theta}) = E\left[\text{etr}\left(i\sum_{j=1}^{N_r}\mathbf{h}_j\mathbf{h}_j^H\mathbf{F}\mathbf{\Theta}\mathbf{F}^H\right)\right] = E\left[\exp\left(i\sum_{j=1}^{N_r}\text{tr}\left(\mathbf{h}_j\mathbf{h}_j^H\mathbf{F}\mathbf{\Theta}\mathbf{F}^H\right)\right)\right]. \tag{7}$$

Applying the cyclic permutation property of the trace function again, the argument of tr() becomes $\mathbf{h}_j^H\mathbf{F}\mathbf{\Theta}\mathbf{F}^H\mathbf{h}_j$ which is a scalar. Hence, (7) can be simplified by omitting the trace operator:

$$\varphi(\mathbf{\Theta}) = E\left[\exp\left(i\sum_{j=1}^{N_r}\mathbf{h}_j^H\mathbf{F}\mathbf{\Theta}\mathbf{F}^H\mathbf{h}_j\right)\right] = E\left[\prod_{j=1}^{N_r}\exp\left(i\mathbf{h}_j^H\mathbf{F}\mathbf{\Theta}\mathbf{F}^H\mathbf{h}_j\right)\right]. \tag{8}$$

Using again the property that the vectors $\mathbf{h}_j$ are i.i.d., we get

$$\varphi(\mathbf{\Theta}) = \prod_{j=1}^{N_r} E\left[\exp\left(i\mathbf{h}_j^H\mathbf{F}\mathbf{\Theta}\mathbf{F}^H\mathbf{h}_j\right)\right] = \left(E\left[\exp(i\mathbf{h}_1^H\mathbf{F}\mathbf{\Theta}\mathbf{F}^H\mathbf{h}_1\right]\right)^{N_r}, \tag{9}$$

in which $\mathbf{h}_1$ has arbitrarily been selected.

Although $\varphi(\boldsymbol{\Theta})$ has been simplified a lot already, it is still not obvious how to take the expectation in (9). Thus, the following steps will focus on substitutions in order to bring $\varphi(\boldsymbol{\Theta})$ into a form which easily allows to solve the $E[\cdot]$ operation.

We know that $\mathbf{h}_1 \sim CN(\mathbf{0}, \mathbf{R}_t)$. We define $\mathbf{w} = \left(\left(\mathbf{R}_t^{1/2}\right)^H\right)^{-1} \mathbf{h}_1$, hence $\mathbf{w} \sim CN(\mathbf{0}, \mathbf{I}_{N_t})$. By substituting $\mathbf{h}_1$ based on $\mathbf{h}_1 = \left(\mathbf{R}_t^{1/2}\right)^H \mathbf{w}$, we obtain

$$\varphi(\boldsymbol{\Theta}) = \left(E\left[\exp\left(i\mathbf{w}^H\mathbf{R}_t^{1/2}\mathbf{F}\boldsymbol{\Theta}\mathbf{F}^H\left(\mathbf{R}_t^{1/2}\right)^H\mathbf{w}\right)\right]\right)^{N_r}. \tag{10}$$

We know by the eigen-value decomposition, that for any Hermitian matrix $\mathbf{B} := \mathbf{R}_t^{1/2}\mathbf{F}\boldsymbol{\Theta}\mathbf{F}^H\left(\mathbf{R}_t^{1/2}\right)^H$, there exists a unitary matrix $\mathbf{V}_{N_t x N_t}$ such that $\mathbf{V}\mathbf{B}\mathbf{V}^H = \boldsymbol{\Lambda}$ where $\boldsymbol{\Lambda}$ denotes a diagonal matrix. Therefore, we define $\mathbf{u} = \mathbf{V}\mathbf{w}$ with $\mathbf{u} \sim CN(\mathbf{0}, \mathbf{I}_{N_t})$. Substituting $\mathbf{w}$ in equation (10), using the Hermitian property of $\mathbf{B}$ and taking into account that the elements of $\mathbf{u}$ are i.i.d., we obtain

$$\begin{aligned}\varphi(\boldsymbol{\Theta}) &= \left(E\left[\exp\left(i\mathbf{u}^H\mathbf{V}\mathbf{R}_t^{1/2}\mathbf{F}\boldsymbol{\Theta}\mathbf{F}^H\left(\mathbf{R}_t^{1/2}\right)^H\mathbf{V}^H\mathbf{u}\right)\right]\right)^{N_r} &(11)\\ &= \left(E\left[\exp\left(i\mathbf{u}^H\mathbf{V}\mathbf{B}\mathbf{V}^H\mathbf{u}\right)\right]\right)^{N_r} = \left(E\left[\exp\left(i\mathbf{u}^H\boldsymbol{\Lambda}\mathbf{u}\right)\right]\right)^{N_r} &(12)\\ &= \left(E\left[\exp\left(i\sum_{j=1}^{N_t}\boldsymbol{\Lambda}_{jj}|u_j|^2\right)\right]\right)^{N_r}. &(13)\end{aligned}$$

We know that the number of eigenvalues $\boldsymbol{\Lambda}_{jj}$ unequal to zero is determined by $\text{rank}(\boldsymbol{\Lambda})$. This rank cannot be greater than the minimum of the ranks of the matrices that $\boldsymbol{\Lambda}$ can be decomposed into [7]. We know that $\text{rank}(\mathbf{V}) = N_t$ and $\text{rank}(\boldsymbol{\Theta}) = N_s$. Assuming that $\mathbf{F}$ and $\mathbf{R}_t$ have full rank, i.e. $\text{rank}(\mathbf{R}_t) = N_t$ and $\text{rank}(\mathbf{F}) = N_s$, the characteristic function is given by

$$\varphi(\boldsymbol{\Theta}) = \left(E\left[\exp\left(i\sum_{j=1}^{N_s}\boldsymbol{\Lambda}_{jj}|u_j|^2\right)\right]\right)^{N_r} = \prod_{j=1}^{N_s} E\left[\exp\left(i\boldsymbol{\Lambda}_{jj}|u_j|^2\right)\right]^{N_r}. \tag{14}$$

Since $\mathbf{u} \sim CN(\mathbf{0}, \mathbf{I}_{N_t})$, $\left|u_j\right|^2$ is a Chi-square distributed random variable with 2 degrees of freedom, i.e. $\left|\mathbf{u}_j\right|^2 \sim X_2^2$. Because the characteristic function of a Chi-square distributed random variable X with k degrees of freedom is given by $E[exp(i\theta X)] = (1 - i\theta)^{-k/2}$, we obtain

$$\varphi(\boldsymbol{\Theta}) = \prod_{j=1}^{N_s}\left(1 - i\boldsymbol{\Lambda}_{jj}\right)^{-N_r} = \det\left(\mathbf{I}_{N_s} - i\boldsymbol{\Lambda}\right)^{-N_r}. \tag{15}$$

As a next step, we substitute back $\boldsymbol{\Lambda} = \mathbf{V}\mathbf{B}\mathbf{V}^H$ and get

$$\begin{aligned}\varphi(\boldsymbol{\Theta}) &= \det\left(\mathbf{I}_{N_s} - i\mathbf{V}\mathbf{B}\mathbf{V}^H\right)^{-N_r} = \det\left(\mathbf{I}_{N_s} - i\mathbf{V}^H\mathbf{V}\mathbf{B}\right)^{-N_r} &(16)\\ &= \det\left(\mathbf{I}_{N_s} - i\mathbf{B}\right)^{-N_r} = \det\left(\mathbf{I}_{N_s} - i\mathbf{F}^H\mathbf{R}_t\mathbf{F}\boldsymbol{\Theta}\right)^{-N_r}. &(17)\end{aligned}$$

Finally, we apply the property $\det(\mathbf{AB}) = \det(\mathbf{A})\det(\mathbf{B})$, leading to

$$\varphi(\mathbf{\Theta}) = \det\left(\left(\mathbf{F}^H\mathbf{R}_t\mathbf{F}\right)\left(\left(\mathbf{F}^H\mathbf{R}_t\mathbf{F}\right)^{-1} - i\mathbf{\Theta}\right)\right)^{-N_r} \quad (18)$$

$$= \det\left(\mathbf{F}^H\mathbf{R}_t\mathbf{F}\right)^{-N_r}\det\left(\left(\mathbf{F}^H\mathbf{R}_t\mathbf{F}\right)^{-1} - i\mathbf{\Theta}\right)^{-N_r}. \quad (19)$$

3.2 Impact of the Precoder Matrix F on the Symbol Error Behavior

In section 3.1, we have developed the characteristic function of $\mathbf{F}^H\mathbf{H}^H\mathbf{HF}$. Now, we will use the obtained results to first identify the pdf of the SNR per stream γ_k, then to state a formula for the average symbol error rate (ASER) for each stream, which will finally allow to determine the impact of the precoder matrix $\mathbf{F}$ on the ASER.

Considering that $\tilde{\varphi}(\mathbf{\Theta}) = \det(\mathbf{R}_t)^{-N_r}\det(\mathbf{R}_t^{-1} - i\mathbf{\Theta})^{-N_r}$ is the characteristic function of the complex Wishart distribution $\mathbf{H}^H\mathbf{H} := \tilde{\mathbf{Z}} \sim W_{N_t}(N_r, \mathbf{R}_t)$ (see [2]) with

$$W_{N_t}(N_r, \mathbf{R}_t) = \frac{\mathrm{etr}\left(-\mathbf{R}_t^{-1}\tilde{\mathbf{Z}}\right)\left(\det\left(\tilde{\mathbf{Z}}\right)\right)^{N_r - N_t}}{\Gamma_{N_s}(N_r)\left(\det\left(\mathbf{R}_t\right)\right)^{N_r}} \quad (20)$$

and Γ_{N_s} being the multivariate Gamma function, it is easy to verify that $\varphi(\mathbf{\Theta}) = \det\left(\mathbf{F}^H\mathbf{R}_t\mathbf{F}\right)^{-N_r}\det\left(\left(\mathbf{F}^H\mathbf{R}_t\mathbf{F}\right)^{-1} - i\mathbf{\Theta}\right)^{-N_r}$ is the characteristic function of the distribution $\mathbf{F}^H\mathbf{H}^H\mathbf{HF} := \mathbf{Z} \sim W_{N_s}(N_r, \mathbf{F}^H\mathbf{R}_t\mathbf{F})$ with

$$W_{N_s}\left(N_r, \mathbf{F}^H\mathbf{R}_t\mathbf{F}\right) = \frac{\mathrm{etr}\left(-\left(\mathbf{F}^H\mathbf{R}_t\mathbf{F}\right)^{-1}\mathbf{Z}\right)\left(\det\left(\mathbf{Z}\right)\right)^{N_r - N_s}}{\Gamma_{N_s}(N_r)\left(\det\left(\mathbf{F}^H\mathbf{R}_t\mathbf{F}\right)\right)^{N_r}}. \quad (21)$$

Following the steps in [2], we first see that the SNR per stream γ_k is a weighted Chi-square distribution with $2(N_r - N_s + 1)$ degrees of freedom, which therefore leads to an upper bound of the ASER given by

$$P_{e,k} \le N_e E\left[e^{-d_{min}^2/4\gamma_k}\right] \quad (22)$$

$$\le N_e \frac{1}{\left(1 + \frac{d_{min}^2\gamma_0}{4[(\mathbf{F}^H\mathbf{R}_t\mathbf{F})^{-1}]_{kk}}\right)^{N_r - N_s + 1}}, \quad (23)$$

where N_e and d_{min} denote the number of nearest neighbors and the minimum distance separating the symbols in the constellation, respectively. If we now consider (3) and let therefore γ_0 in (23) approach infinity, we see that d is given by

$$d = N_r - N_s + 1. \quad (24)$$

We can thus make several conclusions:

- The diversity order of each data stream is $N_r - N_s + 1$.

- The number of transmit antennas N_t does not appear in the exponent in (23), but only influences $\mathbf{R}_t$ and $\mathbf{F}$, respectively. Hence, changing N_t can only provide SNR gain.
- A precoder matrix $\mathbf{F}$ exploiting the knowledge of transmit antenna correlation only affects the covariance matrix of the distribution of the SNR γ_k. This kind of precoding can thus only provide SNR gain but never diversity gain.
- Spatial multiplexing systems with transmit antenna correlation based precoding and systems without precoding have thus the same diversity order. For clarity, assume that $\mathbf{F} = \sqrt{1/N_s}\left[\mathbf{0}_{N_s,(N_t-N_s)}\ \mathbf{I}_{N_s}\right]^T$ with equal power loading $1/N_s$ on each stream is the precoder matrix of the non-precoded case. Clearly, $\mathbf{F}$ does not affect the exponent in (23) and therefore it does not influence the diversity order.

4 Simulations

In our simulations, we present some results confirming the formula for the diversity order stated in (24). We simulate $2 * 10^5$ channel instances per SNR point and $2 * 10^3$ QAM64 symbols per channel instance and data stream. We model a transmit antenna correlation of $\mathbf{R}_{t_{ij}} = 0.7^{|i-j|}$.

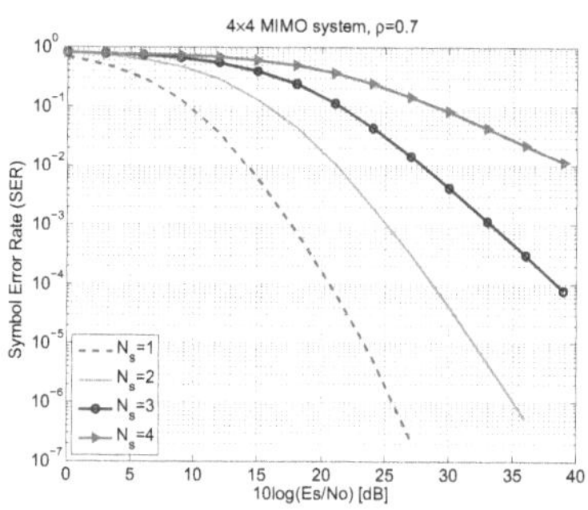

Fig. 1. Fix N_r and N_t, vary N_s

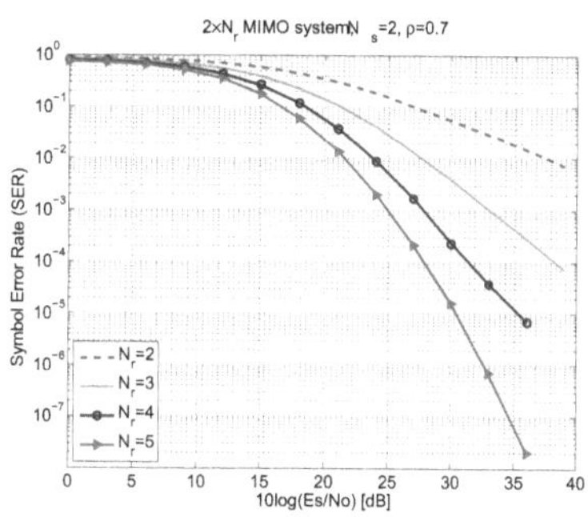

Fig. 2. Fix N_t and N_s, vary N_r

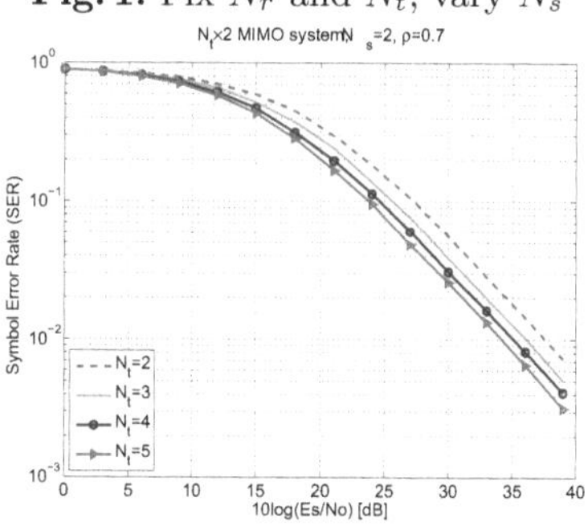

Fig. 3. Fix N_r and N_s, vary N_t

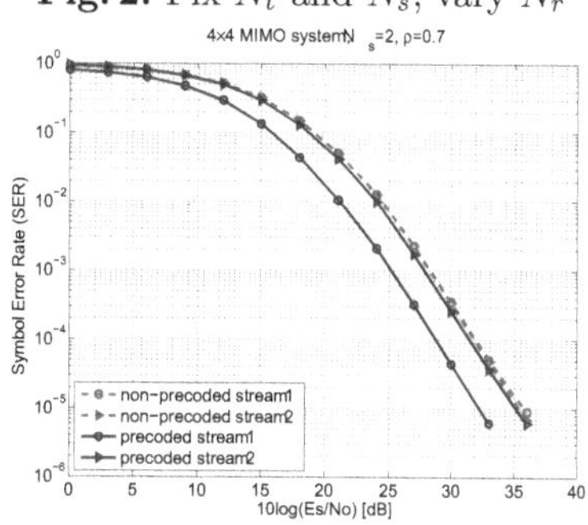

Fig. 4. With vs. without precoding

When looking at the figures, we see that the results obtained from simulations are in accordance with the analytical results derived in section 3.

Figure 1 compares 4×4 ($N_t \times N_r$) MIMO systems with a different number of independent data streams N_s. As all data streams have the same diversity order, it suffices to just show one single stream per system. It is easy to see that each additional data stream causes a diversity loss of 1. In Figure 2, N_t and N_s are fixed and the number of receive antennas N_r is varied. We can observe that each additional receive antenna increases the diversity order by 1. Figure 3 shows that, as predicted in (24), increasing the number of transmit antennas N_t does not affect the diversity order, but it only causes a shift of the SER curve to the left, i.e. an SNR gain. Finally, Figure 4 confirms that precoding based on transmit correlation feedback can never improve the diversity in comparison to systems without precoding, but it can only result in SNR gain. In our case, a system without precoding is a system which just puts each data stream on a separate transmit antenna without performing signal shaping based on the feedback information. In case of $N_s < N_t$, the data streams are put on the outermost antennas.

5 Conclusion

We proved that the diversity order of spatial multiplexing MIMO systems with transmit antenna correlation based precoding is $N_r - N_s + 1$. The number of transmit antennas N_t does not affect the diversity. Furthermore, this kind of precoding will bring no diversity but only an SNR gain in comparison to the case without precoding.

References

1. Oestges, C.: MIMO Wireless Communications. Academic Press, London (2007)
2. Gore, D.A., Heath, R.W., Paulraj, A.J.: Transmit Selection in Spatial Multiplexing Systems. IEEE Communication Letters 6(11), 491–493 (2002)
3. Sampath, H., Paulraj, A.: Linear Precoding for Space-Time Coded Systems With Known Fading Correlations. IEEE Communication Letters 6(6), 239–241 (2002)
4. Bahrami, H.R., Le-Ngoc, T.: Precoder Design Based on Correlation Matrices for MIMO Systems. IEEE Trans. Wireless Commun. 5(12), 3579–3587 (2006)
5. Zheng, L., Tse, D.: Diversity and Multiplexing: A Fundamental Tradeoff in Multiple-Antenna Channels. IEEE Trans. Inform. Theory 49(5), 1073–1096 (2003)
6. Goodman, N.R.: Statistical Analysis Based on a Certain Multivariate Complex Gaussian Distribution (An Introduction). The Annals of Mathematical Statistics 34(1), 152–177 (1963)
7. Horn, R.A., Johnson, C.R.: Matrix Analysis. Cambridge University Press, Cambridge (1985)

Software Simulator to Model an Energy Autonomous System*

Francisco Cabrera, Víctor Araña, Lourdes Suárez, Gonzalo Gutiérrez, and Carlos M. Travieso

Department of Signals and Communications, Technological Centre for Innovation on Communication (CeTIC), University of Las Palmas de Gran Canaria, Campus de Universitario de Tafira, Ed. de Telecomunicación, Pabellón B. 35017, Las Palmas de G.C., Spain
{fcabrera,varana,lsuarez,ggutierrez,ctravieso}@cetic.eu

1 Introduction

Nowadays, applications for marine autonomous systems are increasing due to the interest caused by different issues: global warming, surveillance, monitoring, etc. These systems need to work twenty four hours without interruption. For this reason, it is indispensable to have an efficient energy system. In order to characterize the consumption and energy production is necessary to have a good model [1]. This model must take into account the different elements that work in the system to obtain the maximum power and efficiency.

The autonomous energy systems used mainly in buoy, are biased from renewable energies, more specifically, wind and photovoltaic energies. All these energy systems cannot provide energy continuously. In the case of solar panels, maximum power energy is produced during the day and the available power is lower in a buoy system than land one, because sea movement changes panel orientation and so, the panel efficiency is reduced. In the same way the sea movement affects the proper working of wind generators.

The aim of this software is to model an energy system simulator to evaluate buoy autonomous capacity under different climatic data scenarios and power consumption scenarios. We have called it: AESS – Autonomous Energy System Simulator. In this paper, the different parts of this software are shown.

2 Energy Model

Several publications to model different power autonomous systems have been developed [2, 3]. They have only made an element by element analysis, however, the analysis of the system which includes generators, regulators, accumulators and loads, to evaluate the charge of the battery or SOC (State of Charge), has not been added.

The determination of the battery SOC may be a problem of more complexity depending on the battery type and the aplication in which the battery is used. In this

* This present research was supported with funds from the European Community's Seventh Framework Programme (FP7/2007-2013) under grant agreement nº 218290.

R. Moreno-Díaz et al. (Eds.): EUROCAST 2009, LNCS 5717, pp. 485–491, 2009.

paper, the ampere hour counting method is used to work out the SOC. The ecuation (1) characterizes the energy system through a mathematical model that includes charge and discharge elements.

$$SOC_t = SOC_{t-1} + \sum_{i=1}^{M} \frac{C_{GSi}}{C_B} + \sum_{i=1}^{N} \frac{C_{GWi}}{C_B} - \sum_{i=1}^{Q} \frac{C_{Li}}{C_B} \tag{1}$$

Equation 1 takes into account the last battery state of charge (SOC_{t-1}) and the battery total capacity (C_B), as well as contribution of solar energy (C_{GS}) from M solar panels [3,4,5], wind energy (C_{GW}) from N wind generators [6] and the Q loads of the system (C_{Li}). The state of charge must be considered as an indicator of electrical charge stored by the battery. The value range available is 0<SOC<1. Note that SOC must be understood as the relation between accepted energy and available capacity at all times. When the SOC is the unity the battery cannot accept more energy from the system, because the stored energy fills all the battery capacity. Whereas, when the SOC is zero the battery has no energy. This equation can be expressed as function of currents and sampled each Δt. As a result, the SOC fulfills equation 2.

$$SOC(t) = SOC(t - \Delta t) + \frac{\Delta t}{C_B} \left[\sum_{i=1}^{M} I_{GS_i}(t) + \sum_{j=1}^{N} I_{GW_j}(t) - \sum_{k=1}^{Q} I_{L_K}(t) \right] \tag{2}$$

The terms in the square brackets of equation 2 represent the battery current. When battery current is positive, the battery is charging and otherwise, when battery current is negative, the battery is delivering power toward loads. Therefore, SOC can be simplified and expressed by equation 3.

$$SOC(t) = SOC(t - \Delta t) + \frac{I_B(t) \cdot \Delta t}{C_B} \tag{3}$$

3 Simulator Structure

This simulator tries to give us an idea of the energy balance and therefore, the circuit viability and battery availability. The use of a graphic user interface (GUI) and different wizards have simplified the user interaction. The simulator onion-like structure (Fig 1) where the GUI surrounds the main simulator core and its algorithms has created a good platform for circuit test giving the facilities to create, to access and to load climate scenarios, circuit and power consumption scenarios data files.

This simulator has been developed under Matlab R2007b Linux version but it should work with minor graphical differences under the same MS-Windows version and newer versions. Matlab is a mathematics environment that lets us design graphical interfaces. This environment gives us a powerful way to work with large amounts of data using matrices and standard mathematical functions. Programs can be easily composed down using any text editor and it has have got a good command line interface and error handing.

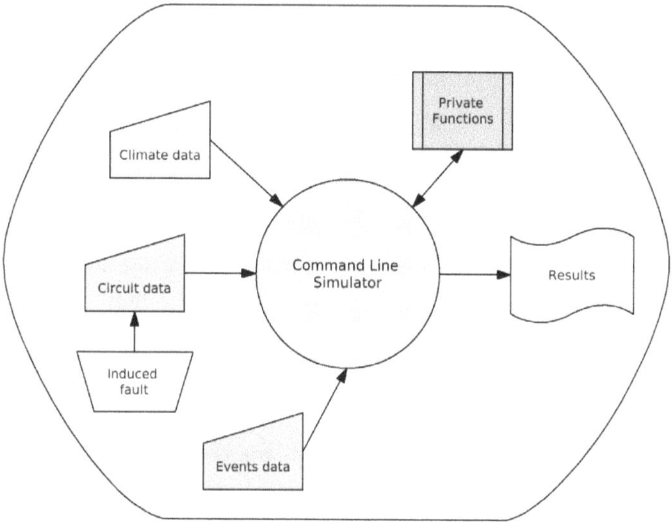

Fig. 1. This figure display the functional blocks of the command line simulator and the different files that feed it

Despite the existence of numerous software tools, it is sometimes difficult to assess the adequacy of these tools for specific tasks. While traditional simulation tools can perform extensive sensitivity analyses, they generally do not permit the user to modify the algorithms that determine the behavior and interactions of the individual components. For this reason, an open architecture is required. In general, software tools can be classified into four categories; pre-feasibility, sizing, simulation and open architecture research tools. This classification is based on the form and purpose of the software. This software covers part of the sizing and simulation roles in order to be a research tool with open and flexible algorithm control architecture as well. The main features of the simulator are [7]:

- The ability to create or load climatic scenarios and determined the efects on a defined circuit, simulating consumptions under predetermined events.
- The ability to represent graphically any parameter versus battery state of charge (SOC) for the simulation time period.
- A wizard-guided data sourcing with a standardized XML data file structure following the W3C's document object model (DOM) standards.
- A flexible file format and internal design that will let users characterize new user-defined elements and scenarios;
- The possibility to induce faults in elements;
- The ability to save and load results for a later visual comparison;
- A settable time step with a minimum resolution of 1 minute.

In short, this software tries to implement a simple graphical user interfacc wizard-based simulator for an autonomous energy system. This system is actually worked in a buoy system.

3.1 Command Line Simulator

The main block is the command line simulator itself that has all necessary logic to achieve the simulation using several private functions. This block is fed by different files with different roles: a climate scenario, a circuit and some consumption events. Moreover, it is possible to induce some circuit faults to test the system's fault-tolerance. The command line simulator function is used to call the simulator from Matlab's command line. This function needs five inputs, the first one is the project name, the second one is the simulation name, and this parameter creates a new simulation into a project's folder. The next ones are a climatic scenario, circuit and scenario filename and the last one is the time step. The project name and simulation must be a string parameter. And time step must be a numeric parameter. After the execution of that function you will receive two results files located inside the project directory 'simulations'. The first one is an *XML* file including a simulator results and the second one it is a *MAT* file with all the internal simulation engine structure. The command line simulator performs the next functions.

- *Internal check*: Take care about simulator tree sanity, make an internal function check and avoid the execution if the elements directory is empty or have the correct syntax ;
- *Input parameters check*: Checks the number and type of input parameters;
- *Input files access check*: Checks if simulation directory and source files exist. Try to open the *XML* files parsing them as a *DOM* nodes and checking at the same time the *XML* structures;
- *Input files syntax check*: Performs a syntax check on each source data files;
- *Input files DOM parsing*: This element has two stages. The first one performs a generic *DOM* parsing of input files. The second one performs a specific parsing that extracts data from this generic *DOM*;
- For every circuit's element an *access check*, *syntax check*, and *DOM parsing* is made;
- *Data integrity check*: Makes a simulation data integrity check before performing the simulation. This stage tries to avoid data inconsistencies. This step will be implemented to improve the software performances. Thus avoiding errors at the time of execution.
- *Simulator engine*: This is the engine that performs the simulation itself. It contains calculi algorithms to evaluate batteries state of charge (*SOC*). This stage could be replaced in the future by other engines with improved behaviour. Nevertheless, at the moment we have implemented a simple algorithm that evaluates a batteries' status of charge starting from an initial charge, taking into account the whole energy balance;

- *DOM output format*: Generates a *DOM* node from the internal results generated by the simulator engine. This *DOM* node gives a well structured document model for final report writing;
- *Result writing*: Creates the project simulations directory if it doesn't exists. Save all internal data used by the simulator engine into a *MAT* file with the simulation name. Copies source files to project directory with a predefined name and save results into a *XML* format.

3.2 Climate Data

The climate data block provides climates data to the simulator during the time that the events are simulated. The first column is indicated the time range, the second column has data from the wind speed. And the third shows the solar radiation data.

3.3 Circuit Data

The circuit data block defines the circuit that the user wants to simulate. This block has the possibility to induce a fail in each of its element. These files contain the technical data and a description of the item. They only contain the technical data necessary to be introduced in the models scheduled. The file circuit has an XML structure. It is formed by three main files, 'battery', 'generator' and 'loads'. Those files contain the manufacturing data and characteristic mathematical model equation as property elements.

3.3 Events Data

Events data block define a group of events that we generate a power consumption scenarios in a period of time. In this case, users have the possibility to create a table of events giving to the simulator a time line in the form of consumption mode changes. This is the period that we are going to simulate starting to the climate data given previously. This block affect to the simulated circuit.

3.4 Private Functions

The private function blocks are functions that reside in subdirectories with the special name private. These functions are called private because they are visible only to M-file functions and M-file scripts.

4 Results

The scenario has been simulated using climate data for a period of three days this data is divided in sequences sixty minutes long because it is more intuitive to have a good representation. The data used in this demo corresponds with historical base data from the Arinaga area in Gran Canaria.

The circuit elements that were created with two solar panels, one wind generator, two batteries and one load (Fig 2), have been loaded and classified with their technical data on the base data of the AESS software.

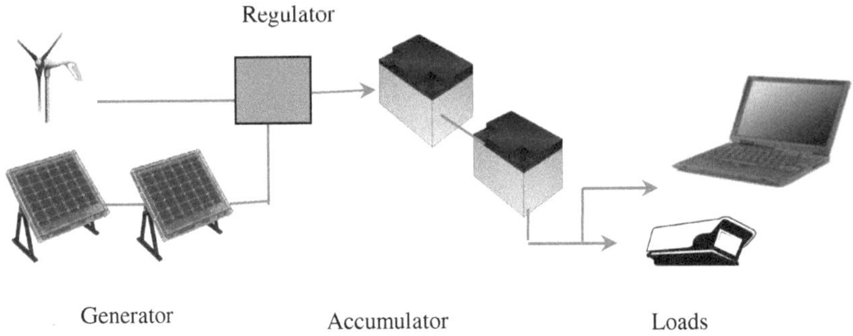

Fig. 2. Scenario example with generators (a wind turbine, two solar panels), accumulator (two batteries in this case), whose energy flow in managed by a regulator and the loads

The AESS viewer is a GUI that lets graphically represents some simulation parameters. The Figure is divided into three axis, the x-axis on the left is represented the variation of the wind velocity (Vs), because it is the parameter that characterises the wind generator behaviour. The graphic shows the evolution of the wind in a period of three days in intervals of 0.5 m/s.

The x-axis on the right represents the variation of the solar radiation, because it is the parameter that characterises the solar panel behaviour. The graphic shows the increases and decreases of the parameter changing with the hour of the day. This parameter influences the representation of the SOC-state of charge (Fig 3 above). This Figure includes three states. The dark blue state indicates that the value of SOC of

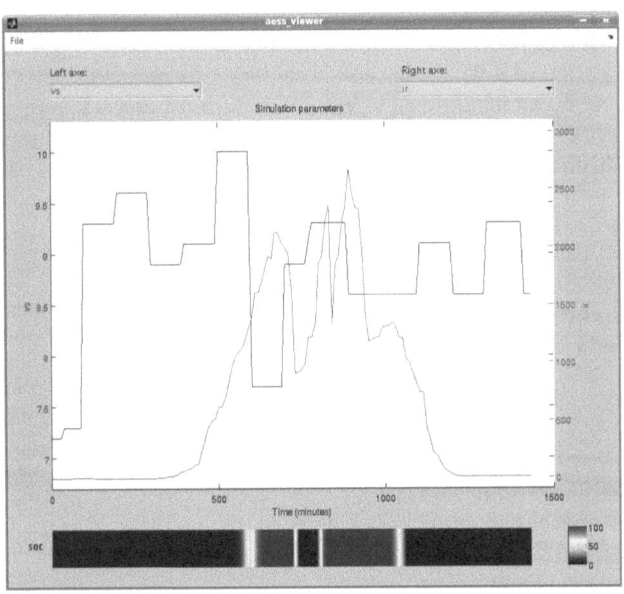

Fig. 3. AEES Viewer represents the three parameters used in this demo in this case the velocity of the wind, radiation and SOC

batteries is between 0-30% of charge, and then light blue state indicates that the value of the SOC of the batteries was between 30-50% of full charge, and finally the red one indicates that the value is on 100% of charge.

In addition to this representation the software has the capacity of represent the curve of characterisation of element (V-I, V-t). The conjunction of those representations gives to the user the opportunity to know in all the time the consumers of their system, the autonomy of the system in different states of work. The user can test the balance energy previously to install their finally system.

5 Conclusions

Therefore, in this present paper, we have showed a new software for autonomus energy system, which includes climatic scenario analysis and the combination of different elements (batteries, solar pannels, wind generator, etc). The results (*SOC*) are shown on a graphical viewer with the combination of the defined previous elements and sceenarios. This software allow the user to have a global idea of the balance energy system. That assement include the iteraction of generator with load and consumer.

This software allows the user to view and represent graphically two o more simulations at the same time. The user can load a simulation carried out previously to compare and actual simulation result with a saved simulation result. Because the user has the possibility to compare the simulation results.

References

1. Strachan, N.P.W., Jovcic, D.: Dynamic Modelling, Simulation and Analysis of an Offshore Variable-Speed Directly-Driven Permanent-Magnet Wind Energy Conversion and Storage System (WECSS). In: OCEANS 2007 – Europe, June 1-6 (2007)
2. Koutroulis, E., Kalaitzakis, K.: Novel battery charging regulation system for photovoltaic applications. IEE Proc. -Electr. Power Appl. 151(2) (March 2004)
3. Markvart, T., Castañer, L.: Practical Handbook of Photovoltaics Fundamentals and Applications. Elsevier, Amsterdam (2003)
4. Hansen, A.D., Sorensen, P., Hansen, L.H., Binder, H.: Models for a Stand-Alone PV System. Rio-R-1219 (EN)/SEC-R-12 (2000)
5. Castañer, L.: Photovoltaic Engineering in Solar Electricity. In: Markvart, T. (ed.), lpp. 74–114. John Wiley, Chichester (1994)
6. Quaschning, V., Hanitsch, R., Zehner, M., Becker, G.: PV simulation and calculation in the internet- the ILSE toolbook. In: 16th European Photovoltaic Solar Energy Conference, Glasgow, May 2000, pp. 2497–2500 (2000)
7. Doumbia, M.L., Agbossou, K., Granger, E.: Simulink Modelling and Simulation of a Hydrogen Based Photovoltaic/Wind Energy System. In: EUROCON 2007. The International Conference on Computer as a Tool, September 2007, pp. 2067–2072 (2007)

On Stochastic Variation in Discrete Time Systems

Yasushi Endow*

Department of Industrial and Systems Engineering
Chuo University
Tokyo 112-8551, Japan
endow@indsys.chuo-u.ac.jp

Abstract. This paper concerns with the variation in discrete time systems driven by a random walk, in contrast with the ordinary Malliavin calculus based on a Brownian motion. A derivative of random functionals with respect to a random walk is introduced and some its fundamental properties are shown. Theories parallel to Malliavin calculus are also discussed in view of applications for discrete time phenomena in signal processing, mathematical finance, and systems science and engineering.

Keyword: Stochastic variation, Random walk, Martingale representation, Walsh functions.

1 Introduction

Malliavin calculus is a celebrated stochastic analysis based on a Brownian motion, and is widely studied and applied in many fields such as system science and financial field, and so on [1]–[5].

Here, we are going to discuss a stochastic analysis based on a random walk. We introduce a derivative (or difference) operator on functionals of a random walk, which exhibits fundamental randomness. Specifically, a functional of a random walk is a function of a discrete-time sample path like a series of outcomes in coin-tossing games and thus, its derivative is defined with respect to the trajectory.

This paper is organized as follows. In 2 a derivative operator on random variables, which are functionals of a random walk, is introduced. The derivative is conducted with respect to the random walk. In 3 some fundamental properties are checked. In 4 the derivative of a functional is also represented in terms of the Walsh functions. In 5 the expectation of a directional derivative is calculated for a special case, and a martingale representation in terms of a random walk is shown. In 6 a stochastic difference equation driven by a random walk is considered, and the derivative of the terminal value of the solution is expressed by the solution of the corresponding stochastic difference equation.

* Supported by the Grant from Chuo University.

R. Moreno-Díaz et al. (Eds.): EUROCAST 2009, LNCS 5717, pp. 492–500, 2009.

2 Definition

2.1 Random Walk

Let $S = \{S_t\}_{t=0}^{T}$ be a random walk defined by

$$S_0 = 0,\ S_t = \sum_{i=1}^{t} \xi_i,\ t = 1, \ldots, T,$$

where $\{\xi_n\}_{n=1}^{T}$ is a sequence of i.i.d. random variables with the distribution $Pr\{\xi_n = \pm 1\} = 1/2,\ n = 1, \ldots, T$. Hereafter, we put $\xi_0 = 0$ for convinience. This random walk moves on the lattice points on $\mathbf{Z}^{T+1}$, where $\mathbf{Z}$ designates the set of integers. Rigorously, it moves randomly on the subset $\Omega \subset \mathbf{Z}^{T+1}$ such that

$$\Omega = \{\omega = (\omega_0, \omega_1, \ldots, \omega_T);\ \omega_0 = 0,\ \omega_t = \sum_{i=1}^{t} z_i,\ z_i = \pm 1,\ t = 0, \ldots, T\}.$$

The random walk is expressed by the coordinate function as $S_t(\omega) = \omega_t,\ t = 0, \ldots, T,\ \omega = (\omega_0, \omega_1, \ldots, \omega_T) \in \Omega$. Let $\mathcal{F}_t = \sigma(S_0, \ldots, S_t)$ be the σ-algebras generated by the random variables $\{S_0, \ldots, S_t\}$, for $t = 0, \ldots, T$, then $\mathcal{F} = \mathcal{F}_T$. A probability measure $\mathbf{P}$ which assumes $\mathbf{P}(\{\omega\}) = 2^{-T}$ for $\omega \in \Omega$ is induced on the space $(\Omega, \mathcal{F})$ naturally. Hence, we obtain the probability space $(\Omega, \mathcal{F}, \mathbf{P})$, on which the random walk S is defined, with the filtration $\{\mathcal{F}_t\}_{t=0}^{T}$. Note that the random walk S is a martingale with respect to the filtration, because S_n is $\mathcal{F}_t$-measurable and $\mathbf{E}[S_{t+1}|\mathcal{F}_t] = S_t,\ t = 0, \ldots, T$. It is also clear that S has independent and identically distributed increments $dS_t = S_t - S_{t-1} = \xi_t$, and $\mathbf{E}[S_t] = 0,\ \mathbf{E}[S_s S_t] = s \wedge t,\ s, t = 0, \ldots, T$.

2.2 Derivative Operator

In the sequel, a random variable $S(h)$ is defined by

$$S(h) = \sum_{t=1}^{T} h(t)\xi_t, \tag{1}$$

for a real-valued function $h = h(t)$ on $\{0, \ldots, T\}$. Let us define the derivative DF of a random variable $F = f(S(h_1), \ldots, S(h_N))$ by the stochastic process $DF = \{D_t F\}_{t=1}^{T}$ such that

$$D_t F = \sum_{i=1}^{N} \partial_i f(S(h_1), \ldots, S(h_N)) h_i(t), \tag{2}$$

where f being partially differentiable with respect to each argument and

$$\partial_i f(S(h_1), \ldots, S(h_N)) = \frac{\partial}{\partial x_i} f(S(h_1), \ldots, x_i, \ldots, S(h_N))\Big|_{x_i = S(h_i)}.$$

Note that the derivative operator D is a linear map transforming a functional of a sample path of the random walk S to a stochastic process. Since $S_i = S(\mathbf{1}_{\{1,\dots i\}})$ and $\xi_i = S(\mathbf{1}_{\{i\}})$, we see that

$$D_t f(S_1, \dots, S_N) = \sum_{i=1}^{N} \partial_i f(S_1, \dots, S_N)\mathbf{1}_{\{1,\dots i\}}(t),$$

$$D_t f(\xi_1, \dots, \xi_N) = \sum_{i=1}^{N} \partial_i f(\xi_1, \dots, \xi_N)\mathbf{1}_{\{i\}}(t).$$

In particular, $D_t\xi_i = \mathbf{1}_{\{i\}}(t)$, $D_t S_k = \mathbf{1}_{\{1,\dots,k\}}(t)$, and $D_t S(h) = h(t)$.

3 Some Fundamental Properties

The following lemmas are immedeate.

Lemma 1. *For differentiable F, G and constants a, b,*

(a) $D(aF + bG) = aDF + bDG,$

(b) $DFG = FDG + GDF.$

Lemma 2.

$$D_t F = \sum_{i=1}^{N} \partial_i f(S(h_1), \dots, S(h_N))D_t S(h_i). \tag{3}$$

We also have the following result.

Lemma 3. *For a predictable process $Z = \{Z_t\}_{t=0}^{T}$,*

$$D_t(\sum_{i=1}^{T} Z_i\xi_i) = Z_t + \sum_{i=1}^{T}(D_t Z_i)\xi_i \tag{4}$$

holds.

Proof: Let

$$Z_i = \sum_{k=0}^{T} \alpha_k \mathbf{1}_{\{k\}}(i),$$

where α_k be $\mathcal{F}_{k-1}$-measurable random variables. Then, by linearity,

$$D_t(\sum_{i=1}^{T} Z_i\xi_i) = D_t(\sum_{i=1}^{T} \alpha_i\xi_i) = \sum_{i=1}^{T} D_t(\alpha_i\xi_i). \tag{5}$$

In view of (b) in Lemma 1,

$$D_t(Z_i\xi_i) = Z_t + (D_t Z_i)\xi_i. \tag{6}$$

On the other hand,

$$\sum_{i=1}^{T}(D_t Z_i)\xi_i = \sum_{i=1}^{T}\sum_{k=1}^{T}(D_t\alpha_k)\mathbf{1}_{\{k\}}(i)\xi_i = \sum_{i=1}^{T}(D_t\alpha_i)\xi_i. \tag{7}$$

Hence, (5)–(7) show (4). □

Lemma 4. *For an ordinary differentiable function f,*

$$D_t\Big(\sum_{u=1}^{T} f(S_u)\xi_u\Big) = \sum_{u=t}^{T} f'(S_u)\xi_u + f(S_t). \tag{8}$$

Proof: It follows from (b) in Lemma 1 and the relation

$$D_t f(S_u) = f'(S_u)\mathbf{1}_{\{1,2\dots,u\}}(t),$$

that

$$\begin{aligned} D_t(f(S_u)\xi_u) &= D_t f(S_u)\xi_u + f(S_u)\mathbf{1}_{\{u\}}(t) \\ &= f'(S_u)\mathbf{1}_{\{1,2\dots,u\}}(t)\xi_u + f(S_u)\mathbf{1}_{\{u\}}(t), \end{aligned}$$

which completes the proof by the linearity of D. □

4 Walsh Functions Representation

In this section we consider the case $T = 2^N$, exclusively. Let H_T be the T-th Hadamard matrix defined recursively by the Kronecker products such that

$$\begin{aligned} H_1 &= [1], \\ H_2 &= \begin{bmatrix} 1 & 1 \\ 1 & -1 \end{bmatrix}, \\ &\vdots \\ H_{2^k} &= \begin{bmatrix} H_{2^{k-1}} & H_{2^{k-1}} \\ H_{2^{k-1}} & -H_{2^{k-1}} \end{bmatrix} = H_2 \otimes H_{2^{k-1}}. \end{aligned}$$

Let $\{w_i,\ i = 1,\dots,T\}$ be the ONS of the discrete Walsh functions on $\{1,\dots,T\}$ defined by the rows in the Hadamard matrix, i.e.,

$$H_T = \begin{bmatrix} w_1 \\ \vdots \\ w_T \end{bmatrix}.$$

Then, every vector $h \subset \mathbf{R}^T$ will be represented uniquely by the ONS as

$$h(t) = T^{-1}\sum_{i=1}^{T}\langle h, w_i\rangle w_i(t),$$

where $\langle h, w_i\rangle = \sum_{j=1}^{T} h_j w_i(t)$.
Since $D_t S(h) = h(t)$, it is also represented by

$$D_t S(h) = T^{-1}\sum_{i=1}^{T}\langle h, w_i\rangle w_i(t).$$

Hence we have the following result.

Theorem 1. *For $F = f(S(h_1), \ldots, S(h_n))$, the derivative DF is expressed by*

$$D_t F = T^{-1}\sum_{j=1}^{T}\Big(\sum_{i=1}^{T}\partial_i f(S(h_1), \ldots, S(h_n))\langle h_i, w_j\rangle\Big) w_j(t),\ n = 0, \ldots, T.$$

5 Directional Derivative and Martingale Representation

For a stochastic process $Z = \{Z_n\}_{n=0}^{T}$, the directional derivative of F is defined by

$$D_Z F = \sum_{t=1}^{T} Z_t(D_t F). \tag{9}$$

Theorem 2. *If Z is predictable, then*

$$\mathbf{E}[D_Z S(h)] = \mathbf{E}[S(h)(D^* Z)] \tag{10}$$

holds, where $D^* Z = \sum_{t=1}^{T} Z_t \xi_t$.

Proof: Since $D_t S(h) = h(t)$, it is clear that

$$\mathbf{E}[D_Z S(h)] = \mathbf{E}[\sum_{t=1}^{T} Z_t(D_t S(h)] = \sum_{t=1}^{T}\mathbf{E}[Z_t]h(t).$$

On the other hand it follows from mutual independence of ξ_i's and the predictability of Z,

$$\mathbf{E}[Z_s\xi_s\xi_t] = \begin{cases}\mathbf{E}[Z_t], & s = t,\\ 0, & s \neq t,\end{cases}$$

and, hence

$$\mathbf{E}[S(h)D^* Z] = \sum_{t=1}^{T}\sum_{s=1}^{T} h(t)\mathbf{E}[Z_s\xi_s\xi_t] = \sum_{t=1}^{T} h(t)\mathbf{E}[Z_t].$$

This completes the proof. □

Remark that the sum D^*Z is a (discrete) stochastic integral of Z with respect to the random walk S, or the martingale transform by Z, since $\xi_t = S_t - S_{t-1}$.

Theorem 3. *Let $M = \{M_t\}_{t=0}^T$ be a process defined by*

$$M_t = M_0 + \sum_{i=1}^t Z_i \xi_i, \ t = 1, \dots, T, \tag{11}$$

and M_0 being a constant. If $Z = \{Z_t\}_{t=0}^T$ is a bounded and predictable process, then M is a square integrable martingale.

Proof: Since Z is predictable, we have that

$$\mathbf{E}[Z_i^2 \xi_i^2] = \mathbf{E}[Z_i^2 \mathbf{E}[\xi_i^2 | \mathcal{F}_{i-1}]] = \mathbf{E}[Z_i^2],$$

and, so

$$\mathbf{E}[M_t^2] = M_0^2 + \sum_{i=1}^t \mathbf{E}[Z_i^2 \xi_i^2] < \infty.$$

By definition we see that for $t = 0, \dots, T-1$,

$$\mathbf{E}[M_{t+1} | \mathcal{F}_t] = M_0 + \sum_{i=1}^t Z_i \xi_i + Z_{t+1} \mathbf{E}[\xi_{t+1} | \mathcal{F}_t] = M_t,$$

since $\mathbf{E}[\xi_{t+1} | \mathcal{F}_t] = 0$. □

Next we consider the reverse to Theorem 3.

Theorem 4. *Let $M = \{M_t\}_{t=0}^T$ be a square integrable martingale. Then there exists a bounded predictable process $Z = \{Z_t\}_{t=0}^T$ such that*

$$M_t = M_0 + \sum_{i=1}^t Z_i \xi_i, \ t = 1, \dots, T. \tag{12}$$

Proof: We prove it by induction. For $t = 1$, putting $A_1 = M_1 - M_0$, we see that it is $\mathcal{F}_1$-measurable, and it will be represented by

$$A_1(\omega) = a \mathbf{1}_{\{\xi_1 = 1\}}(\omega) + b \mathbf{1}_{\{\xi_1 = -1\}}(\omega), \ \omega \in \Omega,$$

where a, b are $\mathcal{F}_0$-measurable, i.e., they are constants. Since

$$0 = \mathbf{E}[A_1 | \mathcal{F}_0] = a \mathbf{P}(\{\xi_1 = 1\}) + b \mathbf{P}(\{\xi_1 = -1\}) = 1/2(a + b),$$

i.e., $b = -a$, we have $A_1 = Z_1 \xi_1$, where $Z_1 = a$.
Assume that (12) holds for t, and consider the case of $t+1$. Since $A_{t+1} = M_{t+1} - M_t$ is $\mathcal{F}_{t+1}$-measurable, it will be expressed by

$$A_{t+1}(\omega) = c(\omega) \mathbf{1}_{\{\xi_{t+1} = 1\}}(\omega) + d(\omega) \mathbf{1}_{\{\xi_{t+1} = -1\}}(\omega), \ \omega \in \Omega,$$

where c, d are $\mathcal{F}_t$-measurable. Similarly as for the case $t = 1$, taking the conditional expectation of A_{t+1} given by $\mathcal{F}_t$, we have that $d(\omega) = -c(\omega)$ for all $\omega \in \Omega$. This show that $A_{t+1} = Z_{t+1}\xi_{t+1}$, where $Z_{t+1} = c$. Hence, the equation (12) holds for $t+1$. This completes the proof. □

Theorem 5. *Let F be an $\mathcal{F}_T$-measurable random variable with $\mathbf{E}[F^2] < \infty$. Then there exists a predictable process $Z = \{Z_t\}_{t=0}^T$ with $\mathbf{E}[Z_t^2] < \infty$ such that*

$$F = \mathbf{E}[F] + \sum_{i=1}^{t} Z_i\xi_i, \ t = 1, \ldots, T. \tag{13}$$

Proof: Define that for $t = 1, \ldots, T$ and $\omega = (\omega_0, \ldots, \omega_{t-1}, *, \cdots, *) \in \Omega$,

$$\begin{aligned} Z_i(\omega) = \frac{1}{2}\Big|&\mathbf{E}[F|\mathcal{F}_t](\omega_0, \ldots, \omega_{t-1}, \omega_{t-1}+1, *, \cdots, *) \\ &-\mathbf{E}[F|\mathcal{F}_t](\omega_0, \ldots, \omega_{t-1}, \omega_{t-1}-1, *, \cdots, *)\Big|, \end{aligned}$$

where the symbol "$*$" denotes any possible value which can be taken. Then, Z_i is $\mathcal{F}_{t-1}$-measurable. By definition of the conditional expectation, we see that for $\omega = (\omega_0, \ldots, \omega_{t-1}, *, \cdots, *) \in \Omega$,

$$\begin{aligned} \mathbf{E}[F|\mathcal{F}_{t-1}](\omega) &= \mathbf{E}[\mathbf{E}[F|\mathcal{F}_t]|\mathcal{F}_{t-1}](\omega) \\ &= \frac{1}{2}\mathbf{E}[F|\mathcal{F}_t](\omega_0, \ldots, \omega_{t-1}, \omega_{t-1}+1, *, \cdots, *) \\ &\quad +\frac{1}{2}\mathbf{E}[F|\mathcal{F}_t](\omega_0, \ldots, \omega_{t-1}, \omega_{t-1}-1, *, \cdots, *). \end{aligned}$$

It follows that

$$\mathbf{E}[F|\mathcal{F}_{t-1}](\omega) + Z_t(\omega)\xi_t(\omega) = \mathbf{E}[F|\mathcal{F}_t](\omega), \tag{14}$$

because of $\xi_t(\omega_0, \ldots, \omega_{t-1}, \omega_t, *, \cdots, *) = \omega_t - \omega_{t-1}$. Applying (14) recursively, we have that

$$\mathbf{E}[F|\mathcal{F}_0] + \sum_{t=1}^{T} Z_t\xi_t = \mathbf{E}[F|\mathcal{F}_T],$$

which means (18). □

6 Stochastic Difference Equation

Let us introduce a stochastic difference equation (SDCE) such that

$$\begin{cases} X_t = X_{t-1} + a(t-1, X_{t-1}) + b(t-1, X_{t-1})\xi_t, \ n = 1, \ldots, T, \\ X_0 = x, \end{cases} \tag{15}$$

where a, b be measurable functions, and x a constant. This is equivalent to

$$X_t = x + \sum_{i=0}^{t-1} a(i, X_i) + \sum_{i=0}^{t-1} b(i, X_i)\xi_{i+1}, \ n = 1, \ldots, T. \tag{16}$$

Now let us consider the partial derivative $Y_t = \partial X_t/\partial x$ of X_t with respect to the initial value x. Formal differentiation of (15) results in

$$\begin{cases} Y_t = Y_{t-1} + a_x(t-1, X_{t-1})Y_{t-1} + b_x(t-1, X_{t-1})Y_{t-1}\xi_t, \ n = 1, \ldots, T, \\ Y_0 = 1, \end{cases} \tag{17}$$

where $a_x = \partial a/\partial x$ and $b_x = \partial b/\partial x$.

Theorem 6. *The derivative $D_t X_T$ is given in terms of the solution of (17) as*

$$D_t X_T = Y_T Y_t^{-1} b(t-1, X_{t-1}), \ n = 1, \ldots, T. \tag{18}$$

Proof: Since, by Lemma 3

$$D_t\Big(\sum_{i=0}^{T-1} b(i, X_i)\xi_{i+1}\Big) = \sum_{i=0}^{T-1} b_x(i, X_i)(D_t X_i)\xi_{i+1} + \sum_{i=0}^{T-1} b(i, X_i)\mathbf{1}_{\{i+1\}}(t),$$

the derivative of X_T is represented by

$$D_t X_T = \sum_{i=0}^{T-1} a_x(i, X_i) D_t X_i + \sum_{i=0}^{T-1} b_x(i, X_i)(D_t X_i)\xi_{i+1}(t) + b(t-1, X_{t-1}). \tag{19}$$

On the other hand, since (17) is equivalent to

$$Y_T = Y_t + \sum_{i=t}^{T-1} a_x(i, X_i)Y_i + \sum_{i=t}^{T-1} b_x(i, X_i)Y_i\xi_{i+1}, \ n = 0, \ldots, T,$$

multiplication of this equation by $Y_t^{-1}b(t-1, X_{t-1})$ results in

$$\begin{aligned} &Y_T Y_t^{-1} b(t-1, X_{t-1})\mathbf{1}_{t\le T} \\ &\quad = b(t-1, X_{t-1})\mathbf{1}_{t\le T} + \sum_{i=t}^{T-1} a_x(i, X_i)Y_i Y_t^{-1} b(t-1, X_{t-1})\mathbf{1}_{t\le T} \\ &\qquad + \sum_{i=t}^{T-1} b_x(i, X_i)Y_i Y_t^{-1} b(t-1, X_{t-1})\xi_{i+1}\mathbf{1}_{t\le T}, \ t = 0, \ldots, T, \end{aligned}$$

which completes the proof in view of (19). □

Remark that $D_T X_T = b(T-1, X_{T-1})$.

References

1. Carmona, R.A., Tehranchi, M.R.: Interest Rate Models: an Infinite Dimensional Stochastic Analysis Perspective. Springer, Berlin (2006)
2. Etherridge, A.: A Course in Financial Calculus. Cambridge University Press, Cambridge (2002)
3. Malliavin, P., Thalmaier, A.: Stochastic Calculus of Variations in Mathematical Finance. Springer, Berlin (2006)
4. Nualart, D.: The Malliavin Calculus and Related Topics. Springer, New York (1995)
5. Di Nunno, G., Øksendal, B., Proske, F.: Malliavin Calculus for Lévy Processes with Applications to Finance. Springer, Heidelberg (2009)

Convolution on Finite Groups and Fixed-Polarity Polynomial Expressions

Radomir S. Stanković, Jaakko T. Astola, and Claudio Moraga*

Dept. of Computer Science, Faculty of Electronics, Niš, Serbia
Dept. of Signal Processing, Tampere University of Technology, Tampere, Finland
European Centre for Soft Computing, 33600 Mieres, Spain &
Technical University of Dortmund, 44149 Dortmund, Germany

Abstract. This paper discusses relationships among convolution matrices and fixed-polarity matrices for polynomial expressions of discrete functions on finite groups. Switching and multiple-valued functions are considered as particular examples of discrete functions on finite groups. It is shown that if the negative literals for variables are defined in terms of the shift operators on domain groups, then there is a relationship between the polarity matrices and convolution matrices. Therefore, the recursive structure of polarity matrices follows from the recursive structure of convolution matrices. This structure is determined by the assumed decomposition of the domain groups for the considered functions.

Keywords: Convolution, Finite groups, Polynomial expressions, Spectral representations.

Discrete functions that are mathematical models of signals encountered in digital systems can be viewed as mappings $f : G \to P$, where G is a finite group of order g, and P is a field that may be a finite (Galois) field, the field of rational numbers Q, or the complex field C. If G is a group decomposable into the product of n subgroups G_i of orders g_i, i.e.,

$$G = \times_{i=1}^{n} G_i, \qquad g = \prod_{i=1}^{n} g_i, \tag{1}$$

then $f(x)$, $x \in G$, can be alternatively viewed as an n variable function $f(x_1, \ldots, x_n)$, $x_i \in G_i$. We denote the space of such functions by $P(G)$.

Example 1. *Switching (or Boolean) functions are a particular class of discrete functions where $G_i = C_2 = (\{0, 1\}, \oplus)$ is the cyclic group of order 2 where $\oplus$ is the addition modulo 2 (EXOR). Switching functions of a given number of variables n under componentwise EXOR and logic AND as the multiplication by a*

* The work leading to this paper was partially supported by the Academy of Finland, Finnish Center of Excellence Programme, Grant No. 213462, and the Foundation for the Advancement of Soft Computing, Mieres, Asturias, Spain.

R. Moreno-Díaz et al. (Eds.): EUROCAST 2009, LNCS 5717, pp. 501–509, 2009.

scalar in scalar form a linear (vector) space $GF_2(C_2^n)$. Functional expressions for switching functions as elements of $GF_2(C_2^n)$ are called bit-level representations.

Alternatively, the logic values 0 and 1 can be interpreted as integers and, then, switching functions are viewed as a subset of elements of the space $C(C_2^n)$ of functions $f : C_2^n \to C$. In this case the corresponding functional expressions are called word-level representations. This approach is especially useful when dealing with binary devices having $k > 1$ outputs, each output represented by a switching function $f_i(x_1, \ldots, x_n)$, $i = 0, 1, \ldots, k-1$, since binary k-tuples $(f_0, f_1, \ldots, f_{k-1})$ can be interpreted as binary representations of integers $z = \sum_{i=0}^{k-1} f_i 2^i$, $z \in \{0, 1, \ldots, 2^k - 1\}$.

Example 2. *In multiple-valued logic, quaternary functions of quaternary-valued variables are viewed as functions on the group $G = C_4^n$, where C_4 is the support group of the field $GF(4)$, into $GF(4)$, thus $f : C_4^n \to GF(4)$. As in the case of binary logic, if elements of $GF(4)$ are identified with integers $0, 1, 2, 3$, with 0 and 1 corresponding to the additive and the multiplicative identity elements of $GF(4)$, these functions can be viewed as integer or complex-valued functions on C_4^n. Therefore, quaternary logic functions can be viewed as elements of spaces $GF_4(C_4^n)$ or $C(C_4^n)$.*

These alternative views to the same class of functions as briefly discussed in examples above can be easily extended to functions on finite not necessarily Abelian groups that are decomposable in the sense expressed by (1) [7].

In this paper, we consider polynomial representations for discrete functions that originate initially from bit-level representations. For the calculation and optimization of these representations we, however, use links to word-level representations and the classical mathematical analysis tools as convolution, spectral transforms, and related fast calculation algorithms. Although the main theoretic results will be formulated for a rather general case, for simplicity of presentation, the examples will be given for polynomial representations of switching functions and quaternary functions.

1 Polynomial Representations

Polynomial representations are a way to represent discrete functions that is widely used in many areas, including digital logic and system design, communication, control, and signal processing, see, for instance [2], [4], and references therein. There are bit-level and and word-level representations depending of the assumed vector spaces. These representations are interesting from both theoretical and practical point of view due to their twofold interpretation. They can be viewed as discrete counterparts of polynomial or Taylor series representations in classical mathematical analysis sharing many of their properties [10] with coefficients related to various classes of differential operators [8]. On the other hand, spectral interpretation permits to view coefficients in polynomial expressions as Fourier-like coefficients, expressing (up to certain measure) good features and properties of the classical Fourier representations. This interpretation also brings possibilities to exploit FFT-like algorithms in their calculations.

Basis functions in terms of which polynomial expressions are defined can be in many cases expressed as product of variables in functions to be represented. A generalization is to use products of polynomials in terms of variables.

Definition 1. *For $f \in P(G)$, the polynomial expression is given by*

$$f = \sum_{i=0}^{g-1} r_i \phi_i(x_1, \ldots, x_n), \quad \phi_i(x_1, \ldots, x_n) = \prod_{k=1}^{n} p^{i_k}(x_k),$$

where i_k is the i-th coordinate in the componentwise representation of g in terms of g_i, $i = 1, \ldots, n$, and $p(x_k)$ are polynomials of order g_i in the k-th variable x_k, including the simplest case when $p(x_k) = x_k$.

In matrix notation we express basis functions in symbolic notation as columns of a row matrix $\mathbf{X} = [\phi_0(x_1, \ldots, x_n), \ldots, \phi_{g-1}(x_1, \ldots, x_n)]$. Since G is a decomposable group, the basis functions $\mathbf{X}$ are the Kronecker product of basis functions on the constituent groups G_i. Thus,

$$\mathbf{X} = \bigotimes_{i=1}^{n} \mathbf{X}_i, \quad \mathbf{X}_i = [\phi_{i,0}, \ldots, \phi_{i,g_i-1}].$$

The coefficients in the representations are defined as entries of a vector $\mathbf{R} = [r(0), \ldots r(g-1)]^T$, determined as

$$\mathbf{R} = \mathbf{TF},$$

where $\mathbf{T} = \mathbf{X}^{-1}$ the inverse of $\mathbf{X}$ over P when its columns ϕ_i are written in expanded form as vectors.

The optimization of polynomial representations is performed by selecting different polarity of literals for variables. For instance, in the case of binary variables, a literal can appear in the positive polarity $x_i \in \{0, 1\}$, $i = 1, \ldots, n$, n-number of variables in a given function f, or the negative polarity defined as $\overline{x}_i = x_i \oplus 1$. In a straightforward way, for a p-valued variable, there are p different polarities, defined as $\overset{c-}{x}_i = x_i \oplus c$, $c = 0, 1, \ldots, p-1$, where $\oplus$ is the addition in the support group of the finite (Galois) field $GF(p)$ where x_i takes its values. We extend the notion of negative literals to functions on decomposable finite groups G as follows. The literal in the polarity k for $x_i \in G_i$ is defined by $\overset{k-}{x}_i = \delta_k(x_i)$, where $\circ$ is the group operation of G_i, and $\delta_k(x) = x \circ k^{-1}$, $k = 0, 1, \ldots, g-1$, is the shift operator on G_i. Notice that for $GF(p)$ we keep the usual way of defining the negative literals, although it should be $\overset{c-}{x}_i \oplus c^{-1}$ for the consistency with the general case. This however, means just a different encoding of all possible negative literals for a p-valued variable x_i.

Since, assignment of polarities for variables is an NP-complete problem, in practice for a given function f all possible fixed polarity expressions are generated and the expression with minimum number of terms is selected. Coefficients in these expressions are conveniently represented as rows of a matrix called the Fixed-polarity matrix for f.

2 Convolution and Fixed-Polarity Matrices

There is a direct relationship between the convolution matrices and fixed-polarity polynomial matrices.

Definition 2. *The convolution product for two functions f_1 and f_2 on a finite group G of order g is defined as*

$$(f_1 * f_2)(\tau) = \sum_{x=0}^{g-1} f_1(x) f_2(x \circ \tau^{-1}), \tau \in G.$$

In matrix notation, $(f_1 * f_2)$ is calculated as the product of the convolution matrix for f_2, $\mathbf{C}_{f_2} = [f_2(\delta_\tau(x))]$, $x, \tau = 0, 1, \ldots, g-1$, and the vector $\mathbf{F}_1 = [f_1(x)]$ of values for f_1.

The fixed-polarity matrix can be calculated as the product of the convolution matrix and the transform matrix as in the following

Remark 1. *The fixed-polarity matrix $\mathbf{Q}_f$ for $f \in P(G)$ with respect to a transform matrix $\mathbf{T}$ is given by*

$$\mathbf{Q}_f = \mathbf{C}_f \mathbf{T}^*, \tag{2}$$

where $\mathbf{T}^*$ is the transpose or transpose complex-conjugate of $\mathbf{T}$ if $P = GF(p)$ and $P = C$, respectively.

Advantages of the convolution based approach to fixed-polarity matrices could be summarized as follows.

1. Fixed-polarity matrices for different polynomial expressions can be studied in a uniform way and there is no need to analyze their structure for each particular case separately to derive rules for construction of polarity matrices [4]. We show that the structure of the fixed-polarity matrices reflects the structure of the domain group, and it is the same for various transforms on the same group.
2. By referring to the convolution product, it is possible to extend the definition of fixed-polarity representations to spectral transforms on a given finite group although in study and applications of these transforms we usually do not assume an underlying structure which would permit definition of an analogue to the notion of logic complement.
3. To determine fixed-polarity expressions we can use various fast convolution algorithms defined for different ways of specification of discrete functions as truth-tables, vectors, cubes, and decision diagrams [3], [6].

The following remark is also possible.

Remark 2. *The recursive structure of the polarity matrix for a given polynomial expression is determined by the structure of the convolution matrices on the considered domain group G for the considered functions. Further, that structure is determined by the assumed decomposition for G and the group operation of G.*

The Remark 1 is a basis to define fast tabular techniques for determination of various fixed polarity expressions for given polarity [1], [9]. It should be noticed that the term tabular does not mean some tables are used for either representation of discrete functions or calculation with them. It refers to the features of the methods consisting of processing sequentially and separately product terms in polynomial expressions of functions considered.

In a general formulation, the method to calculate fixed-polarity matrices consists of the following steps.

1. Use the linearity of the related transform. That means, express the given function f as a sum of characteristic functions expressing values of f at particular points. Thus, write f as

$$f = \sum_{i=0}^{g-1} f(i)J_i(x), \quad J_i(x) = \begin{cases} 1, & i = x, \\ 0, & i \neq x. \end{cases}$$

2. If the given polarity is $H = (h_1, \ldots, h_n)$, then generate the H-th row of the convolution matrix for each $f(i)J_i(x)$. These rows actually are the shifted versions of $f(i)J_i(x)$. That means, the elements of vectors representing values of $f(i)J_i(x)$ are shifted by using $\delta_H(i)$.
3. Perform the related transform T over the generated rows of the convolution matrices for each $f(i)J_i(x)$ used to represent f.
4. Use the linearity of the considered transform in the spectral domain. That means, perform addition of the generated spectra for shifted $f(i)J_i(x)$.

The method is a generalization of the Fast tabular techniques (FTT) for calculation of the Fixed-polarity Reed-Muller expressions for switching functions [9] and a similar method for quaternary functions in [4].

3 Case Study

In this section, we will present examples of fixed polarity expressions and correlation matrices for binary and quaternary logic functions.

3.1 Dyadic Correlation and Arithmetic Expressions

The dyadic convolution is defined as

$$(f * g)(\tau) = \sum_{x=0}^{2^n-1} f(x)g(x \oplus \tau), \quad \tau = 0, 1, \ldots, 2^n - 1.$$

Example 3. *For $n = 2$, the convolution matrix is*

$$\mathbf{C}_f(2) = \begin{bmatrix} f(0) & f(1) & f(2) & f(3) \\ f(1) & f(0) & f(3) & f(2) \\ f(2) & f(3) & f(0) & f(1) \\ f(3) & f(2) & f(1) & f(0) \end{bmatrix}.$$

For a function of n binary-valued variables that is defined by the function vector $\mathbf{F} = [f(0), \ldots, f(2^n-1)]^T$, the arithmetic spectrum $S_f = [S_f(0), \ldots, S_f(2^n-1)]^T$ is defined as $\mathbf{S}_f = \mathbf{A}(n)\mathbf{F}$, where the $(2^n \times 2^n)$ transform matrix $\mathbf{A}(n)$ is defined as

$$\mathbf{A}(n) = \bigotimes_{i=1}^{n} \mathbf{A}(1), \quad \mathbf{A}(1) = \begin{bmatrix} 1 & 0 \\ -1 & 1 \end{bmatrix}, \tag{3}$$

and $\otimes$ denotes the Kronecker product.

The arithmetic expression for an n-variable function is defined as

$$f(x_1, \ldots, x_n) = \left(\bigotimes_{i=1}^{n} \begin{bmatrix} 1 & x_i \end{bmatrix} \right) \left(\bigotimes_{i=1}^{n} \mathbf{A}(1) \right) \mathbf{F}. \tag{4}$$

The optimization of arithmetic expressions in the number of non-zero coefficients count can be performed by selecting different polarity for variables x_i, i.e., the usage of positive x_i and negative $\overline{x}_i = x_i \oplus 1$ literals, but not both for the same variable. In this way, Fixed-polarity arithmetic expressions (FPARs) are defined, see for example, [5].

In matrix notation, FPARs are defined as

$$f(x_1, \ldots, x_n) = \left(\bigotimes_{i=1}^{n} \begin{bmatrix} 1 & x_i^{h_i} \end{bmatrix} \right) \left(\bigotimes_{i=1}^{n} \mathbf{A}^{h_i} \right) \mathbf{F}, \tag{5}$$

where

$$x_i^{h_i} = \begin{cases} x_i, \ h_i = 0, \\ \overline{x}_i, \ h_i = 1, \end{cases} \quad \mathbf{A}^{h_i}(1) = \begin{cases} \begin{bmatrix} 1 & 0 \\ -1 & 1 \end{bmatrix}, \ h_i = 0 \\ \\ \begin{bmatrix} 0 & 1 \\ 1 & -1 \end{bmatrix}, \ h_i = 1, \end{cases}$$

The polarity vector $H = (h_1, \ldots, h_n)$, $h_i \in \{0, 1\}$, $i = 1, \ldots, n$, uniquely specifies each FPRA for a given function f.

Example 4. *For $n = 2$, there are four different FPRA the coefficients of which can be represented by the Fixed-polarity Arithmetic Representation (FPRA)-matrix whose rows represent coefficients $c_{i(H)}$ for different polarities $H(h_1, h_2)$,*

$$\mathbf{Q}_f = \mathbf{FPAR}(2) = \begin{bmatrix} c_{0(0,0)} & c_{1(0,0)} & c_{2(0,0)} & c_{3(0,0)} \\ c_{0(0,1)} & c_{1(0,1)} & c_{2(0,1)} & c_{3(0,1)} \\ c_{0(1,0)} & c_{1(1,0)} & c_{2(1,0)} & c_{3(1,0)} \\ c_{0(1,1)} & c_{1(1,1)} & c_{2(1,1)} & c_{3(1,1)} \end{bmatrix}.$$

If $\mathbf{c}_{i(H)}$ *is the ith column of* $\mathbf{FPAR}(2)$*, then* $\mathbf{c}_{i(H)} = \mathbf{a}_i * \mathbf{F}$*, where* $\mathbf{a}_i$ *is the ith row of the arithmetic transform matrix* $\mathbf{A}(2)$*. In the other words, by Remark 1,*

$$\mathbf{Q}_f = \mathbf{FPAR}(2) = \mathbf{C}_f(2)(\mathbf{A}(1))^T = \begin{bmatrix} f(0) & f(1) & f(2) & f(3) \\ f(1) & f(0) & f(3) & f(2) \\ f(2) & f(3) & f(0) & f(1) \\ f(3) & f(2) & f(1) & f(0) \end{bmatrix} \begin{bmatrix} 1 & -1 & -1 & 1 \\ 0 & 1 & 0 & -1 \\ 0 & 0 & 1 & -1 \\ 0 & 0 & 0 & 1 \end{bmatrix}.$$

3.2 Correlation and Galois Field Expressions in $GF(4)$

The convolution matrix for $f \in GF_4(C_4)$ is given by

$$\mathbf{C}_f(1) = \begin{bmatrix} f(0) & f(1) & f(2) & f(3) \\ f(1) & f(0) & f(3) & f(2) \\ f(2) & f(3) & f(0) & f(1) \\ f(3) & f(2) & f(1) & f(0) \end{bmatrix}.$$

Assume that elements of the Galois field $GF(4)$ are identified with the non-zero integers $0, 1, 2, 3$, with zero assigned to the identity in $GF(4)$. A variable x taking values in $GF(4)$ is given by the vector $\mathbf{x} = [0, 1, 2, 3]^T$.

The set $1, x, x^2, x^3$, where the exponentiation is defined in terms of the multiplication in $GF(4)$, is a basis in $GF_4(C_4)$. In matrix notation, this basis is given by

$$\mathbf{G}_4(1) = \begin{bmatrix} 1\,0\,0\,0 \\ 1\,1\,1\,1 \\ 1\,2\,3\,1 \\ 1\,3\,2\,1 \end{bmatrix}.$$

The positive-polarity Galois field expression in $GF(4)$ is defined by

$$f = \begin{bmatrix} 1 \; x \; x^2 \; x^3 \end{bmatrix} \mathbf{S}_f,$$

where the exponentiation x^i, $i = 2, 3$ is defined in terms of multiplication in $GF(4)$. The vector of spectral coefficients $\mathbf{S}_f = [S_f(0), S_f(1), S_f(2), S_f(3)]^T$ is determined by $\mathbf{S}_f = \mathbf{G}_4^{-1}(1)\mathbf{F}$, where $\mathbf{G}_4^{-1}(1)$ is a matrix inverse to $\mathbf{G}_4$ over $GF(4)$, i.e., $\mathbf{G}_4^{-1}(1) = \begin{bmatrix} 1\,0\,0\,0 \\ 0\,1\,3\,2 \\ 0\,1\,2\,3 \\ 1\,1\,1\,1 \end{bmatrix}$.

By the analogy to the method and borrowing also the notation from the binary case, the negative literals for a quaternary variable x are defined as $\overset{k-}{x} = x \oplus k$, $k = 1, 2, 3$, where $\oplus$ is the addition in $GF(4)$. Thus, $\overset{1-}{\mathbf{x}} = [1, 2, 3, 0]^T$, $\overset{2-}{\mathbf{x}} = [2, 3, 0, 1]^T$, $\overset{3-}{\mathbf{x}} = [3, 0, 1, 2]^T$.

Fixed-polarity Galois field (FPGF) expressions are defined by selecting different polarities for the variables, and can be represented by the Fixed-polarity Galois field (FPGF) matrix in the same way as in the case of binary functions.

The polarity matrix for $f \in GF_4(C_4)$ can be expressed in terms of the convolution matrix as

$$\mathbf{Q}_f = \mathbf{FPGF}(1) = \mathbf{C}_f(1)(\mathbf{G}_4^{-1}(1))^T = \begin{bmatrix} f(0) & f(1) & f(2) & f(3) \\ f(1) & f(0) & f(3) & f(2) \\ f(2) & f(3) & f(0) & f(1) \\ f(3) & f(2) & f(1) & f(0) \end{bmatrix} \begin{bmatrix} 1 & 0 & 0 & 1 \\ 0 & 1 & 1 & 1 \\ 0 & 3 & 2 & 1 \\ 0 & 2 & 3 & 1 \end{bmatrix}$$

$$= \begin{bmatrix} f(0) & f(1)+3f(2)+2f(3) & f(1)+2f(2)+3f(3) & f(0)+f(1)+f(2)+f(3) \\ f(1) & f(0)+3f(3)+2f(2) & f(0)+2f(3)+3f(2) & f(1)+f(0)+f(3)+f(2) \\ f(2) & f(3)+3f(0)+2f(1) & f(3)+2f(0)+3f(1) & f(2)+f(3)+f(0)+f(1) \\ f(3) & f(2)+3f(1)+2f(0) & f(2)+2f(1)+3f(0) & f(3)+f(2)+f(1)+f(0) \end{bmatrix}.$$

Extensions to functions of $n > 1$ variables is done in terms of the Kronecker product for both basis functions $\mathbf{X}(1) = \begin{bmatrix} 1 & x_i & x_i^2 & x_i^3 \end{bmatrix}$, and the transform matrix $\mathbf{GF}^{-1}(1)$ for $n = 1$ in the same way as in the case of binary functions.

4 Closing Remarks

We present relationships between convolution operators on groups and Fixed-polarity matrices used in optimization of bit-level and word-level polynomial expressions for discrete functions including switching and multiple-valued logic functions and particular cases. These relationships are a basis to develop fast calculation algorithms for Fixed-polarity matrices by exploiting fast convolution algorithms based on spectral transforms.

References

1. Almaini, A.E.A., Thomposin, P., Hanson, D.: Tabular techniques for Reed-Muller logic. Int. J. Electronics 70, 23–34 (1991)
2. Jabir, A.M., Pradhan, D., Mathew, J.: GfXpress: A technique for synthesis and optimization of $GF(2^m)$ polynomials. IEEE Trans. CAD 27(4), 698–711 (2008)
3. Falkowski, B.J., Rahardja, S.: Efficient computation of quaternary fixed polarity Reed-Muller expansions. IEE Proc. Computers and Digital Techniques 142(5), 345–352 (1995)
4. Janković, D., Stanković, R.S., Moraga, C.: Optimization of $GF(4)$ expressions using the extended dual polarity property. In: Proc. 33rd Int. Symp. on Multiple-Valued Logic, Tokyo, Japan, pp. 50–56. IEEE Press, Los Alamitos (2003)
5. Malyugin, V.D.: Paralleled Calculations by Means of Arithmetic Polynomials. Physical and Mathematical Publishing Company, Russian Academy of Sciences, Moscow (1997) (in Russian)
6. Stanković, R.S., Karpovsky, M.G.: Remarks on calculation of autocorrelation on finite dyadic groups by local transformations of decision diagrams. In: Moreno Díaz, R., Pichler, F., Quesada Arencibia, A. (eds.) EUROCAST 2005. LNCS, vol. 3643, pp. 301–310. Springer, Heidelberg (2005)
7. Stanković, R.S., Moraga, C., Astola, J.T.: From Fourier expansions to arithmetic-Haar expressions on quaternion groups. Applicable Algebra in Engineering, Communication and Computing AAECC 12, 227–253 (2001)

8. Stanković, R.S., Moraga, C., Astola, J.T.: Derivatives for multiple-valued functions induced by Galois field and Reed-Muller-Fourier expressions. In: Proc. 34th Int. Symp. on Multiple-Valued Logic, Toronto, Canada, May 19-22, pp. 184–189 (2004)
9. Tan, E.C., Yang, H.: Optimization of Fixed-polarity Reed-Muller circuits using dual-polarity property. Circuits Systems Signal Process. 19(6), 535–548 (2000)
10. Yanushkevich, S.N., Miller, D.M., Shmerko, V.P., Stanković, R.S.: Decision Diagram Techniques for Micro- and Nanoelectronic Design Handbook. CRC Press, Taylor & Francis (2006)

Reversible Synthesis through Shared Functional Decision Diagrams

Milena Stanković and Suzana Stojković

Faculty of Electronic Engineering, University of Niš
A. Medvedeva 14, 18000 Niš, Serbia
milena.stankovic@elfak.ni.ac.rs, suzana.stojkovic@elfak.ni.ac.rs

Abstract. Reversible logic synthesis gained much attention recently, primarily due to its applications in low-power computing and quantum computing. There are many synthesis approaches including those using spectral techniques. The algorithm presented in this paper uses the Reed-Muller expansion of a reversible function represented by a Shared Functional Decision Diagram (FDD) to synthesize the function as a network of Toffoli gates. In each step of the algorithm the Toffoli gate with smallest cost is selected, where the cost is defined through complexity of the Shared FDD.

1 Introduction

The synthesis of reversible networks has received much attention in recent years, [2]. In particular, the interest in reversible logic is motivated by its applications in low-power computing and quantum computing. A completely specified n-input n-output Boolean function is called reversible if it maps each input assignment to a unique output assignment and vice versa. A reversible function of n variables can be defined as a truth table or as a permutation on the set of integers $(0, 1, \ldots, 2^n - 1)$. For example, the reversible function in Table 1 can also be specified as the sequence of integers $\{1, 0, 7, 2, 3, 4, 5, 6\}$.

Definition 1. *An n-input n-output gate (or circuit) is* reversible *if it realizes an $n \times n$ reversible function.*

Table 1. A reversible function $f(c, b, a)$

c b a	c' b' a'
0 0 0	0 0 1
0 0 1	0 0 0
0 1 0	1 1 1
0 1 1	0 1 0
1 0 0	0 1 1
1 0 1	1 0 0
1 1 0	1 0 1
1 1 1	1 1 0

R. Moreno-Díaz et al. (Eds.): EUROCAST 2009, LNCS 5717, pp. 510–517, 2009.

Several reversible gates have been proposed in the literature [1,4]. In this paper, we consider realization with generalized Toffoli gates only, witch is often used to construct reversible logic circuits. This gate is defined as follows:

Definition 2. *For the domain variables* $\{x_n, x_{n-1}, \cdots, x_1\}$ *the generalized Toffoli gate has the form* $TOF(C;T)$*, where* $C = \{x_{i_1}, x_{i_2}, \cdots, x_{i_k}\}$, $T = \{x_j\}$, *and* $C \cap T = \Phi$*. It maps a Boolean pattern* $\{x_n, x_{n-1}, \cdots, x_{j+1}, x_j, x_{j-1}, \cdots, x_1\}$ *into* $\{x_n, x_{n-1}, \cdots, x_{j+1}, x_j \oplus x_{i_1} x_{i_2} \cdots x_{i_k}, x_{j-1}, x_1\}$*. Usually, the set* C *is called the control set and the set* T *is called the target.*

The most commonly used Toffoli gates are: the NOT gate ($TOF\{x_j\}$), the CNOT gate, which is also known as the Feynman gate ($TOF\{x_i; x_j\}$), and the original Toffoli gate denoted by ($TOF\{x_{i_1}, x_{x_2}, \cdots, x_{i_k}; x_j\}$), shown in Fig. 1(a), (b) and (c). A reversible network that is constructed by the Toffoli gates only is called the Toffoli network.

The problem of reversible network synthesis has received much attention in recent years and several synthesis methods are proposed [6,2,3]. Some of them are based on transformations in the spectral Reed-Muller domain. There are several synthesis methods for reversible circuits. However, most of them are inefficient in the case of functions with a large number of variables. In this paper we consider a procedure for synthesis of Toffoli networks based on the approach used in [2] and [3]. However, these methods are based on the Reed-Muller expressions for function to be realized, while we use Shared Functional Decision Diagrams (Shared FDD). Since decision diagrams are a data structures convenient for dealing with large functions, the method is suitable for reasonably large functions.

Quantum Cost. The quantum cost of a reversible circuit is the sum of the quantum cost of its gates. The quantum cost of a gate G is the number of elementary quantum operations required to realize the function given by G. These elementary operations are performed by the NOT, CNOT, and three-bit Toffoli gates. NOT and CNOT gates have a quantum cost of one. However, they are not complete because they only realize linear functions. The addition of the three-bit Toffoli gate makes the set of gates complete (i.e., have the ability to implement any reversible function). However, the three-bit Toffoli gate cannot be realized as a single elementary operation. Fortunately, a realization for the

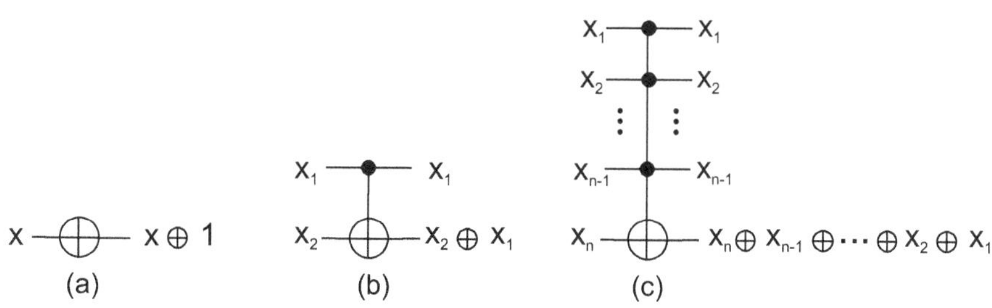

Fig. 1. Main Toffoli gates

three-bit Toffoli gate with a quantum cost of five has been found. Larger Toffoli gates have a higher quantum cost due to the number of elementary quantum operations required for their realizations.

2 Background Theory

In this section we present some basic definitions that will be used later.

2.1 Reed-Muller Spectrum

Definition 3. *Any n-variable Boolean function* $f(x_n, x_{n-1}, \ldots, x_1)$ *can be uniquely represented by the Reed-Muller expression*

$$f(x_n, x_{n-1}, \ldots, x_1) = a_0 \oplus a_1 x_n \oplus \cdots \oplus a_{n+1} x_n x_{n-1} \oplus \cdots \oplus a_{2^n-1} x_n x_{n-1} \cdots x_1$$

with $a_i \in \{0,1\}, i = 1, 2, \ldots 2^n - 1$. *Here* $\oplus$ *denotes the Exclusive OR (XOR) operation. The vector of the coefficients in this expansion is called the Reed-Muller (RM) spectrum of* f.

The RM-spectrum $\mathbf{RM}_f$ of a Boolean function can be efficiently computed by using the Reed-Muller transform defined as:

$$\mathbf{RM}_f = \mathbf{M}(n)\mathbf{F}(n).$$

where

$$\mathbf{M}(n) = \begin{bmatrix} \mathbf{M}(n-1) & 0 \\ \mathbf{M}(n-1) & \mathbf{M}(n-1) \end{bmatrix} \quad \text{and} \quad \mathbf{M}(1) = \begin{bmatrix} 1 & 0 \\ 1 & 1 \end{bmatrix}$$

In this relation the summation is modulo-2, i.e. XOR, and $\mathbf{F}(n)$ is the truth vector of f.

2.2 Decision Diagrams for Reversible Functions

Decision diagrams are a data structure convenient for representation of discrete functions. They are usually used for manipulation with functions of a large number of variables [7]. By the recursive application of the Shannon decomposition to the variables in a Boolean function $f(x_n, x_{n-1}, \ldots, x_1)$ to derive the complete disjunctive normal form, f can be represented by a *Binary Decision Tree* (BDT). The values in the terminal nodes in the BDT are the values of the function represented (elements of the truth-vector $\mathbf{F}(n)$). In a BDT, usually there are some isomorphic subtrees. Due to that, a BDT can be reduced into a *Binary Decision Diagram* (BDD). *Shared DD* (SDD) are used to represent a system of Boolean functions. A BDD representing the Boolean function $f(x_n, x_{n-1}, \ldots, x_1)$ can be transformed into a *Functional Decision Diagram* (FDD), by performing the calculation defined by the transformation matrix $\mathbf{M}(1)$ in each non-terminal node of the BDD. This calculation is performed over the subtrees which ensures the efficiency of this calculation method [7]. For example, Shared FDD for the reversible function in the Table 1 is shown in Fig. 2.

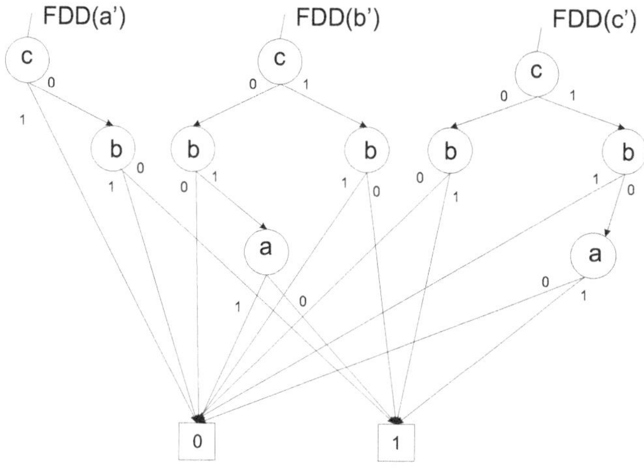

Fig. 2. Shared FDD for the function in Table 1

3 The Procedure for Reversible Synthesis

A naive algorithm for reversible network synthesis based on the ReedMuller expansion of the function, would simply use as many gates as there are terms in the ReedMuller expansion. Clearly, such a method fails to take advantage of any shared functionality that exists between multi-output functions. In the algorithm presented in [2], candidate factors, which are subexpressions common between ReedMuller expansions of multiple outputs, are identified. The factors are then substituted into the ReedMuller expansions to determine if they will be favorable in leading to a solution (i.e., a synthesized network). The primary objective of this algorithm is to minimize the number of gates (i.e., the number of factors needed to convert a ReedMuller expansion into the identity function), whereas its secondary objective is to minimize the size of the individual gates (i.e., the number of literals in the factors).

The input to this algorithm is a ReedMuller expansion of a reversible function $f : (x'_n, x'_{n-1}, \cdots, x'_1) = (x_n, x_{n-1}, \cdots, x_1)$ that is to be synthesized. The output is a network of Toffoli gates that realizes f.

3.1 Description of the Original Algorithm

For implmentation of this algorithm a search tree structure is used for memorizing of parameters of one level in the Teffoli network will be realized. The ReedMuller expansions of all output x'_i in f are obtained in terms of all its input variables x_i and stored in one *Node* of the search tree as *Node.pprm*. *Node.terms* and *Node.elim* contain the total number of terms in the current ReedMuller expansion and the total number of terms that have been eliminated from the original ReedMuller expansion once a substitution is made, respectively, while in *Node.factor*, the factor used in the substitution is stored. Also the *Node.priority* is stored and a priority queue is generated. The priority queue maintains a list of nodes sorted with respect to their priorities.

Variable *bestDepth* stores the number of gates in the best network synthesized for f so far. Variable *bestSolNode* stores a pointer to a leaf node, which represents the last gate of the synthesized circuit.

Initialization:

- Variable *bestDepth* is set to infinity.
- The root node (named *rootNode*) of the search tree is initialized. The depth and factor of this node are set to 0 and NULL, respectively. The ReedMuller expansions of all output variables x_i' in f are obtained in terms of all its input variables x_i and stored as *rootNode.pprm*. Because this is the root node, *rootNode.elim* is set to zero and its priority *rootNode.priority* is set to infinity.
- Finally, the empty priority queue PQ is initialized, and the root node is pushed onto the priority queue. This will be the first node that will be explored during synthesis.

loop

- The most promising node for further exploration is removed from the priority queue and stored in *parentNode*.
- Next step is checking to see whether *parentNode* is worth exploring or not. If the depth of *parentNode* is greater than or equal to $bestDepth - 1$, then *parentNode* can be ignored as it cannot possibly lead to a better solution than the best one seen so far.
- The *parentNode* is explored by examining each output variable x_i' in the RM expansion of f. For each input variable x_i, the RM expansion of x_i' contained in *parentNode.pprm* is searched for factors that do not contain x_i. For example, if $a' = a \oplus 1 \oplus bc \oplus ac$, then the appropriate factors are 1 and bc, as neither contains literal a. For each factor (factor) that has been identified in this manner, the substitution $x_i = x_i \oplus factor$ is made in the ReedMuller expansions of *parentNode*.
- A new node is created, which is a child of *parentNode*. The depth of the child node is incremented by one, and a copy of factor is stored. The Reed-Muller expansion of the child node is set to be the ReedMuller expansion obtained once the substitution has been made. The number of terms in the new ReedMuller expansion and the number of terms eliminated by making the substitution are stored in *childNode.terms* and *childNode.elim*, respectively.
- Finally, *childNode* is analyzed, and one of the following actions is taken.
  1. If the synthesis of f has been completed (i.e., the RM expansions for all x_i' contain only x_i), then the values of $bestDepth$ and $bestSolNode$ are updated if this solution improves upon the best solution found so far.
  2. If the number of terms in the RM expansion has not decreased by making the substitution (i.e., $childNode.elim = 0$), then the node is also disregarded. This guarantees that we only explore those nodes where the

number of terms in the RM expansion is decreasing monotonically with the application of each substitution. Otherwise, the priority of *childNode* is calculated, and the node is inserted into the priority queue.

end loop

The priority of *childNode* is calculated as follows:

$$childNode.priority =$$

$$\alpha childNode.depth + \frac{\beta childNode.elim}{childNode.depth} - \gamma factor.literalCount,$$

where α, β, and γ are weights that sum up to one. The first term gives preference to nodes at a larger depth, the second term addresses the primary objective of minimizing the number of Toffoli gates, while the third term addresses the secondary objective of minimizing the number of control bits of the individual Toffoli gates.

The algorithm repeats the above process until the priority queue becomes empty. This condition implies that there are no more candidate nodes left to explore. Upon termination, *bestSolNode* contains a pointer to the leaf node, which represents the last gate of the synthesized circuit. The path from *rootNode* of the search tree to *bestSolNode* represents the series of Toffoli gates in the synthesized network. The edges of the path represent the substitutions that were made. For each node n along this path, $n.factor$ contains a copy of the substitution $x_i = x_i \oplus factor$. Hence, x_i is the target bit, and the literals in factor represent the control bits of the Toffoli gate.

3.2 Modified Procedure

Our procedure starts from the Shared BDD representing the outputs x'_i of the reversible function f to be realized. Starting BDD is transformed into Shared FDD representing the Reed-Muller expresions of the considered reversible function. During the synthesis a Shared FDDs are stored and nodes in the search tree are the root nodes of the corresponding Shared FDDs. Aslo the new Shared FDD, witch represent ReedMuller expansions after performed transformation is generated by transformation of the parent Shared FDD. Insted of *rootNode.terms* and *rootNode.elim* which contain the total number of terms in the current RM expansion and the total number of terms that have been eliminated from the original ReedMuller expansion once a substitution is made, respectively, we consider the totaal numer of $1-paths$ in the Shared FDD and a number of eliminated $1-paths$, where the $1-path$ is defined as the path from one of the root nodes to the terminal node with the value 1. A path is denoted by $(p_n, p_{n-1}, \ldots, p_1)$, where $p_i \in \{0, 1\}$ are labels at the edges the path consists of.

Step by step, we select Toffoli gates and transform the Shared FDD from the previous step into one new until we get the final Shared FDD representing the Reed Muller spectra of input variables. Each output x'_i in this Shared FDD have only one $1-path$ which correspond to the variable x_i, for the example

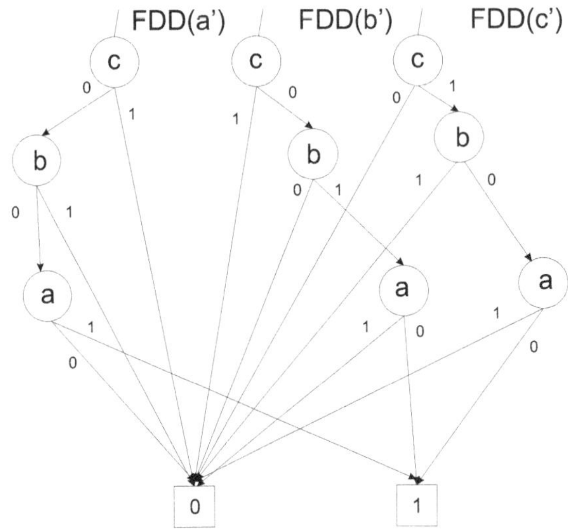

Fig. 3. The target final Shared FDD for the function in Table 1

from Table 1 shown in Fig. 3. The gates are selected by using the following rules:

Rule 1. (Case when the Toffoli gate $TOF\{x_j\}$ is selected and applied.)

We select this gate if there exist 1-path with all $p_i = 0$ and the 1-path with $p_j = 1$ and $p_k = 0$ for all $k \neq j$), (for example the pats $(0, 0, \dots, 0, \dots, 0)$ and $(0, 0, \dots, 1, \dots, 0)$).

When applying this gate, the following transformation of the Shared FDD must be performed: For each 1-path in the Shared FDD with $p_j = 1$ it is necessary to invert the value in the terminal node in the path with $p_j = 0$ and all other p_i equal to these in the considered 1-path.

Rule 2. (Case when the Toffoli gate $TOF\{x_i; x_j\}$ is selected and applied.)

We selected this gate if in the Shared FDD exists the 1-path with $p_j = 1$ and $p_k = 0$ for all $k \neq j$, and the 1-path with $p_i = 1$ and $p_k = 0$ for all $k \neq i$ (for example the 1-paths: $(0, 0, \dots, 1, 0, \dots, 0)$ and $(0, 0, \dots, 0, 1, \dots, 0)$).

In this case the Shared FDD is transformed as: For each 1-path with $p_j = 1$ it is necessary to invert the value in the terminal node in the path with $p_j = 0$, $p_i = 1$ and all other p_k equal to these in the considered 1-path.

Rule 3. (Case when the Toffoli gate $TOF\{x_{i_1} x_{i_2} \cdots x_{i_m}; x_j\}$ is selected and applied.)

We select this gate if there exist 1-path with $p_j = 1$ and $p_k = 0$ for all $k \neq j$, and 1-path with $p_{i_1} = 1, p_{i_2} = 1 \dots p_{i_m} = 1$ and $p_k = 0$ for all $k \neq i_1, i_2, \dots, i_m$, (for example the paths $(0, 0, \dots, 1, 0, 0, \dots, 0)$ and $(0, 0, \dots, 0, 1, 1 \dots, 0)$).

In this case the Shared FDD is transformed as: For each 1-path with $p_j = 1$ it is necessary to invert the value in the terminal node on the path with $p_j = 0$,

$p_{i_1} = 1, p_{i_2} = 1, \ldots, p_{i_m} = 1$, and all other p_k equal to these in the considered 1-path.

Implementation of the proposed procedure is based on several heuristic rules, similar to the rules discussed and proposed in the papers [2,3].

4 Conclusion

In the papaer a modification of the algorithm for reversible function synthesizis as network of Toffoli gates, described in [2] is presented. The algorithm searches the tree of possible factors in priority order to try to find the best possible solution. In our modification the Shared BDD for representing of Reed-Muller expresions and for performing of all necessery transformation is proposed. We expect that this approach will be convenient for working with functions with large number of inputs.

References

1. Toffoli, T.: Reversible computing. Tech. Rep. MIT, Cambridge (1980)
2. Gupta, P., Agrawal, A., Jha, N.K.: An algorithm for synthesis of reversible logic circuits. IEEE Trans. on Computer-aided Design of Integrated Circuits and Systems 25(11), 2317–2329 (2006)
3. Donald, J., Nirajk, J.: Reversible logic synthesis with Fredkin and Peres gates. ACM Journal of Emerging Technologies in Computing Systems 4(1) (2008)
4. Fredkin, E., Toffoli, T.: Conservative logic. Int. J. Theor. Phys. 21(3/4), 219–253 (1982)
5. Sasao, T., Fujita, M. (eds.): Representation of Discrete Functions. Kluwer Academic Publishers, Dordrecht (1996)
6. Zhong, J., Muzio, J.C.: Improved Implementation of a Reed-Muller Spectra Based Reversible Synthesis Algorithm. In: IEEE Pacific Rim Conference on Communication, Computers and Signal Processing (2007)
7. Stanković, R.S., Stanković, M., Janković, D.: Spectral Transforms in Switching Theory: Definitions and Calculations. Nauka, Beograd (1998)

Ternary Haar-Like Transform and Its Application in Spectral Representation of Ternary-Valued Functions

Susanna Minasyan[1], Radomir Stankovic[2], and Jaakko Astola[1]

[1] Tampere University of Technology
Korkeakoulunkatu 1,FI-33720,Tampere Finland
{susanna.minasyan,jaakko.astola}@tut.fi
[2] Department of Computer Science, Faculty of Electronics, 18000 Nis, Serbia
rstankovic@bankerinter.net

Abstract. The paper introduces a signal adaptive Haar-like transform for ternary functions. The term adaptive means, given the signal, we design a transform after a brief analysis of signal features such as appearance of identical patterns or periods of constancy. The proposed transform possesses a fast FFT-like computation algorithm similar to fast algorithms for the classical Haar transform on finite dyadic groups as well as the generalized Haar transforms for multiple-valued functions. The proposed transform is utilized in reduction of the number of nonzero coefficients in spectral representation of ternary functions. The method shows good results, especially, when the given ternary signal contains intervals of constancy or repeated patterns having relatively high frequency of appearance.

Keywords: Haar-like transform,switching functions,adaptive transform, minimization,ternary logic.

1 Introduction

Spectral transforms have found wide applications in many areas of signal processing, such as filtering, pattern recognition, communications, logic design, and some other areas, see, for instance, [13], [10],[12],[14], and references therein. In particular, the Reed-Muller (RM) transform is widely used in solving different problems in logic design, such as AND-EXOR synthesis [1], testability [2],[3], etc. The same is true also for various generalizations of the Reed-Muller transform to multiple-valued functions [7],[8]. In many applications, efficiency of the Reed-Muller transform is based on the property that it is a local transform permitting to focus on particular subareas of the domain of definition of the signal under analysis. The Haar transform is another classical local spectral transform efficiently applied in the same areas [10], [13],[14].

Several generalizations of the Haar transform have been proposed to solve various tasks in signal processing. In particular, the parameterized slant-Haar transforms [4] and parametric Haar-like transforms [5]-[6] have been introduced

R. Moreno-Díaz et al. (Eds.): EUROCAST 2009, LNCS 5717, pp. 518–525, 2009.

in order to permit efficient exploiting of particular features of certain classes of signals that often appear in some signal processing applications. These transforms can be adapted to the properties of signals to be processed by carefully adjusting certain parameters which provides high flexibility in adapting a transform to a given application. In this paper, we define a signal adaptive ternary Haar-like transform over the finite Galois field GF(3). As an illustration of possible applications of this transform, we consider the compactness of the corresponding spectral representations of ternary valued functions in the number of nonzero coefficients. Experimental results show, that the proposed approach to the representation of ternary functions reduces significantly the number of nonzero coefficients in the ternary spectrum, for example, compared to ternary RM and other related transforms. As it should be expected, the results are better when a given signal contains many identical patterns.

The paper is organized as follows. Section 2 presents the basic concepts of the family of parametric Haar-like transforms. Section 3 describes the fast ternary adaptive Haar-like transform. In Section 4 we discuss the application of the proposed ternary Haar-like transform to the reduction of spectral representations of ternary functions. Section 5 brings the experimental results obtained by applying the proposed method to various examples of randomly generated ternary functions.

2 Family of Ternary Parametric Haar-Like Transforms

In practice, orthogonal transforms having fast algorithms are widely used, since this feature ensures their computation efficiency in applications. The Fast Discrete Cosine Transform (FDCT), the Fast Fourier Transform (FFT), the Fast Walsh-Hadamard Transform (FWHT), the Fast Haar Transform (FHT), etc., are examples of fast discrete transforms. Most of these transforms may be represented in a unified decomposition form which involves a set of parameters. In this respect, the common name for such representations is the parametric transforms. The mentioned transforms are mostly used for representation of signals defined in 2^m points. Here we will consider the ternary case, that is, when the dimension of the signal to be represented (and correspondingly of the transform) is a power of 3.

The ternary generalized transform matrix of order $N = 3^m$ (m is an integer) corresponding to various transforms that have fast calculation (FFT-like) algorithms can be uniformly represented by the following product of sparse matrices:

$$H_N = P^{(m+1)} \prod_{j=m}^{1} H^{(j)} P^{(j)} = \prod_{j=m}^{1} \left(\bigoplus_{s=0}^{N/3-1} V^{(j,s)} \right) P^{(j)} \tag{1}$$

where $H^{(j)}, j = 1, \ldots, m$ are block-diagonal matrices of order $(N \times N)$ having 3×3 blocks $V^{(j,s)}$, called butterflies or spectral kernels, located on the main diagonal; $P^{(j)}$ is the permutation matrix of order N. This representation follows

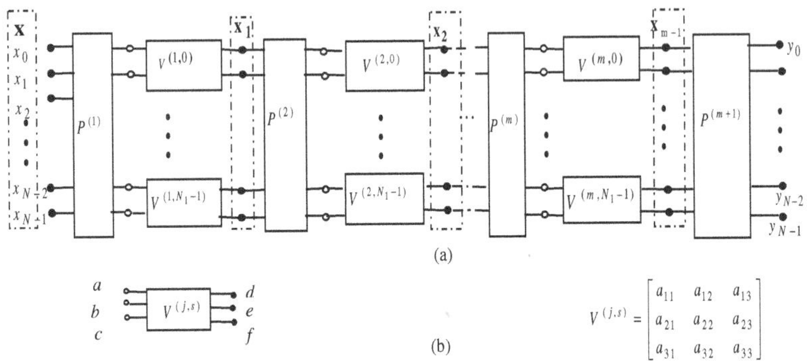

Fig. 1. Unified flow-graph for the fast transform algorithms with ternary kernels

from the decomposition of the transform matrix into submatrices corresponding to steps of the fast algorithms, see for instance [10].

Transforms represented by the transform matrix of the form (1) may be computed iteratively by a fast algorithm through stages:

$$\mathbf{x}_0 = \mathbf{x}; \quad \mathbf{x}_j = H^{(j)} \cdot (P^{(j)}\mathbf{x}_{j-1}), j = 1, ..., m; \quad \mathbf{y} = P^{(m+1)}\mathbf{x}_m \tag{2}$$

where $\mathbf{x}$ is the input signal and $\mathbf{y}$ is the spectrum of it.

According to (2), at the stage $j = 1, \ldots, m$, the input vector $\mathbf{x}_{j-1}$ to that stage is first permuted as specified by the permutation matrix $P^{(j)}$ and then, the resulting vector is multiplied by the block diagonal matrix $H^{(j)}$ which is equivalent to the multiplication of the (3×3) spectral kernels by the corresponding (3×1) subvectors of the permuted vector.

The fast algorithm (2) may be expressed by the fast transform flow-graph in Fig. 1(a).

Fig. 1(b) shows the ternary butterfly which performs calculations specified by the (3×3) parametric matrix $V^{(j,s)}$ which corresponds to the s-th operation of the j-th stage in the fast transform algorithm. Many known fixed transforms and an infinite number of new orthogonal transforms having fast algorithms may be obtained from this unified representation by an appropriate setting of the parameters to certain values. The synthesis of fast orthogonal parametric transforms, in particular, parametric Hadamard-like and Haar-like transforms (PHT) of an arbitrary order based on predefined basis functions called generating vectors may be found in detail in [5].

3 The Family of Ternary Haar-Like Transforms

Recently, the parametric Haar-like transform of the dimension 2^m have found different applications in signal processing [2],[4], as well as in compact representation of binary benchmark functions [6]. In this section, we will consider construction of the ternary-valued parametric Haar-like transform (TPHT) of the dimension 3^m which can be used for compact representations of ternary functions.

The TPHT is a signal adapted transform that has the structure similar to the ternary Haar transform [8]. As it was discussed in [5], the kernels or basic matrices of Haar-like transform are constructed based on a given input signal. The same approach will be used for construction of the ternary Haar-like transforms.

Instead of defining a transform by a transform matrix which can be subsequently decomposed to derive a fast calculation algorithm, we do the opposite. Given a signal, we specify the ternary parametric Haar-like transform through its fast calculation algorithm, which permits an iterative analysis of the signal by the decomposition of it into inputs of steps in the fast calculation algorithm. Due to this decomposition, for each step of the algorithm we select the most suitable parameters of the ternary butterflies, depending on the structure of the input signal to be processed in this step. Therefore, we can generally say that the construction of TPHT is based on a predefined generating vector. The generating vector is a vector based on which the signal adaptive TPHT transform is designed. In other words, the generating vector is an input to fast Haar-like transform algorithm (2). The fast TPHT algorithm in structure is similar to that of fast ternary Haar transform algorithm. The idea of TPHT is close to binary parametric Haar-like transform (BPHT) [2],[4] with the difference that the kernels are ternary and the calculations are over GF(3).

The fast TPHT may be described in the following steps:

1. Assume that $\mathbf{h}$ is the input (or generating) vector to the fast algorithm (2) which has $m = \log_3 N$ stages, where the j-th stage, $j = 1, \ldots, m$, consists of $N/3^j$ ternary butterflies.
2. For the first stage, the permutation matrix is selected as $P^{(1)} = I_N$ (the notation for the permutation at the first stage is used in (1) in order to have a regular form of decomposition). Then, we split the input vector $\mathbf{h}$ into triples (subvectors of length 3). For each triple of entries of the input vector $\mathbf{h}$, we consider the library of all possible combinations of triples containing values 0, 1 and 2. There are altogether ten such combinations: {0 0 0}, {0 0 1},{0 0 2},{0 1 1}, {0 2 2}, {0 1 2}, {1 1 1},{2 2 2}, {1 1 2}, {1 2 2}. For a given triple $[u_{j,s}, v_{j,s}, w_{j,s}]^T$ of an input ternary truth vector of order $N = 3^m$, the transform kernel defining the ternary butterfly to process the triple is chosen as follows:
   - If $u_{j,s} = 0, 1, 2, v_{j,s} = w_{j,s} = 0$ i.e. if we have triples [000], [100],[200], then the corresponding kernel is the one given in Fig.2 (a);
   - If $u_{j,s} = v_{j,s} = w_{j,s} = 1$ or $u_{j,s} = v_{j,s} = w_{j,s} = 2$, i.e. if we have triples [111], [222], then the kernel given in Fig.2 (b) is selected;
   - If $u_{j,s} = 0, v_{j,s} = 1, w_{j,s} = 2$, i.e. if we have triple [012], then the corresponding kernel is the one given in Fig.2 (c);
   - $u_{j,s} = 0, v_{j,s} = w_{j,s} = 1$ or $v_{j,s} = w_{j,s} = 2$, i.e. if we have triples [011], [022], then the kernel given in Fig.2 (d) is used;
   - $u_{j,s} = v_{j,s} = 1, w_{j,s} = 2$, i.e. if we have triple [112], then the corresponding kernel is the one given in Fig.2 (e);

$$\text{(a) } V^{(j,s)} = \begin{bmatrix} 1 & 0 & 0 \\ 0 & 1 & 0 \\ 0 & 0 & 1 \end{bmatrix}; \quad \text{(b) } V^{(j,s)} = \begin{bmatrix} 1 & 0 & 0 \\ 1 & 1 & 1 \\ 2 & 1 & 0 \end{bmatrix}; \quad \text{(c) } V^{(j,s)} = \begin{bmatrix} 0 & 1 & 0 \\ 1 & 1 & 1 \\ 0 & 1 & 1 \end{bmatrix};$$

$$\text{(d) } V^{(j,s)} = \begin{bmatrix} 0 & 1 & 0 \\ 0 & 2 & 1 \\ 1 & 0 & 0 \end{bmatrix}; \quad \text{(e) } V^{(j,s)} = \begin{bmatrix} 1 & 0 & 0 \\ 0 & 1 & 1 \\ 1 & 0 & 1 \end{bmatrix}; \quad \text{(f) } V^{(j,s)} = \begin{bmatrix} 1 & 0 & 0 \\ 1 & 1 & 0 \\ 1 & 0 & 1 \end{bmatrix};$$

Fig. 2. Kernels corresponding to certain input triples

- $u_{j,s} = 1, v_{j,s} = w_{j,s} = 2$, i.e. if we have triple [122], then the corresponding kernel is given in Fig.2 (f).

  The other kernels, corresponding to the remained permuted combinations of the ten triples mentioned, are obtained by permutation of columns in the corresponding kernels in Fig. 2. One can see that the kernel in the case (a) is the identity matrix of order 3, which means no operation, while the kernel in the case (f) is a permuted version of the kernel in the case (e). Therefore, altogether, we have four different kernels for various triples of values 0,1,2.

  It should be noticed that a ternary kernel (butterfly) is selected in such a way that by applying it to an input triple, the first output is always equal to nonzero and the other two outputs are always zeros. This property provides the similarity with the fast Haar transform algorithm.

  In other words, for the jth stage and the sth kernel we have the following:

$$V^{(j,s)} \times [abc]^T = [def]^T, \text{ where } d \neq 0, e = f = 0.$$

  The transform matrix corresponding to the first stage is designed based on kernels specified in Fig. 2. These $V^{(j,s)}$ kernels will be located on the main diagonal of the block-diagonal matrix $H^{(1,s)}$.

3. Then, for $j = 1$, apply the first stage of the flow-graph (2) to the input vector $\mathbf{x}$ to obtain the output $\mathbf{x}_1$.
4. For each $j = 2, \dots n$, define the permutation matrix, so that it passes the first nonzero outputs, obtained from butterflies in the previous stage, to the uppermost butterflies, and pass the zero outputs to the lowest butterflies of the current stage.

   Then, again we define the spectral kernels $V^{(j,s)}$ for the triple $[u_{j,s}, v_{j,s}, w_{j,s}]$ according to the rule above, where $[u_{j,s}, v_{j,s}, w_{j,s}]$ are the corresponding three components of the vector $P^{(j)} x_{j-1}$ that are passed to the s-th operation of the current stage of the flow-graph.

   Apply the j-th stage of the flow-graph to the vector $\mathbf{x}_{j-1}$ to obtain the vector $\mathbf{x}_j$.
5. Select arbitrarily the set of permutations $P^{(m+1)}$ under the restriction that they do not change the position of the first component. Thus, the first row of $P^{(m+1)}$ is a vector $[1, 0, \dots, 0]$.

Since the number of nonzero components reduces to $1/3$ from stage to stage and since the number of stages is $m = \log_3 N$, only the first output of the flow-graph will be nonzero. The desired transform matrix may be computed as the product of block-diagonal and permutation matrices.

4 Application of Ternary Haar-Like Transform: Reduction of Spectra of Ternary Functions

In this section, we present a method based on ternary Haar-like transform for reduction of the number of nonzero coefficients in spectral representations of ternary logic functions. As it was shown in the previous section, the ternary Haar-like transform is constructed based on a given signal, which can be viewed as the generating vector for the transform produced. This was also the case for binary parametric Haar-like transforms [6].

In many cases, the truth vectors of logic functions may contain identical parts (patterns) that may appear with certain frequency. Constant subvectors are included as particular cases. In this respect, the proposed approach may be useful since it can capture the redundant information (i.e., repeating patterns) which occurs in a multiple-valued logic function f, in order to reduce the number of spectral coefficients required to represent f.

The determination of compact representations of ternary-valued functions of ternary variables in terms of coefficients of the ternary parametric Haar-like transform can be described by the following algorithm.

Algorithm: As in the binary case, the ternary Haar-like transform is designed based on a generating vector. In order to find the generating vectors the following procedure is implemented:

1. The original truth vector $\mathbf{F}$ of a ternary function f is subdivided into non-overlapping subvectors $K_i \in K \subseteq f$, $i = 1, \ldots, 3^{n-k}$, of equal lengths 3^k (k is a fixed number), where $K = \{K_i\}_{i=1}^{3^{n-k}}$ is the set of all subvectors in a truth vector.
2. The ternary Haar-like transforms $\mathbf{T}(n, k, K_i)$, $i = 1, \ldots, r, r \leq 3^{n-k}$ are calculated, where r is the number of non-repeated ternary subvectors of length 3^k in a truth vector. The TPHTs of order $N = 3^k$ are constructed based on r subvectors which serve as the generating vectors for the TPHT transform design described in the previous section.
3. The TPHTs are applied to all 3^{n-k} subvectors of the ternary vector in order to determine the spectral representation of a ternary function. Thereafter, the best or optimal transform i.e. the one that gives the minimal number of non-zero spectral coefficients is selected.

It should be noticed that the reconstruction of the original function from its spectral representation requires the information about the generating vector corresponding to the optimal transform.

5 Experimental Results

In this section, the results of testing of the proposed approach based on the TPHT are presented. We performed a series of experiments over the ternary functions which are constructed such that contain 70% of the same repeated pattern that is selected randomly. For example, the vector $[0\ 0\ 0\ 2\ 1\ 1\ 1\ 0\ 1]^T$ was considered as a pattern. The pattern was located at randomly selected place by replacing the content of a subvector in originally synthesized function. Table 1 shows the number of nonzero coefficients in spectra of different transforms, such as TRM, TRMH (ternary Reed-Muller-Haar) (see [8]), TLI [9] (which is the one of the transforms belonging to the family of recursive transforms, Class A), THel [7] transforms and our proposed TPHT. The second column shows the output functions of ternary multi-output functions. We consider here 5-input (N=243) and 3-output ternary functions. The third column gives the number of nonzero elements (1's, 2's) in the original ternary domain. In 4th -8th columns, the number of nonzero coefficients in spectra of different transforms are given. The results in this Table correspond to the application of the transforms to the whole signal (given in brackets) and in a block-wise manner. The last means that the original signal is split into the subvectors of a fixed length and then, the transforms of small order are applied to subvectors. Therefore, the total number of nonzero spectral elements in the original function is the sum of nonzero spectral elements corresponding to each subvector.The proposed ternary Haar-like transform is applied only in the block-wise manner due to the adaptive nature of our transform. In experiments, the length 9 was taken since it's desirable to have a generating vector of small length. The

Table 1. Number of nonzero elements in spectra of TRM, TRMH, THel, TLI, THT transforms of functions with 70% repetitions of an identical pattern

Tern.func.	Outp.	Orig. *#nonz.*	TRM (sub/all)	TRMH (sub/all)	THel. (sub/all)	TLI (sub/all)	THaar-like (sub)
F1	Out1	105	145/146	110/113	100/100	90/86	35
	Out2	100	144/132	105/110	89/74	87/101	31
	Out2	115	156/148	114/124	117/127	103/121	47
F2	Out1	91	127/127	91/91	76/111	76/91	19
	Out2	102	132/149	97/98	81/84	85/90	28
	Out3	135	137/140	110/110	114/118	99/108	55
F3	Out1	126	136/121	107/107	105/98	94/105	48
	Out2	152	183/163	149/152	131/123	131/127	84
	Out3	100	134/138	98/99	85/85	85/109	30
F4	Out1	79	53/132	43/43	74/105	65/72	34
	Out2	91	57/66	47/47	86/105	75/66	36
	Out3	133	170/169	130/135	120/115	120/119	66
F5	Out1	121	133/148	101/102	100/109	94/92	41
	Out2	117	129/140	99/99	99/96	87/94	39
	Out3	72	126/165	114/117	95/146	68/95	37
Av.*#nonz*		109					42

specific feature of our approach is that it always reduces the spectral representation of functions (contained repeated patterns) compared to their original domain representation.

The experiments show that on the average the larger is the number of occurrences of the same pattern in the truth vector, the better is the performance of the proposed approach. In other words, when the same pattern is repeated many times, then the main contribution in zeroing out the spectral coefficients is due to the Haar-like transform based on corresponding pattern-subvectors that may work also good on the rest (distinct) subvectors.

References

1. Sasao, T.: Switching Theory for Logic Synthesis. Kluwer Academic Publishers, Dordrecht (1999)
2. Agaian, S., Astola, J., Egiazarian, K.: Binary Polynomial Transforms and Nonlinear Digital Filters. Marcel Dekker, New York (1995)
3. Stankovic, R., Stankovic, M., Jankovic, D.: Spectral Transforms in Switching Theory, Definitions and Calculations. IP Nauka, Belgrade (1998)
4. Agaian, S., Tourshan, K., Noonan, J.: Parametrisation of slant-Haar transforms. IEE Proc. Vis. Image Signal Process. 150(5), 306–311 (2003)
5. Minasyan, S., Guevorkian, D., Sarukhanyan, H.: On parameterized fast Haar- and Hadamard-like transforms of arbitrary order. In: Proc. of 3rd Int. Conf. On Computer Science and Information Technologies (CSIT 2001), Yerevan, Armenia, pp. 294–298 (2001)
6. Minasyan, S., Astola, J., Stankovic, R., Guevorkian, D.: Reduction of the number of nonzero coefficients in Reed-Muller Haar- like spectrum. In: SMMSP 2008, Moscow (2007)
7. Fu, C., Falkowski, B.: Generation of Multi-polarity Helix Transform over GF(3). IEICE Electron. Express 1(8), 211–216 (2004)
8. Stankovic, R., Stankovic, M., Moraga, C.: Design of Haar Wavelet Transforms and Haar spectral transform decision diagrams for multiple-valued functions. In: Proc. of 31st IEEE Int. Symp. on Multiple-valued Logic (ISMVL 2001), pp. 311–316 (2001)
9. Falkowskiy, B., Fu, C.: Family of Fast Transforms over GF(3). In: 33rd Int. Symp. on Multiple-Valued Logic (ISMVL 2003), pp. 323–329 (2003)
10. Karpovsky, M., Stankovic, R., Astola, J.: Spectral Logic and its Application for the Design of Digital Devices (2008)
11. Grigoryan, A., Agaian, S.: Multidimensional Discrete Unitary Transformations: Representation, Partitioning, and Algorithms. Marcel Dekker Inc., New York (2003)
12. Thornton, M., Dreschler, R., Miller, D.: Spectral Techniques in VLSI CAD. Springer, Heidelberg (2001)
13. Astola, J., Yaroslavsky, L.: Advances in Signal Transforms - Theory and Applications. EURASIP Book Series on Signal Processing and Communications (2007)
14. Yanushkevich, S., Miller, D., Shmerko, V., Stankovic, R.: Decision Diagram Techniques for Electrical Engineers (Handbook). CRC Press/Taylor & Francis (2006)

Complete Sets of Hamiltonian Circuits for Classification of Documents

Bernd Steinbach[1] and Christian Posthoff[2]

[1] Freiberg University of Mining and Technology, Institute of Computer Science, D-09596 Freiberg, Germany
steinb@informatik.tu-freiberg.de

[2] The University of The West Indies, St. Augustine Campus, Trinidad & Tobago
Christian.Posthoff@sta.uwi.edu

Abstract. The calculation of Hamiltonian Circuits is an NP-complete task. This paper uses slightly modified complete sets of Hamiltonian circuits for the classification of documents. The known solution method is based on a SAT-instance with a huge number of clauses which is flattening the knowledge about the problem. We suggest an even more compact model of Boolean equations that preserves the knowledge by summarizing restrictions and requirements. The presented *implicit two-phase SAT-solver* finds efficiently the solution using operations of the XBOOLE library. This solver can be included easily as signal processing unit into the device where the classification of the documents is required.

1 Introduction

A Hamiltonian path in a graph is a sequence of edges that uses each node precisely once. A Hamiltonian circuit [1], also called Hamiltonian cycle, is a cycle in the graph which visits each node exactly once and returns to the starting node. It is well-known that the problem of determining whether such paths or cycles exist is an NP-complete problem [4].

We suggest a slight modification of this problem and apply it to the classification of documents by means of an efficient signal processing unit. Basically our approach can be applied to each document that is transmitted by a sequence of bits. As an example we motivate our approach in the context of FAX transmissions. In spite of several new possibilities the FAX transmission is popular for the fast exchange of textual and graphic data because only simple equipment is required. A disadvantage of the FAX transmission from Alice to Bob is that Eve (an eavesdropper) can see the transmitted information. An easy way to reduce the value of the information seen be Eve is that both *good* and *bad* information are transmitted by FAX. It is necessary that the receiver Bob can select the *good* FAX, and a signal processing unit throws away the *bad* information immediately.

A couple of bits of the transmitted document is used for this detection These bits may form an $n \times n$ adjacency matrix of a graph. Figure 1 illustrates the suggested procedure.

R. Moreno-Díaz et al. (Eds.): EUROCAST 2009, LNCS 5717, pp. 526–533, 2009.

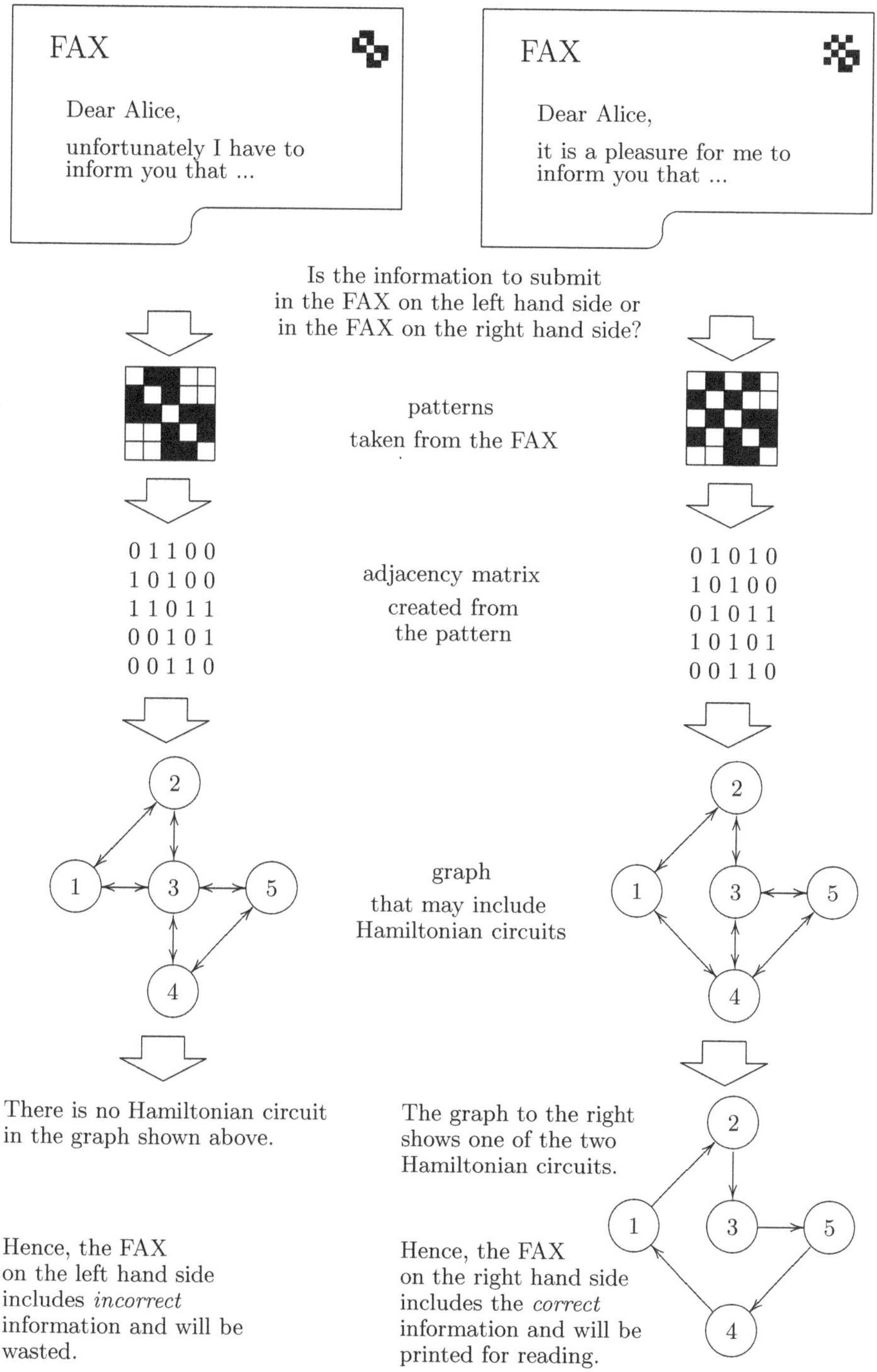

Fig. 1. Principle to select the correct FAX information by Hamiltonian circuits

Reflexive edges that begin and end on the same node are ignored. Instead of a single Hamiltonian circuit a set of Hamiltonian circuits will now be allowed. A complete set of Hamiltonian circuits covers all nodes of the graph. Any received FAX without a complete set of Hamiltonian circuits will be rejected. The number of different complete sets of Hamiltonian circuits is an additional information for the receiver Bob, but hidden for Eve. Due to the missing number of nodes n and the position of the classification bits in the transmitted FAX the eavesdropper Eve is not able to distinguish between *good* and *bad* FAX transmissions.

2 Boolean Model

2.1 Coding the Edges of the Graph by Boolean Variables

Boolean variables are introduced for each edge and each direction as follows:

$$x_{i,j} = \begin{cases} 1 & \text{if the edge is used from node } i \text{ to node } j \\ 0 & \text{otherwise} \end{cases} . \quad (1)$$

Hence, the number of variables needed to express all conditions of complete sets of Hamiltonian circuits is equal to the number of values 1 in the $n \times n$ adjacency matrix of a graph where values 1 on the main diagonal are replaced be values 0. By means of these variables all conditions for complete sets of Hamiltonian circuits can be expressed.

2.2 Simple SAT - Model

The selection of any edge, i.e. $(x_{i,j} = 1)$ determines for Hamiltonian circuits three conditions:

1. The reverse edge is forbidden:

$$x_{i,j} \wedge x_{j,i} = 0 . \quad (2)$$

2. Any edge from the same start node i to any destination node $d_l \neq j$ is forbidden:

$$x_{i,j} \wedge x_{i,d_l} = 0, \ \forall \ d_l \neq j . \quad (3)$$

3. Any edge that ends at the node j and begins at any source node $s_m \neq i$ is forbidden:

$$x_{i,j} \wedge x_{s_m,j} = 0, \ \forall \ s_m \neq i . \quad (4)$$

The final condition for complete sets of Hamiltonian circuits is determined by the nodes of the graph. In each node of the graph one of the outgoing edges must be an element of the complete sets of Hamiltonian circuits:

$$\bigvee_{d_l=1}^{d_{l\,\max}} x_{i,d_l} = 1 . \quad (5)$$

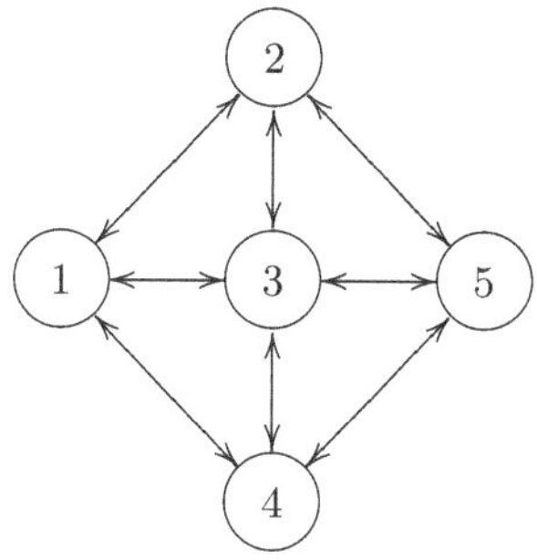

Fig. 2. Simple graph to analyze for complete sets of Hamiltonian circuits

The system of equations defined by the formula given above can be solved directly using XBOOLE [2,3] or transformed into a single equation where a conjunctive form on the left hand side is equal to 1 and solved by a SAT - solver like *zChaff* or *March* . The abbreviation SAT is used for the term *satisfiability*.

For the simple graph of Figure 2 we get the SAT - equation (6) that consists of 93 clauses that include a total of 16 variables.

$$
\begin{aligned}
&(\overline{x}_{1,2} \vee \overline{x}_{2,1})\,(\overline{x}_{1,2} \vee \overline{x}_{1,3})\,(\overline{x}_{1,2} \vee \overline{x}_{1,4})\,(\overline{x}_{1,2} \vee \overline{x}_{3,2})\,(\overline{x}_{1,2} \vee \overline{x}_{5,2}) \wedge \\
&(\overline{x}_{1,3} \vee \overline{x}_{3,1})\,(\overline{x}_{1,3} \vee \overline{x}_{1,2})\,(\overline{x}_{1,3} \vee \overline{x}_{1,4})\,(\overline{x}_{1,3} \vee \overline{x}_{2,3})\,(\overline{x}_{1,3} \vee \overline{x}_{4,3})\,(\overline{x}_{1,3} \vee \overline{x}_{5,3}) \wedge \\
&(\overline{x}_{1,4} \vee \overline{x}_{4,1})\,(\overline{x}_{1,4} \vee \overline{x}_{1,2})\,(\overline{x}_{1,4} \vee \overline{x}_{1,3})\,(\overline{x}_{1,4} \vee \overline{x}_{3,4})\,(\overline{x}_{1,4} \vee \overline{x}_{5,4}) \wedge \\
&(\overline{x}_{2,1} \vee \overline{x}_{1,2})\,(\overline{x}_{2,1} \vee \overline{x}_{2,3})\,(\overline{x}_{2,1} \vee \overline{x}_{2,5})\,(\overline{x}_{2,1} \vee \overline{x}_{3,1})\,(\overline{x}_{2,1} \vee \overline{x}_{4,1}) \wedge \\
&(\overline{x}_{2,3} \vee \overline{x}_{3,2})\,(\overline{x}_{2,3} \vee \overline{x}_{2,1})\,(\overline{x}_{2,3} \vee \overline{x}_{2,5})\,(\overline{x}_{2,3} \vee \overline{x}_{1,3})\,(\overline{x}_{2,3} \vee \overline{x}_{4,3})\,(\overline{x}_{2,3} \vee \overline{x}_{5,3}) \wedge \\
&(\overline{x}_{2,5} \vee \overline{x}_{5,2})\,(\overline{x}_{2,5} \vee \overline{x}_{2,1})\,(\overline{x}_{2,5} \vee \overline{x}_{2,3})\,(\overline{x}_{2,5} \vee \overline{x}_{3,5})\,(\overline{x}_{2,5} \vee \overline{x}_{4,5}) \wedge \\
&(\overline{x}_{3,1} \vee \overline{x}_{1,3})\,(\overline{x}_{3,1} \vee \overline{x}_{3,2})\,(\overline{x}_{3,1} \vee \overline{x}_{3,4})\,(\overline{x}_{3,1} \vee \overline{x}_{3,5})\,(\overline{x}_{3,1} \vee \overline{x}_{2,1})\,(\overline{x}_{3,1} \vee \overline{x}_{4,1}) \wedge \\
&(\overline{x}_{3,2} \vee \overline{x}_{2,3})\,(\overline{x}_{3,2} \vee \overline{x}_{3,1})\,(\overline{x}_{3,2} \vee \overline{x}_{3,4})\,(\overline{x}_{3,2} \vee \overline{x}_{3,5})\,(\overline{x}_{3,2} \vee \overline{x}_{1,2})\,(\overline{x}_{3,2} \vee \overline{x}_{5,2}) \wedge \\
&(\overline{x}_{3,4} \vee \overline{x}_{4,3})\,(\overline{x}_{3,4} \vee \overline{x}_{3,1})\,(\overline{x}_{3,4} \vee \overline{x}_{3,2})\,(\overline{x}_{3,4} \vee \overline{x}_{3,5})\,(\overline{x}_{3,4} \vee \overline{x}_{1,4})\,(\overline{x}_{3,4} \vee \overline{x}_{5,4}) \wedge \\
&(\overline{x}_{3,5} \vee \overline{x}_{5,3})\,(\overline{x}_{3,5} \vee \overline{x}_{3,1})\,(\overline{x}_{3,5} \vee \overline{x}_{3,2})\,(\overline{x}_{3,5} \vee \overline{x}_{3,4})\,(\overline{x}_{3,5} \vee \overline{x}_{2,5})\,(\overline{x}_{3,5} \vee \overline{x}_{4,5}) \wedge \\
&(\overline{x}_{4,1} \vee \overline{x}_{4,1})\,(\overline{x}_{4,1} \vee \overline{x}_{4,3})\,(\overline{x}_{4,1} \vee \overline{x}_{4,5})\,(\overline{x}_{4,1} \vee \overline{x}_{2,1})\,(\overline{x}_{4,1} \vee \overline{x}_{2,1}) \wedge \\
&(\overline{x}_{4,3} \vee \overline{x}_{3,4})\,(\overline{x}_{4,3} \vee \overline{x}_{4,1})\,(\overline{x}_{4,3} \vee \overline{x}_{4,5})\,(\overline{x}_{4,3} \vee \overline{x}_{1,3})\,(\overline{x}_{4,3} \vee \overline{x}_{2,3})\,(\overline{x}_{4,3} \vee \overline{x}_{5,3}) \wedge \\
&(\overline{x}_{4,5} \vee \overline{x}_{4,5})\,(\overline{x}_{4,5} \vee \overline{x}_{4,1})\,(\overline{x}_{4,5} \vee \overline{x}_{4,3})\,(\overline{x}_{4,5} \vee \overline{x}_{2,5})\,(\overline{x}_{4,5} \vee \overline{x}_{3,5}) \wedge \\
&(\overline{x}_{5,2} \vee \overline{x}_{2,5})\,(\overline{x}_{5,2} \vee \overline{x}_{5,3})\,(\overline{x}_{5,2} \vee \overline{x}_{5,4})\,(\overline{x}_{5,2} \vee \overline{x}_{1,2})\,(\overline{x}_{5,2} \vee \overline{x}_{3,2}) \wedge \\
&(\overline{x}_{5,3} \vee \overline{x}_{3,5})\,(\overline{x}_{5,3} \vee \overline{x}_{5,2})\,(\overline{x}_{5,3} \vee \overline{x}_{5,4})\,(\overline{x}_{5,3} \vee \overline{x}_{1,3})\,(\overline{x}_{5,3} \vee \overline{x}_{2,3})\,(\overline{x}_{5,3} \vee \overline{x}_{4,3}) \wedge \\
&(\overline{x}_{5,4} \vee \overline{x}_{4,5})\,(\overline{x}_{5,4} \vee \overline{x}_{5,2})\,(\overline{x}_{5,4} \vee \overline{x}_{5,3})\,(\overline{x}_{5,4} \vee \overline{x}_{1,4})\,(\overline{x}_{5,4} \vee \overline{x}_{3,4}) \wedge \\
&(x_{1,2} \vee x_{1,3} \vee x_{1,4})\,(x_{2,1} \vee x_{2,3} \vee x_{2,5}) \wedge (x_{3,1} \vee x_{3,2} \vee x_{2,4} \vee x_{4,5}) \wedge \\
&(x_{4,1} \vee x_{4,3} \vee x_{4,5})\,(x_{5,2} \vee x_{5,3} \vee x_{5,4}) = 1 \qquad (6)
\end{aligned}
$$

The disadvantages of this SAT – approach are on the one hand the large number of clauses that are required even for small graphs and on the other hand the loss of problem knowledge because the information with regard to single edges is distributed over a large number of clauses.

2.3 Compact Rule-Based Model

We use the same encoding (1) to express which edges are included in a Hamiltonian circuit. As an example we take again the simple graph of Figure 2.

Basically there are two types of laws:

1. *restrictions* that describe prohibited states, and
2. *requirements* that describe necessary choices.

Table 1 show the three general restrictions for complete sets of Hamiltonian circuits and their concrete specification by implications for the edge from node 1 to node 2 in Figure 2.

Table 1. Restrictions for Hamiltonian circuits, expressed for the edge from node 1 to node 2 in Figure 2

Restrictions	Implications
An edge from node i to node j, i.e. $x_{i,j} = 1$, prohibits that the reverse edge is used, i.e $x_{j,i} = 0$.	$(x_{1,2} \Longrightarrow \overline{x}_{2,1}) = 1$
An edge from node i to node j, i.e. $x_{i,j} = 1$, prohibits all edges to other destination nodes d_l, i.e $x_{i,d_l} = 0$.	$(x_{1,2} \Longrightarrow \overline{x}_{1,3}) \wedge$ $(x_{1,2} \Longrightarrow \overline{x}_{1,4}) = 1$
An edge from node i to node j, i.e. $x_{i,j} = 1$, prohibits all edges from other source nodes s_m, i.e $x_{s_m,j} = 0$.	$(x_{1,2} \Longrightarrow \overline{x}_{3,2}) \wedge$ $(x_{1,2} \Longrightarrow \overline{x}_{5,2}) = 1$

The implications can be substituted by:

$$a \Longrightarrow b = \overline{a} \vee b \ , \tag{7}$$

so that the system of equations of Table 1 can be transformed into one single equation:

$$(\overline{x}_{1,2} \vee \overline{x}_{2,1})(\overline{x}_{1,2} \vee \overline{x}_{1,3})(\overline{x}_{1,2} \vee \overline{x}_{1,4})(\overline{x}_{1,2} \vee \overline{x}_{3,2})(\overline{x}_{1,2} \vee \overline{x}_{5,2}) = 1 \ . \tag{8}$$

It can be seen that each clause in the partial SAT formula (8) depends on exactly two negated variables where $\overline{x}_{1,2}$ appear in each clause. Applying the laws of the Boolean algebra to (8) we get:

$$(\overline{x}_{1,2} \vee x_{1,2}\,\overline{x}_{2,1}\,\overline{x}_{1,3}\,\overline{x}_{1,4}\,\overline{x}_{3,2}\,\overline{x}_{5,2}) = 1 \ . \tag{9}$$

The two conjunctions in formula (9) express two constraints with regard to the edge modeled by $x_{1,2}$:

1. if the edge from node 1 to node 2 is not used in a Hamiltonian circuit ($\overline{x}_{1,2}$), then there is no restriction for other edges, and

Table 2. Requirement for Hamiltonian circuits, expressed for the node 1 in Figure 2

Requirement	Clause
There must be an edge starting from a node.	$(x_{1,2} \vee x_{1,3} \vee x_{1,4}) = 1$

2. if the edge from node 1 to node 2 is used in a Hamiltonian circuit ($x_{1,2}$), then five edges indicated by the negated variables of the second conjunction in (9) can not be part of the Hamiltonian circuit.

Table 2 shows the requirement for complete sets of Hamiltonian circuits and their concrete specification by a disjunction for the node 1 in Figure 2.

Putting the restrictive laws and the requirement law together we get for node 1 in Figure 2:

$$\begin{aligned} (\overline{x}_{1,2} \vee x_{1,2}\,\overline{x}_{2,1}\,\overline{x}_{1,3}\,\overline{x}_{1,4}\,\overline{x}_{3,2}\,\overline{x}_{5,2})\wedge \\ (\overline{x}_{1,3} \vee x_{1,3}\,\overline{x}_{3,1}\,\overline{x}_{1,2}\,\overline{x}_{1,4}\,\overline{x}_{2,3}\,\overline{x}_{4,3}\,\overline{x}_{5,3})\wedge \\ (\overline{x}_{1,4} \vee x_{1,4}\,\overline{x}_{4,1}\,\overline{x}_{1,2}\,\overline{x}_{1,3}\,\overline{x}_{3,4}\,\overline{x}_{5,4}) \wedge (x_{1,2} \vee x_{1,3} \vee x_{1,4}) = 1 \ , \end{aligned} \tag{10}$$

which can be simplified (using the distributive law) to

$$\begin{aligned} x_{1,2}\,\overline{x}_{2,1}\,\overline{x}_{1,3}\,\overline{x}_{1,4}\,\overline{x}_{3,2}\,\overline{x}_{5,2} \vee x_{1,3}\,\overline{x}_{3,1}\,\overline{x}_{1,2}\,\overline{x}_{1,4}\,\overline{x}_{2,3}\,\overline{x}_{4,3}\,\overline{x}_{5,3} \\ \vee x_{1,4}\,\overline{x}_{4,1}\,\overline{x}_{1,2}\,\overline{x}_{1,3}\,\overline{x}_{3,4}\,\overline{x}_{5,4} = 1 \ . \end{aligned} \tag{11}$$

Formula (11) describes completely the rule for node 1 in Figure 2. Each conjunction of (11) is a constraint associated to the edge indicated by the non-negated variable and describes completely **all the consequences**. For each edge of any graph such a constraint is defined uniquely.

For the simple graph of Figure 2 we finally get one equation that consists of five rules – one rule for each node, and 16 constraints – one constraint for each edge.

3 Implicit Two-Phase SAT Solver

The idea of our new approach is motivated in Section 2.3. Instead of processing the huge number of clauses (6) we solve an equation built by a much smaller number of rules of type (11).

The **first phase** covers the modeling of the problem by constraints and the calculation of partial solution sets implicitly. Therefore we call this new approach *Implicit Two-Phase SAT-Solver*. Algorithm 1 generates the matrix *mpss* that covers the partial solution sets of complete sets of Hamiltonian circuits of any graph completely.

As result of the first phase we get generally an $m \times m$-matrix for a graph of m edges. This matrix includes m partial solution sets. Each of them is the solution of one constraint and includes exactly one value 1 and so much values 0 as defined by the restrictive laws in Table 1 having the same variable of an edge as premise.

The **second phase** of the *implicit two-phase SAT Solver* is controlled by the requirement laws. In the lines 4 to 8 of Algorithm 2 the union operation UNI of

Algorithm 1. (1. Phase) Create the $m \times m$ matrix of partial solution sets from an $n \times n$ adjacency matrix: $mpss$ CPSS($Sizes\ n, m, AM\ am$)

```
Require: number n of nodes in the graph
    number m of edges in the graph
    adjacency matrix am of the graph
Ensure: m×m matrix mpss that includes for each edge the associated partial solution
    set
mpss ← ∅
for all i such that 1 ≤ i ≤ n do {iterate over all rows}
  for all j such that 1 ≤ j ≤ n do {iterate over all columns}
    if am[i, j] = 1 then {edge exists}
      x ← vector of n^2 dashes
      x[i * n + j] ← 1 selected edge
      x[j * n + i] ← 0 {1. restrictive law: reverse edge}
      for all c such that 1 ≤ c ≤ n do {iterate over the row}
        if c ≠ j then {other columns}
          if am[i, c] = 1 then {2. restrictive law:}
            x[i * n + c] ← 0 {outgoing edge from the start node}
          end if
        end if
      end for
      for all r such that 1 ≤ r ≤ n do {iterate over all rows}
        if r ≠ i then {other rows}
          if am[r, j] = 1 then {3. restrictive law:}
            x[r * n + j] ← 0 {incoming edge to the end node}
          end if
        end if
      end for
      mpss[NumberOfRows(mpss) + 1] ← x
    end if
  end for
end for
remove all columns from mpss that include only dashes
return mpss{the m × m matrix of partial solution sets}
```

XBOOLE [2,3] to construct the solution of the rules for a selected node using the partial solution sets of the constraints generated in the first phase. The solution set s of Algorithm 2 is calculated in line 9 by the intersection operation ISC of XBOOLE, finally. This set s consists of all binary vectors of the length m that solve the SAT problem of complete sets of Hamiltonian circuits. There are 8 Hamiltonian circuits in the example of Figure 2.

4 Experimental Results

A small adjacency matrix of 5×5 bits with 12 values 1, for instance, is sufficient to sign a FAX as *good* or *bad* and deliver additional classification information. The power of such a signal processing unit for classification becomes visible

Algorithm 2. Solve the Problem of Complete Sets of Hamiltonian Circuits: s SCSHC($Size\ n, Vector\ g, MPSS\ mpss$)

Require: number n of nodes,
vector g of the length n that include the numbers of outgoing edges,
matrix $mpss$ of partial solution sets as created in Algorithm 1

Ensure: set of all solution vectors of the length m that indicate the used edges in complete sets of Hamiltonian circuits in the basic graph

$s \leftarrow FULL(mpss)$
$b \leftarrow 0$
for all i such that $1 \leq i \leq n$ **do** {iterate over all nodes}
  $rule \leftarrow \emptyset$
  for all j such that $1 \leq j \leq g[i]$ **do** {iterate over all outgoing edges of a node}
    $constraint \leftarrow mpss[b + j]$
    $rule \leftarrow UNI(rule, constraint)$
  end for
  $s \leftarrow ISC(s, rule)$
  $b \leftarrow b + g[i]$
end for
return s{set of all solutions of complete sets of Hamiltonian circuits over n nodes}

when during the time of receiving a FAX this unit could find all existing 185,868 complete sets of Hamiltonian circuits in an adjacency matrix of 36×36 with 154 values 1.

5 Conclusions

The advantage of the new implicit two-phase approach in comparison with the known traditional SAT-model is that the assignment of one single value does not determine only one value of the solution, but additionally all values of the associated constraint which strongly restricts the remaining search space. For the simple example of Figure 2 the SAT - equation of 93 clauses is replaced by an equation of five rules over 16 constraints!

Basically this approach is not restricted to the detection of Hamiltonian circuits; it can easily be adapted to other tasks such as graph coloring or others.

References

1. Ore, O.: A Note on Hamiltonian Circuits. Amer. Math. Monthly 67, 55 (1960)
2. Posthoff, C., Steinbach, B.: Logic Functions and Equations - Binary Models for Computer Science. Springer, Dordrecht (2004)
3. Steinbach, B., Posthoff, C.: Logic Functions and Equations - Examples and Exercises. Springer Science + Business Media B.V. (2009)
4. Wegener, I.: Complexity Theory - Exploring the Limits of Efficient Algorithms. Springer, Dordrecht (2005)

SPICE Simulation of Analog Filters: A Method for Designing Digital Filters

Corneliu Rusu[1,*], Lacrimioara Grama[1], and Jarmo Takala[2]

[1] Technical University of Cluj-Napoca, FETTI, Signal Processing Group, Baritiu 26-28, RO-400027 Cluj-Napoca, Romania
{corneliu.rusu,lacrimioara.grama}@bel.utcluj.ro
http://www.sp.utcluj.ro

[2] Tampere University of Technology, Department of Computer Systems, P.O. Box 553, FIN-33101 Tampere, Finland
jarmo.takala@tut.fi
http://www.tkt.cs.tut.fi

Abstract. Traditionally IIR digital filters are designed by using analog filters described in time or transform domain, then by converting the analog filters to digital filters using appropriate transformation from s-domain to z-domain. For many engineers analog filters mean certain circuits or a netlist of components, and digital filters are a set of statements in certain software. In this work, we show how to obtain a digital filter from a given netlist of an analog filter by skipping the transfer function description.

Keywords: SPICE, analog filter, digital filter.

1 Introduction

Analog filters have been for many years one of the most important components of every electronic and communications system. Besides one or few active circuits, the analog filter contains components like resistors, capacitors and inductors. Even it is customary to assume that the components are ideal, i.e. they have no parasitic elements, the usually components and active circuits are far from ideal behavior. Consequently, a suitable analysis and synthesis of analog filters is rather difficult without the help of computers. Nowadays computer simulations are used in order to calculate the response of any analog filter [1].

Digital filters have been for many years the most common application of digital signal processors (DSPs). By digitizing any design, we can reproduce it time and time again with exactly the same characteristics. There is also a significant advantage with respects to analog filters. It is possible to reprogram the DSPs and drastically modify the gain or phase response of the filter. Moreover, this can be done without throwing away the existing hardware [2,3].

Traditionally IIR digital filters are designed using analog filters described in time domain or transform domain; then analog filters can be converted to digital

* This research work was supported by CNCSIS project number ID 162/2008.

R. Moreno-Díaz et al. (Eds.): EUROCAST 2009, LNCS 5717, pp. 534–539, 2009.

filters using appropriate transformation from s-domain to z-domain [4]. However many engineers avoid mathematical descriptions like transfer functions and for them analog filters mean certain circuits or a netlist of components, and digital filters are a set of statements in certain software. Although for digital filters hardware implementation is one of main choices, a set of statements in software must be written in designing or implementation phase of any digital filter. In this paper we will show how to design digital IIR filters from analog filters from a given netlist, by skipping the transfer function description.

The paper is organized as follows. First we will present our motivation (Section 2) and then we provide a short description of SPICE simulation program (Section 3). The proposed approach and some examples are presented in Section 4. Few comments are also provided (Section 5).

2 Motivation

In the case of IIR filter, the most often used design method is to convert the digital filter specifications into analog lowpass prototype filter specifications, to determine the analog lowpass filter transfer function $H_a(s)$ meeting these specifications and then to transform it into the desired digital filter transfer function $H(z)$ [5]. This approach has been widely used for several reasons as follows:

- analog approximation techniques are highly advanced;
- they usually yield closed-form solutions;
- extensive tables are available for analog filter design;
- many applications require the digital simulation of analog filters.

The basic idea behind the conversion of an analog prototype transfer function $H_a(s)$ into a digital IIR transfer function $H(z)$ is to apply a mapping from the s-domain to the z-domain so that the essential properties of the analog frequency response are preserved. This implies that the mapping function should be such that the imaginary $j\Omega$ axis in the s-plane be mapped onto the unit circle of the z-plane and a stable analog transfer function be transformed into a stable digital transfer function [6]. To this end, the most widely used methods are forward approximation, backward approximation, and bilinear transform (trapezoidal approximation). Other popular approaches are matched z-transform, impulse invariance, and step invariance [7].

All these methods use a mathematical description of analog filters, either in time or frequency domain, using concepts like transfer function, frequency response, impulse response, step response etc. However, for many engineers analog filters mean certain circuits or a netlist of components, their connections, and maths are used seldom. Although for digital filters hardware implementation is one of main choices, a set of statements in software must be written in designing or implementation phase of any digital filter [8].

It should be mentioned the recent interest in making any analog device available in a digital implementation. Sometimes the design process is time consuming and expensive, as engineers must know and consider the design in both analog and digital domain. Besides, for an accurate design the frequency specifications must be provided, which is not always the case. Sometimes only the analog circuits diagrams are available. Using the proposed approach, one can use a previous implementation in analog domain and a code may be available rapidly. It may happen that this code in not optimal; we just note that after design process, in any implementation the code is optimized according to DSPs facilities and features.

3 Short Description of SPICE

Most of electronic circuit design is carried out with the aid of a computer aided circuit analysis program such as SPICE (Simulation Program with Integrated-Circuit Emphasis). Actually SPICE is considered as industrial standard for computer-aided circuit analysis for microelectronic circuits. SPICE was initially a research project at the University of California at Berkeley in the late 1960s [9]. Today SPICE is synonymous with analogue computer-aided simulation [10] and the major companies offer a properly SPICE version as part of their analogue products.

SPICE is a general-purpose circuit simulator capable of performing the following types of analysis: nonlinear DC analysis, nonlinear transient analysis, linear AC analysis, temperature analysis, noise analysis, sensitivity analysis, Fourier analysis, Monte Carlo analysis and distortion analysis. Various types of circuits can be analyzed by SPICE; they may contain one or more of the following components: independent and/or dependent voltage and/or current sources, resistors, capacitors, inductors, mutual inductors, transmission lines, operational amplifiers, switches, diodes, bipolar junction transistors, function field effect transistors (JFET), metal-oxide semiconductor transistors (MOS), metal Schottky field effect transistors (MESFET) and digital gates.

SPICE solves digitally continuous time differential equations describing circuits. A circuit to be simulated must be described in a sequence of lines. Each line is either a statement, which describes a single element, or a control line, which set model parameters, measurement nodes, or analysis types. SPICE input file format is as follows [9]:

```
Title Statement
      Circuit Description
            Power Supplies/Signal Sources
            Element Descriptions
            Model Statements
      Analysis Requests
      Output Requests
.END
```

The circuit is described as elements connected between nodes. Elements must be uniquely labelled, and the nodes must be distinctly numbered with nonnegative integers between 0, for ground, and 9999 [10].

4 The Proposed Approach

It is the time to answer to the main question: How to convert an analog filter (directly from its netlist or diagram) to a digital filter (software statements) without using explicitly concepts like transfer function or frequency response?

Analog filters are simulated using digital computers and inside the computer we have only discrete implementation. We can use this discrete implementation to obtain a digital filter. If the analog circuit is described using a netlist [7], we just replace every component by its companion model adding to the modified nodal or tableau equations. If the analog circuit is described using a graphical interface, we replace every icon of analog circuit with the corresponding discrete set of companion model.

An example is presented in Figure 1 and the equations which describes the circuit are:

$$\begin{gathered} -Iab(n) + Ibg(n) + Ibc(n) = 0; \\ -Ibc(n) + Icg1(n) + Icg2(n) = 0; \\ Vb(n) - Va(n) = -R_1 Iab(n); \\ Vc(n) = R_2 Icg2(n); \\ Ibg(n+1) = C_1/h[Vb(n+1) - Vb(n)]; \\ Icg1(n+1) = C_2/h[Vc(n+1) - Vc(n)]; \\ Vc(n) - Vb(n) = -L/h[Ibc(n+1) - Ibc(n)]; \end{gathered} \tag{1}$$

We have 7 equations with 7 unknowns, where Va is the input signal and Vc is the output signal. The first two equations from (1) are Kirchoff Current Laws. The other equations are Branch Equations, using companion models. For every statement in the netlist, we have at least one finite difference equation; here h is the integration step-size. The netlist describes the analog filter, the set of difference equations describes the digital filter.

We need some careful manipulation to solve this system. First we solve the first four equations. The unknowns are $Iab(n)$, $Ibg(n)$ and $Icg1(n)$ and $Icg2(n)$.

$$\begin{gathered} -Iab(n) + Ibg(n) + Ibc(n) = 0; \\ -Ibc(n) + Icg1(n) + Icg2(n) = 0; \\ Vb(n) - Va(n) = -R_1 Iab(n); \\ Vc(n) = R_2 Icg2(n). \end{gathered} \tag{2}$$

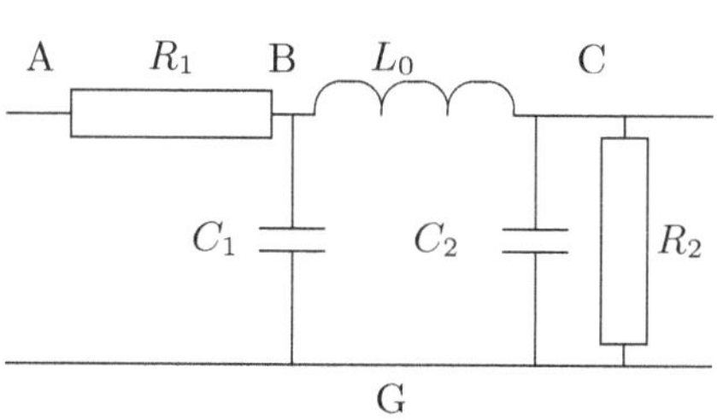

SPICE NETLIST

Va A G
R1 A B
L0 B C
C1 B G
C2 C G
R2 C G
Vc C G

Fig. 1. An example of digital filter obtained from an analog filter

Then we determine $Vb(n+1)$, $Vc(n+1)$ and $Ibc(n+1)$:

$$\begin{aligned} Ibg(n) &= C_1/h[Vb(n+1) - Vb(n)]; \\ Icg1(n) &= C_2/h[Vc(n+1) - Vc(n)]; \\ Vc(n) - Vb(n) &= -L/h[Ibc(n+1) - Ibc(n)]. \end{aligned} \tag{3}$$

One can recognize that the state-space description may be useful in this case. Moreover, this can lead to any canonic structure of digital filter.

Consider now the lowpass filter given in Figure 1 with unit values for the components, and the step-size set at 0.1. The resulting MATLAB code which describes the digital filter is presented in Figure 2. The inputs and the outputs of the digital filter for two frequencies $f_1 = 1/100$ and $f_2 = 51/100$ are shown in Figure 3.

```
clear;close all; N=200;
va=sin(2*pi*1/100*[0:N-1 N]);%va=sin(2*pi*51/100*[0:N-1 N]);
iab=zeros(1,N);ibc=zeros(1,N); icg1=zeros(1,N);icg2=zeros(1,N);
ibg=zeros(1,N);vb=zeros(1,N);vc=zeros(1,N);
R1=1;R2=1;L=1;C1=1;C2=1;h=0.1; A=[-1 1 0 0; 0 0 1 1; 1 0 0 0; 0 0
0 1]; for n=1:N-1
   B=A^(-1)*[-ibc(n) ibc(n) (va(n)-vb(n))/R1 vc(n)/R2]';
   iab(n)=B(1);ibg(n)=B(2);icg1(n)=B(3);icg2(n)=B(4);
   vb(n+1)=h/C1*ibg(n)+vb(n);
   vc(n+1)=h/C2*icg1(n)+vc(n);
   ibc(n+1)=-h/L*(vc(n)-vb(n))+ibc(n);
end
```

Fig. 2. A MATLAB code describing the digital filter

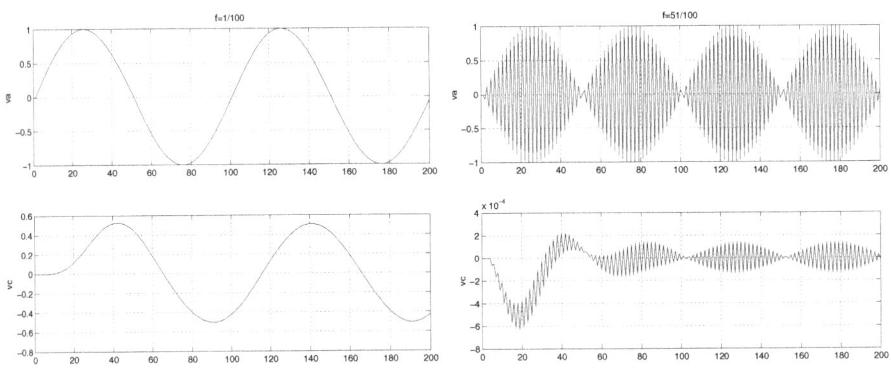

(a) (b)

Fig. 3. The input (top) and the output (bottom) of the digital filter for $f_1 = 1/100$ (a) and $f_2 = 51/100$ (b)

5 Comments and Conclusions

We note that all the process might be performed automatically. A component may have many companion models, resulting in many digital implementations derived from the same analog filter. Usually the simulators of analog filters have sophisticated companion models for components, thus the order of resulting digital filter can be large. As companion models may be nonlinear, digital filter may be a nonlinear in some cases. Since analog filters have infinite impulse response, we obtain only IIR digital filters with this method. The set of difference equations may contain more equations than the canonical form, thus it needs to be simplified. It may happen also that the design process of digital filter is not so straightforward. Moreover, the tuning of parameters may be a problem. However, there are many analog filters in technical literature and they may be speculated to obtain digital filters. The process of deriving the mathematical equations can be overpassed, by replacing any statement in the netlist with corresponding software statement.

To conclude, in this paper we have shown how to obtain a digital filter from a given netlist of an analog filter, by skipping the frequency response or transfer function description. If the analog circuit is described using a netlist, we just replace every component by its companion model. Thus we obtain digital filters directly from analog filters.

References

1. Banzhaf, W.: Computer-Aided Circuit Analysis Using PSpice. Prentice Hall, Englewood Cliffs (1992)
2. Ifeachor, E.C., Jervis, B.W.: Digital Signal Processing - A Practical Approach. Addison-Wesley, Wokingham (1993)
3. Marven, C., Ewers, G.: A Simple Approach to Digital Signal Processing. Texas Instruments (1993)
4. Proakis, J.G., Manolakis, D.G.: Introduction to Digital Signal Processing. Macmillan, New-York (1988)
5. Mitra, S.K.: Digital Signal Processing: A Computer-Based Approach. McGraw-Hill, New York (1998)
6. Oppenheim, A.V., Schafer, R.W.: Discrete-Time Signal Processing. Prentice-Hall, Englewood Cliffs (1989)
7. DeFatta, D.J., Lucas, J.G., Hodgkiss, W.S.: Digital Signal Processing: A System Design Approach. John Wiley & Sons, New-York (1988)
8. Lapsley, P., Bier, J., Shoham, A., Lee, E.A.: DSP Processor Fundamentals - Architectures and Features. Berkeley Design Technology, Inc., Freemont (1994)
9. Vladimirescu, A.: The Spice Book. John Wiley & Sons, Chichester (1993)
10. Tuinenga, P.W.: SPICE A Guide to Circuit Simulation and Analysis Using PSpice. Prentice Hall, Englewood Cliffs (1988)

A Heterogeneous Decision Diagram Package

D. Michael Miller[1] and Radomir S. Stanković[2]

[1] Dept. of Computer Science, Faculty of Engineering
University of Victoria Victoria,
BC, Canada V8W 3P6

[2] Dept. of Computer Science, Faculty of Electronics
University of Niš, 18000 Niš, Serbia

Abstract. This paper describes a decision diagram package for efficient computation with large matrices. The heterogeneous decision diagrams supported by the package do not require the selection variables to have a single domain. Computations can be over a selected field up to the complex numbers including finite fields. Implementation strategies supporting this level of flexibility are presented. Applications are outlined taken from diverse areas such as reversible and quantum logic, spectral transformations and operations over finite fields.

1 Introduction

Reduced ordered binary decision diagrams (ROBDD) were introduced by Bryant [1]. Decision diagram methods are now widely used in conventional logic design [2,3] and very high quality binary decision packages, *e.g.* CUDD [4], are available, Decision diagrams have also been applied in multiple-valued logic [5, 6, 7] and spectral logic [3]. In addition, decision diagrams have been shown to be effective in representing matrices [8] with recent application to reversible and quantum logic [9, 10, 11].

Most decision diagram formulations and implementations assume all selection variables have the same domain and target a particular class of terminal values. Nagayama and Sasao [12] introduced heterogeneous multiple-valued decision diagrams and demonstrated their effectiveness in the representation of logic functions. In their approach, the selection variables can have different domains and by appropriate coding of binary variables as multiple-valued ones they show that this representation is more compact than conventional ROBDD. In this work, we combine this idea with techniques presented in [9] to develop a flexible decision diagram package supporting applications based on matrix operations over selected fields up to the complex numbers.

2 Decision Diagrams

The basic binary decision diagram (BDD) concept was introduced by Lee [13] and further studied and refined by Akers [14]. The full potential of BDD for representation and computation was identified in the seminal work of Bryant [1] who introduced the key concepts of ordering and reduction.

R. Moreno-Díaz et al. (Eds.): EUROCAST 2009, LNCS 5717, pp. 540–547, 2009.

Definition 1. *A* **reduced ordered binary decision diagram** *(ROBDD) is a directed acyclic graph with some number of non-terminal vertices, including a start vertex, and two terminal vertices labeled 0 and 1, respectively. Each non-terminal vertex is labeled by a binary-valued variable and has two outgoing edges labeled 0 and 1, respectively. An ROBDD satisfies the following key criteria:*

- ***ordered:*** *there is a global ordering of the variables such that the variables on every path from the start vertex to a terminal vertex adhere to that ordering and no variable appears more than once on any single path;*
- ***reduced:*** *common subgraphs are shared and redundant nonterminal vertices, those where the 0 and 1 edges point to the same destination, are removed.*

Starting at the root, the value of the Boolean function represented by a ROBDD is found by following the path to a terminal determined by the values of the variables encountered along the path.

Bryant proved the ROBDD representation of a given function is unique up to variable ordering. This unique representation property is critical to the power of ROBDD and extensions to this fundamental structure. A key advantage is that because the representation of each function is unique, common subfunction are represented once and shared in larger representations. Details on the efficient implementation of ROBDD can be found in [15,3,4]. ROBDD can be extended to represent multiple-valued functions [5,6,7]. The major change is that for a p-valued function, each nonterminal vertex has p outgoing edges.

The use of decision diagrams to represent matrices was introduced in [8]. In that work, binary row and column selection variables are used and the matrix becomes a binary input, multi-output function represented by an appropriate decision diagram. The representation can be very compact as common submatrices have shared representations. Recently, this idea has been used in the representation of reversible and quantum functions and circuits [10,11].

The functional behaviour of am n-variable reversible function, gate or circuit can be represented as a $2^n \times 2^n$ permutation matrix. In the quantum case, the matrices are unitary. Quantum multiple-valued decision diagrams (QMDD) [9] were designed to represent these matrices for both the homogeneous binary and homogeneous multiple-valued cases, i.e. where all variables have a common radix. QMDD represent a matrix in a somewhat different manner than the obvious extension of the row and column selection variables used in [8].

Each nonterminal vertex in a QMDD specifies a decomposition of a $p^n \times p^n$ matrix M into p^2 submatrices $M_0, M_1, \ldots, M_{p^2-1}$ of size $p^{n-1} \times p^{n-1}$ as shown in Equation 1.

$$M = \begin{bmatrix} M_0 & M_1 & \ldots & M_{p-1} \\ M_p & M_{p+1} & \ldots & M_{2p-1} \\ \vdots & \vdots & \ddots & \vdots \\ M_{p^2-p} & M_{p^2-p-1} & \ldots & M_{p^2-1} \end{bmatrix} \quad (1)$$

QMDD satisfy Bryant's ordering and reduction criteria (Definition 1).

The key feature of this approach is that each matrix decomposition requires a single vertex, albeit one with p^2 outgoing edges. For the binary case, the vertices have four outgoing edges rather than the two in ROBDD. However as a consequence the depth of a diagram is half of that required by the approach using row and column selection variables with a consequent speedup in operations on the matrices.

In order to take advantage of the structure of the unitary matrices representing quantum functions and circuits, each edge in a QMDD is assigned a complex-valued multiplier α. An edge points to the matrix which is the submatrix rooted by the vertex it points to multiplied by α. The edge multipliers for permutation matrices are all 0 or 1.

The use of edge multipliers requires normalization rules to ensure the uniqueness of the representation of a matrix. The normalization rules used here, Definition 2, were introduced in [9].

Definition 2. *A QMDD is* **normalized** *if (i) there is a single terminal vertex with value 1; (ii) every edge with a multiplier of 0 points directly to the terminal vertex; and (iii) for every nonterminal vertex, the weights* $w_0, w_1, \ldots$ *on the outgoing edges are such that there is a* $w_i = 1$ *where* $w_j = 0, \forall j, 0 \leq j < i$.

3 Heterogeneous Decision Diagrams

The heterogeneous decision diagrams (HDD) introduced in this paper are an extension of QMDD. For HDD, each selection variable, x_i labeling nonterminal vertices has an associated radix r_i. Hence, some nonterminal vertices may partition a matrix into four submatrices, others into nine, and so on. The partitioning rule is the same for all vertices labeled by the same variable. Also while QMDD support arithmetic over the complex-numbers, our implementation of HDD supports operations over various fields including finite fields.

4 Implementation

Our HDD package makes use of standard ROBDD techniques [15, 4] as well as techniques developed for MDD [16] while extending them to account for the heterogeneous nature of the diagrams. We here briefly describe the principal features of the implementation of the HDD package.

Decision diagram packages use dynamic memory allocation for vertices and a tabular approach to ensure the uniqueness of the representation. When a vertex is created, hash-based table lookup is used to see if it already exists in which case the current representation is used. If a vertex is indeed new, it is added to the table. Reference counting for each vertex is used to support garbage collection which can be explicitly invoked by the user or can be applied dynamically by the package when available memory is exhausted. During the garbage collection process, the memory assigned to vertices with zero reference count, and which are thus no longer in use, can be recovered for further use. The difference for HDD is that the memory required for a vertex depends on the radix of the associated variable since the number of outgoing edges is the square of the radix. A separate unique table is used for each variable and each vertex recovered during garbage collection is put on an available space list for the appropriate radix.

The most novel aspect of the HDD package is how it handles number representation and arithmetic. As noted above, each variable has an associated radix. Each variable also has an associated arithmetic mode which can specify arithmetic over a field up to the complex numbers including modular arithmetic over a finite field. The support of arithmetic is modularized so that it can be readily extended to accommodate additional arithmetic modes.

Numbers, including the real and imaginary part of a complex number, are stored in progressing complexity as integers, rationals, quadratic irrationals or floating point (long double) as required. A rational a/b is represented as two integers a and b. A quadratic irrational is a number of the form $\frac{a+b\sqrt{2}}{c}$ where a, b and c are integers. The package automatically chooses the 'best' representation as computation proceeds. The goal is to maintain as accurate a representation as possible. Rationals and quadratic irrationals are kept in canonic form using standard greatest common divisor techniques. Round-off problems only occur when floating point is used, so it is used only when necessary. The standard technique of using a tolerance, which can be set by the user, is used when comparing floating point numbers.

The conversion amongst the representations is straightforward except for conversions from floating point. For example, a quadratic irrational can be expressed as a rational if $b = 0$ and as an integer if $c = 1$ as well. Converting a floating point number to an integer involves a test that the number is no further from the integer than the floating point tolerance. Strictly speaking, every floating point number can be expressed as a rational number. We use a standard continued fraction technique [17] to find the best rational approximation of a floating point value within the floating point tolerance and convert the floating point number to the rational number unless one or both of the integers is too large and would cause future computational problems. Conversion from floating point to a quadratic irrational is not implemented.

5 Applications

We outline a few applications showing the diversity of HDD and the supporting implementation.

Since HDD are an extension of QMDD, they can be applied to all problems suitable for QMDD. This includes the representation and manipulation of the permutation and unitary matrices required for reversible and quantum gates and circuits [9, 7, 18]. A minor performance penalty, typically less than 5%, is incurred when HDD are used because of the overhead in the flexible number and arithmetic support. On the other hand, HDD can directly handle circuits with both binary and multiple-valued values.

QMDD have been used in applications such as equivalence checking of reversible circuits composed of reversible and elementary quantum (square-root-of-NOT) gates [18]. Representations for quite large circuits, including circuits with thousands of gates and a reversible 64-bit adder with 193 variables, are constructed and compared in less than a minute of CPU time and often in just a few seconds.

Miller and Thornton [19] considered the use of QMDD for computing the binary Rademacher-Walsh and Reed-Muller transforms, and the multiple-valued Chrestenson transform. There is an advantage in using the HDD package in this area because it will perform the computations using the most accurate number representation. We here briefly outline the approach and show that HDD can also be used for the non-Kronecker product discrete Haar transform.

The Rademacher-Walsh, Reed-Muller and Chrestenson transforms are Kronecker product-based transforms of the form $S = T^n F$ where $T^n = \otimes_{i=1}^{n} T^1$ with

Rademacher-Walsh	Reed-Muller	Chrestenson
$T^1 = \begin{bmatrix} 1 & 1 \\ 1 & -1 \end{bmatrix}$	$T^1 = \begin{bmatrix} 1 & 0 \\ 1 & 1 \end{bmatrix}$ computation over $GF(2)$	$T^1 = \begin{bmatrix} 1 & 1 & 1 \\ 1 & a^2 & a \\ 1 & a & a^2 \end{bmatrix}, a = e^{-\frac{2}{3}\pi i}$

F is the truth vector of the function which is often coded as $+1$ for logic 0 and -1 for logic 1 for the Rademacher-Walsh case, and is coded as $a^p, 0 \leq p \leq r$, for the Chrestenson case. S is called the spectrum of the function.

The HDD for each of these transform matrices has one vertex for each variable as shown for $n = 3$ in Figure 1. Note that the values shown on the edges are the weights. The edges from each nonterminal vertex are ordered $e_0, e_1, e_2, \ldots$ from left to right. For the Chrestenson case the weights from left to right for each nonterminal vertex are $1, 1, 1, 1, a^2, a, 1, a, a^2$.

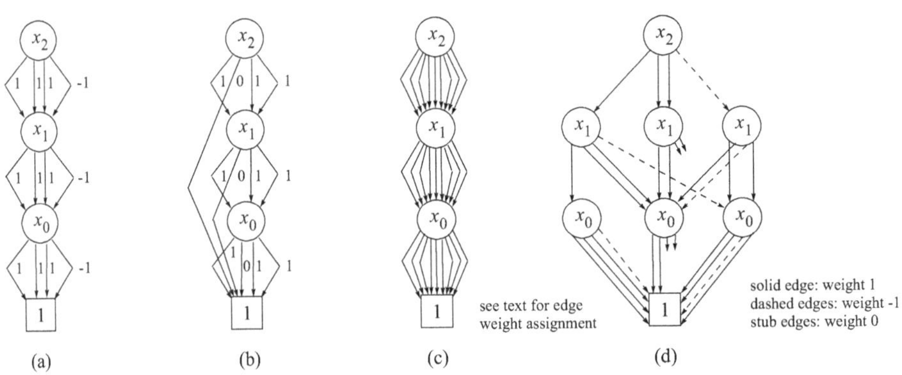

Fig. 1. (a) R-W (b) R-M (c) Chrestenson and (d) discrete Haar transforms ($n = 3$)

The discrete Haar transform is not a Kronecker product-based transform but it can be conveniently decomposed as

$$H^n = \begin{bmatrix} H^{n-1} \otimes \begin{bmatrix} 1 & 1 \end{bmatrix} \\ I^{n-1} \otimes \begin{bmatrix} 1 & -1 \end{bmatrix} \end{bmatrix}, \; H^0 = [1] \tag{2}$$

where I^{n-1} is the $2^{n-1} \times 2^{n-1}$ identity matrix. In general, the HDD for H^n has $3n - 2$ nonterminal vertices. The case for $n = 3$ is shown in Figure 1 (d).

Multiplication of matrices represented as HDD (or QMDD) uses Bryant's apply operation [1] (see [9] for details). This has complexity $O(|G_1| \times |G_2|)$. As shown above, the four transforms considered here have size $O(n)$. The worst case size to represent an p-valued function as a decision diagram is $O(p^n)$. Hence performing the considered spectral transformations using HDD has complexity $O(np^n)$ which is the same as for the well-known fast transform techniques.

The study reported in [19] shows that the decision diagram approach in practice requires far fewer operations than fast transform techniques, typically less than 5%. For well-structured functions, the space requirement is also far less, but the space for a decision diagram becomes much higher than for fast transform techniques when random functions are considered.

As a simple example of using computation over $GF(3)$, consider the matrices:

$$G = \begin{bmatrix} 1 & 0 & 0 \\ 1 & 1 & 1 \\ 1 & 2 & 1 \end{bmatrix}, \quad G^{-1} = \begin{bmatrix} 1 & 0 & 0 \\ 0 & 2 & 1 \\ 2 & 2 & 2 \end{bmatrix} \tag{3}$$

The HDD structure for $G \otimes G$ is straightforward as shown in Figure 2 (a). Figure 2 (b) shows the HDD for $G^{-1} \otimes G^{-1}$. In this case, it is clear how to derive the HDD for an inverse, but no efficient general HDD-based method is known.

Consider matrix D in Equation 4 which is a numeric encoding for the operation table of the group $G = C_2 \times C_3 \times C_3$ where $\times$ denotes Cartesian product and C_p defines addition over $GF(p)$ with the group identity encoded as 0. The HDD for D has 10 nonterminal vertices and rational number edge weights.

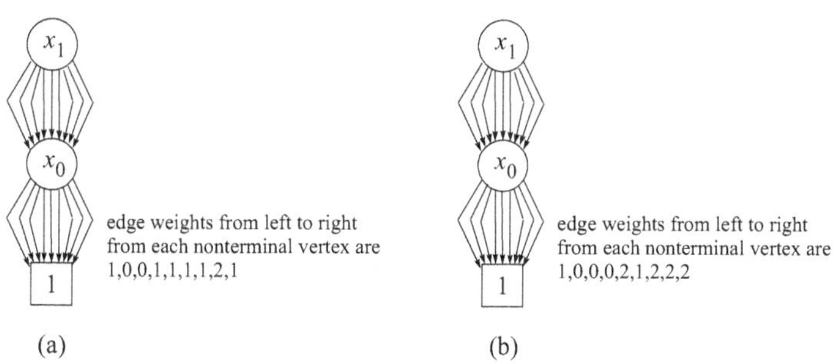

Fig. 2. HDD for (a) $G \otimes G$ and (b) $G^{-1} \otimes G^{-1}$ for G in Equation. 3

$$D = \begin{bmatrix}
0 & 1 & 2 & 3 & 4 & 5 & 6 & 7 & 8 & 9 & 10 & 11 & 12 & 13 & 14 & 15 & 16 & 17 \\
1 & 2 & 0 & 4 & 5 & 3 & 7 & 8 & 6 & 10 & 11 & 9 & 13 & 14 & 12 & 16 & 17 & 15 \\
2 & 0 & 1 & 5 & 3 & 4 & 8 & 6 & 7 & 11 & 9 & 10 & 14 & 12 & 13 & 17 & 15 & 16 \\
3 & 4 & 5 & 6 & 7 & 8 & 0 & 1 & 2 & 12 & 13 & 14 & 15 & 16 & 17 & 9 & 10 & 11 \\
4 & 5 & 3 & 7 & 8 & 6 & 1 & 2 & 0 & 13 & 14 & 12 & 16 & 17 & 15 & 10 & 11 & 9 \\
5 & 3 & 4 & 8 & 6 & 7 & 2 & 0 & 1 & 14 & 12 & 13 & 17 & 15 & 16 & 11 & 9 & 10 \\
6 & 7 & 8 & 0 & 1 & 2 & 3 & 4 & 5 & 15 & 16 & 17 & 9 & 10 & 11 & 12 & 13 & 14 \\
7 & 8 & 6 & 1 & 2 & 0 & 4 & 5 & 3 & 16 & 17 & 15 & 10 & 11 & 9 & 13 & 14 & 12 \\
8 & 6 & 7 & 2 & 0 & 1 & 5 & 3 & 4 & 17 & 15 & 16 & 11 & 9 & 10 & 14 & 12 & 13 \\
9 & 10 & 11 & 12 & 13 & 14 & 15 & 16 & 17 & 0 & 1 & 2 & 3 & 4 & 5 & 6 & 7 & 8 \\
10 & 11 & 9 & 13 & 14 & 12 & 16 & 17 & 15 & 1 & 2 & 0 & 4 & 5 & 3 & 7 & 8 & 6 \\
11 & 9 & 10 & 14 & 12 & 13 & 17 & 15 & 16 & 2 & 0 & 1 & 5 & 3 & 4 & 8 & 6 & 7 \\
12 & 13 & 14 & 15 & 16 & 17 & 9 & 10 & 11 & 3 & 4 & 5 & 6 & 7 & 8 & 0 & 1 & 2 \\
13 & 14 & 12 & 16 & 17 & 15 & 10 & 11 & 9 & 4 & 5 & 3 & 7 & 8 & 6 & 1 & 2 & 0 \\
14 & 12 & 13 & 17 & 15 & 16 & 11 & 9 & 10 & 5 & 3 & 4 & 8 & 6 & 7 & 2 & 0 & 1 \\
15 & 16 & 17 & 9 & 10 & 11 & 12 & 13 & 14 & 6 & 7 & 8 & 0 & 1 & 2 & 3 & 4 & 5 \\
16 & 17 & 15 & 10 & 11 & 9 & 13 & 14 & 12 & 7 & 8 & 6 & 1 & 2 & 0 & 4 & 5 & 3 \\
17 & 15 & 16 & 11 & 9 & 10 & 14 & 12 & 13 & 8 & 6 & 7 & 2 & 0 & 1 & 5 & 3 & 4
\end{bmatrix} \quad (4)$$

Matrix addition, multiplication and Kronecker product for QMDD were discussed in [9]. Those approaches extend directly to HDD. Implementation of Cartesian product is new to this work and is based on the following construction illustrated for the binary case (n_B is the number of variable for B):

$$\begin{aligned} A \times B &= a||B|| + B \;\text{ if }\; A = [a],\;\; ||B|| = 2^{n_B} \\ &= \begin{bmatrix} A_0 \times B & A_1 \times B \\ A_2 \times B & A_3 \times B \end{bmatrix} \;\text{ otherwise} \end{aligned} \quad (5)$$

As with all DD approaches, HDD are quite sensitive to variable order. For example, consider a group composed as the Cartesian product of $\lfloor \frac{n}{2} \rfloor$ occurrences of C_3 and $\lceil \frac{n}{2} \rceil$ occurrences of C_2. Table 1 shows the number of nonterminal vertices for increasing values of n and three plausible variable orderings. Placing the binary (C_2) variables at the top yields the smallest size HDD.

Table 1. Effect of variable ordering on HDD size for various C_2, C_3 compositions

n	C_2 at Top	Ratio	C_2 at Bottom	Ratio	Alternating	Ratio
2	4		5		4	
3	8	2.0	11	2.2	10	2.5
4	20	2.5	32	2.9	22	2.2
5	40	2.0	68	2.1	58	2.6
6	112	2.8	203	3.0	130	2.2
7	224	2.0	419	2.1	346	2.7
8	656	2.9	1256	3.0	778	2.3

6 Conclusions

HDD can support a variety of matrix based techniques in a broad spectrum of potential application areas. The HDD package will support the exploration of new transforms by avoiding time-consuming and often difficult prototype software development.

A full formal analysis of HDD and their implementation is greatly complicated by the use of computed and computation tables. Experience to date, including the results presented above, indicate the approach is quite efficient if there is reasonable structure in the functions and matrices being considered.

Ongoing work includes optimizing the implementation and adding a better user interface along the lines used for the QuIIDPro package [10], We are also interested in characterizing the representative capabilities of HDD, and further studying the effect of variable ordering on HDD size.

References

1. Bryant, R.: Graph-based algorithms for Boolean function manipulation. IEEE Trans. on Computers 35(8), 677–691 (1986)
2. Drechsler, R., Becker, B.: Binary Decision Diagrams - Theory and Implementation. Kluwer Academic Publishers, Dordrecht (1998)
3. Yanushkevich, S., Miller, D.M., Shmerko, V., Stanković, R.: Decision Diagram Techniques for Micro- and Nanoelectronic Design Handbook. CRC/Taylor & Francis (2006)
4. Somenzi, F.: CUDD: CU decision diagram package - release 2.4.2 (2009), `http://vlsi.colorado.edu/~fabio/CUDD/cuddIntro.html`
5. Srinivasan, A., Kam, T., Malik, S., Brayton, R.: Algorithms for discrete function manipulation. In: Proc. Int'l Conf. on CAD, pp. 92–95 (1990)
6. Kam, T., Villa, T., Brayton, R., Sangiovanni-Vincentelli, A.: Multi-valued decision diagrams for logic synthesis and verification. Multiple Valued Logic - An International Journal, 9–24 (1998)
7. Miller, D.M., Thornton, M.A.: Multiple-Valued Logic: Concepts and Representations. Morgan and Claypool (2008)
8. Clarke, E., Fujita, M., McGeer, P., McMillan, K., Yang, J., Zhao, X.: Multi terminal binary decision diagrams: An efficient data structure for matrix representation. In: Proc. Int'l Workshop on Logic Synthesis, pp. P6a:1–15 (1993)
9. Miller, D., Thornton, M.: QMDD: A decision diagram structure for reversible and quantum circuits. In: Proc. Int'l Symp. on Multiple-valued Logic (CD), 6 pages (2006)
10. Viamontes, G., Markov, I., Hayes, J.: QuIDDPro: High-performance quantum circuit simulation V. 3.1 (2007), `http://vlsicad.eecs.umich.edu/Quantum/qp/`
11. Wang, S.A., Lu, C.Y., Tsai, I.M., Kuo, S.Y.: An XQDD-based verification method for quantum circuits. IEICE Trans. on Fundamentals E91-A(2), 584–594 (2008)
12. Nagayama, S., Sasao, T.: Compact representations of logic functions using heterogeneous MDDs. In: Proc. Int'l Symp. on Multiple-valued Logic, pp. 247–252 (2003)
13. Lee, C.: Representation of switching circuits by binary decision diagrams. Bell System Technical Journal 38, 985–999 (1959)
14. Akers, S.: Binary decision diagrams. IEEE Trans. on Computers 27(6), 509–516 (1978)
15. Brace, K., Rudell, R., Bryant, R.: Efficient implementation of a BDD package. In: Proc. Design Automation Conf., pp. 40–45 (1990)
16. Miller, D.M., Drechsler, R.: Implementing a multiple-valued decision diagram package. In: Proc. Int'l Symp. on Multiple-valued Logic, pp. 52–57 (1998)
17. Pheatt, C.: C code snippets (2009), `http://pheatt.emporia.edu/courses/2008/cs260s08/snippets/Code/code.html`
18. Wille, R., Große, D., Miller, D.M., Drechsler, R.: Equivalence checking of reversible circuits. In: Proc. Int'l Symp. on Multiple-valued Logic (CD), pp. 324–330 (2009)
19. Miller, D.M., Thornton, M.A.: QMDD and spectral transformation of binary and multiple-valued functions. In: Workshop on Boolean Problems, pp. 137–144 (2008)

Walsh Matrices in the Design of Industrial Experiments*

Claudio Moraga[1,2] and Héctor Allende[3,4]

[1] European Centre for Soft Computing, 33600 Mieres, Spain
[2] Dortmund University of Technology, 44221 Dortmund, Germany
[3] Technical University Federico Santa María, Valparaíso, Chile
[4] University Adolfo Ibáñez Viña del Mar, Chile

Abstract. Discrete Walsh functions are well known in digital signal processing, telecommunications, and logic design. In this paper we show that they also appear "naturally" in matrices representing forms of interaction among different factors involved in the design of industrial experiments.

1 Introduction

J.L. Walsh introduced in 1923 a set of orthogonal functions [17] that carry his name, as the multiplicative closure of the Rademacher functions [13] disclosed a year earlier. The Walsh functions constitute a complete orthogonal system for the Hilbert space $L_2[0,1)$. Therefore every square-integrable function $f : [0,1) \to \mathbb{R}$ has a Walsh-Fourier Series representation [11]. Furthermore, Walsh functions have been identified as character functions of the dyadic group D, which is defined by the set of 0-1 sequences $\{(x_1, x_2, \ldots)|x_i \in \{0,1\}\}$ with the componentwise addition modulo 2 as group operation [7], [15]. Even though there has been substantial work on the continuous Walsh functions (see e.g. [16], [14]), the discrete Walsh functions have become very well known in the digital world for their applications in telecommunications [8], signal processing [1], [12] and logic design [6], [9]. Discrete Walsh functions have appeared in the literature under different orderings. Among the most important, besides the one originally given by Walsh [17], which is known as "sequency ordering", there is the one introduced by Paley [10], which is directly related to the product of Rademacher functions and the Walsh-Hadamard ordering [9], which is characterized by a Kronecker product structure. Readers looking for more information may like to see [4] for an extended literature review.

F. Pichler reported [12] that discrete Walsh functions in sequency order, appear "quite naturally" in the layout of telephone lines to minimize cross modulation effects. (In Germany, this was known as the "Kreuzungsplan von Pinkert", named after a German telephone engineer around the year 1880.) The present

* Work leading to this paper was partially supported by the Foundation for the Advance of Soft Computing, Mieres, (Spain), and by a Fondecyt 1070220 Research Grant (Chile).

R. Moreno-Díaz et al. (Eds.): EUROCAST 2009, LNCS 5717, pp. 548–554, 2009.

paper shows that they also happen to appear in the design of experiments, according to the method introduced in [3]. An experiment can be defined broadly as an act of observation. This work however, uses a more restrictive definition, and states that an experiment is a series of trials or tests which produce quantifiable outcomes (response variable). In experimental design, the first step is to decide (based on the goals of the experiment) to what factors and alternatives the experimental units are to be exposed, and what project parameters are to be set. Then it should be examined whether any of the parameters cannot be kept at a constant value and account for any undesired variation. Finally, it should be chosen which response variables are to be measured and which should be the subject of complementary experiments.

Experiments generally recommendable in the industrial area are called "factorial" [3], [2]. The goal of these experiments is to determine the effect of variations, interaction and extreme values in the parameters which characterize the industrial process under observation. According to [3], parameters are called "factors" and their types or values are called "levels". For this reason it is usual to speak of "factorial design of experiments". A factorial design uses every possible combination of the alternatives of all the factors. It thus, discovers the effects of each factor and its interactions with the other ones. For instance, if an experiment is to be designed to determine the effect of certain parameters on the hardness of some building blocks, one of the factors could be the kind of cement used in its production and the levels could be "from volcanic ashes" or "from scoria of the iron and steel industry". Notice that even though factors and levels may be nominal categories, their effect will be measured, giving a numerical value. This allows the development of an abstract model, which highlights the interaction among factors and their effect on the system. On the other side, one of the disadvantages of factorial experiments from a practical point of view, is the fact that the number of treatment combinations increases rapidly as the number of factor and/or levels increases. In this case, a fraction of the full factorial design can be used. Fractional factorial designs save time and money but provide less information than the full designs.

2 Design of Industrial Experiments [2], [3]

The simplest experiments consider factors with (only) two levels, which may represent extreme values or selected categories. These experiments allow a fast (preliminary) diagnosis on the effects of the factors upon the performance of the system.

Consider an experiment which involves two factors with two levels each. Let A and B denote the factors and a_1, a_2, b_1, b_2 denote the respective levels. If an experiment is conducted with factor A at level a_1 and factor B at level b_2, this will be denoted as an a_1b_2 run of the experiment. Let r_{ij} denote the *response* of the system under test to the factors A and B with levels a_i and b_j respectively. With this notation the following combined effects may be calculated:

i) the global average effect $= \frac{1}{4}[r_{11} + r_{12} + r_{21} + r_{22}]$
ii) the slope of A $= \frac{1}{2}[(r_{21} + r_{22}) - (r_{11} + r_{12})]$
iii) the slope of B $= \frac{1}{2}[(r_{12} + r_{22}) - (r_{11} + r_{21})]$
iv) the interaction of A and B $= \frac{1}{2}[(r_{22} - r_{21}) - (r_{12} - r_{11})]$

Let X denote a factor with levels x_2 and x_1. It will be agreed that x_2 is *higher* than x_1, where in the case of levels with numerical values, this ordering is taken from $\mathbb{R}$; meanwhile if the levels are nominal categories, the ordering will be defined arbitrarily (but used consistently). The slope of a factor X is a numerical value obtained as the difference between the average response of the system when this factor is at the level x_2 and the average response of the system when this factor is at the level x_1. The higher the value of the slope, the stronger is the effect of the corresponding parameter upon the system. The interaction between two factors is obtained as the difference between the average response when the levels of both factors are of the same type (both high or both low) and the average response when the levels of both factors are of opposite type.

If the following notation is used: 1 for the global average effect; A for the slope of A; B for the slope of B; and AB for the interaction of A and B, the structure of the combined effects may be summarized in Table 1, where the rows correspond to the pairs of levels being considered at the input factors, and the columns contain the signs to calculate the effects based on the corresponding responses. Using the traditional approach of varying one factor at a time, four tests may be performed at the levels indicated in Table 1.

Table 1. Reproduction of Table 2.3 from [3] with rows in reverse order

Combination of level of factors	Effect 1	A	B	AB
a_2b_2	+	+	+	+
a_1b_2	+	-	+	-
a_2b_1	+	+	-	-
a_1b_1	+	-	-	+

The sign table of an experimental design may be directly built as follows: Assign the sign (+) to one of the levels of each factor and the sign (-) to the other. It does not matter which level is chosen for each sign. Build a table with one column for factors and other column per combination of factors. The table rows are obtained as follows. For one factor column, every row corresponds to a given combination of (+) and (-) values for the respective level alternatives. The set of all the rows contains all the combination of the alternatives of all the factors. For the factor-combination columns, the entries correspond to the pointwise multiplication of the signs of the corresponding one-factor columns. For example, each row in column AB will be filled in by multiplying the corresponding signs of columns A and B.

If in the Table 1 the entries with "+" are interpreted as "+1" and those with "-", as "-1", it is easy to recognize the matrix W_4 representing the Walsh functions on 4 points, in Hadamard ordering. It becomes apparent the Kronecker structure showing that $W_4 = W_2 \otimes W_2$. Furthermore, the labels of the columns are obtained as $[1\ B] \otimes [1\ A]$.

Example 1. For the production of wood chipboards, wood shavings are glued with special adhesives. Hardness is a relevant feature of a chipboard. It is assumed that the granularity of wood shavings and the type of adhesive being used affect the hardness of the final product. The following experiment is conducted, where the response is the hardness of the chipboard measured as the weight placed at the center of the board, needed to produce a 5 mm span.

Factors	Levels	
A: Type of adhesive	a_2 :	X.45
	a_1 :	W.75b
B: Granularity of wood savings	b_2 :	Rough
	b_1 :	Fine

Combined Factors Levels	Response (Kg)
a_2b_2	$r_{22} = 23$
a_1b_2	$r_{12} = 10$
a_2b_1	$r_{21} = 17$
a_1b_1	$r_{11} = 16$

Therefore, the global average effect is $\frac{1}{4}[r_{11}+r_{12}+r_{21}+r_{22}] = \frac{1}{4}[23+10+17+16] = 16.5$

The slope of A is given by $\frac{1}{2}[(r_{21}+r_{22})-(r_{11}+r_{12})] = \frac{1}{2}[(17+23)-(16+10)] = 7$. This value indicates that there is an effect of factor A. The fact that this value is positive indicates that the effect increases as the factor changes from level a_1 to level a_2. Similarly, the slope of B, expressed as $\frac{1}{2}[(r_{12}+r_{22})-(r_{11}+r_{21})]$ gives $\frac{1}{2}[(10+23)-(16+17)] = 0$, meaning that this factor -(alone)- does not affect the hardness of the chipboard; however the interaction between A and B, calculated as $\frac{1}{2}[(r_{22}-r_{21})-(r_{12}-r_{11})]$ gives $\frac{1}{2}[(23-17)-(10-16)] = 6$, meaning that B interacts with A. It is easy to see that if B is kept at b_2, a change of A from a_1 to a_2 produces an increment of 13 Kg in the hardness, meanwhile if B is kept at b_1, a change of A from a_1 to a_2 only increases that hardness by 1 Kg.

The fact that the Walsh matrix appears in experiments with two factors and two levels is not a "simple coincidence", since the structure carries on to experiments with 3 factors and two levels. Table 3.2 of [3] -(with reverse ordering of the rows)- shown in Table 2, corresponds to W_8, which is $W_2 \otimes W_2 \otimes W_2$.

Table 2. Table 3.2 of [3] with reverse ordering of the rows

	1	A	B	AB	C	AC	BC	ABC
$a_2b_2c_2$	+	+	+	+	+	+	+	+
$a_1b_2c_2$	+	-	+	-	+	-	+	-
$a_2b_1c_2$	+	+	-	-	+	+	-	-
$a_1b_1c_2$	+	-	-	+	+	-	-	+
$a_2b_2c_1$	+	+	+	+	-	-	-	-
$a_1b_2c_1$	+	-	+	-	-	+	-	+
$a_2b_1c_1$	+	+	-	-	-	-	+	+
$a_1b_1c_1$	+	-	-	+	-	+	+	-

If a more detailed analysis is required, the number of levels may be increased to 3. For every factor, two types of comparisons are made: (i) only the difference of the responses to the extreme levels are considered and the response to the middle level is assigned a weight of 0, and (ii) the difference of twice the response to the middle level is calculated against the sum of the responses with respect to the extreme levels. (Notice that this corresponds to comparing the response to the middle level with the average response to extreme levels, scaled by 2, to operate

Table 3. Design Table for experiments with 2 factors and 3 levels each (The double lines frame the blocks of the Kronecker product)

Factor	Efects								
levels	1	A_1	A_2	B_1	A_1B_1	A_2B_1	B_2	A_1B_2	A_1B_2
a_3b_3	1	1	1	1	1	1	1	1	1
a_2b_3	1	0	-2	1	0	-2	1	0	-2
a_1b_3	1	-1	1	1	-1	1	1	-1	1
a_3b_2	1	1	1	0	0	0	-2	-2	-2
a_2b_2	1	0	-2	0	0	0	-2	0	4
a_1b_2	1	-1	1	0	0	0	-2	2	-2
a_3b_1	1	1	1	-1	-1	-1	1	1	1
a_2b_1	1	0	-2	-1	0	2	1	0	-2
a_1b_1	1	-1	1	-1	1	-1	1	-1	1

with integers.) For the general structure, let the factors be A and B, with levels a_1, a_2, a_3 and b_1, b_2, b_3 , respectively. Order the columns of the Design Table after $[1\ B_1\ B_2] \otimes [1\ A_1\ A_2]$, where B_1 denotes the column for the effect of parameter B following the comparison (i) and B_2 denotes the column for the effect of parameter B following the comparison (ii). Similarly for A_1 and A_2. Under these constraints, the corresponding Design Table is Table 3. It may be seen that the entries of the table, interpreted as a matrix, do not represent a Chrestenson matrix [5], which may be considered as a ternary extension of the Walsh matrix. The Chrestenson matrix has complex-valued entries and this could not adequately be interpreted for the comparison of effects of factors and levels upon the responses. The matrix, however exhibits a Kronecker structure. It is not difficult to recognize that it represents the Kronecker product of the matrix

$$\begin{bmatrix} 1 & 1 & 1 \\ 1 & 0 & -2 \\ 1 & -1 & 1 \end{bmatrix}$$

with itself.

The Kronecker structure of the matrices corresponding to the Design Tables speaks for the scalability of the method. Increasing the number of factors -(with the same number of levels)- being considered, requires only the corresponding increased "Kronecker power" -(products with itself)- of the basic matrix. On the other hand, it becomes apparent that the Design Table will grow exponentially on the number of levels. This complexity growth may be alleviated by using *fractional* experiments, where a subset of factors is kept at some preselected levels meanwhile the others vary. This will lead to corresponding *fractional* Design Tables of tractable size. (See chapter 4 of [3])

The Kronecker structure of the matrices corresponding to the Design Tables also adds flexibility to the design of experiments, to consider factors with different number of levels. For instance, consider an experiment which involves a factor A with levels a_1, a_2, a_3 and a factor B, with levels b_1 and b_2. Then the Design Table will be obtained based on the Kronecker product

Table 4. Design Table for an experiment with a factor with two levels and a factor with three levels

	1	A_1	A_2	B	A_1B	A_2B
a_3b_2	1	1	1	1	1	1
a_2b_2	1	0	-2	1	0	-2
a_1b_2	1	-1	1	1	-1	1
a_3b_1	1	1	1	-1	-1	-1
a_2b_1	1	0	-2	-1	0	2
a_1b_1	1	-1	1	-1	1	-1

$$\begin{bmatrix} 1 & 1 \\ 1 & -1 \end{bmatrix} \otimes \begin{bmatrix} 1 & 1 & 1 \\ 1 & 0 & -2 \\ 1 & -1 & 1 \end{bmatrix}$$

As shown in Table 4.

3 Conclusions

The basic idea for the design of industrial experiments following the method of [3] has been presented. In the case of experiments involving factors with two levels, it is shown that Walsh matrices in Hadamard ordering naturally appear as the structure of the Design Table. Other Kronecker structured matrices may be used to generate Design Tables for experiments involving factors with three levels or experiments where the factors have a different number of levels. The focus of design of experiments, has been the analysis of data from a standard design. It is usual to refer to any factorial design in which one o more level combinations are excluded from the study as a fractional factorial design. To recommend a fraction for a particular problem, is usually hard and it is often impossible to identify the "Best fraction". Kronecker structured design matrices may provide some basic support, since their blocks may be associated to possible fractional factorial experiments.

References

1. Ahmed, N., Rao, K.R.: Orthogonal Trasnsforms for Digital Signal Processing. Springer, Heidelberg (1975)
2. Allende, H., Bravo, D., Canessa, E.: Robust design in multivariate systems using genetic algorithms. Quality and Quantity (to appear, 2009)
3. Allende, H., Canessa, E., Galbiati, J.: Diseño de experimentos industriales. UTFSM, Valparaíso, Chile (2005)
4. Bramhall, J.N.: Bibliography on Walsh and Walsh related functions. IEEE, Los Alamitos (1974)
5. Chrestenson, H.E.: A class of generalized walsh functions. Pacific Jour. Mathematics 5, 17–31 (1955)

6. Edwards, C.: The application of the Rademacher-Walsh transform to digital circuit synthesis. In: Proceedings Theory and Applications of Walsh Functions. The Hatfield Polytechnic (1973)
7. Fine, N.J.: On the Walsh functions. Trans. of the American Mathematical Society 65, 372–414 (1949)
8. Harmuth, H.F.: Radio comunication with orthogonal time functions. Transactions AIEE Comunications and Electronics 79, 221–228 (1960)
9. Moraga, C., Heider, R.: A tutorial review on applications of the Walsh transform in swiching theory. In: Proc. 1st International Workshop on Trasnforms and Filter Banks, pp. 494–512. T.U. Tampere, Findland (1998)
10. Paley, R.E.: On orthogonal matrices. Jr. Math. Phys. 12, 311–320 (1933)
11. Pichler, F.: Walsh functions. Introduction to theory. In: Proc. Nato Advanced Study Institute on Signal Processing. Loughbourough, UK (1972)
12. Pichler, F.: Some historical remarks on the theory of Walsh functions and their applications in information engineering. In: Butzer, P.L., Stanković, R.S. (eds.) Theory and Applications of Gibss Derivatives. Press Matematiki Institut, Belgrade (1989)
13. Rademacher, H.: Einige saetze der reihen von allgemeinen orthogonalfunktionen. Math. Ann. 87, 112–138 (1922)
14. Stanković, R.S.: Walsh and Dyadic Analysis. In: Proceeding of the Workshop. Press Faculty of Electronics, University of Niš, Serbia
15. Vilenkin, N.J.: On a class of complete orthogonal systems (in Russian). Izv. AN, SSSR, Ser. Matem. 11, 363–400 (1947)
16. Wade, W.: L^r inequalities for Walsh series $0 < r < 1$. Acta Sci. Math. 46, 233–241 (1983)
17. Walsh, J.L.: A close set of orthogonal functions. American Jour. of Mathematics 45, 5–24 (1923)

Dynamic Behavior of Time-Domain Features for Prosthesis Control

Stefan Herrmann and Klaus J. Buchenrieder

Institut für Technische Informatik
Universität der Bundeswehr München
D-85579, Neubiberg, Germany
{stefan.herrmann,klaus.buchenrieder}@unibw.de

Abstract. Myoelectric hand-prostheses are used by patients with either above- or below-elbow amputations and actuated with a minimal microvolt-threshold myoelectric signal (MES). Prehensile motions or patterns are deduced from the MES by classification. Current approaches act on the assumption, that MES is adiabatic-invariant and unaffected by fatigue of contributory muscles. However, classifiers fail on the onset of muscle fatigue and cannot distinguish between voluntary-, submaximal-contraction and an intentional release of muscle tension. As a result, patients experience a gradual loss of control over their prostheses. In this contribution we show, that the probability distributions of extracted time- and frequency-domain features are fatigue dependent with regard to locality, skewness and time. Also, we examine over which time-frame, established classifiers provide unambiguous results and how classifiers can be improved by the selection of a proper sampling-window size and an appropriate threshold for select features.

Keywords: Myoelectric Signal Processing, Upper Limb Prosthesis, Muscle Fatigue, Multinormal Distribution, Guilin-Hills Selection.

1 Introduction

Amputees can generate repeatable, but gradually varying MES during muscle contraction or dynamic limb motion, to steer myoelectric hand-prostheses [1]. For this purpose, the MES is collected with a non-invasive skin-surface electrode placed over the muscle. After amplification and filtering, features are extracted from the MES and recognized patterns ranked into predefined categories. These categories represent hand-positions or motions used by the controller to steer the prostheses. Post-processing methods are often applied after classification to smooth the prosthetic motion without unwanted perturbing jumps. Feature selection and the extraction of highly effective features from the MES is most critical, because the use of features over a raw MES not just improves the classification efficiency, but effectively determines the accuracy and performance of the entire control scheme. While many amputees successfully use myoelectric prostheses [2,3], classification schemes fail or degrade on the onset of muscle

R. Moreno-Díaz et al. (Eds.): EUROCAST 2009, LNCS 5717, pp. 555–562, 2009.

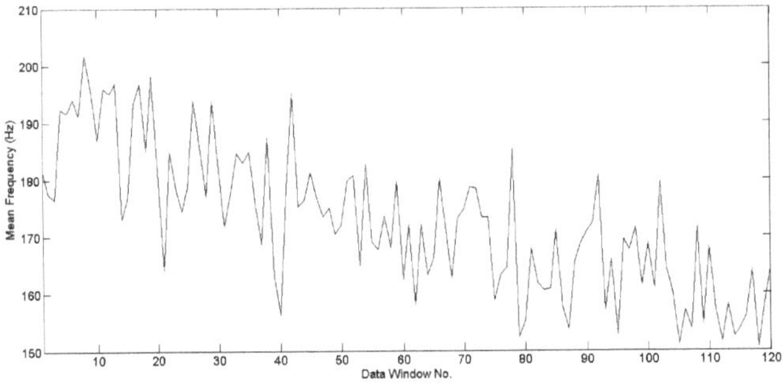

Fig. 1. Shift of the mean-frequency for a strong voluntary contraction (fist)

fatigue [4]. In human biomechanics, physiologists define the point, at which a contraction can no longer be maintained, as the failure point when a muscle fatigues. This accord implies, that fatigue occurs at a specific time and cannot be detected by indicators before failure. It is often viewed as a confounding factor and not as a cumulative effect. The failure point however, is a function of physiological and psychological factors, and it is difficult to accurately know the causal relationship [5]. However, one can exploit the well-known spectral modification property of the MES signal during a sustained contraction [6]. The modification manifests itself as a compression, accompanied with a change of the signal shape, providing the time-course of the fatigue-related physiological and biochemical processes [5,7]. Since the change of the baseline frequency of the MES can be easily calculated, the performance of the individual muscle is generally described with a spectral variable: the fatigue index [8]. The spectral modification advances continuously after five to ten seconds from the onset of an isometric contraction. It provides an indication of the fatigue progress from start. Consider Figure 1, which depicts a 30 seconds interval of a sustained fist-contraction. The frequency gradually starts to shift after 5 to 10 seconds.

2 Features, Fatigue and Classification

Generally, a MES is formed by the superimposition of numerous, individual action potentials, generated by irregular discharges of active motor units in the muscle [1]. Merlo et al. [9] model a MES as:

$$\begin{aligned} s(t) = & \textstyle\sum_j MUAP \cdot T_j(t) + n(t) \\ & \textstyle\sum_j \sum_t k_j \cdot f(\frac{t-\theta_{ij}}{\alpha_j}) + n(t) \end{aligned} \quad (1)$$

k_j is the amplitude factor for the jth motor unit, $f(\frac{t-\theta_{ij}}{\alpha_j})$ the shape of the action potential discharge, θ_{ij} the occurrence time of the motor unit action potential (MUAP), α_j the scaling factor, and $n(t)$ the additive noise. Since many MUAPs are overlapping and the nature of the motor unit discharge is highly irregular, a surface signal is generally considered as a complex, non-stationary stochastic signal. For this reason, an MES must be mapped into a smaller dimension vector, called either a feature-vector or just feature. In calculating features, a recorded MES is segmented into consecutive time frames of, e.g., 200 msec each. For a sustained contraction, the spectral features, mean-frequency (MNF) and median-frequency (MDF), are most commonly used. MNF is defined as the first-order moment for the average frequency of the power spectrum [10], as:

$$MNF = \frac{\int_0^\infty \omega \cdot P(\omega) \cdot d\omega}{\int_0^\infty P(\omega) \cdot d\omega} \tag{2}$$

where $P(\omega)$ is the power spectrum density (PSD) of the MES and ω the frequency variable. The frequency, at which the spectrum is divided into two parts of equal power, is the MDF. With the zero-order moments of the PSD follows:

$$\int_0^{MDF} P(\omega) \cdot d\omega = \int_{MDF}^{\infty} P(\omega) \cdot d\omega = \frac{1}{2} \int_0^{\infty} P(\omega) \cdot d\omega \tag{3}$$

Since fatigue influences amplitude, shape and the scaling factor over time, features in the time-domain are also affected. However, changes in the frequency-domain are most prevailing. Therefore, researchers often neglect the effects in the time- and in the time-frequency-domain [11]. Oskoei, Hu and Gang [10] use the features mean-scale (MNS) and inverse MNS, which is also known as the instantaneous mean-frequency (IMNF), to analyze fatigue in localized dynamic contractions. In their contribution, they use the continuous wavelet transform, defined as:

$$CWT(s,\tau) = \int x(t)\Psi^*_{s,\tau}(t) \cdot dt \tag{4}$$

with the scale parameter s, translation or time-shift parameter τ , and the MES $x(t)$, to obtain the basic function $\Psi_{s,\tau}(t)$ by scaling the mother wavelet $\Psi(t)$ at time τ:

$$\Psi_{s,\tau}(t) = \frac{1}{\sqrt{s}}\Psi(\frac{t-\tau}{s}) \tag{5}$$

The power density function or scalogram (SCAL) is estimated [12] with:

$$SCAL(x,\tau) = |CWT_x(x,\tau)|^2 \tag{6}$$

Like the MNF, the MNS is defined as the first-order moment of the scalogram as follows:

$$MNS = \frac{\int_{ls}^{hs} s \cdot SCAL(s) \cdot ds}{\int_{ls}^{hs} SCAL(s) \cdot ds} \tag{7}$$

Thereby ls is the lowest and hs the highest scale of interest.

3 Experiments

To assess the effects of fatigue, we worked with five right-handed participants, with an average age of 25 years and with no known muscular disorder. Each participant was asked to perform 12 sets of 30 seconds of maximum voluntary fist-contractions, each followed by a 2 minutes relaxation period. Post-tetanic exhaustion was not observed in any of the experiments. For signal acquisition, we utilized a DELSYS Bagnoli EMG System with double differential sensors (DE3.1) placed over the *extensor digitorum* muscle on the right forearm. The data was digitized with a National Instruments PCI 6251 DAQ Card with a sampling rate of 1024Hz. All feature calculations were carried out with MatLab^{TM}V. 7.7. In the feature extraction stage, we derived the standard time-domain elements: mean absolute value (MAV), root mean square (RMS), waveform length (WL), variance of the MES (VAR), Willison Amplitude (WAMP), slope-sign changes (SSC), integrated absolute value (IAV) and zero crossings (ZC). As expected, the analysis confirmed, that fatigue is affecting features in the frequency- (Figure 1) and in the time-domain (Figure 2) by time-variant changes in amplitude and distribution. These changes clearly affect our preferred feature selection and classification method. The Guilin Hills method follows a linear discriminant analysis (LDA) scheme, to select feature combinations and to efficiently compute classification maps [13]. LDA is a method to determine the linear combination of features that best separate two or more classes of objects or events. The resulting combination can be used as a linear classifier or for dimensionality reduction before classification. LDA assumes, that the conditional probability density functions are normally distributed, stationary over time, and that the class covariances are identical and have full rank. Under these assumptions, the Bayes rule for quadratic discriminant analysis allows to practically admeasure points to hand-positions. Even though, we can (almost) perfectly distinguish seven hand-positions with very good repeatability using only three sensors, we noticed a reproducible decline in the performance of our Guilin Hills classi-

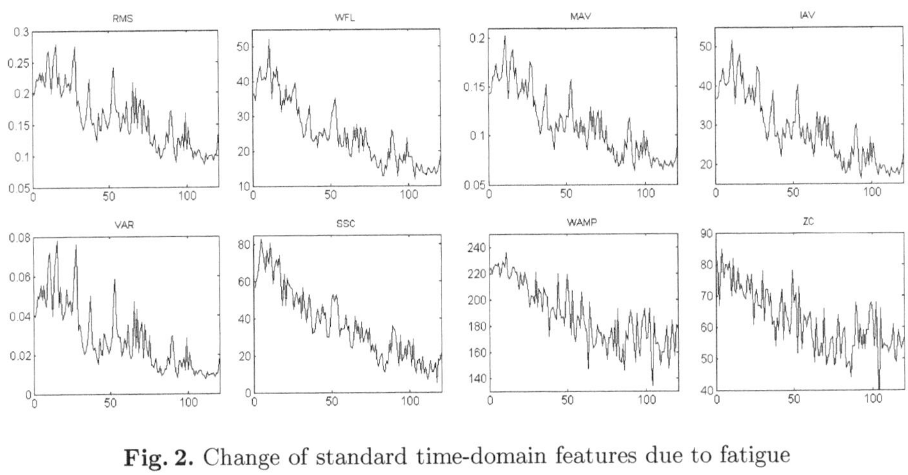

Fig. 2. Change of standard time-domain features due to fatigue

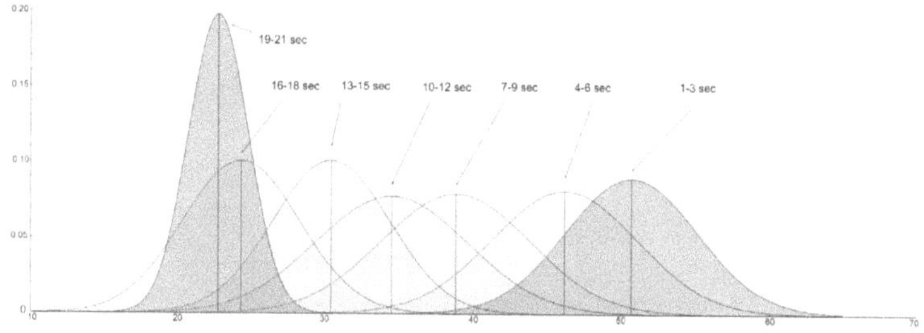

Fig. 3. Fatigue dependent, time-variant change in amplitude and distribution

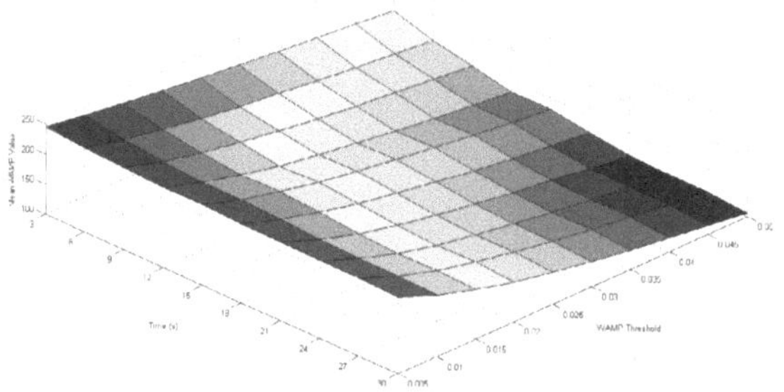

Fig. 4. Effect of different threshold values on the WAMP feature mean value decay

fier. The degradation and ultimate failure is caused by the inevitable fatigue-dependent change. As an example consider Figure 3, which depicts the SSC over a 30 seconds time-frame. It shows the shift, which we observed for all probands, however with a different degree. In the diagram, the bandwidth spans from 10% to 80%, based on the starting value. Similarities for all participants exist, most often the WAMP feature inherits the least, and the VAR feature the most decay. The window-size and window-overlap had no significant impact on the results. In contrast, the threshold value of the WAMP, SSC and ZC feature had a major impact on the significance of the observed feature value. Figure 4 visualizes the WAMP mean-value and standard-deviation over time for different threshold values. The observed fatigue and the decreasing factor is significantly effected by the threshold. As the experiments show, threshold selection is crucial, because low values lead to a saturation. We attribute this to noise under the presence of fatigue. Higher threshold values produce correct values with the drawback of fatigue dependent decay.

Besides the changes of the time-domain features the frequency- and time-frequency-parameter of the acquired signal also changes. Figure 5 presents a 30 seconds trial (top-left), the corresponding scalogram (bottom-left), the mean-scale (top-right),and the averaged feature mean-scale, all showing changes over time.

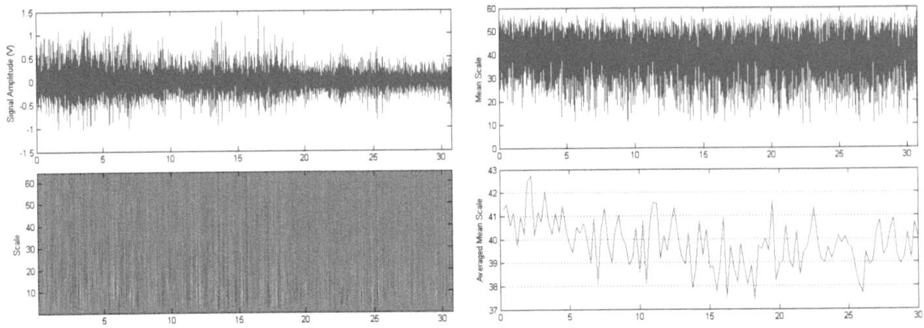

Fig. 5. Visualization of the changes in wavelet scalogram and feature mean scale

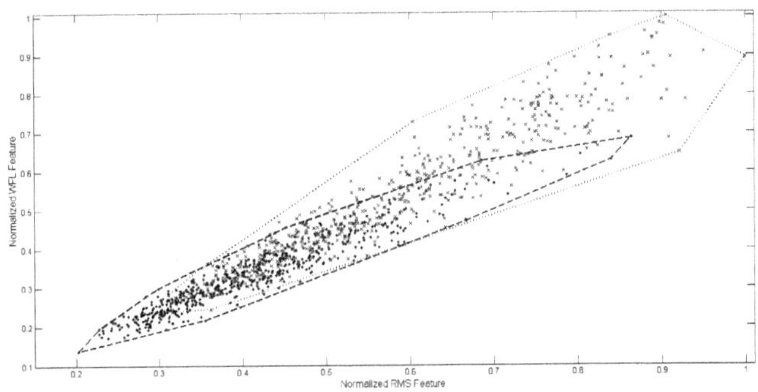

Fig. 6. Change of feature cluster position due to fatigue causing missclassification

Since features in the time-, the frequency-, and the time-frequency-domain are fatigue-time variant, standard classification techniques including neural-nets, degrade and fail with progressive fatigue. Consider Illustration 6, which shows the leeway of a typical classification region, starting at the upper-right and drifting to the lower-left, here contoured after 15 seconds. The time variant probability of the correct classification of a hand pose, based on the initial classification region is expressed by:

$$p_{success,t_x} = \int_{x=-\infty}^{\infty} \int_{y=-\infty}^{\infty} PDF_{t_x,A,B}(x,y) \cdot Class_{A,B}(x,y) \cdot dx \cdot dy$$

whereby the volume intersection between the initial region and the actual region at t_x is the measure for the classification probability at t_x. $PDF_{A,B}(x,y)$ represents the normal distribution for the features A and B and the corresponding mean- (μ_A, μ_B), standard deviation- (σ_A, σ_B) and the correlation coefficient-value (ϱ):

$$PDF_{A,B}(x,y) = \frac{1}{2\pi\sigma_A\sigma_B\sqrt{1-\varrho^2}} \cdot e^{\left(-\frac{1}{2(1-\varrho^2)}\left(\left(\frac{x-\mu_A}{\sigma_A}\right)^2 - 2\varrho\frac{x-\mu_A}{\sigma_A}\frac{y-\mu_B}{\sigma_B} + \left(\frac{y-\mu_B}{\sigma_B}\right)^2\right)\right)}$$

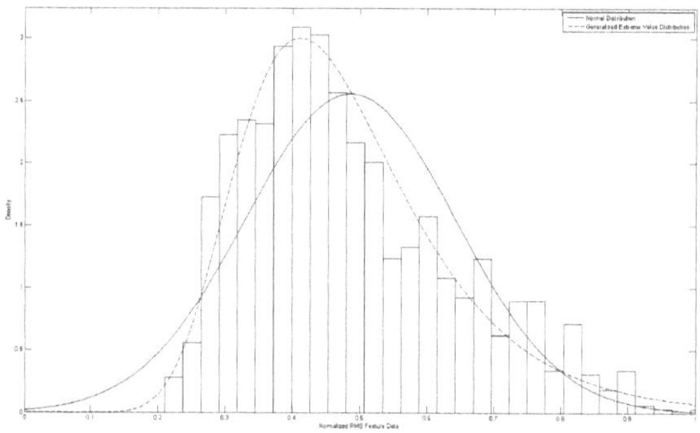

Fig. 7. Histogram of the VAR feature of a 30 second trial with a fitted normal and generalized extreme value distribution

The classification region of one hand pose is thereafter defined through the 5 percent footprint of:

$$Class_{A,B}(x,y) = \begin{cases} 1 & if e^{(-\frac{1}{2(1-\varrho^2)}((\frac{x-\mu_A}{\sigma_A})^2 - 2\varrho\frac{x-\mu_A}{\sigma_A}\frac{y-\mu_B}{\sigma_B} + (\frac{y-\mu_B}{\sigma_B})^2))} \geq 0.05 \\ 0 & otherwise \end{cases}$$

Especially classifiers based on LDA produce false classifications under fatigue, since the method is based on the assumption that the probability density function is normally distributed. Experimentally, we showed that this is not valid for fatigued MES. Considering a whole 30 second trial, the distribution of time-domain features significantly changes over-time. Figure 7 shows the histogram of VAR at the beginning and towards the end of a trial.

4 Conclusions and Current Research

Our previous work, with respect to linear discriminant analysis (LDA-) based prosthetic classifiers, relied on the assumption that myoelectric features are normally distributed. This assumption however, is only valid until the onset of muscle fatigue. Thereafter, calculated features are not normally distributed anymore. The major findings of this ongoing work concerns the general loss in accuracy, which effects all pattern recognition classifiers, especially those relying on Linear Discriminant Analysis (LDA) using features in the time-, frequency- and the time-frequency-domain features. With statistical methods, it is however possible to determine the features being most affected. These can be singled out for exclusion from the classification process. Also we found, that the significance of features varies over time, and good results in the beginning may become inadequate over time. This effect is under investigation and our work will hopefully lead to more reliable classifiers. Since we need to distinguish between voluntary feature decrease and fatigue, this task is not trivial and the mere tracking of features is not sufficient.

References

1. Oskoei, M.A., Hu, H.: Myoelectric Control Systems - A Survey. Biomedical Signal Processing and Control 2(4), 275–294 (2007)
2. Englehart, K., Hudgins, B., Parker, P., Stevenson, M.: Classification of the Myoelectric Signal using Time-Frequency Based Representations. Special Issue of Medical Engineering and Physics on Intelligent Data Analysis in Electromyography and Electroneurography 21, 431–438 (1999)
3. Muzumdar, A.: Powered Upper Limb Prostheses, Control, Implementation and Clinical Application. Springer, Heidelberg (2004)
4. MacIsaac, D., Parker, P., Englehart, K., Rogers, D.: Fatigue Estimation with a Multivariable Myoelectric Mapping Function. IEEE Transactions on Biomedical Engineering 53(4), 694–700 (2006)
5. De Luca, C.J.: The Use of Surface Electromyography in Biomechanics. Journal of Aplied Biomechanics 13, 135–163 (1997)
6. De Luca, C.J.: Myoelectric Manifestations of Localized Muscular Fatigue in Humans. Crit Rev Biomedeical Engineering 11(4), 251–279 (1984)
7. Bischoff, C., Schulte-Mattler, W.J., Conrad, B.: Das EMG-Buch; EMG und periphere Neurologie in Frage und Antwort. Georg Thieme Verlag (2005)
8. Merletti, R., Farina, D.: Myoelectric Manifestations of Muscle Fatigue. In: Wiley Encyclopedia of Biomedical Engineering. John Wiley & Sons, Inc., Chichester (2006)
9. Merlo, A., Farina, D., Merletti, R.: A Fast and Reliable Technique for Muscle Activity Detection from Surface EMG Signals. IEEE Transactions on Biomedical Engineering 50(3), 316–323 (2003)
10. Oskoei, M.A., Hu, H., Gan, J.Q.: Manifestation of Fatigue in Myoelectric Signals of Dynamic Contractions Produced During Playing PC Games. In: Proc. 30th Annual International Conference of the IEEE Engineering in Medicine and Biology Society EMBS 2008, pp. 315–318 (2008)
11. Bonato, P., Roy, S.H., Knaflitz, M., De Luca, C.J.: Time-Frequency Parameters of the Surface Myoelectric Signal for Assessing Muscle Fatigue During Cyclic Dynamic Contractions. IEEE Transactions on Biomedical Engineering 48(7), 745–753 (2001)
12. Radicheva, N., Gerilovsky, L., Gydikow, A.: Changes in the Muscle Fibre Extracellular Action Potentials in Long-Lasting (Fatiguing) Activity. European Journal of Applied Physiology and Occupational Physiology 55, 545–552 (1986)
13. Buchenrieder, K.: Dimensionality Reduction and Classification of Myoelectric Signals for the Control of Upper-Limb Prostheses. In: Proceedings of the IASTED - Human Computer Interaction 2008, pp. 113–119 (2008)

Decomposing Pattern Matching Circuit

Grzegorz Borowik and Tadeusz Łuba

Institute of Telecommunications, Warsaw University of Technology,
Nowowiejska 15/19, 00-665 Warsaw, Poland
{G.Borowik,luba}@tele.pw.edu.pl

Abstract. This paper presents a new cost-efficient realization scheme of pattern matching circuits in FPGA structures with *embedded memory blocks* (EMB). The general idea behind the proposed method is to implement combinational circuits using a net of *finite state machines* (FSM) instead. The application of functional decomposition method reduces the utilization of resources by implementing FSMs using both EMBs and LUT-based programmable logic blocks available in contemporary FPGAs. Experimental results for the proposed method are also shown. A comparison with another dedicated method yields extremely encouraging results: with a comparable number of EMBs, the number of logic cells has been reduced by 95%.

Keywords: Logic synthesis, Pattern matching, Address generator, Finite state machine, Decomposition, FPGA, Embedded memory.

1 Introduction

The paper presents a logic synthesis method targeted at FPGA architectures with specialized embedded memory blocks. The technological advancements in field programmable gate array devices in the past decade have opened new paths for digital system design engineers; unfortunately, existing methods do not ensure effective utilization of the possibilities provided by such modules [3,4].

This paper focuses on the implementation of multi-input logical functions characterized by a huge disproportion between the number of input words yielding a logical response of 1 and the number of words yielding 0 (and vice versa). Such combinational functions are commonly utilized in various types of logical circuits [1,8,10,13,14,16,18].

The general idea behind the proposed method is to implement a combinational circuit using a net of finite state machines. The presented technique works solely on FSMs, and is based on partitioning of the combinational circuit into smaller parts working in parallel; the parts are implemented using FSMs.

An effective implementation of FSMs was obtained by applying a functional decomposition method [7] with internal states encoding aimed at memory based implementation. The memory usage was significantly reduced by using the concept of the address modifier which significantly increases the quality of the finite state machines implemented in FPGA circuits [3].

R. Moreno-Díaz et al. (Eds.): EUROCAST 2009, LNCS 5717, pp. 563–570, 2009.

This way the problem of synthesis of a combinational circuit is transformed into a problem of synthesis of a finite state machine. The scheme applied reduces the logic cells by fully exploiting embedded memories [3,4].

2 Combinational Circuit Synthesis Scheme

An *address generator* (AG) is a typical example of such imbalanced, completely-specified function (Tab. 1a). This function has a long input vector and a relatively small number of registered vectors[1].

Consider a set of k binary registered vectors of n bits. For every registered vector, assign a unique integer from 1 to k. An address generator table shows the relation between the registered vectors and the corresponding integers (Tab. 1b). An address generation function produces the corresponding integer if the input matches a registered vector, otherwise produce 0. k is the weight of the address generation function [16].

Example. The completely-specified Boolean function of 5 inputs from Table 1a has only 5 registered vectors. The number of inputs determines the length of the vector. The list of registered vectors is given in Table 1b.

Table 1. An example of: a) Boolean function, b) address generator table

a)

	0 0 0 0 1 1 1 1	x_2
	0 0 1 1 0 0 1 1	x_1
	0 1 0 1 0 1 0 1	x_0
00	0 0 0 0 1 0 1 0	
01	0 0 0 1 0 0 0 0	
10	0 0 0 0 0 0 0 0	
11	0 0 0 0 1 1 0 0	
x_4, x_3		

b)

x_4	x_3	x_2	x_1	x_0	output
0	0	1	0	0	1
0	0	1	1	0	2
0	1	0	1	1	3
1	1	1	0	0	4
1	1	1	0	1	5

AGs are used in many universal areas, e.g. on the Internet (as IP filtering circuits) or in memory circuits (as memory patching circuits and pattern matching circuits).

The task of building an AG with 40 or more inputs is difficult when the AG is represented as a completely-specified Boolean function. The standard Berkeley's *PLA* file of *.typefr* [6] would have about 10^{12} rows, which makes it hard even to store the *PLA* file on a standard disk drive.

This paper proposes a new implementation of such functions. In the proposed method, a Boolean function is synthesized by splitting it into parts and by implementing those parts using FSMs. Assuming a function with n inputs and creating t state machines, each of the FSMs has n/t bits from the main vector on its input. The state machines produce output indexes of all vectors matched by those n/t bits (Fig. 1). The number of state machines (t) is selected experimentally.

[1] A registered vector is a vector yielding a logical response of 1.

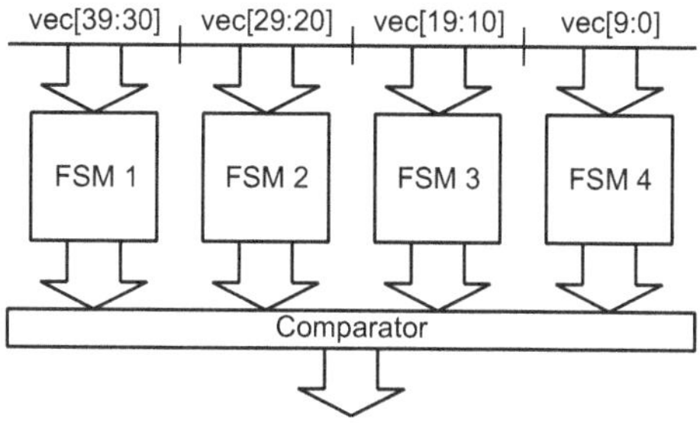

Fig. 1. AG implementation using FSMs ($n = 40$, $t = 4$)

After the above function decomposition, each given input vector yields t sequences of indexes. If an index appears in all t sequences the vector is recognized and that index is its value. If the result does not appear in all t sequences, the vector is not on the list.

The comparator consists of FIFO queues. After the start signal, the FIFOs gather all values yielded by the FSMs. Next, the unit starts searching for a value that repeats in all queues. If the current value is smaller than the largest found so far, a shift is performed. This process is repeated until all of the values match, or until one of them is 0. In the former case the unit produces an index, in the latter – 0. The values from state machines are sorted in ascending order, which speeds up the comparison process.

Example. For the address generator from Table 1b and $t = 2$ (two state machines) 3-input state machines are considered. Setting "001" to the first state-machine input makes it respond with the following values: 1, 2. If the input is "111", the response is: 4, 5.

3 ROM-Based Synthesis

The implementation scheme developed allows the transformation into FSM-description. Using this scheme we can implement FSMs in field programmable gate array devices.

Attempts to solve the problem of finite state machine synthesis resulted in many methods for the structural synthesis of FSMs. Their diversity results from different analysis, different assumptions and, subsequently, designing the methods for specific types of target components. Thus, different methods of the synthesis of FSM for PLA structures, ROM memories, and PLD modules exist [2,9,11]. Although most of the methods gathered and discussed in [12] can be effectively used for synthesis of FSM implemented with gates and flip-flops, they are not efficient for today's programmable structures, particularly for FPGA devices with embedded memory blocks [12]. Such implementations would benefit from a structure with a separate memory block which is common in microprogrammable circuits. However, an advanced apparatus for design of address modifier is required to support the synthesis based directly on the FSM transition table.

A limited size of embedded memory blocks available in FPGA devices is the main argument behind the application of this structure. For example, Altera FLEX family devices have 2048-bit EAB memory blocks. In [15] it is demonstrated that the ROM-based implementation of an example sequential circuit – the *tbk* benchmark – requires 16,384 bits of memory; this considerably exceeds the resources available in the FLEX 10K device. An alternative implementation of this circuit with LUTs requires 895 logic cells (a result from the Altera Quartus II ver. 8.1 system); this also exceeds the resources available in the FLEX 10K device, as it has only 576 cells. Thus, the *tbk* implementation with this device must rely on the a new FSM architecture.

Clearly, a considerably larger number and size of embedded memory blocks in the newer programmable Stratix and Cyclone devices do not eliminate this problem, as there will always be FSMs whose implementation requires more memory than is available in the state-of-the-art programmable devices.

In case when efficient memory utilization is essential, the FSM can be implemented in a structure that includes an address register and ROM memory, in which the reduction of ROM memory size is obtained by the introduction of an additional block for address modification (Fig. 2b).

The address modifier can be synthesized with advanced algorithms of functional decomposition, applied until recently exclusively to synthesis of combinational circuits. Such an approach to address modifier synthesis was proposed in [15,17] (and extended in [3,4]).

The implementation of an FSM shown in Fig. 2b can be seen as a serial decomposition of the memory block included in the structure of Fig. 2a into two blocks: an address modifier and a memory block of smaller capacity than required for the realization of the structure of Fig. 2a. As a result, sequential circuits requiring large-capacity ROM memories (and thus not implementable in the architecture of Fig. 2a) can be implemented using a memory block with a smaller number of inputs and an additional combinational logic block – the address modifier.

Assuming an FSM implementation with an FPGA device, the advantage of the proposed architecture lies in that the address modifier can mapped into a network of LUT cells or into a PAL matrix, while the memory block can be mapped into the built-in EAB matrices. The application of this concept (without the optimization of the state encoding) to the synthesis of the earlier discussed benchmark *tbk* results in a design composed of 333 logic cells and a 4096-bit embedded memory block, which fits entirely in the limited resources of the FLEX structure.

In general, it is possible to treat the address modifier and the memory as separate combinational blocks and implement them independently, with the application of different strategies for decomposition of combinational circuits. Alternating application of serial and parallel decomposition has been shown to be an extremely effective strategy to construct a structure utilizing both logic cells and EMBs. The promising results of other design experiments reported in

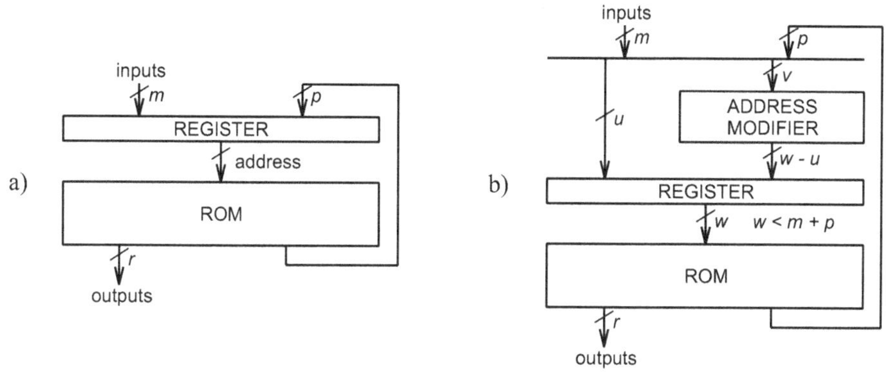

Fig. 2. FSM implementation: a) using ROM memory, b) with the addition of an address modifier

[15] confirm the effectiveness of the architecture of Fig. 2b. The results of the subsequent studies in this area are presented in [3] and [4].

4 Experiments

Several experiments have been performed in order to demonstrate the influence of the application of the proposed method. The experiments have been done using FSM*dec* software [5].

Table 2 presents the results of implementation of pattern recognition circuits using the proposed approach on Stratix EP1S10F484C5 and Stratix III EP3SE50F780C2 devices. Pattern sets with patterns of different length (40 or 260 bits) and different number of patterns to be recognized have been used. Table 2 presents the number of logic cells and EMBs required to implement the recognition circuit in FPGA devices. Noticeably, recognition circuits for large patterns as well as large set of patterns can be implemented with relatively small amount of programmable resources.

In [16], Sasao and Matsuura proposed the super hybrid method as a solution for implementing an address generator function using hash memories and

Table 2. Experimental Results *) EP1S10F484C5, **) EP3SE50F780C2

Vector length (inputs no.)	Number of patterns	Number of logic cells	Number of EMB blocks
40 *	2000	132	36(M4K)
40 *	4000	141	60(M4K)
40 *	8100	167	120(M4K)
260 **	1000	5006	108(M9K)
260 **	4000	6440	218(M9K)

Table 3. Comparison of AG Circuits Implementations in FPGA Structures

Count of:	Super hybrid method	New method	Gain
	a	b	$\frac{a-b}{a} \cdot 100\%$
No. of pattern vectors	1730	2000	
Pattern vector size (bits)	40	40	
No. of outputs	11	11	
M4K memory modules	32	36	-14.0%
M512 memory modules	0	4	
Adaptive Logic Modules	2426	132	94.6%
No. of pattern vectors	3366	4000	
Pattern vector size (bits)	40	40	
No. of outputs	12	12	
M4K memory modules	64	60	5.5%
M512 memory modules	0	4	
Adaptive Logic Modules	4889	141	97.1%
No. of pattern vectors	4705	8100	
Pattern vector size (bits)	40	40	
No. of outputs	13	13	
M4K memory modules	121	120	0.4%
M512 memory modules	0	4	
Adaptive Logic Modules	3560	167	95.3%

reconfigurable PLA. Using this method, an AG with k registered vectors can be implemented with p-input memory instead of n-input memory, where $2^p \geqslant k+1$.

Table 3 shows the comparison of results for AG implementation obtained with the presented in this paper method versus the super hybrid method described in [16]. This comparison uses 3 AG examples of similar complexity to those presented in [16]. Noticeably, implementations obtained with application of the method presented in this paper yield great reduction of logic resources. In the case of first AG example with 2000 pattern vectors 40 bit each, the proposed method yielded over 90% reduction of adaptive logic modules, at the expense of additional memory modules, when compared to the result presented in [16]. For two other examples (4000 vectors and 8100 vectors) the application of the method leads to over 90% reduction of adaptive logic modules as well, but at the same time it also reduces the memory module utilization.

Table 4 presents the comparison of pattern recognition circuits implementations using methods from literature [1,8,14,18] with the method presented in this paper. The method presented by Attig et al. in [1] is based on Bloom filters. The basic idea of the method discussed in [14] is to use hashing to generate a distinct address for each candidate pattern, which is subsequently stored in memory. Simple CRC-style polynomials implemented with XOR gates were used to achieve low cost hashing of the input patterns. In [18] Sourdis et al. presented an approach based on Burkowski's multiterm string comparator and Merkle's hash tree. A brute-force pattern matcher presented in [8] consists of a set of character

Table 4. Comparison of Pattern Matching Circuits in FPGA Structures

Circuit	Through-put [Gbps]	LCs	chars	LCs / char	Memory [kbit]
New method	2.001	6440	130000	0.05	1600
B. Filters [1]	0.502	36720	420000	0.09	629
[14]	2.008	2760	18636	0.15	558
PHmem [18]	2.108	6832	20911	0.32	288
NFAs [8]	7.004	54890	17537	3.13	0

match units that operate in parallel on a buffer of the input data. Noticeably, our implementation of a circuit consisting of a total of 130000 characters achieves similar throughput and uses significantly less logic cells per character compared to the results published earlier. Result from Table 4 prove that the proposed approach allows for the construction of very efficient pattern matching circuits, with the lowest number of logic cells per character.

5 Conclusions

This paper proposes a method to implement pattern matching circuits in FPGA structures with EMBs, based on the application of FSMs. This method applies functional decomposition to yield smaller resource utilization by implementing the combinational parts of the FSMs using both EMBs and LUT-based programmable logic blocks available in contemporary FPGAs. The application of this method requires relatively little FPGA resources when compared to other known methods. The results presented in the paper prove that this approach produces extremely encouraging results compared to other dedicated methods. Using a comparable number of EMBs, the proposed implementation requires over 90% less logic cells than the other methods used in the comparison.

Acknowledgments. This paper was supported by Ministry of Science and Higher Education financial grants: SINGAPUR/31/2006, N517 003 32/0583.

References

1. Attig, M., Dharmapurikar, S., Lockwood, J.: Implementation results of bloom filters for string matching. In: Proc. of 12th Annual IEEE Symposium on Field-Programmable Custom Computing Machines, Napa, CA, April 2004, pp. 322–323 (2004)
2. Barkalov, A., Titarenko, L.: Logic Synthesis for Compositional Microprogram Control Units. Lecture Notes in Electrical Engineering, vol. 22. Springer, Heidelberg (2008)
3. Borowik, G.: Improved state encoding for fsm implementation in fpga structures with embedded memory blocks. Electronics and Telecommunications Quarterly 54(1), 9–28 (2008)

4. Borowik, G., Falkowski, B.J., Łuba, T.: Cost-efficient synthesis for sequential circuits implemented using embedded memory blocks of fpga's. In: Proc. of 10th IEEE Workshop on DDECS, Cracow, Poland, April 2007, pp. 99–104 (2007)
5. Borowik, G.: FSMdec Homepage (2006), `http://gborowik.zpt.tele.pw.edu.pl`
6. Brayton, R., Hachtel, G., McMullen, C., Sangiovanni-Vincentelli, A.: Logic Minimization Algorithms for VLSI Synthesis. Kluwer Academic Publishers, Boston (1985)
7. Brzozowski, J.A., Łuba, T.: Decomposition of boolean functions specified by cubes. Journal of Multi-Valued Logic & Soft Computing 9, 377–417 (2003)
8. Clark, C.R., Schimmel, D.E.: Scalable parallel pattern matching on high speed networks. In: Proc. of 12th Annual IEEE Symposium on Field-Programmable Custom Computing Machines, Napa, CA, April 2004, pp. 249–257 (2004)
9. Cong, J., Yan, K.: Synthesis for fpgas with embedded memory blocks. In: Proc. of ACM/SIGDA 8th Int. Symp. on Field Programmable Gate Arrays, Monterey, California, United States, pp. 75–82 (2000)
10. Dharmapurikar, S., Krishnamurthy, P., Taylor, D.E.: Longest prefix matching using bloom filters. IEEE/ACM Transactions on Networking 14(2), 397–409 (2006)
11. Fuhrer, R.M., Nowick, S.M.: Sequential Optimization of Asynchronous and Synchronous Finite-State Machines: Algorithms and Tools. Kluwer Academic Publishers, Norwell (2001)
12. Łuba, T., Borowik, G., Kraśniewski, A.: Synthesis of finite state machines for implementation with programmable structures. Electronics and Telecommunications Quarterly 55(2) (2009)
13. Nilsen, G., Torresen, J., Sorasen, O.: A variable wordwidth content addressable memory for fast string matching. In: Proc. of Norchip, November 2004, pp. 214–217 (2004)
14. Papadopoulos, G., Pnevmatikatos, D.: Hashing + memory = low cost, exact pattern matching. In: Proc. of 15th Int. Conference on Field Programmable Logic and Applications, pp. 39–44 (2005)
15. Rawski, M., Selvaraj, H., Łuba, T.: An application of functional decomposition in rom-based fsm implementation in fpga devices. Journal of Systems Architecture 51, 424–434 (2005)
16. Sasao, T., Matsuura, M.: An implementation of an address generator using hash memories. In: Proc. of 10th Euromicro Conference on Digital System Design Architectures, Methods and Tools, pp. 69–76 (2007)
17. Selvaraj, H., Rawski, M., Łuba, T.: Fsm implementation in embedded memory blocks of programmable logic devices using functional decomposition. In: Proc. of ITCC, Las Vegas, April 2002, pp. 355–360 (2002)
18. Sourdis, I., Pnevmatikatos, D., Wong, S., Vassiliadis, S.: A reconfigurable perfect-hashing scheme for packet inspection. In: Proc. of 15th Int. Conference on Field Programmable Logic and Applications, pp. 644–647 (2005)

Hardware Approach to Artificial Hand Control Based on Selected DFT Points of Myopotential Signals

Przemyslaw M. Szecówka[1], Jadwiga Pedzińska-Rżany[1], and Andrzej R. Wolczowski[2]

[1] Faculty of Microsystem Electronics and Photonics, Wroclaw University of Technology, Janiszewskiego 11/17, 50-372 Wroclaw, Poland
[2] Institute of Computer Engineering, Control and Robotics, Wroclaw University of Technology, Wybrzeze Wyspianskiego 27, 50-370 Wroclaw, Poland
przemyslaw.szecowka@pwr.wroc.pl

Abstract. Technology of smart hand prosthesis control based on myoelectric signals strongly depends on signal processing algorithms. The application forces specific requirements on computational complexity (for dexterity of prosthesis), processing speed (for fast reaction) and size (for portability). The paper refers to a concept of selected DFT points of EMG signal analysis for patient intention recognition. Signals are acquisited from 6 channels with 1 kHz sampling frequency. Specialized digital hardware is proposed, capable of parallel processing of series of signals. The design was implemented in VHDL, verified and synthesized for FPGA. In-house developed floating point arithmetic was applied. Satisfying processing speed was obtained at a reasonable cost.

1 Introduction

Continuous activity of human organism is reflected in biosignals. Some of them may be relatively easily detected and then applied to control of technical devices. Electrical potentials related with skeleton muscles activity belong to this type of biosignals. They may be measured on the surface of the body in the form of electromyographic signals (EMG). Various movements of a limb are related to the recruitment of distinct motor units of its muscles. Simultaneously different spatial locations of these units, in relation to locations of measuring electrodes, lead to the formation of EMG signals of varying features, in particular with different frequency spectrum (spatial filtration effect). These features depend on the type of executed or imagined (in the case of an amputated limb) movement. Thus they allow discrimination of the EMG signals accompanying individual movements, eventually providing the information about the user's intention. Bioprostheses utilize this information to control the actuators of e.g. artificial hands fingers or knees and foots of artificial legs [1,2,3,4,5].

In general, technology of smart hand bio-prosthesis control strongly depends on signal feature extraction and classification. This encourages two directions of

R. Moreno-Díaz et al. (Eds.): EUROCAST 2009, LNCS 5717, pp. 571–578, 2009.

investigations - the first one focused on simplifying methods of the two steps of myoelectric signals processing and the second one aiming in the efficient hardware implementation of algorithms. The target application forces specific requirements. Diversity of movements and dexterity of prosthesis usually induces high computational complexity. Fast reaction of prosthesis, enabling natural-like appearance and obedience, necessary for human adaptation, requires high processing speed. Portability sets additional boundaries on size and power consumption. Eventually the two issues - algorithm performance and its availability for physical implementation gain similar importance and may be resolved concurrently. In our previous works it was shown that e.g. combination of wavelet transform and Learning Vector Quantization neural networks is quite effective solution [6], which may be effectively implemented in a single FPGA circuit [7]. The other solutions are neural network decision trees and sequential classification [8]. Classic approach to the features extraction from a signal is discrete Fourier transform. It is obvious that DFT/FFT spectra carry the most comprehensive information. The main disadvantage of this approach is high complexity of calculations and the result. The latter one induces adequately high complexity of the classifier, e.g. the neural network. Reasonable solution is to use only selected points of DFT spectrum. Research on this issue is in progress and the results will be published soon. In this paper we focus on dedicated digital hardware providing efficient calculation of this kind of partial DFT. In the next sections a concept of partial DFT is presented, followed by schematics and descriptions of digital architectures, results of synthesis, and finally conclusions.

2 The Concept

In a complex process of handicapped patient intention recognition based on myoelectric signals the DFT/FFT transform appears to be efficient solution for signal features extraction. Simultaneously it was observed that only selected points of DFT spectrum are crucial for further step - hand move classification. These points may be selected using various methods - heuristic and analytic. Observation of signals behaviour, supported by analysis of standard deviations calculated for each DFT point may reveal the most and the least sensitive ones. High correlation ratio between two DFT points may prove redundancy of one of them. Sensitivity analysis applied for neural network model of processing may reveal the most and the least important inputs [9,10]. Growing number of problems may be solved by primitive exhaustive search - comparison of efficiency of various combinations of inputs applied for series of dummy neural networks. However in our case of a few channels with e.g. 256 DFT points in each this approach still remains unreachable for contemporary computers.

Calculation of a single point of DFT spectrum requires series of multiplications and summations performed on the processed signal and the appropriate samples of sinusoid. Computational complexity is proportionally smaller than for primitive calculation of complete DFT spectrum. For reasonably small number of points to be calculated and relatively large number of samples this complexity remains much smaller than smart FFT calculation as well. (It is possible to perform partial FFT calculation omitting redundant operations. However it is very hard to design logic circuit providing free choice of points to be calculated). Thus in this specific context the classic DFT approach is more efficient than FFT.

In our case the intention recognition is based on signal deriving from 6 electrodes placed on the handicapped persons forearm. The sampling frequency is 1 kHz and the time window covers 256 samples. It is presumed that for each channel a few DFT points (Real + Imaginary) shall be calculated. Unfortunately these points may be different for each channel. This induces the need for specialized digital hardware capable of calculation of a series of freely chosen DFT points. This calculation must be fast enough to provide dexter reaction of hand prosthesis. The consecutive windows of samples are delivered every 256 ms. It is presumed however that only 10% of this time interval may be used for signal processing and intention recognition.

3 Architecture

The basic list of input signals contains reset, single clock, sampling strobe, EMG sample input and optionally selector of DFT point to be calculated (it may be embedded in the design as well). The output is series of calculated DFT points (Real and Imaginary), delivered one after another or in parallel.

Key elements of design are data and sinus samples delivery mechanisms, arithmetic units and control. The architectures outlined below were implemented in VHDL (hardware description language) [11,12] and the codes were processed, verified and synthesized with Xilinx ISE tools [13].

Sine and cosine samples generation is based on a look-up table containing 128 samples covering a half of basic sine period. The appropriate frequency may be synthesized by extraction of every p-th sample, where p is division factor between sampling frequency f_0 and the actual frequency f (i.e. $f = p * f_0/N$, where N is the number of samples in a window, in this case 256). The samples from a table are multiplexed, separately for sine and cosine in the two muxes controlled by separate counters, each starting from the appropriate position (sine from 0, cosine from 64) and incrementing by mentioned p. The samples are originally always positive, thus the sign is reversed when appropriate.

Basic idea of synchronous multiply and accumulate architecture is presented in Fig. 1. Consecutive samples of signal meet the appropriate samples of sine/cosine in a multiplier and the results are added to the accumulated sum. The calculation process is controlled by a counter. After reaching 256 it enables the final register to store the result and simultaneously resets the accumulator, enabling calculation of the next DFT point. In-house developed floating point arithmetic modules were applied. The bit vector representation of a number consists of a sign bit, 9 or 18-bit significand (always positive) and 6-bit exponent (2'complement). Octal basis was selected leading to significant savings on complexity when referred to e.g. IEEE format. Adder and multiplier modules designs are based on combinatorial logic containing multiplexed shift operations and fixed point arithmetic modules

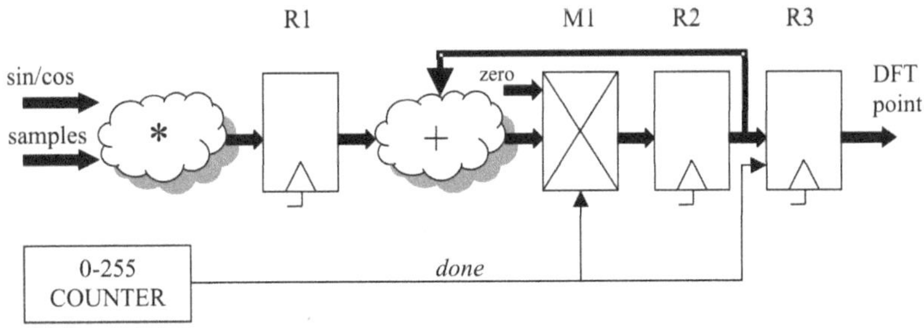

Fig. 1. Basic sequential architecture providing a single DFT point calculation

available via the VHDL *std_logic_signed* and *std_logic_unsigned* packages of IEEE library. In particular the modules may be constructed of very efficient fixed point arithmetic blocks embedded in more advanced FPGA circuits.

The weak point of this concept is unbalanced logic complexity of the two arithmetic modules. Static Timing Analysis (performed with Xilinx tools) revealed that for the floating point arithmetic applied, operation of multiplying unit is almost twice faster than operation of adder. For optimal time allocation the combinatorial logic of adder shall be divided into two parts separated by a register. This would enable twice higher clock frequency, which would decrease the data path processing time from original 4n time units to 3n (3 clock cycles with twice faster clock versus 2 clock cycles with slower clock). Unfortunately this approach disturbs the accumulation mechanism - to provide the data integrity the multiply unit shall deliver new result every second clock cycle, what increases the processing time back to 4n. To avoid gaps in operation of multiplier the authors decided to share the same arithmetic circuitry for both parts of DFT - *Real* and *Imaginary* as it is shown in Fig. 2. The adder logic is divided to two parts separated by register. Every clock cycle the control lines are toggled to deliver either sine or cosine sample to the multiplier unit, then deliver either accumulated *Re* or *Im* parts to the first part of adder and eventually open either *Re* or *Im* register to store the result delivered by the second part of adder. Eventually the two parts of DFT are calculated concurrently taking together 6n time units whilst for the basic concept that would be 8n time units.

Presented arithmetic block providing calculation of a single DFT point may be either replicated or shared for all the EMG channels and frequency points, depending on speed requirements and availability of resources. Various variants were considered and the most reasonable one assumes a little redesign of presented module to provide concurrent calculation of all the desired DFT points of a single channel in a shared processing unit. The two latches dedicated for each DFT point store the cumulated sum and a set of multiplexers provide delivery of the appropriate sine/cosine sample to the input of multiply unit. For each analysed frequency two dedicated counters must be allotted, each of them stepping by its own value p and memorizing which sample of sine/cosine shall be delivered to the multiplier unit. This concept relies on a fact that for 1 kHz

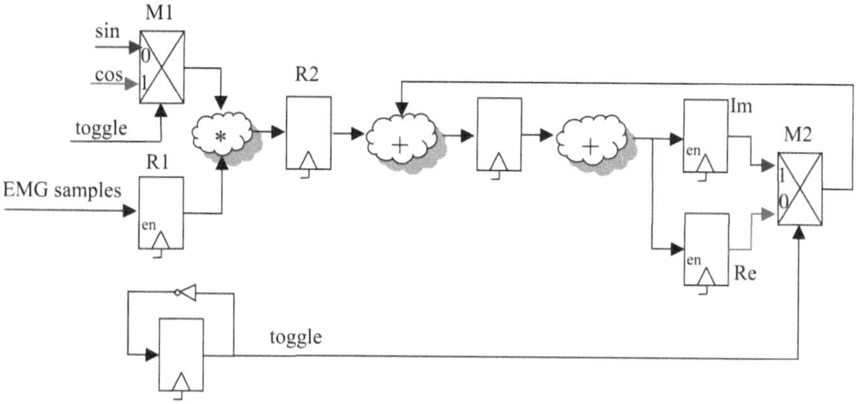

Fig. 2. Concurrent calculation of Real and Imaginary parts of DFT in shared arithmetic block

sampling frequency, the consecutive samples of signal appear quite slowly what encourages to perform all the calculations concerning the present sample in a time gap before the next sample is delivered. Eventually when the last sample of a window comes, calculation of all selected DFT points is almost finished.

These concurrent modules may be either shared again or replicated for each EMG channel. The replication variant is quite easy to implement and reuse, it provides very high processing speed (due to strictly parallel operation) and its overall logic complexity is proportional to the number of channels (if to exclude sine/cosine look-up tables and small amount of control logic elements which produce same signals and thus do not need to be replicated). At the current stage we focus on this solution, and we refer to this variant in the synthesis results analysis given in the next section. It is possible however to share a single arithmetic block for calculation of various DFT points in series of channels. This alternative solution is more complicated from control point of view but brings significant reduction of resources. The idea is presented in Fig. 3. Single arithmetic block is shared for all channels. The appropriate sine/cosine sample is delivered to the multiplier via a complex multiplexer controlled by the signal set up by actual channel, actual frequency to be processed, Real/Imaginary switch and finally by the appropriate dedicated sample counter. The frequency point selector is delivered from circulating shift registers, dedicated for each channel (top-left). Each shift register contains series of specific p factors which represent the list of DFT points to be calculated for the given EMG channel. Eventually this solution mimics operation of a microprocessor, with single arithmetic unit. Nevertheless its data delivery mechanisms definitely outperform microprocessor architectures. The weak points of this solution are a cost of additional logic allocated for control (which could be better used for computation) and difficulties in redesign for varying number of channels and DFT points.

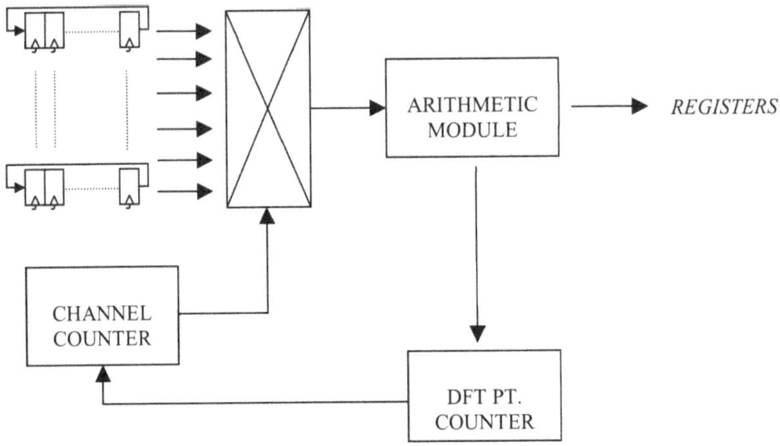

Fig. 3. Concurrent processing of all EMG channels in shared arithmetic module

4 Results

Described architecture variant, designed to process up to 8 DFT points in a single channel was replicated 6 times and synthesized. Synthesis process was run for Xilinx Spartan 3 series *xc3s*1000 FPGA (1 million equivalent gates). The design consumed ca. 5k registers (of 15k available), 6.3k LUTs (of 15k available) 5.3k slices of combinatorial logic (of 7.5k available), leaving some space for placement of classification algorithm. Simultaneously the logic complexity was estimated to 110k equivalent gates. Static Timing Analysis revealed the following key features:

Minimum clock period : 16.211ns
Maximum clock frequency : 61.687MHz
Minimum input arrival time before clock : 9.913ns
Maximum output required time after clock: 12.508ns
Maximum combinational path delay : 21.290ns

The requirement for signal stability on the input for 10 ns before clock edge and the granted stability of output 12.5 ns after clock edge are reasonable and acceptable. Maximum clock frequency, exceeding 60 MHz, enables the following estimation of processing speed:

Concurrent calculation of presumed 8 points of DFT requires ca. 40 clock cycles per signal sample. For 256 samples it takes over 10000 cycles. Nevertheless the calculation starts with the first EMG sample delivered and then it is distributed in all the time gaps between consecutive EMG samples delivered. Eventually the physical delay between the last sample and delivery of DFT points is equal to the mentioned 40 clock cycles. This induces the time of ca. 700 ns (= 40 cycles * 16 ns clock period) effectively consumed for extraction of EMG signal features. The overall acceptable delay between collection of a whole window of samples and reaction of prosthesis is estimated to 25 ms. Thus the delay invoked by the presented features extractor is not significant in the presumed time scale. With this huge speed margin and similar speed of classifier it is possible to apply windows overlapping, e.g. starting collection and processing of new window in the middle of the previous one. This approach, would lead to much more smart and smooth operation of prosthesis.

5 Conclusions

The concept of artificial hand control mechanism based on selected DFT points of multi-channel EMG signal was presented. Two variants of algorithm implementation in dedicated digital hardware were proposed. Both of them met the requirements. The one relying on generic block designed to process a single EMG channel, which may be replicated was found more handy for redesign and reuse.

Typical digital architecture consists of processing blocks (arithmetic, logic), storage (registers, memory) and data delivery mechanisms. When compared with microprocessors, superior performance of dedicated hardware is usually obtained by replication of arithmetic units providing parallel calculation. In presented architectures however the efficient data delivery system played equally important role.

Target application forces specific requirements on processing speed and specific way of its estimation. The 256 ms interval between consecutive windows which would normally be accepted for data processing is unacceptable for prosthesis control. Time allotted for intention recognition process shall not increase the overall delay significantly. Thus it is roughly estimated to 25 ms (10%). The key parameter is not the overall processing time but the delay between the ingress of the last sample of a window and decision about a move to be performed. Proposed hardware provides EMG signal features extraction consuming only 700 ns of the available time slot. Excellent processing speed, together with availability for embedding inside the prosthesis make this solution attractive choice for consideration. Besides the EMG prosthesis control it may be applied in many other fields.

References

1. Zecca, M., Micera, S., Carrozza, M.C., Dario, P.: Control of Multifunctional Prosthetic Hands by Processing the Electromyographic Signal. Critical Reviews in Biomedical Engineering 30(4-6), 459–485 (2002)
2. Englehart, K., Hudgins, B.: A robust, real-time control scheme for multifunction myoelectric control. IEEE Trans. on Biomedical Engineering 50, 848–854 (2003)
3. Chu, J.-U., Moon, I., Kim, S.-K., Mun, M.-S.: Control of multifunction myoelectric hand using a real-time EMG pattern recognition. In: Proc. IEEE/RSJ International Conference on Intelligent Robots and Systems, pp. 3511–3516 (2005)
4. Herrmann, S., Buchenrieder, K.: Dynamic Behavior of Time-Domain Myoelectirc Features for Prosthesis Control. In: Proc. 12th International Conference on Computer Aided Systems Theory, Eurocast, pp. 185–186 (2009)

5. Wolczowski, A.R., Krysztoforski, K.: Control-measurement circuit of myoelectric prosthesis hand. In: Proc. of 13th Conference of the European Society of Biomechanics, Wroclaw, pp. 576–578 (2002); Acta of Bioengineering and Biomechanics 4(supp. 1)
6. Krysztoforski, K., Wolczowski, A.R., Bedzinski, R., Helt, K.: Recognition of palm finger movements on the basis of EMG signals with application of wavelets. TASK Quarterly, 8–18 (2004)
7. Wolczowski, A.R., Szecówka, P.M., Krysztoforski, K., Kowalski, M.: Hardware Approach to the Artificial Hand Control Algorithm Realization. In: Oliveira, J.L., Maojo, V., Martín-Sánchez, F., Pereira, A.S. (eds.) ISBMDA 2005. LNCS (LNBI), vol. 3745, pp. 149–160. Springer, Heidelberg (2005)
8. Kurzynski, M., Wolczowski, A.R.: Control of dexterous hand via recognition of EMG signal using combination of decision-tree and sequential classifier. In: Kurzynski, M., Wozniak, M. (eds.) Computer Recognition Systems, vol. 2, pp. 687–694. Springer, Heidelberg (2007)
9. Żurada, J.M., Malinowski, A., Usui, S.: Perturbation method for deleting redundant inputs of perceptron networks. Neurocomputing 14, 177–193 (1997)
10. Szecówka, P.M., Szczurek, A., Mazurowski, M., Licznerski, B.W., Pichler, F.: Neural Network Sensitivity Analysis Applied for the Reduction of the Sensor Matrix. In: Moreno Díaz, R., Pichler, F., Quesada Arencibia, A. (eds.) EUROCAST 2005. LNCS, vol. 3643, pp. 27–32. Springer, Heidelberg (2005)
11. IEEE Standard 1076 - VHDL Language Reference Manual, 2000 Edition
12. IEEE Standard 1164 - Multivalue Logic System for VHDL Model Interoperability (Std_Logic_1164), 1993 Edition
13. Xilinx ISE Web Pack, ver 9.2 (2008), http://www.xilinx.com

System Approach to Complex Signal Processing Task

Vaclav Gerla[1], Vladana Djordjevic[1], Lenka Lhotska[1], and Vladimir Krajca[2]

[1] Gerstner Laboratory, Czech Technical University in Prague
Technicka 2, 166 27 Prague 6, Czech Republic
[2] Department of Biomedical Technology, Faculty of Biomedical Engineering,
Czech Technical University in Prague
{gerlav,djordv1,lhotska}@fel.cvut.cz, krajcav@fnb.cz

Abstract. This paper describes methods of automatic analysis and classification of biological signals. Polysomnographic (PSG) recordings encompass a set of heterogeneous biological signals (e.g. EEG, EOG, EMG, ECG, PNG) recorded simultaneously. These signals, especially EEG, are very complex and exhibit nonstationarity and stochasticity. Thus their processing represents a challenging multilevel procedure composed of several methods. Used methods are illustrated on examples of PSG recordings of newborns and sleep recordings of adults and can be applied to similar tasks in other problem domains. Analysis was performed using real clinical data.

Keywords: PSG, EEG, neonatal, sleep, segmentation, classification, clustering.

1 Introduction

Electroencephalography (EEG) is the measurement of brain electrical activity by means of electrodes located on the scalp. It has important applications in medicine and in cognitive science. Even since its introduction, electroencephalography has been evolving, both in technical and practical aspects [1]. But visual inspection is still widespread method of EEG signal analysis – neurologists use atlases and appropriate experience obtained in clinical practice for this evaluation. However, this kind of visual approach may not be always appropriate, namely with long-term EEG recordings (such as epileptic, sleep EEG or EEG of newborns). Long-term recordings are very important, because they give us the possibility to follow disorders that are not permanently present but appear incidentally or under certain conditions [2].

With the development of computer technology, the field of EEG got new dimension and the scope of analysis was broadened. Aims of computer assisted processing are to simplify tedious and time consuming work of neurologists, make the evaluation more objective and visualize results and represent them in a convenient form. Electroencephalogram is frequently supplemented with measurements of other biological signals and these measurements altogether are termed polysomnography (PSG). In infants and adults, these usually include electrooculogram (EOG), electromyogram (EMG), electrocardiogram (ECG) and respiration (PNG).

Sleep is a non-uniform biological state that has been divided into several stages. The standard for terminology and scoring of sleep stages is the manual by Rechtschaffen

R. Moreno-Díaz et al. (Eds.): EUROCAST 2009, LNCS 5717, pp. 579–586, 2009.

and Kales [3], which is followed by the vast majority of sleep laboratories, worldwide. Sleep stages classification is one of the diagnostic tools needed for the proper assessment of a number of sleep disorders and other neurological problems. In adults, sleep can be categorized into three states: non-rapid eye movement (NREM), rapid eye movement (REM) and wakefulness. NREM state is divided into four particular stages, reflecting a continuum of lighter to deeper sleep.

Sleep occupies a major portion of the lives of newborns [4]. The ratio of three newborn's behavioral states (wakefulness, active and quiet sleep) is an important indicator of the maturity of the newborn brain in clinical practice. Also, EEG provides useful information reflecting the function of the neonatal brain, may assist in identification of focal or generalized abnormalities, existence of potentially epileptogenic foci or ongoing seizures [5].

This paper focuses on methods appropriate for automated analysis and classification of polysomnographic recordings. Used methods are illustrated on two examples: PSG sleep recordings of adults and PSG recordings of full-term newborns.

2 Material and Methods

2.1 Data

Two data sets were used in this study. Each consisted of 10 subjects, neonates or adults, selected from wider groups. They were used in order to illustrate applied methods and obtained results. Both datasets were scored by experienced neurologists.

The first dataset encompasses PSG recordings of ten healthy fullterm infants, selected based on the similar postconceptional age from a wider group. The EEG activity was recorded from eight referential derivations, namely FP1, FP2, T3, T4, C3, C4, O1 and O2, positioned under the international 10-20 system [6]. Other measured signals were: EOG, EMG, ECG and PNG. The sampling frequency of all measured channels was 128Hz. These data were provided by the Institute for Care of Mother and Child in Prague.

The second used data set consisted of ten sleep PSG recordings of adults. Electrodes were also placed according to standards and EEG activity from nineteen electrodes was recorded. Other recorded PSG channels were EOG, ECG, PNG, EMG and SaO2, all sampled with 250Hz. Recorded time is about 8 hours for each person. These data were provided by The Faculty Hospital Na Bulovce in Prague.

2.2 Methods

The main processing steps in biological signal processing are preprocessing, data representation and classification. Also, optimization and visualization are inseparable parts of a complete processing task. From the data processing process, only processing steps that this study focuses on will be described in more details. Those are: signal segmentation, artifacts detection, feature extraction and classification and clustering. Visualization of processing steps and their intermediate results is also important. It gives the possibility of understanding the results easier, as well as the connection between the obtained results and their interpretation.

Adaptive Segmentation. Segmentation (constant or adaptive) is usual step in preprocessing of the nonstationary, complex signals. Our approach comprises adaptive segmentation. The reason for introducing such algorithm is that the following feature extraction from such relatively homogenous segments would be substantially more effective than the feature extraction from segments of constant length [7].

Adaptive segmentation, opposite to widely used constant segmentation, divides signal into quasi-stationary segments of variable length. Our method utilizes the principle of two joint windows of same length, sliding along the signal. This method is based on calculating the differences of the defined signal parameters of two windows [8]. The change of stationarity of the signal is indicated by local maxima of the difference measure - combined amplitude and frequency difference. When this calculated measure of the difference exceeds predefined threshold, the point is marked as a segment border. The threshold was introduced in order to eliminate the influence of small fluctuations of difference measure. Fig. 1 shows the results of adaptive segmentation applied to parts of EEG recordings. This method can be also used for computing segment boundaries in other recorded channels independently.

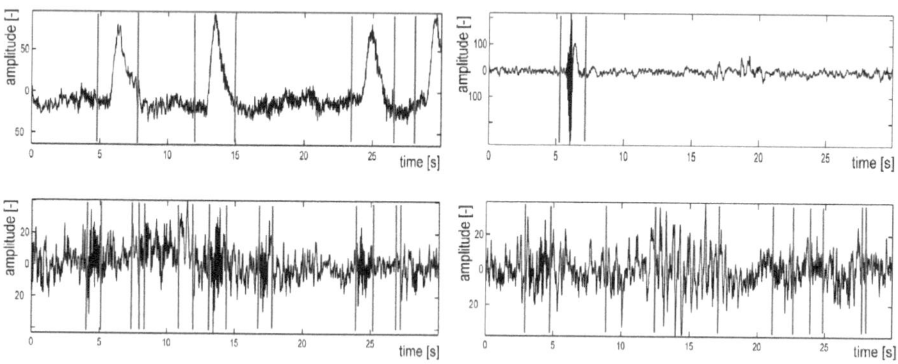

Fig. 1. Results of adaptive segmentation applied to EEG channels

Artifacts Detection. PSG signals, especially EEG signals, naturally contain various artifacts, that may occur at many points during the recording process [9]. The range of physiological and nonphysiological artifacts is very wide (e.g. ocular, muscle, cardiac, electrode, external device artifacts). They increase the difficulty of analyzing EEG in that way that recordings can be unreadable or artifacts can be misinterpreted as pathological activity. Recognition and elimination of artifacts is a complicated task, usually performed by a human expert.

In this study, we decided to use a simple method: signal segments with amplitudes higher than the threshold are marked as artifact segments. The threshold value was computed in two ways, namely based on standard deviation of observed signal (adaptively) or set as predefined constant value. Based on this algorithm, segments containing artifacts were marked and not involved in further analysis. Fig. 2 displays an example of artifacts detection method, applied to EEG FP1 electrode signal. Fig. 3 visualizes comparison of manual and automatic detection of artifacts: results are displayed for detection of artifacts by physician and for automatic detection described above.

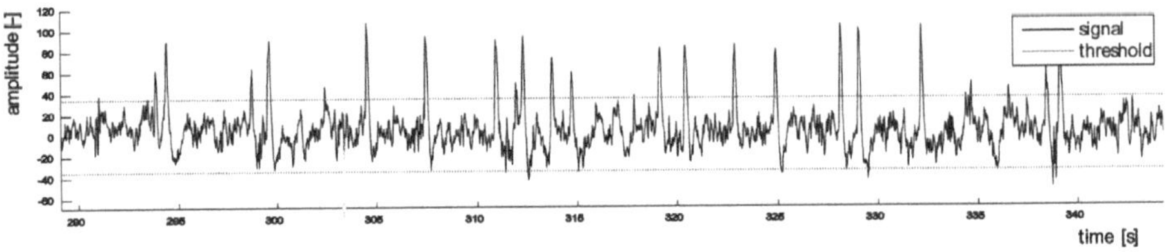

Fig. 2. Detection of artifacts on EEG electrode FP1; the threshold was set to constant value

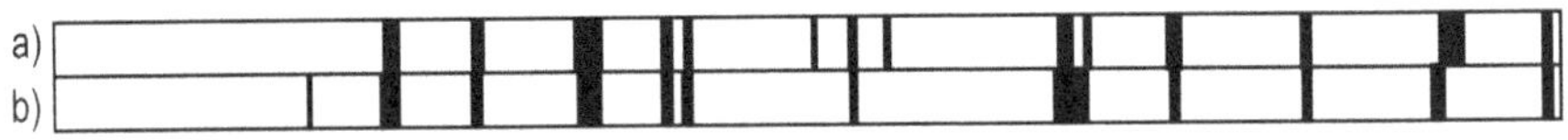

Fig. 3. Comparison of (a) manual and (b) automatic detection of artifacts

Feature Extraction. Feature extraction is an automated recognition of various descriptive features of signals. We extracted features not only from EEG signal, but also from other PSG signals, since this combination helps to enhance precision and robustness of the methods [10].

Each obtained artifact free segment was represented by extracted features. Computed features were: statistical parameters; mean and maximum values of the first and second derivation of the segment samples; Shannon's entropy; absolute and relative power/energy for important EEG frequency bands derived from Fourier and wavelet transforms; statistical values of the wavelet coefficients corresponding to decomposition scales; Shannon's entropy of wavelet transform details and approximations; and mean and maximum values of wavelet coefficients of the first and second derivations. Also, coefficients of correlation and coherence analysis were computed and used as features for appropriate channels. For wavelet analysis, various combinations of mother wavelets and decomposition levels were tested. Features listed above are already known in the literature comprising biological signal processing, e.g. [11], [12]. For ECG and PNG signals, heart rate variability and regularity of respiration were also computed. As the values of the individual features may differ by several orders, feature normalization is performed.

In this way we acquire several hundreds of features, which represent a burden for further processing. There are several different ways in which the dimension of a problem can be reduced. In this work Principal Component Analysis (PCA) approach is used which defines new features (principal components or PCs) as mutually-orthogonal linear combinations of the original features. For many datasets, it is sufficient to consider only the first few PCs, thus reducing the dimension [13]. Fig. 4 illustrates an example of PCA application on newborn data. It can be clearly seen that several first PCs correspond to the expert classification, discriminating quiet sleep stage.

Classification. Generally speaking, classification involves assigning a class to an unknown object. In our case, objects are segments described by vectors of features. For classification, we used both supervised and unsupervised methods.

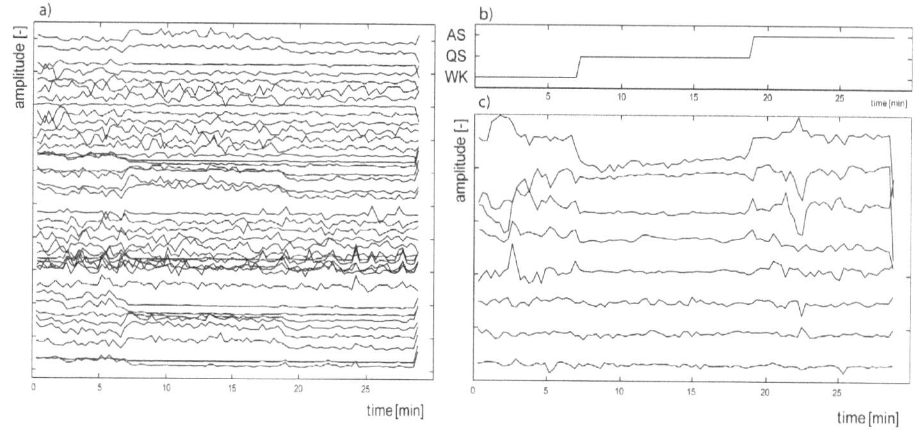

Fig. 4. Example of PCA for newborn data; a) forty original features; b) expert classification (wakefulness, quiet slep, active sleep); c) eight final principal components

Special attention was given to the possible application of Hidden Markov Models (HMMs) [14]. HMMs are a special class of stochastic processes that uniquely determine the future behaviour of the process by its present state. We use the expectation-maximization (EM) algorithm for finding the maximum-likelihood estimate of the parameters of HMMs given a set of observed feature vectors. This algorithm is also known as the Baum-Welch algorithm.

Clustering. Clustering was tested as unsupervised classification method. We applied k-means algorithm and hierarchical clustering as representatives of unsupervised methods [15].

In this study results of hierarchical clustering are presented. This approach allows us to find similar EEG/PSG segments and create clusters. Then assigning the clusters to defined classes (either manually or with computer assistance) is easier than "classical" record evaluation in time domain. In relation to hierarchical clustering we are examining influence of different metrics of distance measurement in state space and influence of choice of cluster generation methods on the number of resulting clusters. The best results were obtained by Ward's clustering. The aim is to reach optimal number of clusters with successive classification having high success rate.

3 Sleep PSG Recordings of Adults

The processing of sleep PSG recordings is a complex process consisting of several steps. In the preprocessing stage, following steps were conducted: resampling of signals with the frequency of 128Hz, 50Hz noise filtering, segmentation and artifacts detection. ECG and PNG signals were segmented into segments of the same fixed length of 20s. Preprocessing stage was followed by feature extraction for individual segments. Further, represented segments were classified. The goal was to obtain final classification of segments to one of six sleep stages, meaning four NREM stages, REM phase and wakefulness.

Fig. 5 presents results of the classification after the use of HMMs. Results are represented by hypnograms, which are the resulting time evolutionary descriptions of sleep in terms of stages. Hypnograms are used by physicians for diagnosis. This kind of representation allows direct comparison of results obtained by automatic analysis with sleep profiles determined visually. Results of clustering analysis are shown in Fig. 6. It can be seen that similar segments are grouped within clusters. Consequently, it is possible to classify each cluster (manually or automatically) to stages.

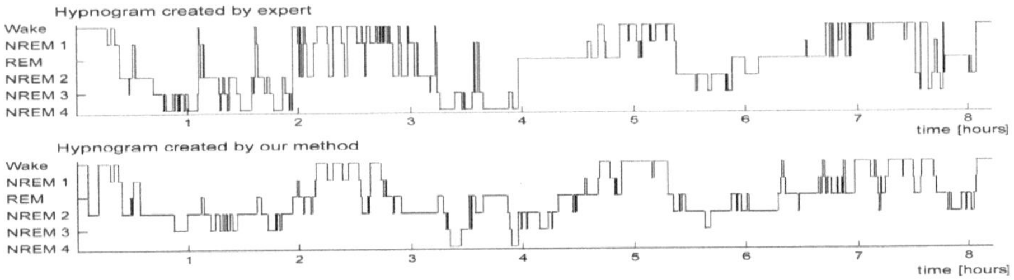

Fig. 5. Comparison of expert (top) and HMM (bottom) classification results

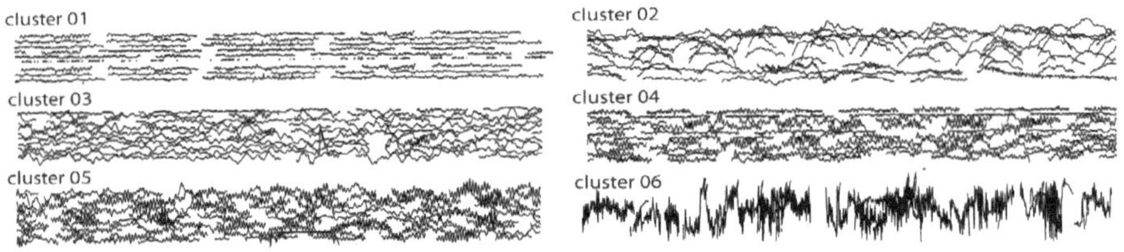

Fig. 6. The example of hierarchical clustering results of C3 electrode signal segments

4 Neonatal PSG Recordings

The main objective in processing of newborn PSG sleep recordings is to design and develop the appropriate combination of feature extraction and classification methods for automatic recognition of behavioral states. Neonatal PSG data from our dataset were first preprocessed: low-pass FIR filter was used to remove frequencies higher than 30Hz, signals were segmented and artifacts were detected. ECG and PNG signals were segmented using constant segmentation algorithm to segments of 20s length, while remaining PSG signals were segmented adaptively.

After obtaining representation of segments by features, classification and clustering were conducted. Concerning HMMs, their application led to results shown in Table 1. With comparison to several other applied classifiers HMMs led to better performance, meaning better classification accuracy. Hierarchical clustering was also tested on this dataset. In Fig. 7 another type of visualization of clustering results is shown.

Table 1. Statistical results. In every group nine newborns was used for training of HMM and remaining one for testing. Classification was made to three classes.

classes	mean success rate [%]	mean TP rate [%]	mean FP rate [%]
AS, QS	87.3 %	79.3 %	6.2 %
Wake , AS, QS	72.5 %	68.0 %	10.3 %

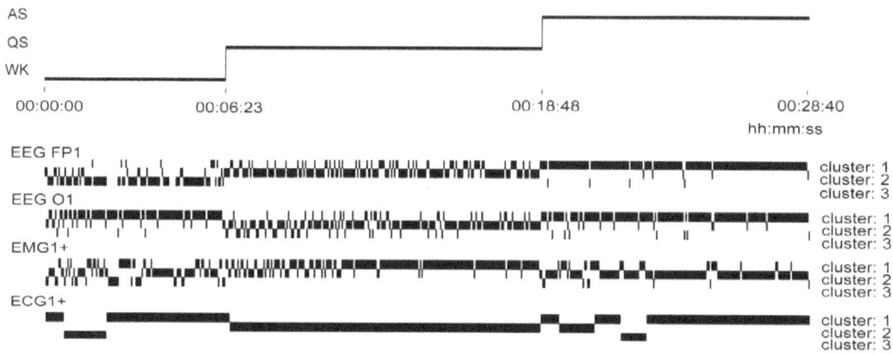

Fig. 7. Clustering results of neonatal data. On the top, expert classification is shown, followed by the cluster structure of two EEG channels, EMG and ECG channels.

5 Discussion and Conclusions

In this paper we have demonstrated methods of PSG data analysis and classification on two examples, namely clinical adult sleep PSG data and sleep data of newborn patients. We have focused on segmentation, feature extraction and classification steps in systems for automatic detection of sleep stages. The results of automatic detection of sleep stages were compared to the sleep profiles determined visually.

In the case of sleep PSG recordings of adults, introduction of features derived form non-EEG channels improved the classification accuracy by enabling discrimination between wakefulness and REM phase. Another problem occurs with the correct classification of transition states. This is related to subjective rating of the expert, because the expert rating was used for training of the system. In order to overcome this problem, unsupervised learning methods were used, namely clustering.

Classification of newborn sleep recordings is a very difficult task. It is necessary to mention that there does not exist any freely accessible annotated base of newborn EEG/PSG recordings. Another problem is that EEG signals are highly dependent on the age. As with the sleep recordings of adults, borders of sleep stages are not strictly defined, which means that the problem with the classification of transition states remains. Concerning the correct identification of behavioral states, the most difficult task was to discriminate wakefulness. This was also improved by supplementing extracted EEG feature set with features derived from other PSG channels.

Attention was also given to the visualization. It is important because it allows doctors to have a clear overview of processing steps and their intermediate results (e.g. spectrograms, coherence maps, trends in individual quantities, results of hierarchical clustering).

The designed methods could be applied to similar tasks in other problem domains as well. In our future work we plan to focus on optimization of parameters of individual processing steps, as well as on identification of other significant features and feature selection methods.

Acknowledgement. This research has been supported by the research program „Information Society“ under grant No. 1ET101210512 „Intelligent methods for evaluation of long-term EEG recordings“, grant IGA MZCR NS 10459 and the research program No. MSM 6840770012 "Transdisciplinary Research in Biomedical Engineering II" of the CTU in Prague.

References

1. Crespel, A., Gelisse, P.: Atlas of Electroencephalography. Awake and sleep EEG, vol. 1. John Libbey Eurotext, Paris (2005)
2. Daube, J.R.: Clinical Neurophysiology, 2nd edn. Mayo Foundation for Medical Education and Research, New York (2002)
3. Rechtschaffen, A., Kales, A. (eds.): A manual of standardized terminology, techniques and scoring system for sleep stages of human subjects. Brain Inform, Los Angeles (1968)
4. Teofilo, L., Lee-Chiong: SLEEP: a comprehensive handbook. John Wiley & Sons, Inc., Hoboken (2006)
5. Mizrahi, E.M., Hrachovy, R.A.: Atlas of Neonatal Electroencephalography, 3rd edn. Walnut St., Philadelphia (2004)
6. Goldensohn, E.S., Legatt, A.D., Koszer, S., Wolf, S.M.: Goldensohn's EEG interpretation: Problems of Overreading and Understanding, 2nd edn. Armonk (1999)
7. Paul, K., Krajca, V., Roth, Z., Melichar, J., Petranek, S.: Comparison of quantitative EEG characteristics of quiet and active sleep in newborns. Sleep Medicine 4, 543–552 (2003)
8. Krajca, V., Petranek, S., Paul, K., Matousek, M., Mohylova, J., Lhotska, L.: Automatic Detection of Sleep Stages in Neonatal EEG Using the Structural Time Profiles. In: IEEE EMBC, Shanghai, China, pp. 6014–6016 (2005)
9. Sterm, M., Engel, J.: Atlas of EEG Patterns. Lippincott Williams & Wilkins (2005)
10. Gerla, V., Paul, K., Lhotska, L., Krajca, V.: Multivariate Analysis of Full-Term Neonatal Polysomnographic Data. IEEE TITB 13(1), 104–110 (2009)
11. Gerla, V., Lhotska, L., Krajca, V.: Utilization of Time-Dependence in EEG Signal Classification. In: EMBEC 2005, SBMILI ČLS JEP, Prague, vol. 11 (2005)
12. Greene, B.R., Faul, S., Marnane, W.P., Lightbody, G., Korotchikova, I., Boylan, G.B.: A comparison of quantitative EEG features for neonatal seizure detection. Clinical Neurophysiology 119, 1248–1261 (2008)
13. Berka, P.J., Rauch, J., Zighed, A.: Data Mining and Medical Knowledge Management. Cases and Applications. Medical Information Science Reference, Hershey (2009)
14. Schlesinger, M.I., Hlavac, V.: Ten lectures on statistical and structural pattern recognition. CTU, Prague (1999)
15. Xu, R., Wunsch, D.C.: Clustering. IEEE Press, Piscataway (2009)

Symbolic Computations on Rings of Rational Functions and Applications in Control Engineering

N.P. Karampetakis[1], E.N. Antoniou[2], A.I.G. Vardulakis[1], and S. Vologiannidis[3]

[1] Aristotle University of Thessaloniki, Department of Mathematics, 54124 - Thessaloniki, Greece
{karampet,avardula}@math.auth.gr

[2] Technological Educational Institute of Thessaloniki, Department of Sciences, 54101 - Thessaloniki, Greece
eantonio@gen.teithe.gr

[3] Technological Educational Institute of Serres, Department of Information and Communication Sciences, 62100 - Serres, Greece
svol@teiser.gr

Abstract. A collection of algorithms implemented in Mathematica 7.0, freely available over the internet, and capable to manipulate rational functions and solve related control problems using polynomial analysis and design methods is presented. The package provides all the necessary functionality and tools in order to use the theory of Ω–stable functions, and is expected to provide the necessary framework for the development of several other algorithms that solve specific control problems.

1 Introduction

Polynomial methods are modern design techniques for complex multi-variable systems, signals and processes based on manipulations of polynomials, polynomial matrices, and other similar objects. The theoretical background of polynomial design techniques for control systems can be traced back to the late fifties. However, their frontal attack to control theory started in the seventies when the first really important results were achieved. One of the most important results is without doubt the parameterization of all controllers that stabilize a given plant, now referred to as Youla-Kucera parameterization. In the eighties, the polynomial methods were used to solve robust control problems and employed also in the field of signal processing. The Youla–Kucera (YK) parametrization of all stabilizing controllers [1], [4] is particularly useful for controller design because all the closed-loop system transfer functions depend affinely in the same parameter that can be optimized over the set of proper stable rational functions [2]. The significance of polynomial methods as a theoretical tool for engineers and applied mathematicians has been proven through the years.

An important research area for control engineering is the development of robust and efficient algorithms solving control problems. The software packages

R. Moreno-Díaz et al. (Eds.): EUROCAST 2009, LNCS 5717, pp. 587–594, 2009.

currently available fall into two categories. The first one includes packages that use numerical methods, having the advantages of high speed and low memory requirements and as a trade-off the loss of numerical accuracy. Well known such packages are "Control Systems Toolbox" [7], "Polynomial toolbox" for MATLAB [5] and "The Control and Systems Library" SLICOT, (see [8], [9]). The second category includes packages using symbolic calculations, that can obtain exact solutions even to parametrical control problems, with higher computational requirements, such as "Control System Professional" for MATHEMATICA [6]. The package presented in the present paper uses the symbolic computation approach.

Many control problems require the design of a compensator satisfying specific requirements, such as the pole placement to a certain subregion Ω of the complex plane $\mathbb{C}$. These kind of problems have already been solved theoretically [10], by the use of $\Omega-$generalized polynomials. Although there exist some toolboxes to manipulate polynomial matrices, such as the "Polynomial toolbox" for MATLAB, there is no software package dealing with the theory and applications of $\Omega-$stable functions/matrices. In this paper we present a collection of algorithms implemented in Mathematica 7.0, freely available over the internet, able to manipulate rational functions and solve related control problems using polynomial analysis and design methods. The package provides all the necessary functionality and tools in order to use the theory of $\Omega-$stable functions and matrices. The existing functions are capable of solving control problems concerning both SISO and MIMO systems, but for brevity reasons the latter is not presented in the following sections.

2 Ω-Stable Rational Functions and the Rings Package

Let Ω be a subset of the extended complex plane $\bar{\mathbb{C}} = \mathbb{C} \cup \infty$, symmetric with respect to the real axis which excludes at least either one point on the real axis $a \in \mathbb{R}$ or ∞. In the following the set Ω will play the role of the forbidden poles region in $\bar{\mathbb{C}}$. Let Ω^C denote the complement of Ω with respect to $\bar{\mathbb{C}}$, i.e. $\Omega \cup \Omega^C = \bar{\mathbb{C}}$. Given a rational function $t(s) \in \mathbb{R}(s)$ we can always factorize it as

$$t(s) = t_\Omega(s)\hat{t}(s),$$

where $t_\Omega(s) = n_\Omega(s)/d_\Omega(s)$, $\hat{t}(s) = \hat{n}(s)/\hat{d}(s)$, with $n_\Omega(s), d_\Omega(s)$ polynomials having zeros in Ω and $\hat{n}(s), \hat{d}(s)$ polynomials with zeros in Ω^C.

A rational function $t(s)$ will be called *Ω-stable* iff it has all its poles in Ω^C. Then $t(s)$ can be written as

$$t(s) = n_\Omega(s)\frac{\hat{n}(s)}{\hat{d}(s)}, \tag{1}$$

where the polynomials $n_\Omega(s), \hat{n}(s), \hat{d}(s)$ have zeros in the regions described above. If we denote by $S_\Omega = \{t(s) : t(s) \text{ is } \Omega\text{-stable}\}$ the set of all Ω-stable rational functions, it can be easily seen that the set S_Ω endowed with the operations of

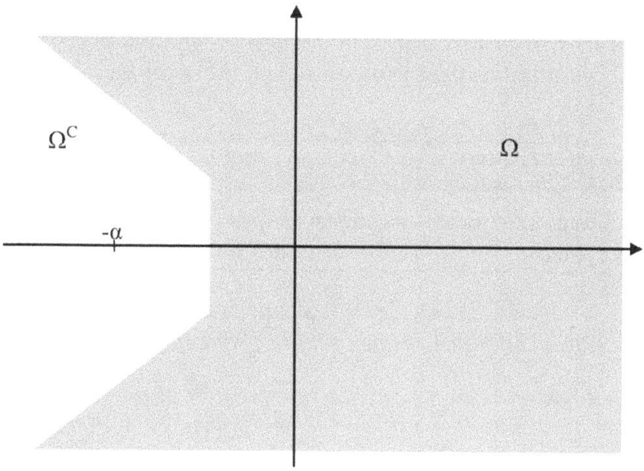

addition and multiplication is a commutative ring, with unity element the real number 1 and no zero divisors (i.e. it is an integral domain). The units of S_Ω are the elements of S_Ω whose inverse also belongs to S_Ω.

In order to be able to manipulate rational functions and solve related control problems, we have implemented the *Rings* package, using Mathematica 7.0. The package provides all the necessary functionality and tools for the experimentation with the theory of Ω−stable functions. In a Mathematica notebook in order to load the *Rings* package, we give

```
<<Rings`
```

The next line provides the global setting for the working ring of proper and Hurwitz stable rational functions:

```
$ForbiddenPolesArea = RightComplexPlane;
```

The `RightComplexPlane` is simply a membership function that gives `True` for values in the extended closed right-half complex plane and false elsewhere. The actual definition of the function is the following:

```
RightComplexPlane[s_]:= (Re[s]>=0) || (Abs[s]==Infinity);
```

Depending on the choice of Ω one may end up with a variety of rings. We give some examples:

- $\Omega_h = \{s : \mathrm{Re}(s) \geq 0\} \cup \{\infty\}$, the ring of proper, Hurwitz stable rational functions.
- $\Omega_s = \{s : |s| \geq 1\} \cup \{\infty\}$, the ring of proper, Schur stable rational functions.
- $\Omega_{pol} = \mathbb{C}$, the ring of polynomials $\mathbb{R}[s]$.
- $\Omega_{pr} = \{\infty\}$, the ring of proper rational functions $\mathbb{R}_p(s)$.

The *Rings* package provides a set of membership functions, to define to most commonly used Ω's, resulting in the above rings of rational functions. The predefined membership functions that can be used as values of **$ForbiddenPolesArea** are

`RightComplexPlane`	$\longrightarrow$	Corresponding to Ω_h
`UnitDiscComplement`	$\longrightarrow$	Corresponding to Ω_s
`FiniteComplexPlane`	$\longrightarrow$	Corresponding to Ω_{pol}
`InfinityPoint`	$\longrightarrow$	Corresponding to Ω_{pr}

In order to check whether a given rational function $t(s)$ belongs to the ring S_Ω defined by the **$ForbiddenPolesArea** setting, we can use

```
RingQ[ts,s]
```

where the argument s indicates the function's indeterminate variable. As in most of the functions of the package one can override the default ring setting provided by **$ForbiddenPolesArea**, using an extra option of the form

```
RingQ[ts,s,ForbiddenPolesArea->UnitDiscComplement]
```

In the latter example the rational function $t(s)$ will be checked against the ring of Schur stable functions. A very important result (see [3]) following from the above discussion, is that the field of rational functions can be considered as a quotient field of S_Ω. Any rational function $t(s) \in \mathbb{R}(s)$ can be written (non - uniquely) as a ratio of two Ω-stable rational functions, i.e.

$$t(s) = \frac{n(s)}{d(s)},$$

where $n(s), d(s) \in S_\Omega$. Given a rational functions $t(s)$ one can obtain the above fractional representation using the function

```
{ns,ds} = RingFraction[ts,s]
```

which returns the numerator $n(s)$ and the denominator $d(s)$ respectively.

Furthermore, one may define the mapping $\delta_\Omega : S_\Omega \to \mathbb{Z}$ via

$$\delta_\Omega(t(s)) = \begin{cases} \# \text{ of zeros of } t(s) \text{ in } \Omega, & \text{if } t(s) \neq 0 \\ -\infty, & \text{if } t(s) = 0 \end{cases}. \tag{2}$$

The mapping $\delta_\Omega(.)$ satisfies

$$\delta_\Omega(t_1(s)t_2(s)) = \delta_\Omega(t_1(s)) + \delta_\Omega(t_2(s)),$$

$$\delta_\Omega(t_1(s) + t_2(s)) \geq \min\{\delta_\Omega(t_1(s)), \delta_\Omega(t_2(s))\},$$

and thus it serves as a *discrete evaluation* or *degree* for the ring S_Ω. For the above mentioned choices of Ω, where $t(s)$ is written as in (1), the discrete evaluation $\delta_\Omega(t(s))$ can be calculated as follows

- For the ring of proper Hurwitz or Schur stable rational functions, $\delta_\Omega(t(s)) = \deg \hat{d}(s) - \deg \hat{n}(s)$.
- For the ring of polynomials $\mathbb{R}[s]$, $\delta_\Omega(t(s)) = \deg n_\Omega(s)$.
- For the ring of proper rational functions $\mathbb{R}_p(s)$, $\delta_\Omega(t(s)) = \deg d(s) - \deg n(s)$.

In the *Rings* package the calculation of the degree of a given Ω-stable rational function $t(s)$, can be done by

```
RingDegree[ts,s]
```

It can be easily seen that the units of S_Ω are exactly the elements $u(s) \in S_\Omega$, satisfying $\delta_\Omega(u(s)) = 0$ i.e. rational functions having no poles and zeros in Ω.

Going one step further, as shown in [3], one may define the euclidean division between two elements of S_Ω. Given $a(s) \neq 0, b(s) \in S_\Omega$ there exist $q(s), r(s) \in S_\Omega$ (called the quotient and remainder respectively), such that

$$b(s) = q(s)a(s) + r(s), \tag{3}$$

satisfying

$$\delta_\Omega(r(s)) < \delta_\Omega(a(s)) \text{ or } r(s) = 0 \tag{4}$$

The integral domain S_Ω equipped with the discrete evaluation $\delta_\Omega(.)$ and the above defined division is thus a euclidean ring. The *Rings* package provides the following division related functionality

`RingDivision[ns,ds,s]`	$\longrightarrow$	Returns the pair {quotient,remainder} of the division of ns by ds
`RingQuotient[ns,ds,s]`	$\longrightarrow$	Returns the quotient of the division of ns by ds
`RingRemainder[ns,ds,s]`	$\longrightarrow$	Returns the remainder of the division of ns by ds
`RingGCD[as,bs,s]`	$\longrightarrow$	Returns the greatest common divisor of as and bs
`RingCoprime[as,bs,s]`	$\longrightarrow$	Checks whether as,bs are coprime rational functions

2.1 Feedback Stabilization of SISO Plants

One of the fundamental problems in control theory is the problem of stabilization of a possibly unstable plant via feedback. Given a linear plant described by its transfer function $P(s)$, the goal of stabilization via feedback, amounts in finding a controller $C(s)$ such that the closed loop system in figure 2.1 has all its poles in a prescribed region of the complex plane.

The stability region may differ depending on the nature of the problem or certain performance requirements imposed by the designer. For instance in continuous time it is desired to move all poles in the open left half complex plane

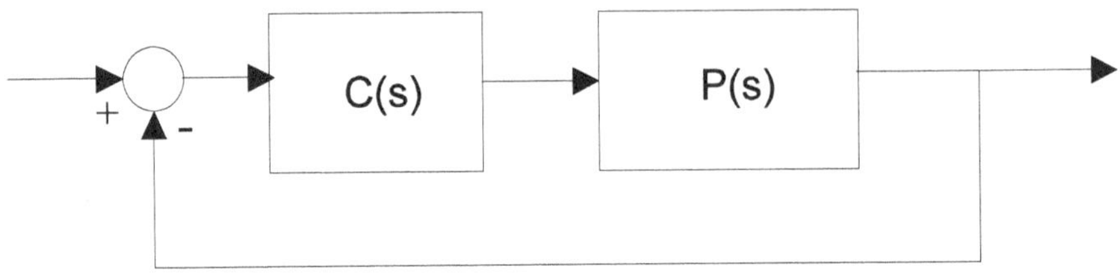

(Re $s < 0$) or even better in some subregion of the left half complex plane symmetrically located with respect to the real axis. Accordingly, in the discrete time case stability is achieved by moving the poles of the closed loop system inside the unit disc or in some subregion of the unit disc.

Let $P(s)$ be a proper rational transfer function describing the plant that we wish to stabilize using the feedback scheme of figure 2.1. We shall restrict our interest to the continuous time case thus it is desired to obtain Hurwitz stability for the closed system. According to the previous section if we set $\Omega = \{s : \mathrm{Re}(s) \geq 0\} \cup \{\infty\}$, we can write $P(s)$ as a ratio of two Ω-stable rational functions, i.e.

$$P(s) = \frac{n(s)}{d(s)},$$

with $n(s), d(s) \in S_\Omega$. We seek to find a proper rational controller $C(s)$, which we assume to be of the form

$$C(s) = \frac{y(s)}{x(s)},$$

where $x(s), y(s) \in S_\Omega$. The closed loop transfer function is given by

$$G(s) = \frac{n(s)y(s)}{d(s)x(s) + n(s)y(s)}. \tag{5}$$

For the stabilization of the closed loop system it is required to determine $x(s)$, $y(s) \in S_\Omega$ such that $G(s)$ is itself Ω-stable. This can be done if the expression $d(s)x(s) + n(s)y(s)$ is a unit of S_Ω or simply 1, i.e.

$$d(s)x(s) + n(s)y(s) = 1. \tag{6}$$

The above equation is *diophantine equation* over the ring of Ω-stable rational functions and can be solved using the euclidean algorithm. If a particular solution of (6) $x_0(s), y_0(s) \in S_\Omega$ is determined, then every other solution can be obtained from the parameterization

$$\begin{aligned} x(s) &= x_0(s) + n(s)t(s) \\ y(s) &= y_0(s) - d(s)t(s) \end{aligned} \tag{7}$$

where $t(s) \in S_\Omega$ is an arbitrary Ω-stable rational function. We illustrate this methodology via the following

Example 1. In order to meet certain performance criteria, the package allows the user to define custom stability areas and hence custom rings of Ω−stable rational functions. Such an area of the complex plane that can guarantee approximately 5% overshoot and 1.5sec settling time of the closed loop system is defined by the function

```
MyArea[s_]:=(Re[s]>=-2.7)||((-2.4<=Arg[s])&&(Arg[s]<=2.4))
```

We also need to provide an arbitrary constant $a \in \mathbb{R}$, such that $-a \in \Omega^C$. Such a choice could be

```
$MyAreaAutomaticAlpha = 3;
```

We can either set `$ForbiddenPolesArea` to `MyArea`, to change the working ring globally, or use the individual option setting for temporary use. We prefer to switch to the custom ring globally, so

```
$ForbiddenPolesArea = MyArea;
```

For the ring corresponding to `MyArea` the transfer function $P(s)$ of a given the plant can be factorized using the `RingFraction` function, which gives

```
Ps=(s-3)/(s^2 +2 s-8);
{Ns, Ds} = RingFraction[Ps, s]
```

$$\left\{\frac{s-3}{s^2+7s+12}, \frac{s-2}{s+3}\right\}$$

In order to compute a stabilizing controller for the given plant, he have to solve the diophantine equation $d(s)x(s) + n(s)y(s) = 1$, where $x(s), y(s)$ are proper and Hurwitz stable rational functions. This can be accomplished by

```
{xo, yo} = RingDiophantineSolve[Ds, Ns, 1, s]
```

$$\left\{\frac{s+33}{s+3}, -\frac{25(s+4)}{s+3}\right\}$$

so a stabilizing controller is given by

```
Cs = yo/xo
```

$$-\frac{25(s+4)}{s+33}$$

while the family of stabilizing controllers is parametrized by the forumla

$$C(s) = \frac{y(s)}{x(s)} = \frac{y_0(s) - d(s)t(s)}{x_0(s) + n(s)t(s)}$$

where $t(s) \in S_\Omega$ is an arbitrary Ω-stable rational function.

This can be computed using

```
Cs = (yo-Ds*t[s])/(xo+Ns*t[s])
```

$$\frac{-\frac{(s-2)t(s)}{s+3} - \frac{25(s+4)}{s+3}}{\frac{(s-3)t(s)}{s^2+7s+12} + \frac{s+33}{s+3}}$$

3 Conclusions

In this paper we have presented a collection of algorithms implemented in Mathematica 7.0, available freely over the internet, able to manipulate rational functions and solve control related problems using polynomial analysis and design methods. The package provides all the necessary functionality and tools in order to use the theory of $\Omega-$stable functions. The user can choose one of the common rings of $\Omega-$stable rational functions, such as the ring of Hurwitz stable, Schur stable, proper rational functions or the ring of polynomials. It is also allowed to introduce user defined rings by describing the set Ω using a simple procedure.

Some commonly used algorithms for the analysis of control systems have been implemented. The package also contains algorithms for the solution of matrix Diophantine equations over a variety of rings, solving several synthesis and design control problems such as stabilization, pole placement, dead-beat control, model matching, disturbance rejection, minimum variance control, LQG or H_2 optimal control, H_∞ optimization, tracking problems etc.The *Rings* package is expected to provide the necessary framework for the implementation of many of the algorithms for the solution of control synthesis and design problems that exist in the literature.

References

1. Kucera, V.: Stability of discrete linear feedback systems. In: Proc. IFAC World Congr., Boston, MA, vol. 1, paper 44.1 (1975)
2. Kucera, V.: Diophantine equations in control - A survey. Automatica 29(6), 1361–1375 (1993)
3. Vardulakis, A.I.G.: Linear Multivariable Control - Algebraic Analysis and Synthesis Methods. John Willey & Sons Ltd., New York (1991)
4. Youla, D.C., Jabr, H.A., Bongiorno, J.J.: Modern Wiener–Hopf design of optimal controllers. IEEE Trans. Autom. Control AC-21(3), 319–338 (1976)
5. Polynomial Toolbox for Matlab, Polyx Ltd., `http://www.polyx.cz/`
6. Control System Professional Version 2, Wolfram Research Inc., `http://www.wolfram.com/products/applications/control/`
7. Control System Toolbox, The MathWorks Inc., `http://www.mathworks.com/products/control/`
8. Sima, V.: SLICOT-based advanced automatic control computations. In: Advances in Automatic Control, pp. 337–350. Kluwer Academic Publishers, Dordrecht (2003)
9. The Control and Systems Library SLICOT, NICONET, `http://www.win.tue.nl/niconet/`
10. Pernebo, L.: An algebraic theory for the design of controllers for linear multivariable systems. I. Structure matrices and feedforward design. IEEE Trans. Automat. Control 26(1), 171–182 (1981)

Nonlinear Systems: A Polynomial Approach

Miroslav Halás

Institute of Control and Industrial Informatics
Faculty of Electrical Engineering and Information Technology
Slovak University of Technology
Ilkovičova 3, 812 19 Bratislava, Slovakia
miroslav.halas@stuba.sk

Abstract. The modern development of nonlinear control theory is related mainly to the use of algebraic methods which show great applicability to solve a number of nonlinear control problems. Recently, the power of such methods were extended by introducing the concept of transfer functions of nonlinear systems. Such a concept represents a generalization, for the transfer functions of nonlinear systems have many analogical properties like those of linear systems. In this chapter some basic properties of the transfer function formalism of nonlinear systems are briefly discussed and references to possible applications are given.

Keywords: nonlinear systems, algebraic approach, polynomial approach, transfer functions.

1 Introduction

Although the Laplace and Z transforms of nonlinear differential and respectively difference equations are not defined transfer functions of nonlinear continuous-, discrete-time and time-delay systems were developed recently. For continuous-time case it was given in [7,10,24], for discrete-time case in [12,14] and for time-delay systems in [8,9]. Such a formalism is equivalent to that of [5] for linear time-varying systems and allow us to associate to a nonlinear system the tangent (or variational) linear system, see for instance [6], over Kähler differentials [17] except that now the time-varying coefficients of the polynomials are not necessarily independent [18]. The nonlinear transfer function formalism is, in principle, similar to the linear theory, except that the polynomial description relates now the differentials of system inputs and outputs, and the resulting polynomial rings are non-commutative. Such a formalism has been already employed in [22] to investigate some structural properties of nonlinear systems, in [15] to study the nonlinear model matching, in [11] to study the observer design and in [13] to study the realization problem of nonlinear systems. In this chapter we discuss some of basic properties of transfer functions of nonlinear systems with a possible application to a few control problems, namely nonlinear model matching [15,16]. To make the text easy-to-follow, many technicalities are rather avoided and the reader is referred to the works listed above.

R. Moreno-Díaz et al. (Eds.): EUROCAST 2009, LNCS 5717, pp. 595–602, 2009.

2 Algebraic Background

We will use the algebraic formalism of [3,4,26] which employ the concept of differential one-forms to study nonlinear continuous-, discrete-time and, respectively, time-delay systems and that of [7,8,14] which introduces the transfer functions of such systems. For the sake of simplicity, here our attention is restricted to the case of SISO systems.

To make notations compact we for any variable $\xi(t)$ write only ξ. In case of differential equations, e.g. continuous-time nonlinear systems, we use the symbols $\dot{\xi}$, $\ddot{\xi}$ and $\xi^{(k)}$ to denote the first, second and k-th time derivation of $\xi(t)$ respectively. Similarly, for discrete-time case we use ξ^+, ξ^{++} and $\xi^{[k]}$ to denote the first, second and, respectively, k-th forward time shift. Finally, for nonlinear time-delay systems we mix up the two previous notations. In particular, the time delays are denoted as backward time shifts ξ^-, ξ^{--} and $\xi^{[-k]}$ respectively.

Using the introduced notation, nonlinear continuous-time systems are objects of the form

$$y^{(n)} = \varphi(y, \dot{y}, \ldots, y^{(n-1)}, u, \dot{u}, \ldots, u^{(m)}) \tag{1}$$

discrete-time systems are objects of the form

$$y^{[n]} = \varphi(y, y^+, \ldots, y^{[n-1]}, u, u^+, \ldots, u^{[m]}) \tag{2}$$

and finally nonlinear time-delay systems are objects of the form

$$y^{(n)} = \varphi(\{y^{[-k]}, \dot{y}^{[-k]}, \ldots, y^{(n-1)[-k]}, u^{[-k]}, \dot{u}^{[-k]}, \ldots, u^{(m)[-k]}; k \geq 0\}) \tag{3}$$

In all φ is assumed to be an element of the field of meromorphic functions $\mathcal{K}$ and $y, u \in \mathbf{R}$ and $m < n$.

Remark 1. In the case of systems without delays even if one starts with the usual state-space representation it is, at least locally, always possible to eliminate the state variables to get an input-output equation of the form (1) or (2) respectively, see for instance [3]. In the time-delay case the state elimination algorithm [1] might, however, results in an input-output equation representing a system of a neutral type. For the sake of simplicity, here we consider only systems that admit input-output equations of the form (3), see also [16].

We define three separate left Ore rings (algebras) $\mathcal{K}[s]$, $\mathcal{K}[\delta]$ and $\mathcal{K}[\delta, s]$ of polynomials over $\mathcal{K}$ with the usual addition and the (non-commutative) multiplications given by the commutation rules:

$$sa = as + \dot{a} \qquad \text{for } \mathcal{K}[s] \tag{4}$$

$$\delta a = a^+\delta \qquad \text{for } \mathcal{K}[\delta] \tag{5}$$

$$\begin{aligned} sa &= as + \dot{a} \\ \delta a &= a^-\delta \qquad \text{for } \mathcal{K}[\delta, s] \\ \delta s &= s\delta \end{aligned} \tag{6}$$

In all $a \in \mathcal{K}$. The rings $\mathcal{K}[s]$, $\mathcal{K}[\delta]$ and $\mathcal{K}[\delta, s]$ thus represent the rings of linear ordinary differential, shift and, respectively, differential time-delay operators that act over the vector space of one-forms $\mathcal{E} = \text{span}_{\mathcal{K}}\{\text{d}\xi; \xi \in \mathcal{K}\}$ in the following ways:

$$* : \mathcal{K}[s] \times \mathcal{E} \to \mathcal{E}; \quad \left(\sum_i a_i s^i\right) * v = \sum_i a_i v^{(i)}$$

$$* : \mathcal{K}[\delta] \times \mathcal{E} \to \mathcal{E}; \quad \left(\sum_i a_i \delta^i\right) * v = \sum_i a_i v^{[i]}$$

$$* : \mathcal{K}[\delta, s] \times \mathcal{E} \to \mathcal{E}; \quad \left(\sum_{i,j} a_{ij} \delta^j s^i\right) * v = \sum_{i,j} a_{ij} v^{(i)[-j]}$$

for any $v \in \mathcal{E}$. For the sake of simplicity the symbols $*$ are usually dropped.

Note that the commutation rules (4), (5) and, respectively, (6) actually represent the rules for differentiating, shifting and differential time-delaying respectively.

Note also that we use the same symbol δ to denote both a forward shift operator in the discrete-time case and a delay operator in the time-delay case, as it is a convention in both. Whether we think of δ as a forward shift operator (5) or a delay operator (6) will be clear from the context and the class of systems we work with.

Lemma 1 (Ore condition). *For all non-zero $a, b \in \mathcal{K}[s]$ ($\mathcal{K}[\delta]$ or, respectively, $\mathcal{K}[\delta, s]$), there exist non-zero $a_1, b_1 \in \mathcal{K}[s]$ ($\mathcal{K}[\delta]$ or, respectively, $\mathcal{K}[\delta, s]$) such that $a_1 b = b_1 a$.*

Thus, the ring $\mathcal{K}[s]$ ($\mathcal{K}[\delta]$ or, respectively, $\mathcal{K}[\delta, s]$) can be embedded to the non-commutative quotient field $\mathcal{K}\langle s\rangle$ ($\mathcal{K}\langle\delta\rangle$ or, respectively, $\mathcal{K}\langle\delta, s\rangle$) by defining quotients [21] as

$$\frac{a}{b} = b^{-1} \cdot a$$

The addition and multiplication are defined as

$$\frac{a_1}{b_1} + \frac{a_2}{b_2} = \frac{\beta_2 a_1 + \beta_1 a_2}{\beta_2 b_1}$$

where $\beta_2 b_1 = \beta_1 b_2$ by Ore condition and

$$\frac{a_1}{b_1} \cdot \frac{a_2}{b_2} = \frac{\alpha_1 a_2}{\beta_2 b_1} \tag{7}$$

where $\beta_2 a_1 = \alpha_1 b_2$ again by Ore condition.

Due to the non-commutative multiplication (4) ((5) or, respectively, (6)) they, of course, differ from the usual rules. In particular, in case of the multiplication (7) we, in general, cannot simply multiply numerators and denominators, nor cancel them in a usual manner. We neither can commute them as the multiplication of quotients is non-commutative as well.

Example 1 ([15]). Consider two quotients

$$\frac{1}{s-y}, \frac{1}{s}$$

from $\mathcal{K}\langle s\rangle$. Then

$$\frac{1}{s-y}+\frac{1}{s}=\frac{2s-y-2\dot{y}/y}{s^2-(y+\dot{y}/y)s}$$

and

$$\frac{1}{s-y}\cdot\frac{1}{s}=\frac{1}{s^2-ys-\dot{y}}\neq\frac{1}{s}\cdot\frac{1}{s-y}=\frac{1}{s^2-ys}$$

Once the fraction of two skew polynomials is defined we can introduce the transfer function of the nonlinear systems (1), (2) and (3) respectively as elements $F(s)\in\mathcal{K}\langle s\rangle$, $F(\delta)\in\mathcal{K}\langle\delta\rangle$ and, respectively, $F(\delta,s)\in\mathcal{K}\langle\delta,s\rangle$ such that $\mathrm{d}y=F(s)\mathrm{d}u$, $\mathrm{d}y=F(\delta)\mathrm{d}u$ and $\mathrm{d}y=F(\delta,s)\mathrm{d}u$ respectively.
For instance, after differentiating (3) we get

$$\mathrm{d}y^{(n)}-\sum_{i,j}\frac{\partial\varphi}{\partial y^{(i)[-j]}}\mathrm{d}y^{(i)[-j]}=\sum_{i,j}\frac{\partial\varphi}{\partial u^{(i)[-j]}}\mathrm{d}u^{(i)[-j]}$$

or alternatively

$$a(\delta,s)\mathrm{d}y=b(\delta,s)\mathrm{d}u$$

where $a(\delta,s)=s^n-\sum_{i,j}\frac{\partial\varphi}{\partial y^{(i)[-j]}}\delta^j s^i$ and $b(\delta,s)=\sum_{i,j}\frac{\partial\varphi}{\partial u^{(i)[-j]}}\delta^j s^i$ are in $\mathcal{K}[\delta,s]$.
Then

$$F(\delta,s)=\frac{b(\delta,s)}{a(\delta,s)}$$

Example 2. Consider the nonlinear time-delay system $\ddot{y}=\dot{y}^-u^-$. Then

$$\begin{aligned}\mathrm{d}\ddot{y}&=u^-\mathrm{d}\dot{y}^-+\dot{y}^-\mathrm{d}u^-\\ s^2\mathrm{d}y&=u^-\delta s\mathrm{d}y+\dot{y}^-\delta\mathrm{d}u\end{aligned}$$

and the transfer function

$$F(\delta,s)=\frac{\dot{y}^-\delta}{s^2-u^-\delta s}$$

Notice that here s and δ stand for differential and, respectively, time-delay operator (6).

3 Properties of Transfer Functions of Nonlinear Systems

Transfer functions of nonlinear systems have many properties we expect from transfer functions. They are invariant with respect to state-transformations. They provide input-output description and are related to the accessibility and observability of a nonlinear system. Finally, they also allow us to use the transfer function algebra when combining systems in series, parallel and feedback connection.

3.1 Invariance of Transfer Functions

Consider for instance the discrete-time case [14] and a system described by the following state-space representation

$$\begin{aligned} x^+ &= f(x,u) \\ y &= g(x) \end{aligned} \tag{8}$$

where the entries of f and g are from the field of meromorphic functions $\mathcal{K}$ and $x \in \mathbf{R}^n$, $u \in \mathbf{R}$ and $y \in \mathbf{R}$.

The transfer function can be computed as

$$F(\delta) = C(\delta I - A)^{-1}B \tag{9}$$

where $A = (\partial f/\partial x)$, $B = (\partial f/\partial u)$ and $C = (\partial g/\partial x)$.

Note that the entries of $(\delta I - A)$ are (non-commutative) skew polynomials and thus the inversion is not trivial, see [7], and leads to solving linear equations over non-commutative fields [20].

Proposition 1 ([14]). *Transfer function (9) of nonlinear discrete-time system (8) is invariant with respect to the state transformation $\xi = \phi(x)$.*

Proof. For any state transformation $\xi = \phi(x)$ one has $\operatorname{rank}_{\mathcal{K}} T = n$ where $T = (\partial\phi/\partial x)$. Since $\mathrm{d}\xi = T\mathrm{d}x$, in the new coordinates we have

$$\begin{aligned} \mathrm{d}\xi^+ &= T^+AT^{-1}\mathrm{d}\xi + T^+B\mathrm{d}u \\ \mathrm{d}y &= CT^{-1}\mathrm{d}\xi \end{aligned}$$

where T^+ means δ is applied entrywise to T. Thus, the transfer function reads

$$F(\delta) = CT^{-1}(\delta I - T^+AT^{-1})^{-1}T^+B = C({T^+}^{-1}\delta T - A)^{-1}B + D$$

After applying the commutation rule $\delta T = T^+\delta$ we get $F(\delta) = C(\delta I - A)^{-1}B$.

Example 3. Consider the system

$$\begin{aligned} x^+ &= \frac{e^u}{x} \\ y &= \ln x \end{aligned}$$

The input-output map of the system is linear $y^+ = -y + u$. Note that $A = \partial f/\partial x = -e^u/x^2$, $B = \partial f/\partial u = e^u/x$ and $C = \partial g/\partial x = 1/x$ and following the non-commutative multiplication rules (5) and (7) the transfer function can be computed as

$$\begin{aligned} F(\delta) = C(\delta I - A)^{-1}B &= \frac{1}{x} \cdot \frac{1}{\delta + e^u/x^2} \cdot \frac{e^u}{x} = \frac{1}{\delta x + e^u/x} \cdot \frac{e^u}{x} = \\ &= \frac{1}{e^u/x\delta + e^u/x} \cdot \frac{e^u}{x} = \frac{e^u}{e^u\delta + e^u} = \frac{1}{\delta + 1} \end{aligned}$$

For continuous-time and time-delay counterparts see [7] and [8] respectively.

Remark 2. The invariance of transfer functions can be viewed as a consequence of the change of basis formula of the pseudo-linear map [2] and also a consequence of the fact that the tangent linear systems of two equivalent nonlinear systems are almost identical [6] with that sense that certain associated left modules are isomorphic.

3.2 Accessibility and Observability

The transfer functions as those of linear systems are strongly related to the notions of accessibility (controllability) and observability of nonlinear systems.

For instance for continuous-time case we can use the concept of autonomous elements introduced in [3] which in terms of the polynomial approach leads to the presence or non-presence of common left factors of polynomials derived from input-output equation of the system [25].

Proposition 2. *Let $F(s) = \frac{b(s)}{a(s)}$ be the transfer function of the nonlinear system (1). Then the system is accessible if and only if the polynomials $a(s)$ and $b(s)$ have no common left factors.*

Also the observability condition can be stated in terms of polynomials.

Proposition 3 ([22]). *Let $F(s) = \frac{b(s)}{a(s)}$ be the transfer function of the nonlinear system*

$$\dot{x} = f(x, u)$$
$$y = g(x)$$

where $x \in \mathbf{R}^n$, $u \in \mathbf{R}$ and $y \in \mathbf{R}$. Then the nonlinear system is observable if and only if

$$\deg a(s) = n$$

Proof (Sketch). If the system is not observable one obtains, by eliminating the state variables, an i/o equation $y^{(r)} = \varphi(y, \dot{y}, \dots, y^{(r-1)}, u, \dot{u}, \dots, u^{(r-1)})$ where $r < n$ and then $\deg a(s) = r$.

For discrete-time and time-delay counterparts see [14] and [8,16] respectively.

Remark 3. In the module approach the presence of common left factors of the polynomials is equivalent to that whether the module associated to the system is torsion free or not, see for instance [23].

3.3 Transfer Function Algebra and Model Matching

When the nonlinear systems are being combined in series, parallel and feedback connection we can even use the algebra of transfer functions of nonlinear systems [7,14,8]. Following example demonstrates how to handle a series connection of two nonlinear time-delay systems.

Example 4 ([8]). Consider the two systems $\dot{y}_A = y_A u_A^-$ and $y_B = \ln u_B$ with the transfer functions

$$F_A(\delta, s) = \frac{y_A \delta}{s - u_A^-} \qquad F_B(\delta, s) = \frac{1}{u_B}$$

The systems are combined together in a series connection. For the connection A→B, when $u_B = y_A$, the resulting transfer function is

$$F(\delta, s) = F_B(\delta, s) F_A(\delta, s) = \frac{1}{u_B} \cdot \frac{y_A \delta}{s - u_A^-} = \frac{y_A \delta}{y_A s + \dot{y}_A - u_A^- y_A} = \frac{\delta}{s}$$

Hence, the combination A→B is linear from an input-output point of view which is easy to check $\dot{y}_B = u_A^-$. However, when the systems are connected as B→A, that is $u_A = y_B$, the result is different

$$F(\delta, s) = F_A(\delta, s) F_B(\delta, s) = \frac{y_A \delta}{s - u_A^-} \cdot \frac{1}{u_B} = \frac{y_A \delta}{u_B^- s - u_B^- \ln u_B^-}$$

This time, it does not yield a linear system. Note that here we used the multiplication rules (6) and (7).

The transfer function algebra can be, to advantage, employed in the nonlinear model matching problem, following the same ideas as in the linear case. The previous example serves as a stepping stone, and a typical control problem can be formulated as for given model $G(s)$ and a system $F(s)$ find a compensator $R(s)$ such that $G(s) = F(s) \cdot R(s)$. Then clearly, solution to the problem consists of computing $R(s) = F^{-1}(s) \cdot G(s)$. Preliminary problem statements and solutions to nonlinear model matching following the above mentioned ideas were given in [15,16] and have contact points to that of [19] for linear time-varying systems except that now the time-varying coefficients of the polynomials are not necessarily independent [18] and the resulting input-output differential forms of compensators might not be integrable. Other applications of the transfer function formalism of nonlinear systems can be found in [11,13].

References

1. Anguelova, M., Wennberg, B.: State elimination and identifiability of the delay parameter for nonlinear time-delay systems. Automatica 44, 1373–1378 (2008)
2. Bronstein, M., Petkovšek, M.: An introduction to pseudo-linear algebra. Theoretical Computer Science 157, 3–33 (1996)
3. Conte, G., Moog, C.H., Perdon, A.M.: Algebraic Methods for Nonlinear Control Systems, 2nd edn. Theory and Applications. Springer, London (2007)
4. Aranda-Bricaire, E., Kotta, Ü., Moog, C.H.: Linearization of discrete-time systems. SIAM J. Cont. Opt. 34, 1999–2023 (1996)
5. Fliess, M.: Une interprétation algébrique de la transformation de Laplace et des matrices de transfert. Linear Algebra and its Applications 203, 429–442 (1994)
6. Fliess, M., Lévine, J., Martin, P., Rouchon, P.: Flatness and defect of non-linear systems: introductory theory and examples. Int. J. Control 61, 1327–1361 (1995)

7. Halás, M.: An algebraic framework generalizing the concept of transfer functions to nonlinear systems. Automatica 44, 1181–1190 (2008)
8. Halás, M.: Nonlinear time-delay systems: a polynomial approach using Ore algebras. In: Loiseau, J.J., et al. (eds.) Topics in Time-Delay Systems. LNCIS, vol. 388, pp. 109–119. Springer, Heidelberg (2009)
9. Halás, M.: Ore algebras: a polynomial approach to nonlinear time-delay systems. In: 9th IFAC Workshop Time-Delay Systems, Nantes, France (2007)
10. Halás, M., Huba, M.: Symbolic computation for nonlinear systems using quotients over skew polynomial ring. In: 14th Mediterranean Conference on Control and Automation, Ancona, Italy (2006)
11. Halás, M., Kotta, Ü.: A polynomial approach to the synthesis of observers for nonlinear systems. In: 47th Conference on Decision and Control, Cancun, Mexico (2008)
12. Halás, M., Kotta, Ü.: Extension of the concept of transfer function to discrete-time nonlinear control systems. In: European Control Conference, Kos, Greece (2007)
13. Halás, M., Kotta, Ü.: Realization problem of SISO nonlinear systems: a transfer function approach. In: 7th IEEE International Conference on Control & Automation, Christchurch, New Zealand (2009)
14. Halás, M., Kotta, Ü.: Transfer Functions of Discrete-time Nonlinear Control Systems, Proc. Estonian Acad. Sci. Phys. Math. 56, 322–335 (2007)
15. Halás, M., Kotta, Ü., Moog, C.H.: Transfer function approach to the model matching problem of nonlinear systems. In: 17th IFAC World Congress, Seoul, Korea (2008)
16. Halás, M., Moog, C.: A Polynomial Solution to the Model Matching Problem of Nonlinear Time-delay Systems. In: 10th European Control Conference, Budapest, Hungary (2009)
17. Johnson, J.: Kähler differentials and differential algebra. Annals of Mathematics 89, 92–98 (1969)
18. Li, Z., Ondera, M., Wang, H.: Simplifying skew fractions modulo differential and difference relations. In: International Symposium on Symbolic and Algebraic Computation ISSAC, Linz, Austria (2008)
19. Marinescu, B., Bourlès, H.: The exact model-matching problem for linear time-varying systems: an algebraic approach. IEEE Transactions on Automatic Control 48, 166–169 (2003)
20. Ore, O.: Linear equations in non-commutative fields. Annals of Mathematics 32, 463–477 (1931)
21. Ore, O.: Theory of non-commutative polynomials. Annals of Mathematics 34, 480–508 (1933)
22. Perdon, A.M., Moog, C.H., Conte, G.: The pole-zero structure of nonlinear control systems. In: 7th IFAC Symposium NOLCOS, Pretoria, South Africa (2007)
23. Pommaret, J.F., Quadrat, A.: Localization and parametrization of linear multidimensional control systems. Systems and control letters 37, 247–260 (1999)
24. Zheng, Y., Cao, L.: Transfer function description for nonlinear systems. Journal of East China Normal University (Natural Science) 2, 15–26 (1995)
25. Zheng, Y., Willems, J., Zhang, C.: A polynomial approach to nonlinear system controllability. IEEE Transactions on Automatic Control 46, 1782–1788 (2001)
26. Xia, X., Márquez-Martínez, L.A., Zagalak, P., Moog, C.H.: Analysis of nonlinear time-delay systems using modules over non-commutative rings. Automatica 38, 1549–1555 (2002)

Robust Control of a Two Tank System Using Algebraic Approach

Marek Dlapa*, Roman Prokop, and Monika Bakosova

Tomas Bata University in Zlin
Nad Stranemi 4511, 760 05 Zlin, Czech Republic
{dlapa,prokop}@fai.utb.cz, bakosova@ka.chtf.stuba.sk

Abstract. The paper deals with design of a robust controller via algebraic μ-synthesis for a two tank system, which is a well known benchmark problem. The controller is obtained by decoupling two-input two-output system into two identical SISO (Single-Input Single-Output) plants. The task of robust controller design is then performed by finding a suitable pole placement for the SISO systems. The robustness is measured by the structured singular value denoted μ. The final controller is verified through simulation for plants perturbed by worst case perturbations.

1 Introduction

Algebraic methods ([7], [11]) are well known and easy to use for SISO (single-input single-output systems) systems described by continuous or discrete transfer functions. However, if applied to MIMO (multi-input multi-output) systems computational difficulties are increasing. In this paper, the problem of MIMO system design is treated via decoupling the MIMO system into two identical SISO plants, which are approximated by transfer functions with simple structure. This guarantees decoupled result control, and simplifies derivation of pole placement formulae.

In order to measure the robust stability and performance, the structured singular value denoted μ is used ([5], [6]). The algebraic μ-synthesis [2] overcomes some difficulties connected with *D-K* iteration, namely the fact that it does not guarantee convergence to a global or even local minimum, which leads to non-optimality of the resulting controller [10]. Moreover, controllers obtained via algebraic approach can have simpler structure as there is no need of absorbing the scaling matrices into generalized plant, and hence no need of further simplification causing deterioration of frequency properties of the controller.

In this paper, the algebraic μ-synthesis is applied to the control of a two tank system [1], which is a well known benchmark for the robust control design.

The following notation is used: $\|\cdot\|_\infty$ denotes $\mathbf{H}_\infty$ norm, $\mathbf{R}_{ps}$ is the ring of Hurwitz-stable and proper rational functions, and $\mathbf{I}_n$ is the unit matrix of dimension n.

2 Model Description

Detailed description of the two tank system is in [1] and related papers (e.g. [8] and [9]). Here only points important for the proposed method are presented.

* Corresponding author.

R. Moreno-Díaz et al. (Eds.): EUROCAST 2009, LNCS 5717, pp. 603–609, 2009.

The system consists of two water tanks in cascade depicted in Fig. 1. The upper tank (tank 1) is fed by hot and cold water via computer controllable valves. The lower tank (tank 2) is fed by water from an exit at the bottom of tank 1. A constant level is maintained in tank 2 by means of overflow. A cold water bias stream also feeds tank 2 and enables the tanks to have different steady-state temperatures.

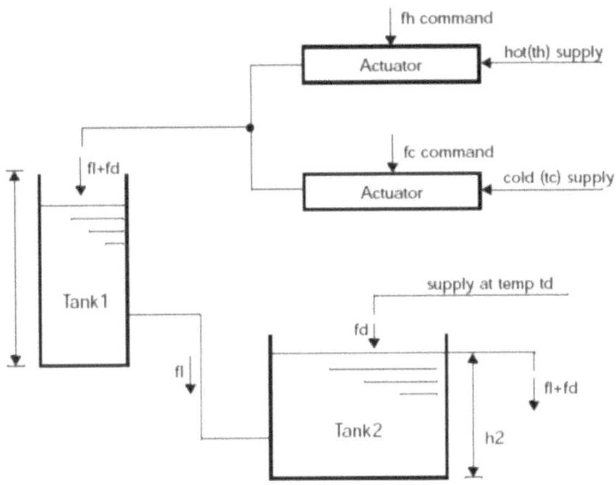

Fig. 1. Schematic diagram of the two tank system

From the brief description follows that the two tank system is a MIMO plant with two measured signals: t_1, t_2 and two inputs f_h and f_c. The quantities t_1, t_2 represent temperatures of tank 1 and 2 respectively. The input signals are commands to hot flow (f_{hc}) and cold flow (f_{cc}) actuators, which are transformed to hot water flow f_h and cold water flow f_c. Third measured signal is water level in tank 1 (h_1), which is not controlled. This quantity is, however, important for assessment of controller performance.

Due to linearization quantities h_1 and t_1 should be in a prescribed range:

$$0.25 \le h_1 \le 0.75 \tag{1}$$

$$0.25 \le t_1 \le 0.75 \tag{2}$$

Nominal model from the inputs f_h, f_c to the outputs t_1, t_2 can be written in two dimensional transfer matrix

$$\mathbf{P}_{nom}(s) \equiv \begin{bmatrix} P_{11}(s) \, P_{12}(s) \\ P_{21}(s) \, P_{22}(s) \end{bmatrix} \equiv$$

$$\equiv \begin{bmatrix} \dfrac{0.0036s^2 + 0.0001s + 7.8157 \cdot 10^{-7}}{s^3 + 0.0491s^2 + 0.0007s + 3.0684 \cdot 10^{-6}} & \dfrac{0.0004s + 4.6643 \cdot 10^{-6}}{s^3 + 0.0491s^2 + 0.0007s + 3.0684 \cdot 10^{-6}} \\ \dfrac{-0.0109s^2 - 0.0004s - 2.3447 \cdot 10^{-6}}{s^3 + 0.0491s^2 + 0.0007s + 3.0684 \cdot 10^{-6}} & \dfrac{0.1562 \cdot 10^{-4} s + 0.0187 \cdot 10^{-4}}{s^3 + 0.0491s^2 + 0.0007s + 3.0684 \cdot 10^{-6}} \end{bmatrix} \tag{3}$$

The differences from the nominal model are treated by multiplicative perturbations at the outputs of measured quantities (Fig. 2).

Perturbation weights are transfer functions of the form:

$$W_{h1} = 0.01 + \frac{0.5s}{0.25s+1} \tag{4}$$

$$W_{t1} = 0.1 + \frac{20\hat{h}_1 s}{0.2s+1} \tag{5}$$

$$W_{t2} = 0.1 + \frac{100s}{s+21} \tag{6}$$

here $\hat{h}_1 = 0.75$ is steady state value of h_1. For details on nominal models of tank 1 and 2 see [1].

Sensor noise is modelled by adding weighted unknown input to h_1, t_1, and t_2. The appropriate weights are

$$W_{t1noise} = 0.03 \tag{7}$$

$$W_{t2noise} = 0.03 \tag{8}$$

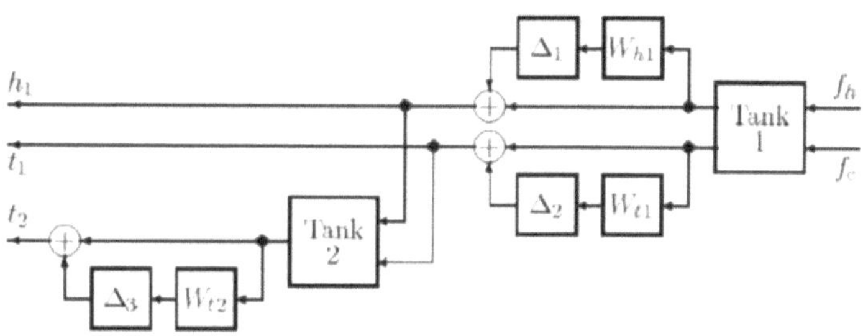

Fig. 2. Schematic representation of the perturbed, linear, two tank model

3 Algebraic μ-Synthesis

For the purposes of algebraic design the MIMO system is decoupled into two identical SISO plants. For decoupling of the nominal plant $\mathbf{P}_{nom}$ it is satisfactory to have the controller in the form

$$\mathbf{K}(s) = K(s)\mathbf{I}_2 \det[\mathbf{P}_{nom}(s)] \frac{1}{P_{11}(s)} [\mathbf{P}_{nom}(s)]^{-1} \tag{9}$$

The choice of decoupling matrix prevents controller from cancelling any poles or zeroes in the right half-plane so that the internal stability of the nominal feedback loop is held. The MIMO problem is now transformed into finding a controller $K(s)$, which is tuned via setting the poles of the nominal feedback loop with the plant

$$\mathbf{P}_{dec}(s) = \frac{1}{P_{11}(s)} \det[\mathbf{P}_{nom}(s)][\mathbf{P}_{nom}(s)]^{-1}\mathbf{P}_{nom}(s) = \frac{1}{P_{11}(s)} \det[\mathbf{P}_{nom}(s)]\mathbf{I}_2 \tag{10}$$

Let

$$P_{dec} \equiv \frac{1}{P_{11}(s)} \det[\mathbf{P}_{nom}(s)] \tag{11}$$

Then the transfer function P_{dec} can be approximated by 2nd order system

$$P^*_{dec}(s) = \frac{b_{dec}(s)}{a_{dec}(s)} = \frac{0.0027s^2 - 0.0017s + 0.0001}{s^2 + 0.0164s + 0.0001} \tag{12}$$

and the controller $K = N/M$ is obtained by solving the Diophantine equation

$$A_{dec}M + B_{dec}N = 1 \tag{13}$$

with A_{dec}, B_{dec}, $M, N \in \mathbf{R}_{ps}$.

It follows from the algebraic theory that the asymptotic tracking of the reference signal is achieved if and only if $A_{dec}M$ is divisible by F_r and disturbance is suppressed if $A_{dec}M$ is divisible by F_d. Here F_r and F_d are Laplace transforms of the reference and disturbance respectively. By the analysis of polynomial degrees of a_{dec} and b_{dec} transfer functions A_{dec}, B_{dec}, M and N are chosen so that the number of closed loop poles is minimal and the asymptotic tracking for step reference signal is achieved:

$$A_{dec} = \frac{a_{dec}}{\prod_{i=1}^{2}(s+\alpha_i)}, \quad B_{dec} = \frac{b_{dec}}{\prod_{i=1}^{2}(s+\alpha_i)} \tag{14}$$

$$M = \frac{sm}{\prod_{i=3}^{4}(s+\alpha_i)}, \quad N = \frac{n}{\prod_{i=3}^{4}(s+\alpha_i)} \tag{15}$$

Degrees of polynomials m and n are:

$$\partial m = 1, \partial n = 2 \tag{16}$$

Thus the characteristic polynomial of the nominal closed loop has 4 pairs of poles $-\alpha_i$, which represent the tuning parameters. The resulting controller is

$$K(s) = \frac{n}{sm} \tag{17}$$

The open-loop interconnection is in Fig. 3 with performance and reference weights

$$W_{t1perf} = \frac{130s+1}{400s+1}, \; W_{t2perf} = \frac{130s+0.5}{800s+1} \tag{18}$$

$$W_{t_{1cmd}} = 1, \; W_{t_{diffcmd}} = 1 \tag{19}$$

$$\begin{bmatrix} t_{1err} \\ t_{2err} \end{bmatrix} = \begin{bmatrix} 1 & 0 \\ 1 & 1 \end{bmatrix} \begin{bmatrix} W_{t_{1cmd}} & 0 \\ 0 & W_{t_{diffcmd}} \end{bmatrix} \begin{bmatrix} w_{11} \\ w_{12} \end{bmatrix} - \begin{bmatrix} t_1 \\ t_2 \end{bmatrix} - \begin{bmatrix} W_{t1noise} & 0 \\ 0 & W_{t2noise} \end{bmatrix} \begin{bmatrix} w_{21} \\ w_{22} \end{bmatrix} \tag{20}$$

Controller inputs are tracking errors of measured temperatures t_{1err}, t_{2err}. The relaxed performance weight is justified by additional postulate of decoupled result control for the nominal loop, which is not common in standard design, and which makes the task of achieving the robust performance and stability more difficult. This modification of the interconnection does not degrade the uncertainty model. The resulting performance can be observed by simulations for nominal and perturbed plants.

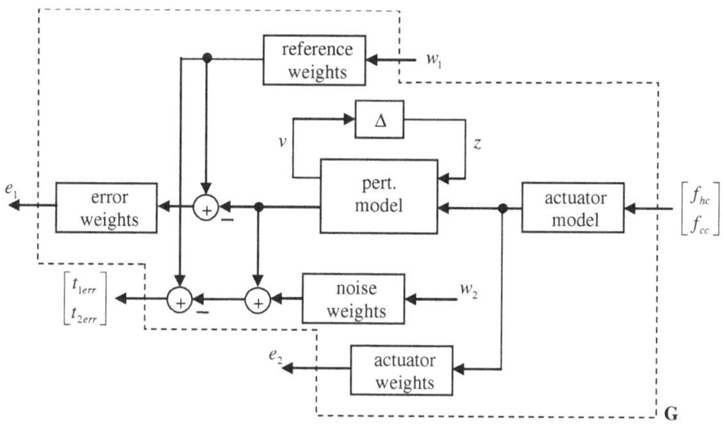

Fig. 3. Interconnection structure for algebraic approach

The controller design is then reduced to minimization of the peak of the μ-function. The cost function is defined as

$$\sup_{\omega \in \mathbf{R}} \mu_{\tilde{\Delta}}[\mathbf{F}_l(\mathbf{G},\mathbf{K})] \tag{21}$$

Here $\tilde{\Delta}$ denotes augmented perturbation including performance weights in feedback loop.

In order to overcome the multimodality of the cost function an evolutionary algorithm Differential Migration (e.g. [3]) was used for searching optimal values of α_i. Rough results obtained from DM were tuned up by the Nelder-Mead simplex method. Poles were constrained to the interval of -300 to 0.

Simulations for worst case perturbations are in Fig. 4 and 5. It can be observed that no steady state error is present, and response to ramp reference signal is monotonous, which is not true for the *D-K* iteration.

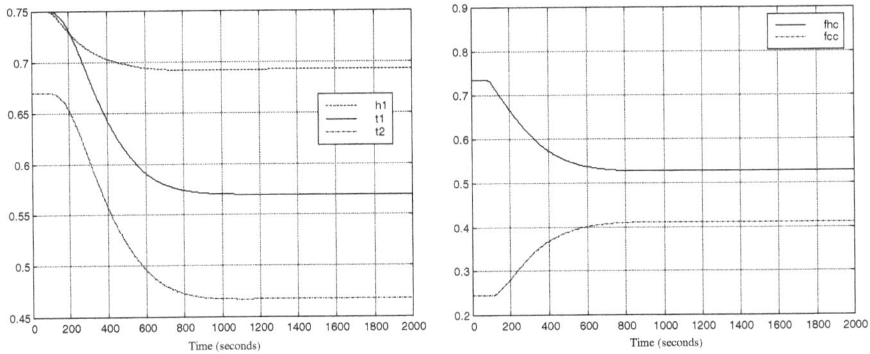

Fig. 4. Response to reference signal which ramps (from 80 to 100 seconds) t_1 from 0.75 to 0.57, and t_2 from 0.67 to 0.47 for algebraic approach and perturbed plant

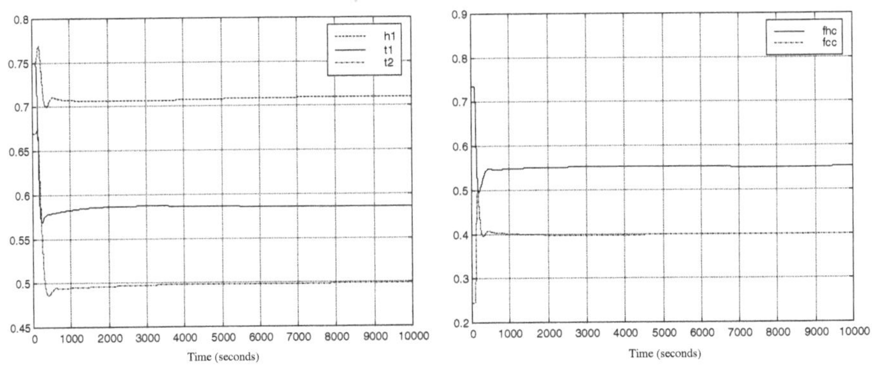

Fig. 5. Response to reference signal which ramps (from 80 to 100 seconds) t_1 from 0.75 to 0.57, and t_2 from 0.67 to 0.47 for *D-K* iteration and perturbed plant

4 Conclusion

The paper has presented another application of the algebraic approach to a MIMO system. The plant has been decoupled into two identical SISO systems and the controller has been designed via optimization of nominal plant poles. The performance and robustness have been measured by supremum of the μ-function in frequency domain. Besides its simpler structure, the resulting controller satisfies robust performance condition and guarantees the robust stability. Simulation proved that the algebraic approach has monotonous response and fast set point tracking for ramp reference signal. Moreover, the asymptotic tracking is achieved, which is not held for the reference method. The better performance of the controller is due to the fact that the algebraic method implements decoupled control for the nominal closed loop. This scheme cannot be used in the scope of the *D-K* iteration. The *D-K* iteration yields higher robustness as it fully utilizes the MIMO structure of the controller, and makes a trade-off between robust stability and performance. However, the higher stability is achieved at the expense of worse performance.

Although it is apparent that generally the presented approach cannot substitute the standard method, in this case, the algebraic design proves better performance than the standard procedure, and confirms the results obtained in [4].

Acknowledgment

This work was supported by the project MSM7088352102.

References

1. Balas, G.J., Doyle, J.C., Glover, K., Packard, A., Smith, R.: μ- Analysis and Synthesis Toolbox for Use with MATLAB. The MathWorks, Inc. (1998)
2. Dlapa, M., Prokop, R.: μ-Synthesis: Simple Controllers for Time Delay Systems. In: Proceedings of 11th Mediterranean Conference on Control and Automation, Rhodes, Greece (2003)
3. Dlapa, M., Prokop, R.: Differential Migration: a new algorithm for global optimization. In: Proceedings of the 6th Asian Control Conference, pp. 618–626 (2006)
4. Dlapa, M., Prokop, R.: Control of the HIMAT Aircraft via Algebraic μ-Synthesis. In: Proceedings of the 7th Portuguese Conference on Automatic Control (2006)
5. Doyle, J.C.: Analysis of Feedback Systems with Structured Uncertainties. Proceedings of IEEE, Part-D 129, 242–250 (1982)
6. Doyle, J.C.: Structure Uncertainty in Control System Design. In: Proceedings of 24th IEEE Conference on Decision and Control, pp. 260–265 (1985)
7. Kučera, V.: Discrete Linear Control: The Polynomial Equation Approach. Wiley, New York (1972)
8. Smith, R.S., Doyle, J., Morari, M., Skjellum, A.: A case study using μ: Laboratory process control problem. In: Proc. Int. Fed. Auto. Control, vol. 8, pp. 403–415 (1987)
9. Smith, R.S., Doyle, J.: The two tank experiment: A benchmark control problem. In: Proc. Amer. Control Conf., vol. 3, pp. 403–415 (1988)
10. Stein, G., Doyle, J.: Beyond Singular Values and Loopshapes. AIAA Journal of Guidance and Control 14(1), 5–16 (1991)
11. Vidyasagar, M.: Control Systems Synthesis: A Factorization Approach. MIT Press, Cambridge (1985)

Comparing Algebraic and Constrained Pole Assignment Controllers for a Thermal System

Mikuláš Huba[1,2], František Jelenčiak[2], and Peter Ťapák[1]

[1] Institute of Control and Industrial Informatics
Faculty of Electrical Engineering and IT, Slovak University of Technology
Ilkovičova 3, 812 19 Bratislava, Slovakia
[2] MI/PRT, FernUniversität in Hagen, Universitätsstr. 27,
D-58084 Hagen, Germany
{peter.tapak,mikulas.huba}@stuba.sk

Abstract. This paper compares two approaches to the control design for a thermal plant. This is dominant by two modes of the heat transfer - the fast mode (radiation) and the slow mode (conduction), by the nonlinear state dependent dynamics and by necessity to apply robust control design approaches. The algebraic and the constrained pole assignment controllers are compared both from the point of view of the nominal tuning required for achieving fastest possible monotonic transient responses and from the availability and easy of the controller robustification. The real experiment is used to verify the results.

Keywords: algebraic approach, constrained control, pole assignment control, thermal process.

1 Introduction

The temperature control represents one of the basic tasks in many areas of control applications. At the same time, by its specific dynamics with the fast and slow mode [1], [2], by the nonlinear character, available sensors, easy construction and maintenance and an easy physical interpretation of running processes, control of scaled laboratory thermal plant models is appropriate also for demonstrating and comparing different methods and tools of identification and control [6], [8], [14], [15], [16], [18]. The two modes of control bring several challenges for the designer. As e.g. reported by Åström and coworkers [1], properties important for control are hidden in the step response that is dominated by the slow time constant, but for the controller tuning the faster mode is critical. These "tricky" properties showed to be not easy to deal with by the traditional approaches of the PID control. Furthermore, since the dynamics is typically nonlinear, it varies with the operating point and so the controller is typically designed by robust approaches. Besides of different "advanced" approaches as e.g. [14], or [16], recently this problem was approached by algebraic design [15] and showed to give promising results. As it was shown by simulation [9] for the batch example with the fast and slow mod presented by Åström and coworkers [1], [2] the constrained pole assignment controller gives reasonably better control quality than

R. Moreno-Díaz et al. (Eds.): EUROCAST 2009, LNCS 5717, pp. 610–617, 2009.

the "best" PI controller [1]. When based on the Integral Square Error (ISE), the disturbance step response ISE value falls down to approximately 45% of the value corresponding to the PI control and for the step response it simultaneously achieves just 82% of the PI controller ISE value! Furthermore, while the traditional PI controller is practically useless without additional anti-windup circuitry, for the new solution no additional circuitry is required. This paper enables immediate comparison of the algebraic and the constrained pole assignment control approaches by controlling the same plant. It shows weak and strong points of both approaches and outlines several possible directions for future research. One possibility is to evaluate influence of the nonlinear terms in the thermal plant dynamics, or the robustness analyses of the discrepancies between the linear model and local dynamical properties. It would be also interesting to evaluate and compare different types of the approximation of small time delays that are not considered within the controller structure but finally determine the achievable rate of the output changes under monotonic dynamics and the best possible achievable control quality.

2 Thermo-Optical Plant

The thermo-optical plant laboratory model (see Fig. 1) offers measurement of 8 process variables: controlled temperature, its filtered value, ambient temperature, controlled light intensity, its derivative and filtered value, the fan speed of rotation and current. The temperature and the light intensity control channels are interconnected by 3 manipulated voltage variables influencing the bulb (heat & light source), the light-diode (the light source) and the fan (the system cooling). Besides these, it is possible to adjust two parameters of the light intensity differentiator. Within Matlab/Simulink or Scilab/Scicos schemes [11] the plant is represented as a single block and so limiting needs on costly and complicated software packages for real time control. The (supported) external

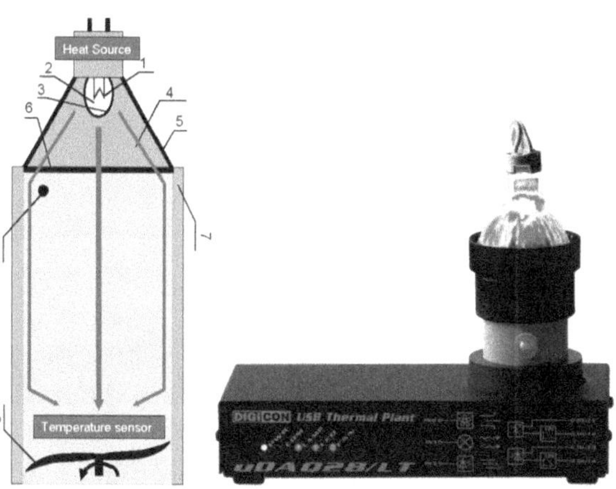

Fig. 1. The thermo-optical plant and scheme of its thermal channel

converter cards are necessary just for sampling periods below 50ms. Currently, more than 40 such plants are used in labs of several EU universities.

The thermal plant consists of a halogen bulb 12V DC/20W (elements 1-6), of a plastic pipe wall (element 7), of its internal air column (element 8) containing the temperature sensor PT100, and of a fan 12V DC/0,6W (element 9 that can be used for producing disturbances, but also for control). Next, we will consider temperature control by the bulb voltage. In paper [10] the plant dominant dynamics was analytically described by the 2nd order nonlinear model. However, it is well known that the heat is usually transferred by three different modes - conduction, convection and radiation. In this paper we will experimentally analyze this physical problem from the control point of view.

3 The Algebraic Approach

The first order model with additional accumulative delay (time constant $T_d << T_1$ approximating the nonmodelled dynamics) is one of the simplest models (1) one can choose to describe the plant's dynamics. It is simple to obtain the model parameters and it allows designing a PI controller.

$$G_1(s) = \frac{K_1}{T_1 s + 1} \frac{1}{T_d s + 1} = \frac{K_s}{s + a} \frac{1}{T_d s + 1} \tag{1}$$

There are many ways to design a controller for plant (1). We would like to compare the algebraic approach and the constrained pole assignment using this simple model. Let us start with algebraic approach.

The following controller design can be found in [15]. All the stabilizing controllers are given through a solution of a diophanitne (Bézout) equation

$$AP_0 + BQ_0 = 1 \tag{2}$$

in parametric form

$$\frac{Q}{P} = \frac{Q_0 - AT}{P_0 + BT} \tag{3}$$

for the system (1) without the accumulative delay

$$G(s) = \frac{b_0}{s + a_0} = \frac{B}{A} \tag{4}$$

it leads to 1DOF controller [15]

$$\frac{Q}{P} = \frac{\ddot{q}_1 s + \ddot{q}_0}{s}; \ddot{q}_1 = \frac{2m - a_0}{b_0}; \ddot{q}_0 = \frac{m^2}{b_0} \tag{5}$$

which is a traditional PI control law. The proper tuning of parameter m is analyzed in [15]. The table of the control loop characteristics corresponding to particular m is introduced in [15], as well.

4 PI_1-Controller

Let us consider a piecewise constant reference signal $w(t)$, the controlled output variable $y(t)$, the control (manipulated) variable $u(t)$ and the required closed loop dynamics

$$F_w(s) = Y(s)/W(s) = -\alpha/(s-\alpha) \tag{6}$$

characterized by the closed loop pole $\alpha < 0$. In the presence of input disturbances v, the required dynamics (6) can be achieved by the P-controller

$$u_r = sat\{K_R e + u_w - v\}; e = w - y; K_R = \frac{\alpha T_1 + 1}{K_1}; u_w = \frac{w}{K_1} \tag{7}$$

In order to get monotonic transients without overshooting in the presence of the nonmodelled dynamics approximated by the time constant T_d, the closed loop pole should be restricted to the interval

$$\alpha \in < -\frac{(1+aT_d)^2}{4T_d}, 0 > \tag{8}$$

whereby the limit admissible pole value

$$\alpha_e = -\frac{(1+aT_d)^2}{4T_d} \tag{9}$$

corresponds to the double real dominant pole of the closed loop system with the P-controller (7) and the plant (1). The disturbance observer (DOB) based I-action can be introduced by reconstructing the plant input disturbance v by means of an inverse plant model as

$$v = \frac{1}{K_1}\frac{1+T_1 s}{1+T_f s}Y(s) - \frac{1}{1+T_f s}U_r(s) \tag{10}$$

5 $PI_1 - P$ Controller

Constrained pole assignment control aproach can be used also, when considering the nominal dynamics of the plant with two different modes - the slow and fast one [6], [8], and [18], it could correspond to thermal plant with two ways of heat transfer [16] - e.g. with the heat radiation (fast mode) and the heat conduction via body of the plant (slow mode). Such a situation could be characterized by the plant transfer function

$$G_2(s) = \left(\frac{K_1}{T_1 s + 1} + \frac{K_2}{T_2 s + 1}\right)\frac{1}{T_d s + 1} \tag{11}$$

Without the time delay T_d that characterizes the non-modelled loop dynamics, the output of both channels can be described as

$$\dot{y}_1 = (K_1 u - y_1)/T_1; \dot{y}_2 = (K_2 u - y_2)/T_2 \tag{12}$$

For $T_d = 0$ and the system output

$$y = y_1 + y_2 \quad (13)$$

one can write

$$\dot{y} = \dot{y}_1 + \dot{y}_2 = \bar{K}u - \frac{1}{T_1}y - \tau y_2; \bar{K} = \left(\frac{K_1}{T_1} + \frac{K_2}{T_2}\right) \quad (14)$$

$$e = w - y; \dot{e} = -\dot{y} \quad (15)$$

The control signal u_w that maintains the system output at $w = const$ is

$$u_w = w/\left(K_1 + K_2\right) \quad (16)$$

Using this control the steady state outputs of the particular channels are

$$y_{1\infty} = K_1 u_w = wK_1/\left(K_1 + K_2\right); y_{2\infty} = K_2 u_w = wK_2/\left(K_1 + K_2\right) = w_2 \quad (17)$$

The pole assignment control requires control error dynamics described by

$$\dot{e} = \alpha e \quad (18)$$

whereby α is a chosen closed loop pole. Substituting into (15) one gets

$$-\bar{K}u + \left(y - w + w\right)/T_1 + \tau\left(y_2 - w_2 + w_2\right) = \alpha(w - y) \quad (19)$$

that corresponds to parallel structure of the P-P controller

$$u = \frac{1}{\bar{K}}w + K_R e + K_{R2} e_2 \quad (20)$$

$$e_2 = w_2 - y_2; K_R = \frac{\alpha + 1/T_1}{\bar{K}}; K_{R2} = \frac{T_1 - T_2}{K_1 T_2 - K_2 T_1} \quad (21)$$

This $P - P$ controller can be expanded to $PI_1 - P$ controller by the I-action designed as the disturbance reconstruction and compensation.

6 Real Experiment

All three designed controllers are compared using real experiment. The first order model's (1) parameters are $T_1 = 505; K_1 = 7.8; T_d = 20$. The model with two first order channels (11) parameters are $T_1 = 989; T_2 = 66; K_1 = 4.7; K_2 = 3.2; T_d = 20$. The PI-controller was tuned according to [15], where the transients whithout overshooting have been chosen as required. The PI_1-controller's closed loop pole $\alpha = \alpha_e/1.3 = -0.009$, corresponds to equivalent pole (9) which has been 'slowed down' to obtain the fastest posible monotonic transients when the disturbance observer I-action is present. The $PI_1 - P$-controller's closed loop pole $\alpha = \alpha_e/1.3 = -0.016$ was chosen to obtain the fastest posible monotonic transients, as well. The results of the experiments are in Fig. 2.

It is obvious that the simplest constrained pole assignment controller, PI_1-controller, gives better results than the classic PI-controller. The $PI_1 - P$-controller based on regular control error decrease of the system (11) gives much faster transients with small overshoot caused by the thermal plant nonlinearities, it gives best results in the step response and the disturbance rejection.

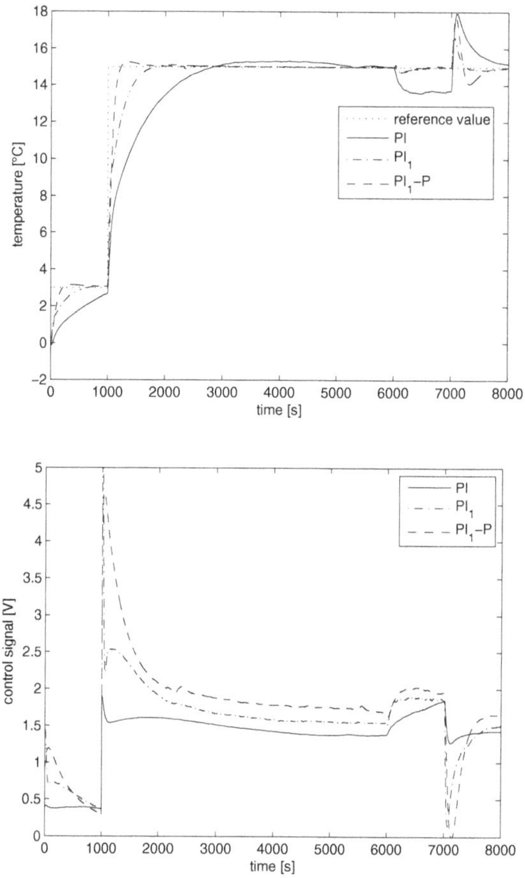

Fig. 2. Real experiments comparison

7 Conclusions

Several control design methods have been applied to a laboratory model of the thermal plant. It is hard to decide which one is better, however, not all solutions of algebraic design have attractive features. Pole assignment control enables modular treatment of properties for a broader group of control problems, it gives dynamics scalable by the closed loop pole from the fully linear one up to the minimum time control. Constrained control requires to use special structures and to keep specific rules. The aim was to present the advantage of transparent interpretation of the constrained pole assignment control. The algebraic approach can lead to control loops containing the disturbance observer circuitry, as well (see e.g. [23]).

References

1. Åström, K.J., Panagopoulos, H., Hägglund, T.: Design of PI Controllers based on Non-Convex Optimization. Automatica 34, 585–601 (1998)
2. Åström, K.J., Hägglund, T.: Benchmark Systems for PID Control. In: IFAC Workshop on Digital Control - Past, present and future of PID Control, Terassa, Spain, pp. 181–182 (2000)
3. Huba, M.: Constrained pole assignment control. In: Menini, L., Zaccarian, L., Abdallah, Ch.T. (eds.) Current Trends in Nonlinear Systems and Control, pp. 163–183. Boston Birkhäuser, Basel (2006)
4. Huba, M., Kurčík, P., Kamenský, M.: Thermo-Optical Plant uDAQ28/LT Technical and User Manual, STU Bratislava (2006)
5. Huba, M., Šimunek, M.: Modular Approach to Teaching PID Control. IEEE Trans. on Industrial Electronics 54(6), 3112–3121 (2007)
6. Huba, M., Vrančič, D.: Constrained control of the plant with two different modes. In: Int. Conf. Process Control 2007, Strbske Pleso, Hight Tatras (2007)
7. Jelenčiak, F., Mikleš, J.: Process identification: Nonlinear systems. In: Proceeding of the 13th Int. Conf. Process Control 2001, Strbske Pleso, pp. 80–89 (2001)
8. Jelenčiak, F., Kurčík, P., Huba, M.: Thermal plant for education and training. In: IEEE Int. Conf. ERK 2007, Portorož, Slovenia (2007)
9. Jelenčiak, F., Huba, M., Masár, I., Gerke, M.: Constrained pole assignement control for nonlinear plant. In: Buletinul institutului politehnic Iasi Tomul Liv (LVIII), FASC, Electrotechnică, energetică, electronică, vol. 4, pp. 547–554 (2008)
10. Jelenčiak, F.: Identification of nonlinear plants by modified method of consecutive integrals. PhD Thesis, STU Bratislava (2008) (in Slovak)
11. Kamenský, M., Kurčík, P., Huba, M.: Scicos possibilities for communication with real processes. In: IEEE Int. Conf. ERK2007, Portorož, Slovenia, pp. 31–33 (2007)
12. Kulhavy, R., Karny, M.: Tracking of slowly varying parameters by directional forgetting. In: Proc. 9th IFAC World Congr., Budapest, Hungary, pp. 79–83 (1984)
13. Ljung, L.: System Identification: Theory for User. Prentice-Hall, Englewood Cliffs (1987)
14. Matko, D., Kavsek-Biasizzo, K., Skrjanc, I., Music, G.: Generalized predictive control of a thermal plant using fuzzy model. In: Proc. American Control Conf., vol. 3, pp. 2053–2057 (2000)
15. Matuš, R., Prokop, R.: Single-Parameter Tuning of PI Controllers: From Theory to Practice. In: Proc. 17th IFAC World Congress Seoul, pp. 4964–4969 (2008)
16. Milanic, S., Karba, R.: Modelling the Thermal Plant with Artificial Neural Networks. In: Proc. 4th European Congress on Intelligent Techniques and Soft Computing, Aachen, Germany, vol. 1, pp. 289–293 (1996)
17. Strejc, V.: Auswertung der dynamischen Eigenschaften von Regelstrecken bei gemessenen Ein- und Ausgangssignalen allgemeiner Art, Messen, Steuern, Regeln, vol. 3, pp. 7–11 (1960)
18. Ťapák, P., Huba, M., Žáková, K.: Constrained Control for Systems with Relative Degree One. In: Proc. 17th IFAC World Congress, Seoul, Korea, July 6-11, 2008, pp. 5814–5819 (2008)
19. Unbehaunen, H., Rao, G.P.: Identification of Continuous Systems. North Holland Systems and Control Series, vol. 10 (1987)
20. Vidyasagar, M.: Nonlinear Systems Analysis. Prentice Hall, Englewood Cliffs (1993)

21. Welty, J., Wicks, C.E., Wilson, R.: Fundamentals of Momentum, Heat, and Mass Transfer. John Wiley, New York (1984)
22. Zhou, K., Doyle, J.C., Glover, K.: Robust and Optimal Control. Prentice Hall, Englewood Cliffs (1996)
23. Umeno, T.Y.H.: Robust speed control of dc servomotors using modern two-degree-of-freedom controller design. IEEE Trans. Ind. Electr. 38, 363–368 (1991)

Nonlinear Controllers for a Fluid Tank System

Vladimír Žilka, Miroslav Halás, and Mikuláš Huba

Institute of Control and Industrial Informatics
Faculty of Electrical Engineering and IT, Slovak University of Technology
Ilkovičova 3, 812 19 Bratislava, Slovakia
{vladimir.zilka,miroslav.halas,mikulas.huba}@stuba.sk

Abstract. This chapter deals with an application of the algebraic formalism in nonlinear control systems on the nonlinear controller design for a fluid tank system. A nonlinear discrete-time model of a fluid tank system and its transfer function are derived. Then, nonlinear continuous- and discrete-time controllers are designed using transfer function formalism for nonlinear systems which was developed recently. Verification on the real plant is also included and it suggested the modification of the original model of one-tank system to achieve better performance on the real plant.

Keywords: nonlinear systems, polynomial approach, transfer functions, fluid tank system.

1 Introduction

An algebraic formalism in nonlinear control systems, both continuous- [1] and discrete-time [2], shows great applicability in solving a number of control problems, like decompositions to canonical forms, feedback linearization, disturbance decoupling problem, to name a few possibilities. A power of such a formalism was recently extended by introducing transfer functions of nonlinear systems [3,4,5,6,7]. Such objects show many properties we expect from transfer functions. One of them is the possibility to use the transfer function algebra when combining systems in series, parallel and feedback connection. In this paper, we depict how such a transfer function formalism can be adopted to design a nonlinear controller, both continuous- and discrete-time, for a fluid tank system. Our attention is restricted to a controller which linearizes closed loop, eliminates an input disturbance and deals with the controller output constraint. Results are demonstrated on a real fluid tank system.

2 Fluid Tank System

Algebraic methods in nonlinear control theory shows great applicability to solve a number of nonlinear control problems. However, nice theoretical results are only occasionally carried over and implemented in practice where also the problems with noises, model inaccuracies, etc., have to be taken into account. Thus, it is

R. Moreno-Díaz et al. (Eds.): EUROCAST 2009, LNCS 5717, pp. 618–625, 2009.

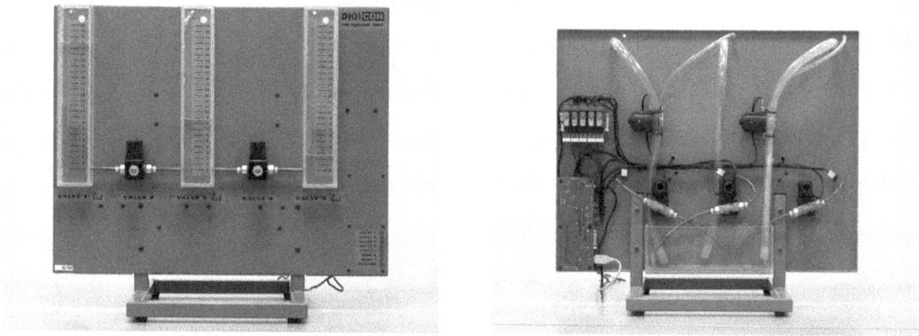

Fig. 1. Three-tank system - front and back view

important to try to do so at least on real laboratory plants. In this work the well known fluid tank system serves to that purpose. In Fig. 1 we can see front and back view of such a laboratory plant. This system can be used as one-, two- or three-tank system, SISO or even MIMO, as it has two pumps. It is quite easy to use, for the communication is established via USB interface and there is no need of additional data acquisition boards and the plant does no require even any special maintenance. In addition, a liquid flow process makes it ideal for use in teleexperiments with a web camera visual feedback.

3 Nonlinear Models of the Fluid Tank System

Consider the fluid tank system described by the state-space equations

$$\begin{aligned} \dot{x} &= \frac{1}{A}u - c\sqrt{x} \\ y &= x \end{aligned} \tag{1}$$

where x denote a level of a liquid, A denote a tank area and c is a flow coefficient.

Due to the nonlinear relations we are not able to find any solution to (1). Hence, to find a nonlinear discrete-time model of (1) we use Euler's approximation

$$\dot{x} = \frac{\mathrm{d}x(t)}{\mathrm{d}t} \approx \frac{\Delta x(t)}{\Delta t}$$

Clearly, we can think of Δt as a sampling period T which implies that $\Delta x(t) = x(t+1) - x(t)$. That is

$$\dot{x} \approx \frac{x^+ - x}{T}$$

Hence, from (1) we get the following nonlinear discrete-time approximation

$$\begin{aligned} x^+ &= x + \frac{T}{A}u - cT\sqrt{x} \\ y &= x \end{aligned} \tag{2}$$

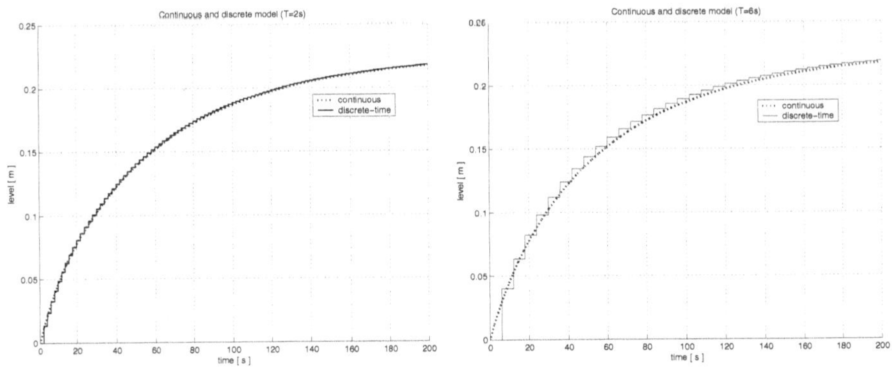

Fig. 2. Transient responses: nonlinear discrete-time approximation of the fluid tank system with $A = 0.001, c = 0.0249$ and $T = 2s$ (on the left) and $T = 6s$ (on the right)

where x^+ stands for $x(t+1)$. Obviously, the less sampling period T we choose the more accurate approximation we get. In Fig. 2 we can see the simulation results for the system with T chosen to be $2s$ and, respectively, $6s$.

Now, following the lines of [6], the transfer function of (2) can be computed as

$$\mathrm{d}x^+ = \mathrm{d}x + \frac{T}{A}\mathrm{d}u - \frac{cT}{2\sqrt{x}}\mathrm{d}x$$

$$(z - 1 + \frac{cT}{2\sqrt{x}})\mathrm{d}x = \frac{T}{A}\mathrm{d}u$$

and

$$F(z) = \frac{K}{z - D} \tag{3}$$

where $K = \frac{T}{A}$ and $D = 1 - \frac{cT}{2\sqrt{x}}$.

4 Constrained Discrete-Time Nonlinear Controller

Now, we use the discrete-time model of the fluid tank system (2) and its transfer function (3) to design a nonlinear discrete-time controller. The aim is to design a controller which:

- satisfies a linearity of the closed loop,
- eliminates an input disturbance and
- deals with a control signal constraint.

The requirement of the closed loop linearity can be satisfied easily by a feedback linearization. If we want the closed loop dynamics to be determined by a time constant T_1 we obtain regular static state feedback

$$u = [(1 - D_1)w + cT\sqrt{x} - (1 - D_1)x]\frac{A}{T} \tag{4}$$

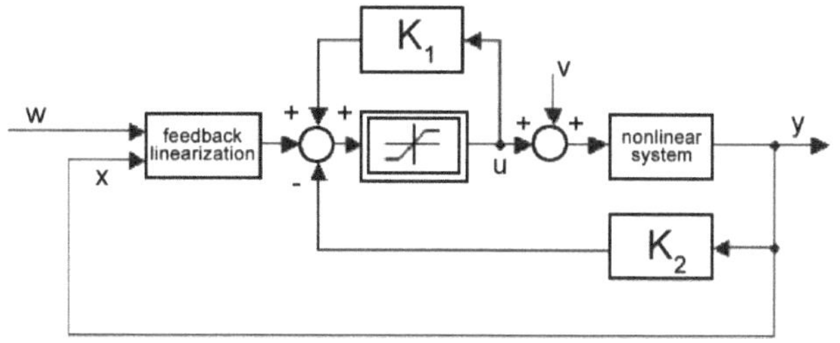

Fig. 3. Closed loop structure

where w denotes new input and $D_1 = e^{-T/T_1}$. Under this feedback the input-output description of the closed loop is linear

$$dy = \frac{1 - D_1}{z - D_1} dw$$

Clearly, to linearize the system one did not need the introduced transfer function formalisms. However, the situation is different in eliminating an input disturbance and dealing with a control signal constraint, where a use of transfer functions is unavoidable.

To satisfy the remaining design requirements we will consider the control structure [8] depicted in Fig. 3. The input disturbance v is eliminated via the feedback filter K_2 which tries to reconstruct v and subtracts it from controller output. The block K_1 only removes the impact of K_2 while controlling the system (via the feedback linearization). It is worth noting that the entire control structure has the properties of a PI controller. An important difference consists in a fact that there is no problem with the wind-up effect, in contrast to the classical PI controller with the control signal constraint.

Of course, the ideal filter $K_2(z) = \frac{1}{F(z)}$ is not realizable. Hence, we use

$$K_2(z) = \frac{(1 - \lambda_f)(z - D)}{(z - \lambda_f)K} \tag{5}$$

where $\lambda_f = e^{-T/T_f}$ and T_f is a time constant which characterizes how fast the disturbance elimination will be.

Transfer function (5) corresponds to the input-output difference equation

$$\begin{aligned} K(z - \lambda_f)dy_{K_2} &= (1 - \lambda_f)(z - D)du_{K_2} \\ K(y_{K_2}^+ - \lambda_f y_{K_2}) &= (1 - \lambda_f)(u_{K_2}^+ + cT\sqrt{u_{K_2}} - u_{K_2}) \end{aligned}$$

where u_{K_2} and y_{K_2} denote input and output respectively to the filter. The realization (state-space description) can be found as

$$\begin{aligned} x_{K_2}^+ &= \lambda_f x_{K_2} + \frac{1 - \lambda_f}{K}(\lambda_f u_{K_2} + cT\sqrt{u_{K_2}} - u_{K_2}) \\ y_{K_2} &= x_{K_2} + \frac{1 - \lambda_f}{K} u_{K_2} \end{aligned}$$

The filter $K_1(z)$ is linear system with the transfer function

$$K_1(z) = \frac{1 - \lambda_f}{z - \lambda_f} \tag{6}$$

Remark 1. Note that the continuous-time version can be easily computed from (4) where we let T tend to 0. That is

$$u = \frac{wA}{T_1} + Ac\sqrt{x} - \frac{xA}{T_1}$$

However, the real laboratory plant allows us to work with the minimum sample time of $250ms$. Hence, only the discrete-time version will be implemented.

4.1 Simulation Results

It is quite important to note that all controllers were designed usign the nonlinear discrete-time approximation (2) of the fluid tank system (1). But as a matter of fact the continuous-time system (1) is to be controlled. And since (2) is only its approximation (more or less accurate) the quality of the closed loop transient responses highly depends on the chosen sampling period T. This is something which constitutes a fundamental difference with respect to the linear case and it is due to the fact that we were not able to find the discrete-time model of the system (1), only its approximation.

In simulations we used the system (1) with $A = 0.001$, $c = 0.0249$. We found these constants by identification of the tank system. The flow coefficient c of the valve can be easily obtained from (1). Note that if the pump is off, the state-space representation (1) reduces to

$$\dot{x} = -c\sqrt{x}$$
$$y = x$$

and in this case, c can be, by integrating the equations, computed as

$$c = \frac{2\sqrt{x_{\text{init}}} - 2\sqrt{x_{\text{final}}}}{\Delta t} \tag{7}$$

where x_{init} is the initial level of a liquid before opening the valve and x_{final} its final level when we close the valve. Finally, Δt represents the duration of the tank getting empty.

However, there is also another approach to identify the flow coefficient c. If there is a non-zero constant flow into the tank the water level x approaches its steady state. That is

$$\frac{1}{A}u = c\sqrt{x}$$
$$c = \frac{u}{A\sqrt{x}} \tag{8}$$

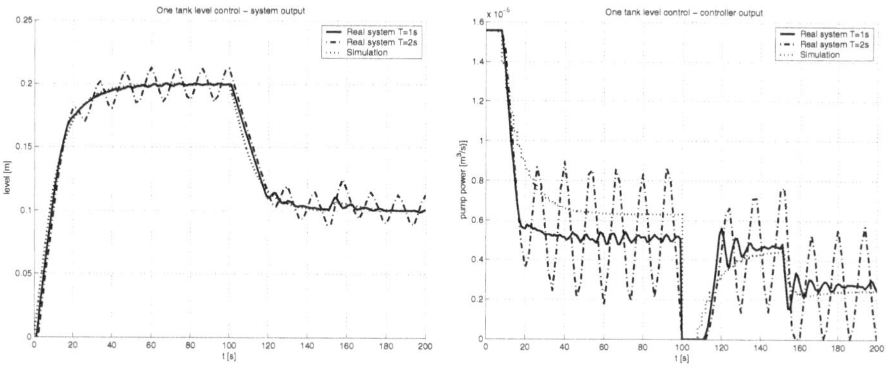

Fig. 4. System output (left) and controller output (right). Sampling period $T = 1s$ and $T = 2s$, disturbance step from 0 to $2e^{-6}m^3s^{-1}$ in time $150s$.

Note that the full pump power at $12V$ is about $1.6e^{-5}m^3s^{-1}$ and we expect that the input-output characteristic of the pump is linear.

In simulations and practical implementations, Fig. 4, the time constants T_1 and T_f, which determine the dynamics of the control and of the disturbance elimination, were chosen to be $4s$ and $10s$ respectively. The control signal was constraint to the interval $\langle 0; 1.6e^{-5} \rangle$. Transient responses for different sampling periods T are depicted in Fig. 4. As we can see, the period $T = 1s$ produces proper quality of the control, but that with $T = 2s$ has too high oscillations. However, in simulations even the sampling period $T = 4s$ still satisfies the quality of the control. One can thus conclude that the problem is not caused by the nonlinear discrete-time approximation (2) of the continuous-time model (1) but we have to change the original model (1) if the better performance is required.

4.2 Improved Identification and Model

Firstly, we estimate the input-output characteristic of the pump. The characteristic, measured with step $0.36V$, is depicted in Fig. 5. As it can be seen, it is nonlinear and thus the approximation by a polynomial of the 4-th degree was chosen. In designing a controller this nonlinearity will be eliminate by the inverse, also approximated by a polynomial of the 4-th degree.

However, the biggest source of inaccuracy is the valve. Therefore, we estimate the flow coefficient c separately for a couple of intervals that differ each other by $2cm$ of the liquid level in the tank. The procedure remains unchanged (7) and the results are depicted in Fig. 5 with the comparison to the values of c obtained by using (8). As can be seen in Fig. 5 (on the right), the flow coefficient c has a nonlinear characteristic. Firstly, we tried to approximate it with a curve

$$c = b + \frac{k}{\sqrt{x}} \tag{9}$$

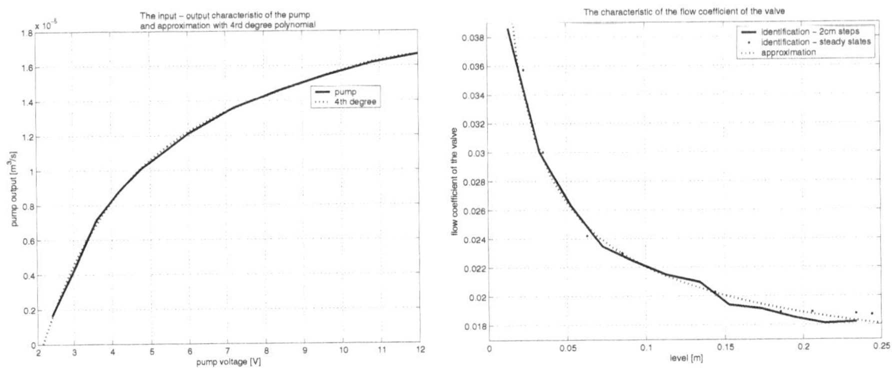

Fig. 5. Pump approximation with 4th degree polynomial is on the left, there is 2 approaches in valve identification on the right

where b and k are parameters to find. This means that (2) easily modifies to

$$\begin{aligned} x^+ &= x + \frac{T}{A}u - kT - bT\sqrt{x} \\ y &= x \end{aligned} \tag{10}$$

However, such an approximation does not have correct physical meaning - we might have problems to explain what k really means. Physically, the problem is caused by offset between the drain hole and the valve. In fact, the valve is situated a few cm under the drain hole of the tank. Thus, a better approximation of c is based on adding an offset to the liquid level which results in the new model

$$\begin{aligned} x^+ &= x + \frac{T}{A}u - cT\sqrt{x + \text{offset}} \\ y &= x \end{aligned} \tag{11}$$

Now, the improved controllers can be design.

Results. Using the improved model (11) with more accurate approximation of the plant dynamics one can repeat the controller design. Results are depicted in Fig. 6. As we can see, there are obvious differences between the former results and the new ones. Even the sampling period $T = 3s$ produces now the better control performance than the former controller with $T = 2s$. The improved controller, when tested, had a critical sampling period about $T = 7s$. With this period the system has oscillations as the original one, Fig. 4. The results when employing the model (10) are pretty much the same. Hence, we can say that, in this case, the accuracy of the model play a key role in the quality of control, while the quality of the nonlinear discrete-time approximation of the continuous-time system does not that big.

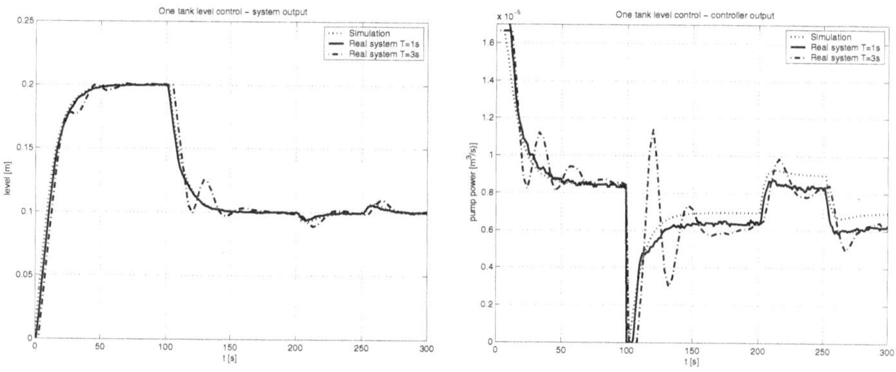

Fig. 6. System output (left) and controller output (right). Sampling period $T = 1s$ and $T = 3s$,disturbance step from 0 to $2e^{-6}m^3s^{-1}$ in time $150s$.

References

1. Conte, G., Moog, C.H., Perdon, A.M.: Algebraic Methods for Nonlinear Control Systems, 2nd edn. Theory and Applications. Springer, London (2007)
2. Aranda-Bricaire, E., Kotta, Ü., Moog, C.H.: Linearization of discrete-time systems. SIAM J. Cont. Opt. 34, 1999–2023 (1996)
3. Halás, M.: An algebraic framework generalizing the concept of transfer functions to nonlinear systems. Automatica 44, 1181–1190 (2008)
4. Halás, M., Huba, M.: Symbolic computation for nonlinear systems using quotients over skew polynomial ring. In: 14th Mediterranean Conference on Control and Automation, Ancona, Italy (2006)
5. Halás, M., Kotta, Ü.: Extension of the concept of transfer function to discrete-time nonlinear control systems. In: European Control Conference, Kos, Greece (2007)
6. Halás, M., Kotta, Ü.: Transfer Functions of Discrete-time Nonlinear Control Systems, Proc. Estonian Acad. Sci. Phys. Math. 56, 322–335 (2007)
7. Halás, M., Kotta, Ü., Moog, C.H.: Transfer function approach to the model matching problem of nonlinear systems. In: 17th IFAC World Congress, Seoul, Korea (2008)
8. Huba, M.: Gain scheduled PI level control of a tank with variable cross section. In: 2nd IFAC Conference on Control Systems Design, Bratislava, Slovakia (2003)

Pre-identification for Real-Time Control

Karel Perutka

Faculty of Applied Informatics, Tomas Bata University in Zlin,
Nam. T.G.M. 5555, CZ-76001 Zlin, European Union
KPerutka@fai.utb.cz

Abstract. The paper deals with the algorithm named as pre-identification, which denotes the simple general identification algorithm used for the system identification. The identification is realized before the system is controlled. It can be used in case the controlled system is time-invariant or slightly time-variant. Furthermore, the identified system might be nonlinear. Pre-identification provides a priori system description which is necessary for switching self-tuning control or useful for nonlinear control. The verification of the pre-identification usefulness was realized on several laboratory apparatuses in real-time using PC.

Keywords: Closed-loop identification, identification algorithms, least-squares identification, nonlinear control, self-tuning control, switching algorithms.

1 Introduction

First of all, let us provide introduction into identification and control. During last years, there were reached many interesting results in the area of system identification theory and practice.

Campi and Weyer derived the confidence regions for the parameters of identified models [2]. Reinelt et al. compared different approaches to model error modeling in robust identification [15]. Gerencsér evaluated the rate of convergence of the least mean square method [4]. Mišković et al. discussed the role of excitation of all references in closed-loop identification [13]. Chapman et al. provided the necessary results for structural identifiability for nonlinear compartment systems [3]. Goodwin et al. presented the robust identification of process models [5]. Hildebrand and Solari mentioned that it is possible to optimize the input in an identification experiment with respect to a performance cost function of a closed-loop system involving explicitly the dependence of the designed controller on the identified model [7]. Hjalmarsson and Ninness published the expression for the variance of scalar frequency functions estimated using the least-squares method [8]. Mahata an Garnier introduced the identification method for continuous-time linear dynamic errors-in-variables models [12]. Jaulin presented a study on the application of interval analysis and consistency techniques to state estimation of continuous-time systems described by nonlinear ordinary differential equations [10]. Yang et al. used the method of iterative global separable nonlinear

R. Moreno-Díaz et al. (Eds.): EUROCAST 2009, LNCS 5717, pp. 626–632, 2009.

least-squares (GSEPNLS) for identification of multi input single output (MISO) systems with unknown time delays of the input [20]. Wang and Yin were interested in system identification with the usage of quantized output observations only and they introduced the identification algorithm for system gains [19]. Ling and Ljung presented a structured way of using the tool analysis of variance and it is used for nonlinear autoregressive model with exogenous input identification with many candidate regressors [11]. Söderström published the survey of errors-in-variables methods in system identification [18]. Soukens et al. described the generation of initial estimates for the dynamic part of a Hammerstein model and it is shown that ARMAX or Box-Jenkins models result in better initial estimates than ARX or output-error (OE) models even in the absence of disturbing noise [17]. Hsu et al. was dealing with the parametric and nonparametric curve fitting [9]. Hassaine et al. introduced a new approach to increase the robustness of the classical L2-norm state estimation [6]. Romijn et al. proposed novel model reduction methodology to approximate large-scale nonlinear dynamical systems [16].

The paper is organized in the following way. Firstly, the pre-identification is formulated. This is followed by the results obtained at the apparatus in the laboratory.

2 Pre-identification

First of all, let us provide the general description of the presented approach called pre-identification.

The purpose of pre-identification is that the controlled system is supposing to be completely identified before the control task starts. Therefore it is denoted as a pre-identification.

During the pre-identification, the system to be controlled is viewed as "a black box" model and it is identified by direct and/or indirect continuous-time algorithms. The identification is divided into following steps:

1. There is closed-loop feedback system without controller. The difference of reference and output signal is sent to the input of the system model. The reference signal values are same as they are used during control.
2. The whole interval of control is divided into intervals based on the change of some of reference signals, each interval is identified separately.
3. Each interval is identified several times, every time by different method of identification. Model obtained by identification method is separately compared with measured response. The parameters of model that are nearest to the measured data are used in control.

3 Laboratory Experiment

The verification was realized firstly by simulation in MATLAB-Simulink at the Hammerstein model of the system (nonlinearity was in the form of root funtion).

Fig. 1. Photo of laboratory apparatus connected to the PC

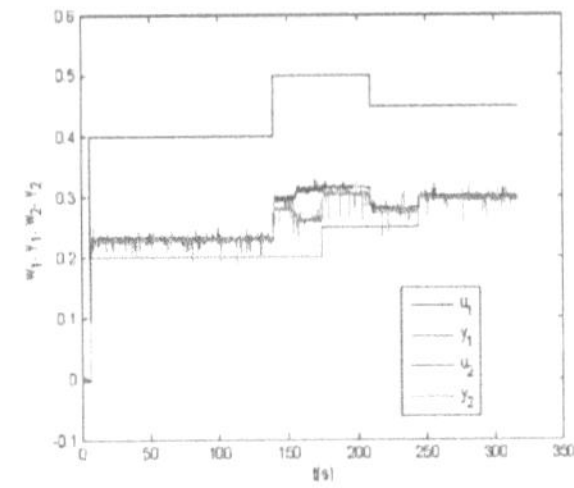

Fig. 2. System responses measurement

This was followed by the verification in laboratory at laboratory apparatus - rewinding machine (see Fig. 1) and at the helicopter model, at both models several times. This paper contains the results from one randomly selected experiment at the rewinding machine only.

Based on the model analysis done by Perutka [14], the rewinding machine is a non-linear system. Laboratory apparatus simulates several practical tasks of tension and speed of material during continuous processes. Examples of these processes can be found during the manufacturing of paper, wires, plastic foil etc. During these processes, the material goes through the workstation, where the speed and tension of the material are measured. The apparatus enables similar conditions. The flexible belt is mounted on three shafts. Two of them are connected with amplifiers of two servomotors and these wheels are fixed. Third one, on the top of the apparatus, is mounted on the jib that is connected to the spring. This wheel simulates the workstation. Two servomotors control the speed of all wheels and the belt tension.

The control voltages of the servo motors amplifiers are two inputs, both drives are bidirectional. There are four outputs, the voltage corresponding to the speed of servomotors and the voltage corresponding to tension and speed of belt. Two outputs were chosen in purpose to verify the pre-identification at nonlinear system. The chosen outputs are the voltages corresponding to the speed of top wheel and belt tension.

The verification was done in the following way. Firstly, the pre-identification was realized (see Fig. 2–8), identification was realized by instrumental variable method and least-squares method with directional forgetting. Those methods are well know, their description is available for instance in Perutka [14]. After pre-identification, the obtained data (see Fig. 7, 8) were used for decoupled self-tuning control by a set of single input single output (SISO) controllers (see Fig. 9). Self-tuning controllers are described for instance in Bobal et al. [1]. In purpose of comparison, the decoupled self-tuning control without data from pre-identification was performed (see Fig. 12).

The quantities in the following figures are following. Fig. 2: t - time, u_1 - input of first subsystem, y_1 - output of first subsystem, u_2 - input of second subsystem, y_2 - output of second subsystem; Fig. 3, 5, 7: t - time, $b0_1, a0_1, a1_1$ -parameters of first subsystem model; Fig. 4, 6, 8: t - time, $b0_2, a0_2, a1_2$ - parameters of second

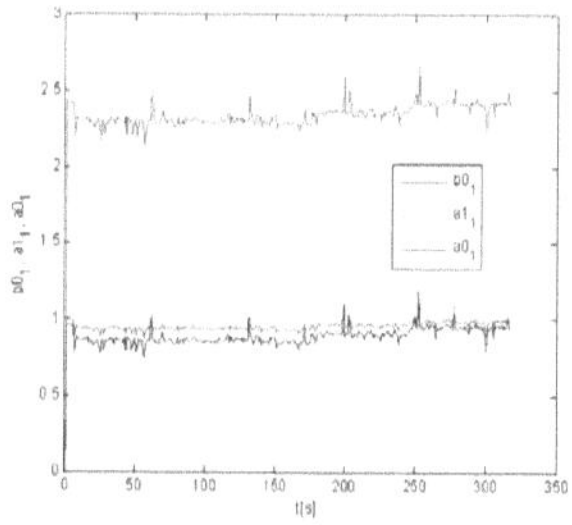

Fig. 3. History of system parameters of first subsystem obtained by instrumental variable method

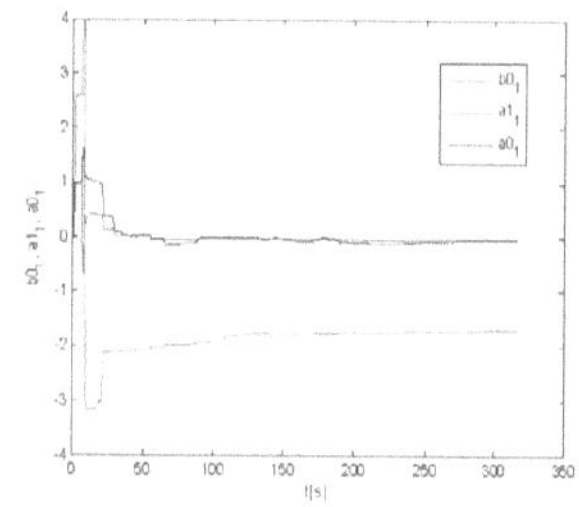

Fig. 4. History of system parameters of first subsystem obtained by least-squares method with directional forgetting

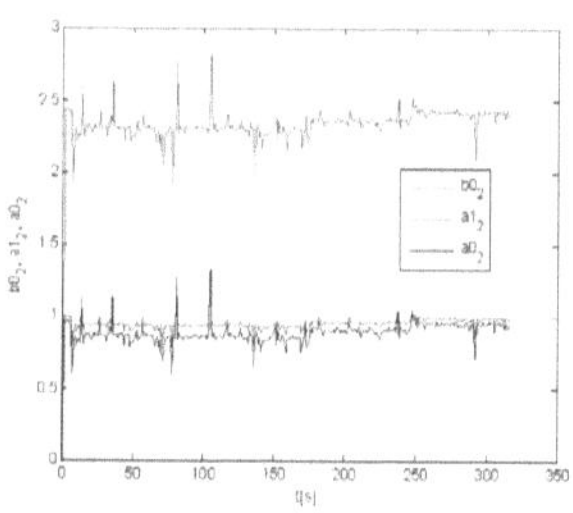

Fig. 5. History of system parameters of second subsystem obtained by instrumental variable method

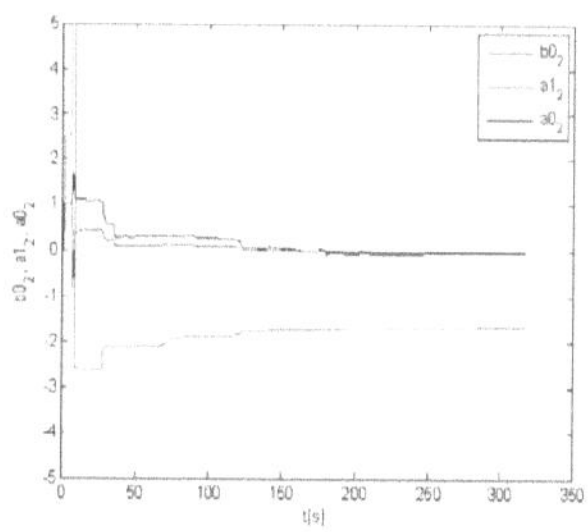

Fig. 6. History of system parameters of second subsystem obtained by least-squares method with directional forgetting

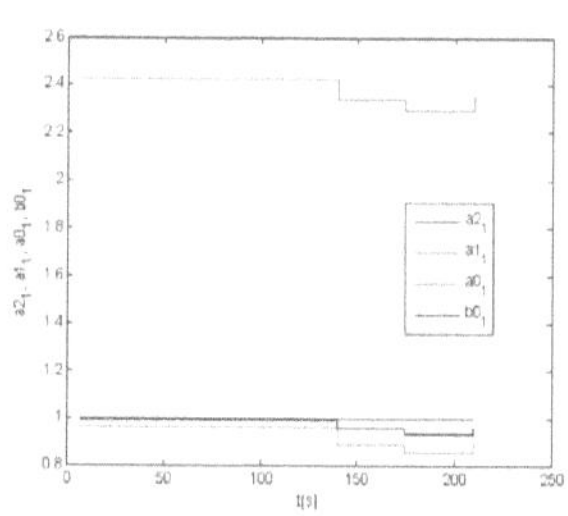

Fig. 7. History of model system parameters for first subsystem obtained by pre-identification

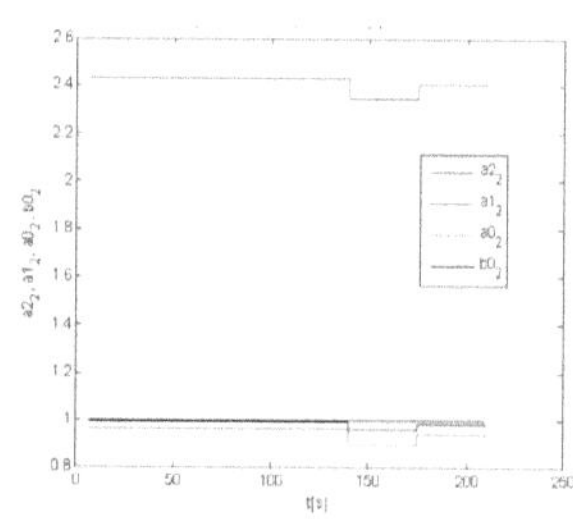

Fig. 8. History of model system parameters for second subsystem obtained by pre-identification

subsystem model; Fig. 7: $a2_1$ - additional parameter of first subsystem model; Fig. 8: $a2_2$ - additional parameter of second subsystem model; Fig. 9, 12: t - time; u_1 - action signal of first sub-controller; y_1 - output of first subsystem; w_1 - reference signal of first subsystem; u_2 - action signal of second sub-controller;

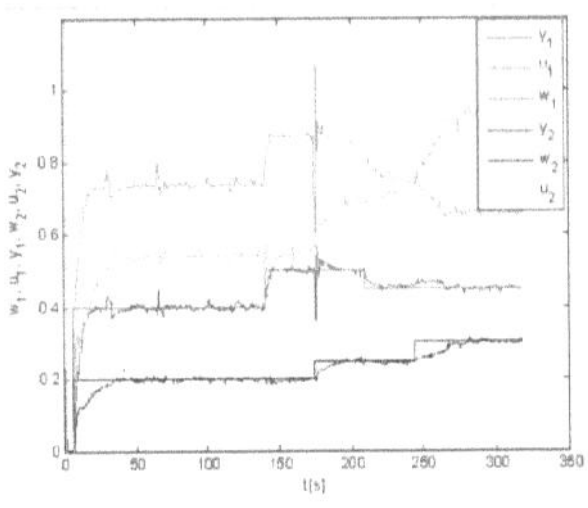

Fig. 9. History of apparatus control using data obtained by pre-identification

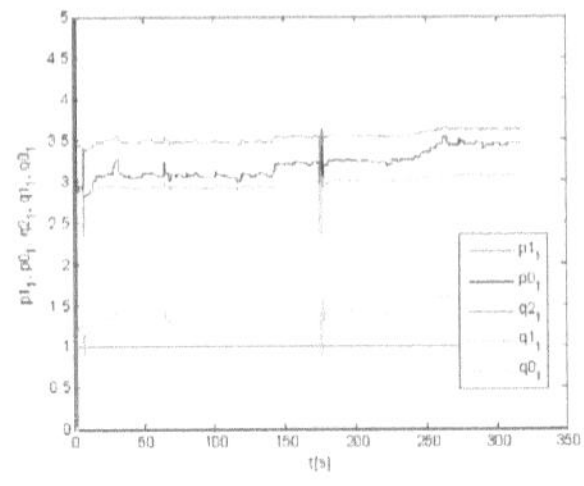

Fig. 10. History of first sub-controller parameters

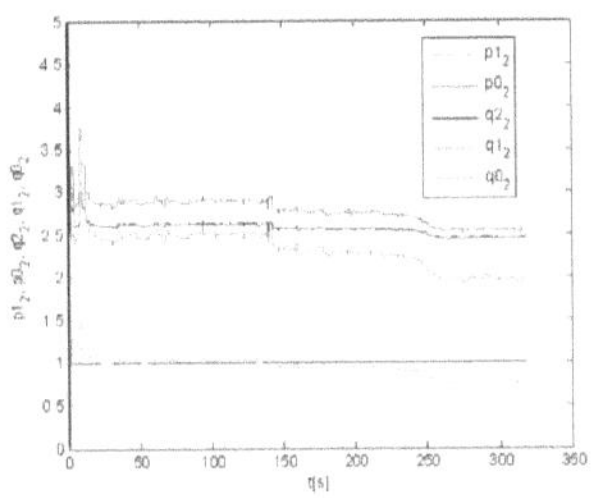

Fig. 11. History of second sub-controller parameters

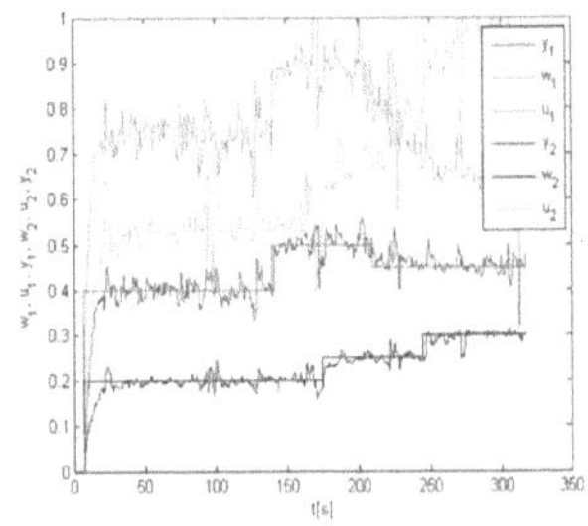

Fig. 12. History of apparatus control without pre–identification

y_2 - output of second subsystem; w_2 - reference signal of second subsystem; Fig. 10, 11 - all quantities in the label box or y–axis are parameters of the sub-controller; t - time.

4 Conclusions

The paper presented the approach denoted as pre-identification. This approach was verified in the following way. Firstly, the measurement of the response of the system was done. The measurement was followed by the identification using recursive instrumental variable method. Then, the identification using recursive least squares with directional forgetting was performed. After that, the simulated response of system using the parameters obtained by these two methods was done. It was compared with the measured response for each method, the correlation coefficients were counted. The better coefficients determined which identified parameters were used. Finally, the real-time control in MATLAB using Humusoft Real-Time Toolbox of the laboratory apparatus was realized. The main advantage of pre-identification is clear from the comparison of the history of control (Fig. 9, 12), the control with pre-identification is distinctly less biased. The future work will pay attention to the modification of pre-identification in purpose to be relevant for more controlled systems, because nowadays it is used only for linearized system model.

Acknowledgement

The author would like to mention that the paper was supported by the research programme of the Ministry of Education, Youth and Sport (MSMT) of the Czech Republic, grant No. MSM 7088352101.

References

1. Bobal, V., Böhm, J., Fessl, J., Machacek, J.: Digital Self-tuning Controllers. Springer, London (2005)
2. Campi, M.C., Weyer, E.: Guaranteed non-asymptotic confidence regions in system identification. Automatica 41, 1751–1764 (2005)
3. Chapman, M.J., Godfrey, K.R., Chappell, M.J., Evans, E.D.: Structural identifiability for a class of non-linear compartmental systems using linear/non-linear splitting and symbolic computation. Mathematical Biosciences 183, 1–14 (2003)
4. Gerencsér, L.: Rate of convergence of the LMS method. Systems & Control Letters 24, 385–388 (1998)
5. Goodwin, G.C., Agüero, J.C., Welsh, J.S., Yuz, J.I., Adams, G.J.: Robust identification of process models from data. Journal of Process Control 18, 810–820 (2008)
6. Hassaine, Y., Delourme, B., Panciatici, P., Walter, E.: M-Arctan estimator based on the trust-region method. International Journal of Electrical Power and Energy Systems 28, 590–598 (2006)
7. Hildebrand, R., Solari, G.: Identification for control: Optimal input intended to identify a minimum variance controller. Automatica 43, 758–767 (2007)
8. Hjalmarsson, H., Ninness, B.: Least-squares estimation of a class of frequency functions: A finite sample variance expression. Automatica 42, 589–600 (2006)
9. Hsu, K., Novara, C., Vincent, T., Milanese, M., Poolla, K.: Parametric and non-parametric curve fitting. Automatica 42, 1869–1873 (2006)
10. Jaulin, L.: Nonlinear bounded-error state estimation of continuous-time systems. Automatica 38, 1079–1082 (2002)
11. Lind, I., Ljung, L.: Regressor and structure selection in NARX models using a structured ANOVA approach. Automatica 44, 383–395 (2008)
12. Mahata, K., Garnier, H.: Identification of continuous-time errors-in-variables models. Automatica 42, 1477–1490 (2006)
13. Mišković, L., Karimi, A., Bonvin, D., Gevers, M.: Closed-loop identification of multivariable systems: With or without excitation of all references? Automatica 44, 2048–2056 (2008)
14. Perutka, K.: Decentralized Adaptive Control, thesis. Tomas Bata University, Zlin, Czech Republic (2007)
15. Reinelt, W., Garulli, A., Ljung, L.: Comparing different approaches to model error modeling in robust identification. Automatica 38, 787–803 (2002)
16. Romijn, R., Özkan, L., Weiland, S., Ludlage, J., Marquardt, W.: A grey-box modeling approach for the reduction of nonlinear systems. Journal of Process Control 18, 906–914 (2008)
17. Schoukens, J., Widanage, W.D., Godfrey, K.R., Pintelon, R.: Initial estimates for the dynamics of a Hammerstein system. Automatica 43, 1296–1301 (2007)

18. Söderström, T.: Errors-in-variables methods in system identification. Automatica 43, 939–958 (2008)
19. Wang, L.Y., Yin, G.G.: Asymptotically efficient parameter estimation using quantized output observations. Automatica 43, 1178–1191 (2007)
20. Yang, Z.-J., Iemura, H., Kanae, S., Wada, K.: Identification of continuous-time systems with multiple unknown time delays by global nonlinear least-squares and instrumental variable methods. Automatica 43, 1257–1264 (2007)

Realization of Continuous–Time Nonlinear Input–Output Equations: Polynomial Approach

Maris Tõnso and Ülle Kotta

Institute of Cybernetics at Tallinn University of Technology
Akadeemia tee 21, 12618, Tallinn, Estonia
{maris,kotta}@cc.ioc.ee

Abstract. The aim of the paper is to apply the polynomial methods to nonlinear realization problem. A new formula is presented which allows to compute the differentials of the state coordinates directly from the polynomial description of the nonlinear system, yielding a shorter and more compact program code in *Mathematica* implementation.

Keywords: nonlinear control system, continuous-time system, input-output models, polynomial methods, state space realization.

1 Introduction

Most results on nonlinear identification are achieved for systems, described by input-output (i/o) differential equations. At the same time the majority of techniques for nonlinear system analysis and control design are based on state-space description. The problem that we deal with in this paper is that of recovering the state-space model, whenever possible, starting from an arbitrary nonlinear higher order i/o differential equation.

In [1] the algebraic formalism, based on differential one-forms, has been applied for studying the realization problem. The coordinate-free necessary and sufficient realizability conditions were formulated in terms of the integrability of certain $\mathcal{H}_k$ subspaces of one-forms and the differentials of the state coordinates were defined as the basis elements of the last subspace. The algorithm to calculate the subspaces was given. Slightly different point of view in the studies of nonlinear control systems is provided by the polynomial approach in which the system is described by two polynomials from the non-commutative ring of left skew polynomials that act on input and output differentials. Polynomial approach has been used so far to study the reduction of the (set of nonlinear) i/o equations [2,3], the transfer equivalence [2] and used also in extending the concept of transfer function into the nonlinear domain [4,5].

The aim of the present paper is to apply the polynomial approach to the realization problem. This allows to simplify the step-by-step algorithm for computation of $\mathcal{H}_k$ subspaces given in [1]. A new formula is presented which allows to compute the $\mathcal{H}_k$ subspaces of one-forms directly from the polynomial description of the nonlinear system. The new method is more direct and therefore better

R. Moreno-Díaz et al. (Eds.): EUROCAST 2009, LNCS 5717, pp. 633–640, 2009.

suited for implementation in computer algebra packages like Mathematica or Maple. Note that the realization problem for discrete-time nonlinear systems was addressed within polynomial approach in [6], extending the linear case, discussed in [7].

The paper is organized as follows. Section 2 describes the realization problem studied in this paper and recalls its solution in terms of $\mathcal{H}_k$ subspaces. Section 3 introduces the polynomial framework and Section 4 presents the solution of the realization problem in terms of polynomials, describing the system. In Section 5 two examples and discussion are provided.

2 Problem Statement and the Algebraic Framework

Consider a nonlinear system Σ, described by a higher order i/o differential equation, relating the input u, the output y and a finite number of their time derivatives,

$$y^{(n)} = \phi(y, \ldots, y^{(n-1)}, u, \ldots, u^{(n-1)}). \tag{1}$$

In (1), $u \in U \subset I\!R$ is the scalar input variable, $y \in Y \subset I\!R$ is the scalar output variable, n is a nonnegative integer, and ϕ is a real analytic function, defined on $Y^n \times U^n$.

The realization problem is defined as follows. Given a nonlinear system, described by the i/o equation of the form (1), find, if possible, the state coordinates $x \in I\!R^n$, $x = \psi(y, \ldots, y^{(n-1)}, u, \ldots, u^{(n-1)})$ such that in these coordinates the system takes the classical state space form

$$\dot{x} = f(x, u), \quad y = h(x), \tag{2}$$

called the realization of (1).

Below we briefly recall the algebraic formalism, described in [1]. Let $\mathcal{K}$ denote the field of meromorphic functions in a finite number of the independent system variables $\{y, \ldots, y^{(n-1)}, u^{(k)}, \ k \geq 0\}$ and $s : \mathcal{K} \to \mathcal{K}$ denote the time derivative operator $\mathrm{d}/\mathrm{d}t$. Then the pair $(\mathcal{K}, s)$ is differential field [8]. Over the field $\mathcal{K}$ one can define a differential vector space, $\mathcal{E} := \mathrm{span}_{\mathcal{K}}\{\mathrm{d}\varphi \mid \varphi \in \mathcal{K}\}$ spanned by the differentials of the elements of $\mathcal{K}$. Consider a one-form $\omega \in \mathcal{E} : \omega = \sum_i \alpha_i \mathrm{d}\varphi_i$, $\alpha_i, \varphi_i \in \mathcal{K}$. Its derivative $\dot{\omega}$ is defined by $\dot{\omega} = \sum_i \dot{\alpha}_i \mathrm{d}\varphi_i + \alpha_i \mathrm{d}\dot{\varphi}_i$.

The relative degree r of an one-form $\omega \in \mathcal{E}$ is defined to be the least integer such that $s^r\omega \notin \mathrm{span}_{\mathcal{K}}\{\mathrm{d}y, \ldots, \mathrm{d}y^{(n-1)}, \mathrm{d}u, \ldots, \mathrm{d}u^{(n-1)}\}$. If such an integer does not exist, we set $r = \infty$. A sequence of subspaces $\{\mathcal{H}_k\}$ of $\mathcal{E}$ is defined by

$$\begin{aligned} \mathcal{H}_1 &= \mathrm{span}_{\mathcal{K}}\{\mathrm{d}y, \ldots, \mathrm{d}y^{(n-1)}, \mathrm{d}u, \ldots, \mathrm{d}u^{(n-1)}\} \\ \mathcal{H}_{k+1} &= \{\omega \in \mathcal{H}_k \mid \dot{\omega} \in \mathcal{H}_k\}, \ k \geq 1. \end{aligned} \tag{3}$$

Note that $\mathcal{H}_k$ contains the one-forms whose relative degree is equal to k or higher than k. It is clear that the sequence (3) is decreasing. Denote by k^* the least integer such that $\mathcal{H}_1 \supset \mathcal{H}_2 \supset \cdots \supset \mathcal{H}_{k^*} \supset \mathcal{H}_{k^*+1} = \mathcal{H}_{k^*+2} = \cdots =: \mathcal{H}_\infty$.

In what follows we assume that the i/o differential equation (1) is in the irreducible form, that is, $\mathcal{H}_\infty$ is trivial. A nth-order realization of equation (1) will be accessible if and only if system (1) is irreducible. If we find a nth-order realization for an i/o equation (1) which is in fact "reducible", the realization will be non-accessible.

System (2) is said to be single-experiment observable if the observability matrix has generically full rank $\mathrm{rank}_{\mathcal{K}}\left[\partial(h(x), sh(x), \ldots, s^{n-1}h(x))/\partial x\right] = n$.

Theorem 1. *The nonlinear system Σ, described by the irreducible i/o differential equation (1), has an observable and accessible state-space realization iff for $1 \leq k \leq n+1$ the subspaces $\mathcal{H}_k$, defined by (3), are completely integrable. Moreover, the state coordinates can be obtained by integrating the basis vectors of $\mathcal{H}_{n+1}$.*

We say that $\omega \in \mathcal{E}$ is exact, if there exists $\zeta \in \mathcal{K}$ such that $\mathrm{d}\zeta = \omega$. A subspace is integrable or closed, if it has a basis which consists only of closed one-forms. Note that closed one-forms are locally exact. Integrability of the subspace of one-forms can be checked by the Frobenius theorem.

Theorem 2. *[9] Let $\mathcal{V} = \mathrm{span}_{\mathcal{K}}\{\omega_1, \ldots, \omega_r\}$ be a subspace of $\mathcal{E}$. $\mathcal{V}$ is closed iff $\mathrm{d}\omega_k \wedge \omega_1 \wedge \ldots \wedge \omega_r = 0$, for all $k = 1, \ldots, r$.*

3 Polynomial Framework

Polynomial framework is built upon the linear algebraic framework. The differential field $(\mathcal{K}, s)$ induces a ring of left polynomials $\mathcal{K}[\partial, s]$. The elements of $\mathcal{K}[\partial, s]$ can be uniquely written in the form $a(\partial) = \sum_{i=0}^{n} a_i \partial^{n-i}$, $a_i \in \mathcal{K}$ where ∂ is a polynomial indeterminate and $a(\partial) \neq 0$ if and only if at least one of the functions a_i, $i = 0, \ldots, n$ is nonzero. If $a_0 \not\equiv 0$, then the positive integer n is called the degree of the left polynomial $a(\partial)$ and denoted by $\deg a(\partial)$. In addition, we set $\deg 0 = -\infty$. For $a \in \mathcal{K}$ let us define the multiplication

$$\partial \cdot a = a \cdot \partial + s(a). \tag{4}$$

If the multiplication is defined by (4), the ring $\mathcal{K}[\partial, s]$ is proved to satisfy left Ore condition [10], and $\partial^n \cdot a \in \mathcal{K}[\partial, s]$, for $n \geq 1$, and $\partial^n \cdot a = \sum_{i=0}^{n} C_n^i s^{n-i}(a)\partial^i$. A ring D is called an integral domain, if it does not contain any zero divisors. This means that if a and b are two elements of D such that $ab = 0$, then $a = 0$ or $b = 0$.

Lemma 1. *[10]*

(i) The ring $\mathcal{K}[\partial, s]$ is an integral domain.
(ii) If a and b are nonzero left polynomials, then $\deg(a\,b) = \deg a + \deg b$.

For $\Phi \in \mathcal{K}$ we define $\mathrm{d} : \mathcal{K} \to \mathcal{E}$ as follows: $\mathrm{d}\Phi := \sum_{i=0}^{n-1} \partial\Phi/\partial y^{(i)}\, \mathrm{d}y^{(i)} + \sum_{l=0}^{k} \partial\Phi/\partial u^{(l)}\, \mathrm{d}u^{(l)}$. $\mathrm{d}\Phi$ is said to be the total differential (or simply the differential) of the function Φ and it is a differential one-form. It is proved in [1] that $s(\mathrm{d}\Phi) = \mathrm{d}(s\Phi)$. Let us define $\partial^k \mathrm{d}y := \mathrm{d}(s^k y)$ and $\partial^l \mathrm{d}u := \mathrm{d}(s^l u)$, for $k, l \geq 0$ in the vector space $\mathcal{E}$. Since every one-form $\omega \in \mathcal{E}$ has the form $\omega = \sum_{i=0}^{n-1} a_i \mathrm{d}y^{(i)} + \sum_{j=0}^{k} b_j \mathrm{d}u^{(j)}$, where $a_i, b_j \in \mathcal{K}$, so ω can be expressed in terms of the left polynomials $\omega = \left(\sum_{i=0}^{n-1} a_i \partial^i\right) \mathrm{d}y + \left(\sum_{j=0}^{k} b_j \partial^j\right) \mathrm{d}u$. A left polynomial can be considered as an operator acting on the elements of $\mathcal{E}$:

$$\left(\sum_{i=0}^{k} a_i \partial^i\right)(\alpha \mathrm{d}\nu) := \sum_{i=0}^{k} a_i \left(\partial^i \cdot \alpha\right) \mathrm{d}\nu,$$

with $a_i, \alpha \in \mathcal{K}$ and $\mathrm{d}\nu \in \{\mathrm{d}y, \mathrm{d}u\}$. It is easy to notice that $\partial(\omega) = s(\omega)$, for $\omega \in \mathcal{E}$. Additionally, using the induction principle, one can show that $\partial^n(\mathrm{d}\Phi) = \mathrm{d}(s^n \Phi)$.

Instead of working with equation (1), describing the control system, we can work with its differential

$$\mathrm{d}y^{(n)} - \sum_{i=0}^{n-1} \frac{\partial \phi}{\partial y^{(i)}} \mathrm{d}y^{(i)} - \sum_{j=0}^{n-1} \frac{\partial \phi}{\partial u^{(j)}} \mathrm{d}u^{(j)} = 0 \tag{5}$$

that can be rewritten as

$$p(\partial)\mathrm{d}y = q(\partial)\mathrm{d}u, \tag{6}$$

with $p(\partial) = \partial^n - \sum_{i=0}^{n-1} p_i \partial^i$, $q(\partial) = \sum_{j=0}^{n-1} q_j \partial^j$ and $p_i = \partial\phi/\partial y^{(i)} \in \mathcal{K}$, $q_i = \partial\phi/\partial u^{(j)} \in \mathcal{K}$. Equation (6) describes the nonlinear system behavior in terms of two polynomials $\{p(\partial), q(\partial)\}$ in derivative operator $\partial := s$ over the differential field $\mathcal{K}$.

Note that in the polynomial description, the systems $\phi(\cdot) = 0$ and $\phi(\cdot) +$ constant $= 0$ are not distinguished for arbitrary constant value. In order to overcome such situations one has to fix the constant and assume it to be defined by the equilibrium point of the system, around which the one forms will be integrated to get the state coordinates. Note that in many papers the assumption $\phi(0, \ldots, 0) = 0$ was made which however is sometimes restrictive if the i/o equation does not admit a zero equilibrium point. Though we make the same assumption for simplicity, note that this assumption can be relaxed.

4 Problem Solution: Polynomial Approach

We introduce certain one-forms in terms of which our main result (Theorem 3) will be formulated. The one-forms

$$\omega_{k,l} := \bar{p}_{k,l}(\partial)\mathrm{d}y + \bar{q}_{k,l}(\partial)\mathrm{d}u, \tag{7}$$

for $k = 1, \ldots, n+1$, $l = 1, \ldots, n$, where $\bar{p}_{k,l}(\partial)$ and $\bar{q}_{k,l}(\partial)$ are Ore polynomials, which can be recursively calculated as left quotients from equalities

$$\begin{aligned} \bar{p}_{k,l-1}(\partial) &= s\bar{p}_{k,l}(\partial) + r_{k,l}, \deg r_{k,l} = 0, \\ \bar{q}_{k,l-1}(\partial) &= s\bar{q}_{k,l}(\partial) + \rho_{k,l}, \deg \rho_{k,l} = 0, \end{aligned} \tag{8}$$

with initial polynomials $\bar{p}_{10} = -s$, $\bar{q}_{10} = 0$, for $k = 1$ and

$$\bar{p}_{k,0}(\partial) = \sum_{i=n-k+1}^{n-1} p_i \partial^i - \partial^n, \quad \bar{q}_{k,0}(\partial) = \sum_{i=n-k+1}^{n-1} q_i \partial^i \tag{9}$$

for $2 \leq k \leq n+1$. The following lemmas are necessary to prove the main result. The proofs are omitted due to lack of space.

Lemma 2. *The one-forms $\omega_{k,l}$, for $k = 1, \ldots, n$, defined by (7), satisfy the relationships*

(i) for $l = 1, \ldots, n-k$

$$\omega_{k+1,l} = \omega_{k,l} + \sum_{i=0}^{n-k-l} \binom{-l}{n-k-l-i} \left[p_{n-k}^{(n-k-l-i)} \mathrm{d}y^{(i)} + q_{n-k}^{(n-k-l-i)} \mathrm{d}u^{(i)} \right],$$

(ii) for $l = n-k+1, \ldots, n$ $\omega_{k+1,l} = \omega_{k,l}$.

Lemma 3. *For the one-forms $\omega_{k,l}$, defined by (7), the property $\dot{\omega}_{k,l} = \omega_{k,l-1} - r_{k,l}\mathrm{d}y - \rho_{k,l}\mathrm{d}u$, holds, where $\deg r_{k,l} = \deg \rho_{k,l} = 0$ for $l = 2, \ldots, n$ and for $k = 1, \ldots, n+1$.*

Theorem 3. *For the i/o model (1), the subspaces $\mathcal{H}_k$ can be calculated as*

$$\mathcal{H}_k = \mathrm{span}_{\mathcal{K}}\{\omega_{k,l}, \mathrm{d}u, \ldots, \mathrm{d}u^{(n-k)}\}, \quad k = 1, \ldots, n \tag{10}$$

and

$$\mathcal{H}_{n+1} = \mathrm{span}_{\mathcal{K}}\{\omega_{n+1,l}\}, \tag{11}$$

where $\omega_{k,l}$ for $l = 1, \ldots, n$ are defined by (7).

Proof. The proof is by induction. We first show that formula (3) holds for $k = 1$. From (7), the quotient polynomials $\bar{p}_{1,l}(\partial) = -\partial^{n-l}$ and $\bar{q}_{1,l}(\partial) = 0$ for $l = 1, \ldots, n$. Consequently, $\omega_l = -s^{n-l}\mathrm{d}y = -\mathrm{d}y^{(n-l)}$ and $\mathcal{H}_1 = \mathrm{span}_{\mathcal{K}}\{-\mathrm{d}y^{(n-1)}, \ldots, -\mathrm{d}y, \mathrm{d}u, \ldots, \mathrm{d}u^{(n-1)}\}$ that agrees with the definition of $\mathcal{H}_1$.

Assume next that formula (10) holds for k and we prove it to be valid for $k+1$. The proof is based on definition of the subspaces $\mathcal{H}_k$. We have to prove that $\mathcal{H}_{k+1} = \mathrm{span}_{\mathcal{K}}\{\omega_{k+1,l}, \mathrm{d}u, \ldots, \mathrm{d}u^{(n-k-1)}\}$, calculated according to formula (10), satisfies the condition (3).

First, we show that basis one-forms $\omega_{k+1,l}, \mathrm{d}u, \ldots, \mathrm{d}u^{(n-k-1)}$ are in $\mathcal{H}_k$. It is obvious that $\mathrm{d}u, \ldots, \mathrm{d}u^{(n-k-1)} \in \mathcal{H}_k$. Lemma 2 represents the one-forms $\omega_{k+1,l}$ as a linear combination of vectors $\omega_{k,l}$, $\mathrm{d}y$, ..., $\mathrm{d}y^{(n-k)}$, $\mathrm{d}u$, ..., $\mathrm{d}u^{(n-k)}$. Though $\mathrm{d}y$, ..., $\mathrm{d}y^{(n-k)}$ are not listed explicitly among the basis vectors of $\mathcal{H}_k$ in (10), they can be expressed as a linear combination of the other basis vectors. From (8) and (9) follows that the coefficients of the higher order terms of polynomials $\bar{p}_{k,l}(\partial)$ are always 1 as well as $\deg \bar{p}_{k,l}(\partial) = n - l$ and $\deg \bar{q}_{k,l}(\partial) = n - l - 1$ for $l = 1, \ldots, n-1$. It means that $\bar{p}_{k,l}(\partial)$ and $\bar{q}_{k,l}(\partial)$ for $l = 1, \ldots, n-1$ have the form $\bar{p}_{k,l}(\partial) = \sum_{j=0}^{n-l-1} p_{k,l,j}\partial^j - \partial^{n-l}$, $\bar{q}_{k,l}(\partial) = \sum_{j=0}^{n-l-1} q_{k,l,j}\partial^j$. For $l = n$

we get $\bar{p}_{k,n}(\partial) = 1$ and $\bar{q}_{k,n}(\partial) = 0$. Consequently, $\omega_{k,n} = \mathrm{d}y$. The rest of the differentials $\mathrm{d}\dot{y}, \ldots, \mathrm{d}y^{(n-k)}$ can be recursively computed from (7) as follows: $\mathrm{d}y^{(l)} = \omega_{k,n-l} - \sum_{j=0}^{l-1} \bar{p}_{k,n-l,j}\mathrm{d}y^{(j)} - \sum_{j=0}^{l-1} \bar{q}_{k,n-l,j}\mathrm{d}u^{(j)}$ for $l = 1, \ldots, n-k$.

Second, we show that the derivatives of the one-forms, computed according to (10) and (7), also belong to $\mathcal{H}_k$. Again, it is obvious that $\mathrm{d}\dot{u}, \ldots, \mathrm{d}u^{(n-k)} \in \mathcal{H}_k$. We have to prove that $\dot{\omega}_{k+1,l} \in \mathcal{H}_k$. According to Lemma 3, $\dot{\omega}_{k+1,l} = \omega_{k+1,l-1} - r_{k+1,l}\mathrm{d}y - \rho_{k+1,l}\mathrm{d}u$, where $\deg r_{k+1,l} = \deg \rho_{k+1,l} = 0$ for $l = 2, \ldots, n$. It was proved in the previous step that $\omega_{k+1,l}$ for $l = 2, \ldots, n$ and $\mathrm{d}y$ are in $\mathcal{H}_k$. For $l = 1$ we have to show separately that $\dot{\omega}_{k+1,1} \in \mathcal{H}_k$. From (7) we have: $\dot{\omega}_{k+1,1} = s\bar{p}_{k+1,1}(\partial)\mathrm{d}y + s\bar{q}_{k+1,1}(\partial)\mathrm{d}u$. Increasing k by 1 and taking $l = 1$ in (8) allows us to express $s\bar{p}_{k+1,1}$ and $s\bar{q}_{k+1,1}$ and substitute them into the previous equality. $\dot{\omega}_{k+1,1} = (\bar{p}_{k+1,0}(\partial) - r_{k+1,1})\mathrm{d}y + (\bar{q}_{k+1,0}(\partial) - \rho_{k+1,1})\mathrm{d}u$. Replacing in the above equality the initial polynomials $\bar{p}_{k+1,0}(\partial)$ and $\bar{q}_{k+1,0}(\partial)$ by their expressions (9) and using the relations $\partial^i \mathrm{d}y = \mathrm{d}y^{(i)}$ for $i = n-k, \ldots, n$ and $\partial^j \mathrm{d}u = \mathrm{d}u^{(j)}$, for $j = n-k, \ldots, n-1$, we obtain $\dot{\omega}_{k+1,1} = \sum_{i=n-k}^{n-1} p_i \mathrm{d}y^{(i)} - \mathrm{d}y^{(n)} + \sum_{j=n-k}^{n-1} q_j \mathrm{d}u^{(j)} - r_{k+1,1}\mathrm{d}y - \rho_{k+1,1}\mathrm{d}u$. Finally, replacing $\mathrm{d}y^{(n)}$ in the above equality by the right-hand side of (5), we get: $\dot{\omega}_{k+1,1} = -\sum_{i=0}^{n-k-1} p_i \mathrm{d}y^{(i)} - \sum_{j=0}^{n-k-1} q_j \mathrm{d}u^{(j)} - r_{k+1,1}\mathrm{d}y - \rho_{k+1,1}\mathrm{d}u$. The latter means that the one-forms $\dot{\omega}_{k+1,1}$ can be expressed as a linear combination of the basis vectors of $\mathcal{H}_k$. This completes the proof.

The differentials of the state coordinates can be found from the subspace $\mathcal{H}_{n+1}$, see Theorem 1. Though in case of realizable i/o equation, $\mathcal{H}_{n+1}$, defined by (11), is completely integrable, the one-forms $\omega_{n+1,l}$ for $l = 1, \ldots, n$, are not necessarily always exact. Therefore, one has to find for $\mathcal{H}_{n+1}$ a new integrable bases, using linear transformations. From Theorem 3 the next corollary can be concluded[1].

Corollary 1. *For realizable i/o equation (1), the differentials of the state coordinates can be calculated as the integrable linear combinations of the one-forms*

$$\omega_l = \bar{p}_l(\partial)\mathrm{d}y + \bar{q}_l(\partial)\mathrm{d}u, \quad l = 1, \ldots, n \tag{12}$$

where $\bar{p}_l(\partial)$ and $\bar{q}_l(\partial)$ can be computed recursively from

$$\bar{p}_{l-1}(\partial) = s\bar{p}_l(\partial) + r_l, \quad \bar{q}_{l-1}(\partial) = s\bar{q}_l(\partial) + \rho_l, \tag{13}$$

with the initial polynomials

$$\bar{p}_0(\partial) = p(\partial), \quad \bar{q}_0(\partial) = q(\partial). \tag{14}$$

Remark 1. Note that in the linear case integrability aspect does not come into the play since all the one-forms $\omega_1, \ldots, \omega_n$ are integrable. This is so because all the polynomial coefficients p_i, q_i, $i = 0, \ldots, n-1$ are real numbers and not the functions, depending on control system variables.

[1] The first index $n + 1$ is omitted in Corollary.

Remark 2. The results of [3] allow to check in terms of polynomials $p(\partial)$ and $q(\partial)$ if the i/o differential equation is irreducible or not, and in case if it is reducible, to find a reduced system description. Namely, equation (1) is irreducible if and only if $p(\partial)$ and $q(\partial)$ have no common left divisors. So, our results complement those of [3], allowing to find the minimal state space realization, no matter whether one starts from irreducible or reducible system description.

5 Examples and Discussion

Example 1. Consider the control system $\ddot{y} = 3uy + 2u\dot{u} + \dot{y}^2$ that can be described by two polynomials $p(\partial) = -\partial^2 + 2\dot{y}\partial + 3u$, $q(\partial) = 2u\partial + (3y + 2\dot{u})$.

To find the state coordinates, one has to, according to Corollary 1, compute the one-forms ω_1, ω_2, defined by (12). From equalities $\bar{p}_0(\partial) := p(\partial) = s\bar{p}_1(\partial) + r_1$, $\bar{q}_0(\partial) := q(\partial) = s\bar{q}_1(\partial) + \rho_1$ one can find the left quotients $\bar{p}_1(\partial) = -\partial + 2\dot{y}$, $\bar{q}_1(\partial) = 2u$. Furthermore, from equalities $\bar{p}_1(\partial) = s\bar{p}_2(\partial) + r_2$, $\bar{q}_1(\partial) = s\bar{q}_2(\partial) + \rho_2$ one can find the left quotients $\bar{p}_2(\partial) = -1$, $\bar{q}_2(\partial) = 0$. Finally, the one-forms that define the differentials of the state coordinates, can be computed according to (12) as follows

$$\begin{aligned}\omega_1 &= \bar{p}_1(\partial)\mathrm{d}y + \bar{q}_1(\partial)\mathrm{d}u = -\mathrm{d}\dot{y} + 2\dot{y}\mathrm{d}y + 2u\mathrm{d}u\\ \omega_2 &= \bar{p}_2(\partial)\mathrm{d}y + \bar{q}_2(\partial)\mathrm{d}u = -\mathrm{d}y.\end{aligned}$$

Though the subspace $\mathrm{span}_{\mathcal{K}}\{\omega_1, \omega_2\}$ is completely integrable, ω_1 is not exact and we have to replace ω_1 by an integrable linear combination of ω_1 and ω_2 to obtain the differentials of the state coordinates $\mathrm{d}x_1 = -\omega_2 = \mathrm{d}y$, $\mathrm{d}x_2 = -\omega_1 - 2\dot{y}\,\omega_2 = \mathrm{d}(\dot{y} - u^2)$ yielding the state equations $\dot{x}_1 = u^2 + x_2$, $\dot{x}_2 = 3ux_1 + (u + x_2)^2$.

Example 2. Consider the control system $y^{(3)} = \sqrt{\dot{y}\ddot{u}}$. We use again Corollary 1. Initial polynomials in (14) are for this example

$$\bar{p}_0(\partial) := p(\partial) = -\partial^3 + \frac{\ddot{u}}{2\sqrt{\dot{y}\ddot{u}}}\partial, \quad \bar{q}_0(\partial) := q(\partial) = \frac{\dot{y}}{2\sqrt{\dot{y}\ddot{u}}}\partial^2.$$

Recursive computation of the left quotients, according to (13), yields

$$\begin{aligned}&\bar{p}_1(\partial) = -\partial^2 + \frac{\sqrt{\ddot{u}}}{2\sqrt{\dot{y}}}, && \bar{p}_2(\partial) = -\partial, && \bar{p}_3(\partial) = -1,\\ &\bar{q}_1(\partial) = \frac{\sqrt{\dot{y}}}{\sqrt{\ddot{u}}} - \frac{\ddot{u}\ddot{y} - \dot{y}u^{(3)}}{4\sqrt{\dot{y}\ddot{u}^3}}, && \bar{q}_2(\partial) = \frac{\sqrt{\dot{y}}}{\sqrt{2\ddot{u}}}, && \bar{q}_3(\partial) = 0.\end{aligned}$$

Finally, by (12),

$$\omega_1 = \frac{\sqrt{\ddot{u}}}{2\sqrt{\dot{y}}}\mathrm{d}y - \mathrm{d}\ddot{y} - \frac{\ddot{u}\ddot{y} - \dot{y}u^{(3)}}{4\sqrt{\dot{y}\ddot{u}^3}}\mathrm{d}u + \frac{\sqrt{\dot{y}}}{\sqrt{2\ddot{u}}}, \quad \omega_2 = -\mathrm{d}\dot{y} + \frac{\sqrt{\dot{y}}}{\sqrt{2\ddot{u}}}, \quad \omega_3 = -\mathrm{d}y.$$

Unfortunately, the subspace $\mathrm{span}_{\mathcal{K}}\{\omega_1, \omega_2, \omega_3\}$ is not integrable and therefore, the i/o equation does not admit a state space form.

Theorem 3 provides an alternative, polynomial method for computing the bases vectors for the subspaces $\mathcal{H}_k$. Note that the polynomial method has some advantages in computer implementation. First, it is direct, meaning that there is no need to compute step-by-step all the $\mathcal{H}_k$ subspaces in order to find $\mathcal{H}_{n+1}$. Second, its program code is shorter and more compact. Algebraic method requires to solve a pseudolinear system of equations, which is linear with respect to unknowns, but those coefficients are nonlinear functions, and not real numbers. If the expressions, found on previous steps, have been not enough simplified, there is a chance that *Mathematica* may be unable to solve the pseudolinear system of equations and the computation fails. Polynomial method does not require solving any system of equations. In case of the second example from Section 5, the algebraic method requires to insert the additional simplification commands into the program code, while the polynomial method was able to produce the result without any intermediate simplification. The main disadvantage of the polynomial method is computation time. However, for discrete-time systems, polynomial method is also less time-consuming, if compared to the algebraic method. Unfortunately, this is not the case for continuous-time systems. The reason lies in the complex multiplication rule (4), compared to that one needs in the discrete-time case: $\partial \cdot a = \delta(a) \cdot \partial$, where δ is the shift operator.

Acknowledgements

This work has been supported by the Estonian Science Foundation grant N6922.

References

1. Conte, G., Moog, C., Perdon, A.: Algebraic Methods for Nonlinear Control Systems. Springer, London (2007)
2. Kotta, Ü., Kotta, P., Nõmm, S., Tõnso, M.: Irreducibility conditions for continuous-time multi-input multi-output nonlinear systems. In: 9th International Conference on Control, Automation, Robotics and Vision, Singapore, pp. 807–811 (2006)
3. Zheng, Y., Willems, J., Zhang, C.: A polynomial approach to nonlinear system controllability. IEEE Transaction on Automatic Control 46, 1782–1788 (2001)
4. Halás, M.: An algebraic framework generalizing the concept of transfer functions to nonlinear systems. Automatica 44(5), 1181–1190 (2008)
5. Zheng, Y., Cao, L.: Transfer function description for nonlinear systems. J. East China Normal Univ. (Nat. Sci.) 2, 15–26 (1995)
6. Kotta, Ü., Tõnso, M.: Realization of discrete-time nonlinear input-output equations: polynomial approach. In: Proc. of the 7th World Congress on Intelligent Control and Automation, Chongqing, China, pp. 529–534 (2008)
7. Rapisarda, P., Willems, J.: State maps for linear systems. SIAM J. Control. Optim. 35(3), 1053–1091 (1997)
8. Kolchin, E.: Differential algebra and algebraic groups. Academic Press, London (1973)
9. Choquet-Bruhat, Y., DeWitt-Morette, C., Dillard-Bleichi, M.: Analysis, manifolds and physics: Part I: Basics. North-Holland, Amsterdam (1982)
10. McConnel, J., Robson, J.: Noncommutative Noetherian Rings. Birkhäuser, Basel (1987)

Using Heuristic Optimization for Segmentation of Symbolic Music

Brigitte Rafael[1], Stefan Oertl[1], Michael Affenzeller[2], and Stefan Wagner[2]

[1] Re-Compose GmbH
Vienna, Austria
{brigitte.rafael,stefan.oertl}@re-compose.com
[2] Upper Austrian University of Applied Sciences,
School of Informatics, Communications and Media
Heuristic and Evolutionary Algorithms Laboratory
Hagenberg, Austria
{michael.affenzeller,stefan.wagner}@fh-hagenberg.at

Abstract. Solving the segmentation problem for music is a key issue in music information retrieval (MIR). Structural information about a composition achieved by music segmentation can improve several tasks related to MIR such as searching and browsing large music collections, visualizing musical structure, lyric alignment, and music summarization. Various approaches using genetic algorithms have already been introduced to the field of media segmentation including image and video segmentation as segmentation problems usually have complex fitness landscapes. The authors of this paper present an approach to apply genetic algorithms to the music segmentation problem.

1 Introduction

During the last years plenty of research has been done in the field of media segmentation. Evolutionary techniques were introduced for image segmentation [8,1] as well as video segmentation [3]. For MIR, segmentation is also an important issue as it provides an insight into the internal structure of a composition.

Music segmentation targets at the identification of boundaries between structurally relevant parts of a composition to enable or improve several MIR-related tasks. Current approaches are either the similarity matrix [10,11,12], hidden Markov models [7], or the application of the shortest path algorithm [5]. The most common approach aims at detecting structure boundaries with the aid of a novelty score [6,12]. Methods drawing on that score are limited to compositions following certain rules and principles as they require the existence of domain knowledge (extraopus). However, the authors' method is not based on this kind of a priori knowledge but focuses on the information provided within the piece itself (intraopus). In consequence, it can be applied to a broader musical spectrum. Music is analyzed by its degree of self-similarity and repetitions are used to detect segments. The algorithm then clusters similar segments to create segment groups (i.e., collections of nonoverlapping segments that fulfill a similarity condition).

R. Moreno-Díaz et al. (Eds.): EUROCAST 2009, LNCS 5717, pp. 641–648, 2009.

This paper presents the idea of applying a genetic algorithm to achieve the optimal segmentation of a composition. The first section demonstrates the suitability of genetic algorithms for music segmentation. The second section describes the mapping of music features to the components of the genetic algorithm. A discussion of the results and an outlook on future work concludes the paper.

2 Suitability of Genetic Algorithms for Music Segmentation

A segmentation of a composition consists of nonoverlapping segments that represent the internal structure of the music. Similar segments form segment groups through clustering. There are several reasons why music segmentation is not a trivial problem:

- Segments can vary in duration and in the number of notes they contain.
- Segments within a segment group usually are not exact repetitions but approximate ones.
- Distances between segments do not have to be regular; gaps of undefined duration may exist.
- Clustering similar segments may lead to ambiguities if overlapping segments are detected and one of them has to be chosen.
- Single notes between detected segments pose a problem if they cannot be unambiguously assigned to neither of the neighbouring segments. A decision has then to be made if they should form their own segments or if they belong to any of the existing ones.

As segments can start at any arbitrary position of the composition, the runtime for the evaluation increases exponentially for longer compositions. Therefore it is not possible to evaluate all potential segmentations but a solution of sufficient quality has to be found in reasonable time. Given all these circumstances, the problem domain of music segmentation turns out to be highly suited for applying genetic algorithms.

3 Implementation of the Genetic Algorithm

3.1 Problem Domain

The music data is represented in the MIDI (Musical Instrument Digital Interface) format. MIDI is a protocol for the exchange of digital music and for the communication and synchronization of software instruments. In contrast to audio formats like MP3 or WAV, MIDI does not transmit audio signals but only information on how these audio signals should be created. Instead of saving a note recorded with a specific instrument, MIDI sends events describing pitch, duration, etc. of the respective note.

An advantage of MIDI data compared to audio data is the availability of separate tracks. Instead of having all instruments merged into one common data

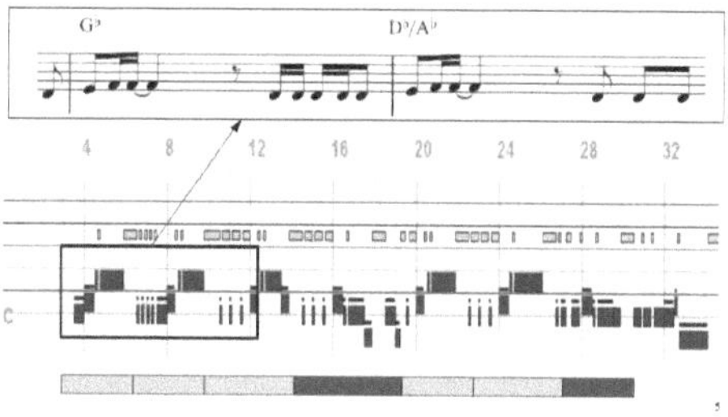

Fig. 1. Graphical representation of a MIDI track

stream, the information in MIDI files is separated into single tracks. As a consequence, each track can be analyzed isolated from the remaining composition and its optimal solution can be found independently from the other tracks.

Fig. 1 illustrates an extract from the notes of one track. The upper part shows a common graphical representation of notes. The lower part is similar, also containing five staff lines and an additional line for Middle C. Notes are displayed as blue boxes and the box widths show note durations. Lines above notes indicate an increment of the pitch value by one semitone. Rests are represented as grey boxes. Vertical lines correspond to the vertical lines in the upper picture and represent bar changes. The colored rectangles below the notes give a sample segmentation. Segments of the same color belong to the same segment groups.

3.2 Problem Encoding

The success of a genetic algorithm mainly depends on a suitable problem representation [9]. The authors have decided to use simple bit vector individuals allowing the application of existing operators. To encode a segmentation of a track one beat of the track is mapped to one bit of the individual. The genome size thus depends on the beat length of the track. Bits of value 1 indicate the start of a new segment. A sample encoding can be seen in Fig. 2. A new segment starts with each 1 in the bit vector, apart from the first one. This results from the precondition that a segment must have a minimum duration of one bar (which equals four beats in this case). As a consequence of the mapping explained above, segments can start on full beats only. However, this is not considered a problem because segment changes are unlikely to appear between full beats.

As a drawback of the chosen representation it is not possible to define regions without segments. However, a representation also allowing empty regions needs an additional marker for the end of a segment and therefore requires more complicated operators. For such a representation the crossover as well as the mutation operator have to ensure that each segment start bit has a corresponding end bit, thus increasing computational costs and slowing down the evolution process. For the chosen representation, empty segments can be simulated with an appropriate evaluation function that ignores single segments (= segments without any similar segments in the corresponding segment group).

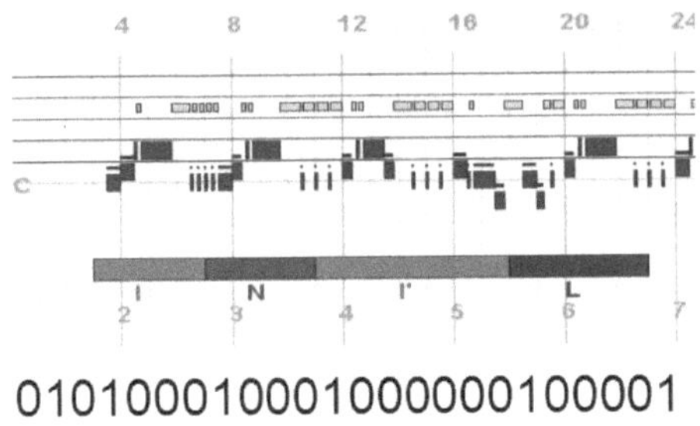

Fig. 2. Encoding of a segment as a bit vector

To produce the inital population segments are created randomly by assigning 0 or 1 to each bit in an individual. Since segments must be of a minimum length, the probability of 0 is higher than the probability of 1. Various probability ratios are used for different individuals leading to a wider variation of segment durations.

3.3 Evaluation

The quality of a segmentation increases if

- many segments within a segment group are identical
- segments within a segment group that are not identical are at least very similar
- segment starts coincide with note starts
- segments start at first beats of bars
- distances between segments are regular

The quality of a segmentation decreases if

- segments within segment groups differ in duration
- segments belonging to different groups differ in duration
- segment boundaries clip notes
- there is a high diversity of segments within one group
- segments do not contain enough notes to form a valid segment

All these features are weighed and summed up to form the evaluation function. The parameters for weighing the factors of the evaluation function have been tested by music experts to find the optimal values.

To define the similarity between two segments the music sequences within the segments are aligned using dynamic programming (compare [4]) and a similarity score is calculated. If the score exceeds a defined threshold, the segments belong to the same segment group.

3.4 Parameters

For the test cases the authors chose a track of a duration of 312 beats resulting in a bit vector length of 312 bits. Population size (P) ranged between 100 and 500 individuals. As a selection operator, tournament selection with size 3 and 4 was tested as well as roulette selection. Single point crossover and bit flip mutation were applied for reproduction. For all test cases 1-elitism was employed to evolve 3000 generations (G). Table 1 gives the settings of some sample test runs as well as the total and average best results.

Table 1. Summary of parameter settings and results without offspring selection

Selection operator	P	G	M rate	Best fitness	Avg fitness	Avg evaluations
Tournament (3)	100	3000	0.01	138.75	128.79	297000
Tournament (3)	500	3000	0.01	142.63	135.90	1497000
Tournament (4)	500	3000	0.01	145.51	137.43	1497000
Tournament (3)	100	3000	0.10	67.47	57.30	297000
Roulette	100	3000	0.01	115.58	109.32	297000
Roulette	500	3000	0.01	119.93	111.85	1497000
Roulette	100	500	0.10	62.65	51.97	297000

3.5 Introducing Offspring Selection

To increase the solution quality offspring selection [2] was introduced for some settings. To get a comparable number of evaluations 500 generations were evolved in the tests using offspring selection instead of 3000 generations in the earlier tests. Table 2 gives results of some sample tests with offspring selection.

Table 2. Summary of parameter settings and results with offspring selection

Selection operator	G	M rate	MaxSelPress	Best fitness	Avg fitness	Avg evaluations
Tournament (3)	500	0.01	100	149.12	139.54	868409
Tournament (3)	500	0.01	25	142.14	135.25	130125
Roulette	500	0.01	100	144.85	132.06	374458
Roulette	500	0.01	25	142.59	136.46	308527

4 Results

Comparing the settings described above tournament selection turned out to be more successful than roulette selection. Higher mutation rates caused an early stagnation of the best fitness within a population (see Fig. 3). Better results were achieved with population sizes also increasing the number of necessary evaluations and thus the total runtime of the algorithm. Fig. 4 shows the progress of the best, average, and worst fitness within a population during its evolution process using the test setting of line three in Table 1.

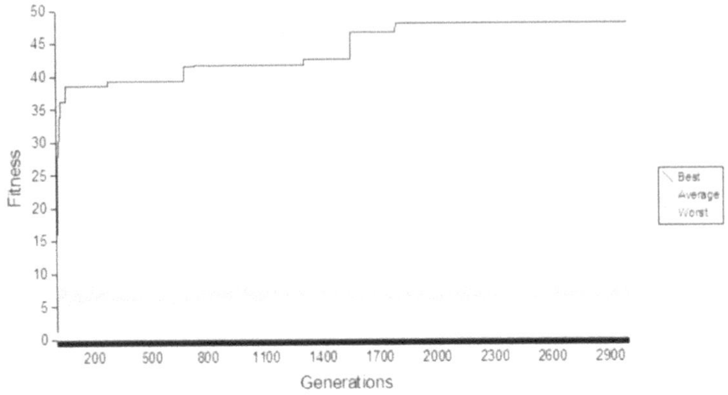

Fig. 3. Tournament selection with a high mutation rate

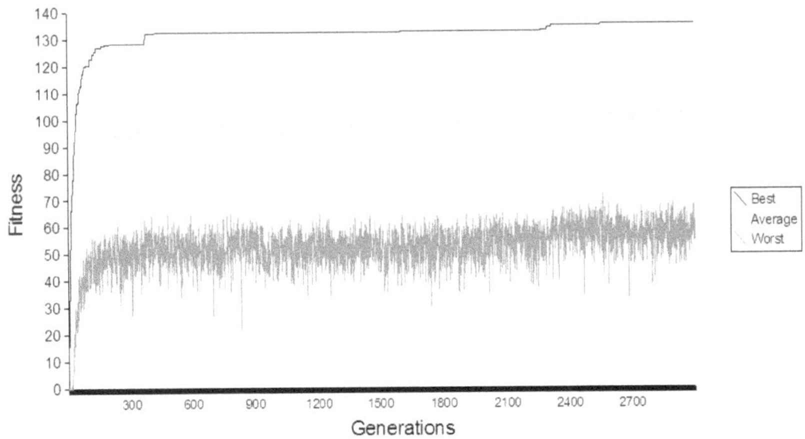

Fig. 4. Tournament selection without offspring selection

Offspring selection generally led to better results with a lower number of evaluations than tests without offspring selection. To get comparable results five times more evaluations were necessary in tests without offspring selection. Consequently, offspring selection emerged as a very important parameter for this genetic algorithm. The progress of a sample test run using the settings of line two in Table 2 is given in Fig. 5.

Figures 6 and 7 show example segmentations in different phases of the evolution. While the segments seem to be random in the beginning, clear segments and segment groups can be detected towards the end of the evolution process.

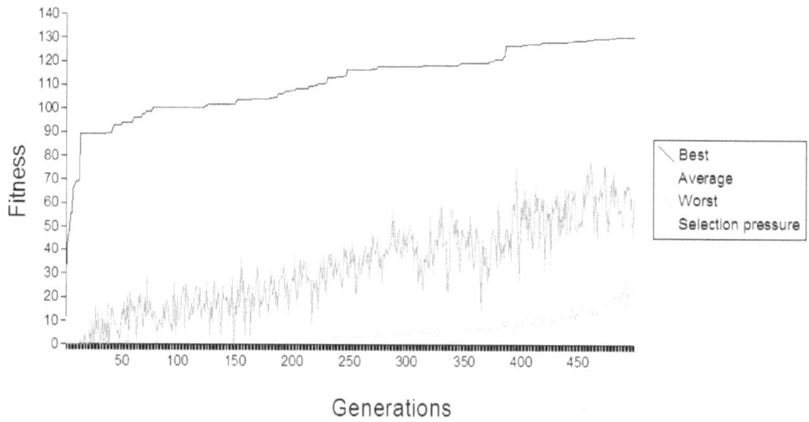

Fig. 5. Tournament selection with offspring selection

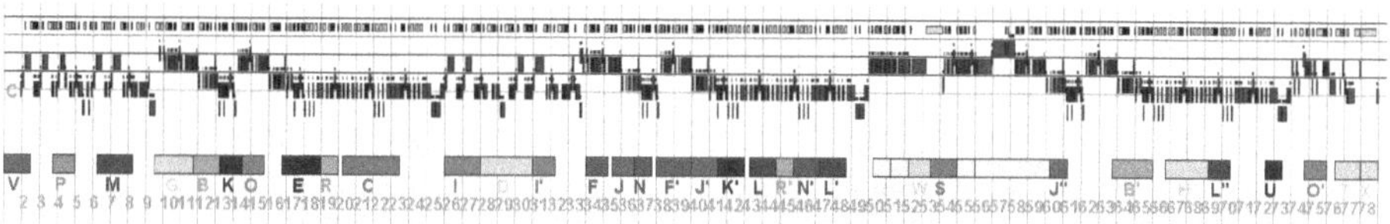

Fig. 6. Segmentation in an early generation

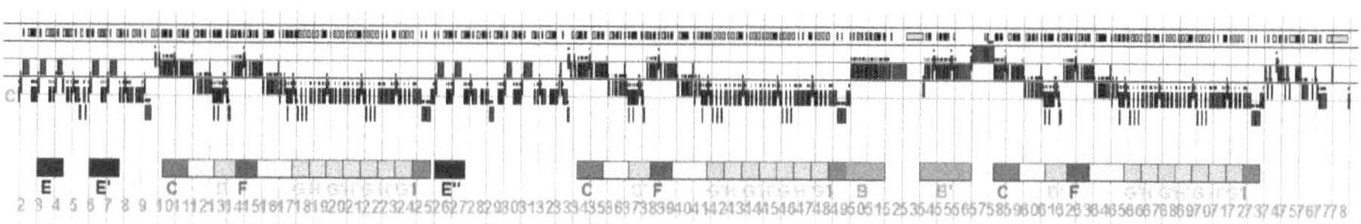

Fig. 7. Segmentation in a later generation

5 Conclusion

In this paper the reader has been introduced to the problem domain of music segmentation as well as to the application of genetic algorithms to the segmentation problem. Good results have been achieved with the first test settings but more test cases have to follow. New settings will be introduced and existing ones are to be tested further, especially focusing on the settings of offspring selection.

The present algorithm performs a segmentation for each track separately. Future work will also approach the combination of segmentations of different tracks to form one global segmentation. Smaller segments found by the genetic algorithm will be further structured hierarchically to form longer segments until the longest structural parts of a composition are reached (e.g., chorus and verse). Building such a hierarchy might also be done using a genetic algorithm.

Currently the genetic algorithm is only applied to segmenting MIDI data. The feasibility to adapt the approach for segmenting audio data will be examined during future research.

References

1. Abdulghafour, M.: Image segmentation using fuzzy logic and genetic algorithms. In: WSCG (2003)
2. Affenzeller, M., Wagner, S.: Offspring selection: A new self-adaptive selection scheme for genetic algorithms. In: Adaptive and Natural Computing Algorithms, pp. 218–221 (2005)
3. Chiu, P., Girgensohn, A., Wolf, P., Rieffel, E., Wilcox, L.: A genetic algorithm for video segmentation and summarization. In: IEEE International Conference on Multimedia and Expo, pp. 1329–1332 (2000)
4. Jehan, T.: Hierarchical multi-class self similarities. In: IEEE Workshop on Applications of Signal Processing to Audio and Acoustics, pp. 311–314 (2005)
5. Jensen, K.: Multiple scale music segmentation using rhythm, timbre, and harmony. EURASIP Journal on Applied Signal Processing 2007(1) (2007)
6. Lee, K., Cremer, M.: Segmentation-based lyrics-audio alignment using dynamic programming. In: Proceedings of the 9th International Conference on Music Information Retrieval, pp. 395–400 (2008)
7. Levy, M., Noland, K., Sandler, M.: A comparison of timbral and harmonic music segmentation algorithms. In: Proceedings of the Acoustics, Speech, and Signal Processing, vol. 4, pp. 1433–1436 (2007)
8. Maulik, U.: Medical image segmentation using genetic algorithms. IEEE Transactions on Information Technology in Biomedicine 13(2), 166–173 (2009)
9. Michalewicz, Z.: Genetic Algorithms + Data Structures = Evolution Programs. Springer, Heidelberg (1992)
10. Mueller, M., Ewert, S.: Joint structure analysis with applications to music annotation and synchronization. In: Proceedings of the 9th International Conference on Music Information Retrieval, pp. 389–394 (2008)
11. Paulus, J., Klapuri, A.: Music structure analysis by finding repeated parts. In: AMCMM 2006: Proceedings of the 1st ACM workshop on Audio and music computing multimedia, p. 5968. ACM Press, New York (2006)
12. Peiszer, E.: Automatic audio segmentation: Segment boundary and structure detection in popular music. Master's thesis, Vienna University of Technology, Vienna, Austria (2007)

Fitting Rectangular Signals to Time Series Data by Metaheuristic Algorithms

Andreas M. Chwatal and Günther R. Raidl

Vienna University of Technology, Vienna, Austria
{chwatal,raidl}@ads.tuwien.ac.at

Abstract. In this work we consider the application of metaheuristic algorithms to the problem of fitting rectangular signals to time-data series. The application background is to search for transit signals of exoplanets in stellar photometric observation data. The presented algorithms include an Evolution Strategy and Differential Evolution; both algorithms use an efficient reduction of the search space by exactly solving a subproblem. The presented results affirm the presented methods to be promising and effective tools for the discovery of the first multi-transit planetary system.

1 Introduction

Fitting parametrized models to data series is a frequently performed task in scientific computing. Nevertheless, finding (near-)optimal fits of superposed periodical signals to time-series data becomes a non-trivial problem when non-sinusoidal models are considered. In this case it is not always possible to derive good model parameters from the Fourier spectrum. Noisy data may further complicate this task. Finding good fits, which is in fact a continuous parameter optimization problem, is a computationally challenging task under these circumstances. In this work, we consider the problem of fitting rectangular signals to time-data series, and present metaheuristic algorithms to solve the problem.

2 Problem Description

The particular application background comes from the field of astronomy, in particular the problem of finding signals from transiting exoplanets in stellar photometric light-curves. For a comprehensive overview on exoplanets and detection methods see [1]. A transiting planet periodically shadows some of the light from its host star for a short time when it moves into our line of sight to the star. During the transit the luminosity of the star is marginally reduced. By neglecting the in- and egress phases, the transit-lightcurve can be well approximated by a periodic rectangular signal. The corresponding parameters are the period p the transit occurs with, a phase offset τ, the length l of the transit, and finally the transit depth d. The latter parameter corresponds to the percentage of light from the star being shadowed by the transiting planet. Figure 1 depicts

R. Moreno-Díaz et al. (Eds.): EUROCAST 2009, LNCS 5717, pp. 649–656, 2009.

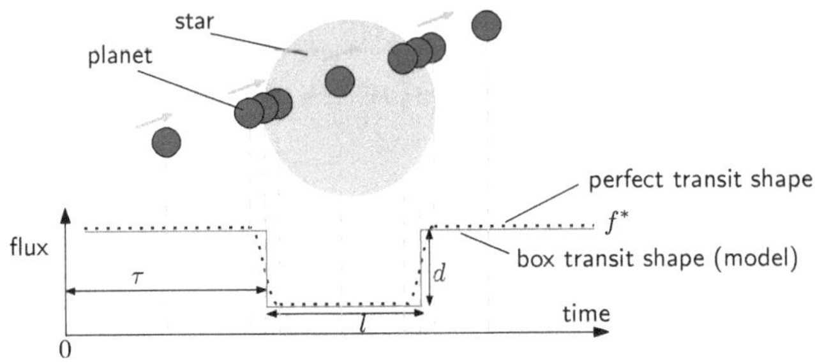

Fig. 1. Transiting planet and corresponding lightcurve

the situation for a single planet. Assuming M planets, the signal of the model at time t is given by

$$\phi(t) = f^* - \sum_{j=1}^{M} \chi_j^t d_j, \tag{1}$$

where f^* denotes a further parameter describing the regular flux (luminosity) of the host star; χ_j^t indicates if planet j is transiting at time t and is given by

$$\chi_j^t = \begin{cases} 1 & \text{if } \tau_j < t \bmod p_j \leq \tau_j + l_j \\ 0 & \text{otherwise.} \end{cases} \tag{2}$$

The observed data series is given by a list of $\{(t_i, f_i)\}, 1 \leq i \leq N$, where t_i denotes a particular observation time and f_i the observed photon flux (i.e. luminosity) at that given time. Let further $m_j = (p_j, l_j, d_j, \tau_j)$ and hence $\boldsymbol{m}$ be the vector of all model parameters (except f^*). The overall quality of the fit can be characterized by the root mean square error

$$f(\boldsymbol{m}, f^*, \boldsymbol{t}, \boldsymbol{f}) = \sqrt{\frac{1}{N}\sum_{i=1}^{N}(f_i - \phi(t_i))^2}. \tag{3}$$

The overall objective is to find a parameter setup for $\boldsymbol{m}$ and f^* minimizing Eq. (3), i.e. to find a model with minimal deviation from the observations. Due to stellar fluctuations and measurement errors real-world instances contain noisy signals. The signal-to-noise ratios can be expected to be very low, i.e. the respective values of d_j will be in the same order of magnitude as the standard deviation σ_f of the input values.

3 Previous Work

Several applications of genetic algorithms in astronomy are outlined in [2], and have since then been successfully applied for many purposes. In particular for the detection of exoplanets, evolutionary algorithms have been used with some success. For instance, an evolution strategy for fitting Keplerian models to radial velocity data is described in [3].

The development of efficient transit detection algorithms has recently gained more interest in the scientific community, as space-based missions like CoRoT[1] provide a great amount of observational data. One of the most popular approaches is the *box fitting least-square algorithm* [4]. This approach, as well as *phase dispersion minimization* [5] have the main drawback, that they are only directly applicable for finding single planet transits.

So far, no multi-planet system could be discovered by the transit method, which is possibly due to the difficulty of detecting their signals in (existing) observational data. More efficient techniques to tackle this numeric optimization problem would thus be a valuable contribution to exoplanet research.

4 Improvement and Evaluation of Candidate Solutions

The overall search process becomes more efficient when optimal values of depths d_j are derived from p_j, l_j, τ_j for each planet j. For this purpose we introduce binary flags $(b_1, \ldots, b_M)$ for each observation point $o_i = (t_i, f_i), i = 1, \ldots, N$, indicating which planet is transiting at the given time. These flags can be interpreted as integer number with binary representation $b_1 b_2 \ldots b_M \in [0, 2^M - 1]$, implying a partitioning of the set $O = \{o_1, \ldots, o_N\}$ of all observation points $O = O_0 \cup O_1 \cup \ldots \cup O_{2^M-1}$. Assuming two planets $M = 2$ we obtain the set of out-of-transit observations O_0, the sets O_1, O_2 of transit events of planets one and two respectively, and the set O_3 where planets one and two are transiting simultaneously. Optimal transit depths can be derived by minimizing

$$f(\boldsymbol{d}) = \sum_{i=1}^{N} (f_i - (f^* - \sum_{j=1}^{M} \chi_j^i d_j))^2, \tag{4}$$

which can be achieved by solving the system of linear equations resulting from $\frac{\partial f(\boldsymbol{d})}{\partial d_k} = 2 \sum_{i=1}^{N} \left(f_i - f^* + \sum_{j=1}^{M} \chi_j^i d_j \right) \cdot \chi_k^i = 0$ for all $k = 1, \ldots, M$. Let $\hat{f}^K = \sum_{i \in \bigcup_{k \in K} O_k} f_i, K \subseteq \{0, \ldots, 2^M - 1\}$ denote the sum of the observed photon fluxes from groups $\bigcup_{k \in K} O_k$, and $\hat{f} = \sum_{i=1}^{N} f_i$ analogously. Let further $n_K = |\bigcup_{k \in K} O_k|$ and $\tilde{\chi}_j^i, j = 1, \ldots, 2^M - 1, \ i = 1, \ldots, N$ indicate if observation i belongs to group j. For the case $M = 2$ we can now derive a direct expression by the partial derivative $\frac{\partial f(\boldsymbol{d})}{\partial d_1} = 2 \cdot \sum_{i=1}^{N} \left(f_i - 2f^* + 2 \sum_{j=1}^{2^M-1} \tilde{\chi}_j^i d_j \right) \cdot (\tilde{\chi}_1^i + \tilde{\chi}_3^i) = 0$ from which we obtain

$$d_1 = f^* - \frac{\hat{f}^{1,3}}{n_{1,3}} - \frac{n_3}{n_{1,3}} \cdot d_2, \quad \text{and} \quad d_2 = f^* - \frac{\hat{f}^{2,3}}{n_{2,3}} - \frac{n_3}{n_{2,3}} \cdot d_1, \tag{5}$$

where $f^* = \hat{f}^0 / n_0$. By inserting d_2 into the equation for d_1 we obtain

$$d_1 = \left((1 - \frac{n_3}{n_{1,3}}) \hat{f}^0 - \frac{1}{n_{2,3}} \hat{f}^{2,3} + \frac{n_3}{n_{1,3} \cdot n_{2,3}} \hat{f}^{2,3} \right) \cdot \left(1 - \frac{n_3^2}{n_{1,3} \cdot n_{2,3}} \right)^{-1}, \tag{6}$$

and a corresponding equation for d_2 by inserting d_1 into the equation for d_2.

[1] CoRoT: **Co**nvection **Ro**tation and planetary **T**ransits; European space telescope.

5 Fitness-Landscape Analysis

In order to evaluate the applicability of metaheuristics for solving this problem, we performed a comprehensive fitness-landscape analysis. For this purpose we created numerous test instances containing signals from two planets. For each configuration we created multiple instances with different signal-to-noise ratios. Figure 2 shows the fitness-distance correlation for one typical instance. For the measure of the distances to the global optimum we used simple Euclidean distances. The left plot shows the view of the whole parameter space. One can see that there is almost no correlation of fitness values to the distances to the global optimum. All points have roughly the same value, which corresponds to the level of noise of the input instance. This effect is due to adjustment of the model transit-depths and the out-of-transit stellar flux according to the other (randomly created) parameter values, as described in Section 4. As a consequence the depths are set to zero for most configurations, and the out-of-transit stellar flux is set to the average value of all data points. Values higher than this average seldomly occur when the out-of-transit average (due to the model) is lower than the in-transit average value.

The right plot of Fig. 2 shows a closer view to the global optimum. Here, parameter values have been enumerated in a discretized way such that all distances are smaller than 0.42. This plot clearly shows that a strong correlation of distances to fitness values appears when coming close to the global optimum. These results indicate that it is very hard to find the region of the global optimum, but if that region has been found, it is relatively easy to find the global optimum itself. For problems with these properties metaheuristic algorithms are known to be a good choice. For the particular case, they must facilitate effective mechanisms for self-adaptation, i.e. to facilitate an explorative search process until the region of the global optimum is found, and then change their behavior to a fine grained exploitative search.

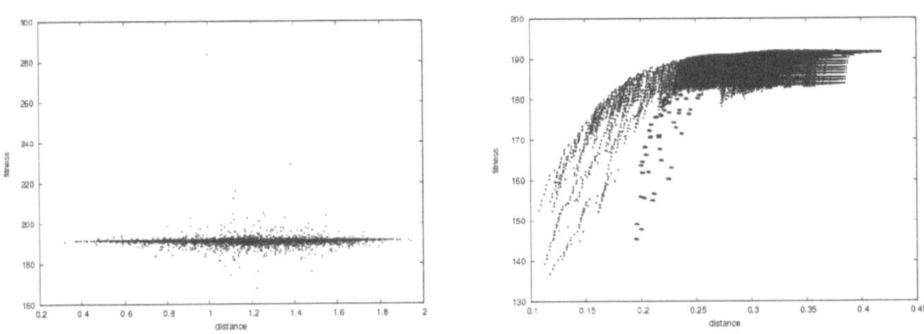

Fig. 2. Fitness-distance correlation diagrams, for the whole parameter-space (left), and a restricted parameter space close to the global optimum solution (right)

6 Metaheuristic Algorithms

There exists a variety of algorithms for heuristically solving difficult continuous parameter optimization problems like *Evolution Strategies* (ES), *Differential Evolution* (DE) and *Continuous Scatter Search* (CSS) [6,7,8]. These algorithms are population-based approaches, iteratively modifying and evaluating a set of candidate solutions. In order to find strengths and drawbacks of these methods w.r.t. this particular problem we implemented them without major modifications. Preceding experiments showed that concerning our problem ES and DE are clearly superior to CSS, as the latter one suffers from relatively time consuming subset generation. Hence, we now focus on ES and DE, which are briefly described in the following.

Individuals are directly encoded as vectors of real values in both approaches. Both algorithms do not use any local optimization method, except the techniques described in Section 4. General purpose local optimizers as for instance the Nelder-Mead method [9] turned out to be too time-consuming, and are, as indicated by the Figure 2, only beneficial if already very close to the global optimum. They are therefore not used in our evaluation.

6.1 Evolution Strategy

The ES can be classified as a (μ, λ)-ES with self-adaptation of strategy parameters [10], where μ denotes the size of the population and λ the number of offsprings created in each generation. It turned out to be advantageous to use a variant of elitist-selection which creates the new population by deterministically taking the best μ individuals from the μ parents and λ offsprings, but taking at most $\hat{\mu}$ individuals from the parents. Hence, our selection is in fact in-between $(\mu + \lambda)$-selection and (μ, λ)-selection.

Mutation is considered to be the primary operator and is performed by adding Gaussian random values to the parameters x_k (see Eq. (7)), where the standard deviation is given by a strategy parameter σ_k, associated with each parameter.

$$x'_k = x_k + N_k(0, \sigma'_k) \tag{7}$$

These strategy parameters are also modified by the evolutionary operators, which facilitates self-adaption of the search process.

$$\sigma'_k = \sigma_k \cdot e^{N(0,\tau_0)+N_k(0,\tau)} \tag{8}$$

After the application of the evolutionary operators, the optimal transit-depths $d_j, j = 1, \ldots, M$ are calculated before fitness function evaluation. If some depth is set to 0.0 – implying that this particular planet-model does not improve the quality of the fit at all – a new random planet is created on this position, which might increase diversity among the population. Prior to mutation recombination operators might be applied with some probability. We use the intermediate recombination, given by

$$x'_k = \alpha_k \cdot x^1_k + (1 - \alpha_k)x^2_k, \tag{9}$$

where x_k^1 and x_k^2 denote the parameters of the parents and α_k is a uniform random number from the interval $[-\beta, 1+\beta]$ for each parameter k, where $\beta = \frac{1}{2}$ turned out to be most successful.

6.2 Differential Evolution

Differential Evolution (DE) is a particular variant of an evolutionary algorithm, operating on a population of individuals which are encoded by a vector of real parameter values. Mutation is performed by combining three randomly selected individuals with indices (r_1, r_2 and r_3) to a new individual $v_{i,t+1}$ by

$$v_{i,t+1}^j = x_{r_1,t}^j + F \cdot (x_{r_2,t}^j - x_{r_3,t}^j), \tag{10}$$

where $F \in [0, 2]$. Using the notation $u_{i,t+1} = (u_{i,t+1}^1, u_{i,t+1}^2, \ldots, u_{i,t+1}^{3 \cdot M})$ for a particular individual, crossover is given by

$$u_{i,t+1}^j = \begin{cases} v_{i,t+1}^j & \text{if } r_j \leq C_R \vee j = r_i \\ x_{i,t}^j & \text{if } r_j > C_R \wedge j \neq r_i \end{cases} \tag{11}$$

where $C_R \in [0, 1]$ denotes the crossoverrate and $r_j, r_i \in [0, 1]$ random numbers. The new individual $x_{i,t+1}$ is obtained by

$$x_{i,t+1} = \begin{cases} u_{i,t} & \text{if } f(u_{i,t}) < f(x_{i,t}) \\ x_{i,t} & \text{otherwise.} \end{cases} \tag{12}$$

7 Results

For an extensive evaluation of our algorithms we created artificial test-instances. Real stellar signals typically do not only contain the rough (nearly) rectangular signals from the transiting planet, but also portions of stellar jitter and measurement errors. We take this into account by adding Gaussian random variables to each data point in the artificial signal. We thus create three instances for each configuration: one strictly rectangular signal and two noisy ones with different standard deviations.

Table 1 shows the results of 50 independent runs for various test instances. The first part shows the results for single signals, whereas the second part contains two-planet signals. For each algorithm we report the percentage of times where optimal solutions have been obtained and the average running times. Each column contains three values corresponding to signals without noise and with noise of $\sigma = 100$ and $\sigma = 300$ for the particular instances respectively. For some instances no results are available (indicated by "n/a"), as the algorithm stopped prematuerly because of many solutions having lower fitness values than the solution of the artificial signal.

For both algorithms we set the number of maximum iterations to 1000. We did not impose a time limit, but runs have been stopped when the global optimum

Table 1. Test-instances and corresponding success ratios of evolution strategy and differential evolution and average running times

Instance-name	Parameters p	l	d	τ	(50/100,500)-ES (% opt.)	$t_{avg}[s]$	DE ($\|P\| = 200$) (% opt.)	$t_{avg}[s]$
art-100	1.0	0.10	100.0	0.5	100,100, 62	34, 34,311	100,100, 86	181,196,207
art-101	1.0	0.10	500.0	0.5	100,100,100	26, 19, 31	100,100,100	182,192,217
art-102	2.0	0.10	100.0	0.5	100,100, 28	12, 13,313	100,100, 28	129,135, 65
art-103	2.0	0.10	500.0	0.5	100,100,100	15, 12, 13	100,100,100	127,159,132
art-104	2.0	0.10	100.0	0.5	100,100, 94	12, 14, 48	100,100,100	108,109,140
art-105	2.0	0.05	500.0	0.5	100, 90,n/a	13, 49,n/a	100,100,n/a	103,115,137
art-106	1.0	0.05	500.0	0.5	98, 96, 70	22, 37,134	100,100,100	264,164,161
art-107	1.0	0.05	300.0	0.5	100,100, 46	23, 24,296	100,100, 86	159,155,180
art-108	1.0	0.05	100.0	0.5	94, 96,n/a	37, 27,n/a	100,100,n/a	162,164, 52
art-109	1.0	0.02	500.0	0.5	70, 72, 4	46, 60,576	100,100, 0	136,141,192
art-110	1.0	0.02	300.0	0.5	86, 76, 4	30, 61,498	100,100, 8	137,149,291
art-111	1.0	0.02	100.0	0.5	82, 26,n/a	38,208,n/a	100, 44,n/a	149,147,163
art-210	1.0/2.2	0.10/0.10	500.0/500.0	0.5/1.0	92, 90, 94	205,322,252	12, 12,n/a	526,531,602
art-211	1.0/2.2	0.10/0.10	500.0/300.0	0.5/1.0	98, 92, 80	314,318,422	56, 36, 8	518,527,616
art-212	1.0/2.2	0.10/0.05	300.0/500.0	0.5/1.0	78, 78, 48	322,440,658	20, 12, 28	479,453,475
art-213	1.0/2.2	0.05/0.05	500.0/500.0	0.5/1.0	88, 88, 46	374,601,663	0, 0, 0	481,500,474
art-214	1.0/7.5	0.05/0.20	400.0/500.0	0.5/1.0	54, 52, 40	316,306,396	32, 28, 14	217,328,340
art-215	1.0/7.5	0.10/0.20	400.0/500.0	0.5/1.0	72, 78, 72	348,339,312	100,100, 98	607,597,446
art-216	1.0/3.1	0.05/0.10	400.0/500.0	0.5/1.0	56, 60, 4	316,483,735	12, 6, 2	351,355,426
art-217	1.0/3.1	0.05/0.10	500.0/400.0	0.5/1.0	66, 76, 18	382,525,720	0, 0, 4	447,466,418
art-218	1.0/3.1	0.05/0.10	500.0/300.0	0.5/1.0	82, 72, 22	594,770,831	0, 0, 6	451,477,472
art-219	1.0/3.1	0.05/0.10	500.0/200.0	0.5/1.0	74, 76, 4	744,647,807	0, 2, 0	481,479,484

was found. With "global optimum" we refer to a solution which is close to the artificial signal and has the same (or lower) objective function value. Altough unlikely, better solutions might exist, i.e. solutions where arbitrary fitting of the noise yields lower deviations to the observations than the original imposed signal. Such situations are indicated by "n/a" in Table 1, as the algorithm is prematurely stopped in these cases.

For ES we used the parameter setting $\mu = 100, \lambda = 500$, and $\hat{\mu} = 50$. Prior to mutation we performed intermediate recombination for the strategy parameters and parameters with a probability of 0.8. For the DE algorithm we used $F = 1, C_R = 0.5$ and a population size of 200.

For both algorithms we used the parameter-space reduction as described in Section 4 and an advanced method to speed up the fitness-function evaluation which is beyond the scope of this paper. The optimal calculation of the depths significantly improves the ability of the algorithm to improve existing solutions quickly. All tests have been performed on a heterogenous cluster mostly consisting of recent hardware like Intel Core2 Quad, Intel Xeon and Dual-Core AMD Opteron processors.

The results show that optimal solutions can be obtained with high probability and acceptable running times for these data instances. Although not part of this work, we want to emphasize that the algorithms have comparable performance on real data-instances obtained from the CoRoT space telescope, which are known to contain planetary signals. For this purpose we added additional artificial signals to selected data instances, as so far no multi-planet signals have been found in this data.

8 Conclusions and Future Work

Both algorithms, ES and DE, exhibit a good performance and robustness on the test-instances presented in Section 7. Generally the ES converges faster which is mainly due to the effectiveness of the self-adaptation mechanism regarding the particular structure of the solution space. Although the DE algorithm generally requires longer running times, for some instances higher success ratios are obtained. Hence, both approaches have a justification to be used in practice.

An important part of transit detection algorithms, not considered in this work, is to compute a value indicating the statistic significance of the resulting fit. Such a measure enables to distinguish real signals from signals containing just noise and non-periodic signals and obviously should keep the false-alarm probability to a reasonably small rate. Techniques, currently used for single-planet signals (e.g. see [4]) are not directly applicable to multi-planet fits obtained by this approach. Hence we currently simply use the ratio between the standard deviation of the obtained fit to the standard deviation of the raw data, or alternatively Student's t-test in order to test if the in-transit levels have significantly different values in comparison to the out-of-transit levels. More extensive (blind) testing needs to be performed to asses the reliability of these approaches, but also more elaborate techniques might be necessary. Nevertheless, it is likely that current indicators are already able to select a reasonable subset of candidates from the huge amount of real-world input-data being worth analyzed in more detail subsequently. The application of the presented algorithms to yet publicly available CoRoT data is ongoing.

References

1. Deeg, H., Belmonte, J.A., Aparicio, A. (eds.): Extrasolar Planets. Cambridge University Press, Cambridge (2008)
2. Charbonneau, P.: Genetic Algorithms in Astronomy and Astrophysics. Astrophysical Journal Supplement 101, 309–334 (1995)
3. Chwatal, A.M., Raidl, G.R.: Determining orbital elements of extrasolar planets by evolution strategies. In: Moreno Díaz, R., Pichler, F., Quesada Arencibia, A. (eds.) EUROCAST 2007. LNCS, vol. 4739, pp. 870–877. Springer, Heidelberg (2007)
4. Kovács, G., Zucker, S., Mazeh, T.: A box-fitting algorithm in the search for periodic transits. Astronomy and Astrophysics 391, 369–377 (2002)
5. Stellingwerf, R.F.: Period determination using phase dispersion minimization. Astrophysical Journal 224, 953–960 (1978)
6. Bäck, T.: Evolutionary Algorithms in Theory and Practice. Oxford University Press, New York (1996)
7. Storn, R., Price, K.: Differential evolution – a simple and efficient heuristic for global optimization over continuous spaces. Journal of Global Optimization 11, 341–359 (1997)
8. Glover, F., Laguna, M., Marti, R.: Fundamentals of scatter search and path relinking. Control and Cybernetics 29(3), 653–684 (2000)
9. Nelder, J., Mead, R.: A simplex method for function minimization. The Computer Journal (7), 308–313 (1964)
10. Schwefel, H.P.: Numerical Optimization of Computer Models. Wiley, Chichester (1981)

Virtual Sensors for Emissions of a Diesel Engine Produced by Evolutionary System Identification

Stephan M. Winkler[1], Markus Hirsch[2], Michael Affenzeller[1], Luigi del Re[3], and Stefan Wagner[1]

[1] Heuristic and Evolutionary Algorithms Laboratory
Upper Austria University of Applied Sciences
School of Informatics, Communications and Media
Softwarepark 11, 4232 Hagenberg, Austria
{stephan.winkler,michael.affenzeller,stefan.wagner}@fh-hagenberg.at
[2] Linz Center of Mechatronics, Altenbergerstraße 69, 4040 Linz, Austria
markus.hirsch@lcm.at
[3] Institute for Design and Control of Mechatronical Systems
Johannes Kepler University Linz, Altenbergerstraße 69, 4040 Linz, Austria
luigi.delre@jku.at

Abstract. In this paper we discuss the generation of models for emissions of a Diesel engine, produced by genetic programming based evolutionary system identification: Models for the formation of NO_x and particulate matter emissions are identified and analyzed. We compare these models to models designed by experts applying variables section and the identification of local polynomial models; analyzing the results summarized in the empirical part of this paper we see that the use of enhanced genetic programming yields models for emissions that are valid not only in certain parts of the parameter space but can be used as global virtual sensors.

1 Introduction and Experimental Data

Virtual sensors are in general simulation models that can be used instead of physical sensors. If there are no appropriate models available to the required precision, virtual sensor design must be based on data; we are in this context speaking of data based system identification [6]. In this paper we concentrate on the discussion of models for emissions of a common rail direct injection 2 liter 4 cylinder production Diesel engine, produced by local polynomial modeling as well as genetic programming based evolutionary system identification; models for the formation of nitrogen oxides (NO_x) and particulate matter (PM), the two most demanding emissions of Diesel engines, are identified and analyzed.

In Section 2 we summarize the identification methods that have been applied for identifying emission models for the investigated engine, namely local in parameter linear regression using a restricted set of input parameters (as described in Section 2.1) and enhanced evolutionary system identification [9] based on genetic programming (GP, see [5], e.g.) as discussed in Section 2.2. The results of

R. Moreno-Díaz et al. (Eds.): EUROCAST 2009, LNCS 5717, pp. 657–664, 2009.

these empirical test studies are summarized in Section 3; a discussion of these results (given in Section 4) concludes this paper.

We have used data recorded at a dynamic engine test bench at the Institute for Design and Control of Mechatronical Systems at Johannes Kepler University (JKU) Linz, Austria. The data was composed of selected engine states relevant for defining the combustion process, given as measured or calculated quantities of the engine control unit (ECU), as well as target values for NO_x and PM emissions which have been measured with fast emission sensors. Seven different measurements with a sampling time of 50 ms have been recorded:

- Six measurements $M_{1...6}$ (each recorded over approximately 10–15 minutes) have been recorded for different engine speeds and amounts of injected fuel. These measurements were done as a result of optimal design of experiment strategies for achieving short but well exciting signals for nonlinear identification (for details on this see [4]). The average engine speed (N) was set to 1000 and 2000 revolutions per minute, and the average amount of total injected fuel (q) to 5, 10 and 20 mg per cycle; thus, combining these parameter settings we get 6 conditions under which the engine has been tested. Figure 1 shows a visualization of the parameters N and q of the samples that are included in these six sets of measurements.
- The engine was also tested following the New European Driving Cycle (NEDC), a standardized driving cycle which is used for evaluating the fulfillment of emission standards for passenger cars. The NEDC data ($NEDC$) is within the operating range covered by the measurements $M_{1...6}$.

Based on these data sets our goal is to identify models for NO_x and PM emissions using the data sets $M_{1...6}$ and testing these models on $NEDC$ data.

2 Identification Methods Used for Designing Virtual Sensors

2.1 Variables Selection and Local Polynomial Regression

Given a data collection including m input features storing the information about n samples, a in parameter linear model is defined by the vector of coefficients $\theta_{1...m}$. For calculating the vector of modeled values $Y \in \mathbb{R}^n$ using the given input values matrix $U \in \mathbb{R}^{m \times n}$, these input values are multiplied with the corresponding coefficients and added: $Y = U \cdot \theta$; the coefficients vector θ can be computed by standard least square error minimization according to $\theta = \left(U^T U\right)^{-1} U^T Y$. Theoretical background of this approach can be found in [6].

For identifying models for NO_x and PM emissions in the context of our research project we have restricted the set of input features to the following four variables: The total amount of injected fuel per cycle (q), the engine's speed (N), the manifold air pressure (MAP), and the concentration of oxygen in the exhaust (O_2exh). This selection of input variables, all available in the ECU, has been made on the basis of physical knowledge about combustion engines;

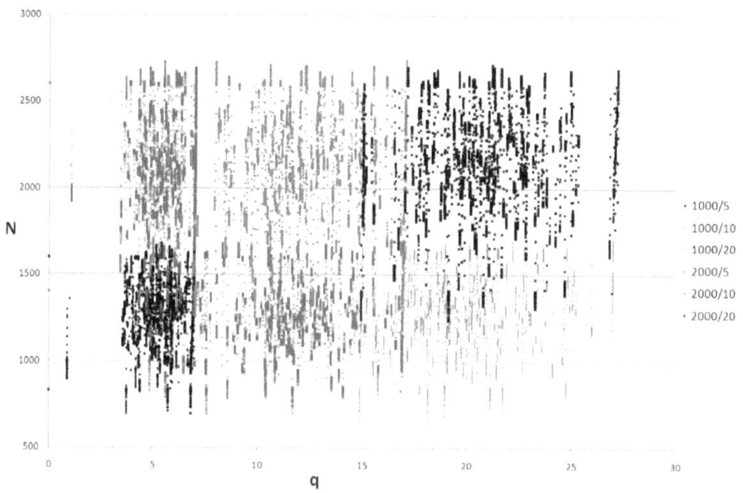

Fig. 1. Visualization of engine parameter values of the training measurements: Each spot represents the engine speed (N) and the amount of injected fuel (q) of one data sample. Each sample belongs to one of the six available sets of measurements; spots representing samples belonging to the same set are displayed using the same visualization style (as indicated in the legend of the chart).

further information can for example be found in [3] and [8]. As the process of emission formation is a nonlinear one, even in the six restricted data sets, second order multiplications of these input features are also used. Though the formulation now includes nonlinearities, the structure is linear in parameters and we are looking for is a set of (in total 30) parameters $\theta_{i,j}$ so that NO_x and PM can be described as $NO_x(t) = \theta_{1,1} + \theta_{1,2} \cdot q(t) + \theta_{1,3} \cdot N(t) + \theta_{1,4} \cdot MAP(t) + \theta_{1,5} \cdot O_2exh(t) + \theta_{1,6} \cdot q(t)^2 + \theta_{1,7} \cdot q(t) \cdot N(t) + \ \ldots \ + \theta_{1,15} \cdot O_2exh(t)^2$ and $PM(t) = \theta_{2,1} + \theta_{2,2} \cdot q(t) + \theta_{2,3} \cdot N(t) + \theta_{2,4} \cdot MAP(t) + \theta_{2,5} \cdot O_2exh(t) + \theta_{2,6} \cdot q(t)^2 + \theta_{2,7} \cdot q(t) \cdot N(t) + \ \ldots \ + \theta_{2,15} \cdot O_2exh(t)^2$, respectively.

The fact that static models (present output is defined only by present input values without time offsets) are used for representing a dynamic system can be reasoned by the fact that not only directly settable input quantities such as $q(t)$ but also internal engine states such as $MAP(t)$ or $O_2exh(t)$, which already contain dynamic effects, have been used here.

2.2 Evolutionary System Identification

Basically, genetic programming (GP) is based on the theory of genetic algorithms (GAs) and utilizes a population of solution candidates which evolve through many generations towards a solution using certain evolutionary operators and a selection scheme increasing better solutions' probability of passing on genetic information; the goal of a GP process is to produce a computer program solving the optimization problem at hand. In the case of structure identification, solution

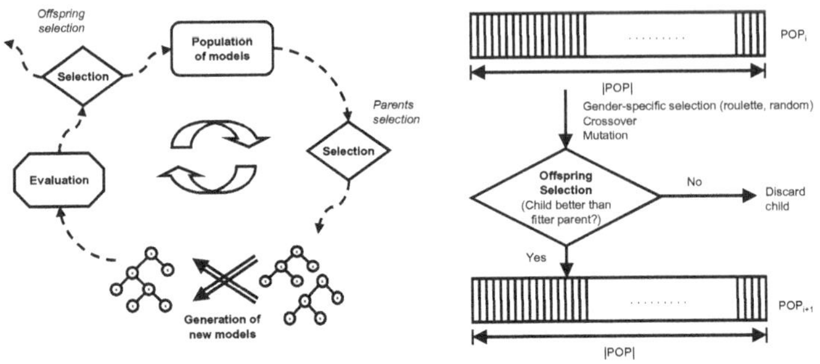

Fig. 2. Left: The extended genetic programming cycle including offspring selection. **Right:** Strict offspring selection as used here within the GP process.

candidates represent mathematical models; these models are applied to the given training data and the so generated output values are compared to the original target data. The left part of Figure 2 visualizes how the GP cycle works: As in every evolutionary process, new individuals (in GP's case, new programs) are created and tested, and the fitter ones in the population succeed in creating children of their own; unfit ones die and are removed from the population [5].

Within the last years we have set up an enhanced and problem domain independent GP based structure identification framework that has been successfully used in the context of various different kinds of identification problems for example in mechatronics, medical data analysis, and the analysis of steel production processes; please see [9] for an extensive overview the authors' research activities in these fields. One of the most important problem independent concepts used in our implementation of GP-based structure identification is offspring selection [1], an enhanced selection model that has enabled genetic algorithms and genetic programming implementations to produce superior results for various kinds of optimization problems. As in the case of conventional GAs or GP, offspring are generated by parent selection, crossover, and mutation. In a second (offspring) selection step (as it is used in our GP implementation), only those children become members of the next generation's population that outperform their own parents; the algorithm repeats the process of creating new children until the number of successful offspring is sufficient to create the next generation's population. In [2] and [9] interested readers can find several examples and analyses of the effects of OS in GAs.

In our research project in the context of the identification of virtual sensors we have used all available (or rather "allowed") variables, i.e., not only those 4 variables that have already been described (q, N, MAP, and O_2exh), but also several other ones (in total 31 variables) including temperatures and parameters of the ECU. These variables are allowed in this context simply because these values are also available in the context of standard commercial automobiles; information about other emissions (as for example CO_2) has not been used.

3 Results Documentation: Analysis of Identified Models

3.1 Using in Parameter Linear Models

The approach applied using in parameter linear regression models for representing local behaviors was the following one: All six available measurement data sets $M_{1...6}$ have been used for identifying 2nd order polynomial models using MATLAB®. When estimating the target values for test data X all models $m_{1...6}$ are applied and the resulting target value for sample i ($t(i)$) is calculated using the following procedure for smooth switching:

$$
\begin{aligned}
x_k &:= \texttt{apply}(m_k, X(i))(\forall k \in [1\ldots 6]) \\
c &:= (N-1400)/200 \quad (c<0 \Rightarrow c:=0; c>1 \Rightarrow c:=1) \\
y_1 &:= x_1 \cdot c + x_2 \cdot (1-c) \\
y_2 &:= x_3 \cdot c + x_4 \cdot (1-c) \\
y_3 &:= x_5 \cdot c + x_6 \cdot (1-c) \\
c &:= (q-4)/2 \quad (c<0 \Rightarrow c:=0; c>1 \Rightarrow c:=1) \\
y_4 &:= y_2 \cdot c + y_3 \cdot (1-c) \\
c &:= (q-14)/2 \quad (c<0 \Rightarrow c:=0; c>1 \Rightarrow c:=1) \\
t(i) &:= y_1 \cdot c + y_4 \cdot (1-c)
\end{aligned}
$$

This procedure is of course applied for estimating NO_x as well as PM emissions (separately); for estimating NO_x values we use the models identified using NO_x training data (${m^{NO_x}}_{1...6}$) as prediction models $m_{1...6}$, and for estimating PM values we use the prediction models (${m^{PM}}_{1...6}$).

We here give the qualities of the so resulting combined estimation models as the estimated values' mean squared error (*mse*) on NEDC data: The *mse* of the estimated NO_x values for the NEDC data set is $\mathbf{1.4 \cdot 10^4}$, for the particulate matter emissions (which are given in terms of opacity) the *mse* is **2.635**.

3.2 Using Evolutionary System Identification

Using GP we have trained models on the basis of all six data sets $M_{1...6}$, i.e., we have collected all samples in one big training data collection and applied enhanced GP using strict OS for learning models; the GP implementation in HeuristicLab (as described for example in [7]) has been used as described for example in [9]. We have not defined any model structures that are fixed for the NO_x and PM identification tasks, all available arithmetic and logical functions (as given in Table 1 and discussed in detail in [9]) have been used; the maximum hierarchy depth of the produced formulas has been set to 12. As the reader can see in Table 1, mathematical functions and terminal nodes are used as well as Boolean operators for building complex arithmetic expressions. There are in fact no structural restrictions for the use of Boolean blocks in formulae; of course, [Then/Else] and Boolean expressions have to be connected to [IF] nodes, but there are no other restrictions regarding the use of Boolean blocks within mathematical expressions.

Table 1. Set of function and terminal definitions for GP based system identification

Functions		
Name	**Arity**	**Description**
+	2	Addition
-	2	Subtraction
*	2	Multiplication
/	2	Division
e^x	1	Exponential Function
IF	3	If [Arg0] then return [Then] branch ([Arg1]), otherwise return [Else] branch ([Arg2])
$\leq$, $\geq$	2	Less or equal, greater or equal
&&, ||	2	Logical AND, logical OR
Terminals		
Name	**Parameters**	**Description**
var	x, c	Value of attribute x multiplied with coefficient c
const	d	A constant double value d

The population size was set to 1000, we applied standard one-point crossover and mutation operators (the mutation rate was set to 15%), and the maximum selection pressure (defining the algorithms' termination criterion) was set to 500. We have applied static modeling (i.e., no time offsets were allowed for model inputs – as also done in the case of constrained polynomial modeling). The GP approach has been tested 5 times independently; the quality of these models is estimated by testing them on the NEDC data set. Using this static GP approach we retrieved models for NO_x that show an average test *mse* of **9351.1742** ($\sigma = 1624.25$); for PM we retrieved models with average test *mse* of **1.8455** ($\sigma = 0.1418$).

3.3 Discussion

Figure 3 shows the evaluation of the models for NO_x (produced by polynomial regression as well as GP) that performed best on training data, evaluated on a part of the given test (NEDC) data; Figure 4 shows the residuals of these models when evaluated on this part of the NEDC data. The mean squared error of the model produced by GP (when evaluated on test data) is **7847.43**. As we can see in these figures, there are significant errors that occur when the amount of injected fuel becomes very low (or even zero). This is because the data for the identification of this area was not available in sufficient detail, and therefore we here extrapolate by applying the models to parameter regions not available in the training data. Very high local peaks of the regression model can be explained: Small delays, not important for slow changes, given by the geometrical location of the lambda sensor in the exhaust, which is used for measuring O_2exh, cause a mismatch of the inputs which causes the high peaks. Slightly filtering q could reduce this effect.

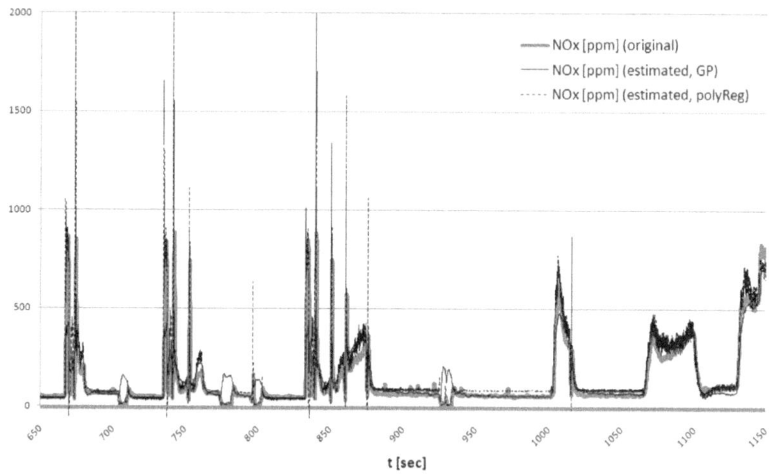

Fig. 3. Evaluation (on a part of the NEDC data) of models produced by GP and polynomial regression for the engine's NO_x emissions

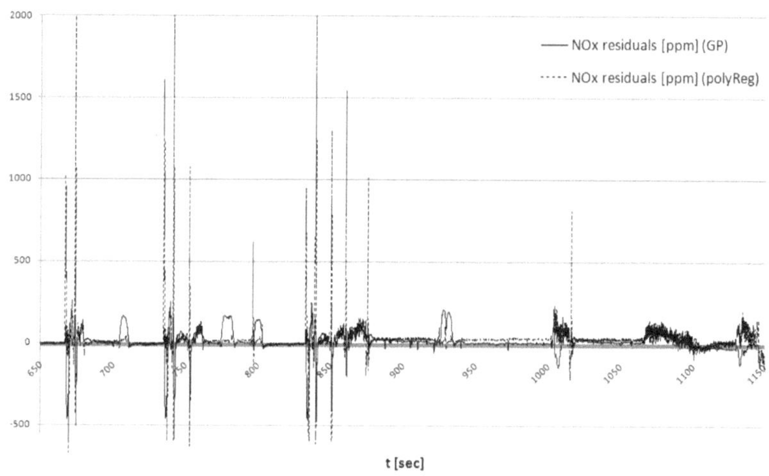

Fig. 4. Residuals (on a part of the NEDC data) of models produced by GP and polynomial regression for the engine's NO_x emissions

4 Conclusion

Differently from classical optimization methods, genetic algorithms address the optimization issue in a biologically inspired way trying to recognize dominant "parts of the solution" and build the final result around these parts. Conceptually, this holds also for genetic programming: GP is able to provide analytical expressions, a frequent wish of designers, especially if these can be interpreted physically. In the course of the project which is the background of the condensed information presented in this paper, GP based structure identification has been compared to a much better established method, the NARX polynomial modelling approach, and has shown to be a viable alternative, with validation results absolutely comparable, if not superior. Differently from classes like the polynomial ARX models, GP is able to build new function kernels which can allow a much better insight into the system. Of course, as for every heuristic method, there is no guarantee for it and the computational effort can become rather large, but it can provide a very interesting and probably unique approach to the system model, in this case to the emissions.

Another important discovery of this work was that, at the end, the choice of the data has a paramount importance, much more than in the linear case.

Acknowledgements

The work described in this paper was done within the Translational Research Program project L284-N04 "GP-Based Techniques for the Design of Virtual Sensors" sponsored by the Austrian Science Fund (FWF). The involved research organizations are the Heuristic and Evolutionary Algorithms Laboratory at the Upper Austria University of Applied Sciences, Faculty of Informatics, Communications and Media, and the Linz Center of Mechatronics.

References

1. Affenzeller, M., Wagner, S., Winkler, S.: Goal-oriented preservation of essential genetic information by offspring selection. In: Proceedings of the Genetic and Evolutionary Computation Conference (GECCO) 2005, vol. 2, pp. 1595–1596. Association for Computing Machinery (ACM), New York (2005)
2. Affenzeller, M., Winkler, S., Wagner, S., Beham, A.: Genetic Algorithms and Genetic Programming – Modern Concepts and Practical Applications. Chapman & Hall/CRC (2008)
3. Hirsch, M., Alberer, D., del Re, L.: Grey-box control oriented emissions models. In: Proceedings of IFAC World Congress 2008, pp. 8514–8519 (2008)
4. Hirsch, M., del Re, L.: Adaped D-optimal experimental design for transient emission models of diesel engines. In: Proceedings of SAE Congress 2009 (2009)
5. Langdon, W.B., Poli, R.: Foundations of Genetic Programming. Springer, Heidelberg (2002)
6. Ljung, L.: System Identification – Theory For the User, 2nd edn. PTR Prentice Hall, Upper Saddle River (1999)
7. Wagner, S.: Heuristic Optimization Software Systems – Modeling of Heuristic Optimization Algorithms in the HeuristicLab Software Environment. PhD thesis, Johannes Kepler University Linz (2009)
8. Warnatz, J., Maas, U., Dibble, R.W.: Combustion - Physical and Chemical Fundamentals, Modeling and Simulation, Experiments, Pollutant Formation. Springer, Heidelberg (1996)
9. Winkler, S.: Evolutionary System Identification - Modern Concepts and Practical Applications. PhD thesis, Johannes Kepler University Linz (2008)

Solving the Euclidean Bounded Diameter Minimum Spanning Tree Problem by Clustering–Based (Meta–)Heuristics

Martin Gruber and Günther R. Raidl

Institute of Computer Graphics and Algorithms
Vienna University of Technology, Vienna, Austria
{gruber,raidl}@ads.tuwien.ac.at

Abstract. The bounded diameter minimum spanning tree problem is an $\mathcal{NP}$-hard combinatorial optimization problem arising in particular in network design. There exist various exact and metaheuristic approaches addressing this problem, whereas fast construction heuristics are primarily based on Prim's minimum spanning tree algorithm and fail to produce reasonable solutions in particular on large Euclidean instances.

In this work we present a method based on hierarchical clustering to guide the construction process of a diameter constrained tree. Solutions obtained are further refined using a greedy randomized adaptive search procedure. Especially on large Euclidean instances with a tight diameter bound the results are excellent. In this case the solution quality can also compete with that of a leading metaheuristic.

1 Introduction

The *bounded diameter minimum spanning tree* (BDMST) problem is a combinatorial optimization problem appearing in applications such as wire-based communication network design when quality of service is of concern, in ad-hoc wireless networks, and also in the areas of data compression and distributed mutual exclusion algorithms.

The goal is to identify a tree-structured network of minimum costs in which the number of links between any pair of nodes is restricted by a constant D, the diameter. More formally, we are given an undirected connected graph $G = (V, E)$ with node set V and edge set E and associated costs $c_e \geq 0$, $\forall e \in E$. We seek a spanning tree $T = (V, E_T)$ with edge set $E_T \subseteq E$ whose diameter does not exceed $D \geq 2$, and whose total costs $c(T) = \sum_{e \in E_T} c_e$ are minimal. This task can also be seen as choosing a *center* – one single node if D is even or an edge in the odd-diameter case – and building a height-restricted tree where the unique path from this center to any node of the tree consists of no more than $H = \lfloor \frac{D}{2} \rfloor$ edges. The BDMST problem is known to be $\mathcal{NP}$-hard for $4 \leq D < |V| - 1$ [1].

R. Moreno-Díaz et al. (Eds.): EUROCAST 2009, LNCS 5717, pp. 665–672, 2009.

2 Previous Work

To solve this problem to proven optimality there exist various integer linear programming (ILP) approaches like hop-indexed multi-commodity network flow models [2,3] or a Branch&Cut algorithm based on a more compact model but strengthened by a special class of cutting planes [1]. They all have in common that they are only applicable to relatively small instances, i.e. significantly less than 100 nodes when dealing with complete graphs. For larger instances, metaheuristics have been developed, including evolutionary algorithms [4,5], a variable neighborhood search, and an ant colony optimization [6] which is currently the leading metaheuristic to obtain high-quality solutions.

In contrast to the large variety of metaheuristic approaches the number of simple and fast construction heuristics applicable to very large instances is limited. They are primarily based on Prim's minimum spanning tree (MST) algorithm and grow a height-restricted tree from a chosen center. One such example is the *center based tree construction* (CBTC) [7]. This approach works reasonably well on instances with random edge costs, but on Euclidean instances this leads to a backbone (the edges near the center) of relatively short edges. The majority of nodes have to be connected to this backbone via rather long edges, see Fig. 1(a). On the contrary, a reasonable solution for this instance, shown in Fig. 1(c), contains a backbone that consists of a few longer edges to span the whole area and allows the majority of nodes to be connected as leaves by much cheaper edges. In a pure greedy construction heuristic this structure is difficult to realize. In the *randomized tree construction approach* (RTC, Fig. 1(b)) from [7] not the overall cheapest unconnected node is always added to the partial spanning tree but a random node is selected and connected by the cheapest feasible edge. Thus at least the possibility to include longer edges into the backbone at the beginning of the algorithm is increased. For Euclidean instances RTC has been so far the best choice to quickly create a first solution as basis for exact or metaheuristic approaches.

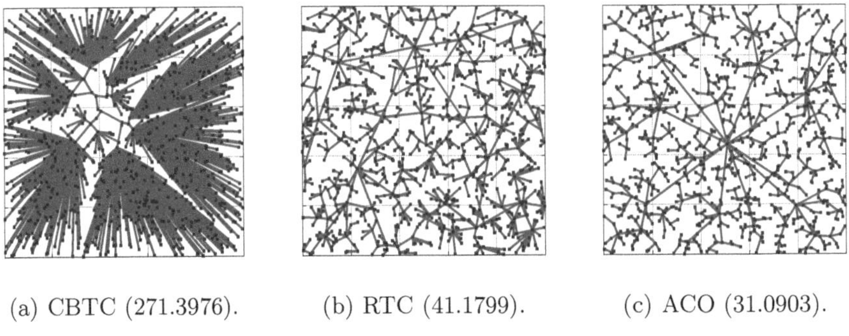

(a) CBTC (271.3976). (b) RTC (41.1799). (c) ACO (31.0903).

Fig. 1. A BDMST with $D = 10$ constructed using (a) CBTC, compared to (b) RTC and (c) a solution obtained by an ACO (complete, Euclidean graph with 1000 nodes distributed randomly in the unit square; objective values are given in parentheses)

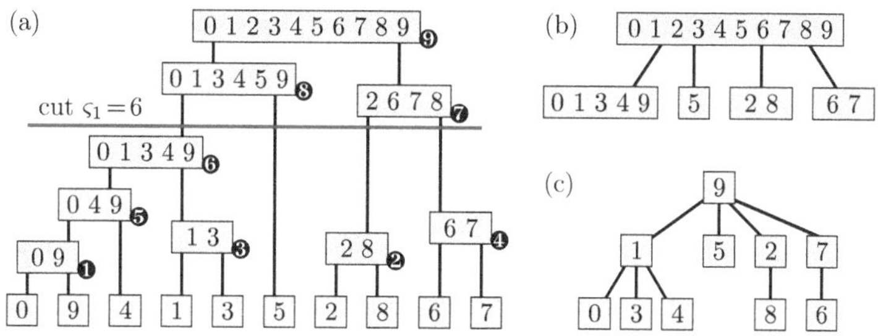

Fig. 2. Hierarchical clustering (a), height-restricted clustering (b), and the resulting BDMST with $D = 4$ (c) after choosing a root for each cluster in (b). In (a) ❶...❾ denote the merge numbers, cf. Section 3.2.

3 Clustering-Based Construction Heuristic

The clustering-based construction heuristic is especially designed for very large Euclidean instances and is based on a hierarchical clustering that guides the algorithm to find a good backbone. It can be divided into three steps: Creating a hierarchical clustering (*dendrogram*) of all nodes based on the edge costs, deriving a height-restricted clustering (HRC) from this dendrogram, and finding for each cluster in the HRC a root (center) node. In the following we will concentrate on the even-diameter case.

3.1 Hierarchical Clustering

For the purpose of creating a good backbone especially for Euclidean instances agglomerative hierarchical clustering provides a reasonable guidance. To get spatially confined areas, two clusters A and B are merged when $\max\{c_{a,b} : a \in A,\ b \in B\}$ is minimal over all pairs of clusters (complete linkage clustering).

The agglomeration starts with each node being an individual cluster, and stops when all nodes are merged within one single cluster. The resulting hierarchical clustering can be illustrated as a binary tree, also referred to as a *dendrogram*, with $|V| - 1$ inner nodes, each representing one merging operation during clustering, and $|V|$ leaves; see Fig. 2(a) for an example with $|V| = 10$.

3.2 Height-Restricted Clustering

After performing the agglomerative hierarchical clustering, the resulting dendrogram has to be transformed into a height-restricted clustering (HRC). In general, the dendrogram itself will violate the height (diameter) constraint, see Fig. 2(a). Therefore, some of the nodes in the dendrogram have to be merged to finally get a HRC of height $H - 1$; see Fig. 2(b).

For the quality of the resulting tree this merging of dendrogram nodes is a crucial step. It can be described by $H - 1$ *cuts* through the dendrogram. Preliminary tests revealed that the information at which iteration two clusters have been merged in the agglomeration process, the *merge number*, allows a fine-grained control of the cutting positions (leaves are assigned a merge number of zero). Based on these merge numbers cutting positions ς are computed as

$$\varsigma_i = (|V|-1) - 2^{i \cdot \frac{\log_2 x}{H-1}} \qquad i = 1, \ldots, H-1\,, \tag{1}$$

where x is a strategy parameter that can be interpreted as the number of nodes that shall form the backbone. An experimental evaluation showed that for $D \geq 6$ promising values for x can be found close to $|V|$. Only in case of the smallest possible even diameter of four x should be chosen near $\frac{|V|}{10}$. In practice, a good value for x for a specific Euclidean instance and D can be determined by applying binary search for $x \in \left[\frac{|V|}{10}, |V|\right]$. These cutting positions can now be utilized to build the HRC for the BDMST using a simple tree traversal algorithm.

3.3 Determining Root Nodes

Finally, from the height-restricted clustering a BDMST has to be derived by identifying for each (sub-)cluster an appropriate root; cf. Figs. 2(b) and (c). This can be done heuristically in a greedy fashion based on rough cost estimations for each cluster followed by a local improvement step, or by a more sophisticated approach based on dynamic programming.

In the following we will require a more formal and in some points augmented definition of a height-restricted hierarchical clustering. Let $C^0 = \{C_1^0, \ldots, C_{|V|}^0\}$ be the set of clusters at the lowest level 0, where each node of V forms an individual cluster. Moreover, let $C^k = \{C_1^k, \ldots, C_{i_k}^k\}$ be the clustering at the higher levels $k = 1, \ldots, H$. All C_i^k, $i = 1, \ldots, i_k$, are pairwise disjoint, and $C_1^k \cup C_2^k \cup \ldots \cup C_{i_k}^k = C^{k-1}$. C^H is the highest level, and it is singleton, i.e. $C^H = \{C_1^H\}$; it refers to all nodes in V aggregated within one cluster. Furthermore, by $V(C_i^k)$ we denote the set of nodes in V represented by the cluster C_i^k, i.e. the nodes part of this cluster and all its sub-clusters at lower levels; $V(C^k) = V(C_1^k) \cup \ldots \cup V(C_{i_k}^k) = V$, and $V(C_1^k) \cap \ldots \cap V(C_{i_k}^k) = \emptyset$, for all $k = 0, \ldots, H$. This definition corresponds to the simple height-restricted clustering introduced before and shown in Fig. 2(b) with the exception that clusters at level zero corresponding to the individual nodes have not been realized explicitly.

Greedy Heuristic with Local Improvement: A simple greedy heuristic to find an initial root for each cluster C_i^k can be based on *stars*, i.e. trees with a diameter of two where a single node v of the cluster acts as center while the remaining nodes $V(C_i^k) \setminus \{v\}$ are connected directly to it. Such a star can be computed for every node $v \in V(C_i^k)$, the center leading to a star of minimal costs for C_i^k is chosen as root for this cluster. The heuristic starts at cluster C^H and assigns roots to clusters top-down until reaching the leaves of the simple height-restricted clustering. Note that a node already selected as root at a level l no longer has to be considered in levels less than l, which can also cause an empty cluster in case all nodes of it are already used as roots at higher levels. This heuristic runs in time $\mathcal{O}(H \cdot |V|^2)$.

In a following local improvement step the selection of root nodes is refined. In case a cluster C_i^k with chosen root v is no leaf of the simple HRC not all nodes of $V(C_i^k) \setminus \{v\}$ will straightly connect to v in the final tree but only the roots of the direct sub-clusters of C_i^k at level $k-1$, cf. Fig. 2(c). This sub-cluster root information was not available in the greedy construction process but now can be used to adapt for each cluster the chosen root node iteratively. This refinement of assigned roots to clusters requires for one iteration time $\mathcal{O}(H \cdot \delta^{\max} \cdot |V|)$, where $\delta^{\max}$ is the maximal branching factor in the HRC.

Dynamic Programming: There are multiple effects on the tree when choosing a specific node v as root for a cluster, e.g. v no longer can act as root for one of the sub-clusters, but it also has not be connected as leaf to the tree. These effects increase the complexity of deriving an optimal BDMST for a given hierarchical clustering to such an extent that it is in general computationally unattractive. Nevertheless, when making certain assumptions it is possible to formulate an efficient approximate dynamic programming approach for this problem.

Let $c(C_i^k, v)$ denote the minimum costs of the subtree of the BDMST defined by the cluster C_i^k if it is rooted at node $v \in V(C_i^k)$, i.e. node v has been chosen as root for cluster C_i^k. Beside other implications one major point when choosing a node v as root is that it no longer has to be connected elsewhere in the tree. When computing $c(C_i^k, v)$ and selecting another node w from the same sub-cluster $C_{j'}^{k-1}$ that v is also part of, then the costs $c(C_{j'}^{k-1}, w)$ also contain the costs to connect node v (perhaps as root of one of the sub-clusters, more likely as a leaf of the BDMST). To exactly compute the contribution of v to the costs of $c(C_{j'}^{k-1}, w)$ is in practice usually not worth the (huge) effort, in particular when considering the costs of edges between root nodes in relation to the costs of connecting a leaf to the tree via a short edge, which is the goal of the whole clustering heuristic.

This observation can be used to formulate an approximate dynamic programming approach utilizing a correction value κ_v for each node $v \in V$ which estimates the costs arising when v has to be connected as leaf to the BDMST. There are various possibilities to define these correction values, preliminary tests showed that a simple choice usually is sufficient: For each cluster at level one the cheapest star is determined, and for a node v of such a cluster, κ_v are the costs to connect it to the center of the best star. The costs $c(C_i^k, v)$ can now be recursively defined for each level and node of a cluster as follows:

$$c(C^0_{\mathrm{ord}(v)}, v) = 0 \qquad \forall v \in V \tag{2}$$

$$\phi(C_i^k, v) = \sum_{C_j^{k-1} \in C_i^k \setminus \{C_{j'}^{k-1}\}} \min_{u \in V(C_j^{k-1})} \left(c_{v,u} + c(C_j^{k-1}, u) \right)$$

$$c(C_i^k, v) = \min\left(c(C_{j'}^{k-1}, v),\ c_{v,w} + c(C_{j'}^{k-1}, w) - \kappa_v \right) + \phi(C_i^k, v)$$

$$\forall k = 1, \ldots, H;\ \forall v \in V(C_i^k);\ C_{j'}^{k-1} \in C_i^k \mid v \in V(C_{j'}^{k-1});\ w \in V(C_{j'}^{k-1}) \mid w \neq v \tag{3}$$

Algorithm 1. refineCuts(ς)

```
input  : cutting positions ς_i, i = 1,...,H−1
output: improved cutting positions
T* ← buildTree(ς) ;                                  // currently best BDMST T*
ς* ← ς ;                                  // currently best cutting positions ς*
clear cache for sets of cutting positions and insert ς;
lwi ← 0 ;                                           // loops without improvement
repeat
    for i = 1,...,H − 1 do
        if i = 1 then Δ ← (|V| − 1) − ς*_1 else Δ ← ς*_{i−1} − ς*_i ;
        repeat ς_i ← ⌊ς*_i + Δ · N(μ, σ²) + 0.5⌋; until check(ς_i) is ok
    if ς ∈ cache then lwi ← lwi + 1 and continue;
    insert ς into cache;
    T ← buildTree(ς);
    if c(T) < c(T*) then T* ← T; ς* ← ς; lwi ← 0; else lwi ← lwi + 1;
until lwi ≥ l_max ;
```

At level zero each node is a single cluster. Therefore, in (2) the costs of the corresponding subtrees can be initialized with zero (ord(v) assigns each node $v \in V$ a unique index within 1 and $|V|$). Then the costs $c(C_i^k, v)$ are composed of two parts: The minimal costs of either using directly the subtree rooted at v from level $k-1$ or another node w from the same sub-cluster $C_{j'}^{k-1}$, plus for all remaining direct sub-clusters the minimal costs to connect a node u of a sub-cluster with its subtree to v, referred to as $\phi(C_i^k, v)$ in (3). After deriving all these costs in a bottom-up fashion, optimal root nodes leading to these costs can be chosen top-down in a second pass. This dynamic programming approach computes roots for clusters within time $\mathcal{O}(H \cdot |V|^2)$.

Connecting Leaf Nodes: When strictly following the clustering the leaves of the BDMST have to connect to the root nodes of their respective clusters. However, this strategy neglects the fact that there are in general much cheaper opportunities since a leaf node can be attached to any root node of a cluster without violating the height and therefore the diameter restriction. Thus, releasing the leaves from their strict membership to a specific cluster and to allow them to establish the cheapest possible connection to an arbitrary root can improve the solution quality substantially.

4 Refining Cutting Positions

In Section 3.2 the computation of initial cutting positions ς_i, $i = 1, \ldots, H-1$, through the dendrogram to derive a height-restricted clustering has been presented. Since these ς_i have a formidable impact on solution quality we additionally implemented an approach similar to a greedy randomized adaptive search procedure (GRASP) [8] to further refine them, see Algorithm 1. In each iteration all cutting positions of the currently best solution are perturbated using the difference Δ to the next lower indexed cutting position (for ς_1 the value $(|V|-1)-\varsigma_1$ is used), multiplied with a Gaussian distributed random value $N(\mu, \sigma^2)$.

Table 1. Averaged objective values over all 15 Euclidean Steiner tree instances from [9] with 1000 nodes for various even diameter bounds and (meta-) heuristics, the standard deviations are given parentheses

	without VND				with VND			
D	CBTC	RTC	CL	t(CL) [s]	RTC	CL	ACO	t(CL) [s]
4	329.026 (6.02)	146.492 (3.88)	**68.323** (0.70)	2.54 (0.09)	65.206 (0.55)	**65.160** (0.56)	65.801 (0.48)	5.56 (1.01)
6	306.266 (9.02)	80.864 (2.40)	**47.170** (4.61)	4.55 (0.49)	41.458 (0.36)	**41.313** (0.50)	42.117 (0.29)	9.94 (1.52)
8	288.384 (7.52)	53.253 (1.33)	**36.941** (1.34)	5.92 (0.42)	35.051 (0.35)	**34.217** (0.29)	34.749 (0.21)	11.61 (1.61)
10	266.366 (9.01)	41.120 (0.68)	**33.341** (0.66)	6.79 (0.42)	32.118 (0.31)	**30.970** (0.24)	31.039 (0.22)	13.43 (2.16)
12	250.002 (8.01)	35.759 (0.47)	**31.956** (0.44)	7.11 (0.33)	30.290 (0.29)	29.180 (0.26)	**28.636** (0.22)	14.68 (2.49)
14	237.140 (6.28)	33.364 (0.30)	**31.018** (0.33)	7.00 (0.64)	29.094 (0.28)	28.009 (0.23)	**26.652** (0.30)	15.05 (3.00)
16	224.312 (5.72)	32.196 (0.24)	**30.429** (0.29)	7.20 (0.72)	28.243 (0.28)	27.136 (0.19)	**25.576** (0.15)	15.63 (2.89)
18	210.987 (7.63)	31.583 (0.24)	**30.135** (0.27)	7.32 (0.81)	27.601 (0.27)	26.560 (0.20)	**24.881** (0.18)	16.78 (3.61)
20	197.177 (7.99)	31.268 (0.22)	**30.038** (0.28)	7.57 (0.76)	27.109 (0.26)	26.108 (0.23)	**24.370** (0.14)	18.54 (3.89)
22	183.016 (8.03)	31.086 (0.22)	**30.074** (0.28)	8.56 (0.98)	26.698 (0.28)	25.805 (0.21)	**24.013** (0.16)	21.39 (5.19)
24	172.825 (10.59)	30.992 (0.23)	**30.160** (0.27)	8.28 (1.41)	26.365 (0.27)	25.452 (0.24)	**23.772** (0.19)	21.36 (6.42)

To derive an actual BDMST from the cutting positions ς in `buildTree(`ς`)` a fast construction heuristic should be applied like the greedy heuristic with local search presented in the previous Section 3.3. To avoid redundant computations a cache is used to identify sets of cutting positions ς already evaluated. Furthermore, a new cutting position ς_i is only accepted if it lies within the interval $[|V|-2, 1]$ and if it differs from all ς_j, $j < i$, which is tested in `check(`ς_i`)`. The whole refinement process is stopped when $l_{\max}$ iterations without improvement have been performed, or no sets of new cutting positions could be found.

5 Computational Results

The experiments have been performed on an AMD Opteron 2214 (2.2GHz) utilizing benchmark instances already used in the corresponding literature from Beasley's OR-Library [9] originally proposed for the Euclidean Steiner tree problem. These complete instances contain point coordinates in the unit square, and the Euclidean distances between each pair of points are taken as edge costs. As performance differences are more significant on larger instances, we restrict our attention here to the 15 largest instances with 1000 nodes.

Table 1 summarizes the results obtained for various heuristics. Given are the objective values averaged over all 15 instances (30 independent runs per instance), together with the standard deviations in parentheses. Considered are the two established construction heuristics CBTC and RTC [7] as well as the clustering heuristic CL. In GRASP a mean μ of 0 and, after preliminary tests, a variance σ^2 of 0.25 was used, and the procedure was aborted after $l_{\max} = 100$ iterations without improvement. The time (in seconds) listed is the over all instances averaged running time of the clustering heuristic, which was also used as time limit for the corresponding executions of CBTC and RTC. To verify statistical significance paired Wilcoxon signed rank tests have been performed.

Clearly, CBTC is not suited for this type of instances; its strength lies in problems with random edge costs. CL outperforms RTC for every diameter bound significantly, where the gap in solution quality is huge when D is small. When applying a strong variable neighborhood descend (VND) as proposed in [6] to the best solutions of the various construction heuristics the differences are flattened. Nevertheless, the BDMSTs derived from CL solutions are of higher quality in general. On instances with a small diameter bound these trees – computed in a few seconds – can also compete with results from the leading metaheuristic, the ACO from [6], obtained after one hour of computation.

6 Conclusions

On the more difficult to solve Euclidean BDMST instances fast construction heuristics proposed so far were primarily based on Prim's MST algorithm and were too greedy to compute reasonable results. We presented a constructive heuristic that exploits a hierarchical clustering to guide the process of building a diameter-constrained tree.

In particular on large Euclidean instances the BDMSTs obtained by the clustering heuristic are in general of high quality and outperform the other construction heuristics significantly, especially when the diameter bound is tight. When using a strong VND to further improve these solutions they can also compete with results from an ACO, currently the leading metaheuristic for this problem.

References

1. Gruber, M., Raidl, G.R. (Meta-)heuristic separation of jump cuts in a branch&cut approach for the bounded diameter minimum spanning tree problem (2008); special issue on Matheuristics of Operations Research, Computer Science Interface Series. Springer, Heidelberg (to appear, 2009)
2. Gouveia, L., Magnanti, T.L.: Network flow models for designing diameter-constrained minimum spanning and Steiner trees. Networks 41(3), 159–173 (2003)
3. Gouveia, L., Magnanti, T.L., Requejo, C.: A 2-path approach for odd-diameter-constrained minimum spanning and Steiner trees. Networks 44(4), 254–265 (2004)
4. Raidl, G.R., Julstrom, B.A.: Greedy heuristics and an evolutionary algorithm for the bounded-diameter minimum spanning tree problem. In: Lamont, G., et al. (eds.) Proc. of the ACM Symposium on Applied Computing, pp. 747–752. ACM Press, New York (2003)
5. Singh, A., Gupta, A.K.: Improved heuristics for the bounded-diameter minimum spanning tree problem. Soft Computing – A Fusion of Foundations, Methodologies and Applications 11(10), 911–921 (2007)
6. Gruber, M., van Hemert, J., Raidl, G.R.: Neighborhood searches for the bounded diameter minimum spanning tree problem embedded in a VNS, EA, and ACO. In: Keijzer, M., et al. (eds.) Proc. of the Genetic and Evolutionary Computation Conference 2006, vol. 2, pp. 1187–1194 (2006)
7. Julstrom, B.A.: Greedy heuristics for the bounded diameter minimum spanning tree problem. Journal of Experimental Algorithmics (JEA) 14, 1.1:1–1.1:14 (2009)
8. Feo, T., Resende, M.: Greedy randomized adaptive search procedures. Journal of Global Optimization 6, 109–133 (1995)
9. Beasley, J.: OR-Library: Capacitated MST (2005), `http://people.brunel.ac.uk/~mastjjb/jeb/orlib/capmstinfo.html`

Solving the Rectangle Packing Problem by an Iterative Hybrid Heuristic

David Beltrán-Cano, Belén Melián-Batista, and J. Marcos Moreno-Vega*

Dpto. Estadística, I.O. y Computación, University of La Laguna,
Avda. Astrofísico Francisco Sánchez s/n, 38271 S.C. de Tenerife, Spain
{jbeltran,mbmelian,jmmoreno}@ull.es

Abstract. In this paper we propose an iterative hybrid heuristic approach consisting of two phases to solve the Rectangle Packing Problem. In the first phase, a strip width value W is fixed and the corresponding Strip Packing Problem is solved using an efficient hybrid GRASP-VNS heuristic. In the second one, a new value W is determined. The above phases are repeated until the stopping condition is met. Then, the results obtained by this iterated heuristic are compared with the results given by a Simulated Annealing given in the literature. The comparative analysis corroborates the effectiveness of the proposed hybrid approach.

Keywords: Rectangle Packing Problem, Strip Packing Problem, GRASP, VNS, SA.

1 Introduction

Given a set of rectangular pieces, $\mathcal{R}$, where each rectangle $R_i \in \mathcal{R}$ has fixed width w_i and height h_i, a packing of $\mathcal{R}$ is a non-overlapping placement of the rectangular pieces in the plane. The Rectangle Packing Problem (RPP) consists in finding the packing with minimal area. This problem can be found in some areas of business and industry (design of chips, textile or leather industries ...).

Some researches have tackled the rectangular packing problem. Murata et al. [7] proposed the sequence-pair representation which encodes the *left-right* and *up-down* topological relations between rectangles on the plane in two permutations. Using this representation some approaches have been proposed. A Simulated Annealing is proposed in [7]. Imahori et al. [5] proposed three metaheuristic algorithms for solving the RPP with General Spatial Costs (random multi-start local search (MLS), iterated local search (ILS) and a random walk (WALK)). The computational results showed that ILS found better solutions than the other two algorithms for many instances. Recently, Imahori et al. [6] designed more efficient techniques to evaluate solutions and incorporated them into ILS. Then, ILS was compared with two Simulated Annealing algorithms (SA-BSG and SA-SP) that use the coding schemes by Nakatake et at. [8] and

* This research has been partially supported by the projects TIN2008-06872-C04-01 and PI 2007/019.

R. Moreno-Díaz et al. (Eds.): EUROCAST 2009, LNCS 5717, pp. 673–680, 2009.

Murata et al [7], respectively. ILS performed better than these two algorithms for the RPP with General Spatial Cost. However, the SA procedure proposed in [8] provides better results for the RPP of minimizing the area, since the algorithm is specially tailored for this problem, which is the problem tackled in this paper. Drakidis at al. [2] presented a genetic algorithm with sequence-pair representation (GA-SP). In the computational experience, they compared the GA-SP with SA-SP and with another Genetic Algorithms previously proposed in the literature. The principal conclusion is that GA-SP is comparable to the rest of procedures.

With the purpose of solving the RPP, we propose an iterative hybrid heuristic approach which consists of two phases. In the first phase, a strip width value, W, is fixed and the corresponding Strip Packing Problem (SPP) is solved using the efficient hybrid heuristic GRASP+VNS described in [1]. In the second one, a new value W is determined. The Strip Packing Problem is defined as follows. Consider a strip of fixed width W and infinite height, and a finite set of rectangles, $\mathcal{R}$, where each rectangle $R_i \in \mathcal{R}$ has fixed width w_i and height h_i and at least one of their sides is smaller than W. The strip packing problem consists in packing the rectangles in the strip minimizing the height of the packing.

The rest of the paper is organized as follows. Section 2 describes the iterative hybrid approach proposed to solve de rectangle packing problem. The computational experience is reported in Section 3. Finally, the conclusions are summarized in Section 4.

2 Solution Approach: Iterative Hybrid Heuristic

In order to solve the rectangle packing problem, we propose an iterative hybrid heuristic that iteratively solves a strip packing problem. Since in the strip packing problem the strip width is fixed, the iterative heuristic fixes a strip width value W and then solves the corresponding SPP using the efficient hybrid heuristic GRASP+VNS described in [1]. Once this problem is solved, a new strip width value is selected.

The hybrid heuristic used to solve the SPP combines GRASP as constructive method and Variable Neighborhood Search (VNS) as a postprocessing step. GRASP (Greedy Randomized Adaptive Search Procedure) [3] is a constructive metaheuristic that consists of two phases; a constructive phase and a post-processing phase. In the constructive phase, a solution is iteratively constructed by selecting at random one element of a restricted candidate list. The goal of the post-processing phase is to improve this solution by running an improvement method. These steps are repeated until a stopping criterion is met. The best solution found along the search procedure is provided as the result.

The GRASP procedure uses the notion of *contour*. Once a new rectangle is added to the partial solution, it determines a new upper contour and new wasted areas. Given a partial solution, the upper contour consists of a series of horizontal segments taken from left to right. The contour C is represented by a series of values as follows:

$$C = [(y^1, x_1^1, x_2^1), (y^2, x_1^2, x_2^2), ..., (y^c, x_1^c, x_2^c)],$$

where y^i is the y-coordinate (height) of the i-segment and $[x_1^i, x_2^i]$ is the interval of x-coordinates of its points, for $i = 1, 2, ..., c, c \geq 1$. The contour C verifies $x_1^1 = 0$, $x_2^c = W$ and, for $1 \leq i < c$, $x_2^i = x_1^{i+1}$. It also verifies $y^i \neq y^{i+1}, for 1 \leq i < c$.

Let $C(t)$ be the contour at iteration t that is determined by the partial solution at iteration t. All the rectangles that do not belong to the solution are evaluated by means of their adjustment to the lowest segment of $C(t)$; i.e. the segment of the contour with minimum height. Given a parameter $\alpha \in [0, 1]$, the Restricted Candidate List (RCL), which consists of a set of rectangles, is constructed as follows.

$$RCL = \{R(w_i, h_i) \in R_2 : w_i \leq l \leq w_i + \alpha\},$$

where R_2 is the set of rectangles not in the partial solution and l is the length of the lowest segment. Only the rectangles that fit the best to the length of the lowest segment of the contour are in the restricted candidate list. If the RCL is empty, the rectangles that fit the best to the lowest segment are selected. If there are not rectangles that fit in the lowest segment, then the contour has to be rebuilt.

In order to carry out the post-processing phase of the above mentioned GRASP procedure, a Variable Neighborhood Search (VNS) [4] is considered. The k (a parameter) last rectangles of the solution are extracted and the VNS is applied over the solution space obtained by all possible permutations of these k rectangles.

The iterative procedure begins with a strip width and increases it until a maximum value using a step parameter. For each strip width, the corresponding SPP is solved by running the proposed heuristic.

3 Computational Experience

This section summarizes the computational experience carried out in this paper. The results provided by the iterative hybrid heuristic proposed in this paper to solve the RPP are compared with the best algorithm provided in the literature for this problem. All the codes were implemented in C++ and run in a Pentium processor with 2GB of memory RAM.

In order to corroborate the effectiveness of the proposed approach, we have implemented a Simulated Annealing as proposed by Nakatake, et al. [8], which, to the best of our knowledge, is the best algorithm proposed in the literature to solve the RPP. Nakatake, et al. proposed a new method of packing the rectangles based on the bounded-sliceline grid (BSG) structure. This structure has been used in this paper to implement the Simulated Annealing as proposed by these authors. According to these authors, the grid size, r, should not be close to the square root of the number of objects in order to guarantee the good performance of the method. Moreover, different grid sizes provide different results.

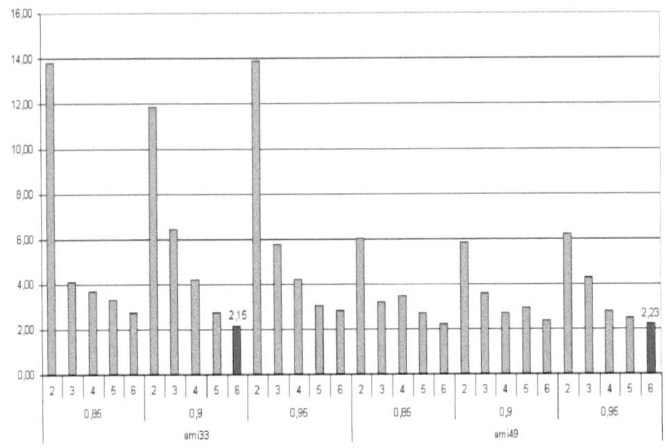

Fig. 1. Minimun Deviation for the BSG-SA algorithm: ami33, ami49

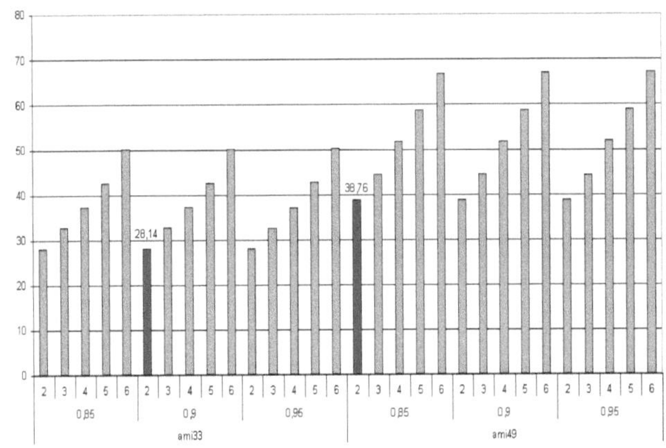

Fig. 2. Minimun CPU Time for the BSG-SA algorithm: ami33, ami49

With the purpose of evaluating the effect of the parameters α and r involved in the Simulated Annealing procedure, the problems *ami*33, *ami*49, *rp*100, *pcb*146 and *rp*200 were solved using different combinations of these two parameters. The algorithm was run 10 times for each parameters combination. Figures 1 and 2 show the minimum relative percentage deviations with respect to the lower bound, which is obtained by adding up the areas of all the rectangles to be packed, and the CPU times for the problems *ami*33 and *ami*49, respectively. Figures 3 and 4 show the same information for the problems *rp*100, *pcb*146 and *rp*200.

For the same value of α, the deviation values decrease as the grid size increases. In the small problems, the best deviation values are obtained for the grid size value $r = square\_root(number\_of\_objects) + 6$.

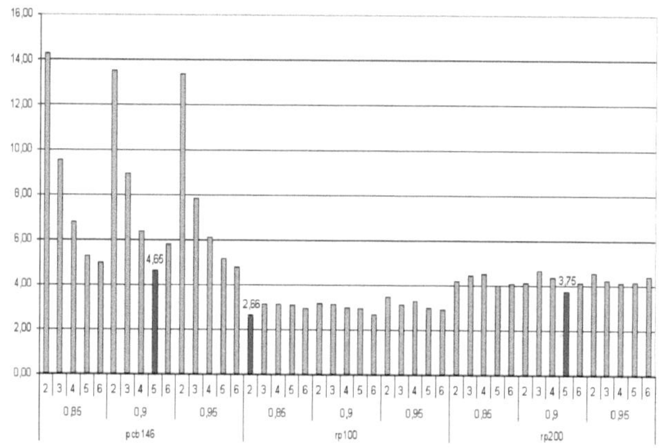

Fig. 3. Minimun Deviation for the BSG-SA algorithm: pcb146, rp100, rp200

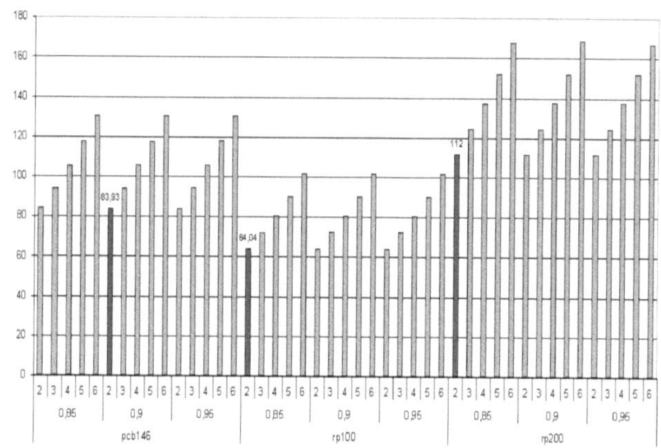

Fig. 4. Minimun CPU Time for the BSG-SA algorithm: pcb146, rp100, rp200

For the hybrid algorithm GRASP-VNS proposed in this paper, the computational experience was carried out in two phases. In the first phase we analyze the behavior that the value of the different parameters have on the efficiency and effectiveness of the proposed method. Different values for these parameters were set and the CPU time and the relative percentage deviation between the objective value and the lower bound were obtained. The small problems (*ami*33, *ami*49 and *rp*100) were used during this phase.

The parameters used to obtain these graphs are the following in reverse order of their appearance on the x-axis of the graphs: *W_initial*, that determines the width of the lowest strip used in the iterative process (it is a percentage on the side of the lowest square that would ideally contain all the boxes that you want to pack); *W_final*, that determines the width of the widest strip used in the iterative process; *Step*, that represents the step value with which the width of the strip iteratively increases; N_iter_GRASP, that indicates the number of iterations for the GRASP procedure; and N_iter_VNS, that indicates the number of iterations for the VNS procedure (if this value is equal to zero, VNS is not used as a postprocessing phase of the GRASP). These results are shown in Figures 5 and 6.

For the *ami*33 problem, the best objective values are obtained with a wider window in which the width of the strip iteratively varies ($w\_inicial$ = 5, w_final = 5). For this window, both the hybrid approach and the GRASP procedure achieve the best results. There is not a clear conclusion about the effect of using VNS as a postprocessing method. In some cases, increasing the number of iterations of VNS let us obtain better results, but in other cases, the

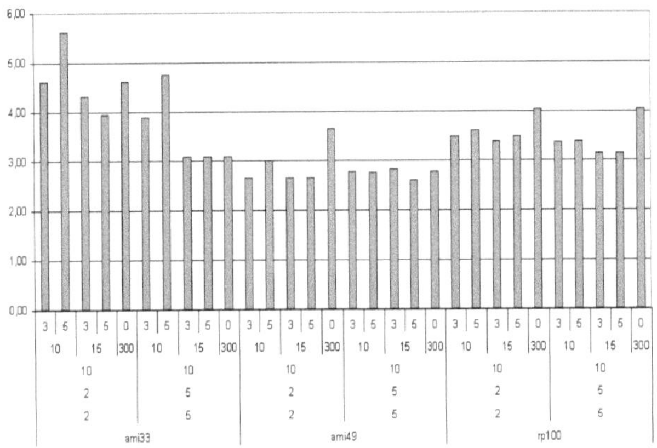

Fig. 5. Minimun Deviation for the GRASP-VNS algorithm: ami33, ami49

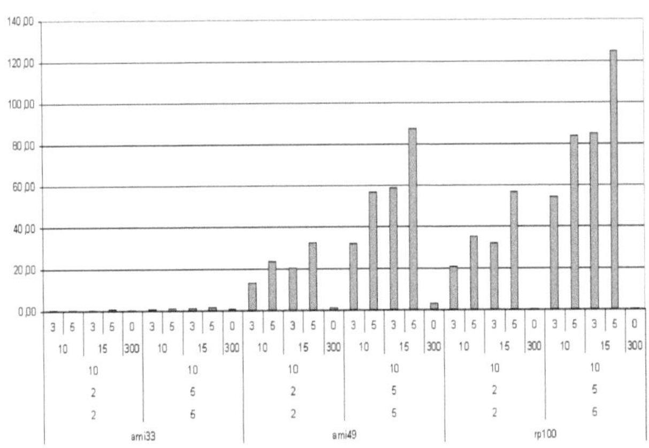

Fig. 6. Minimun CPU Time for the GRASP-VNS algorithm: ami33, ami49

Table 1. BSG-SA and GRASP-VNS best results

	BSG-SA		GRASP-VNS	
Problem	Deviation	CPU Time	Deviation	CPU Time
*ami*33	2.15	50.15	3.10	0.49
*ami*49	2.23	67.23	2.61	87.36
*pcb*146	4.79	130.69	1.88	308.59
*rp*100	2.66	64.04	3.14	85.13
*rp*200	3.75	152.12	3.01	433.64

opposite happens. There is a clear trend in the CPU time; if the number of iterations of VNS increases, the time of the hybrid procedure also increases.

For the *ami*49 problem, the best objective values are obtained with a wider window in which the width of the strip iteratively changes ($w\_inicial = 5$, $w\_final = 5$). There are not significant differences in the best deviations when the values of the parameters are changed. However, there is one combination of parameters that makes the hybrid approach GRASP + VNS perform better the the GRASP. The CPU time trends are the same as the observed in the ami33 problem. GRASP is more efficient than GRASP + VNS. If N_iter_GRASP is kept constant, the best values are obtained with the widest window.

For the *rp*100 problem, the best objective values are achieved with a wide window with $N\_iter\_GRASP = 15$ and $N\_iter\_VNS = 3$. Note that increasing the number of iterations of VNS has no effect on the quality of the solutions obtained. Therefore, it is required that the GRASP procedure uses VNS as a postprocessing method to improve the solutions, but it is not necessary to perform too many iterations of VNS to achieve good quality solutions. GRASP is the most efficient method. It runs 300 iterations in less than a second.

From the previous observations, we conclude that it is recommended to use a wide window in which to vary the width of the strip ($w\_inicial = w\_final = 5$). Despite the efficiency reached by the GRASP procedure, this method alone is not able to find the best objective function values in all instances. Therefore, VNS has to be used with a small value of iterations, as the postprocessing method in GRASP.

In the second phase of the computational experience carried out in this paper for the GRASP-VNS method, the larger problems (*pcb*146 and *rp*200) were solved following the parameters setting obtained from the above analysis. Each problem was solved 10 times.

Table 1 shows the best objective function values obtained by BSG-SA and GRASP-VNS. From the results reported in this table, we observe that the iterative hybrid method proposed in this paper performs better than the SA over the instances *pcb*146 and *rp*200, while the SA is better over the other three instances. Moreover, for some instances, the GRASP, which is most efficient than the SA, is able to reach the best results without the use of the VNS as a postprocessing step.

4 Conclusions

In this work, we propose an iterative hybrid heuristic to solve the Rectangle Packing Problem, which combines GRASP and VNS as postprocessing step. The results obtained by this method are compared with the results given by a Simulated Annealing proposed in the literature. The hybrid method performs better than the SA over two instances, while the SA is better over three instances. However, for some instances, the GRASP, which is more efficient than the SA, is able to reach the best results without the use of the VNS as a postprocessing step in some instances.

References

1. Beltrán, J.D., Calderón, J.E., Jorge, R., Moreno-Pérez, J.A., Moreno-Vega, J.M.: GRASP-VNS hybrid for the Strip Packing Problem. In: Proceedings of the First International Workshop in Hybrid Metaheuristics (HM 2004) at ECCAI, pp. 79–90 (2004)
2. Drakidis, A., Mack, R.J., Massara, R.E.: Packing-based VLSI module placement using genetic algorithm with sequence-pair representation. In: IEE Proc.-Circuits Devices Syst., vol. 153, pp. 545–551 (2006)
3. Feo, T.A., Resende, M.G.C.: Greedy randomized adaptive search procedures. Journal of Global Optimization 6, 109–133 (1995)
4. Hansen, P., Mladenovic, N.: Variable neighborhood search: principles and applications. European Journal of Operations Research 130, 449–467 (2001)
5. Imahori, S., Yagiura, M., Ibaraki, T.: Local search algorithms for the rectangle packing problem with general spatial costs. Mathematical Programming 97, 543–569 (2003)
6. Imahori, S., Yagiura, M., Ibaraki, T.: Improved local search algorithms for the rectangle packing problem with general spatial costs. European Journal of Operational Research 167, 48–67 (2005)
7. Murata, H., Fujiyoshi, K., Nakatake, S., Kajitani, Y.: VLSI module placement based on rectangle packing by the sequence pair. IEEE Transactions on Camputer Aided Design 15, 1518–1524 (1996)
8. Nakatake, S., Fujiyoshi, K., Murata, H., Kajitani, Y.: Module placement on BSG-structure and IC layout applications. In: Proc. Intl. Conf. Comput. Adided Des., pp. 484–491 (1996)

New Approximation-Based Local Search Algorithms for the Probabilistic Traveling Salesman Problem

Dennis Weyland, Leonora Bianchi, and Luca Maria Gambardella

IDSIA, Dalle Molle Institute for Artificial Intelligence
{dennis,leonora,luca}@idsia.ch
www.idsia.ch

Abstract. In this paper we present new local search algorithms for the Probabilistic Traveling Salesman Problem (PTSP) using sampling and ad-hoc approximation. These algorithms improve both runtime and solution quality of state-of-the-art local search algorithms for the PTSP.

1 Introduction

The field of combinatorial optimization under uncertainty has received increasing attention within the last years. Combinatorial optimization problems containing uncertain and dynamic information can be used for more realistic models of real world problems. One common way of representing the uncertainty is to express input parameters by probability distributions instead of single values. Combinatorial optimization problems using this kind of stochastic information are called Stochastic Combinatorial Optimization Problems (SCOPs). Among the SCOPs the Probabilistic Traveling Salesman Problem (PTSP) is currently one of the most significant problems. A broad overview about recent developments in this field is given in [4].

Since most of the Stochastic Combinatorial Optimization Problems are generalizations of combinatorial optimization problems it is not surprising that they are in practice usually harder to solve than the underlying (deterministic) combinatorial optimization problem. For problem sizes of practical relevance exact approaches are computationally too expensive and therefore heuristics are mostly used to tackle these kind of problems.

Currently local search algorithms play an important role in this field and a lot of algorithms for the PTSP are either itself local search algorithms or use local search algorithms as subroutines. In this paper we introduce new local search algorithms for the Probabilistic Traveling Salesman Problem and show that these new algorithms outperform the state-of-the-art local search algorithms for the PTSP.

2 Probabilistic Traveling Salesman Problem

The Probabilistic Traveling Salesman Problem (PTSP) is a generalization of the well known Traveling Salesman Problem (TSP). In contrast to the TSP each city

R. Moreno-Díaz et al. (Eds.): EUROCAST 2009, LNCS 5717, pp. 681–688, 2009.

in the PTSP has to be visited only with a certain probability, thus allowing more realistic models and scenarios. The goal here is to find a so called a-priori tour that visits all cities exactly once, minimizing the expected cost over all possible a-posteriori tours, where cities which do not require a visit are just skipped without chaning the order of the a-priori tour.

As a generalization of the TSP the PTSP is NP-hard and therefore algorithms computing near optimal solutions in a reasonable amount of time are of great interest. Especially local search algorithms play an important role in this field.

Formally we can define the PTSP over a complete undirected edge- and node-weighted graph $G = \{V, c, p\}$. $V = \{1, 2, \ldots, n\}$ is the set of nodes which represent the customers, $p : V \longrightarrow [0, 1]$ is the probability function that assigns to each node the probability that the node requires a visit and $c : V \times V \longrightarrow \mathbb{R}^+$ is the symmetric cost function that represents the non-negative travel costs between any two nodes. For real world problems c usually obeys the triangle inequality or even represents distances in an Euclidean space. The goal is to find a permutation $\tau : V \rightarrow V$ (the a-priori solution) which minimizes the expected cost over all a-posteriori solutions. A closed-form expression for this cost due to [8] is given in section 3.2.

3 Local Search Algorithms

We propose five new local search algorithms for the PTSP. Starting point for our work is a state-of-the-art local search algorithm for the PTSP which uses the 2.5-opt neighbourhood and a sampling-based approximation for the difference between the expected costs of two neighbouring solutions. It is called *2.5-opt-EEs* and a detailed description can be found in [7] with some extensions in [3].

A description of the 2.5-opt neighbourhood operator, which is a combination of the famous 2-opt and 1-shift neighbourhood operators, can be found in [9]. A lot of common optimization techniques for local search algorithms that are used in our new algorithms are also discussed in that work in the context of the TSP.

After introducing the general local search framework we show how the exact cost of a solution can be computed. Then we give a detailled presentation of our new algorithms.

3.1 Local Search Framework

Algorithm 1 describes the general local search framework for the PTSP. At the beginning an initial solution is created ignoring the additional customer probabilities and using the well known nearest neighbour heuristic for the PTSP. As long as our current solution is not a local optimum, we replace the current solution with an improving one. The way in which we explore the neighbourhood and in which the improving solution is chosen differs among our algorithms and is described in the corresponding chapters.

Algorithm 1. *Local Search Framework for the PTSP*

1. *Create an initial solution S using the nearest neighbour heuristic*
2. *While S is not a local optimum regarding the chosen neighbourhood:*
   Replace S by an improving solution
3. *Return S*

For a better runtime behaviour we use neighbourhood lists and don't look bits which are both common optimization techniques for the PTSP and which are also described in [9]. Additionaly we use delta-evaluation for the comparison of two solutions. Instead of comparing the solution costs directly only the difference of the solution costs is calculated. For the neighbourhood operators used in this work delta-evaluation leads to another major runtime improvement.

3.2 Exact Evaluation

Here we show two ways of computing the exact cost of a solution. The first approach calculates the sum of the a-posteriori costs over all possible combinations of realizations of the random variables each multiplied with the according probability that such realizations occur. Since there are two different possibilities for each customer we have in total 2^n different combinations. This naive computation requires exponential runtime and is therefore not useful for any practical implementations, but can be used as a starting point for an approximation with Monte-Carlo-Sampling (cf. 3.3).

Another way to calculate the exact solution cost analytically is to sum over all edges and multiply their costs with the probability that they occur in the a-posteriori tour. The probability that a certain edge occurs in the a-posteriori tour is the product of the probabilities, that both of its vertices require a visit and that all the vertices that are between them in the a-priori tour do not require a visit. Let $\tau : V \longrightarrow V$ be the permutation that represents our current solution and let $\tau_i = \tau(i) \ \forall i \in V$. Then the expected cost of the a-posteriori tour can be written as

$$E(\tau) = \sum_{i=1}^{n} \sum_{j=i+1}^{n} c(\tau_i, \tau_j)\, p(\tau_i)\, p(\tau_j) \prod_{k=i+1}^{j-1} (1 - p(\tau_k))$$
$$+ \sum_{i=1}^{n} \sum_{j=1}^{i-1} c(\tau_i, \tau_j)\, p(\tau_i)\, p(\tau_j) \prod_{k=i+1}^{n} (1 - p(\tau_k)) \prod_{k=1}^{j-1} (1 - p(\tau_k))$$

Using this formulation the expected cost of a solution can be calculated in runtime $\mathcal{O}(n^2)$ by adding the summands in a certain order. Although this is a lot better than the exponential runtime of the first approach, it is still too slow for input instances of reasonable sizes. This is why there is a huge need for fast and accurate approximations of the expected cost of a solution. We introduce such approximations in the next chapters in the context of the algorithms where they are used.

3.3 2.5-Opt-Optimized

This algorithm is based on a state-of-the-art algorithm in [7]. It uses the 2.5-opt neighbourhood within our local search framework. The solutions are explored in a random order in each iteration and the first improving solution is used to replace the current solution.

For the comparison of two solutions we use Monte-Carlo-Sampling with the naive exact computation from 3.2, but instead of summing over all possible scenarios, we sample s scenarios at the beginning of the algorithm using the know probabilities and take the average over the a-posteriori costs regarding the sampled scenarios as an approximation for the expected costs. Using the same samples during the whole run of the algorithm is a well known variance reduction technique. With delta-evaluation it is even possible to calculate the difference of two solutions in the 2.5-opt neighbourhood regarding a particular sample in constant time. This leads to a total runtime of $\mathcal{O}(s)$ for the comparison of two solutions using s samples.

For this algorithm it is necessary to calculate for a customer i and a sample s the first customer in the a-priori tour prior to customer i that requires a visit regarding sample s as well as the first customer after customer i that requires a visit regarding sample s. The runtime of the algorithm can be improved by computing these values at the beginning of the algorithm for the initial solution and by updating them after each improvement step. We call this algorithm *2.5-opt-sampling*.

3.4 1-Shift-Delta

The special structure of the 1-shift neighbourhood makes it possible to push this precalculation process to an extreme. Here we compute for the initial solution the delta-values (differences between the solutions) for all possible 1-shift moves. After each improvement step these values are then updated. For the 1-shift neighbourhood the number of values that really change and require an update stays small in relation to the number of all possible moves. Unfortunately this does not hold for the 2-opt neighbourhood, since large segments of the a-priori tour are reversed in some cases and this makes it impossible to benefit from the precalculations for this neighbourhood operator.

The precalculation of the delta-values makes it possible to explore the whole neighbourhood in each step and to replace the current solution with the best solution in its neighbourhood. We call this algorithm *1-shift-delta*.

3.5 2.5-Opt-Depth

Like *2.5-opt-sampling* this algorithm uses the 2.5-opt neighbourhood within our local search framework. The solutions are also explored in a random order in each iteration and the first improving solution is used to replace the current solution.

The calculation of the solution costs is based directly on the formula given in the section about the exact evaluation of the solution costs. Instead of summing over all edges, here we sum only over those edges whose vertices have a distance of at most d in the a-priori tour, where d is called the depth of the approximation. The formula changes to

$$E_{depth}(\tau) = \sum_{i=1}^{n} \sum_{j=i+1}^{\min\{i+d,n\}} c(\tau_i, \tau_j)\, p(\tau_i)\, p(\tau_j) \prod_{k=i+1}^{j-1} (1 - p(\tau_k))$$
$$+ \sum_{i=1}^{n} \sum_{j=1}^{i+d-n} c(\tau_i, \tau_j)\, p(\tau_i)\, p(\tau_j) \prod_{k=i+1}^{n} (1 - p(\tau_k)) \prod_{k=1}^{j-1} (1 - p(\tau_k))$$

The computational time required for this approach is $\mathcal{O}(dn)$. Usually d is a constant between 10 and 50, which leads to a good tradeoff between runtime and approximation accuracy. Another important speedup can be achieved by computing the difference of the costs of two solutions in the 2.5-opt neighbourhood. In this case only the $\mathcal{O}(d^2)$ edges that are at a distance of at most d to at least one of the removed/added edges have to be considered, leading to a runtime of $\mathcal{O}(d^2)$. The resulting local search algorithm is called *2.5-opt-depth.*

3.6 2.5-Opt-Threshold

This algorithm is similar to *2.5-opt-depth* and also uses the 2.5-opt neighbourhood. The solutions are explored in a random order in each iteration and the first improving solution is used to replace the current solution. But instead of summing over all edges whose vertices have a distance of at most d in the a-priori tour, here we sum over all edges that occur in the a-posteriori solution with a probability of at least t, where t, with $0 < t < 1$, is called the threshold.

The computational time required for this approach can be bounded by $\mathcal{O}(un)$ where u depends on t and the problem instance and is usually a lot smaller than n for reasonable values of t. By calculating the difference of the costs of two solutions in the 2.5-opt neighbourhood an important speedup can be achieved. The computational time in this case decreases to $\mathcal{O}(u^2)$. The resulting local search algorithm is called *2.5-opt-threshold.*

3.7 2.5-Opt-Combined

Since the local search algorithms *2.5-opt-depth*, *2.5-opt-threshold* and *2.5-opt-sampling* use different perturbed variants of the exact solution cost function, we tried to combine these algorithms. Especially the following combination of *2.5-opt-threshold* and *2.5-opt-sampling* turned out to be useful.

Algorithm 2. *Combined Local Search Algorithm for the PTSP*

1. *Create an initial solution S using the nearest neighbour heuristic*
2. *Repeat i times:*
   Use 2.5-opt-threshold *with S as the initial solution, store the result in S*
   Use 2.5-opt-sampling *with S as the initial solution, store the result in S*
3. *Return S*

We will refer to this combined local search approach with *2.5-opt-combined.*

4 Experiments and Results

We tested different parameterizations of the algorithms on instances from the TSPLIB benchmark [10] supplemented with the probabilities for the customers (tsplib instances), on Euclidean instances in which customers are distributed uniformly at random in a square (uniform instances) and Euclidean instances in which customers are distributed around a certain number of centers which themselves are distributed uniformly at random in a square (clustered instances). We either have selected the same probability for each city with typical values of $0.05, 0.1, 0.2, \ldots, 0.5$ or we have selected the probabilities uniformly from a fixed interval. For each algorithm and each parameterization of the algorithm we performed 50 independent runs and calculated the average runtime and the average solution quality. Complete numerical results of our experiments will be soon available at [1].

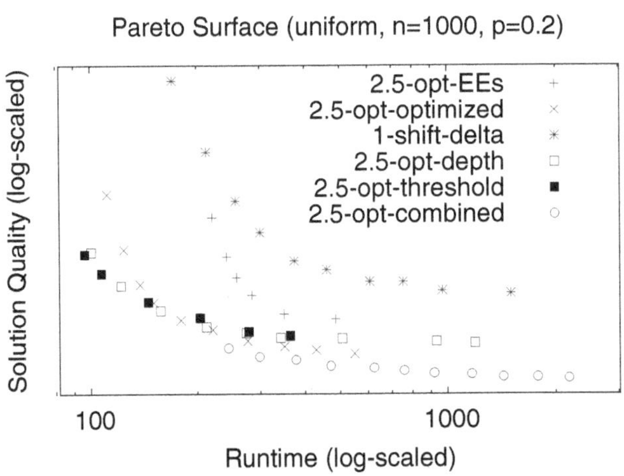

Fig. 1. Pareto Surface for different parameterizations of the algorithms *2.5-opt-EEs*, *2.5-opt-optimized*, *1-shift-delta*, *2.5-opt-depth*, *2.5-opt-threshold* and *2.5-opt-combined* on uniform instances with $n = 1000$ and homogeneous probabilities of $p = 0.2$. Since the PTSP is a minimization problem, better solutions are plotted with a lower solution quality.

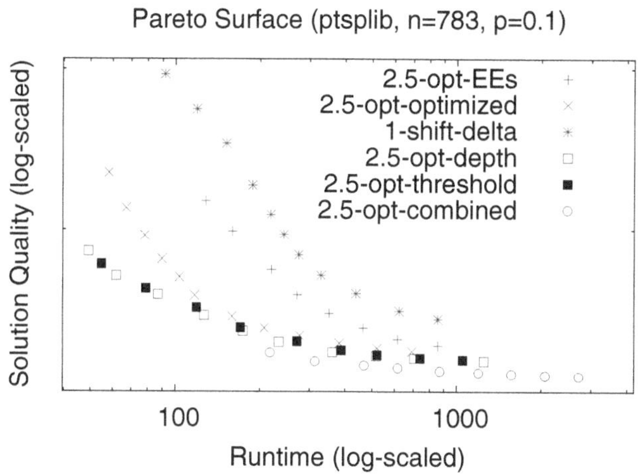

Fig. 2. Pareto Surface for different parameterizations of the algorithms *2.5-opt-EEs*, *2.5-opt-optimized*, *1-shift-delta*, *2.5-opt-depth*, *2.5-opt-threshold* and *2.5-opt-combined* on a 783-city TSPLIB instance with homogeneous probabilities of $p = 0.1$. Since the PTSP is a minimization problem, better solutions are plotted with a lower solution quality.

Since heuristics for the PTSP should both be fast and produce good solutions, the development of these heuristics can be seen as a multi-objective optimization problem. That means the goal here is to find the so called Pareto front of non-dominated algorithms. The practitioner then can pick the proper algorithm concerning the required solution quality or the available runtime. Therefore we used runtime/solution quality graphs for the visualization of the results.

In our experiments we found out that the algorithms *2.5-opt-optimized*, *2.5-opt-depth*, *2.5-opt-threshold* and *2.5-opt-combined* always dominate *2.5-opt-EEs* which itself dominates *1-shift-delta* for most of the problem instances. In all our tests *2.5-opt-combined* is part of the Pareto front on its whole runtime range, whereas for lower runtimes the situation is not clearly decided between *2.5-opt-optimized*, *2.5-opt-depth* and *2.5-opt-threshold*.

Figure 1 illustrates the results for a uniform instance with $n = 1000$ customers and homogeneous probabilities of $p = 0.2$ and figure 2 illustrates the results for a 783-city TSPLIB instance with homogeneous probabilities of $p = 0.1$. These results are typical for our experiments and representative for the results mentioned above.

5 Conclusions and Outlook

In this work we have presented new sampling- and ad-hoc approximation-based local search algorithms for the PTSP with significant improvements over a state-of-the-art local search algorithm. These improvements could be achieved on the basis of the following three ideas: Precomputation combined with more elaborate data structures for the sampling-based algorithms, the use of ad-hoc approximation and the alternation of local search with sampling and ad-hoc approximation.

The new algorithms now build the Pareto front for small runtimes (i.e. at most some seconds) and in principle they could also be used in iterated local search algorithms or hybrid heuristics to create new state of the art algorithms in higher runtime regions. Ant Colony Optimization combined with a local search algorithm is a widely used hybrid heuristic for the PTSP, eg. in [5], [6] and [2]. It seems very promising to use the new local search algorithms within an Ant Colony Optimization algorithm.

Acknowledgements. Dennis Weyland admits support from the Swiss National Science Foundation, grant 36RBVRPSAM.

References

1. Complete Numerical Results, `http://www.idsia.ch/~weyland`
2. Balaprakash, P., Birattari, M., Stützle, T., Dorigo, M.: Adaptive sample size and importance sampling in estimation-based local search for stochastic combinatorial optimization: A complete analysis. Technical Report TR/IRIDIA/2007-015, IRIDIA, Université Libre de Bruxelles, Brussels, Belgium (September 2007)
3. Balaprakash, P., Birattari, M., Stützle, T., Yuan, Z., Dorigo, M.: Estimation-based ant colony optimization and local search for the probabilistic traveling salesman problem. Technical report, Brussels, Belgium (September 2008)
4. Bianchi, L., Dorigo, M., Gambardella, L.M., Gutjahr, W.J.: A survey on metaheuristics for stochastic combinatorial optimization problems. Accepted for publication at Natural Computing (2008), doi. 10.1007/s11047-008-9098-4
5. Bianchi, L., Gambardella, L.M., Dorigo, M.: An ant colony optimization approach to the probabilistic traveling salesman problem. In: PPSN VII: Proceedings of the 7th International Conference on Parallel Problem Solving from Nature, pp. 883–892. Springer, Heidelberg (2002)
6. Bianchi, L., Gambardella, L.M., Dorigo, M.: Solving the homogeneous probabilistic traveling salesman problem by the aco metaheuristic. In: ANTS 2002, pp. 176–187. Springer, Heidelberg (2002)
7. Birattari, M., Balaprakash, P., Stützle, T., Dorigo, M.: Estimation-based local search for stochastic combinatorial optimization. Technical Report TR/IRIDIA/2007-003, IRIDIA, Université Libre de Bruxelles, Brussels, Belgium (February 2007)
8. Jaillet, P.: Probabilistic Traveling Salesman Problems. PhD thesis, Massachusetts Institute of Technology, Cambridge, MA (1985)
9. Johnson, D.S., McGeoch, L.A.: The traveling salesman problem: A case study. In: Aarts, E.H.L., Lenstra, J.K. (eds.) Local Search in Combinatorial Optimization, pp. 215–310. Wiley, Chichester (1997)
10. TSPLIB, `http://www.iwr.uni-heidelberg.de/groups/comopt/software/TSPLIB95/`

Evolving 6-State Automata for Optimal Behaviors of Creatures Compared to Exhaustive Search

Patrick Ediger, Rolf Hoffmann, and Mathias Halbach

Technische Universität Darmstadt
FB Informatik, FG Rechnerarchitektur
Hochschulstr. 10, 64289 Darmstadt, Germany
{ediger,hoffmann,halbach}@ra.informatik.tu-darmstadt.de

Abstract. We applied an Island Model Genetic Algorithm (GA) to a Multi-Agent System (MAS) modeled in Cellular Automata (CA) in order to find the optimal behavior of the agents. The agents' task is to visit all free cells in a cellular grid containing obstacles as fast as possible. For this investigation we used a previously defined set of five different environments. The agents are controlled by a finite state machine with a restricted number of states and outputs (actions of the agents). Finite state machines with 4 to 7 states have been evolved by the GA. We compared the effectiveness (quality of solutions) and efficiency of the GA to an exhaustive search of all possible solutions. A special hardware (FPGA logic) has been used to enumerate and evaluate all 6-state finite state machines. The results show that the GA is much faster but almost as effective as the exhaustive search.

1 Introduction

The general goal of our investigation is to find optimal local behaviors (algorithms) for agents (also called creatures here) that have to fulfill a certain task in a multi-agent-system (MAS). For this purpose we are developing efficient methods that allow to find automatically the optimal local algorithm or an algorithm with a feasible performance. In this investigation the agents' task is the "Creature's Exploration Problem" (CEP) [1]: several agents (creatures) are moving around in a two-dimensional cellular automata (CA) grid, which serves as a fixed environment for moving agents and consists of empty cells and cells with obstacles. The agents have to visit all the empty cells at least once in shortest time, i. e., with a minimum number of discrete time steps. The local algorithm we are looking for here is an algorithm that defines the movement of the agents. An algorithm will be defined by a finite state machine (FSM) that controls the actions of the agents.

For a specific problem set comprising of five environments with eight agents (Fig. 1), that are initially equally distributed at the borders, a full search of all possible FSMs aided by special hardware [2] was performed in [3], where

R. Moreno-Díaz et al. (Eds.): EUROCAST 2009, LNCS 5717, pp. 689–696, 2009.

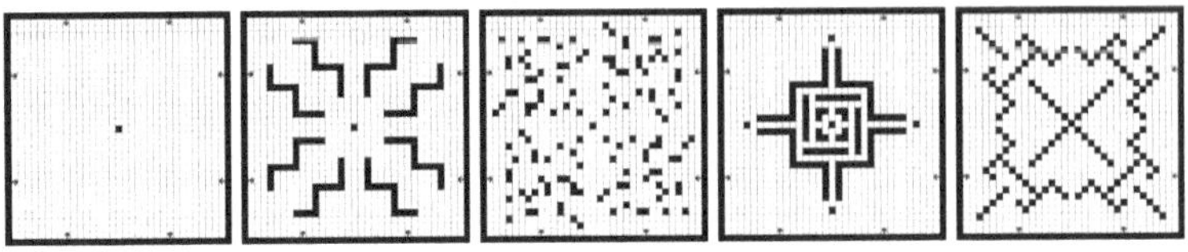

Fig. 1. The five initial configurations (33 × 33). 8 agents are placed at the borders.

the maximum number of states was restricted to $s = 6$ in order to keep the search space small enough for enumeration and simulation within a reasonable time limit. We used the number of time steps needed to complete the task as a metric to compare the quality of algorithms because there exist 6-state-FSMs that are sufficiently complex to fulfill the task completely on all environments. The results of the enumeration were then compared to the algorithms found by genetic algorithms (GA). The question behind this investigation is whether GA is an appropriate instrument to evolve state based algorithms and if the results of the enumeration and GA differ considerably when the search space is varied, i. e., the number of states s is varied.

In former investigations we have experimented with different optimization techniques like time-shuffling [4], and genetic programming [5] and we studied the CEP in [6] with respect to multi-agent efficiency. Modeling the behavior with a state machine with a restricted number of states and evaluation by enumerations was also undertaken in SOS [7,8]. Additional work was done by these authors using genetic algorithms. The CEP based on our model was further investigated in [9].

The remainder of this paper is organized as follows. Section 2 describes how the MAS is modeled in CA. In Section 3 the details and results of the exhaustive search are presented. In Section 4 the genetic algorithm is described in detail and its results are compared to the optimal solutions found by the full search. The paper ends with a conclusion in Section 5.

2 Modeling the MAS in CA

The whole system is modeled by cellular automata. It consists of a two-dimensional environment ($n \times m$ grid) with borders or without borders (wrap-around) and $k = 8$ uniform agents. Each cell is either *empty*, an *obstacle* or an *agent*, where an obstacle is a cell which the agents cannot visit, e. g., the border. If the cell is an *agent* an additional information about its direction (N, E, S, W) is needed. Thus the cell state is a combination of the type (agent, obstacle or empty) and the direction of the agent. An agent can only read the information from one cell ahead (*target cell, front cell*). If it detects a border cell or an agent in front or a conflict, it will turn right (R) or left (L) 90 degrees staying on the same cell. A conflict occurs when two or more agents want to move to the same front cell (crossing point, cell in conflict, mediator), because there can only be one

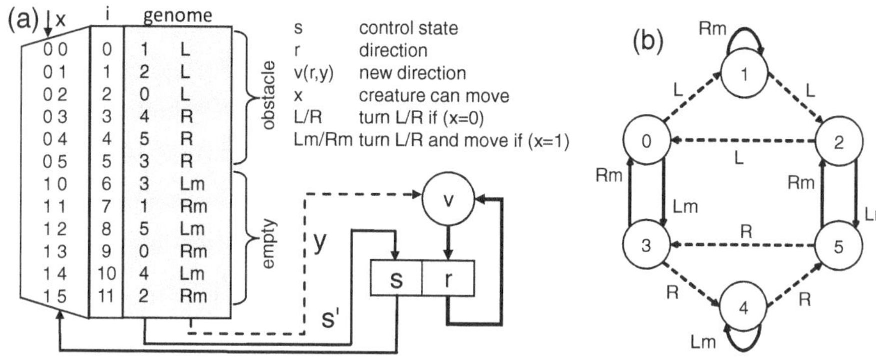

Fig. 2. A state machine (a) models an agent's behavior. Corresponding 6-state graph (b), dashed line for $x = 0$, solid line for $x = 1$.

agent on one cell at the same time. The conflict detection is either realized by an extended neighborhood (Manhattan distance 2), or by an arbitration logic [10] which is available in each cell. The arbitration logic evaluates the move requests coming from the agents and replies asynchronously by an acknowledge signal in the same clock cycle. In the case that the agent does not detect a conflict or an obstacle or another agent in the front cell it moves forward to the front cell and simultaneously turns to the right (*Rm*) or to the left (*Lm*). Thus an agent performs the rule:

1. (*Evaluate move condition* x): If (front cell $==$ OBSTACLE $\vee$ AGENT $\vee$ CONFLICT) then $x = 0$ else $x = 1$
2. (*React*): If (x) then Rm/Lm else R/L

Modeling the movement in CA actually means that the target cell, which must be in state *empty*, changes to the state *agent* while the cell that was in state *agent* changes to *empty*. The cells in state *obstacle* never change their state.

The decision which of the actions R/L or Rm/Lm will be performed depends on the behavior of the agent. The behavior (algorithm) of an agent is defined by a finite state machine (FSM). Input of the FSM is the move condition x. Output of the FSM is the signal y. The action R/L is performed if $y = 1/0$ and $x = 0$. The action Rm/Lm is performed if $y = 1/0$ and $x = 1$. The actions were defined in this way in order to keep the control automaton as simple as possible.

A state machine is defined by a state transition table (Fig. 2) with the current input x, the current state s, the next state s' and the current output y. It is represented by concatenating its components to a string or a simplified string, namely the values of s' and the actions induced by the values of y, e. g.:

1R5L3L4R5R3L-1Lm2Rm3Rm4Lm5Rm0Lm
= 1R5L3L4R5R3L-1L2R3R4L5R0L *//simplified representation*

3 Exhaustive Search

The number of state machines which can be coded using a state table is $M = (\#s\#y)^{(\#s\#x)}$ where $\#s$ is the number of states, $\#x$ is the number of different input values and $\#y$ is the number of different output actions. Thus 12^{12} state machines can be coded with $\#s = 6$ states. Note that not all of these behaviors are distinct (e. g., permutation of the states leads to equivalent behaviors) or useful (e. g., state graphs which make little use of the inputs or which are weakly connected). First the equivalent automata (under state permutation) and automata in which the states are not strongly connected are disregarded by a sophisticated "intelligent enumeration" technique [3]. The resulting amount of relevant automata (Table 1) with 4 to 6 states was simulated on the problem set (Fig. 1) for 4 and 5 states in software (dual core PC, 2.4GHz) and for 6 states by the aid of configurable logic (FPGA).

The optimal algorithms were selected using the fitness function f, being the sum of all fitnesses f_i of the environment i, defined as follows:

$$f_i = \begin{cases} r + n_i + p & \text{if (not successful}_i \ \& \ n_i > q) \\ r + n_i & \text{if (not successful}_i \ \& \ n_i \leq q) \\ g_i & \text{if (successful}_i) \end{cases}$$

where n_i is the number of not visited cells, p is a penalty, r is another penalty for not successful FSMs, and g_i is the number of generations needed in case of success. The goal was to find only such algorithms which are "completely" successful (visit all empty cells) on *all* 5 environments. The values $p = 450$, $r = 20,000$ and $q = 500$ were chosen, because the relatively high value for r gives us the desired dominance relation, because after 10,000 time steps the simulation is aborted if not successful. The simulation (fitness evaluation) is also stopped if there is no improvement within 2,000 simulation steps relatively counted from the last improvement.

The results (Table 2) show that there are no 4-state FSMs capable of solving all environments and that the best 5-state FSMs surprisingly outperform the best 6-state FSMs. Note that the FSMs with 5 states are not included in the set of 6-state FSMs due to the enumeration technique.

Additionally 100,000 simulations with random walk were performed on the whole problem set. 60,704 were completely successful. Taking only these into

Table 1. The number of possible state machines compared to the number of relevant automata and the simulation time in software (SW) or in hardware (HW) with 4, 5, 6 and 7 states

#s	M	Relevant Automata	Percentage	Simulation Time
4	8^8	798,384	4.8%	≈ 3.7 hours (SW)
5	10^{10}	98,869,740	1.0%	≈ 7 days (SW)
6	12^{12}	14,762,149,668	0.2%	≈ 30 days (HW)
7	14^{14}	2,581,401,130,308	0.02%	-

Table 2. The optimal algorithms for 4, 5 and 6 states and their fitness values compared to random walk. The column "Rank in set" indicates the rank within the set of FSMs that use exactly #s states.

RANK	STRING REPRESENTATION	RANK IN SET	FITNESS	SUCCESSES
1	**1L3L2R0L4R-2L0R4L3R1R**	**1 (#s = 5)**	**3,284**	**5**
2	1R3R2L0R4L-2R0L4R3L1L	2 (#s = 5)	3,566	5
3	**5L0R2L5R4L1R-2R1L4R2R5L3L**	**1 (#s = 6)**	**3,618**	**5**
4	0L1L3R4R2R-1R2L4L3L0R	3 (#s = 5)	3,857	5
5	1R4L1L3R0L5R-3L2R3L5L4R1R	2 (#s = 6)	3,904	5
⋮	⋮	⋮	⋮	⋮
	0L1L2R3R-1R2L3L0R	**1 (#s = 4)**	**23,712**	**4**
-	Best Random Walk	-	9,545	5
-	Random Walk Average	-	20,525	5 (60.7%)

account, their fitness value was $f = 20,525$ on average, which is significantly worse than the best algorithms defined by FSMs with 5 or more states.

4 Evolving Automata

"Intelligent enumeration" is not applicable anymore when the number of states increases above a certain level (Table 1). As a first alternative approach we tried a random search in the set of 6-state FSMs. 2,000,000 FSMs were randomly generated and simulated in software. None of these were completely successful and only 4 algorithms could solve 4 out of 5 environments. It can be concluded that the density of feasible solutions is too low and that random search is not an adequate method.

Thus we evaluated another more promising heuristic approach. The used method to evolve the FSMs is an Island Model GA. The components (s'_i, y_i) of the state table (Fig. 2) define the genome (individual, possible solution). P populations of N individuals are updated in each generation (optimization cycle). During each cycle M offsprings are produced in each population. The union of the current N individuals and the M offsprings are sorted according to their fitness (function f as described in Section 3) and the N best are selected to form the next population. An offspring is produced as follows:

1. (GET PARENTS) Two parents are chosen for each offspring. Each parent is chosen from the own population with a probability of $(1 - p_1)$ and from an arbitrary other population with the probability of p_1.
2. (CROSSOVER) Each new component (s'_i, y_i) of the genome string is taken from either the first parent or the second parent with a probability of 50%. This means that the tuple (next state, output) for the position i=(input, state) is inherited from any parent.
3. (MUTATION) The string being modified by the crossover is afterwards mutated with a probability of p_2. If a mutation shall be performed, an arbitrary

Table 3. The best evolved solutions. The column "Rank in set union" denotes the rank of an algorithm within a set of FSMs including all sets with less states. "Detections" indicates the number of runs in which this solution was found. The "detection range" denotes the first and the last generation in which the best algorithm was found.

#s	Best Fitness	Rank in set union	Detections	Detection range
≤ 4	23,712	1 ($\#s \leq 4$)	4/6	2,551 - 4,237
≤ 5	3,566	2 ($\#s \leq 5$)	3/6	16,100 - 40,034
≤ 6	3,284	1 ($\#s \leq 6$)	5/6	17,076 - 32,040
≤ 7	2,932	unknown	3/6	12,548 - 21,098

Table 4. The number of feasible solutions found in each run of the GA after 4,500/40,800 generations

#s	Generations	Computation time	Run 1	Run 2	Run 3	Run 4	Run 5	Run 6
≤ 4	4,500	0.5 hours per run	-	-	-	-	-	-
≤ 5	40,800	5.0 hours per run	3	5	7	4	6	5
≤ 6	40,800	4.8 hours per run	67	42	53	83	42	55
≤ 7	40,800	4.7 hours per run	>100	>100	>100	>100	>100	>100

position i is chosen and a new value (randomly chosen from the set of valid values) is replacing the existing one. Thereby the next state and the output is changed at position i.

We performed six independent runs on a dual core PC (2.4GHz) for each number of states $\#s = 4 \ldots 7$ with the following parameters: $P = 7$, $N = 100$, $M = 10$, $p_1 = 2\%$, $p_2 = 9\%$. Note that unlike the "intelligent enumeration" the GA does not exclude solutions that are reducable to an equivalent solution with less than $\#s$ states. Furthermore $(\#s - 1)!$ equivalent algorithms exist which only differ in permutations of the states 1 to $(\#s - 1)$, assuming state 0 to be always the initial state. When evolving the 4-state FSMs each run was stopped after 4,500 generations. In the other cases each run was stopped after 40,800 generations with 70 offsprings each (2,856,000 simulations).

The results of the GA are very satisfying. In more than half of the runs one of the two optimal solutions was found (Table 3). Feasible solutions (completely successful algorithms) were found multiple times in all of the runs (Table 4), except the runs evolving 4-state FSMs, because there are no feasible solutions. It is also shown here that the computation time for approx. 40,000 generations (dual core, 2.4GHz) is much lower than the time for the intelligent enumeration in hardware. The numbers suggest that the GA is a very fast and reliable instrument to optimize the agents' behavior if it is performed multiple times or if one does not require the absolute optimum, but a solution close to it.

Taking a closer look at the state graphs of successful FSMs (Fig. 3) reveals certain cycles that induce strategies like "move straight" or "move diagonal". In a sequence of the simulation of the best 5-state and the best (evolved) 7-state automaton on one of the 5 environments (Fig. 4) it can be seen that the

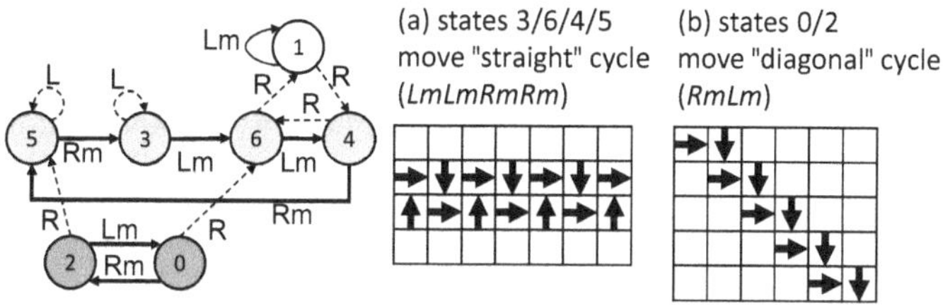

Fig. 3. The best evolved 7-state FSM comprises of the two main moving cycles (a) "straight" and (b) "diagonal"

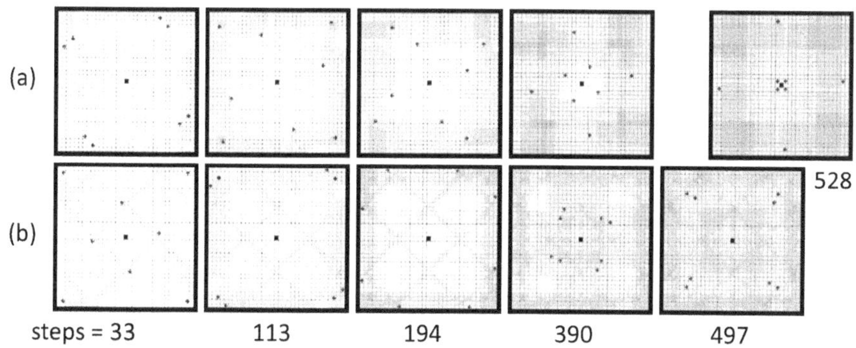

Fig. 4. A simulation sequence of the first environment comparing (a) the best 5-state algorithm and (b) the best evolved 7-state algorithm showing the typical visiting patterns. The darker the shading the more often the cells were visited.

5-state automaton produces "straight" patterns and that the 7-state automaton first produces "diagonal" patterns and then changes its strategy to "straight". Obviously it is an advantage to change this strategy at a certain point of time. We have not yet systematically investigated this phenomenon, but it seems to be an interesting approach to construct automata by hand or to use this information in order to optimize heuristic methods.

5 Conclusion

The effectiveness of a GA compared to exhaustive search was shown for the CEP with 8 agents. We have evolved state machines with up to 7 states for the agents' behavior with an Island Model GA and we compared them to all possible FSMs (4 to 6 states) found by an intelligent enumeration technique [3].

In spite of the relatively inconsistent and huge search space [5], in more than half of the runs the GA was able to find the optimal or a near optimal algorithm within less than 40,800 generations, i. e., less than 5 hours computation time. The enumeration of all possible solutions was much slower and did not offer better results. We conclude that the proposed Island Model GA is an appropriate instrument to evolve state based algorithms even if one requires a solution close to the optimum. Additional studies with Random Walk and Random Search revealed that random walkers need much more time steps to solve the task compared to FSM-controlled agents and that it is unlikely to find a feasible FSM by chance.

In our future work we want to compare the effectiveness and efficiency with other heuristic methods such as Simulated Annealing. We will also investigate more complex agents and tasks and use hardware support for the time consuming and parallelizable simulations. Among others we will try to answer the following questions that arise from this paper: (1) Is there an optimal number of states, i. e., does the effectiveness and/or the efficiency of the GA decrease above a certain number of states? (2) How do the parameters of the GA influence its performance?

References

1. Halbach, M., Heenes, W., Hoffmann, R., Tisje, J.: Optimizing the behavior of a moving creature in software and in hardware. In: Sloot, P.M.A., Chopard, B., Hoekstra, A.G. (eds.) ACRI 2004. LNCS, vol. 3305, pp. 841–850. Springer, Heidelberg (2004)
2. Halbach, M., Hoffmann, R.: Parallel hardware architecture to simulate movable creatures in the CA model. In: Malyshkin, V.E. (ed.) PaCT 2007. LNCS, vol. 4671, pp. 418–431. Springer, Heidelberg (2007)
3. Halbach, M.: Algorithmen und Hardwarearchitekturen zur optimierten Aufzählung von Automaten und deren Einsatz bei der Simulation künstlicher Kreaturen. PhD thesis, Technische Universität Darmstadt (2008)
4. Ediger, P., Hoffmann, R., Halbach, M.: How efficient are creatures with time-shuffled behaviors? In: PASA. LNI, vol. 124, pp. 93–103. GI (2008)
5. Komann, M., Ediger, P., Fey, D., Hoffmann, R.: On the effectivity of genetic programming compared to the time-consuming full search of optimal 6-state automata. In: Vanneschi, L., Gustafson, S., Ebner, M. (eds.) EuroGP 2009. LNCS. Springer, Heidelberg (2009)
6. Halbach, M., Hoffmann, R.: Solving the exploration's problem with several creatures more efficiently. In: Moreno Díaz, R., Pichler, F., Quesada Arencibia, A. (eds.) EUROCAST 2007. LNCS, vol. 4739, pp. 596–603. Springer, Heidelberg (2007)
7. Mesot, B., Sanchez, E., Peña, C.-A., Pérez-Uribe, A.: SOS++: Finding smart behaviors using learning and evolution. In: Standish, R., Bedau, M., Abbass, H. (eds.) Artificial Life VIII, pp. 264–273. MIT Press, Cambridge (2002)
8. Sanchez, E., Pérez-Uribe, A., Mesot, B.: Solving partially observable problems by evolution and learning of finite state machines. In: Liu, Y., Tanaka, K., Iwata, M., Higuchi, T., Yasunaga, M. (eds.) ICES 2001. LNCS, vol. 2210, pp. 267–278. Springer, Heidelberg (2001)
9. Di Stefano, B.N., Lawniczak, A.T.: Autonomous roving object's coverage of its universe. In: CCECE, pp. 1591–1594. IEEE, Los Alamitos (2006)
10. Halbach, M., Hoffmann, R., Both, L.: Optimal 6-state algorithms for the behavior of several moving creatures. In: El Yacoubi, S., Chopard, B., Bandini, S. (eds.) ACRI 2006. LNCS, vol. 4173, pp. 571–581. Springer, Heidelberg (2006)

Analysis of the Properties of the Harmony Search Algorithm Carried Out on the One Dimensional Binary Knapsack Problem

Jerzy Greblicki and Jerzy Kotowski

Institute of Computer Engineering, Control and Robotics
Wrocław University of Technology
11/17 Janiszewskiego St., 50-372 Wrocław, Poland
{jerzy.greblicki,jerzy.kotowski}@pwr.wroc.pl

Abstract. In the paper we carried out the analysis of the properties of the Harmony Search Algorithm (*HSA*) on a well known one-dimensional binary knapsack problem. Binary knapsack problems are among the most widely studied problems in discrete optimization. Since the optimization versions of these problems are nP-hard, practical solution techniques do not ask for optimality, but are heuristics that generate feasible, suboptimal solutions. In this paper we describe the 0-1 knapsack problem itself, the backgrounds of the *HSA*, Baldwin and Lamarck Effects and the numerical tests. The result of the tests performed is surprised a bit.

Keywords: knapsack problem, HSA, Baldwin Effect, Lamarck Effect.

1 Introduction

Knapsack problem is among some of the most widely studied problems in discrete optimization. Theoretical interest arises from the fact that, although it is among the simplest discrete optimization problems to state, it is often quite difficult to solve. Exact algorithms to solve it are based on branch and bound, dynamic programming, or a hybrid of the two. Since the optimization versions of these problems are nP-hard, practical solution techniques do not ask for optimality, but are heuristics that generate feasible, suboptimal solutions.

Existing meta-heuristic algorithms are based on ideas found in the paradigm of natural or artificial phenomena. The newest method in his area is the Harmony Search Algorithm (*HSA*) [6,7]. It was conceptualized from the musical process of searching for a perfect state of harmony, such as jazz improvisation.

Although the *HSA* is a comparatively simple method, it has been successfully applied to various optimization problems. The main aim of this paper is to check an effectiveness of the Harmony Search approach to the titled extremely simple knapsack problem. In details, in this paper, we describe the 0-1 knapsack problem itself, the backgrounds of the Harmony Search Algorithm, Baldwin and Lamarck Effects as like as the idea of the numerical tests conducted.

R. Moreno-Díaz et al. (Eds.): EUROCAST 2009, LNCS 5717, pp. 697–704, 2009.

Algorithm 1. (determining the relaxation for Problem 1)

Step 1 Put $f_u = 0$, $i = 1$.
Step 2 If $a_i \leq b$ then put $b = b - a_i$ and $f_u = f_u + c_i$. Otherwise go to Step 4.
Step 3 If $i = n$ then end. Otherwise substitute $i = i + 1$ and go to Step 2.
Step 4 Put $f_u = f_u + d_i b$.

2 Knapsack Problem

The 0-1 (binary) knapsack problem can be described as follows:

Problem 1. (binary knapsack problem)

Let $x \epsilon R^n$ be a vector of binary variables containing x_i. It is necessary to determine such values of x_i for which

$$f(x) = \sum_{i=1}^{n} c_i x_i \rightarrow max \tag{1}$$

$$\sum_{i=1}^{n} a_i x_i \leq b. \tag{2}$$

The parameters which appear in Problem 1 i.e.: c_i, $i = 1, \ldots, n$, a_i, $i = 1, \ldots, n$ and b, are positive. The problem lies in finding the most appropriate way of packing our knapsack such that the sum of the price values of the elements in the knapsack will be maximal. It is not possible to take a fraction of the object.

Problem 1 is nP-hard. One can obtain its greedy solution by prelude enumeration of the decision variables to fulfill the condition $d_i \geq d_{i+1}$, $i = 1, \ldots, n-1$ where $d_i = {c_i}/{a_i}$. Now, we can also determine the upper bound (relaxation) f_u for Problem 1 as a solution to the following linear programming problem (3)-(5).

Problem 2. (linear packing problem, continuous knapsack problem)

$$f(x) = \sum_{i=1}^{n} c_i x_i \rightarrow max \tag{3}$$

$$\sum_{i=1}^{n} a_i x_i \leq b \tag{4}$$

$$0 \leq x_i \leq 1, \quad i = 1, \ldots, n. \tag{5}$$

After the enumeration of the decision variables, as stated above, the value of f_u in Problem 2 can be obtained by the few steps as in the Algorithm 1.

After enumerating the variables one can determine an effective greedy solution with the value of the goal function (1) equals f_g (see Algorithm 2).

Both approaches in Algorithms 1 and 2 are of the n-complexity and they differ only in Step 4 of Procedure 1. Let f_o denotes the optimal value of the goal function for the Problem 1. The feasible set of solutions for Problem 2 is a super set of the feasible solutions for Problem 2. The greedy solution is an acceptable solution for Problem 1. So, in light of this $0 \leq f_g \leq f_o \leq f_u$.

Algorithm 2. (determining the value of greedy solution for Problem 1)

Step 1 Put $f_g = 0$, $i = 1$.
Step 2 If $a_i \leq b$ substitute $b = b - a_i$ and $f_g = f_g + c_i$.
Step 3 If $i = n$ then end, otherwise substitute $i = i + 1$. Go to Step 2.

Algorithm 3. (Harmony Search Algorithm)

Step 1 Initialize the problem and algorithm parameters.
Step 2 Initialize the harmony memory.
Step 3 Improvise a new harmony.
Step 4 Update the harmony memory.
Step 5 Check the stopping criterion.

3 Harmony Search Algorithm

Harmony Search Algorithm (*HSA*) is a meta heuristic algorithm mimicking the improvisation process of musicians. In the process, each musician plays a note for finding a best harmony all together. Likewise, each decision variable in optimization process has a value for finding a best vector all together.

HSA has several advantages when compared with traditional gradient-based optimization techniques. First, *HSA* does not require complex calculus. More, *HSA* does not require initial value settings for the decision variables, thus it may escape local optima. What's more, *HSA* can handle discrete as well as continuous variables. The steps in the *HSA* are as presented in Algorithm 3 [5,6,7]:

In **Step 1**, the optimization problem is specified as follows: *minimize* $f(x)$, $x = (x_1, x_2, \ldots, x_n)$, where $l_i \leq x_i \leq u_i$ for each decision variable x_i. The values l_i and u_i are the lower and upper bounds for each decision variable. The *HSA* parameters are also specified in this step. These are:

- HMS harmony memory size i.e. the number of solutions in the harmony memory *HM*;
- $HMCR$ harmony memory considering rate;
- PAR pitch adjusting rate; and
- NI the number of improvisations, or a stopping criterion.

In **Step 2**, the *HM* matrix is filled with as many randomly generated solution vectors as the HMS. In **Step 3**, a new harmony vector $x^{'}$ is generated based on three procedures: memory consideration, pitch adjustment and random selection. In the memory consideration, the value of the first decision variable for the new vector is chosen from any of the values in the specified *HM* range. The $HMCR$, which varies between 0 and 1, is the rate of choosing one value from the historical values stored in the *HM*, while $(1 - HMCR)$ is the rate of randomly selecting one value from the possible range of values. Every component obtained by the memory consideration is examined to determine whether it should be pitch-adjusted. This operation uses the PAR parameter, which is the rate of pitch adjustment. If the pitch adjustment decision is *YES*, an appropriate new x_i is replaced by the random value from $[l_i, u_i]$. In **Step 3**, *HM* consideration, pitch adjustment or random selection is applied to each variable of the new harmony.

In **Step 4**, if the new harmony vector is better than the worst harmony, the new harmony is included in the *HM* and the existing worst harmony is excluded from it. In **Step 5**, if the stopping criterion (maximum number of improvisations) is satisfied, computation is terminated. Otherwise, **Steps 3** and **4** are repeated.

Nowadays a new version of the *HSA*, named an Improved Harmony Search Algorithm (*IHSA*), is proposed [3]. In *IHSA* two parameters, namely PAR and BW are changing suitable during calculations.

4 Baldwin Effect, Memetic Algorithms and Lamarckian Evolution

The Baldwin Effect [1,2,4,8,9,10,11,12] was proposed by J. Mark Baldwin and others about 100 years ago. The Baldwin Effect refers to the notion that learning can change the environment for a species in such a way as to influence the selective environment for the learned behavior or some closely related character. The Baldwin Effect is sometimes referred to as the simple notion that, through evolution, unlearned can replace learned behavior [12].

The genotype is the genetic constitution of an individual. In a living organism, this is typically the organism's DNA. In evolutionary computation, it is typically a string of bits. The phenotype is the set of observable characteristics of an organism, as determined by the organism's genotype and environment. The distinction between genotype and phenotype is clear in biological evolution, but the distinction does not exist in many of the simpler examples of evolutionary computation.

The Baldwin Effect is a very useful tool in constrained optimization. Its idea lies in a self-dependent approach to the process of the population evolution and to the process of seeking the feasible solution to the studied problem. It is made by projecting I of the genotype g on the cover of the set of the feasible solutions (a phenotype p) and by assuming that the adjusting of the chromosome (used for instance by the procedure of selection) is the measure of its projection quality: $f(g) = f(p) = f(I(g))$. In many cases we may assume that there is no special coding, so $G = X$. The idea of the method lies now in accepting the following rule of calculation of the fitness function value for every chromosome: $f(x) = f(I(x))$.

Memetic algorithms [1] are a population-based approach in optimization problems. It was shown that they are orders of magnitude faster than traditional genetic algorithms for some problem domains. In general, memetic algorithms combine local search heuristics with crossover operators.

Lamarckian evolution is another approach to combining evolutionary search and local search. It is based on the inheritance of acquired characteristics an individual can pass the characteristics acquired through learning to its offspring genetically (encoded in the genotype). Many researches agree that Lamarckian evolution (Lamarck Effect) and memetic algorithms mean almost the same.

Due to the notation presented in the previous sub-paragraph we may state that an idea of the Lamarckian evolution (or memetic algorithms) leads to the one additional operator in the general scheme of the classic evolutionary based algorithm: after the projection $y = I(x)$ (or$p = I(g)$ we must state $x = y$.

5 Choosing the Quality Measure of the Algorithm

The following characteristics are for the already considered binary instances of packing problems:

f_g	greedy solution;
f_o	optimal solution;
f	solution of the problem obtained by way of the analyzed algorithms.

For obvious reasons we have $0 < f_g \leq f \leq f_o$. The values of f_g and f_o are relatively large but close to each other meaning that the fraction f_g/f_o has a value close to one. The evaluation of position of f within the range $[f_g, f_o]$ can be used in order to increase the sensitivity of the solution quality measure. That is why the following quality measure w is proposed for each test carried out:

$$w = \frac{f - f_g}{f_o - f_g}. \tag{6}$$

Take note that $0 \leq w \leq 1$.The measure w takes the value 0 if it was not possible to improve the greedy solution and the value 1 if an optimum solution was found. It is therefore a very comfortable and intuitive way of assessing the effectiveness of the analyzed calculation algorithm. Obviously, the value of the quality measure will change with respect to a particular instance of a solved optimization problem. In order to eradicate this effect it is essential to calculate the average W after obtaining a large amount of instances P of the problem.

$$W = \frac{1}{|P|} \sum_{p \epsilon P} w_p. \tag{7}$$

To check empirically an influence of the $|P|$ on the value W we solved near 30000 of the problem instances with $n = 50$. Data were randomly generated due to the following rule: $d_{i+1} = r \cdot d_i$, where r is a uniformly distributed random variable: $r \epsilon R[0.99, 0.9999]$. We stated that $|P| = 5000$ is an adequate size of the population tests. The quantities of W differ from each other by less than $0,01$.

No, we may find the best values of the *HSA* parameters for the considered Problem 1. At the literature the proposed values of the four HSA parameters are usually as follows: $NI = 5000$, $HMS = 100$, $HMCR = 0.74$ and $PAR = 0.1$. It is of no use at this moment to consider the effect of the NI parameter on the value of measure W. This dependence is obviously increasing. It is enough to assume $NI = 5000$. Finally, the problem may be formulated as follows:

Problem 3. (optimization of HSA parameters)
Let $W = W(HMS, HMCR, PAR)$ for a certain set of instances P for Problem 1. It is necessary to determine such a set of parameters HMS, $HMCR$ and PAR, for which:

$$W\,(HMS, HMCR, PAR) \rightarrow max \tag{8}$$

The Gauss-Siedel technique was used to determine the best parameters of the HS algorithm independently for pure HS, HS+Baldwin and HS+Lamarck. The resulting optimizing process of parameters for the HS algorithm supported by the Baldwin effect (n=50, NI=5000) is presented in Table 1.

Table 1. Optimizing of parameters for the HSA with Baldwin effect

HMS	HMCR	PAR	W	Comments
100	0.74	0.1	0.8636	Starting point
100	0.74	**0.4**	0.8670	argmax PAR
100	**0.78**	0.4	0.8707	argmax HMCR
80	0.78	0.4	0.8738	argmax HMS
80	0.78	**0.5**	0.8754	argmax PAR
80	**0.82**	0.5	0.8775	argmax HMCR
150	0.82	0.5	0.8851	argmax HMS
150	**0.82**	**0.5**	**0.8851**	Optimum

Calculating the value W, presented at the Table 1, required the solving of 100000 instances of the knapsack problem with $n = 50$ and $NI = 5000$. We did the same in the aim of obtaining optimal values of parameters for the *HSA* supported by the Lamarck effect and for the pure *HSA*. To conduct these tests we solved 30000 instances of knapsack Problem 1. The optimal parameters for the *HSA* method in all three cases are practically the same.

As the first results of these tests we formulated the following conclusions:

1. The $W(HMS, HMCR, PAR)$ function has the maximum value for finite values of its three parameters.
2. Optimal value of parameter PAR differs five times from the starting value which is similar to the proposed value in literature concerning this subject.
3. Optimal values of parameters $HMCR$ and PAR are significantly larger than the proposed values in the literature. It may be a result of the specifications of the knapsack problem (binary variables).
4. The quality of obtained results decreases when the value of parameter *HMS* is too large. This means (as in normal life) that sometimes having a good memory could be a curse if not worse.

6 Numerical Tests

The last test carried out was aimed at determining the relation between the size of problem n and the value of the measure W for the all considered versions of

Table 2. An influence of the size of the problem n on the efficiency of the *HSA*

n	Pure HSA	Lamarck	Baldwin
5	0.7405	0.9702	0.9982
10	0.7600	0.9631	0.9825
20	0.7432	0.9692	0.9711
30	0.6808	0.9558	0.9573
50	0.5553	0.9239	0.9044
70	0.3200	0.8473	0.8106
100	0.0595	0.6290	0.6477
120	0.0177	0.4198	0.5172
150	0.0026	0.1789	0.3200
200	0.0001	0.0229	0.1008

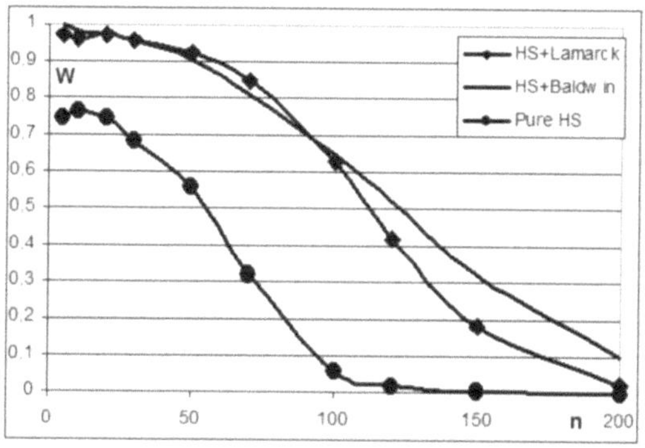

Fig. 1. The influence of the size of the problem n on the efficiency of the *HSA*

the *HSA*. The test was carried out using optimal parameters presented in the Table 1. The resulting data of the test performed were presented in Table 2. A graphical interpretation of the fitting dependencies we present on Figure 1.

The results of the test are surprisingly negative. For a problem with $n = 200$ variables, the greedy solutions are rarely in need of change. The value of the quality measure W is around 4%. Determining an optimal solution to Problem 1 using the B&B method lasts only a fraction of a second.

7 Conclusion

The algorithm for solving the packing problem based on the Harmony Search algorithm was presented in this article. There were 3 versions presented, that is, the original and two others supported by the Baldwin and Lamarck effects.

The tests carried out showed a definite advantage in the algorithms which were supported by the Baldwin and Lamarck effects as compared to the classic HSA. Unfortunately, the tests also showed beyond reasonable doubt that the efficiency of all these algorithms is relatively small. An optimal solution can definitely be found only for problems that are small in size (practically only up to $n = 200$). For this problem size, the classic algorithm based on the idea of partial enumeration (B&B method), always finds an optimum, and faster.

An interesting result obtained during the course of the tests was the statement that, for an intuitionally introduced optimization quality measure of calculation procedure, the HMS parameter assumes a finite optimal value. This means that, for the considered binary optimization problem, a large harmony memory can be unfavorable. This observation can suggest that – as in real life – a memory that's too good can be the cause of many problems. It can be an interesting trial to determine the class of an optimization problem for which this phenomenon can be observed.

References

1. Arabas, J.: Lectures on Evolutionary Algorithms. WNT, Warsaw (2001) (in Polish)
2. Baldwin, J.M.: A new Factor in Evolution. American Naturalist 30, 441–451 (1896)
3. Cheng, Y.M., Lansivaara, L., Li, T., Chi, S.C., Sun, Y.J.: An improved harmony search minimization algorithm using different slip surface generation methods for slope stability analysis. Engineering Optimization 40, 95–115 (2008)
4. French, R.M., Messinger, A.: Genes, Phenes and the Baldwin Effect: Learning and Evolution in a Simulated Population, pp. 277–282. The MIT Press, Cambridge (1994)
5. Geem, Z.W., Kim, J.H., Loganathan, G.V.: A new heuristic optimization algorithm: harmony search. Simulation 76(2), 60–68 (2001)
6. GGeem, Z.W., Tseng, C.L.: New methodology, harmony search and its robustness. In: 2002 Genetic and Evolutionary Computation Conference, pp. 174–178 (2002)
7. Geem, Z.W.: Optimal cost design of water distribution networks using harmony search. In: Environmental Planning and Management Program, pp. 1–49. Johns Hopkins University (2005)
8. Greblicki, J., Kotowski, J.: Optimal RNS Moduli Set for DSP Applications. In: Proc. of MMAR 2005, 11th IEEE Conference on Methods and Models in Automation and Robotics (2005)
9. Greblicki, J., Kotowski, J.: The Greedy Solution of the OPTIMAL RNS Moduli Set Problem. In: Proc. of MMAR 2005, 11th IEEE Conference on Methods and Models in Automation and Robotics (2005)
10. Kotowski, J.: The use of the method of illusion to optimizing the simple cutting stock problem. In: Proc. MMAR 2001, 7th IEEE Conference on Methods and Models in Automation and Robotics, vol. 1, pp. 149–154 (2001)
11. Turney, P.: Myths and Legends of the Baldwin Effect. In: Proc. GECCO 1999, Genetic and Evolutionary Computation Conference (1999)
12. Weber, B.H., Depew, D.J.: Evolution and Learning: The Baldwin Effect Reconsidered. MIT Press, Cambridge (2003)

An Algorithm of Schedule Planning for Tanker Drivers

Jerzy Greblicki and Jerzy Kotowski

Institute of Computer Engineering, Control and Robotics
Wrocław University of Technology
11/17 Janiszewskiego St., 50-372 Wrocław, Poland
{jerzy.greblicki,jerzy.kotowski}@pwr.wroc.pl

Abstract. In this paper a certain modification of an evolutionary algorithm is characterized, which is intended for planning a schedule for tanker drivers working for a petrol base. The description of a computational algorithm is presented, starting from the description of a particular genotype and phenotype. The most important element of the algorithm is the procedure of projection of the genotype on the set of phenotypes based on the Baldwin Effect. The final part of the paper presents computational tests and plans for the future.

Keywords: cutting stock, vehicle routing problem, genetic algorithm.

1 Introduction

In this study a certain modification of an evolutionary algorithm is characterized, which is intended for planning a schedule for tanker drivers working for a petrol base. The modification is based on the Baldwin Effect [2,3,4].

Each tanker has several separate chambers which, for safety reasons, can only be full or empty. Such tankers are designed in response to market needs. One tanker with many chambers can serve many customers and supply each of them with many different sorts of petrol. The existing systems enable continuous loading and unloading of the vehicles. You can expect great interest in this kind of software among large petrol concerns and in consequence noticeable financial benefits in distributing this tool. The problem of schedule planning was transformed into a certain cutting stock problem. Tankers are subordinated to the cutting process, and their chambers are the elements of this process [6]. A specific feature of the problem is that the cost of each cutting stock scheme is set as a travelling salesman problem. It is the cost of the cheapest closed route, connecting customers of petroleum based products, by tanker delivery.

In this study the description of a computational algorithm is presented. The most important element of the algorithm is the procedure of projection of the genotype on the set of phenotypes, i.e. feasible solutions [6].

2 Genesis of the Problem

To tackle the problem of schedule planning for tanker drivers one should take into consideration at least the following parameters and characteristics: number of

R. Moreno-Díaz et al. (Eds.): EUROCAST 2009, LNCS 5717, pp. 705–712, 2009.

drivers available at any given moment, number of vehicles and their types, with full information about the capacity of each particular tank chamber, consumers' demands made on the amounts of products, data about each client: capacity of their tanks, minimal available petrol level in the tanks and current level in the containers, names of clients, their addresses and priorities, suitable time of delivery, type of pouring containers and available auxiliary equipment, unloading regulations, kind of product being transported in a chamber, parking site for each tanker, location of petrol base, tank loading and unloading time, permissible speed of the vehicle and other traffic restrictions.

The foregoing analysis has revealed the complexity of the title problem. In this paper an algorithm for solving a very simplified version of the problem above will be presented:

For the transport base under analysis the following is known: number of tankers, number and capacity of chambers in each tank, number of petrol types delivered to client, number of clients, current demands with reference to particular petrol types and matrix of the cost of transport between clients. We assume one petrol station and one type of tank. All chambers in the tank have the same capacity. The chambers must be empty or full during transport.

Problem 1. Arrange delivery in such a way that consumer demands are fulfilled, and total transport cost is minimal.

3 Mathematical Model of the Optimization Problem

In this part we introduce all the symbols referring to the output data for the optimization problem. Hence,

p	type of product: $p\epsilon\{1,2,\ldots,lp\}$, where lp is the number of products,
lk	number of chambers in one tanker,
vk	capacity of a single chamber.

The symbols referring to clients are as follows:

r	receivers' number $r\epsilon\{1,2,\ldots,lr\}$, where lr is the number of clients,
z_r	order placed by $r-th$ client, $z_r\epsilon R^{lp}$,
M	matrix of distances, $M = M_{(lr+1)x(lr+1)}$.

The petrol station is in node $r = 0$. The needs of clients can be written as:

$$Z = \{z_{ij}\}_{lp\,x\,xlr} = [z_1, z_2, \ldots, z_{lr}]\,. \tag{1}$$

For the safety reasons mentioned, it can be assumed that clients needs can be expressed in terms of the number of petrol chambers ordered:

$$Y = \left\{y_{ij} = \frac{z_{ij}}{vk}\right\}_{lp\,x\,lr}. \tag{2}$$

Any particular order can be written as the lk-elemental sequence of pairs: $w = \{(p_i, r_i)\,,\ i = 1,2,\ldots,lk\}$, where p_i stands for the petrol type being in the $i-th$

tank chamber and r_i is the number of the receiver of the content of the $i-th$ chamber. It can also be assumed that the order in which the pairs appear in the w responds to the visiting plan of the tank. Let m stand for the number of orders and let $W = \{w_1, w_2, \ldots, w_m\}$ denote the set of all orders.

Definition 1. *Set W is a feasible solution to the problem of schedule planning for tanker drivers, if the execution of the orders belonging to it satisfies all the receivers' needs.*

Let $f(w)$ be the cost of a given order belonging to W. You can, in a natural way, describe the cost of each feasible solution W: The routes must be designed in such a way that each point is visited only once by exactly one vehicle, all routes start and end at the depot, and the total demands of all points on one particular route must not exceed the capacity of the vehicle

$$F(W) = \sum_{w \epsilon W} f(w). \tag{3}$$

Let Ψ denote the family of the feasible solutions for the discussed problem: $W \epsilon \Psi$.

Definition 2. *The Set $W_o \epsilon \Psi$ is the optimal solution to the problem of schedule planning for tanker drivers, if:*

$$F(W_o) = min_{w \epsilon W} F(W). \tag{4}$$

4 Analysis of the Problem

The optimization problem considered in this paper belongs to the wide group of discrete tasks well known in literature as the Vehicle Routing Problem (VRP). The VRP is a central problem in the fields of transportation, distribution and logistics. The utilization of computerized methods in this area often results in significant savings ranging from 5% to 20% in total costs, as reported in [10].

The classic VRP is the problem of designing low cost routes from one depot to a set of customers, each requiring a specified weight of goods to be delivered. The problem is to find a set of delivery routes satisfying the particular requirements and giving minimal total cost [7,8,9]. Some of the most important restrictions, which appear in VRP problems, are presented in [9] as: CVRP, VRPTW, MDVVRP, VRPPD, SDVRP, SVRP, PVRP, etc.

No exact algorithm can be guaranteed to find optimal tours within reasonable computing time when the number of cities is large. Classification of the solution techniques is considered in [1,9]: exact approaches (branch and bound approach and branch and cut approach), heuristic methods (constructive methods and the 2-Phase Algorithm) and meta-heuristics: (ant algorithms, genetic algorithms, simulated annealing, tabu search, the adaptive memory procedure, etc).

In our study it was proposed to solve the Problem 1 with a method based on the idea of an evolutionary algorithm. Below is presented an implementation of an evolutionary algorithm for the problem being considered. The important element of this algorithm is the procedure of setting the fitness function value on the basis of the Baldwin Effect.

5 The Baldwin Effect

The Baldwin Effect [2,3,4,11,12] was proposed by J. Mark Baldwin and others about 100 years ago. It refers to the notion that learning can change the environment for a species in such a way as to influence the selective environment for the learned behavior or some closely related character [12].

Since the end of the 1980s, several researchers [11] have observed a synergistic effect in evolutionary computation when there is an evolving population of learning individuals. Learning has benefits (the first aspect of the Baldwin Effect) but it also has costs (the second aspect).

For our purposes some necessary definitions should be stated now. The *genotype* is the genetic constitution of an individual. In evolutionary computation, it is typically a string of bits. The *phenotype* is the set of observable characteristics of an organism, as determined by the organism's genotype and environment.

The distinction between genotype and phenotype is clear in biological evolution, but the distinction does not exist in many of the simpler examples of evolutionary computation. *Phenotypic plasticity* is the ability of an organism to adapt to its environment. In its most general sense, the Baldwin Effect deals with the impact of phenotypic plasticity on evolution.

6 Idea of the Optimization Algorithm

Let the following static, constrained optimization problem be considered:

$$f(x) \rightarrow max \tag{5}$$

$$x \epsilon D \subset X \tag{6}$$

The idea of the suggested method lies in accepting the following rule of calculation of the fitness function value for every chromosome:

$$p(x) = f(I(x)) \tag{7}$$

In (7) p stands for the fitness function, f is an original goal function (5) and I is a projection on the feasible solutions. In evolutionary computation the idea presented above stands for the Baldwin Effect described in the chapter [6].

7 Features of the Optimization Problem

For convenience, we formulate Problem 1 in the form proposed by Gilmore and Gomory [5]. The A is a matrix built on the base of all orders. Each matrix column responds to the particular order. Each matrix row responds to one, non-zero demand from matrix Y (2). So, an element a_{ij} of such a matrix indicates the number of chambers with particular petrol ordered by an individual client as a part of the $j-th$ order. We can limit the number of columns to those orders that fulfill the following condition:

$$\sum_i a_{ij} = lk. \tag{8}$$

Let $A = A_{k\,x\,s} = [a_1, a_2, \ldots, a_s]$, where k is a the number of non-zero demands in matrix Y and s is the number of orders which fulfill the condition (8). Let $b = (b_1, b_2, \ldots, b_k)^T \epsilon R^k$ be a vector of orders built of all non-zero elements of the Y matrix, and let $c = (c_1, c_2, \ldots, c_s) \epsilon R^s$ denote the vector of prices of individual strategies. We have to solve the traveling salesman problem to obtain any element of the c vector. As a consequence, the title problem can be reduced to solving the integer programming problem:

$$c \cdot x \rightarrow min \tag{9}$$
$$A \cdot x \geq b \tag{10}$$
$$x \geq 0 \; and \; integer. \tag{11}$$

In a real situation a number of the A matrix columns can reach hundreds of millions. Setting only the cost of each strategy demands a huge amount of time, so it is impossible to find a solution to the problem (9)-(11) with the technique described in the Gilmore-Gomory's two-level algorithm. The implementation of the evolutionary algorithm leads to following:

- The cutting strategy (an order for a driver) plays the role of a single gene.
- The subset of a given number of the A matrix columns constructed from the coefficients described with formula (8) acts as the genotype.
- The role of the phenotype is played by another subset of the A matrix columns with an adequate subset of a vector x. The phenotype should be the feasible solution to the problem: $A \cdot x \geq b$.
- The relation between phenotype and genotype is described by the projection procedure. The value for the fitness function of the genotype responds to the cost of the phenotype, by virtue of (7).

So, the implementation of the evolutionary algorithm consists of:

- Creating the proper initial population. Each element of the population is a genotype. The genotype is a single matrix built of feasible strategies.
- Establishing proper procedures that will act as the basic evolutionary operators: reproduction, crossover, mutation and selection.
- Working out a procedure for setting the values of the fitness function for the genotype. An algorithm based on the Baldwin Effect, it is necessary to work out a projection which assigns the genotype to a proper phenotype.

Algorithm 1. Constructional algorithm

Step 1 Set $j = 0$ (strategy number). Go to step 2.
Step 2 Assume $j = j + 1$, $nk = lk$, $a_{ij} = 0$ for $i = 1, \ldots, k$. Go to step 3.
Step 3 According to the M matrix of distances find a client who is the longest distance away from the petrol base and such that in vector b there are components responding to it, which fulfill the condition $b_i > 0$. Go to step 4.
Step 4 Set $a_{ij} = max(nk, b_i)$ and $nk = nk - a_{ij}$, $b_i = b_i - a_{ij}$. Go to step 5.
Step 5 If $b_i = 0$, for $i = 1, 2, \ldots, k$ then go to step 8. If $nk = 0$, go to step 7. Otherwise go to step 6.
Step 6 Find the client closest to the previous one, for whom $b_i > 0$. Go to step 4.
Step 6 The next tank was filled. Return to step 2.
Step 8 The end of the algorithm. If $j < s$, generate an appropriate number of columns in the A matrix using random generation to attain a full genotype.

- Choosing an evolutionary strategy and describing its parameters such as probability of mutation and crossover, size of populations, etc.

More details are found below.

An evolutionary strategy: The described algorithm uses the evolutionary scheme known as Strategy (μ,λ). In this strategy every population consists of μ elements. During evolution each new base population is created by the selection from a λ elementary transitory population. A transitory population is created via appropriate operators of reproduction, crossover and mutation.

An individual: In the described problem an individual is represented by the A matrix made of feasible strategies. For the needs of the projection procedure, we assumed a redundancy of A matrix columns: $s \ggg lc_{min}$, where

$$lc_{min} = \frac{\sum_{i=1}^{k} b_i}{lk}. \tag{12}$$

The relation of (12) describes the minimal number of tankers needed.

Initial population: Two procedures: random generation and constructional algorithm, are implemented. A constructional algorithm (see Algorithm 1) uses the idea of a greedy solution.

Reproduction: According to Strategy (μ,λ) the new base population is created by selection of μ elements from a λ elementary transitory population.

Crossover: For the recombination process we cross an appropriate number of columns in two matrices selected from the transitory population. The recombination method is a classic one-point crossover.

Mutation: Mutation occurs for any matrix with the given probability p_m. The established algorithm requires that we replace a randomly chosen gene (column) by another gene in the matrix being considered.

Projection: In terms of the Baldwin Effect the projection operator is an essential part of the implemented algorithm, which enables us to obtain a fitness value of the genotype, as stated in formula (7). In the presented optimization problem this idea leads to the calculation of the the greedy solution of the problem (9)-(11). More detail of this procedure is presented in Algorithm 2.

Algorithm 2. Projection

Step 1 Calculate c_j, $j = 1, \ldots, s$, by virtue of (9), for all strategies (orders) included in the A matrix. Set $F = 0$ and $x_j = 0$, $j = 1, \ldots, s$. Go to step 2.
Step 2 If $\{j : a_j \leq b,\ j = 1, 2, \ldots, s\} = \emptyset$, then go to step 5. Otherwise go to step 3.
Step 3 Find j_0 for which $c_{j_0} = \min_{a_i \leq b,\, i=1,\ldots,s} c_j$.
Step 4 Calculate $x_{j_0} = \min_{a_{ij}>0,\, i=1,\ldots,k} \frac{b_i}{a_{ij}}$. Set $F = F + c_{j_0} \cdot x_{j_0}$ and $b_i = b_i - a_{ij_0} \cdot x_{j_0}$, $i = 1, 2, \ldots, k$. Return to step 2.
Step 5 If $b_i = 0$, for $i = 1, 2, \ldots, k$, then go to step 7. Otherwise go to step 6.
Step 6 Find $j_0 = \min_{x_j=0,\, j=1,2,\ldots,s} j$. Replace the strategy a_{j_0} in the A matrix by a new strategy obtained with the help of steps 2-7 in Algorithm 1 for the current values of the elements in vector b.
Step 7 The end of the procedure. A new phenotype was found. Its description is stored in the A matrix and x vector. The cost of the phenotype equals F.

Selection: To put a selection operator into motion we have, first of all, to know the fitness values of all genotypes being members of the base population. In the presented problem the assessment of the chromosome is associated with the solution of an appropriate minimization problem (9)-(11). A standard roulette scheme was used for the selection procedure. This scheme is oriented to the maximization process. Therefore, the normalized fitness value G_m of the given genotype m, where $m = 1, 2, \ldots, \mu$ is set as

$$G_m = \frac{F_m - F_{max}}{F_{min} - F_{max}} \tag{13}$$

In the relation of (13) F_m denotes a fitness value for the phenotype, F_{min} (F_{max}) is the minimal (maximal) value of the fitness function over the base population. When use is made of Strategy (μ,λ) the new transitory population is created by selecting λ elements from a μ elementary base population.

8 Numerical Tests

To check the numerical properties of the approach, a special program in the C++ language was prepared. The idea of the object oriented approach was implemented during the programming process.

The tests were carried out with an IBM PC. Two kinds of tests were conducted: with random generation of the consumers' needs and the matrix of distances between them, and with real data attained with the help of Microsoft AutoRoute Express. In the latter case, the scenario was chosen in the proximity of Wroclaw, Poland. This region is divided into two separate sub-areas due to the Odra River passing through this district. The small number of bridges connecting the river banks makes the problem difficult even for a small number of clients. These tests enabled us to examine the numerical properties of the developed tool in difficult cases.

The total number of tests in both cases exceeded 1000. On the basis of these tests, we postulate that the best parameters of the optimization algorithm are:

- the total number of generations: 4000,
- the size of the base population: $\mu = 20$,
- the size of the transitory population: $\lambda = 30$,
- the probability of mutation: $p_m = 0.9$,
- the probability of crossover: $p_c = 0.8$.

9 Final Conclusions

The conclusion resulting from all the tests performed is as follows: In evolutionary computation the Baldwin Effect leads to effective optimization algorithms.

The optimization problem presented here describes a very simple case and does not include many important assumptions or limitations which must be taken into account in practice. They are, for example, road traffic regulations, types of vehicle, specific situations concerning particular clients, etc. In the near future, our research will focus on the development of comprehensive and far more sophisticated software. As already mentioned in the Introduction section, we expect an enhanced interest in such software on the stock market.

References

1. Anderson, D., Anderson, E., Lesh, N., Marks, J., Mirtich, B., Ratajczak, D., Ryall, K.: Human-Guided Simple Search, Working Paper, Mitsubishi Electric Research Laboratory, Cambridge, USA (2000)
2. Arabas, J.: Lectures on Evolutionary Algorithms. WNT, Warsaw (2001) (in Polish)
3. Baldwin, J.M.: A new Factor in Evolution. American Naturalist 30, 441–451 (1896)
4. French, R.M., Messinger, A.: Genes, Phenes and the Baldwin Effect: Learning and Evolution in a Simulated Population. In: Artificial Live IV, pp. 277–282. MIT Press, Cambridge (1994)
5. Klempous, R., Kotowski, J., Szlachcic, E.: Interactive procedures in large scale two-dimensional cutting stock problems. Journal of CAM 66, 323–332 (1996)
6. Kotowski, J.: The use of the method of illusion to optimizing the simple cutting stock problem. In: Proc. MMAR 2001, 7th IEEE Conference on Methods and Models in Automation and Robotics, vol. 1, pp. 149–154 (2001)
7. Laporte, G., Semet, F.: Classical Heuristics for the Vehicle Routing Problem, Les Cahiers de GERAD, G-98-54, 1-19 (1999)
8. Potvin, J.-Y., Robillard, C.: Clustering for Vehicle Routing with a Competitive Neural Network. Neuro-computing 8, 125–139 (1995)
9. Sokołowski, M., Szlachcic, E.: A New Heuristic Algorithm for the Vehicle Routing Problem with Time Windows. In: Proc. MMAR 2001, 7th IEEE Conference on Methods and Models in Automation and Robotics, vol. 1, pp. 1201–1206 (2004)
10. Toth, P., Vigo, D.: The Vehicle Routing Problem. Monographs on Discrete Mathematics and Applications. SIAM, Philadelphia (2001)
11. Turney, P.D.: Myths and Legends of the Baldwin Effect. In: Proc. GECCO 1999, Genetic and Evolutionary Computation Conference (1999)
12. Weber, B.H., Depew, D.J.: Evolution and Learning: The Baldwin Effect Reconsidered. MIT Press, Cambridge (2003)

A Kruskal-Based Heuristic for the Rooted Delay-Constrained Minimum Spanning Tree Problem

Mario Ruthmair and Günther R. Raidl

Institute of Computer Graphics and Algorithms
Vienna University of Technology, Vienna, Austria
{ruthmair,raidl}@ads.tuwien.ac.at
http://www.ads.tuwien.ac.at

Abstract. The rooted delay-constrained minimum spanning tree problem is an NP-hard combinatorial optimization problem arising for example in the design of centralized broadcasting networks where quality of service constraints are of concern. We present a construction heuristic based on Kruskal's algorithm for finding a minimum cost spanning tree which eliminates some drawbacks of existing heuristic methods. To improve the solution we introduce a greedy randomized adaptive search procedure (GRASP) and a variable neighborhood descent (VND) using two different neighborhood structures. Experimental results indicate that our approach produces solutions of better quality in shorter runtime when having strict delay-bounds compared to an existing centralized construction method based on Prim's algorithm. Especially when testing on Euclidian instances our Kruskal-based heuristic outperforms the Prim-based approach in all scenarios. Moreover our construction heuristic seems to be a better starting point for subsequent improvement methods.

1 Introduction

When designing a network with a single central server broadcasting information to all the participants of the network some applications, e.g. video conferences, require a limitation of the maximal delay from the server to each client. Beside this delay-constraint minimizing the total cost of establishing the network is in most cases an important design criterium. In another example we consider a package shipment organization with a central depot guaranteeing its customers a delivery within a specified time horizon. Naturally the organization wants to minimize the transportation costs but at the same time has to hold its promise of being in time.

These network design problems can be modeled using a combinatorial optimization problem called *rooted delay-constrained minimum spanning tree (RDCMST) problem*. The objective is to find a minimum cost spanning tree of a given graph with the additional constraint that the sum of delays along the paths from a specified root node to any other node must not exceed a given delay-bound.

R. Moreno-Díaz et al. (Eds.): EUROCAST 2009, LNCS 5717, pp. 713–720, 2009.

More formally, we are given a graph $G = (V, E)$ with a set of n nodes V, a set of m edges E, a cost function $c : E \rightarrow \mathbb{R}^+$, a delay function $d : E \rightarrow \mathbb{R}^+$, a fixed root node $s \in V$ and a delay-bound $B > 0$. An optimal solution to the RDCMST problem is a spanning tree $T = (V, E')$, $E' \subseteq E$, with minimum cost $c(T) = \sum_{e \in E'} c(e)$, satisfying the constraints: $\sum_{e \in P(s,v)} d(e) \leq B$, $\forall v \in V$. $P(s, v)$ denotes the unique path from the specified root node s to a node $v \in V$.

The RDCMST problem is $\mathcal{NP}$-hard because a special case called *hop-constrained minimum spanning tree problem*, where $d(e) = 1$, $\forall e \in E$, is shown to be $\mathcal{NP}$-hard in [1], so all more general variants of this problem are $\mathcal{NP}$-hard too.

2 Previous Work

Exact approaches to the RDCMST problem have been examined by Gouveia et al. in [2], but these methods can only solve small graphs with significantly less than 100 nodes to proven optimality in reasonable time if considering complete instances.

A heuristic approach was presented by Salama et al. in [3], where a construction method based on Prim's algorithm to find a minimum spanning tree [4] is described. This Prim-based heuristic starts from the root node and iteratively connects the node which can be reached in the cheapest way without violating the delay-constraint. If at some point no node can be connected anymore, the delays in the existing tree are reduced by replacing edges. These steps are repeated until a feasible RDCMST is obtained. A second phase improves the solution by local search using the edge-exchange neighborhood structure.

There are many recent publications dedicated to the *rooted delay-constrained minimum Steiner tree problem* which is a generalization of the RDCMST problem. In this variant only a subset of the nodes has to be reached within the given delay-bound, the other nodes can optionally be used as intermediate (Steiner) nodes. Several metaheuristics have been applied to this variant, e.g. a tabu-search in [5], a GRASP in [6] and a path-relinking approach in [7].

3 Kruskal-Based Construction Heuristic

A general problem of the Prim-based heuristic especially on Euclidian instances is the fact that the nodes in the close surrounding of the root node are connected rather cheaply, but at the same time delay is wasted, and so the distant nodes can later only be linked by many, often expensive edges, see Fig. 1. The stricter the delay-bound the more this drawback will affect the costs negatively. This fact led us to a more de-centralized approach by applying the idea of Kruskal's minimum spanning tree algorithm [8] to the RDCMST problem.

3.1 Stage 1: Merging Components

In the beginning of stage one of the construction heuristic all edges are sorted by ascending costs and then iteratively added to the solution preventing cycles

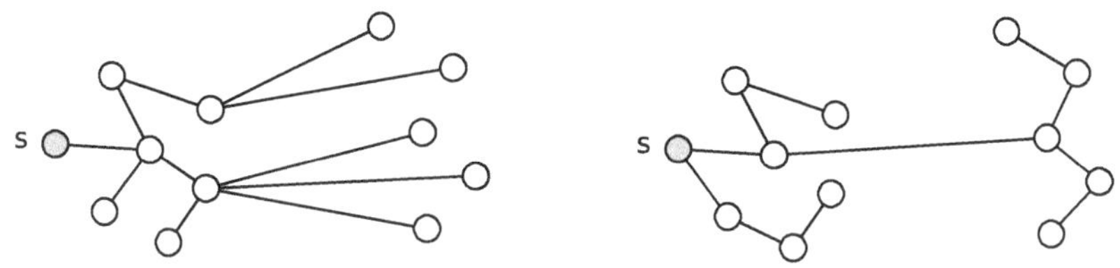

(a) Prim-based (b) Kruskal-based

Fig. 1. Prim-based heuristic (a) compared to Kruskal-based heuristic (b)

until a feasible spanning tree is formed. In other words, components initially consisting of single nodes are merged by adding edges to result in one connected tree. The challenge is to maintain the feasibility of the partial solutions, i.e. to satisfy the delay-constraint to the root node throughout the whole merging process. In the Prim-based approach in [3] checking the feasibility of adding an edge to the existing tree naturally runs in constant time whereas our decentralized algorithm needs more effort to achieve this. We have to store and update additional information for each node $v \in V$:

- the path-delay $d_s(v) := \sum_{e \in P(s,v)} d(e)$ from the root node to node v
- the maximum delay $d_{max}(v)$ to any other node in the same component
- the predecessor $pred(v)$ on the path $P(s, v)$, initialized with node v

To initialize $d_s(v)$ Dijkstra's algorithm [9] calculates the path $P(s, v)$, $\forall\ v \in V$ with the shortest path-delay. The paths themselves are not added to the solution, we just keep them in mind to always have a possible feasible connection to the root node available. This fallback paths are essential for stage two of the heuristic (see Section 3.2).

At this time we are able to decide if a solution exists or not because if the shortest-delay-path exceeds the specified delay-bound for any node then we cannot build a feasible tree and therefore stop here.

Initially we have a set of components $C = \{C_1, ..., C_k\}$, $k = n$. Everytime we add an edge to the solution two components are merged and thereby k is decreased by 1 until set C only contains one component. For each component C_i we specify one node v_{C_i} which is nearest to the root node – it can be seen as the local root node of the subtree C_i. As mentioned above the path $P(s, v_{C_i})$, $v_{C_i} \neq s$, is not part of the tree, we just use it for testing the feasibility of a partial solution.

Now we start iterating over the sorted edge-list. Let $e = (u, v) \in E$ be the next edge on the list and $C_u \ni u$, $C_v \ni v$ be the components incident to e. The decision of adding e to the tree and thereby merging the two components C_u and C_v is based upon fulfilling at least one of the following two conditions:

1. $d_s(u) + d(e) + d_{max}(v) \leq B$
2. $d_s(v) + d(e) + d_{max}(u) \leq B$

So if it is allowed to add edge e to the solution the node information of all nodes in the newely created component C_{uv} has to be updated. First of all we have to specify the new $v_{C_{uv}}$. There are many possibilities of choosing this node with the only constraint that $d_s(v_{C_{uv}})$ plus the delay of path $P(v_{C_{uv}}, w)$ has to satisfy the delay-bound for all $w \in C_{uv}$. A very simple and fast method turned out to be the most successful one: if only condition 1 is met then $v_{C_{uv}} = v_{C_u}$, when condition 2 holds, we choose v_{C_v}, and if both conditions are satisfied we prefer the v_{C_i} where the corresponding inequality has a larger gap to the delay-bound.

Beginning from this chosen local root node for component C_{uv} we perform a depth-first search to modify $pred(w)$ and $d_s(w)$, $\forall\ w \in C_{uv}$ using $d_s(v_{C_{uv}})$ as the starting delay. The maximal extents $d_{max}(w)$ can be determined in linear time profiting from the tree structure of the component.

The iterations stop if the solution only consists of one component, which means that it is already feasible, or there are more than one components but no more edges left in the list. The latter case is handled in stage two.

To conclude, stage one consists of sorting all edges of the graph in $\mathcal{O}(m \log m)$ time, testing each one for feasibility in constant time and updating the node information in $\mathcal{O}(n)$ time if an edge is added which can happen at most $n-1$ times due to the properties of a tree. So the total runtime is in $\mathcal{O}(m \log m + n^2)$.

3.2 Stage 2: Extension to a Feasible Solution

At the end of stage one the graph needs not to be connected, so in stage two the remaining subtrees are attached to the component which contains the root node by adding the shortest-delay-path $P(s, v_{C_i})$, $\forall\ C_i \in C$. At least one of the edges of a path $P(s, v_{C_i})$ creates a cycle when adding it to the solution, otherwise all edges of $P(s, v_{C_i})$ would have been included in stage one. So the main task in this stage is to dissolve resulting cycles to form a tree without violating the delay-constraint.

Paths are added by backtracking the shortest-delay-path starting from node v_{C_i} until a node u with minimal delay $d_s(u)$ is reached. We can be sure that path $P(s, u)$ is already the shortest-delay-path and do not have to go further – in the worst case however we end up at the root node. Now we add the missing edges along path $P(u, v_{C_i})$ until we are back at v_{C_i}. Cycles can occur if edge $e = (v, w)$ is added and $pred(w) \neq w \neq v$, indicating that two different paths $P(s, w)$ exist in the tree. Removing edge $(pred(w), w)$ dissolves this cycle and at the same time maintains feasibility because the delay d_s of any node in component C_w can only get smaller or stay equal since $d_s(w)$ now is the smallest possible delay and all other nodes depend on that. In $C_{pred(w)}$ no delays are affected by the removal of edge $(pred(w), w)$ since all nodes are connected to the root node through path $P(s, v_{C_{pred(w)}})$.

Since the dissolving of cycles can be done in constant time and each node is examined at most once, stage two runs in $\mathcal{O}(n)$.

3.3 Modifications

Two modifications in stage one usually lead to better results when applying a subsequent improvement method (see Section 5):

1. A delay-factor $df \geq 1$ is introduced and multiplied with the left side of the inequalities when checking the feasibility of adding an edge. In other words, the delay-bound is lowered by the factor $\frac{1}{df}$.
2. If stage one has added a predefined number of edges $< (n-1)$ it is aborted and stage two uses shortest-delay-paths to attach the left components.

Both modifications provide a solution where the gap between the node-delays $d_s(v)$ and the delay-bound is larger than in the spanning tree of the standard implementation. This higher "residual delay" leads to more possibilites in a following improvement phase and therefore often results in solutions with less total cost.

4 GRASP

To provide many different feasible starting solutions for a subsequent improvement phase we extended stage one of the Kruskal-based construction heuristic with a *greedy randomized adaptive search procedure* (GRASP) [10]. In each iteration of stage one do:

1. store all feasible edges in a candidate list (CL)
2. select a subset of least-cost edges of CL with
$$c(e) \leq \min_{e \in CL} c(e) + \alpha \cdot (\max_{e \in CL} c(e) - \min_{e \in CL} c(e))$$
for a predefined parameter $\alpha \in [0, 1]$ and insert them into a restricted candidate list (RCL)
3. randomly choose an edge from the RCL
4. merge components by adding this edge

5 Variable Neighborhood Descent

We introduce a *variable neighborhood descent* (VND) [11] for improving a constructed solution by performing a local search switching between two neighborhood structures: *Edge-Replace* (ER) and *Component-Renew* (CR). The standard implementation of a VND as it is described in [11] was modified to provide here better results in a shorter runtime: A neighborhood structure is searched by next-improvement until a local optimum is reached; then we switch to the other one continuing until no better solution can be found anymore.

A move in the Edge-Replace neighborhood removes the most expensive edge and connects the resulting two components in the cheapest possible way. A complete neighborhood search is done in $\mathcal{O}(nm)$ time.

A Component-Renew move also deletes the most expensive edge, but completely dissolves the component which is now separated from the root node; it then re-adds the individual nodes by applying a Prim-based algorithm. As before in some cases not all single nodes can be added due to the delay-bound. These remaining nodes are again joined to the root component by shortest-delay-paths, dissolving created cycles. A complete neighborhood search is done in $\mathcal{O}(n^3)$ time.

Table 1. Comparison of Prim- and Kruskal-based heuristics, applied on random instance sets with 500 and 1000 nodes (B: delay-bound, C: only construction, CV: construction and VND, CGV: construction with GRASP and VND, $\bar{c}$: average final objective values, σ: standard deviations, $t[s]$: running times in seconds)

		R500						R1000					
		Prim-based			Kruskal-based			Prim-based			Kruskal-based		
B	Test	$\bar{c}$	σ	$t[s]$	$\bar{c}$	σ	$t[s]$	$\bar{c}$	σ	$t[s]$	$\bar{c}$	σ	$t[s]$
6	C	19651	1583	0.1	**10785**	643	0.0	24053	3065	0.5	**14717**	710	0.0
	CV	9624	624	0.8	**9177**	633	0.5	11691	845	4.0	**10123**	544	3.0
	CGV	9340	578	12.2	**9067**	643	9.2	10858	558	64.6	**9942**	505	57.5
8	C	13020	1709	0.0	**8285**	428	0.0	15291	1826	0.0	**11779**	575	0.0
	CV	6795	546	0.8	**6035**	292	0.5	9433	1163	4.2	**6796**	322	3.2
	CGV	6352	368	13.8	**5871**	293	12.8	7719	471	68.8	**6610**	284	60.3
10	C	9555	1666	0.0	**7071**	328	0.0	11275	2051	0.0	**10277**	500	0.0
	CV	5914	686	0.8	**4554**	210	0.8	7299	747	4.3	**5172**	219	3.3
	CGV	4975	274	14.7	**4421**	200	13.5	5715	408	72.7	**5040**	202	70.3
15	C	5793	1037	0.0	**5565**	401	0.0	**6945**	1113	0.1	7996	533	0.0
	CV	3941	432	1.1	**2939**	142	0.8	4726	562	4.7	**3402**	158	3.6
	CGV	3102	238	15.9	**2811**	117	16.0	3459	205	79.8	**3291**	121	86.4
20	C	**4235**	861	0.0	4733	379	0.0	**4972**	892	0.1	6788	437	0.1
	CV	2947	378	1.1	**2215**	117	0.9	3410	415	5.0	**2603**	108	5.1
	CGV	2247	192	15.0	**2124**	87	18.9	2579	112	84.9	**2517**	83	98.7
30	C	**2783**	400	0.0	3757	359	0.0	**3382**	502	0.2	5062	475	0.2
	CV	2011	245	1.2	**1553**	87	1.0	2314	204	7.5	**1888**	67	6.4
	CGV	1501	88	19.2	**1468**	69	21.7	1825	61	111.3	**1812**	56	134.3
40	C	**2070**	318	0.0	3353	353	0.0	**2540**	358	0.5	3979	416	0.5
	CV	1496	194	1.4	**1221**	52	1.1	1894	212	7.4	**1562**	55	7.4
	CGV	1167	56	20.8	**1155**	52	25.4	1491	45	134.1	**1486**	42	189.1

6 Experimental Results

Our testing environment consists of Intel quad-core processors with 2.83 GHz and 8 Gigabytes of RAM. Three kinds of tests are performed to compare the Kruskal-based to the Prim-based heuristic [3]:

1. only the deterministic construction heuristic (in the result tables this test is abbreviated with "C")
2. the deterministic construction followed by the VND, using $df = 1.5$ ("CV")
3. the construction with the GRASP extension followed by the VND, using $\alpha = 0.25$, stopping after ten starts without gain and taking the average values of 30 runs ("CGV")

The instance sets R500 and R1000 each contain 30 complete instances with 500 and 1000 nodes and random integer edge-costs and -delays uniformly distributed in $[1, 99]$. The root node is set to node 0 in all tests. The comparison of only one constructed solution (test "C" in Table 1) indicates that our Kruskal-based heuristic produces usually significantly better solutions than the Prim-inspired

Table 2. Comparison of Prim- and Kruskal-based heuristics, applied on Euclidian instance sets with 500 and 1000 nodes (B: delay-bound, C: only construction, CV: construction and VND, $\bar{c}$: average final objective values, σ: standard deviations, $t[s]$: running times in seconds)

		E500						E1000					
		Prim-based			Kruskal-based			Prim-based			Kruskal-based		
B	Test	$\bar{c}$	σ	$t[s]$	$\bar{c}$	σ	$t[s]$	$\bar{c}$	σ	$t[s]$	$\bar{c}$	σ	$t[s]$
0.8	C	19.12	0.44	0.1	**18.03**	0.40	0.1	27.56	0.43	0.7	**25.40**	0.32	0.3
	CV	19.00	0.47	1.4	**17.53**	0.40	2.1	27.15	0.65	22.0	**24.81**	0.32	15.6
0.9	C	19.11	0.41	0.1	**18.04**	0.38	0.1	27.48	0.44	0.7	**25.36**	0.32	0.4
	CV	19.02	0.37	1.6	**17.41**	0.36	2.2	26.97	0.76	20.9	**24.65**	0.29	16.3
1.0	C	19.17	0.49	0.1	**17.83**	0.43	0.1	27.38	0.49	0.8	**25.32**	0.29	0.4
	CV	18.97	0.49	1.9	**17.26**	0.34	2.1	26.80	0.93	16.7	**24.51**	0.31	15.4
1.5	C	18.92	0.48	0.2	**17.46**	0.52	0.1	27.30	0.50	1.0	**24.78**	0.32	0.4
	CV	18.75	0.56	2.9	**16.79**	0.36	2.4	26.71	1.07	23.9	**23.85**	0.26	19.4
2.0	C	18.87	0.60	0.2	**17.37**	0.49	0.1	27.29	0.46	1.1	**24.54**	0.37	0.5
	CV	18.69	0.67	3.3	**16.51**	0.33	2.6	26.33	1.29	34.6	**23.49**	0.23	16.6
3.0	C	18.53	0.59	0.2	**17.02**	0.49	0.1	27.04	0.43	1.2	**24.17**	0.29	0.6
	CV	18.09	0.80	4.0	**16.22**	0.30	2.3	25.69	1.43	48.9	**23.14**	0.24	14.0

algorithm, especially if the delay-constraint is strict. Only in tests with high delay-bounds the Prim-based solution exceeds the Kruskal-based one, but this advantage disappears when also applying the VND. In this test and also when using the GRASP extension ("CV" and "CGV") our heuristic outperforms the Prim-based approach with clear statistical significance. In addition we can observe a higher dependence of the Prim-based heuristic on the specific edge-costs and -delays of the instances noticeable in the higher standard deviation values.

Concerning the runtime the Kruskal-based approach can compete with the Prim-based one and often even beats it, although the administration effort is higher when updating the node information in each step of stage one. We can observe that the runtime is nearly independent of the specified delay-bound B in contrast to the Prim-based heuristic, where tight bounds lead to longer runtimes due to the repeated delay-relaxation process, see Table 1. The general slight increase of the runtime when raising the bound is caused by the fact that in a preprocessing step all edges with $d(e) > B$ are discarded since no feasible solution can include these edges. So tests with lower delay-bounds have to handle less edges.

Additionally we tested our construction heuristic on two sets each consisting of 15 Euclidian instances from the OR-Library originally used for the Euclidian Steiner tree problem [12]. These instances consist of 500 respectively 1000 points randomly distributed in the unit square and the edge-costs correspond to the Euclidian distances between these points. We extended these input data by edge-delays normally distributed around the associated costs and chose a point near the center as root node. The results shown in Table 2 clearly demonstrate the superiority of the Kruskal-based heuristic even if using high delay-bounds. At no time even the VND-improved Prim-based solution reaches the quality of our just constructed spanning tree.

7 Conclusions and Future Work

We introduced a Kruskal-based construction heuristic for the rooted delay-constrained minimum spanning tree problem which produces faster and better results especially for tight delay-bounds and Euclidian edge-costs compared to the Prim-based approach. The runtime is almost independent of the delay-constraint and the cost- and delay-values of the instances. Furthermore the Kruskal-based heuristic seems to be a better starting point for improvement with the presented VND and GRASP.

In the future we want to extend the VND with more neighborhoods maybe based on new solution representations to better diversify the search and therefore find new feasible solutions. Furthermore, we try to apply a modified version of our de-centralized construction heuristic on the rooted delay-constrained minimum Steiner tree problem and compare it to existing approaches.

References

1. Dahl, G., Gouveia, L., Requejo, C.: On formulations and methods for the hop-constrained minimum spanning tree problem. In: Handbook of Optimization in Telecommunications, pp. 493–515. Springer, Heidelberg (2006)
2. Gouveia, L., Paias, A., Sharma, D.: Modeling and Solving the Rooted Distance-Constrained Minimum Spanning Tree Problem. Computers and Operations Research 35(2), 600–613 (2008)
3. Salama, H.F., Reeves, D.S., Viniotis, Y.: An Efficient Delay-Constrained Minimum Spanning Tree Heuristic. In: Proceedings of the 5th International Conference on Computer Communications and Networks (1996)
4. Prim, R.C.: Shortest connection networks and some generalizations. Bell System Technical Journal 36, 1389–1401 (1957)
5. Skorin-Kapov, N., Kos, M.: The application of Steiner trees to delay constrained multicast routing: a tabu search approach. In: Proceedings of the 7th International Conference on Telecommunications, vol. 2, pp. 443–448 (2003)
6. Skorin-Kapov, N., Kos, M.: A GRASP heuristic for the delay-constrained multicast routing problem. Telecommunication Systems 32(1), 55–69 (2006)
7. Ghaboosi, N., Haghighat, A.T.: A Path Relinking Approach for Delay-Constrained Least-Cost Multicast Routing Problem. In: 19th IEEE International Conference on Tools with Artificial Intelligence, pp. 383–390 (2007)
8. Kruskal, J.B.: On the shortest spanning subtree of a graph and the traveling salesman problem. Proceedings of the American Mathematics Society 7(1), 48–50 (1956)
9. Dijkstra, E.W.: A note on two problems in connexion with graphs. Numerische Mathematik 1(1), 269–271 (1959)
10. Feo, T., Resende, M.: Greedy Randomized Adaptive Search Procedures. Journal of Global Optimization 6(2), 109–133 (1995)
11. Hansen, P., Mladenović, N.: Variable neighborhood search: Principles and applications. European Journal of Operational Research 130(3), 449–467 (2001)
12. Beasley, J.E.: A heuristic for Euclidean and rectilinear Steiner problems. European Journal of Operational Research 58, 284–292 (1992)

Applying Ant Colony Optimisation to Dynamic Pickup and Delivery

Martin Ankerl and Alexander Hämmerle

Profactor GmbH, Im Stadtgut A2, 4407 Steyr-Gleink, Austria

Abstract. We present an optimisation algorithm called "King of The Hill" ACO (KoTH-ACO) based on the MAX-MIN Ant System for a TSP problem extended for the dynamic pickup and delivery problem. The KoTH algorithm shows faster convergence and better solution qualities than the MAX-MIN Ant System in our benchmark instances. In addition, the runtime performance of ACO systems could be improved with approximate probability calculation.

Keywords: ant colony optimisation, TSP, pickup delivery, MAX-MIN AS.

1 Motivation

Solving dynamic pickup and delivery (P&D) problems is daily business for dispatchers in forwarding companies. The challenge is to find optimal assignments of shipments to freight vehicles and to calculate optimal routes for the vehicles. "Optimality" refers to criteria like transport costs, travel time, or distance. In general a shipment is characterised through pickup and delivery locations, quantity, and maximum tour duration.

The dynamics of the problem arises from new shipments dispatched to the optimisation algorithm. This may happen during runtime of the algorithm (when calculating transport plans) or during the plan execution phase, whilst the vehicles are actually on the road - the latter requiring re-routing of vehicles.

In a research project with an Austrian logistics company we defined dynamic P&D as the problem to be solved, being subject to industrial requirements like good runtime performance of the algorithm with a response time in few seconds. The Ant Colony Optimisation (ACO) Metaheuristic, with self-adaptivity and scalability as core features, was the prime candidate for solving a realistic, dynamic P&D problem. In the following we describe the core features of our ACO implementation, show its applicability to dynamic P&D, and how the performance requirements have been achieved.

2 Ant Colony Optimisation for Pickup and Delivery

The ACO Metaheuristic [1] iteratively constructs tours until the termination criteria is met (e.g. the desired fitness is achieved). Good tours are allowed to put down pheromones along their path, and this information is used probabilistically in future tour constructions. Listing 1 shows the pseudo code.

R. Moreno-Díaz et al. (Eds.): EUROCAST 2009, LNCS 5717, pp. 721–728, 2009.

```
function ACO_Metaheuristic {
    while (termination_criteria_not_met()) {
        for (number_of_tours_per_iteration) {
            construct_tour()
        }
        update_pheromones()
    }
}
```

Listing 1. ACO pseudo code

2.1 TSP Tour Construction

Tour construction itself is an iterative process, where successive travel destinations are selected based on a predefined problem specific heuristic function and the already learnt pheromone information.

```
function construct_tsp_tour() {
    visitable_locations = all_locations()
    while (visitable_location not empty) {
        location = choose_next(visitable_locations)
        path.add(location)
        visitable_locations.remove(location)
    }
    return path
}
```

Listing 2. TSP tour construction pseudo code

Whenever a location has to be chosen in `choose_next`, for each of the n visitable locations a probability is calculated which is used in a roulette wheel selection. The probability calculation is based on a heuristic function $\eta_{i,j}$ and pheromone levels $\tau_{i,j}$, both values are weighted with constants α and β and then multiplied together, as shown in equation 1. In short, an ant will choose node j to move from node i with a probability $p_{i,j}$.

$$p_{i,j} = \frac{\left(\tau_{i,j}^{\alpha}\right)\left(\eta_{i,j}^{\beta}\right)}{\sum_{j=1}^{n}\left(\tau_{i,j}^{\alpha}\right)\left(\eta_{i,j}^{\beta}\right)} \qquad (1)$$

After a tour has been constructed this way, usually a local optimiser like n-Opt is applied to further improve it.

2.2 Pickup and Delivery Tour Construction

The tour construction mechanism in Listing 2 has to be modified for the pickup and delivery problem, because it adds a precedence constraint to the locations: A pickup has to be a part of the tour before the delivery. Since the construction is iteratively and in successive order, this precedence constraint can be easily implemented by allowing a visit of a corresponding delivery only when a pickup has already occurred.

In addition, the capacitated pickup & delivery problem adds another constraint that allows a pickup only when the transport resources capacity is not exceeded. The modified algorithm that accommodates both constraints can be seen in Listing 3, the added parts are highlighted.

```
function construct_p&d_tour() {
    visitables = all_pickup_locations()
    while (visitables not empty) {
        repeat {
            location = choose_next(visitable_locations)
        } until (transportresource has enough space)
        if (location is pickup) {
            visitables.add(delivery for location)
        }
        path.add(location)
        visitables.remove(location)
    }
    return path
}
```

Listing 3. Pickup & Delivery tour construction

The modifications to the TSP tour construction algorithm are simple, and it removes all invalid tours from the search space while the feasible solution space does not get restricted. This behaviour has parallels to constraint implementations in a Constraint Solver where the propagation is strong enough to fully resolve the problem.

3 The KoTH-ACO Algorithm

The "King of The Hill" (KoTH) ACO is a modified MAX-MIN Ant System (MMAS) [7]. In MMAS, T. Stützle et al. have extended the Ant System in three key aspects based on an analysis of the search space of Traveling Salesman Problem (TSP) and Quadratic Assignment Problem (QAP). These changes are:

1. After each iteration only one ant is allowed to add pheromones, either the iteration-best or the global-best.
2. Pheromones are limited to an interval $[\tau_{min}, \tau_{max}]$ which prevents stagnation due to overfitting.
3. Pheromones are initialized to τ_{max} to start the tour creation process with a more global search, and iteratively evaporate to be able to gradually move to a more local search.

This MMAS algorithm was shown to produce good solutions for the TSP and QAP problem. We have therefore used MMAS as the baseline in our research and applied it to multiple test instances from our industry partner. Based on our observations we have developed several modifications:

1. Elitist population: After each iteration the iteration-best ant is added to a population of fixed size (e.g. 20) containing the *n* global best ants. Every ant of this elitist population is allowed to add pheromones, in relation to their own fitness and the number of ants in the population.

2. Adaptive α and β: Every ant of the elitist population has configuration parameters α and β attached. When an ant constructs a tour, the average α and β values of the population is calculated, and settings that randomly slightly deviate from these settings are used for a new ant.

3.1 Elitist Population

In MMAS only a single ant is allowed to update the pheromones after each iteration. The trail update rule is given by Equation 2

$$\boldsymbol{\tau_{i,j}(t+1) = \rho \cdot \tau_{i,j}(t) + \Delta\tau_{i,j}{}^{best}} \tag{2}$$

where $\Delta\boldsymbol{\tau}_{\mathbf{i,j}}{}^{\mathbf{best}} = \mathbf{1/f(s^{best})}$ and $\mathbf{f(s^{best})}$ is the tour cost of either the iteration-best or the global-best solution. In KoTH we use an elitist population that is updated after each iteration according to the algorithm in Listing 4.

```
function update_population() {
    if (!population.contains(iteration_best) {
        if (population.size > n) {
            population.remove(population.worst)
        }
        population.add(iteration_best)
    }
}
```

Listing 4. Population update algorithm in KoTH

The population therefore contains the n-1 unique best solutions and the iteration best. For performance reason equality check is simply performed by comparing the fitness of two solutions. After each iteration the whole population is used to update pheromone trails using the update rule from Equation 3.

$$\boldsymbol{\tau_{i,j}(t+1) = \rho \cdot \tau_{i,j}(t) + \frac{1}{n}\sum_{k=1}^{n} \Delta\tau_{i,j}{}^{best_k}} \tag{3}$$

The rationale behind using a population is to enforce the recombination of different good solutions. A similar approach to using an elitist population in ACO is realized in the Omicron ACO [10]. In our experiments a population size of 20 was experimentally determined to produce good results over a wide range of test instances.

3.2 Adaptive α and β

Before a new tour is generated a new α and β value is calculated and used for this tour creation. These configuration parameters are attached to the solution when it is added to the population. The adaptive α and β parameters are calculated using the populations average α and β as described in Equation 2. r is a Gaussian distributed double value with mean 1.0 and standard deviation 0.1. r is regenerated until it is greater than zero.

$$\alpha = \frac{1}{n}\sum_{i=1}^{n} \boldsymbol{\alpha_i \cdot r} \tag{4}$$

Whenever an ant uses the new parameters to produce a solution that is good enough to replace another tour from the elitist population, it will become part of the population and therefore influence the average α and β values in successive iterations. This behaviour gradually moves the parameters into a range that increases the probability for creating better solutions. The adaption allows for interesting observations, as can be seen in a typical optimisation run in Figure 1.

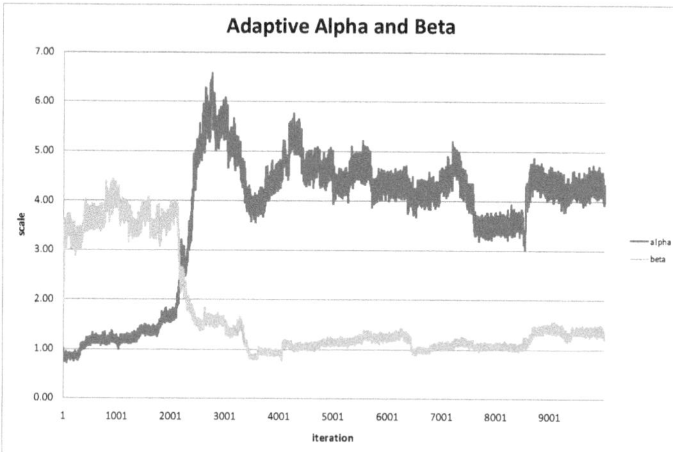

Fig. 1. Average α and β values during one optimisation run

Here the optimisation is initialized with α=1, and β=2. Shortly after optimisation starts, the β parameter increases to a value between 3.0 and 4.0, which means that the heuristic is the main driving force in finding improved solutions.

After several iterations α starts to increase, until at a tipping point of around 2000 iterations the heuristic is not able to provide any more helpful information for tour construction, and the Metaheuristic switches to recombining the pheromone information to construct improved solutions from it.

The behaviour of the parameters can vary greatly between different scenarios. For example, in some of our benchmarks the β value immediately falls down to a low value while the pheromones gradually increase. This is an indication that the heuristic function is not well suited for this particular problem. In other scenarios the α parameter always stays below β which means that the heuristic is very good and pheromones are not necessary.

3.3 Dynamics

One important aspect for our customer is that when transportation problem changes frequently by e.g. adding or removing shipments. Reasonably good solutions should be found quickly despite these changes. Our approach to support this dynamic updates is this: 1. clear the whole population, 2. remove or add the shipments to the problem in a way that indices for tour locations are preserved, 3. initialize pheromones to and from the new shipment locations with τ_{max}. This way we can use the previously learnt information of the pheromones in the updated problem. Figure 2 shows an excerpt of a typical update during an optimisation run.

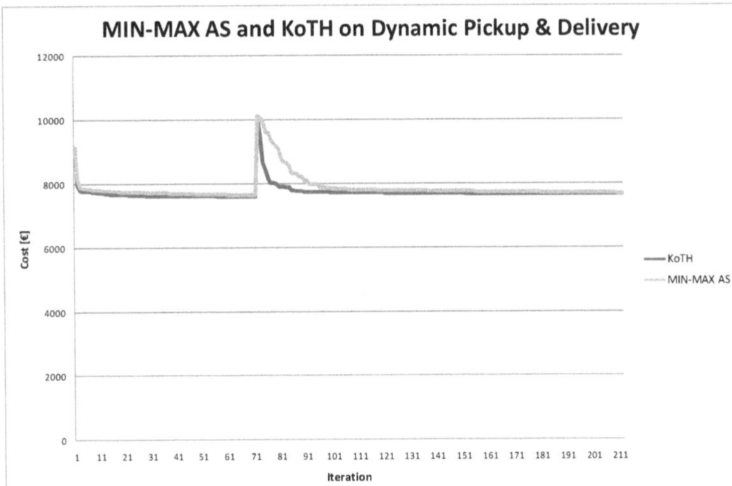

Fig. 2. Change of fitness when an additional shipment is added to the problem. The graphs show the transportation costs averaged over 20 simulations of the same scenario.

Here, at some time during the optimisation, the problem is updated by adding another shipment. The population is cleared, and the pheromones are updated so that the links to the new pickup and delivery locations are update with τ_{max}. The KoTH-ACO quickly reaches a fitness close to the previous solution while the MIN-MAX AS takes much longer for this process.

4 Performance Optimisation

M. Dorigo et. al. have shown that local optimisation is important to produce good results. Unfortunately, the widely used local search algorithms from TSP problems like 2-Opt can not be directly applied to the P&D problem since the precedence constraint and capacity constraint can drastically limit the possible exchange operations. Although a good local search can improve solution quality, we have decided not to develop a customized P&D local search algorithm because it is too tightly intervened with the problem at hand. Since we extend the algorithm with time windows and other constraints, every time constraints are added to the problem the local search would have to be rewritten. Therefore we have decided to use a brute force approach and instead improve the base performance of the core algorithm.

When profiling an ACO algorithm it can clearly be seen that the most performance relevant part is the probability calculation of Equation 1, the slow part is the *pow* calculation. The slowness stems from the accuracy required from the IEEE 754 floating point standard. Since this accuracy is by far not needed in the probabilistic selection scheme, there is the possibility to use an approximation. One solution would be to have α=1 and β=2 constant so that probability can be calculated with $\tau_{i,j} \cdot \eta_{i,j} \cdot \eta_{i,j}$ but that would lose the adaptive behavior. Therefore we have developed an approximation for *pow* based on [9] shown in Listing 5 that can be more than 10 times faster the original *pow*.

```
public static double pow(double a, double b) {
    int x = (int) (Double.doubleToLongBits(a) >> 32);
    int y = (int) (b * (x - 1072632447) + 1072632447);
    return Double.longBitsToDouble(((long) y) << 32);
}
```

Listing 5. Fast *pow* approximation for Java. Ports were also benchmarked in C, C++ and C#

The approximation takes advantage of the IEEE 754 floating point representation of data to effectively use a lookup table with 2048 entries. Although the approximation can be highly inaccurate especially for large values of b, the overall optimisation performance of the KoTH algorithm is not noticably impared in all our benchmarks. Using this approximation translates into a 3 to 4 times more solutions per second, which further dimishes the need for a local optimiser. The performance can be further improved when extended to full $a^b \cdot c^d$ calculation and the fact that the exponents do not change within one tour creation.

5 Conclusion

We have extended the classical MAX-MIN AS with a population based approach similar to the Omicron ACO [10]. In contrast to MAX-MIN AS, we allow multiple solutions to contribute to the pheromone reinforcement which leads to a better combination of different solutions and improves performance and solution quality.

The use of a population introduces the possibility to use a self-adaptive configuration for the ACO. Specifically, we found that an adaptive configuration of the weighting between pheromones and heuristic leads to better performance than fixed values for the weightings. This self-adaptive configuration can also be used as an indicator for the quality of the heuristic function: when the heuristic is well suited for the problem at hand, it will finally get a high weighting.

We have conducted experiments on the basis of P&D problems provided by an Austrian logistics company. We found that for static problems with 50 shipments and 3 vehicles, and 500 shipments and 10 vehicles, respectively, our ACO implementation is fast enough to meet the performance requirements. The flexibility of our implementation with respect to dynamics was tested with 55 shipments and 2 vehicles, where batches from 1-8 shipments were periodically added to the problem. We found that the solution quality recovered more quickly than MMAS from the disturbances caused by the batches, demonstrating the ability of our ACO implementation to cope with dynamic P&D problems.

We are currently working on adding many more constraints to the P&D problem like time windows, shifts, prohibition of mixed loadings, multiple capacity types, multiple inhomogeneous transport resources, etc. Initial experiments with different constraint handling mechanisms indicate that stochastic ranking combined with a good heuristic function leads to fast results and high quality tours.

References

1. Dorigo, M., Di Caro, G.: The ant colony meta-heuristic. In: Corne, D., Dorigo, M., Glover, F. (eds.) New Ideas in Optimization, pp. 11–32. McGraw Hill, London (1999)
2. Gambardella, L.M., Taillard, E., Agazzi, G.: MACS-VRPTW: A Multiple Ant Colony System for Vehicle Routing Problems with Time Windows. In: Corne, D., Dorigo, M., Glover, F. (eds.) New Ideas in Optimization, pp. 63–76. McGraw Hill, London (1999)
3. Held, M.: Analysis and Improvement of Constraint Handling in Ant Colony Algorithms. Bachelor Thesis, Clayton School of Information Technology, Monash University (2005)
4. Meyer, B., Ernst, A.: Integrating ACO and constraint propagation. In: Dorigo, M., Birattari, M., Blum, C., Gambardella, L.M., Mondada, F., Stützle, T. (eds.) ANTS 2004. LNCS, vol. 3172, pp. 166–177. Springer, Heidelberg (2004)
5. Montemanni, R., Gambardella, L.M., Rizzoli, A.E., Donati, A.V.: A new algorithm for a Dynamic Vehicle Routing Problem based on Ant Colony System. In: Second International Workshop on Freight Transportation and Logistics (2003)
6. Runarsson, T.P., Yao, X.: Stochastic ranking for constrained evolutionary optimization. IEEE Transactions on Evolutionary Computation 4(3), 284–294 (2000)
7. Stützle, T., Hoos, H.H.: MAX-MIN Ant System. Future Generation Computer Systems 16(9), 889–914 (2000)
8. Rizzoli, A.E., Montemanni, R., Lucibello, E., Gambardella, L.M.: Ant colony optimization for real-world vehicle routing problems. Istituto Dalle Molle di Studi sull'Intelligenza Artificiale (IDSIA), Galleria 2, 6928 Manno, Switzerland, Ant Optima, via Fusoni 4, 6900 Lugano, Switzerland (May 2007)
9. Schraudolph, N.N.: A Fast, Compact Approximation of the Exponential Function, number IDSIA-07-98 (1998)
10. Barán, B., Gómez, O.: Omicron ACO. A New Ant Colony Optimization Algorithm. CLEI Electron. J. 8(1) (2005)

Model Driven Rapid Prototyping of Heuristic Optimization Algorithms*

Stefan Wagner, Gabriel Kronberger, Andreas Beham, Stephan Winkler, and Michael Affenzeller

Heuristic and Evolutionary Algorithms Laboratory
School of Informatics, Communications and Media - Hagenberg
Upper Austria University of Applied Sciences
Softwarepark 11, A-4232 Hagenberg, Austria
{swagner,gkronber,abeham,swinkler,maffenze}@heuristiclab.com

Abstract. In this paper the authors describe a model driven approach for the development of heuristic optimization algorithms. Based on a generic algorithm model, several operators are presented which can be used as algorithm building blocks. In combination with a graphical user interface, this approach provides an interactive and declarative way of engineering complex optimization heuristics. By this means, it also enables users with little programming experience to develop, tune, test, and analyze heuristic optimization techniques.

1 Introduction

Since the beginning of the 1990s heuristic optimization is a very active field of research. Many different algorithms have been developed that were applied to optimization problems of numerous domains. Thereby, especially the development and application of metaheuristic algorithms is a favored approach, as such algorithms can be easily reused for many different problems [2].

However, according to the No Free Lunch theorem for heuristic search (see for example [5]), no single heuristic optimization algorithm is able to outperform all other algorithms for all possible problems. This simple fact led to a magnitude of different heuristic optimization paradigms, which often adapt strategies found in nature to solve technical optimization problems (e.g., evolutionary algorithms, simulated annealing, ant colony optimization, particle swarm optimization). All these algorithms have specific characteristics and are suitable for different problem scenarios and solution space topologies.

As a consequence, selecting a suitable algorithm for a given optimization problem is a cumbersome task. Extensive empirical testing is required to compare different optimization paradigms and to select appropriate parameter values. Furthermore, in many cases of hard real-world optimization problems it is not

* The work described in this paper was done within HEUREKA!, the Josef Ressel centre for heuristic optimization sponsored by the Austrian Research Promotion Agency (FFG). Visit http://heureka.heuristiclab.com for more details.

R. Moreno-Díaz et al. (Eds.): EUROCAST 2009, LNCS 5717, pp. 729–736, 2009.

enough to tune parameters in order to achieve satisfactory results. Approaches of different paradigms have to be combined, leading to hybrid algorithms or even new classes of metaheuristic optimization techniques.

For this reason, heuristic optimization software systems have to support rapid prototyping and testing of algorithms. By providing a set of ready-to-use components and by supporting data analysis and visualization, the process of assembling heuristic optimization algorithms can be simplified. However, for combining these components profound knowledge of the underlying framework and good programming skills are required in most cases. Therefore, algorithm development is usually restricted to software engineers who have to bridge the gap between heuristic optimization experts and experts in the problem domains.

In order to overcome this drawback, the authors propose a model driven approach to heuristic algorithm engineering. Based on a generic algorithm model developed by the authors in a previous work [4], several operators for representing heuristic optimization algorithms are discussed in this paper. In contrast to similar ideas in the heuristic optimization community (for example EASEA [1]), modeling is thereby abstracted from a concrete programming language as far as possible. Additionally, by providing a graphical user interface, development and testing of heuristic optimization algorithms becomes possible even for users with little experience or interest in programming and software development.

2 Generic Algorithm Model

In [4] the authors introduced a generic model for representing arbitrary algorithms. In this model algorithms are described as operator graphs, whereby each operator represents a single operation. The data manipulated by an algorithm is structured in hierarchical data structures called scopes. When an algorithm is executed by an engine, its operators are applied on the scopes in order to access and manipulate variables or sub-scopes. For additional details about the model and for a detailed description of its benefits with respect to the representation of parallel algorithms, the reader is referred to [4].

3 Modeling Heuristic Optimization Algorithms

Based on the generic algorithm model, several operators are presented in the following which can be used as building blocks for defining heuristic optimization algorithms.

3.1 Basic Operators

VariableInjector. Each algorithm execution starts with applying an initial operator on an empty global scope. Therefore, an operator is required for adding user-defined variables to a scope. This task is fulfilled by the operator *VariableInjector* which can be used for example to add global variables such as the population or neighborhood size, the mutation rate, the tabu tenure, the maximum number of iterations, or the random number generator. Each succeeding operator will then be able to access these values.

Manipulation Operators. Manipulation operators are basic operators that change the variables of a scope. For example, a *Counter* operator can be used for incrementing variables. Furthermore, custom manipulations operators can be defined which represent problem-specific or encoding-specific manipulation operations.

Comparator. The *Comparator* operator is responsible for comparing the values of two variables. It expects two input variables which should be compared and a comparison operation specifying the type of comparison (e.g., less, equal, greater or equal). After retrieving both variable values from the scope, it creates a new Boolean variable containing the result of the comparison and adds it to the scope.

SubScopesCreater. The operator *SubScopesCreator* is used to extend the scope tree. It expects an input variable specifying how many new sub-scopes should be appended to scope it is applied on.

3.2 Control Operators

Operators have to define which operations an engine has to execute next (i.e., which operators are applied on which scopes). Therefore, the execution flow of an algorithm can be defined using control operators that do not manipulate variables or scopes but return successor operations. In the following, basic control operators are presented to model sequences and branches. Any other control structure known from classical programming languages (for example switch statements or loops) can be realized by combining these operators.

SequentialProcessor. *SequentialProcessor* represents a sequence of operations. It tells the engine to apply all its sub-operators on the current scope.

UniformSequentialSubScopesProcessor. This operator can be used to navigate through the hierarchy levels of the scope tree. It tells the engine to apply its sub-operator on each sub-scope of the current scope. As each solution is represented as a scope, this operator can for example be used for evaluating or manipulating a set of solutions.

SequentialSubScopesProcessor. *SequentialSubScopesProcessor* can also be used to apply operators to sub-scopes. However, in contrast to the UniformSequentialSubScopesProcessor, it provides a sub-operator for each sub-scope which enables individual processing of all sub-scopes.

ConditionalBranch. The operator *ConditionalBranch* can be used to model simple binary branches. It retrieves a Boolean input variable from the scope tree.

Depending on the value of this variable, it tells the engine to apply either its first sub-operator (true branch) or its second sub-operator (false branch) on the current scope.

3.3 Selection and Reduction

Seen from an abstract point of view, a large group of heuristic optimization algorithms (improvement heuristics) follows a common strategy: In an initialization step one or more solutions are generated either randomly or using construction heuristics. Then these solutions are iteratively manipulated in order to navigate through the solution space and to reach promising regions. In this process manipulated solutions are usually compared with existing ones to control the movement in the solution space depending on solution qualities. Selection splits solutions into different groups either by copying or moving them from one group to another; replacement merges solutions into a single group again and overwrites the ones that should not be considered anymore.

As each solution is represented as a scope and scopes are hierarchically organized, selection and replacement operations can be realized in a straight forward way: On the one hand, selection operators split sub-scopes of a scope into two groups by introducing a new hierarchical layer of two sub-scopes in between, one representing the group of remaining solutions and one holding the selected ones. Thereby solutions are either copied or moved depending on the type of the selection operator. On the other hand, reduction operators represent the reverse operation. A reduction operator removes the two sub-scopes again and reunites the contained sub-scopes. Depending on the type of the reduction operator, this reunification step may also include elimination of some sub-scopes that are no longer required. The general principle of selection and reduction operators is schematically shown in Figure 1.

According to this simple principle of selection and reduction of solutions, a set of selection and reduction operators can be defined which can be used as a basis for realizing complex selection and replacement schemes. These operators are described in the following.

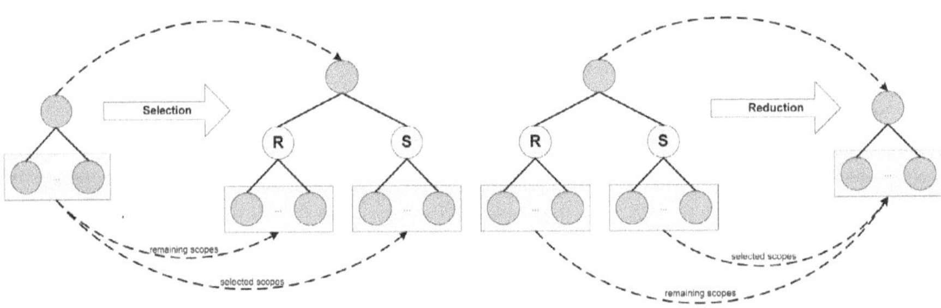

Fig. 1. General principle of selection and reduction operators

Selection Operators. The most trivial form of selection operators are the two operators *LeftSelector* and *RightSelector* which select sub-scopes either starting from the leftmost or the rightmost sub-scope. If the sub-scopes are ordered for example with respect to solution quality, these operators can be used to select the best or the worst solutions of a group. If random selection of sub-scopes is required, a *RandomSelector* can be used which additionally expects a random number generator as an input variable.

In order to realize more sophisticated ways of selection, *ConditionalSelector* can be used which selects sub-scopes depending on the value of a Boolean variable contained in each sub-scope. This operator can be combined with a selection pre-processing step to inject this Boolean variable into each scope depending on some other conditions.

Furthermore, several classical quality-based selection schemes well-known from the area of evolutionary algorithms can be realized as well, as for example fitness proportional selection (*ProportionalSelector*), linear rank selection (*LinearRankSelector*), or tournament selection with variable tournament group sizes (*TournamentSelector*). Additionally, other individual selection schemes can be integrated easily by implementing custom selection operators.

Reduction Operators. Corresponding reverse operations to LeftSelector and RightSelector are provided by the two reduction operators *LeftReducer* and *RightReducer*. Both operators do not reunite sub-scopes but discard either the scope group containing the selected or the group containing the remaining scopes. LeftReducer performs a reduction to the left and picks the scopes contained in the left sub-scope (remaining scopes) and RightReducer does the same with the right sub-scopes (selected scopes). Additionally, another reduction operator called *MergingReducer* reunites both scope groups by merging all sub-scopes.

Sorting Operators. Many selection operators consider solution quality as the main property affecting selection. However, as there are many different ways how solution qualities can be represented, selection operators should be abstracted from quality values as much as possible. For the operators which just expect an ordering of sub-scopes and do not need to consider exact quality values (for example best selection, worst selection, linear rank selection, tournament selection), operators are required for reordering sub-scopes regarding some property. *SubScopesSorter* is a representative of this class of operators and reorders sub-scopes depending on the value of a double variable contained in each scope which usually represents the quality value. Additionally, in other cases, as for example multi-objective optimization problems, custom sorting operators realizing other ways of ordering can be implemented.

3.4 Modularity

All operators introduced so far represent simple functionality required as a basis for building complex heuristic optimization algorithms. However, working directly with these simple operators can become quite cumbersome for users.

Even for simple algorithms, such as a hill climber or a canonical genetic algorithm, operator graphs are quite large and complex as the level of abstraction of these operators is rather low. As a consequence, developing algorithms is a complex and error-prone task.

To overcome these difficulties, a concept of modularization is needed. Users have to be able to define new operators fulfilling more complex tasks by combining already existing operators, either simple or of course also combined ones. For example, it is reasonable to define an operator for picking the best n solutions out of a set of solutions or one for processing all solutions of a set with an evaluation operator. These operations are very common in different kinds of heuristic optimization algorithms, so it is not suitable to define them again and again when creating a new algorithm. Instead, using combined operators enables reuse of complex operators in different algorithms.

The two operators *CombinedOperator* and *OperatorExtractor* are responsible for modularization of operator graphs and are described in detail in the following. An important aspect is that combined operators can be created by users. Therefore, it is possible for users to develop own or share existing operator libraries containing various combined operators for specific tasks. By this means, the level of abstraction is not determined by the algorithm model, but depends only on its users.

CombinedOperator. A *CombinedOperator* contains a whole operator graph. When executed it tells the engine to apply the initial operator of its graph on the current scope. Therefore, the whole operator graph is executed by the engine subsequently.

In order to have a clearly defined interface and to enable parameterization of combined operators, the user has to declare which variables are read, manipulated or produced by the operators contained in the graph. However, parameterization is not restricted to data elements. As operators are also considered as data, operators can be injected into the scope tree that are used somewhere in the operator graph which enables a functional style of programming. For example, a combined operator might be needed that encapsulates manipulation and evaluation of all solutions in a solution set. Such an operator can be defined by using an UniformSequentialSubScopesProcessor to iterate over all solutions (sub-scopes) in a set (scope) applying a SequentialProcessor on each solution. The sequential processor contains three sub-operators, one for manipulating solutions, one for evaluating them, and one for counting the number of evaluated solutions. However, the two operators for manipulating and evaluating a solution do not have to be defined in the operator graph directly but can be retrieved from the scope tree. Consequently, the combined operator can be parameterized with the required manipulation and evaluation operator, but the contained operator graph does not have to be changed in any way. In other words, combined operators enable the definition of reusable algorithm building blocks that are independent of specific optimization problems.

OperatorExtractor. Another operator called *OperatorExtractor* can be used to retrieve operators from the scope tree. OperatorExtractor is a placeholder that can be added anywhere in an operator graph of a combined operator and expects an input variable containing an operator. The operator looks for that variable in the scope tree recursively and tells the engine to apply the contained operator on the current scope.

4 Interactive Algorithm Engineering

All operators discussed in the previous section have been implemented by the authors in the HeuristicLab[1] optimization environment [3]. In combination with the graphical user interface of HeuristicLab (see Figure 2), this enables a declarative and interactive definition of complex heuristic optimization algorithms. By this means, also users with little or no programming experience can easily develop, tune, test, and analyze sequential and parallel metaheuristic optimization techniques.

Therefore, the generic algorithm model and the operators discussed in this paper provide a basis for bridging the gap between heuristic optimization experts,

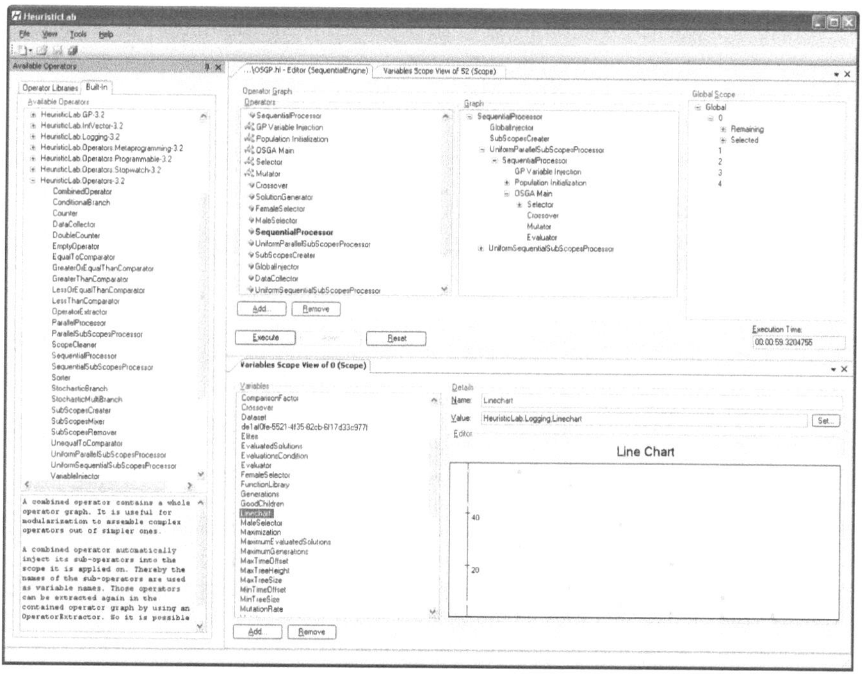

Fig. 2. Screenshot of HeuristicLab 3.0

[1] http://www.heuristiclab.com

practitioners of different application domains, and software engineers. On the one hand, experts can use basic combined operators representing fundamental concepts required in a broad spectrum of heuristic optimization algorithms or can just use the provided simple operators for developing complex algorithms; they are able to work on a low level of abstraction to benefit from genericity and flexibility to a large extend. On the other hand, practitioners can use complex combined operators representing whole heuristic optimization algorithms as black box solvers to attack problems of their interest. However, if necessary each user can take a glance at the definition of a combined operator in order to explore the internal functionality of an algorithm.

5 Conclusion

Based on the generic algorithm model described in [4], several operators have been discussed in this paper that can be used to define heuristic optimization algorithms. Due to the page restriction of this contribution, problem specific concepts such as different solution encodings or evaluation and manipulation operators have not been covered in detail. However, these aspects will be described in subsequent publications and it will be shown that they can also be realized easily as additional data elements and operators. Furthermore, concrete case studies will also be presented in the future to demonstrate how different metaheuristic algorithms such as genetic algorithms or simulated annealing can be modeled.

References

1. Collet, P., Lutton, E., Schoenauer, M., Louchet, J.: Take it EASEA. In: Deb, K., Rudolph, G., Lutton, E., Merelo, J.J., Schoenauer, M., Schwefel, H.-P., Yao, X. (eds.) PPSN 2000. LNCS, vol. 1917, pp. 891–901. Springer, Heidelberg (2000)
2. Doerner, K.F., Gendreau, M., Greistorfer, P., Gutjahr, W., Hartl, R.F., Reimann, M. (eds.): Metaheuristics: Progress in Complex Systems Optimization. Operations Research/Computer Science Interfaces Series. Springer, Heidelberg (2007)
3. Wagner, S.: Heuristic Optimization Software Systems - Modeling of Heuristic Optimization Algorithms in the HeuristicLab Software Environment. PhD thesis, Johannes Kepler University, Linz, Austria (2009)
4. Wagner, S., Kronberger, G., Beham, A., Winkler, S., Affenzeller, M.: Modeling of heuristic optimization algorithms. In: Bruzzone, A., Longo, F., Piera, M.A., Aguilar, R.M., Frydman, C. (eds.) Proceedings of the 20th European Modeling and Simulation Symposium, pp. 106–111. DIPTEM University of Genova (2008)
5. Wolpert, D.H., Macready, W.G.: No free lunch theorems for optimization. IEEE Transactions on Evolutionary Computation 1(1), 67–82 (1997)

Heuristic Methods for Searching and Clustering Hierarchical Workflows

Michael Kastner, Mohamed Wagdy Saleh, Stefan Wagner, Michael Affenzeller, and Witold Jacak

Heuristic and Evolutionary Algorithms Laboratory
School of Informatics, Communications and Media - Hagenberg
Upper Austria University of Applied Sciences
Softwarepark 11, A-4232 Hagenberg, Austria
mkastner@fh-hagenberg.at, mohamed-wagdy.saleh@isi-hagenberg.at,
{swagner,maffenze,jacak}@fh-hagenberg.at

Abstract. Workflows are used nowadays in different areas of application. Emergency services are one of these areas where explicitly defined workflows help to increase traceability, control, efficiency, and quality of rescue missions. In this paper, we introduce a generic workflow model for describing fire fighting operations in different scenarios. Based on this model we also describe heuristics for calculating the similarity of workflows which can be used for searching and clustering.

1 Introduction

In the last years, process models were frequently used to describe workflows of business processes in the economic field [10,7]. Following the idea of process-centric management, the explicit definition of workflows helps to increase traceability, control, efficiency, and quality. However, describing tasks by using formal process models is not restricted to the area of business process modeling. Especially in the area of emergency services, explicit process definitions are also of major importance to reduce the risk of human errors and to improve effectiveness [9].

In this paper, we focus on the scenario of a decentralized fire fighting organization. In this organization, hierarchical workflows (i.e. workflows consisting of actions, transitions, and sub-workflows) are used to describe actions to be executed in different kinds of emergency situations. As emergency situations are hardly ever exactly the same, most workflows have to take specific characteristics of a concrete emergency scenario into account. Therefore, individual workflows are defined by many users and a priori generalization is hard to achieve.

In order to keep a large number of workflows manageable, two strategies may be applied: First, performing a fuzzy search on all existing workflows is helpful for defining a new workflow. By this means, the user is enabled to check, if similar workflows have been created already, and to compare and adapt the new workflow iteratively. Second, clustering algorithms can be used to group all workflows in order to identify which workflows describe similar tasks, which

R. Moreno-Díaz et al. (Eds.): EUROCAST 2009, LNCS 5717, pp. 737–744, 2009.

workflows can act as representatives for a whole cluster, and to perform an a posteriori simplification or standardization.

As a foundation for both approaches (searching and clustering), we introduce a generalized workflow model. Based on the work of Jung and Bae [8], we describe several heuristics for calculating the similarity of workflows on the semantic (actions) as well as on the structural (transitions) level. Furthermore, we develop a new method of combining different similarity measurements at a time. This approach allows us to take advantage of the individual characteristics of the proposed measurements and to apply them in combination in order to achieve satisfying clustering results. Finally, the suitability of this approach is evaluated by clustering artificially generated sets of workflows using classical clustering algorithms (k-Means, DBSCAN, and Expectation Maximization).

2 Workflow Model

In cooperation with fire fighters of different fire departments we defined a generalized workflow model. Each activity/task carried out by a fire fighter in an operation is called an action. Each action contains an ID, a name, a status, a description, and a set of keywords. The concept of keywords was introduced to facilitate the comparison of actions. Similar keywords in multiple actions indicate that these actions might deal with a similar topic. The status of an action can either be *to do*, *in progress*, or *done* and is set by a fire fighter when carrying out a workflow. Furthermore, an additional status *irrelevant* was also defined to mark actions which are not applicable in a concrete emergency situation.

In order to facilitate different combinations of actions, four action subtypes were identified: The first type is a *single action* which represents a single and atomic activity. The second type is called *checklist* and describes a collection of single actions that can be carried out in any order. For modeling multiple alternatives, the type *conditional action* was introduced which contains multiple

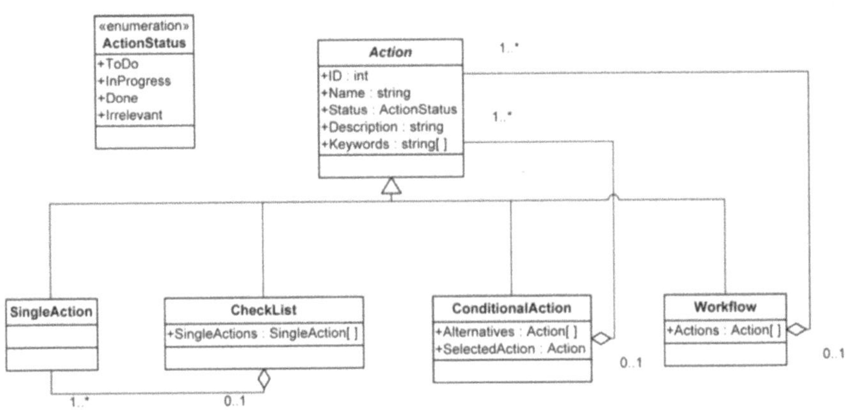

Fig. 1. UML chart of the workflow model

actions of any type from which one action has to be chosen. Finally, the last type called *workflow* contains arbitrary many actions of any type which have to be executed in a predefined linear sequence. Obviously, as a whole workflow is considered as a type of action, workflows can be structured hierarchically. An UML representation of the workflow model is shown in Figure 1.

3 Similarity Measures

Based on the workflow model described above, we introduce several heuristics to calculate the similarity of actions in this section. At first, basic similarity measures are presented that turned out to be suitable for comparing workflows in the domain of fire fighting operations. Then it is described how these similarity measures can be combined in order to apply different similarity measures at a time depending on the types of actions that are compared. This approach allows the flexible definition of complex comparison heuristics that benefit from the individual characteristics of the basic measurement values.

3.1 Activity Similarity Measure (ASM)

As presented by Jung and Bae in [8], the activity similarity measure calculates how many activities are commonly shared between two actions by using the Cosine measure. It is assumed that the degree of similarity of two actions increases, as the number of shared activities increases. The transitions between activities are not taken into account. Furthermore, the ASM only considers top level actions; actions contained on lower hierarchy levels as for example the actions of a sub-workflow are not compared.

3.2 Transition Similarity Measure (TSM)

The transition similarity measure [8] is a more strict measure than the ASM, as it considers the transitions of activities in each action (i.e. the sequence in which activities are performed). It is assumed that the degree of similarity of two actions only increases, if activities in both actions are carried out in the same order. For both actions a transition vector is calculated which indicates each pairwise sequence of activities included in an action. Then the similarity of these transition vectors is again calculated by using the Cosine measure. Similarly to the ASM, also the TSM only considers activities on the top level of each action.

3.3 Single Action Activity Similarity Measure (SAASM)

As the two similarity measures described above (ASM and TSM) do not take multiple hierarchy levels into account, we propose another similarity measure called single action activity similarity measure. It is assumed that the similarity of actions can be described by the similarity of all contained single actions regardless of the hierarchy level on which they occur. Therefore, all single actions of both actions are compared pairwise and the maximum similarity value for each single action is calculated. Then all maximum similarity values are averaged to get the similarity value for the two actions.

3.4 Single Action Transition Similarity Measure (SATSM)

The single action transition similarity measure is based on the SAASM but additionally considers the sequence in which single actions appear on all hierarchy levels of an action. It calculates the average similarity of all transitions contained in two actions, whereby the similarity of a transition is defined as the similarity of its source and destination single action.

3.5 Keyword Similarity Measure (KSM)

The keyword similarity measure is a simple way of measuring the similarity of two actions just by considering their keywords. It is defined as the relative number of identical keywords contained in both actions. For example, if three keywords out of seven are contained in both actions, the KSM gives a similarity value of 3/7.

3.6 Combination of Similarity Measures

The basic similarity measures described above can be applied to calculate the similarity of arbitrary actions contained in a workflow. However, it is reasonable to apply different similarity measures depending on the type of actions that are compared. For example, for comparing two checklists ASM can be applied as the single actions contained in a checklist can be executed in any order. Consequently, the sequence of these single actions (i.e. the transitions) do not have to be considered. In contrast, for comparing conditional actions or sub-workflows a similarity measure that also takes the sequence of actions into account might be more reasonable.

Therefore, a specific combination of similarity measures can be defined in a matrix as shown exemplarily in Table 1. When comparing the actions contained in a workflow, the respective similarity measure is selected depending on the compared actions. This approach allows for a high degree of flexibility. By this means, workflow comparison can be easily tuned according to the structure and content of workflows in order to reflect different application scenarios.

Table 1. Combination matrix of similarity measures

	Single Action	Checklist	Conditional Action	Workflow
Single Action	KSM	ASM	ASM	ASM
Checklist		ASM	SAASM	SAASM
Conditional Action			TSM	SATSM
Workflow				SATSM

4 Experiments

4.1 Setup

In order to evaluate the similarity measures proposed above with respect to their suitability for clustering workflows, we applied them on a set of artificially generated workflows. The creation of these benchmark workflows was performed in two steps: At first, 49 workflows were randomly generated. Thereby, for each workflow the maximum depth was randomly chosen between 3 and 6 and the maximum number of child actions per hierarchy level was chosen randomly between 3 and 10. In the second step, 8 workflows were randomly chosen out of the 49 generated workflows. For each of these selected workflows, 10 new workflows were created by applying few modifications. These modifications were done either by changing the keywords, adding additional child actions, swapping the order of child actions, or removing child actions. As a result, we obtained a set of 88 workflows representing 8 clusters, each containing 11 workflows. The remaining 41 workflows generated in the first step were kept as noise. This led to a total of 129 workflows. For all these workflows the pairwise similarity was calculated, resulting in a matrix of 16641 similarity values, where each line represents the similarities of a workflow to all other workflows.

Clustering of the similarity data set was performed using WEKA[1] which contains tools for data pre-processing, classification, regression, clustering, association rules, and visualization [11]. In order to compare the results of different classical clustering algorithms, we applied k-Means [2,3], DBSCAN (Density Based Spatial Clustering of Applications with Noise) [6,1], and EM (Expectation Maximization) [4,5]. As discussed in Section 3, different combinations of similarity measures can be used. In order to show the different characteristics of the proposed similarity measures, we carried out four experiments: in the first experiment we applied only SAASM, in the second only SATSM, in the third only ASM, and in the last experiment we used a combination of similarity measures as described in Table 1. For example, a typical result of SAASM and k-Means is shown in Figure 2.

4.2 Results

Table 2 shows the clustering results. For each experiment and for each clustering algorithm the number of workflows contained in each cluster is listed. Analyzing the results it can be noticed that the identified workflows vary significantly depending on which kind of similarity measure is used. This highlights the suitability of combined similarity measures in order to be able to tune workflow comparison for different application scenarios. k-Means showed the most robust results as the clusters had almost a fair distribution of workflows. After experimenting with different parameters, DBSCAN with $epsilon = 1.4$ and $minPts = 2$ showed ideal results in experiment 3 and experiment 4 as the workflows were clustered almost exactly in the way they were intended to be due to the generation procedure. The results of EM were not satisfying, as the number of clusters ranged from 2 to 5 clusters.

[1] `http://www.cs.waikato.ac.nz/ml/weka/`

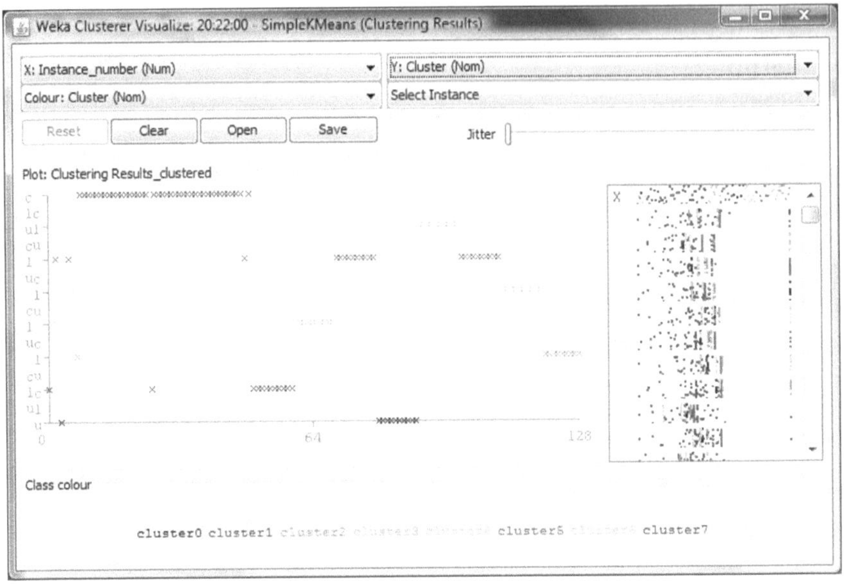

Fig. 2. Typical clustering result of SAASM and k-Means visualized in WEKA

Table 2. Clustering results

	k-Means	DBSCAN	EM		k-Means	DBSCAN	EM
Exp. 1	62	73	54	Exp. 2	30	99	10
	7	4	38		7	30	40
	19	2	24		17		30
	12	2	13		16		19
	6	3			23		30
	6	2			22		
	6	2			9		
	11	2			5		
Exp. 3	11	11	11	Exp. 4	11	11	54
	7	11	95		6	11	75
	10	11	11		5	11	
	4	11	11		5	11	
	64	10	1		5	10	
	11	11			11	11	
	11	11			75	11	
	11	11			11	11	

5 Conclusion and Future Work

In this paper we described a generic workflow model for representing different sequences of activities that have to be carried out in fire fighting operations. In order to support clear structuring and reuse, workflows are modeled as hierarchical structures. Each workflow contains a sequence of actions, whereby each action can either be a single activity, a checklist containing multiple single activities, a conditional action representing multiple alternatives, or a workflow itself.

Based on this workflow model, we presented several measures for calculating the similarity of workflows on the semantic (actions) as well as on the structural (transitions) level. These similarity measures represent the foundation for searching and clustering large numbers of workflows in order to identify workflows representing similar tasks and to enable simplification and standardization in the domain of fire fighting. Furthermore, we developed a new approach for combining different similarity measures depending on the type of actions that are compared. By this means, specialized similarity measures can be easily defined in a flexible way in order to reflect different application scenarios.

The suitability of the proposed similarity measures has been evaluated by applying classical clustering algorithms (k-Means, DBSCAN, Expectation Maximization) on a set of artificially generated workflows.

Future work will concentrate on the application of workflow similarity measures on different workflows in the domain of fire fighting operations as well as in the area of emergency services in general. As the experiments described in this paper were done using randomly generated workflows, the evaluation of the proposed similarity values on real workflows is still an open task. Furthermore, we will also continue experimenting with different combinations of similarity measures and clustering algorithms in order to identify a configuration that is best suited for clustering and analyzing workflows in the fire fighting domain.

Acknowledgements

The work described in this paper was done within the project "emergency mission control center" (emc^2) sponsored by the Austrian Research Promotion Agency (FFG).

References

1. Arlia, D., Coppola, M.: Experiments in parallel clustering with dbscan. In: Proceedings of the 7th International Euro-Par Conference Manchester on Parallel Processing, pp. 326–331. Springer, Heidelberg (2001)
2. Arthur, D., Vassilvitskii, S.: How Slow is the k-Means Method? In: Proceedings of the twenty-second annual symposium on Computational geometry, pp. 144–153. ACM Press, New York (2006)
3. Arthur, D., Vassilvitskii, S.: k-means++: The advantages of careful seeding. In: Proceedings of the eighteenth annual ACM-SIAM symposium on Discrete algorithms, pp. 1027–1035. Society for Industrial and Applied Mathematics (2007)
4. Borman, S.: The expectation maximization algorithm – a short tutorial (July 2004), `http://www.seanborman.com/publications/`
5. Dellaert, F.: The expectation maximization algorithm. Technical Report GIT-GVU-02-20, Georgia Institute of Technology (February 2002)

6. Ester, M., Kriegel, H.P., Sander, J., Xu, X.: A density-based algorithm for discovering clusters in large spatial databases with noise. In: Proceedings of 2nd International Conference on Knowledge Discovery and Data Mining, pp. 226–231. AAAI Press, Menlo Park (1996)
7. Fischer, L. (ed.): 2008 BPM & Workflow Handbook - Spotlight on Homan-Centric BPM, 1st edn. Future Strategies (2008)
8. Jung, J.-Y., Bae, J.: Workflow clustering method based on process similarity. In: Gavrilova, M.L., Gervasi, O., Kumar, V., Tan, C.J.K., Taniar, D., Laganá, A., Mun, Y., Choo, H. (eds.) ICCSA 2006. LNCS, vol. 3981, pp. 379–389. Springer, Heidelberg (2006)
9. Lundberg, J.: Principles of Workflow Support in Life Critical Situations. Blekinge Institute of Technology Licentiate Dissertion Series, vol. 2. Blekinge Institute of Technology (2007)
10. van der Aalst, W., van Hee, K.: Workflow Management: Models, Methods, and Systems. Cooperative Information Systems. MIT Press, Cambridge (2004)
11. Witten, I.H., Frank, E.: Data Mining: Practical machine learning tools and techniques, 2nd edn. Morgan Kaufmann, San Francisco (2005)

Model Instability in Microarray Gene Expression Class Prediction Studies

Henryk Maciejewski and Piotr Twaróg

Institute of Computer Engineering, Control and Robotics,
Wroclaw University of Technology,
ul. Wybrzeże Wyspiańskiego 27, 50-370 Wrocław, Poland
Henryk.Maciejewski@pwr.wroc.pl,
Piotr.Twarog@gmail.com

Abstract. This work is devoted to the problem of building a sample classifier based on data from microarray gene expression experiments. Two specific issues related to this are tackled in this paper: (a) selection of parameters of a classification model to ensure best generalization power, and (b) variability of expected prediction error (EPE) for new data as a function of the model parameters. A method is presented for selection of model parameters minimizing the EPE in studies where the number of samples (n) is much smaller then the number of attributes (d). Due to very unstable behaviour of the EPE in the space of model parameters, it seems essential that microarray studies involve systematic search for the right model parameters, as shown in this work.

Keywords: Class prediction, gene expression microarrays, expected prediction error, cross-validation.

1 Introduction

One of most promising although challenging applications of gene expression microarrays is related to class prediction. This involves building patterns or profiles of gene expression for predictive of prognostic purposes. Although still not widely used in clinics, microarray gene expression profiling has been recently reported as the basis for the first FDA-approved clinical prognostic test - the MammaPrint [4], used to determine the likelihood of breast cancer returning after a woman's initial cancer.

More wide-spread use of microarrays for diagnostic class prediction requires that several issues related to microarray technology itself (accuracy and reproducibility of results) as well as analysis of microarray data are resolved. This paper contributes to the latter issue - it is motivated by some shortcomings observed in literature on class prediction from microarray data. Analysis of sample microarray datasets shows [8,9,11] that slight changes in parameters of a predictive model can produce very different estimates of expected prediction error for new data (EPE). This implies that the task of building a model minimizing the EPE requires that careful search for the *right* model's parameters is performed,

R. Moreno-Díaz et al. (Eds.): EUROCAST 2009, LNCS 5717, pp. 745–752, 2009.

including such important parameters as e.g., number of hidden neurons, number of training iterations, or parameters of a support vector machine (SVM) (such as cost C and nonlinear kernel parameter γ), depth of decision trees, etc. This issue is however commonly overlooked in most microarray studies (e.g., [6], [10], etc.) where authors pay little consideration to how they obtain the actual model parameters.

In this work, we propose a generic approach to selection of model parameters minimizing the EPE. We discuss the challenges related to this step caused by high dimensionality of microarray data d and small number of samples n (a $d \gg n$ problem). We then demonstrate, by analysis of a number of microarray datasets, the instable behaviour of the EPE observed in the space of model parameters. This is supposed to justify the conclusion, that the approach proposed should be rigorously used for fine-tuning class predictors built from microarray data.

2 Choice of Classifier Parameters in Microarray Studies

An important step in building a classification model from training data involves setting the parameters of the model to achieve the best generalization property, ie., the smallest prediction error for new data. This is essentially related to deciding on the right complexity of the model, hitting the balance between over-simplifying and overfitting. This step requires that predictive performance for new data can be estimated. Two approaches to this are available:

- For simple models, EPE can be estimated based on the number of the training cases, the number of parameters of the model and the error observed on the training set. An example of this is the Schwarz's Bayesian Criterion. The right complexity of the model can be selected by minimizing this criterion.
- In a general case of more complex models (such as e.g., neural nets or decision trees), the EPE can be estimated by the split-sample or related method, where the model is trained on a subset of the available training data, with the EPE estimated based on remaining cases. However as shown in [2], in order to use this estimate for selection of the model parameters minimizing EPE (ie., model fine-tuning), one should split the data into three subsets: the *training* subset, the *validation* subset (used to compare different models and to select the one generalizing best), and finally, the *test* subset (used to estimate the EPE for new data). This is known as the hold-out method.

An attempt to fine-tune a classifier built from microarray data using the hold-out method poses a major challenge. Typically, the number of training cases is relatively small ($n \sim 10^2$), and it is much smaller then the dimensionality of the cases $d \gg n$. For such data, estimation of the EPE can be done by data reuse methods, such as e.g., leave-out-one cross-validation (CV) where the data is repeatedly split into $n-1$ training cases versus 1 test case. The EPE is then estimated as ([7]): $EPE = \frac{1}{n}\sum_{i=1}^{n} L\left(y_i, f^{-i}\left(x_i\right)\right)$, where $f^{-i}\left(x_i\right)$ denotes the class of the case x_i as predicted by the classifier f^{-i} built with the case x_i left out from the training data, and y_i is the true class label of the case x_i. L assumes

1 for an erroneous prediction (i.e., for $y_i \neq f^{-i}(x_i)$), and 0 otherwise. This approach is now commonly used in microarray literature. However, an important question arises here about how the *right* parameters of the model should be determined to guarantee the smallest EPE. Typically, microarray studies overlook this problem by either (a) setting the parameters of the model arbitrarily, without any justification, or (b) fine-tuning the model on the training CV subset, or (c) fine-tuning the model on the left-out (testing) split. Clearly, considering very unstable nature of the EPE in the space of model parameters, as shown in Sect. 4, (a) can produce a model far from optimal, while (b) seems to be prone to overfitting and (c) does not produce reliable estimates of predictive performance for new data (thus yielding overoptimistic estimates of the EPE).

In order to overcome this shortcoming, an approach is proposed to essentially incorporate the hold-out method with the cross-validation procedure. This method involves two nested cross-validation loops: an internal loop used to fine tune the model (working as the *validation* hold-out split), and an external loop used to estimate the EPE for new data (working as the *test* hold-out split). This approach is similar to the procedure outlined in [3] (developed to select the right parameters of an SVM), and essentially allows to implement some heuristic search in the space of model parameters. This is covered in detail in the following section.

3 Nested CV Loops for Model Selection

The procedure will be explained using the following notation. Let us represent the whole data set from microarray experiment as $D = \{(x_1, y_1), \ldots, (x_n, y_n)\}$, where $x_i \in R^d$ denote vectors of gene expression of the n samples tested, and y_i their corresponding class labels. The function $L(y, y^*)$ equals 1 for different class labels $y \neq y^*$, and 0 otherwise.

The proposed algorithm of nested cross-validation loops can be summarized in the following steps.

1. Split the set D into the training Z_i and test T_i subsets: $T_i = \{(x_i, y_i)\}$, $Z_i = D - T_i$.
2. Based on the set Z_i perform a feature selection procedure, i.e., leave only relevant $d^* \ll d$ features in samples in Z_i and T_i. The reduced dimensionality feature vectors will be denoted x', and the corresponding training and test sets Z'_i, T'_i .
3. Select an initial value of the vector of classifier parameters $P = P_0, ip = 0$.
4. Split the set Z'_i into the actual training U'_j and validation V'_j subsets: $V'_j = \{(x'_j, y_j)\}, j \neq i$, and $U'_j = Z'_i - V'_j$.
5. Build a classifier with parameters P based on the training set U'_j. Denote the model as $f_P^{-i,j}$.
6. Test the model on validation set, i.e., compute the misclassification rate $mr_{i,j}(P) = ps_j \cdot L\left(y_j, f_P^{-i,j}(x'_j)\right)$, where ps_j denotes a model specific *prediction strength* in favour of the predicted class $f_P^{-i,j}(x'_j)$ (such as e.g., neuron

activation in artificial neural networks, normalized distance from the separating hyperplane in linear separation algorithms or purity of the leaf in decision tree).
7. Repeat Steps 4 through 6 for $j = 1, 2, \ldots, i-1, i+1, \ldots, n$.
8. Calculate the overall misclassification rate using all subsequent samples from Z_i' treated as validation data: $vmr_i(P) = \sum_{j=1,2,\ldots,i-1,i+1,\ldots,n} mr_{i,j}(P)$.
9. If a condition for stopping the search in the space of model parameters is not fulfilled, assign $ip = ip + 1$, $P = P_{ip}$ and repeat steps 4 through 8. The stopping criterion and the sequence of vectors $P_0, P_1, P_2, \ldots$ should be established with a search algorithm.
10. Else find $\hat{P}$ such that: $vmr_i(\hat{P}) = \min_P vmr_i(P)$.
11. Train the model with the vector of parameters $\hat{P}$ using the training data set Z_i, denote the model as $f_{\hat{P}}^{-i}$.
12. Test the model on the set T_i, compute: $e_i = L\left(y_i, f_{\hat{P}}^{-i}(x_i')\right)$.
13. Repeat Steps 1 through 12 for $1 = 1, 2, \ldots, n$.
14. Calculate the estimate of EPE for new data as: $EPE = \frac{1}{n}\sum_{i=1}^{n} e_i$.

Although both CV loops in this procedure were organized as leave-out-one CV loops, it is quite straightforward to express them as k-fold CV. However, leave-out-one CV seems more widely used in microarray studies, due to its smaller bias ([7]).

As argued in Sect. 2, the outer loop of this CV procedure allows to estimate the EPE based on data never used for model training and fine-tuning. Similarly, the inner CV loop is used to select the parameters $\hat{P}$ of the model, based on how the model performs on validation data, disjoint from training data. This clearly brings benefits of the Bishop's hold-out method [2] to studies with small number of training cases.

The method requires that some search algorithm is used in step 9 to decide on how subsequent vectors of model parameters $P_0, P_1, \ldots$ are determined and when the process should stop. Heuristic search methods can be applied here, such as:

- The grid search method, possibly making the grid interpoint distances smaller as the algorithm progresses. This approach was used in [3].
- The tabu search method [5], which looks for the best not yet checked candidates in the neighbourhood of the current point. This yields significantly faster performance of the process as compared with the grid search.

4 Numerical Examples of Model Instability

The purpose of the numerical examples given in this section is to (a) demonstrate how instable the EPE seems to be in the space of parameters of the classification models, and (b) show how the double CV loop improves the quality of models, as compared with the current common practice where models are either fine-tuned based on training data or this step is omitted at all. To illustrate these concepts, a comprehensive study was performed involving different microarray datasets (e.g., data first published in [1], [6], [10]), as well as different classifiers (ANN, SVM, decision tree and random forests). Although here we present only a small portion of these results, the conclusions we draw are consistently confirmed by all the datasets/models tested.

The first example shown in Fig. 1 and 2 is based on Alon's colon data [1] classified with a neural network (MLP). For the purpose of this example, two selected parameters of the model (the number of training iterations and the seed of a random number generator used to assign initial weights) were fine-tuned based on the validation data (Fig. 1), or based on the training data (Fig. 2). In both figures, the EPE is depicted as a function of remaining two parameters of the model: the number of neurons in the hidden layer of the MLP and the number of genes selected by the feature selection procedure (denoted d^* in step 2 in Sect. 3).

It is clear from both figures that the EPE realized by the model proves very sensitive with respect to the model's parameters. This observation seems to hold for different data sets and different classifiers. This implies that it is unlikely to achieve optimal generalization of models unless a systematic search in the space of model parameters is performed.

Comparing Figs. 1 and 2 gives some indication how this search should (should not) be done. It is clear that fine-tuning the model (with respect to the number of training iterations and initial weights) based on training data (Fig. 2) leads to strong overfitting in the central, top and right part of the graph. By fine-tuning the model on validation data (double CV loop, Fig. 1), the best result was obtained (5 misclassified samples, with 14 neurons in the hidden layer). The model fine-tuned on training data (double CV not used – Fig. 2) would overfit for this big number of hidden neurons. The best results obtained in Fig. 2 was 7 errors, obtained under 8 hidden neurons.

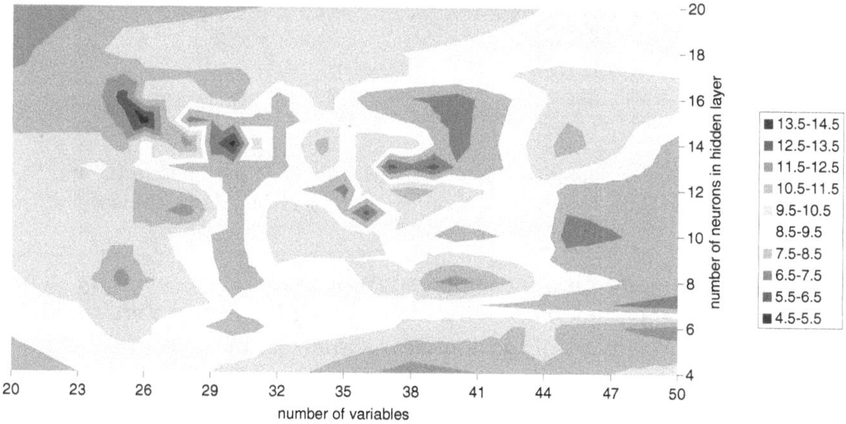

Fig. 1. Total number of classification errors with the ANN parameters (initial weights and number of training iterations) fine-tuned on validation data (double CV loop). Results for the Alon's colon data [1].

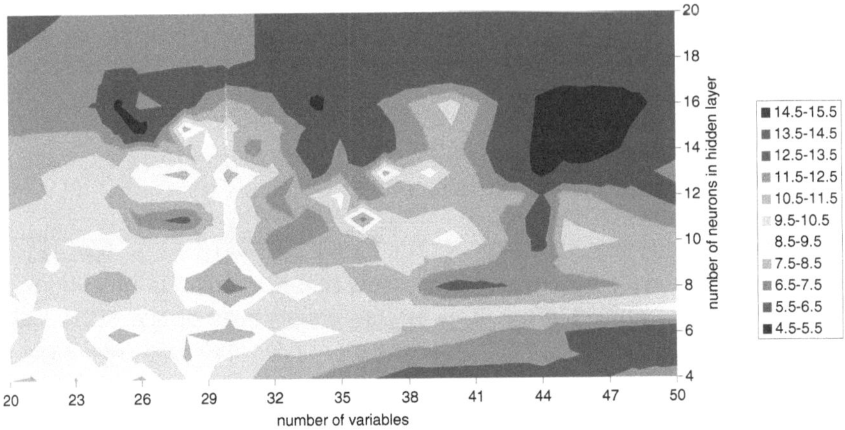

Fig. 2. Total number of classification errors with the ANN parameters (initial weights and number of training iterations) fine-tuned on training data. Double CV loop not used. Results for the Alon's colon data [1].

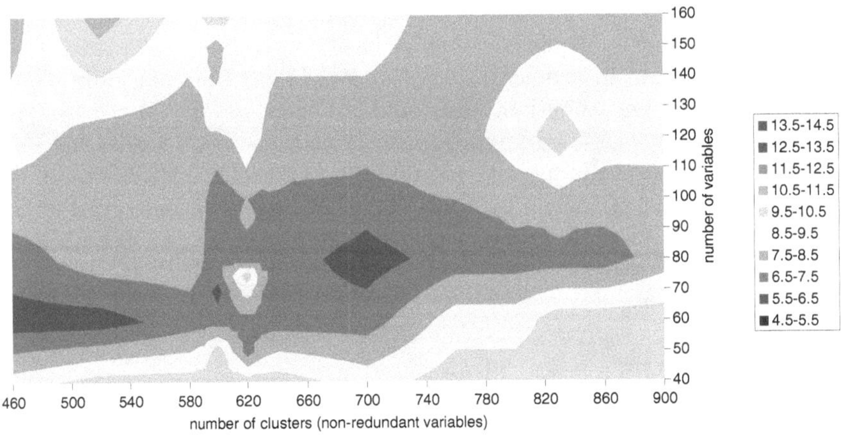

Fig. 3. Total number of classification errors with the SVM parameters (C and γ) fine-tuned on validation data (double CV loop). Results for the Alon's colon data [1].

Similar results pertaining the support vector machine classifier are shown in Figs. 3 and 4. The parameters of this model being fine-tuned were the cost parameter (C) and parameter of the nonlinear kernel (γ). In the study illustrated in Fig. 3 these parameters were fine-tuned based on the validation data (double CV loop used), while in Fig. 4 - based on the training data. The EPE measured for new data is shown in these pictures as a function of remaining parameters of the model (in this example these were: the number (d^*) of features used, and a specific parameter related to the feature selection procedure, which was based on results of clustering in the space of d features, similarly to the idea presented in [12]).

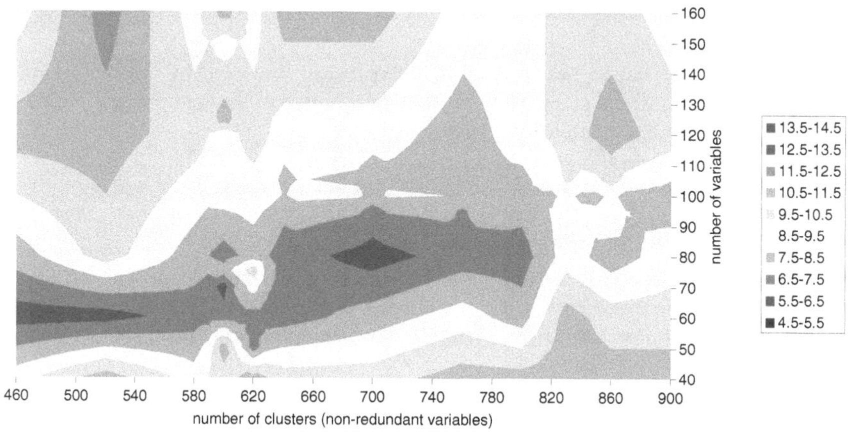

Fig. 4. Total number of classification errors with the SVM parameters (C and γ) fine-tuned on training data. Double CV loop not used. Results for the Alon's colon data [1].

Although the SVM model is considered less likely to overfit then the ANN, we still observe the effect of overfitting along the number of variables axis in Fig. 4. This is clearly due to fine-tuning the model based on training data. The SVM model seems more stable then the ANN, as the best results are not contained in isolated points but rather create larger areas in the space of parameters. Interestingly, the nested CV loop seems to further smoothen the EPE observed in the space of model parameters. This increases the chance of obtaining correct prediction results on really new data.

5 Conclusions

In this work we demonstrated, by analysis of a number of microarray datasets, the nonlinear nature of the EPE observed in the space of model parameters. This study indicates that there are some areas in the space of model parameters that yield significantly better generalization behaviour of the model, however the model is very sensitive in this respect (thus instable).

Conclusions from this come in the form of a recommendation: microarray studies should consciously apply search techniques in the model parameters in order to achieve best possible generalization behaviour (grid/heuristic, such as tabu search techniques). Due to small number of training cases available in microarray studies, this search is challenging. However, it is doable using the double cross-validation loop as shown in this paper.

Finally, considering the highly nonlinear nature of the EPE (or instability of prediction models), a question arises whether the best EPE obtained by elaborate fine-tuning of the model will guarantee the same predictive performance for the *actually new data*? Further analysis is clearly needed to get insight into the reason of the EPE nonlinearity/instability (data related? model related?).

References

1. Alon, U., et al.: Broad patterns of gene expression revealed by clustering analysis of tumor and normal colon tissues probed by oligonucleotide arrays. Proc. Natl. Acad. Sci. USA 96, 6745–6750 (1999)
2. Bishop, C.M.: Neural Networks for Pattern Recognition. Oxford University Press, Oxford (1995)
3. Hsu, C.-W., Chang, C.-C., Lin, C.-J.: A Practical Guide to Support Vector Classification. National Taiwan University, Taipei (2008)
4. Glas, A.M., et al.: Converting a breast cancer microarray signature into a high-throughput diagnostic test. BMC Genomics 7, 278 (2006)
5. Glover, F., Laguna, M.: Taboo search. Kluwer Academic Publishers, Dordrecht (1997)
6. Golub, T., et al.: Molecular classification of cancer: Class discovery and class prediction by gene expression monitoring. Science 286, 531–537 (1999)
7. Hastie, T., Tibshirani, R., Friedman, J.: The Elements of Statistical Learning. Data Mining, Inference and Prediction. Springer, Heidelberg (2002)
8. Maciejewski, H.: Adaptive selection of feature set dimensionality for classification of DNA microarray samples. In: Computer recognition systems. Springer Advances in Soft Computing (2007)
9. Maciejewski, H.: Quality of feature selection based on microarray gene expression data. In: Bubak, M., van Albada, G.D., Dongarra, J., Sloot, P.M.A. (eds.) ICCS 2008, Part III. LNCS, vol. 5103, pp. 140–147. Springer, Heidelberg (2008)
10. Singh, D., et al.: Gene expression correlates of clinical prostate cancer behavior. Cancer Cell 1, 203–209 (2002)
11. Twaróg, P.: Sample classification based on DNA microarray assays. MSc Thesis. Wroclaw University of Technology (2008) (in Polish)
12. Yu, L., Liu, H.: Redundancy Based Feature Selection for Microarray Data. In: Proc. 10th ACM SIGKDD Conference on Knowledge Discovery and Data Mining, pp. 737–742 (2004)

Conflict Resolution in Multiagent Systems Based on Wireless Sensor Networks

Witold Jacak and Karin Pröll

Dept. of Software Engineering at Hagenberg
Upper Austrian University of Applied Sciences
Softwarepark 11, A 4232 Hagenberg, Austria
jacak@fh-hagenberg.at, proell@fh-hagenberg.at

Abstract. The design of intelligent and sensor-based autonomous agents learning by themselves to perform complex real-world tasks is a still-open challenge for artificial and computational intelligence. In this paper a concept of a framework for conflict resolution in an autonomous robotic agent system is presented. The structure of an intelligent robotic agent consists of two independent subsystems: the action and motion planning system and the action and motion reactive control system with integrated conflict resolution methods.

1 Introduction

We present the concept of a framework for an autonomous robotic agent that is capable of showing both local sensor-based reactive behavior and global action planning based on external sensor network. In multi-agent robotic systems, one is primarily interested in the behavior and interactions of a group of agents and a dynamic surrounded world, based on the models of the agents themselves and environment stimuli. Reactive systems have no internal history or long-term plans, but calculate or choose their next action based solely upon the current perceptual situation. On the other hand, machine learning-based models are motivated by the representation of the system's knowledge. Such models permit to use more powerful and more general methods than reactive models; this, however, makes them inadequate for many real-time applications where a dynamic change in the environment occurs. Usually an agent has only partial information about the world state obtained from own perception system (sensors system). Machine learning aims to overcome the limitations such as knowledge bottleneck, engineering and tractability bottleneck, by enabling an agent to collect its knowledge on-the-fly, through real-world experimentation. Processing, storing and using information obtained during several task executions is called lifelong learning. For this reason it is necessary to extend the reactive control system of sensor-based global preplanning system by omitting the knowledge bottleneck in classical machine learning approach.

R. Moreno-Díaz et al. (Eds.): EUROCAST 2009, LNCS 5717, pp. 753–760, 2009.

2 Conflict Resolution for Robotic Agent System

In our concept, the structure of an intelligent robotic agent consists of two independent subsystems: the action and motion planning system and the action and motion reactive control system with integrated conflict resolution method [1,2,3,4]. The action planning system uses an aggregated world model storing knowledge about all conflicts, which occurred in the past. Conflicts occur when a collision between a robotic agent and an unknown dynamical object or another agent in workspace is possible. The action controller is able to solve the conflict situation (space conflict for autonomous robotic agent) in a reactive manner. The general conflict resolution part of the agent makes use of knowledge not only from sensors mounted on active agent (local sensor network) but also from distributed global sensor network (see Fig 1). A sensor network is a collection of sensor nodes deployed in the workspace.

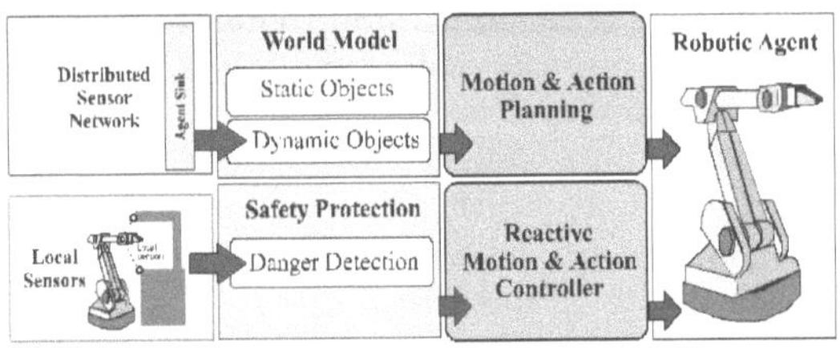

Fig. 1. Structure of sensor-based conflict resolution system

Being battery powered and deployed in remote areas they have limited energy resource and hence limited lifetime. Other constraints include limited memory, processing power, and bandwidth. The accuracy of information is location dependent. Due to these limitations data aggregation is an important consideration for sensor networks. The idea is to combine the data coming from different sources and reroute it further, after eliminating redundancy, minimizing number of transmissions and thus saving energy. Sensors will be used for inventory maintenance and unknown object recognition and motion tracking. The agent requests a monitoring of some segment of space in which the next parts of motion are supposed to take place from the global sensor network. If the global sensor network identifies unknown objects in these segments then this information can be used by the agents safety system to preplan a reaction prior to recognition of these objects by the local perception units. This can help to avoid possible conflict situations in advance.

3 Sensor Network Based Conflict-Free-Motion Planning System

3.1 Static and Dynamic World Model

The knowledge represented here is the geometrical model of the robotic agent environment. Many different methods can be used for geometrical representation of the agent service space. One of them is the triangle approximation another is cubic approximation. In a model with triangle approximation, the points in the triangle net (lying on service space border) are coupled in triangle walls. The walls represent the data objects of the world model. The other model describes the service space of the robot manipulator as cubic approximation. The service space of robotic agent can be discretized in the form of the cubic raster (voxels). The number of voxels depends of the accuracy of approximation. If an obstacle occupies the voxel then it obtains the value 1 of the space occupancy function. The level of fullness of knowledge leads to two different methods of resolving of conflict situation. The both models are convenient to represent geometrical environment of known objects in agent workspace. The model is used for collision-freeness testing between agents and surrounding world. We can say that the position of an agent is collision free if it does not collide with any static or dynamic obstacle in its workspace. To test such conditions we should have full knowledge about the surrounding static and dynamic world that means that the geometrical model of agents environment should be completely known. To obtain fast and fully computerized methods for collision detection, we use additional geometric representation of each object on the scene. We introduce the ellipsoidal representation of 3D objects, which uses ellipsoids for filling the volume. The ellipsoidal representation of an object is convenient to test collision freeness of agent positions. The 3D models of objects represent the static part of agents world model. To construct the model of unknown objects, which penetrate the agents environment, it is necessary to continuously modify the geometrical representation. The dynamic part of model is modified based on sensor data coming from wireless sensors network.

3.2 Sensor Network

A sensor network is composed of a large number of tiny autonomous devices, called sensor nodes. Each sensor node has four basic components: a sensing unit, a processing unit, a radio unit, and a power unit. Since a sensor node has limited sensing and computational capabilities and can communicate only within short distances. The nodes are deployed densely and coordinate amongst themselves to achieve common information [7].Some examples of sensor network applications are as follows:

Intrusion detection and tracking. Sensors are deployed along the border to detect, classify, and track intruding objects.[8]

Environmental monitoring. Specialized sensor nodes that are able to detect changes in environment.[6]

These sensor networks applications differ significantly. However, the tasks performed by the sensors are similar: sensing the environment, processing the information, and sending information to the base station(s). A node in a sensor network has essentially three different tasks [5]: Sensing: detecting changes of environment, Communicating: forwarding information, acting as an intermediate relay in a path, Computing: data aggregation, processing, and compression.

Routing techniques are needed to send data between sensor nodes and the base station. Several routing protocols are proposed for sensor networks. These protocols can be divided into the following categories: data-centric protocols, hierarchical protocols, location based protocols, and some QoS-aware protocols [6]. For monitoring the surrounding environment of the robotic agent we propose a sensor network based on a virtual grid representation of the monitored area and for data delivery a combination of event driven and query-driven data delivery protocol between sensor network and a mobile sink component (AS) of the robotic agent is used [8]. The mobility of AS results from movements of the robotic agent.

Virtual Grid Structure of Sensor Network. The monitored area is divided into virtual grids. We notate $G(x,y)$ as the grid coordinates, where x is grid x-coordinate and y is grid y-coordinate in Cartesian space. After sensors have been deployed, a node is selected to act as grid head. The grid heads task is to record information about events and disseminate it to other nodes for collaborative signal and information processing. At first, each node obtains neighboring information by start message. Utilizing this information, a node that is closest to the center of the grid is selected as head. If a head has not enough resources, one of other nodes in the same grid will be selected to replace it. We assume that the side length of grids to guarantee that the grid heads can communicate with neighboring grids directly. Here, we also assume that the communication range of node is able to communicate with the neighboring grids. The grids are grouped in sub-networks in hierarchical way. Each sub-network obtains one head selected from the set of grids heads. The sub-network head collects the messages from grids heads within the sub-network. The agent sink (AS) stores the topology of the virtual grid and uses this information to perform the queries to the sensor network. The task of AS is to find the virtual cell of grid where the intrusion of a new dynamic object is registered. An AS expects to obtain the event information instantly when such event occurred. An event usually happens unexpected. Therefore, a sensor has to signal an event after it was detected. This sensor is called source node. The source node propagates a register packet to all grid heads. The format of register packet is $< P_{type}, Src_{id}, Src\ G(x,y), hc, event\ type, time\ to\ expire >$, where P_{type} is packet type, Src_{id} is the identifier of the source node, $Src\ G(x,y)$ is the sources grid location, and hc is the hop count. When heads receive this packet, they store the register packet in their register table and route it to the sub-network head. This information is kept within a certain period of time (*time to expire*). If a head does not receive any further register packets and *time to expire* is elapsed, it removes the information from the register table. The information of an event is distributed to the grid heads and summarized to the sub-network head in the following way: The grid heads store the x, y positions of active sensor nodes (an event has been signaled) whereas in the subnet-work heads only grid numbers with active sensor nodes are registered (active grids). For the next query the AS maintains a list containing all sub-network heads surrounding the next segment of its motion path of robotic agent and all sub-networks with active grids. AS can now decide to obtain detailed information about all active sensor nodes by querying the grid heads of all active grids or summarized information about all active grids by querying only the sub-network heads. For reconstructing the geometric model of the intruding object both - detailed or summarized - information can be used. Using detailed information a more accurate shape of the base of the intruding object can be deduced. The use of summarized information leads to a very approximate representation based on shapes of active grids. (see Fig. 2)

By maintaining a dynamic list of sub-network heads for querying, a general broadcast to all sub-network heads in the virtual grid can be avoided. The possibility to query either sub-network heads or grid heads further reduces transmission activities in the network.

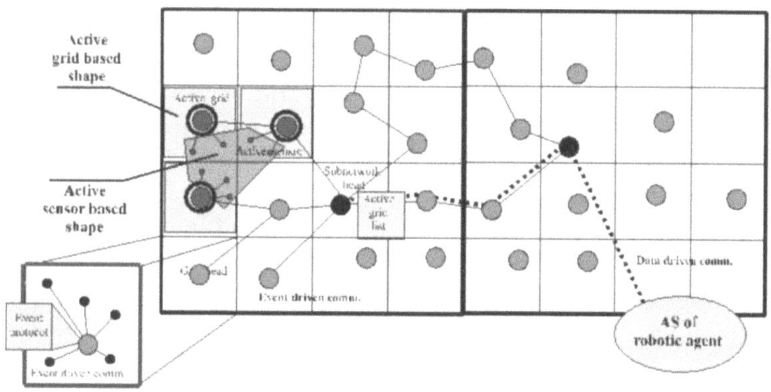

Fig. 2. Use of summarized or detailed information from virtual grid for reconstructing shape of intruding object

World Model Updating. The path planner of robotic agent sends a query to AS to check if any grid cell in sensor network has an active signal of a new object intrusion. AS uses the sub-network heads and grid heads and hierarchical protocol to collect the $G(x, y)$ position of grids to be active. The $G(x, y)$ and side length d will be used to approximation of object shape and volume to be intruded in the work space of robotic agent. The 3D geometrical model of an intruding object is constructed in a 2,5D representation of the shape registered by sensors.

3.3 Global Motion Planner

Agent Model. The most suitable model of hardware component of robotic agent is a discrete dynamic system. One of the ways to construct such model of robot's kinematics is based on an arbitrary discretization of angle increments of the agent mechanical joints [3].Using the fact that all the angles can change only by a defined increment, we define the input set U of the model as: $U = \times\{u_i | i = 1,..,n\}$ where $u_i = \{-\delta q, 0, \delta q\}$ is the set of possible (admissible) directions of changes of the i-th joint angle. Having defined the set U, it is possible to describe the changes of successive configurations of the agent's link as discrete linear system with the state transition function as:

$$q(k+1) = q(k) + \lambda u(k) \tag{1}$$

Checking for the collision-freeness of the agent configuration (skeleton) can be reduced to the "broken line - ellipsoid" intersection detection problem, which has an easy analytical solution. The complete formal explanation of the FSM model of agent kinematics is presented in [3,4].

Motion Planner. For the robotic agent we can define the problem of achievement the goal position as the problem of reachability of the final state set from the agent's current state (current position). In order to solve this problem we apply graph searching in the state transition graph. The process of applying the transition function to a current state we term expanding the graph node. Expanding current state q_c, successors of q_c etc. ad infinitum, makes explicit the graph that implicitly is defined by current state and transition function. The way of expanding the graph will depend on the form of the cost function using to evaluate each node. As the evaluation function we can use the sum of the cost function $c(q_c, q)$ and a cost estimate function $h(q, q_f)$, for example the rectilinear distance between agent position and terminal position q_f. Using the standard A* procedure we can find the state trajectory (if exists) $q* = (q_c, q(2), .., q(k), .., q_f)$ from current state to final state qf which includes only feasible states. In order to solve the path-planning problem we apply the graph searching procedure to the agent's state transition graph. The development of the search graph will start from the node (configuration) q_c, by the action for all possible inputs from the set U. Thus, it becomes essential to quickly check for the non-repeatability of the nodes generated, and their feasibility. A configuration q is said to be feasible if it does not collide with any object in the surrounding world.

3.4 One Step Motion Planning

The continuously adapted world model is used to test the collision freeness. Each dynamic object registered from sensor network field is inserted into the world model as a new obstacle in 2,5D form. The task of the motion planning and execution component is to plan the collision free configuration of the robot's manipulator based on information coming from the world model from knowledge base. To realize this task, the on line motion planner calculates the changes of robot configuration to avoid the obstacle in the best possible way. This knowledge is represented as geometrical model of work scene and mathematical model of agent's actor direct and inverse kinematics. The one step action planner of the agent generates the new movement in following steps (see Fig. 3):

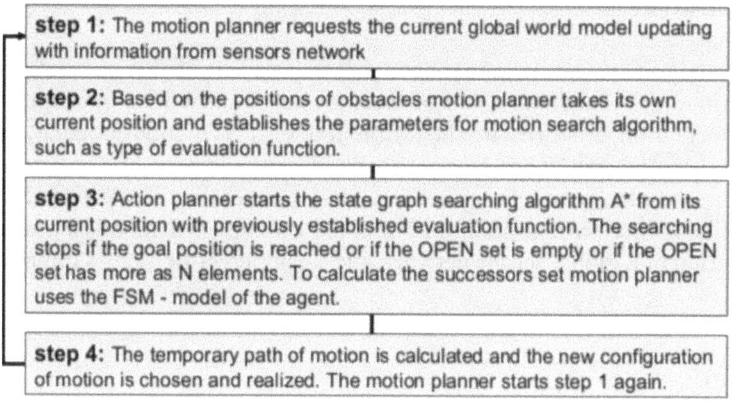

Fig. 3. Conflict resolving algorithm

3.5 Example

The following example (see Fig. 4) shows the trajectories of the robotic agent at times $T_1, T_2, \ldots T_8$ avoiding the cylindrical object crossing its initial path. At each point of time T_i the motion planner uses the information about the obstacles current position from the sensor network for recalculating a new collision free motion as a temporary path from current to final position. When the object leaves, the robotic agent does not return to its initial trajectory at time T_1 but takes the shortest way from the current to the final position.

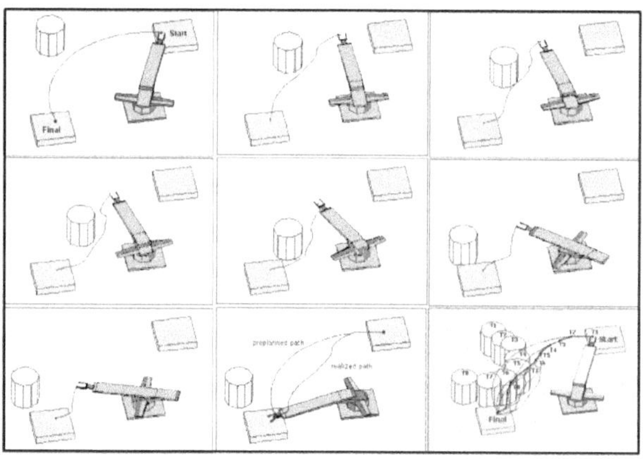

Fig. 4. One step motion planning to avoid collision with cylindrical object

4 Remarks

Using global sensor network unknown objects in a robotic agents workspace can be identified prior to recognition of these objects by the local perception units. This information is stored in the world model of the robotic agent, which is continuously updated. The aggregated world model is used by the agents safety system to preplan a reaction to avoid possible conflict situations. Especially for mobile robots with a manipulator fixed on top of a mobile base unit the combination of a global sensor network and local perception units improves their mobility.

References

1. Jacak, W., Pröll, K.: Heuristic Approach to Conflict Problem Solving in an Intelligent Multiagent System. In: Moreno Díaz, R., Pichler, F., Quesada Arencibia, A. (eds.) EUROCAST 2007. LNCS, vol. 4739, pp. 772–779. Springer, Heidelberg (2007)
2. Jacak, W., Proell, K., Dreiseitl, S.: Conflict Management in Intelligent Robotic System based on FSM Approach. In: Moreno-Díaz Jr., R., Buchberger, B., Freire, J.-L. (eds.) EUROCAST 2001. LNCS, vol. 2178, pp. 359–386. Springer, Heidelberg (2001)
3. Jacak, W.: Intelligent Robotic Systems. Kluwer Academic/Plenum Publishers (1999)
4. Pröll, K.: Intelligent Multi-Agent Robotic Systems: Contract and Conflict Management, PhD Thesis, Johannes Kepler University Linz /Austria (2002)
5. Al-Karaki, J.N., Kamal, A.E.: Routing techniques in wireless sensor networks: a survey. IEEE Wireless Communications 11(6), 6–28 (2004)
6. Dousse, O., Tavoularis, C., Thiran, P.: Delay of intrusion detection in wireless sensor networks. In: Proceedings of the 7th ACM international Symposium on Mobile ad hoc Networking and Computing, Florence, Italy, May 22-25, pp. 155–165 (2006)
7. Karl, H., Willig, A.: Protocols and Architectures for Wireless Sensor Networks. Wiley, USA (2005)
8. Chen, T.-S., Chang, Y.-S., Tsai, H.-W., Chu, C.-P.: Data Aggregation for Range Query in Wireless Sensor Networks. Journal of Information Science and Engineering 23(4), 1103–1121 (2007)

Evolutionary Selection in Simulation-Based Optimization

Andreas Beham, Monika Kofler, Michael Affenzeller, and Stefan Wagner

Josef Ressel Centre for heuristic optimization - Heureka!
School of Informatics, Communications and Media - Hagenberg
Upper Austria University of Applied Sciences
Softwarepark 11, A-4232 Hagenberg, Austria
{abeham,mkofler,maffenze,swagner}@heuristiclab.com
http://www.heuristiclab.com

Abstract. In this work we examine the effect of elitist and non-elitist selection on a supply chain problem. The problem is characterized by an output constraint which in turn separates the search space in a feasible and a non-feasible region. Additionally the simulation output is noisy due to a stochastic demand model. We will show analyze which strategy is able to perform a walk on the boundary between the feasible and infeasible space. Additionally a new selection scheme is introduced based on a statistical test to evaluate the difference between two solutions given a number of noisy quality values. This selection scheme is described and evaluated on the problem situation.

Keywords: simulation-based optimization, selection, evolutionary algorithms.

1 Introduction

Simulation plays a very important role in the analysis of today's business processes. A company can improve the qualities of its decisions with the help of accurate simulation models. The parameters of these models represent the decisions a company can take. Finding the optimal decision variables quickly and in an automated way is the goal of simulation-based optimization.

More technically simulation-based optimization treats the optimization of one or multiple output value(s) resulting from the run of a simulation model. Different algorithms from gradient based methods to complex memory-based strategies are developed to perform these tasks. Among these, metaheuristics are viewed as having a very good potential to deal with more and more complex models, and still find good solutions in acceptable time [4]. However, the stochastic nature of the simulation model creates a difficulty that many metaheuristics are still not completely prepared to deal with.

In deterministic optimization the evaluation of a single configuration results in a value that represents the quality, also called fitness, of that configuration. In simulation-based optimization, using a stochastic simulation model, this quality

R. Moreno-Díaz et al. (Eds.): EUROCAST 2009, LNCS 5717, pp. 761–768, 2009.

varies from evaluation to evaluation so the value cannot be completely trusted. The question arises whether the averaged quality is enough or whether there is need to take the variance of the quality samples into account. Additionally constraints on the parameter configuration and especially on the simulation output are a further difficulty. Reevaluating a configuration sometimes brings a big burden on the computational resources and possibly allows fewer iterations in which the algorithm can find a good solution. All these problems together create new challenges and new opportunities for metaheuristic developments.

2 Selection under Uncertain Conditions

The problem of selecting a solution as part of the next generation has been a big topic of research. The problem of choosing the right selection and replacement operator is not always clear: The amount of selection pressure to be exerted depends not only on the problem situation, but also on the algorithm configuration. In the traveling salesperson problem it has been empirically shown that crossover operators such as the edge recombination crossover, and the maximal preservative crossover do not work well together with fitness proportional selection, but require tournament selection with a reasonably high group size [1]. The term selection pressure denotes the strength of the preference of good solutions over bad ones. The higher the selection pressure, the less effect a bad solution has on the development of the evolutionary search path. A good balance is often required to obtain good solutions as too much selection pressure will lead to premature convergence and a quick loss of diversity, while too little pressure may not lead to convergence in acceptable time.

In evolution strategies [7] [3] two selection, also called replacement, schemes exist; they are denoted as *Plus* and *Comma*. Plus selection is elitist which means that individuals may survive as many generations as no better solutions have been found. Comma selection is purely generational so that individuals live exactly one generation. There, the selection pressure is scalable in the form of a ratio between the offspring population size λ and the parent population size μ: $\frac{\lambda}{\mu}$. The larger this ratio becomes, the higher the selection pressure. If the ratio is 1 then no selection pressure exists and the algorithm performs a random walk in the search space.

Simulation-based optimization provides an additional difficulty to the problem of exerting the right amount of selection pressure. Because of the stochastic output, the average quality value is just an estimation of the unknown true quality. In this work we have looked into using replicated simulation runs, but not relying on the average quality in all cases. Rather, we analyze the fitness samples from the simulation model with statistical hypothesis tests. A number of different statistical tests exist to determine if two stochastic sizes are equal. Typically the hypothesis H_0 that two populations p_1 and p_2 are equal is tried to be rejected given a certain significance level α. These tests may or not make assumptions on the distribution of p_1 and p_2, e.g. that they are of equal variance such as in the F-test, and others that can cope with unequal variances such as the popular

Student's t-test. In this work we have used the Mann-Whitney-Wilcoxon rank sum test to compare two solutions [6]. This test is similar to Student's t-test, but has the advantage that the samples need not stem from a normal distribution. Performing such a comparison may be useful to determine whether two configurations are equal with respect to the given samples and significance level. When the average would likely mislead the optimizer the test might return a non-significant result, while the test may reject H_0 when the average can be seen as a more trusted predictor. Additionally we want to evaluate the performance when using variance reduction techniques such as common random numbers (CRN) [5]. There, the same random seed is used in e.g. all first simulation replication. Then a different, random seed is used in all second replicated runs and so on. The output of two runs can thus be compared directly replication by replication. Naturally, the number of replications and different random seeds depends on the variance of the model output.

2.1 Uncertain Best Selection

The uncertain selection step that we are introducing performs a pair-wise two-tailed rank sum test between all solution candidates. If a significant difference can be detected the average of the two candidates is used to decide on the winner of this comparison and the winner's rank is increased by one. When all solutions have been ranked that way, selection starts by picking the highest ranking solutions and proceeding to lower ranking solutions until the required number of solutions to be selected has been reached. In case multiple solutions have the same rank, random selection is performed on those solutions. Since a random selection is unbiased it is able to keep genetic diversity when a clear decision on a search direction is not feasible. The decision is prolonged to the point when the test suggests that a solution is better than another.

3 Supply Chain Simulation

The simulation model is an inventory system as described in [5] and available within AnyLogic^{TM} 6. It consists of three components: A retailer, a wholesaler, and a factory. The factory produces goods out of an endless supply of raw materials, the retailer sells these goods to customers, while the wholesaler acts as buffer between the factory and the retailer. Each of these components manages an inventory. Two decision variables, representing a lower and an upper bound, model the order behavior in this supply chain. At the beginning of a day orders are placed to refill the inventory to the upper bound given the current level, backlog, and expected arrival of items from previous orders. Each action, such as ordering items, manufacturing, and holding them, as well as backlogged orders increase the cost value which in turn has to be minimized. Additionally there is a constraint that marks a solution as feasible if the maximum customer waiting time at the retailer is below a certain threshold.

3.1 Fitness Function and Design of Penalty

The output constraint separates the solution space into a feasible and an infeasible region. The optimization approach thus needs to handle infeasible solutions such that they are either discarded or a penalty is added to the fitness value so that these solutions are of less importance to the evolutionary optimization process. We have decided to use such a penalty, describe three different possibilities, and evaluate their effect on the optimization approach. The fitness function in the feasible case is:

$$\text{minimize } cost = \frac{\sum \text{OderingCosts} + \sum \text{HoldingCosts} + \sum \text{BacklogCosts}}{\text{simulation days}}$$

In the infeasible case following penalties are compared. Exemplary quality progress curves are given in Figure 1.

- **Penalty 1** sets *cost* to a constant, but high value which is well above the cost values obtained in the feasible region. So the infeasible region is modeled as a plateau. However, this creates a difficulty for an optimization algorithm as a search direction is not given. The optimizer will have to evaluate configurations randomly until a feasible one is found and a search direction may be present in its neighborhood. It may thus have more trouble finding a solution initially and it may be possible that the optimizer gets lost in the plateau if it is large enough.
- **Penalty 2** adds a constant, but high value to the cost so that the infeasible region is no longer a plateau, and a search direction is available everywhere in the fitness landscape. This allows the optimizer to make decisions towards better solutions anywhere in the solution space. But as can be seen in Figure 1 the direction can also lead away from the feasible region, resulting in an undesirable optimization behavior.
- **Penalty 3** adds the maximum waiting time to a constant, but high value. Thus, in the infeasible region the goal becomes to minimize the maximum waiting time, instead of the costs. We found this to be best suited for a neighborhood based optimizer such as the evolution strategy. The infeasible region's slope points towards the feasible region and thus quickly leads the optimizer back.

We'd also like to note that the undesirable situation in Penalty 2 is hardly possible when the optimizer is elitist. In Plus selection a solution can only be replaced when a better one is found. The problem thus appeared only in evolution strategies with Comma selection.

4 Experiment Setup

The investigation has been conducted with HeuristicLab [8], a flexible optimization environment. Previous work and results [2] in HeuristicLab have shown the

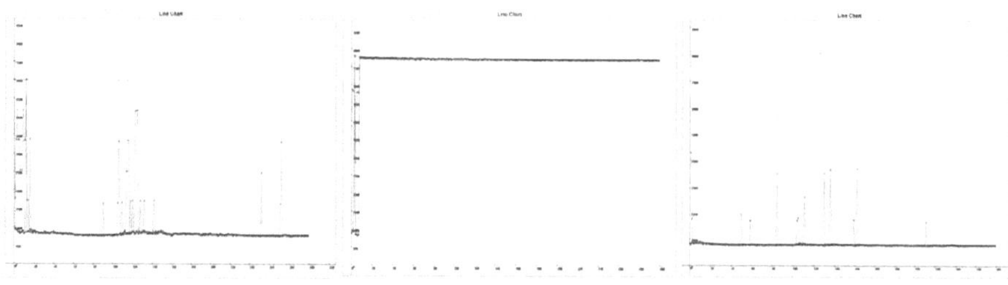

Fig. 1. Optimization behavior under three different penalties, from left to right: Penalty 1, 2, and 3. The dark line represents the evolution of the best quality over 300 generations, lighter lines represent average and worst qualities. Using Penalty 2 (middle image) the optimizer converges in the infeasible region.

possibilities of this generic environment for heuristic simulation-based optimization. The evolution strategies which have been applied make use of Plus as well as Comma selection, intermediate recombination and modify the parameter vector by adding a normal distributed random number $N(0, \sigma_i)$ to the parameter and rounding the result to the nearest integer. The mutation strength vector σ is initialized randomly. The basic layout of the algorithm is given in [3]. The evolution strategy applied can be described as $(2/2_i,20)$-ES, $(2/2_i+20)$-ES, $(5/5_i,20)$-ES, and $(5/5_i+20)$-ES. There have been 1, 5, and 10 simulation replications performed in tests with and without CRN and with and without the uncertain best selector. The significance level α in the uncertain best selector has been set to 0.05. The maximum number of evaluations, including replications, has been set to 20,000 in each configuration. Thus using less replications allows the search to proceed longer. In total 40 different configurations have been created; each configuration has been tested 30 times. The best solutions obtained in each configuration have been tested a 100 times on the simulation model in order to decide about the robustness of a certain solution. A solution is considered 100% robust if in all 100 cases the maximum customer waiting time remained below the threshold.

5 Results

In Figure 2 the average results of all 40 configurations are displayed in a scatter plot where the x axis represents the average robustness, and the y axis represents the average quality. The interesting solutions have been marked in black. They represent the pareto front above 70% robustness and have been obtained with the (2/2,20)-ES and (5/5,20)-ES with 5 or 10 replications, and without CRN. These configurations thus performed best on average.

In Figure 3 and in Figure 4 the Comma and Plus strategy are compared with and without CRN. It can be seen that the choice of the selection scheme becomes less important when CRN is used. Both the Plus and the Comma ES variant produce solutions of about equal quality and robustness, the Comma strategy

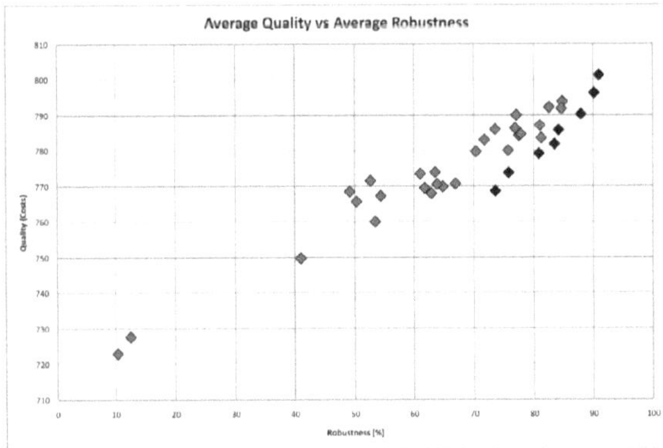

Fig. 2. Average results found in 40 different configurations

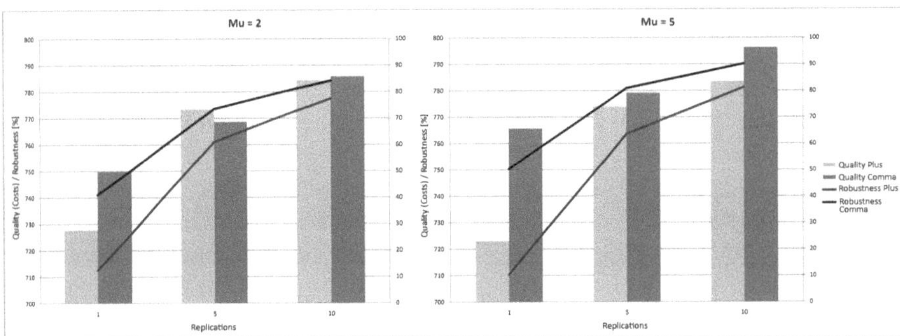

Fig. 3. Comparison of Plus and Comma selection without CRN. The lighter bar represents the average quality obtained in Plus selection, while the darker bar stands for Comma selection. Similarly the lines, which are fit to the secondary axis, represent robustness in Plus selection (lighter) and Comma selection (darker).

performs slightly better with regard to robustness when fewer replications are made, while Plus selection becomes better in the case of 10 replications. When using only a single "common" random number the optimization treats a deterministic version of the simulation model. As can be seen this does not lead to robust results.

When not using CRN more robust results can be achieved. Also it can be seen that using Comma selection results in very robust solutions already with a few replications, 5 in this case, while Plus selection is not able to achieve an average robustness larger than 70% until 10 replications are made. Plus selection is leading the search deeper into the infeasible region when still a few feasible solutions can be found.

We can also see that in both cases the algorithms benefit from a larger population size as the results are generally better when μ is set to a higher value.

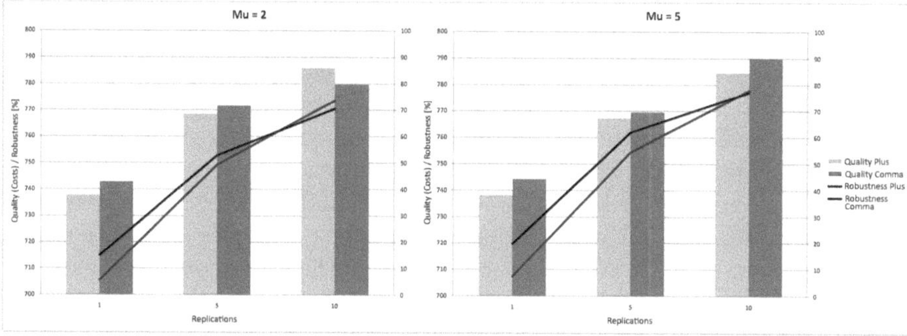

Fig. 4. Comparison of Plus and Comma selection with CRN. The lighter bar represents the average quality obtained in Plus selection, while the darker bar stands for Comma selection. Similarly the lines, which are fit to the secondary axis, represent robustness in Plus selection (lighter) and Comma selection (darker).

Finally in Figure 5 we would also like to show some results using the uncertain best selector. Unfortunately the results without CRN do not show much benefit from using this selection step. Robustness is slightly higher than in those results where the average value has always been used, but the differences are not pronounced. When using CRN however, two samples are more directly comparable and thus also the statistical test benefits. It is visible how the squares and the "x" which mark those results obtained using the uncertain best selector are among the most robust solutions obtained. Also it can be seen again that the bigger population size positively effected the search as the results are located closer to each other and more to the right.

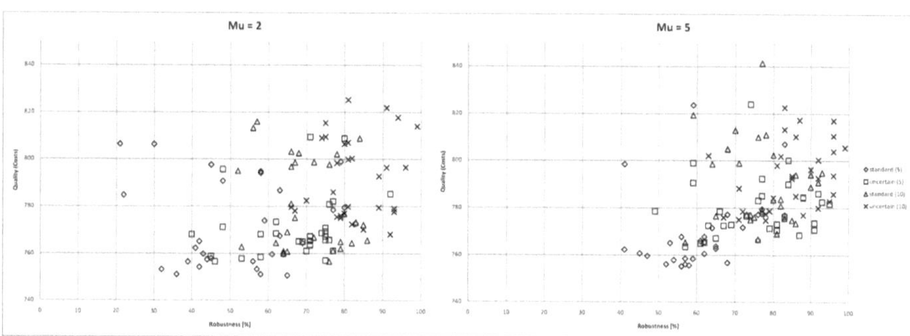

Fig. 5. Results obtained without and without the uncertain best selector when using CRN. Spades and Triangles represent solutions found with standard Comma selection, while those marked by x and squares were found using the uncertain best selector.

6 Conclusions

We have discussed an approach to optimizing a stochastic simulation model and analyzed it with respect to elitist and non-elitist selection and shown how the fitness function and the design of the penalty may influence the search. To optimize stochastic simulation models we have introduced a new selection method based on the Mann-Whitney-Wilcoxon rank sum test and shown that it performed quite well in combination with variance reduction techniques such as common random numbers (CRN). We can conclude that in the case of output constraints elitist selection may require more replications and if used in conjunction with CRN results in good and robust solutions. If solution evaluation takes a lot of time, it is an option not to use replications, or only a few and employ non-elitist selection as it achieves good and robust solutions in those cases as well.

For future work it would certainly be interesting to look into dynamically adapting the amount of replications that are necessary for significant decisions and applying to and evaluating the performance of the introduced concepts on further simulation models. Additionally it might be interesting to apply multiobjective optimization and fully explore the pareto front of quality and robustness.

Acknowledgements. The work described in this paper was done within the Josef Ressel Centre for heuristic optimization sponsored by the Austrian Research Promotion Agency (FFG).

References

1. Affenzeller, M., Winkler, S., Wagner, S., Beham, A.: Genetic Algorithms and Genetic Programming - Modern Concepts and Practical Applications. Numerical Insights. CRC Press, Boca Raton (2009)
2. Beham, A., Affenzeller, M., Wagner, S., Kronberger, G.: Simulation optimization with heuristiclab. In: Bruzzone, A., Longo, F., Piera, M.A., Aguilar, R.M., Frydman, C. (eds.) Proceedings of the 20th European Modeling and Simulation Symposium (EMSS 2008), pp. 75–80. DIPTEM University of Genova (2008)
3. Beyer, H.-G., Schwefel, H.-P.: Evolution strategies - a comprehensive introduction. Natural Computing 1, 3–52 (2002)
4. Fu, M.C., Glover, F., April, J.: Simulation optimization: A review, new developments, and applications. In: Proceedings of the 2005 Winter Simulation Conference, pp. 83–95 (2005)
5. Law, A.M.: Simulation Modeling and Analysis, 4th edn. McGraw-Hill, New York (2007)
6. Mann, H.B., Whitney, D.R.: On a test of whether one of two random variables is stochastically larger than the other. Annals of Mathematical Statistics 18, 50–60 (1947)
7. Rechenberg, I.: Evolutionsstrategie. Friedrich Frommann (1973)
8. Wagner, S.: Heuristic Optimization Software Systems - Modeling of Heuristic Optimization Algorithms in the HeuristicLab Software Environment. PhD thesis, Johannes Kepler University, Linz, Austria (2009)

Feature Selection Based on Pairwise Classification Performance

Stephan Dreiseitl[1] and Melanie Osl[2]

[1] Dept. of Software Engineering
Upper Austria University of Applied Sciences
A-4232 Hagenberg, Austria
[2] Dept. of Biomedical Engineering
University for Health Sciences, Medical Informatics and Technology
A-6060 Hall, Austria

Abstract. The process of feature selection is an important first step in building machine learning models. Feature selection algorithms can be grouped into wrappers and filters; the former use machine learning models to evaluate feature sets, the latter use other criteria to evaluate features individually. We present a new approach to feature selection that combines advantages of both wrapper as well as filter approaches, by using logistic regression and the area under the ROC curve (AUC) to evaluate *pairs* of features. After choosing as starting feature the one with the highest individual discriminatory power, we incrementally rank features by choosing as next feature the one that achieves the highest AUC in combination with an already chosen feature. To evaluate our approach, we compared it to standard filter and wrapper algorithms. Using two data sets from the biomedical domain, we are able to demonstrate that the performance of our approach exceeds that of filter methods, while being comparable to wrapper methods at smaller computational cost.

Keywords: Feature selection, feature ranking, pairwise evaluation.

1 Introduction

In machine learning and data mining, *feature selection* is the process of choosing, from a list of available features, those that are best suited for a particular problem [1]. Choosing appropriate features is beneficial from two points of view: From the point of view of machine learning, all algorithms can benefit from considering only features of interest without having to deal with noise (irrelevant features) in the data. From the point of view of particular application areas, the identification of relevant features may provide interesting information for the applications area itself. This, most notably, is the case in bioinformatics, where *biomarker discovery* is important for both medical as well as economical reasons, because biomarkers constitute valuable intellectual property for pharmaceutical companies.

R. Moreno-Díaz et al. (Eds.): EUROCAST 2009, LNCS 5717, pp. 769–776, 2009.

There are two competing methodologies for selecting suitable feature subsets for supervised machine learning algorithms: wrappers and filters [2]. *Wrapper methods* [3] make use of machine learning algorithms to evaluate the quality of feature subsets. Because the number of feature subsets is exponential in the number of features, features subsets are usually evaluated incrementally or decrementally, i.e., by adding or removing single features from subsets. An alternative to this form of greedy feature inclusion/exclusion strategy is to use more advanced heuristic search methods, e.g. as provided by genetic algorithms [4]. Wrappers thus always evaluate and judge features in combination with other features in a subset. In contrast, *filter methods* [5] work independently of machine learning algorithms by providing their own criteria for judging the merit of a feature, such as correlation with the value to be predicted. The large majority of filter methods evaluate each feature individually. Using filters, it is thus possible to rank-order the features independent of any particular machine learning algorithm.

Given the choice between wrapper and filter approaches, wrappers are often preferred when a learning algorithm has already been selected, because wrappers make use of an algorithm to select feature subsets that work well with this particular algorithm. A drawback of wrapper approaches is their high computational complexity, which can only partially be alleviated by heuristic search through subset space. Filter approaches, by evaluating features one by one, do not share this disadvantage, and are thus generally much faster than wrappers. There is, however, a drawback to considering every feature by itself: By not evaluating combinations of features, filters cannot take correlations and redundancies among features into account.

In this paper, we propose a hybrid filter-wrapper approach to feature selection. We use a machine learning algorithm to evaluate the performance of *pairs* of features. This constitutes the wrapper component of our approach. We then rank the features based on the matrix of pairwise performance numbers; this constitutes the filter part of our approach.

Various algorithms for considering feature selection based on pairs of features were recently proposed in the literature. The works of Bo and Jonassen [6], Pekalska *et al.* [7] and Harol *et al.* [8] all incrementally add *pairs* of features to a feature subset. In contrast, we only *evaluate* features in pairs, but rank-order them individually. The work of Michalak and Kwasnicka [9] extends on the papers listed above by considering feature correlations before evaluating them either in pairs or individually.

2 Method

The basis of our approach to feature ranking is the evaluation of the discriminatory ability of pairs of features. In our algorithm, the discriminatory ability is measured by training a logistic regression model [10], and measuring the discriminatory power of this model by ROC analysis. The area under the ROC curve (AUC) is a one-number summary of classifier performance [11], and is generally considered to be a better measure than accuracy [12].

2.1 Pairwise Feature Selection

Our pairwise feature ranking algorithm is a greedy search through feature space, at each step choosing a new feature that is optimal for the already chosen features. A ranking is obtained from the order in which the features are added to the set of already chosen features. This process of ranking features is done within a ten-fold cross-validation setup. This means that ten rankings are produced (one for each of the ten cross-validation runs), and the final ranking is determined as an average ranking (by assigning points to positions in a ranking, and averaging these points).

A more precise description of one cross-validation iteration of the algorithm is given by the following pseudo-code:

1. Rank features individually based on their discriminatory power, as measured by the AUC of a logistic regression model trained on only this feature. Choose as first feature f^* the one with highest AUC. Initialize the set S of already chosen features as $S \leftarrow \{f^*\}$, and the set of remaining features R as all other features.
2. Iterate the following until $R = \emptyset$:
   (a) Use a logistic regression model and the AUC to calculate the discriminatory power of all pairs in $\{f^*\} \times R$. Thus, only the combination of the last selected feature with all remaining features has to be calculated.
   (b) The next selected feature f^* is the one that achieves the highest AUC in combination with a feature in S. Set $S \leftarrow S \cup \{f^*\}$. Set $R \leftarrow R \setminus \{f^*\}$.

If one is interested in only a small number n of features, the algorithm can be modified to terminate after ranking the first (best) n features.

In our algorithm, we evaluate a classifier on the same data that it was trained on. This is generally discouraged in machine learning. Here, however, it is necessitated by the need for an independent test set to evaluate the quality of feature rankings (and not to evaluate the quality of an individual logistic regression model). This setup is used because the goal of classifier training is, in this case, not model building with a subsequent evaluation of model quality, but the relative performance of feature pairs. This relative performance is unaffected by evaluating the classifiers on their training sets. The evaluation of feature rankings is described in more detail in Section 2.3.

2.2 Other Feature Ranking Algorithms

To validate the feature ranking algorithm described above, we compare its performance with that of filter and wrapper approaches. These approaches are described briefly in the following.

Information gain. This filter method is suitable for discrete features; thus, continuous features first have to be discretized before their information gain can be computed. The information gain IG for the class label C given a feature f is defined as the decrease in entropy H of the class label when subtracting the conditional entropy of C given f, i.e. as $\mathrm{IG}(C|f) = H(C) - H(C|f)$.

ReliefF. This filter method repeatedly samples the data set and finds, for each sampled data point, the k nearest neighbors in both classes. A feature's importance is calculated as the mean distance (over all k neighbors and all sampled points) to points in the other class, minus the mean distance to points in the same class [13].

Greedy forward selection. This wrapper method uses logistic regression and the AUC to, one by one, greedily add the best of the remaining features to a solution set. The feature ranking is given by the order in which the features are added to the solution set.

2.3 Evaluation of Feature Ranking Algorithms

In every iteration of the ten-fold cross-validation loop, one tenth of the data set is set aside to evaluate the feature rankings. This evaluation is done after cross-validation, i.e., using the final (averaged) ranking. The quality of a feature ranking is determined as the discriminatory power of the first k features in the ranking, for $k = 1, \ldots, 15$. The discriminatory power is measured as the average AUC for ten logistic regression models, each trained on the training part (nine tenths) of the data set in a cross-validation iteration. The models were evaluated on the remaining part of the data set that had not previously been used in determining the ranking.

3 Experiments

Our algorithm was evaluated and compared on two data sets from the biomedical domain. The first is a data set containing 107 morphometric features extracted from pigmented skin lesion images obtained by dermoscopy at the Department of Dermatology of the Medical University of Vienna [15]. There are 105 melanoma images in the data set, and 1514 images of benign lesions. The second is a publically available micro-array data set for lung cancer diagnosis [14], with 97 cancer cases and 90 controls expressed in 22215 genes (features).

The feature rankings obtained by our algorithm (and the filters and wrapper we compare it with) were evaluated as described above. The rankings of information gain (IG), ReliefF (RF) and the greedy forward selection (GFS) were also obtained as averages over ten cross-validation iterations using the same portions of the data set we used for determining the pairwise feature ranking (PFR).

Fig. 1 summarizes the results of evaluating the four feature ranking algorithms on the melanoma data set. It can be seen that the performance of our method quickly exceeds that of the other methods, with GFS requiring nine features to exceed our method (and reaching its highest AUC of 0.906 at 13 features). The best performance of PFR (AUC of 0.882) is achieved with only four features, whereas IG requires 11 features to achieve its maximum AUC value of 0.859. RF is consistently worse than the others, with its highest AUC of 0.814 at eight features.

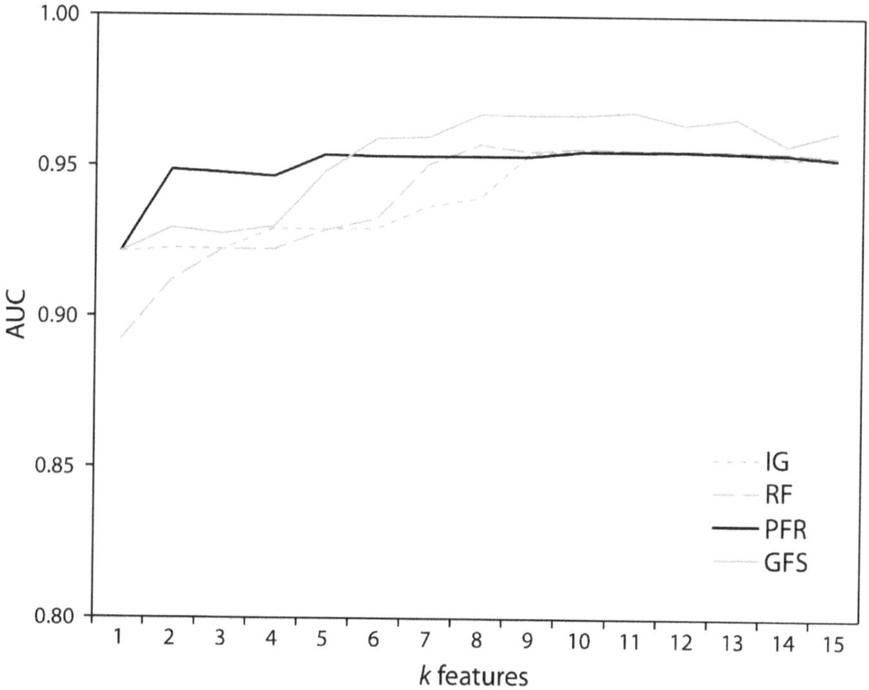

Fig. 1. The discriminatory power of the first k features obtained by information gain, ReliefF, greedy forward selection, and our pairwise ranking method on the melanoma image data set.

The performance comparison of the four feature ranking methods on the melanoma image data set is shown in Fig. 1. One can observe that IG, GFS and PFR all choose the same best individual feature, while RF lags behind. PFR quickly achieves its highest discriminatory power of 0.954 with only five features, whereas IG and RF require eight and nine features, respectively, to achieve this level (with RF slightly exceeding PFR). GFS performs best, with a maximum AUC value of 0.968 using eight features. Using six or more features, the performance of GFS exceeds that of PFR.

4 Discussion

The experiments presented above show that our new method of pairwise feature ranking compares favorably with the two filter methods information gain and ReliefF, while exceeding the performance of a greedy wrapper approach for small numbers of features. This is not surprising. By considering the interaction of features (if only in pairs), it is to be expected that pairwise feature ranking performs better than filter methods that evaluate features individually. On the other hand, it is also to be expected that wrapper methods eventually (using

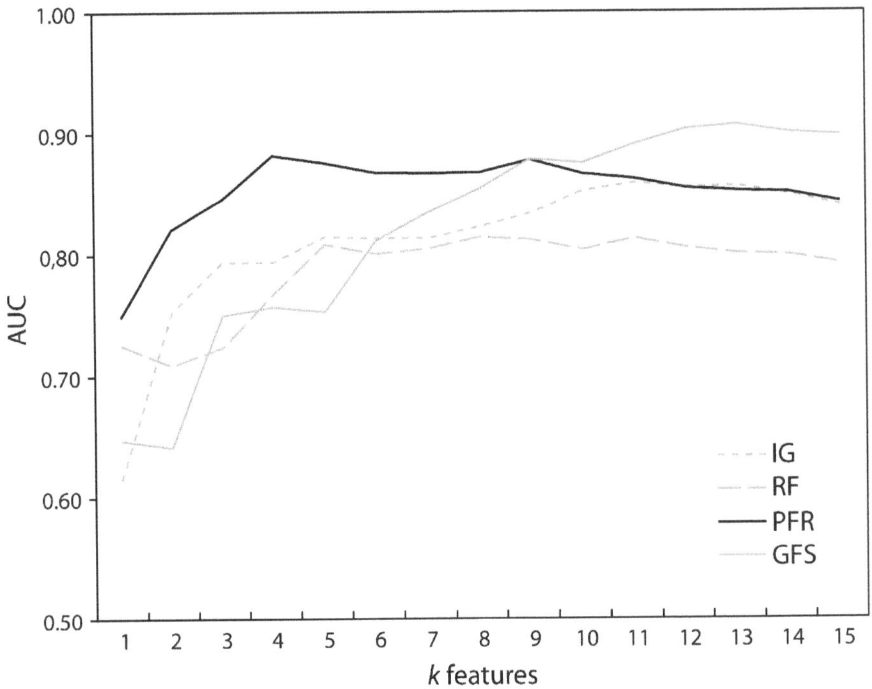

Fig. 2. The discriminatory power of the first k features obtained by information gain, ReliefF, greedy forward selection, and our pairwise ranking method on the lung cancer set.

more and more features) will detect beneficial feature interactions that cannot be found looking only at pairs of features.

It is, however, interesting that pairwise feature ranking outperforms the wrapper method for small numbers of features. A large part of this can be attributed to the rather high variability of the ten rankings obtained by the wrapper approach in the cross-validation loop. This means that features with high positions in one ranking may have lower positions in other rankings. This effect is caused by redundancy in the features—if one feature is chosen by the wrapper, it may happen that an equally good feature is not added to the solution set, because it does not contribute (in combination) more than another feature that is not as relevant when considered by itself. Thus, for small numbers of features, the ranking is dominated not by *great* features (that are sometimes ranked high, sometimes lower), but by *good* features that have, on average, high positions in the ranking.

Considering features individually (or only in pairs) can therefore be beneficial. As one can observe from the large regions in Fig. 2 in which the performance stays constant, there is also a drawback to considering features individually or in pairs: Redundancy between features is either not considered at all (for the filters), or only to a limited degree (for pairwise feature selection). Thus, features that do not contribute to the performance of a feature *set* may appear high in a feature *ranking*. Therefore, the performance of a set does not change when such features are included. This effect is more prominent in the lung cancer data set, and not as visible in the melanoma data set. A possible approach to counteract this phenomenon is to re-order feature rankings in such a way that redundant features are demoted to positions in the ranking in which they *do* contribute to discriminatory power (because a complementary feature is ahead in the ranking). This idea is further investigated in a recent paper [16].

Wrappers *select* features (by evaluating subsets of features), whereas filters *rank* features (by evaluating them individually). As used in this paper, wrappers can also produce feature rankings as the order in which features are added to a subset. Similarily, filters can also select feature subsets by taking the k best features in a ranking. The choice of k, however, is usually done empirically, by choosing the number of features that maximizes discriminatory power. As can be seen in Fig. 2 for PFR and RF, and in Fig. 1 for GFS, performance curves are not necessarily monotonically increasing functions of feature numbers. The decrease in performance with more features can be attributed to the fact that a feature further down in the ranking may still be the best of all remaining features, even if it does not increase the discriminatory power of a feature set. Rather than merely *not increase*, such a feature may actually *decrease* performance by overfitting a model to a spurious feature.

Although the run-time of our algorithm is quadratic in the number of features, and thus comparable to a greedy best-first wrapper approach, it is still preferable to wrappers because individual models are much smaller and thus faster to calculate. Furthermore, by evaluating only two features at a time, our approach is not as prone to overfitting and other phenomena associated with high-dimensional data in machine learning tasks.

Acknowledgements. This work was funded in part by the Austrian genome research program (GEN-AU), project *Bioinformatics Integration Network* (BIN III).

References

1. Guyon, I., Elisseeff, A.: An introduction to variable and feature selection. The Journal of Machine Learning Research 3, 1157–1182 (2003)
2. Hall, M.A., Holmes, G.: Benchmarking attribute selection techniques for discrete class data mining. IEEE Transactions on Knowledge and Data Engineering 15, 1437–1447 (2003)
3. Kohavi, R., John, G.H.: The Wrapper Approach. In: Feature Selection for Knowledge Discovery and Data Mining, pp. 33–50. Kluwer Academic Publishers, Dordrecht (1998)
4. Yang, J., Honavar, V.: Feature subset selection using a genetic algorithm. IEEE Intelligent Systems and their Applications 13, 44–49 (1998)
5. John, G.H., Kohavi, R., Pfleger, K.: Irrelevant features and the subset selection problem. In: Proceedings of the 11th International Conference on Machine Learning (1994)

6. Bo, T.H., Jonassen, I.: New feature subset selection procedures for classification of expression profiles. Genome Biology 3 (2002); research0017.1–0017.11
7. Pekalska, E., Harol, A., Lai, C., Duin, R.P.W.: Pairwise selection of features and prototypes. In: Proceedings of the 4th International Conference on Computer Recognition Systems, pp. 271–278 (2005)
8. Harol, A., Lai, C., Pekalska, E., Duin, R.P.W.: Pairwise feature evaluation for constructing reduced representations. Pattern Analysis & Applications 10, 1433–7541 (2007)
9. Michalak, K., Kwasnicka, H.: Correlation-based feature selection strategy in classification problems. International Journal of Applied Mathematics and Computer Science 16, 503–511 (2006)
10. Hosmer, D.W., Lemeshow, S.: Applied Logistic Regression, 2nd edn. Wiley-Interscience Publication, Hoboken (2000)
11. Hanley, J.A., McNeil, B.J.: The meaning and use of the area under a receiver operating characteristic (ROC) curve. Radiology 143, 29–36 (1982)
12. Provost, F., Fawcett, T., Kohavi, R.: The case against accuracy estimation for comparing induction algorithms. In: Proceedings of the 15th International Conference on Machine Learning, pp. 445–453 (1998)
13. Kononenko, I.: Estimating attributes: Analysis and extensions of RELIEF. In: Bergadano, F., De Raedt, L. (eds.) ECML 1994. LNCS, vol. 784, pp. 171–182. Springer, Heidelberg (1994)
14. Spira, A., Beane, J.E., Shah, V., Steiling, K., Liu, G., Schembri, F., Gilman, S., Dumas, Y.M., Calner, P., Sebastiani, P., Sridhar, S., Beamis, J., Lamb, C., Anderson, T., Gerry, N., Keane, J., Lunburg, M.E., Brody, J.S.: Airway epithelial gene expression in the diagnostic evaluation of smokers with suspect lung cancer. Nature Medicine 13, 361–366 (2007)
15. Dreiseitl, S., Ohno-Machado, L., Kittler, H., Vinterbo, S., Billhardt, H., Binder, M.: A comparison of machine learning methods for diagnosis of pigmented skin lesions. Journal of Biomedical Informatics 34, 28–36 (2001)
16. Osl, M., Dreiseitl, S., Cerqueira, F., Netzer, M., Baumgartner, C.: Improving feature ranking algorithms by demoting redundant features. J. Biomed. Inform. 42(4), 721–725 (2009)

On the Influence of Selection Schemes on the Genetic Diversity in Genetic Algorithms*

Michael Affenzeller, Stephan Winkler, Andreas Beham, and Stefan Wagner

Heuristic and Evolutionary Algorithms Laboratory
School of Informatics, Communications and Media
Upper Austrian University of Applied Sciences, Campus Hagenberg
Softwarepark 11, 4232 Hagenberg, Austria
{maffenze,swinkler,abeham,swagner}@heuristiclab.com

Abstract. This paper discusses some aspects of the general convergence behavior of genetic algorithms. Careful attention is given to how different selection strategies influence the progress of genetic diversity in populations. For being able to observe genetic diversity over time measures are introduced for estimating pairwise similarities as well as similarities among populations; these measures allow different perspectives to the similarity distribution of a genetic algorithm's population during its execution. The similarity distribution of populations is illustrated exemplarily on the basis of some routing problem instances.

1 Introduction

In the theory of genetic algorithms (GAs) population diversity and premature convergence are often considered as closely related topics. The loss of genetic diversity is usually identified as the primary reason for a genetic algorithm to prematurely converge. However, the reduction of genetic diversity is also needed in order to end up with a directed search process towards more promising regions of the search space.

What we would expect from an ideal genetic algorithm is that it loses the alleles of rather poor solutions on the one hand, and on the other hand that it slowly fixes the alleles of highly qualified solutions with respect to a given fitness function. In natural evolution the maintenance of high genetic diversity is important for the ability to adopt to changing environmental conditions. In contrast to this, in artificial evolution, where usually constant optimization goals are to be solved, the reduction of genetic diversity is even necessary for target-oriented search.

In this paper the dynamics of population diversity is documented and discussed in detail on the basis of typical and well known benchmark problem instances of routing problems like the travelling salesman problem (TSP) [3] or the capacitated vehicle routing problem [7] with (CVRPTW) or without time windows (CVRP). Different problem specific similarity measures $similarity(s_1, s_2)$

* The work described in this paper was done within the Josef Ressel centre for heuristic optimization sponsored by the Austrian Research Promotion Agency (FFG).

R. Moreno-Díaz et al. (Eds.): EUROCAST 2009, LNCS 5717, pp. 777–784, 2009.

are introduced for the respective problem representations on the basis of which some test showcases are described. In these tests we examine different parent selection strategies for standard GAs as well as GAs using offspring selection [1].

The rest of the paper is organized as follows: Section 2 discusses some general aspects of selection in evolutionary algorithms and explains the basic principles of offspring selection, and Section 3 introduces the similarity measures used for the comparison of solutions candidates; in Section 4 we document empirical results.

2 Selection Schemes

Concerning guidance of search corresponding to the given fitness function, selection is the driving force of GAs. In contrast to crossover and mutation, selection is completely generic, i.e. independent of the actually employed problem and its representation. A fitness function assigns a score to each individual in a population that indicates the 'quality' of the solution the individual represents. The fitness function is often given as part of the problem description or based upon the objective function.

In the standard genetic algorithm the probability that a chromosome in the current population is selected for reproduction is proportional to its fitness (roulette wheel selection). However, there are also many other ways of accomplishing selection. These include for example linear-rank selection or tournament selection [4], [6]. However, all evenly mentioned GA-selection principles have one thing in common: They all just consider the aspect of sexual selection, i.e. mechanisms of selection only come into play for the selection of parents for reproduction. Offspring selection [1] defies this limitation by considering selection in a more general sense.

2.1 Selection and Selection Pressure

In the population genetics view, especially in the case of not so highly developed species, sexual selection covers only a rather small aspect of selection which appears when individuals have to compete to attract mates for reproduction. The population genetics basic selection model basically considers the selection process in the following way:

random mating → selection → random mating → selection → ......

In other words this means that selection is considered to depend mainly on the probability of surviving of newborn individuals until they reach pubescence which is called viability in the terminology of population genetics. The essential aspect of offspring selection in the interpretation of selection is rarely considered in conventional GA selection.

2.2 Offspring Selection

In principle, offspring selection (OS) acts in the following way: the first selection step chooses the parents for crossover either randomly or in the well-known way of genetic algorithms by proportional, linear-rank, or some kind of tournament selection strategy. After having performed crossover and mutation with the selected parents, offspring selection is inserted: For this purpose, we check the success of the apparently applied reproduction in order to assure the proceeding of genetic search mainly with successful offspring in that way that the used crossover and mutation operators were able to create a child that surpasses its parents' fitness. Therefore, a new parameter, called success ratio ($SuccRatio \in [0,1]$), is introduced. The success ratio gives the quotient of the next population members that have to be generated by successful mating in relation to the total population size. Our adaptation of Rechenberg's success rule, originally stated for the $(1+1)-ES$ [5] for genetic algorithms says that a child is successful if its fitness is better than the fitness of its parents, whereby the meaning of 'better' has to be explained in more detail: is a child better than its parents, if it surpasses the fitness of the weaker, the better, or is it in fact some kind of mean value of both?

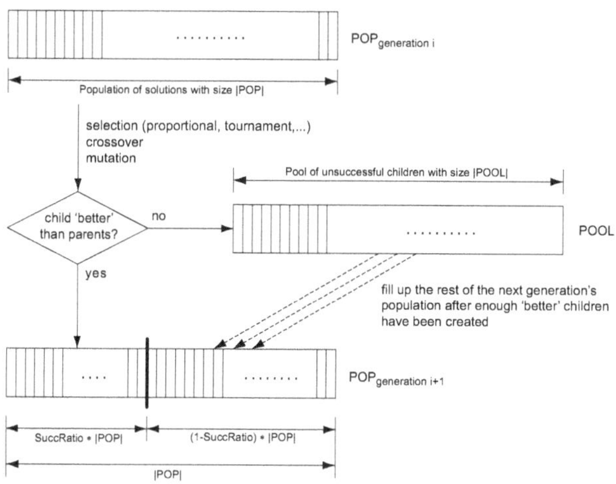

Fig. 1. Flowchart of the embedding of offspring selection into a genetic algorithm

As an answer to this question we claim that an offspring only has to surpass the fitness value of the worst parent in order to be considered as "successful" in the beginning, while as evolution proceeds the child has to be better than a fitness value continuously increasing between the fitness of the weaker and the better parent. As in the case of simulated annealing, this strategy gives a broader search at the beginning, whereas at the end of the search process this operator acts in a more and more directed way. Having filled up the claimed ratio ($SuccRatio$) of the next generation with successful individuals using the success criterion defined above, the rest of the next generation ($(1-SuccRatio)\cdot \mid POP \mid$) is simply filled up with individuals randomly chosen from the pool of individuals that were also created by crossover, but did not reach the success criterion. The actual selection pressure $ActSelPress$ at the end of a single generation is defined by the quotient of individuals that had to be considered until the success ratio was reached and the number of individuals in the population in the following way:

$$ActSelPress = \frac{|POP_{i+1}| + |POOL|}{|POP|}$$

Fig. 1 shows the operating sequence of the above described concepts. With an upper limit of selection pressure $MaxSelPress$ defining the maximum number of children considered for the next generation (as a multiple of the actual population size) that may be produced in order to fulfill the success ratio, this new model also functions as a precise detector of premature convergence:

If it is no longer possible to find a sufficient number of $(SuccRatio \cdot |POP|)$ offspring outperforming their own parents even if $(MaxSelPress \cdot |POP|)$ candidates have been generated, premature convergence has occurred.

As a basic principle of this selection model a higher success ratio causes higher selection pressure. Nevertheless, higher settings of success ratio and therefore of selection pressure do not necessarily cause premature convergence as the preservation of fitter alleles is additionally supported and not only the preservation of fitter individuals. Also it becomes possible within this model to state selection pressure in a very intuitive way that is quite similar to the notation of selection pressure in evolution strategies. Concretely, we define the actual selection pressure as the ratio of individuals that had to be generated in order to fulfill the success ratio to the population size. For example, if we work with a population size of say 100 and it would be necessary to generate 2000 individuals in order to fulfill the success ratio, the actual selection pressure would have a value of 20. Via these means we are in a position to attack several reasons for premature convergence as illustrated in the following sections. Furthermore, this strategy has proven to act as a precise mechanism for self-adaptive selection pressure steering, which is of major importance in the migration phases of parallel evolutionary algorithms.

3 Similarity Measures

The observance of genetic diversity over time is the main objective of this paper. For this reason we apply specific similarity measures in order to monitor and to analyze the diversity and population dynamics. According to the definitions stated in [2] we will use the following problem independent definitions where the concrete definition of $similarity(s_1, s_2)$ has to be stated separately for a certain problem representation:

- **Similarity between two solutions**
  As similarity measures do not have to be symmetric, we use the mean value of the two possible similarity calls and so define a symmetric similarity measure.

$$sim(s_1, s_2) = \frac{similarity(s_1, s_2) + similarity(s_2, s_1)}{2} \qquad (1)$$

– **Similarity of a solution s within a population P**
In order to have a measure for the similarity of a certain solution s within a population P at a certain iteration we calculate the average and the maximum similarity of s to all other population members in the following way:

$$meanSim(s,P) = \frac{1}{|P|-1} \sum_{s_* \in P, s_* \neq s} sim(s, s_*) \tag{2}$$

$$maxSim(s,P) = max_{(s_* \in P, s_* \neq s)}(sim(s, s_*)) \tag{3}$$

– **Similarity within a population P**

$$meanSim(P) = \frac{1}{|P|} \sum_{s \in P} meanSim(s,P) \tag{4}$$

$$maxSim(P) = \frac{1}{|P|} \sum_{s \in P} maxSim(s,P) \tag{5}$$

Whereas the similarity definitions stated in the formulae 1 – 5 do not depend on a concrete problem representation, the similarity itself has to be defined according to the problem representation at hand.

The similarity measure between two TSP-solutions t_1 and t_2 used here is defined as a similarity value sim between 0 and 1:

$$sim(t_1, t_2) = \frac{| e : e \in E(t_1) \wedge e \in E(t_2) |}{| E(t_1) |} \in [0, 1] \tag{6}$$

giving the quotient of the number of common edges in the TSP solutions t_1 and t_2 and the total number of edges. E here denotes the set of edges in a tour. The according distance measure can then be defined as

$$d(t_1, t_2) = 1 - sim(t_1, t_2) \in [0, 1] \tag{7}$$

Thus, the similarity or the distance of two concrete TSP solutions can be measured on a linear scale between the values 0 and 1.

The similarity measure for two VRP solutions t_1 and t_2 is calculated in analogy to the TSP similarity using edgewise comparisons. However, as big routes in the VRP are subdivided into smaller routes, a maximum similarity sim_{max} is calculated for each route $r \in t_1$ to all routes $s \in t_2$. These values are summed for all routes r_i and finally divided by the number of routes.

4 Results

The results shown in this section are aimed to just show the basic principle of dynamic diversity analysis for genetic algorithms on the basis of two different GA selection paradigms which are very characteristically. For more sophisticated analyses with tests for more benchmark instances of different combinatorial, real-valued and genetic programming problems performing a sufficient number of test runs for each parameter setting with a sophisticated discussion of the achieved results the interested reader is referred to the book [2].

A very detailed representation of genetic diversity in a population is the statement of pairwise similarities or distances for all members of a population. An appropriate measure, which is provided in the HeuristicLab framework, is to illustrate the similarity as a $n \times n$ matrix where each entry indicates the similarity in form of a grey scaled value. Fig. 2 shows an example: The darker the $(i, j) - th$ entry in the $n \times n$ grid is, the more similar are the two solutions i and j. Not surprisingly, the diagonal entries, which stand for the similarity of solution candidates with themselves, are black indicating maximum similarity.

Fig. 2. Degree of similarity/distance for all pairs of solutions in a SGA's population of 120 solution candidates after 10 generations

Unfortunately, this representation is not very well suited for a static monochrome figure. Therefore, the dynamics of this $n \times n$ color grid over the generations is shown in numerous colored animations available at the website of the book [2][1].

For a meaningful figure representation of genetic diversity over time it is necessary to summarize the similarity/distance information of the entire population in a single value. An average value of all n^2 combinations of solution pairs in form of a mean/max similarity value of the entire population as a value between 0 and 1 can be calculated according to the Formulas 2 to 5 stated in section 2. This form of representation allows to display genetic diversity over the generations in a single curve. Small values around 0 indicate low average similarity, i.e., high genetic diversity and vice versa high similarity values of almost 1 indicate little genetic diversity (high similarity) in the population. In the following we

[1] `http://gagp2009.heuristiclab.com`

Table 1. Overview of standard GA and offspring selection GA parameters

Parameters for the standard GA		**Parameters for the offspring selection GA**	
Generations	100,000	Population Size	500
Population Size	120	Elitism Rate	1
Elitism Rate	1	Mutation Rate	0.05
Mutation Rate	0.05	Selection Operator	Roulette
Selection Operator	Roulette	Mutation Operator	Simple Inversion
Mutation Operator	Simple Inversion	Success Ratio	0.7
		Maximum Selection Pressure	250

show results of exemplary test runs of GAs applied to the *kroA200* 200 city TSP instance taken from the TSPLib using the parameter settings given in Table 1 and OX crossover.

Fig. 3 shows the genetic diversity curves over the generations for a conventional standard genetic algorithm as well as for a typical offspring selection GA. The gray scaled values of Fig. 3 show the progress of mean similarity values of each individual (compared to all others in the population); average similarity values are represented by solid black lines.

For the standard GA it is observable that the similarity among the solution candidates of a population increases very rapidly causing little genetic diversity already after a couple of generations; it is only mutation which is responsible for reintroducing some new diversity keeping the evolutionary process going. Without mutation the algorithm converges as soon as the genetic diversity of the population is lost, which happens very soon in case of the standard GA. In terms of global solution quality, the finally achieved results with an offspring selection GA are slightly superior to the standard GA and quite close (about 0,5% to 5%) to the global optimum. But for the standard GA this property only holds for well adjusted mutation rates of about 5%. Without mutation the standard GA fails drastically whereas the GA with offspring selection is still able to achieve quite the same solution qualities (about 0,5% to 2% off the global optimum) [2]. The explanation for this behavior is quite simple when we take

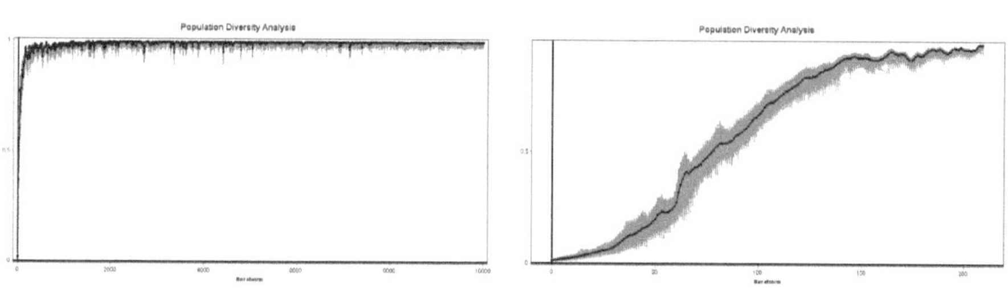

Fig. 3. Left: Genetic diversity over time in the population of a conventional GA (left figure), **right:** genetic diversity over time in the population of a GA with offspring selection

a look at the genetic diversity over time: in case of offspring selection diversity disappears slowly and controlled whereas in the case of the standard GA diversity is lost very soon and from that time on it is only mutation that keeps evolution running.

Summarizing these results it can be stated for the TSP experiments that the illustration in form of a static figure is certainly some kind of restriction when the dynamics of a system should be observed. For that reason the website of the book [2] contains some additional material showing the dynamics of pairwise similarities for all members of the population (as indicated in Fig. 2) in the form of short motion pictures.

References

1. Affenzeller, M., Wagner, S.: Offspring selection: A new self-adaptive selection scheme for genetic algorithms. In: Ribeiro, B., Albrecht, R.F., Dobnikar, A., Pearson, D.W., Steele, N.C. (eds.) Adaptive and Natural Computing Algorithms. Springer Computer Science, pp. 218–221. Springer, Heidelberg (2005)
2. Affenzeller, M., Winkler, S., Wagner, S., Beham, A.: Genetic Algorithms and Genetic Programming: Modern Concepts and Practical Applications. CRC Press, Boca Raton (2009)
3. Larranaga, P., Kuijpers, C.M.H., Murga, R.H., Inza, I., Dizdarevic, D.: Genetic algorithms for the travelling salesman problem: A review of representations and operators. Artificial Intelligence Review 13, 129–170 (1999)
4. Michalewicz, Z.: Genetic Algorithms + Data Structures = Evolution Programs. Springer, Heidelberg (1992)
5. Rechenberg, I.: Evolutionsstrategie. Friedrich Frommann Verlag (1973)
6. Schöneburg, E., Heinzmann, F., Feddersen, S.: Genetische Algorithmen und Evolutionsstrategien. Addison-Wesley, Reading (1994)
7. Thangiah, S.R., Potvin, J.-Y., Sun, T.: Heuristic approaches to vehicle routing with backhauls and time windows. International Journal on Computers and Operations Research 23(11), 1043–1057 (1996)

Solving a Real–World FAP Using the Scatter Search Metaheuristic

José M. Chaves-González, Miguel A. Vega-Rodríguez, Juan A. Gómez-Pulido, and Juan M. Sánchez-Pérez

Univ. Extremadura. Dept. Technologies of Computers and Communications, Escuela Politécnica. Campus Universitario s/n. 10071. Cáceres, Spain
{jm,mavega,jangomez,sanperez}@unex.es

Abstract. Frequency planning is a very important task in the design and operation of current GSM networks. For this reason, the frequency assignment problem (FAP) is a well-known problem in the Operations Research which includes different mathematical models depending on the specific conditions of the application which is being designed. However, most of these models are not close from considering current technologies aspects which are deployed in GSM networks. In this work, we use a formulation of FAP, developed in published work, which focuses on aspects which are used in real-word GSM networks. We focus on solving this problem for a realistic-sized, real-world GSM network, using the Scatter Search algorithm. We have analyzed and fixed the SS algorithm to the FAP problem and, after a detailed statistical study, the obtained results prove that this approach can compute accurate frequency plans for real-world instances in an optimum way. In fact, our results surpass all the results previously published in the literature.

Keywords: FAP, Frequency Planning, SS, real-world GSM network.

1 Introduction

Frequency planning is an optimization NP-hard problem, so its resolution using metaheuristics, such as the Scatter Search (SS) algorithm, is very appropriate [2]. But moreover, the SS metaheuristic has proved to be very effective in the resolution of optimization problems (such as the FAP is) and consequently the usage of this algorithm has been significantly increased in the last few years [3]. For this reason, we have studied, adapted and evaluated SS to a real-world GSM network with 2612 transceivers that currently operates in a quite large U.S. city (Denver city).

It is important to point that GSM (global system for mobile) is the most successful mobile communication technology nowadays. In fact, at the end of 2007 GSM services were in use by more than 3 billion subscribers [4] across 220 countries, representing approximately 85% of the world's cellular market. One of the most relevant and significant problems that it can be found in the GSM technology is the frequency assignment problem (FAP), because frequency planning is a very important and critical task for current (and future) mobile communication operators.

R. Moreno-Díaz et al. (Eds.): EUROCAST 2009, LNCS 5717, pp. 785–792, 2009.

The two most significant elements we find in the frequency planning problem are the transceivers (TRXs) which give support to the communication and the frequencies which make possible the communication. A mobile communication antenna includes several TRXs placed in several sectors of the antenna and each TRX has to have assigned a specific frequency in the most optimum way to provide the widest coverage and minimizing the interferences produced in the network. The problem is that there are not enough frequencies (there are not more than a few dozens) to give support to each transceiver (there are usually thousands of them) without causing interferences. It is completely necessary to repeat frequencies in different TRXs, so, a good planning to minimize the number of interferences is highly required.

The rest of the paper is structured as follows: In section 2 we present the background of our frequency assignment problem and the mathematical formulation we have used for its resolution. Section 3 describes the metaheuristic used in this study (SS). In section 4 we explain the experiments and the results obtained in the adjustment and optimization of the algorithm to the FAP problem. Finally, the conclusions and future work of the research are discussed in the last section.

2 Frequency Planning Problem in GSM Networks

The two most relevant components which refer to frequency planning in GSM systems are the *antennas* or, as they are more known, base transceiver stations (BTSs) and the *TRX*. The TRXs of a network are installed in the BTSs where they are grouped in sectors, oriented to different points to cover different areas. The instance we use in our experiments is quite large (it covers the city of Denver, USA, with more than 500,000 inhabitants) and the GSM network includes 2612 TRXs, grouped in 711 sectors, distributed in 334 BTSs. We are not going to extend the explanation of the GSM network architecture but the reader interested in it can consult reference [6].

There are several ways of quantifying the interferences produced in a telecommunication network, but the most extended one (and the method we use) is using what is called the *interference matrix* [9], denoted by M. Each element $M(i,j)$ of this matrix contains two types of interferences: the *co-channel interference*, which represents the degradation of the network quality if the cells i and j operate on the same frequency; and the *adjacent-channel interference*, which occurs when two TRXs operate on adjacent channels (e.g., one TRX operates on channel f and the other on channel $f+1$ or $f-1$). Therefore, an accurate interference matrix is an essential requirement for frequency planning because the ultimate goal of any frequency assignment algorithm will be to minimize the sum of all the interferences (1), or in other words, to find a solution p which minimizes the cost function (C). For a deeper explanation of the mathematical concepts of the problem, please, consult [1].

$$C(p) = \sum_{t \in T} \sum_{u \in T, u \neq t} C_{sig}(p,t,u) \tag{1}$$

The smaller the value of C is, the lower the interference will be, and thus the better the communication quality. In order to define the function $C_{sig}(p,t,u)$, let s_t and s_u be the sectors (from $S = \{s_1, s_2, \ldots, s_m\}$) in which the transceivers t and u are installed, which are $s_t = s(t)$ and $s_u = s(u)$ respectively. Moreover, let $\mu s_t s_u$ and $\sigma s_t s_u$ be the two elements of the corresponding matrix entry $M(s_t,s_u)$ of the interference matrix with respect to sectors s_t and s_u. Then, $C_{sig}(p,t,u)$ is equal to the following expression:

$$\begin{cases} K & \text{if } s_t = s_u, |p(t)-p(u)| < 2 \\ C_{co}(\mu_{s_t s_u}, \sigma_{s_t s_u}) & \text{if } s_t \neq s_u, \mu_{s_t s_u} > 0, |p(t)-p(u)| = 0 \\ C_{adj}(\mu_{s_t s_u}, \sigma_{s_t s_u}) & \text{if } s_t \neq s_u, \mu_{s_t s_u} > 0, |p(t)-p(u)| = 1 \\ 0 & \text{otherwise} \end{cases} \tag{2}$$

$K>>0$ is a very large constant defined by the network designer to make undesirable allocating the same or adjacent frequencies to TRXs serving the same area (e.g., placed in the same sector). $C_{co}(\mu,\sigma)$ is the cost due to *co-channel interferences*, and $C_{adj}(\mu,\sigma)$ represents the cost in the case of *adjacent-channel interferences* [8].

3 Scatter Search

Scatter Search (SS) [12, 13] is a metaheuristic oriented to solve optimization problems (such as the FAP). The algorithm works with a quite small set of solutions (called *RefSet*, which usually includes no more than 10 solutions). The proposed solutions are encoded as arrays of integer values, p; where $p(t_i) \in F_i$ is the frequency assigned to the transceiver t_i. Therefore, the solutions encoded are tentative frequency plans which solve the given FAP problem instance.

One of the most interesting features of SS is the double criteria of quality and diversity that the algorithm uses to work with the individuals. According to this criteria, the solutions held in the *RefSet* are divided into quality solutions (the best frequency plans for the FAP problem) and diverse solutions (the most different ones, which provide a wider exploration of the space search –in our case, the set of frequency plans that solve the problem–). A brief description of the algorithm we have used can be seen in figure 1.

Algorithm 1

```
initialize (population, PopulationSize)
population = improvementMethod (population)
best_solutions = selectBestSolutions (population, RefSetSize/2)
diverse_solutions = selectDivSolutions (population, RefSetSize/2)
RefSet = generateRefSet (best_solutions, diverse_solutions)
while (not time-limit) do
    subSets = subsetGenerationMethod (RefSet)
    while (subset not examined in subSets) do
        Solution = solutionCombinationMethod (subset)
        Solution = improvementMethod (Solution)
        if (cost (Solution) < worstCost (RefSet) AND Solution ∉ RefSet) then
            RefSet = updateRefSet (Solution)
        end if
    end while
end while
```

Fig. 1. Pseudocode for Scatter Search

The algorithm starts with the generation of the population using a *Diversification Generation Method* which creates random individuals, so that all TRXs included in each individual are assigned with one of its random valid frequencies (line 1). Then (line 2), an *Improvement Method* fixed to the FAP problem is applied to each population individual to try to improve it (obtaining a better frequency plan). This *Improvement Method* will be applied again (in line 10) with the same aim over the solution obtained as a result of the *Combination Method* (line 9). As a general rule, the *Improvement Method* will be used each time that a new solution is generated (as a result of either the *Diversification Generation Method* or the *Solution Combination Method*) and it is basically a local search method which tries to improve each solution making a search through the individual and trying different frequencies in all the TRXs of each sector in a solution. After generating the *RefSet* (line 5) we use a *Subset Generation Method* (line 7) to create all possible subsets from the *RefSet*. The next step is to apply the *Solution Combination Method* (line 9) to the solutions in each subset. The solutions are combined in a pair-wise way and after that, the local search will be applied again (line 10). Frequency plannings are replaced in the *RefSet* so that the best solutions to the FAP problem keep in there (lines 11 and 12). Finally, when all combinations have been made, an iteration of the algorithm has been completed. Then, the *RefSetSize/2* best solutions are saved in the *RefSet* and a new population is generated to select the *RefSetSize/2* most diverse solutions. The distance used to measure the diversity among frequency plans is the sum of the absolute values of the difference between all the frequencies assigned to TRX. With this new *RefSet* the algorithm restarts a new iteration from line 6 until the time limit expires.

4 Experimental Evaluation

In this section we present the different experiments we performed with the standard version of the algorithm to adjust and improve it with the purpose of obtaining the best frequency plans which give a solution to the FAP problem. As we said in the Introduction section, we have used a real-world instance to perform all our experiments. We considered that this is much more representative than using theoretical instances [14], because our requirements have come directly imported from the industry (which makes more practical the tests performed). In fact, due to the real approach used (section 2) our solutions do not consider only the computation of high performance frequency plans, but also the prediction of QoS, which is very important for the industry. To specify, we have used a real GSM network with 2,612 TRXs distributed in 711 sectors and with only 18 available channels (from 134 to 151) which is currently operating in a U.S. 400 km^2 city with more than 500,000 people, so its solution is of great practical interest.

In order to provide the results with statistical confidence and check the improvements provided by each version of the algorithm within short and long periods of time, we have considered 30 independent runs for each experiment taking in consideration three different time limits (120, 600 and 1800 seconds).

Besides, according to the mathematical formulation of the problem (section 2.1), the results are given in function of the cost that a frequency plan is able to obtain. The smaller the value of the cost is, the better the frequency plan (and the result) will be.

4.1 Empirical Results

Our starting points were the previous works published in [15] and [16] where our best result with SS was: 93820.4 cost units on average (after 30 executions, see table 1). From that state we have performed several improvements and fine adjustments over the algorithm until we have obtained optimum results with SS when solving a realistic FAP problem. In fact, to our best knowledge, we have obtained the best results tackling this problem, if we compare our results with the ones found in the bibliography which work with the same problem instance ([1], [11], [15], and [16]).

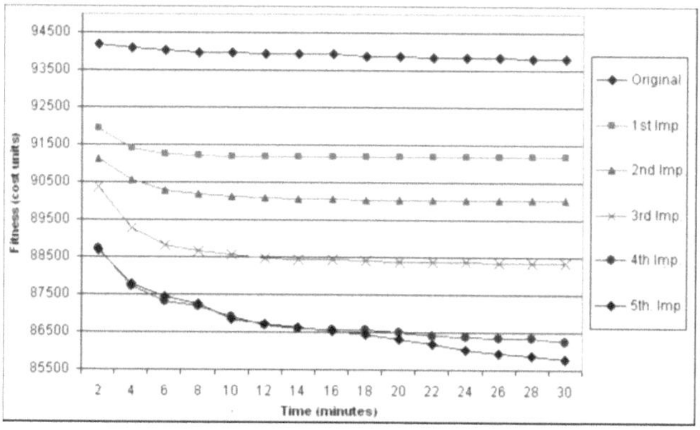

Fig. 2. Evolution of the results in different improvement stages

Figure 2 summarizes the main improvement stages applied in our study. The starting result was the one obtained in [16] (*Original* line in the chart of figure 2), where the algorithm had not been thoroughly tuned to the problem yet and the frequency plans we were able to obtain with SS had a cost of over 93800 cost units on average. The values of the algorithm parameters here were 40 for the *population size*; 10 for the *RefSet size*; a *balanced RefSet* (with 5 quality solutions and 5 diversity solutions); and uniform crossover at sector level (every 20-40 sectors) for the *Solution Combination Method* of the algorithm (see section 3). As we can see in figure 2, we have highlighted 5 different stages in the improvement of the algorithm which match with important advances in the algorithm tuning. These improvements are clearly represented in the graph of figure 2 with different lines (from *1st Imp.* to *5th Imp.*) and they are going to be explained in the following paragraphs.

The first improvement we got (which corresponds with the line *1st Imp.* in fig. 2) was motivated thanks to the study we performed about the crossover of the solutions. In the initial version of the algorithm (represented by the line *Original* in fig. 2) we used a uniform crossover at sector level (every 20-40 sectors) and after doing the study we discovered that configuring the crossover at TRX level was much better because results were improved significantly. But even more, we studied the frequency in which the crossover had to be done and our final conclusion was that the *uniform crossover at TRX level* exchanging the parent in every TRX of the solution was the optimum one (we performed several test as well with different random crossovers, but the uniform one gave us the best results). The possible explanation to the great improvement achieved with this crossover configuration is that the solutions obtained include a high diversity level, and this is very positive since the search space is increased if the solutions are very different among them. Therefore, with the study performed with the crossover we could reduce the cost of the frequency plans obtained from 93820 to 91186 (on average after 30 executions), which represents a quite good improvement.

The second improvement in the algorithm (represented in figure 2 by the *2nd Imp.* line) was obtained thanks to the improvement made in the *Diversification Generation Method* of the algorithm (see section 3). This method is very important, since it creates the initial population of individuals that the algorithm will use to work with, and as a general rule, we can state that the better the solutions are from the beginning the best results will be obtained at the end. In fact, this can be observed in figure 2, where the difference between the *1st Imp.* line and the *2nd Imp.* line is just the starting point of both lines (2nd Imp. starts with a lower -a better- cost than 1st Imp. thanks to the improvement developed in the *Diversification Generation Method* –this improvement reduces the cost at the end from 91186 to 90032–). The pseudo-code for this method can be observed in figure 3. Basically, this method avoids the highest-cost interferences (with *K* cost, see equation 2).

The third improvement made over the algorithm presented in section 3 was developed thanks to the study performed in the *RefSet proportions*. The *RefSet* includes quality and diversity solutions, and before this study was completed, we supposed that the half of the solutions of the *RefSet* had to be in the quality set and the other half in the diversity set. Therefore, if the *RefSet* size was equal to 10, both quality and diversity sets included 5 solutions each one. However, after doing a complete study with all the different proportions that the *RefSet* could take with 10 solutions ([1,9], [2,8],..., [9,1]) we concluded that the best configuration was [1,9]. This means that with one single solution in the quality set is enough to preserve the information of the best solution held in the *RefSet*. Line *3rd Imp.* in figure 2 shows us the evolution in the results with this improvement (at the end, the reduction of the cost goes from 90032 to 88330, which is a considerable reduction). As we can see, in the first 10 minutes the evolution of the algorithm with this improvement included is very strong, so the *RefSet proportions* is a very important parameter to consider because it determines the amount of solutions which will be saved through each iteration and the ones which will be updated from the population to provide diversity to the search.

The fourth improvement is shown in figure 2 through the line *4th Imp.* As we can see there, the evolution of the algorithm is improved in all the slices of time (from 2 to 30 minutes, reducing the final cost planning to 86263). These positive results are achieved thanks to the spread in the usage of the method of generation of a new solution which was explained in the pseudo-code of figure 3. In the *4th Imp.*, we discovered that it was very appropriate that each time a new solution was created, the *Efficient Generation Method* (fig. 3) is used to improve the solution, or in other words, this method is used not only in the creation of the initial population, but also each time that the population is regenerated, etc.

Finally, the evolution in the last stage of our study, and the present result we have, is represented in figure 2 with the line *5th Imp.* As we can observe, this line is the one with the best evolution in the whole graph because it keeps a good evolution not only in the first minutes of execution, but also in the last ones. The study developed for this

```
Algorithm 2
Solution = ∅
while ( current_sector <= last_sector ) do
  for (all TRX ∈ current_sector ) do
    freq_TRX = random (freq_range)
    if ( cost (current_sector, TRX, freq_TRX) < K_Cost ) then
      Solution = insert (Solution, TRX, freq_TRX )
    else
      freq_TRX = search_freq_without_K_cost (current_sector, freq_TRX )
      Solution = insert ( Solution, TRX, freq_TRX )
    end if
  end for
end while
```

Fig. 3. Pseudocode for Efficient Diversification Generation Method

Table 1. Empirical results (in cost units) for 3 different time limits

	120 seconds			600 seconds			1800 seconds		
	Best	Avg.	Std.	Best	Avg.	Std.	Best	Avg.	Std
Initial SS	91216.7	94199.6	1172.3	91069.8	93953.9	1178.6	91069.8	93820.4	1192.3
Final fixed SS	86169.4	**88692.7**	1124.9	84570.6	**86843.8**	950.5	84234.5	**85767.6**	686.3
ACO	90736.3	93439.5	1318.9	89946.3	92325.4	1092.8	89305.9	90649.9	727.5
LSHR	88543.0	92061.7	585.3	88031.0	89430.9	704.2	87743.0	88550.3	497.0
PBIL	-	-	-	-	-	-	162810	163394	346.78

improvement was about the size of the *RefSet*. We performed a complete set of tests with different sizes (from 3 to 15, keeping the proportions fixed in the *3rd Imp.* of this study) and the results told us that the best *RefSet size* to solve the FAP problem with the real instance used is 9, with the proportions [1,8]. This new size reduces a bit the set of work that the algorithm uses, allowing that SS performs more iterations and consequently improving the algorithm evolution. Results obtained with this improvement are detailed in the row *Final fixed SS* of table 1 (they are compared with the results obtained using the original version of the algorithm -row *Initial SS*- and the results found in the bibliography for the same problem but different metaheuristics).

5 Conclusions and Future Work

In this paper we present a complete study about the configuration and fine tuning of the SS algorithm to solve the FAP problem in a real-world GSM network composed of 2612 transceivers. We have improved the results obtained in all the other works found in the bibliography (using very different metaheuristics, [1], [11], [15], and [16], table 1) tackling the FAP problem with the same real-world instance, but moreover, we have improved significantly the best results that SS was able to obtain, such as we can see in table 1 (where the frequency planning costs are reduced to 85767.6 cost units on average after 30 independent executions). To carry out this task we have performed a complete study with several important parameters, such as the size and proportions of the *RefSet*, the crossover type and the usage of a fine tuned solution generation method. In conclusion, the results obtained in our experiments show that the SS algorithm is able to obtain optimum frequency plans (the best ones, to our best knowledge) in different periods of time when it is properly fixed.

Future work includes the evaluation of the algorithms using additional real-world instances, we will also use cluster and grid computing in order to speedup all our experiments, and we will try to extend the mathematical model to deal with more advanced issues in GSM frequency planning.

Acknowledgments. This work was partially funded by the Spanish Ministry of Science and Innovation and FEDER under the contract TIN2008-06491-C04-04 (the MSTAR project). José M. Chaves-González is supported by the research grant PRE06003 from Junta de Extremadura (Spain).

References

1. Luna, F., et al.: ACO vs EAs for Solving a Real-World Frequency Assignment Problem in GSM Networks. In: GECCO 2007, London, UK, pp. 94–101 (2007)
2. Glover, F.W., Kochenberger, G.A.: Handbook of Metaheuristics. Int. Series in Operations Research and Management Sciences. Kluwer A. P., Norwell (2003)
3. Glover, F., et al.: Scatter search. In: Ghosh, A., Tsutsui, S. (eds.) Advances in Evolutionary Computing: theory and Applications. Natural Computing Series, pp. 519–537. Springer, New York (2003)
4. GSM World (2007), http://www.gsmworld.com/news/statistics/index.shtml
5. Aardal, K.I., et al.: Models and Solution Techniques for Frequency Assignment Problems. 4OR 1(4), 261–317 (2003)
6. Eisenblätter, A.: Frequency Assignment in GSM Networks: Models, Heuristics, and Lower Bounds. PhD thesis, Technische Universität Berlin (2001)
7. Mouly, M., Paulet, M.B.: The GSM System for Mobile Comm. Telecom Publishing (1992)
8. Mishra, A.R.: Radio Network Planning and Opt. In: Fundamentals of Cellular Network Planning and Optimisation: 2G/2.5G/3G.. Evolution to 4G, pp. 21–54. Wiley, Chichester (2004)
9. Kuurne, A.M.J.: On GSM mobile measurement based interference matrix generation. In: IEEE 55th Vehicular Technology Conference, VTC Spring 2002, pp. 1965–1969 (2002)
10. Walke, B.H.: Mobile Radio Networks: Networking, Protocols and Traffic Performance. Wiley, Chichester (2002)
11. Domínguez, D., et al.: Using PBIL for Solving a Real-World Frequency Assignment Problem in GSM Networks. In: EPIA 2007, Guimarães, Portugal, pp. 207–218 (2007)
12. Martí, R., et al.: Principles of Scatter Search. European Journal of Operational Research 169, 359–372 (2006)
13. Laguna, M., et al.: Scatter Search: Methodology and Implementation in C. Kluwer Academic Publishers, Norwell (2002)
14. FAP Web (2009), http://fap.zib.de/
15. Chaves-González, J.M., Vega-Rodríguez, M.Á., Domínguez-González, D., Gómez-Pulido, J.A., Sánchez-Pérez, J.M.: SS vs PBIL to Solve a Real-World Frequency Assignment Problem in GSM Networks. In: Giacobini, M., Brabazon, A., Cagnoni, S., Di Caro, G.A., Drechsler, R., Ekárt, A., Esparcia-Alcázar, A.I., Farooq, M., Fink, A., McCormack, J., O'Neill, M., Romero, J., Rothlauf, F., Squillero, G., Uyar, A.Ş., Yang, S. (eds.) EvoWorkshops 2008. LNCS, vol. 4974, pp. 21–30. Springer, Heidelberg (2008)
16. Luna, F., et al.: Metaheuristics for Solving a Real-World Frequency Assignment Problem in GSM Networks. In: GECCO 2008, Atlanta, GE, USA, pp. 1579–1586 (2008)

On the Success Rate of Crossover Operators for Genetic Programming with Offspring Selection*

Gabriel Kronberger, Stephan Winkler, Michael Affenzeller, Andreas Beham, and Stefan Wagner

Heuristic and Evolutionary Algorithms Laboratory
School of Informatics, Communications and Media - Hagenberg
Upper Austria University of Applied Sciences
Softwarepark 11, A-4232 Hagenberg, Austria
{gkronber,swinkler,maffenze,abeham,swagner}@heuristiclab.com

Abstract. Genetic programming is a powerful heuristic search technique that is used for a number of real world applications to solve amongst others regression, classification, and time-series forecasting problems. A lot of progress towards a theoretic description of genetic programming in form of schema theorems has been made, but the internal dynamics and success factors of genetic programming are still not fully understood. In particular, the effects of different crossover operators in combination with offspring selection are largely unknown.

This contribution sheds light on the ability of well-known GP crossover operators to create better offspring when applied to benchmark problems. We conclude that standard (sub-tree swapping) crossover is a good default choice in combination with offspring selection, and that GP with offspring selection and random selection of crossover operators can improve the performance of the algorithm in terms of best solution quality when no solution size constraints are applied.

1 Genetic Programming

Genetic programming (GP) is a generalization of genetic algorithms first studied at length by John Koza [5]. Whereas the goal of genetic algorithms is to find a fixed length vector of symbols that encodes a solution to the problem, the goal of genetic programming is to find a variable-length program that solves the original problem when executed. Common practice is to use a tree-based representation of computer programs similar to so called symbolic expressions of functional programming languages such as LISP.

Genetic programming is a powerful heuristic search method that has been used successfully to solve real world problems from various application domains, including classification, regression, and forecasting of time-series [9,16].

* The work described in this paper was done within HEUREKA!, the Josef Ressel center for heuristic optimization sponsored by the Austrian Research Promotion Agency (FFG).

R. Moreno-Díaz et al. (Eds.): EUROCAST 2009, LNCS 5717, pp. 793–800, 2009.

Offspring selection [1] is a generic selection concept for evolutionary algorithms that aims to reduce the effect of premature convergence often observed with traditional selection operators by preservation of important alleles [2]. The main difference to the usual definition of evolutionary algorithms is that after parent selection, recombination and optional mutation, offspring selection filters the newly generated solutions. Only solutions that have a better quality than their best parent are added to the next generation of the population. In this aspect offspring selection is similar to non-destructive crossover [21], soft brood selection [3], and hill-climbing crossover [13]. Non-destructive crossover compares the quality of one child to the quality of the parent and adds the better one to the next generation, whereas offspring selection generates new children until a successful offspring is found. Soft brood selection generates n offspring and uses tournament selection to determine the individual that is added to the next generation, but in comparison to offspring selection the children do not compete against the parents. Hill-climbing crossover generates new offspring from the parents as long as better solutions can be found. The best solution found by this hill-climbing scheme is added to the next generation. The recently described hereditary selection concept [11,12] also uses a similar offspring selection scheme in combination with parent selection that is biased to select solutions with few common ancestors.

2 Motivation

Since the very first experiments with genetic programming a lot of effort has been put into the definition of a theoretic foundation for GP in order to gain a better understanding of its internal dynamics. A lot of progress [9,17,18,20] towards the definition of schema theorems for variable length genetic programming and sub-tree swapping crossover, as well as homologous crossover operators [19] has been made. Still, an overall understanding of the internal dynamics and the success factors of genetic programming is still missing. The effects of mixed or variable arity function sets or different mutation operators in combination with more advanced selection schemes are still not fully understood. In particular, the effects of different crossover operators on the tree size and solution quality in combination with offspring selection are largely unknown.

In this research we aim to shed light on the effects of GP crossover operators regarding their ability to create improved solutions in the context of offspring selection. We apply GP with offspring selection to three benchmark problems: symbolic regression (Poly-10), time series prediction (Mackey-Glass) and classification (Wisconsin diagnostic breast cancer). The same set of experiments was also executed for the 4-bit even parity problem, but because of space constraints the results of those experiments are not reported in this paper.

Recently we have analyzed the success rate of GP crossover operators with offspring selection with strict solution size constraints [6]. In the paper at hand we report results of similar experiments with the same set of crossover operators and benchmark problems, but without strict solution size constraints.

3 Configuration of Experiments

The crossover operators used in the experiments are: standard (sub-tree swapping) [5] [20], one-point [9], uniform [15], size-fair, homologous, and size-fair [7]. Additionally, the same experiments were also executed with a crossover variant that chooses one of the five crossover operators randomly for each crossover event [6]. Except for the crossover operator, the problem specific evaluation operator, and the function set all other parameters of the algorithm were the same for all experiments. The random initial population was generated with probabilistic tree creation (PTC2) [10] and uniform distribution of tree sizes in the interval $[3; 50]$. A single-point mutation operator was used to manipulate 15% of the solution candidates by exchanging either a function symbol (50%) or a terminal symbol (50%). See Table 1 for a summary of all GP parameters.

To analyze the results, the quality of the best solution, average tree size in the whole population as well as offspring selection pressure were logged at each generation step together with the number of solutions that have been evaluated so far. Each run was stopped as soon as the maximal offspring selection pressure or the maximal number of solution evaluations was reached.

Offspring selection pressure of a population is defined as the ratio of the number of solution evaluations that were necessary to fill the population to the population size [1]. High offspring selection pressure means that the chance that crossover generates better children is very small, whereas low offspring selection pressure means that the crossover operator can easily generate better children.

3.1 Symbolic Regression – Poly-10

The Poly-10 symbolic regression benchmark problem uses ten input variables $x_1, \ldots, x_{10}$. The function for the target variable y is defined as $y = x_1x_2 + x_3x_4 + x_5x_6 + x_1x_7x_9 + x_3x_6x_{10}$ [8,14]. For our experiments 100 training samples were generated randomly by sampling the values for the input variables uniformly in the range $[-1, 1[$. The usual function set of +,-,*, % (protected division) and the terminal set of $x_1 \ldots, x_{10}$ without constants was used. The mean squared errors function (MSE) over all 100 training samples was used as fitness function.

3.2 Time Series Prediction – Mackey-Glass

The Mackey-Glass ($\tau = 17$)[1] chaotic time series is an artificial benchmark data set sometimes used as a representative time series for medical or financial data sets [8]. We used the first 928 samples as training set, the terminal set for the prediction of $x(t)$ consisted of past observations $x_{128}, x_{64}, x_{32}, x_{16}, x_8, x_4, x_2, x_1$ and integer constants in the interval $[1; 127]$. The function set and the fitness function (MSE) were the same as in the experiments for Poly-10.

---

1 Data set available from: http://neural.cs.nthu.edu.tw/jang/benchmark/

3.3 Classification – Wisconsin Diagnostic Breast Cancer

The Wisconsin diagnostic breast cancer data set from the UCI Machine Learning Repository [4] is a well known data set for binary classification. Only a part (400 samples) of the whole data set was used and the values of the target variable were transformed to values 2 and 4. Before each genetic programming run the whole data set was shuffled, thus the training set was different for each run.

Again the mean squared errors function for the whole training set was used as fitness function. In contrast to the previous experiments a rather large function set was used that included functions with different arities and types (see Table 1). The terminal set consisted of all ten input variables and real-valued constants in the interval $[-20; 20]$.

Table 1. General parameters for all experiments and specific parameters for each benchmark problem

General parameters for all experiments	Population size	1000
	Initialization	PTC2 (uniform [3..50])
	Parent selection	fitness-proportional (50%), random (50%)
		strict offspring selection, 1-elitism
	Mutation rate	15% single point (50% functions, 50% terminals)
	constraints	unlimited tree size and depth
Poly-10	Function set	ADD, SUB, MUL, DIV (protected)
	Terminal set	$x_1 \dots x_{10}$
	Fitness function	Mean squared errors
	Max. evaluations	1.000.000
Mackey-Glass	Function set	ADD, SUB, MUL, DIV (protected)
	Terminal set	$x_{128}, x_{64}, \dots, x_2, x_1$, constants: 1..127
	Fitness function	Mean squared errors
	Max. evaluations	5.000.000
Wisconsin	Function set	ADD, MUL, SUB, DIV (protected), LOG, EXP, SIGNUM, SIN, COS, TAN, IF-THEN-ELSE, LESS-THAN, GREATER-THAN, EQUAL, NOT, AND, OR, XOR
	Terminal set	$x_1, \dots, x_{10}$, constants: [−20..20]
	Fitness function	Mean squared errors
	Max. evaluations	2.000.000

4 Results

Figure 1 shows the quality progress (MSE, note log scale), average tree size, and offspring selection pressure for each of the six crossover operators over time (number of evaluated solutions). The first row shows the best solution quality, the second row shows average tree size over the whole population and the third row shows offspring selection pressure.

Size-fair, homologous, and mixed crossover are the most successful operators, whereas onepoint and uniform crossover show rather bad performance. The average tree size grows exponentially in the experiments with standard and mixed crossover, whereas with onepoint, uniform, size-fair and homologous crossover the average tree size stays at a low level. The most interesting result is that offspring selection pressure stays at a low level over the whole run when standard

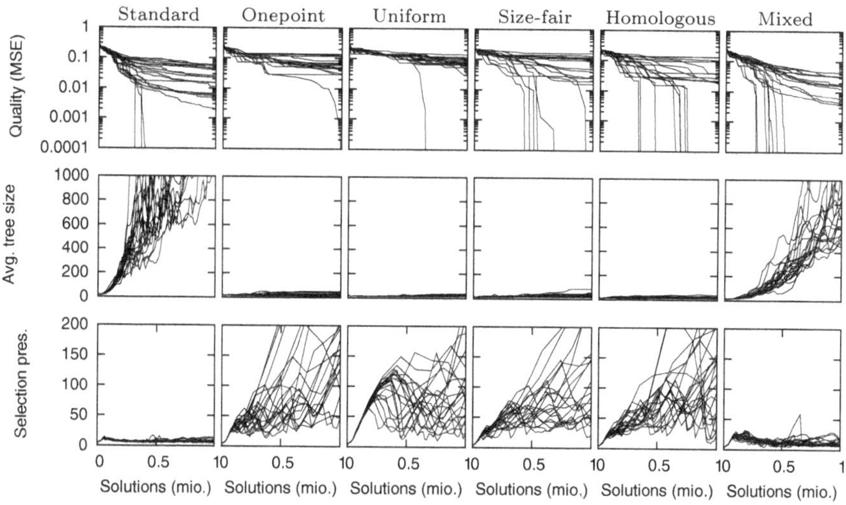

Fig. 1. Best solution quality (MSE, note log scale), average tree size, and offspring selection pressure for 20 runs with each crossover operator for the Poly-10 problem

or mixed crossover are used. Offspring selection pressure rises gradually over the whole run when standard crossover is used with size constraints [6]. The different behavior when no size constraints are applied indicates that larger offspring solutions are more likely to be better than their parent solutions than offspring solutions of equal or smaller size. The offspring selection pressure charts for onepoint, uniform, size-fair and homologous crossover show the usual effect, namely that it becomes increasingly harder for crossover to produce successful children.

Figure 2 shows the results for the Mackey-Glass problem. Standard crossover and mixed crossover show good performance in terms of solution quality and the expected exponential growth of solution size. Size-fair crossover had similar behavior as homologous crossover. Onepoint and uniform crossover are the least effective operators. The offspring selection pressure charts show that with onepoint and uniform crossover the offspring selection pressure rises quickly. The runs with standard crossover and mixed crossover again have low offspring selection pressure over the whole run.

Figure 3 shows the results for the Wisconsin classification problem. Mixed crossover performs better than standard crossover for this problem. Onepoint, uniform, size-fair, and homologous crossover reached similar solution quality, except for one outlier with homologous crossover. The offspring selection pressure curves of onepoint and uniform crossover show that offspring selection pressure remains at a low level until a point of convergence is reached where the offspring selection pressure rapidly increases to the upper limit. The explanation for this is that onepoint and uniform crossover cause convergence to a fixed tree shape. When all solutions have the same tree shape it becomes very hard to find better solutions. Only the runs with size-fair crossover show the usual pattern of gradually increasing offspring selection pressure. An interesting result is that

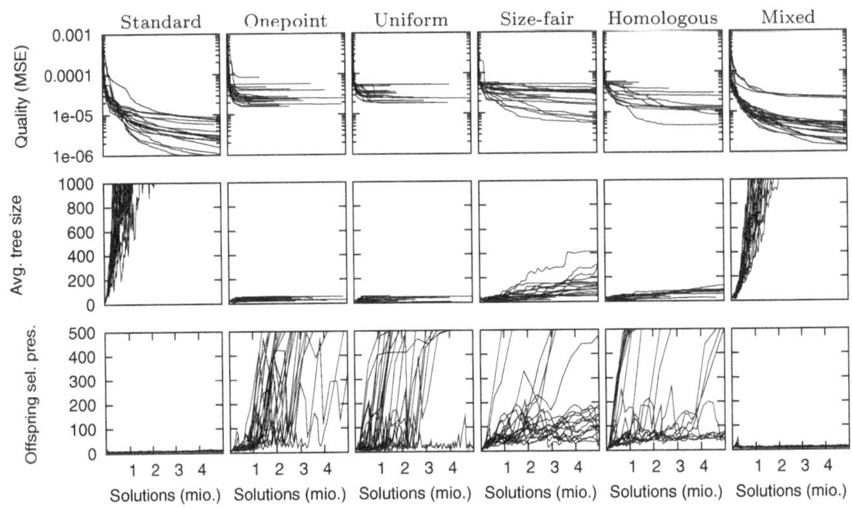

Fig. 2. Best solution quality (MSE, note log scale), average tree size, and offspring selection pressure (note different scale) for 20 runs with each crossover operator for the Mackey-Glass problem

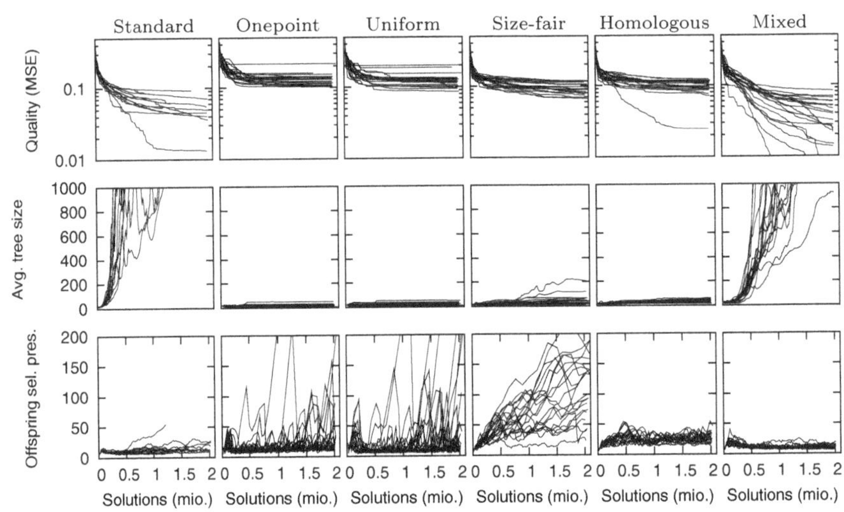

Fig. 3. Best solution quality (MSE, note log scale), average tree size, and offspring selection pressure for 20 runs with each crossover operator for the Wisconsin classification problem

offspring selection pressure also remains low for homologous crossover even though it doesn't show the exponential growth in solution size as standard and mixed crossover. The flat offspring selection pressure curve could be caused by either the extended function set or the structure of the data set. Further investigations are necessary to fully explain this observation.

5 Conclusion

Based on the results for the benchmark problems it can be concluded that standard (sub-tree swapping) crossover is a good default choice. The results also show that onepoint and uniform crossover operators do not perform very well on their own. They also have the tendency to quickly freeze the tree shape, and should be combined with mutation operators which manipulate tree shape.

The aim of the experiments with the mixed-crossover variant was to find out if a combination of all five crossover operators in one GP run has a beneficial effect either in terms of achievable solution quality or efficiency. For two of the three benchmark problems the runs with mixed crossover found better solutions than runs with standard crossover. This result is in contrast to the results of experiments with strict size constraints where runs with mixed crossover did not find better solutions than runs with standard crossover [6].

Acknowledgment

G.K. thanks the participants of EuroGP 2009 for the insightful comments and discussions.

References

1. Affenzeller, M., Wagner, S.: Offspring selection: A new self-adaptive selection scheme for genetic algorithms. In: Adaptive and Natural Computing Algorithms. Springer Computer Series, pp. 218–221. Springer, Heidelberg (2005)
2. Affenzeller, M., Winkler, S.M., Wagner, S.: Effective allele preservation by offspring selection: An empirical study for the TSP. International Journal of Simulation and Process Modelling (2009) (accepted to appear)
3. Altenberg, L.: The evolution of evolvability in genetic programming. In: Kinnear Jr., K.E. (ed.) Advances in Genetic Programming, ch. 3, pp. 47–74. MIT Press, Cambridge (1994)
4. Asuncion, A., Newman, D.J.: UCI machine learning repository (2007)
5. Koza, J.R.: Genetic Programming. MIT Press, Cambridge (1992)
6. Kronberger, G., Winkler, S., Affenzeller, M., Wagner, S.: On crossover success rate in genetic programming with offspring selection. In: Vanneschi, L., Gustafson, S., Moraglio, A., De Falco, I., Ebner, M. (eds.) Genetic Programming, pp. 232–243. Springer, Heidelberg (2009)
7. Langdon, W.B.: Size fair and homologous tree genetic programming crossovers. Genetic Programming and Evolvable Machines 1(1/2), 95–119 (2000)
8. Langdon, W.B., Banzhaf, W.: Repeated patterns in genetic programming. Natural Computing (2008); Published online: May 26, 2007
9. Langdon, W.B., Poli, R.: Foundations of Genetic Programming. Springer, Heidelberg (2002)

10. Luke, S.: Two fast tree-creation algorithms for genetic programming. IEEE Trans. Evolutionary Computation 4(3), 274–283 (2000)
11. Murphy, G., Ryan, C.: Exploiting the path of least resistance in evolution. In: GECCO 2008: Proceedings of the 10th annual conference on Genetic and evolutionary computation, Atlanta, GA, USA, July 12-16, pp. 1251–1258. ACM, New York (2008)
12. Murphy, G., Ryan, C.: A simple powerful constraint for genetic programming. In: O'Neill, M., Vanneschi, L., Gustafson, S., Esparcia Alcázar, A.I., De Falco, I., Della Cioppa, A., Tarantino, E. (eds.) EuroGP 2008. LNCS, vol. 4971, pp. 146–157. Springer, Heidelberg (2008)
13. O'Reilly, U.-M., Oppacher, F.: Hybridized crossover-based search techniques for program discovery. In: Proceedings of the 1995 World Conference on Evolutionary Computation, Perth, Australia, 29 November -1 December, vol. 2, pp. 573–578. IEEE Press, Los Alamitos (1995)
14. Poli, R.: A simple but theoretically-motivated method to control bloat in genetic programming. In: Ryan, C., Soule, T., Keijzer, M., Tsang, E.P.K., Poli, R., Costa, E. (eds.) EuroGP 2003. LNCS, vol. 2610, pp. 204–217. Springer, Heidelberg (2003)
15. Poli, R., Langdon, W.B.: On the search properties of different crossover operators in genetic programming. In: Genetic Programming 1998: Proceedings of the Third Annual Conference, University of Wisconsin, Madison, Wisconsin, USA, July 22-25, pp. 293–301. Morgan Kaufmann, San Francisco (1998)
16. Poli, R., Langdon, W.B., McPhee, N.F.: A Field Guide to Genetic Programming. Lulu.com (2008)
17. Poli, R., McPhee, N.F.: General schema theory for genetic programming with subtree-swapping crossover: part I. Evol. Comput. 11(1), 53–66 (2003)
18. Poli, R., McPhee, N.F.: General schema theory for genetic programming with subtree-swapping crossover: part II. Evol. Comput. 11(2), 169–206 (2003)
19. Poli, R., McPhee, N.F., Rowe, J.E.: Exact schema theory and markov chain models for genetic programming and variable-length genetic algorithms with homologous crossover. Genetic Programming and Evolvable Machines 5(1), 31–70 (2004)
20. Poli, R., Rowe, J.E., Stephens, C.R., Wright, A.H.: Allele diffusion in linear genetic programming and variable-length genetic algorithms with subtree crossover. In: Foster, J.A., Lutton, E., Miller, J., Ryan, C., Tettamanzi, A.G.B. (eds.) EuroGP 2002. LNCS, vol. 2278, pp. 212–227. Springer, Heidelberg (2002)
21. Soule, T., Foster, J.A.: Code size and depth flows in genetic programming. In: Koza, J.R., Deb, K., Dorigo, M., Fogel, D.B., Garzon, M., Iba, H., Riolo, R.L. (eds.) Genetic Programming 1997: Proceedings of the Second Annual Conference, Stanford University, CA, USA, July 13-16, pp. 313–320. Morgan Kaufmann, San Francisco (1997)

On Structural Identification of 2D Regression Functions for Indoor Bluetooth Localization

Rene Mayrhofer[1], Stephan Winkler[2], Helmut Hlavacs[1], Michael Affenzeller[2], and Stefan Schneider[1]

[1] University of Vienna, Faculty of Computer Science, Dr.-Karl-Renner Ring 1, A-1010 Wien, Austria
{rene.mayrhofer,helmut.hlavacs}@univie.ac.at
[2] Heuristic and Evolutionary Algorithms Laboratory, Upper Austrian University of Applied Sciences, Campus Hagenberg, Softwarepark 11, A-4232 Hagenberg, Austria
{stephan.winkler,michael.affenzeller}@heuristiclab.com

Abstract. In-door localization of mobile devices is a common problem for many current and future applications, for example to control infrastructure services or for personalized in-building navigation systems. Sufficiently capable Bluetooth support is often available in off-the-shelf mobile devices such as mobile phones, which makes Bluetooth an attractive technology for cheap and widely available in-door localization systems. However, Bluetooth has been optimized to deal with effects of radio frequency transmission such as reflection and multi-path propagation. It therefore produces highly non-linear relationships between the distance of devices and their perceived signal strength. In this paper, we aim to identify these relationships for a specific dataset of 2D device positions using structural identification methods. Driven by an extended genetic algorithm, we aim to find optimal mappings in form of non-linear equations for x and y coordinates, thus producing formal regression functions.

1 Introduction

Bluetooth localization has many interesting applications in the area of ubiquitous and pervasive computing, both out-doors and in-doors. Its main advantage is broad availability in devices — most of today's mobile phones already include sufficiently capable Bluetooth chipsets with low power consumption that can continuously remain in visible mode without significantly impairing battery life. However, Bluetooth has not been optimized for localization, as we show based on collected real-world data. Depending on the accuracy and precision reached by a specific method, certain application areas may not be supportable (such as selecting a printer to use from an array of side-by-side ones when the localization accuracy is only in the meter range). We therefore aim for highest possible accuracy given off-the-shelf mobile hardware.

Localization methods are usually distinguished between infrastructure-based (public) and client-based (private) ones. Bluetooth localization has the potential to support both, that is, to allow the infrastructure to localize mobile clients as

R. Moreno-Díaz et al. (Eds.): EUROCAST 2009, LNCS 5717, pp. 801–808, 2009.

well as to support self-localization of these clients based on the environment. This potentially supports a wide range of applications, for example: to use a mobile phone to interact with infrastructure services such as printers, displays, or doors based on their location; to track users (that is, the mobile phones they carry) as part of an in-building navigation system that displays personalized arrows on infrastructure displays; or as an in-building tour guide executed locally on the mobile phone to display localized content.

The major problems of Bluetooth localization are accuracy and update speed. Accuracy is impaired by radio frequency (RF) effects such as reflection, absorption, and multi-path propagation of signals between a sender and a receiver. Bluetooth has been developed to deal with – and partially make use of – such effects to provide noise-resilient communication in the free 2.4 GHz range. On the other hand, these effect along with building elements (e.g. doors, walls, windows, etc. all have different RF characteristics) lead to non-linear relationships between the measured signal strength and the real distance between two devices. Our aim is to model these relationships with highest possible accuracy.

2 Related Work

Indoor localization in general and Bluetooth-based localization in particular have recently seen increasing research interest. Apart from outdoor applications with GPS as the standard localization method, the most commonly employed indoor sensing technologies are 802.11 WLAN (e.g. [1,2]), 802.15 Bluetooth (e.g. [3]), various ultra-wide band (UWB) implementations (e.g. the commercial Ubisense system), and ultrasound (e.g. [4]). In comparison to UWB and ultrasonic localization systems, Bluetooth and WLAN typically provide significantly worse accuracy (in the area of meters compared to cm-range for UWB and ultrasound), but support off-the-shelf mobile devices such as laptops or smart phones with their built-in wireless networking hardware. The major advantage of mobile device-based methods is that self-localization does not necessarily reveal this highly sensitive information to third parties. Users can benefit from localized services while still safeguarding their own location tracks. While UWB systems are typically infrastructure-based, Bluetooth, WLAN, and ultrasound support both.

In this paper, we assume that not only the visibility of Bluetooth devices (the most commonly analyzed information for simple Bluetooth localization systems, e.g. [3,5,6]), but the relative signal strength readings between the fixed base stations and the mobile device are used as the basis for location estimation. This specific method has already been studied before, and the following three publications were especially inspiring for our work. Kotanen et al. [7] present a client-based localization system that tries to explicitly estimate the distance to each of the base stations (whose positions are assumed to be known to the mobile device) from Bluetooth RSSI readings and a consecutive Kalman filtering step. The reported average absolute error is 3,76 m, which indicates the difficulty of directly estimating the distance between two Bluetooth devices from their signal strength. In contrast, we use machine learning methods to map from trained signal strength readings to absolute positions and therefore implicitly support more complex, location-dependent models for these relationships. By using structure identification, we lessen the disadvantage of black-box behavior that typical machine learning methods such as neural networks exhibit. Ye [8] presents an infrastructure-based Bluetooth signal strength sensing system, but omitted details on the localization heuristic due to a claimed patent application. The average absolute error is reported in the range of 7 m. Genco et al. [9] were first to apply genetic algorithms to infrastructure-based Bluetooth localization, albeit not for the actual location estimation but for minimizing the number of required base stations. The best reported accuracy is 37,5 cm.

3 Problem Specification

We assume multiple Bluetooth devices (typically USB "dongles") to be distributed over a limited environment, for example an office consisting of multiple rooms. These infrastructure devices are placed strategically and fixed in their location. A mobile Bluetooth device then roams freely within this environment and should be localized based on Bluetooth readings. Both the infrastructure devices (either connected to the same host or to multiple networked hosts) and the mobile device can perform so-called Bluetooth inquiries to query which other Bluetooth devices are in range and can create direct connections to those devices that have been found. When a connection has been opened, both communication partners can locally determine an estimate of the signal strength. These measurements are specific to the respective device and generally not comparable. We assume a calibration phase involving multiple different mobile devices but the same set of infrastructure "sensors" during which signal strengths are systematically recorded for multiple locations (of the mobile device).

The problem is thus, given a set of n signal strength estimates (between the mobile device and each of the infrastructure devices), to determine the approximate 2D position in the form of x and y values. In our current implementation, the signal strength estimate is based on RSSI readings as provided by the Bluetooth HCI link-level API. Potential positions at which the mobile device should be localized may not have been included in the calibration set, and some form of interpolation or regression is therefore required.

4 Location Estimation

This mapping problem can be addressed with different approaches. Practical experience shows that simple mappings are not sufficient (cf. Fig. 2b). We therefore focus on two machine learning methods that have already been applied successfully to related problems: standard neural networks and genetic algorithms.

4.1 Neural Network Approximation

Multi-Layer Perceptrons (MLPs) are feed-forward neural networks composed of multiple layers of neurons. The first layer (also called input layer) directly encodes the inputs of the mapping problem, in our case the 4 signal strength estimates. The last layer (also called output layer) computes the required outputs, in this case the approximation of x and y coordinates. We use the standard approach of smoothing input values (to lessen the influence of noise) by averaging over a sliding time window and scaling to $[0.1; 0.9]$. Out of the available measurements, a random selection of 60% is used for training the MLP using backpropagation learning, 20% for validation during training, and the remaining 20% for testing and computing the accuracy. The training process is manually stopped when the error rate on the test set converges to a perceived minimum.

4.2 Evolutionary System Structure Identification

Genetic programming (GP) is based on the theory of genetic algorithms (GAs) and utilizes a population of solution candidates which evolves through many generations towards a solution using certain evolutionary operators and a selection scheme increasing better solutions' probability of passing on genetic information; the goal of a GP process is to produce a computer program solving the optimization problem at hand. In the case of *structure identification*, solution candidates represent mathematical models; these models are evaluated by applying the formulae to the given training data and comparing the generated output to the original target data. Figure 1 visualizes the GP cycle: As in every evolutionary process, new individuals (in GP's case, new programs) are created and tested, and the fitter ones in the population succeed in creating children of their own; unfit ones die and are removed from the population [10].

Within the last years we have set up a GP based structure identification framework that has been successfully used in the context of various different kinds of identification problems (e.g. mechatronics, medical data analysis, and the analysis of steel production processes [11]). One of the most important problem independent concepts used in our implementation of GP-based structure identification is offspring selection [12], an enhanced selection model that has enabled genetic algorithms and genetic programming implementations to produce superior results for various kinds of optimization problems. As in the case of conventional GAs or GP, offspring are generated by parent selection, crossover, and mutation. In a second (offspring) selection step, only those children become members of the next generation population that outperform their own parents. This process of creating new children is repeated until the number of successful offspring is sufficient to create the next generation's population.

Genetic programming can be used for data based modeling. A given system is to be analyzed and its behavior is to be modeled formally. This process is (especially in the context of modeling dynamic physical systems) called *system identification* [13]. The main goal here is to determine the relationship of a dependent (target) variable t to a set of specified independent (input) variables z. Thus, we search for a function f that uses z and a set of coefficients w such that $t = f(z, w) + \epsilon$ where e represents the error (noise) term. The structure of f is not pre-defined – it is part of the GP based identification process to identify

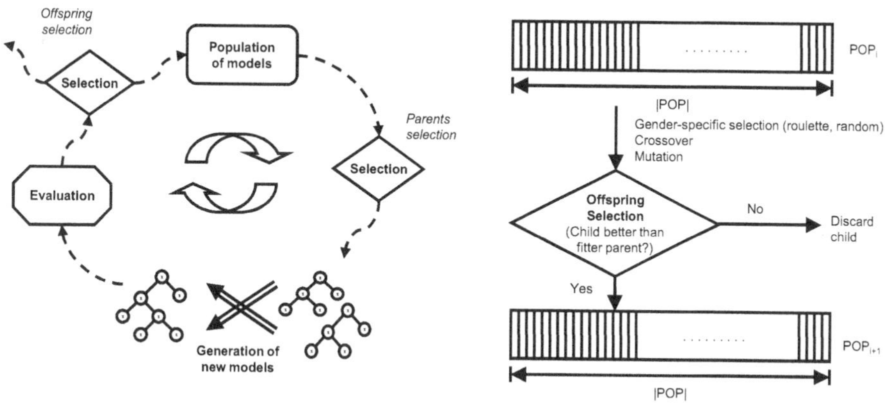

Fig. 1. Left: The extended genetic programming cycle including offspring selection. **Right:** Strict offspring selection as used here within the GP process.

the optimal structure for a formula, to find a set of relevant variables, and to optimize the terminals' parameters.

Applying this procedure we assume that a model can be created with which it will also be possible to predict correct outputs for other data examples (test sample); from the training data we want to generalize to situations not known (or allowed to be analyzed) during the training phase.

5 Experimental Evaluation

Initial data collection was carried out using four base stations distributed over different rooms of a small flat and recording RSSI estimates to two different mobile devices: a laptop (Bluetooth class 2) and a mobile phone (Bluetooth class 3). Systematic training data was then collected by placing the mobile devices on a 7x7 grid with 50 cm field width, resulting in an overall area of 3x3 m^2 (see Fig. 2a) and collecting roughly 15 minutes of measurements for each grid point at 0.5 Hz. Figure 2b gives an example of the recorded signal strengths from a single sensor (base station 4) to the mobile phone at all grid points and shows clearly that we can not assume a simple linear dependency between distance and Bluetooth signal strength.

5.1 Results Using Neural Network Approximation

For MLPs, the number of so-called "hidden" layers and numbers of neurons in each of these layers have a significant influence on the overall performance. Unfortunately, these are specific to the problem and data set and therefore need to be optimized alongside the actual neural weights. For this data set, a fully connected 4-60-30-2 network, i.e., the required 4 input and 2 output neurons with 2 hidden layers with 60 and 30 neurons, respectively, performed best.

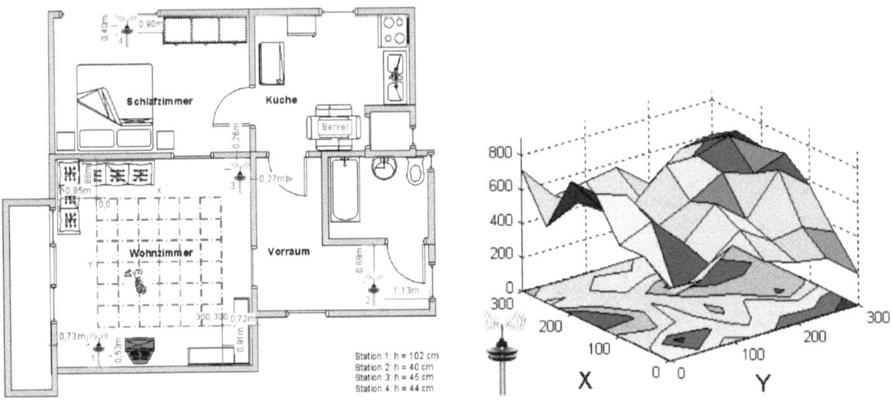

(a) 7x7 grid with 50 cm field width with base station sensors in different rooms

(b) Measurements for mobile phone signal strength estimates (sensor 4)

Fig. 2. Experimental setup

After smoothing, 67 measurements were available for each grid point. Randomly chosen sets for training and testing resulted in an averaged absolute error of 9.6 cm (median 5.4 cm) with the class 2 device (laptop) and 3.4 cm (median 1.9 cm) with the class 3 device (mobile phone).

5.2 Results Using Evolutionary System Structure Identification

We have also applied enhanced genetic programming (i.e., GP with strict offspring selection) for identifying mathematical models that are able to predict the x and y coordinates of the mobile devices. Based on manual optimization, we used a population size of 500 and maximum model complexity of 10 levels (i.e., the maximum height of the evolved structure trees was set to 10) with single point crossover and single point mutation (mutation probability: 10%) as genetic operators, mean squared error (mse) as evaluation function, strict offspring selection (i.e., success ratio and comparison factor were both set to 1.0 and maximum selection pressure set to 300), and the maximum selection pressure as termination criterion.

Using these algorithmic settings we have executed 5 independent test runs each for identifying models for the x and y coordinates. The identification data available for the algorithm have been split into a set of training data (containing 80% of the identification data) and a validation set (containing the remaining 20% of the identification data); the algorithms were configured to use the set of training samples for optimizing the evolved models, and eventually those models were presented as results that performed best (with respect to mse) on the validation samples. The resulting models have been evaluated on test data not seen by the identification algorithm and not smoothed to obtain more realistic error estimates. Due to the independent training of models for x and y coordinates, we also analyze them separately:

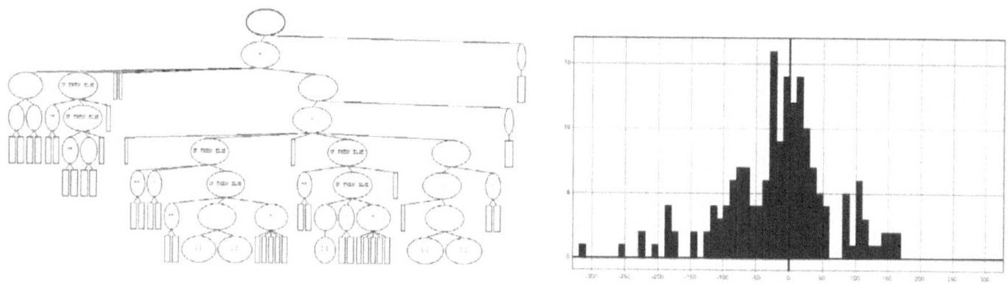

Fig. 3. Left: Structure tree representation of the best model identified by GP for x coordinates. **Right:** Resulting error distribution of this model.

- The models for x coordinates performing best on validation data show a *mse* of 41.87 cm ($\sigma = 5.92$) on training data, and 44.15 cm ($\sigma = 7.24$) on the test data set. Figure 3 shows a structure tree representation of the evaluation of the best model (with respect to training quality) and the distribution of the errors made using this model on test data.
- The models performing best on validation data show a *mse* of 55.68 cm ($\sigma = 5.43$) on training data, 55.5 cm ($\sigma = 8.16$) on test data.

These results are promising, but also show potential for further improvement of the quality of the results achievable using GP. We see that the test quality of the results is close to their training quality, i.e., overfitting does not seem to be major issue in this context. We are confident that the use of filtered data as well as more complex models should lead to significantly better results, potentially even comparable to those achieved by the MLP but with the advantage of explicit models for obtaining the coordinate estimates.

6 Conclusions and Outlook

The general approach of Bluetooth localization based on a learned model of systematic signal strength readings is usable for both infrastructure-based and client-based localization, but needs extensive training. Because this training step only needs to be performed once for each location, it can still be practical for in-door settings that require high localization accuracy and can afford certain fixed infrastructure components. The learned and potentially manually optimized model (which is small and can be used efficiently at run-time) can then be transmitted automatically to all clients (for example using Bluetooth OBEX push to any new devices entering the area).

Compared to other approaches, accuracy is potentially better, because the complex models constructed by the two machine learning methods studies in this paper can potentially accommodate difficult settings (e.g. with metal and different surfaces in in-door scenarios) where linear or other simple relationships between distance and signal strength readings can not be assumed.

Current results achieved using a neural network are impressive, but still preliminary. Instead of training a certain percentage of the measurements from *all* grid points (i.e., all the data that is being tested has already been provided to the network during training), it would be more practical to use only a smaller number of grid points for training and then testing with points for which *no* measurements were trained. Our GP models were trained with this approach and show that, in principle, errors should not increase significantly. Work is underway to make the results of both methods more comparable.

A significant advantage of structure identification is the creation of a "white-box" model, offering introspectability and the possibility for manual tuning as well as small models (which consist of only two explicit mathematical formulae for x and y). In the future, we aim to further optimize the GP structure identification approach towards the results currently only achieved with "black-box" neural networks, as well as a live implementation.

References

1. Bahl, P., Padmanabhan, V.N.: RADAR: An in-building rf-based user location and tracking system. In: Proc. IEEE Infocom 2000, pp. 775–784. IEEE CS Press, Los Alamitos (2000)
2. LaMarca, A., Chawathe, Y., Consolvo, S., Hightower, J., Smith, I., Scott, J., Sohn, T., Howard, J., Hughes, J., Potter, F., Tabert, J., Powledge, P., Borriello, G., Schilit, B.: Place lab: Device positioning using radio beacons in the wild. In: Gellersen, H.-W., Want, R., Schmidt, A. (eds.) PERVASIVE 2005. LNCS, vol. 3468, pp. 116–133. Springer, Heidelberg (2005)
3. Bruno, R., Delmastro, F.: Design and analysis of a Bluetooth-based indoor localization system. In: Conti, M., Giordano, S., Gregori, E., Olariu, S. (eds.) PWC 2003. LNCS, vol. 2775, pp. 711–725. Springer, Heidelberg (2003)
4. Hazas, M., Kray, C., Gellersen, H., Agbota, H., Kortuem, G., Krohn, A.: A relative positioning system for co-located mobile devices. In: Proc. MobiSys 2005, pp. 177–190. ACM Press, New York (2005)
5. Bargh, M., de Groote, R.: Indoor localization based on response rate of Bluetooth inquiries. In: Proc. MELT 2008. ACM Press, New York (2008)
6. Jevring, M., de Groote, R., Hesselman, C.: Dynamic optimization of Bluetooth networks for indoor localization. In: Proc. AASN 2008 (October 2008)
7. Kotanen, A., Hännikäinen, M., Leppäkoski, H., Hämäläinen, T.D.: Experiments on local positioning with Bluetooth. In: Proc. ITCC 2003, April 2003, pp. 297–303 (2003)
8. Ye, J.Y.: Atlantis: Location based services with Bluetooth. Department of Computer Science, Brown University (2005)
9. Genco, A., Sorce, S., Scelfo, G.: Bluetooth base station minimal deployment for high definition positioning. Dipartimento di Ingegneria Informatica, Universitá di Palermo (2005)
10. Langdon, W.B., Poli, R.: Foundations of Genetic Programming. Springer, Heidelberg (2002)
11. Winkler, S.: Evolutionary System Identification - Modern Concepts and Practical Applications. PhD thesis, Institute for Formal Models and Verification, Johannes Kepler University Linz (2008)
12. Affenzeller, M., Wagner, S., Winkler, S.: Goal-oriented preservation of essential genetic information by offspring selection. In: Proceedings of the Genetic and Evolutionary Computation Conference (GECCO) 2005, vol. 2, pp. 1595–1596. Association for Computing Machinery, ACM (2005)
13. Ljung, L.: System Identification – Theory For the User, 2nd edn. PTR Prentice Hall, Upper Saddle River (1999)

Grid-Enabled Mutation-Based Genetic Algorithm to Optimise Nuclear Fusion Devices

Antonio Gómez-Iglesias[1], Miguel A. Vega-Rodríguez[2], Francisco Castejón-Magaña[1], Miguel Cárdenas-Montes[3], and Enrique Morales-Ramos[2]

[1] National Fusion Laboratory, CIEMAT, Madrid, Spain
{antonio.gomez,francisco.castejon}@ciemat.es
http://www.ciemat.es
[2] Dep. of Technologies of Computers and Communications, University of Extremadura, Cáceres, Spain
mavega@unex.es, enmorales@alumnos.unex.es
[3] Dep. of Basic Research, CIEMAT, Madrid, Spain
miguel.cardenas@ciemat.es

Abstract. Fusion community is becoming more important as long as fusion energy is considered the next generation of energy. However, many problems are presented in fusion devices. One of these problems consists of improving the equilibrium of confined plasma. Some modelling tools can be used to improve the equilibrium, but the computational cost of these tools and the number of different configurations to simulate make impossible to perform the required tests to obtain optimal designs. With grid computing we have the computational resources needed for running all these tests and with genetic algorithms (GAs) we can look for an approximate result without exploring all the solution space. This work joins all these ideas. The obtained results are very encouraging.

Keywords: Grid Computing, Genetic Algorithm, Mutation-Based Algorithm, Nuclear Fusion.

1 Introduction

Nowadays there is a global concern about the problem of global warming and the use of fossil fuels like our main energy supply. Several possibilities are growing as future energy sources, being the nuclear fusion among the most promising ones.

Nuclear fusion is the process by which multiple atomic nucleus join together to form a heavier nucleus. It is accompanied by the release or absorption of energy. The fusion of two light nuclei generally releases energy while the fusion of heavy nuclei absorbs energy [6]. To get these fusion reactions we need special devices to confine the particles, in plasma state, which will fuse. This confinement is done by means of magnetic fields generated by external coils.

There are some modelling tools which can simulate the behaviour of these devices. The computational resources needed for these tools are not only extremely high, but also take long to finish a single simulation. The number of

R. Moreno-Díaz et al. (Eds.): EUROCAST 2009, LNCS 5717, pp. 809–816, 2009.

possible configurations for these devices is also high, so the number of different tests to be performed requires a long execution time. These tools use configuration files with many parameters to get all the possible configurations of these devices.

Here grid computing is a good solution, because using the computational resources and distributed paradigm of the grid [3][5], many tests can be performed using different configurations. So, with grid computing and modelling tools scientists can predict the behaviour of plasma confinement devices and can look for optimal configurations in order to improve the equilibrium of confined plasma.

To explore the solution space, grid computing is useful but, even using this paradigm, to get approximate configurations can be challenging if we use a brute-force algorithm. For this reason we present a distributed GA [7] to find out these configurations inner this huge solution space. Every individual in the population is the encoded version of a tentative solution.

The rest of the paper is organised as follows: section 2 introduces the workflow to measure the equilibrium (fitness function) for a single configuration. Section 3 shows the mutation-based GA designed, whereas section 4 describes how this GA can be executed using the grid. In section 5 we present the results obtained grid computing and, finally, section 6 displays the conclusions and future work.

2 Equilibrium. How to Measure It

In any nuclear fusion device is important to get the best equilibrium for confined plasma as possible to avoid transport of particles. Transport (neoclassical one) can be considered as the trajectory of a particle and the problem consists of particles drifting apart their trajectories determined by the magnetic fields so they are not finally useful to obtain the fusion reaction [2].The probability of getting an interaction with other particles is not only lower, but also to obtain reactions in the core of the plasma is more difficult. This decreases the efficiency of the device.

2.1 Designing the Workflow

After many expressions involving concepts of magnetic fields and plasma physics [2][11], we get the target function given by

$$F_{targetfunction} = \sum_{i=1}^{N} \left\langle \left| \frac{\vec{B} \times \vec{\nabla} |B|}{B^3} \right| \right\rangle_i \tag{1}$$

In this equation, i represents the different magnetic surfaces in TJ-II, and B, the magnetic field. Our aim is to minimise the value given by this function. The computational cost of this equation is really high due to the number of operations to perform to calculate the different values for B_i and the number of iterations i.

The workflow we need to execute to measure the equilibrium in the TJ-II, a magnetic confinement device localized in Madrid, is widely explained in the related work [8]. This workflow takes more than one hour on average to perform its calculations.

Another problem we have to face, apart from the high execution time, when trying to improve the design of a nuclear fusion device is the number of parameters of the VMEC configuration file, being this value over 300. Furthermore, there are many interrelations among these parameters, so they cannot be modified without taking into account these interrelations. Such considerations - the long number of parameters and the interrelations among them - are critical to implement a GA to get an optimised configuration in a reasonable time.

3 Mutation-Based Genetic Algorithm

This mutation-based algorithm uses the sample standard deviation of each chromosome in the whole population to perform the mutation. This function assures some level of convergence in the values of chromosomes, even though this convergence is only noticed after a long number of generations, as well as a large diversity of the population. Each selected chromosome is modified by a value between 0 and the standard deviation for that chromosome. We are using a floating point representation of the solution domain [10] as well as some binary chromosomes that, instead of being mutated, are crossed using the binary chromosomes of the two parents of the new offspring.

We have developed two different versions of a genetic algorithm and both are described in this section. The two versions can be classified as master-slave parallel GA once they are running using the grid [4]. As long as in these algorithms, selection and crossover consider the entire population they are also known as global parallel GAs [1].

3.1 Random Selection and Replacement

The way to proceed of this algorithm is explained here, showing the different stages of a genetic algorithm and the implementation proposed.

Selection. All the individuals of the population are selected in pairs using a tournament selection with a tournament size of two. We have a list representing the entire population in which one index in the list is randomly selected and the individual with this index is checked. If the individual has not been previously selected, it is chosen. In case it would have been used in a previous confrontation, we increase the index until and individual not previously selected is found. If we reach the end of the list, the process follows from the beginning of the list until the initial random index is reached. In this case, the process finishes. The other individual is selected using the same procedure.

Evaluation. For evaluation, we use the fitness value calculated for each individual. The individual, in each pair, with the highest value, this is, the worse individual, will be selected for mutation. As explained in section 2 our target is to minimise transport levels and this is reached by minimising the value of the fitness function (eq. 1).

Mutation. The chromosomes of the worst individual in the pair of parents are randomly selected and mutated using the standard deviation, as mentioned. The number of chromosomes to be mutated is randomly selected by a number in the range $\{1 - number\_of\_chromosomes\}$. Once we have the number of chromosomes to mutate, they are randomly selected in a similar way to the previously explained for selection of the individuals. The binary chromosomes are also randomly crossed.

Replacement. The new individual will replace the worse one in the pair of parents and will be used in the following generation instead of the previous individual. The main characteristic of this implementation is the high dispersion of the chromosomes in the population. This can be desirable in some situations, although it has the problem of extremely dispersed results even when we could have a subset of individuals with optimised fitness values. For this reason we propose a new approach by not changing the algorithm but using other replacement procedure in order to get better results.

3.2 Worst Individual Replacement

The main problem of the previously described version is its high dispersion. As long as all the selections are randomly performed, the dispersion of the chromosomes in the individuals of the population becomes extremely high. Even though the individual used as reference to perform the mutation is always the best of the two parents, the random selection of individuals makes that the best one could be confronted to the best-but-one individual and this will be replaced by the new offspring, and at the same time we could have a confrontation of the worst individual with the worst-but-one, being the first one replaced. In this situation, we would have lost a good one individual and would have kept a bad one, being the dispersion very high.

To avoid this, we have introduced a sorted list of all the elements in the population based on the fitness value of all of them. This approach has been introduced previously in GA providing good results [12] and it constitutes an elitism replacement policy.

All the individuals are also selected for breeding in pairs. The chromosomes are also mutated randomly. Thereby we ensure diversity within final population. The sorted list is, finally, used to create a file with the best results obtained during the execution of the algorithm.

4 The Grid. Non-supervised System

The designed process would work perfectly without human supervision but the problem will appear after getting the new generation, because the user should resend the jobs. Also, in case of failure, the system could not recover its original status.

To get a non-supervised system we have developed a set of python scripts which interact with the metascheduler and the proxy to manage all the required

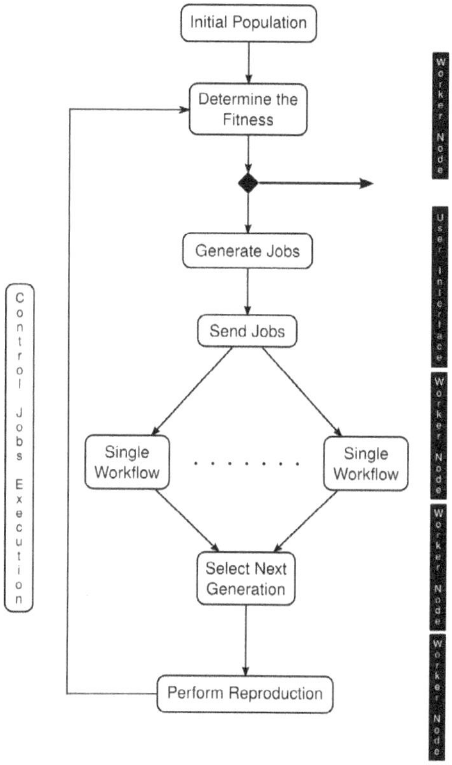

Fig. 1. Final workflow of the mutation-based GA using the grid

processes in a proper way. With these scripts, introduced in this section, the resulting workflow is as shown in Fig. 1. In this figure we can see where is executed each component of the workflow. The final algorithm is a distributed GA following a master-slave model, where reproduction, selection and mutation occur just as they would on a single computer -being this single computer also a *Worker Node* (WN) of the grid- and fitness evaluation is spread across a network of computers. Even more, due to the paradigm of the grid, this is a geographically distributed genetic algorithm.

4.1 Looking Up for Failed Jobs

Sometimes, because of failures in the grid infrastructure or other non-desired situations, some jobs fail and these failures are not managed by the metascheduler to resend them. For this reason one of the functionalities implemented by our scripts is to look up for these jobs, which are not included into the population for the following generation. Once all the jobs have finished, and the new generation has been created, the failed jobs are resubmitted.

This way to proceed is also used in case of jobs with high execution time. Under some conditions a job can have an execution time extremely long, being better to cancel this job and send it within the next generation. We are considering that a job is taking much time than needed when its execution time is twice the average time of finished jobs. Jobs are resubmitted only once.

4.2 Failure Recovery

In case of failure in the *User Interface* (UI) or any other resource in the grid making the python script to stop, it can be reloaded with the proper flag to recover the execution at the same time it was finished.

To recover from general failures in the grid infrastructure, after each generation the system performs a full backup of the individuals of the population as well as all the configuration files, report files and intermediate files. This allows recovering the system at any required iteration automatically.

5 Results

In this section we present some results obtained with the grid. Firstly we show some results related to the time required to evaluate some generations in the GA for both two versions. After that, we present some results focused on the evolution of the fitness function within the population along the generations. In these results, RSRA means *Random Selection and Replacement Algorithm* while WIRA, *Worst Individual Replacement Algorithm*. The tests have been performed several times in order to compare both algorithms.

5.1 Execution Time

The size of the population used for these tests has been set to 200. For our tests we are using the desired ideal stop condition where the fitness value is 0 -or a value close to 0- and a maximum number of iterations of 40.

The execution times obtained for both algorithms are shown in Table 1. This table shows the number of generations used like reference to get some relevant results, the number of single workflows execution (fitness evaluations) and the execution time required to perform all these calculations and the average time for these calculations. All the times are in the *hh:mm:ss* format.

Table 1. Results for both two versions of the algorithm after 40 generations

Test	RSRA	WIRA
Number of Fitness Evaluations	4,245	4,745
Total Execution Time	6,748:07:01	6,786:23:12
Average Execution Time	01:23:27	01:19:53

The required time to obtain these results, for RSRA, was 481:36:10 (more than 20 days). Considering the total execution time being more than 6,748 hours, which is 271 days, the advantage of using grid computing becomes clear.

5.2 Fitness Values Evolution

Fig. 2 shows the average fitness value for the elements within the population after an iteration for both algorithms.

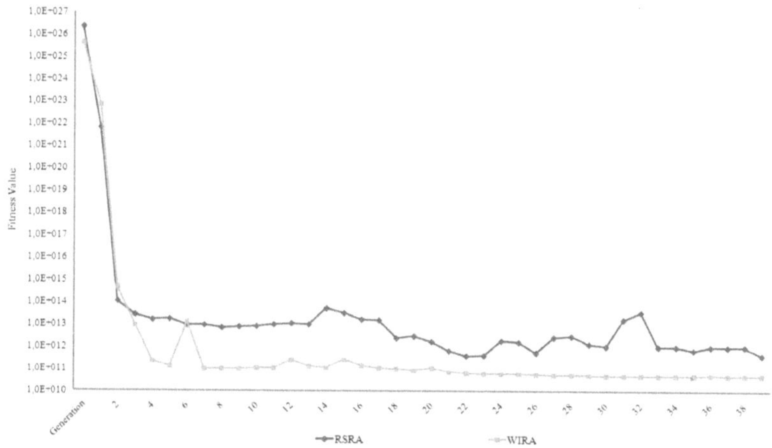

Fig. 2. Evolution of fitness values

In this figure we can see how the results for WIRA are better and this is due to the lower dispersion within the elements in the population, which, in fact, allows us to get a better convergence. And we also have some levels of dispersion within the population, which are useful to avoid local minima values. Table 2 shows the best fitness values obtained at certain generations for both algorithms. In this table we can see how WIRA gets better results than RSRA.

Table 2. Best fitness values

Algorithm	Generat. 1	Generat. 10	Generat. 20	Generat. 30	Generat. 40
RSRA	9.9435e+10	9.9435e+10	9.9435e+10	6.8491E+10	6.8491e+10
WIRA	9.7164e+10	6.0778e+10	5.1402e+10	4.9952E+10	4.9912e+10

6 Conclusions and Future Work

In this paper we have shown the work and results obtained with a mutation-based GA which allows to optimise the equilibrium of a magnetic confinement device for fusion reactions. Grid computing offers an optimal paradigm for GAs because of its high number of computational resources and its distributed behaviour.

With GAs a problem occurs when the dispersion becomes higher, because we could be close to an optimal solution but looking for other ones which are not related to this configuration. A method for replacement of worst individuals must be implemented in order to reduce the dispersion. Once this method has been developed, the results become better and the convergence among the fitness of the individuals within the entire population increases.

As future work we think of developing an evolutionary algorithm (EA) such as Scatter Search [9] to compare results of these two GAs with EA. Many other functions to improve the configuration of nuclear fusion devices can be implemented and introduced in our system, having the possibility to implement a multi-objective system. New selection methods should be developed to study how this can change the results.

References

1. Adamidis, P.: Review of Parallel Genetic Algorithms Bibliography. Tech. rep. vers. 1, Aristotle University of Thessaloniki, Thessaloniki, Greece (1994)
2. Bellan, P.M.: Fundamentals of Plasma Physics. Cambridge University Press, Cambridge (2006)
3. Berman, F., Hey, A., Fox, G.C.: Grid Computing. Making the Global Infrastructure a Reality. John Wiley & Sons, Chichester (2003)
4. Chipperfield, A., Fleming, P.: Parallel and Distributed Computing. In: Zomaya, A.Y.H. (ed.) Handbook - Parallel Genetic Algorithms. McGraw-Hill, New York (1996)
5. Foster, I., Kesselman, C.: The Grid: Blueprint for a New Computing Infrastructure. Morgan Kaufmann, San Francisco (1999)
6. Freidberg, J.: Plasma Physics and Fusion Energy. Cambridge University Press, Cambridge (2007)
7. Golberg, D.E.: The Design of Innovation: Lessons from and for Competent Genetic Algorithms. Addison-Wesley, Reading (2002)
8. Gómez-Iglesias, A., et al.: Grid Computing in order to Implement a Three-Dimensional Magnetohydrodynamic Equilibrium Solver for Plasma Confinement. In: PDP 2008 (2008)
9. Laguna, M., Martí, R.: Scatter Search Methodology and Implementations. Kluwer Academic Publishers, Boston (2003)
10. Michalewicz, Z.: Genetic Algorithms + Data Structures = Evolution Programs. Springer, Heidelberg (1999)
11. Miyamoto, K.: Fundamentals of Plasma Physics and Controlled Fusion. Iwanami Book Service Center, Tokyo (1997)
12. Sarma, J., De Jong, K.A.: An Analysis of the Effect of the Neighborhood Size and Shape on Local Selection Algorithms. In: Voigt, H.M., Ebelinlg, W., Rechenberg, I., Schewefel, H.P. (eds.) PPSN 1996. LNCS, vol. 1141, pp. 236–244. Springer, Heidelberg (1996)

Priority Rule Generation with a Genetic Algorithm to Minimize Sequence Dependent Setup Costs

Monika Kofler, Stefan Wagner, Andreas Beham, Gabriel Kronberger, and Michael Affenzeller

Heuristic and Evolutionary Algorithms Laboratory
School of Informatics, Communications and Media - Hagenberg
Upper Austria University of Applied Sciences
Softwarepark 11, 4232 Hagenberg, Austria
{mkofler,swagner,abeham,gkronber,maffenze}@heuristiclab.com

Abstract. Setup costs are a crucial factor in many branches of industry and frequently sequence dependent. However, the empirical acquisition of setup costs is inaccurate and not practicable for companies with large product portfolios operating in volatile markets. We therefore propose an abstract model for the estimation of such sequence dependent setup costs and subsequently apply dispatching and scheduling strategies to generate optimized production sequences. Both approaches are tested on randomly generated test instances and a real-world production scenario.

Keywords: scheduling, dispatching, setup costs, genetic algorithms.

1 Introduction

In the last decades, many branches of industry saw a shift from mass production to flexible manufacturing of customized products in small lot sizes. This is often coupled with more frequent setups to prepare machines or tooling for the subsequent product to be manufactured. As noted by Allahverdi et al. [3] the majority of scheduling research since the mid-1950s considered setup time as negligible or part of the processing time. However, in a more recent follow-up survey [4] a significant increase in interest in scheduling problems involving setup times/costs was detected, with more than 300 papers published between 1999 and 2006 compared to only 190 papers that were published prior to 1999.

Setup times vary vastly between industries, but there are many fields where it is a major component and can amount up to 40% - 70% of the total manufacturing time. Setup times and costs are frequently coupled but not necessarily directly proportional. In this paper, setup costs comprise monetary, temporal, or personnel costs. Following the classification in [4], we present an approach to optimize the special case of non-batch, sequence-dependent setup costs.

Many algorithms in the scheduling literature that aim to minimize sequence-dependent setup cost rely on the availability of a matrix for each machine, which

R. Moreno-Díaz et al. (Eds.): EUROCAST 2009, LNCS 5717, pp. 817–824, 2009.

contains the required setup costs for all possible ordered pairs of consecutive products. As proposed by Gilmore and Gomory [7], the problem to minimize the total setup costs can then be reformulated as an asymmetrical Traveling Salesman Problem (aTSP), where each city represents a product and distances between cities correspond to the required setup times.

In real life applications a complete setup costs matrix is frequently not available and empirical values from the production plant are usually averages that conceal the sequence dependency of setups. White and Wilson [11] address this issue by collecting data, namely setup times and characteristics, in a production facility over several months and by developing a prediction equation for the required setup times for new products, which can be incorporated in scheduling decisions. The developed equation is specifically targeted to the production scenario observed in their case study and highly dependent on the parts manufactured in the company as well as the available manufacturing equipment. We investigated this problem from the point of view of a manufacturer who produces machines for the metalworking industry. In this scenario, the automatic optimization of job sequences with regard to setup costs is meant to be an additional feature for their customers. Visits to selected customers confirmed that setup cost estimation needs to be treated case-by-case. However, it would not be practical to generate a custom prediction equation for each customer. As a consequence, we discarded the empirical measurements entirely and replaced the complicated real-world setup process by an abstract model developed in cooperation with our partner. This also involved the analysis of setup operations on their machines and a subsequent definition of a set of characteristics, that can be used to define custom setup cost functions for different production scenarios.

The rest of this paper is organized as follows: Section 2 presents an abstract model for the setup problem and states the optimization objective. In Section 3 two optimization approaches, which are based on scheduling and dispatching, are introduced. Results obtained with both approaches are listed and analyzed in Section 4. The paper is concluded with an outlook on future work and a summary in Section 5.

2 An Abstract Model for Machine Setup Problems

We designed a generic optimization system, which can be adapted for different industries where sequence dependent setup costs are relevant, such as chemical compounds manufacturing, metal processing, food processing, or paper industries. The overall optimization concept for a single machine environment is illustrated in Figure 1, with industry, machine, or customer specific elements marked by numbers.

Each optimization algorithm receives a set of production jobs as input, which should be arranged such that the total setup costs are minimized. We believe that machine manufacturers are best suited to model and characterize setup requirements, while the customers can better assess the relevance of these model parameters for their production scenarios. To illustrate this point, setup - as defined by

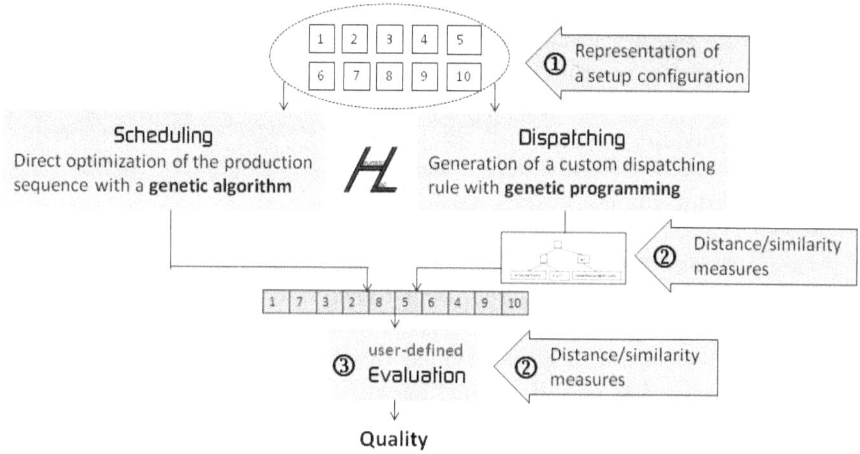

Fig. 1. Schematic overview of the generic optimization system and required inputs that must be supplied to optimize job processing plans with respect to setup costs. Two alternative optimization approaches based on scheduling and dispatching were implemented.

a manufacturer - may include full or partial dismantling of the old setup, cleaning, additional setup work, transport of tools to and from the machine etc., and all or none of the above might be relevant for a certain manufacturers optimization problem [12,2]. For example, transport times might be negligible if the tools are stored in a trolley beside the machine. In this paper we therefore propose an optimization approach that can integrate a custom, customer-defined objective function. A default objective function could of course be defined by the manufacturer as well. In any case, we suggest that the manufacturer supplies

1. a definition of a model to represent setup configurations and
2. a definition of a set of characteristics, that either assess a certain configuration or evaluate similarities/differences between pairs of configurations.

In order to evaluate generated job sequences the customer should

3. define an estimation function for the setup costs.

The characteristics defined in step 2 can be incorporated into the estimation function in step 3. In addition, customer-specific data could be used to further improve the scheduling objective function. For example, data extracted from an enterprise resource planning system, such as job due dates or material availability, could be coupled with setup costs to realize full production planning optimization.

2.1 Example: Modeling a Construction Set

To demonstrate the functionality of the system, we chose to model a construction set that consists of different building blocks, namely baseplates of various dimensions and 72 different bricks, which can be assembled to construct 3-dimensional setup configurations, as illustrated by Figure 2.

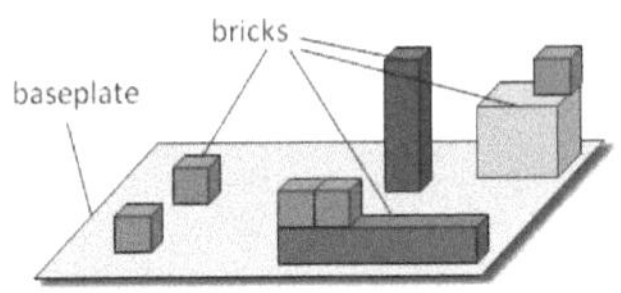

Fig. 2. Model of a setup configuration

In this abstraction, a rectangular baseplate represents a machine and bricks represent tools. A setup configuration consists of one base plate of pre-defined width and length plus a set of n bricks, which are defined by their width, length, and height. Each building block has an index, starting with zero for the baseplate and then enumerating the bricks. In addition, bricks in a configuration have a position (x,y) and four possible rotation states. If bricks overlap, they are stacked on each other, a brick with lower index having precedence over one with a higher index. Given a single baseplate and a set of configuration definitions, we wish to find a building sequence that entails the least setup effort. A number of simple similarity measures is defined for the model, which can later be incorporated into the objective function and the generated dispatching rule. The set of simple similarity measures includes

- **NumSameBricks** - number of brick types present in both configurations
- **NumBrickPairs** - the number of identical bricks, meaning bricks that only have to be moved and/or rotated between two configurations
- **NumRotatedBricks** - the number of bricks that are assembled in a non-standard rotation state
- **NumBricksSource** - the total number of bricks in the source configuration
- **NumBricksTarget** - the total number of bricks in the target configuration
- **MinTotalMovementCosts** - sum of the minimal distance that identical bricks must be moved between two configurations plus distances of added/removed bricks to the nearest edge of the baseplate
- **MinWeightedMovementCosts** - same as *MinTotalMovementCosts*, but the distances are weighted by the brick volume, assuming that larger bricks are harder to move
- **SurfaceSimilarity** - similarity measure for surface similarity, looking at the height profile from above
- **VolumeSimilarity** - volume similarity of two configurations, counting overlapping areas and weighting them in relation to the total volume difference

Based on these simple setup characteristics, we define the objective function as

$$Costs = MinWeightedMovementCosts + \alpha \cdot TransportTools + \beta \cdot RotationTools,$$

where *TransportTools* returns the number of added/removed bricks and *RotationTools* denotes the number of rotated tools. Parameters α and β specify the costs per individual transport or rotation. For the optimization runs in Section 4 we set $\alpha = 20$ and $\beta = 3$.

3 Scheduling vs. Dispatching

Scheduling and dispatching are two ways of solving production planning problems that peruse fundamentally different optimization strategies. Scheduling is a global approach, which creates a schedule - i.e. a sequence of jobs per machine - that is intended to be optimal with respect to some criteria, such as minimal accumulated setup costs. It involves planning of the whole job processing sequence in advance and is not directly suited for volatile environments where jobs arrive dynamically during the day. In contrast, dispatching is a local approach, which chooses the next most eligible candidate for processing from a set of pending jobs every time the machine becomes idle, and is thus more convenient for dynamic environments. Both scheduling and dispatching rules strive to optimize some criterion, but dispatching is necessarily greedy with respect to the employed dispatching rule, while scheduling may sacrifice short term optimality for long term optimality on the whole schedule.

3.1 Scheduling: Job Sequence Optimization with Genetic Algorithms

Genetic algorithms (GAs) are randomized, population-based search techniques, which were pioneered by John Holland [8] in the early 70s. They have been widely studied, experimented and applied to practical problem situations. GAs draw inspiration from concepts of evolutionary biology, such as mutation, selection, and crossover, to evolve a population of candidate solutions. In this study, solutions are represented as permutation lists, where each position encodes one of n jobs. The employed genetic algorithm is a variant of the so-called standard genetic algorithm, albeit equipped with an enhanced selection scheme that was first introduced in [1]. The selection scheme, which is called *offspring selection*, only accepts children that could outperform their parents. Genetic algorithms are well suited to solve production scheduling problems, both for single and multi-machine environments [6].

3.2 Dispatching: Generation of Complex Rules with Genetic Programming

We also propose the generation of customized dispatching rules, that rely on product or machine configuration similarities to reduce total setup costs on a machine for a given number of jobs. Genetic programming (GP) can be seen as a special GA variant that traditionally operates on tree structures, which can for instance represent computer programs or mathematical formulas [9]. As shown in the authors' previous work [5] the use of GP to generate dispatching rules has a couple of advantages: First and foremost the time-consuming generation of the dispatching rule can be done offline on a training data set, but the actual job sequence decisions during production are made in real-time with the rule. Moreover, if the production scenario at a plant changes, for example due to an increase in workload, the rule generation step can be repeated to adapt to the new situation.

4 Results

We tested our approach with the toy model from Section 2.1 and on a real-world production scenario. Two test instances were generated: Dataset 1 consists of a 50x50 base plate and 100 randomly generated setup configurations with 0-20 bricks, uniformly drawn from 72 different brick types and rotated with a probability of 5%. Dataset 2 was generated with the same settings as Dataset 1, albeit with a larger problem dimension of 200 setup configurations that must be scheduled. Lacking a known best solution, we compare the results to the best performing simple global dispatching rule, as listed in Table 1. All test runs were conducted with HeuristicLab, a generic and extensible framework for heuristic and evolutionary optimization [10].

It can be seen that the *Max(MinWeightedMovementCosts)* dispatching strategy yielded the best results on both test instances, which is not surprising, since the chosen objective function is a sum of the *MinWeightedMovementCosts* plus transport and rotation costs, as detailed in Section 2.1. To generate custom, composite dispatching rules, a GP was configured with 1-elitism, 15% mutation rate, maximum formula tree height of 10, and maximum tree size of 40. A population size of 100 for Dataset 1 and 200 for Dataset 2 was used. Likewise, a GA with 1-elitism, population size of 100, proportional selection, order crossover, inversion mutation, and 5% mutation rate was employed to directly optimize the job sequences. In addition, both approaches used offspring selection to steer the algorithm.

Results averaged over 10 optimization runs are listed in Table 2. Composite dispatching rules improved the results by 5% at most, compared to the best simple priority rule, while the direct optimization of job sequences achieved a maximal improvement of 10%. These meager results can be attributed to the random generation of configurations in the test instances, which produced very diverse setups. In real-world problems, part similarity is much more pronounced and setup configurations therefore exhibit a stronger resemblance to eachother.

Table 1. Results for the global application of a single, simple dispatching rule on two test instances. The respective best results are in bold face.

ID	Sort	Dispatching Rule	Dataset 1	Dataset 2
0		NumSameBricks	663,716	1,342,512
1		NumBrickPairs	664,248	1,329,912
2		NumRotatedBricks	736,494	1,490,724
3	MIN	NumBricksTarget	734,852	1,487,702
3		MinTotalMovementCosts	764,144	1,589,214
4		MinWeightedMovementCosts	758,926	1,593,271
5		SurfaceSimilarity	723,684	1,503,179
6		VolumeSimilarity	728,723	1,506,459
7		NumSameBricks	762,535	1,581,838
8		NumBrickPairs	763,171	1,581,838
9		NumRotatedBricks	746,946	1,491,929
3	MAX	NumBricksTarget	744,064	1,488,374
10		MinTotalMovementCosts	661,304	1,285,762
11		MinWeightedMovementCosts	**608,175**	**1,201,339**
12		SurfaceSimilarity	742,862	1,496,264
13		VolumeSimilarity	733,587	1,505,583

Table 2. Results obtained for direct sequence optimization with GA and the generation of complex dispatching rules with GP

Problem	Dispatching Variant	Algorithm	Average	Std.	Runtime
Dataset 1	Scheduling: Direct sequence optimization	SGA	562,056	6,837	15 min
	Dispatching: Generation of complex rule	GP	588,959	6,967	10 hrs
Dataset 2	Direct optimization of configuration sequences	SGA	1,162,544	12,782	30 min
	Generation of complex rule	GP	1,165,626	8,797	1 day

For the real-world industry application, we were indeed able to reduce setup costs by 15% with generated dispatching rules and by 25% with direct sequence optimization. In both applications, direct sequence optimization with GA ouperforms GP generated rules, both with respect to setup cost reduction and runtime.

5 Conclusions and Future Work

In this section we will interpret the obtained results and point out areas of interest for future research.

For the problem dimensions investigated in this study, runtime and solution quality were in favor of the scheduling approach. However, with growing complexity the monthly or weekly generation of a custom dispatching rule and subsequent application in real-time during production becomes more and more attractive. Whether the effort for the generation of a dispatching rule is worthwhile ultimately depends on the short and long term stability of the rule. We believe that composite rules with conditionals have the potential to express such complex sequencing decisions and wish to focus on rule stability analyses in a future extension of our work.

Secondly, we emphasized the importance of integrating user-defined objective functions. In this study only one objective function has been employed both for the toy problem and the real-world scenario. Therefore the influence of different objective functions on the algorithmic performance is still a pending question and will have to be looked at in more detail.

Finally, different optimization approaches were only compared with each other. The logical next step would be to directly match them against a human expert and validate the generated job sequences in field tests in the production facility. Structural analyses of the generated dispatching rules could reveal similarities - or differences - with strategies employed by the human planner. This step is essential, not only from a validation perspective, but also to perform acceptance testing and establish trust in the software.

Acknowledgments

The work described in this paper was carried out as part of the IPS (Intelligent Production Steering) project and partially funded by the Mechatronics Cluster and the regional government of Upper Austria.

References

1. Affenzeller, M., Wagner, S.: Offspring selection: A new self-adaptive selection scheme for genetic algorithms. In: Ribeiro, B., Albrecht, R.F., Dobnikar, A., Pearson, D.W., Steele, N.C. (eds.) Adaptive and Natural Computing Algorithms. Springer Computer Series, pp. 218–221. Springer, Heidelberg (2005)
2. Aggarwal, S.C.: A note on "The influence of setup time on job shop performance" by Wilbrecht and Prescott. Management Science 19(11), 1332–1333 (1972)
3. Allahverdi, A., Gupta, J.N.D., Aldowaisan, T.: A review of scheduling research involving setup considerations. Omega International Journal of Management Science 27, 219–239 (1999)
4. Allahverdi, A., Ng, C.T., Cheng, T.C.E., Kovalyov, M.Y.: A survey of scheduling problems with setup times or costs. European Journal of Operational Research 187, 985–1032 (2008)
5. Beham, A., Winkler, S., Wagner, S., Affenzeller, M.: A genetic programming approach to solve scheduling problems with parallel simulation. In: Proceedings of the 22nd IEEE International Parallel & Distributed Processing Symposium (IPDPS 2008). IEEE, Los Alamitos (2008)
6. Cheng, R., Gen, M., Tsujimura, Y.: A tutorial survey of job-shop scheduling problems using genetic algorithms, part ii: hybrid genetic search strategies. Computers & Industrial Engineering 36, 343–364 (1999)
7. Gilmore, P.C., Gomory, R.E.: Sequencing a one-state variable machine; a solvable case of the traveling salesman problem. Operations Research 12, 655–674 (1964)
8. Holland, J.H.: Adaption in Natural and Artificial Systems. University of Michigan Press (1975)
9. Koza, J.R.: Genetic Programming: On the Programming of Computers by Means of Natural Selection. MIT Press, Cambridge (1992)
10. Wagner, S.: Heuristic Optimization Software Systems - Modeling of Heuristic Optimization Algorithms in the HeuristicLab Software Environment. PhD thesis, Johannes Kepler University, Linz, Austria (2009)
11. White, C.H., Wilson, R.C.: Sequence dependent set-up times and job sequencing. International Journal of Production Research 15(2), 191–202 (1977)
12. Wilbrecht, J.K., Prescott, W.B.: The influence of setup time on job shop performance. Management Science 16(4), B274–B280 (1969)

A GRASP–VNS Hybrid for the Fuzzy Vehicle Routing Problem with Time Windows

J. Brito[1], F.J. Martínez[1], J.A. Moreno[1], and J.L. Verdegay[2]

[1] Group of Intelligent Computing, I.U.D.R, University of La Laguna, 38271, La Laguna, Spain
{jbrito,fmartinz,jamoreno}@ull.es

[2] Department of Computer Science and Artificial Intelligence, University of Granada, 18071, Granada, Spain
verdegay@decsai.ugr.es

Abstract. We consider the Vehicle Routing Problem with time windows where travel times are triangular fuzzy numbers. The weighted possibility and necessity measure of fuzzy relations is used to specify a confidence level at which it is desired that the travel times to reach each customer fall into their time windows. In this paper we propose and analyze a solution procedure consisting in hybridizing a Variable Neighborhood Search (VNS) and a Greedy Randomize Adaptive Search Procedure (GRASP) for the corresponding optimization problem.

Keywords: Vehicle Routing Problem, GRASP, VNS, Fuzzy travel time.

1 Introduction

Global transportation and logistics environments require computational methods for planning that reduce operating costs, optimize available resources and improve performance and customer service. These environments also demand methods and tools that provide more control, capacity for adaptation and flexibility in their operations, such as location and planning, warehousing, distribution, and transportation. Transportation and logistics organizations often face combinatorial problems on operational and strategic levels, specifically in the case of the design strategies for planning route distributions.

A standard objective in a distribution system for geographically dispersed customers is to determine the set of routes for the available vehicles which satisfy several constraints and minimize total fleet operating cost. Vehicle Routing Problems (VRP) are concerned with finding the optimal set of routes, beginning and ending at a depot, for a fleet of vehicles to serve customers with demands for some commodity. Constraints such as capacity limitations on the amount of demand of the customers served by each vehicle, or service time or time window to serve each customer are included among the real-life requirements in VRP.

In many practical problems it is necessary to realize that the available knowledge about some data and parameters of the model are imprecise or uncertain. Exact travel time between two locations in routing problems is frequently unknown in advance

R. Moreno-Díaz et al. (Eds.): EUROCAST 2009, LNCS 5717, pp. 825–832, 2009.

because of road, traffic conditions, etc.; consequently, travel time is represented by fuzzy numbers which can more accurately model this imprecision.

The inability of exact approaches to solve medium and large scale vehicle routing problems, as well as the difficulty in evaluating the objective function in real-life complex problems are two important reasons why heuristics and metaheuristics are mainly employed for the solution of VRP. Different methods have been developed to solve the Vehicle Routing Problem with Time Windows (VRPTW), including the Greedy Randomize Search Procedure (GRASP) and Variable Neighborhood Search (VNS) for most of these problems [1].

In this paper we use the Fuzzy Sets and Possibility Theory proposed by Zadeh [9] and the ideas developed by Dubois and Prade [4] and others, to manage the optimistic and pessimistic view on the comparison of fuzzy number. We also propose the combination of this fuzzy approach and a hybrid GRASP-VNS heuristics to obtain solutions of the Fuzzy Vehicle Route Problem with Time Windows (FVRPTW).

This paper is organized as follows. Section 2 describes the approach to operate with fuzzy travel times. Section 3 explains the FVRPTW and the codification proposed for its solutions. Section 4 details the hybrid metaheuristic implementation used to find solutions. In Section 5, computational results are described and compared. Finally, conclusions are included in the last section.

2 Fuzzy Travel Time

An ordinary set can be described, among other ways, by using the characteristic function on an universe, in which 1 indicates membership and 0 non-membership. However, in many cases, the membership is not clear when the sets are imprecisely described. In order to deal with them, Zadeh [8] introduced the concept of a fuzzy set given by a membership function from the universe to the real interval [0,1]. The value zero is used to represent complete non-membership, the value one is used to represent complete membership, and values in between are used to represent intermediate degrees of membership. Fuzzy sets have been well developed and applied in a wide variety of real problems. For each $\alpha \in [0,1]$, the α-cut of a fuzzy set is the ordinary set of values where the membership is equal or greater than α. The support of a fuzzy set is the set of values whose membership is positive and its mode is the value with maximum membership. A fuzzy number is usually defined as a fuzzy set of the real line whose α-cuts are closed and bounded intervals, with compact support and unique mode [3].

In real-life applications of routing problems, it is often the case that the exact travel time between two locations cannot be known in advance. A vague or imprecise quantity can be modeled by making use of the concept of fuzzy number. The introduction of fuzzy numbers will allow the approximate numeric values of travel time to be manipulated.

A simple model for these fuzzy numbers is the triangular fuzzy numbers. A triangular fuzzy number is given by its support $[a_1,a_3]$ (the set of possible values) and its mode $a_2 \in [a_1,a_3]$ (the most plausible value). This triangular fuzzy number is denoted by $Tr(a_1,a_2,a_3)$ and has the following membership function:

$$\mu_{Tr}(x)=\begin{cases}0 & x<a_1\\ \dfrac{x-a_1}{a_2-a_1} & a_1\le x\le a_2\\ \dfrac{x-a_3}{a_2-a_3} & a_2\le x\le a_3\\ 0 & a_3<x\end{cases} \tag{1}$$

We use triangular fuzzy numbers to model the imprecision in travel times along each edge of the underlying network. A little knowledge can be used to assign intervals $[T_{min},T_{max}]$ to represent the travel times and define fuzzy numbers. Precise distributions would require a deeper knowledge of the instance and usually yield very complex calculations. If the knowledge can be used to specify the value T_{mod} that appears to be more plausible than others, a natural extension is to use fuzzy numbers. Thus, the user or decision maker of the problem can subjectively estimate, based on his experience and intuition and/or available data, that travel times at the vehicle correspond to fuzzy triangular numbers $Tr(T_{min},T_{mod},T_{max})$.

The usual arithmetic operators for real numbers are extended to fuzzy numbers by the *Extension Principle*. Then the sum of two triangular numbers is a triangular number and the maximum between two triangular numbers is approximated by a triangular number. We use the formulas:

$$Tr(a_1,a_2,a_3)+Tr(b_1,b_2,b_3)=Tr(a_1+b_2,a_2+b_2,a_3+b_3) \tag{2}$$
$$Tr(a_1,a_2,a_3)\vee Tr(b_1,b_2,b_3)\cong Tr(a_1\vee b_2,a_2\vee b_2,a_3\vee b_3)$$

Several methods for comparisons of fuzzy numbers are available [11]. We use the Possibility and Necessity for comparing the time a vehicle reaches a customer with the upper limits of his time window.

If $\tilde{t}$ is this fuzzy time and b is the upper bound then the possibility is

$$Pos(\tilde{t}\le b)=\sup_{x\le b}\mu_{\tilde{t}}(x) \tag{3}$$

and the necessity is $Nec(\tilde{t}\le b)=1-Pos(\tilde{t}\ge b)$. Therefore if $\tilde{t}=Tr(a_1,a_2,a_3)$ then:

$$Pos(\tilde{t}\le b)=\begin{cases}1 & a_2<b\\ \dfrac{b-a_1}{a_2-a_1} & a_1\le b\le a_2\\ 0 & b<a_1\end{cases}\quad Nec(\tilde{t}\le b)=\begin{cases}1 & a_3<b\\ \dfrac{b-a_3}{a_2-a_3} & a_2\le b\le a_3\\ 0 & b<a_2\end{cases} \tag{4}$$

The possibility measures the best case while the necessity measures the worst case then the possibility shows the most optimist attitude toward the events and the necessity shows the most pessimist attitude. We consider a linear combination of the possibility and the necessity to modulate the optimistic/pessimistic attitude of the decision-maker. The parameter $\lambda\in[0,1]$ in the Weighted Possibility and Necessity given by $W(A) = \lambda Pos(A) + (1-\lambda)Nec(A)$ measures the degree of optimism of the decision maker.

3 Model Description

The VRPTW, normal extension of VRP, is given by a set of k identical vehicles to serve a set of n customers within given time windows departing from a depot. Each vehicle goes by a route visiting a number of nodes satisfying their own demand.

The assumptions of the FVRPTW model are:

- Each vehicle is assigned to only one route on which there may be more than one customer.
- Each customer is visited by one and only one vehicle.
- Each route begins and ends at the depot.
- Each vehicle has a container with a capacity limitation and the total loading of each vehicle cannot exceed its capacity.
- Each customer is served within its time window.
- The travel times between customers are triangular fuzzy number.

The objective of the optimization problem is to minimize the total traveled distance by all the vehicles in their routes.

We consider the following indices and model parameters:

- The vehicles indexes are: $k = 1, 2, ..., m$;
- The customers indexes are: $i = 1, 2, ..., n$;
- The depot index is $i = 0$;
- The amount of demand of customer i is $q[i]$, $i = 1, 2, ..., n$;
- The capacity of vehicle k is $C[k]$, $k = 1, 2, ..., m$;
- The distance from customer i (or depot 0) to j is $d[i,j]$, $i, j = 0, 1, ..., n$;
- The triangular fuzzy travel time from customer i to j is $T[i,j]$, $i, j = 0, 1, ..., n$;
- The unloading time at customer i is $U[i]$, $i = 1, 2, ..., n$;
- The time window of customer i is $[a[i],b[i]]$; $i = 1, 2, ..., n$;
  $a[i]$ and $b[i]$ are the respective beginning and end of the time window.

The operational plan consisting of the routes for the m vehicles to serve the n customers is given by a single decision vectors x of size n+m+1 denoted by $x = (x_0, x_1, x_2, ..., x_{n+m})$ that is a rearrangement of $(0, 1, 2, ..., n+m)$ such that $x_0 = 0$ and $x_{n+m} > n$. Each element of the solution greater than n represents a new vehicle that begins at depot in such a way that $x_r = n+k$ represents that vehicle k returns to the depot. The customers visited for this vehicle are those of the indexes that appeared in the solution from the previous vehicle or from $x_0 = 0$ if it is the first vehicle. In this way, the vector arrangement x ensures that that each vehicle is used at most once, all tours begin and end at the depot, each customer is visited by one and only one vehicle, and there is no sub-tour.

The objective function to be minimized is the total traveled distance of the solution x that is computed by:

$$F = \sum_{r=0}^{n} d[x_r, x_{r+1}] \tag{5}$$

where for every $i > n$, $d[i,j] = d[j,i] = d[0,j]$, for all $j = 1,2, \ldots, n$.

The constraints of the problem are the capacity constraints of the vehicles and the time windows constraints of the customers. The capacity constraints are constraints of real values and are tested for each solution considered by computing the load Q of each vehicle to serve the assigned customers.

The amount $Q[.]$ of demand served by each vehicle is computed by the following recurrent formula on the vector x. Take $Q[x_0] = 0$ and then for $r = 1, 2, \ldots, n+m$, apply: $Q[x_r] = Q[x_{r-1}] + q[x_r]$. The capacity constraint of each vehicle is verified each time that $x_r > n$ by $Q[x_r] \leq C[x_r - n]$. To compute the load of the next vehicle $Q[x_r]$ is set again to 0 and the recurrent formula $Q[x_i] = Q[x_{i-1}] + q[x_i]$ is again applied for $i = r+1, r+2, \ldots,$ until the next i such that $x_i > n$.

The service time $S[.]$ for each customer is a fuzzy number obtained by similar recurrent formula on the vector x. To compute the fuzzy service time for each customer we apply the formula: $S[x_r] = a[x_r] \vee (S[x_{r-1}] + U[x_{r-1}] + T[x_{r-1}, x_r])$ with the following conventions. The initial values at the depot are $S[x_0] = 0$ and $U[x_0] = 0$. If x_r corresponds to a vehicle ($x_r > n$) the service time is set to 0 to compute the service time for the next customer x_{r+1} by the above recurrent formula. For these cases, the values of $U[x_r] = U[j]$ for $j > n$ are also null, so that the service time for the next customer x_{r+1} (that is the first customer of the next vehicle) is: $S[x_{r+1}] = a[x_{r+1}] \vee T[0, x_{r+1}]$. A positive value for $U[j]$ for these cases ($j > n$) would correspond to some time-consuming preprocessing operation at the depot such as the loading of the vehicles. The time windows for these indexes ($[a[j], b[j]]$ for $j > n$) do not exist unless the vehicle has to return to the depot within a given time interval.

Note that since the times between customers $T[i,j]$ are triangular fuzzy numbers then each service time $S[x_r]$ is also a triangular fuzzy number. The unload times $U[x_r]$, the beginning $a[x_r]$ and the end $b[x_r]$ of the time windows are real numbers which can be considered triangular fuzzy numbers where the three defining numbers are equal. In addition, the arithmetic operator in the constraints refers to an addition of fuzzy numbers. This determines that, the comparison between quantities is also ambiguous and therefore are fuzzy constraints. We consider an approach proposal by Delgado et al. [2] which considering fuzzy solutions for *models with fuzzy coefficients in the constraints.* The particular method used to verify if a solution satisfies the time window constraints is the Weighted Possibility and Necessity. Therefore, at confidence value α, the fuzzy service time of customer x_r is within the corresponding time window if $W(S[x_r] \leq b_r) \geq \alpha$. Therefore, at this confidence value α, the solution x verifies the time window constraints if: $min_r\, W(S[x_r] \leq b_r) \geq \alpha$.

4 Solution Approach

We propose a hybrid solution algorithm based on Greedy Randomized Adaptive Search Procedure (GRASP) and Variable Neighborhood Search (VNS) metaheuristics for the FVRPTW. The total distance traversed by the vehicles are minimized while the capacity constraints are satisfied and the service times fall within time windows at a given confidence level. GRASP is a metaheuristic proposed by Feo and Resende [7]. GRASP consists of two phases: a construction phase, in which a randomized greedy function is used to produce a feasible solution, and an improvement phase, in which a local search replacing the constructed solution by a better one. Variable

Neighborhood Search (VNS) is a metaheuristic proposed in 1997 by Hansen and Mladenovic based upon systematic change of neighborhoods in a search [5]. VNS has rapidly developed in both methods and applications.

```
CONSTRUC_GRASP Method
  1.Let x ={} be an initial empty partial solution. Set t ← 0.
  2.Repeat the following sequence until there is not element to
    include in the partial solution:
    (a)Construct the restricted candidate list RCL.
    (b)Choose at random an element e_{t+1} of RCL.
    (c)Update the partial solution x_{t+1} ← x_t + {e_{t+1}}.
    (d)Set t ← t + 1.
```

Fig. 1. Construct Greedy Randomized Adaptive Search Procedure (GRASP) pseudo-code

The *solution construction GRASP* mechanism builds a solution step-by-step by adding a random new element from a candidate list (the restricted candidate list RCL) to the current partial solution. The elements are initially ordered in a candidate list with respect to a previously defined greedy function. The RCL consists of a bounded set of the best new elements. The probabilistic component of a GRASP is characterized by a random choice of the element that is not necessarily the top candidate of the RCL. There is no guarantee that the solution generated in the construction phase is locally optimal. Subsequently, a local search is applied to the current solution as an *improvement* phase.

```
GENERAL_VNS Method
  1. Take an initial solution x.
  2. Repeat the following steps, until a stop condition:
   (i)  Set k ← 1;
   (ii) Repeat the following steps, until k = k_max:
      a) Shaking: Take a random neighbour x' of x; x'∈N_k(x)
      b) Improving:
        b1) Set k' ← 1;
        b2) Repeat the following steps until k' = k'_max:
            Find the best neighbour x'' of x' in N_k'(x):
              If f(x'') < f(x') then set x' ← x'' and k' ←1;
              otherwise, set k' ←k'+ 1.
      c) Move or not:
        If the local minimum x'' is better than x
          then set  x ← x'',  and  k ← 1;
          otherwise, set k ← k + 1;
```

Fig. 2. Variable Neighbourhood Search (VNS) pseudo-code

The basic schemes of VNS can be described by the combination of series of random and improving (local) searches based on changing neighborhood structures. The main phases of basic VNS are *Shaking* and *Local Search* (*LS*). A *LS* consists in applying an improving move to the current solution while possible. When the search stops at a local minimum, a *shake procedure* performs a random search for a new starting point for a new local search. The improving local search and the random shake procedure are usually based on a standard move that determines the neighborhood structures.

We use a neighborhood structure based on a mechanism that moves a subset of consecutive elements of the solution and can be described as follows: Let $N_k(x)$ be the set of solutions obtained from x by a k-chain move. A k-chain move consists in taking a chain or segment of the solution with length k and moving it to another part of the solution. Given that k is fixed, we choose two positions i and j of the solution vector and insert the k elements which are in positions j, j+1, j+2, ..., j+k–1 after position i. The service times and level of satisfying the times windows constraints need to be updated only in modified routes.

The proposed hybrid GRASP-VNS approach is obtained by using GRASP for the generation of the initial solution for VNS or using the VNS in the post-processing phase of GRASP.

5 Experiments

The GRASP-VNS solution approach has tested by comparing it with GRASP and VNS with the instances of the papers by Zheng and Liu [10] and Peng et al. [6] that deal also with a FVRPTW with triangular travel times. These instances have 18 and 20 customers respectively. We solved the instances for $\alpha = 0.6$ and table 1 shows the objective function F and the computational time in seconds with several number of iterations.

Table 1. Experimental results with 18 and 20 customers instances

18 customers	5000		10000		20000	
	F	Time	F	Time	F	Time
GRASP	349.70	25	349.50	42	349.50	54
VNS	360.50	110	339.50	269	339.50	271
GRASP-VNS	345.70	87	345.70	88	345.70	91

20 customers	5000		10000		20000	
	F	Time	F	Time	F	Time
GRASP	710	2.21	710	2.21	710	2.21
VNS	920	8.58	920	8.58	920	8.58
GRASP-VNS	685	492	670	641	660	1588

In both cases the best results are obtained with GRAS-VNS hybrid approach.

6 Conclusions

Uncertainty needs to be managed in real logistic and transportation systems. Fuzzy Logic systems and the Possibility and Necessity measures can be used to manage the uncertainty in time travel: Triangular fuzzy travel time can be assigned with a little knowledge and efficiently managed with metaheuristics. The Weighted Possibility and Necessity modulate the optimistic/pessimistic attitude of the Decision Maker. GRASP-VNS hybrid algorithm provides an interesting way to find good solutions to FVRPTW, improving the results of both GRASP and VNS, and providing flexibility and adaptability.

Future research will include a deeper analysis of the fuzzy concepts needed to formalize the optimization problem and computational analysis of the application of these metaheuristics to real instances of the FVRPTW and related problems.

Acknowledgement

Supported by project TIN2008-06872-C04-01/TIN and TIN2008-06872-C04-04//TIN of the Spanish Government (70% are FEDER funds) and TIC-02970-JA.

References

1. Bräysy, O., Gendreau, M.: Vehicle Routing Problem with Time Windows. Part I: Route Construction and Local Search Algorithms. Part II: Metaheuristics. Transportation Science 39(1), 104–139 (2005)
2. Delgado, M., Verdegay, J.L., Vila, M.A.: A General Model for Fuzzy Linear Programming. Fuzzy Sets and Systems 29, 21–29 (1989)
3. Dubois, D., Prade, H.: Fuzzy Sets and Systems, Theory and Applications. Academic, New York (1980)
4. Dubois, D., Prade, H.: Possibility Theory: An Approach to Computerized Processing of Uncertainty. Kluwer, Dordrecht (1988)
5. Hansen, P., Mladenović, N., Moreno Pérez, J.A.: Variable Neighbourhood Search: Methods and Applications. 4OR: A Quarterly Journal of Operations Research (2008) (in press), doi:10.1007/s10288-008-0089-1
6. Peng, J., Shang, G., Liu, H.: A Hybrid Intelligent Algorithm for Vehicle Routing Models with Fuzzy Travel Times. In: Huang, D.-S., Li, K., Irwin, G.W. (eds.) ICIC 2006. LNCS (LNAI), vol. 4114, pp. 965–976. Springer, Heidelberg (2006)
7. Resende, M.G.C., Ribeiro, C.C.: Greedy Randomized Adaptive Search Procedure. In: Glover, F., Kochenberger, G. (eds.) Handbook in Metaheuristics, pp. 219–249. Kluwer, Dordrecht (2003)
8. Zadeh, L.A.: Fuzzy Sets. Information and Control 8(3), 338–353 (1965)
9. Zadeh, L.A.: Fuzzy Sets as a Basis for a Theory of Possibility. Fuzzy Sets and Systems 1(1), 3–28 (1978)
10. Zheng, Y., Liu, B.: Fuzzy Vehicle Routing Model with Credibility Measure and its Hybrid Intelligent Algorithm. Applied Mathematics and Computation 176(2), 673–683 (2006)
11. Zhu, Q., Lee, E.S.: Comparison and Ranking of Fuzzy Number. In: Kacyprzyk, J., Fedrizzi, M. (eds.) Fuzzy Regression Analysis, pp. 21–44. Omnitech, Heidelberg (1992)

Performance Modelling for Avionics Systems

Visar Januzaj[1], Ralf Mauersberger[2], and Florian Biechele[3]

[1] Technische Universität Darmstadt, Fachbereich Informatik,
FG Formal Methods in Systems Engineering - FORSYTE
Hochschulstr. 10, 64289 Darmstadt, Germany
januzaj@forsyte.de

[2] EADS Innovation Works, 85521 Ottobrunn, Germany
ralf.mauersberger@eads.net

[3] EADS Defence & Security, 85077 Manching, Germany
florian.biechele@eads.com

Abstract. The new paradigm of Integrated Modular Avionics (IMA) [1] necessitates the analysis and validation of non-functional requirements for IMA systems. This includes the analysis of their performability. In this paper we present an initial approach of a performance modelling framework, based on the SAE standardised modelling and analysis language AADL [2, 3], to integrate performance analysis in the toolchain of this practical context. The proposed framework is a hybrid of static and dynamic systems analysis and includes aspects of performance evaluation.

1 Introduction

Due to the increasing demands and complexity of avionics systems emerges the need of a methodology to cope with system development issues, such as spatial demands, power resources onboard the aircraft and high maintenance costs. To overcome these issues the *Integrated Modular Avionics* (IMA) [1] defines avionics system as an integrated system with multiple functions hosted on a cabinet of processors. In consequence IMA systems are comprised as distributed embedded systems where software components interact with each other and the hosting physical architecture by a set of standardised interfaces, similar in function to operating system calls in general-purpose computer systems. This abstraction enables dynamically configurable systems where software components are deployed on host processors according to criteria such as criticality, system load and run-time faults. The deployment/binding between hardware and software components is defined in so called system configuration tables or *blueprints*. System blueprints including an initial system configuration as well as reconfigurations for identified system failure types are loaded at runtime and conditions are monitored by designated system components. Despite its advantages, IMA demands sophisticated analysis of avionics systems. Thus a blueprint generation requires foregoing system analysis, e.g. schedulability analysis, to assure a correct operation of avionics software so that deadlines are met and memory resources are sufficient. To achieve a successful application of the IMA concept in the

R. Moreno-Díaz et al. (Eds.): EUROCAST 2009, LNCS 5717, pp. 833–840, 2009.

avionics domain, supporting techniques are essential for blueprints generation and their integration into the system development process.

We present a performance modelling framework which facilitates the automatic generation of blueprints for IMA systems. Our framework not only proposes a method for execution time prediction in modern avionics systems, more importantly, it shows how those predictions can be integrated and used for the determination of stable and feasible system configurations (blueprints). In order to support the IMA technology we use the SAE standardised *Architecture & Analysis Description Language* (AADL) [2,3] which offers an abstract but precise description of the components of a system architecture and facilitates the application of performance analyses [4].

2 The Performance Modelling Framework

Figure 1 describes our modelling and analysis framework. Our approach is divided into two major stages. Within the first stage, *component analysis*, we separately analyse the hardware and software components. For each analysis a corresponding profile is generated. The analysis is performed separately for the following reasons:

- once the hardware is analysed, we do not need to repeat its analysis for possible software changes, or if new software is available
- we do not need to run tests for each software on each available hardware

This approach can also be applied for execution time prediction of software being still under development. In the second stage, *modelling and system analysis*, we model, according to the collected profile data, and analyse the generated system. The purpose of the latter analysis is to determine possible configurations in order to finally generate feasible and stable system blueprints.

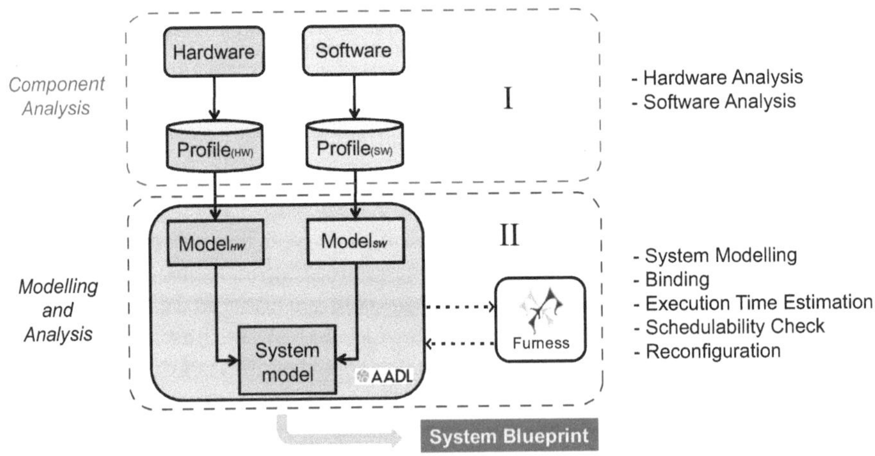

Fig. 1. The performance modelling framework

2.1 Component Analysis

This analysis concerns the execution time prediction and aims at generating corresponding *profiles* for each component type. These profiles hold measurement information about the estimated execution time for each instruction/operation (on a hardware) and their occurrence during an execution run (of a software), respectively.

Hardware Analysis. For the sake of simplicity the hardware is analysed on a test-based manner, i. e. carrying out a number of benchmarks. For each benchmark we collect the execution time and perform an operation counting (cf. Software Analysis), i. e. we count during the execution of a benchmark the occurrence of each operation type, e. g. `ADD`, `MUL` and `ALLOC`. The benchmarks are C programs that try to reflect as good as possible the behaviour and nature of the software used in avionics systems.

We represent each benchmark as a *linear inequation* built by the number of the operations occurred and the observed execution time, e. g. for a benchmark b we get : $200 \cdot \texttt{ADD} + 300 \cdot \texttt{MUL} + 50 \cdot \texttt{DIV} + 9 \cdot \texttt{ALLOC} + \ \ldots \ \leq \ 33,9 \ ms.$

All resulting linear inequations are put together creating a linear inequation system. Mathematically we represent such a linear inequation system as follows : $O \ \underline{v} \ \leq \ \underline{t}$, where O is a $n \times m$ matrix built out of the number of the occurrences for each operation (called in the following *variable*) in each benchmark. The number of rows n corresponds to the number of benchmarks and the number of columns m to the number of variables. The $1 \times m$ vector v represents the variable vector. t represents the $1 \times n$ vector of execution times for each benchmark.

In order to get more accurate execution time estimations (or even to get a solution at all) we add in both sides of our linear inequations a variation vector δ, representing potential differences on execution times that a benchmark might have (during different runs). Thus we get the following representation : $\underline{t} - \underline{\delta} \leq O \ \underline{v} \leq \underline{t} + \underline{\delta}$.

To avoid arbitrary solutions we need to add further constraints to our linear inequation system. These constraints are based on processor data sheets and represent a weighting for each variable. The varying weights evolve from the fact that, e. g. a floating-point division takes much longer than a simple address load. The constraints play an important role since they not only reflect the processor's architecture but also lead to more qualitative and better solutions.

The estimated execution time of the variables is calculated by minimising over the sum of all variations δ (subject to the linear inequation system and the corresponding constraints): $min \ \sum_{i=1}^{n} \ \delta_i$, where n is the number of the benchmarks. The minimisation problem is solved using linear programming solvers. The solutions represent the estimated execution time for each variable and are stored in the *hardware profile*.

Software Analysis. In this step the source code structure is extracted and a rather symbolic execution time estimation is calculated. To achieve this we perform *code instrumentation* using the LLVM compiler framework [5], i. e. the source code is translated into a LLVM bytecode intermediate representation

```
property set New_Properties is
    Variables : list of aadlstring applies to (thread, processor);
    ExecTime  : list of aadlreal applies to (processor);
    Occurrence: list of aadlinteger applies to (thread);
end New_Properties;
```

Fig. 2. New properties

```
processor CPU
  properties
    New_Properties::Variables => ("MUL", "ADD", "LOAD");
    New_Properties::ExecTime  => (0.0035, 0.0013, 0.0007);
end CPU;

thread T
  properties
    New_Properties::Variables  => ("MUL", "ADD", "LOAD");
    New_Properties::Occurrence => (309, 53, 137);
end T;
```

Fig. 3. Applying new properties

(IR). The language- and platform-independent IR is represented by a reduced set of RISC-like instructions. Furthermore, by providing a modified `gcc` compiler the LLVM framework facilitates the analysis of optimised code, thus increasing the software analysis accuracy. To keep track of each instruction type execution we add to the IR corresponding *counters*. These counters are incremented each time an instruction is executed. A program run is symbolically represented as an equation of the following form:

$$3579 \cdot \texttt{ADD} + 759 \cdot \texttt{MUL} + 53 \cdot \texttt{LOAD} + \;\ldots\; = \; exT, \tag{1}$$

where exT is the execution time which at this point is unknown and will be estimated later on. The code structure (at the function level) and the symbolic execution time estimation form together the *software profile.*

2.2 Modelling and System Analysis

We model the component profiles and the resulting system using AADL. A detailed overview of AADL can be found in [2, 3]. The AADL components are divided into three categories: (i) *Hardware components* : `processor`, `memory`, `bus`, `device`, (ii) *Software components* : `process`, `thread`, `subprogram`, `data`, `thread group` and (iii) *System components* : `system` (representing the composition of all components).

The hardware profile is mapped to hardware components and the software profile is mapped to software components, respectively. However, most of the profiles data cannot be mapped to AADL components. Using own property definitions one can add supplementary attributes to each AADL component.

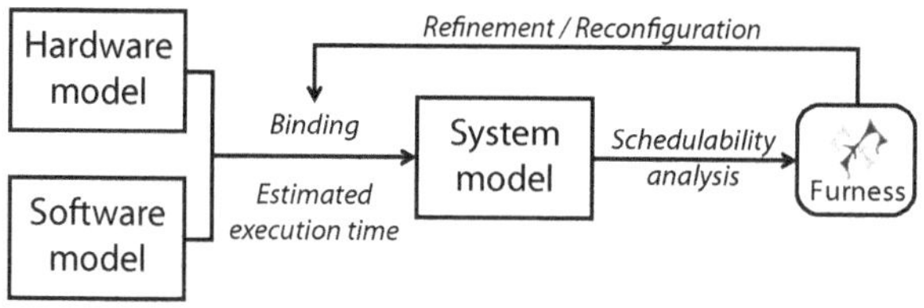

Fig. 4. System analysis

The newly defined properties (cf. Fig. 2) for both hardware (`processor`) and software components (`thread`) represent the variables, their occurrence and their estimated execution time. The application of the new properties is illustrated in Fig. 3. Using the data stored in the property entries we determine the estimated execution time for each thread depending on a particular binding.

System Analysis. In this section we introduce the final step of our approach, which involves the following actions:

– calculate a set of possible bindings for hardware and software components
– determine the execution time estimation for each chosen binding
– find feasible system configurations by performing schedulability analysis

These actions are repeated until a feasible and sufficient schedule is found. Hence, the automated system deployment process is completed. Figure 4 depicts the three steps introduced above.

In order to include possible hardware failures we perform *reconfigurations*, i. e. for each found feasible system configuration we exclude a hardware component and repeat the actions above. We keep excluding hardware components until no binding can be found. All computed reconfigurations are stored and integrated into the AADL system model as *failure modes*. We will however not discuss in this work the reconfiguration process.

Binding. This step is one of the important ones, since the quality of determined bindings influences plainly the quality of the system deployment and the generated blueprint. Possible bindings are determined by cross combining the available hardware and software in consideration of various important criteria, e. g. taking into account the speed of the processor in order to find the more appropriate hardware to handle the specified software and by clustering processes that have dense intercommunication to achieve an efficient deployment.

Execution Time Estimation. The estimated execution time of a software bound to a particular hardware is calculated by simply building equations of the form shown in equation (1) and evaluating the generated equation. Each chosen binding is specified by the `Allowed_Processor_Binding` property of the `system` component. Such an entry could be for instance: `Allowed_Processor_Binding => reference c1 applies to p1.t1;`, indicating that thread `T` of the process

P is bound to the processor CPU, i. e. T will run on CPU. Note that t1, p1 and c1 are references to the implementations of T, P and CPU, respectively.

To calculate the estimated execution time of T running on CPU (cf. Fig. 3), as mentioned above, we create out of the properties (Variables and Occurrence) of thread T the following equation:

$$309 \cdot \texttt{MUL} + 53 \cdot \texttt{ADD} + 137 \cdot \texttt{LOAD} = \; exT$$

In order to determine exT we replace each variable (ADD, MUL and LOAD) with the values found in the execution time properties of processor CPU, namely in ExecTime. Thus we get the following:

$$309 \cdot 0.0035 + 53 \cdot 0.0013 + 137 \cdot 0.0007 = \; exT$$

The value of exT ($= 1.2463$) is then added in the specified timing property of thread T as follows: `Compute_Execution_Time => 1.2463 Ms .. 1.2463 Ms;` , indicating that T needs 1.2463 milliseconds for its execution.

Schedulability Analysis. Having defined the binding and calculated the corresponding estimated execution time we perform the schedulability analysis. The schedulability analysis checks the compliance of thread scheduling constraints using the Furness toolset [6]. Furness applies the ACSR [7] process algebra to determine schedulability. If a model is not schedulable Furness displays a failing trace (a timed system trace). Otherwise, it shows the analysis of best-case and worst-case time responses [6], thus facilitating the validation of various system configurations and the establishment of a thoroughly feasible system deployment.

Depending on these schedulability analysis results one can decide if it is required to repeat the previous actions, i. e. to choose a different binding, calculate the corresponding estimated execution time and run the schedulability analysis again. The received results are evaluated by the system developer. Nonetheless, one can automatically run the schedulability analysis for a set of chosen bindings and identify a feasible binding.

3 Related Work

Execution time predictions support in particular the development of embedded systems, enabling various system analysis regarding specified performance, such as schedulability analysis to examine if task deadlines are met [8]. A detailed overview of techniques and tools that deal with the execution time analysis is given in [9] and a comparison of different timing analysis methods and approaches can be found in [10, 11]. A modular architecture for a timing analysis approach combining different analysis methods is introduced in [12], making possible the exchange of results and the comparison of such methods. In [13] an approach is presented for computing tight bounds for WCET using *static* analysis. Wang et al. [14] present an approach based on *source code instrumentation* concerning binary translation and microarchitecture-related issues. A similar approach is presented in [15]. The authors investigate different methods and propose a statistical analysis-based approach which improves estimates of execution time. Both latter approaches have similar aspects to our component analysis approach. However, the main distinction is that we keep the analysis of hardware and software strictly separated. As a result of research on timing analysis a number of tools evolved, such as aiT [16], SWEET [17], Chronos [18] and OTAWA [19]. The importance, challenges and the evolution of the embedded system design is discussed in [20]. Different methodologies and development environments, such as SCADE [21] and Matlab/Simulink [22], have been applied on the design and development of embedded systems. We use in our framework the SAE standardised language AADL [2, 3]. Powerful tools such as Furness [6] and Cheddar [23] allow the schedulability analysis of systems modelled in AADL.

4 Conclusion

We introduced in this work an approach applicable to the performance modelling of modern avionics systems. Following the IMA ideology, a separate analysis of hardware and software ensures a decoupling of any possible analysis dependencies/impact in case of an update, be it a hardware or a software update. Furthermore, the application of AADL facilitates a direct support of the IMA concept. The execution time estimation method which we present in this paper is well suited for observing differences between various hardware. However, the quality of these observations is very closely bound to the quality of the benchmarks used and the constraints built. Due to the clearly defined interface between component analysis and the system modelling and analysis, one can smoothly integrate any other method for execution time prediction. We have tested our method with a set of benchmarks and the results look promising with regard to future activities, achieving an accuracy up to 90%. Since the work is still in progress, the current results have rather an experimental character. Nevertheless this activity is extensively fed by an industrial context.

The distinctiveness of our approach is based on the fact that it not only tackles important non-functional properties in modern avionics system design and development, such as execution time estimation, resource awareness composition and reusability of function units, but it also shows how those particular observations can be integrated and used for the deployment of such systems.

Acknowledgements

We would like to thank Andreas Holzer and Stefan Kugele for their fruitful comments in finalising this paper.

References

1. Garside, R., Joe Pighetti Jr., F.: Integrating modular avionics: A new role emerges. In: IEEE/AIAA 26th Conference on Digital Avionics Systems, DASC 2007 (2007)
2. Society of Automotive Engineers: SAE Standards: Architecture Analysis & Design Language (AADL) - AS5506 (November 2004) and AS5506/1 (June 2006)

3. Feiler, P.H., Gluch, D.P., Hudak, J.J.: The architecture analysis & design language (aadl): An introduction. Technical report, SEI, Carnegie Mellon (2006)
4. Feiler, P.H.: Modeling of system families. Technical report, Software Engineering Institute, Carnegie Mellon (2007)
5. The LLVM Framework, `http://www.llvm.org`
6. Furness Toolset v.1.6., User Guide, `http://www.furnesstoolset.com/files/Furness%20Toolset%20User%20Guide.pdf`
7. Brémond-Grégoire, P., Lee, I., Gerber, R.: ACSR: An Algebra of Communicating Shared Resources with Dense Time and Priorities. In: Best, E. (ed.) CONCUR 1993. LNCS, vol. 715, pp. 417–431. Springer, Heidelberg (1993)
8. Engblom, J., Ermedahl, A., Sjödin, M., Gustafsson, J., Hansson, H.: Worst-case execution-time analysis for embedded real-time sys. STTT 4(4), 437–455 (2003)
9. Wilhelm, R., Engblom, J., Ermedahl, A., Holsti, N., Thesing, S., Whalley, D., Bernat, G., Ferdinand, C., Heckmann, R., Mitra, T., Muller, F., Puaut, I., Puschner, P., Staschulat, J., Stenström, P.: The worst-case execution-time proble overview of methods and survey of tools. ACM Transactions on Embedded Computing Systems (TECS) 7(3), 1–53 (2008)
10. Engblom, J., Ermedahl, A., Stappert, F.: Comparing Different Worst-Case Execution Time Analysis Methods. In: The Work-in-Progress session of the 21st IEEE Real-Time Systems Symposium (RTSS 2000) (November 2000)
11. Kirner, R., Knoop, J., Prantl, A., Schordan, M., Wenzel, I.: Wcet analysis: The annotation language challenge. In: WCET (2007)
12. Ermedahl, A.: A Modular Tool Architecture for Worst-Case Execution Time Analysis. PhD thesis, Uppsala University: Acta Universitatis Upsaliensis (June 2003)
13. Ferdinand, C., Heckmann, R., Langenbach, M., Martin, F., Schmidt, M., Theiling, H., Thesing, S., Wilhelm, R.: Reliable and precise WCET determination for a real-life processor. In: Henzinger, T.A., Kirsch, C.M. (eds.) EMSOFT 2001. LNCS, vol. 2211, pp. 469–485. Springer, Heidelberg (2001)
14. Wang, Z., Sanchez, A., Herkersdorf, A., Stechele, W.: Fast and accurate software performance estimation during high-level embedded system design. In: edaworkshop, Hannover, Germany (May 2008)
15. Giusto, P., Martin, G., Harcourt, E.: Reliable estimation of execution time of embedded software. In: DATE 2001: Proceedings of the conference on Design, automation and test in Europe, pp. 580–589. IEEE Press, Los Alamitos (2001)
16. AbsInt, `http://www.absint.com/ait/`
17. SWEET, `http://www.mrtc.mdh.se/projects/wcet/sweet.html`
18. Chronos, `http://www.comp.nus.edu.sg/~rpembed/chronos/`
19. Otawa, `http://www.otawa.fr/`
20. Henzinger, T.A., Sifakis, J.: The discipline of embedded systems design. IEEE Computer 40(10), 32–40 (2007)
21. SCADE, `http://www.esterel-technologies.com/products/scade-suite/`
22. The MathWorks Inc.: Using Simulink (2000)
23. Singhoff, F., Legrand, J., Nana, L., Marcé, L.: Cheddar: A flexible real time scheduling framework. In: Proceedings of the 2004 Annual ACM SIGAda International Conference on Ada, Atlanta, GA, USA (2004)

Object-Oriented Petri Nets-Based Modeling of Resources in Project Engineering

Vladimír Janoušek and Šárka Květoňová

Brno University of Technology,
Bozetechova 2, 61266 Brno, Czech Republic
{janousek,kvetona}@fit.vutbr.cz

Abstract. Project engineering is a domain where suitable formal models and tools are still needed. The paper presents a way how the dynamically instantiable, multilevel Petri nets can be applied in all significant processes of project engineering. The main emphasis is put on the resources modeling, simulation, and scheduling. We use the *Object oriented Petri nets* (OOPN) formalism which is a kind of *multi-level Petri nets* with a possibility to synchronize events at different levels. In the case of the project modeling domain, the first level corresponds to the *project plans* and the second level corresponds to the *resources*.

Keywords: Object oriented Petri net, modeling, simulation, monitoring, optimization.

1 Introduction

Project engineering is a very actual, broad, and competitive domain. It concerns conceptualisation, development, integration, implementation and management of projects in a variety of fields. An important part of project engineering is resources management domain on which this paper is focused. This domain is the most crucial part of project management because it affects success and/or failure of the whole project. We keep in mind not only human resources, but material, financial etc., too. The main motivation of our underlying research is to simplify the whole process of projects engineering by means of appropriate models, techniques, and tools. The main accent is put on resources allocation optimization an adaptation of the whole process to changing requirements and conditions during project life time.

Basic Notions. In the following, basic terms and relations of project management and planning/scheduling domains are given. *Project* is a temporary effort undertaken to create a unique product or service, or result conforming to certain specifications and applicable standards [1]. *A process* is a series of actions bringing about a result. It is a complex of mutually connected resources and activities, which changes inputs to outputs. At present, activities and resources under the project are managed almost entirely like processes. A *resource* processes the individual operations, eventually, it serves as means an operation realization. *Project*

R. Moreno-Díaz et al. (Eds.): EUROCAST 2009, LNCS 5717, pp. 841–848, 2009.

management is a procedure of managing and directing time, material, personnel and costs to complete a particular project in an orderly and economical manner; and to meet established objectives in time, costs, and technical results. *Project Portfolio Management* [8] is about more than running multiple projects. Each portfolio of projects needs to be assessed in terms of its business value and adherence to strategy. The portfolio should be designed to achieve a defined business objective or benefit. *Planning* is a process of a set of proper activities creation to gain the predefined goals. *Scheduling* is a process of converting a general or outline plan for a project into a time-based representation given information on available resources and time constraints [10]. *Static (Off-line) scheduling* requires a knowledge about all the resources, their parameters, all the requirements, constraints and all criteria for the scheduling process in advance, to complete the schedule before the system starts to run. *On-line scheduling* [6] represents a process of creating a schedule in run-time. The schedule is re-created each time the conditions in the environment are changed or modified. The scheduling method has to be sufficiently fast in this case.

Related Work. Nowadays there exist many approaches, methods and formalisms which are using formal models, simulation and optimization in the project management domain. Manfred Mauerkirchner is focused on human resource allocation, in conrete, Resource Constrained Project Scheduling Problem (RCPSP) and uses a specific non analytical multiobjective function for allocation of qualified human resources [9]. In general, solving this problem has been a challenge for researches for many years. The basic reviews can be found e.g. in [7], [12]. Several approaches use Petri nets for project modeling, e.g. [5,3]. But, an interesting possibility which is still neglected nowadays is object oriented principles together with Petri nets use in the project planning, scheduling and monitoring. It is necessary to implement it by a suitable way to realize not only off-line scheduling, but also on-line optimization or dynamic modification of project parameters depending on the actual external conditions which are evolving in time (changes in resources structure, in project plans etc.).

We use the *Object oriented Petri nets* (OOPN) formalism [2], which is a kind of the *multi-level Petri nets* [11] with a possibility to synchronize events at different levels. The OOPN formalism is very interesting for the project portfolio modelling domain because it offers the concept of dynamically instantiable Petri nets and shared places. This feature allows for straightforward modelling of resources shared among a set of running projects (processes). Apart from obvious approaches in the area of project modeling, our model is well structured and allows for dynamic instantiations of project plans or sub-plans.

2 Object Oriented Petri Nets

Object Oriented Petri Nets (OOPN) consist of Petri nets organised in classes. Each class consists of an object net and a set of dynamically instantiable method nets. Places of the object net are accessible for the transitions of the method nets.

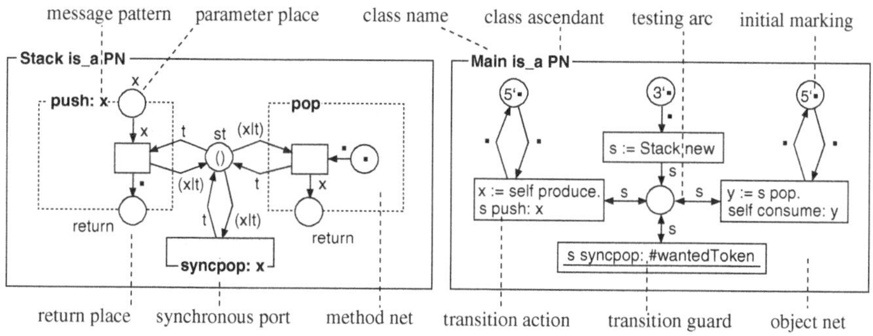

Fig. 1. An OOPN example

Object nets as well as the method nets can be inherited. Inherited transitions and places of the object nets (identified by their names) can be redefined and new places and/or transitions can be added in subclasses. Inherited methods can be redefined and new methods can be added in subclasses. Classes can also define special methods called synchronous ports, which allow for synchronous interactions of objects. Message sendings and object creations are specified as actions attached to transitions. The transition execution is polymorphic — the methods which has to be invoked is chosen according to the class of the message receiver that is unknown at the compile time. A token in OOPN represents either a trivial object (e.g., a number or a string) or an instance of a Petri net-described class consisting of an instance of the appropriate object net and possibly several concurrently running instances of the invoked method nets.

An example illustrating the OOPN formalism is shown in Figure 1. As it is depicted in Figure 1, a place can be inscribed by an initial marking (a multiset of objects) and an initial action (allowing a creation and initialization of the objects to be initially stored in the place; not shown in the Figure 1). A transition can have a guard restricting its firability, and an action to be performed whenever the transition is fired. Finally, arcs are inscribed by multiset expressions specifying multisets of tokens to be moved from/to certain places by the arcs associated with a transition being fired.

The OOPN on the Figure 1 demonstrates that the method nets of a given class can share access to the appropriate object net – the places of the object net are accessible from the transitions belonging to the method nets. In this way the execution of the methods can modify the state of the object. Class Main describes an active object, which can instantiate (and communicate with) a passive object – stack (an object is passive if its object net contains no transitions). Each method net has parameter places and a return place. These places are used for passing data (object references) between the calling transition and the method net. Apart from method nets, classes can also define special methods called synchronous ports that allow for synchronous interactions of objects. This form of communication (together with execution of the appropriate transition and synchronous port) is possible when the calling transition (which calls a synchronous port from its guard) and the called synchronous port are executable simultaneously.

The transition guards and actions can send messages to objects. An object can be either primitive (such as a number or a string), or non-primitive (defined by Petri nets). The way how transitions are executed depends on the transition actions. A message that is sent to a primitive object is evaluated atomically (thus the transition is also executed as a single event), contrary to a message that is sent to a non-primitive object. In the latter case, the input part of the transition is performed and, at the same time, the transition sends the message. Then it waits for the result. When the result is available, the output part of the transition can be performed. In the case of the transition guard, the message sending has to be evaluable atomically. Thus, the message sending is restricted only to primitive objects, or to non-primitive objects with appropriate synchronous ports.

PNtalk. The OOPN formalism is implemented by a tool called PNtalk. PNtalk allows for a specification and simulation of OOPN-based models. In PNtalk, it is possible to specify delayed transition execution (in simulation time). Similarly to Simula-67, the delay is accomplished by sending *hold:* to *self* from a transition action. It is also possible to specify the message sending from a transition guard. In the former case, the execution of the output part of the transition is delayed, in the later case, it specifies the required enabling time for the transition. Note that it is possible to synchornize the simulation time with real time. A more detailed explanation of the PNtalk system and the PNtalk project can be found e.g. in [2,4].

3 Projects and Resources Modeling

The concept of OOPN-based modeling of the projects distinguishes two-levels in the model. The first level corresponds to the *project plans* and the second level corresponds to the *resources*. In the following, we explain the concept in more detail.

Actually, an OOPN object can model multiple projects (project portfolio [8]). Tokens in the object net's places represent the shared resources. They are distributed in the places according to their roles. Method nets correspond to the individual project plan templates. Their instances can be dynamically created and destroyed (it corresponds to a start and a finish of a project) and they share an access to the object net's places containing pointers to the resource objects. Project start is modelled by the appropriate message sent to the project portfolio object, possibly with parameters. At the same time a new instance of the method net (i.e. project plan) is created and starts to run. Note that it is possible to invoke a method (i.e. instantiate a project template) several times (with specific parameters) and the invocations can overlap in time. It corresponds to the situation where the project templates are instantiated.

An example of the approach is shown in Figure 2. There are described three different project templates (OOPN methods) which share the same resources.

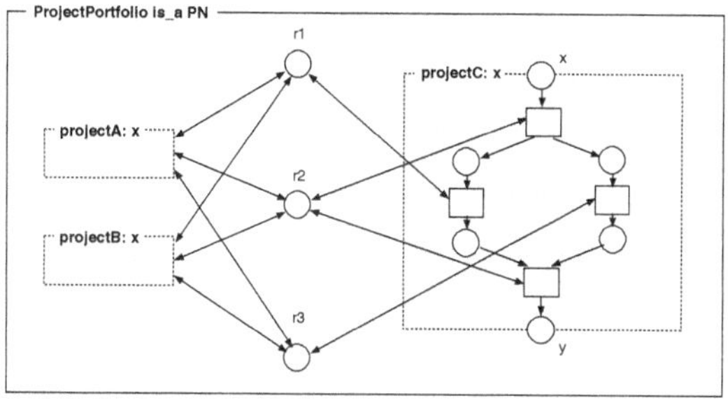

Fig. 2. Project portfolio model (basic idea)

Actually, the resource types (their roles) are modelled by places of an object net). Project templates A and B are collapsed (their structure is not shown). Resources are modelled by individual objects which are available as tokens in the places. Actually, the tokens are references to the objects. So it is possible to have a resource available under two and more roles modelled by the places. Each *activity* (modeled as a transition of a method net which models a project) attempts to allocate all the resources it needs (by means of the corresponding synchronous ports calling) and if it succeeds, it uses the resources for some time (by means of an invocation of the corresponding method of the resource). Figure 3 depicts a simple example showing how a resource having two roles is used by two activities.

The above sketched model demonstrates only the core idea. Actually, it is necessary to model the resources in the context of some constraints, such as actual availability, skills, compatibility with the activities etc. In the case of multiple resources alocated to an activity, we must take in account also the quality of their team collaboration. All these attributes can be expressed using OOPN formalism effectively. Figure 4(a) depicts a definition of a resource with skills specified in a form of a set of tokens (*activity*, *skills*) in the correponding place. Note that the shadow parts of the class specification are inherited from the superclass

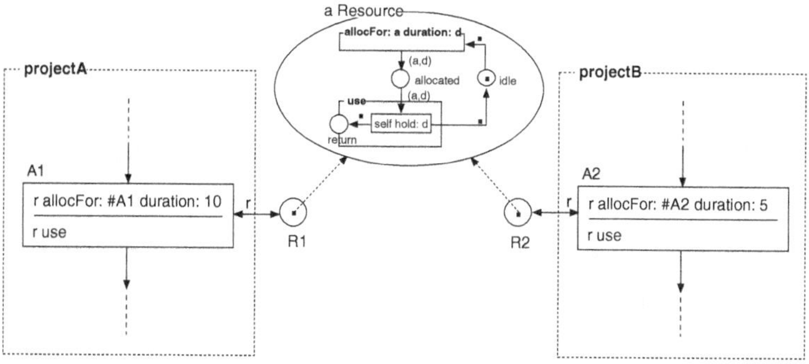

Fig. 3. Conflict of two activities requesting one resource having two roles

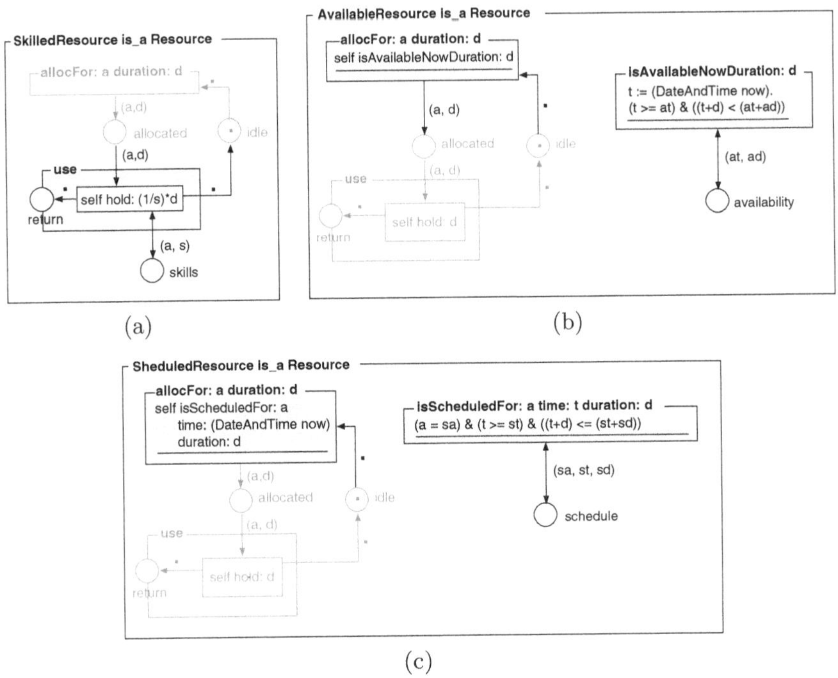

Fig. 4. Resource with skills (a), availability (b), and schedule (c) specified

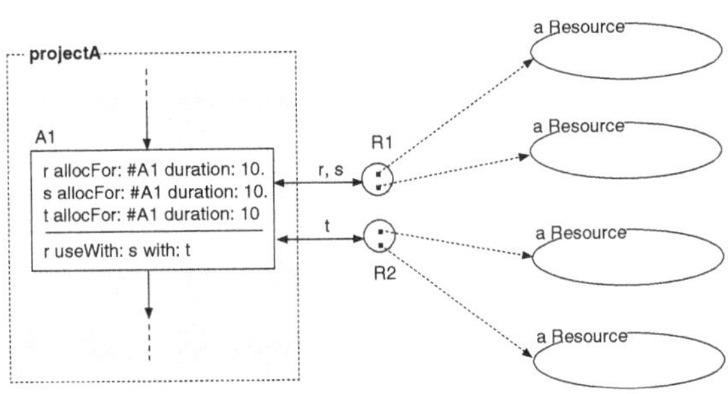

Fig. 5. Multiple resource allocation

Resource. Figure 4(b) depicts a definition of a resource with availability specified in a form of a set of tokens (*time*, *duration*) in the correponding place. Figure 5 depicts an example of multiple resources allocation.

4 Scheduling and Monitoring

The above sketched model can serve as a basis for scheduling [10]. A *scheduling process* generates a schedule satisfying the specified criteria, e.g. availability, skills, experience, and costs of the resources. We use genetic algorithm (GA) for scheduling. As part of the fitness function, a simulation is performed in order to check the feasibility of each candidate schedule and for gaining the essential performance results such as time and costs.[1] The resulting *schedule* is then attached to the corresponding resource as a set of triples $(activity, time, duration)$. When a resource with a schedule is being allocated (by the appropiate synchronous port calling), it checks whether it is scheduled for the requested activity, current time and the requested duration (see Fig. 4(c)).

A model with the scheduled resources can be used in the *monitoring* process. In that case, the model is simulated in real time and is being dynamically updated according to the actual state of the reality. If necessary, a repeated scheduling is performed on-line. In that case, a clone of the model is used as a basis for the scheduling process. When a new schedule is found, it is incorporated back to the model as a set of tokens in the place *schedule*, as defined by ScheduledResource (see Fig. 4(c)).

Implementation. The above sketched way of OOPN use requires a simulator with a reflective interface (metaobject protocol, MOP) allowing for dynamical modifications of the model being simulated. This enhancement of the OOPN simulator (PNtalk) has already been proposed and experimentally implemented [4]. OOPN/PNtalk MOP allows for inspection and edition of particular nets which define classes, as well as individual net instances implementing the actually living objects and actually running method invocations. It also allows to make a clone of the whole running system and to store/restore it to/from a database in a serialized form.

5 Conclusion

We have presented a concept of OOPN-based modeling of resources in project engineering. The main advantages of using OOPN in project engineering include (1) formal nature of the model enabling an analysis of projects using mathematical methods, (2) intuitively understandable model representation thanks to the visual nature of the formalism, (3) well structured model with clear mapping to the notions of the domain, and (4) *model continuity* – an OOPN-based model is used as the main model which it is being dynamically updated continuously during the monitoring process, while its transformation to/from other views (models) is possible at any time. The model evolution is possible thanks to the PNtalk meta-object protocol, allowing for inspection end edition of the model at run time.

[1] During the scheduling process, the resource models are augmented in such a way that they use the candidate schedule representation compatible with the scheduling mechanism based on GA. Actual scheduling mechanism is not discussed here in detail due to the space limitation.

Acknowledgments. This work was supported by Czech Grant Agency and Ministry of Education, Youth and Sports under the contracts GA102/07/0322 and MSM 0021630528.

References

1. Anbari, F.: Q & As for the PMBOK Guide. Project Management Institute (2005) ISBN 1930699395
2. Ceska, M., Janousek, V., Vojnar, T.: PNtalk — a computerized tool for object oriented petri nets modelling. In: Moreno-Díaz, R., Pichler, F. (eds.) EUROCAST 1997. LNCS, vol. 1333, pp. 591–610. Springer, Heidelberg (1997)
3. Jiao, X.C., Yan, H.S., Wang, Z.: Modeling, scheduling and simulation of product development process by extended stochastic high-level evaluation Petri nets. Robotics and Computer-Integrated Manufacturing 19(4), 329–342 (2003)
4. Janousek, V., Koci, R.: Towards an Open Implementation of the PNtalk System. In: Proceedings of the 5th EUROSIM Congress on Modeling and Simulation. EUROSIM-FRANCOSIM-ARGESIM, Paris, FR (2004)
5. Kim, J., Desrochers, A.A., Sanderson, A.C.: Task planning and project management using petri nets. In: IEEE International Symposium on Assembly and Task Planning, p. 265 (1995)
6. Kocjan, W.: Dynamic scheduling state of the art report. Technical Report T2002:28, Dynamic scheduling state of the art report (2002)
7. Kolisch, R., Hartmann, S.: Experimental investigation of heuristics for resource-constrained project scheduling: An update. European Journal of Operational Research 174(1), 23–37 (2006)
8. Lee, B.: Multi-project management in software engineering using simulation modeling. Software Quality Journal 12(1), 59–82 (2004)
9. Mauerkirchner, M.: A general planning method for allocation of human resource groups. In: Moreno-Díaz, R., Buchberger, B., Freire, J.L. (eds.) EUROCAST 2001. LNCS, vol. 2178, pp. 172–181. Springer, Heidelberg (2001)
10. Pinedo, M.: Scheduling: Theory, Algorithms, and Systems, 2nd edn. Prentice Hall, Englewood Cliffs (2002)
11. Valk, R.: Petri Nets as Token Objects: An Introduction to Elementary Object Nets. In: Desel, J., Silva, M. (eds.) Application and Theory of Petri Nets. LNCS, vol. 120, Springer, Heidelberg (1998)
12. Weglarz, J.: Project scheduling. Recent models, algorithms and applications. Kluwer Academic, Dordrecht (1998)

Simulation Based Design of Control Systems Using DEVS and Petri Nets

Radek Kočí and Vladimír Janoušek

Faculty of Information Technology, Brno University of Technology,
Bozetechova 2, 616 00 Brno, Czech Republic
{koci,janousek}@fit.vutbr.cz

Abstract. Current model-based design methodologies use executable semi-formal models allowing for transformations including code generation. Nevertheless, the code should be finalized manually and further development or debugging by means of prime models is impossible. The paper introduces an approach to the system design called Simulation Based Design which uses fromalisms of DEVS (Discrete-Event Systems Specification) and Object Oriented Petri Nets (OOPN) allowing for clear modeling, a possibility to check correctness by means of simulation as well as by formal verification. The approach is based on techniques such as incremental development in the simulation, reality-in-the-loop simulation, and model-continuity. The model is understood as an executable program valid through all development stages including the deployment (the target system).

1 Introduction

Current trend in systems design methodologies aims at an efficiency and safety of development processes as well as at the quality of resulted systems. The classic methodologies define the development process as a sequence of analysis, design, implementation, testing, and deployment. These methodologies are simple for its realization, but is not sufficient for design of complex systems, because the system requirements change during the development process and the classic methodologies do not support it very much. As a response to this the incremental and iterative design methodologies were developed, e.g., Unified Process or Agile Methodologies. They allow for concurrent execution of design phases, the changes are done by small steps, lays stress on testing, etc. These methodologies are more successful for ordinary applications, but have their limitations. The main problem is that the implementation and testing are separated from the design phase. For instance, the models are usually used in the design phases, but when we wish to test dynamic properties of developed system or to get the target system, we have to implement an executable prototype or system according to these models.

Another kind of methodologies which were investigated in the last decade are commonly known as *Model-Driven Software Development* or *Model-Based Design* (MBD). The most popular methodology is *Object Management Group's Model*

R. Moreno-Díaz et al. (Eds.): EUROCAST 2009, LNCS 5717, pp. 849–856, 2009.

Driven Architecture (MDA) [7] based on Executable UML [11]. An important feature of these methods is that they use executable models. In these methodologies of new generation, the designer creates models and tests their correctness by simulation, so that there is no need to make a prototype. They also allow for model transformations including code generation. Nevertheless, the code has to be usually finalized manually and these changes are not incorporated back to the models. It entails a possibility of semantic mistakes or imprecision between models and transformed code. Moreover, the further development, debugging and investigating of target application by means of prime models is impossible, so the significance of model has declined.

The paper introduces an approach to the software system design called *Simulation Based Design*. We understand it as an approach which combines the concepts of *model-continuity*, *incremental development in the simulation*, and *reality-in-the-loop simulation*. The model continuity makes a tendency towards an elimination of generating the source code from models [3,5]. The system is developed incrementally, models are being improved and are simulated in each design step (*incremental development in the simulation*). It is possible to simulate external components to test the system functionality. The next step is to exchange simulated components for their real realization and to test developed system in the real conditions (*reality-in-the-loop simulation)*. Finally, the developed models, in particular the control of the system logic, is deployed into the target system. The next important idea is multi-paradigm modeling. In the design, it is useful to benefit from special properties of different formalisms, which can be combined. The paper is aimed to the DEVS and Petri Nets formalisms which are used in a complement way. While OOPN is the object based formalism, DEVS serves as a component based framework for embedding other formalisms.

The presented concepts are supported by the tool named PNtalk/SmallDEVS [5,6] based on both formalisms. Its architecture results from principles of meta-level and reflective architectures [3,4] allowing models to reciprocal cooperates with the tool environment, to define and use special statements which are not directly included in the used formalism, etc., which enables model continuity and reality-in-the-loop simulation techniques.

2 DEVS

DEVS stands for Discrete Event System Specification. DEVS specifies a system hierarchically. A model can be specified as an atomic or as a coupled model. Coupled models consist of interconnected atomic and coupled models. This results in a model hierarchy of loosely coupled models.

Atomic model. An atomic model represents a simple, indivisible entity. The atomic models is defined as a structure:

$$M = (X, Y, S, ta, \delta_{int}, \delta_{ext}, \lambda)$$

where

- X (resp. Y) is the set of input events (resp. the set of output events);
- S is the set of states;
- $ta : S \longrightarrow R^{+}_{0,\infty}$ is the time advance function, that returns the lifetime of a state;
- $\delta_{ext} : Q \times X \longrightarrow S$ is the external input transition function, where $Q = \{(s, t_e) | s \in S, t_e \in [0, ta(s)]\}$ is the total state set and t_e is the time elapsed since last transition;
- $\delta_{int} : S \longrightarrow S$ is the internal state transition function;
- $\lambda : S \longrightarrow Y^{\varepsilon}$ is the output function where $Y^{\varepsilon} = Y \cup \{\varepsilon\}$ and ε denotes the silent event.

The system is always in some state $s \in S$. If no external event occurs, the system is staying in state s for $ta(s)$ time. If the elapsed time t_e reaches $ta(s)$, then the value of $\lambda(s)$ is propagated to the output and the system changes to state $\delta_{int}(s)$. If an external event $x \in X$ occurs on the input in time $s \leq ta(s)$, then the system changes its state to $\delta_{ext}(s, t_e, x)$.

The definition of the atomic model can be simply modified in such a way that it allows to use input and output ports. This makes the composition of models significantly easier from practical point of view.

Coupled model. Coupled models describe the system as a network of coupled components. These components can be atomic or coupled models. This is in fact a recursive hierarchical definition. In this way, output events of a component can become input events of another component. Zeigler [13] showed, that the DEVS formalism is closed under coupling, in other words, for every coupled model a equivalent atomic model can be constructed.

3 Object Oriented Petri Nets

Several attempts to combine Petri nets and objects has been done in the nineteens, for instance Object Petri Nets [10], Cooperative Nets [12], Nets-in-nets formalism [1]. They are supported by specialized tools like, e.g., Renew [9]. One of the formalisms covering advantages of Petri nets and object orientation is a formalism of Object Oriented Petri Nets (OOPN) [2]. Petri nets allow to describe properties of the modeled system in a proper formal way and the object-orientation brings structuring and a possibility of net instantiation.

The formalism of OOPN consists of Petri nets organized in classes. Every class consists of an object net and a set of dynamically instantiable method nets. Object nets as well as method nets can be inherited. Inherited transitions and places of object nets (identified by their names) can be redefined and new places and/or transitions can be added in subclasses. Inherited methods can be redefined and new methods can be added in subclasses. A token in OOPN represents either a trivial object (e.g., a number or a string) or an instance of some Petri net. A *class* is specified by an object net, a set of method nets, a set of synchronous ports, a set of negative predicates, and a set of message selectors corresponding to its method nets and ports.

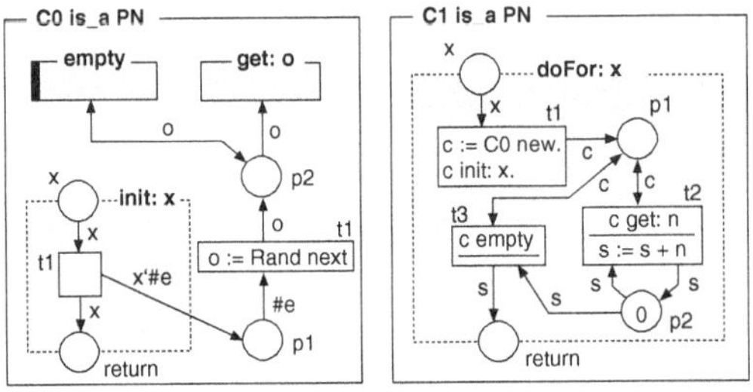

Fig. 1. An OOPN example

Object nets consist of places and transitions. Every place has its initial marking. Every transition has conditions (i.e. inscribed testing arcs), preconditions (i.e. inscribed input arcs), a guard, an action, and postconditions (i.e. inscribed output arcs).

Method nets are similar to object nets but, in addition, each of them has a set of parameter places and a return place. Method nets can access places of the appropriate object nets in order to allow running methods to modify states of objects which they are running in. Method nets are *dynamically instantiated* by message passing specified by *transition actions.*

Synchronous ports are intended for synchronous interaction of objects. Synchronous port is a hybrid of method and transition that allow for synchronous interactions of objects. In order to be fired, the synchronous port has to be called (a method concept) and has to be firable (a transition concept). Synchronous port can be called only from a transition guard. The transition can be fired only if all called synchronous ports are able to fire. A special variant of synchronous port is *negative predicate.* Its semantics is inverted – the calling transition is firable if the negative predicate is not firable.

An example illustrating the important elements of the OOPN formalism is shown in Figure 1. There are depicted two classes *C0* and *C1*. The object net of the class *C0* consists of places *p1* and *p2* and one transition *t1*. The object net of the class *C1* is empty. The class *C0* has a method *init:*, a synchronous port *get:*, and a negative predicate *empty*. The class *C1* has the method *doFor:*. The semantics of the method *doFor:* execution is to randomly generate x numbers and return their sum.

4 Simulation Based Design

This chapter introduces basic principles of the simulation based design. For better understanding, the presented principles will be demonstrate on the simple robotic example. It is a part of the *leader-follower* example, which is one of classic robotic tasks consisting of at least two robots which are moving along the defined space. If two or more robots meet, they come to an agreement who will be the leader and the others follow him (reproduce his moving). If some robot hits, e.g., the wall and is not able to continue in motion reproducing, he leaves the formation. The example uses Player/Stage software [8]. Player provides a network interface to a variety of robot and sensor hardware. Stage simulates a population of mobile robots moving in and sensing a two-dimensional bitmapped environment.

4.1 The Design Framework Based on OOPN and DEVS

As we have mentioned, the DEVS formalism serves as a component based framework for embedding other formalisms. Each DEVS component has the input and output ports which serve for communication between components. The component defined by the OOPN formalism delegates the communication responsibility to chosen OOPN object. This object (delegate) is just one in the component and its time-life is adherent to the component. The communication is provided via places of the object net—the port named `x` is mapped onto the place `x`. Let us show it on the example of the robotic system. The control mechanism is realized by OOPN and is wrapped onto a DEVS component named `PNAgent` (see the Figure 2 on the left). This component has three input ports which receive data from robot's sensors (`sonars`, `bumpers`, and `position`) and two output ports which reproduce generated commands (`rotateTo` and `move`). The delegate object is represented by the OOPN class `RCPlatform` (see the Figure 2 on the right). There are three special input places `sonars`, `bumpers`, and `position` corresponding with the component's ports. As soon as the data has been received to the input ports, they are placed to the matching places and the transition `createEvent` is fired. If the control mechanism makes a decision that, e.g., the robot has to rotate, the method `rotateTo:` is called, and the value representing a rotate degree is placed into the place `rotateTo`. This place is a special output place corresponding with the component's output port `rotateTo`. As soon as the value is placed to this port, it is accessible for another components which can read this port.

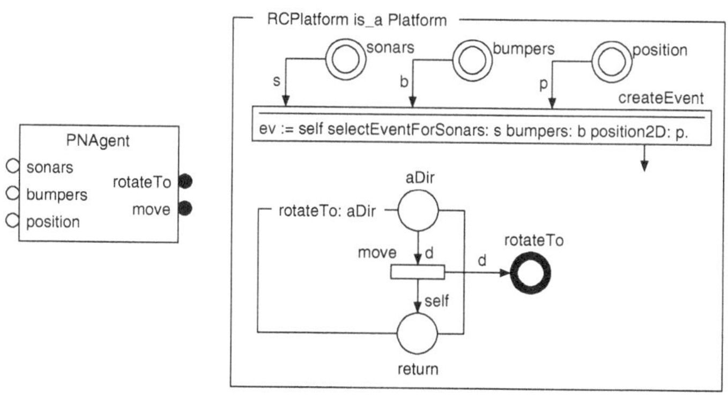

Fig. 2. DEVS ports and OOPN places mapping

4.2 System Design in Simulation

In the presented approach to the system development in simulation, we have to distinguish the system run and the simulation run. The simulation run allows to define and use special elements, e.g., we can use a special kind of places collecting statistics data. The implementation of that places is less efficient that the ordinary one, but it is no problem in the simulation run. We can also use simulated components instead of its real (software) variant. There are also three different kinds of time in which the model can execute—model time, real time, and quasi-real time. If the model is running in the model time, it runs in a pure simulation. If it is running in quasi-real time, the real time is a base for clock unit which is multiplied by defined constant. If it is running in real time, no time manipulation is supported—it is used for the deployed model (system).

Let us have a DEVS component of the robotic control. To test its correctness, we need either to implement the real communication between this component and the robot's sensors and actuators or to design their simulated variant. The second variant is much easier to realization so that the designer can quick and automatic test the component. The principle is shown in the Figure 3. The component `Test` simulates the robot (its state, position, etc.) and its neighborhood. It reacts to the `PNAgent` outputs and generates test data of sensors. The next component named `Stat` collects output data from other components and is able to trace the simulation run and to generate statistic data. Both `Test` and `Stat` components can be realized in the OOPN, DEVS, or other formalism.

The example of the `Stat` realization is shown in the Figure 4. It is implemented using the OOPN formalism and shows only a part of the class. Each input is stored in the place including the time stamp (see `self currentTime`)—the time can be real or model, it depends on the selected mode. Let us investigate the time which is consumed by `PNAgent` to make decision about the next step. If a new sensor data is generated (inject into the place `position`), the time stamp is stored to the place `p1`. The `PNAgent` indicates that the decision is made via the port `req`. Then the new time stamp is generated (the variable `t2`) and the consumed time is a difference between these two stamps (`t2-t1`).

To get statistic data we have to implement methods or make use the meta-level architecture of the PNtalk/SmallDEVS framework. It enables to define macros which are transformed into the relevant methods. The example is shown in the Figure 4. There is defined a macro `getMax` generating a method `getMax` which iterates for each object stored in the place `s2` (a pair `(t2,t1)` in our example)

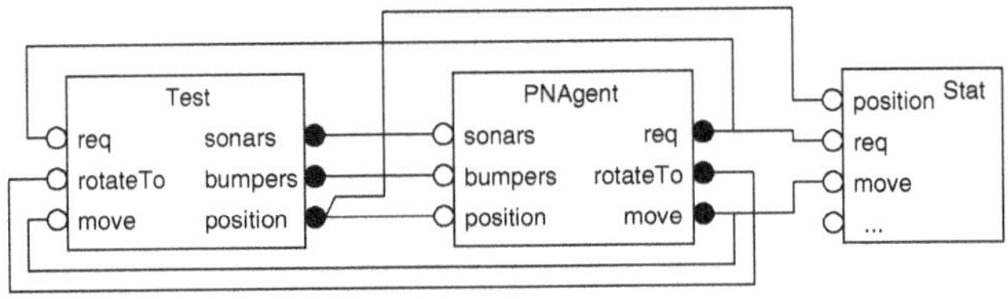

Fig. 3. Simulation Based Testing

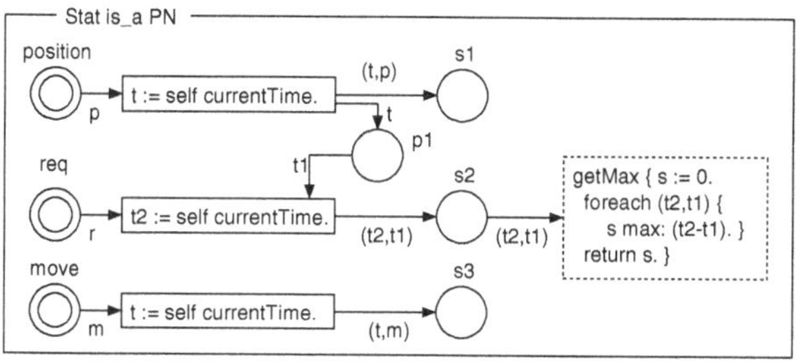

Fig. 4. Stat class in OOPN

and returns the maximum value of the consumed time (`t2-t1`). It very eases the design of such a kind of methods.

4.3 Model Continuity

The model continuity, together with reality-in-the-loop simulation, allows to use models in each development stages including the system deployment. Let us continue in our example. When we have tested the component `PNAgent`, we change a simulated communication (the component `Test`) for a real communication with the player/stage software (another DEVS component with the same interface). The components coupling will look the same like in the previous case (the Figure 3), just the component `Test` will have different implementation. Now we are able to test our system in real conditions using the same way (e.g., the component `Stat`). Finally, we remove all testing and simulated components, change the mode to the real time, and are able to deploy models to the system as is.

5 Conclusions

The paper has presented basic concepts of the *Simulation Based Design* technique. It uses formal models which can be simulated as well as deployed into the target system, makes a tendency towards an elimination of generating source codes from models, and uses various simulation techniques which ease the design and testing of particular components. A possibility to deploy models into the target system as is allows to debug system on the model basis—the system is always seen as a set of models which ease the debugging and further development of systems. The Simulation Based Design allows for high-quality and rapid development and results in more effective design process as well as in less numbers of mistakes and errors in the designed system. The further research will be aimed at the efficiency of the model execution in real time, the advancement of the tool supporting the presented design technique, and the usage of it in complex systems.

Acknowledgement. This work was supported by the Czech Grant Agency under the contracts GA102/07/0322, GP102/07/P306, and Ministry of Education, Youth and Sports under the contract MSM 0021630528.

References

1. Cabac, L., Duvigneau, M., Moldt, D., Rölke, H.: Modeling dynamic architectures using nets-within-nets. In: Ciardo, G., Darondeau, P. (eds.) ICATPN 2005. LNCS, vol. 3536, pp. 148–167. Springer, Heidelberg (2005)
2. Češka, M., Janoušek, V., Vojnar, T.: PNtalk - A Computerized Tool for Object Oriented Petri Nets Modelling. In: Proceedings of the 5th International Conference on Computer Aided Systems Theory and Technology – EUROCAST 1997, pp. 229–231. Las Palmas de Gran Canaria, ES (1997)
3. Janoušek, V., Kočí, R.: PNtalk: Concurrent Language with MOP. In: Proceedings of the CS&P 2003 Workshop. Warsaw University, Warsawa (2003)
4. Janoušek, V., Kočí, R.: Towards an Open Implementation of the PNtalk System. In: Proceedings of the 5th EUROSIM Congress on Modeling and Simulation. EUROSIM-FRANCOSIM-ARGESIM, Paris, FR (2004)
5. Janoušek, V., Kočí, R.: Towards Model-Based Design with PNtalk. In: Proceedings of the International Workshop MOSMIC 2005. Faculty of management science and Informatics of Zilina University, SK (2005)
6. Janoušek, V., Kočí, R.: Simulation and design of systems with object oriented petri nets. In: Proceedings of the 6th EUROSIM Congress on Modelling and Simulation, p. 9. ARGE Simulation News (2007)
7. Kleppe, A., Warmer, J., Bast, W.: MDA Explained: The Model Driven Architecture – Practice and Promise, 1st edn. Addison-Wesley Professional, Reading (2003)
8. Kranz, M., Rusu, R.B., Maldonado, A., Beetz, M., Schmidt, A.: A player/stage system for context-aware intelligent environments. In: Proceedings of the System Support for Ubiquitous Computing Workshop, the 8th Annual Conference on Ubiquitous Computing (2006)
9. Kummer, O., Wienberg, F., Duvigneau, M., Schumacher, J., Köhler, M., Moldt, D., Rölke, H., Valk, R.: An extensible editor and simulation engine for Petri nets: Renew. In: Cortadella, J., Reisig, W. (eds.) ICATPN 2004. LNCS, vol. 3099, pp. 484–493. Springer, Heidelberg (2004)
10. Lakos, C.A.: From Coloured Petri Nets to Object Petri Nets. In: DeMichelis, G., Díaz, M. (eds.) ICATPN 1995. LNCS, vol. 935. Springer, Heidelberg (1995)
11. Raistrick, C., Francis, P., Wright, J., Carter, C., Wilkie, I.: Model Driven Architecture with Executable UML. Cambridge University Press, Cambridge (2004)
12. Sibertin-Blanc, C.: Cooperative Nets. In: Valette, R. (ed.) ICATPN 1994. LNCS, vol. 815. Springer, Heidelberg (1994)
13. Zeigler, B., Kim, T., Praehofer, H.: Theory of Modeling and Simulation. Academic Press, Inc., London (2000)

Transforming UML-Based System Descriptions into Simulation Models as Part of System Development Frameworks*

Andreas W. Liehr and Klaus J. Buchenrieder

Institut für Technische Informatik
Universität der Bundeswehr München
D-85577, Neubiberg, Germany
{andreas.liehr,klaus.buchenrieder}@unibw.de

Abstract. The support of the engineering process for computer-based systems by Co-Design frameworks is the key to lower production cost and the time-to-market for complex devices. Early in the design cycle, system models must be rated, concerning design constraints and decisions must be made. To enable system engineers to gauge concurring system-realizations approaches, such frameworks must allow the prediction of key characteristics for the system under development, like performance, battery consumption and heat production. Hereby, the prediction of the system performance constitutes the vital constraint.

In this work, we demonstrate, how an EQN-based performance simulation method can be seamlessly embedded into a Hardware / Software Co-Design framework based on UML system descriptions. The presented approach yields a performance simulation model, ready for the automated simulation process, and therefore fosters the performance evaluation of the system under development.

Keywords: Hardware / Software Co-Design, Performance Simulation, System Modeling, Unified Modeling Language, Queuing Network Models.

1 Introduction

In the recent decade, computer based systems became ubiquitous in everydays live, as the prevailing existence of Moores law resulted in a still ongoing decay of production and development costs for mobile computer devices. Hence, specialized devices for a multitude of applications, like navigation or multimedia playback, emerged as well as affordable multi-purpose systems, like smartphones or netbooks. As this mass market attracted numerous established IT suppliers and startup-companies during its phase of constitution and rapid growth, this

* This work has been supported within a subcontract between Infineon Technologies AG and the Universität der Bundeswehr München. This contract is part of the Project "Verteilte integrierte Systeme und Netzwerkarchitekturen für die Applikationsdomänen Automobil und Mobilkommunikation", (VISION), 01 M 3078.

R. Moreno-Díaz et al. (Eds.): EUROCAST 2009, LNCS 5717, pp. 857–864, 2009.

saturized market segment is characterized by a high competition between existing providers and a rapid technological evolution. To survive in such a market, it is imperative, to quickly react to changes in the consumers' taste and demands when introducing novel products. Such short time-to-market prerequisites can be supported by Hardware / Software Co-Design frameworks, especially if these allow to determine whether consumer demands, like performance or battery life time are fulfilled or not. Such frameworks bestow a competitive advantage to companies, employing this technology. The ability to predict the expected performance of a system under development constitutes one vital component of such a framework. Therefore, a performance estimation model must be one of its integral parts. Recognizing this demand, numerous approaches for performance estimation early in the design cycle have been brought forward [1,2,3].

A promising specification language for Hardware / Software systems within a Co-Design framework is the UML profile for Modeling and Analysis of Real-Time and Embedded systems (MARTE) [4]. It adds capabilities for model-driven development of Real Time and Embedded Systems to UML. MARTE supports the specification, design, and verification / validation of systems in different stages of the development process and is therefore a fitting system description language for hardware, software and the mapping between them. Therefore, MARTE constitutes system models with three design views: 1) an application model, specifying the system functionality; 2) a resource model, representing the execution platform; and 3) an allocation model, that maps the function to the architecture [5]. Consequently, first approaches with MARTE as system description language have been disclosed in [6,7,8].

In this contribution, we show how MARTE-based system descriptions can be exploited to derive performance simulation models in the form of Extended Queuing Network Models (EQN). To realize this objective, we employ a component-based transformation on the hardware model to deduce the EQN network structure. Consequently, we derive the initialization network components that influence job behavior from the software model and the allocation model of the initial system description, so that the jobs of the EQN emulate the application of the computer based system under development. As we show, our approach automatically yields EQNs, ready for simulation of performance constraints, from MARTE based system models.

2 Deriving Simulation Models from MARTE System Descriptions

Our transformation method consists of a two-stage process, in which a UML MARTE-based system description, consisting of a hardware, a software and an allocation model, is input and yields an EQN representing this system ready for performance simulation. A brief introduction to EQNs with respect to performance simulation can be found in [9] and [10].

2.1 The Traffic Sign Recognition System

For illustration of our method, we provide an example from our industrial partner Infineon Technologies AG. The example constitutes a recognition system for traffic signs developed for the automotive domain.

The functionality of the recognition system is depicted in Figure 1 and consists of three phases. First, a video stream is captured by a camera and processed in real-time to determine images, that contain traffic signs. A frame that contains the image of a traffic sign is unhinged immediately and forwarded to the classification routine.

During classification, the depicted sign is compared with characteristic features, stored in a database. If positive identification prevails, the third phase is executed.

In the visualization phase, traffic sign specific information is presented to the driver. Additionally, information about the environment is evaluated and serves to augment the visualization. Such information comprises factors, like the speed of the car or weather data. The recognition system is executed continuously and does not halt after a traffic sign is recognized.

The functional description is serialized as UML activity diagram, extended with real-time and performance annotations according to the Real-Time Execution (RTE) Model of Computation and Computing, which is part of the MARTE profile.

Figure 2 shows the architectural realization of the traffic sign recognition system. Its core system is able to serve as stand-alone system, as it provides all functionality. The System Interface (SIF) is the bus interface for the camera module. The data-stream must be processed imediately, otherwise frames will be dropped as soon as new input data arrive. The FlexRay Controller enables the communication of the core system with peripheral modules, connected to

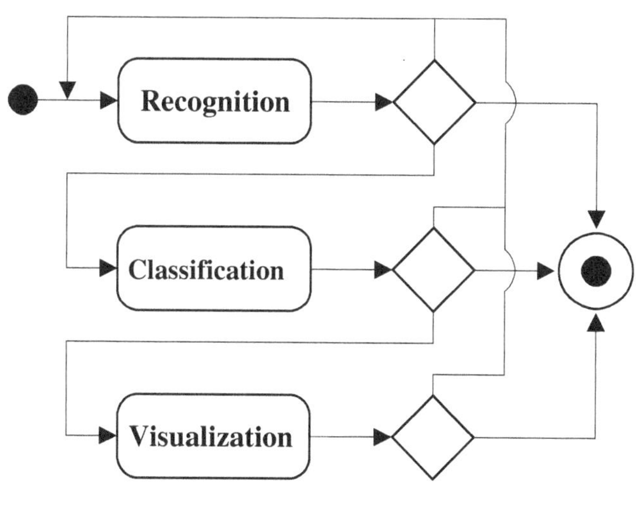

Fig. 1. Functional Description of the Traffic Sign Recognition System

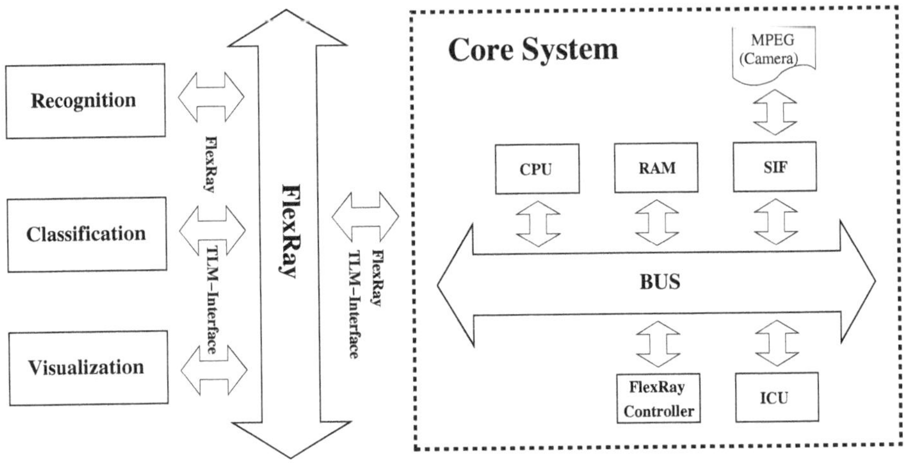

Fig. 2. Architecture of the Traffic Sign Recognition System

the FlexRay bus. Additional modules enhance the performance of the system. Such modules exist for the recognition, the classification and the visualization of scene data and are executed in parallel.

In our system development framework, the hardware model is specified as UML composite structure diagram, enriched with the Hardware Resource Modeling (HRM) specification, defined in the MARTE profile. The interrelation between hardware components is modeled within the composite structure diagram and the MARTE-representation of each component holds a reference to a hardware pattern within a pattern database, which characterizes this component in terms of performance. Such patterns are composed in EQN syntax and can be combined to the final EQN structure of the holistic system. To give an impression of such EQN patterns, we depict in Figure 3 an exemplar system simulation model, containing a CPU with a round robin scheduler, memory and bus.

The allocation diagram, which is the third component of our system model, specifies the architectural components that are utilized for every activity of the activity diagram, representing the functionality. Allocation diagrams are introduced by the MARTE profile and take the form of UML composite diagrams. Further information to allocation diagrams and examples of their usage can be found in [4] and [8].

2.2 Generating Performance Simulation Models

To obtain the performance simulation model from existing system descriptions in UML MARTE, as described in the preceding subsection, the following procedure is applied:

First, a list is compiled, which contains only the hardware components of the system hardware model that is utilized during the execution of the system under development. Only architectural components contributing to the functionality

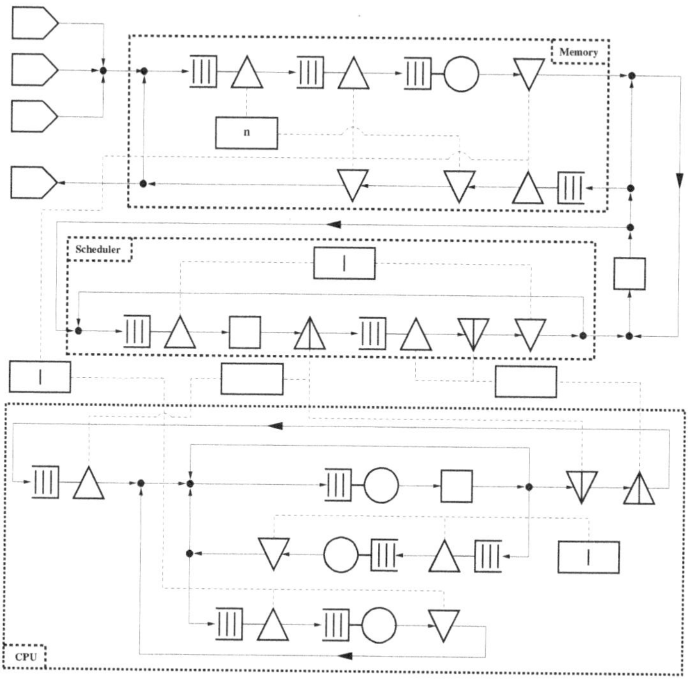

Fig. 3. EQN of a System Consisting of CPU with Scheduler, Memory and Bus

of the system are considered as part of the performance simulation model to build. As the hardware model might contain components, not utilized by any functional part of the system, the resulting set of architectural components in the simulation model can be a subset of the hardware components specified in the system model. Which hardware components are not utilized can be derived from the allocation model. If no mapping within the allocation model exists for a component, it is not involved in the execution of any application.

From the resulting list of hardware components, a simplified EQN, representing the system architecture is compiled. This simplified EQN holds placeholders for each hardware component. How the interconnection between the components is realized, can be inferred from the composite structure diagram serving as hardware model compliant to the MARTE HRM specification. Figure 4 depicts the representation of the traffic sign recognition system as simplified EQN model. As shown by this example, jobs can be routed between all hardware components but buses. Buses are realized as passive queues and their timing behavior is modeled by servers with appropriate service time, dependent on the amount of data to transfer and of the bus protocol.

Next, the component placeholders in the simplified EQN are replaced by EQN patterns. These patterns are taken from a pattern database. Thereby, a reference, relayed for every component of the composite structure diagram, indicates the appropriate pattern for each component. Consequently, a specific EQN

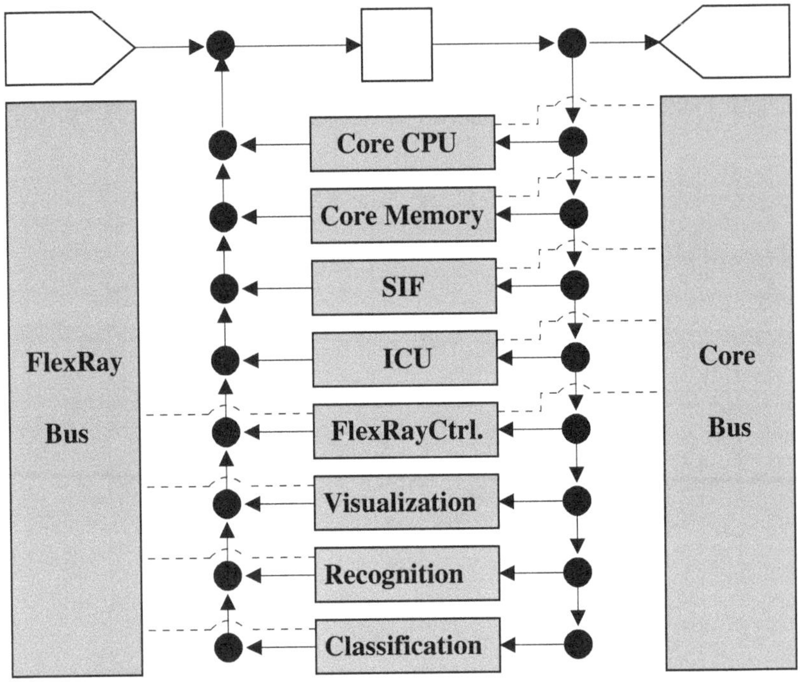

Fig. 4. Simplified EQN representing the Traffic Sign Recognition System

pattern must exist for every hardware component used in the system development process. Such patterns have to be produced by the developer of the design framework or must be provided by the vendor of the component.

At this point, the first stage of the generation method yields a valid EQN. In the second stage, we map the behavior of the system, defined with the activity diagram of the system model, to the EQN. As a result, the behavior of the jobs within the EQN meet the defined behavior of the applications of the system. This is established as follows:

The behavior of application of the system is translated to jobs within the simulation model. Initialization times for applications within the activity diagram are translated to building rules for sources of the EQN. Hereby, every type of application, constitutes an independent job class. This distinction into job classes is mandatory, to allow for the simulation of different types of applications executing in parallel and competing for hardware resources.

Next, the behavior of the application is modeled on the granularity level of single activities. The allocation diagram defines, which hardware resources are utilized by each activity, and the activity diagram contains the utilization pattern of these resources. Information about resources is stored in global variables of the EQN and influence the behavior of the jobs in the network during simulation.

Finally, the transition between single activities of an application during runtime are included in the simulation model. This is realized with global variables, forming an execution schedule for each application. These global variables are processed by a central steering instance within the EQN, a set node, invoked every time a job passes the corresponding set node.

The simplified EQN, depicted in Figure 4, illustrates the mechanism of the central steering instance. The set node is depicted as square in the upper part of the Figure. Jobs enter the network through the source in the upper left corner of the Figure and are routed to the central steering instance. From this point, jobs are routed through the network to the first component, where work is incurred by the application. For this purpose, the routing target is determined by a function of the set node depending on the class of the job and the global variables representing the appropriate schedule. The utilization pattern, the job incurs to this component is determined by the global variables, defining the behavior of the applications activity. When the processing is finished, the job returns to the central steering instance. From here, the job is directed to the next hardware component, it is processed from. This procedure is repeated, until the job finishes and leaves the network through the sink in the upper right-hand corner of Figure 4.

3 Conclusion and Outlook

In this work, we have shown how performance simulation can be seamlessly integrated into a UML MARTE-based system development framework for Hardware / Software Co-Design. Therefore, we detail a transformation method that derives system simulation models from UML system descriptions, consisting of a hardware model, a software model and a mapping from functionality to architecture. The resulting simulation model is an Extended Queuing Network Model, which can be directly evaluated on an EQN simulator, as introduced in [11]. The method is characterized by its autonomy, as it can be implemented for execution without user interaction. Additionally, it fosters a high degree of reusability, as a hardware component pattern database is employed to reuse pre-specified architectural parts.

Future work comprises the utilization of model-to-model transformation techniques to let users benefit from a higher degree of automation in the performance engineering process as part of system development frameworks.

Since the visualization of performance-simulation results is inherently important in the Hardware / Software Co-Design process, we are striving to provide advanced visual feedback as integral part for such frameworks. First results of this ongoing work have been presented in [12].

References

1. Balsamo, S., Di Marco, A., Inverardi, P., Simeoni, M.: Model-based performance prediction in software development: a survey. IEEE Transactions on Software Engineering 30(5), 295–310 (2004)
2. Wagh, R., Bellur, U., Menezes, B.: Transformation of uml design model into performance model - a model-driven framework. In: Manolopoulos, Y., et al. (eds.) ICEIS 2006. LNBIP, vol. 3, pp. 576–580. Springer, Heidelberg (2006)

3. Estefan, J.A.: Survey of model-based systems engineering (mbse) methodologies. Technical report, California Institute of Technology, Pasadena, CA, USA (May 2007)
4. OMG: Uml profile for modeling and analysis of real-time and embedded systems (marte). OMG Document 07-08-04, Object Management Group (August 2007)
5. Taha, S., Radermacher, A., Gerard, S., Dekeyser, J.L.: An open framework for detailed hardware modeling. In: The IEEE Second International Symposium on Industrial Embedded Systems (SIES), Lisbon, Portugal, vol. 1, pp. 118–125. IEEE Computer Society, Los Alamitos (2007)
6. Demathieu, S., Thomas, F., Andre, C., Gerard, S., Terrier, F.: First experiments using the uml profile for marte. In: 11th IEEE International Symposium on Object Oriented Real-Time Distributed Computing (ISORC 2008), Orlando, FL, USA, May 2008, pp. 50–57 (2008)
7. Mraidha, C., Tanguy, Y., Jouvray, C., Terrier, F., Gerard, S.: An execution framework for marte-based models. In: 13th IEEE International Conference on Engineering of Complex Computer Systems (ICECCS 2008), Belfast, Ireland, pp. 222–227. IEEE Computer Society, Los Alamitos (2008)
8. Liehr, A.W., Buchenrieder, K.J., Rolfs, H.S., Nageldinger, U.: Generation of MARTE Allocation Models from Activity Threads. In: Languages for Embedded Systems and their Applications. Lecture Notes in Electrical Engineering, vol. 36. Springer, Heidelberg (2009)
9. Sauer, C.H., Chandy, K.M.: Computer Systems Performance Modeling. Prentice Hall, Englewood Cliffs (1981)
10. Liehr, A.W., Buchenrieder, K.J.: Simulating inter-process communication with extended queuing networks. In: Asia Simulation Conference 2008 / The 7th Internation Conference on System Simulation and Scientific Computing, Beijing, China, pp. 1027–1031. IEEE Computer Society, Los Alamitos (2008)
11. Liehr, A.W., Buchenrieder, K.J.: An xml based simulation method for extended queuing networks. In: Louca, L.S. (ed.) 22nd European Conference on Modelling and Simulation, Nicosia, Cyprus, European Council for Modelling and Simulation, June 2008, pp. 322–328 (2008)
12. Liehr, A.W., Buchenrieder, K.J., Nageldinger, U.: Visual feedback for design-space exploration with uml marte. In: The Fifth International Conference on Innovations in Information Technology, Al Ain, UAE, pp. 44–48. IEEE Computer Society, Los Alamitos (2008)

Model-Based Design and Verification of Reactive Systems

Jiří Hýsek, Milan Češka, and Vladimír Janoušek

Brno University of Technology, Faculty of Information Technology
Božetěchova 2, 612 66 Brno, Czech Republic
{ihysek,ceska,janousek}@fit.vutbr.cz

Abstract. The article is focused on a model-based design and verification of reactive systems. In opposite to common approaches to developing reliable systems, at first a model of a system is constructed in high level visual language, then it is verified on this level of abstraction and consequently a low level code for target platform is generated. In approach discussed in this article an UML statechart formalism is used for construction of the model. This model is translated into Promela model and verified by the SPIN model checker.

Keywords: model-based design, formal verification, state charts, reactive systems.

1 Introduction

Many approaches of developing reliable computer systems focus on formal methods of analysis and verification of systems implemented in common computer languages. They have to fight many obstacles caused by using low-level programming language to specify the system. This obstacles could be avoided or at least simplified by using different system specification techniques. In our approach we are using an UML statecharts [7] as specification language. Systems specified by the statechart are reasonably readable for humans, it is also used for documenting purposes so UML models are self-documenting. Statecharts operate on a high level of abstraction, which makes models smaller and there is less probability of errors caused by a programmer. An application of the statecharts has been already used before, e.g. in NASA's Deep Space 1 mission [1], or at the Jet Propulsion Laboratory for specification of communication protocols [2]. Our approach differs mainly in the choice of target language and domain of use.

1.1 Model-Based Design

Common approaches of software system development are using a manually written code in a programming language for an implementation of the system.

A model-based design approaches are using a high level modeling language to specify the system. Target code is automatically generated from this model. High level of abstraction used by this approach brings several advantages.

R. Moreno-Díaz et al. (Eds.): EUROCAST 2009, LNCS 5717, pp. 865–872, 2009.

- Readability and maintainability.
- Avoiding of "low-level problems" like pointer operations.
- High-level features could be hopefully exploited in order to reduce a state space used during a process of verification.

1.2 Reactive Systems

By a reactive system we understand any system that responds to an external stimuli (inputs) and triggers events (outputs) that may be perceived by its observer (Fig. 1). An environment of the system is not explicitly defined in a model of the system, nevertheless we will have to consider it during the process of verification.

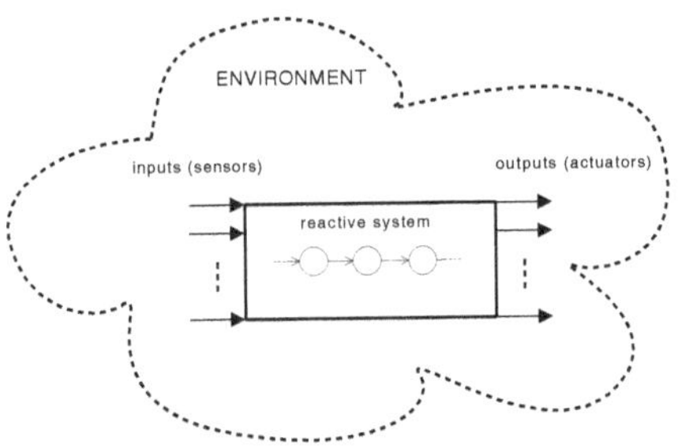

Fig. 1. Reactive system surrounded by its environment

2 Approach of Model-Based Design

In our approach, which is schematically illustrated in Fig. 2, the UML statechart formalism is used as a modelling language. UML is frequently used visual specification language. The model is analyzed and verified on this level of abstraction. It's transformed into a corresponding model in Promela and verified by the SPIN model checker [3]. When the model satisfies given correctness specifications it's translated into a code for a target platform which is simulation framework SmallDEVS [6] in our case. It's an implementation of Zeigler's DEVS [4] formalism, written in Smalltalk. Nevertheless, a phase of generating a SmallDEVS code is not discussed in this article.

2.1 Statecharts

UML is widely used formalism, hence there are a lot of CASE tools which can be used for constructing a statechart models. Many of them can export models into a standardized XML file, an open format, that can be easily processed. Statechart is a formalism developed in order to have a visual specification language for complex systems [4]. It's convenient for developers, because it supports a rich set of features. For that reason it could be difficult to analyze it. Hence we are

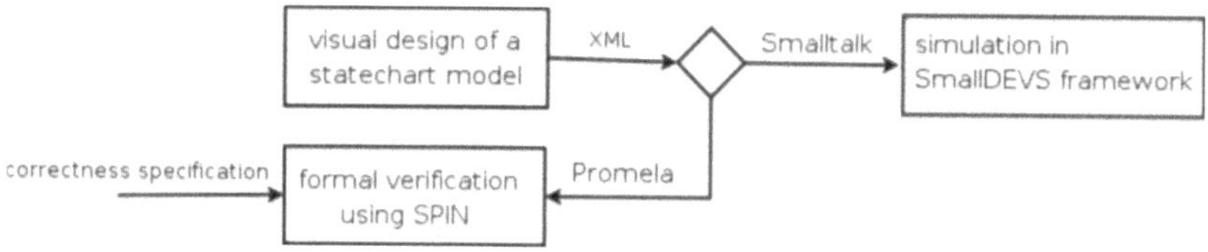

Fig. 2. A schema of model-based approach for design of computer systems

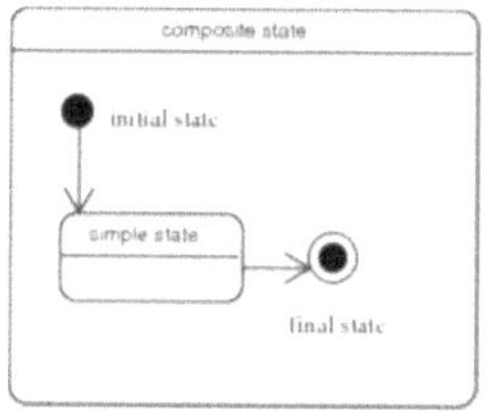

Fig. 3. Syntax of initial pseudostate, composite, simple and final state of statechart

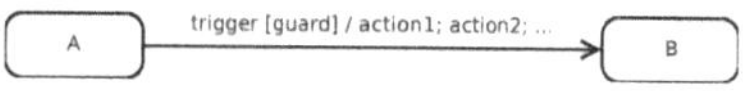

Fig. 4. Syntax of transition

using a restricted version of statecharts, some unsupported constructions are automatically transformed into semantically identical construction using only supported features, some are not allowed at all.

The most significant features of statecharts are hierarchically ordered states, concurrent regions, labelled transitions and inscription language. A syntax of statechart states is showed at Fig. 3. The label of the transition contains *a trigger* (name of input signal), *a guard* (boolean condition specified by inscription language of the model) and a sequence of *actions* (statements of the inscription language). A Fig. 4 shows a syntax of transition. A detailed information about syntax and semantics of statecharts can be find in UML specification [7].

3 Verification of Statecharts

For the verification of the statechart model is used the LTL model checker SPIN. It analyzes models described by the modeling language Promela. In order to verify the model we have implemented a translator from the statechart model to the Promela code.

Since our research focuses on reactive systems, i.e. systems that respond to external signals and react by sending signal to their environment, we have to bear in mind an environment which is implicitly a part of the modeled system, but it is not included in the statechart model. We are assuming that every event produced by system's environment can occur in any time and in arbitrary order, and the output signals of the system are handled by the environment every time.

However the environment has to be modeled by the Promela code. Promela representation of the environment basically consists of two processes. One of them simulates external events according that the model acts (input signals of the model). It repeatedly sends random signals on model's input channel. The second one represents a part of the environment which accepts model's output signals. This process repeatedly picks up the signals from model's output channel.

3.1 Translation into Promela

Each composite state and parallel region which is included in the statechart model is modeled as single process in Promela. Initially there are always 3 active processes.

- A process which represents a top-level composite state. This state is always implicitly included in statechart model. Each other state is a child state of this top-level state.
- The signal producer – a process which simulates that part of a model's environment which generates input signals on which the model reacts.
- The signal consumer – a process which simulates that part of the environment which reacts on signals that are send by the model to the environment.

For each process of the statechart which represents a composite state a global variable is defined where an identifier of active sub-state is stored. This variable is initialized with the value of identifier of initial sub-state of the composite state. Input signals are handled by an infinite loop. System reacts only to input signals which could trigger any transition enabled in current active state. Other input signals are dropped. An example of a composite state and it's translation into a Promela is described in the Fig. 5 and a snippet of it's model in Promela is as follows:

```
#define initial_1 = 0;
#define state_A = 1;
#define state_B = 2;
#define final_1 = 3

/* this variable contains active state's identifier */
int composite_state_1;
int x = 0;
...
```

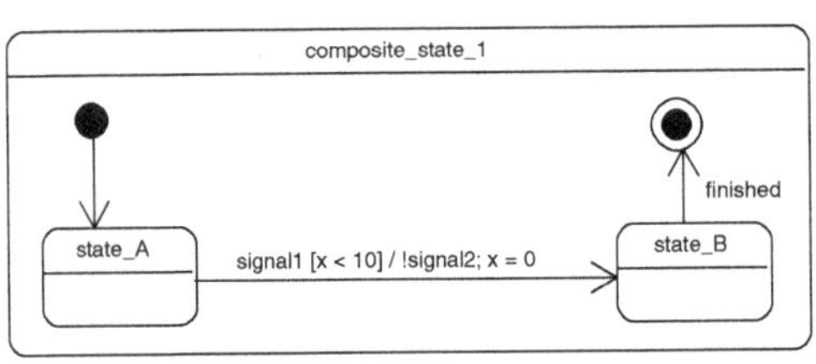

Fig. 5. An example of composite state

```
proctype proc_composite_state_1() {
composite_state_1 = initial_1;     /* initial state is active */

do
:: composite_state_1 == initial_1 ->
   if
   :: true -> composite_state_1 = state_A
   fi
:: composite_state_1 == state_A ->
   if
   :: (input?signal1) && (x < 10) ->   output!signal2;
                                       x = 0;
                                       composite_state_1 = state_B
   :: input?x -> skip
   fi
:: composite_state_1 == state_B ->
   if
   :: input?finished -> composite_state_1 = final_1
   :: input?x -> skip
   fi
:: composite_state_1 == final_1 -> break
od
}
...
```

Let's briefly describe a function this code. A variable which carry an identifier of active substate of the composite state from Fig. 5 is declared as global variable *composite_state_1*. Immediately after the process *proc_composite_state_1*() representing this state is activated, the variable is set to the identifier of initial state. In the following cycle is periodically checked whether any transition could be fired. A transition can be fired when a source state of a transition is active, transition's guard condition is satisfied and a trigger signal has been received and not dispatched yet. When a transition is taken following sequence of commands is performed.

1. An exit action of a source state of the transition,
2. a list of transition's actions,
3. setting an active state to the target state of transition,
4. an entry action of the target state.

When there are more than one transitions which can be fired, the system can choose arbitrary one. Semantics of Promela is the same – if more than one branch can be taken, then a random one is taken. The SPIN model checker will verify all possible runs.

3.2 Signals

The model itself is not an isolated system, it needs to communicate with it's environment and another components. Therefore a source of events could be somewhere beyond the model and the model can emits signals that are not addressed to any part of the model itself. It is necessary to handle this situation by the verification model.

Events used in the statechart model are represented by signals in the Promela. Signals are transmitted through *channels*. From the view of the model, three types of channels are distinguished.

- An *input* channels which are used for sending external input signals to the model. The signal producer process emits signals into this channel. The set of input channels contains by default a single channel called *input*. If the channel with signal in trigger action of transitions is not defined, assumed channel is *input*. If some translation is triggered by signals at different channel and no translation in model is sending a signal on this channel, this channel is also assumed as input channel. If there are some transitions triggered by a time event in the model a special channel *time_out* is inserted into the set of input channels.
- An *internal* channels which are used by the model for sending an internal signals. This channels are used for communication between parallel regions.
- An *output* channels which are used for sending output signals from the model to it's environment. A set of output channels by default contain a channel called *output*. If any transition sends a signal within it's action and channel is not specified, it's send into a channel *output*. When any transition sends a signal through different channel and no transition is triggered by any signal from this channel, this channel is also assumed as an output channel.

Promela representation of the signal producer and signal consumer looks as follows:

```
active proctype produce_input() {
  do
  :: input!signal1
  :: input!signal2
  :: input!signal3
  :: ...
  od
}
```

```
active proctype consume_output(){
        mtype x;
        do
        ::output?x -> skip
        od
    }
```

The producer process is repeatedly sending random signals into an *input* channel (or all input channels if there are more of them), the consumer process is repeatedly picking up all signals appearing on the *output* channel (or more output channels if there are more of them).

3.3 Example of Verification

Assume a simple model of vending machine (Fig. 6). Immediately after the machine starts it goes to the state called *idle*. There the machine waits until a coin is inserted (i.e. until an environment sends signal called *coinIn*). After that the machine waits 15 second for users choice. The choice is signalized by a signal

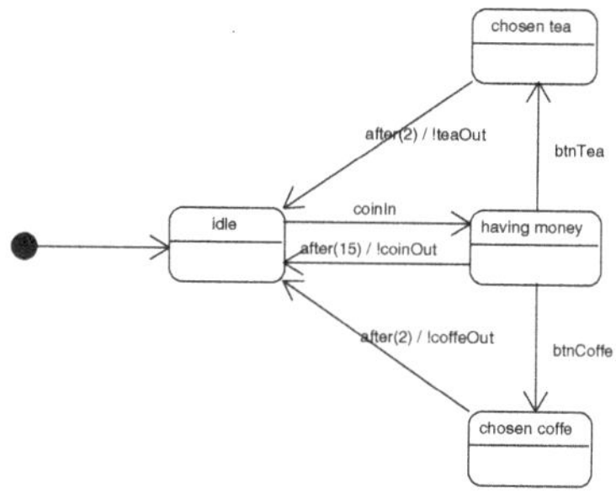

Fig. 6. Simple model of vending machine

from environment called *btnTea* or *btnCoffe*. If no drink is chosen, the machine sends signal coinOut, to say to environment to release back the coin. If a drink is chosen, the system goes to an appropriate state, waits two seconds and release the drink. This delay is only for simulation purposes.

We would like to verify whether the machine cannot release a drink without a coin being inserted. We can specify this by properties written as LTL formulae. For sake of readability of formula, let's define several predicates:

- *is_idle* is true iff the system is in the *initial state* or in the state *idle.*
- *inserted_coin* is true iff the system is in the state *having money.*
- *release_drink* is true iff the system is in the state *chosenTea* or *chosenCoffe.*

In the rest of the example we will talk about these shortcuts as about single states. For example *release_drink* will denote abstract state covering states *chosenTea* and *chosenCoffe*. Verification properties could be as follows:

$$\Box(is\_idle \Rightarrow is\_idle \ \mathbf{U} \ inserted\_coin)$$

$$\Box(is\_idle \Rightarrow \neg(is\_idle \ \mathbf{U} \ release\_drink))$$

The first one says that always holds a condition "when the system is in *is_idle* then it stays in this state until the coin is inserted". It means there are no way to leave idle state until user of the vending machine puts a coin into it.

The second formula says that always holds a condition "when the system is in *is_idle*, then there is no path leading directly from *is_idle* into *release_drink*.

After a statechart model is exported into .xmi file, it is automatically translated into a Promela by a tool we are working on. For experimenting we are using graphical interface of SPIN model checker called jSpin. All we need is to open Promela model, define predicates, define the property as LTL formula and run a verification.

4 Conclusion

This article has described some specific features of our approach of model-based design and verification of reactive systems. The described method is already implemented, but the translation tool is a prototype still in construction. In addition to discussed method we are able to generate a DEVS models. One of the application domains we take in account is a design and simulation of control systems of autonomous reactive robots. For the simulation of robots is used a SmallDEVS framework. Because of Smalltalk's dynamism we are able to inspect and modify models during the simulation, so we can experiment with them interactively in dynamic environment. This feature could be useful e.g. for simulating hardware malfunctions appearing during the simulation, and study the behavior of the model in such situations. In spite of the target code is Smalltalk, there are no obstacles complicating translation of models into a mainstream production programming language like C or Java.

Acknowledgements

This work has been supported by the Grant Agency Czech Republic, the grant No. GA 102/07/0322 "Advanced Formal Approaches in the Design and Verification of Computer-Based Systems" and the Research Plan Np. MSM 0021630528 "Security-Oriented Research in Information Technology".

References

1. Rouquette, N.F., Neilson, T., Chen, G.: The 13th technology of Deep Space One. In: Proceedings of the 1999 IEEE Aerospace Conference, pp. 477–487 (1999)
2. Wagstaff, L.K., Peters, K., Scharenbroich, L.: From Protocol Specification to Statechart to Implementation. California Institute of Technology (2008)
3. Holzmann, G.J.: The SPIN Model Checker: Primer and Reference Manual. Addison-Wesley, Reading (2004)
4. Zeigler, B.P., Praehofer, H., Kim, T.G.: Theory of Modeling and Simulation, 2nd edn. Academic Press, London (2000)
5. Harel, D.: Statecharts: a visual formalism for complex systems. Science of Computer Programming (1987)
6. Homepage of SmallDEVS project, `http://perchta.fit.vutbr.cz/projekty/10`
7. OMG. The Unified Modeling Language (UML) Specification - Version 1.4, Joint submission to the Object Management Group (OMG) (September 2001), `http://www.omg.org/technology/uml/index.htm`

Resonant Tunnelling Diode-Based Circuits: Simulation and Synthesis

Marek A. Bawiec

The Institute of Computer Engineering, Control and Robotics
Wrocław University of Technology, Poland
marek.bawiec@pwr.wroc.pl

Abstract. In this present how to use SPICE for transient analysis of Boolean logic circuits based on monostable–bistable transition logic element (MOBILE). MOBILE circuit is composed of negative differential resistance (NDR) and takes advantages of negative resistance in $I-V$ characteristic of NDR. We also propose and verify a method how to construct NDRs library for SPICE. Our method generates GTG for a given Boolean function of up to n variables.

Keywords: Nanoelectronic Devices, NDR Modelling, SPICE Simulation, Boolean Logic Synthesis.

1 Introduction

Scaling electronic devices to the nanometer dimensions brings quantum effects in the focus of attention. For many years this was due to an adverse effect of thermionic emission, tunnelling and Coulomb blockades on the operation of the classical electronic devices. Although this phenomena are usually considered parasitic it is possible to construct new nanoscale devices that will benefit from this non-classical behaviour achieving greater performance.

Before constructing complex circuits based on nanoscale devices it is required to model the behaviour of a single device as well as the whole circuit. Moreover, this has to be done in a manner that enables circuit simulation and verification in a reasonable amount of time. Later on, circuit structure and synthesis algorithms have to be constructed so that one can use nanoelectronic devices to build a complex circuit implementing desired functionality and achieving specific properties.

Negative differential resistance (NDR) devices take advantages of quantum effects. In our work we focus on resonant tunnelling diodes (RTDs) which are among one of well studied and analysed nanoelectronic devices [1,2,3,4,5,6]. RTDs have quantum feature which leads to negative resistance in current versus voltage ($I-V$) characteristic. We focus on RTDs since they are well analysed, however, results of our work can be applied to other NDR devices. Moreover, our method of GTG circuit synthesis can by applied to any commercially available SPICE like environments, e.g.: Cadence SPECTRE, PSPICE or HSPICE.

R. Moreno-Díaz et al. (Eds.): EUROCAST 2009, LNCS 5717, pp. 873–880, 2009.

2 Related Work

It is not a new concept that nanoelectronic devices negative differential resistance (NDR) property can be used to build logic circuits. However, today's knowledge doesn't clearly define how to do it. Circuits proposed by Avedillo [1,7] and Berezowski [2] are both based on the monostable-bistable transition logic element (MOBILE) concept [8]. MOBILE was invented to exploit negative differential resistance which is an inherent property of the resonant tunnelling diodes. Basic circuit operating in MOBILE regime consists of two NDR elements: load (NDR_l) and driver (NDR_d) connected in series (Fig. 1–B). By applying bias voltage V_{bias}, which varies between $0[V]$ and V_{DD} (Fig. 1–A) the circuit switches from a monostable OUT_m state to one of the bistable states: OUT_{b1} — low voltage (logic "0"), or OUT_{b2} — high voltage (logic "1"). The output state depends on the relation between peak currents I_{pl} and I_{pd} (see I-V characteristic on Fig. 1–C). Since the NDR with smaller value of peak current always switches to high resistance state when V_{bias} increases, therefore the circuit will be in OUT_{b1} state if $I_{pl} < I_{pd}$ and OUT_{b2} if $I_{pl} > I_{pd}$. Precisely there are four phases in the operation of MOBILE circuit that are determined by the V_{bias} voltage that changes according to 4-phase clocking scheme (Fig. 1–A):

- evaluation phase (V_{bias} increases) in which circuit evaluates the output state by switching from monostable to one of the bistable states,
- hold phase (V_{bias} high) in which circuit remains in the state selected in previous phase independently of the actual relation of I_{pl} and I_{pd};
- reset phase (V_{bias} drops) when circuit returns to the initial monostable state, and
- wait phase (V_{bias} low) when the circuit awaits for adjusting of I_{pl} and I_{pd} relation.

Adjusting relation between the peak currents I_{pl} and I_{pd} allow to control the output of the circuit and this enables to implement Boolean functions. Method how to build NDR-based circuits that implement Boolean function was presented by

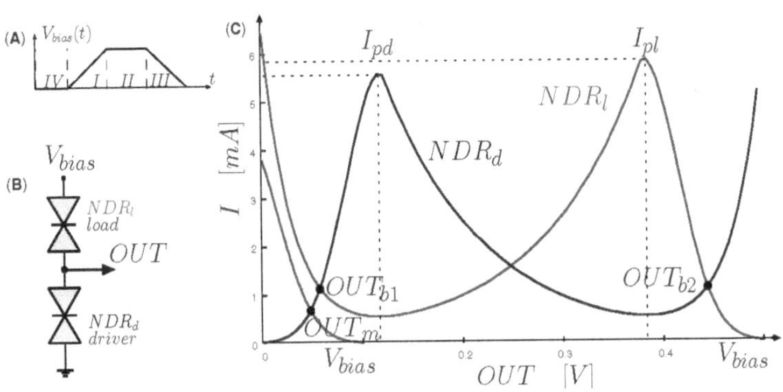

Fig. 1. MOBILE Idea — (A) V_{bias} clocking scheme, (B) circuit structure, (C) $I(V)$ characteristic with output states marked with bullets for Monostable as OUT_m, OUT_{b1} and OUT_{b2} as Bistable states and peak currents of both NDRs

Avedillo et al. [7] and Berezowski [2]. They both proposed to build circuits from the series of NDRs (NDR_l and NDR_d) paralleled by branches of either NDR/-transistor pair or NDR and serially connected serial–parallel (SP) transistor network.

2.1 Structures of NDR-Based Logic Circuits

Avedillo et al. [1] assumed that all NDRs in the circuit may have different parameters (i.e. peak currents) and the transistor network consist of exactly one transistor. As presented in [1] the circuit consisting of $n+1$ branches of this type computes weighted sum of n binary inputs and quantises its result. This enables to implement some boolean functions, of up to n inputs (e.g. any 2–input boolean function, simple AND and OR functions of $n > 2$ variables). This concept was extended in [7] where a branch of serially connected transistors was used instead of a single transistor. This modification enables to implement all n–input boolean functions with at most $2^n + 1$ branches.

Generalised threshold gate proposed by Avedilo et al. [7] was extended by Berezowski [2], who propose to use a series of identical NDRs, serially connected with a serial-parallel transistor network. Such an approach results in reduced number of branches and makes which become conditional on the complexity of the circuit rather then the number of inputs. He also gave an iterative formula that describes circuit behaviour and used an exhaustive search to show that all boolean functions of up to 4 variables can be implemented in a circuit of at most 6 branches.

2.2 RTD Models

As is was mentioned in previous sections before NDR based logic circuits will become widely available it is necessary to construct methods of NDR circuits synthesis and simulation. To do this we need to build a software model of single NDR device and circuits resulting from the synthesis. There are several NDR device but resonant tunnelling diode (RTD) is among one of the well studied. Moreover, there are several real-live implementations of such device with different parameters and properties. Therefore, in our simulations we have focused on resonant tunnelling diodes as an representative NDR device.

Previous work on simulation and circuit synthesis for non-classical devices used several different models of a RTD [3,6,9,10,13,11,12,15]. One of the earliest model proposed by Bhattacharya and Mazumder [3] utilises equations derived from physical parameters of the RTD. This method enabled for fast and robust simulation while being difficult in implementation and having reduced flexibility. On the other hand, Yan and Deen [6] proposed an empirical interpreted model that provides flexibility and ease of implementation while increasing simulation time. This model describes an RTD with a set of nine different parameters that allow to form the $I-V$ characteristic of the device. Another approach was proposed by Rose et al. [4] who used a universal device model (UDM) that captures nanoelectronic device behaviour by qualitatively representing the fundamental quantum effects. This approach enables to model different nanoscale devices without compact physical models. However, implementing new and robust UDM models on SPICE requires modifying the internal source code of the simulator [4], which is infeasible.

3 Implementation of Yan-Deen Model

Since our work was focused on RTDs thus we have analysed different models of RTDs and verified possibility to implement them in SPICE software. Despite the fact that there are many models proposed in the literature, we found that some of them are either not publicly available or are simply theoretical and their simulation is either not flexible or slow. Since we haven't found any suitable model we constructed our own SPICE implementation (List. 1) of the resonant tunnelling diode. Our implementation is based on the model proposed by Yan-Deen [6] and uses all nine parameters to profile the $I-V$ characteristic of the device. As proposed in the original paper by Yan-Deen, we implemented RTD using three voltage–controlled current sources "G" connected in parallel.

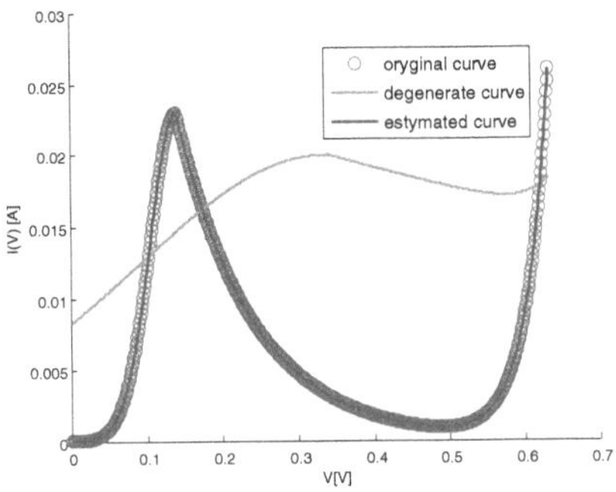

Fig. 2. Calling MATLAB nlinfit function result

```
.SUBCKT RTDYD N00104 N00119 PARAMS:
+Ip1 = 0.28e-3 Is = 1.4e-11
+N = 1.085 Vp = 0.14 Vt = 0.0273
+rop = 0.035 ron = 0.05 R = 0 M = 2e4

.FUNC Idiode(VCC)    { Is*(exp(VCC/(N*Vt))-1) }

.FUNC Itp_exp1(VCC) { exp(-((VCC-Vp)^2) / (2*rop^2)) }
.FUNC Itp_expInt(VCC)    { exp( M*(VCC-Vp) ) }
.FUNC Itp_exp2(VCC) { exp( (1-VCC/Vp)*Itp_expInt(VCC) ) }
.FUNC Itp(VCC)       { Ip1*Itp_exp1(VCC)*Itp_exp2(VCC) }

.FUNC Itne_exp1(VCC)     { exp(-(VCC-Vp)/(2*ron)) }
.FUNC Itne_expInt(VCC)   { exp(M*(Vp-VCC)) }
.FUNC Itne_exp2(VCC)     { exp((VCC/Vp -1)*Itne_expInt(VCC)) }
.FUNC Itne(VCC) { ((Ip1*Itne_exp1(VCC))-Itp(VCC))
                *Itne_exp2(VCC) }

G_G1    N00104 N00119 VALUE { Idiode( V(N00104, N00119) ) }
G_G2    N00104 N00119 VALUE { Itp( V(N00104, N00119) ) }
G_G3    N00104 N00119 VALUE { Itne( V(N00104, N00119) ) }
.ENDS RTDYD
```

Listing 1. Yan and Deen SPICE model implementation

However, implementation of the RTD is only half of the job since nine parameters of the model has to be setup. This was done with MATLAB using nonlinear regression (by nlinfit function) [14] and real $I-V$ characteristic of the resonant tunnelling diode (Fig. 2) from [5].

Using Yan-Deen model allows to create numerical model of the RTD that is flexible and gives ability to model almost any $I-V$ characteristic by simply adjusting parameters with no need to recompile anything. Convergence and ease of integration gives additional advantage over previous, publicly available models.

4 Synthesis

With implemented model of a single RTD diode we were able to construct and simulate larger circuits. We have verified two existing structures of circuits that are composed of RTDs [1,2]. Circuit structure proposed by Avedilo et al.[1] is build from the series of two RTDs parallelled by branches of RTD/transistor pair. It is assumed that any RTD in the network can have different parameters and is serially connected to a single transistor. On the other hand Berezowski [2] used identical RTDs serially connected to serial-parallel transistor network. Papers by Avedilo and Berezowski showed that circuit of both structures can implement some boolean functions, but no effective synthesis algorithm was given – in [2] exhaustive search was used to verify that any 4 intput boolean function can be implemented. There was also no proof that either circuit structure is capable of implementing Boolean function of an arbitrary number of variables.

We have analysed both structures and focused our attention on [2]-like circuits for which circuit synthesis requires to determine the number of branches and structure of SP networks for each branch. Moreover circuits proposed by Berezowski [2] introduce less issues concerning their physical implementation comparing to the circuits proposed by Avedillo [7]. This is mainly because circuits proposed by Berezowski use less RTD elements thus reducing the following issues: (i) inaccuracies in physical design of NDRs may accumulate and may result in erroneous quantisation; (ii) power consumption increases with the number of NRD in the circuit.

We have proposed an effective, iterative algorithm that given any boolean function outputs information required to implement [2]-like circuits. To the best of our knowledge this is the first effective algorithm for NDRs based circuit synthesis. According to [2] the GTG circuit can be described in a recursive fashion:

$$Y_l(\mathbb{X}^n) = \begin{cases} 0 & l = 0 \\ Y_{l-1}(\mathbb{X}^n) + N_l(\mathbb{X}^n) & l = 2k-1 \\ Y_{l-1}(\mathbb{X}^n)\overline{N_l(\mathbb{X}^n)} & l = 2k, \end{cases} \quad (1)$$

where $N_l(\mathbb{X}^n)$ represents a serial-parallel network of transistors in l-th branch. Nevertheless this formula is relatively simple it is suitable for formal analysis and makes it difficult to synthesise the circuit for given Boolean function. In fact this formula miss some important observations that allow to simplify it further. Missing assumptions can be derived from the analysis of the operation of GTG circuits in which upper and lower branches of the circuit switch on in turns. In terms of $N_l(\mathbb{X}^n)$ functions such operation means that any two functions $N_i(\mathbb{X}^n)$ and $N_j(\mathbb{X}^n)$ for $i \neq j$ meet an absorption rule, that is

$$\texttt{either } N_i(\mathbb{X}^n)N_j(\mathbb{X}^n) = N_i(\mathbb{X}^n) \quad \texttt{or } N_i(\mathbb{X}^n)N_j(\mathbb{X}^n) = N_i(\mathbb{X}^n). \tag{2}$$

Moreover we can use absorption rule to number function $N_l(\mathbb{X}^n)$ in such a way that $N_i(\mathbb{X}^n) \propto N_j(\mathbb{X}^n)$ which means that $N_i(X)$ absorbs $N_j(\mathbb{X}^n)$. As a consequence of this fact any three functions $N_i(\mathbb{X}^n) \propto N_j(\mathbb{X}^n) \propto N_k(\mathbb{X}^n)$ satisfy the following equalities:

$$N_i(\mathbb{X}^n)\overline{N_j(\mathbb{X}^n)} = N_i(\mathbb{X}^n) \oplus N_j(\mathbb{X}^n), \tag{3}$$

$$N_i(\mathbb{X}^n)\overline{N_j(\mathbb{X}^n)} + N_k(\mathbb{X}^n) = N_i(\mathbb{X}^n) \oplus N_j(\mathbb{X}^n) \oplus N_k(\mathbb{X}^n). \tag{4}$$

The above observations lead to the following observation that all $N_l(\mathbb{X}^n)$ functions from (1) can be numbered based on absorption rule, and the model can be simplified to an EXOR formula:

$$Y(\mathbb{X}^n) = \bigoplus_{i=1}^{n} N_i(\mathbb{X}^n). \tag{5}$$

There are several advantages of the new GTG model despite the fact that it is much simpler then the previous one. The biggest advantage of the new model is the existence of a prove that there exist a GTG circuit consisting of at most $n+2$ branches and NDR elements that can implement any n-variable Boolean function (due to the lack of space please refer to the full version of this paper for details). Apart from showing that GTG can implement any Boolean function we have construct an synthesis algorithm (Algorithm 1) that given Boolean function, in Reed-Muller canonical form, outputs $N_l(\mathbb{X}^n)$ functions for the GTG circuit.

Algorithm 1. Reed-Muller based GTG circuit synthesis

Require: n–variable Boolean function $Y(\mathbb{X}^n)$
Ensure: NDR_l vs. NDR_d relation, and $N_i(\mathbb{X}^n)$ functions

1: Transform $Y(\mathbb{X}^n)$ to Reed-Muller canonical form, i.e.
$$Y(\mathbb{X}^n) = a_0 \oplus \bigoplus_i N_i(\mathbb{X}^n),$$
2: **if** $Y(0^n) = 0$ **then** $\text{NDR}_l > \text{NDR}_d$
3: **else** $\text{NDR}_l < \text{NDR}_d$,
4: $Y(\mathbb{X}^n) = \text{Sort}(Y(\mathbb{X}^n))$,
5: set $i = 1$, $j = 2$,
6: **if** $N_i(\mathbb{X}^n)N_j(\mathbb{X}^n) \neq N_k(\mathbb{X}^n)$ for $k = i, j$ **then**
7: set $N_i(\mathbb{X}^n) \leftarrow N_i(\mathbb{X}^n) + N_j(\mathbb{X}^n)$,
8: set $N_j(\mathbb{X}^n) \leftarrow N_i(\mathbb{X}^n)N_j(\mathbb{X}^n)$,
9: $Y(\mathbb{X}^n) = \text{Sort}(Y(\mathbb{X}^n))$,
10: **else** set $j = j + 1$,
11: **if** $j > \text{Count}(Y(\mathbb{X}^n))$ **then** $i = i + 1$, $j = i + 1$,
12: **if** $i < \text{Count}(Y(\mathbb{X}^n))$ **then** goto 6-th step,

Proposed algorithm utilises two auxiliary functions: $Sort(\mathbb{X}^n)$ and $Count(\mathbb{X}^n)$, where Sort($Y(\mathbb{X}^n)$) is a function that given an EXOR sum of $N_i(\mathbb{X}^n)$ terms, outputs $Y(\mathbb{X}^n)$ with EXOR terms ordered according to the smallest number of variables in products and the biggest number of terms in a sum. Count($Y(\mathbb{X}^n)$) is a function that outputs the number of EXOR terms in $Y(\mathbb{X}^n)$ expression. After above considerations we prepared transient simulation in PSPICE for Boolean function: $Y(\mathbb{X}^4) = \overline{x_2}x_3x_4 + x_1\overline{x_2 x_3 x_4} + x_1x_2\overline{x_3}x_4 + x_1x_2x_3\overline{x_4}$. Such GTG structure have 6 branches and 4 of them are controlled by activation functions:

- $N_1(\mathbb{X}^4) = x_1x_2 + x_3$,
- $N_2(\mathbb{X}^4) = x_1x_3 + x_2x_3 + x_3x_4 + x_1x_2x_4$,
- $N_3(\mathbb{X}^4) = x_1x_3x_4 + x_2x_3x_4$,
- $N_4(\mathbb{X}^4) = x_1x_2x_3x_4$.

Simulation fully verified our synthesis method, we analysed many different function and presents one as an example (Fig. 3).

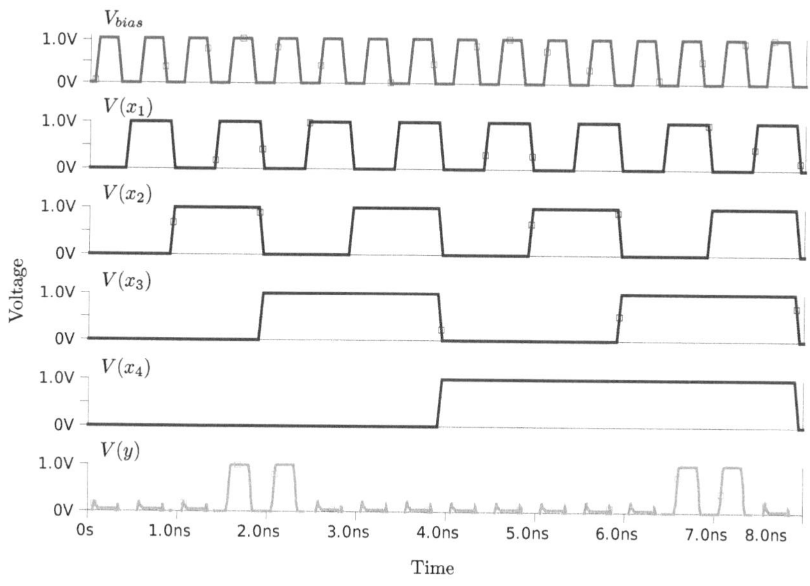

Fig. 3. Transient simulation for $Y(\mathbb{X}^4) = \overline{x_2}x_3x_4 + x_1\overline{x_2 x_3 x_4} + x_1x_2\overline{x_3}x_4 + x_1x_2x_3\overline{x_4}$ function in GTG structure

5 Conclusion

This paper presents a method how to carry on SPICE compatible transient analysis for the RTDs based circuits. Proposed synthesis method allows to build SPICE library of RTD devices and is alternative to one proposed in [13]. We have proposed new model of the GTG circuit that simplifies the analysis and simulation of the GTG circuit. Simplified model of the GTG circuit allowed us to construct a synthesis algorithms that outputs GTG circuit structure and $N_l(\mathbb{X}^n)$ functions for a Boolean function of an arbitrary number of variables. To the best of our knowledge this the first synthesis algorithm for GTG circuit. Because there are at most n unate functions $N_l(\mathbb{X}^n)$ required to implement n variable Boolean function therefore GTG circuit constructed using our synthesis algorithm consist of at most $n+2$ branches.

References

1. Avedillo, M.J., Quintana, J.M., Pettenghi, H., Kelly, P.M., Thompson, C.J.: Multi-threshold threshold logic circuit design using resonant tunnelling devices. IET Electr. Lett. 39(21), 1502–1504 (2003)
2. Berezowski, K.S.: Compact binary logic circuits design using negative differential resistance devices. IET Electr. Lett. 42(16), 902–903 (2006)
3. Bhattacharya, M., Mazumder, P.: Augmentation of spice for simulation of circuits containing resonant tunneling diodes. IEEE Trans. Computer-Aided Design 20, 39–50 (2001)
4. Rose, G.S., Ziegler, M.M., Stan, M.R.: Large-Signal Two-Terminal Device Model for Nanoelectronic Circuit Analysis. IEEE Trans. V. Large Sc. Int. (VLSI) Syst. 12(11), 1201–1208 (2004)
5. Schulman, J.N., De Los Santos, H.J., Chow, D.H.: Physics-based RTD current-voltage equation. IEEE Electr. Dev. Lett. 17(5), 220–222 (1996)
6. Yan, Z., Deen, M.J.: New RTD Large-Signal DC Model Suitable for PSPICE. IEEE Trans. on Comp.-Aided Des. 14(2), 167–172 (1995)
7. Avedillo, M.J., Quintana, J.M., Pettenghi, H.: Logic Models Supporting the Design of MOBILE-based RTD Circuits. In: Proc. Int'l Conf. Appl. Spec. Syst., Arch. & Process. (ASAP 2005), Samos, Greece, pp. 254–259 (2005)
8. Akeyoshi, T., Maezawa, K., Mizutani, T.: Weighted sum threshold logic operation of MOBILE (monostable-bistable transition logic element) using resonant-tunneling transistors. IEEE Electr. Dev. Lett. 14(10), 475–477 (1993)
9. Long, M., Ying-Long, H., Yang, Z., Liang-Chen, W., Fu-Hua, Y., Yi-Ping, Z.: A Small Signal Equivalent Circuit Model for Resonant Tunnelling Diode. Chinese Physics Letters 23(8), 2292–2295 (2006)
10. Sellai, A., Al-Hadhrami, H., Al-Harthy, S., Henini, M.: Resonant tunneling diode circuits using Pspice. Microelectronics Journal 34(5), 741–745 (2003)
11. Lake, R., Junjie, Y.: A physics based model for the RTD quantum capacitance. IEEE Electr. Dev. Tran. on 50(3), 785–789 (2003)
12. Wang, J.M., Sukhwani, B., Padmanabhan, U., Dongsheng, M., Sinha, K.: Simulation and Design of Nanocircuits With Resonant Tunneling Devices. IEEE Tran. on Cir. and Sys. 54(6), 1293–1304 (2007)
13. Chen, K.J., Guofu, N.: Logic synthesis and circuit modeling of a programmable logic gate based on controlled quenching of series-connected negative differential resistance devices. IEEE Journal of Solid-State Cir. 38(2), 312–318 (2003)
14. Seber, G.A.F., Wild, C.J.: Nonlinear Regression. John Wiley & Sons Inc., Chichester (1989)
15. Schulman, J.N.: Extension of Tsu-Esaki model for effective mass effects in resonant tunneling. Appl. Phys. Lett. 72, 2829–2831 (1998)

A Practical Methodology for Integration Testing*

Laura M. Castro[1], Miguel A. Francisco[2], and Víctor M. Gulías[1]

[1] MADS Group, University of A Coruña, Spain
{lcastro,gulias}@udc.es
http://www.madsgroup.org
[2] LambdaStream S.L., Spain
miguel.francisco@lambdastream.com
http://www.lambdastream.com

Abstract. The recognition of the importance of verification and validation tasks, within the software development process or life cycle, is growing significantly. Still, its unarguably complexity and the great amount of time and resources needed to perform testing properly, together with the industry's unawareness of the most powerful and versatile testing tools, makes that, in practise, these activities are often underestimated and diminished, or just simply ignored and skipped, sometimes due to client's demands or hard time-to-market constraints.

Integration testing is a specific kind of testing, which is gathering more and more attention within a software engineering industry that has been for quite some time already relying in structuring application and systems in different modules and components. In this paper, we propose a generic and re-usable model-based methodology for testing integration between different components, and illustrate it using a real case study, LiveScheduler, a scheduler and control tool for transmissions on live broadcast channels through the Internet.

1 Introduction

Far away from those monolithic systems from the early days of computer programming, nowadays component-based applications are probably the most common architectural structure to be found in software applications. As part of their development life cycle, different kinds of testing activities can be performed, also taking into account how they are internally arranged. When dealing with systems which are composed by different modules, these elements should not only be tested on their own (to ensure they perform correctly) but also in combination (to ensure they interact properly). The latter is called *integration testing.*

In the following pages we present a testing methodology to deal with integration testing, grounded on model-based testing. The testing procedure we propose is illustrated on the basis of a real application, LiveScheduler. LiveScheduler is a software component that enables scheduling of events which represent live

* Partially supported by MEyC TIN2005-08986, XUGA PGIDIT07TIC005105PR.

R. Moreno-Díaz et al. (Eds.): EUROCAST 2009, LNCS 5717, pp. 881–888, 2009.

broadcast data to be transmitted through the Internet. However the LiveScheduler component does not really start or stop any event, it just stores information about their periodicity and configuration, and instead of physically setting up the actual transmissions, it interacts with an external entity which does so: VoDKA [1], a powerful streaming server. It is the VoDKA server which handles multicast channels using a wide variety of different contents: local files, HTTP sources, other multicast sources,... So LiveScheduler relies in VoDKA for the actual creation of the multicast channels with the different contents, according to the user preferences, and it just has the responsibility of properly invoking VoDKA when it is due. Hence, seamless integration between LiveScheduler and VoDKA is a key factor for the good operation of the entire service, as happens regularly in component-based applications. Without diminishing all the other aspects to be tested in a software product, this integration testing is among the most important, since the global functionality of the system may heavily depend on these interactions to be virtually error-free.

The proposed method for integration testing has been implemented using an automatic test case generation tool called Quviq QuickCheck[2,3]. QuickCheck is a very powerful and versatile testing tool that, on the one hand, can be used to write customised data generators and system properties specifications[4], and on the other hand, provides a mechanism to define a state machine that can be used to test state-full system behaviour in a very simple and structured manner[5]. Using QuickCheck state-machine facilities, we have put in practise our methodology. Nevertheless, the testing procedure we explain here is not bounded to this tool, so it can be easily exported, translated and re-used to any other integration testing environment or scenario.

2 State Machine-Based Methodology

Our proposal is based on the use of a state-machine as a way of modelling testing case execution: for a system in a certain initial state, state transitions are triggered as a result of invocations to the different operations or services (i.e. use cases) to be tested, resulting in the system state modification as a sort of side effect. Very complex system behaviour can be modelled using a state machine, since for each operation to be executable at a certain state, some conditions might have to be met, and some other may have to be true after the state transition is completed. Those conditions that need to be true before a certain operation can be executed are called *preconditions*, and those conditions that should be fulfilled once the execution is finished are named *postconditions*.

Using a testing state machine, test case generation and execution means generating random sequences of state transitions (representing the operations to test) from a given initial state, each of them according to their preconditions (i.e. if a precondition is not true, that transition will not be eligible to be the next step in the testing sequence), updating the internal state, and then checking their postconditions.

Therefore, the steps for applying a state machine-based integration testing methodology involve first defining the internal state structure of the testing

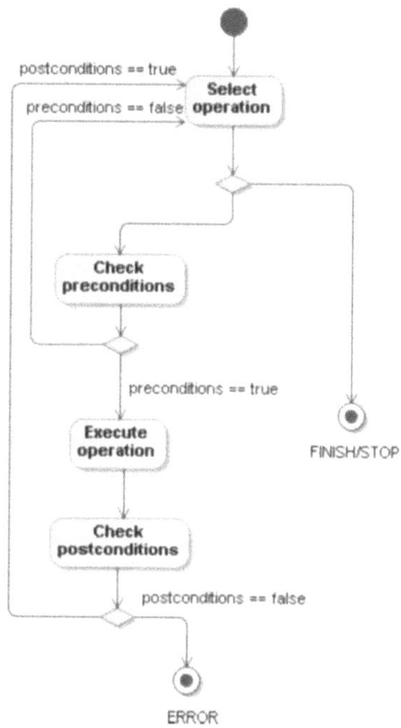

Fig. 1. Testing state machine

state machine, its contents as initial state, and then the set of operations to test and their preconditions and postconditions, and how the execution of each operation may modify the content of the internal state.

The internal state must hold the data that may be needed to check relevant constraints at postconditions, and that could also be used at the preconditions, if necessary. The internal state shall also serve as an additional argument for the state machine transitions (i.e. the relevant operations from the integration perspective), that are responsible for modifying it.

In a good software design, an application provides one or more facades [6] which make public and available the list of functions for invoking each use case. These are the most convenient operations to be called as the transitions of the state machine in our proposed methodology, since this state-machine based approach is aligned with the principles of *black box testing*, rather than white box testing or code inspection. It might be the case that the actual operations in the system or module to be tested receive or return additional information that is not relevant from the testing point of view, so that we would rather not need to provide or handle it. In these cases, we can keep the state machine definition clean and simple by defining wrapper functions for each operation to test. Wrapper functions can adapt both parameters and/or returned values for/from the original functions, and be replace them as the state transition operations.

Finally, for each operation we need to define preconditions and postconditions, as well as the stage change, which specifies how the state transition (execution of the operation) affects the internal state. Of course, preconditions, postconditions, and state change must reflect the properties of the system. Domain constraints should be identified and translated into these check points, but always bearing in mind that actual fulfilment of those constraints must already be responsibility of the code or component to be checked. That is, we need to confirm that those conditions are being assured, avoiding by all means restricting our test-cases more than the normal use of the application would.

Regardless of the external elements our components under test will interact with, adopting a black box perspective it is very likely that we need to assume that the state-machine operations can be invoked providing any argument with any value. In other words, specially when performing integration testing, we should not rely on the component's caller to any extent, and instead test our components' behaviour against all possible situations. Thus, this will keep the preconditions as small and simple (i.e. ideally, empty, `true`) as possible.

As for the postconditions, this is the place where important tests can be performed to check whether or not each operation result is what should be expected. Thus, for the integration testing aim, the postconditions must check if the component actually invokes the proper operations from other component and if it uses the returned values in the right way. For that, we propose an additional technique, explained in the next section.

Be it as it may, the best way to produce an exhaustive commands list that includes all relevant operations is by adding them gradually, so we should start testing a small state machine with few transitions and check it error-free before proceeding and considering more possibilities. It is very important to remember that each state machine operation should resemble an exact operation/use case/functionality in our system, so that the state machine performs no extra work, or bears no extra responsibility, but performing the actual transitions.

2.1 The Dummy Component Technique

When dealing with integration testing, the main goal is checking that two (or more) components work properly together. In other words, we want to be sure that when a service is requested from one component, which relies in other component to perform the operation, the former invokes the right methods from the latter. For such matter we don't actually need the second component, because *we are not testing functionality but* only *interaction*. What we do need is a replacement for such component that offers the same (or a subset of) interface and provides the same kind of answers. We don't care if the service is actually provided, if any operation at all is performed inside that *dummy* replacement component, or if all the process of building up a request result is just simulated.

Each *dummy* function in the introduced *dummy* component must, of course, receive the same parameters and return the same values than the original functions in the replaced component. For external applications, the dummy component must *pretend to work* in the same way than the original one. Thus, even though the new component needs not make any real action, for other components it shall seem to work correctly. The same procedure will be used in all the functions, the only constraint being that the dummy implementation must mimic the behaviour of the original component in the best possible way. Implementing such a dummy component should be a very easy task, and its advantages include faster response times and collateral effects avoidance.

Still, to be really able to check the proper connection between the first component requests and the called functions in the external component (which, in fact, is precisely the key aspect to integration testing), we require something more. We suggest that the dummy component registers all access to its interface functions, and also that it provides a way to retrieve that information. This data trace can be later on recovered to verify that the correct interactions are taking place in all possible scenarios. So, for each operation to test, there will be a way of checking which functions were called in the external component.

So when executing test cases, generated as sequences of state-machine transitions, the set of interactions actually produced with the external module can be retrieved after calling each function under test. This set of operations should be then inspected, to check that the proper calls are being made for each particular request or sequence of requests. This kind of check is not to be made as part of the implementation of the wrapper; instead, it represents exactly the postcondition that the operations need to satisfy in order to be considered correct. Thus, it is the postconditions where we have all the information about the operations invoked on the external component, the information about the current system state, and of course the arguments to the function itself. Postcondition implementation should hence consist in checking, among other possible constraints, the list of invoked functions to make sure that all (and only) the appropriate dummy component functions have been called. Storing as exhaustive as possible information about calls (function name, arguments, result) is preferred, so should an operation depend on the results of previous operations, or on the values provided as function arguments, all that information is available.

Successful execution of tests produced with this methodology provides confidence on that for each operation in the tested component, the correct operations in the external component are called, and in the correct order. Of course, since we are testing against a dummy external component, this testing does not imply that the entire system works properly in every sense. But remember that we are here testing components integration, and that individual component functionality testing should be previously done separately.

2.2 Negative Testing

In the testing procedure we have just described, we approach the integration between two components or applications. The methodology is based on the implementation of dummy components which offer the same interface and emulate the same behaviour than the original components, in order to avoid any changes in the component to test.

However, there is a number of failure scenarios that fall outside of the scope of the state-machine integration testing as we have stated it so far. When testing interaction between components, relevant questions such as what happens if the external component fails to respond to a request need to be taken into account. The kind of testing we have suggested with our state-machine based methodology only focuses on *positive testing* right now, but what about those situations in which the functions in the external component may fail due to external circumstances (such as network congestion or similar)? If we do not consider these situations, then we will never be able to control how these errors affect the behaviour of our system.

To deal with *negative integration testing*, a good option is to create new transitions in the state machine which force a failure situation, such as delaying or dropping the answers of the dummy external component. Realize that, by doing so, we are adding a new communication scenario, which obviously needs to be properly handled at the corresponding postconditions.

If we generate and execute test sequences with these new transitions, i.e. representing waiting times and exceptions in the communication between the components, the tests may fail because the expected trace is not the same one which is obtained when some other exception or error occurs. So, a key aspect here is to properly deal with this in the postcondition function, where we can act accordingly, comparing the traces taking into account anomalous situations, such as delay and/or even network congestion parameters. With this basic approach, we can handle very precise situation models. For example, the delay can be a particular parameter of each operation, hence reproducing very different and changing testing scenarios.

Last but not least, the same procedure explained here for timeouts can be re-used for any other type of possible exceptions or errors which can occur in the communication between two components, like dynamic granting of permissions or privileges, for instance.

3 LiveScheduler and QuickCheck: Case Study and Methodology Implementation

Using the integration testing methodology we have proposed here, we have tested interaction between LiveScheduler and VoDKA. As we outlined in section 1, LiveScheduler is an application used for programming events over broadcast multimedia channels, which are actually handled by the VoDKA server. With regard to integration testing, the key pieces of functionality to be tested are the use cases that involve starting or stopping events, activities that require interaction between LiveScheduler and the VoDKA server.

In our case study, applying the dummy component technique meant that we could do without a working VoDKA server as long as we replaced it with a dummy component offering the same API and answers to the LiveScheduler component (so neither the real VoDKA server nor the LiveScheduler module need to be modified, of course, see figure 2).

Also, applying the ideas in the previous section about negative testing, we have been able to test LiveScheduler's behaviour whenever the communication channels get slow, or even if there are errors such as timeouts and so on.

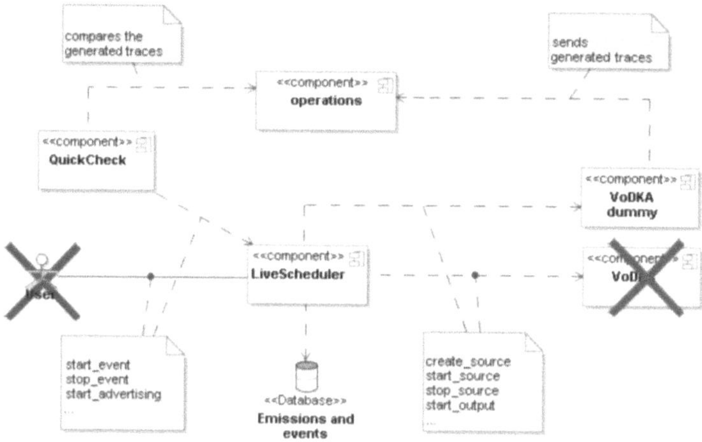

Fig. 2. Introduction of dummy component

With regard to the use of the automatic testing tool QuickCheck, its suitability to implement our methodology is based on the mechanisms that provides, to easily define a state machine model: a series of call-back functions are used to defined the system initial state, the set of operations (transitions) and corresponding pre- and postconditions, and the effect of the transitions on the internal state. Besides, QuickCheck allows not only the generation, but also the execution of test cases, applying a very convenient shrinking process to failure cases whenever a problem is found. Last but not least, QuickCheck adds some other interesting features such as the possibility of granting each transition with an associated probability of occurrence.

As one can easily see, all these properties introduce very important advantages with respect to other testing tools when it comes to implementing our integration testing methodology.

4 Conclusions

In this paper we have introduced and explained a process to test integration between different components of an application or system. This particular testing scenario is very common in modern systems, which are usually structured in different separate modules that work together to offer full services or provide complex functionalities. In all those situations, integration is an essential aspect to test. Thus, we consider that the experience we present here can be of interest as an example of how to deal with this particular testing task.

The methodology we propose, which is inspired on model-based testing and uses a state machine as central element, has important advantages. First of all, the use of a state machine allows us to define a way to generate not only the kind of operation sequences that we will usually test, but all sequences that adjust to the set of constraints specified in the state machine definition, no matter how strange or improbable they might be. Having the possibility of testing all sort of even unlikely cases is very useful to find the kind of errors that are more expensive to fix, when found after releasing or sending the product to a production environment, and may potentially save a lot of time and effort.

The criteria to determine if a testing case has been successful or not is based on replacing external components by API-equivalent dummy ones, in combination with the inspection of their communication (invocation) traces. By using dummy components, collateral effects on testing components or applications integration can be avoided, while still being able to test their interaction to a full extent. Of course, the new dummy external component does not need to replicate the original external component functionality: it only has to provide the same interface and emulate the external component replies, so its development time should not be significant. Designing a wide-purpose framework for specifying a possible trace and comparing two traces with several possibilities and branches of operations would facilitate all the procedure, and is one of the open research lines in which we are working at the moment.

Finally, even though this methodology has been put to practise using a specific testing tool, we consider that these concepts can be translated straight forward to other environments and that the ideas we have exposed here can be re-used to be applied when testing integration of applications with comparable tools/development platforms.

References

1. Gulías, V.M., Barreiro, M., Freire, J.L.: VoDKA: Developing a video-on-demand server using distributed functional programming. Journal on Functional Programming 15(3), 403–430 (2005)
2. QuviQ: QuviQ QuickCheck (2008), `http://www.quviq.com`
3. Claessen, K., Hughes, J.: Quickcheck: a lightweight tool for random testing of haskell programs. In: ACM SIGPLAN Notices, pp. 268–279. ACM Press, New York (2000)
4. Arts, T., Castro, L.M., Hughes, J.: Testing erlang data types with quviq quickcheck. In: ERLANG 2008: Proceedings of the 2008 ACM SIGPLAN workshop on Erlang, pp. 1–8. ACM, New York (2008)
5. Arts, T., Hughes, J., Johansson, J., Wiger, U.: Testing telecoms software with quviq quickcheck. In: ERLANG 2006: Proceedings of the 2006 ACM SIGPLAN workshop on Erlang, pp. 2–10. ACM Press, New York (2006)
6. Gamma, E., Helm, R., Johnson, R., Vlissides, J.: Design patterns: elements of reusable object-oriented software. Addison-Wesley Professional, Reading (1995)

Safety Oriented Laparoscopic Surgery Training System

Andrzej Wytyczak-Partyka[1], Jan Nikodem[1], Ryszard Klempous[1], Jerzy Rozenblit[2], Radoslaw Klempous[3], and Imre Rudas[4]

[1] Wroclaw University of Technology, Institute of Computer Engineering, Control and Robotics, 27 Wybrzeze Wyspianskiego St., 50-370 Wroclaw, Poland
[2] Electrical and Comp. Eng. Dept., The University of Arizona, Tucson, USA
[3] Marciniak Hospital, Traugutta St. 116, 50-420 Wrocaw, Poland
[4] Budapest Tech, Becsi ut 96/B H-1034, Budapest, Hungary

Abstract. The discussed training system employs several means for encouraging safe behavior during laparoscopic surgery procedures. The elements of no-fly zones, magnetic position sensing and expert systems are tied together to form a complex system that provides guidance and performance assessment. A 3D reconstruction algorithm is used for the purpose of defining no-fly zones and has been tested in a simulator developed for the purpose of this work. An expert system has been built in cooperation with surgeons that, based on simple rules, can assess the risk of the trainee's actions. Despite the shortcomings of the 3D reconstruction process, the training system performed as expected during experiments. Simple exercises like touching points in 3D space were performed and scored appropriately to whether a no-fly zone has been breached or not. Also simple advice could be provided to the trainee in order to help improve the results.

Keywords: laparoscopic surgery, training, image processing, expert system, fuzzy logic.

1 Introduction

Laparoscopy is a surgical technique, where operating sites within the human body are accessed under the guidance of an endoscopic camera, through special long tools, that are introduced into the body via small incisions - as opposed to open surgery, where one large incision is made and the surgeon has direct access to the tissue and is viewing the operating site with his bare eye. Because of that change of perception of visual and haptic sensations laparoscopy is much more demanding and difficult to master than traditional surgery.

Soon after laparoscopic surgery has been adopted in the 1980's the medical society discovered the need of specificaly designed training programs with objective performance assessment. This need has been addressed by applying simulators, an idea that came from the aviation industry [1,2]. Several such systems [3,4] are available with emphasis on the Virtually Asissted Surgical Trainer (VAST)

R. Moreno-Díaz et al. (Eds.): EUROCAST 2009, LNCS 5717, pp. 889–896, 2009.

[4] system, which serves as a basis for the training system built for the purpose of this work.

Currently [5,6] the main part of laparoscopic surgeon training is learning to operate the endoscopic camera [7] and instruments by performing excercises on physical models. The main purpose of the system described in this paper is to aid in that process but also enforce safe behaviour during procedures - by informing the trainee about approaching certain predefined zones where the operating instruments shouldn't appear [8,9] as well as providing information about certain features of movements, like speed or roughness. That way the trainee is forced not only to master the basic set of skills but also to operate in a safe manner.

An expert system application has been proposed earlier [9] but has been extended to offer not only advice on improving the skills but also can now provide risk assessment. The expert system is based on a set of rules created in cooperation with surgeons and a fuzzy classification of movements.

2 System

The system contains of a standard laparoscopic setting - a surgical instrument and a camera, which are insterted in to a obscured box. The instrument also has an embedded position sensor, which together with the images from the endoscopic camera provide information for the computer system - as outlined in Figure 1.

An important part of the system is the image processing component, which is reponsible for creating a 3D model of the operating field. It is possible to obtain such a model through application of a certain set of image processing algorithms (an approach called structure-from-motion 3D recovery) directly from a set of 2D images of the operating field, coming from the laparoscopic camera.

2.1 Training

The training is based on simple tasks that trainees have to complete within a specified time. Several types of exercises have been proposed by others [14, 15], the common goal is to practice dexterity, coordination and depth perception. Example exercises are:

- knot tying,
- cutting and suturing,
- picking up objects,
- touching different points on a model.

The trainee's score is calculated with respect to :

- elapsed time,
- length of the path of the instrument tip,
- accuracy,
- hazard,

where the hazard score is related to the events of approaching no-fly zones - hazardous regions, defined at the beginning of each exercise.

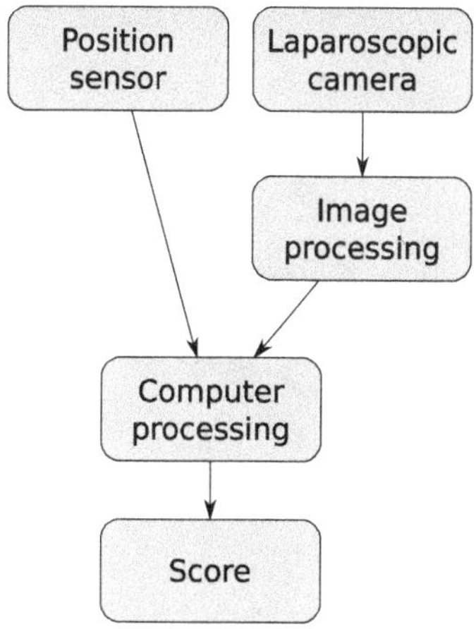

Fig. 1. Components of the training system

2.2 No-Fly Zones

The concept of no-fly zones is an important part of the described system, which just like the idea of applying simulation in surgical training has been borrowed from the aviation industry. In the system presented here the no-fly zones are used to focus the training on the skill of avoiding certain regions of the operating field - just like during normal procedures the surgeon has to avoid irritating or cutting certain anatomical structures (i.e. nerves, vessels).

$$S = \left(\frac{k_t}{t}, \frac{k_s}{s}, k_A \cdot A, k_H \cdot H\right) \tag{1}$$

with

$$H = -\frac{2R_H}{|d(H_c, T)| + R_H} + 1, \tag{2}$$

and $d(H_c, T)$ being the distance between the center of the no-fly zone (approximated by a sphere) H_c and the instrument tip T.

A simulator of the system has been developed and used to evaluate the performance of the proposed concepts. As it is shown in Figure 2(a) the simulator contains several elements - an instrument, image of the operating field, no-fly zone proximity indicator and the H measure indicator. An example of applying the modified scoring function in a training drill is shown in Figure 2.

The no-fly zones are located in the 3D space of the operating field. To achieve that a 3D model of the operating field has to be obtained first. It has been shown [10] that an application of structure-from-motion computer vision algorithms (also outlined in Figure 2(b)) can be used for that purpose.

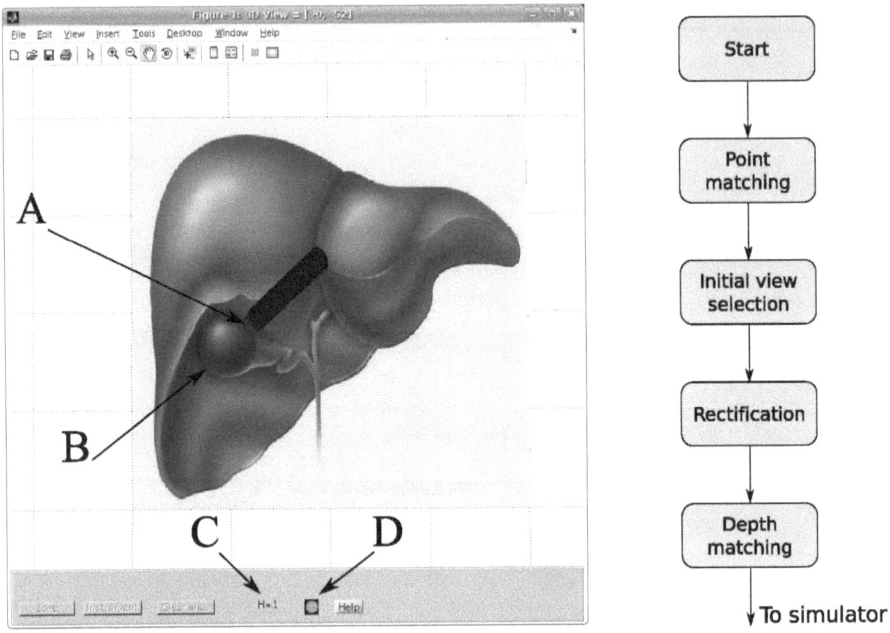

(a) Simulator of the system, running in Matlab. A - tip of the instrument, B - no-fly zone location, C - H measure indicator, D - no-fly zone proximity indicator

(b) Image processing steps

Fig. 2. Simulator and the image processing pipeline

2.3 Expert System

In order to build a usable training system some expert knowledge and fuzzy set theory has been embedded into the described system [11,12]. This knowledge is expressed in a number of rules, similar to the following:

- if (speed is F-) and (smoothness is S+) and (distance is D+) then (Risk is R-),
- if (speed is F+) and (smoothness is S-) and (distance is D-) then (Risk is R+),
- if (speed is F+) or (smoothness is S+) or (distance is D+) then (Risk is R+),
- if (speed is F+) then (Risk is R+),
- if (smoothness is S-) then (Risk is R+),
- if (distance is D-) then (Risk is R+),

where F^+, S^+ and D^+ are fuzzy sets containing movements that are, respectively, fast, smooth and distant from no-fly zones.

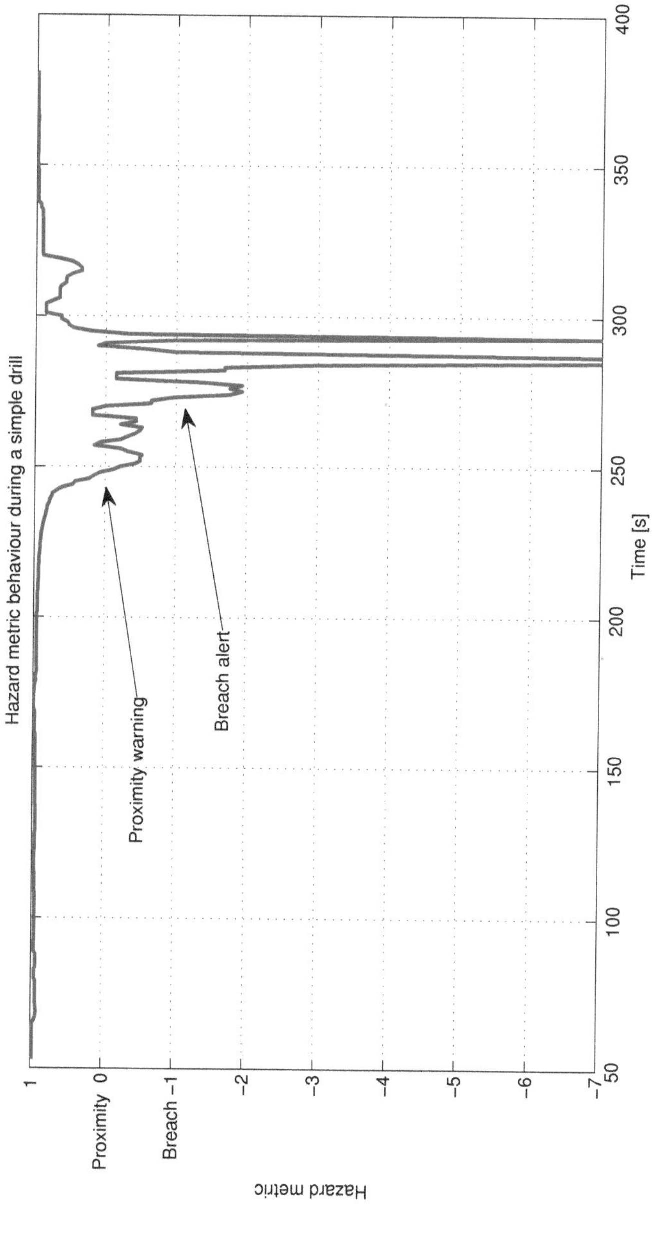

Fig. 3. Graves Disease workflow in Taverna

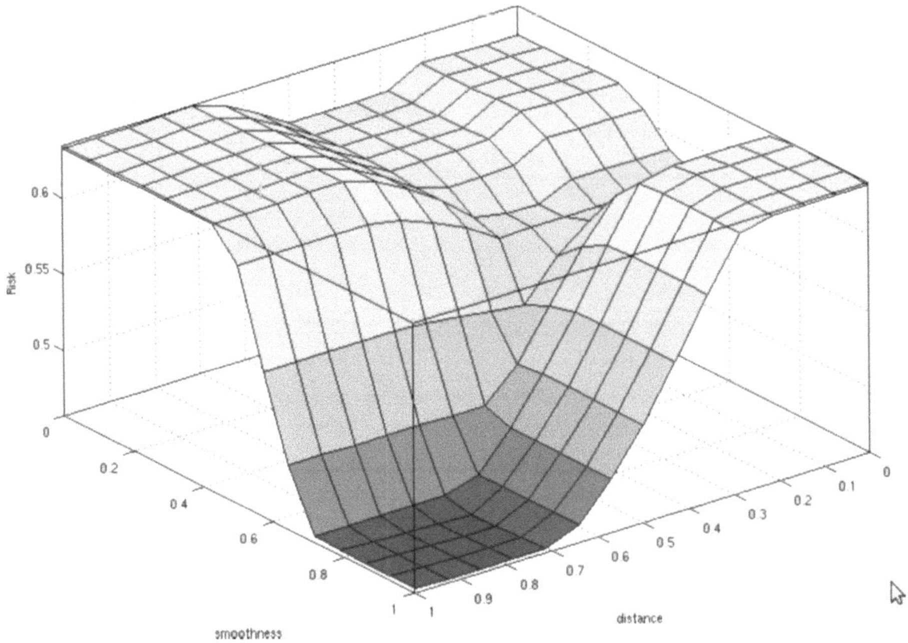

(a) Smoothness vs. distance

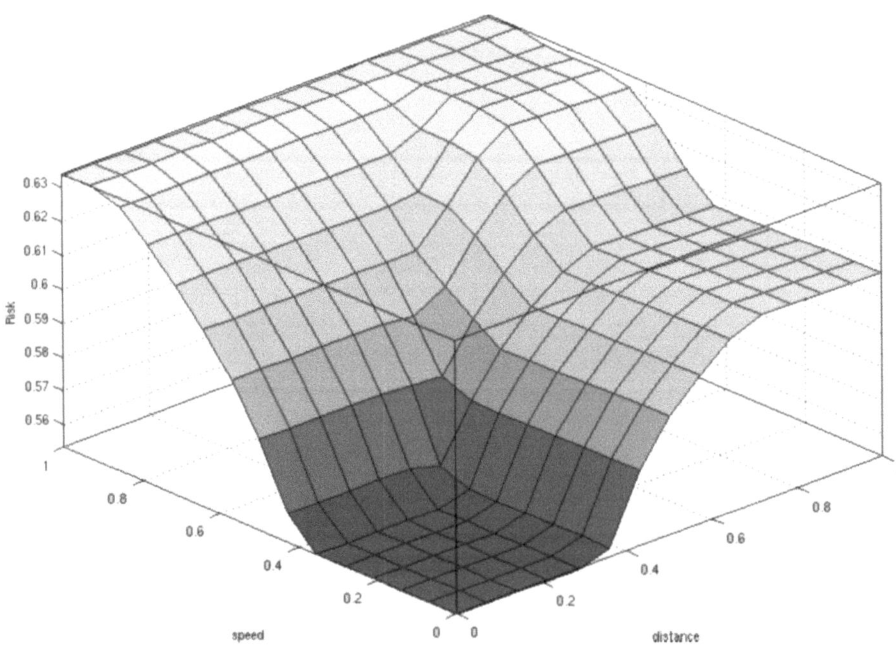

(b) Speed vs. distance

Fig. 4. Risk function in terms of smoothness, speed and distance

The main goal of the expert system is assessment of the risk involved with the actions taken by the trainee, an example of the risk output in terms of movement speed, smoothness and no-fly zone distance is shown in Figure 4. Additionally the system can offer advice (i.e. can tell the trainee that the movements are too fast, too rough, etc.), based on the movement classification, to help the trainee in improving missing skills.

Further detail on classification of movements and expert advice offered by the system has been given in [9].

3 Conclusions

The main focus of the system described in this paper is enforcing safe behaviour during laparoscopic procedures. Several means are used to achieve that goal - application of computer simulation, no-fly zones, expert system. The trainee's performance can be scored adequately and advice on how to improve can be offered. The concepts brought together in this system have been tested and can serve as a tool in laparoscopic training programs.

The image processing part (used mainly for the purpose of locating the no-fly zones in the 3D space of the operating field) of the system needs further attention, but already shows promising results. Especially models that are poorly textured produce 3D representations biased with a significant error. Stereo matching algorithms that deal with such textureless objects are available and they will be examined for application in future versions of the proposed system.

In future versions of the system it will be interesting to add a parameter describing the visibility of the instrument, as well as it's distance from anatomical structures, not only from the no-fly zones. That way better guidance and risk assessment will be offered.

References

1. Kneebone, R.: Simulation in surgical training: educational issues and practical implications. Medical Education 37(3), 267–277 (2003)
2. Fried, G.M., Feldman, L.S., Vassiliou, M.C., Fraser, S.A., Stanbridge, D., Ghitulescu, G., Andrew, C.G.: Proving the Value of Simulation in Laparoscopic Surgery. Annals of Surgery 240(3), 518–528 (2004)
3. Simbionix WWW page
4. Feng, C., Rozenblit, J.W., Hamilton, A.J.: A Hybrid View in a Laparoscopic Surgery Training System. In: Proceedings of the 14th Annual IEEE International Conference and Workshops on the Engineering of Computer-Based Systems, pp. 339–348 (2007)
5. Gorman, P.J., Meier, A.H., Krummel, T.M.: Computer-assisted training and learning in surgery. Computer Aided Surgery 5(2), 120–130 (2000)
6. Cosman, P.H., Cregan, P.C., Martin, C.J., Cartmill, J.A.: Virtual reality simulators: Current status in acquisition and assessment of surgical skills. ANZ Journal of Surgery 72(1), 30–34 (2002)

7. Fraser, S.A., Klassen, D.R., Feldman, L.S., Chitulescu, G.A., Stanbridge, D., Fried, G.M.: Evaluating laparoscopic skills. Surgical Endoscopy 17(6), 964–967 (2003)
8. Wytyczak-Partyka, A., Nikodem, J., Klempous, R., Rozenblit, J.: A novel interaction method for laparoscopic surgery training. In: Proceedings of the 8th Annual IEEE International Conference and Workshops on Human Systems Interactions, pp. 858–861 (2008)
9. Wytyczak-Partyka, A., Nikodem, J., Klempous, R., Rozenblit, J., Feng, C.: Computer-guided laparoscopic training with application of a fuzzy expert system. In: Gelbukh, A., Morales, E.F. (eds.) MICAI 2008. LNCS (LNAI), vol. 5317, pp. 965–972. Springer, Heidelberg (2008)
10. Wytyczak-Partyka, A.: Image processing in computer-guided surgical training. Master's thesis, Wroclaw University of Technology (2008)
11. Kandel, A.: Fuzzy Expert Systems. CRC Press, Boca Raton (1992)
12. Zadeh, L.A.: Fuzzy sets. In: Advances In Fuzzy Systems. World Scientific Series, pp. 394–432 (1996)

Co-operative Extended Kohonen Mapping (EKM) for Wireless Sensor Networks

Zenon Chaczko, Perez Moses, and Christopher Chiu

Faculty of Engineering & IT, University of Technology, Sydney, Australia
zenon.chaczko@uts.edu.au, zindpm@gmail.com, christopher.chiu@uts.edu.au

Abstract. This paper discusses a methodology to manage wireless sensor networks (WSN) with self-organising feature maps, using co-operative Extended Kohonen Maps (EKMs). EKMs have been successfully demonstrated in other machine-learning contexts such as learning sensori-motor control and feedback tasks. Through a quantitative analysis of the algorithmic process, an indirect-mapping EKM can self-organise from a given input space, such as the WSN's external factors, to administer the WSN's routing and clustering functions with a control parameter space.

Preliminary results demonstrate indirect mapping with EKMs provide an economical control and feedback mechanism by operating in a continuous sensory control space when compared with direct mapping techniques. By training the control parameter, a faster convergence is made with processes such as the recursive least squares method. The management of a WSN's clustering and routing procedures are enhanced by the co-operation of multiple self-organising EKMs to adapt to actively changing conditions in the environment.

Keywords: Software Engineering, Extended Kohonen Wireless Sensor Networks (WSN), Sensor Actor Networks (SANET).

1 Introduction

The governance of wireless networks, particularly with WSNs, is important in managing the routing and clustering mechanisms in a dynamic and volatile environment [2]. Unlike static wired networks, the reliability and guarantees on communication stability can never be assured. Furthermore, the dependency on nodes to follow through on message relaying means that the failure of one node may lead to an entire branch losing total connectivity. Another main aspect is security, as the security within the network can be prone to denial-of-service attacks or technological espionage such as 'packet sniffing'. These particular domain concerns, among many more which are inherent with wireless networking, require the wireless network structure to adapt to changing environmental concerns to ensure the network's continual stability and robustness [10].

As with all open systems, wireless sensor network communications is inherently dynamic, such that the focus in achieving optimum conditions in a given environment is to determine the efficient routing path through the network.

R. Moreno-Díaz et al. (Eds.): EUROCAST 2009, LNCS 5717, pp. 897–904, 2009.

Modelling the domain-level conditions in a WSN is complex in nature, with a multitude of variables that influence the optimum routing condition. Transforming the multi-dimensional concerns to a single map is a formative approach to reduce the problem space into a single-dimensional map, in which the routing conditions can be established for the network.

The proposition for an alternate feature map approach is to be used for wireless sensor network management concerns, through co-operative EKMs with indirect mapping [8]. An indirect-mapping EKM approach is novel to existing direct-mapping methods by utilising the following techniques [11]:

- **Direct-Mapping Approaches**
  Direct-mapping involves creating a Self-Organising Feature Map (SOM) that maps a sensory input directly to the node's stimuli in the wireless network. This would include parameters such as signal-to-noise ratios, battery charge levels and available bandwidth. Therefore, Direct-mapping methods map the continuous sensory stimuli space to discrete clustering or routing directives for each node, as seen from a different perspective.
- **Indirect-Mapping Approaches**
  Indirect-mapping requires mapping the continuous sensory stimuli space to the node clustering or routing directive as an end result. The determination of the quality of clustering and routing approaches can be achieved through evaluating the cluster set data with validity matrices such as Dunn's Separation Index, Calinski-Harabasz and Davies-Bouldin's indexing methodologies [3]. Hence, the indirect-mapping approach maps sensory stimuli indirectly to a node clustering or routing directive with the utilisation of control parameters.

2 Examining Co-operative EKMs

The directives given to a node in a wireless sensor network's control space is formed as a discrete set of commands to be used by reinforcement learning algorithms [7, 11], or at the minimum level, pre-defined static rules. In recent years, autonomous agent research in dynamic systems theory and reinforcement learning propose the operation of such directives in a continuous control space [7], to allow the indirect-mapping method to provide finer directive decisions than in direct mapping. Focussing on the flexibility and precision in sensory stimuli control is imperative in a wireless network domain where external environmental factors directly affect the network's robustness and reliability.

In traditional contexts, SOM or EKM in passive learning control is established and documented by Ritter [11]. However, the inconsequential problem of combining multiple SOMs or EKMs for sophisticated system control is a potential area of study. If the sensory control is insufficiently clarified, the routing or clustering decision made by a wireless node may be unexpected or undesirable, leading to a potential inertia to route data back to the sink or wireless gateway [7, 8]. When SOMs or EKMs are established in the weighted-sum ensemble, a similar problem of inertia also takes place.

To solve the problem of routing inertia, the combinational approach with co-operative EKMs will be applied to wireless sensor networks [8]. The co-operation and competition of multiple EKMs that similarly self-organise can enable a non-holonomic wireless node to optimise its routing and clustering choices in unexpected changes in its environment. In contrast, a node managed by the weighted-sum ensemble method will approach a routing inertia, even though the wireless network also implements a continuous sensory control space.

The environmental concern for a given wireless network domain can be summarised in the following statement tasks shown in Figure 1 on the following page; of which the environmental concerns deal with external factors such as sensory stimuli and node conditions [9]:

- For an initial state described by the input vector $u\,(0)$ in input space U;
- Adapt a new clustering or routing sequence of control vectors $c\,(t)\,, t = 0, \ldots, T-1$ in the sensory control space C;
- With the resultant goal state elaborated by $u\,(T) \in U$ that adapts the network structure for a desired objective or target state.

The following algorithmic categories have been considered for an adaptive wireless sensor network environment which suits the above statement tasks:

- **Feature Mapping**
  By using a SOM proposed by Kohonen [5], such as the Extended Kohonen Map (EKM) [12], the map self-organises to partition continuous sensory space into discrete regions. The feature map's generalisation capability arises from its self-organisation during training [6], such as when every node in the WSN is effectively trained to map a localised sensor region. This approach increases the sensory representation's resolution in the frequently encountered stimuli regions [6]. This conduct reflects biological sensory perceptions where frequent practice leads to better predictive capability of common, anticipated events.
- **Multivariate Regression**
  An alternative approach formulates the statement task as a non-linear multivariate regression problem. Uninterrupted mapping from U to C is done by training a multilayer perceptron (MLP), which offers possible generalisation capability [3, 10, 12]. The main disadvantage prior to training the network is that training samples must be collected for each time step 't' to define quantitative error signals. As this sampling process can be very tedious and computationally difficult, it is solved with the reinforcement learning approach by providing a qualitative success or failure feedback only at the end of the executing control sequence [4].

3 Experimental Framework

The experiment to evaluate the effectiveness of co-operative EKMs on WSNs is conducted using the MATLAB environment originally developed by

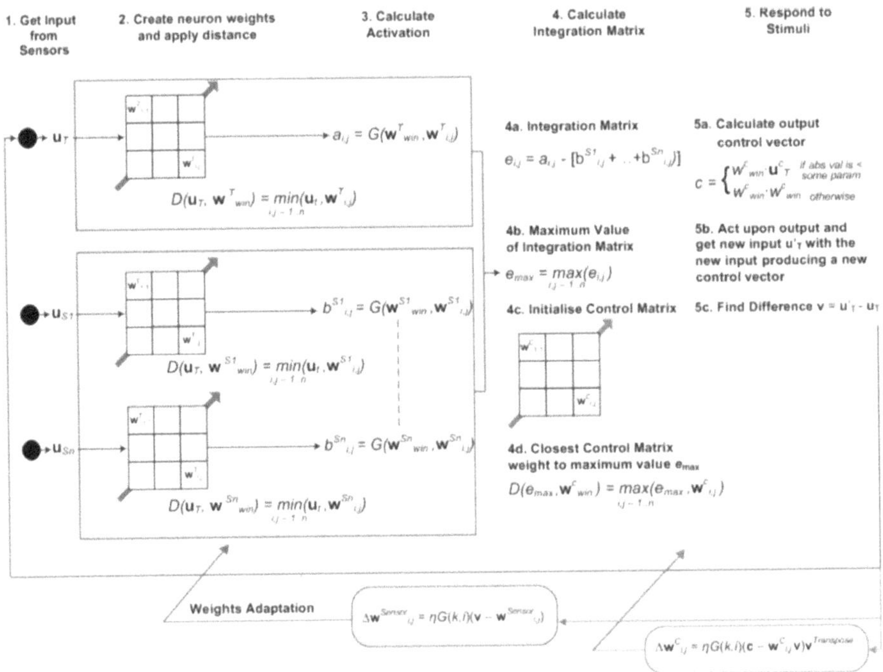

Fig. 1. Diagram of Co-operative EKM Functions

Chaczko, et al in 2003 [1]. The simulation framework, depicted in Figure 2 on the current page, allows for convenient monitoring and tracking of WSN events by programing event trajectories in the network field. The experiment is completed with the following methodology:

1. A population of n nodes is distributed randomly using the Fast Mersenne-Twister method, in a two-dimensional network area of 100m x 100m. The total node population sampled include:

   (a) 100;
   (b) 250;
   (c) 500;
   (d) 1000;
   (e) 2000;
   (f) 3000;
   (g) 4000;
   (h) 5000 nodes.

2. An event trajectory is executed from a point in the network area; of which the test path can take the following courses:

   (a) **Linear Path**
   A linear path consists of an event trajectory where the entry and exit point from the network area consists of a constant gradient level.

   (b) **Arc-formation Path**
   An arc-formation path consists of an event trajectory where the entry

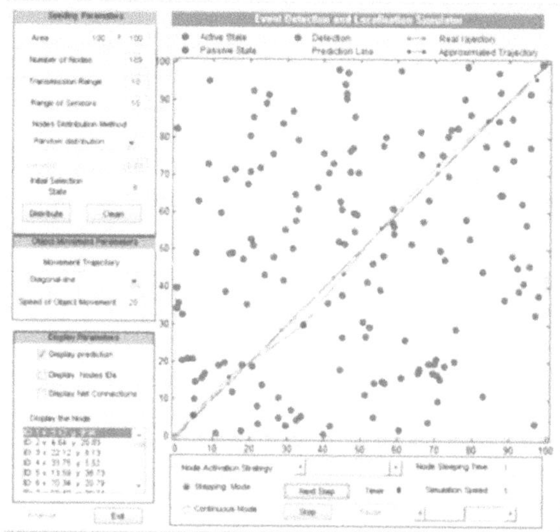

Fig. 2. Event Detection and Localisation Simulator Screenshot

and exit point of the network area will either be increasing or decreasing in the level of gradient, such that it forms a segment of a circle.

(c) **Pseudo-random Path**
A pseudo-random path using the Mersenne-Twister method combines elements of 2(a) and 2(b) at various points throughout the trajectory, until it reaches the exit point of the network area.

3. The algorithm selects the route from the node in range of the approximate trajectory to be established to the sink; such that the closer the algorithm is to calculating the event path, the more optimum the route will be to establish successful communications to the sink. The two algorithms that are assessed:

   (a) **Linear Approximation**
   Linear approximation estimates the approximate trajectory to determine the route path using general Euclidean geometry to calculate the final point of the event, based on the current nodes that are in contact with the event and tracing a path between the previous and current nodes.

   (b) **Co-operative EKMs**
   Co-operative EKMs use an indirect-mapping SOM map to train the control parameters in which to converge at the final trajectory point, in such a fashion to actively train the neural network to seek positive outcomes in reaching a final route from the trajectory's path to the sink.

4. The experiment is executed for 1000 iterations to calculate the mean rate of successful identification of the trajectory's target point.

(a) A successful identification is where the final point is within a 98% confidence interval of the entire network area.
(b) Therefore, a maximum margin of error is determined to be an area of 2m x 2m of where the final approximate point is calculated.

4 Results

The results shown in Figure 3 on this page demonstrate that in comparison to linear approximation and co-operative EKMs, the greatest promise in the final results can be achieved when pseudo-random trajectory tracking is required. While linear approximation of pseudo-random trajectories are expected to perform inadequately due to the inflexibility of the algorithm to accept large degrees in variation or change in the final result, co-operative EKMs demonstrate a noticeable improvement in the identification rate over linear approximation.

The analysis of co-operative EKMs in Table 1 on the next page, when assessed in terms of performance of pseudo-random tracking, requires more analysis into the algorithmic process. In particular, the thresholds established for determining positive or negative learning reinforcement is an issue that needs to be evaluated for an in-depth assessment. The tolerance levels used to calculate the thresholds is a delicate concern, as subtle variations in tolerance may yield undesirable results. As a case in point, reducing tolerance levels too far will result in the inflexibility of the algorithm to adapt to changes the event trajectory; the corollary is that generous tolerance levels will yield undesirable tracking results when noise or faulty nodes produce invalid sensory data.

As a consequence, the potential of Co-operative EKMs to identify events within a wireless sensor network show great promise; but as with all passive learning heuristic methods, a heuristic ensemble approach is required to train

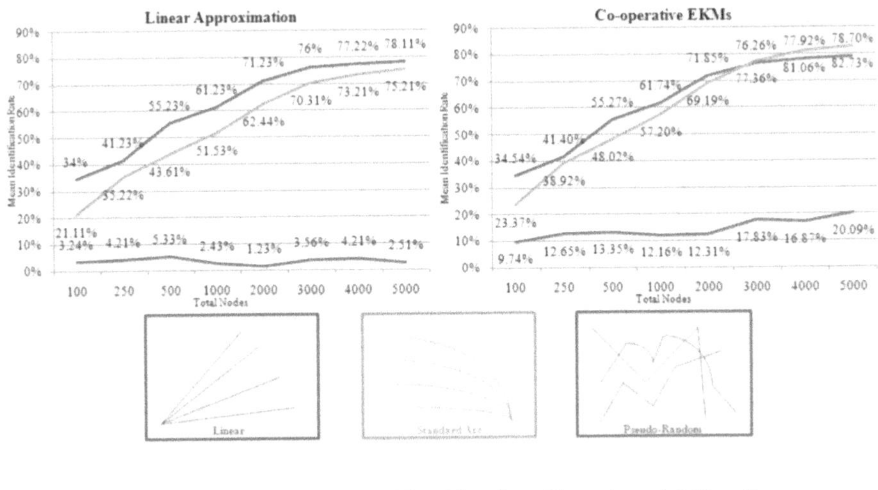

Fig. 3. Accuracy of Target Identification: Experimental Results

Table 1. Comparative Assessment and Evaluation

	Linear Approximation	Co-operative EKMs
Quantitative Assessment	Linear approximation is limited with small node numbers as the sparse distribution is a poor fit to the estimated euclidean geometry path	Demonstrate an indirect-mapping EKM can provide detection to optimise for local (obstruction) and global (target seeking) concerns
Qualitative Assessment	Linear approximation cannot compensate for path obstruction or unpredictable movement without further algorithmic improvement	Preliminary results show a smoother and finer tracking mechanism to monitor events in real-time, compensating for random events [8, 10]

the algorithm to evaluate and determine the tolerance thresholds that are most suitable for the current conditions. The implementation of co-operative EKMs with alternative heuristic algorithms such as genetic programming will be considered to evaluate improvement in the mean identification rate.

5 Conclusion

An innovative method of adaptive wireless sensor network governance responsibilities with co-operative EKMs has been established through evaluation; the preliminary results demonstrate indirect-mapping EKM generates more proficient wireless network governance decisions than other local learning methods like direct-mapping EKM. With recursive least squares, the control parameters of the indirect-mapping EKM can be trained to allow rapid convergence and improved optimisation when compared to the gradient descent.

The experimental results show a positive determination of the positive capability of co-operative EKMs in wireless sensor network routing. The notable variability in the identification rate is a result in the need to improve the quality of training mechanisms to reinforce positive selection processes, so the aggregation of the final routing selection is optimal for the given scenario. In addition, the current data-sets in future experimental assessments will be based on physical test-bed environments, in order to create a real-world scenario for quantitative evaluation of target identification and routing.

While linear and arc trajectories can be easily predicted using co-operative EKMs, further experimental study is required to improve the routing identification results of pseudo-random tracking using the technique of unsupervised learning. There are limitations to the degree of success that can be achieved with passive learning methods, before active supervision is necessary to train the system to seek patterns in target behaviour and act accordingly using weighting functions.

References

[1] Chaczko, Z., Klempous, R., Nikodem, J., Nikodem, M.: Methods of Sensors Localization in Wireless Sensor Networks. In: Proceedings of the 14th Annual IEEE International Conference and Workshops on the Engineering of Computer-Based Systems, March 2007, pp. 145–152 (2007)
[2] Ghosh, J., Nag, A., Howlett, R.J., Jain, L.C.: An Overview of Radial Basis Function Networks: Radial Basis Function Networks: New Advances in Design, pp. 1–36. Physica-Verlag, New York (2001)
[3] Halkidi, M., Batistakis, Y., Vazirgiannis, M.: Clustering Validation Techniques. Intelligent Information Systems Journal 17(2-3), 107–145 (2001)
[4] Kaelbling, L.P., Littman, M.L., Moore, A.W.: Reinforcement Learning: A Survey. Journal of Artificial Intelligence Research, 237–285 (1996)
[5] Kohonen, T.: Self-Organizing Maps, 3rd edn. Springer, New York (2000)
[6] Martinetz, T.M., Ritter, H.J., Schulten, K.J.: Three-dimensional Neural Net for Learning Visuomotor Co-ordination. IEEE Transactions on. Neural Networks, 131–136 (1990)
[7] Millán, J.D.R., Posenato, D., Dedieu, E.: Continuous-Action Q-Learning. Machine Learning, 249–265 (2002)
[8] Low, K.H., Leow, W.K., Ang Jr., M.H.: Task Allocation via Self-organizing Swarm Coalitions in Distributed Mobile Sensor Network. In: 19th National Conference on Artificial Intelligence, pp. 28–33 (2004)
[9] Low, K.H., Leow, W.K., Ang Jr., M.H.: An Ensemble of Cooperative Extended Kohonen Maps for Complex Motion Tasks. Neural Computation 17(6), 1411–1445 (2005)
[10] Pomerleau, D.A.: Efficient Training of Artificial Neural Networks for Autonomous Navigation. Neural Computing, 88–97 (1991)
[11] Ritter, H., Schulten, K., Denker, J.S.: Topology Conserving Mappings for Learning Motor Tasks: Neural Networks for Computing. In: American Institute of Physics: 151st Conference Publication Proceedings, Snowbird, Utah, pp. 376–380 (1986)
[12] Sharkey, A.J.C., Sharkey, N.E.: Combining Diverse Neural Nets. Knowledge Engineering Review, 231–247 (1997)

Morphotronic System Applications

Zenon Chaczko[1] and Germano Resconi[2]

[1] Faculty of Engineering & IT, University of Technology, Sydney, Australia
[2] Dept. of Mathematics and Physics, Catholic University Brescia, I-25121, Italy
zenon.chaczko@uts.edu.au, resconi@numerica.it

Abstract. This paper discusses a newly proposed Morphotronic System paradigm that can be used as a general computation model for construction of software. The Morphotronic System allows for a definition of very flexible software prototypes, in which a processing path can be interpreted as an extremum principle. This approach offers a significant improvement to traditional software practices. The system purpose is stated as a conceptual input to the computer system at a starting point of the computation process while the local machine state is completely ignored. The system context and its rules are generated as resources to allocate the computational components. The morphotronic system applies non-Euclidean geometry which allows to shape the context and to define the projection operators for an ideal network of forms.

Keywords: Software, Turing Machine, Morphotronic System.

1 Introduction

Software contains a set of instructions which are implemented in a specific context for a specific purpose. Traditionally designed software use a standard programming language which contains syntactic rules which aggregate a set of instructions. The Turing Machine (TM) is used as a conceptual model and the system architecture is based on the same concepts to realise the purpose. However, observations in nature show that the purpose is obtained without taking into account of the Turing Machine. In the proposed approach, we use biomimetics for the construction of software prototypes as a general computation model. This model can be used to decide how to take actions in a given moment and in one specific location of the memory. Therefore, in one table we have to decide on all possible actions, as well as the trajectory of actions resulting in the implementation of a required task. Considering that no external assistance exists in defining a purpose, nor a predefined plan for a desired path of actions to generate the purpose, only conceptual work can give a definition of the path. Each element of the path is independent from other elements, and the connection is made at the conceptual level with user intervention. The TM offers the connection between any element of the path and the actions only, thus complex actions have to be defined by a path of elementary actions. The semantic part of the purpose is forgotten and only remains in the mind of the developer.

R. Moreno-Díaz et al. (Eds.): EUROCAST 2009, LNCS 5717, pp. 905–912, 2009.

Recent work in the domains of autopoietic systems, biomimetic middleware [2], constructal theory [1], immuno-computing [4], holographic computing, morphic computing [5], quantum computers[6]. Neural Networks and similar others suggest a change is required to the traditional approaches based on the original Turing Machine model [3]. The first point is the new computation model which offers a total change of perspective in development. By beginning to state the purpose as the starting point for defining the computation process, the purpose becomes the conceptual input to a computer or a machine. Secondly, one should ignore the local machine state and generate the context and its rules as resources to locate or allocate the computational component(s). In proposing the new computational model, we extend the definition of *self* to a set of entities that are strongly inter-connected or correlated. Hence, the first definition of the global *self* entity is the context. The context is not created by just connecting individual entities with individual *self*. *Self* is associated with all elements of the context at once. The same stands for the definition of *purpose*, where *purpose* is defined by a task or a goal made by a set of conceptual entities that cannot be separated one from another, and that are associated with only one *self*. In constructal theory, allometric rules describe the *self* and *self-shaping* as a description of the purpose [2]. Constraint is a *purpose* in dissipative thermodynamics, where the variational and extremum principle realises the context of purpose or constraint.

2 Constructal Theory, Dissipative Phenomena and the Projection Operator

The search for extremisation (or variational) principles in physical systems is often quite fruitful. These can be applied to find the state of the system to describe fluctuations to find dynamic laws, discovering solutions of these equations of motion, solving constraints on the direction of processes and evolutions and so forth. In mechanics, one has the Lagrangian and Hamiltonian formulae with the principle of least action to find the equations of motion. In equilibrium thermodynamics, the principle of maximum entropy or minimum free energy is used to solve the equilibrium state; while in near-equilibrium, thermodynamics is used for the entropy production which it is minimized in order to find the stationary state. The situation of thermodynamic equilibrium, and with nonlinear dynamics is more difficult. A general variational principle is not known to exist, but can one at least define and describe regions where some principles do apply. The entropy production is used frequently to study physical systems; from simple electrical networks to complex chemical reaction systems, fluid systems, ecological and climate systems. Entropy production is related to the construct theory as postulated by A. Bejan [1]. This means that force E generates the flux J the distribution of which minimises the dissipation of energy in the system, and maximises the entropy production, otherwise we have the least imperfection shape of the flux in the system. It is assumed that forces and fluxes are conjugate variables governed by the set of force E and sources of flux J which form

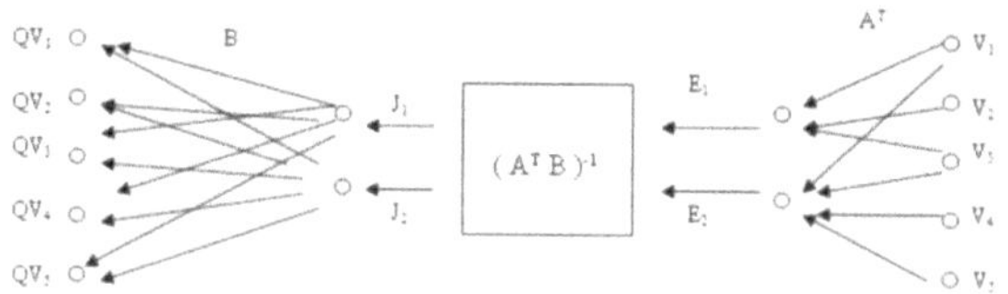

Fig. 1. The projection operator

a relation in a linear case such as: $E = ZJ$. Force E can be computed from the forces V according to $E = A^TV$ equation; where the E entities are the External Sources of the system and E_i collects external information and introduces the external information in a compressed form. The relation $E = ZJ$ is called the convergent *WRITE* process (right side in Fig. 1). The external information enters the system and generates sources of internal fluxes of information J. The fluxes by the internal network B reproduce the best model of the external values V or QV thus from the sources of the flux, the sum of all fluxes in the system can be calculated as:$I = QV = BJ$, where the propagator matrix B represents the internal network in which fluxes are propagated in all the internal parts of the system. The action of operator B is a divergent *READ* (autopoietic) process from the sources J to I elements (left side in Fig. 1). In the morphotronic system the projection operator can use the MIMO representation (see Fig. 1).

3 Biomimetic Model of Morphotronic System

Let's consider a cell's membrane with embedded macromolecules C that are able to absorb and diffuse molecules from the environment. For the molecules of A and B type, there are two possible independent fluxes, flowing from the inside and outside of the cell. The flux J_1 belongs to molecule A, and the fluxJ_2 to both A and B molecules. Only some of molecules A join with B, thus we have: $J_3 = J_1 - J_2 = \emptyset$. The flux of A generates E_1 by the macromolecule M that actively transports A inwards and outwards of the cell. The M macromolecule's generator E_2 transports the A and B molecules from/into the environment. Since fluxes J1 and J2

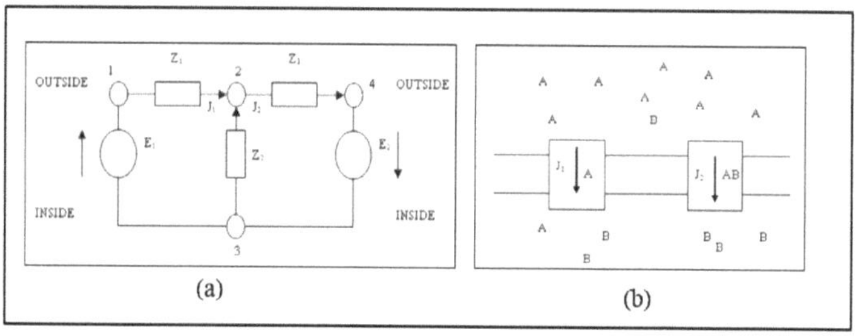

Fig. 2. (a) Electrical circuit stimulating levels of molecule concentrations; (b) of the same electrical circuit represented as the biomimetic model of morphotronic system

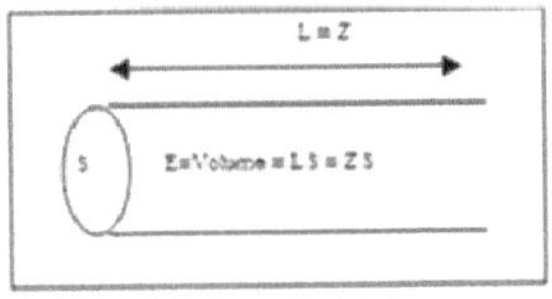

Fig. 3. A vessel with cross sectional area S, volume E, length L

are bounded by flux J3 thus the biological process can be represented as the circuit (Fig. 2a). The flux can be defined as:$J_{i,j} = \frac{p_j - p_i}{R_{i,j}}$, where p_j and p_i are the density of the molecules at different points of the cell. In the circuit (Fig.2), for the generator E_1 the molecule A concentration outside the cell is greater then the concentration inside the cell. Since $E_2 = -E_1 1$, the concentration of B inside is greater than outside and since $J_1 > J_2$, the concentration of A outside is greater than the concentration of B inside (Fig. 2b).

3.1 Biological Allometry

Murray's law that is observed in certain biological structures (blood arteries) can be eplained using the Morphotronic system theory. In $E = LS = ZS$ equation, E is a vessel's volume that represent the Force), S is the vessel's cross-surface with direction that represents Flux and the vessel's length L is the impedance or the cross matrix Z. From forces$V$ or volume of an elementary vessel, sources E of the volumes can be calculated as: $E = A^T V$.

4 Binding in the Morphotronic System

Let us consider two independent or non-coupled systems (electric circuits) seen as two separate meshes or loops (Fig. 4).

There are tow meshes $C_1 = (1, 2, 3, 1)$ and $C_2 = (4, 2, 3, 4)$ and there are two currents, one in each mesh. the samples for two currents are:

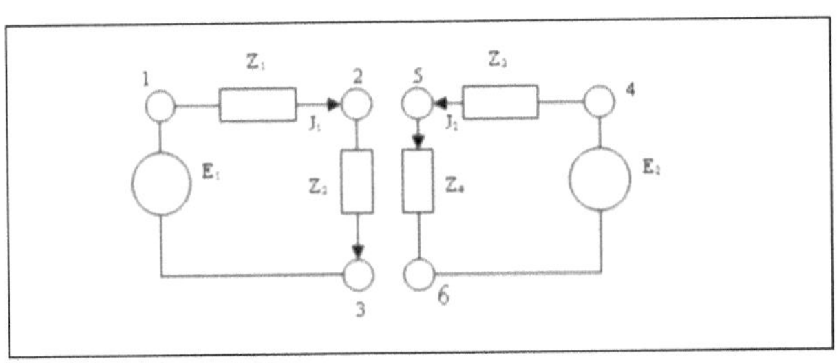

Fig. 4. Independent, non-coupled electric circuits

$$A_{c_1} = \begin{bmatrix} & C_1 \\ (1,2) & J_1 \\ (2,3) & J_1 \\ (4,2) & 0 \end{bmatrix}, \ A_{C_2} = \begin{bmatrix} & C_1 \\ (1,2) & 0 \\ (2,3) & J_2 \\ (4,2) & J_2 \end{bmatrix} = \begin{bmatrix} 1 & 0 \\ 1 & 0 \\ 0 & 1 \\ 0 & 1 \end{bmatrix},$$

For the two samples $J_1 = J_2 = 1$, hence from two meshes fluxes can be collected into one matrix such as:

$$A = \begin{bmatrix} & C_1 & C_1 \\ (1,2) & 1 & 0 \\ (2,3) & 1 & 1 \\ (4,2) & 0 & 1 \end{bmatrix} \text{or simply } A = \begin{bmatrix} 1 & 0 \\ 1 & 1 \\ 0 & 1 \end{bmatrix},$$

Impedances can be calculated as: $z = \begin{bmatrix} Z_1 & 0 & 0 \\ 0 & Z_2 & 0 \\ 0 & 0 & Z_3 \end{bmatrix}$ and the sources cross matrix is given as:

$$Z_c = A^T Z A = \begin{bmatrix} 1 & 0 \\ 1 & 1 \\ 0 & 1 \end{bmatrix}^T \times \begin{bmatrix} Z_1 & 0 & 0 \\ 0 & Z_2 & 0 \\ 0 & 0 & Z_3 \end{bmatrix} \begin{bmatrix} 1 & 0 \\ 1 & 1 \\ 0 & 1 \end{bmatrix} = \begin{bmatrix} Z_1 + Z_2 & Z_2 \\ Z_2 & Z_2 + Z_3 \end{bmatrix}$$

the sources of force or the voltages E_1 and E_2 are defined as: $\begin{bmatrix} E_1 \\ E_2 \end{bmatrix} \begin{bmatrix} 1 & 0 \\ 1 & 1 \\ 0 & 1 \end{bmatrix}^T \times \begin{bmatrix} V_{12} & 0 \\ V_{23} & 0 \\ V_{45} & 0 \end{bmatrix} = \begin{bmatrix} V_1 + V_2 \\ V_2 + V_3 \end{bmatrix}$

and the source of currents are: $\begin{bmatrix} J_1 \\ J_2 \end{bmatrix} = (A^T Z A)^{-1} E = \begin{bmatrix} \frac{E_1(Z_1+Z_3) - E_2 Z_2}{Z_1 Z_2 + Z_1 Z_3 + Z_2 Z_3} \\ \frac{E_2(Z_1+Z_2) - E_1 Z_2}{Z_1 Z_2 + Z_1 Z_3 + Z_2 Z_3} \end{bmatrix}$,

since J_1 and J_2 are bound this they can be expressed as: $\begin{bmatrix} J_1 \\ J_2 \end{bmatrix} = \begin{bmatrix} \frac{E_1 - J_2 Z_2}{Z_1 + Z_2} \\ J_2 \end{bmatrix}$,

where the invariant L or power can be defined as:

$L = J^T Z J = \begin{bmatrix} J_1 \\ J_2 \end{bmatrix}^T \begin{bmatrix} Z_1 + Z_2 & Z_2 \\ Z_2 & Z_2 + Z_3 \end{bmatrix} \begin{bmatrix} J_1 \\ J_2 \end{bmatrix} = (Z_1 + Z_2) J^2 + (Z_2 + Z_3)^2 + 2 Z_2 J_1 J_2$

$$QV = ZA(A^T Z A)^{-1} E$$

$$QV = \begin{bmatrix} QV_1 \\ QV_2 \\ QV_3 \end{bmatrix}^T \begin{bmatrix} [(Z_1 + Z_3) Z_1] E_1 - [Z_1 Z_2] E_2 \\ [(Z_2 Z_3)] E_1 + [Z_1 Z_2] E_2 \\ [(Z_1 + Z_2) Z_3] E_2 - [Z_1 Z_3] E_1 \end{bmatrix} \frac{1}{Z_1 Z_2 + Z_1 Z_3 + Z_2 Z_3}$$

5 Morphotronic Sensornets

A sensornet is a system made of many sensors distributed at different locations in the 3D space. Inside a predefined cluster of sensors there is an infinite number of coordinate points at which information can be received from any sensor that is located in the cluster. Among the infinite number of points there is a special point C called the cluster barycentre, where the weighted sum of transmitted energy Σ from all the sensors reaches the minimum value. The C and Σ are context dependent, where the context is represented by the sensor's position and weights. The energy Σ is not additive, as the sum of all energies $\Sigma_1, \Sigma_2, \ldots, \Sigma_n$ that come from all parts of the cluster is less or equal to the total energy Σ. It is convenient

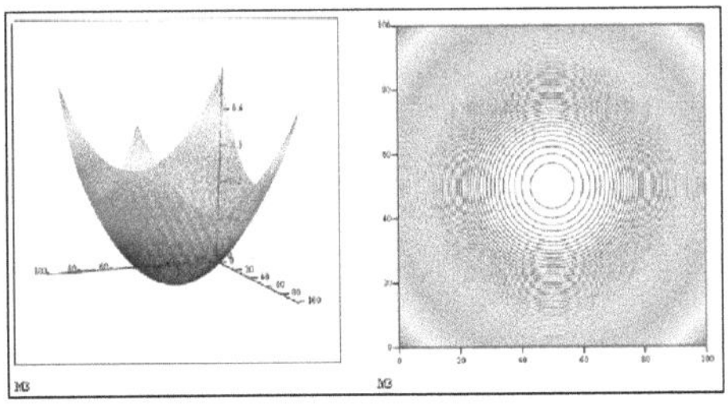

Fig. 5. The surface and the silhouette plots of the $F(x, y, z)$ with the min value at the point $P(x_0, y_0, z_0)$

to consider parts of the cluster in separation from the the observed cluster. The sensors that transmit information generate around itself the field F_s; while the field F is the weighted superposition of all the fields $F_1, F_2, ..., F_n$ generated by the individual sensors in the cluster. Applying the projection operator for the field X, the sensor weights can be calculated in order to construct the best approximation of X using the superposition function. Sensors positioned further from the C centre have small weights and higher energy levels while sensors with larger weights that are closer to the C centre have lower energy. For the sensor S located at $P(x_0, y_0, z_0)$ the field$F$ represents the energy loss by the sensor. The Euclidean distance D is given by:

$$D^2 = (x–x0)^2 + (y–y0)^2 + (z–z0)^2 = F, \tag{1}$$

and can be represented by the parabolic form: $(x–x0)^2 + (y–y0)^2 + (z–z0)^2 - F = 0$, where the minimum (0) point is located in the point (x_0, y_0, z_0). The field F is graphically depicted in the two dimensional space (see 5). The superposition of two or more parabolic functions at different point locations is also a parabolic function thus the minimum value is the average of the points defined by the parabolic function where F_{cl} is denoted as:

$$F_{cl} = w_1 D_1^2(x, y, z) + w_2 D_2^2(x, y, z) + ... + w_n D_n^2(x, y, z) \tag{2}$$

where F_{cl} is given as:

$$F_{cl} = w_1 \left[(x - x_1)^2 + (y - y_1)^2 + (z - z_1)^2\right] + w_2 \left[(x - x_2)^2 + (y - y_2)^2 + (z - z_2)^2\right] + \ldots + w_n \left[(x - x_n)^2 + (y - y_n)^2 + (z - z_n)^2\right] =$$

$$\sum w_k \left[(x - \frac{\sum_{k=1}^{n} w_k x_k}{\sum_k w_k})^2 + (y - \frac{\sum_{k=1}^{n} w_k y_k}{\sum_k w_k})^2 + (z - \frac{\sum_{k=1}^{n} w_k z_k}{\sum_k w_k})^2 \right] +$$

$$\sum_{k=1}^{n} w_k x_k^2 - \frac{(\sum_{k=1}^{n} w_k x_k)^2}{\sum_k w_k} + \sum_{k=1}^{n} w_k y_k^2 - \frac{(\sum_{k=1}^{n} w_k y_k)^2}{\sum_k w_k} + \sum_{k=1}^{n} w_k z_k^2 - \frac{(\sum_{k=1}^{n} w_k z_k)^2}{\sum_k w_k} =$$

$$\sum w_k \left[(x - \frac{\sum_{k=1}^{n} w_k x_k}{\sum_k w_k})^2 + (y - \frac{\sum_{k=1}^{n} w_k y_k}{\sum_k w_k})^2 + (z - \frac{\sum_{k=1}^{n} w_k z_k}{\sum_k w_k})^2 \right] + \Sigma \,,$$

thus the cluster's min Σ is located at:

$$x_B = \frac{\sum_{k=1}^{n} w_k x_k}{\sum_k w_k} =< x >, \; y_B = \frac{\sum_{k=1}^{n} w_k y_k}{\sum_k w_k} =< y >, \; z = \frac{\sum_{k=1}^{n} w_k z_k}{\sum_k w_k} =< z >, \tag{3}$$

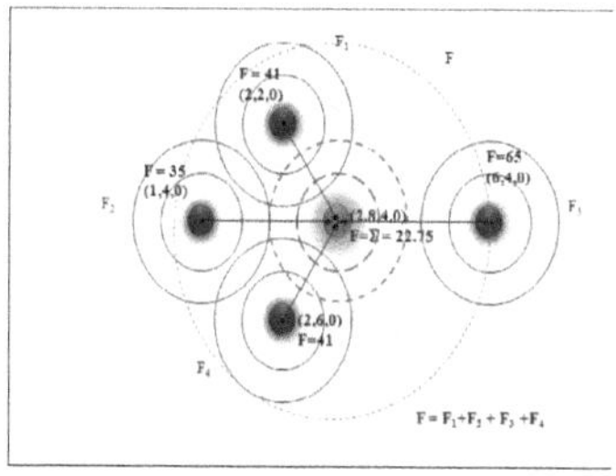

Fig. 6. Superposition of the field F the minimum value $\Sigma = 22.75$ is in the barycentre point $C = (2.8,\ 4,\ 0)$

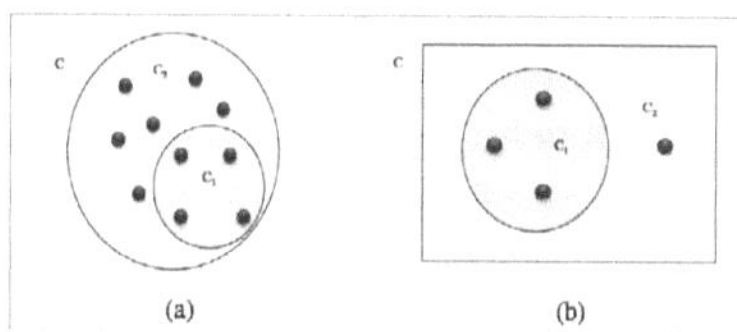

Fig. 7. Bifurcation of sub-clusters $C_1$and $C_2$where $C_1 \cap C_2 = \emptyset$ *and* $C = C_1 \cup C_2$

where the (x_B, y_B, z_B) is the barycenter of the cluster with w_j weights (masses). For a given set of four sensors at the positions (2,2,0), (1,4,0), (6,4,0) and (2,6,0) and with the weights (Fig. 6) the 5 formula can be instantly applied to compute the values of Σ and C. With a simple computation Σ can be obtained as:

$$\sum = w_1(x_1^2 + y_1^2 + z_1^2) + w_2(x_2^2 + y_2^2 + z_2^2) + ... + w_n(x_n^2 + y_n^2 + z_n^2) - \frac{1}{\sum\limits_{k=1}^{n} w_k}\left[(\sum\limits_{k=1}^{n} x_k)^2 + (\sum\limits_{k=1}^{n} y_k)^2 + (\sum\limits_{k=1}^{n} z_k)^2\right] \quad (4)$$

5.1 Cluster Decomposition and Fusion

Decomposition (bifurcation) of a sensor cluster into sub-clusters can be expressed algebraically as:

$$\sum_1 = w_1(x_1^2 + y_1^2 + z_1^2) + w_2(x_2^2 + y_2^2 + z_2^2) + ... + w_{n1}(x_{n1}^2 + y_{n1}^2 + z_{n1}^2) - \frac{1}{\sum\limits_{k=1}^{n1} w_k}\left[(\sum\limits_{k=1}^{n1} x_k)^2 + (\sum\limits_{k=1}^{n1} y_k)^2 + (\sum\limits_{k=1}^{n1} z_k)^2\right]$$

$$\sum_2 = w_1(x_1^2 + y_1^2 + z_1^2) + w_2(x_2^2 + y_2^2 + z_2^2) + ... + w_{n2}(x_{n2}^2 + y_{n2}^2 + z_{n2}^2) - \frac{1}{\sum\limits_{k=1}^{n2} w_k}\left[(\sum\limits_{k=1}^{n2} x_k)^2 + (\sum\limits_{k=1}^{n2} y_k)^2 + (\sum\limits_{k=1}^{n2} z_k)^2\right]$$

where $n = n_1 + n_2$ thus

$$\Sigma_1 + \Sigma_2 = w_1(x_1^2 + y_1^2 + z_1^2) + w_2(x_2^2 + y_2^2 + z_2^2) + ... + w_n(x_n^2 + y_n^2 + z_n^2) - \left(\frac{1}{\sum\limits_{k=1}^{n1} w_k}\left[(\sum\limits_{k=1}^{n1} x_k)^2 + (\sum\limits_{k=1}^{n1} y_k)^2 + (\sum\limits_{k=1}^{n1} z_k)^2\right] + \frac{1}{\sum\limits_{k=1}^{n2} w_k}\left[(\sum\limits_{k=1}^{n2} x_k)^2 + (\sum\limits_{k=1}^{n2} y_k)^2 + (\sum\limits_{k=1}^{n2} z_k)^2\right]\right)$$

The difference between the original cluster C and the sum of C_1 and C_2 sub-clusters can be expressed as:

$$\Delta - \sum(1,2,\ldots,n) - (\sum_1(1,2,\ldots,n_1) + \sum_2(n_1+1, n_2+2,\ldots,n) =$$
$$\left(\frac{1}{\sum_{k=1}^{n1} w_k}\left[(\sum_{k=1}^{n1} x_k)^2 + (\sum_{k=1}^{n1} y_k)^2 + (\sum_{k=1}^{n1} z_k)^2\right] + \frac{1}{\sum_{k=1}^{n2} w_k}\left[(\sum_{k=1}^{n2} x_k)^2 + (\sum_{k=1}^{n2} y_k)^2 + (\sum_{k=1}^{n2} z_k)^2\right]\right) -$$
$$\frac{1}{\sum_{k=1}^{n} w_k}\left[(\sum_{k=1}^{n} x_k)^2 + (\sum_{k=1}^{n} y_k)^2 + (\sum_{k=1}^{n} z_k)^2\right] \tag{5}$$

The separation of the cluster of n sensors results in a positive gain of energy Δ where$\Delta \geq 0$ while a fusion of sub-clusters has the opposite effect and a negative energy ($\Delta \leq 0$) is gained.

6 Conclusion

The world of biology suffers from inadequacy of mathematical instrumentation and a shortage of useful computational models helping to describe biological phenomena. Construction of man-made systems would greatly benefit if better models and techniques to describe biology are found. By no means perfect, the theory of Morphotronic Systems aims to advance from the point where traditional computation models such as: von Neuman's architecture, Turing Machine or Cellular Automata can no longer be of help when dealing with the challenges of biology. The underlying factor that somewhat limits these classical approaches is the fact that they miss to place the Purpose as a central aspect of these models. Morphotronics perceives a complex system as a network of forms for which the Extreme principle can be applied to compute the length of the bounding links to obtain the minimum path. In such a network, the morphotronic mechanism substitutes the local description of forms with the global description. The global rule (allometric) can be defined as minimum velocity of entropy production. An ideal communication is the network of forms embedded in a context where global rules are present as a minimum condition or an invariant. The morphotronic theory offers a type of computation for searching an optimal solution to communication that is coherent with the ideal model inside a context and described by its rules. The theory of Morphotronic Systems applies the Non-Euclidean geometry to model the shape of the context and to define the projection operators for an ideal network of forms.

References

1. Bejan, A.: Shape and Structure, From Engineering to Nature. University Press, Cambridge (2000)
2. Chaczko, Z.: Biomimetic Middleware for WSN. In: EMSS 2008, Italy, September 17-19 (2008)
3. Newman, M.E.J.: The Structure & Function of Complex Networks. Santa Fe Inst. Pub. (2004)
4. Tarakanov, et al.: Immunocomputing: Principles and Applications. Springer, Heidelberg (2003)
5. Resconi, G., Nikravesh, M.: Morphic Computing. Applied Soft Computing Journal (July 2007)
6. Resconi, G., Nikravesh, M.: Morphic Computing: Quantum and Field. In: Nikravesh, M., Zadeh, L.A., Kacprzyk, J. (eds.) Forging the New Frontiers: Fuzzy Pioneers II, July 2007. Studies in Fuzziness and Soft Computing. Springer, Heidelberg (2007)

SNIPER: A Wireless Sensor Network Simulator

Sourendra Sinha, Zenon Chaczko, and Ryszard Klempous

Faculty of Engineering, University of Technology, Sydney
Wroclaw University of Technology, Wroclaw
souren.sinha@gmail.com, zenon@eng.uts.edu.au, ryszard.klempous@pwr.wroc.pl

Abstract. Wireless Sensor Networks (WSN) is fast becoming the holy grail of digital surveillance, data monitoring and analysis. Being relatively cheap, low-powered and easy to deploy WSNs is being adopted in a range of different fields. Currently, the focus is towards optimizing techniques used to form sensor clusters and to route data within a network. In order to satisfy these goals a significant amount of research is being done globally and what tends to be lacking at times is the right tools to make such research work less time consuming and inexpensive. The use of simulation tools is one such adaptation that can help researchers closely analyse a particular aspect of WSN while sticking with a known environment that is applicable to other scenarios in a similar way. This paper presents the performance and features of WSNSim, a WSN Simulator and the immediate advantages that can be experienced by researchers working on various realms in this area.

1 Introduction

In this digital age WSNs are rapidly finding their place in a wide range of applications, for e.g. patient care, precision agriculture, building monitoring, and many more. However, being inherently limited in their resource availability a major requirement still is to find new ways to optimize data dissemination in a reliable and timely manner from the nodes.

In order to effectively design and develop new protocols and applications targeted towards WSNs, there is a strong need to use an equally effective and reliable simulation environment. Sensor nodes are, by design, tiny and inexpensive, but due to their need to work in large numbers make full-scale testing a fairly expensive process to be carried out using real hardware. Analytical modelling using tools such as MATLAB help in acquiring a quick insight, but they fail to provide anything close to realistic results that can be acquired through proper implementations and good simulation tools.

The proposed WSN Simulator employs a framework that benefits several potential areas of research in this field. The primary contributions from the simulation framework can be summed up as follows:

1. Enables researchers to easily target particular sectors of research work, e.g. sensor node distribution, routing, and clustering.

R. Moreno-Díaz et al. (Eds.): EUROCAST 2009, LNCS 5717, pp. 913–920, 2009.

2. Imitate the behaviour of real-world sensors such that the internal modules can be studied in detail, for e.g., the impact of intense usage on the radio, battery and CPU.
3. Load testing through repetitive runs of a particular scenario to analyse node behaviour and power consumption. The simulator also contains integrated analysis tools to assess various conditions and their impact on the network as a whole.
4. The mode of communication between the nodes is based on a virtual wireless network where each node communicates with its own radio (module). The nodes do not need to be aware of the topology or their exact geographic location, but this depends on the type of algorithm used to form clusters or route data between nodes.

2 Related Work

By far the most popular and widely used simulator currently in use for WSN is NS-2. It has a fairly rich set of IP network focussed protocols for communication and has been written in C++. However, because it was never designed to specifically target WSNs it does not scale very well and has trouble simulating when the node count exceeds over 100. Also, the way in which NS-2 has been designed makes it inherently difficult for a researcher to enrich it with a new protocol or node component.

J-Sim is another general purpose WSN simulator written completely in Java, thus making it platform independent and easier for any researcher familiar with Object-Oriented Programming to be able to alter it and use for their specific set of test cases. However, J-Sim too was initially designed for wired networks and thus, like NS-2, suffers from a few inherent design issues in terms of extensibility. It does integrate 802.11 MAC scheme and models the node's battery, CPU, radio and sensor.

The proposed framework is targeted specifically for WSN, and thus care has been taken from a design perspective to ensure ease of integration of other protocols in the future. It has been developed completely in Java and can be recompiled easily under any IDE that supports Java. Although it has been developed on the Windows Platform, the simulator is expected to run on other Linux distributions as well. The Simulator is almost feature complete in terms of the integration of the key essential elements, and can be used to monitor clustering and routing algorithms of sensor nodes. This paper will first present the primary components of the framework and the overall structure before discussing a WSN cluster formation and routing model called SNIPER, and how the framework accommodates it fro simulation purposes. Finally, the last section will present the conclusions that have been drawn from this development effort and the future aspects of it.

3 WSN Simulation Framework

In order to truly simulate a WSN it is not enough to simply distribute random nodes and study their interaction through direct peer-to-peer like communication. The simulator ought to imitate the real world by not only offering a range of wireless protocol message exchange, but also accommodate factors such as timing, power consumption, environmental factors, etc.

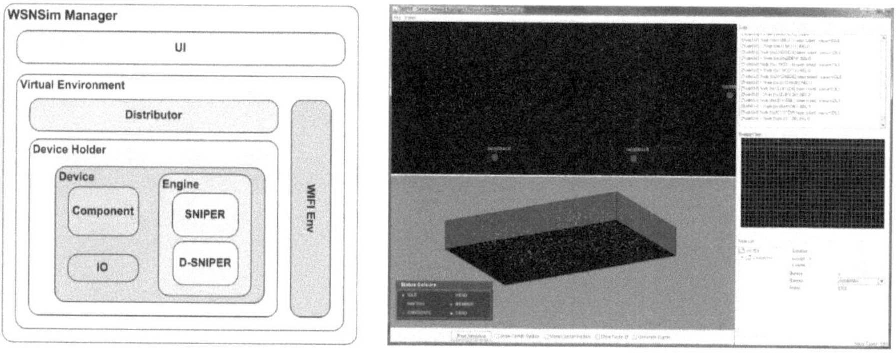

Fig. 1. Simulator Structure and UI

The primary goal of the WSNSim is to analyse the behaviour of nodes in a typical WSN, and to understand the process of cluster formation and data routing. However, in order to truly simulate inter-node communications various factors need to be accounted for, of which some are critical while the others are complimentary. In the real world, each node is designed to monitor the immediate environment and consists of a sensor, battery, radio and a CPU. The WSNSim has been designed so that each of these components is accounted for as best as possible, but tradeoffs have been made to achieve the balance between simulation details and execution performance. An overall structure of the simulator is presented in Fig. 1.

3.1 User Interface

The WSNSim supports a GUI to the user to view and analyse the behaviour of a number of sensor nodes. As shown in Fig. 1 the interface is fairly simple and allows users to specify the maximum number of nodes to be used for testing and analysis. The interface also allows the user to either start a single run of the simulator or sequence a series of test runs to study various aspects, such as Cluster Formation and/or data routing against number of nodes, time taken per test run and estimated power level remaining. Since, the WSNSim attempts to mimic the real-world inter-node communication through the use of a virtual environment the nodes need not be location aware, and instead rely on the transmission and receive power for RF messages. This necessitated the environment to use a scaling factor to separate the nodes by a suitable distance – the default scaling factor used is 25:1.

3.2 Virtual Environment

In any environment where a WSN is setup the sensor nodes tend to simply consist of the very basic components such as a battery, radio and one or more sensors. In order for the nodes to determine their relative location and to communicate with other nodes, they rely on the environment and its associated elements. The proposed framework has adopted a similar construct by allowing the nodes to be deployed in a virtual environment that allows nodes to be placed at any coordinate. In doing so, the environment is aware of the exact coordinate of the node and is able to provide the medium of communication between the nodes by creating a virtual WIFI bubble, as shown in Fig. 2.

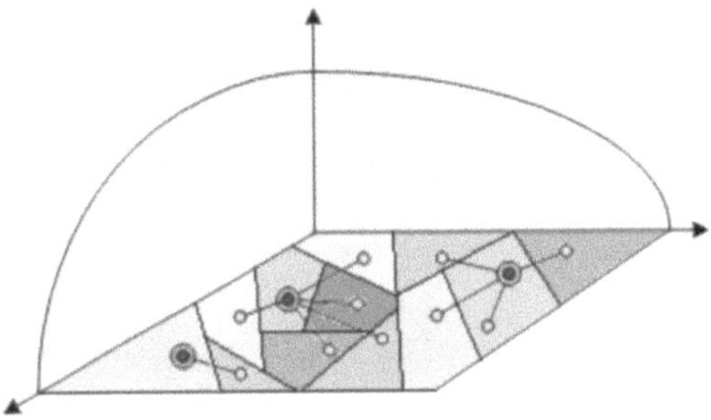

Fig. 2. Virtual WIFI Environment

The WIFI bubble adopts the Voronoi Diagram based methodology, whereby, it assumes that for every point of transmission there is a set *S* of potential recipients $\{p_1, p_2, p_3, \ldots, p_n\}$ that is nearest to it than any other recipient in the environment.

$$V(p_i) = \{x: |p_i - x| \leq |p_j - x|, \forall j \neq i\} \quad (1)$$

Where, V (p_i) is the Voronoi region of the point p_i. When a sensor node transmits a message the virtual WIFI environment is able to determine the point of origin, p_i, of that node and then determine the respective Voronoi Area to calculate the set of potential recipients. The pseudo code of this algorithm is shown below:

```
INPUT: point of transmission – transmitting node
OUTPUT: set of recipient node(s)
Calculate Voronoi();
Do begin
  for every node begin
     calculate Voronoi_Area();
  end
  get CoverageArea();
  if (coverage_area < max_area) then
     begin
        NotifyNode();
        outside_coverage = false;
     end
  else
     outside_coverage = true;
  end
 while (!outside_coverage)
```

Given that for each radio transmission the above algorithm is to be applied implies that the complexity of this for the overall WSN is O (n log n).

The location attribute of each of the nodes are placed in the virtual environment identifies the respective x, y and z values on a 3-D plane. This allows the WIFI bubble to calculate the estimated receive signal at individual nodes for any given transmission packet from a particular source. The receive signal strength is determined in dBm using the equation shown below:

$$E_{Rx} = E_{Tx} * \left(\frac{\lambda}{4 * \pi * d^2}\right) \tag{2}$$

Where E_{Rx} is received signal power, E_{Tx} is transmitted signal power, λ is wavelength of light, and d is the distance of node from source. As the distance increases between the source node and a potential recipient node, the receive signal strength also begins to fall. While the WIFI bubble continues to relay the data for fairly low signal strength, once the threshold preset for each node is crossed, the nodes from that sector and beyond will continue to ignore the message should it arrive. Also, the framework estimates the probable energy consumption in transmitting a packet by radio to be proportional to $1/d^2$ for short distances, and $1/d^4$ for longer distances, as per the following equation 3:

$$E_{Tx} = \begin{cases} lE_{elec} + l\varepsilon_{fs}d^2 & d < d_0 \\ lE_{elec} + l\varepsilon_{amp}d^4 & d \geq d_0 \end{cases} \tag{3}$$

Where, E_{elec} = energy of electronic signals, l = size of a message, d = distance between transmitter and receiver, ε_{amp} = amplification factor, d_0 = limit distance over which d is affected, M = size of the area, D_{toBS} = distance to Base Station.

As well as supporting the underlying communication mechanism of the sensor nodes and being responsible for maintaining their actual locations on the plane, the virtual environment also supports a collection of distributors. Each distributor is responsible for generating a set of co-ordinates that can be used to randomly distribute a number of sensor nodes. The supported types of distributors include:

- Grid Distributor – the whole 2-D plane is broken up into N x M grid and each co-ordinate of the grid is randomly populated by a node.
- Spiral Distributor – the plane is divided into 4 quadrants and starting at the origin (centre of the plane); nodes are placed at random distances from it in a spiral manner.
- Random Distributor – using a Poisson process of distribution nodes are randomly distributed on the 2-D plane.

The WIFI bubble also allows the use of dedicated channels for communication. Each device in the virtual environment that is equipped with a radio may open one or more channels to transmit data.

3.3 Simulation Engine – SNIPER

The simulation framework can simultaneously support more than one engine, and depending on the engine chosen at run-time the clustering and routing algorithm used for the nodes would change. Currently, the following assumptions have been made in designing the framework:

- All nodes are stationary
- The nodes may be equipped with one or more sensors
- The virtual environment provided by the framework can support more than one device – and this includes sensor node and sink alike

- A sink is considered to be a much more powerful device in comparison to a node in terms of energy and CPU
- Communication between the nodes is done in an extremely ideal environment with no collisions – however, the framework is well capable of supporting packet loss and collisions if necessary

The SNIPER engine does not require the nodes to be location aware and assumes the nodes to be randomly distributed in an environment, with no sink or beacon necessarily in sight. It also assumes that the nodes in the network are fairly simple, low-powered entities equipped with a radio that allows transmission range control, thus allowing control over the level of energy consumption.

SNIPER uses a rating based mechanism in forming cluster among a group of nodes. These ratings primarily include leadership, capability and reliability and are calculated using the following equation:

$$rating = \left|\left(e^{(-K1*time)} - K2\right) * d\right| \quad (4)$$

Where, `K1` is a temporal value dependent on how often a node transmits data, `K2` is a constant, and `d` is a value calculated randomly in the range 1-50. The values of `K1` and `K2` are coefficients used to assign the initial leadership and trust rating of the nodes, while `d` determines a reliability factor, and the three values together determine the node's actual rating. The higher the value of the rating, the better suited a node is to be declared as a Cluster Head. Based on tests conducted the most optimal values for `K1` and `K2` have been selected to be 2.3 and 0.1, respectively for leadership rating, and 1.1 and 0.4 respectively for trust rating.

Once the nodes have grouped themselves into clusters, SNIPER assigns each node to be a Cluster Head, Broker or a basic node. The Cluster Head is responsible for monitoring the cluster and determine routing paths based on energy level heuristics determined based on the following equation:

$$\eta(r,s) = \frac{(I - e_r)^{-1}}{\sum_{n \in R}(I - e_n)^{-1}} \quad (5)$$

Where, `I` is the initial energy, e_r is the current energy level of receiver node r. In order to setup routing paths the SNIPER algorithm attempts to find nodes that are placed in range of more than one Cluster Heads and get those nodes assigned as BROKERs. Once elected the BROKERs are intended to transmit messages between one or more cluster heads, and/or the sink.

Unlike many traditional algorithms, SNIPER is based more on affinity aspects of neighbouring nodes, rather than the proximity and other attributes. This characteristic is employed by Cluster Heads in routing data to other clusters. Each broker is able to advertise its reliability and capability ratings based on its current energy level and location relative to nearby Cluster Heads. However, while the capability rating is directly related to current energy level, the reliability rating comes from the number of successful data transfer it participates in. As a result, over time the reliability rating rises and capability drops, and at the point where the two reach the most optimal point, the Cluster Head tries to make decisions to ensure that the Broker is not over-exercised and thus be available to serve as the next Cluster Head, if necessary.

All data packets that are transmitted using the SNIPER algorithm have a lifetime of 10 hops, and do not need for a sink to be present initially for cluster formation to occur.

4 Evaluation of the Simulation Framework

In this section, the results of a performance study of the simulation framework against the J-Sim simulator framework have been presented. The scenario involves doing several test runs of the SNIPER clustering algorithm and incrementing the number of sensors by 250, from an initial value of 250, in each cycle. The maximum number of test cycles is limited to 5 and the response timeout of each of the nodes is fixed at 5secs – implying that after transmitting a message a node will wait up to 5 secs for the node to send a response back.

The charts presented in Fig. 3 show that the number of cluster formed (shown in (a)), is not directly related to the number of nodes, but rather their relative location to each other and also the inherent attributes. Similarly, in (b) we can see that the load per Cluster Head does depend on the number of sensors – it increments in each cycle. Finally, similar to the simulation tests run under J-Sim [7] the time taken to form clusters has some correlation to both the number of sensors and the proximity to the nodes to each other.

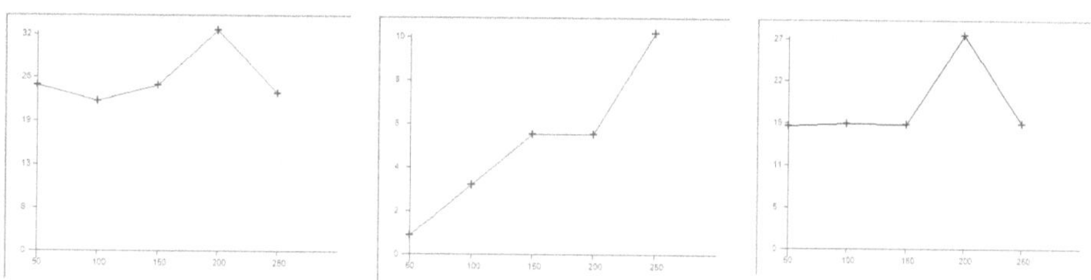

Fig. 3. Plot of sensor against (a) sink count, (b) load/sink and (c) time taken

5 Conclusion and Future Work

In this paper, a simulation framework for Wireless Sensor Networks has been presented. We have focussed on the aspects of the framework that allows users to target specific areas of research and be able to make assumptions about the rest and yet get acceptable results in a sandbox environment. We have found that with the creation of a virtual environment in which the sensor devices are deployed a WIFI bubble is also created, allowing the nodes to communicate with each other using their RF antennas. The paper has also presented one of the algorithms that have been integrated with the simulator to study clustering and routing aspects of WSN nodes.

There are still several aspects of the framework that can be further investigated. For instance, the inter-node communication currently is assumed to occur in an ideal environment and thus issues such as collision, impact of air, obstacles, etc. and have not been fully evaluated. In the future, there is still more work to be done towards a D-SNIPER model, whereby the engine is more dynamic in taking actions, and further reducing energy requirements.

Further analysis with more algorithms integrated in the framework is necessary to fully assess the various aspects like ease of integration, dynamic binding to the selected algorithm, and more data analysis.

References

1. Egea-López, E., Vales-Alonso, J., Martínez-Sala, A., Pavón-Marñio, P., García-Haro, J.: Simulation tools for wireless sensor networks. In: Proceedings of the International Symposium on Performance Evaluation of Computer and Telecommunication Systems (SPECTS 2005), Philadelphia, Pa, USA (July 2005)
2. Wikipedia (May 6, 2009), `http://www.comnets.uni-bremen.de/~mab/cruise/simulation-tool-comparison-matrix.html`
3. Fan, Y., Tao, W., Subir, B.: Toward In-Band Self-Organization in Energy-Efficient MAC Protocols for Sensor Networks (2008)
4. Chen, G., Branch, J., Pflug, M., Zhu, L., Szymanski, B.K.: Sense: a wireless sensor network simulator. In: Advances in Pervasive Computing and Networking, ch. 1, pp. 249–267. Springer, New York (2006)
5. Miller, J.A., Nair, R.S., Zhang, Z., Zhao, H.: JSIM: a Java based simulation and animation environment. In: Proceedings of the 30th Annual Simulation Symposium, Atlanta, Ga, USA, April 1997, pp. 31–42 (1997)
6. Viera, M., Ruiz, B., Loureiro, A.F., Fernandes, A., Nogueira, S.: Scheduling Nodes in Wireless Sensor Networks: A Voronoi Approach
7. Sobeih, A., Chen, W., Hou, J., Kung, L., Li, N., Lim, H., Tyan, H., Zhang, H.: J-Sim: A Simulation and Emulation Environment for Wireless Sensor Networks

Embedded Fortress –Software Environment for Intellectual Property Protection in Embedded Systems

Adam Handzlik[1], Tomasz Englert[1], and Andrzej Jablonski[2]

[1] Microtech International S.A., 20 Wolowska str., 51-116 Wroclaw, Poland
a.handzlik@microtech.com.pl, t.englert@microtech.com.pl
http://www.microtech.com.pl
[2] Faculty of Electronics the Wroclaw University of Technology
27 Wybrzeze Wyspianskiego str., 50-370 Wroclaw, Poland
andrzej.jablonski@pwr.wroc.pl
http://eka.pwr.wroc.pl

Abstract. Embedded Fortress system is the tool designed for electronics engineers and designers, who wish to secure the Intellectual Property (IP) in their embedded systems. The main tasks of Embedded Fortress software include providing designer with a ready-made security.

The solutions in the system will allow the user to select the security that meets his/her expectations from the point of view of security and hardware requirements.

Keywords: embedded systems, Intellectual Property, protection, Field Programmable Gate Arrays, IP Core.

1 Introduction

Embedded Fortress (EF) system is the tool designed for electronics engineers and designers, who wish to secure the Intellectual Property (IP) in their embedded systems. The main tasks of Embedded Fortress software include providing designer with a ready-made security. The solutions in the system will allow the user to select the security that meets his/her expectations from the point of view of security and hardware requirements.

Analyzing the market offer, it is worth noticing that the quickly-growing electronics suppliers market is increasingly seeking to secure their product and Intellectual Property (IP). Securing a product is currently a laborious and time-consuming task [1]. It requires from a designer getting to know many offers of electronics components and deep security methods knowledge. As a consequence, the designer devotes a considerable amount of his/her time on issues which do not impact the functionality of the designed system itself, just its security.

As the above analysis shows, there is no tool that would offer security solutions, adequate to the needs of a designer, available in the market. This statement provided a basis for the development of the Embedded Fortress system, which is mainly meant to provide security solutions fitted to the needs of a design.

R. Moreno-Díaz et al. (Eds.): EUROCAST 2009, LNCS 5717, pp. 921–929, 2009.

The Embedded Fortress system is designed for a wide range of electronic engineers and manufacturers, considering the varying levels and knowledge on the security methods.

2 Project

Additional design guideline, related to the dynamic nature of the electronics market, is the flexible System design which enables frequent updates and changes to the database. An important function is performed by an intuitive user interface, which helps to provide the EF System with the data necessary to specify the solution and display it clearly to the user.
Below is a summary of EF System features:

- Scalable database design,
- Interactive user interface,
- Answer provided in possibly the least number of attempts, maximum of 10 attempts,
- Understandable and simple questions asked to the user,
- Unambiguous expert system process result,
- User response processed by expert system within less than 3 seconds,
- Windows environment operation,
- Database design that allows using web applications.

The basic assumption of Embedded Fortress System architecture is its modularity, which greatly simplifies the implementation, testing as well as later modifications. The target system platform for the Embedded Fortress System is Microsoft Windows operating system, version XP or later. System implementation was performed in Microsoft Visual Studio.Net programming environment, using C, XML technologies. The modules making up the Embedded Fortress System are:

- Embedded Fortress Application the main application,
- Embedded Fortress Expert System,
- Embedded Fortress Solution Browser,
- Embedded Fortress Library solution libraries.

This system structure enabled dividing the process of implementation, testing and the whole system integration into stages. The master module is Embedded Fortress Application, which manages the content displayed on the screen and allows user to communicate with the remaining system modules. An important component of the module is the part responsible for user interaction. Use of various controls and forms allows the dialogue between the module and the user. The module is also responsible for gathering and storing the information on the project on which the user is working.

This is related, among others, to accessing the interface that allows to save and load the project.

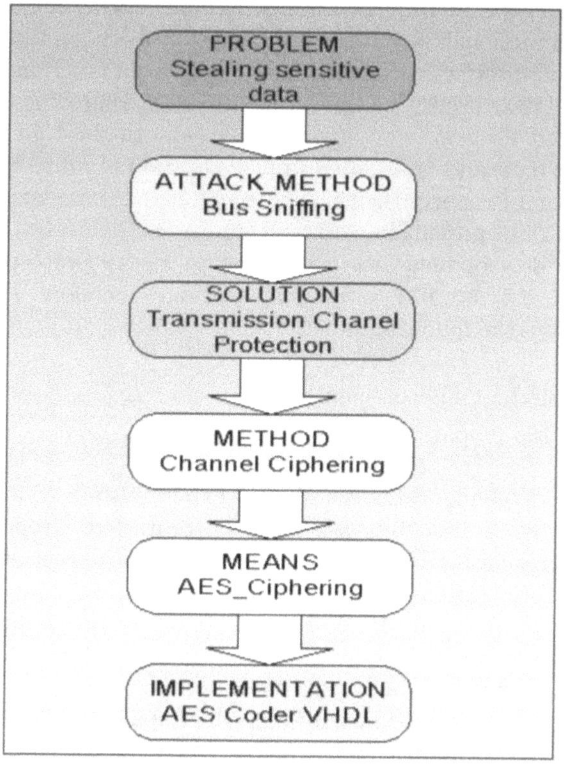

Fig. 1. An example of knowledge structure

Embedded Fortress Expert System is a practical embodiment of expert system. Its task is to propose the best security solution based on the knowledge gathered in the knowledge base and the information obtained from the user.

The Embedded Fortress Solution Browser module is responsible for presenting solutions from the solution library. The module also provides an interface used by user to communicate with code generators which help to set the parameters of a selected generator and then to generate the code.

Embedded Fortress Library is a collection of all solutions proposed by Embedded Fortress System. This module includes an XML file with a list of all security solutions available and a directory which for each individual security contains a file with the solution description and the solution code generator (if a specific solution has the generator).

Our solution is therefore a functional and technical innovation.

3 Tools for Solution

We used advanced development tools for project realization. According to the design guidelines, the system was implemented in Visual Studio 2005 environment, in C language, using XML technology.

Descriptions of modules implementation provide information on the method in which a specific functionality was executed, relations between the modules as well as information on interfaces of classes employing given functionality and on the structure of XML files used. The aim of the project is to provide reliable software tool for embedded electronics designers for advising the best methods of protection for their systems and automated implementation of protection mechanisms against Intellectual Property theft. The project area of interests is protection of software and/or data present in electronic embedded devices/systems.

The solution development has been done to assure protection of designed equipment from copying and reverse engineering. Therefore, the EF software environment offers the following functions:

- Protection method advisor/configurator,
- Protection chip preparation.

Protection advisor/configurator is a kind of expert system which offers a set of protection methods limited by system (to be protected) properties size, calculation power, pins availability, application importance, cost of protection etc. Protection chip preparation covers generation of the code1 for protecting chip (configuration bitstream for CPLD or similar one time programmable chip) and generation of code2 (an IP core for FPGA or library for uC / processor) which performs corresponding action in protected system.

This new approach to embedded systems development and protection confers several benefits by enabling the following capabilities:

- radical reduction of time spent on implementation of security mechanisms during embedded systems development,
- reduction of all lifecycle management costs of the project,
- shortening the time-to-market for embedded systems,
- bringing online expert deign-for-security knowledge to the company at the point of need,
- increasing the Intellectual Property security by selecting the most suitable and multi-level protection mechanisms,
- providing higher product flexibility through the possibility of implementing different protection mechanisms.

The task of Embedded Fortress Expert System is to provide advice to the user on widely-understood security issues in embedded systems. Complex structure of embedded systems makes description of such systems highly difficult. Thus, to assist in system description, as well as to facilitate providing more accurate solutions by expert, we decided to choose a context-type expert system model. The context in our expert system may be the embedded system, as well as its all components, such as processor, memory etc. The more precise the system description, the more accurate expert system replies.

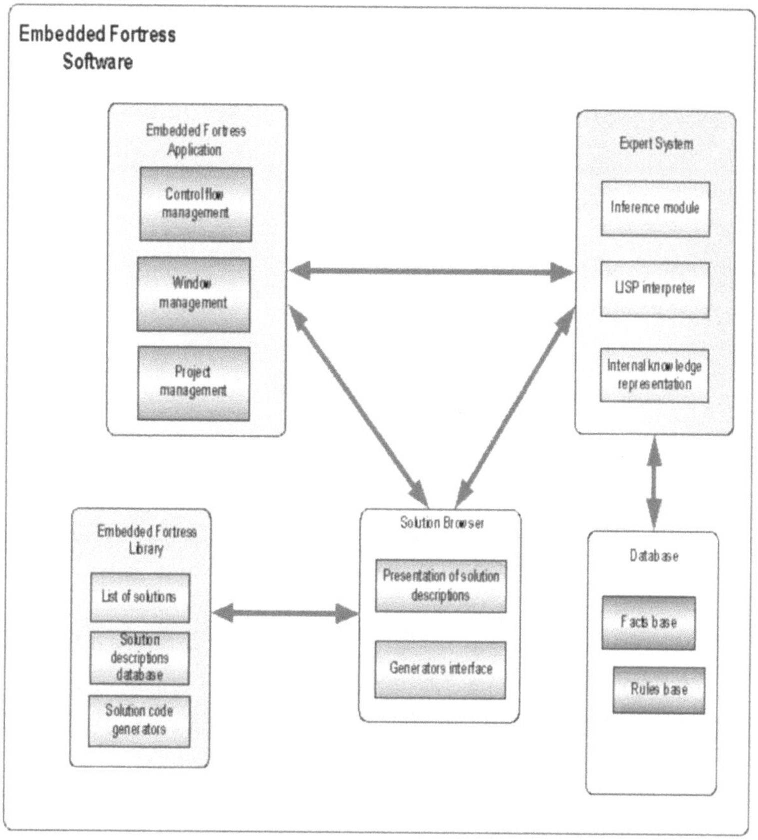

Fig. 2. Embedded Fortress System physical structure the conceptual level

Each context includes two tables:

- ANSWERS contains user information on a given context,
- IMPLICATIONS contains implications generated by expert system with reference to the context. In this table, the expert systems solution to the user question will be ultimately placed as well.

Below is a chart of such context model.

The primary context will always be SYSTEM type context. Within this context, there are two tables ANSWERS and IMPLICATIONS. Expert system performs reasoning using data from both tables, and then places the generated solution in IMPLICATIONS table. COMPONENT type contexts also feature two tables and based on them the expert system performs reasoning, but additionally, on the component level, expert system may use the information in the tables belonging to the primary SYSTEM type context. The ultimate expert system response is a sum of answers to users questions generated for all contexts.

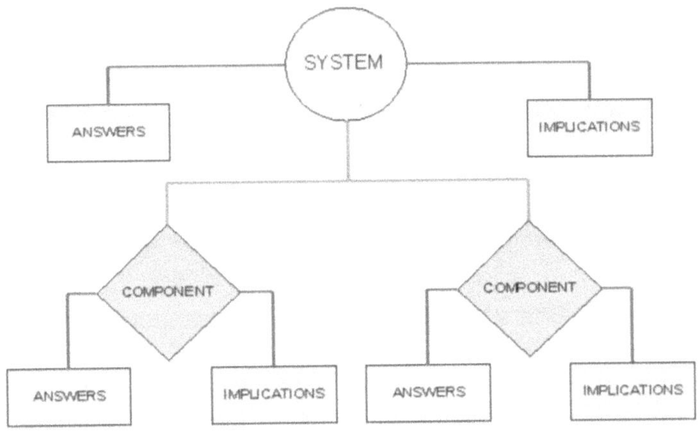

Fig. 3. Context expert system model

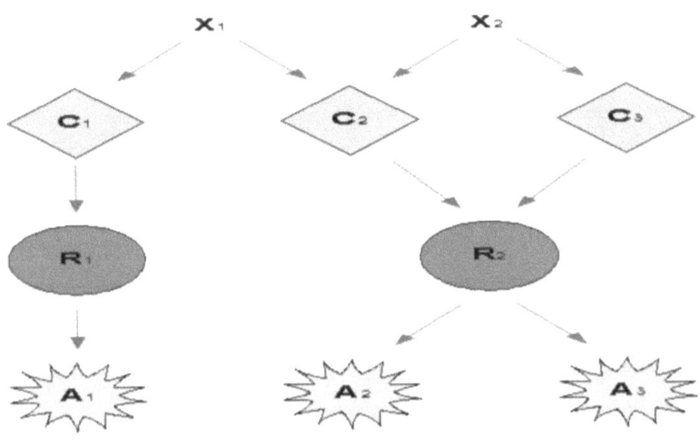

Fig. 4. Symbol representation of knowledge base

Knowledge in expert system has 2 representations:

- External –in XML file, containing a set of attributes, facts and rules,
- Internal –consisting of information loaded in the application from the external file together with IMPLICATIONS and ANSWERS.

Knowledge is represented by a set of rules in condition-action form. The system uses a set of X attributes tested in conditions C. R - inference rule - is run when the relevant set of conations is fulfilled. Then, an appropriate set of A actions is executed (Fig. 4). Sets of attributes, conditions, rules and actions are stored in individual tables indexed with their names. They create two-directional lists among one another, which enable a convenient passage in such a graph.

In Embedded Fortress expert system, we use backward chaining inference type. This means that tips are searched by setting a goal and seeking rules that may solve it. Rules are run if their conditions are met (Fig. 4).

4 Embedded Fortress Application Module Implementation

The implementation of Embedded Fortress Application was divided into 3 steps:

1. Creating application GUI
2. Creating GUI management class
3. Creating project management class

The basic assumption in GUI implementation was separating interface visualization mechanisms from mechanisms responsible for filling the windows with content. To achieve this, two classes responsible for user interface were created:

- EmbeddedFortressApplication –responsible for visualization interface,
- ProjectView –responsible for controlling the windows and willing them with content.

The task of EmbeddedFortressApplication is to create Embedded Fortress application window and ProjectView class instance, and then to provide handles to individual elements of GUI to ProjectView class. ProjectView class creates instances of project class (ProjectClass), expert system (ExpertSupervisorClass) and solution browser (SolutionBrowserClass), and then fills the windows with appropriate content. ProjectView class captures events generated from the application window and then modifies the content in windows, in line with users commands. This class invokes methods from ExpertSupervisor, SolutionBrowserClass classes to change the displayed content or to hand over the control over a part of the windows to these classes. This solution allowed to design one universal, simple and clear GUI used by all application modules to communicate with the user.

Work with Embedded Fortress system is based on a project concept. During each system session, user works on a project which contains the information on the directory name and project name, as well as on solutions generated by the user, added to favourites and user responses in expert system. To be able to store this information effectively, ProjectClass, GeneratedSolutionsClass and FavouriteSolutionClass were created. ProjectClass object contains project description fields (name, project directory) as well as instances of GeneratedSolutionsClass, FavouriteSolutionClass and ExportSupervisor classes. This class must be able to save the values of its fields in a file, to allow their subsequent loading in the project. We used XML serialization for this purpose. Thanks to this method, the project files are saved on the disk as XML files.

The expert system was implemented as a separate functional module. Creating the expert system consisted in implementing LISP interpreter, inference module and creating knowledge base. Additionally, the intermediate class between GUI and expert system needed to be created, allowing expert and user interface communication.

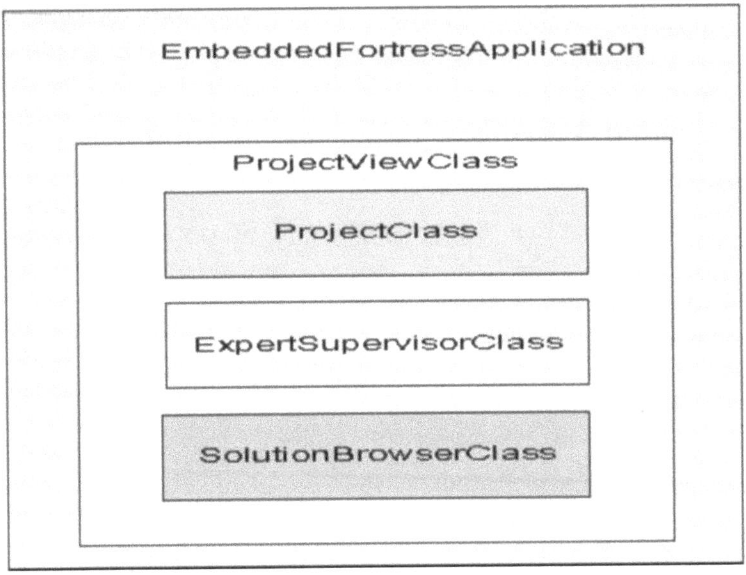

Fig. 5. Relations of classes in Embedded Fortress application

5 Conclusion

Thanks to this form of presentation, knowledge is gathered in a structured, easily editable way and, most importantly, relations between individual attributes are presented clearly, which considerably facilitates creating rules for expert system. This structure of knowledge acquisition points out the new ways in which Embedded systems may be attacked, and then enables design of security means that would protect systems against such attacks. Tree knowledge base structure allows to add easily lower and higher levels, creating a logical complete picture. This document presents the software architecture concept and implementation method of Embedded Fortress system. It shows the division into functional modules together with their implementation. Additionally, it contains expert system operation description, knowledge gathering methods and the methodology of creating code generator for solutions available in Embedded Fortress system.

Many industry manufacturers in Europe develop and manufacture embedded systems every day. They have to still improve their solutions as well as production process to be competitive on European and world markets, always a step ahead towards their rivals. They have to also protect their Intellectual Property adequately to avoid financial losses. Embedded Fortress makes European manufacturers more competitive in continuing contest on markets by increasing their efficiency and productivity, reducing financial losses, reducing manufacturing costs and enlarging quality of their products.

The project Embedded Fortress System was made together with University of Reading (UK) under patronage of the International Initiative Committee EUREKA!

References

1. Keating, M., Bricaud, P.: Reuse methodology manual for system-on-a-chip design. Kluwer Academic Publishers, Boston (2001)
2. Skotarczyk, A., Englert, T., Handzlik, A.: Embedded Systems State of the art and trends. Internal Technology Report, Microtech International Ltd., Wroclaw (2005)
3. Short, S.: Building XML Web Services for Microsoft.NET Platform. Polish edition Microsoft Press, Warszawa (2003)
4. Microsoft Official Course: 2557A - Building COM+ Applications Using Microsoft.NET Enterprise Services (2002)
5. Microsoft Official Course: 2349B-Programming with the Microsoft.NET Framework (Microsoft Visual C.NET) (2002)

Collaborative XML Document Versioning

Sebastian Rönnau and Uwe M. Borghoff

Institute for Software Technology,
Universität der Bundeswehr München,
Werner-Heisenberg-Weg 39,
85577 Neubiberg, Germany
Sebastian.Roennau@unibw.de

Abstract. Document formats based on XML are widely used in today's office collaboration. However, most supporting tools like version control systems, or business process systems do not handle these documents adequately. Parallel editing of documents across network and system borders is almost impossible.

Our recent research showed that versions of XML documents can be merged in a reliable way using deltas with context fingerprints. In this paper, we present a collaboration strategy based on these deltas that allows for a highly dynamic distributed collaboration among XML documents.

Keywords: XML, documents, merge, version control, context fingerprints, distributed collaboration.

1 Overview

Today's office work demands a highly dynamic collaboration. Teams and projects are built on top of classic hierarchies. Inter-project, inter-team cooperation and collaboration are everyday tasks. Since most people are part of different projects and different teams, the human beings are more important than their function. This kind of dynamic environment needs highly adaptive tools to support the collaboration of teams and their members, respectively. However, most existent business process tools require fixed process rules and workflows, as shown e.g. in [1]. Version control sytems, however, usually require a central server hosting the actual state of all documents related to the process [2]. These preconditions do not hold in situations as described above: organizational, infrastructural and security-related issues prevent that everyone has equal access to the central system. Because of these constraints it has become accepted practice to send documents via e-mail, and to merge document versions manually, which is both time-consuming and error-prone.

We offer a solution for this problem by providing a toolkit for versioning and merging XML documents. Our approach is device- and media-independent. Its main goal is to empower people to share documents without forcing them to use a specific document workflow or to be bound to a restricted server system. In

R. Moreno-Díaz et al. (Eds.): EUROCAST 2009, LNCS 5717, pp. 930–937, 2009.

contrast to most workflow-based systems, we do not require a linear evolution of the document, but rely on a distributed collaboration strategy. This strategy is based on a collaborative editing model for distributed application. A key prerequisite to effective distributed collaboration is a reliable merge capability. Our delta-based versioning model used as backbone of the collaboration model allows for reliable merging using context fingerprints. This technique has been introduced in previous work [3], and turned out to be highly reliable.

The remainder of this paper is organized as follows. In Section 2, we present our collaboration strategy based on collaborative editing using an XML-aware delta-model. We compare our approach to other techniques in Section 3.

2 Collaboration Strategy

Traditionally, documents are regarded as evolving in a linear way. However, especially in highly dynamic and creative environments, this precondition does not hold. Therefore, we introduce our view towards the document evolution model in Section 2.1. In Section 2.2, we present our delta model used as backbone of our collaboration model, including an empirical evaluation of the reliability of our approach.

2.1 Linear Evolution vs. Collaborative Editing

Business processes organize tasks and activities to achieve a certain goal. Usually, processes can be visualized by a flowchart. Documents can be used as central information carrier within business processes [4]. Therefore, the evolution of such a document obeys the same fixed rules as the process on the whole. Usually, each participant in the workflow contributes a well-defined and unambiguous part to the document. On the one hand, conflicts are nearly excluded as the business process places the responsibility for each part of the document to different users. On the other hand, the strict workflow hinders a creative collaboration. Any discussion about the document to create has to be held outside the business process, thus leading to a second layer of communication relationships not supported by the business process.

Version control systems offer a precise tracking of the changes performed on a document. They are based on a centralistic view, where the server hosts the complete evolution of a document [2]. A user requesting access to a document has to check it out from the repository. Changes have to be performed on the local copy of the document, and propagated to the server afterwards. Fig. 1 shows a document A being edited in parallel. A is checked out by three different persons, called Alice, Bob and Carol. Alice creates the version A' on her local working copy, and commits the updated version to the repository. In the meantime, Bob has modified his working copy of A to A''. When he tries to commit his changes to the repository, the document versions A' and A'' have to be merged. Two major drawbacks arise in the sceanrio presented here. First, Carol cannot be notified about the actual changes on the document. She can only become aware of the ongoing work on the document after the commitment of A'. This prevents

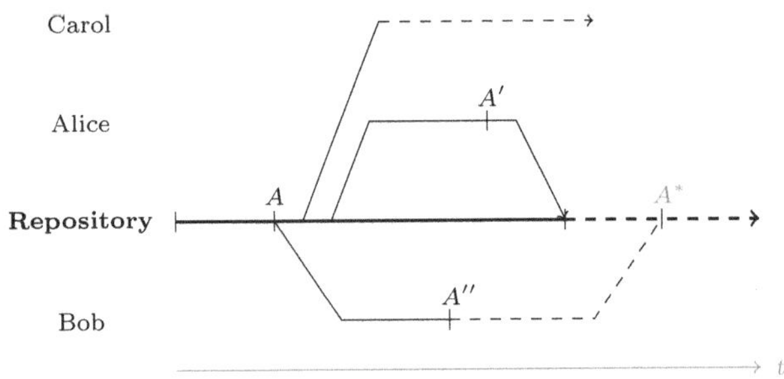

Fig. 1. In a linear document evolution model, changes have to be propagated to the central repository. A common awareness of the ongoing work is not possible.

a common awareness crucial to collaborative work [5]. Second, the repository has the only valid time axis in terms of the document evolution model. Even if Bob has performed his changes earlier on his working copy, Alice's version A' is entitled to come first in the document evolution due to the earlier commitment. These drawbacks require strong organizational measures to enforce an effective collaboration, thus restraining creative work.

In the last years, distributed version control systems emerged that aimed to solve these issues. They rely on a collaborative cooperation model, highly inspired by the development processes used in OpenSource software systems [6]. Due to the growing success of OpenSource collaboration processes, their methods are more and more adopted by traditional industries [7]. The main ideas of distributed collaboration environments is to allow the users to freely exchange their versions of the documents, without organizational restrictions.

Fig. 2 shows an example scenario. All participants have their own timeline of the document, a linear evolution line cannot be constructed any more. All

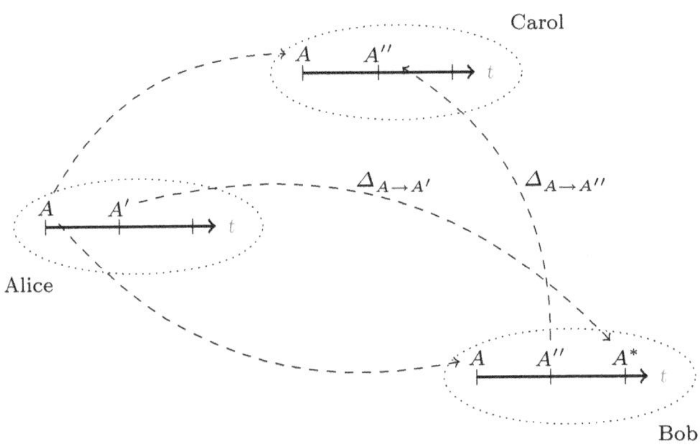

Fig. 2. In a distributed, highly collaborative environment, changes can be arbitrarily exchanged between all participants

changes are propagated through deltas. Note that we do not define, whether the deltas have to be pushed to the other participants or pulled by them. We believe that in most collaboration scenarios, a hybrid approach will be appropriate. Deltas can be distributed, e.g. using automatically generated e-mails, or fetched in case they are requested. This collaboration scenario is highly suitable for dynamically changing groups and for a bazaar-style collaboration [6].

2.2 Delta Model

The backbone of our collaboration strategy is the XML delta model. We rely on a delta model presented in previous work [3,8], which is based on context-oriented fingerprints for the reliable identification of edit operations within a document.

Two versions of one document are compared using a so-called *diff* tool. It detects all changes performed on the document, and stores them within a *delta*. This delta contains only the edit operations needed to construct the new document version out of the former one. The reconstruction process is performed by the *patch* tool. As the delta is usually much smaller than the corresponding documents, it will likely be distributed instead of the whole document.

If different persons update a document independently, their changes have to merged afterwards. In our model, the delta that describes the changes between the original document, and one of the updated documents is applied to the other document version, as shown in Fig. 3. Thus, the *merge* is actually performed by the patch tool. Note that this approach is a common solution in the domain of line-based documents, e.g. for source code [9].

Applying a delta to a document version it was not computed can cause severe problems. Any insert or delete can affect the paths to all subsequent elements within the document. Therefore, absolute paths cannot be used to identify an edit operation. Alternatively, the node names could be used as identifier. However, this is as well inappropriate as the use of absolute paths. The reason for that is that office document do not have unambiguous node names in general. E.g., considering an ODF text document, each paragraph is stored as a text node being child of a `text:p` node [10]. Therefore, an operation addressing a paragraph could not ensure the correctness of the address.

To reliably identify an edit operation, it is required to enrich edit operations with additional information. This is a precondition to meaningful merges. We assume the context of an edit operation to be a reliable indicator, whether an edit operation should be applied or not. In our approach, the context of each edit operation is stored within a so-called *context fingerprint*, which is a normalized representation of the surrounding nodes of an edit operation. We use a normalization technique based on hash values. For details, please refer to [3,8].

During the merge process, the context fingerprint of each edit operation to apply is compared to the according context within the document to patch. If the delta is applied to the version of the document it has been computed for, the fingerprints match completely, as no merge is necessary in this case. In all other cases, the best match w.r.t. the given context fingerprint is searched within the document. A match is evaluated on the basis of the amount of matching nodes

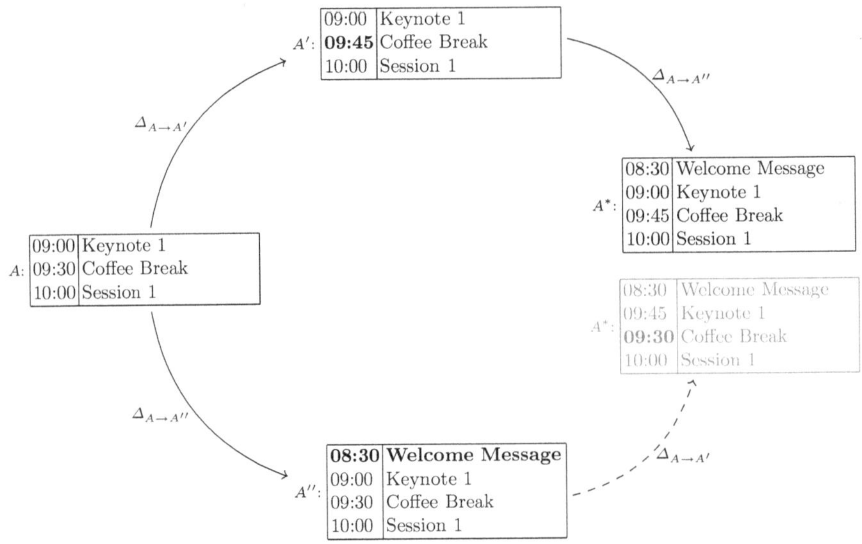

Fig. 3. Applying a delta to a document version it was not computed for leads to unwanted results easily. Performing the delta marked with a dashed line would create a wrong document version.

within the context, as well as on the distance of the match to the edit operation itself. Our delta model allows for detecting conflicting edit operations, as the context fingerprint also contains a recursively computed hash over the subtree to delete or to update.

To evaluate the reliability of our approach, we constructed a merge scenario based on ODF text documents. Different versions of one base document have been created, and compared. Afterwards, the deltas have been applied to other versions of the document. The resulting documents have been compared with a manually merged version of the document. Over 1,000 edit operations have been merged that way. Table 1 shows the results. In total, 95% of the edit operations are applied as expected. This results from the fact that some edit operations could not be applied, as the corresponding nodes had been deleted in the meantime, thus denoting true negatives. Other ones have been rejected as the corresponding nodes had been updated in the meantime, thus denoting a true conflict. False negatives and false conflicts occur in case that our merge procedure was not able to find the correct position of the edit operation, even

Table 1. The merge quality is very high, over 95% of the operations are performed as expected. No false positives occur, which means wrongly applied edit operations.

	positives		negatives		conflicts	
	true	false	true	false	true	false
edit operations	87.54%	0.0%	1.12%	0.37%	6.56%	4.41%

though it was still present in the document to patch, thus leading to rejected edit operations. Note that no false positives occured. False positives are edit operations which have been wrongly applied to the document.

The evaluation shows that deltas using contextfingerprints allow for a meaningful and reliable merge of document versions.

3 Related Work

The question of collaborative editing of XML documents has been discussed extensively with different focuses.

The operational transformation (OT) approach can be used to immediately reconcile a common document version across a shared document [11]. Using frequent reconciliation steps, the probability of errors is minimzed. The applicability of OT to the XML domain has been shown [12]. Additionally, there exist approaches to repair inconsistencies resulting from operations conflicting w.r.t. the DTD of a document [13]. The OT approach is mostly used in collaborative editors, where a synchronous collaboration is demanded, thus requiring a permanent connection of all participants. Furthermore, most approaches need a central server for dispatching, which is an organizational barrier to efficient collaboration over organizational bounds.

XML-based databases have become more and more popular in the last years. They offer an intuitive way of accessing XML-based data, disregarding their physical representation. A common way to ensure a parallel editing on an XML document is to use locking techniques, which try to isolaten the area affected by the editing process of a user [14,15]. These approaches show two major drawbacks concerning our use-case. First, the database approach needs a central server hosting the data. Second, most database-oriented transaction models rely on an unordered tree model not suitable to the domain of documents.

A document-centric view towards collaboration is used by approaches that annotate the original document with editing information representing the changes. Annotations can be applied to XML documents in general [16], or in domain-specific manner, e.g., for office documents [17,10]. On the one hand, the annotation approach avoids the problems of updated paths and ambiguous node names. On the other hand, the differences cannot be extracted but have to be remain within the annotated document. This leads to two major issues. First, the whole document has to be transmitted by each participant at each synchronization step, which leads to a dramatical overhead in transmitted data[1]. Second, it might be unwanted to keep all editing information within the document for privacy reasons.

To avoid the complete retransmission of the whole document, several XML-aware diff tools have been proposed [18,19,20]. All of these diff tools are able to compute the delta between two XML documents efficiently. However, none of them is able to merge the generated deltas afterwards. To avoid erroneous merge

[1] Especially large documents edited by different users can easily reach several megabytes in space.

results, it is useful to verify whether the document which should be patched was updated before [19].

In order to offer a merge capability for documents, a three-way diff approach can be used [21]. In this case, two new versions of the document are compared w.r.t. their nearest common ancestor in terms of the document evolution. An XML-aware three-way diff has already been proposed [22]. In loosely-coupled environments it appears to be nearly impossible to determine the nearest common ancestor of a document without a central repository.

4 Conclusions and Future Work

In this paper, we have presented a collaboration strategy empowering a collaborative versioning of XML documents. The backbone of our approach is an XML-aware delta model using context fingerprints for a reliable merge capability.

The ability for arbitrary communication relationships is both a blessing and a curse. The fast exchange of document versions allows for a highly dynamic collaboration, fostering creative working. However, it is possible to lose the overview of the different versions. The complexity of the communication relationships increases dramatically in large teams, and in long-term collaboration scenarios. This issue has been studied in the context of OpenSource software development [23], where the need for a group awarness supported by different organizational and technical measures has been stressed.

In future work, we will focus on the creation of an interface of our collaboration tools to a distributed version control system. This way, we try to bring together the domains of software development and of document editing.

References

1. Boyer, J.M.: Interactive office documents: a new face for web 2.0 applications. In: DocEng 2008: Proceeding of the eighth ACM symposium on Document engineering, pp. 8–17. ACM, New York (2008)
2. Tichy, W.F.: Design, implementation, and evaluation of a revision control system. In: ICSE 1982: Proceedings of the 6th international conference on Software engineering, pp. 58–67. IEEE Computer Society Press, Los Alamitos (1982)
3. Rönnau, S., Pauli, C., Borghoff, U.M.: Merging changes in xml documents using reliable context fingerprints. In: DocEng 2008: Proceeding of the eighth ACM symposium on Document engineering, pp. 52–61. ACM, New York (2008)
4. Boyer, J.M., Dunn, E., Kraft, M., Liu, J.S., Shah, M.R., Su, H.F., Tiwari, S.: An office document mashup for document-centric business processes. In: DocEng 2008: Proceeding of the eighth ACM symposium on Document engineering, pp. 100–101. ACM, New York (2008)
5. Dourish, P., Bellotti, V.: Awareness and coordination in shared workspaces. In: CSCW 1992: Proceedings of the 1992 ACM conference on Computer-supported cooperative work, pp. 107–114. ACM, New York (1992)
6. Raymond, E.S.: The Cathedral and the Bazaar: Musings on Linux and Open Source by an Accidental Revolutionary. O'Reilly & Associates, Inc., Sebastopol (2001)

7. Neus, A., Scherf, P.: Opening minds: cultural change with the introduction of open-source collaboration methods. IBM Syst. J. 44(2), 215–225 (2005)
8. Rönnau, S., Borghoff, U.M.: Versioning XML-based office documents. Multimedia Tools and Applications 43(3), 253–274 (2009)
9. FSF: Comparing and Merging Files. Free Software Foundation (2002)
10. Brauer, M., Weir, R., McRae, M.: OpenDocument v1.1 specification (2007)
11. Sun, C., Ellis, C.: Operational transformation in real-time group editors: issues, algorithms, and achievements. In: CSCW 1998: Proceedings of the 1998 ACM conference on Computer supported cooperative work, pp. 59–68. ACM, New York (1998)
12. Davis, A.H., Sun, C., Lu, J.: Generalizing operational transformation to the standard general markup language. In: CSCW 2002: Proc. of the 2002 ACM conf. on Computer supported cooperative work, pp. 58–67. ACM, New York (2002)
13. Skaf-Molli, H., Molli, P., Rahhal, C., Naja-Jazzar, H.: Collaborative writing of xml documents. In: 3rd Int. Conf. on Information and Communication Technologies: From Theory to Applications, ICTTA 2008, April 2008, pp. 1–6 (2008)
14. Dekeyser, S., Hidders, J.: Path locks for xml document collaboration. In: WISE 2002: Proceedings of the 3rd International Conference on Web Information Systems Engineering, Washington, DC, USA, pp. 105–114. IEEE Computer Society, Los Alamitos (2002)
15. Haustein, M., Härder, T.: An efficient infrastructure for native transactional xml processing. Data Knowl. Eng. 61(3), 500–523 (2007)
16. Fontaine, R.L.: Merging xml files: a new approach providing intelligent merge of XML data sets. In: Proceedings of XML Europe 2002 (2002)
17. Paoli, J., Valet-Harper, I., Farquhar, A., Sebestyen, I.: ECMA-376 Office Open XML File Formats (2006)
18. Zhang, K., Shasha, D.: Simple fast algorithms for the editing distance between trees and related problems. SIAM J. Comput. 18(6), 1245–1262 (1989)
19. Cobéna, G., Abiteboul, S., Marian, A.: Detecting Changes in XML Documents. In: Proceedings of the 18th International Conference on Data Engineering, San Jose, CA, 26 February - 1 March, pp. 41–52. IEEE Computer Society, Los Alamitos (2002)
20. Lindholm, T., Kangasharju, J., Tarkoma, S.: Fast and simple xml tree differencing by sequence alignment. In: DocEng 2006: Proceedings of the 2006 ACM symposium on Document engineering, pp. 75–84. ACM, New York (2006)
21. Khanna, S., Kunal, K., Pierce, B.C.: A formal investigation of diff3. In: Arvind, V., Prasad, S. (eds.) FSTTCS 2007. LNCS, vol. 4855, pp. 485–496. Springer, Heidelberg (2007)
22. Lindholm, T.: A three-way merge for XML documents. In: DocEng 2004: Proceedings of the 2004 ACM symposium on Document engineering, pp. 1–10. ACM, New York (2004)
23. Gutwin, C., Penner, R., Schneider, K.: Group awareness in distributed software development. In: CSCW 2004: Proceedings of the 2004 ACM conference on Computer supported cooperative work, pp. 72–81. ACM, New York (2004)

Parallel Distributed Genetic Algorithm for Expensive Multi-Objective Optimization Problems

Ewa Szlachcic[1] and Waldemar Zubik[2]

[1] Institute of Computer Engineering, Control and Robotics
Wrocław University of Technology
11/17 Janiszewskiego St., 50-372 Wrocław, Poland
ewa.szlachcic@pwr.wroc.pl
[2] TETA SA Al. Wisniowa 1, 53-137 Wroclaw, Poland
w.zubik@teta.com.pl

Abstract. In many Multi-Objective Optimization Problems it is required to evaluate a great number of objective functions and constraints and the calculation effort is very high. The use of parallelism in Multi-Objective Genetic Algorithms is one of the solutions of this problem. In this work we propose an algorithm, based on parallelization scheme using island model with spatially isolated populations. The intent of the proposed paper is to illustrate that modifications made to a selection and resolution processes and to a migration scheme have further improved the efficiency of the algorithm and good distribution of Pareto front.

Keywords: multi-objective optimization, parallel genetic algorithm, multiple resolution, island model, migration strategy.

1 Introduction

In many Multi-Objective Optimization Problems it is necessary to evaluate a large number of objective functions and constraints then the calculation process is very time-consuming. Some of the great disadvantages of meta-heuristics methods concerning Multi-Objective Optimization Problems (MOPs) are dealing with large search spaces and with an extremely high calculation cost. These problems named expensive Multi-Objective Optimization Problems (eMOPs) are nowadays in great interests [4,8,11,10].

In MOPs three typical approaches are solved by optimising one objective at a time while taking other objectives as a changing constraints or combining all objectives into a single objective with some weight coefficients and last one – optimizing all objectives simultaneously [3]. In the first case, it can be very difficult to precisely and accurately select these weights, even for someone familiar with the problem domain [2,8]. Even small perturbations in the weight coefficients can sometimes lead to different solution. This approach returns single solution. The changing constraints approach have also many disadvantages. The

R. Moreno-Díaz et al. (Eds.): EUROCAST 2009, LNCS 5717, pp. 938–946, 2009.

third general approach is to determine an entire Pareto optimal solution set. A Pareto optimal set is a set of solutions that are non-dominated with respect to each other. Pareto optimal sets can be of varied sizes, but the size of the Pareto set usually increases with the increase of objective functions number [4,11].

It is important to emphasize that the computational efficiency of MOGA depends on the complexity of Pareto ranking ($O(kL_p^2)$), where k is the number of objective functions and L_p is the population size [11,12]. The use of parallelism in Multi-Objective Genetic Algorithms is one of the solutions for expensive Multi-Objective Optimization Problems. There are many studies that are related to parallel genetic algorithms [1,7,8,10,11,9] for MOPs. These algorithms allow the use of larger population size, they can use more memory to cope with more difficult problems and they improve the population diversity [5]. Some more reasons to justify their parallelization are that they reduce the probability of finding suboptimal solutions and they can cooperate in parallel with another search technique.

In this work we propose alternate migration strategies with modified replacement scheme used in parallel Multi-Objective Genetic Algorithm. The island model is used with spatially isolated subpopulations and multiple resolution idea.

2 Problem Formulation

The Multi-Objective Optimization Problem is to find a set of optimal vectors $x^* \in X \subset R^n$, which optimizes a set of objective functions and satisfies a set of constraint functions such that:

$$F(x^*) = \min_{x \in X} F(x) \ = \min_{x \in X} \ [f_1(x), ..., f_k(x)] \tag{1}$$

F(x) denoted as $F(x): \ X \subset R^n \rightarrow R^k$ is a given vector of component functions $f_i(x) \ : \ X \subset R^n \rightarrow R$ for $i \in \{1, 2, ..., k\}$. The minimum of $F(x^*)(1)$is taken in the sense of the standard Pareto order on an objective functions space. Pareto optimal solution x* is understood as such that there exists no feasible vector x, which would decrease some criterion without causing a simultaneous increase in at least one other criterion. The set of Pareto optimal solutions named Pareto front [4] consists of non-dominated points, satisfying Pareto dominance definition. Minimum is stated in the sense of standard Pareto order as [12]:

If there exist two vectors $u = (u_1, u_2, ..., u_k)$and $v = (v_1, v_2, ..., v_k)$ in R^k,then

$$u \preceq v \iff u_i \leq v_i \ \forall i \in 1, ..., k. \tag{2}$$

The relation $u \prec v$ gives us an order in the sense: $u \preceq v$ and $u \neq v$.

Definition 1. *A vector $x^* \in X$ is called Pareto optimal solution for multi-objective optimization problem if there is no $x \in X$ such that $F(x) \prec F(x^*)$.*

The set $P^* = \{x \in X; \, and \, X \subset R^n\}$is defined as Pareto optimal set and its image under $F(x)$ is called as the Pareto front PF:*

$$PF^* := \{F(x) = (f_1(x), f_2(x), ..., f_k(x) : x \in P^*\}. \quad (3)$$

Pareto front determines the space in R^k formed by the objective solutions of the Pareto optimal set. As mentioned in [2] we try to obtain a good representation of the Pareto front that has to achieve the main requirement: precision and diversity. At each step of the optimization process in a multi-objective genetic approach a set of solutions is constructed and we try to choose non-dominated points among all dominated solutions. It allows to classify these solutions to two subsets according to the Pareto optimality and Pareto dominance concepts.

3 The Principle of Parallel Distributed Multi-Objective Genetic Algorithm

In the paper we propose a parallel Distributed Multi-Objective Genetic Algorithm (pDMOGA), based on parallelization scheme using island model with isolated subpopulations. Each island can apply a serial multi-objective genetic algorithm with different parameters. The algorithm combines the parallel MOGA [11] and the Strenght Pareto Evolutionary Algorithm 2 [15] with some modifications. In discussed numerical optimization problems variables are described with real number representation. Each node can use different resolution scheme it means that the number of precision digits after decimal point can not be equal among islands.

As mentioned in [11] the Pareto optimal solutions are found in fewer iterations using low resolution representations than using higher resolution representations. The search space is smaller as the resolution decreases and we need to explore fewer solutions to produce approximation of PF*. The low resolution islands approach faster to Pareto optimal front PF* than high resolution nodes. The islands can use different parameters setting. At each node modified Strenght Pareto Evolutionary Algorithm 2 (MSPEA2) as a serial algorithm is executed for a number of generations called an epoch. At the end of each epoch, individuals migrate between islands [16]. Then a resolution modification procedure is needed to incoming individuals to adapt its chromosomes to the new resolution parameter. Finally the root process combines the non-dominated front of each island to receive the optimal PF^* .

3.1 Migration Topology Procedure

The population is divided into small subpopulations, which work independently. Each island encodes the solutions with its own resolution parameter. The migration topology depends on resolution parameters. The individuals migrate between neighboring islands using two migration schemes: strict topology when an individuals migrate to the near island where the resolution is greater by one and complete topology alternatively [16]. Each islands migrate solutions to all of its children in the hierarchy. The second migration topology allow for migration to the islands with greater resolution then by one. When the migration process starts each node takes some individuals from its archive set and send them to the basic subpopulation of its neighbors.

3.2 Selection and Clustering Scheme

On each island the serial MSPEA2 algorithm is applied. The fitness assignment is based on the rank of an individual at generation t, which is dominated by individuals [13] in current and archive subpopulations. The archive subpopulation consists of unic individuals. The proposed temporary set consists of duplicated individuals, which are used to fulfill the archive set using selection operators. In MSPEA2 six different crossover operators are used for different islands with special attention to reciprocal translocation operator. In standard selection process non-dominated individuals are copied to archive set. The individuals are sorted with along to the fitness functions values. If the number of elements overcomes the maximum number of Pareto archive set we remove the redundancy of elements according to the distance of fitness fuctions values between neighborhood individuals. The method of the archive set reduction can influence on the convergence of the whole pDMOGA algorithm. In this case the changes of crossover or mutation types or changes of the algorithm parameters as crossover or mutations probabilities are insufficient solutions. The proposed clustering procedure, based on removing an element which has a minimum distance to another individual in an objective function space gives a required diversity of elements. In the case of lack of chromosomes we have to take the rest of individuals from the primary population. In this moment the problem of individuals diversity arrives so a special attention was carried out to a selection procedure for archive set elements. In the new island the incomming individuals are converted to the resolution, which is in the node. At the end of the search procedure – the root process combines the non-dominated front of each island regarding for good spread and distribution of non-dominated solutions along the Pareto optimal front.

4 Numerical Results

In order to investigate the performance of pDMOGA according to the MOPs many tests have been made with MSPEA2 as the serial algorithm. The numerical tests for pDMOGA were carried out using six multi-objective problems [15] named ZDTx, which cause difficulties to any MOGA algorithms. These problems have different types of Pareto optimal front: problem ZDT1 – presents convex Pareto front, problem ZDT2 – has a non-convex Pareto front and problem ZDT3– a Pareto front composed of some discontinuous convex regions and ZDT6 - convex Pareto front, In the paper we show numerical results only for three ZDTx test problems: ZDT1, ZDT3 and ZDT6, described in Table 1.

4.1 Evaluation Methods

The quality of the Pareto front obtained by MOGA can be evaluated considering three goals: to achieve a good distribution of the non-dominated solutions

Table 1. Benchmark test problems

Type of MOP	Minimization of objective functions	Constraints
ZDT1	$f_1(x) = x_1$ $f_2(x) = g(x)\left[1 - (f_1(x)/g(x))^{1/2}\right]$ $g(x) = 1 + 9 \times \sum_{i=2}^{m} x_i/(n-1)$	$0 \leq x_i \leq 1$ $i = 1, ..., n$ $n = 30$
ZDT3	$f_1(x) = x_1$ $f_2(x) = g(x)\left[1 - (\frac{f_1(x)}{g(x)})^{1/2} - \frac{f_1(x)}{g(x)} sin(10\pi f_1(x)\right]$ $g(x) = 1 + 9 \times \sum_{i=2}^{m} x_i/(n-1)$	$0 \leq x_i \leq 1$ $i = 1, ..., n$ $n = 30$
ZDT6	$f_1(x) = 1 - e^{-4x_1} sin(6\pi x_1)$ $f_2(x) = 1 - \left(\frac{f_1(x)}{g(x)}\right)^2$ $g(x) = 1 + 9 \times (\sum_{i=2}^{m} x_i/(n-1))^{0,25}$	$0 \leq x_i \leq 1$ $i = 1, ..., n$ $n = 30$

obtained, to minimize the distance between the approximation obtained and the true Pareto front. The last one concerns maximization of the spread of the Pareto front obtained.

The effectiveness of the pDMOGA was evaluated with the help of three metrics: Success Counting (SC), Inverted Generational Distance (IGD) and Spacing metric (SP) [11]. SC evaluates the closeness to the Pareto-optimal front calculating the number of non-dominated vectors that belong to Pareto front [%]. IGD measures the overall progress of optimal Pareto front calculating the Euclidian distance d_i between each vector of Pareto optimal front and the nearest member of non-dominated solutions set. For this metric the lower metric values are preferred so the best value will be for IGD equal to zero. It means that the whole non-dominated set includes in the Pareto optimal front. The last one SP measures the distance of r neighbouring vectors in Pareto front according to search space:

$$SP = \sqrt{\frac{1}{r-1} \sum_{i=1}^{r} \left(\overline{d} - d_i^f\right)^2} \tag{4}$$

A way of measuring the range variance of neighboring vectors in Pareto front d_i^f looks as below:

$$d_i^f = \min_j \left(\mid f_1^i(x) - f_1^j(x) \mid + \mid f_2^i(x) - f_2^j(x) \mid \right) \; i,j = 1, ..., r. \tag{5}$$

where $\overline{d}$ denotes the average of all d_i^f. A value of SP indicates how PF set is spreadthroughout the whole front in a decision function space.

We also try to evaluate the efficiency of the parallel algorithm using the well-known metrics [1]: Speedup S_p, Efficiency E_p and Serial Fraction F_p. Speedup compares the average execution time since the algorithm MSPEA2 and pDMOGA are stopped when they have reached the same solution. The Efficiency E_p determines the ratio between Speedup S_p and the number of islands used in the process.

The metric F_p measures the performance of an island model algorithm on a fixed-size problem with p heterogeneous nodes.

$$F_p = \frac{1/S_p - 1/p}{1 - 1/p} \tag{6}$$

The pDMOGA algorithm was run with the following parameters: 400 individuals in one population, the archive size is equal to 10-200. All runs were performed for 1-20 island size with 5-50 epochs. Six different crossover operators were used and the best value of crossover probability was fixed at 0.87, the mutation probability was changed among 0.01-0.02 values. In the experiments 15 independent runs were performed varying the number of islands, number of epochs and number of generations performed. The convergence to an optimal Pareto front was received in all cases.

4.2 Comparison of Results for Test Problems

The simulation for test problems ZDTx were done to study the effectiveness of the pDMOGA and the efficiency of pseudo parallel algorithm. The different types of migration scheme were tested. In this paper we show only the influnce of increasing number of processes (islands) for the effectiveness of pDMOGA. The SC metric shown on Fig.1 indicates that our algorithm for all three test problems converges very closely to the Pareto optimal set. Pareto-optimal solutions are achieved as non-dominated solutions existing on feasible set of constraints. The convergence of an algorithm considering the SC metric reached its maximum with certain number of islands and then dropped. For the Inverted Generational Distance and Spacing metrics the quality of non-dominated solutions and distribution along Pareto front ameliorate with the increase of islands number (Fig.2 and Fig. 3). Note that the non-dominated solutions are not sufficient for a serial MOGA.

The efficiency of the proposed pDMOGA algorithm was checked according to the Speedup, Efficiency and Serial Fraction metrics also according to the increasing number of processes.

The results are presented on Fig. 4 only for ZDT3 problem, which has Pareto front composed of some discontinuous convex regions.

It is clear that there is sufficient number of islands when the parallel idea is useful and speedup metrics increase. For the great number of islands the communications and synchronisation of the migration processes take an important

Fig. 1. SC metric according to the number of islands for ZDT1, ZDT3 and ZDT6

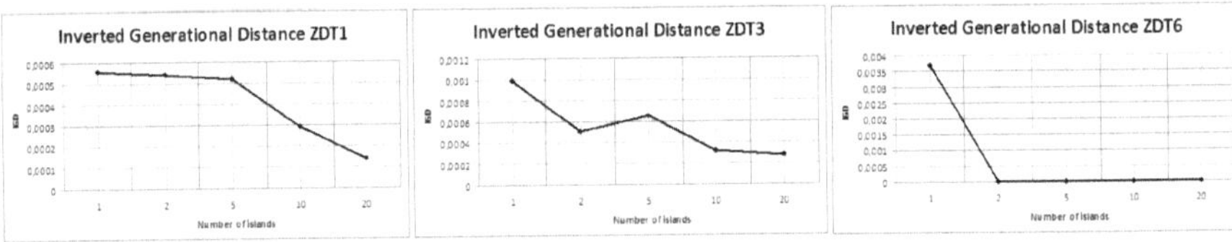

Fig. 2. IGD metric to the number of islands for ZDT1, ZDT3 and ZDT6

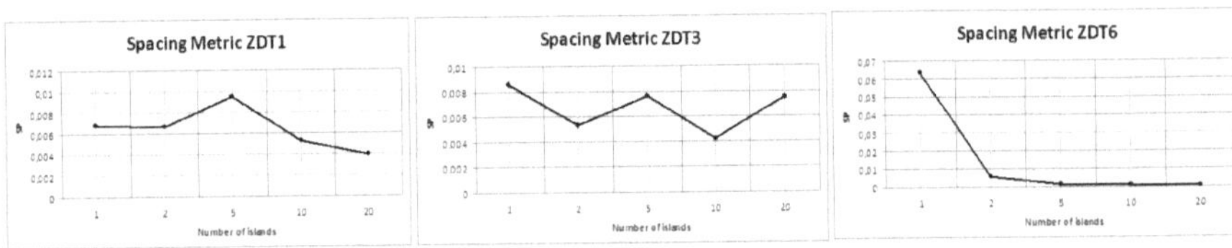

Fig. 3. SP metric to the number of islands for ZDT1, ZDT3 and ZDT6

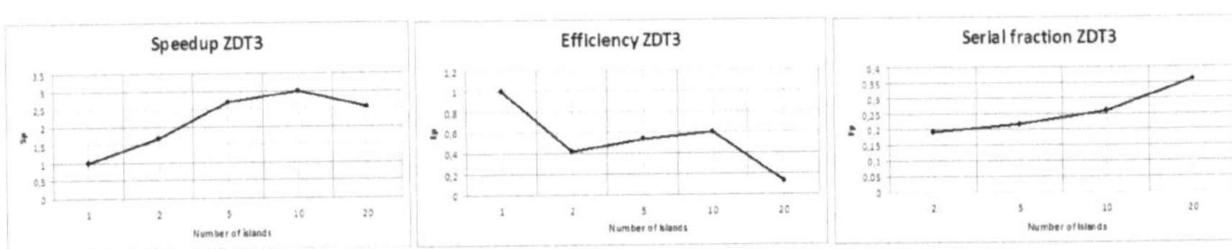

Fig. 4. Efficiency analysis of pDMOGA for ZDT3 problem

influence and the efficiency of the algorithm decreases. Taking under consideration the widespread of $PF^*$the Pareto optimal front consists of non-dominated Pareto optimal vectors, which are equidistantly spaced along the PF*.

5 Conclusions

In the paper we propose an algorithm for expensive Multi-Objective Optimization Problem. The approach is based on parallelization scheme using island model with isolated subpopulations. The idea of multiple resolutions concerning different islands with alternate migration strategies allows to maintain the diversity of individuals. The presented algorithm combines the parallel Distributed Multi-Objective Genetic Algorithm and modified Strenght Pareto Evolutionary Algorithm 2 as a serial part. The convergence of the pDMOGA algorithm has been accelerated by incorporating some modifications in SPEA2 algorithm.The modified SPEA2 are employed to enhance the pDMOGA performance in terms of convergence behavior.

The computational results for ZDTx tests problems indicate that pDMOGA is a promising approach and demonstrate the effectiveness of the proposed algorithm. The non-dominated solutions have the diversity in objective space and in design variable space. The approach characterized by dividing the population for the heterogenuous islands with different parameter settings strongly improves the convergence of pDMOGA. High parallel efficiency was derived with proposed pDMOGA. The changing migration strategy allows to receive the Pareto optimal solutions well distributed along the whole Pareto front.

References

1. Alba Torres, E., Troya Linero, J.M.: A survey of parallel distributed genetic algorithms. Complexity 4(4), 31–51 (1999)
2. Augusto, O.B., Rabeau, S., Depince, P., Bennis, F.: Multi-objective genetic algorithms: A way to improve the convergence rate. Eng. Application of Artificial Inteligence 19, 501–510 (2006)
3. Burke, E.K., Landa Silva, J.D.: The influence of the fitness evaluation method on the performance of multiobjective search algorithm. Eur. Journal of Operational Research 169, 875–897 (2006)
4. Coello Coello, C.A., Lamont, G.B., Van Veldhuiz en, D.A.: Evolutionary algorithms for solving multi-objective problems, 2nd edn. Springer, Heidelberg (2007)
5. David, A., Zydallis, J.B., Lamont, B.: Considerations in engineering parallel multiobjective evolutionary algoritms. IEEE Trans. on Evoutionary Computation 7(3), 144–173 (2003)
6. Deb, K.: Multi-objective genetic algorithms: problem difficulties and construction of test problems. Evolutionary Computation 7(3), 205–230 (1999)
7. Hiroyasu, T., Miki, M., Watanabe, S.: The new model of parallel genetic algorithm in multi-objective optimization problems. In: IEEE Proc. of the 2000 Congress on Evol. Comp., pp. 333–340 (2000)
8. Kamiura, J., Hiroyasu, T., Miki, M.: MOGADES: Multi-objective genetic algorithm with distributed environment scheme. In: Computational Intelligence and Applications (Proceedings of the 2nd International Workshop on Intelligent Systems Design and Applications), pp. 143–148 (2002)
9. Kaneko, M., Hiroyasu, T., Miki, M.: A parallel genetic algorithm with distributed environment scheme. In: Proceedings of the International Conference on Parallel and Distributed Processing Techniques and Applications, vol. 2, pp. 619–625 (2000)
10. Konak, A., Coit, D., Smith, A.: Multi-objective optimization using genetic algorithms: A tutorial. Reliability Engineering and System Safety 91, 992–1007 (2006)
11. Lopez, J.A., Coello, C.A.: MRMOGA- A new parallel multi-objective evolutionary algorithm based on the use of multiple resolutions. Concurrency Computation: Pract. Exper. 7 (2006)
12. Mitchell, M.: An introduction to genetic algorithms. MIT Press, Cambridge (1996)
13. Villalobos-Arias, M.A., Pulido, G.T., Coello Coello, C.A.: A proposal to use stripes to maintain diversity in a multi-objective particle swarm optimizer. IEEE Trans. on Evolutionary Computation 7 (2005)

14. Zitzler, E., Laumanns, M., Thiele, L.: SPEA2: improving the strength Pareto evolutionary algorithm. Technical Report 103, Computer Engineering and Networks Laboratory (TIK), Swiss Federal Institute of Technology (ETH) Zurich, Gloriastrasse 35, CH-8092 Zurich, Switzerland (2000)
15. Zitler, E., Deb, K., Thiele, L.: Comparison of multi-objective evolutionary algorithms: empirical results. Technical Report 70,Computer Engin. and Networks Lab., Swiss Federal Inst. of Technology (ETH) Zurich, Zurich (December 1999)
16. Zubik, W.: Meta-heuristic algorithms for multi-objective optimization. Praca magisterska, M.Sc. Eng. Thesis. Wroclaw University of Technology (2008) (in Polish)

Author Index

Kolibri 360 GmbH
Ferdinand-Jühlke-Straße 7
99095 Erfurt, Deutschland
produktsicherheit@kolibri360.de